Lecture Notes in Mechanical Engineering

Lecture Notes in Mechanical Engineering (LNME) publishes the latest developments in Mechanical Engineering—quickly, informally and with high quality. Original research reported in proceedings and post-proceedings represents the core of LNME. Volumes published in LNME embrace all aspects, subfields and new challenges of mechanical engineering. Topics in the series include:

- Engineering Design
- Machinery and Machine Elements
- Mechanical Structures and Stress Analysis
- Automotive Engineering
- Engine Technology
- Aerospace Technology and Astronautics
- Nanotechnology and Microengineering
- Control, Robotics, Mechatronics
- MEMS
- Theoretical and Applied Mechanics
- Dynamical Systems, Control
- Fluid Mechanics
- Engineering Thermodynamics, Heat and Mass Transfer
- Manufacturing
- Precision Engineering, Instrumentation, Measurement
- Materials Engineering
- Tribology and Surface Technology

To submit a proposal or request further information, please contact the Springer Editor of your location:

China: Dr. Mengchu Huang at mengchu.huang@springer.com
India: Priya Vyas at priya.vyas@springer.com
Rest of Asia, Australia, New Zealand: Swati Meherishi at swati.meherishi@springer.com
All other countries: Dr. Leontina Di Cecco at Leontina.dicecco@springer.com

To submit a proposal for a monograph, please check our Springer Tracts in Mechanical Engineering at http://www.springer.com/series/11693 or contact Leontina.dicecco@springer.com

Indexed by SCOPUS.

All books published in the series are submitted for consideration in Web of Science.

More information about this series at http://www.springer.com/series/11236

Volodymyr Tonkonogyi ·
Vitalii Ivanov · Justyna Trojanowska ·
Gennadii Oborskyi · Anatolii Grabchenko ·
Ivan Pavlenko · Milan Edl ·
Ivan Kuric · Predrag Dasic
Editors

Advanced Manufacturing Processes II

Selected Papers from the 2nd Grabchenko's International Conference on Advanced Manufacturing Processes (InterPartner-2020), September 8–11, 2020, Odessa, Ukraine

Springer

Editors
Volodymyr Tonkonogyi
Odessa National Polytechnic University
Odessa, Ukraine

Justyna Trojanowska
Poznan University of Technology
Poznan, Poland

Anatolii Grabchenko
National Technical University
"Kharkiv Polytechnic Institute"
Kharkiv, Ukraine

Milan Edl
University of West Bohemia
Pilsen, Czech Republic

Predrag Dasic
Department Trstenik
Academy of Professional Studies Šumadij
Trstenik, Serbia

Vitalii Ivanov
Sumy State University
Sumy, Ukraine

Gennadii Oborskyi
Odessa National Polytechnic University
Odessa, Ukraine

Ivan Pavlenko
Sumy State University
Sumy, Ukraine

Ivan Kuric
University of Zilina
Zilina, Slovakia

ISSN 2195-4356 ISSN 2195-4364 (electronic)
Lecture Notes in Mechanical Engineering
ISBN 978-3-030-68013-8 ISBN 978-3-030-68014-5 (eBook)
https://doi.org/10.1007/978-3-030-68014-5

This Springer imprint is published by the registered company Springer Nature Switzerland AG
The registered company address is: Gewerbestrasse 11, 6330 Cham, Switzerland

Preface

This volume of Lecture Notes in Mechanical Engineering contains selected papers presented at the 2nd Grabchenko's International Conference on Advanced Manufacturing Processes (InterPartner-2020), held in Odessa, Ukraine, during September 8–11, 2020. The conference was organized by Odessa National Polytechnic University, National Technical University "Kharkiv Polytechnic Institute", Sumy State University, and International Association for Technological Development and Innovations.

InterPartner-2020 focuses on promoting research and developmental activities, intensifying scientific information interchange between researchers, developers, and engineers.

InterPartner-2020 received 133 contributions from 13 countries around the world. After a thorough peer-review process, the program committee accepted 82 papers written by authors from 13 countries. Thank you very much to the authors for their contribution. These papers are published in the present book, achieving an acceptance rate of about 62%.

We want to take this opportunity to thank members of the program committee and invited external reviewers for their efforts and expertise in contribution to reviewing, without which it would be impossible to maintain the high standards of peer-reviewed papers.

Thank you very much to keynote speakers: Vitalii Ivanov (Ukraine), Dagmar Caganova (Slovak Republic), Justyna Trojanowska (Poland), Jan Pitel (Slovak Republic), Slawomir Luscinski (Poland) and Milan Edl (Czech Republic) for sharing their knowledge and experience.

The book "Advanced Manufacturing Processes II" was organized into seven parts according to the main conference topics: Part 1—Production Planning, Part 2—Design Engineering, Part 3—Advanced Materials, Part 4—Manufacturing Technology, Part 5—Machining Processes, Part 6—Process Engineering, and Part 7—Quality Assurance.

The first part "Production Planning" includes studies in the intelligent predictive decision support system, the use of augmented reality solutions in enterprises, ways for production line balancing and improvement of assembling technologies at production lines, and application of lean analysis in complex product manufacturing process. This part also includes studies in the field of the improvement of manufacturing processes, printing industry, logistical issues, as well as ecological activities in the use and recycling of products.

The second part "Design Engineering" includes studies in mathematical modeling and numerical simulation of the operating processes, including mixing, delamination, and oscillations. Notably, elastic characteristics, deflected modes, dynamic state, and structures' failure are analyzed in this part. This part also consists of kinematics and dynamics of a particle, modeling of hydraulic systems, and design optimization of structures. The possibilities of using engineering methodology for modeling the system "spindle assembly–cutting system" during machining are additionally included in this part.

The third part "Advanced Materials" presents studies in the field of thermo-mechanical properties of composites, deformation of alloys, mass balance in the deposition of ionic-plasma coatings, and ultrasonic treatment of polymers. This part also includes studies aimed at improving properties for structurally inhomogeneous materials, carbide coatings, and polyurethane foam molding. Problems related to the quality assessment, quantitative metallography, and radial isostatic compression of porous materials are also solved in this part.

The fourth part "Manufacturing Technology" is based on the manufacture of parts, their treatment, and technological support, as well as on numerical simulation of the technological processes. The environmental impact of additive manufacturing, minimization of roughness, and errors are also included in this part. Notably, improvement of characteristics for cutting tools, the use of solid lubricants, the influence of technological factors on the reliability, wear resistance, stability, and quality parameters are analyzed in this part.

The fifth part "Machining Processes" aims to develop and implement methods for multi-criteria optimization of machining and finishing processes and improvement of the quality for cutting tools, modeling of tool surface treatment, and simulation of shaping processes. Additionally, issues related to the development of dynamic models for the machining processes using wear-resistant and self-adaptive coatings are analyzed in this part. Ways for simulation of vibrational turning, tapered thread machining, fine boring, and double-sided grinding are also developed in this part.

The sixth part "Process Engineering" includes studies in the fields of parameter identification of heat supply systems and refrigeration capacity of air-conditioning systems, as well as the optimal design of combustion engines and the improvement of characteristics for atomizers. Substantiation of pressure compensator construction for nuclear power plants is also included in this part. Notably, the part analyzes petroleum and bio-components in fuel compositions, dynamic processes of mechatronic systems, the temperature in elements of wind power installations, thermal loading and air cooling of engines, as well as cavitation and output characteristics of hydraulic devices.

The seventh part "Quality Assurance" includes issues related to ensuring the quality of training engineers using a virtual environment, information quality in energy management systems, and auditing sustainable development for organizations. This part presents ways to improve measuring tools and test methods for nanomaterials, chemical parameters of mineral waters, and the selection of lithological layers. Additionally, ensuring the software assistance for the smart factory production line and industrial robots certification are also analyzed.

We appreciate the partnership with Springer, StrikePlagiarism, and EasyChair for their support during the preparation of InterPartner-2020.

Thank you very much to InterPartner Team. Their involvement and hard work were crucial to the success of the conference.

InterPartner's motto is "Science unites people together".

September 2020

Volodymyr Tonkonogyi
Vitalii Ivanov
Justyna Trojanowska
Gennadii Oborskyi
Anatolii Grabchenko
Ivan Pavlenko
Milan Edl
Ivan Kuric
Predrag Dasic

Organization

Honorary Chair

Anatolii Grabchenko	National Technical University "Kharkiv Polytechnic Institute," Ukraine

General Chair

Volodymyr Tonkonogyi	Odessa National Polytechnic University, Ukraine

Co-chair

Vitalii Ivanov	Sumy State University, Ukraine

Steering Committee (in alphabetical order)

Vitalii Ivanov	Sumy State University, Ukraine
Ievhen Ostroverkh	National Technical University "Kharkiv Polytechnic Institute," Ukraine
Volodymyr Tonkonogyi	Odessa National Polytechnic University, Ukraine

Program Committee (in alphabetical order)

Gabriel Abba	University de Lorraine, France
Viktor Antonyuk	National Technical University of Ukraine "KPI named after I. Sikorskyi," Ukraine
Katarzyna Antosz	Rzeszow University of Technology, Poland
Michal Balog	Technical University of Kosice, Slovak Republic
Kristina Berladir	Sumy State University, Ukraine
Sergiy Byelikov	Zaporozhye National Technical University, Ukraine

Yurii Petrakov	National Technical University of Ukraine “KPI named after I. Sikorskyi,” Ukraine
Jan Pitel	Technical University of Kosice, Slovak Republic
Leonid Polonskyi	Zhytomyr State Technological University, Ukraine
Grigore M. Pop	Technical University of Cluj-Napoca, Romania
Oleksandr Povstyanoy	Lutsk National Technical University, Ukraine
Ivan Pyzhov	National Technical University “Kharkiv Polytechnic Institute,” Ukraine
Erwin Rauch	Free University of Bozen-Bolzano, Italy
Michal Rogalewicz	Poznan University of Technology, Poland
Andrii Rogovyi	Kharkiv National Automobile and Highway University, Ukraine
Oleksandr Salenko	National Technical University of Ukraine “KPI named after I. Sikorskyi,”, Ukraine
Michal Sasiadek	University of Zielona Gora, Poland
Vira Shendryk	Sumy State University, Ukraine
Robert Sika	Poznan University of Technology, Poland
Volodymyr Sokolov	Volodymyr Dahl East Ukrainian National University, Ukraine
Vadym Stupnytskyy	Lviv Polytechnic National University, Ukraine
Volodymyr Tonkonogyi	Odessa National Polytechnic University, Ukraine
Justyna Trojanowska	Poznan University of Technology, Poland
Raul Turmanidze	Georgian Technical University, Georgia
Valentyn Tikhenko	Odessa National Polytechnic University, Ukraine
Nicolae Ungureanu	Technical University of Cluj-Napoca, Romania
Anatolii Usov	Odessa National Polytechnic University, Ukraine
Yana Vasylchenko	Donbass State Machine-Building Academy, Ukraine
George Vosniakos	National Technical University of Athens, Greece
Jerzy Winczek	Czestochowa University of Technology, Poland
Oleg Zabolotnyi	Lutsk National Technical University, Ukraine
Jozef Zajac	Technical University of Kosice, Slovak Republic
Viliam Zaloga	Sumy State University, Ukraine
Przemyslaw Zawadzki	Poznan University of Technology, Poland

Invited External Reviewers (in alphabetical order)

Yevheniia Basova	National Technical University “Kharkiv Polytechnic Institute,” Ukraine
Yelizaveta Chernysh	Sumy State University, Ukraine
Maryna Demianenko	Sumy State University, Ukraine
Yuliia Denysenko	Sumy State University, Ukraine
Nikolaos Galanis	National Technical University of Athens, Greece

Jan Hrdlicka	Czech Technical University in Prague, Czech Republic
Maryna Ivanova	National Technical University "Kharkiv Polytechnic Institute," Ukraine
Nikolaos Karkalos	National Technical University of Athens, Greece
Serhii Khovanskyi	Sumy State University, Ukraine
Kateryna Kostyk	National Technical University "Kharkiv Polytechnic Institute," Ukraine
Dmytro Levchenko	Innovative Solutions, LLC, Ukraine
Dmitriy Muzylyov	Kharkiv Petro Vasylenko National Technical University of Agriculture, Ukraine
Panagiotis Karmiris Obratański	National Technical University of Athens, Greece
Emmanouil Papazoglou	National Technical University of Athens, Greece
Serhii Sharapov	Sumy State University, Ukraine
Malgorzata Sokala	Kielce University of Technology, Poland
Yuliia Tarasevych	AGH University of Science and Technology, Poland
Roman Vaskin	Sumy State University, Ukraine
Tatiana Volina	Sumy National Agrarian University, Ukraine
Tetiana Zhylenko	Sumy State University, Ukraine

InterPartner Team (in alphabetical order)

Anna Balaniuk	Odessa National Polytechnic University, Ukraine
Kristina Berladir	Sumy State University, Ukraine
Maryna Demianenko	Sumy State University, Ukraine
Vitalii Ivanov	Sumy State University, Ukraine
Oleksandr Liaposhchenko	Sumy State University, Ukraine
Gennadii Oborskyi	Odessa National Polytechnic University, Ukraine
Ievhen Ostroverkh	National Technical University "Kharkiv Polytechnic Institute," Ukraine
Ivan Pavlenko	Sumy State University, Ukraine
Andrey Pavlyshko	Odessa National Polytechnic University, Ukraine
Volodymyr Tonkonogyi	Odessa National Polytechnic University, Ukraine

Contents

Production Planning

Intelligent Predictive Decision Support System for the Maintenance Service Provider

Katarzyna Antosz[1(✉)] and Małgorzata Jasiulewicz-Kaczmarek[2]

[1] Faculty of Mechanical Engineering and Aeronautics, Rzeszow University of Technology, 12, Al. Powstancow Warszawy, 35-959 Rzeszów, Poland
katarzyna.antosz@prz.edu.pl

[2] Faculty of Management Engineering, Poznan University of Technology, Prof. Rychlewskiego 2, 60-965 Poznań, Poland

Abstract. A maintenance process of internal vehicle transport is important from the production companies and also service providers. Failures of transport vehicles for a production company mean difficulties in the realization of production and auxiliary processes. Failures of transport vehicles for a service organization means those employees of the organization are assigned to carry out service activities for a longer time. This issue can be a particular problem, especially for small service organizations. Hence, in order to plan the maintenance activities, it is appropriate to predict failures to prevent them by undertaking adequate preventive actions. In this work, the failure risk was calculated based on the data from the maintenance processes collected. Additionally, it is proposed the solution, especially for small maintenance service providers, which can be used for maintenance activities taken under consideration the criticality of internal vehicles. The method presented in the article, which supports decision making regarding service planning, can help companies providing maintenance outsourcing services.

Keywords: Intelligent decision making tool · Fuzzy logic · Outsourcing of maintenance processes

1 Introduction

Companies specializing in technical infrastructure service (machinery and equipment, installations, means of transport) currently operate in almost all possible areas of the market, providing services for both commercial and manufacturing enterprises. Along with the ubiquitous automation and computerization, the demand for services related to, among others, repair, maintenance, and periodic inspection of machines and devices is growing. These needs create a huge field of activity for maintenance companies. Concerning manufacturing enterprises, maintenance service providers support: (i) non-functional areas, such are office heating, air conditioning, lighting; (ii) auxiliary areas to production, such as water treatment for production facilities, air compression, transport vehicles and storage facilities; (iii) production area such as production equipment.

The maintenance tasks realized by service providers are extremely important because the continuity of operation of the processes carried out by the clients of the maintenance company and their success depends on efficiently functioning machines,

V. Tonkonogyi et al. (Eds.): InterPartner 2020, LNME, pp. 3–13, 2021.
https://doi.org/10.1007/978-3-030-68014-5_1

devices, and means of transport. Also, regular maintenance of devices extends their life cycle and increases the efficiency of their operation and also has a positive effect on production efficiency [1, 2].

Due to the wide spectrum of maintenance services, running a maintenance company is a difficult task. The main problems of managers of such a company include the need to coordinate the dates of planned maintenance activities (e.g., inspections) and unplanned activities related to fixing the failures at customers' facilities located in different places. On the other hand, service technicians often struggle with maintaining machines from different manufacturers, which forces them to comply with various control procedures and service principles.

Enterprises implementing outsourcing maintenance contracts tend to achieve the "perfect balance" between costs and the availability and technical condition of equipment covered by the service contract. It is mainly done through: adaptation of exploitation strategies to machines depending on their criticality, adjustment of the frequency of preventive works on machines to their criticality, and technical condition.

Machine criticality is a key element for both the company and the maintenance service provider. The same machine can be critical in one enterprise, but not in another. An enterprise providing maintenance services should categorize machines together with client employees to include crucial and specific factors. The results of the categorization will be the basis for determining the scope of maintenance activities that should be taken to keep the risk of failure at the level required by the customer.

Each failure resulting in a stoppage of the machine, production line or even the entire plant may have several consequences, such as costs of removing the failure (e.g., cost of spare parts, cost of work of employees), reduction of product quality or difficulties in the entire supply chain. In addition to purely financial consequences, the occurrence of failure may cause safety risks for machine operators and third parties in the vicinity of the machine; also, it may harm the natural environment [3, 4]. Conducting a criticality analysis enables the selection of an appropriate maintenance strategy [5]. It should be emphasized, however, that selecting the strategy is a complex technical, economic and organizational task, requiring knowledge of market needs (as a recipient of maintenance services), the balance of total costs as well as profits, and the technical capabilities of the used equipment. When selecting the strategy, the specificity of tasks performed by a given company and key problems, that are generated in connection with maintenance processes, are of great importance.

Another element after determining the maintenance strategy, important for the implementation of the service company's activities, is the adjustment of the frequency of preventive work on machines to their criticality and technical condition. Planning and scheduling of maintenance are perceived as the "center" of maintenance management [6]. What and when it is to be performed directly affects production (availability of machinery and equipment), safety (tasks planned are generally safer than unplanned), environment (compliance with legal and sustainable requirements), costs (additional working hours) and indirectly marketing (availability of products for the client) [7]. At the same time, planning and scheduling of maintenance are influenced by the availability of financial resources, human resources (e.g., availability of competences) [8], information (e.g., historical data on operational events) and material and technical resources (internal availability and the possibility of acquiring them from a market environment).

Planning is a decision-making process that deals with the allocation of resources to jobs, in given time intervals, and its task is to optimize one or more goals [9]. Resources and tasks in any organization can take different forms. They can be machines in a repair workshop or people carrying out tasks. The decision to implement many tasks exploiting shared resources (and this situation occurs in enterprises providing services) is very complicated.

Different types of machines are assigned to different periods of service, depending on the frequency of failures and their criticality. One of the challenges is to determine the right time interval between services and their ranges. It is necessary to ensure that:

- costly and unnecessary maintenance activities were not performed long before the actual occurrence of the failure;
- costly activities caused by the failure as a consequence of too long time intervals between planned preventive actions were not necessary.

The need to plan maintenance and repair activities, the variety of planning methods used, and the resulting benefits are the subject of many publications [10–14].

2 Literature Review

The review presents some solutions supporting maintenance transport. In research [15], the case study takes its starting point in the perspective of an actor considering how to develop vehicle maintenance services for its customers and points at the need to enable understanding of the conditions for vehicle maintenance, which necessitates identification and analysis of the variety across transport service settings. The main objective of the work [16] is to design a system for diagnosis, measurement, and improvement of productivity aimed at quality of service. The result in the study [17] is researched on how to achieve an effective fleet maintenance planning in transport companies, which contributes to increasing the fleet energy efficiency and in achieving the companies' goal. Some decision systems are also presented in the works [18–20].

Although the literature on the subject presents some solutions to decision support systems, in the field of vehicle transport maintenance, there is still a lack of dedicated, simple, low-cost, and intelligent solutions, especially for small service providers. That is why the main research problem in this paper is: How to support maintenance service providers in planning their maintenances tasks based on the criticality of vehicles and predict adequate maintenance activities, especially in a small service provider?

This paper is organized as follows. First, a short literature review related to maintenance service providers is presented. Moreover, the research problem is posed. Next, the analysis of maintenance activities frequency based on case study service provider is introduced. In the third chapter, the intelligent predictive decision support system for the services provider is presented. Finally, the main conclusion and added value of this work are discussed.

3 Research Methodology

3.1 Structure of Internal Vehicle Transport Classification

The paper presents a practical analysis of the process of maintaining internal transport, the main purpose of which is to keep vehicles in readiness for work. The analysis was carried out based on the principles of organization and implementation of machine maintenance activities for various vehicles and used in various locations. The planning of maintenance activities of the service provider must anticipate failures to take them and take appropriate preventive action. In this analysis, the data is based on data collected in maintenance processes performed by the service provider.

The subject of the following analyses is Internal Vehicle Transport (*IVT*). Different criteria were used to classify IVT, such: type of IVT (internal combustion vehicle, electric), a model (m_1, m_2,…,m_n), age, and working time (WT) (Table 1). Ranges for criteria were identified during the data analysis from the case company and concerning requirements from internal vehicle producers. For analysis, the data delivered by the service provider were used. The boxplot in Fig. 1 shows the value distribution of Age (A) and working time (WT). Only the boxplot of working time shows the outliers, which do not influence further analyses.

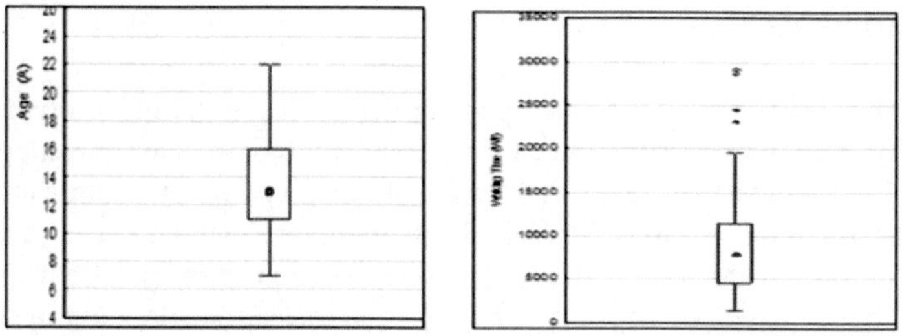

Fig. 1. The boxplot of value distribution of Age (A) and Working Time (WT).

3.2 Maintenance Activities Analyses

The first of the study was to identify the type and the frequency of maintenance activities (*MA*). Mainly the type of maintenance activities was divided into two types planned and unplanned activities. To identify the frequency of maintenance activities to criteria were analyzed: age (A) and Working Time (WT). The analysis showed that the age of IVT is in the range of 7–22 years, and the maximum value of the working time of IVT is almost 30000 h. Three categories of analyzed criteria have been established: *High (H), Medium (M))* and *Low (L)* (Table 1). The criteria categories have been established based on authors and service organization experience.

Table 1. Classification criteria of IVT.

Categories	Criteria	Age of IVT (A) [years]	Working time (WT) of IVT [hours]
	L - Low	<8	<8000
	M - Medium	8–16	8000–16000
	H - High	>16	>16000

To analyze the frequency of maintenance activities of each IVT category, the maintenance activities frequency *(MAF)* indicator, according to Eq. (1), was calculated.

$$MAF = \frac{NoMA_C}{\sum NoMA} \tag{1}$$

where: *MAF* – the maintenance activities frequency, *NoMAc* – many maintenance activities in every category (considering *A* and *WT*),

$\sum NoMAF$– the total number of maintenance activities in the analyzed time.

Table 2 shows the obtained results, the value of *the MAF* indicator in each category.

Table 2. The value of the *MAF* indicator in each category.

Age (A) of IVT [years]		L	M	H
Working time (WT) [hours]	L	0.08	0.28	0.16
	M	0.05	0.22	0.08
	H	0.02	0.11	0.00

The experience of authors and personnel of service organizations let to identify the risk level of the *MAF* indicator. The level of risk of *the MAF* indicator was divided into three categories: Low, when the value of *MAF* is less than 0.10, Medium when the value of *MAF* is in the range 0.10 to 0.20, High when the value of the *MAF* indicator is more than 0.20 (Table 3). According to these assumptions, the risk level of *MAF* for every category was specified and presented in Table 4.

The possible use of the presented matrix of risk has been proved with workers in the analyzed service organization. In order to provide any circumstances (especially if the values of criteria are near to the border of the ranges), it is important to use an approach based on fuzzy logic to support the *MAF* indicator.

Table 3. The *MAF* categories.

The categories	Low	Medium	High
The value	<0.10	0.10–0.20	>0.20

Table 4. The risk level of *MAF*.

Criteria	Age (A) [years]			
		Low	Medium	Hight
Working time (WT) [hours]	Low	Low	High	Medium
	Medium	Low	High	Low
	High	Low	Medium	Low

4 Results

4.1 Process of Fuzzy Interference

The shown case study described in the paper, a Mamdani-type process of fuzzy inference is utilized. The membership functions were formulated, taking into account the values of the A and WT. However, it might be changed membership functions according to maintenance experts. Figure 2 presents the total view of the suggested fuzzy system of risk importance assessment. To calculate the risk rank, there is the option to choose A and WT, as the inputs to the fuzzy system. Because the Mamdani type fuzzy inference process is easy to understand and the most intuitive, so it was used to present that analyzed case study.

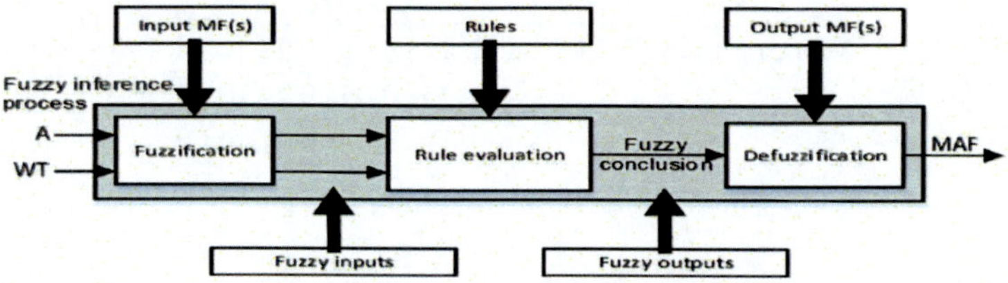

Fig. 2. Fuzzy Interference Process.

In the proposed Mamdani Fuzzy Interference System, two quantitative inputs were used *(A)* and *(WT)*, and output is the *MAF* indicator. The MATLAB (R2019b) was used to implement the proposed fuzzy inference process [21] (Fig. 3).

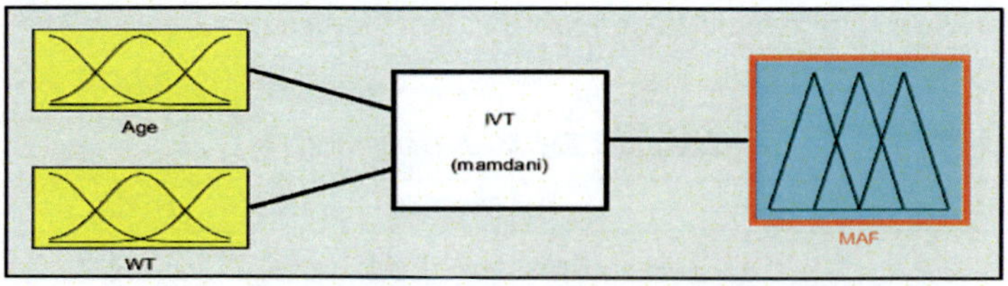

Fig. 3. Designer of Fuzzy Logic (Matlab R2019b).

4.2 Parameters of Fuzzy Interference System

For the inputs (*A, WT*) and output (*MAF*), the membership functions *MFs* were specified. In this paper, the functions of Gaussian membership available in MATLAB (R2019b) were incorporated by the authors for input *A* and *WT*, were used [21, 22] The curve of Gaussian membership function is described by the Eq. (2):

$$f(x) = exp^{\frac{-0.5(x-c)^2}{\sigma^2}} \tag{2}$$

where σ is the standard deviation, and *c* is the mean. Figure 4 presents the Gaussian membership function (GMF) for inputs *A* and *WT*.

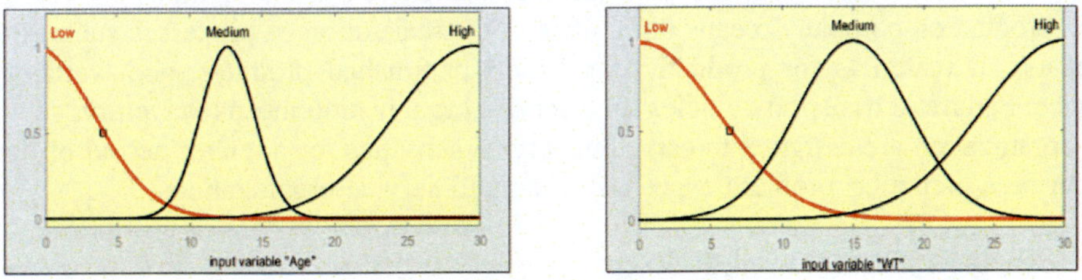

Fig. 4. GMF of *A* and *WT*.

Based on the work [23], the authors' and service organization experience, and in order to reduce the difference regarding the mathematical modeling and the practical implementation, the triangular membership functions for output *MAF* were used (3).

$$\mathrm{y} = \mathrm{trimf}\,(\mathrm{x}, [\mathrm{x\ b\ c}]) \tag{3}$$

The curve of the triangular membership function is a vector x, and it depends on three parameters a, b, and c. The b parameter is the triangle peak, and the a and c parameters determine the "feet". The membership functions for output *MAF* was determined based on the ranges presented in Table 3 (see Fig. 5). Based on Table 4, the fuzzy rule base was established (see Fig. 6).

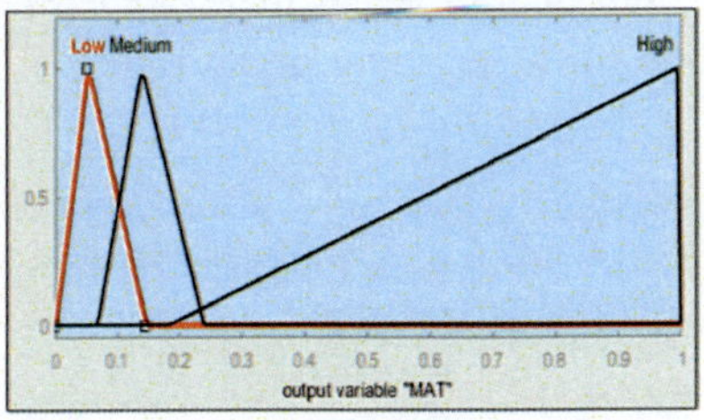

Fig. 5. A rule base for proposed FIS.

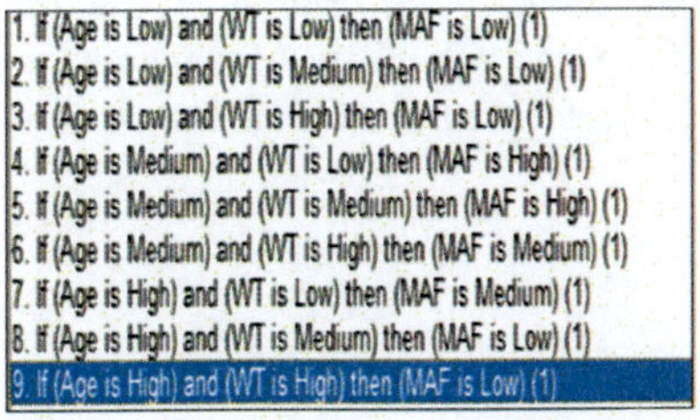

Fig. 6. A rule base for proposed FIS.

4.3 Analysis of the Results and Discussion

Figure 7 and Fig. 8 present the calculation value of *MAF* (3D) three-dimensional risk (*MAT*) profiles relative to *A* and *WT*. The calculation was made for the $A = 7$ and $WT = 7.000$ h. The value of *the MAF* indicator estimated by the FIS is 0.396. According to the corresponding linguistic value presented in Table 3 is H. This linguistic value H means that for this *IVT* group means a high frequency of performing maintenance activities. It is very important information from the service company's point of view. In the next step, the service company should analyze the type of maintenance activities performed for this *IVT* group (*Age* – Medium and *Working Time* - Low). In the analyzed group of *IVT*, a significant part of 80% was corrective maintenance, and preventive maintenance constituted only 20%. It means that the failure rate for the analyzed transport vehicles was high. Failure of transport vehicles for a production company means difficulties in the realization of processes such as the transport of materials for production, transport of finished products, and warehouse service. Failure of transport vehicles for a service organization means that employees of the organization are assigned to carry out service activities for a longer period of time. It can be a particular problem especially for small service organizations.

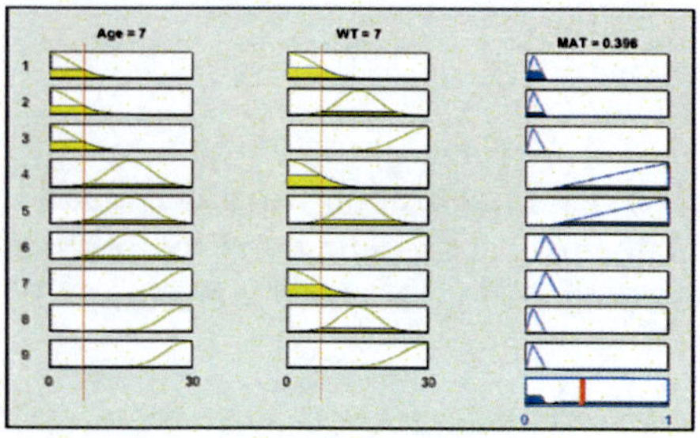

Fig. 7. Calculations [$MAT = 0.396$ for $A = 7$ and $WT = 7.000$] and rule view.

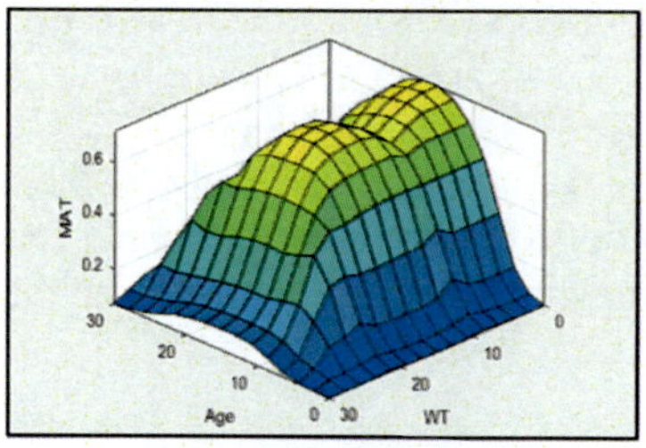

Fig. 8. 3D profile.

Figure 9 shows the type of maintenance activities implemented for the *IVT* group (percentage of Preventive Maintenance (PM) and Corrective Maintenance (CM)).

The proposed solution to this situation is to increase the number of preventive maintenance activities. For a production company, this would mean increasing the availability and reliability of transport vehicles, while for a servicing organization, it would help to plan the maintenance activities.

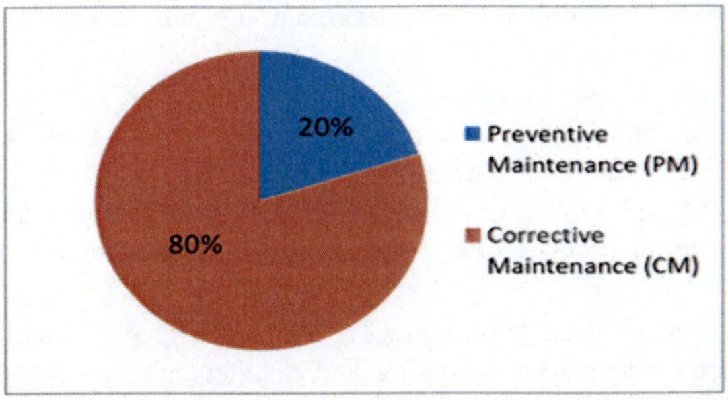

Fig. 9. Percentage of Preventive Maintenance (PM) and Corrective Maintenance (CM).

5 Conclusions

The method presented in the article, which supports decision making regarding service planning, can help companies providing maintenance outsourcing services. However, recently, along with new technologies such as RFID, various sensors, and wireless telecommunications as well as data control and acquisition, both the quantity and quality of *IVT* data during the period of their use is very high. Modern *IVT* with installed sensors and connected thanks to IoT technology can provide a very large amount of data describing their work and technical condition. It will enable companies that provide maintenance services to better plan their work, and thus to better manage human resources. The use of new technologies for planning service work seems to be an excellent strategy for any company providing maintenance services, looking for ways to perform services at the level of satisfying customers while effectively using their resources. It is also the direction of future research of the authors of this work.

Acknowledgements. The authors would like to acknowledge the employees of the service provider for delivered data and participation in this research.

References

1. Araújo, A.F., Varela, M.L.R., Gomes, M.S., Barreto, R.C.C., Trojanowska, J.: Development of an intelligent and automated system for lean industrial production adding maximum productivity and efficiency in the production process. In: Hamrol, A., Ciszak, O., Legutko, S., Jurczyk, M. (eds.) Advances in Manufacturing. LNME, pp.131–140 (2018)
2. Jasiulewicz-Kaczmarek, M.: Practical aspects of the application of RCM to select optimal maintenance policy of the production line. In: Nowakowski, T., et al. (eds.) Safety and Reliability: Methodology and Applications - Proceedings of the European Safety and Reliability Conference, ESREL 2014, pp. 1187–1195. Taylor & Francis Group, London (2015)
3. Burduk, A.: An attempt to adapt serial reliability structures for the needs of analyses and assessments of the risk in production systems. Eksploatacja i Niezawodnosc-Maintenance Reliab. **3**, 85–96 (2010)

4. Garg, A., Deshmukh, S.: Maintenance management: literature review and directions. J. Qual. Maintenance Eng. **12**(3), 205–238 (2006)
5. Jasiulewicz-Kaczmarek, M., Żywica, P.: The concept of maintenance sustainability performance assessment by integrating balanced scorecard with non-additive fuzzy integral. Eksploatacja i Niezawodnosc – Maintenance Reliab. **20**(4), 650–661 (2018)
6. Gopalakrishnan, M., Skoogh, A.: Machine criticality based maintenance prioritization: identifying productivity improvement potential. Int. J. Prod. Perform. Manag. **67**(4), 654–672 (2018)
7. Koochaki, J., Bokhorst, J.A.C., Wortmann, H., Klingenberg, W.: The influence of condition-based maintenance on workforce planning and maintenance scheduling. Int. J. Prod. Res. **51**(8), 2339–2351 (2013)
8. Allahverdi, A.: The third comprehensive survey on scheduling problems with setup times/costs. Eur. J. Oper. Res. **246**(2), 345–378 (2015)
9. Basri, E., Abdul Razak, I., Ab-Samat, H., Kamaruddin, S.: Preventive maintenance (PM) planning: a review. J. Qual. Maintenance Eng. **23**(2), 114–143 (2017)
10 Baptista, M., Sankararaman, S., Medeiros, I.P., Nascimento, C., Jr., Prendinger, H., Henriques, E.M.P.: Forecasting fault events for predictive maintenance using data-driven techniques and ARMA modeling. Comput. Ind. Eng. **115**, 41–53 (2018)
11. Gopalakrishnan, M., Skoogh, A., Laroque, C.: Simulation-based planning of maintenance activities in the automotive industry. In: Proceedings of the 2013 Winter Simulation Conference (2013)
12. Kujawińska, A., Diering, M., Żywicki, K., Rogalewicz, M., Hamrol, A., Hoffmann, P., Konstańczak, M.: Methodology supporting the planning of machining allowances in the wood industry. In: International Joint Conference SOCO 2017-CISIS 2017-ICEUTE 2017, León, Spain, 6–8 September 2017. Advances in Intelligent Systems and Computing, pp. 338–347. Springer (2018)
13. Hedvall, K., Lind, F.: Analysing an activity in context: a case study of the conditions for vehicle maintenance. Ind. Mark. Manag. **58**, 69–82 (2016)
14. Kujawińska, A., Diering, M., Rogalewicz, M., Żywicki, K., Hetman Ł.: Soft modelling-based methodology of raw material waste estimation. In: Intelligent Systems in Production Engineering and Maintenance – ISPEM 2017. Advances in Intelligent Systems and Computing, vol. 637, pp. 407–417 (2017)
15. Parada, M., Madriz-Rodriguez, M.D., Castillo-Pedraza, M.: Productivity management system for the service sector in the San Cristobal municipality of Tachira state, Venezuela. Ciencia Unemi **11**(26), 63–78 (2018)
16. Moyseenko, S.S., Moroz, E.O.: Optimization of the transport service system of fishing vessels. Mar. Intellect. Technol. **3**(4), 168–176 (2018)
17. Vujanovic, D.B., Momcilovic, V.M.: A hybrid multi-criteria decision making model for the vehicle service center selection with the aim to increase the vehicle fleet energy efficiency. Therm. Sci. **22**(3), 1549–1561 (2018)
18. Ratnayake, R.M.C., Antosz, K.: Risk based maintenance assessment in the manufacturing industry: minimization of suboptimal prioritization. Manag. Prod. Eng. Rev. **8**(1), 38–45 (2017)
19. Burduk, A., Grzybowska, K., Safonyk, A.: The use of a hybrid model of the expert system for assessing the potentiality manufacturing the assumed quantity of wire harnesses. LogForum **15**(4), 459–473 (2019)

20. Antosz, K., Stadnicka, D.: An intelligent system supporting a forklifts maintenance process. Advances in Intelligent Systems and Computing, vol. 637, pp. 13–22 (2018)
21. Matlab, MATLAB 7.12.0 (R2019b): Fuzzy logic Toolbox, 1984–2019. The MathWorks, Inc. (2019)
22. Mathworks, Fuzzy inference system modelling: Gaussian combination membership function. www.mathworks.se/help/fuzzy/gauss2mf.html. Accessed 10 Oct 2019
23. Ross, T.: Fuzzy Logic in Engineering Applications, 3rd edn. Wiley, Singapore (2010)

Readiness to Use Augmented Reality Solutions in Small and Medium Enterprises in Poland: A Survey

Pawel Bun(✉), Damian Grajewski, and Filip Gorski

Poznan University of Technology, Poznan, Poland
pawel.bun@put.poznan.pl

Abstract. The article discusses the results of work to assess the readiness of Polish industrial companies to apply augmented reality technology in practice. As part of the work, a tool in the form of a survey was prepared, available online as a digital survey, and a mobile application. The survey contains several dozen questions divided into sections, and not only the technical readiness of a given company is assessed, but also the potential area of application of the Augmented Reality technique in its activities such as training, support for production and production-related processes and support for machine operation and maintenance. In addition to the development and implementation of the tool, it was also verified at the enterprises from the Wielkopolska region and neighboring regions. Survey results can be used by those enterprises to help them get ready to introduce Augmented Reality in their work. Gathered data alongside its analysis are presented in this paper.

Keywords: Smart factories · Maintenance · Industry 4.0

1 Introduction

Industry 4.0 is a paradigm for the increasing digitization and automation of the manufacturing environment [1]. One of the pillars of Industry 4.0 is the Smart Factories [2] in which the human workforce is integrated into manufacturing systems, and it must be flexible and adaptive [3]. In order to reach that goal, Virtual and Augmented Reality (AR) are used for employee training and work on the production line [4]. The AR technology is used for displaying dynamic spatial and flat visualizations overlaid on real-world objects, Azuma's [5] definition of the AR technology includes constant bond between real-world and digital objects as well as an interaction between the user and the virtual objects. The AR solutions can be based on cellphones, tablets [6], or headsets [7], such as Microsoft HoloLens.

AR is not yet widely used in most Polish manufacturing companies. In order to determine the reasons for that, the survey was taken among companies included in small and medium-sized enterprises aimed at determining the degree of technological readiness of companies to implement this technology. The proposed tool in the form of a survey enables both the assessment of readiness itself and obtaining

V. Tonkonogyi et al. (Eds.): InterPartner 2020, LNME, pp. 14–23, 2021.
https://doi.org/10.1007/978-3-030-68014-5_2

recommendations aimed at enabling the introduction of AR in the company if not all the requirements have been met.

2 Literature Review

Augmented and Virtual Reality are widely used in Industry 4.0. In most of the production companies that utilize the power of AR it is used for supporting work in:

- operation and maintenance processes of machines and devices [8, 9],
- interactive training and training [10],
- product and process design [11].

Unlike virtual reality (VR), AR is not well known to Polish companies. It is despite the undoubted potential of their applications in the production and operation of machines and technical devices and their important place in the concept of Industry 4.0 (the concept of the so-called augmented operator, mutual communication of machines, the Internet of Things, etc.). It is also reflected in industry reports on new technologies – for example, the Gartner Hype Cycle for Emerging Technologies. It is the annually updated progress chart of new technologies, in version 2018 (Fig. 1) no longer includes a point representing VR – Gartner analysts have recognized that this technology has already reached a plateau. Meanwhile, augmented reality is on this curve in the least favorable position, i.e., the so-called "bottom of disappointment".

It means that the technology has already lost its so-called "Wow effect" resulting from her novelty, and yet she has not yet convinced her recipients enough that the prognosis for her life was fully positive.

There is a clear trend and need to use AR and VR in numerous literary publications. Stoltz [13] described possibilities of using AR in the warehouse, Gorski [14] showed the possibility of using it to train workers in 3D Printing. Gattullo [15] proposed a methodology for AR manuals for Industry 4.0. Palmarini [16] did a systematic review of AR applications in maintenance. Bun [17] and Ivanov [18] described how to introduce AR and VR in engineering education. Despite that, Polish small and medium enterprises are reluctant to introduce such novelties, and there are several reasons [19]. The most important seems to be problems and costs resulting from initial long-term analyzes leading to the assessment of the potential of a given enterprise. Most companies are also aware of the need to take actions that must be taken to be able to smoothly implement virtual technologies (Augmented/Mixed or Virtual Reality) in their daily production activities. Hence the need to develop a tool that quickly and in a simple form will allow initially answering whether a given company is technically ready for the implementation of AR/MR technology.

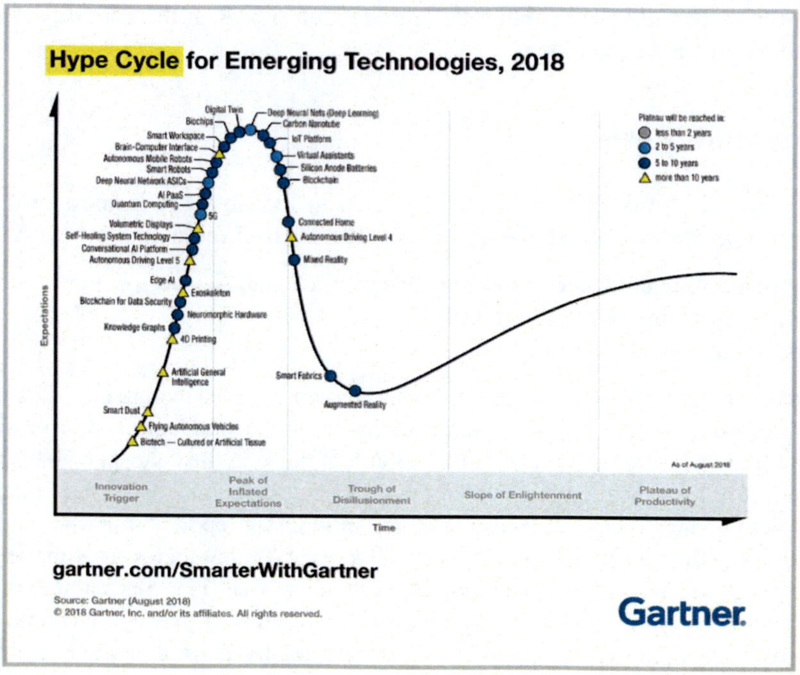

Fig. 1. 5 trends emerge in the gartner hype cycle for emerging technologies, 2018 [12].

3 Research Methodology

3.1 Survey Structure

During the preparation of the survey, the following assumptions were made:

- the survey consists entirely of closed questions,
- some questions appear only if the previous questions are correctly answered,
- some of the questions are single and part of multiple-choice – this information was included in the content of the question,
- dozens of questions were developed and divided into sections.

The following sections were adopted:

G1. Nature of production activities and working conditions (11 questions)

This section contains basic questions about company size and production activity (production volume, range, materials, etc.). This section also includes questions about the possibility of modifying products and production stations by applying markers on them to support image display in AR.

1. Specify the company size
2. Specify the type of production
3. Specify the type of products manufactured (max. 2)
4. Specify the dimensions of manufactured products (max. 2)

5. Specify materials used in production (main 2)
6. Specify dominant technological processes (max. 3)
7. Is it possible to apply additional markings on products/semi-finished products in the form of sticky/engraved markers?
8. Is it possible to apply additional markings on production machines in the form of sticky/engraved markers?
9. Is it possible to modify the design of products to include markers or RFID/NFC sensors that are part of the product?
10. Determine the risk to the health and life of employees during the implementation of production processes in the company
11. Specify the conditions in the production space (any number of responses)

G2. Mechanization and automation of production (5 questions)

Section 2 contains questions about the participation of manual, mechanized activities and robotic in the company's manufacturing activities.

1. Specify the share of manual activities in production
2. Specify the number of CNC machines and automated stations:
3. Specify the degree of robotization of production
4. Specify the flexibility and reconfigurability of production
5. Mark what machines and devices are used in the company to carry out production and production-related processes (any number of answers)

G3. Digitization of production and operating processes (12 questions)

The third section contains questions about the computer-aided design methods used in the company, production preparation, and operation as well as production data flow.

1. Specify the use of CAx systems in your company (select one or more answers)
2. Specify the degree of digital integration of production processes
3. Specify the degree of digitization of production stations
4. Specify the degree of computerization of engineering works
5. What are 2D/3D data formats about products used by employees and engineering staff? (tick max. 3)
6. What part of the products manufactured by the company has its full 3D representation?
7. What proportion of production positions/machines in the company has its 3D representation?
8. How is the conceptual design of new products (max. 2 answers)?
9. How is the design of new processes and production stations carried out?
10. How do production department employees report problems to engineers and maintenance teams?
11. Do production employees have access to KPI?
12. Is it possible to display current production data in the company? (information on machines or process flow)

G4. Staff (11 questions)

The questions of the fourth section concern the company's personnel, both production workers and engineering staff. The questions relate to, among others, age, internship, rotation, and training.

1. Specify the average age of production employees
2. Specify the standard deviation/age distribution of production workers
3. Specify the required qualifications of production employees (max. 2)
4. Specify the time to introduce the production employee to full efficiency as:
5. Specify the average employment time for production workers:
6. Specify the engineering and specialist staff present and dominating in the company (max. 4):
7. Specify the average age of engineering staff:
8. How often does production staff take part in various thematic training?
9. How often does engineering staff take part in various thematic training?
10. What are the most common topics for training production workers? (max. 2)
11. What are the most common topics in engineering training? (max. 2)

G5. IT infrastructure, security (5 questions)

This section contains basic questions about IT equipment and network traffic security introduced in the enterprise.

1. Specify the number of desktops/laptops used in the enterprise per employee:
2. Specify the number of mobile devices (tablets, smartphones, phablets) used for work in the enterprise, per employee
3. Mark the AR/VR devices that are included in the company (any number of answers)
4. Specify how to transfer data between computers in your company
5. Specify the level of security associated with the communication between computers and devices

G6. Maintenance (6 questions)

The sixth section is the questions about the operation, servicing, and inspection of machines and equipment used for the implementation of production and auxiliary processes.

1. Specify the number of machines (machines, robots, and production stations):
2. Specify the frequency of failure of production and other machines (in general):
3. Specify the frequency of inspections of production and other machines (in general):
4. Specify the dominant method of removing failures of production machines and other:
5. Specify the dominant method of inspecting production and other machinery:
6. Specify the dominant way to service your company's products

G7. Research and development, use of new technologies (9 questions)

The last section contains questions about R&D staff and projects, as well as new technologies used in the company, with particular emphasis on VR/AR.

1. How often does the company use virtual reality technology?
2. Specify the applications of VR technology in the company (select one or more answers):
3. How often does the company use augmented and/or mixed reality technologies?
4. Specify the applications of AR technology in the company (select one or more answers):
5. Specify how content development for AR/VR technology in the company:
6. Select the technologies and concepts that are used in the company (any number of answers):
7. How many people do research and development of new technologies in the company?
8. How many R&D projects are currently carried out in the company?
9. What own budget (without co-financing) is the company able to spend on the development and implementation of new digital technologies that can help in the production area?

The answer to each question is associated with the granting of a specific point value – from −1 to +5 points, with the points divided into the so-called assessment of general readiness (almost every question has an impact on this sum) and assessment of the potential of AR implementation in three areas: training (T), support for production and production-related processes (P) and support for machine operation and maintenance (E).

3.2 Data Gathering

Data from companies were collected using a survey implemented in two ways. The first is an online survey placed on Google's servers. The screenshot of the survey is shown in Fig. 2. The survey does not allow an immediate result - the answers to the questions are collected and placed in a supporting spreadsheet, where the algorithm sums them up and calculates a recommendation.

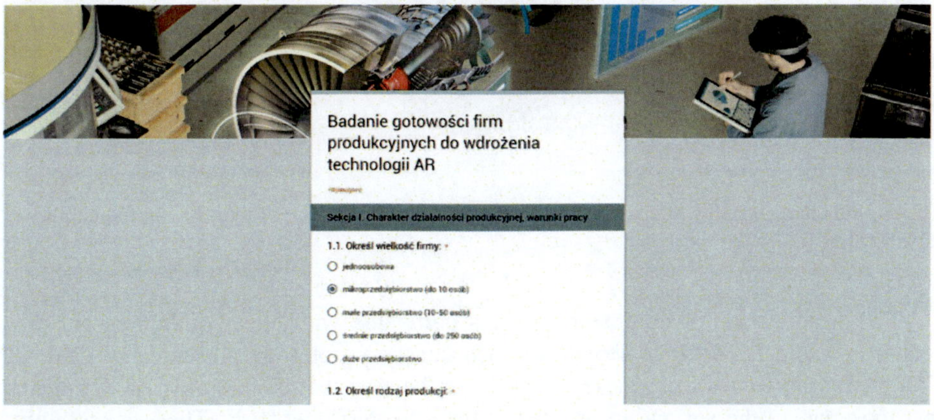

Fig. 2. An online survey used to gather data from companies. The screenshot shows the beginning of the survey - the first question from the first section. All questions are marked as mandatory.

The second way to implement the tool is a mobile application - it was prepared in the version for the Internet browser (WebGL) and the Android system, using the Unity software version 2018.2 and the Visual Studio 2017 compiler, the programming of the functionality was done using the C # language. The application automatically calculates the results and sends an email informing the user about the results.

4 Results

The survey was conducted online from May to August 2018. The answers from each of the questions were presented in the form of pie charts; they were also correlated with expert judgment (specific point value – from −1 to +5 points) to determine whether the company is ready to implement AR. For example, for question G4.4 presented at Fig. 3, the expert judgment was:

- less than 2 business days (0)
- 2–5 business days (+1; +1P; +1T)
- 5–10 business days (+1; +1P; +2T)
- 10–15 business days (+1; +2P; +2T)
- a month or more (+1; +3P; +1T)

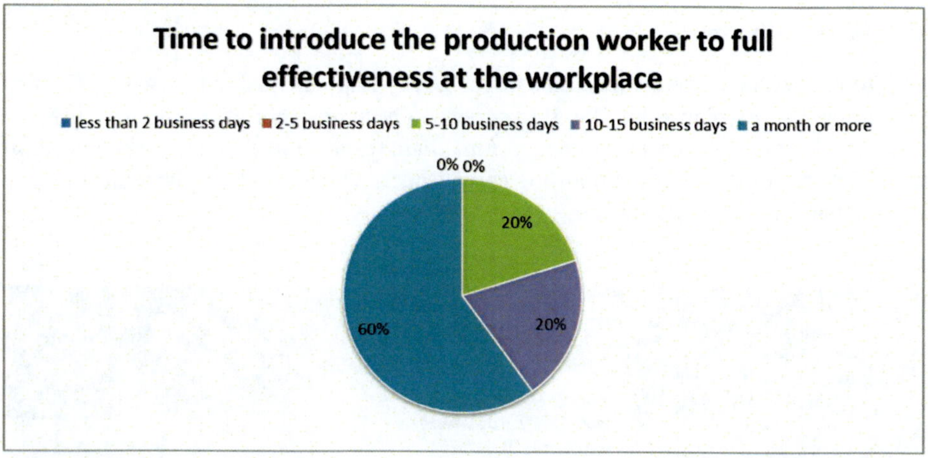

Fig. 3. Pie chart – results of answers to the question about the time of introducing the employee to full efficiency at the workplace.

The mean values of companies' scores correlated with expert judgment are presented in Table 1. Maximal scores for each question group are presented in Table 2. Comparing the values from these tables, it can be seen that the majority of companies that took part in the survey are not ready for it to implement AR. The weakest point score falls on the G5 category.

Table 1. Mean values from surveys. G – general score, T – training recommendation, P – production recommendation, E – maintenance recommendation.

	G1	G2	G3	G4	G5	G6	G7
G	9.78	2.67	9.89	6.67	0.56	1.00	3.56
T	0.00	2.56	0.00	3.11	0.00	0.00	0.00
P	0.00	0.67	0.00	0.00	0.00	0.00	0.00
E	0.00	2.00	0.00	0.00	0.00	2.11	0.44

Table 2. Maximum values from for each category.

	G1	G2	G3	G4	G5	G6	G7
G	30.00	4	30	21	20	2	29
T	1	3	0	9	0	0	1
P	0	3	0	6	0	0	1
E	0	3	0	0	0	11	1

5 Conclusions

A functional survey tool was obtained to test the readiness of manufacturing companies to implement AR technology. The mobile application was tested on a dozen or so different devices, including a PC, laptop, and Android mobile phones - the correct and proper operation of the application was reported everywhere without errors and freezes. The tool in both versions is online - requires access to the Internet.

While data was being collected from companies, surveys were completed by 20 company representatives. The companies that completed the survey positively evaluated the very fact of the creation of such a tool, as well as the simplicity and speed of use. No technical difficulties were encountered when using the tool, and the range of questions was assessed as detailed enough. There were slight doubts as to who is to use such a tool in companies - it was stated that this does not have to be a decision-making person, but only an employee who has full knowledge of the nature of production, human resources and other activities of the enterprise or can acquire such knowledge.

Four out of five surveyed enterprises were classified as not ready for implementation of AR technology - the companies scored below 50% points, and additionally, in three out of four cases, the sum of points in two sections was below 2. It was mainly due to the low level of digitization of production and operational processes, a relatively low level of robotization of production, as well as a small number of mobile devices used during work, the lack of any AR/VR devices on the enterprise's equipment and virtually no funds for R&D. Four companies received over 50% of the points (except category G5) – the companies were recommended to incur additional investments before the AR solution could be implemented - primarily as support in operation and service.

The expert assessment overlapped fully with the assessment of the algorithm, which means that it was properly prepared for this sample of enterprises. Further work should involve more extensive research and perhaps refine the scope of questions and modify the scale of points. The authors plan to create an interactive assistant who will inform the company what investments are necessary to introduce the necessary IT infrastructure successfully. It should be borne in mind; however, that in order to successfully introduce AR and VR technologies to companies, it is necessary to change the management approach to investments in new technologies and R&D.

Acknowledgments. The research work had the partial financial support of Ministry of Science and Higher Education, Republic of Poland, under the project 02/23/DSPB/8716 and European Regional Development Fund under the project 0613/SBAD/8727.

References

1. Oesterreich, T.D., Teuteberg, F.: Understanding the implications of digitisation and automation in the context of industry 4.0: a triangulation approach and elements of a research agenda for the construction industry. Comput. Ind. **83**(Suppl. C), 121–139 (2016)
2. Żywicki, K., Zawadzki, P., Górski, F.: Virtual reality production training system in the scope of intelligent factory. Advances in Intelligent Systems and Computing, vol. 637, pp. 450–458 (2018)
3. Yew, A.W.W., et al.: Towards a griddable distributed manufacturing system with augmented reality interfaces. Robot. Comput.-Integr. Manuf. **39**, 43–55 (2016)
4. Azuma, R.T.: A survey of augmented reality. Presence Teleoperators Virtual Environ. **6**(4), 355–385 (1997)
5. Damiani, L., et al.: Augmented and virtual reality applications in industrial systems: a qualitative review towards the industry 4.0 era. IFAC-PapersOnLine **51**(11), 624–630 (2018)
6. Rumiński, D., Walczak, K.: Creation of interactive AR content on mobile devices. Lecture Notes in Business Information Processing: International Conference on Business Information Systems, vol. 160, pp. 258–269 (2013)
7. Hamacher, A., et al.: Application of virtual, augmented, and mixed reality to urology. Int. Neurourol. J. **20**(3), 172–181 (2016)
8. Flatt, H., et al.: A context-aware assistance system for maintenance applications in smart factories based on augmented reality and indoor localization. In: 2015 IEEE 20th Conference on Emerging Technologies & Factory Automation (ETFA), pp. 1–4. IEEE (2015)
9. Quint, F., Loch, F., Bertram, P.: The challenge of introducing AR in industry-results of a participative process involving maintenance engineers. Procedia Manuf. **11**, 1319–1323 (2017)
10. Limbu, B.H., Jarodzka, H., Klemke, R., Wild, F., Specht, M.: From AR to expertise: a user study of an augmented reality training to support expertise development. J. Univ. Comput. Sci. **24**(2), 108–128 (2018)
11 Mourtzis, D., Zogopoulos, V., Katagis, I., Lagios, P.: Augmented reality based visualization of CAM instructions towards industry 4.0 paradigm: a CNC bending machine case study. Procedia CIRP **70**, 368–373 (2018)
12. Gartner analysis of trends in Emerging Technologies. https://www.gartner.com/smarterwithgartner/5-trends-emerge-in-gartner-hype-cycle-for-emerging-technologies-2018/. Accessed 28 Dec 2019

13. Stoltz, M.-H., et al.: Augmented reality in warehouse operations: opportunities and barriers. IFAC-PapersOnLine **50**(1), 12979–12984 (2017)
14. Gorski, F., Wichniarek, R., Kuczko, W., Bun, P., Erkoyuncu, J.A.: Augmented reality in training of fused deposition modelling process. In: Hamrol, A., Ciszak, O., Legutko, S., Jurczyk, M. (eds.) Advances in Manufacturing. Lecture Notes in Mechanical Engineering, pp. 565–574 (2018)
15 Gattullo, M., Scurati, G.W., Fiorentino, M., Uva, A.M., Ferrise, F., Bordegoni, M.: Towards augmented reality manuals for industry 4.0. A methodology. Robot. Comput.-Integr. Manuf. **56**, 276–286 (2019)
16. Palmarini, R., Erkoyuncu, J.A., Roy, R., Torabmostaedi, H.: A systematic review of augmented reality applications in maintenance. Robot. Comput.-Integr. Manuf. **49**, 215–228 (2018)
17. Bun, P., Trojanowska, J., Ivanov, V., Pavlenko, I.: The use of virtual reality training application to increase the effectiveness of workshops in the field of lean manufacturing. In: Bruzzone, A.G., Ginters, E., Mendivil, E.G., et al. (eds.) 4th International Conference of the Virtual and Augmented Reality in Education, VARE 2018, pp. 65–71 (2018)
18. Ivanov, V., Pavlenko, I., Trojanowska, J., Zuban, Y., Samokhvalov, D., Bun, P.: Using the augmented reality for training engineering students. In: Bruzzone, A.G., Ginters, E., Mendivil, E.G., et al. (eds.) 4th International Conference of the Virtual and Augmented Reality in Education, VARE 2018, pp. 57–64 (2018)
19. Grzelczak, A., Kosacka, M., Werner-Lewandowska, K.: Employees competences for Industry 4.0 in Poland–preliminary research results. DEStech Trans. Eng. Technol. Res. (ICPR) 139–144 (2017)

Production Line Balancing in a Mixed-Model Production System: A Case Study

Michał Grześkowiak and Justyna Trojanowska(✉)

Poznan University of Technology, 3, Piotrowo Street, 60-965 Poznan, Poland
justyna.trojanowska@put.poznan.pl

Abstract. This paper looks at the balancing of a coating line in an automotive manufacturing company. A well-balanced production line is standardized one, with jobs performed in the same sequence, evenly distributed workload and a constant takt time. The project discussed in this paper consisted of the development of a balanced work standard in a mixed model production system, with additional measures taken to boost the manufacturing capacity. The key challenge in balancing a mixed-model production system are differences in time and labor consumption across operations. The focus should be put on the elimination of any wastes and errors observed during the balancing tests through continuous monitoring and root cause analysis. Optimal distribution of workload can be efficient only if the company develops solutions to eliminate and prevent issues, causing abnormalities. Owing to the measures implemented in the company under analysis, a constant takt time has been introduced, and the value-added in the process boosted.

Keywords: Production management · Manufacturing capacity · Takt time

1 Introduction

Nowadays, an increasing number of manufacturing companies produce customized goods that can be configured end-users in a way that best meets their requirements. Each order is different, made specifically to the customer's specification, while products vary not only in terms of design but also the time and labor required to make them.

The challenges faced when managing mixed-model production processes stem from little standardization of procedures. Aiming to improve the flow of materials in mixed-model production, it is essential to identify common elements of the process and standardize the maximum possible number of jobs, with many options within them.

Practice shows that each production process involves a variety of factors that impede the performance [1–5]. There are being put forward many different solutions and tools, improving the manufacturing management process [6–10]. However, some are very difficult to put into practice.

This paper addresses the question of how to balance a coating line in an automotive manufacturing company. A well-balanced, mixed-model production system is based on standardization of labor, consisting in the performance of certain jobs in the same sequence by staff during the same shift or various shifts, even workload distribution among staff working in the same area, and maintenance of a constant takt time. As part

V. Tonkonogyi et al. (Eds.): InterPartner 2020, LNME, pp. 24–32, 2021.
https://doi.org/10.1007/978-3-030-68014-5_3

of the project under analysis, a new, balanced work standard in a mixed-model production system has been developed, and additional measures have been initiated to boost manufacturing capacity.

2 Literature Review

An assembly line is a special production system, which consists of workstations lined up in the order concerning each other, where work packages are implemented on a product while it passes through these stations. Assembly lines are the places where several products and components are combined and processed [11].

Mixed-model assembly lines, most commonly used in practice, support simultaneous assembly of many similar product models, launched in any sequence. As the trend for current markets is to have a wider product range and variability, mixed-model assembly lines are preferred over the traditional single-model assembly lines [12].

According to the literature, the balancing of a production line is highly beneficial for mass production, where products have been manufactured for a longer period of time in the same production line [13]. However, many production environments do not meet these requirements.

In an automotive industry production line, balancing poses a significant challenge due to the trade-off between machine idle time and work-in-process accumulation between different machines [14].

Assembly line balancing is a process which is used for assigning tasks and work packages to workstations, while adapting to technology-based capacity constraints and priority relations, keeping the loss of time at a minimum, and increasing line efficiency [15].

The assembly line balancing problem arises when designing or redesigning an assembly line, and it consists of finding a feasible assignment of tasks to workstations in such a way that the assembly costs are minimized, the demand is met, and the constraints of the assembly process are satisfied. The assembly line balancing problem has been extensively researched, and comprehensive literature reviews addressing it include the works of Ghosh, Scholl, and Becker [16–19].

3 Research Methodology

In this study, a coating line has been balanced in line with the methodology developed for an automotive manufacturing company. The methodology is shown in Fig. 1.

The first step of the methodology consists of the calculation of the takt time, which should be maintained across the plant in order to ensure that a high-quality product is delivered on time. In step two, the manufacturing process is monitored – a process map is developed, or photographs are taken during a normal working day to identify the operations performed by staff and mudas (wastes) in the process. Next, technology routing is verified to identify operations performed according to the existing standards. Measurements of the duration of operations provide data necessary to balance the line. The timing method is recommended here, and time-keeping records should be

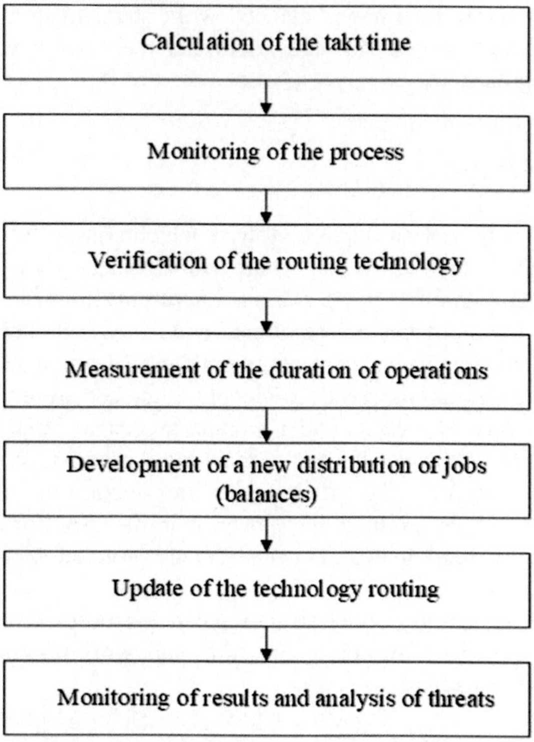

Fig. 1. The methodology of the study.

appended with information whether a particular operation adds value to the process or is a muda (waste). Next, a new distribution of jobs (balances) is developed for each workstation, followed by updating the technology routing and training of the staff. Finally, the line is monitored for the impact of the implemented changes and analysis of potential threats.

4 Results

4.1 The Production Process

In line with the methodology, the balancing of the coating line in the automotive manufacturing company under analysis started with the calculation of the takt time, which should be maintained across the plant to ensure that the customer receives a high-quality product on time. For this purpose, the 'effective working time' was determined with the formula (1).

$$\begin{aligned} &\textit{effective working time} = \textit{number of working days} \times \textit{duration of one shift}, \\ &\textit{including break} \times \textit{number of shifts} \end{aligned} \tag{1}$$

The calculation was made for the following variables:

- number of working days, excluding public holidays = 242 days,
- duration of one shift, including a break = 7.5 h,
- number of shifts = 2,

The effective working time obtained = 3630 h/year.

Assuming the number of customer orders at 1450 items, a takt time of TT = 2.5h was determined with formula (2), which means that every 2.5 h, one ready-made product should leave the production line.

$$takt\ time(TT) = \frac{effective\ working\ time}{number\ of\ orders} \tag{2}$$

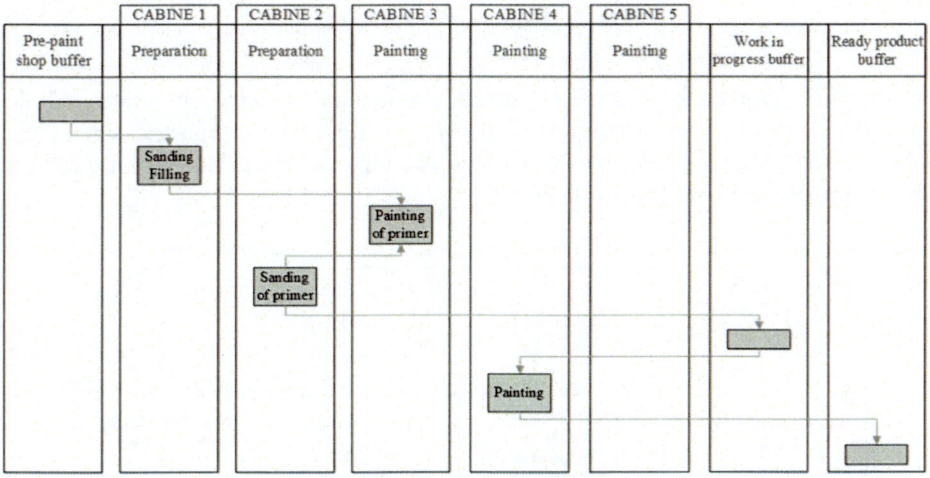

Fig. 2. Map of processes.

Figure 2 Shows a map of manufacturing processes in the company under analysis. A detailed flowchart of operations is shown in Fig. 3.

As shown in Fig. 3, the paint shop has two preparation booths and three spray booths. Sanding and filling operations are performed in the preparation zones. One spray booth is used for the application of primer. Two layers of primer are applied — a layer of reactive primer and a layer of primer. Colored coatings are sprayed in the two other booths, according to the design.

4.2 Balancing of the Coating Line

The process was photographed during a routine working day to identify any disruptions and wastes. Each operator was assigned a production support staff member, who

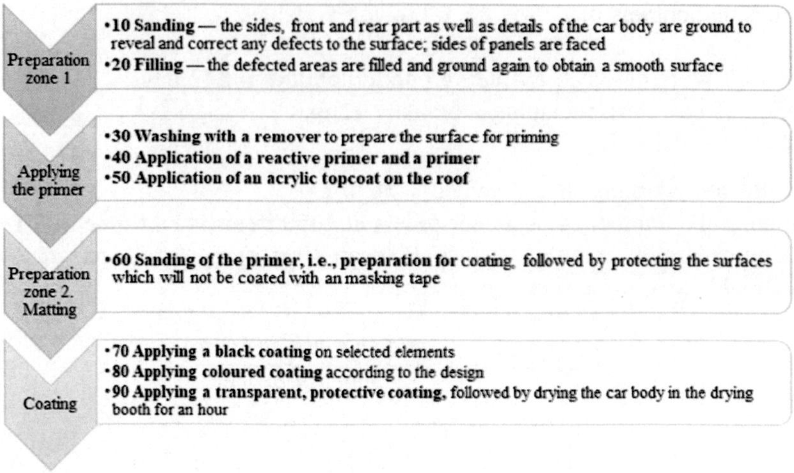

Fig. 3. Flowchart of operations.

monitored the process to draw up a spaghetti diagram and identify any wastes. All the operations performed were assigned to one of the following categories: VA (Value Adding), NNVA (Necessary, but Non-Value Adding), or NVA (Non-Value Adding). The results obtained are shown in Fig. 4.

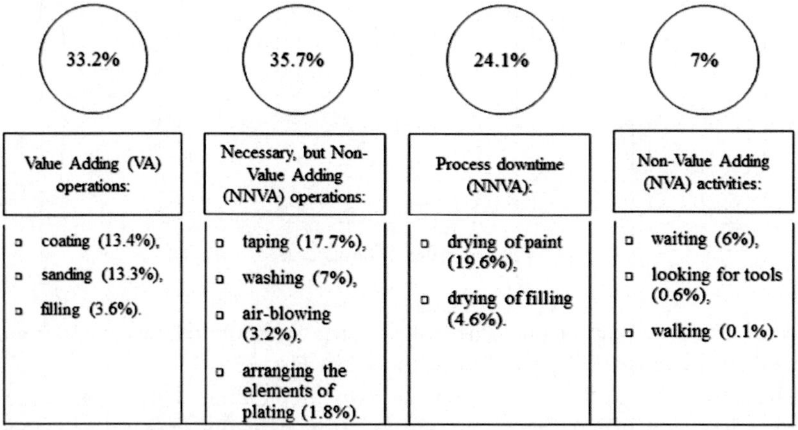

Fig. 4. Results of photographs of a working day.

It was found based on the results that value-adding (VA) operations accounted for 33.2% of the process, while 35.7% of the process was made up of necessary but non-value-adding (NNVA) operations. Aimed at preparing the car body for the application of primers and the main coating, they were an integral part of the process flow. Process downtime was another category of operations that did not add value to the process. Non-value adding (NVA) activities accounted for 7% of the process.

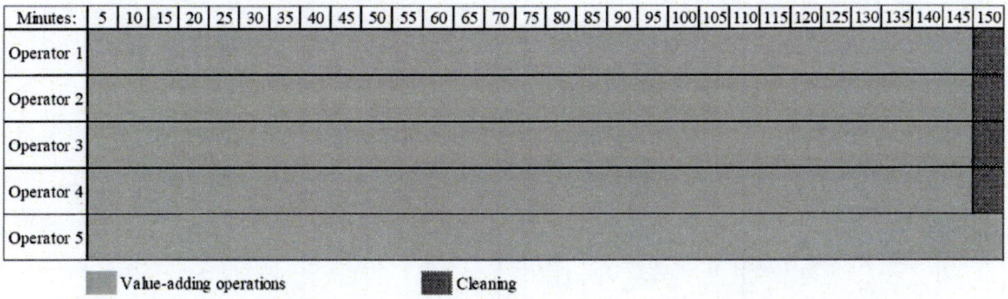

Fig. 5. Diagram of work in the booth where vehicles are prepared for coating.

In the next step, timing measurements were taken, and the results were shown on a timescale for analysis and assessment (see Fig. 5).

As part of the improvement process, a special table was put up at the entrance to each booth to note down balances, time spent by each vehicle in the booth, and the number of staff required to perform a given operation. For each order, leaders noted down any disruptions which occurred during the operation. The notes would be collected daily for analysis, and the disruptions—discussed at a weekly meeting of the project team.

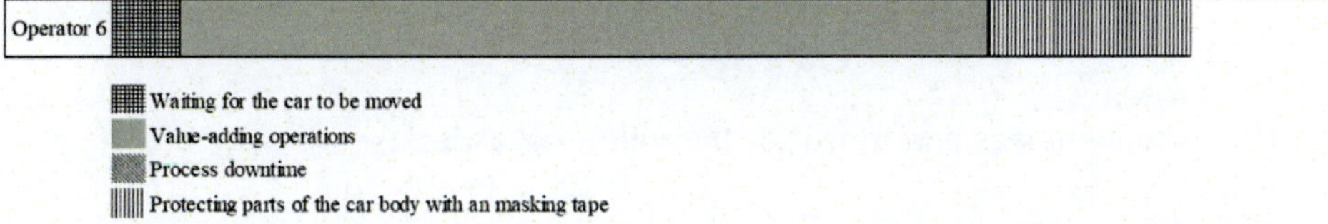

Fig. 6. Diagram of work in the booth where vehicles are prepared for coating after changes.

An additional measure, taken as part of the balancing of the preparation zones, was the putting up of a shadow box for grinders. A diagram of the organization of work in the preparation booth after the implementation of changes is shown in Fig. 6.

The same balances were implemented in other booths. Having tested the balances, the authors once again took photographs during a normal working day (see Fig. 7).

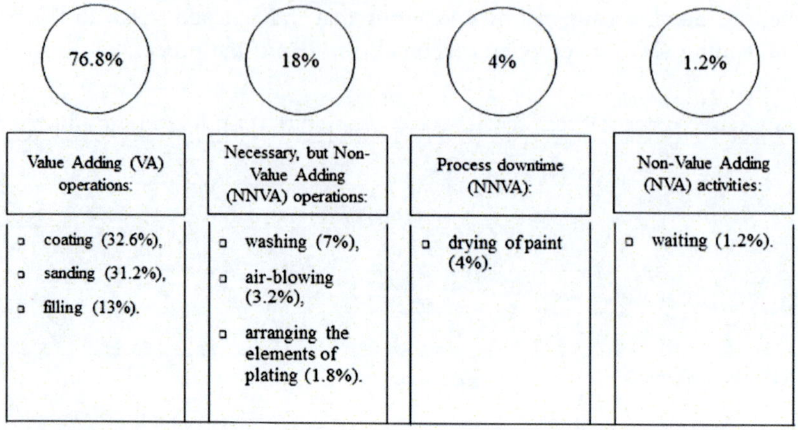

Fig. 7. Results of photographs of a working day, after balancing.

A comparison of Figs. 4 and 7 shows that the share of value-adding operations soared from 33.2% to 76.8%, mainly owing to better utilization of spare time during the process downtime.

4.3 Results

The assessment of the balanced coating line was based on the calculated line efficiency and smoothness coefficient.

Line efficiency, determined with the formula (3), represents the proportion of the meantime of utilization of the balanced line to the takt time multiplied by the number of staff operating analysis, multiplied by 100%. The closer the result is to 100%, the better the outcome of balancing (in theory).

$$Mean\ time\ of\ utilisation = \frac{TT}{number\ of\ staff} \tag{3}$$

$$LE = \frac{4625}{150 \times 32} \times 100\% = 96,35\%$$

The calculation was performed for the following variables:

- 4625 [min] – the value taken from an example routing for the selected product,
- 150 [min] – takt time,
- 32 – number of staff.

The smoothness coefficient (formula 4) represents downtime. A zero-value smoothness coefficient represents no downtime.

$$SI = \sqrt{\sum\nolimits_{i=1}^{n} (4800 - 4625)^2} = 175[\mathrm{min}] \quad (4)$$

The calculation was performed for the following variables:

- 4800 [min] – the product of takt time and number of staff,
- 4625 [min] – mean line utilization time.

5 Conclusions

As a result of the balancing, even distribution of workload among the staff was obtained, and a constant takt time of 2.5 h was introduced.

Improvements to the process, such as appointment of a booth leader who gives early notice of readiness of the car to be moved to the next booth, and reduction of process stoppages, has resulted in shortening the duration of the entire operation by as much as ca. 15%. The photographs taken during a working day before and after the balancing show that the share of value-adding operations rose from 33.2% to as much as 76.8%.

In order to boost the production capacity in the balanced area, the operation of coating in black has been eliminated, what has brought savings of EUR 40 per unit (in the costs of materials and labour only) and reduced the total time of handling a single item by 154 min throughout the process.

The key challenge in balancing a mixed model production system is the differences in the time and labor consumption in particular operations. Therefore, the focus should be put on the elimination of any wastes and errors observed during the balancing tests through continuous monitoring and root cause analysis. Optimal distribution of workload can be efficient only if the company develops solutions to eliminate and prevent issues, causing abnormalities.

References

1. Liaposhchenko, O., Khukhryanskiy, O., Moiseev, V., Ochowiak, M., Manoilo, E.: Intensification of foam layered apparatus by foam stabilization. J. Eng. Sci. **5**(2), F13–F18 (2018). https://doi.org/10.21272/jes.2018.5(2).f3
2. Ivanov, V., Dehtiarov, I., Pavlenko, I., Kosov, I., Kosov, M.: Technology for complex parts machining in multiproduct manufacturing. Manag. Prod. Eng. Rev. **10**(2), 25–36 (2019). https://doi.org/10.24425/mper.2019.129566
3. Antosz, K.: Maintenance – identification and analysis of the competency gap. Eksploatacja i Niezawodnosc Maintenance Reliab. **20**(3), 484–494 (2018). https://doi.org/10.17531/ein.2018.3.19.
4. Sobaszek, Ł., Gola, A., Kozłowski, E.: Application of survival function in robust scheduling of production jobs. In: Ganzha, M., Maciaszek, M., Paprzycki, M. (eds.) Proceedings of the 2017 Federated Conference on Computer Science and Information Systems (FEDCSIS), pp. 575–578. IEEE, New York (2017)

5. Shramenko, N., Muzylyov, D.: Forecasting of overloading volumes in transport systems based on the fuzzy-neural model. In: Ivanov, V., et al. (eds) Advances in Design, Simulation and Manufacturing II. DSMIE 2019. Lecture Notes in Mechanical Engineering, pp. 311–320. Springer, Cham (2020).
6. Varela, M.L.R., Putnik, G.D., Cruz-Cunha, M.M.: Web-based technologies integration for distributed manufacturing scheduling in a virtual enterprise. Int. J. Web Portals **4**(2), 19–34 (2012). https://doi.org/10.4018/jwp.2012040102
7. Varela, M.L.R., Ribeiro, R.A.: Distributed manufacturing scheduling based on a dynamic multi-criteria decision model. In: Recent Developments and New Directions in Soft Computing, Studies in Fuzziness and Soft Computing, vol. 317, pp. 81–93. Springer, Switzerland (2014)
8. Arrais-Castro, A., Varela, M.L.R., Putnik, G., Ribeiro, R., Dargam, F.: Collaborative negotiation platform using a dynamic multi-criteria decision model. Int. J. Decis. Supp. Syst. Technol. **7**(1), 1–14 (2015). https://doi.org/10.4018/ijdsst.2015010101
9. Ivanov, V., Vashchenko, S., Rong, Y.: Information support of the computer-aided fixture design system. In: Paper presented at the CEUR Workshop Proceedings, vol. 1614, pp. 73–86 (2016)
10. Ivanov, V., Dehtiarov, I., Pavlenko, I., Liaposhchenko, O., Zaloga, V.: Parametric optimization of fixtures for multiaxis machining of parts. In: Hamrol A., Kujawińska A., Barraza M. (eds) Advances in Manufacturing II, MANUFACTURING 2019, Lecture Notes in Mechanical Engineering, pp. 335–347 (2019). https://doi.org/10.1007/978-3-030-18789-7_28.
11. Asar, K., Andrew, J.D.: A knowledge based design methodology for manufaturing assembly lines. J. Comput. Ind. Eng. **41**(4), 441–467 (2001)
12. Simaria, A.S., Xambre, A.R., Filipe, N.A., Vilarinho, P.M.: A decision support system for assembly and production line balancing. Handb. Bus. Inf. Syst. (2010). https://doi.org/10.1142/9789812836069_0016
13. Sivasankaran, P., Shahabudeen, P.: Literature review of assembly line balancing problems. Int. J. Adv. Manuf. Technol. **73**(9–12), 1665–1694 (2014). https://doi.org/10.1007/s00170-014-5944-y
14. Naik, G.R., Raikar, V.A., Naik, P.G.: Fuzzy non-linear optimization model for production line balancing of Jadhav industries using genetic algorithm. Int. J. Adv. Res. Comput. Sci. **8** (8), 315–326 (2017)
15. Aysun, T., Yalcin, Y., Mahmut, K.: Heuristic production line balancing problem solution with MATLAB software programming. Int. J. Clothing Sci. Technol. **28**(6), 750–779 (2016). https://doi.org/10.1108/IJCST-01-2016-0002
16. Ghosh, S., Gagnon, R.J.: A comprehensive literature review and analysis of the design, balancing and scheduling of assembly systems. Int. J. Prod. Res. **27**, 637–670 (1989)
17. Scholl, A.: Balancing and Sequencing of Assembly Lines. PhysicaVerlag, Heidelberg (1999)
18. Becker, C., Scholl, A.: A survey on problems and methods in generalized assembly line balancing. Eur. J. Oper. Res. **168**, 694–715 (2006)
19. Scholl, A., Becker, C.: State-of-the-art exact and heuristic solution procedures for simple assembly line balancing. Eur. J. Oper. Res. **168**, 666–693 (2006)

Ecological Activities of Manufacturing Companies in the Use and Recycling of Products

Ewa Dostatni, Jacek Diakun(✉), Jolanta Jurga, and Łukasz Kowalski

Poznań University of Technology, 5, Maria Skłodowska-Curie Square, 60-965 Poznań, Poland
jacek.diakun@put.poznan.pl

Abstract. In the paper, the research results on ecological activities in manufacturing companies are described. The research was conducted in selected small and medium (SME) and large enterprises in Wielkopolska (Greater Poland) Region (Poland). The main method used in the study was a questionnaire survey. The questionnaire inquired the ecological solutions in selected areas of enterprises' activities. The questionnaire was divided into groups according to product lifecycle. Thus comparison of ecological maturity in all stages was possible. The questionnaires were focused on the level of implemented ecological activities at the use and recycling phases of the product lifecycle about company size. The article discusses the negative impact of companies' products on the environment, arguments for eco-design, considering companies as being "green", handling of products withdrawn from the market, methods of segregation of used products and ways of disassembly of used products. The results showed differences between ecological practices in Polish small-and-medium and large enterprises and revealed appearing tendencies in enterprises' activities.

Keywords: Questionnaire survey · Small and medium enterprise · Large enterprise · Product lifecycle

1 Introduction

Environmental protection has been widely discussed in scientific papers. However, more importantly, it is being put into practice in many manufacturing companies. Ignored for many years, negative environmental footprints are finally being reduced. Numerous preventive measures minimizing the environmental impact of economic development have been developed. This paper presents an analysis of environmental solutions implemented in select manufacturing companies. The analysis was conducted in the Greater Poland Region, Poland, which has one of the highest rates of gross domestic product (GDP) per capita and is Poland's second-largest province. The province is economically varied, with agriculture and manufacturing as the leading sectors. About 30% of its population is employed in the manufacturing industry. Almost 97% of businesses are small and medium-sized enterprises [1].

V. Tonkonogyi et al. (Eds.): InterPartner 2020, LNME, pp. 33–41, 2021.
https://doi.org/10.1007/978-3-030-68014-5_4

The research aim of this paper is to analyze the environmental activities and solutions implemented in selected manufacturing companies in Wielkopolska (Greater Poland) region in Poland. The companies being surveyed operate in various industries and vary in terms of manufacturing methods, the complexity of processes, scale of activity, and organization of work. The analyzed companies were from various industry sectors and of various sizes, divided into two groups: small and medium-sized enterprises (SMEs), and large-sized enterprises (LEs). Selected aspects of the environmental activity of the two groups of enterprises, with a focus on the differences, were analyzed. The study was based on a questionnaire survey. The success of the study depended on the collaboration of the respondents and their willingness to provide information to the authors that would help them understand how the enterprises function. The results of the survey were analyzed. The results were also broken down by the size of enterprises under analysis and used for a comparison of pro-environmental behavior in enterprises of various sizes.

2 Literature Review

Up to date, little analysis has been carried out on the environmental awareness of businesses. Many entrepreneurs underestimate the environmental impact of their business, especially those operating in the small business sector. Running a small business, they believe they leave little or no environmental footprint. Statistical data shows that SMEs form a significant part of the world's economies, including Poland's economy. Environmental awareness manifests itself in taking responsibility for one's actions in the context of environmental protection and preservation of natural resources, and active participation in environmental protection. An eco-minded attitude is built through processing the knowledge gained on environmental protection and converting it into appropriate daily habits [2]. Environmental awareness is the responsibility not only of individuals but also businesses. The growth of social-environmental awareness and environmental organizations in the 1980s led to the origin of corporate social responsibility. The European Commission has defined corporate social responsibility as the responsibility of enterprises for their impact on society [3]. Corporate social responsibility includes, without limitation, purposeful activity undertaken by enterprises to reduce the negative impact on the environment [4]. The results of studies presented in the Innovation and CRS at SMEs report (Report, 2012), commissioned by the Innovation Development Agency, show that many businesses fail to realize what beneficial impact their activity could have on the environment if it were conducted on a large scale. Employees have little knowledge of environmental protection. They do not know the tools to incorporate the good environmental practice into their business [2]. Also, regional circumstances in environmental regulations lead to differences in the activity of companies in the implementation of environmental issues [5].

Many studies have been conducted to examine environmental awareness in enterprises. Van Hemel and Cramer [6] analyzed motivation and barriers in the implementation of ecodesign in 77 SMEs in the Netherlands. Cote´ et al. [7] examined eco-efficiency in 25 SMEs in Canada. Erkko et al. [8] looked at the use of eco-efficiency strategies in 40 British businesses based on the Eco-Management and

Audit Scheme (EMAS) reports. Netregs surveyed 5,554 SMEs in Scotland by phone to examine their eco-minded practices [9]. Fernandez-Vine et al. [10] studied eco-efficiency in SMEs in Venezuela.

The growing demand for pro-environmental solutions in companies is the driving force behind the development of pro-environmental business solutions [11]. Enterprises face increasing requirements from the modern economy concerning environmental protection. Seeking to meet the requirements, companies implement solutions that minimize their environmental impact. Pro-environmental behavior has captured a great amount of attention in recent years. In order to build their image and competitive edge, companies follow new trends [12–15]. Pro-environmental solutions are often implemented in companies as eco-innovations at the strategic level of management. It is enforced by increasing pressure on minimizing the impact of industry on the environment [16, 17]. Properly implemented environmental solutions should cover all company departments. In general, changes implemented in manufacturing have the most significance [18]. However, tangible results can be obtained only through action covering all areas of operational activity. Environmental awareness must be raised among all employees to achieve the objective of minimizing negative environmental impact. Only a thorough understanding of the actions taken at all organizational levels will translate into real financial profit and improvement of the overall company image.

Pro-environmental practice in business should be analyzed and improved across the product development process, as only then it brings the desired results [19]. Approximately 80% of a product's sustainability performance is defined during the early stages of its development [20]. In order to include environmental factors in the design process, it is necessary to identify the product-related environmental aspects and incorporate them into the design process at the early stages of product development [21]. This approach is referred to as ecodesign, design for the environment, or design compliant with the principles of sustainable development [22–27].

3 Research Methodology

The study was based on a questionnaire survey. At stage one, the study problem was specified in detail. One of the important decisions concerned with the subject area. The questionnaire aimed to collect data on environmental solutions applied in selected manufacturing companies. The implementation of environmental solutions gives the best results when it covers the entire product life cycle. Therefore, the questionnaire was divided into areas [28], with each area covering a selected stage of the product life cycle.

The questionnaire consists of an introduction and a set of questions concerning the enterprise (company name, the industry of operation, a type of company, market of operation, type of production).

The main body of the questionnaire contains questions grouped by the stages of the product life cycle. The "Product use" set of questions is aimed to collect information on the environmental performance of the manufactured products and the company's approach to eco-design (negative impact on the environment, arguments for eco-design, the competitiveness of products). The "Product recycling" group contains questions regarding handling products after the end of the life cycle, i.e., disassembly or segregation of used products.

A total of 16 open, closed, semi-closed, yes/no, and filtering questions were used, arranged in tables, including 9 questions on the pro-ecological activities of companies and 7 general questions on company characteristics. The total number of respondents was 29, where 13 of them were large enterprises (LE), and 16 of them were small-and-medium (SMEs) enterprises.

The surveys were conducted in person in a company, by email and by phone. Direct surveys provided the best results – 100% of answers were obtained during the first contact, and they contained the least inconsistent data. Regarding inconsistent data, the company, if possible, was asked to clarify answers in additional contact. Moreover, direct contact with the respondent was an opportunity to obtain a great amount of additional information that provides insight into the company's operation and its approach to environmental issues.

Finally, sixteen companies classified as SMEs, and thirteen as LEs, were queried. The SMEs represented the following industries: clothing, building, household appliances, heating, medical, cooling, tire manufacturing, rubber and rubber-and-metal products, plastic packaging, injection molds, corrugated packages, rubber products, and aluminum profile systems. The LEs represented such industries as household appliances, automobile, agricultural machinery, transport equipment, medical equipment, flexible packaging, and window fittings.

4 Results

What concerns product use, a majority of the SMEs and LEs state that the most common negative impact of their products is energy consumption. This opinion is especially popular among LEs. Only a small number of LEs indicate that their products generate greenhouse gases, radiation, or acidification of air. No SMEs or LEs indicated the generation of smog by their products. Toxicity to aquatic life was indicated by one SME (Fig. 1).

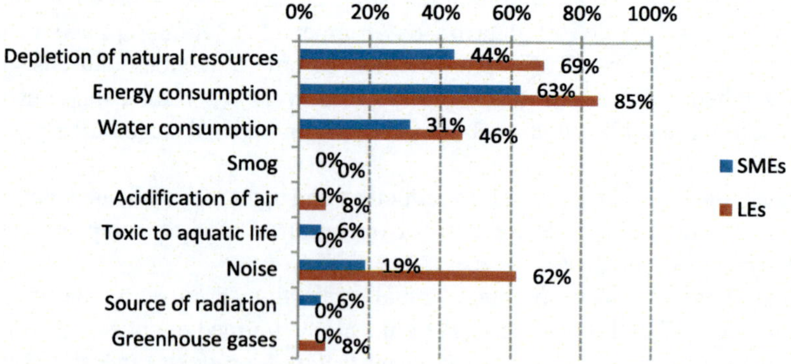

Fig. 1. The environmentally negative impact of products.

The most common arguments for eco-design in LEs include the improvement of products and technologies and the reduction of material and energy consumption by-products at all stages of the life cycle. The latter, together with competitiveness, is the most common for SMEs (Fig. 2).

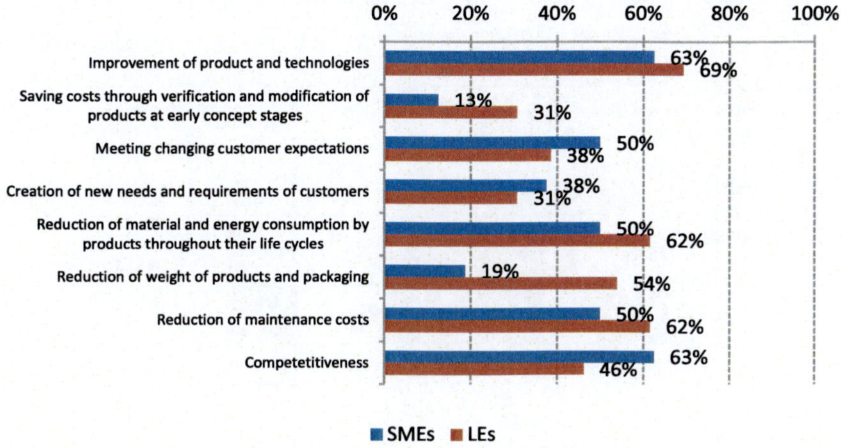

Fig. 2. Arguments for ecodesign.

Most of the reviewed companies consider themselves "green" ones (Fig. 3). Many factors can be a motivation to become a "green company". A highly promising result was obtained in the group of LEs – both environmental awareness and compliance with legal standards motivate them equally. Legal standards are a slightly stronger motivator for SMEs than environmental awareness. Surprisingly, the possibility of being granted EU subsidies is not the prevailing reason for maintaining a sound environmental policy (Fig. 4).

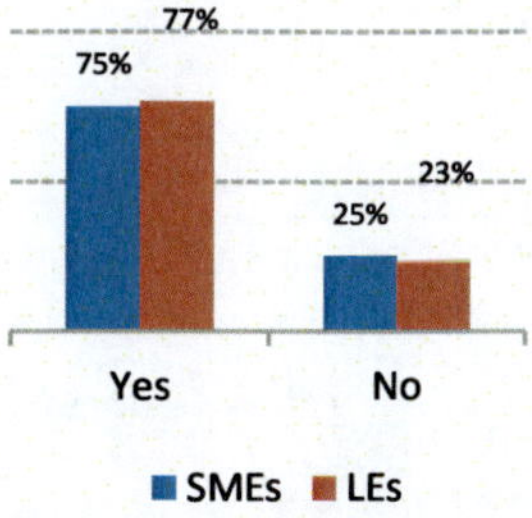

Fig. 3. Having the "green company" image.

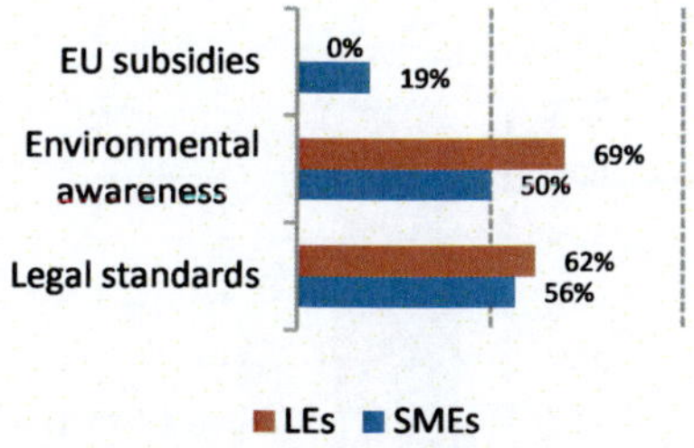

Fig. 4. Motivation to become a "green company".

What concerns product recycling, from the analysis, it seems that at the end of the product lifecycle 100% of SMEs and LEs recycle or recover products.

At the end of the life cycle, products of both SMEs and LEs are recycled or processed for re-use. SMEs and LEs have different policies for handling products withdrawn from the market. Most LEs have a product handling scheme in place which provides for the manufacturing of recycled material. This practice is the least common among SMEs, which typically re-use components of used products. Re-use of products is the least common practice among both LEs and SMEs (Fig. 5).

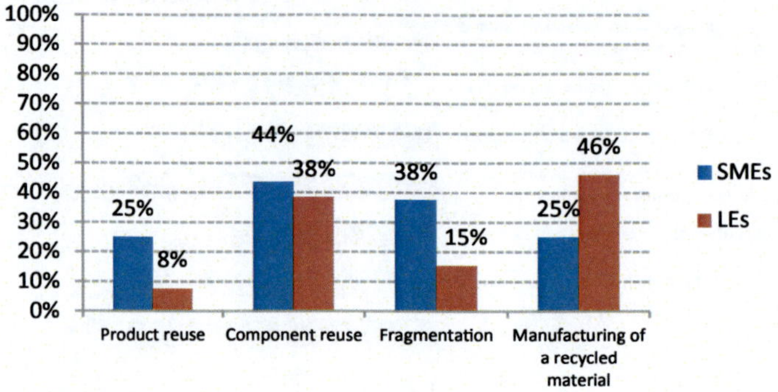

Fig. 5. Handling products withdrawn from the market.

The type of segregation of used products depends on the character of products. Most SMEs and LEs segregate used products by geometric dimensions and shape. The second most common answer was "Other". The respondents defined the segregation as "manual disassembly" or provided information on commissioning external companies to perform disassembly. Magnetic segregation or segregation by electrical conductivity is rare and carried out only by LEs (Fig. 6).

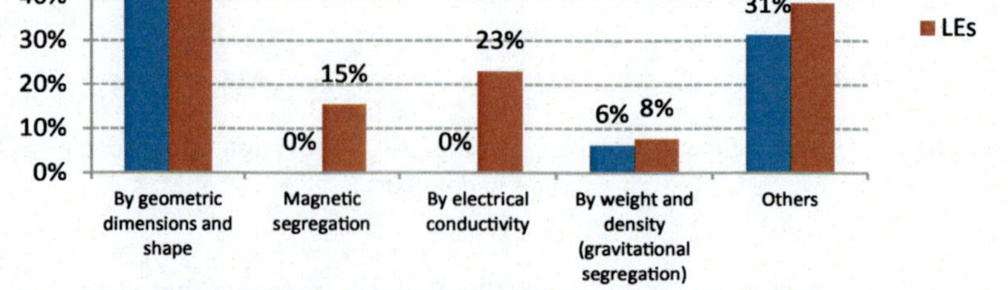

Fig. 6. Segregation of used products.

The most common form of disassembly is manual, with the LEs leading in this category. Mechanical disassembly is less frequent, but also more common at LEs. A small number of LEs use electronic or automated disassembly, but the methods are non-existent at the reviewed SMEs (Fig. 7).

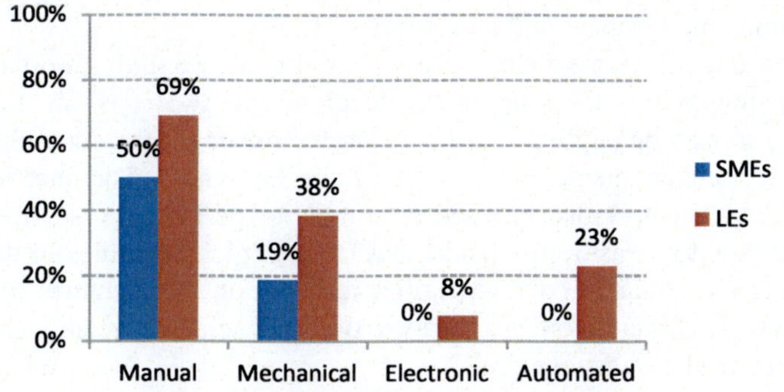

Fig. 7. Disassembly of used products.

5 Conclusions

The analysis presented above reveals many similarities in the environmental policies pursued by enterprises. The similarities at the stage of product design concern criteria of great importance, such as a selection of materials, consideration of the compatibility of materials, the type of joints, and the conducting of preliminary analyses.

At the manufacturing stage, the similarities in the environmental activity of enterprises cover waste management, segregation of products, use of eco-labeling (most of the reviewed enterprises do not use eco-labeling), compliance with applicable legal standards, and the energy-saving packages held. In the research conducted in 2004–2008 concerning eco-awareness of SMEs in Poland [30], it was proved that awareness of the negative influence of energy consumption and waste amount on the environment is increasing. The results of this research confirmed a high awareness of a negative influence of energy consumption and waste amount on the environment.

The implementation of environmental solutions at the stages of product manufacturing and use also varies across the two sizes of enterprises. At the recycling stage, differences were noticeable in the manner of handling products withdrawn from the market. SMEs tend to re-use product components, whereas LEs tend to produce recycled material.

The questions regarding recycling provided information on awareness concerning the handling of products after they have reached their lifetime. Providing an answer was often difficult since external companies are commissioned to handle the process, and their practices may not be known to the respondents. Many LEs stated that they received reports from external companies commissioned to handle products at the end of their lifetime, which means a higher level of awareness can be expected from them than from the reviewed SMEs.

The study leads to the conclusion that SMEs opt to implement relatively simple solutions that do not require specialist knowledge or substantial capital expenditures. They often fail to understand the concept of the product life cycle or accept their perspectives. SMEs have limited interest in the environmental activity and often do an absolute minimum in this respect. LEs, considering their reputation and position, are more willing to undertake environmental activity, which is also often related to their level of environmental impact and awareness of threats.

Authorities and policymakers can use the results of this study in order to prepare strategies and regulations to improve the level of eco-awareness in manufacturing companies. It can also be applied to industry professionals for the determination of the current state of their companies in the area of eco-awareness. The questionnaire prepared for this study could also be used as a tool for periodical assessment within a company, and thus to measure the tendencies of ecological activities in the company.

The authors are going to conduct further research on the activities of other companies based on the survey in order to analyze the tendencies and relationships between factors included in the questionnaire.

Acknowledgements. The work presented in the paper has been financed under 0613/SBAD/4677 grant for Poznan University of Technology.

References

1. General characteristics of the Wielkopolska province. https://wzk.poznan.uw.gov.pl/sites/default/files/zalaczniki/ogolna_charakterystyka_cz2.pdf, Accessed 01 May 2020
2. Domańska, A.: Environmental awareness as part of corporate social responsibility of the pharmaceutical and medical industry. Zeszyty Naukowe Politechniki Częstochowskiej, Zarządzanie **10**, 138–145 (2013)
3. Communication From The Commission To The European Parliament, The Council, The European Economic And Social Committee And The Committee Of The Regions A Renewed EU Strategy 2011–14 For Corporate Social Responsibility, Brussels (2011)
4. Innovation and CRS at SMEs Innowacjei CSR w MSP. https://odpowiedzialnybiznes.pl, Accessed 01 June 2015
5. Kong, D., Xiandong, Y., Chen, L., Wei, Y.: Business strategy and firm efforts on environmental protection: evidence from China. Bus. Strategy Environ. **28**(6), 1–20 (2019)
6. van Hemel, C., Cramer, J.: Barriers and stimuli for ecodesign in SMEs. J. Cleaner Prod. **10**, 439–453 (2002)
7. Coté, R., Booth, A., Louis, B.: Eco-efficiency and SMEs in Nova Scotia, Canada. J. Cleaner Prod. **14**, 542–550 (2006)
8. Erkko, S., Melanen, M., Mickwitz, P.: Eco-efficiency in the Finnish EMAS reports – a buzz word? J. Cleaner Prod. **13**, 799–813 (2005)
9. A review of changing environmental attitudes and behaviours among small and medium-sized businesses in England, https://www.netregs.gov.uk/netregs/1169119/?lang=_e, Accessed 01 June 2015
10. Fernández-Viñé, M.B., Gómez-Navarro, T., Capuz-Rizo, S.F.: Eco-efficiency in the SMEs of Venezuela: current status and future perspectives. J. Cleaner Prod. **18**, 736–746 (2010)
11. Baumann, H., Boons, F., Bragd, A.: Mapping the green product development field: engineering, policy and business perspectives. J. Cleaner Prod. **10**, 409–425 (2002)

12. Hart, S.L., Dowell, G.: A natural-resource-based view of the firm: fifteen years after. J. Manag. **37**, 1464–1479 (2011)
13. Kara, S., Ibbotson, S., Kayis, B.: Sustainable product development in practice: an international survey. J. Manuf. Technol. Manag. **25**, 848–872 (2014)
14. Bryke, M., Starzyńska, B.: Human lean green conception as the instrument of sustainability of organizational development oriented towards the increase of its effectiveness. Res. Pap. Wrocław Univ. Econ. **77**, 119–136 (2015)
15. Jasiulewicz-Kaczmarek, M.: Integrating lean and green paradigms in maintenance management. In: 19th IFAC World Congress, pp. 4471–4476. Cape Town (2014)
16. Del Río, P., Peñasco, C., Romero-Jordán, D.: Distinctive features of environmental innovators: an econometric analysis. Bus. Strategy Environ. **24**(6), 361–438 (2015)
17. Demirel, P., Kesidou, E.: Sustainabilityoriented capabilities for eco-innovation: meeting the regulatory, technology, and market demands. Bus. Strategy Environ. **28**(5), 847–857 (2019)
18. Gangala, C., Modi, M., Manupati, V.K., Varela, M.L.R., Machado, J., Trojanowska, J.: Cycle time reduction in deck roller assembly production unit with value stream mapping analysis. In: Rocha, Á., Correia, A., Adeli, H., Reis, L., Costanzo, S. (eds.) Recent Advances in Information Systems and Technologies. WorldCIST 2017. Advances in Intelligent Systems and Computing, vol. 571, pp. 509–518. Springer, Cham (2017).
19. Rodrigues, V.P., Pigosso, D.C.A., McAloone, T.C.: Measuring the implementation of ecodesign management practices: a review and consolidation of process-oriented performance indicators. J. Cleaner Prod. **156**, 293–309 (2017)
20. McAloone, T., Bey, N.: Environmental Improvement through Product Development: a Guide. Danish Environmental Protection Agency, Copenhagen (2009)
21. Grajewski, D., Diakun, J., Wichniarek, R., Dostatni, E., Buń, P., Górski, F., Karwasz, A.: Improving the skills and knowledge of future designers in the field of ecodesign using virtual reality technologies. Procedia Comput. Sci. **75**, 348–358 (2015)
22. Dostatni, E., Diakun, J., Grajewski, D., Wichniarek, R., Karwasz, A.: Multi-agent system to support decision-making process in design for recycling. Soft. Comput. **20**, 4347–4361 (2016)
23. Kujawińska, A., Vogt, K., Wachowiak, F.: Ergonomics as significant factor of sustainable production. In: Golińska, P., Kawa, A. (eds.) Technology Management for Sustainable Production and Logistics, Book Series: EcoProduction, 2015, pp. 193–203. Springer, Heidelberg (2015)
24. Pigosso, D.C.A., Rozenfeld, H., McAloone, T.C.: Ecodesign maturity model: a management framework to support ecodesign implementation into manufacturing companies. J. Cleaner Prod. **59**, 160–173 (2013)
25. Pigosso, D.C., McAloone, T.C., Rozenfeld, H.: Characterization of the state-of the- art and identification of main trends for ecodesign tools and methods: classifying three decades of research and implementation. J. Indian Inst. Sci. **95**, 405–427 (2015)
26. Diakun, J., Dostatni, E., Grajewski, D., Wichniarek, R., Karwasz, A., Brzezinski, W., Ciechanowicz, B.:Modelling and recycling-oriented assessment of household appliances. In: Burduk A., Mazurkiewicz D. (eds.) Intelligent Systems in Production Engineering and Maintenance – ISPEM 2017. ISPEM 2017. Advances in Intelligent Systems and Computing, vol. 637, pp. 306–315. Springer, Cham (2017).
27. Dostatni, E., Rojek, I., Hamrol, A.: The use of machine learning method in concurrent ecodesign of products and technological processes. In: Hamrol, A, Ciszak, O., Legutko, S., Jurczyk, M. (eds.) Advances in Manufacturing. Lecture Notes in Mechanical Engineering, Issue19, pp. 321–330. Springer, Cham (2018).
28. Selech, J., Joachimiak-Lechman, K., Klos, Z., Kulczycka, J., Kurczewski, P.: Life cycle thinking in small and medium enterprises: the results of research on the implementation of life cycle tools in Polish SME-s Part 3: LCC-related aspects. Int. J. Life Cycle Assess. **19**, 1119–1128 (2014)

Application of the RFID Technology at a Production and Assembly Line

Anna Karwasz(✉) and Łukasz Paczeŝny

Poznan University of Technology, 3, Piotrowo Street, 61-138 Poznan, Poland
anna.karwasz@put.poznan.pl

Abstract. This paper looks at modern wireless communication technologies commonly used in the manufacturing industry. The Radio Frequency Identification (RFID) technology, applied in warehousing and transport logistics as well as in product manufacturing and assembly of products, is discussed in more detail. The authors consider standards that must be complied with when using the RFID technology. The study has been designed to build and implement an electronic system at the Smart Factory production process improvement laboratory for the improvement of production systems. The task of the completed system is to calculate how long a given machine or device works, how often it is started, and with what frequency. By calculating the uptime and frequency of machine start-ups, the implemented RFID technology-based system provides data based on which optimal uptime, as well as inspection and maintenance intervals, can be established. It allows you to react when the machine is running for too long without inspection or maintenance. What can save a company from losses when a machine breaks down.

Keywords: Chips · Smart Factory · 4.0 Industry

1 Introduction

The manufacturing output is growing year on year in all industries. As a consequence, manufacturers, sales staff, and freight forwarders face the challenge of processing huge amounts of data in logistics, manufacturing, and warehousing processes [1]. Many digital technologies available in the 21st-century economy are useful in handling this task. Concepts like big data and related analytical tools, social media, computing clouds, and mobile technologies are growing in importance.

The industry is striving to maximize automation and place less reliance on the human factor for value-added data collection and design. However, automated applications are not able to entirely replace the work of human hands [2]. We are witnessing the development of new professions with new areas of competence. The focus in the engineering industry is on a systemic approach, new technology applications, and improvement of the existing solutions in line with the Lean Manufacturing methodologies [3]. The flow of information and the flow of goods occur simultaneously, and any disruptions generate additional costs and cause delays. The supporting technologies can mitigate challenges resulting from the misalignment.

V. Tonkonogyi et al. (Eds.): InterPartner 2020, LNME, pp. 42–51, 2021.
https://doi.org/10.1007/978-3-030-68014-5_5

The ever-increasing importance of logistic processes drives the development of automatic identification technologies, such as optical character recognition (OCR), barcode systems, and radio frequency identification (RFID), which help minimize the costs of logistics and improve the flow of information. They also eliminate human errors, intentional or unintentional mistakes resulting from work fatigue. Implementing new technologies in manufacturing companies also means building a competitive advantage [4–9].

The research aimed to show the need to adapt the production line to changing industrial conditions and reducing production costs by preventing machine failures in time.

2 Literature Review

RFID technology is widely used in many areas of life. It stands behind many of our daily activities without us realizing it [10]. Most commonly used in retail and logistics, it has many other applications. The barcode technology speeds up the work of shop personnel, as prices and quantities of products can be read directly from barcodes [11]. A contactless card system can open a door without the need to enter a code, as well as keep a record of working hours and control whether an employee is at their workplace [12]. The technology also has many commercial applications, such as e.g., automated borrowing and returning of books at the university library and protection of books against theft, or record-keeping and archiving of files and documents as well as facilitating their localization, search of contents and traceability. In the 1990s, RFID technology began to be applied in sports. Today, it facilitates the measurement of lap times for particular runners in a large group of athletes [9, 12]. One of the earliest applications of the RFID technology was the identification and record-keeping of livestock, with microchips embedded in ear tags [13]. An automated motorway toll collection system was introduced in Poland in 2011. Other European countries have also implemented similar systems. The solution in Poland uses transponders mounted in cars and a system of overhead gantries which register the passage of cars. Configured in the 5.8 GHz band, the devices are located on almost 4 thousand kilometers of national motorways [14]. Modern cars use RFID technology for keyless entry and start. The door unlocks once the car detects the key in the vicinity.

The authors undertook to build and implement a system that would use the RFID technology to detect tools moving along a manufacturing or assembly line and keep a tally of how many times each tool passed the line. This data is required for scheduled maintenance – having completed a certain number of cycles; the tools need to undergo an inspection to prevent failure or seizure of the machine.

Built to the author's original design, the system consists of the following components:

- an Arduino UNO minicontroller (a replacement),
- an RFID MF RC522 13.56 MHz module,
- RFID wireless chips,
- a casing,

- signal cables, and
- a power supply.

The underlying assumption was to obtain a ready-to-use version of the system at a minimum cost. Therefore, a replacement minicontroler, as well as a casing printed in 3D, were used [15].

The Smart Factory laboratory for the improvement of production systems at the Poznan University of Technology (Fig. 1) has been designed to demonstrate the operation of an assembly or production line and give an insight into the product flow management [16]. It helps young engineers to understand the ever-more popular concepts of Lean Management and 4.0 Industry.

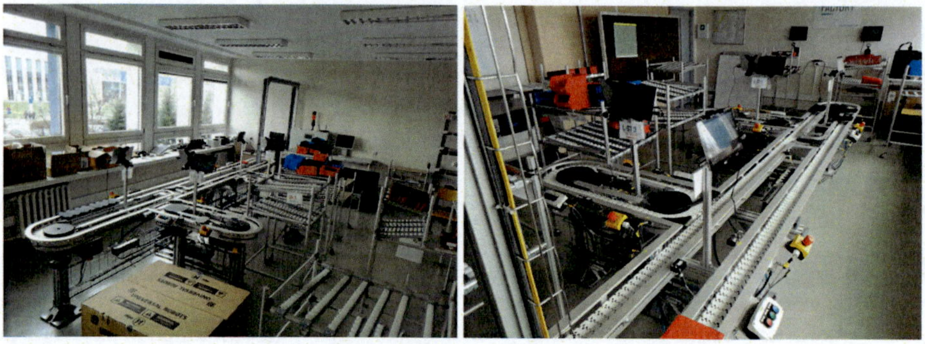

Fig. 1. The Smart Factory laboratory [16].

The production system model implemented in the laboratory consists of:

- an automated assembly and production line,
- assembly stations,
- additive manufacturing stations, and
- a warehouse for the storage of raw materials and end products.

The automated production line is composed of four manufacturing loops with assembly workstations. The loops are fitted with switch points where pallets carrying the product can be diverted to any transporting loop. An RFID antenna that detects RFID-tagged items is mounted in front of every switch point and assembly workstation. The production line is controlled from a control box containing power supply units, protection circuits, and a programmable logic controller (PLC). The status of particular devices is displayed in the corresponding columns. The communications system which controls the devices uses the AS-interface and ProfiNet industrial networks.

The assembly stations are self-contained organizational units for the manufacturing of components for end products. The components are stored in flow racks fitted with heads which scan RFID-tagged transport containers. The RFID technology is used here for detecting transport containers. End products are also stored in flow racks. Each storage space (for components and end products) has a barcode. Management of the entire production system is supported by the 4Factory IT system, while the flow of materials and communications are controlled based on the RFID technology and the Internet of Things (IoT) concept.

The assembly line is equipped with trays (see Fig. 2) on which products are assembled and transported. For the study, a transmitter has been stuck to each tray, and a receiver fitted at the assembly line keeps a tally of how many times the tray has passed through the line [5, 6].

a) 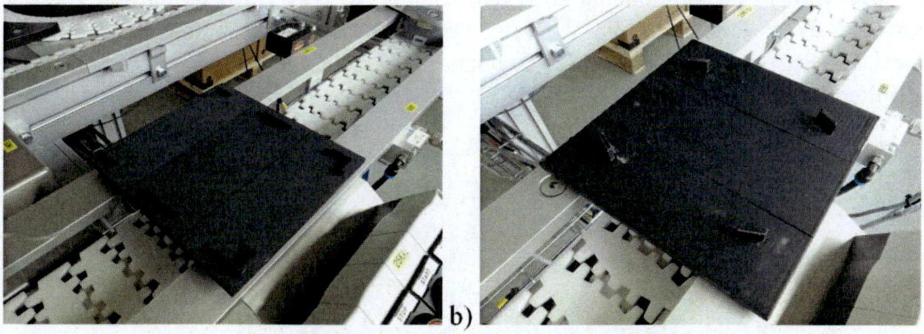

Fig. 2. Assembly tray: a) top, b) bottom.

As mentioned before, the RFID technology is used here to keep a tally of the number of work cycles and transmit, e.g., a sound alert to prevent a failure or seizure of the machine (in this case - the tray).

3 Research Methodology

The RFID technology-based system relies on an Arduino UNO minicomputer. Having considered its purchase price of ca. PLN 90, a decision was taken to replace the original machine with a replacement product made in China. A printed circuit board with an ATmega328 microcontroller (see Fig. 3), which is equivalent to that used in Arduino UNO, was bought for PLN 23, i.e., almost a quarter of the price of the original product.

Fig. 3. The replacement for Arduino UNO [17].

The ATmega328 microcontroller was selected not only for its price, but also for its number of digital inputs and outputs, required to connect other system components.

An indispensable part of the system, which enables contactless communication between the minicontroller and the RFID tags, is an MF RC522 13.56 MHz module (see Fig. 4). The module, purchased for a price of PLN 12, enables contactless communication between devices using the RFID standard. It is an integrated component that records and reads data from transponders and does not require any additional power supply. Its range of up to 50 mm depends on the size of the RFID coil, as well as the arrangement of components [18]. Based on an RFID RC522 circuit, the module is powered by a voltage of 3.3 V. Data are transferred through the serial peripheral interface (SPI) protocol, often used for data interchange between microprocessors and peripheral systems [20].

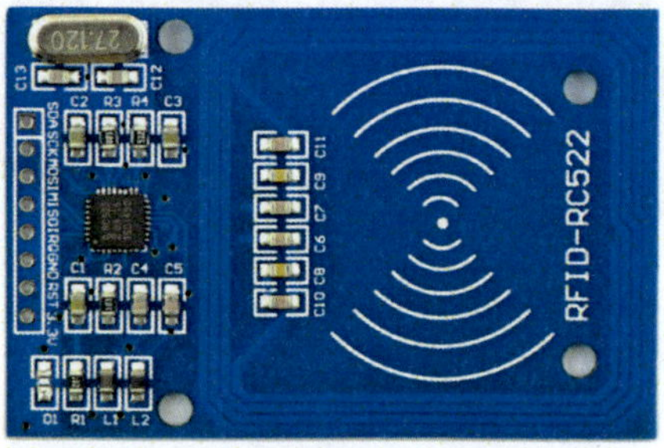

Fig. 4. The MFRC522 module [19].

Two sets of the components referred to above were bought, because the number of work cycles is recorded in the RFID tag's memory. One system was mounted at the assembly line, and the other – at the 'theoretical' maintenance unit in the laboratory, where the maintenance staff keeps track of the machine's work cycles to know exactly when the machine needs maintenance.

The system is also equipped with an LCD display (Fig. 5) from the Nokia 5110 mobile phone model. The display has a resolution of 48 × 84 pixels and communicates via the SPI. It has been implemented into the system for its small size, good readability, and low price.

A few RFID chips were bought to tag the assembly trays. The RFID chip models were selected for their small size. The system includes several other components which enhance its functionality, such as:

- cables,
- a sound generator,
- a step-down voltage regulator to reduce the voltage generated by the minicomputer from 5 V to 3.3 V,
- a 12 V 2 A power supply,
- a casing printed by the authors using the additive technology (3D printing).

Fig. 5. LCD Nokia 5110 display [21].

The total cost of the system amounts to PLN 142.55.

The proposition of the maintenance plan is based on production line workers at one specific place where the system is located. They need to observe the system LCD screen with information about the number of cycles. There are prepared special stickers: yellow, orange, and red dots. They used it according to instruction (Fig. 6).

1. Reading the counter value when tool come to the assembly place.
2. Choosing a right sticker (n- number from LCD screen):
   - **4<=n<10 – YELLOW STICKER**
   - **10<=n<15- ORANGE STICKER**
   - **15<=n<255 – RED STICKER**
3. Sticking a sticker at right down corner of tool (only at upper surface)

Fig. 6. Instruction with maintenance scheme.

Sticker is information for maintenance workers because it specifies kinds of inspection steps (they are specified for the specific tool). Colors represent the accuracy level of maintenance (yellow – basic, orange – advanced, red – most advanced).

Schematic diagram of code representing a few important steps (Fig. 7).

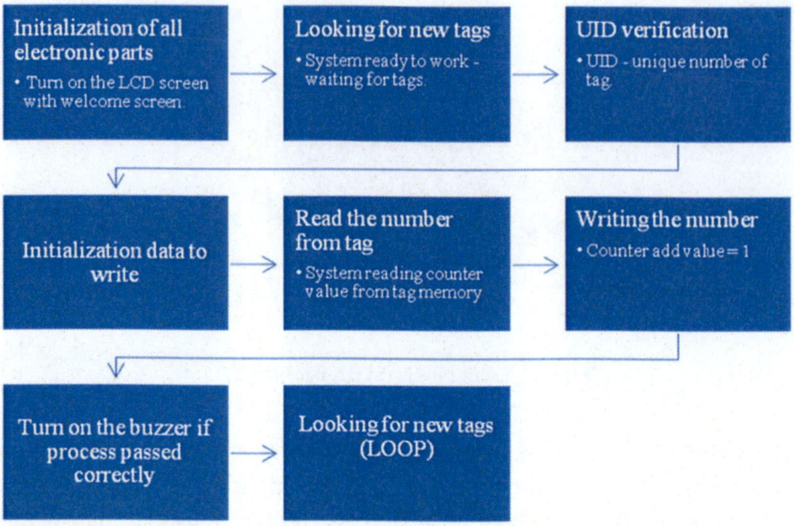

Fig. 7. Schematic diagram of code representing important steps.

4 Results

In order to connect all the components into one system, it was necessary to write code and install it on the minicomputer. The Arduino Software (IDE) was downloaded, free of charge, from the developer's website [22]. In the next step, libraries controlling the electronic components of the system were compiled using the Arduino Software (IDE). Finally, the cables were connected. For that purpose, pins were welded to the LCD display and the RFID module. The arrangement of pins was established based on the downloaded libraries.

Once the system had been put together, the software was installed to enable it. The Arduino IDE application is written in the Arduino programming language, a derivative of C/C++. The start-up went smoothly, and the system worked as anticipated. There were some modifications necessary only to the text display on the LCD panel. The system has been programmed to read the ID and number of passages a tool makes along the production line once the tool's RFID tag reaches the reader's range. The number of passages read is automatically increased by one, and the new number transmitted and saved in the tag's memory. The maximum number on the counter is 255, determined by the RFID tag's memory. Once it is reached, the counter is automatically reset to zero. The counter may also be reset manually by pressing a button, which – by default – is not mounted on the system located at the production line to prevent the results against corruption. The system located in the theoretical maintenance unit is equipped with that button.

In the next step, the casing for the system was printed. To mount the system beside the assembly line properly and aesthetically, it was necessary to design a casing that would hold the system components and position them correctly to read the data. The model, created in the Inventor software program (Fig. 8), was printed on a 3D printer.

Fig. 8. Casing design created in the Inventor software program.

The maintenance unit designated within the laboratory is equipped with a device for reading and resetting the counter of the RFID tags. The Arduino IDE software installed on the computer at the station enables control of the maintenance device. To read the counter, the minicomputer needs to be connected to the computer at the maintenance unit, and the live view function enabled in the Arduino IDE software. The maintenance intervals of the assembly trays have been determined in the designed maintenance procedure.

With the power supply connected, the system starts, and the LCD shows the home screen. Once the first assembly tray with an RFID tag passes the antenna of the reader module, the display shows a corresponding message and displays the transponder's unique identification number (UID). The counter, which is the key component of the software, is displayed at the bottom of the LCD panel in a larger font to facilitate the read.

The counter is read, and the value saved within less than a second. For the process to run smoothly, it is enough to bring the RFID tag within the range of the antenna. The read values are sufficient to ensure proper operation of the system at the production line (Fig. 9).

An antenna is based on the circular coil, and the maximum distance between RFID tag and coil is 50 mm (if there are not any metal parts between). 3D printing casing is not an obstacle for electromagnetic waves.

After dozens of attempts, a decision was made to locate the antenna possibly nearest to moving tools with RFID tags (see Table 1). A distance of 10 mm is the best choice between signal stability and non-collision moving. There is also one important condition – the antenna needs to be located parallel to the RFID tag.

The system ensures 100% of correctly read and write operations with that kind of mounting.

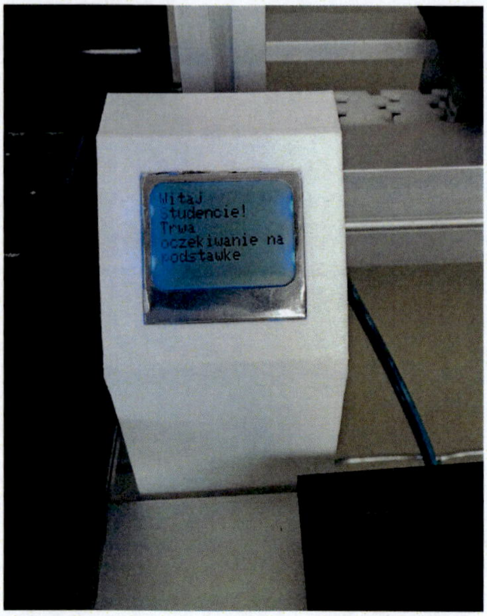

Fig. 9. The distance of mounting between antenna and RFID tag.

Table 1. The distance of mounting between antenna and RFID tag.

Distance [mm]	Number of tries	Correct	No correct	% of correct reads and wrights
30	10	3	7	30
20	10	6	4	60
10	10	10	–	100

5 Conclusions

The RFID technology-based system designed and built by the authors enables the determination of maintenance intervals at the production and assembly line.

With the RFID technology implemented in the laboratory in the way described above, the system can be easily developed and modified. Using the C/C++ programming language, the program code can be modified freely, or the program can be even replaced with a newer, more functional version.

The presented system could also be compliant with industrial standards, but in the assumption, it was created to be a solution for a specific problem. In the authors' opinion, it is a very useful device for any who wants to learn about RFID technology, programming, and modifying ready to use code. The next step will be the implementation of RFID technology into the production company and testing it in industrial conditions.

Acknowledgment. The presented results of the research, carried out under the theme No. 0613/SBAD/8727, were funded with a grant to science granted by the Ministry of Science and Higher Education.

References

1. Ocicka, B.: Mobile Technologies in Logistics and Supply Chain Management. PWN, Warszawa (2017). (in Polish)
2. Automation of Polish industry. https://www.magazynprzemyslowy.pl/komentarze/Automatyzacja-polskiego-przemyslu-W-jakim-miejscu-jestesmy,10559,1. Accessed 10 Jan 2019
3. Pająk, E.: Production Management, PRODUCT, Technology, Organization. PWN, Warszawa (2019). (in Polish)
4. Landt, J.: The history of RFID. IEEE Potent. **24**(4), 8–11 (2005)
5. Gładysz, B., Grabia, M., Santarek, K.: RFID from design to Implementation. Polish Perspective. PWN, Warszawa (2017). (in Polish)
6. RFID. https://pl.wikipedia.org/wiki/RFID. Accessed 10 Jan 2019
7. RFID technologies introduction. https://automatykaonline.pl/Artykuly/Montaz-i-transport/technologie-rfid-wprowadzenie. Accessed 10 Jan 2019
8. Kujawińska, A., Vogt, K., Diering, M., Rogalewicz, M., Waigaonkar, S.D.: Organization of visual inspection and its impact on the effectiveness of inspection. In: Advances in Manufacturing. Lecture Notes in Mechanical Engineering, pp. 899–909 (2018)
9. Starzyńska, B., Klembalska, A.: A digital repository of science resources of research institute as a source of knowledge from the area of production engineering for SMES. In: 24th International Conference on Production Research, pp. 525–530 (2017)
10. RFID application. https://www.rfidpolska.pl/zastosowanie-rfid/. Accessed 10 Jan 2019
11. Długosz, J.: Modern Technologies in Logistics. PWE, Warszawa (2009). (in Polish)
12. RFID application. https://www.pwsk.pl/rfid/zastosowania-rfid/https://www.pwsk.pl/rfid/zastosowania-rfid/. Accessed 10 Jan 2019
13. Examples of RFID system applications in animal identification. https://rfid-lab.pl/tag/identyfikacja-zwierzat. (in Polish)
14. viaTOLL. https://pl.wikipedia.org/wiki/ViaTOLL. Accessed 10 Jan 2019
15. Pacześny, Ł.: The use of RFID technology in signal recognition on the assembly line. Engineering work, Poznan University of Technology, Poznań (2019). (in Polish)
16. Żywicki, K., Zawadzki, P.: Fulfilling individual requirements of customers in smart factory model. In: Advances in Manufacturing, 2019 LNME, pp. 185–194. Springer, Heidelberg (2019)
17. Arduino Uno equivalent. https://kamami.pl/plytki-zgodne-z-arduino-inne/180240-arduino-uno-r3-odpowiednik-plytka-z-mikrokontrolerem-atmega328.html. Accessed 10 Jan 2019
18. MFRC522 Datasheet. https://www.nxp.com/docs/en/data-sheet/MFRC522.pdf. Accessed 10 Jan 2019
19. RFID module MFRC522. https://botland.com.pl/pl/rfid/6765-modul-rfid-mf-rc522-1356mhz-spi-karta-i-brelok.html. Accessed 10 Jan 2019
20. Serial Peripheral Interface. https://pl.wikipedia.org/wiki/Serial_Peripheral_Interface. Accessed 10 Jan 2019
21. Graphic LCD display. https://botland.com.pl/7214-thickbox_default/wyswietlacz-lcd-graficzny-84x48px-nokia-5110-sparkfun.jpg. Accessed 10 Jan 2019
22. https://www.arduino.cc/en/guide/windows. Accessed 10 Jan 2019

Check for updates

Improvement of the Assembling Technology for Precision Joints Using the Dimensional Information

Oleksandr Kupriyanov(✉) and Nataliia Lamnauer

Ukrainian Engineering Pedagogics Academy, 16, Universitetskaya Street, Kharkiv 61003, Ukraine
a_kupriyanov@uipa.edu.ua

Abstract. In the paper, it is proposed to memorize and use data on actual dimensions to improve the technology of manufacturing and assembly of machine-building products. It is proposed to store dimension information in a database, and use Direct Part Marking Identification to identify parts. Information on the values of the fit dimensions of parts allows us to apply the kitting algorithm for the actual dimensions, based on the ranking of parts. The created program that implements the proposed algorithm is illustrated. The proposed kitting technology of machine-building parts for assembly has been tested on the example of a two-element joint Ø50H7/f7. As a result, for a batch of 30 kits, the gap scatter decreased from 0.027 mm to 0.007 mm. Thus, the use of dimension information can significantly increase the dimensional accuracy of joints without increasing the accuracy of manufacturing parts. The proposed approach is the development of the ideas of selective assembly in the conditions of Industry 4.0. At the same time, a database and Direct Part Marking Identification are used to store dimension information. Instead of a static kitting algorithm for selective assembly – parts assigned to the same group are collected – a dynamic algorithm is used in which the number of groups is equal to the number of kits of parts.

Keywords: Assembly · Direct part marking identification · Kitting · Dimensional accuracy of joints

1 Introduction

Part dimension data is information that allows you to improve the manufacturing process. Unfortunately, despite the significant successes of information technology, at the shop level of a machine-building enterprise, information on parts is not used enough.

In most cases, the dimensions of parts obtained for control purposes are not saved after the rejection of parts. The received information is lost, but it could, with subsequent use, have a significant positive effect.

Various types of control charts can be attributed to the methods of extended use of dimension information. For building control charts, the parameters of the parts are measured, but then this information does not belong to specific parts. Control charts are

V. Tonkonogyi et al. (Eds.): InterPartner 2020, LNME, pp. 52–60, 2021.
https://doi.org/10.1007/978-3-030-68014-5_6

used to reject existing or resize parts manufactured in the future, but not to improve subsequent assembly and operation.

Loss of already obtained information about dimensions is all the more unreasonable in our time of intensive development of information technologies. Storing significant amounts of information is cheap, and the speed of access to such information far exceeds the needs of engineering.

In order to further use the information about the dimensions, it is necessary to unambiguously associate the data on the dimensions of the parts with specific instances of these parts, and the specific nodes in which they are used. Such tasks can be solved by marking. Marking systems have been significantly developed in recent years, but in trade, and not in engineering: barcoding, Direct Part Marking Identification (DPMI), Radio Frequency Identification (RFID).

In the paper, a complex optimization problem of the manufacturing of high-precision joints is investigated with the aim of getting the required quality and reducing the overall production-related costs through the actual size assembly.

2 Literature Review

Databases can effectively store a large amount of information about parts and products, accumulated as a result of measurement, fixing technological parameters, etc. For associating information in a computer with specific details, their marking is necessary. In addition to directly identifying parts instances, some marking systems can store sufficient data for practice.

Various marking systems differ in the amount of information stored, cost, range, speed of application, and reading, the ability to overwrite [1, 2]. The most common are barcoding. Barcoding technology has been widely used for decades; codes are linear and two-dimensional [3]. A linear bar code contains a limited amount of information and is intended primarily for product identification. There are a large number of types of two-dimensional codes that allow you to store much more information, up to several mega-bytes.

Radio Frequency Identification is a method of automatic identification of objects in which data stored in so-called RFID tags are read or written by means of radio signals. Compared to barcoding, this technology has advanced capabilities. In particular, it can be used in the absence of a line of sight [4]. The use of such marks is limited by their dimension and weight, as well as the fact that most engineering parts have magnetic properties.

In trade, labels are often used for bar codes. In mechanical engineering, however, external labels are not sufficiently resistant to the harsh conditions of manufacture and use. For machine-building parts, the image must be applied in a manner resistant to coolant, oil, moisture, and abrasion. These requirements are met by direct component marking, Direct Part Marking Identification [5]. This is the technology of applying the image directly to the components. The technology of impact marking, etching marking, laser marking is applied [6].

In production conditions, it is possible to use specialized software and hardware for information support of assembly technology. Of interest is equipment that allows you to transfer measurement results to ordinary personal computers, the cost of which is several times lower than specialized solutions. Manufacturers of measuring tools offer a wide range of such solutions. The first group allows you to pair measuring instruments with a computer via wires. The other group uses radio or Wi-Fi [7] to transmit measurement results, which gives more freedom of work. In this case, the transfer of numerical data to a computer is emulated by pressing a keyboard, which guarantees compatibility with all programs.

The measurement results are entered into a pre-prepared MS Excel table and other programs, while they are converted into keyboard codes. In the table, the user selects a cell and presses a button on the interface of the measuring device. The result is entered in this cell. When you click the button next time, the result is entered into the lower cell. Thus, it is possible to enter the measurement results into a computer quickly, and then carry out their processing using specialized computer programs.

The traditional technology of manufacturing products in mechanical engineering uses information about the actual dimensions of parts only for the operational control of the technological process [8]. Ultimate suitability control gauges are characterized by high speed and accuracy of control. However, they provide little information for long-term process control. If you get the numerical values of the dimensions, when measured with universal measuring tools with high accuracy, these data are not stored and are very limitedly used to control the technology [9].

The reason for such an organization of traditional technology in the past is easily explainable – it is not easy to store the dimensions obtained by linking them to parts and products without using computer technologies. Therefore, this was rarely done; we give only one well-known example of such a technology organization - selective assembly [10]. During selective assembly, a range is recorded in which the measured parameter of the part falls, and this information is used for the following technological operations, in particular, assembly ones. The full and operational use of part dimension data, which is possible with Industry 4.0, allows you to organize the assembly process in a new way, effectively using part dimension information [11].

Despite the various approaches proposed and realized in the industrial practice for the assembly phase of the product development process, finding the best production plan efficiency is becoming more challenging. The paper [12] proposed a novel method for integrating and optimizing product design and production planning to maximize investment efficiency and reduce the overall production cost for high-precision assembled products. A novel optimization problem was formulated that combines tolerance design with assembly resource configuration.

The assembly process has been the subject of scientific work, mainly due to the multiple aspects involved from geometrical matters. Assembly technique selection and geometrical allocation tolerances accordingly are also critical to precision joints. The potential benefits of the method were studied, representing the assembly of a simple mechanical structure [13].

The part dimension distribution plays a significant role in the assembly. The normal distribution to include skewed distributions as members, and hence the method is applicable for quality improvement, either the distribution of quality measurements follow us to use the symmetric or skewed distributions [14]. Paper [15] presents a new geometric model for representing statistically-based tolerance regions.

In addition to the assembly technology, the precise equipment is needed. A detailed manipulator design analysis is conducted for the customized different inside actuation mechanisms [16]. The manipulator overall stiffness evaluation is needed and can be conducted by simulation [17] or special generic design method [18]. The manipulator trajectories can be generated to satisfy widely varying tasks towards a fully reconfigurable control without any hardware or software adjustment [19].

3 Research Methodology

It is proposed to receive, store, and use information about the dimensions of manufactured parts and assembly units. Information is proposed to be stored in the database, and the record in the database should be associated with the dimensions of the real part using Direct Part Marking Identification. Thus, DPMI will have a key role that associates part dimension data in the database with the part itself.

Measuring tools are interfaced with storage devices based on a computer. The actual dimensions obtained are saved and associated with marked parts. This allows the future use of information on details both to improve previous and subsequent operations along the route, as well as to justify corrective technological actions.

Acquisition methods based on an individual selection are proposed. The essence of the proposed methods of acquisition consists in ranking the parts before assembly and selecting in the kit parts of each type with the same rank. The kitting position should contain the same number of parts of each type included in the assembly unit. Kitting sequence with ranking:

1) measurement of fit dimensions of all parts, dimensions are saved, parts are marked;
2) dimensions are ordered for each part in ascending order – ranking;
3) parts with the same serial number - rank are selected in the kits.

Assembly is carried out in the usual way for this joint.

In the discrete kitting process, all parts from the ranked lots are completed and then sent to the assembly. Thus, all the parts received at the kitting position will be assembled, and the quality of selection, determined by the deviation of the dimension of the closing link of the dimensional chain from the best value, is determined by the worst of the resulting kits.

The following organization of the assembly site is proposed. Fit dimensions of parts are measured to obtain a numerical result, the dimensions of each part are entered into the program, and parts are marked using Direct Part Marking Identification. The computer program gives out what details you need to take in each kit to obtain the best quality compounds and gives numerical results of the effectiveness of kitting. After assembling a batch of parts, the process is repeated for a new batch.

Automatic kitting area 1 (Fig. 1) includes a hopper 1 of shaft-type parts, a hopper of 2 sleeve-type parts, measuring and marking position 3 for shaft-type parts, measuring and marking position 4 for shaft-type parts, drive 5 for shaft-type parts, drive 6 for sleeve-type parts, transport device 7, assembly position 8, hopper of assembled joints 9, the control device 10.

Lots of parts of the shaft type and sleeve type in the same quantity are fed from the bins 1, 2 to the measuring and marking positions 3, 4, where measurement and marking are performed, information about the dimensions of each marked part is stored in the control device, and the parts are fed to drives 5, 6. After measuring all the details of the batch, automatic kitting is performed, then the completed joint is sent to the assembly position, which is performed in the usual way for this product. The cycle can be repeated at arbitrary intervals.

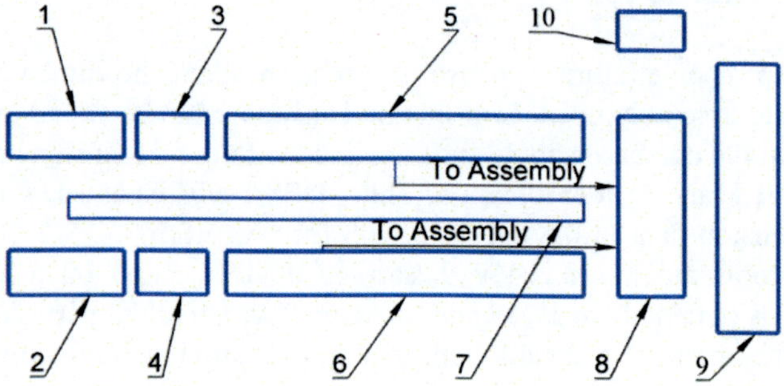

Fig. 1. Assembly area diagram.

4 Results

For the automation of the kitting process, a computer program was developed on VBA in MS Excel. Figure 2 shows the interface, and the program can be adapted to a specific product and production.

A computer program has the following fields, options, and settings. The allowed structurally minimum and maximum fit-gap, as well as its optimal value, are entered into the "Demand" fields (Fig. 2, pos. 1). "Demand" parameters do not change and are entered once for every joint.

In the fields "Shaft dimension" (Fig. 2, pos. 2) and "Bore dimension" (Fig. 2, pos. 3), the measured dimensions are entered. The "Before kitting" fields (Fig. 2, pos. 4) show the minimum and maximum gap, which can be obtained without kitting when assembling with full interchangeability. In the "After" fields (Fig. 2, pos. 5) – the same as a result of kitting with ranking.

Selected part numbers in kits are displayed in the "To kitting-up" fields (Fig. 2, pos. 6), the "Fit dimension" of each kit is also displayed there (Fig. 2, pos. 8). "Demand" dimensions may not be achieved for all kits, the number of satisfying requirements is displayed in "Have kitt", and in the "To kitting-up" fields only suitable ones are displayed. The "Start kitting-up" button (Fig. 2, pos. 9) starts the selection of parts in kits The "Clear / New" button (Fig. 2, pos. 10) prepares data for entering the dimensions of the next batch of parts.

The technical result that is achieved is to ensure full collectability with a significant reduction in the variation in the dimension of the closing link. For example, the program was tested on a shaft-sleeve joint, fit Ø50H7/f7, lot size 30 pcs., with the dimension distribution according to the normal law. In Fig. 3 a) shows the values of the fit dimensions and the sifting field before kitting (the case of complete interchangeability), in Fig. 3 b) - kitting with ranking. Inclined lines show the resulting scatter in the dimension of the closing link. As a result, for a batch of 30 kits, the gap scatters decreased from 0.027 to 0.007 mm. Thus, the use of dimensional information can significantly increase the dimensional accuracy of joints without increasing the accuracy of manufacturing parts.

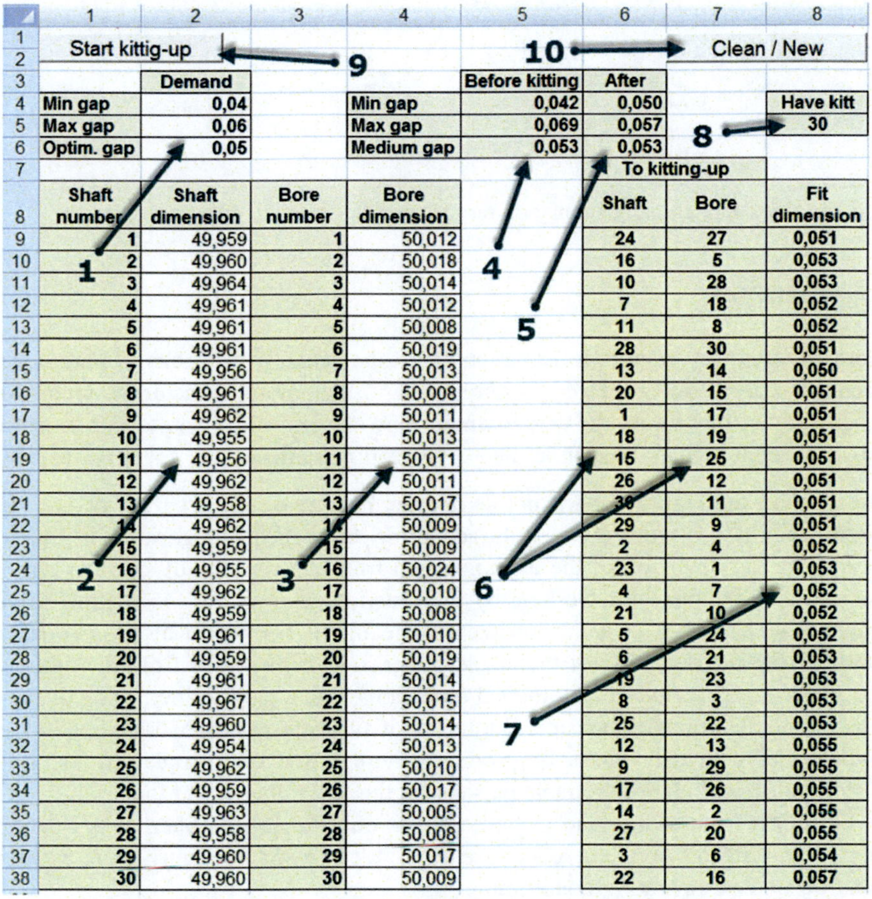

	1	2	3	4	5	6	7	8
1	Start kittig-up						Clean / New	
2								
3		Demand			Before kitting	After		
4	Min gap	0,04		Min gap	0,042	0,050		Have kitt
5	Max gap	0,06		Max gap	0,069	0,057		30
6	Optim. gap	0,05		Medium gap	0,053	0,053		
7						To kitting-up		
8	Shaft number	Shaft dimension	Bore number	Bore dimension		Shaft	Bore	Fit dimension
9	1	49,959	1	50,012		24	27	0,051
10	2	49,960	2	50,018		16	5	0,053
11	3	49,964	3	50,014		10	28	0,053
12	4	49,961	4	50,012		7	18	0,052
13	5	49,961	5	50,008		11	8	0,052
14	6	49,961	6	50,019		28	30	0,051
15	7	49,956	7	50,013		13	14	0,050
16	8	49,961	8	50,008		20	15	0,051
17	9	49,962	9	50,011		1	17	0,051
18	10	49,955	10	50,013		18	19	0,051
19	11	49,956	11	50,011		15	25	0,051
20	12	49,962	12	50,011		26	12	0,051
21	13	49,958	13	50,017		30	11	0,051
22	14	49,962	14	50,009		29	9	0,051
23	15	49,959	15	50,009		2	4	0,052
24	16	49,955	16	50,024		23	1	0,053
25	17	49,962	17	50,010		4	7	0,052
26	18	49,959	18	50,008		21	10	0,052
27	19	49,961	19	50,010		5	24	0,052
28	20	49,959	20	50,019		6	21	0,053
29	21	49,961	21	50,014		19	23	0,053
30	22	49,967	22	50,015		8	3	0,053
31	23	49,960	23	50,014		25	22	0,053
32	24	49,954	24	50,013		12	13	0,055
33	25	49,962	25	50,010		9	29	0,055
34	26	49,959	26	50,017		17	26	0,055
35	27	49,963	27	50,005		14	2	0,055
36	28	49,958	28	50,008		27	20	0,055
37	29	49,960	29	50,017		3	6	0,054
38	30	49,960	30	50,009		22	16	0,057

Fig. 2. The program interface for the organization of acquisition.

The more kits to sort, the better the results of kitting with ranking. In fact, with one kit, there are no options for selection, and the kitting efficiency, defined as the difference between the smallest and largest resulting gaps, corresponds to that with complete interchangeability. The more kits, the wider the scatter of resulting gaps already.

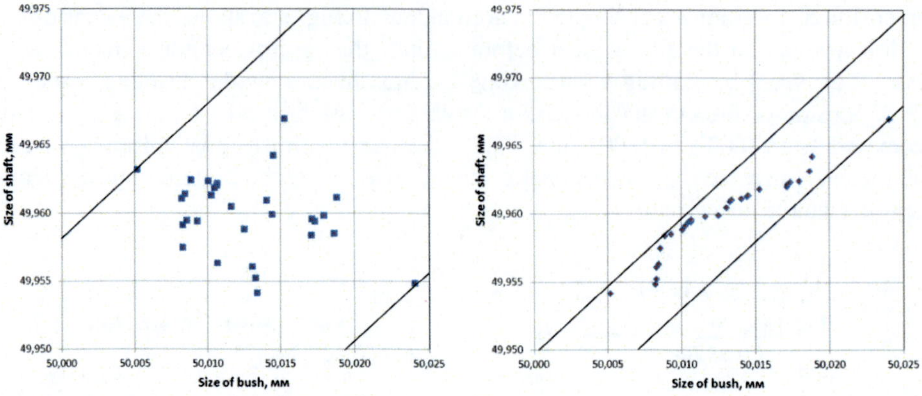

a) b)

Fig. 3. Assembly simulation results with the selection of parts in kits.

5 Conclusions

The article proposes to store information about the actual dimensions of parts for the subsequent improvement of the technological results of manufacturing production. Storage and use of data on the actual dimensions of parts allow you to implement the following technological, economic, and organizational effects:

- to organize the kitting of parts in kits in order to increase the accuracy of assembly. Parts should be selected into kits according to a special algorithm in order to minimize the deviation of the assembly gap from the structurally optimal one.
- to organize payment for workers by the criterion of proximity to optimal dimensions. The tolerance principle of interchangeability is replaced by an approach of minimizing the deviation of the actual dimensions of the parts from the optimum.
- to analyze the previous technological operations with a view to improving them – to determine the tuning dimension, frequency of sub-adjustment, overhaul period, etc.
- when analyzing the causes of product failure based on complaints, the stored information will identify reliable reasons for the low reliability of the products. This will improve the design and technology not only based on experimental data but based on statistics on the practical use of products, which are much more extensive and depend on operating conditions.
- select the best suppliers of components, task performers.

Technologically, an approach using marking and a special selection of parts into kits can be arranged for all dimensions and parts, but it is economically feasible to start with precision and responsible ones. Thus, the result is a significant increase in dimensional accuracy of joints without changing the manufacturing technology of parts.

It should be noted that in the case of kitting with ranking, requirements for the stability of the statistical parameters of the technology increase. It is noteworthy that the dimensions of the parts do not need to be within tolerance. This means, for example, for a two-element joint, that from the shaft and hole parts made with a defect to a minimum, a suitable joint will be assembled.

The proposed approach is the development of the ideas of selective assembly in the conditions of Industry 4.0. At the same time, a database and Direct Part Marking Identification are used to store dimension information. Instead of a static kitting algorithm for selective assembly—parts assigned to the same group are collected—a dynamic algorithm is used in which the number of groups is equal to the number of kits of parts.

New characteristic features of the proposed method of kitting parts are:

– the lack of constant dimensions of groups of parts during kitting compared with selective assembly;
– a simple algorithm for processing information about the dimension of parts.
– The advantages of the proposed method in comparison with selective assembly are:
– reduction of work in progress;
– effective reduction in the dimension variation of the closing link of the dimensional chain;
– the simplicity of the computer implementation of the kitting algorithm.

The proposed acquisition technology can be used both in serial and mass production.

References

1. Fusko, M., Rakyta, M., Binasova V.: Data collection for technical services. In: Transcom 2015, Zilina, Slovak Republic (2015)
2. Athanasios, K., Farhad, N.: Linear barcodes, direct part marking and RFID for material handling applications. Int. J. Manuf. Technol. Manage. **21**(3), 208–224 (2010). https://doi.org/10.1504/IJMTM.2010.035432
3. Gaubatz, M., Ulichney, R.: Barcodes on non-flat surfaces. In: NIP & Digital Fabrication Conference, Printing for Fabrication, pp. 115–119, San Francisco (2019). https://doi.org/10.2352/ISSN.2169-4451.2019.35.115
4. Mithu, B., Chao, C., Tracy, M.: RFID implementation in retail industry: current status, issues, and challenges. In: Decision Science Institute Conference, Phoenix AZ (2007)
5. Lingling, L., Yaoquan, Y., Tao, G.: Direct part mark bar code image preprocessing. Int. J. Adv. Pervasive Ubiquit. Comput. **7**(3), 13–24 (2015)
6. Cao, R., Qiu, H., Li, J., Lu. C.: Quality assessment and technical parameter optimization of laser direct part marking. Appl. Laser Technol. **31**(2), 151–156 (2011). https://doi.org/10.3788/AL20113102.0151
7. Ding, H., YL Xiong, YL.: Computational manufacturing. Progress in Natural Science 12(9), 641–648 (2002)
8. Kilic, H.S.: Design of kitting system in lean-based assembly lines. Assembly Autom. **32**(3), 226–234 (2012)
9. Kost'al, P., Mudrikova, A., Vaclav S., Diaz, R.: Manufacturing component base broadening in the flexible manufacturing system by using a group technology. Mater. Sci. Forum, 45–54 (2019). https://doi.org/10.4028/www.scientific.net/MSF.952.45

10. Babu, R., Asha, A.: Minimising assembly loss for a complex assembly using Taguchi's concept in selective assembly. Int. J. Prod. Qual. Manage. **15**(3) (2015)
11. Kupriyanov, A.: Kitting-up of two-element assemblage on the base of ranking. Technologia i Automatyzacja Montazu **1**(95), 15–20 (2017)
12. Tsutsumiab, D., Gyulaic, D., Kovacsc, A., Tiparyc, B., Uenoa, Y., Nonakaa, Y., Fujitab, K.: Joint optimization of product tolerance design, process plan, and production plan in high-precision multi-product assembly. J. Manuf. Syst. **54**, 336–347 (2020). https://doi.org/10.1016/j.jmsy.2020.01.004
13. Andolfatto, L., Thiebaut, F., Lartigue, C., Douilly, M.: Quality- and cost-driven assembly technique selection and geometrical tolerance allocation for mechanical structure assembly. J. Manuf. Syst. **33**(1), 103–115 (2014). https://doi.org/10.1016/j.jmsy.2013.03.003
14. Chiang, J.Y., Tsai, T.R., Lio, Y.L., Lu, W.B., Shi, D.M.: An integrated approach for the optimization of tolerance design and quality cost. Comput. Ind. Eng. **87**, 186–192 (2015). https://doi.org/10.1016/j.cie.2015.05.003
15. Xu, S.G., Keyser, J.: Statistical geometric computation on tolerances for dimensioning. Comput. Aided Des. **70**, 193–201 (2016). https://doi.org/10.1016/j.cad.2015.06.012
16. Khalid, A., Mekid, S.: Intelligent spherical joints based tri-actuated spatial parallel manipulator for precision applications. Rob. Comput. Integr. Manuf. **54**, 173–184 (2018). https://doi.org/10.1016/j.rcim.2017.11.005
17. Kupriyanov, O., Romanov, S.: Simulation of induction heating for railway wheel set elements during assembly and disassembly. In: Ivanov, V., et al. (eds.) Advances in Design, Simulation and Manufacturing II. DSMIE-2019. Lecture Notes in Mechanical Engineering, pp. 25–34. Springer, Cham. (2020). https://doi.org/10.1007/978-3-030-22365-6_16
18. Andersen, AL., Brunoe, T., Ditlev., Nielsen, K.: Towards a generic design method for reconfigurable manufacturing systems Analysis and synthesis of current design methods and evaluation of supportive tools. J. Manuf. Syst. **42**, 179–195 (2017), https://doi.org/10.1016/j.jmsy.2016.11.006
19. Avram, O., Valente, A.: Trajectory planning for reconfigurable industrial robots designed to operate in a high precision manufacturing industry. Fact. Future Digital Environ. **57**, 461–466 (2016). https://doi.org/10.1016/j.procir.2016.11.080

Improvement of the Warehouse Functioning: A Study Based on an Enterprise in the Printing Industry

Jan Lipiak[1] and Mariusz Salwin[2](✉)

[1] Etigraf Printing House, 52, Głowackiego Street, Sulejówek 05-071, Poland
[2] Faculty of Production Engineering, Warsaw University of Technology, 85, Narbutta Street, Warsaw, Poland
mariusz.salwin@onet.pl

Abstract. A warehouse is a component of a logistical system. It appears as a link in both the supply chain and supply network. Goods are temporarily stored in a warehouse and then directed to other links. As far as a warehouse is concerned, the flows of goods are considered as the objects of both delivery and reception; those may be concentrated or separated. The roles of a warehouse have an impact on its organization of work, used technology as well as location. The article aims to present a method of improving the functioning of a warehouse concerning a chosen enterprise in the printing sector. A variety of recurring errors was found in the functioning of the warehouse of the analyzed enterprise. Those were disrupting the enterprise's activity (mostly in terms of logistics). The irregularities were connected with the structure of the information system as well as with the organization of work. The method of improvement of the warehousing process used by the chosen enterprise was presented. The changes in terms of the information system were observed to have a positive impact on solving the problems reported by the employees.

Keywords: Printing Industry · Warehouse · Logistical System

1 Introduction

Warehouse management is an important part of the whole logistics of companies. It covers operational activities in storage, internal and external transport, production and distribution. Year after year, companies develop more and more efficiently and therefore very good management of warehouse logistics can open their doors to new markets and new customers [1–3].

Warehouse management is constantly looking for innovative solutions that will generate added value for the customer with their functionality. The implementation of such solutions is the way to the company's success and, at the same time, allows them to be competitive in a difficult national and international market. Modern companies in the field of warehouse management use integrated IT systems. Their use eliminates unnecessary manipulation activities, affects the speed of information transfer between the company's departments, and thus contributes to the acceleration of the processes of receiving, storing, issuing, and documenting all activities. However, despite the use of

V. Tonkonogyi et al. (Eds.): InterPartner 2020, LNME, pp. 61–71, 2021.
https://doi.org/10.1007/978-3-030-68014-5_7

the latest technologies and qualified staff, it is not always possible to avoid problems in this area [4, 5].

Warehouse management combines various aspects of business management and transport. This expertise includes a thorough knowledge of the necessary processes, their technical and operational feasibility, and successful implementation into an overall functioning system. Due to the different requirements of the customer's order placement, production requirements, attractiveness, and size of products and many other factors, there are no universal and generally applicable rules to achieve this [3, 6, 7].

Warehouse management aspects are often a problem for companies. These problems are often a factor in generating financial losses. Companies need efficient warehouses regardless of what they produce. Mainly it depends on the nature of production, which type of warehouse is the best solution for the organization. For ensuring smooth production, the materials and raw used materials should be collected in sufficient quantity in a place from where they can be quickly delivered to the production line. This applies to every production process. Finished products coming off the production line must be stored on the factory premises until the appropriate quantity is sent to the recipients. The warehouse is, therefore the factory's gateway to and from the factory [8–10].

A characteristic industry that uses storage systems is printing. Printing houses need well-functioning warehouse management due to a number of materials (paints, inks, substrates, adhesives, varnishes) and tools used in production (printing rollers, dies, blanking dies), as well as the specificity of finished products (packaging, labels, book, newspapers). Efficient warehouse management is a way to organize the processes taking place inside the printing house, regain space for storing tool materials and finished products, and thus reduce operating costs [11, 12].

In the paper, we focused on the analysis of the functioning of the warehouse in a company from the printing industry. The conducted consultations with the management of the company and the employees of the warehouse allowed determining the emerging problems and their elimination, before which we managed to improve the functioning of the warehouse.

2 Literature Review

A warehouse is defined as a functional-organizational entity intended to the storage of tangible goods (stocks) in a distinguished, according to the established technology, area of a storage construction, equipped with proper technical measures, managed and serviced by the team of people [13–15]. It is a part of a logistic system, and it constitutes a junction in which the goods are temporarily stored and then directed to other phases of the process. The streams of goods occurring in a warehouse may be of both a delivery and a reception nature; those may either concentrate or separate. The functions performed by a warehouse are determined by factors such as work organization, location, and technology [5, 6, 16].

The functions performed by a warehouse in a logistic system are as follows [3, 4, 8]:

- coordination of the size of supply and demand caused by their fluctuations,
- support of the marketing processes through creating sets or gathering stocks necessary for launching a promotional campaign,
- reduction of the costs of transport, possible due to the decrease in the frequency of deliveries with a simultaneous increase of their size,
- support of the production processes that provide a continuous supply of the production in the necessary resources and packages on the one hand and a current reception of the finished product on the other hand.

Warehousing is a set of activities connected with reception, storage, stockpiling, completing, movement, maintenance, registration, controlling, and launching tangible goods – stocks on a temporary basis [13, 14]. Warehouses in the producing enterprises can be divided according to the criterion such as purpose, the character of the stored goods, storing conditions as well as technical-organizational solutions [1, 17]. Depending on the relative positions of zones, there are three crucial technological systems of warehouses [3, 4, 8]:

- through the warehouse (input and output zones are located at the opposite sides of the storage area),
- angled warehouse (input and output zones are located at the neighboring walls of the storage area),
- sack warehouse (input and output zones are located at the same wall of the storage area).

As far as the warehousing process in a service providing enterprise is concerned, both a character of piece loading units and a possibility of accumulation on stacks have an impact on a method of storing [13, 14]. Considering a physical form and a peculiar technological similarity, the following types of load units can be distinguished [3, 4, 8, 18]:

- micro-units (non-palletized – based on the logistics containers and packages),
- palletized units (formed on flat, bar, box or special pallets),
- package units (made from goods which length is considerably longer than other measurements and amounts to more than 1.2 m),
- container units (used in transport).

Currently, pallet cargo units are widely used in transport and warehousing. Among the cargo units, one can distinguish homogenous units which load includes only one part of one assortment position and heterogeneous units, which are made of at least two parts or two assortment positions [17, 19, 20].

A warehouse process is a set of operational actions connected with reception, storage, completion, and launching of tangible goods in properly adapted places while complying with the determined organizational and technological conditions. A warehouse process consists of material and information flows [17, 19]. According to the definition presented above, a warehouse process is divided into four basic phases: receiving, storing, completing (completion), and finally, launching. Reception of goods from the supplier relates to the acknowledgement of receipt and giving the recipient (a warehouse) the responsibility for goods.

The essential operations implemented during the reception of goods, as far as a warehouse of a service enterprise is concerned, include as follows [3, 4, 8]:

- unloading, using the owned measures of internal transport and handling devices,
- sorting through dividing the goods into groups according to determining similarities,
- identification connected with the identification measures or the unequivocal identification of goods,
- qualitative and quantitative inspection,
- preparation of goods for storage which can be based on a development of load units or decreasing their height, transshipping to containers, proper provision of signs etc.,
- submission of the delivery to the storage area if the goods are transported by the employees completing the reception.

The next stage consists of storage, meaning a set of actions connected with both arrangement and placement of stock either on the surface or within the forming space of a warehouse (e.g., in the storage equipment) in an organized way, under the characteristics of stock as well as the existing conditions. Storage of goods is the most basic function of a warehouse connected with temporary warehousing [1, 17, 19].

Among the basic operations implemented during the storage of goods one may distinguish (in terms of the examined warehouse): receipt of goods from the receipt zone if the goods are not delivered to the storage area by the employees responsible for reception, an arrangement of goods in the storage area, storage of goods by maintaining required conditions, periodical inspection and finally, submission of goods to the completion zone.

Completing is defined as an operation that takes place in the warehouse process, which consists of a collection of stocks from the equipment used for storage or from stacks to complete the order of stocks in accordance with the assortment and quantity specifications for a particular recipient. Completing may take place inside or outside the warehousing place [17, 20, 21].

The basic operations executed in terms of the completion phase include as follows [3, 4, 8]:

- proper preparation of load units for the needs of completion,
- completion of orders implemented in the completion zone under the order,
- quantity inspection which confirms the completeness of the created load unit and the compliance with the order concerning both assortment and quantity,
- packaging and forming of transport units to protect the goods against damage.

Redemption of goods is understood as physical actions connected with the delivery of goods from a warehouse for the determined recipient, including confirmation of handover. The crucial operations executed during the redemption of goods in an enterprise of a service sector include [9]:

- packaging of goods in parcels and courier shipment to a proper recipient,
- control of redemption concerning checking the conformity of prepared goods with delivery documentation; control of completeness of prepared transport units and compliance of the way of forming and marking them with the requirements of the recipient.

Particular phases of warehousing, as well as the entire subsystem of a warehouse occurring in a holistic approach in a logistics system of an enterprise, are complicated. Those include numerous elements that are capable of causing mistakes, disrupting the activities, or completely disorganizing and, consequently, paralyzing the functioning of the organization [22, 23]. One of those factors is the human factor, meaning employees. The mistakes occurring during warehousing caused by the human factor include as follows [3, 24]:

- improper marking of goods and storing them in an inappropriate place,
- lack of respect to the determined procedures (procedural violations may also result from unconsciousness),
- defective communication of a warehouse with different departments or sections of an enterprise,
- incorrect set up of the information system,
- incorrectly filled documents or lack of them,
- irregularities during an inventorying,
- supplying an incorrect amount of goods from a warehouse,
- wrong address of shipment of goods from a warehouse,
- improper organization of work of a warehouse,
- insufficient training of workers,
- lack of procedures for emergencies,
- varying amounts of goods in terms of documentation and physical quantities,
- shelves not being adjusted to the gathered goods.

The future of warehousing is primarily related to data analysis, digitalization, and automation of storage systems towards intelligent solutions. A well-functioning warehouse system reduces costs and improves operational efficiency and the overall level of services provided by the company. Reducing warehousing costs is a major priority for many companies to enhance economic efficiency. Implementation of modern warehouse solutions is a high cost but also involves subsequent operating costs. For this reason, some companies are still using a conventional warehouse system to improve and modernize it with human resources [10, 24–26].

The implementation and service costs mentioned above may lead to interest in PSS-based solutions [27, 28]. Innovative warehouse solutions for PSS will eliminate the problem of investment and the lack of warehouses in printing houses. They will provide printing houses with individual solutions, necessary services, replacement and regeneration of worn components, flexibility and efficiency of the warehouse system in terms of adjusting product changes. In addition, they will help to eliminate and solve a number of problems related to storage. PSS can reduce storage costs and improve economic efficiency, which is currently one of the priorities of many printing houses as well as companies from other industries [25, 29, 30].

3 Research Methodology

3.1 Research Aim and Methodology

The article aims to present a method of improving the functioning of the warehouse concerning a selected company from the printing industry. A case study method was used - a company from the printing industry. The research was conducted in the period from August 2019 to October 2019.

In the paper we formulated the following research questions:

- What mistakes made by employees adversely affect warehouse processes and have a significant impact on the functioning of the entire company?
- What methods can be used to improve the functioning of the warehouse in a printing company?

The research methodology used in this article included the following steps:

1. Identification of the causes of the problem – personal observations of the management of the department and the company to collect information about the warehouse process and problems occurring in the process.
2. In-depth interviews with employees - an expert questionnaire method was chosen to conduct the research. This method makes it possible to collect data to reveal the actual situation that takes place in the storage process and warehouse of the printing company.
3. Analysis of solutions – the brainstorming method was chosen, in which all employees of the storage department and the management of the company participated. At this stage, solutions for warehouse improvement were sought and analyzed.
4. Choice of the solution – based on the brainstorming, the head of the department in consultation with the president, which decides on the best solution.
5. Implementation and verification - the last stage includes the implementation of the best solution and assessment of the effectiveness of the project.

3.2 Research Problem

A printing company's warehouse was analyzed. The warehouse is used to store semi-finished products that are used in production and finished products that go to it from the production department. Concerning the warehouse of a selected company from the printing industry, the authors of this article have observed errors made by employees, which cause damage to the warehouse processes, having a significant impact on the functioning of the entire company.

The observations, interviews conducted with the employees, and the research using an expert questionnaire allowed determining the central irregularities in the functioning of the warehouse in the investigated printing company. These include:

- IT system structure – during the reception of e.g., ten pallets of one assortment with one batch of product, the system in possession required a 10-fold introduction of a batch of product, due to the lack of possibility to receive the whole delivered batch at the same time,
- recommendations for the return of the goods during the collection of the delivery were generated automatically, the system allocated the place where the goods were placed according to the shelves until they were finished, which resulted in placing the pallet with the goods on the upper level, even though the shelves standing next to it had free space on the lower levels,
- individual insertion of the placement space in the scanner for each finished pallet,
- one assortment in many places far away from each other.

For the above reasons, the changes were implemented what resulted in the improvement of warehousing processes.

4 Results

For solving the problems presented earlier, it was proposed to implement several improvements to the warehouse system in the analyzed company.

The employee, while receiving the determined amount of one assortment with one batch, does not have to enter each palette separately. The way of generating the placing location was also changed. The receiving employee, using a system, can choose a shelf in which a certain delivery should be placed. It requires choosing the number of a shelf. The system automatically finds empty locations and provides indications regarding placing a particular delivery. The ABC rotation scheme on shelves was created to generate locations for placing certain assortments, including their percentage in an overall sale. It was assumed that A products include those articles which constitute 60% of overall sold goods, B products – up to 30%, and C products are slow-moving items with only 10% of the overall sale. The ABC rotation scheme includes placing the items from A group in the lowest location, nearest to the release line, B group in the middle locations on the shelves, and the items from the C group are placed in the furthest locations on the shelves.

The completing employee can enter the location place only once for each sale order, without having to do it for each palette separately. What is more, a request for generating sustainable places in a warehouse was reported to arrange the goods there properly. However, after introducing the changes, each item is assigned a permanent place in the location (locations of cardboard/ piece collections). This solution contributed to the fact that the items of one type are located next to each other, instead of being placed all over the warehouse. The sustainable places can be modified using a computer, which can be done only by the store manager. The IT department improved also printing of pallet labels - after the implementation of the improvement concept those are printed after the completion of the entire order, instead of being printed after the completion of one palette as it took place before.

It should be underlined that due to the implementation of a variety of changes in the warehouse of the described enterprise, the quality of provided services has improved. What is more, it has had a positive impact on the quality of working conditions. In Fig. 1, the average unloading time, the average systemic application time, as well as the average time of placing the delivery into shelves, are presented. Implementation of the systemic changes contributed to the decrease of those times, as far as each of those processes is concerned. Those changes are presented for one delivery. However, by multiplying them by the number of monthly or yearly deliveries, a considerable difference can be observed – the saved time is even larger (Fig. 1).

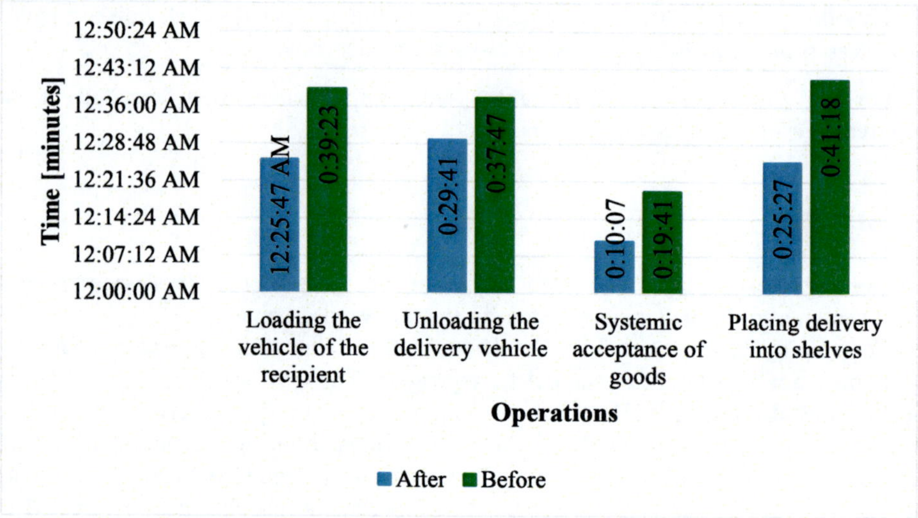

Fig. 1. Results of the changes in terms of the functioning of a warehouse – time of handling activities.

Due to the changes in terms of the informatics system, the problems reported by the employees were solved. As far as printed documents are concerned, those contain a place for barcodes regarding the numbers of orders. What is more, the possibility of printing the sales orders according to the direction of deliveries was introduced. The printed sale orders also include the information about the number of mixed and completed palettes.

5 Conclusions

Among the fundamental tasks performed by a warehouse, one may include the storage of goods and handling activities. Storage of items takes place whenever the item stays unmoved. Above all, it is connected with its storage in determining conditions of storing, such as moisture, temperature, air quality, or protection against theft. As far as handling activities are concerned, those always take place during the receipt of goods as well as releasing goods. Those also include any movements inside a warehouse and, finally, the activities causing the change of the character of the load. The efficient management of the entire warehouse process is a major challenge for a large group of companies.

In the paper, a five-stage process of improving the functioning of the warehouse in a printing company was conducted. The process was based on observations, questionnaire research, and consultation with employees. The research revealed, first of all, problems with the proper placement of goods in the warehouse. The solutions to the problems were generated based on the brainstorming process, which involved the employees of the warehouse department and the management board of the company. The introduced changes contributed to improving warehouse processes. The analysis carried out underlines the importance of effective warehouse management. Thanks to this, the quality and effectiveness of services have improved.

The approach proposed in the paper is addressed to micro, small and medium-sized enterprises. This is since companies of this size have mostly small warehouses in which the presented solutions bring significant organizational and economic effects. Additionally, they have a small base of suppliers and customers, and in the storage department, where only a few employees work, is not their main source of income. It should be remembered that in this size of enterprises, there is a great potential to implement modern warehouse solutions. However, their implementation may prove unattractive due to limited financial resources. With this in mind, an appropriate warehouse organization based on well-known solutions is very attractive for companies of this size.

The paper was created in close cooperation with the printing company and represents an added value for the analyzed company. The conducted research has shown that effective improvement of the warehouse operation depends on a number of factors. The main emphasis was placed on the cooperation of the department's employees with the company's top management in order to find optimal solutions. An important element in improving the functioning of the warehouse for the described company was also to make the employees aware of the importance of generating unnecessary costs for the company, and thus for themselves. What is more, the emphasis was also placed on implementing the basis for changing the organizational culture of the workplace. The employees understood that the newly created order should be maintained, and attention should be paid to taking care of tools that are necessary for everyday work. It should be emphasized that significant success is connected not only with the involvement of the whole team in the search for further problems to be solved but also with their joint elimination. Therefore, the team identifies itself even more with the organization, having the feeling that all the changes are also introduced to their benefit.

The article shows an area where printing house experience can be used to design PSS. The article highlights problems related to storage in a printing company and how to solve them. Using this experience will allow you to design PSS that reflect the real industrial situation in the printing house. The problems and needs of the printing house will support the design of services that are to be an answer to them. Thanks to this, PSS will contain what the enterprise needs. PSS designed in this way can support the printing house and eliminate emerging problems. It is worth noting that an important aspect in the design of PSS will be communication and trust between the printing house and the PSS provider.

References

1. Chan, F.T.S., Chan, H.K.: Improving the productivity of order picking of a manual-pick and multi-level rack distribution warehouse through the implementation of class-based storage. Expert Syst. Appl. **38**, 2686–2700 (2011)
2. Burinskiene, A.: Order picking process at warehouses. IJLSM **6**, 162 (2010)
3. Tompkins, J. (ed.): The Warehouse Management Handbook. Tompkins, Raleigh, NC (1998)
4. Ten Hompel, M., Schmidt, T.: Warehouse Management: Automation and Organisation of Warehouse and Order Picking Systems. Springer, Berlin; New York (2007)
5. Palšaitis, R., Čižiūnienė, K., Vaičiūtė, K.: Improvement of warehouse operations management by considering competencies of human resources. Procedia Eng. **187**, 604–613 (2017)
6. Khanzode, V., Shah, B.: A comprehensive review of warehouse operational issues. IJLSM **26**, 346 (2017)
7. Pyza, D., Jachimowski, R., Jacyna-Gołda, I., Lewczuk, K.: Performance of Equipment and Means of Internal Transport and Efficiency of Implementation of Warehouse Processes. Procedia Engineering **187**, 706–711 (2017)
8. Frazelle, E.: World-class warehousing and material handling. McGraw-Hill, New York (2002)
9. Gattorna, J.: Dynamic supply chains: how to design, build and manage people-centric value networks. Pearson Education, Harlow, England (2015)
10. Bartolini, M., Bottani, E., Grosse, E.H.: Green warehousing: Systematic literature review and bibliometric analysis. J. Clean. Prod. **226**, 242–258 (2019)
11. Lipiak, J.: Methods of Improving the Flexographic Printing Process Using the Process Approach and Structural and Technological Changes. Publishing House of the Warsaw University of Technology, Warsaw (2018)
12. Lipiak, J., Salwin, M.: The improvement of sustainability with reference to the printing industry – case study. In: Hamrol, A., Grabowska, M., Maletic, D., Woll, R. (eds.) Advances in Manufacturing II, pp. 254–266. Springer International Publishing, Cham (2019)
13. De Koster, M.B.M., Van der Poort, E.S., Wolters, M.: Efficient orderbatching methods in warehouses. Int. J. Prod. Res. **37**, 1479–1504 (1999)
14. Eben-Chaime, M.: Operations sequencing in automated warehousing systems. Int. J. Prod. Res. **30**, 2401–2409 (1992)
15. Zhang, J., Onal, S., Das, S.: The dynamic stocking location problem – dispersing inventory in fulfillment warehouses with explosive storage. Int. J. Prod. Econ. **224**, 107550 (2020)
16. You, M., Xiao, Y., Zhang, S., Yang, P., Zhou, S.: Optimal mathematical programming for the warehouse location problem with Euclidean distance linearization. Comput. Ind. Eng. **136**, 70–79 (2019)
17. Guthrie, B., Parikh, P.J., Kong, N.: Evaluating warehouse strategies for two-product class distribution planning. Int. J. Prod. Res. **55**, 6470–6484 (2017)
18. Sun, D., Meng, Y., Tang, L., Liu, J., Huang, B., Yang, J.: Storage space allocation problem at inland bulk material stockyard. Transp. Res. Part E: Logistics Transp. Rev. **134**, 101856 (2020)

19. Mourtzis, D., Samothrakis, V., Zogopoulos, V., Vlachou, E.: Warehouse design and operation using augmented reality technology: a papermaking industry case study. Procedia CIRP **79**, 574–579 (2019)
20. Chen, F., Wang, H., Xie, Y., Qi, C.: An ACO-based online routing method for multiple order pickers with congestion consideration in warehouse. J. Intell. Manuf. **27**, 389–408 (2016)
21. Faber, N., De Koster, R.B.M., Smidts, A.: Survival of the fittest: the impact of fit between warehouse management structure and warehouse context on warehouse performance. Int. J. Prod. Res. **56**, 120–139 (2018)
22. Wyrwicka, M.K., Kliber, M.C., Brzeziński, Ł.: Social climate management in enterprises. Res. Logistics Prod. **5**, 163–179 (2015)
23. Lototsky, V., Sabitov, R., Smirnova, G., Sirazetdinov, B., Elizarova, N., Sabitov, Sh.: Model of the automated warehouse management and forecasting system in the conditions of transition to industry 4.0. IFAC-PapersOnLine **52**, 78–82 (2019)
24. Gu, J., Goetschalckx, M., McGinnis, L.F.: Research on warehouse operation: a comprehensive review. Eur. J. Oper. Res. **177**, 1–21 (2007)
25. Leng, J., Yan, D., Liu, Q., Zhang, H., Zhao, G., Wei, L., Zhang, D., Yu, A., Chen, X.: Digital twin-driven joint optimisation of packing and storage assignment in large-scale automated high-rise warehouse product-service system. Int. J. Comput. Integrated Manuf. **32**, 1–18 (2019)
26. Yener, F., Yazgan, H.R.: Optimal warehouse design: literature review and case study application. Comput. Ind. Eng. **129**, 1–13 (2019)
27. Salwin, M., Kraslawski, A., Lipiak, J.: State-of-the-art in product-service system design. In: Panuwatwanich, K., Ko, C.-H. (eds.) The 10th International Conference on Engineering, Project, and Production Management, pp. 645–658 (2020)
28 Salwin, M., Kraslawski, A., Lipiak, J.: State-of-the-art in product-service system design. In: Ivanov, V., Trojanowska, J., Pavlenko, I., Zajac, J., Perakovic, D. (eds.) Advances in Design, Simulation and Manufacturing III, pp. 187–200. Springer (2020)
29. Salwin, M., Gladysz, B., Santarek, K.: Technical product-service systems—a business opportunity for machine industry. In: Hamrol, A., Ciszak, O., Legutko, S., and Jurczyk, M. (eds.) Advances in Manufacturing, pp. 269–278. Springer, Cham (2018)
30. Salwin, M., Kraslawski, A., Lipiak, J., Golebiewski, D., Andrzejewski, M.: Product-service system business model for printing houses. J. Clean. Prod. **274**, 122939 (2020)

Determination of the Production Frequency and Batch Size for the Manufacturing Process

Paulina Rewers(✉), Marta Czaja, Kamila Janczura, and Jacek Diakun

Poznan University of Technology, 5, M. Sklodowskiej-Curie Sq., Poznan 60-965, Poland
paulina.rewers@put.poznan.pl

Abstract. Production leveling (jap. Heijunka) is an effective method of balancing production and controlling stocks. It also provides an appropriate level of flexibility. This article presented an analysis of the literature on production leveling. In particular, the focus was on definitions, key goals, and methods for implementing leveled production. Also presented is a fragment of research on developing a method for planning the flow of finished products from the production process. The focus was on determining such a variant of the production batch size and production frequency for the selected model of the production process, which, which the given input parameters will get the best results in terms of the degree of order processing and machine load. To achieve the assumed goal, three variants of the production plan were proposed, a computer simulation was created in FlexSim, and the results of the simulation were analyzed.

Keywords: Production planning · Production leveling · Computer simulation

1 Introduction

Production companies that want to remain competitive on the market must constantly adapt to changing demand conditions. Fluctuations in demand can be leveled by increasing the level of stocks of finished products or increasing the flexibility of the production system [1]. In the traditional approach, the production system is separated from the demand. Therefore, the stabilization of production is achieved due to increased stocks. However, it leads to a longer cycle, an increase in the production batch size, and thus to a reduction in production flexibility [2, 3]. On the other hand, increasing the flexibility of the production system, consisting of the rapid implementation of current customer demand, forces the ongoing transfer of demand to the production system. In this case, production usually takes the form of customer order production. It determines the creation of variable production schedules, which results in a periodic increase in overtime or idle times [4]. Usually, the organization of a flexible system is very difficult or even impossible. Therefore, an intermediate solution is required, which may become production leveling [5].

Heijunka has gained popularity among manufacturing companies as a method for leveling production and better stock control [2]. Heijunka's main goal is to reduce

V. Tonkonogyi et al. (Eds.): InterPartner 2020, LNME, pp. 72–82, 2021.
https://doi.org/10.1007/978-3-030-68014-5_8

production spikes by creating cyclical production schedules. Such schedules to help in the stabilization of production, while not increasing the stock [1]. The cyclical production schedule can also contribute to reducing the number of shortages, better use of production resources, and increasing production efficiency while reducing the consumption of raw materials, i.e., material, energy, etc. As a result, the company can reduce the amount of pollution and waste generation, which is part of a trend of sustainable development of enterprises [6].

Unfortunately, in many companies, production planning according to the rules Heijunka is very difficult. It is especially true in cases where production is not stable, i.e., machines are fail-safe, there is no repeatability of production, etc. Therefore, it is important to prioritize planning before planning according to the principles of heijunka, to systematize production by implementing Lean Manufacturing tools, such as SMED, 5S, and others. The main premise of leveling production is to produce products in small batches, as often as possible, and in the largest possible mix. However, it is very difficult to correctly determine the volume and frequency of production of individual products to meet the current customer demand and not exceed the permissible stock levels.

In connection with the above, it was proposed to conduct research aimed at developing a method of planning the flow of products from production, which in its intention, will indicate the most favorable variant of the production plan. The article presents a fragment of research concerning the determination of the production batch size and production frequency for a selected model of the production process. Three variants of the production plan were proposed. A computer simulation of the process was created in the FlexSim software. Achieved results were analyzed considering the degree of order fulfillment (on time – when the order is placed by the client, after the time and lack implementation) and the load on the machines. The analysis aims to indicate the best production plan in terms of the adopted evaluation criteria.

2 Literature Review

Production leveling is a planning technique-oriented at even inflow of goods from the production process. It requires the determination of sequence and rate of inflow of products from the production process in a way to sell directly from the warehouse and avoid sudden changes in the production schedule. The production plan must be repeatable and not cause sudden peaks or delays. With production leveling, enterprises can get the inflow of products from the production process and stock levels under control [3, 4, 7].

The key objectives of leveled production, as described in the literature of the subject, include: ensuring smooth flow throughout the supply chain [8], smoothing out production peaks and troughs [9], reducing stocks levels [10], avoiding excessive workload [11].

In the literature, descriptions of the implementation of production leveling in enterprises can be found. One of them was described in Araujo and Queiroz [12]. According to the authors, the implementation involves two stages: product prioritization, i.e., the selection of products most often bought by the customer, and the creation of a leveled production plan. The authors also showed that as a result of undertaken actions, average monthly inventories decreased by 23%, average production batch size decreased by 18%, which resulted in an increase in monthly production by 10%.

The methodology of implementing production leveling in an enterprise with low production volume and high product diversity was described in their works by Bohnen et al. [13, 14]. The procedure starts with the preparation and analysis of the value stream map and demand analysis. Based on the analysis, the so-called leveling model. The second stage is the division of products into families. This division depends on the similarity of technological processes or product design. The third step is to create a leveling formula. This formula is reflected in the value of the EPEI (Every Part Every Interval) indicator. It illustrates the frequency of production of products within a set schedule on a specific time scale. In the last stage, it is assumed that the previously created level production pattern will be launched and its continuous improvement. In this case, improvement is defined as narrowing the product families until a time schedule is created for individual product pieces.

The methodology of implementing leveled production with a large variety of products was presented by Liker and Meier [15]. The authors define their method as "cut and chop", and it includes such stages as a division of products into product families with common characteristics and production steps, scheduling, launching production.

A different methodology is presented in the papers [7, 16, 17]. The methodology consists of five stages: determining a group of products, division of selected products into families, determining the frequency of repeating, determining stock levels, establishing a leveled production plan.

The most important and the most difficult to implement is the last stage - creating a leveled production plan. There are three factors to consider when creating a leveled production plan: production batch size, the frequency of batch production (intervals), and the order of manufactured product families.

3 Research Methodology

3.1 Main Research Steps

The research methodology was adopted, consisting of six basic steps:

- Creating a manufacturing process model,
- Assumption the size and frequency of customer orders,
- Creating variants of the production plan,
- Creating a model in a simulation program,
- Starting the simulation and generating simulation results,
- Analyzing the results and developing conclusions.

The first stage is creating a model of the manufacturing process. At this stage, the number of positions is assumed, links between them are established, products are determined, etc. It is also simulated data such as order history, operation time, setup time,, the work calendar, etc.

The next stage is assuming the size and frequency of customer orders. It was assumed that non-seasonal products are produced and that all products are produced according to a leveled plan.

After analyzing the orders, variants of the production plan are created. The described example assumes three variants of the plan, which will be described later in the article.

The next step is to create a model in the simulation program. At this stage, the designed manufacturing process is mapped in a simulation environment, the model is programmed, and numerical data is implemented. The presented study used the FlexSim simulation program. After running the simulation, results are generated, which can then be analyzed. For this study, the results were developed in terms of order processing and machine load.

3.2 Input Data for the Simulation Model

It was assumed that the production process model would consist of six work stations (Fig. 1).

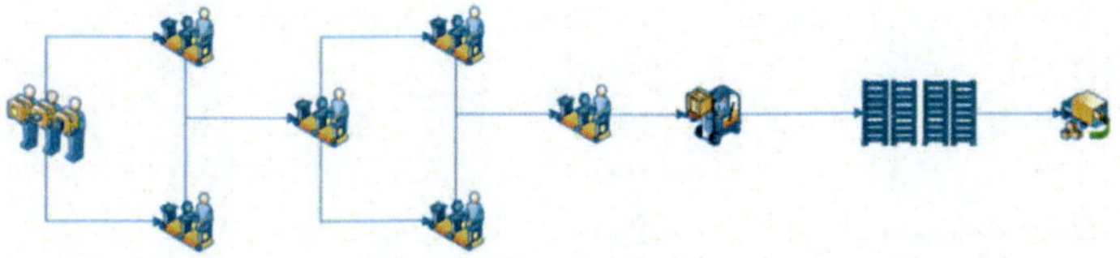

Fig. 1. Model of the manufacturing process.

For the study, the average daily volume of customer orders (n), operation time (t_j), setup time (T_{pz}), available production time (production 5 days a week, one eight-hour shift with a 30-min. break) and the number of products (10 different products, structurally similar and technologically) (Table 1).

Based on the data defined above, three variants of the production plan were adopted. Option 1 assumes the production of the average daily volume of customer orders in one lot. It means that the production batch has as many pieces as the average daily customer demand. Option 2 assumes the production of products twice a day, in production batches reduced by half. Option 3 assumes production three times a day, in production batches reduced three times. Unlimited storage capacities are also assumed. Production volumes and frequencies in individual variants are presented in Table 2.

It has been assumed that the simulation will be carried out for 180 working days. Customer daily orders were generated for each business day. Orders were generated randomly according to the normal distribution with a specified average (equal to the average daily customer demand) and standard deviation (equal to 5 pieces). Such conditions are most similar to conditions in reality. Production leveling should ensure the availability of products within the specified standard deviation range. A total of 1800 orders were generated. A fragment of the order table is presented in Table 3.

Table 1. Simulation data.

Product No.	M1 t_j [min.]	M2 t_j [min.]	M3 t_j [min.]	M4 t_j [min.]	n [pcs./day]	T_{pz} [min.]
1	0,43	0,2	0,4	0,23	220	2
2	0,38	0,18	0,38	0,2	210	2
3	0,34	0,18	0,35	0,19	200	2
4	0,66	0,28	0,6	0,28	160	2
5	0,7	0,3	0,65	0,3	140	2
6	0,72	0,31	0,7	0,31	120	2
7	0,9	0,45	0,88	0,43	120	2
8	0,89	0,4	0,85	0,41	80	2
9	0,92	0,49	0,9	0,47	75	2
10	1	0,55	0,82	0,54	60	2

Table 2. Production volumes and frequency in variants.

Product No.	Variant 1		Variant 2		Variant 3	
	Lot size [psc.]	Frequency	Lot size [psc.]	Frequency	Lot size [psc.]	Frequency
1	220	1 x day	110	2 x day	60	4 x day
2	210		100		50	
3	200		100		50	
4	160		80		40	
5	140		70		36	
6	120		60		30	
7	120		60		30	
8	80		40		20	
9	75		40		20	
10	60		30		16	

3.3 Simulation Model

The next stage of research was the preparation of the model of the manufacturing process. The model was implemented in FlexSim simulation environment and consists of (Fig. 2):

- sources of raw materials (Entrance),
- four buffers (Buffer 1, 2, 3, 4),
- six machines (Machine 1a, 1b, 2, 3a, 3b, 4),
- finished goods warehouse (Supermarket),
- an element enabling leaving products from the supermarket (Supermarket_exit),
- customer order acceptor (Orders)
- element issuing production orders (Orders_exit).

Table 3. Fragment of the order table, source: own work.

Product No,	Order quantity [pcs.]	Order time [min.]
1	200	450
2	210	450
3	220	450
4	170	450
5	150	450
6	105	450
7	135	450
8	70	450
9	95	450
10	40	450

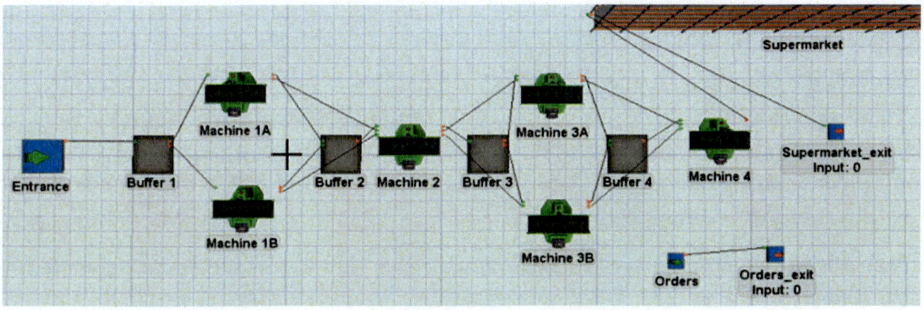

Fig. 2. Model of the manufacturing process.

The production plan, with operation times, changeovers times, the order in which the products are made, and the number of items to be manufactured, is implemented in the "Entrance" element. It is a source of raw materials that allows you to determine the specific number of products produced, and also to determine the time when the raw materials needed for production are to appear on the buffer. "Buffer" is an interstate magazine. When the machine finishes producing one piece of product, it takes another from the buffer. When the last machine in the production line finishes producing the product, the finished product goes to the supermarket shelves, where the FIFO principle applies.

The element defined as "Supermarket_exit" allows you to release products from the supermarket and control of the number of pieces of all products that were collected to carry out the customer's order. The "Orders" element specifies the size of the order, as well as the time after which they are carried out (orders appear once a day, at the end of the business day). The last element, "Orders_exit" allows you to complete orders and control the number of completed orders.

A table ("Order history") has also been implemented in the simulation model in which the simulation results will be automatically saved. The table contains the following elements (Fig. 3):

no_orders	name_product	item_type_product	order_quantity	submit_order	realization_order	time_of_realization_order	status_order	number_of_currentl	number_pieces_
1.00	Product1	1.00	200.00	450.00	489.84	39.84	zrealizowane	200.00	0.00
2.00	Product2	2.00	210.00	450.00	534.39	84.39	zrealizowane	210.00	0.00
3.00	Product3	3.00	220.00	450.00	981.91	531.91	zrealizowane	220.00	0.00
4.00	Product4	4.00	170.00	450.00	1018.76	568.76	zrealizowane	170.00	0.00

Fig. 3. Simulation results table ("Order history") example view.

- no_orders – number of orders,
- name_product and item_type_product – allows checking if another 10 orders concern 10 different products.,
- order_quantity – number of items ordered by the customer,
- submit_order – a time of order placement by the customer,
- realization_order – time during which the customer received the product,
- time_of_realization_order – if the order was completed after time, it allows checking after what time it was fully completed,
- status_order – in this case, the degree of order fulfillment is considered in two cases: completed and in progress, however, the latter after the end of the simulation is considered as an unrealized order,
- number_of_currently_realized_pieces – if the order has been completed, the value of this column is equal to the value from the "order_quantity" column,
- number_pieces_to_realize – if the order has the status in progress, i.e., it was not completed during the simulation, you can check in this column how many pieces were missing to complete the order.

The general scheme of order fulfillment in the simulation model is presented as a BPMN diagram in Fig. 4.

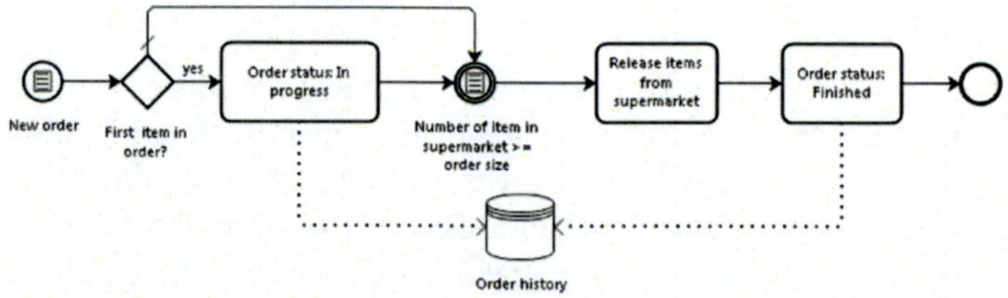

Fig. 4. Order fulfillment implemented in the simulation model.

4 Results

4.1 Realization of Orders

Figure 5 shows the implementation of orders for variant 1. In this option, most orders (1796 orders) were realized, with 1235 of them realized on time.

Figure 6 shows the execution of orders for variant 2. In this option, 55 orders were unrealized.

Realized on time [pcs.]	1235
Realized after time [pcs.]	561
Unrealized [pcs.]	4

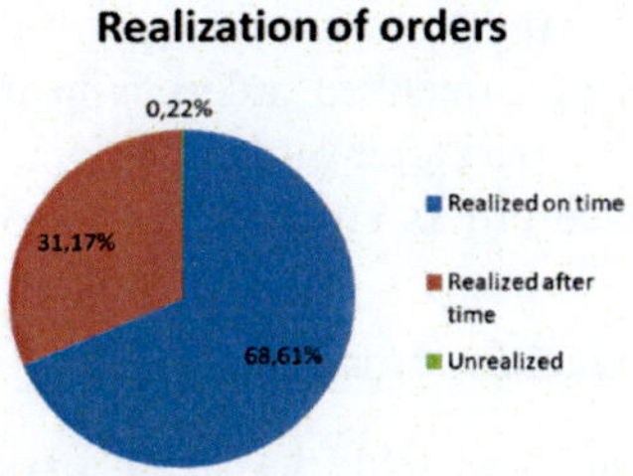

Fig. 5. Realization of orders for variant 1.

Realized on time [pcs.]	236
Realized after time [pcs.]	1509
Unrealized [pcs.]	55

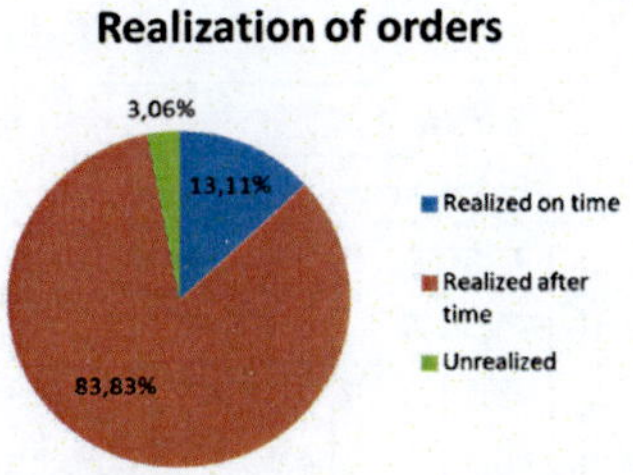

Fig. 6. Realization of orders for variant 2.

Figure 7 shows the execution of orders for variant 3. In variant 3, the highest number of unrealized orders was recorded. It is worth noting, however, that the percentage of orders realized after time is very similar to variant 2.

Realized on time [pcs.]	18
Realized after time [pcs.]	1594
Unrealized [pcs.]	188

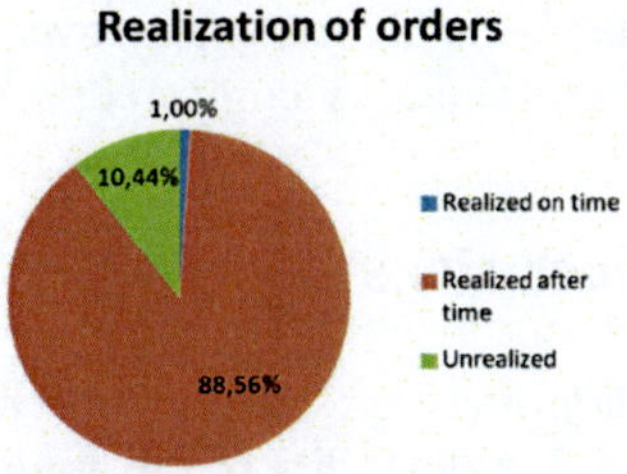

Fig. 7. Realization of orders for variant 3.

The summary of simulation results concerning the order fulfillment is presented in Table 4.

Table 4. Orders in different variants.

	Realized on time [pcs.]	Realized after time [pcs.]	Unrealized [pcs.]
Variant 1	1235	561	4
Variant 2	236	1509	55
Variant 3	18	1594	188

The largest number of orders realized on time was in variant 1. Thus the lowest number of unrealized orders is in this option. It is worth noting that the percentage of orders realized after time in variant 3 is very similar to variant 2. The largest number of unrealized orders arose in variant 3.

4.2 Machine Load

Machines load table has been exported from FlexSim (Table 5). The table contains information about changeover time (Setup), idle time (Idle), and work (Processing).

Table 5. Machines load.

	Variant 1			Variant 2			Variant 3		
	Processing	Setup	Idle	Processing	Setup	Idle	Processing	Setup	Idle
Machine 1a	95,50%	4,50%	0,00%	91,30%	8,70%	0,00%	84,30%	15,70%	0,00%
Machine 1b	95,50%	4,50%	0,00%	91,30%	8,70%	0,00%	84,30%	15,70%	0,00%
Machine 2	88,10%	4,50%	7,30%	81,90%	8,70%	9,40%	79,80%	15,70%	4,50%
Machine 3a	90,80%	4,50%	4,60%	86,80%	8,70%	4,50%	80,00%	15,70%	4,20%
Machine 3b	90,60%	4,50%	4,80%	86,80%	8,70%	4,50%	80,00%	15,70%	4,20%
Machine 4	85,60%	4,50%	9,80%	81,90%	8,70%	9,40%	81,60%	15,70%	2,70%

In variant 1, Machine 4 has the highest idle time value, while machines 1a and 1b have been in operation set up all the time. In variant 2, machine 4 and machine 2 have idle time value, while machines 1a and 1b have been in operation or set up all the time. In variant 3, the highest changeover times can be seen, which are caused by the highest production frequency compared to other variants.

5 Conclusions

Production leveling stabilizes the production schedule, as well as ensures an appropriate level of flexibility of the production system while maintaining the lowest level of stocks. The presented research is a fragment of research on the development of a method for planning the flow of products from the production process. This method is to enable the creation of a leveled production plan.

The paper focuses on determining the variant of the production batch size and production frequency for the selected model of the production process, which with the given input parameters, will obtain the best results in terms of the degree of order processing and machine load. Three variants of the production plan were proposed, a computer simulation of the process was created in FlexSim, and the results of the simulation were analyzed. The analysis aims to indicate the best variant of the production plan.

Variant 1 assumed the production of the average daily volume of customer orders in one batch. Variant 2 assumed the production of products twice a day, in production batches reduced by half. Variant 3 assumed production three times a day, in production batches reduced three times.

From the results presented, it can be seen that the largest number of orders completed on time were in variant 1. Thus the lowest number of unrealized orders is in this option. It may be because, in this variant, the least time of all available working time was spent on setup or idle. It is worth mentioning that the percentage of orders processed over time in variant 3 is very similar to variant 2. Most unrealized orders arose in variant 3. Therefore, for the presented case, the most favorable batch size is one in which each product is produced in one batch. It means that the batch should be as large as possible and produced with the lowest frequency possible.

The article presents a fragment of conducted research. Simulations are currently being carried out for different input data and different manufacturing process models. Work is also underway to refine the simulator in the FlexSim simulation program, which will enable even more accurate reflection of the actual elements of the production system.

References

1. Liker, J.K.: The Toyota Way, 14 Management Principles from the Worlds Greatest Manufacturer, 1st edn. McGraw-Hill, London (2004)
2. Korytkowski, P., Grimaud, F., Dolgiu, A.: Exponential smoothing for multi-product lot-sizing with heijunka and varying demand. Manage. Prod. Eng. Rev. **5**(2), 20–26 (2014)
3. Dennis, P.: Lean Production Simplified, 2nd edn. Productivity Press, New York (2007)
4. Dolgui, A., Proth, J.M.: Supply chain engineering: useful methods and techniques. Springer (2010)
5. Rother, M.: Toyota Kata: Managing People for Continuous Improvement and Superior Results. McGraw-Hill, London (2009)
6. Veleva, V., Hart, M., Greiner, T., Crumbley, C.: Indicators of sustainable production. J. Clean. Prod. **9**(5), 447–452 (2001)
7. Rewers, P., Hamrol, A., Żywicki, K., Kulus, W., Bożek, M.: Production leveling as an effective method for production flow control - experience of polish enterprises. Procedia Eng. **182**, 619–626 (2017)
8. Monden, Y.: Toyota Management System. Productivity Press, New York (1993)
9. Andel, T.: Accentuate heijunka, eliminate junk, supply chain flow. Mater. Handling Eng. **54** (8), 77 (1999)
10. Coleman, J.B., Vaghefi, M.: Heijunka: a key to the Toyota production system. Prod. Inventory Manage. J. **34**(4), 31–35 (1994)
11. Rinehart, J.: After lean production: evolving employment practices in the world auto industry. Am. J. Sociol. **104**(4), 1212–1214 (1997)
12. Araujo, L.F.D., Queiroz, A.A.D.: Production Leveling (Heijunka) implementation in a batch production system: a case study. International Federation for Information Processing, pp. 105–112 (2010)
13. Bohnen, F., Maschek, T., Deuse, J.: Leveling of low volume and high mix production based on a Group Technology approach. J. Manuf. Sci. Technol. **4**, 247–251 (2011)
14. Bohnen, F., Buhl, M., Deuse, J., Schneider, R.: Effiziente Kleinserienfertigung durch Produktionsnivellierung. Productivity Management **14**, 19–22 (2009)

15. Liker, J.K., Meier, D.: The Toyota Way Fieldbook, A Practical Guide for Implementing Toyota's 4Ps, 1st edn. McGraw-Hill, London (2011)
16. Rewers, P., Trojanowska, J., Diakun, J., Rocha, A., Reis, L.P.: A study of priority rules for a levelled production plan. In: Advances in Manufacturing. Lecture Notes in Mechanical Engineering, pp. 111–120, Springer (2018)
17. Rewers, P., Żywicki, K., Diakun, J.: Comparison study of different production control policies in condition of various demand for final products. In: Advances in Manufacturing II Volume 2 - Production Engineering and Management, Lecture Notes in Mechanical Engineering, pp. 268–280. Springer (2019)

Product-Service System: A New Opportunity for the Printing Industry

Mariusz Salwin[1(✉)], Krzysztof Santarek[1], Andrzej Kraslawski[2], and Jan Lipiak[3]

[1] Warsaw University of Technology, 85, Narbutta Street, Warsaw, Poland
mariusz.salwin@onet.pl
[2] Lappeenranta University of Technology, P.O. Box 20, 53581 Lappeenranta, Finland
[3] Etigraf Printing House, 52, Głowackiego Street, 05-071 Sulejówek, Poland

Abstract. Product-Service System (PSS) is a combination of products and services to meet specific customer needs. It is a concept that allows companies to build a competitive edge and supports sustainable development. The area of PSS design methods, as well as industrial practice, do not show examples of how this approach can be used in the printing industry. This article aims to develop a conceptual model of PSS for the printing industry. The presented example was developed during a workshop conducted in a printing company. Creating a conceptual model of PSS based on real problems, needs, and service expectations of the company, we hope to draw attention to a number of important issues. The created PSS model provides the printing house with tools in the form of services that will eliminate production problems, improve production efficiency, minimize the adverse impact of printing production on the environment and affect the extension of the machine life cycle.

Keywords: Product-Service System (PSS) · Business model · Printing machines

1 Introduction

Changes in the economic structures in which manufacturing companies move away from production only, towards the provision of products to which services are added. This paradigm shift is related to the global trend of increasing the share of the services sector in national economies [1–3]. In 2018, the industry accounted for over 65% of global GDP. In the service economy, the satisfaction of individual customer needs plays a key role. Customers themselves are not interested in purchasing from owning a product, but in the opportunities, it offers [4]. All this makes the research in the area of the Product-Service System intensify in recent years [5, 6].

Product-Service System is a concept developed in Scandinavia. It is a special case of servicing [7]. The strategy based on the Product-Service System focuses on meeting the needs of customers and not on the product itself, which is a response to changes in the economy and society. Although this concept may bring more profit to the company, the real goal of its implementation is to achieve the best possible product and service

V. Tonkonogyi et al. (Eds.): InterPartner 2020, LNME, pp. 83–95, 2021.
https://doi.org/10.1007/978-3-030-68014-5_9

configuration. This configuration is to meet customer needs as much as possible, be environmentally friendly, and enable the manufacturer to achieve maximum profit at the lowest cost [8–11].

The Product-Service System is closely linked to sustainability [12]. This is primarily related to the more efficient use of materials and reduction of waste [13]. This is very important for increasing public awareness and environmental policy [14].

Environmental and sustainability issues are significant for the printing industry, which generates different types of waste. Each printing house is committed to pro-environmental measures and responsible waste management. Waste management creates high costs for companies and forces them to take very thoughtful action [15].

Printing is a characteristic production sector, which includes not only printing and bookbinding services but also machines and equipment and materials used in the production process. Printing has played a key role in the development of the Renaissance and the scientific revolution and laid the foundations for a modern economy based on knowledge and its dissemination. Over the centuries, printing has undergone significant changes. It has entered an industrial phase, whose development is taking place at an even faster pace. In modern printing, we distinguish several basic printing methods and machines used in each of them [16–18].

Printing company owners, analysts, and researchers gathered at the 22nd Polygraphic Symposium in 2019 identified ten trends that will affect the printing industry. One of the issues discussed in detail was the implementation of solutions based on the Product-Service System in printing houses, where the customer does not pay for the printing machines and equipment, but for using them. This issue was considered to be the way forward for the printing industry.

The article aims to develop the Product-Service System model for the printing sector. The presented example was developed during a workshop conducted in a printing company.

2 Literature Review

2.1 Product-Service System in Industrial Practice

This stage of analysis concerned PSS models used in the industry. The analyzed PSS models have been developed in large companies and are addressed to various industries. Another characteristic feature of the products offered in this model is their high value, technological level, and long life cycle. Among the listed PSS models, it was not possible to find a model used in the printing industry.

2.2 Product-Service System Design

The literature analyzed gives a number of examples of case studies in which given PSS design methods were applied in practice. Not all of the methods examined were precisely assigned to a specific industry, therefore, for the purpose of this study, we have assigned the techniques to particular economic sectors (Fig. 1).

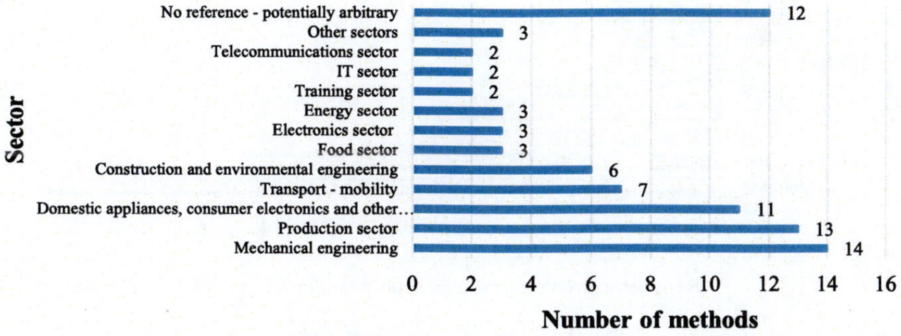

Fig. 1. Classification of PSS design methods by sector [19].

At this moment, it should be noted that some methods are addressed to several completely different industries. Most of the methods, 14, are aimed at mechanical engineering [19]. As many as 12 design methods for PSSs could not be allocated to any industry, which proves that they can be used in any sector of the economy [19]. It is worth noting that the available literature does not provide PSS design methods addressed to the printing industry.

2.3 Printing Industry

The printing sector is responsible for producing newspapers, books, magazines, packaging, brochures, labels, advertising catalogs, direct marketing materials, and other promotional materials. The unique feature distinguishing them from other products are in the possession and transmission of information. They are printed with the appropriate text and illustrations [20–22].

The largest printing markets include China, USA, India, Brazil, and EU countries. In Europe, the printing industry is an important employer and has a long tradition. This sector is linked to other industries.

The revenues of printing companies operating in the EU countries in 2009–2015 amounted to an average of 87,985,000,000 EUR (Figs. 2 and 3). In the EU countries, 725,800 people work in this sector in 119,591 companies [20–22].

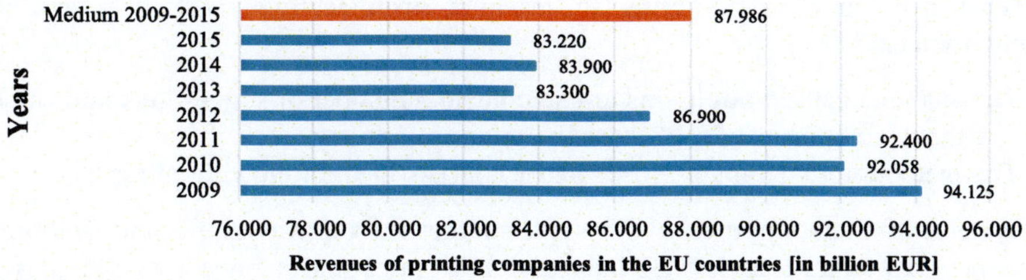

Fig. 2. Revenues of printing companies in the EU countries [in billion EUR] [20–22].

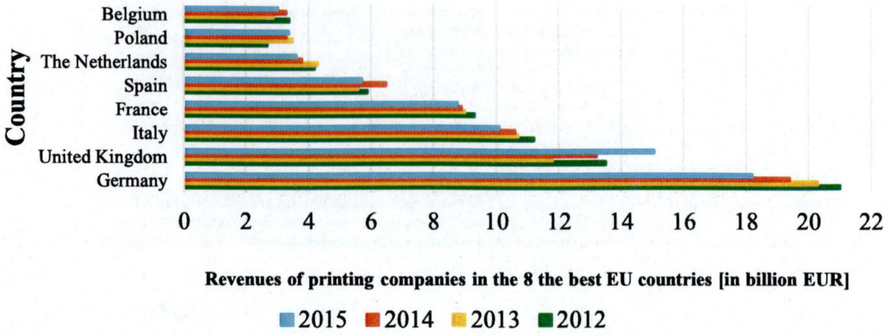

Fig. 3. Revenues of printing companies in the EU countries [in billion EUR] [20–22].

2.4 Printing Machines

Printing machines are a strategic sector for world economies. They are key capital goods because the products they produce are essential for the functioning of other branches of the economy.

Printing machines are the main element of the printing house equipment. They are used to make prints on an industrial scale. The machines differ in construction, the way the fonts are assembled, and the method of printing on a given substrate. They are durable, technically advanced devices that, when used properly, are efficient and fast in operation. The essential features of printing machines from the customer's point of view include roll and print width, roll diameter, and print speed.

At present, the main manufacturers offer on the market specialized printing machines in various variants of equipment together with basic services (installation, maintenance, training).

3 Research Methodology

The article aims to develop the Product-Service System model for the printing sector. The article formulates the following research questions:

- What are the possibilities of using PSS in the printing industry?
- What benefits can PSS bring to machine manufacturers, customers and the environment?

The analyses carried out in this article contain a number of suggestions that can be used in the development of a PSS for the printing industry.

The methodology adopted in this paper consists of the following phases:

1. Systematic literature review. This phase focuses on two equivalent activities, namely a literature review of industrial PSS cases and a review of PSS design methods. The authors searched for the term "Product-Service System in industry" or its synonyms. Databases were used in the search (ProQuest, Springer Link, Science Direct, Taylor & Francis Online, EBSCOhost, Scopus, Emerald, Insight, Web of

Science, Ingenta, Dimensions, Wilma, IEEE Xplore Digital Library and Google Scholar). The next step was to define the selection criteria. The authors focused on works written in English (articles published in magazines, conference materials, book chapters, reports, and white papers). The result was 120 works, in which PSS models operating in the industry were characterized. The works covered the period 2001–2019. Using the same databases and time frames, the authors searched for the term "Product-Service System design" or its synonyms. The result was 64 articles (including scientific articles, conference materials, monographic chapters, and books), in which 60 PSS design methods were found.

2. Analysis of the printing industry. This stage focuses on the analysis of users and manufacturers of printing machines. In this stage, industry reports were used for analysis.
3. Company survey. They were investigating problems and needs related to the industrial printing machine, demand for additional services dedicated to printing machines in a flexographic printing house. The Pareto–Lorentz analysis was used in the analyzed company to identify problems that generate significant losses and costs for the company. Then, using brainstorming, the primary service needs of the enterprise were identified. Based on this information, a questionnaire was developed containing proposals of services that were to respond to the problems and needs of the enterprise. The services in the questionnaire were divided into three groups. From each group, the company's employees selected the essential services from their point of view.
4. Designing the Product-Service System model. Based on a literature review and company research, the PSS model for industrial printing machines was developed. The new PSS model contains the services that were the most popular and the most beneficial for the company.

4 Results

4.1 Characteristics of the Analyzed Company

The analyzed company for flexographic printing is with a long tradition and experience. The company specializes in the production of labels and laminates. The printing house has a professional graphic studio, modern machinery. The printing house employees are a creative team with many years of experience. Thanks to these resources, the company can meet many tasks, starting from graphic design, through printing, finishing, and ending with delivering the finished product to the customer. Throughout its activity, the company enjoys the trust of a large group of customers from many industries (meat, food, pharmaceutical, cosmetic, and chemical). Among the wide range of labels produced by the company, there are five main types of products: self-adhesive labels, heat-shrinkable labels, OPP films, tea tags, laminates.

Thanks to continuous development, we are able to print any type of label up to a maximum width of 430 mm and a repeatable length of 600 mm on all types of paper and film. In the analyzed flexographic printing house, there are printing machines which are built in a modular system (BOBST M5). The modern machine park allows us to print materials of up to 12 colors and to be additionally refined with UV varnish, foil, gilding with cold-stamping and hot-stamping and in-line embossing. There is also a possibility of multi-layered label printing from the adhesive side.

4.2 Company Problems and Needs

This phase aimed to identify the problems and needs of the printing house, which is a user of printing machines. The analysis took place in the form of workshops carried out in the analyzed company.

4.2.1 Company Problems

As mentioned above, the company is well established on the market. However, I am still struggling with a number of problems. In the analyzed company, use Pareto-Lorentz analysis to identify them (Fig. 4). These problems are still being reduced but still generate high costs for the company. In 2018, the key ones were the rebuilding of the printing machine, printer failures and stoppages, finished product incompatibilities, waste disposal, material losses, and training. All these factors are linked to each other and result in losses for the company.

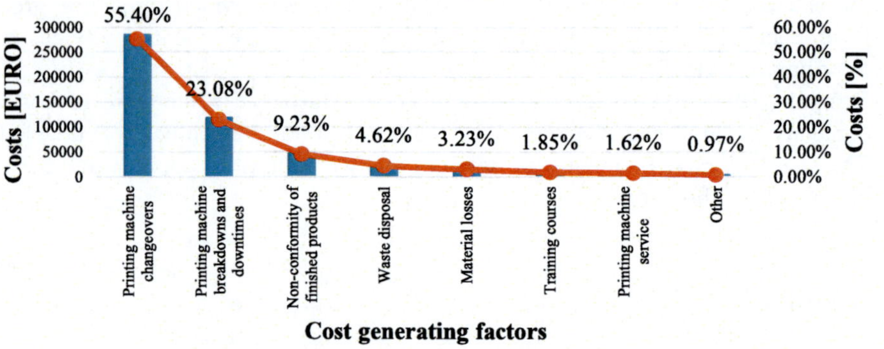

Fig. 4. Analysis of losses of a printing company in 2018.

The most significant losses for the printing house are caused by machine changeovers, breakdowns and downtimes is in total 77.8%. The factor worth looking at is waste disposal. All this creates the effective working time of a printing machine during a working day is 55%. This is also related to the low Overall Equipment Effectiveness, which averaged 50% in 2018.

The high machine changeover time is specific for the printing industry, especially for flexography. This type of production takes place on order, so a dozen or so different orders can be carried out daily, between which there is always a changeover and a number of activities related to it. An important role in the whole printing process is played by operators who, based on their own knowledge, determine the order of orders themselves. The order of orders outside the execution date is influenced by the changeover time, color or raw material width. Therefore, the training of operators in which the analyzed company invests becomes a significant issue. The cost of waste disposal is a noteworthy fact.

It is worth noting that almost every printing company faces similar problems.

4.2.2 Company Needs

The workshops have identified the main needs that are related to the following areas:

- training - purchasing, upgrading, or retrofitting a machine with available options and tools that were not purchased early, requires re-training of the machine operators.
- the printing process - selection and mixing of ink, cleaning of printing equipment, changeover, and development of new printing forms, diagnosis of printing errors.
- operation and service of the machine – replacement of lamps, filters, UV mirrors, and other machine parts do not take place immediately, but after a certain time indicated by the service.
- waste management – despite the fact that the company has international certificates and standards and is constantly improving, this area of activity is very troublesome and, at the same time, involves high financial costs.

These points have a significant impact on the machine's performance and, therefore, on financial losses. There is also a burden on the environment due to unnecessary use of energy, water, and disposal of paint and products that do not meet quality requirements.

4.2.3 The Service Needs of the Company

In the next stage of the workshop, service proposals for the PSS model for printing machines were developed. The services were developed in order to respond to the problems and needs of the company so that the new PSS model meets its strict requirements. Three groups of services were created: universal services related to the printing machine, services related to the printing process, additional services.

From each of the groups of services, representatives of the management board, middle management, and production employees selected the most essential services from their point of view (Table 1).

Table 1. Services selected by the company's employees from the broadened areas.

	Universal services relating to printing machine	Print process services	Additional services
The Management Board	Financial services	Selection, delivery, and optimization of paint consumption for a specific order	Take-back
	Delivery, installation, and start-up of the machine	Optimization of printing machine conversions	Optimizing the use of equipment
	Health and safety services	Optimal time of preparation and realization of printing forms	Optimizing the use of materials
	Training courses	Preparation of CTP discs	Disposal of production waste
		Disposal of paint and substrates	Update (Reconstruction, upgrade)
		Exchange of mirrors, filters and UV lamps	Quality control
Middle management	Guarantee	Choice of inks and printing substrate	Diagnostics and error recovery
	Training courses	Technological process design	Lean tools
	Supply of spare parts	Exchange of mirrors, filters and UV lamps	Production planning and monitoring
	Regeneration, overhaul, repair, maintenance	Selection, delivery, and optimization of paint consumption for a specific order	Monitoring, testing, and diagnosis of machine operation
		Optimization of printing machine conversions	Quality control
		Optimal time of preparation and realization of printing forms	
		Preparation of CTP discs	
Production workers	Cleaning	Washing rollers, aniloxes	Optimizing the use of equipment
	Noise reduction	Selection, delivery, and optimization of paint consumption for a specific order	Optimizing the use of materials
	Training courses	Optimization of printing machine conversions	Disposal of production waste
	Health and safety services	Optimal time of preparation and realization of printing forms	Quality control
		Preparation of CTP discs	

At this stage, the company has confirmed its interest in hiring a machine and services that will improve the machine and the printing process. Additionally, the analyzed company was interested in a monthly subscription for using the machine more than in buying it.

4.3 Product-Service System for Printing Industry

Using an approach based on the PSS philosophy, as well as the knowledge gained on the problems and needs of the company, the next stage of the workshop was to build a PSS model. In the analyzed system, the main sides of the system, in this case include the machine manufacturer and the customer. In the model, the main components will be the printing machine and services the company is interested in (Table 2).

Table 2. Service packages in Product-Service System for the printing industry.

Area	Service package 1 – elementary	Service package 2 – intermediate	Service package 3 – advance
Training	Training courses	Health and safety services	Technological process design
Printing process	Cleaning	Selection, delivery, and optimization of paint consumption for a specific order	Optimizing the use of equipment
	Noise reduction	Optimization of printing machine conversions	Optimizing the use of materials
		Optimal time of preparation and realization of printing forms	Production planning and monitoring
		Preparation of CTP discs	Quality control
		Exchange of mirrors, filters and UV lamps	Lean tools
Operation and service of the machine	Delivery, installation, and start-up of the machine	Choice of inks and printing substrate	Monitoring, testing, and diagnosis of machine operation
	Supply of spare parts	Washing rollers, aniloxes	Take-back
	Guarantee		Diagnostics and error recovery
	Regeneration, overhaul, repair, maintenance		Update (Reconstruction, upgrade)
Waste management		Disposal of paint and substrates	Disposal of production waste
Additionally	Financial services	Financial services	Financial services

A company producing flexographic printing machines delivers them to the customer. Together with the machine, the customer receives a service package (Table 2, Fig. 5). The manufacturer charges a fixed fee per printing hour and a fixed monthly subscription fee. It is worth noting that the ownership will not be transferred to the user but remains with the machine manufacturer. The customer only needs to use the machine, i.e., to print as many orders as possible, and not to own the machine itself and deal with a number of issues related to its maintenance. The machine itself can be exchanged for a new one after a fixed period of time.

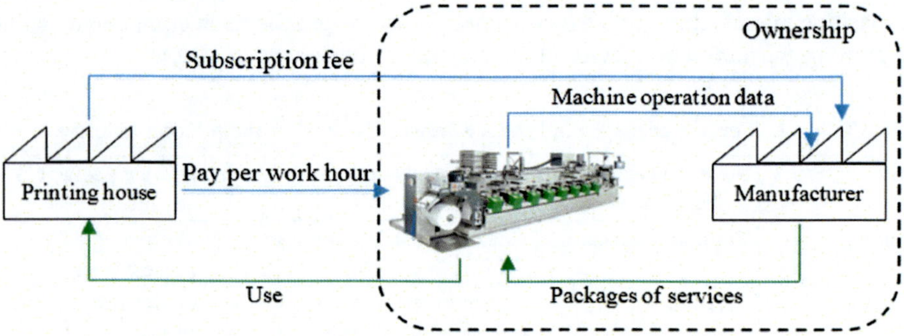

Fig. 5. Product-Service System for the printing industry —a concept.

An approach based on the philosophy of product and service systems brings a lot of benefits to all parties to the transaction. The client focuses primarily on his core business, i.e., printing and expanding the printing house. It is also not interested in service and repair activities as well as cleaning. An additional benefit for him is that he does not have to train people and buy the equipment needed for the mentioned activities himself, which saves time and resources. Also very important here are situations in which it can be enough to break down. The manufacturer is also obliged to remove it quickly because it is also in his interest to keep the machine running continuously. This approach allows the customer to increase the number of orders (Table 3).

A manufacturer of a printing machine earns money at the same time from the production of machines and services that are related to this machine. It is primarily in his interest to use the best possible materials and technological solutions to build from a high-quality machine that will have a long life cycle. This will ensure the possibility of data collection so that the manufacturer will be able to improve their parameters in new generations of machines and improve the printing process. It is essential that the elements of the printing machine which have a short life cycle can be quickly replaced. This is all to guarantee the reliability, minimize expenses on possible repairs and maintenance of the machine.

Table 3. Product-Service System for the printing industry —a concept.

Ownership	Sale	Services	Advantages for the manufacturer	Customer benefits
Ownership is retained by the manufacturer of the printing machine	Monthly subscription	Service package 1 - elements	Longer and more lasting relationships with customers	Focus on core business
		Service package 2 - intermediate	Control and monitoring of machines	Elimination of unnecessary costs
	Fee per worked hour	Service package 3 - advance	Improving the performance of printing machines	Increase in production
		Additional services	Less impact on the environment	

5 Conclusion

The article analyses the printing market. This sector is characterized by a large number of small and micro-enterprises. Such a large number of this size of enterprises proves a great application potential, a wide network of clients, and a large sales market. The specificity and size of this sector give a wide range of new business models for manufacturers of printing machines. It is worth noting that the manufacturers' offer for new offers based on PSS is very limited.

Product-Service System is a comprehensive solution that meets the needs and requirements of customers. An essential element of worldwide research is the use of all theoretical aspects of PSS and its implementation in industrial practice. In the article, there were conducted PSS design workshops for the printing industry. On this basis, the main problems and needs of the company and services that meet them were integrated. The printing machine, integrated with the services that the company really needs, has been integrated into the whole PSS system.

The following conclusions can be drawn from the design workshops conducted:

- in generating new PSS models, it is necessary to examine customer needs and problems. The active participation of representatives of the organization using the product is indicated in these activities.
- by properly adjusting services to the customer's problems and needs, it can radically minimize the adverse impact of printing production on the environment and at the same time extend the machine life cycle,
- the selected services should be divided into packages, so that the customer is not forced to use all the services at once, but can only choose the ones he needs,
- the package of services related to the printing process is significant support for users of printing machines, which has not been offered on the market so far,

- the developed concept of the PSS model for printing machines can directly or indirectly increase the production capacity of a printing house by eliminating time losses related to changeovers and breakdowns,
- the subscription fee does not generate costs associated with the depreciation of the machine for the customer. Additionally, the customer does not have to invest large amounts of money to purchase the machine,
- retaining ownership on the manufacturer's side allows him to continuously control the operation of the machine so that he sees the weakest links and components of the machine, which he will improve in future generations,
- this model may be addressed in particular to micro, small and medium-sized enterprises and people wishing to start-up in the printing industry,
- in the model in question, there are practically no restrictions on adding services related to printing machines,
- an important challenge for the manufacturer may be to create a service network and to standardize and standardize the production of machines.

Besides, the design workshops provide a solid basis for developing and implementing a PSS model for printing machines. It shows which aspects need to be addressed and which elements need to be addressed when creating a fully customized PSS.

References

1. Salwin, M., Kraslawski, A., Lipiak, J., Golebiewski, D., Andrzejewski, M.: Product-service system business model for printing houses. J. Clean. Prod. **274**, 122939 (2020)
2. Salwin, M., Kraslawski, A., Lipiak, J.: State-of-the-art in product-service system classification, In: Ivanov, V., Trojanowska, J., Pavlenko, I., Zajac, J., Perakovic, D. (eds.) Advances in Design, Simulation and Manufacturing III, pp. 187–200. Springer (2020)
3. Salwin, M., Gladysz, B., Santarek, K.: Technical product-service systems—a business opportunity for machine industry. In: Hamrol, A., Ciszak, O., Legutko, S., Jurczyk, M. (eds.) Advances in Manufacturing, pp. 269–278. Springer International Publishing, Cham (2018)
4. Baines, T.S., Lightfoot, H., Benedettini, O., Whitney, D., Kay, J.M.: The adoption of servitization strategies by UK-based manufacturers. Proc. Inst. Mech. Eng. Part B J. Eng. Manuf. **224**, 815–829 (2010)
5. Beuren, F.H., Gomes Ferreira, M.G., Cauchick Miguel, P.A.: Product-service systems: a literature review on integrated products and services. J. Clean. Prod. **47**, 222–231 (2013)
6. Maussang, N., Zwolinski, P., Brissaud, D.: Product-service system design methodology: from the PSS architecture design to the products specifications. J. Eng. Design **20**, 349–366 (2009)
7. Goedkoop, M., van Haler, C., te Riele, H., Rommers, P.: Product Service-Systems, ecological and economic basics (1999)
8. Manzini, E., Vezzoli, C.: A strategic design approach to develop sustainable product service systems: examples taken from the 'environmentally friendly innovation' Italian prize. J. Clean. Prod. **11**, 851–857 (2003)
9. Mont, O.K.: Clarifying the concept of product–service system. J. Clean. Prod. **10**, 237–245 (2002)

10. Morelli, N.: Designing product/service systems: a methodological exploration. Des. Issues **18**, 3–17 (2002)
11. Neely, A.: Exploring the financial consequences of the servitization of manufacturing. Oper. Manage. Res. **1**, 103–118 (2008)
12. Boons, F., Lüdeke-Freund, F.: Business models for sustainable innovation: state-of-the-art and steps towards a research agenda. J. Clean. Prod. **45**, 9–19 (2013)
13. Sundin, E., Nässlander, E., Lelah, A.: Sustainability indicators for small and medium-sized enterprises (SMEs) in the transition to provide product-service systems (PSS). Procedia CIRP **30**, 149–154 (2015)
14. Roy, R.: Sustainable product-service systems. Futures **32**, 289–299 (2000)
15. Lipiak, J., Salwin, M.: The improvement of sustainability with reference to the printing industry – case study. In: Hamrol, A., Grabowska, M., Maletic, D., Woll, R. (eds.) Advances in Manufacturing II, pp. 254–266. Springer International Publishing, Cham (2019)
16. Rees, F.: Johannes Gutenberg: inventor of the printing press. Compass Point Books, Minneapolis, Minn (2006)
17. Eisenstein, E.L.: The Printing Revolution in Early Modern EUROPE. Cambridge University Press, Cambridge (2005)
18. Kelley, M.R., Sorce, P.: Printing as an Industry, Commodity, and Activity: an Economic Analysis of Growth and Inter-industry Transactions. RIT Printing Industry Center, Rochester (2006)
19. Salwin, M., Kraslawski, A., Lipiak, J.: State-of-the-art in Product-Service System design, In: Panuwatwanich, K., Ko, C-H. (eds.) The 10th International Conference on Engineering, Project, and Production Management, pp. 645–658. Springer (2020)
20. KPMG, Polish Brotherhood of Gutenberg Knights, Faculty of Journalism, Information and Bibliology, University of Warsaw: The printing and printed packaging market in Poland (2018)
21. KPMG, Polish Brotherhood of Gutenberg Knights, Faculty of Journalism, Information and Bibliology, University of Warsaw: The printing and printed packaging market in Poland (2016)
22. KPMG, Polish Brotherhood of Gutenberg Knights, Faculty of Journalism, Information and Bibliology, University of Warsaw: The printing and printed packaging market in Poland (2015)

Rationalization of Grain Cargoes Transshipment in Containers at Port Terminals: Technology Analysis and Mathematical Formalization

Natalya Shramenko[1,2](✉), Dmitriy Muzylyov[1], and Vladyslav Shramenko[1,3]

[1] Kharkiv Petro Vasylenko National Technical University of Agriculture, 44, Alchevskyh Street, Kharkiv 61002, Ukraine
nshramenko@gmail.com

[2] Ukrainian State University of Railway Transport, 7, Feierbakh Square, Kharkiv 61050, Ukraine

[3] V. N. Karazin Kharkiv National University, 4, Svobody Sq., Kharkiv 61022, Ukraine

Abstract. The grain market is actively developing in Ukraine. The grain is exported mainly through ports. Since current scientific regulations and practice requirements indicate the relevance of logistics principles, an efficient organization and operation of delivery systems require a detailed study of transportation processes according to their compliance with logistics principles. The study aim is the mathematical formalization of grain cargoes transshipment process in containers at port terminals using a multimodal delivery system. Research tasks can be determined and solved when the system approach, methods of analysis and synthesis, logistics, tools of economic and mathematical modeling were used. This can help to achieve stated goals in a short-time research period. Marine terminal complexes of Ukraine are at the stage of active reform. This fact was revealed from general researches for the practical state of organization issue of transportation in combined railway and water transport. Almost all indicators show a significant increase in volumes of goods transshipment, and above all, cereals, which is due to the quick growth of agricultural production. The efficiency criterion was designed for the transshipment processes of containers with cargo at the port terminal. This is total specific costs associated with cargo transshipment in containers at port terminals. Analytical dependencies were obtained for certain components of costs related to containers overloading according to "railway carriage-ship" technology and containers transshipment of through warehouses as a result of the analysis of technologies.

Keywords: Grain · Transportation · Multimodal systems · Costs · Technology of cargo transshipment · Seaport · Terminal · Warehouse · Railway carriage · Ship · Efficiency criterion

V. Tonkonogyi et al. (Eds.): InterPartner 2020, LNME, pp. 96–105, 2021.
https://doi.org/10.1007/978-3-030-68014-5_10

1 Introduction

The grain market is actively developing in Ukraine. The grain is exported mainly through ports. The grain is supplied to port terminals by road [1], and rail [2] transports. However, there is an insufficient efficient organization of cereals cargo turnover in grain supply chains from grain elevators to ports. One of the reasons for inefficient logistics is insufficient load-carrying possibilities that are caused by limited quantities of specialized rolling stock (railway wagon for grain, trucks for grain delivery) [3] and also its irrational use [4, 5]. Excessive rolling stock idle at terminals leads not only to increased costs over the entire logistics chain [6] but also to increased time of vehicle turnover [7].

Since current scientific regulations and practice requirements indicate the relevance of logistics principles, the efficient organization and operation of delivery systems require a detailed study of transportation processes according to their compliance with logistics principles. Technological processes associated with the handling of containers in seaports require additional research. Optimization of the container transshipment process at the port will reduce the cost of grain delivery.

2 Literature Review

The recent emphasis on organization [8] and efficient operations of multimodal cargo delivery systems [9] is due to reductions in production and trade cycle length [10] and cooperation of participants [11], increases in storage costs of cargoes [12] and necessities to accelerate responses to consumer demand [13].

Ukraine is quickly developing grain infrastructure at terminals [14]. Thus, grain business is a promising direction for Ukraine and requires further development both for export supplies and for processing complexes inside the country.

The projected further growth of grain delivery [15] has led to various transport technologies applications [16] for both export and domestic transportations. Alternative conventional methods of grain transportation by road [17] and railway [18, 19] are piggyback [20, 21], bimodal [22, 23] and container [24]. There is an imbalance between container cargoes turnover in opposite directions along corresponding routes. The economic factor has no significant impact on delivery options choosing [25] in this case.

The growing trend will continue for grain transportation by container according to expert opinion. These conclusions based on world experience (Japan, USA, Canada, Argentina, Australia) [25–27], as well as national peculiarities [28].

Studies have shown that failures at seaports can cause undesirable wave effects that negatively affect transport network operation in wholes, as well as economic and social aspects [29, 30]. Therefore, effective cooperation among all subjects is a prerequisite for the effective delivery of integrated logistics services during supply, especially in terms of transport network requirements [31, 32], environmental protection, and general safety [33]. This aspect will also give result in expected benefits incoming from complex transport [34] and logistics operations [35].

Part of containers is transshipment at ports directly from land modes of transport to ships according to direct options [36], and part is going into warehouses [37] for short-time storage. The study results [38] show that direct overload mode can significantly reduce operating costs.

Thus, models and approaches require further development, especially in sphere related to operation organizing in interaction modules (including port terminals).

3 Research Methodology

Port infrastructure [35] and management structure [39] should also facilitate effective use of port resources, railway carriage [40, 41], and vehicles. Besides, excessive downtime of railway rolling stock arriving at the port for container loading onto ships should be avoided to reduce the cost of its downtime and its turnaround time.

It is necessary to optimize the interaction of different transport kinds at marine terminals, which will allow achieving minimum transportation costs [42].

The study aim is the mathematical formalization of grain cargoes transshipment process in containers at port terminals using a multimodal delivery system.

Research problems:

- To analyze container cargo supply dynamics in Ukraine in railway and water transport, including grain transportation;
- To analyze alternative technologies for transshipment of containerized cargo;
- To create an efficiency criterion for process assessment of container cargoes transshipment in ports;
- To formalize port terminal costs associated with container cargo transshipment using alternative technologies.

Research tasks can be determined and solved when the system approach, methods of analysis and synthesis, logistics, tools of economic and mathematical modeling were used. This can help to achieve stated goals in a short-time research period.

4 Results

4.1 Analysis of Transshipment Trends in Ukrainian Seaports

The annual volume of cargo transshipment exceeded 160 million tons for the first time [43] at 13 active Ukrainian ports in 2019.

Maximum indicators were recorded for grain cargoes transshipment in Ukrainian seaports (Fig. 1) in 2019.

Container transportation took third place among cargo in 2019 [43, 44], and first time during the last ten years exceeded the volume of 1 million TEU transshipment (Fig. 2).

Container cargo transshipment shows a significant growth during 2016–2019 in Ukraine (Fig. 2).

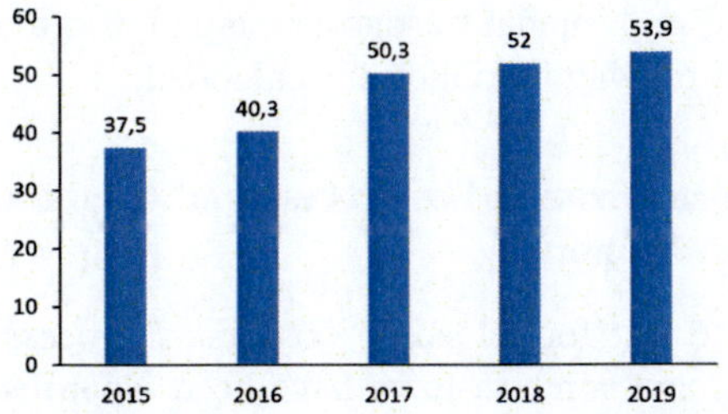

Fig. 1. Dynamics of grain cargo transshipment in Ukrainian seaports, million tons. Source: Created by the authors based on [43, 44].

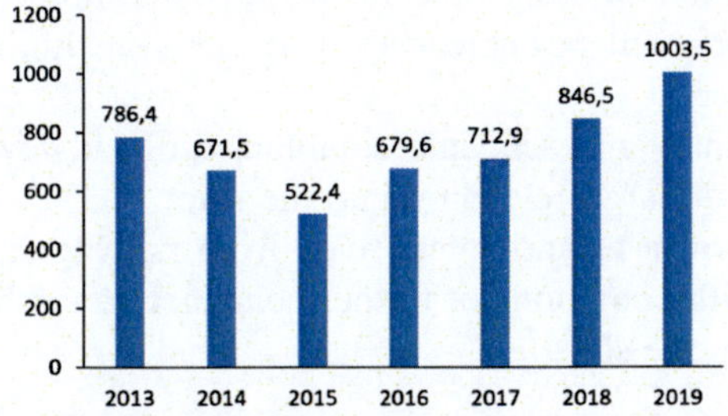

Fig. 2. Container turnover through Ukrainian seaports during 2013–2019, thousands TEU. Source: Created by the authors based on [43, 44].

The largest influence was provided by the Ukrainian agricultural sector (Fig. 3) into record transshipment of cargo at seaports [43, 44] in 2019.

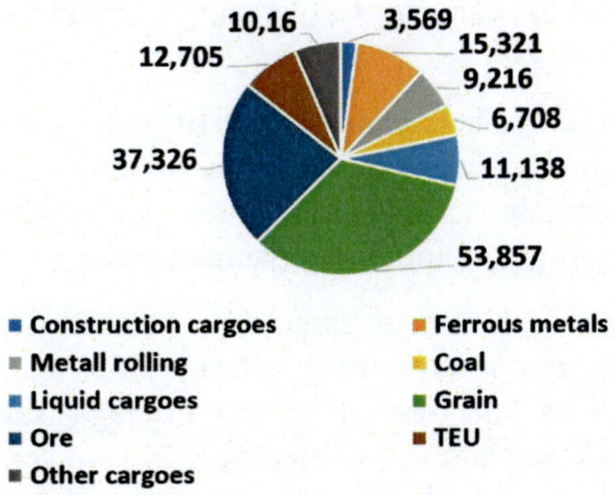

Fig. 3. Transshipment volumes by cargo categories through Ukrainian seaports in 2019, million tons. [43, 44]. Source: Created by the authors based on

However, quickly growth rate of agricultural production does not correspond to efficiency levels of logistics. Coordination between grain elevators, terminals, carriers, ports, and all participants must be improved during cargo delivery. This will help to increase the efficiency of multimodal transportation and also to optimize technological processes in ports when railway carriages are unloaded.

4.2 Mathematical Formalization for Processes of Container Transshipment in Seaports

The analysis was carried out for reloading containers process at ports according to alternative technologies, and some features have been identified.

Technology peculiarities are the next for container cargoes transshipment according to the direct option ‘railway carriage-ship’:

- Long idle time of railway carriage waiting for cargo transshipment to ship;
- Excessive of normative waiting time for cargo transshipment, as a consequence, time wagon turnover will be increasing, and therefore it’s necessary to increase railway carriages fleet;
- The need to coordinate and schedule compliance of railway carriage to arrive (a group of wagons) for cargo reloading into the ship;
- The necessity to form a technological route from railway stations of departure to destination ports with conditions of further containers transshipment from arriving railway carriages to one ship.

Technology peculiarities are the next for container cargoes transshipment according to option ‘railway carriage-warehouse-ship’:

- limited container terminal area;
- probability of waiting in queue for transshipment to the warehouse;
- if containers are temporary storage of containers at port terminals then cargo owners will get additional supply chain costs;
- additional loading and unloading related to storage and sorting of containers;
- increased probability of damage to containers during handling and storage operations.

Total specific costs of containers reloading are selected as an efficiency criterion for process assessments during container cargoes transshipment in ports.

Efficiency criterion includes in case of container transshipment according to direct option ‘railway carriage-ship’ the following parameters:

- costs for arriving and departure of cargo mini-train to (from) the transshipment place according to direct option ‘railway carriage-ship’ $S_{ar/dep}$;
- costs associated with an idle time of railway carriage, shunting locomotive and container waiting for transshipment to ship S_{idle};
- costs of container transshipment from railway carriages to ship at departure ports $S_{transsh}$.

The efficiency criterion includes the next parameters during containers transshipment through warehouses at port:

- costs for arriving and departure of cargo mini-train to the place of container transshipment from railway carriage to warehouse $R_{ar/dep}$;

- costs associated with an idle time of railway carriage, shunting locomotive and container waiting for transshipment to warehouses at port R_{idle};
- costs associated with container storing in warehouses at the departure port R_{stor};
- costs for transshipment through warehouses to ship at the departure port $R_{transsh}$;
- costs associated with 'dead capital' R_{dcap}.

Analysis of alternative technologies for processing containers at port terminals realizes as follows. Analytical dependencies have been obtained for certain components of costs for transshipment containers in ports according to direct option 'railway carriage-ship' and reloading containers through warehouses as an analysis result of technologies.

Costs for arriving and departure of cargo mini-train to (from) the transshipment place in ports:

$$S_{ar/dep} = R_{ar/dep} = C_{oploc} \cdot t_{loc} + C_{spwag} \cdot L \cdot k, \tag{1}$$

where C_{oploc} – specific costs per 1-h operation by shunting locomotive with freight traffic during arriving cargo mini-train to transshipment front in the port, UAH/h; t_{loc} – operating time of shunting diesel locomotive at arriving and departure of cargo mini-train to transshipment front in port, h; C_{spwag} – specific costs per 1 wagon-kilometer of total mileages for railway carriage from cargo fleet when mini-train arrive at transshipment front in port, UAH/wag-km; L – additional mileage for arriving or departure of cargo mini-train to (from) the transshipment front in port, km; k – quantity of railway carriage in one time arriving at the port, wag.

It should be noted that when shunting diesel locomotive is operating during arriving-departure of cargo mini-train to transshipment front at ports, the time required for direct arriving-departure of cargo mini-train to transshipment front is included. This criterion also includes times that need for registration of documents for containers import into the port and their transshipment to ships (warehouse), as well as the idle time when railway carriages are waiting for cargo reloading.

Costs associated with an idle time of railway carriage, shunting locomotive and container waiting for transshipment and directly during cargo reloading through transshipment front at the port:

$$S_{idle} = R_{idle} = \left(t_{idle} + \frac{n_{ship} \cdot Q}{q_c} \right) \cdot \left(C_{cont} \cdot Q + C_{wag} \cdot k + C_{spwag} \right), \tag{2}$$

where t_{idle} – railway carriage idle time waiting for transshipment to ship, UAH/h; n_{ship} – coefficient of parallel operations during containers transshipment; Q - quantity of containers in one time arriving at the port, unit; q_c - crane capacity, cont./h; C_{cont} – container usage cost, UAH/h; C_{wag} – railway carriage idle hour cost, UAH/wag-h.

Costs of container transshipment from railway carriages to ship:

$$S_{transsh} = C_{load/unl} \cdot \frac{n_{ship} \cdot Q}{q_c}, \tag{3}$$

where $C_{load/unl}$ – cost of loading and unloading at port terminals, UAH/h.

Costs associated with container storing in warehouses at the departure port

$$R_{stor} = t_{stor} \cdot (C_{cont} + C_{stor}) \cdot Q, \quad (4)$$

where t_{stor} – storage time of containers with cargo in the warehouse, h; C_{stor} – one-hour cost of containers storage with cargo in the warehouse, h.

Costs for transshipment through warehouses to ship at the departure port:

$$R_{transsh} = 2 \cdot \left(C_{load/unl} \cdot \frac{n_{stor} \cdot Q}{q_c} \right), \quad (5)$$

where n_{stor} – coefficient of parallel operations during containers transshipment according to technology 'railway carriage-warehouse-ship'.

Costs associated with 'dead capital':

$$R_{dcap} = \frac{M_{cont} \cdot [1 - P(Q)] \cdot t_{stor} \cdot i}{24 \cdot 360}, \quad (6)$$

where M_{cont} – goods market price carried in the container, UAH/cont.; $P(Q)$ – the probability of cargo damage during transshipment through warehouses; i – the interest rate on deposits, proportion; 360 – days period per year accepted in calculations.

Values of idle time of railway wagons waiting for transshipment according to 'railway carriage-ship' and 'railway carriage-warehouse-ship' technologies, as well as a time of container reloading from cargo mini-trains to warehouse (to ship), time of container transshipment from warehouses to ships are random values. Therefore, it is necessary to carry out time observations and analysis of statistical data to determine the distribution character of these random values.

5 Conclusions

Marine terminal complexes of Ukraine are at the stage of active reform. This fact was revealed from general researches for the practical state of organization issue of transportation in combined railway and water transport. Almost all indicators show a significant increase in volumes of goods transshipment, and above all, cereals, which is due to the quick growth of agricultural production. However, there are disadvantages in ports, such as inconsistency in operation between different transport kinds during goods transshipment under the direct option, failure to achieve a normative idle time under the performance of cargo operations according to direct technology and through warehouses.

The efficiency criterion was designed for the transshipment processes of containers with cargo at the port terminal. This is total specific costs associated with cargo transshipment in containers at port terminals.

Analytical dependencies were obtained for certain components of costs related to containers overloading according to "railway carriage-ship" technology and containers transshipment of through warehouses as a result of the analysis of technologies.

The mathematical formalization of the process of grain cargoes transshipment in containers at the seaport, which is proposed, will contribute to designing mathematical models and on formations of rational technologies for container transshipment in the port.

References

1. Shramenko, N., Muzylyov, D.: Forecasting of overloading volumes in transport systems based on the fuzzy-neural model. In: Ivanov, V., et al. (eds.) Advances in Design, Simulation and Manufacturing II. DSMIE-2019, Lecture Notes in Mechanical Engineering, pp. 311–320. Springer, Cham (2020)
2. Zubkov, V.N., Mamaev, E.A., Chislov, O.N., Ivanchenko, V.N., Ryazanova, E.V., Chebotareva, E.A.: Perspective technologies for the transportation of agricultural goods in rail and sea traffic. Polit. Internet Electron. Sci. J. Kuban State Agrarian Univ. **124**, 275–297 (2016)
3. Rustamov, R.: Assessment of the prospects for the development of grain logistics in Ukraine. Transp. Syst. Transp. Technol. **8**, 127–133 (2014)
4. Medvedev, E.P.: Factor analysis of the organization of transportation support for grain harvesting in Ukraine. Project Manage. Syst. Anal. Logist.. **18**(1), 86–93 (2016)
5. Medvedev, E.P.: Current status and prospects of transport provision for wheat harvesting. Proc. State Univ. Infrastruct. Technol. **31**, 236–244 (2018)
6. Shramenko, N., Shramenko, V.: Optimization of technological specifications and methodology of estimating the efficiency of the bulk cargoes delivery process. Sci. Bull. Nat. Min. Univ. **3**, 146–151 (2019)
7. Shramenko, N., Shramenko, V.: Mathematical model of the logistics chain for the delivery of bulk cargo by rail transport. Sci. Bull. Nat. Min. Univ. **5**(167), 136–141 (2018)
8. Petrashevskyi, O.L., Kyrychenko, A.I.: Ways of improving effective management of goods delivery under multimodal transportation. Probl. Transp. Coll. Sci. Works **9**, 3–16 (2012)
9. Shramenko, N.Y.: Methodology for evaluation of synergy effect in terminal cargo delivery system. Actual Probl. Econ. **8**(182), 439–444 (2016)
10. Heiets, I., Spivakovskyy, S., Spivakovska, T.: Innovative business models for full cycle operating airlines. Int. J. Bus. Perform. Manage. **20**(4), 356–377 (2019). Special Issue on: TBM 2019, Transformative Business Models – Disruptive Innovation in Finance, Logistics and Tourism
11. Spivakovskyy, S., Spivakovska, T, Al-Gazou, A.: Cooperation between Middle East countries and Ukraine in aerospace industry. Int. J. Bus. Perform. Manage. **20**(4), 78–399 (2019). Special Issue on: TBM 2019, Transformative Business Models – Disruptive Innovation in Finance, Logistics and Tourism
12. Shramenko, N.: Effect of process-dependent parameters of the handling-and-storage facility operation on the cargo handling cost. Eastern-Eur. J. Enterp. Technol. **5/3**(77), 43–47 (2015)
13. Shramenko, N., Muzylyov, D., Shramenko, V.: Methodology of costs assessment for customer transportation service of small perishable cargoes. Int. J. Bus. Perform. Manage. **21** (1/2), pp. 132–148 (2020). Special Issue on: TBM 2019, Transformative Business Models – Disruptive Innovation in Finance, Logistics and Tourism
14. Vernigora, R., Rustamov, R.: Analysis of the storage system of Ukrainian grain. Transp. Syst. Transp. Technol. **13**, 10–18 (2017)

15. Kingwell, R., Elliott, P., White, P., Carter, C.: An emerging challenge for Australian wheat exports. Australian Export Grains Innovation Centre. https://aegic.org.au/wp-content/uploads/2016/04/Ukraine-Supply-Chain-Full-Report.pdf. Accessed 30 Apr 2019
16. Hyland, M.F., Mahmassani, H.S., Mjahed, L.B.: Analytical models of rail transportation service in the grain supply chain: deconstructing the operational and economic advantages of shuttle train service. Transp. Res. Part E Logist. Transp. Rev. **93**, 294–315 (2016)
17. Muzylyov, D., Shramenko, N., Shramenko, V.: Integrated business-criterion to choose a rational supply chain for perishable agricultural goods at automobile transportations. Int. J. Bus. Perform. Manage. **21**(1/2), 166–183 (2020). Special Issue on: TBM 2019, Transformative Business Models – Disruptive Innovation in Finance, Logistics and Tourism
18. Pasichnaja, E., Trapenov, V., Khan, V.: Prospects for the development of railway transportation to port grain terminals. Eng. Vestnik of Don: Electron. Sci. J. **4** (2017). https://www.ivdon.ru/ru/magazine/archive/n4y2017/445. Accessed 30 Nov 2019
19. Lomot'ko, D.V., Veyisov, T.Z.: Improvement of technology of transfer of cargo flow at interaction of railway and sea transport. Collect. Sci. Works Ukrainian State Acad. Railway Transp. **150**, 91–97 (2014)
20. Shramenko, N.: The methodological aspect of the study feasibility of intermodal technology of cargo delivery in international traffic. Sci. Bull. Nat. Min. Univ. **4**(160), 145–150 (2017)
21. Shramenko, N.: Evaluation of the effectiveness of piggyback traffic in the context of creating transport and logistics clusters. Sci. Bull. Nat. Min. Univ. **6**(162), 151–155 (2017)
22. Myamlin, S.V., Korobieva, R.G., Malashkin, V.V., Besarab, D.A.: Improving grain logistics through the introduction of bimodal technologies. Collect. Sci. Works DNUZT Acad. V. V. Lazaryan **14**, 69–77 (2017)
23. Korobyova, R.G., Rustamov, R.W., Grevtsov, S.V.: Introduction of bimodal technologies of grain cargo transportation in Ukraine. Transp. Syst. Transp. Technol. **9**, 29–34 (2015)
24. Kimberly, V.: Marketing U.S. Grain and Oilseed by Container. DP-272. North Dakota State University. Fargo: Upper Great Plains Transportation Institute (2014). https://www.ugpti.org/resources/reports/downloads/dp-272.pdf. Accessed 11 Jan 2020
25. Kawasaki, T., Matsuda, T.: Containerization of bulk trades: a case study of US–Asia wood pulp transport. Marit. Econ. Logist. **17**(2), 179–197 (2015)
26. Lawrence, R., Nolan, J., Schoney, R.: Simulating contestability in freight transportation: the Canadian grain handling and transportation system. J. Transp. Econ. Pol. (JTEP) **50**(4), 325–349 (2016)
27. Kingwell, R., White, P.: Argentina's grains industry: implications for Australia (2018). https://www.aegic.org.au/wp-content/uploads/2018/11/AEGIC-Argentina-Report_18_LR.pdf. Accessed 28 Jan 2020
28. Levitsky, I.E., Korobieva, R.G., Rudenko, N.V.: Container transportation in the Odessa region. Transp. Syst. Transp. Technol. **4**, 61–64 (2012)
29. Kurapati, S., Lukosch, H., Verbraeck, A., et al.: Improving resilience in intermodal transport operations in seaports: a gaming approach. EURO J. Decis. Process. **3**, 375–396 (2015)
30. Panasenko, N.N., Yakovlev, P.V.: Containerization of the international transport system. Bull. ASTU. Ser. Marine Eng. Technol. **4**, 103–116 (2016)
31. Gleim, S., Nolan, J.: Canada's grain handling and transportation system: a GIS-based evaluation of potential policy changes. J. Transp. Res. Forum **54**(3), 70–78 (2015)
32. Vachal, K., Button, K.J.: Impact of shuttle rates on local grain flows. Int. J. Transp. Econ. **7**, 91–102 (2003)
33. Semenov, I., Filina-Dawidowicz, L., Trojanowski, P.: Integrated approach to information analysis for planning the transport of sensitive cargo. Arch. Transp. **51**(3), 65–76 (2019)

34. Jacyna, M., Semenov, I.N., Trojanowski, P.: The research directions of increase effectiveness of the functioning of the RSA with regard to specialized transport. Arch. Transp. **35**(3), 27–39 (2015)
35. Juhel, M.: Globalisation, privatisation and restructuring of ports. Int. J. Maritime Econ. **3**, 139–174 (2001)
36. Liang, C., Hwang, H., Gen, M.: A berth allocation planning problem with direct transshipment consideration. J. Intell. Manuf. **23**, 2207–2214 (2012)
37. Panasenko, N.N., Yakovlev, P.V.: Containerization of the international transport system. Bull. Astrakhan State Techn. Univ. Series Marine Technol. **4**, 103–116 (2016)
38. Zeng, Q., Feng, Y., Chen, Z.: Optimizing berth allocation and storage space in direct transshipment operations at container terminals. Maritime Econ. Logist. **19**(3), 474–503 (2017)
39. Tyulenev, K.G.: Management of Container Transportation in Foreign Economic Activity. Institute of Transport Problems Russian Academy of Sciences, St. Petersburg (2017)
40. Bojovic, N., Milenkovic, M.: The best rail fleet mix problem. Oper. Res. Int. J. **8**, 77–87 (2008)
41. Valkova, S.S., Stepanets, A.V., Veryutina, V.E.: The formation of complex of tasks on a management by cargo handling of railway wagons in a seaport. Sci. Tech. Bull. Volga Region **1**, 146–150 (2013)
42. Zabors'kyy, L.O.: Methodical bases of organization of transport and technological processes in cargo delivery systems. Ph.D. thesis, Odessa (2008)
43. Ports of Ukraine. Analytical reports online. https://ports.com.ua/uk/analitics/gruzooborot-portov-2019-infografika. Accessed 18 Feb 2020
44. Official site. State Statistics Service of Ukraine. Transport. www.ukrstat.gov.ua. Accessed 20 Feb 2020

Application of Lean Analyses and Computer Simulation in Complex Product Manufacturing Process

Dorota Stadnicka(✉) and Maksymilian Mądziel

Rzeszow University of Technology,
12, Al. Powstancow Warszawy, 35-959 Rzeszow, Poland
dorota.stadnicka@prz.edu.pl

Abstract. The paper presents an application of lean analyses and computer simulations (CSs) in a case study of a manufacturing line that is used to manufacture complex products. The complexity of the manufacturing process (MP) is increased by realizing the MPs of components in parallel. Moreover, there is a necessity of cooperation with external service providers, to which the components/products are sent four times on different stages of the MP. In the first analyses, such factors as machine/workstation availability, takt time, and the number of employees are considered. While, in CS the following factors were used: machine availability, batch size, and way of cooperation with external service providers. Several recommendations which can be useful in the development of simulation models for complex product manufacturing system are also presented. The model was developed in Flexim. Then, simulations were performed based on a simulation plan. The simulation results were analyzed and discussed. Finally, conclusions were presented.

Keywords: Value stream mapping · Discrete event simulations · Lean concept · Machines availability

1 Introduction

The performance of the manufacturing process (MP) depends on different factors. Among others, the following can be distinguished: row material quality, MP quality, workers availability, machine availability, work organization, etc. To guarantee that a customer will receive the products on time, a company, first of all, has to ensure that the MPs will be stable [1]. It enables to create a correct plan and schedule of an MP. If a company wants to improve the performance, it has to undertake adequate actions, for example, to realize a six sigma project [2, 3]. It can help with the identification of problems and their possible further elimination. Sometimes, such methods as quality analysis with the use of A3 report [4] or material and information flow with the use of value stream mapping [5, 6] as well as other lean tools application [7] can be enough. Sometimes, more sophisticated tools as simulations are indispensable [8].

In this work, a CS-VSMap being a result of value stream mapping, is used to create a simulation model. The presented MP realizes a complex product; therefore, the MS is

V. Tonkonogyi et al. (Eds.): InterPartner 2020, LNME, pp. 106–116, 2021.
https://doi.org/10.1007/978-3-030-68014-5_11

complex as well. On the base of data concerning the MS, a simulation model was developed. The problems which appeared in the development process and the ways of their elimination are presented in the paper. The simulation model was used to perform simulations with chosen sets of data. The results of simulations were compared with the results of the lean analysis. The data introduced to the model have to be changed to perform simulations in different scenarios. The factors, which have been taken into consideration in the research, in both initial analysis and CSs are as follow: machine availability, takt time, batch size, number of employees, and way of cooperation with external service providers. The research goal was to discuss how the simulations can help in the analysis of an MS, where a complex product is realized.

2 Literature Review

VSM allows seeing the whole material and information flow. It enables discovering problems that affect MPs [9] or warehouse operations [10]. The implementation of different solutions to eliminate the problems can improve the manufacturing line (ML) organization and performance [11]. However, in complex MSs, VSM analysis cannot be sufficient. To see how the proposed solutions will affect the ML performance, a set of simulations can help to choose the best solutions [12]. In MSs different simulation methods such as system dynamics simulations [8], discrete event simulation [12, 13], as well as different simulation software, can be used. In literature, it is presented how different methods and software can be applied in MSs analyses. For example, in the study [14], the author presents how Enterprise Dynamics can be used in storage subsystem design in a flexible MS. In the research work [15] with the use of Tecnomatix Plant Simulation, the authors analyze how buffer capacity can influence the throughput. In work [16], two simulation tools: FlexSim and Tecnomatix Plant Simulation are compared. The paper [17] presents a process of building a simulation model in Vensim and the benefits of its application. In the paper [18], Witness software is used to maximize efficiency and prevent waste in the MS design phase. At the same time, the paper [19] presents a combination of lean concept and simulations. In the publications, the problems concerning simulations of the complex product manufacturing system are not sufficiently discussed.

3 Research Methodology

In the frame of this work, two research problems are discussed. The first problem concerns the low performance of an ML. A case study of ML, where a complex product is manufactured, was analyzed. Different factors influence ML performance. In this analysis, the following factors are analyzed: machine availability, takt time, and many employees. The case study company would like to deliver to a customer 20 pcs of ready products per day. Therefore, a set of analysis needed to be performed to find out under which circumstances the goal can be achieved.

The second research goal concerns development of the simulation model, which can be useful in the analyses and which can be applied in performing simulations for the case study of ML. In CSs, the following factors are analyzed: machine availability, batch size, and way of cooperation with external service providers.

In the study, the MP presented in work [20] was used. Also, a CS-VSMap, developed in work [20], was input to current studies.

3.1 A Manufacturing Line Description

The simplified structure of processes in VSMap, presented in work [15], and the duration of them, are presented in Table 1. The analyzed MP consists of 19 processes which are realized in the company as well as by cooperators. Availability (AVA) of the workstations/machines realizing the MPs are presented in Table 1 together with information concerning processing time (*PT*) and the number of employees (NE) realizing the processes. Each workstation works for two shifts. Available working time (AWT) for a shift equals 27 000 s. Real available working time is lower because the chosen workstations/machines availability is decreased. The processes: 3E, 6E, 15E, and 18E are realized by cooperators (external service providers). Transport between the manufacturing company and cooperators is currently realized once a day. However, because the collaborators are situated close to the manufacturing company, the transports can be realized more often. Possible takt time (*TT*), in the case when AVA equals 100%, which should be achieved, is 2 700 s.

Table 1. Data concerning the MPs; AVA – current process availability, AWT – available working time, *PT* – processing time, *PT-Batch* – processing time for a batch, NE – number of employees engaged in the process, CT_{max} – maximal cycle time; TT = 2 700 s (for AVA = 100% and AWT = 54 000 s/day).

Process	AVA [%]	AWT [s]	*PT* [s]	*PT-Batch* batch = 4pcs [s]	NE	CT_{max} [s]
1. Deburring I	80	43 200	180	720	0.25	2 160
2. Bending	90	48 600	60	240	0.25	2 430
4. Deburring II	100	54 000	120	480	0.25	2 700
5. Shipment to a subcontractor	80	43 200	3 600	14 400	1	2 160
7. Turning I	90	48 600	360	1 440	0.25	2 430
8. Cutting I	70	37 800	78	312	0.25	1 890
9. Turning II	90	48 600	2 073	8 292	1	2 430
10. Cutting	100	54 000	2 040	8 160	1	2 700
11. Turning III	90	48 600	2 190	8 760	1	2 430
12. Turning IV	100	54 000	90	360	0.25	2 700
13. Milling	90	48 600	960	3 840	0.5	2 430
14. Deburring III	80	43 200	510	2 040	1	2 160
16. Welding	90	48 600	2 100	8 400	1	2 430
17. Assembly of components	100	54 000	7 500	30 000	2	2 700
19. Final assembly + Pressure test	100	54 000	6 300	25 200	2	2 700

Takt time was calculated with the use of Eq. (1). The calculations are presented by Eq. (2).

$$TT\ [s/pcs] = \text{Number of shifts} \cdot AWT\ [s/shift] / \text{Customer order}\ [pcs/day] \quad (1)$$

$$TT\ [s/pcs] = 2 \cdot 27\,000\ [s/shift]/20\ [pcs/day] = 2\,700\ [s/pcs] \quad (2)$$

Two employees are involved in each of the following processes: P17 and P19. In each of the other processes, one employee works.

3.2 Identification of Bottleneck Processes

Because of the decreased availability, the processes do not work at the same place, which is expected to be equaled 2 700 s if the customer orders 20 products per day, and the company works 2 shifts per day. Current *CTs* of the processes were compared to the maximum cycle times (CT_{max}) for the processes with decreased availability, expected to ensure on-time deliveries (Fig. 1). CT_{max} was calculated with the use of the formula (3). From the data presented in Fig. 1, it can be concluded that the problem exists for three processes: P5, P17, and P19. In the simple analyses, the company decided to add additional employees to process 17 and 19 and implement training and standardization in process 5. Thus, the situation on the ML has changed (Fig. 2). From Fig. 2 it can be seen that the problem was solved. After changes *CT* for P5 equals 600 s, *CT* for P17 equals 2 500 s, and *CT* for P19 equals 2 100 s. There are no bottleneck processes anymore. Therefore, according to the performed analysis, the goal of 20 pcs per day should be achieved.

$$CTmax\ [s/pcs] = TT\ [s/pcs] \cdot AVA\ [\%]/100 \quad (3)$$

In the performed analyses the following factors were taken into consideration: (1) process availability (AVA), which equaled from 70% to 100%, (2) takt time (*TT*) which was calculated on the base of the possible number of products ordered by a customer, (3) number of employees engaged in the process. From the performed analyses, it can be concluded that each process can manufacture 20 pcs per day.

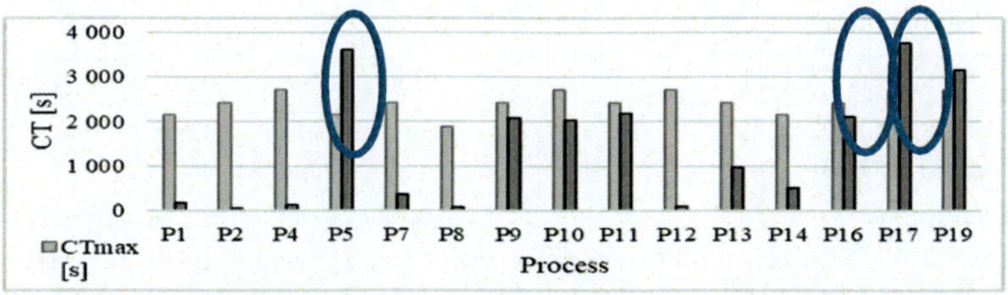

Fig. 1. Cycle time (*CT*) comparison with maximal cycle times (CT_{max}) for the processes – initial state (12 employees) – for demand equaled 20 pcs per day.

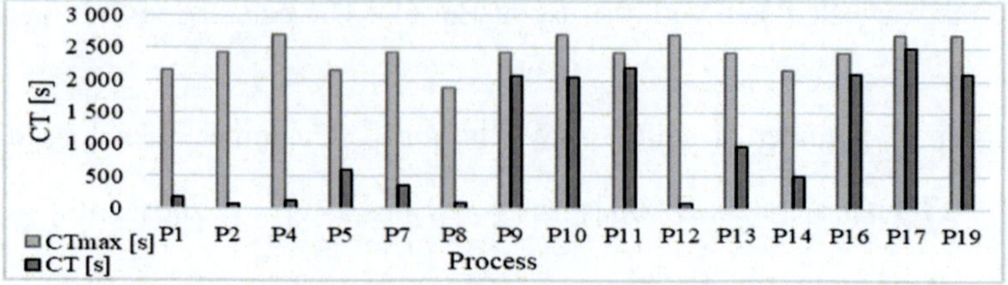

Fig. 2. Cycle time (*CT*) comparison with maximal *CTs* (CT_{max}) for the processes – after the implementation of the proposed changes – for demand equaled 20 pcs per day.

4 Results

4.1 Simulation Plan

In the simulation experiment, the following factors were applied: (1) process availability (AVA), which could be 100% or lower, (2) batch size (BS), which equaled 1 pcs or 4 pcs, (3) way of cooperation with external service providers: transport between the company and external service providers might be realized few times a day; the cooperators can provide services for up to 60 pcs per day; the factories of external service providers are situated close to the manufacturing company.

Tables 2 and 3 present a simulation plan and data. In Scenario 1, if only one component is ready is sent to collaborators. Therefore, the transport between the manufacturing company and collaborators is realized many times per day. In some cases, it might not be reasonable. Therefore, in Scenario 2 the number of pieces sent in one transport to collaborators is 4 pcs. In Scenarios 3 and 4, the number of pieces increased to 16 pcs. It seems that the most reasonable solution is sending at once 20 pcs of products, which is the target number for the process to manufacture (Scenarios 5, 6, and 7). In other scenarios (Scenario 8), the number of pieces sent to external service providers is even higher. In the planned scenarios, the number of products sent to collaborators ranged from 1 to 26. The authors intended to see if increasing the number of pcs would affect ML performance. In Scenarios 1, 6, and 8, one-piece flow is

Table 2. Simulation plan; AVA – Availability, BS – batch size, NPC – Maximal number of products in cooperation.

	AVA [%]	BS [pcs]	Max NPC [pcs]
Scenario 1	100	1	1
Scenario 2	100	4 pcs (1 batch)	4 (1 batch)
Scenario 3	100	4 pcs (1 batch)	16 (4 batches)
Scenario 4	Decreased for chosen processes	4 pcs (1 batch)	16 (4 batches)
Scenario 5	100	4 pcs (1 batch)	20 (5 batches)
Scenario 6	100	1	20
Scenario 7	Decreased for chosen processes	4 pcs (1 batch)	20 (5 batches)
Scenario 8	Decreased for chosen processes	1	26

simulated. In other scenarios, the batch size was increased to 4 pcs. In Scenarios 1–3 and 5 it was assumed that AVA equals 100% for all processes. In other scenarios, the AVA was decreased for chosen processes.

Table 3. Details for scenarios with decreased availability; AVA – current process availability, AWT – available working time for machines/workstations with different availabilities.

	Availability of machines/workstations in the following processes [%]						
	P5	P9	P11	P13	P16	P17	P19
Scenarios: 4, 7, 8	80	90	90	90	90	100	100
AWT [s]	43 200	48 600	48 600	48 600	48 600	54 000	54 000

Details, concerning scenarios, in which AVA is decreased, are presented in Table 3 together with available working time (AWT) for the workstations/machines with different levels of availability. In all scenarios, the inventories between processes have no limits. Moreover, the materials and components are always available. Furthermore, a simplification concerning assigning one employee to each process (apart from processes: P17 and P19 where three operators work in each process) was applied.

Simulations were performed for the time period equaled 4 weeks (1 080 000 s) and 6 weeks (1 620 000 s). The results of the simulations are presented later in work.

4.2 Simulation Model Development

The simulation model was developed in the Flexsim software based on the MP structure, which presents material flow and is shown in Fig. 3. The processes which are realized by collaborators (3E, 6E, 15E, 18E) were treated as 1 day of waiting. It means that every time products are sent to cooperators and spend there 54 000 s before being sent back to the company (after two working shifts). Data from Table 1 were used in the simulation model. Only data for processes P5, P17, and P19 are different because the company made some changes. For P5, processing time equals 600 s. P17 processing time equals 7 500 s, but they are three workers working on the three products in parallel. A similar situation is for process P19, but processing time for this process is 6 300 s.

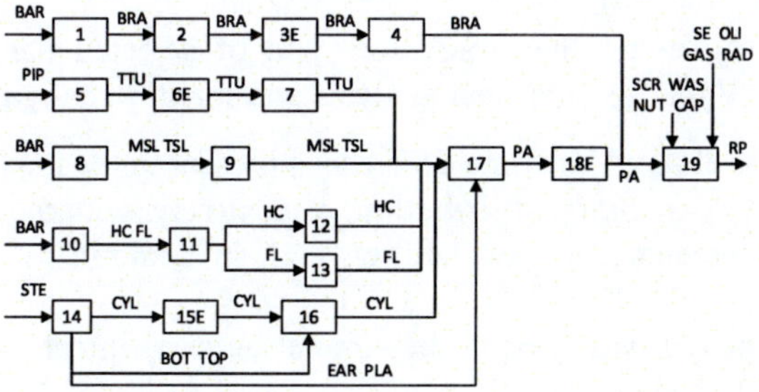

Fig. 3. MP structure with material flow.

Moreover, it was assumed that input material/component are always available, which means that the problem of delivery delays was not analyzed in the frame of this work.

Based on process development of simulation models, the following problems were discovered:

Problem 1. Batching of products. This problem was solved by setting *Target Batch Size* values in the queue preceding the external process.

Problem 2. Sending the components to different processes. The problem appeared when, after being manufactured in one process, the products went to different processes to be further manufactured. This problem was solved by the appropriate connection of objects in the model and choosing the appropriate output port for a given product by the *Flow* function.

Problem 3. Creating sets of components for an assembly process. The problem appeared before assembly processes when a set of components had to be prepared for the assembly process. This problem was solved by using the *Combiner* object to assemble components to make a product.

Problem 4. External processes implemented by service providers. In the case of four processes, they had to be realized externally, by service providers. The products can be sent to cooperators in batch, and the simulation model has to be ready to create such batches. This problem was solved by using the *Batch* function in queues before the process and setting the appropriate *Content* value for the external process.

Problem 5. In one process, two types of components are manufactured. In the simulation model, the total processing time for both components was introduced, and the process content was 2 pcs.

Therefore, the following recommendations are suggested to be applied in the development of a computer model of a complex product manufacturing system:

Recommendation 1. Checking the model for bottleneck formation after starting the simulation to check the correctness of the connections of individual objects.

Recommendation 2. Setting the right number of products at the entrance to the model to eliminate unnecessary waste of time by the *Source* object.

Recommendation 3. Controlling the work of individual model objects using the *Dashboard* object, which enables easier diagnosis of the model's work.

Recommendation 4. Saving the simulation data of selected parameters of model objects, e.g., to a *CSV file* for later data analysis and making the required calculations.

Recommendation 5. If the AVA time decrease is present in the model, it is possible to use the *Sent to port by percentage* distribution, respectively, with the available time and sending a specific percentage of availability, e.g., in addition to the queue or sink objects.

The simulation model was used in the simulation experiment in which results are presented in the next section.

4.3 Results of Simulations and Discussion

The results of the performed simulations are presented in Table 4. The simulations were continued for 6 weeks (30 days). Only in Scenario 1, additionally, 4 weeks (20 days) period was applied. Times for *ML-TT* and *LT* for scenarios where batching was considered refers for a whole batch, and *TH* was multiplied by the number of products in batch. Figure 4 is a graphical presentation of *ML-TTs* comparison for different scenarios and with *TT*. The *TT* depends on customer requirements. If the customer expects 20 products per day, it means that *TT* will equal 2 700 s (see Eq. 1).

$$\mathrm{TT} = 54\,000\ \mathrm{s}/20\ \mathrm{pcs} = 2\,700\ \mathrm{s} \tag{4}$$

Table 4. Simulation results; *TH* – Throughput, *LT* – lead time, ML – manufacturing line.

	Working time	*ML-TT*	*LT*	*TH* [pcs]
Scenario 1(4)	4 weeks	41 min 34 s	34 h 6 min 19 s	384
Scenario 1(6)	6 weeks	41 min 34 s	34 h 6 min 19 s	600
Scenario 2	6 weeks	2 h 45 min 45 s	46 h 35 min 22 s	588
Scenario 3	6 weeks	2 h 44 min 20 s	54 h 44 min 38 s	576
Scenario 4	6 weeks	2 h 59 min 21 s	54 h 44 min 38 s	524
Scenario 5	6 weeks	2 h 45 min 19 s	57 h 48 min 38 s	572
Scenario 6	6 weeks	41 min 27 s	49 h 52 min 19 s	580
Scenario 7	6 weeks	2 h 59 min 26 s	58 h 28 min 38 s	520
Scenario 8	6 weeks	43 min 19 s	54 h 52 min 19 s	538

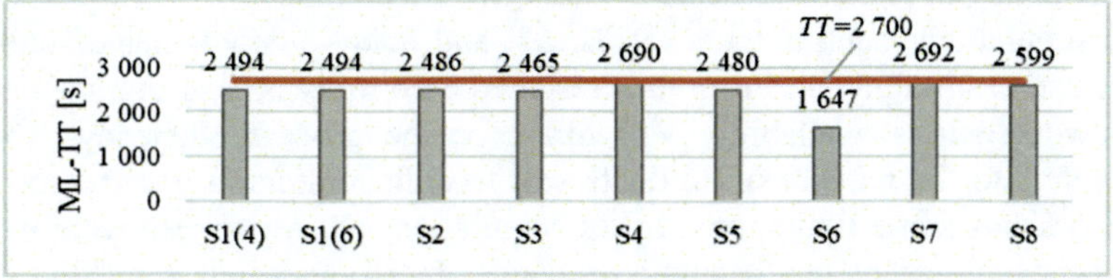

Fig. 4. ML-TT comparison for different scenarios.

If the process of *CT* or *ML-TT* is lower than *TT*, it means that the process can manufacture the expected number of products. In all presented scenarios, *ML-TTs* are lower than *TT*. Therefore, the simulations confirmed the conclusions drawn based on the analysis presented earlier in work. However, when *TH* is taken under analyses, it can be seen that only in case of Scenario 1, it was possible to manufacture the expected number of products, i.e., 600 pcs in 6 weeks, what gives 20 pcs per day. In this case, the one-piece flow has been implemented. It is the most expected material flow organization in a lean manufacturing system. It can also be noticed that in 4 weeks it was impossible to manufacture the expected number of products. It is connected with the time needed to fill the ML with products. In the presented case study, the customer has to send an order 6 weeks before delivery time to get the 600 products.

However, it can be troublesome to organize a one-piece flow between the manufacturing company and external collaborators. Therefore, it can be indispensable to increase the number of products sent in one transport to a collaborator. In Scenario 2, the number of products sent to collaborators increased to 4 pcs. Moreover, in this scenario, the products are moving through the ML in batches consisted of 4 pcs. Actually, in this case, it is the most expected material flow organization since the customer requires to deliver products in containers with 4 products insight. However, this solution caused a reduction in the number of manufactured products by 12 pcs.

Moreover, *LT* increased from 34 h 6 min 19 s to 46 h 35 min 22 s, i.e., by 37%. Further increasing of many products sent to collaborators caused the further reduction of manufactured products number and increasing *LT*. However, it was possible to achieve the presented results only if AVA equaled 100%. In Scenarios 4, 7, and 8, the AVA parameter was decreased. It caused a further reduction in the number of manufactured products. Furthermore, it was impossible to manufacture the expected number of products even if one-piece flow system was implemented.

To summarize, it has to be said that the analyses which are presented in the first part of the paper are not enough to understand whether the manufacturing system can deliver a certain number of products.

5 Conclusions

On the results of the presented research, it can be concluded that for analysis of a complex product manufacturing system, value stream mapping and simple analyzes can be not enough. Therefore, in such cases, it is recommended to implement CSs. From the simulations, such outputs as *TH*, *LT*, and many products manufactured in a fixed time can easily find out. From the presented case study is clear that the decreased machines/workstations availability will influence the process efficiency. The same situation refers to the number of products sent to collaborators in one transport.

The work has some limitations. In the simulation, all inventories were treated as FIFO lanes, which means that if only a product (a batch) appeared in a FIFO lane, the next workstation took a product and started MP. While, in reality, in the ML supermarkets are situated in some places. In the simulations also workers' assignments to the workstations were not analyzed. Moreover, setup times were not applied in the model.

In future work, the simulation model will be improved, and supermarkets will be introduced for further simulations. Future work will also analyze setup times and workers utilization, taking into account that some employees realize more than one process, and in some processes, more than one employee is needed. Moreover, in future research, a simulation experiment according to a plan which will allow identifying the influence of predetermined factors on *ML-TT*, *LT*, and *TH* will be performed.

References

1. Burduk, A.: Stability analysis of the production system using simulation models. In: Process Simulation and Optimization in Sustainable Logistics and Manufacturing, pp. 69–83. Springer, Cham (2014)
2. Bożek, M., Hamrol, A.: Analysis of efficiency of Lean Manufacturing and Six Sigma in a production enterprise. Manage. Prod. Eng. Rev. **3**, 14–25 (2012)
3. Cohen, A., Alhuraish, I., Robledo, C., Kobi, A.: A statistical analysis of critical quality tools and companies' performance. J. Clean. Prod. **255**, 120221 (2020)
4. Pérez-Pucheta, C.E., Olivares-Benitez, E., Minor-Popocatl, H., Pacheco-García, P.F., Pérez-Pucheta, M.F.: Implementation of Lean Manufacturing to reduce the delivery time of a replacement part to dealers: a case study. Appl. Sci. **9**(18), 3932 (2019)
5. Bhamu, J., Shailendra Kumar, J.V., Sangwan, K.S.: Productivity and quality improvement through value stream mapping: a case study of Indian automotive industry. Int. J. Prod. Qual. Manag. **10**(3), 288–306 (2012)
6. Stadnicka, D., Antonelli, D.: Application of value stream mapping and possibilities of manufacturing processes simulations in automotive. FME Trans. **43**, 279–286 (2015)
7. Pereira, A.C., Dinis-Carvalho, J., Alves, A.C., Arezes, P.: How Industry 4.0 can enhance Lean practices. FME Trans. **47**(4), 810–822 (2019)
8. Antonelli, D., Litwin, P., Stadnicka, D.: Multiple System Dynamics and Discrete Event Simulation for manufacturing system performance evaluation. Procedia CIRP **78**, 178–183 (2018)
9. Singh, J., Singh, H.: Application of lean manufacturing in automotive manufacturing unit. Int. J. Lean Six Sigma **11**(1), 171–210 (2020)
10. Purba, H.H., Aisyah, S.: Productivity improvement picking order by appropriate method, value stream mapping analysis, and storage design: a case study in automotive part center. Manage. Prod. Eng. Rev. **9**, 71–81 (2018)
11. Masuti, P.M., Dabade, U.A.: Lean manufacturing implementation using value stream mapping at excavator manufacturing company. Mater. Today Proc. **19**, 606–610 (2019)
12. de Sousa Junior, W.T., Montevechi, J.A.B., de Carvalho Miranda, R., Campos, A.T.: Discrete simulation-based optimization methods for industrial engineering problems: a systematic literature review. Comput. Ind. Eng. **128**, 526–540 (2019)
13. Jinrong, Y.U., Dali, H.U.: Production line balancing and optimization based on discrete event system simulation. Acad. J. Manuf. Eng. **17**(2), 112–117 (2019)
14. Gola, A.: Reliability analysis of reconfigurable manufacturing system structures using computer simulation methods. Eksploatacja i Niezawodnosc – Maint. Reliab. **21**(1), 90–102 (2019
15. Kłos, S., Stadnicka, D.: An analysis of the impact of buffer allocation and maintenance on the effectiveness of a manufacturing system using computer simulation. In: International Conference on Information Systems Architecture and Technology, pp. 355–368. Springer, Cham (2017)
16. Grabowik, C., Ćwikła, G., Kalinowski, K., Kuc, M.A.: Comparison analysis of the computer simulation results of a real production system. In: International Workshop on Soft Computing Models in Industrial and Environmental Applications, pp. 344–354. Springer, Cham (2019)
17. Malec, E.: The benefits of using computer simulation models to support decision-making. In: Advances in Manufacturing, pp. 205–214. Springer, Cham (2018)

18. Chramcov, B., Bucki, R.: Lean manufacturing system design based on computer simulation: case study for manufacturing of automotive engine control units. In: Handbook of Research on Design and Management of Lean Production Systems. IGI Global (2014)
19. Al-Fandi, L., Lam, S.S., Ramakrishnan, S.: A framework to reduce problem complexity using lean concepts with simulation. In: Doolen, T., Van Aken, E. (eds.) IIE Annual Conference. Proceedings, pp. 1–7. Institute of Industrial and Systems Engineers (IISE) (2011)
20. Bukowska, B., Stadnicka, D.: Value stream mapping of unique complex product manufacturing process. Technologia i Automatyzacja Montażu **1**, 36–43 (2020)

Design Engineering

Modeling of the Mixing Process in the Gravitational Mixer Using the Theory of Markov Chains

Igor Dudarev(✉), Serhii Holiachuk, Yurii Hunko, and Svitlana Panasyuk

Lutsk National Technical University, 75, Lvivska Street, Lutsk 43018, Ukraine
i_dudarev@ukr.net

Abstract. Modeling of the mixing process of four bulk materials in the gravitational mixer using the Markov chains theory can be used to determine the design parameters of the gravitational mixer. These mixer parameters will provide a good quality of a mixture. Modeling of the mixing process requires a detailed study of the components flow pattern in the gravitational mixer. The proposed mathematical model of the mixing process can be used to describe the process of mixing a different number of components in the gravitational mixer. The results of experimental studies of candy-dragees mixing in the gravitational mixer confirmed the adequacy of the proposed mathematical model of the mixing process. The gravitational mixer is a piece of beneficial equipment for the mixing process, and producers in the process industries can use it for mixing different raw materials and finished products. The mixing process of bulk materials in the gravitational mixer occurs without energy consumption and damage to bulk materials. Besides, the gravitational mixer is compact and easy to maintain.

Keywords: Mixer parameters · Mixing model · Mixture quality · Mixture flows

1 Introduction

The mixing process of bulk materials (components) is prevalent in the process industries [1]. During the mixing of components, it is essential to ensure their uniform distribution by the volume of the mixture, since it is the uniform distribution of components in the mixture that determines its quality. Also, for individual bulk and granular prescription components and finished food products are unacceptable damage during mixing, because they lose their properties. In the food industry, mixers of various types are used to mix bulk materials [2]. The most common are gravitational, drum, vibration, and centrifugal mixers [3]. Most mixers are equipped with individual working bodies that provide intensive mixing of the mixture components, which is unacceptable for original materials because it entails their damage or deterioration of quality properties. Thus, the determining factor in choosing the mixer design is the physical and mechanical properties of the mixture components. Among the known

V. Tonkonogyi et al. (Eds.): InterPartner 2020, LNME, pp. 119–128, 2021.
https://doi.org/10.1007/978-3-030-68014-5_12

mixer designs, drum and gravitational mixers cause the slightest damage to the components of the mixture. Besides, gravitational mixers without moving working bodies have the lowest energy costs. At the same time, drum mixers have a low degree of filling the volume of the working chamber and, as a rule, periodic action; respectively, the consumption of time for loading and unloading the mixer reduces its productivity. Only two components can be mixed at the same time in most gravitational mixers. Considering the analysis of mixer designs, the new mixer should be continuous, provide mixing of more than two components, and should not contain active working bodies.

The research aim is to simulate the mixing process of bulk materials in the gravitational mixer using the Markov chains theory and justify the mixer design parameters.

2 Literature Review

The determination of mixer parameters that would provide the necessary quality of the mixture is carried out by modeling the mixing process [4]. The mixing process is complicated to simulate because it depends on many factors: the properties and number of the components [5]; the particle size ratio and proportion of components in the mixture [6]; the design of the mixer [7]; the way the components are loaded [8]. The stochastic methods of mixing process modeling that include models based on Markov chains theory are more common than traditional methods involving the use of deterministic models [9]. Markov chains have been mainly employed in the literature on particle technology for determining particle flow pattern in mixers [10]. As drum mixers are the most common in the food industry, respectively, most models of the mixing process are designed for drum mixers [11]. These models use the discrete element method (DEM) [12]. DEM-method could successfully simulate the complex mixing dynamics of monodisperse particles [13] and gives information about the positions of components particles in the mixture [14]. Most models of the mixing process are designed to describe the mixing of the two components [15]; there are also models describing the mixing of three or more components [16]. Individual models of the mixing process allow considering the size and density of bulk materials [17]. Also, models of the mixing process based on the Markov chains theory are widely used to describe the mixing process in the gravitational mixers [18]. The development of such models requires a detailed study of the component flow pattern in the mixer.

3 Research Methodology

The modeling of the mixing process of four bulk materials was carried out using mathematical equations that describe the Markov random process with discrete states and time. The modeling was carried out on the assumption that during the fall of the particle of material A from the plate of the mixer combiner, three cases of its motion are possible [19]: the particle of material A does not fly through the plane V, and it does not contact with the particles of material B (Fig. 1a); the particle of material A does not fly through the plane V, as a result of contact with the particle of material B, it changes the trajectory (Fig. 1b); the particle of material *A* flies through the plane *V* (Fig. 1c). Each of these cases of particle movement due to a number of factors: the properties of bulk materials; the design parameters of mixer; the interaction of the particles of different materials. Also, in the study, it was assumed that all three cases of motion of the particles of material *A* are equally probable. Accordingly, the probability that the particle of material *A* does not fly through the plane *V* is $P = 0.667$, and the probability that the particle of material *A* will fly through the plane *V* is $P = 0.333$. Similar assumptions can be made for the particle of material *B*.

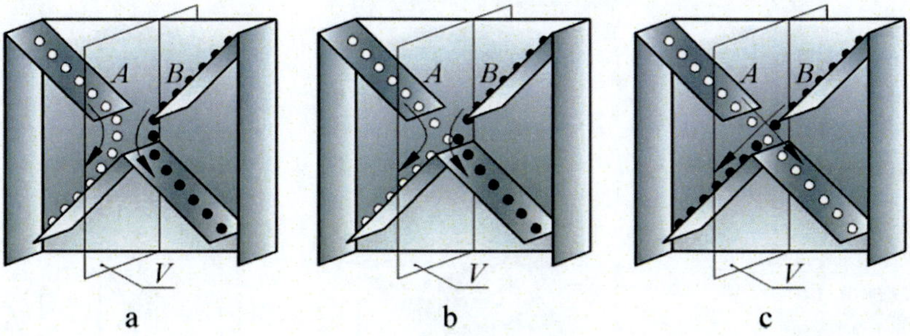

Fig. 1. Cases of movement of the particles of bulk materials *A* and *B* in the mixer section: the particles do not fly through the plane *V* (a, b); the particles fly through the plane *V* (c).

Verification of the applicability of the mathematical model of the mixing process was carried out experimentally in the gravitational mixer (Fig. 2a, and Fig. 2b). The gravitational mixer consists of sections of the same design installed one above the other. The section is formed by the body, inside which two vertical partitions are fixed, delineating the section into three parts. From the operational parts of the section, the combiner and the separator are attached to each vertical partition. Each combiner and separator is formed by two plates.

The research was carried out on candy-dragees of four colors: yellow, green, blue, and red. Candy-dragees were indicated: yellow – component *A*; green – component *B*; blue – component *C*; red – component *D*. The components *A*, *B*, *C*, and *D* were mixed in a ratio of 1:1:1:1 ($c_{0A} = c_{0B} = c_{0C} = c_{0D} = 25\%$, where c_{0u} – the base content of component *u* in the mixture, which should be in the ideal mixture). The study was carried out in the gravitational mixer with five and seven sections. Four flows of the mixture were sent to four containers after the mixing process in the mixer. The mixture from each container was divided into components *A*, *B*, *C*, and *D*, and the number *N* of each component in the mixture was determined. The average number of components in the mixture was calculated separately from each container:

$$N_{mu} = \frac{\sum_{z=1}^{k} N_{uz}}{k}, \quad (1)$$

where N_{mu} – the average number of component u in the mixture, dragees; N_u – the number of component u in the mixture from each container, dragees; k – the number of experiments.

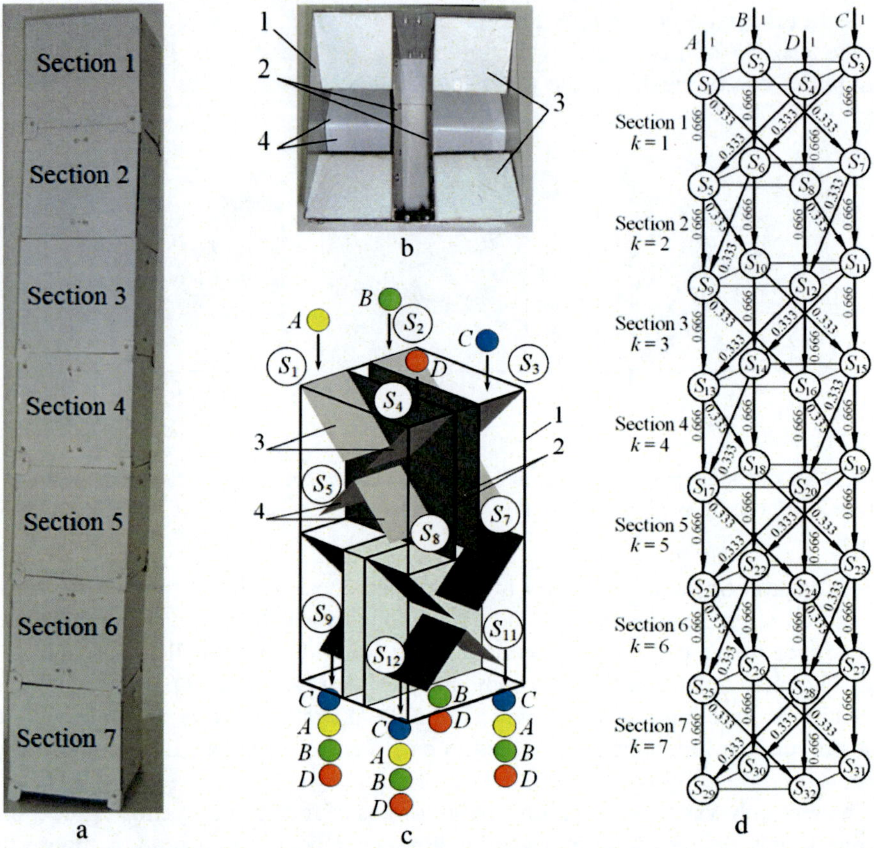

Fig. 2. The gravitational mixer (a) and its section (b), and diagram (c) explaining movement of the components particles in the top two sections, and graph (d) of possible states S_i of the components particles with the indication of probabilities P of the particles transition from the state to the state at each stage k during movement by mixer sections: 1 – the section body; 2 – the vertical partitions; 3 – the plates of the mixer combiner; 4 – the plates of the mixer separator.

The average content of components in percentage was calculated separately for the mixture from each container:

$$c_u = \frac{N_{mu}}{\sum_{u=1}^{r} N_{mu}} \cdot 100\%, \tag{2}$$

where c_u – the average content of component u in the mixture, %; $\sum_{u=1}^{r} N_{mu}$ – the total average number of all components in the mixture, dragees; r – the number of components in the mixture ($r = 4$).

The quality index of the mixture characterizes the quality of the candy-dragee mixture. To determine the quality index of the mixture, the average content c_u of components in the mixture was compared with the base content c_{0u} of components:

$$\Delta_u = \frac{|c_{0u} - c_u|}{c_{0u}} \cdot 100\%, \tag{3}$$

where Δ_u – the quality index of the mixture on the content of component u, %.

The quality index Δ_u of the mixture characterizes the quality of the mixture separately for each component. The quality of the mixture can be divided into the following groups: $\Delta_u \leq 5\%$ – excellent; $5\% < \Delta_u \leq 10\%$ – good; $10\% < \Delta_u \leq 20\%$ – satisfactory; $\Delta_u > 20\%$ – unsatisfactory. Another important index of the quality of mixing is the degree of damage to candy-dragees. The damage of candy-dragees is determined organoleptically after mixing in the gravitational mixer. Also, the distribution of each component in all containers was determined. The content of component u in container l was calculated by the following equation:

$$c_{cul} = \frac{N_{mul}}{\sum_{j=1}^{l} N_{muj}} \cdot 100\%, \tag{4}$$

where

c_{cul} – the content of component u in container l, %;

$\sum_{j=1}^{l} N_{muj}$ – the total number of component u in all containers, dragees;

l – the number of containers ($l = 4$).

Each container corresponds to a particular state S_i of the mathematical model of the mixing process. Accordingly, the probability p_i of staying component u in the state S_i multiplied by 100% calculated by the equations of the mathematical model can characterize the probable content c_{pu} as a percentage of the component u in the mixture for the state S_i. Comparison of the probable content c_{pu} of the component u in the mixture with the content c_{cul} of the same component allows to determine how accurately the mathematical model describes the mixing process in the mixer.

4 Results

The Fig. 2c presents the motion of flows of components A, B, C, and D, and their mixtures in the gravitational mixer. In each of the four initial flows of the components that are fed to the top section of the mixer, a single particle (dragee) is selected, and the probability of staying this particle on the plates of the separators of each mixer section is determined. The locations of the component particle on the plates of the mixer separators are named states S_i (where S_i – the possible state i of the component particle). Each plate of the separators of all mixer sections and the plates of the combiners of the top section of the gravitational mixer corresponds to a particular state S_i of the components particles.

All states S_i of the particles of components A, B, C, and D are numbered as shown in Fig. 2d. The number of possible states for the particles of all four components will be $n = 32$. The particles of the four components, moving down the mixer sections, can be in states $S_1, S_2, \ldots, S_n$. Possible transitions of the components particles from the state to the state in the graph are marked by arrows (Fig. 2d). According to the accepted assumptions, there are known probabilities P of possible transitions of the components particles from the state to the state, which are marked next to the arrows in Fig. 2d. The probability of a particle transition from the state S_i to the state S_j is denoted by P_{ij}. If individual transitions are impossible, then their probabilities are zero. The probability $p_1(k), p_2(k), \ldots, p_n(k)$ of states of the component particle after an arbitrary stage k [19]:

$$p_i(k) = \sum\nolimits_{j=1}^{n} p_j(k-1) \cdot P_{ji}, \left(i = \overline{1,n}\right), \tag{5}$$

where P_{ij} – the probability of particle transition from the state S_i to the state S_j.

During the mixture motion, the components particles can be in the states (Fig. 2d): the particles of component A – $S_1, S_5, S_8, S_9, S_{10}, \ldots, S_{32}$; the particles of component B – $S_2, S_6, S_7, S_9, S_{10}, \ldots, S_{32}$; the particles of component C – $S_3, S_6, S_7, S_9, S_{10}, \ldots, S_{32}$; the particles of component D – $S_4, S_5, S_8, S_9, S_{10}, \ldots, S_{32}$.

Figure 3 presents the results of experimental studies of the mixing process of candy-dragees of four colors (components A, B, C, and D) in the gravitational mixer containing five and seven sections. The diagram in Fig. 3a shows the average number N_{mu} of components in the mixtures from four containers corresponding to the states S_{21}, S_{22}, S_{23}, and S_{24} after mixing in five sections. The diagram in Fig. 3b shows the average number N_{mu} of the components in the mixtures from four containers corresponding to the states S_{29}, S_{30}, S_{31}, and S_{32} after mixing in seven sections. The average content c_u of each component in the mixtures from four mixer containers was calculated by the Eq. (2). Analysis of mixtures from four containers after mixing the components in five sections indicates that the average content c_u of components in the mixtures varies widely (Fig. 3c): for component A – c_A = 19.5–29.4%; for component B – c_B = 20.5–28.5%; for component C – c_C = 20.5–30.2%; for component D – c_D = 21.1–30.3%. Such a significant difference between the average content c_u of components in the mixtures from different containers indicates a lack of components mixing in the five mixer sections. For this case, the quality index Δ_u of the mixture for each component, with a base content c_{0u} = 25% of these components in the mixture, is within (Fig. 3e): for component A – Δ_A = 7.2–22.0%; for component B – Δ_B = 6.0–18.0%; for component C – Δ_C = 12.0–20.8%; for component D – Δ_D = 6.4–21.2%. The following quality index Δ_u of the mixture indicates the poor quality of the mixture after components mixing in the five mixer sections. Thus, five sections of the gravitational mixer are not enough to get a good quality of the mixture.

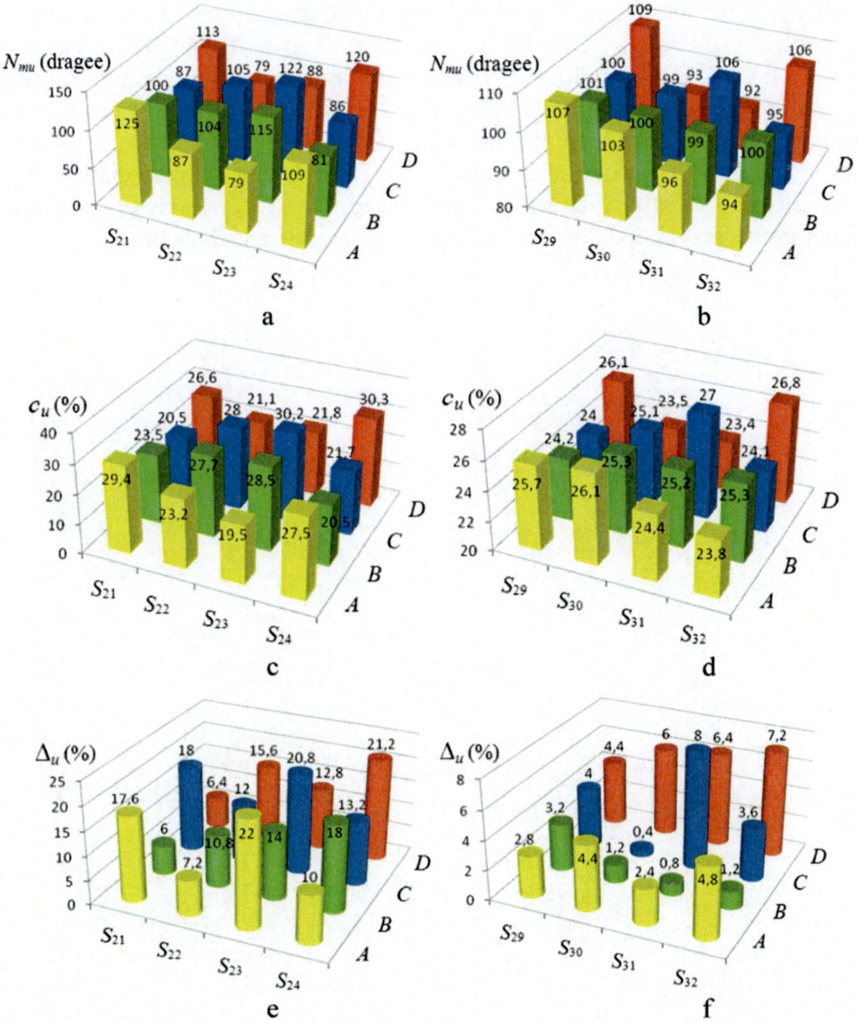

Fig. 3. The average number N_{mu} (a, b) and content c_u (c, d) of each component in the mixtures and the quality index Δ_u (e, f) of the mixture from the containers corresponding to the states S_{21}, S_{22}, S_{23}, and S_{24}, after mixing in five sections (a, c, e) and from the containers corresponding to the states S_{29}, S_{30}, S_{31}, and S_{32}, after mixing in seven sections (b, d, f): ■ – component A; ■ – component B; ■ – component C; ■ – component D.

Analysis of mixtures from four containers after components mixing in the seven mixer sections indicates that the average content c_u of components in the mixtures is within (Fig. 3d): for component A – c_A = 23.8–26.1%; for component B – c_B = 24.2–25.3%; for component C – c_C = 24.0–27.0%; for component D – c_D = 23.4–26.8%. This difference between the average content c_u of components in the mixtures from different containers indicates a good degree of mixing. For this case, the quality index Δ_u of the mixture for each component, with a base content c_{0u} = 25% of these components in the mixture, is within (Fig. 3f): for component A – Δ_A = 2.4–4.8%; for component B – Δ_B = 0.8–3.2%; for component C – Δ_C = 0.4–8.0%; for component D – Δ_D = 4.4–7.2%. The following values of the quality of the mixture indicate the good quality of the mixture after components mixing in the seven mixer sections. According to the content of components A and B, the quality of the mixture is excellent. The candy-dragees inspection after mixing in the gravitational mixer showed that the dragees surfaces were not damaged and cracked. Thus, the candy-dragees are not damaged in the gravitational mixer and retain their quality and shape.

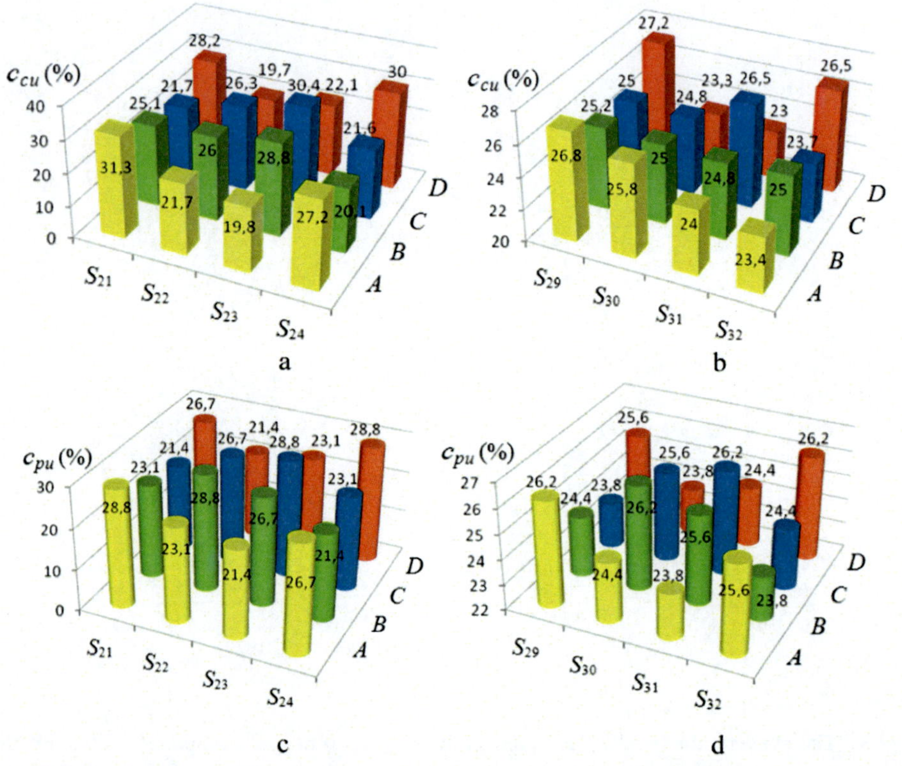

Fig. 4. The content c_{cu} (a, b) and probable content c_{pu} (c, d) of components in the containers corresponding to the states S_{21}, S_{22}, S_{23}, and S_{24}, after mixing in five sections (a, c) and from the containers corresponding to the states S_{29}, S_{30}, S_{31}, and S_{32}, after mixing in seven sections (b, d): ■ – component A; ■ – component B; ■ – component C; ■ – component D.

The study of the distribution of each component in the four flows of the mixture, which come out of the fifth section of the gravitational mixer, shows that the experimental values of the content c_{cu} of each component in the containers (Fig. 4a) do not differ significantly from the probable content c_{pu} of these components in the containers (Fig. 4c), as determined by the modeling of the mixing process. Thus, among all components, the largest difference between the experimental and model data is for component B in the mixture from the container, which corresponds to the state S_{22}. In this case, the experimental content of the component is c_{cB} = 26.0%, and the probable content of this component obtained from the modeling is c_{pB} = 28.8%. Thus, the model of mixing process accurately describes the process of four bulk materials mixing in the gravitational mixer, which consists of five sections.

The study of the distribution of each component in the four flows of the mixture, which come out of the seventh section of the gravitational mixer, shows that the experimental values of the content c_{cu} of each component in the containers (Fig. 4b) do not differ significantly from the probable content c_{pu} of these components in the containers (Fig. 4d), as determined by the modeling of the mixing process. Thus, among all components, the most substantial difference between the experimental and model data is for component A in the mixture from the container, which corresponds to the state S_{32}. In this case, the empirical content of the component is c_{cA} = 23.4%, and the probable content of this component obtained from the modeling is c_{pA} = 25.6%. Thus, the model of mixing process accurately describes the process of four bulk materials mixing in the gravitational mixer, which consists of seven sections.

5 Conclusions

The proposed design of the gravitational mixer has the advantages: the mixer does not contain active working bodies, so the components are not damaged; in the mixer, there are no energy costs for the mixing process; the mixer is compact and easy to maintain. Modeling of the mixing process in the gravitational mixer using Markov chains theory can be used to describe mathematically the component flows in the mixer sections and to determine the probable content of each component in the mixture at each stage of mixing. Also, the number of mixer sections can be determined by the proposed mathematical model of the mixing process. Such mixer sections should be not less than seven. A comparison of the experimental study results of candy-dragees mixing with the data obtained during the modeling indicates that the mathematical model of bulk materials mixing with sufficient accuracy describes the mixing process. Candy-dragees mixing in the mixer containing seven sections provides a good quality of the mixture since the quality indexes of the mixture are within the range Δ_u = 0.4–8.0%. Studies involving the mixing of components with different properties and size of particles in the proposed gravitational mixer should be realized in the future.

References

1. Thakur, R.K., Vial, C.H., Nigam, K.D.P., Nauman, E.B., Djelveh, G.: Static mixers in the process industries – a review. Chem. Eng. Res. Des. **81**(7), 787–826 (2003). https://doi.org/10.1205/026387603322302968
2. Ghanem, A., Lemenand, T., Della Valle, D., Peerhossaini, H.: Static mixers: mechanisms, applications, and characterization methods – a review. Chem. Eng. Res. Des. **92**(2), 205–228 (2014). https://doi.org/10.1016/j.cherd.2013.07.013

3. Ivanec, V.N., Borodulin, D.M., Popov, A.M., Tikhonov, V.V.: Design of drum type apparatus for processing of bulk materials. Procedia Chem. **10**, 391–399 (2014). https://doi.org/10.1016/j.proche.2014.10.066
4. Arntz, M.M.H.D., Den Otter, W.K., Briels, W.J., Bussmann, P.J.T., Beeftink, H.H., Boom, R.M.: Granular mixing and segregation in a horizontal rotating drum: a simulation study on the impact of rotational speed and fill level. AIChE J. **54**(12), 3133–3146 (2008). https://doi.org/10.1002/aic.11622
5. Zhu, H.P., Zhou, Z.Y., Yang, R.Y., Yu, A.B.: Discrete particle simulation of particulate systems – a review of major applications and findings. Chem. Eng. Sci. **63**(23), 5728–5770 (2008). https://doi.org/10.1016/j.ces.2008.08.006
6. Jain, N., Ottino, J.M., Lueptow, R.M.: Regimes of segregation and mixing in combined size and density granular systems: an experimental study. Granular Matter **7**(2–3), 69–81 (2005). https://doi.org/10.1007/s10035-005-0198-x
7. Ottino, J.M., Khakhar, D.V.: Mixing and segregation of granular materials. Annu. Rev. Fluid Mech. **32**, 55–91 (2000). https://doi.org/10.1146/annurev.fluid.32.1.55
8. Ferraris, C.F.: Concrete mixing methods and concrete mixers: state of the art. J. Res. NIST **106**(2), 391–399 (2001). https://doi.org/10.6028/jres.106.016
9. Bakin, M., Kapranova, A., Verloka, I.: Modern methods of mathematical description of bulk materials mixing process. Fundam. Res. **5**, 923–927 (2014)
10. Berthiaux, H., Mizonov, V., Zhukov, V.: Application of the theory of Markov chains to model different processes in particle technology. Powder Technol. **157**(1–3), 128–137 (2005). https://doi.org/10.1016/j.powtec.2005.05.019
11. Santomaso, A.C., Ding, Y.L., Lickiss, J.R., York, D.W.: Investigation of the granular behavior in a rotating drum operated over a wide range of rotational speed. Chem. Eng. Res. Des. **81**(8), 936–945 (2003). https://doi.org/10.1205/026387603322482176
12. Godlieb, W., Deen, N.G., Kuipers, J.A.M.: Characterizing solids mixing in DEM simulations. In: 6th International Conference on Multiphase Flow, ICMF, Leipzig, Germany (2007)
13. Doucet, J., Hudon, N., Bertrand, F., Chaouki, J.: Modeling of the mixing of monodisperse particles using a stationary DEM-based Markov process. Comput. Chem. Eng. **32**(6), 1334–1341 (2008). https://doi.org/10.1016/j.compchemeng.2007.06.017
14. Wen, Y., Liu, M., Liu, B., Shao, Y.: Comparative study on the characterization method of particle mixing index using DEM method. Proc. Eng. **102**, 1630–1642 (2015). https://doi.org/10.1016/j.proeng.2015.01.299
15. Tasirin, S.M., Kamarudin, S.K., Hweage, A.M.A.: Mixing process of binary polymer particles in different type of mixers. Mod. Appl. Sci. **3**(6), 88–95 (2009). https://doi.org/10.5539/mas.v3n6p88
16. Mizonov, V., Balagurov, I., Berthiaux, H., Gatumel, C.: A Markov chain model of mixing kinetics for ternary mixture of dissimilar particulate solids. Particuology **31**, 80–86 (2017). https://doi.org/10.1016/j.partic.2016.05.006
17. Porion, P., Sommier, N., Faugere, A.M., Evesque, P.: Dynamics of size segregation and mixing of granular materials in a 3D-blender by NMR imaging investigation. Powder Technol. **141**(1–2), 55–68 (2004). https://doi.org/10.1016/j.powtec.2004.02.015
18. Göbel, F., Golshan, S., Norouzi, H.R., Zarghami, R., Mostoufi, N.: Simulation of granular mixing in a static mixer by the discrete element method. Powder Technol. **346**, 171–179 (2019). https://doi.org/10.1016/j.powtec.2019.02.014
19. Dudarev, I.: Simulation of bulk materials mixing process in gravitational mixer. Food Ind. **23**, 67–73 (2018). https://doi.org/10.24263/2225-2916-2018-23-12

Simulation of Delamination Processes of Multilayer Mechanical Engineering Structures

Aleksandr Gondlyakh[1(✉)], Andrey Chemeris[1], Aleksandr Kolosov[1], Aleksandr Sokolskiy[1], and Sergiy Antonyuk[2]

[1] National Technical University of Ukraine "Igor Sikorsky Kyiv Polytechnic Institute", 37, Prospect Peremohy, Kyiv 03056, Ukraine
avg_hotm@hotmail.com
[2] Technical University of Kaiserslautern, 67653 Kaiserslautern, Germany

Abstract. The article is devoted to the problems of numerical simulation of the delamination processes of multilayer spatial systems under static loading. The iterative-analytical theory of spatial multilayer structures is used. A particular multilayer eight-node finite element has been developed and numerically implemented. The solution of nonlinear problems is carried out based on the Newton-Kantorovich algorithm, supplemented by a block that implements an iterative-analytical method of variable approximations. A comparison of the results of numerical solutions with analytical solutions shows their good agreement. As an example of a semi-rigid coupling, results of the numerical modeling of elastic disks package deformations depending on changes in the coefficient of friction between the disks are given. The developed methods allow one to estimate reliably of the deformed state of multilayer elements of mechanical engineering equipment, depending on the change of physical and mechanical characteristics and parameters of material strength during the exploitation of its elements. Research results can be used to develop systems for supporting the lifecycle of mechanical engineering equipment.

Keywords: Multilayer structures · Iterative-analytic theory · Fracture · Delamination zone · Numerical simulation · Finite element method

1 Introduction

Multilayer structures are widely used in mechanical engineering, chemical, and petrochemical industries. These include metal-composite vessels of internal pressure, the shells of reaction columns, heat exchangers, separators, couplings, springs, bimetallic distillation columns, and apparatuses.

The peculiarity of deformation of spatial multilayer structures of structural and composite materials is the possibility of the occurrence of both through cracks and the formation in the process of their distortion of the delamination zones. Therefore, even the initially uniform in thickness structure becomes a multilayer system. Since the thicknesses of exfoliated portions are often small in comparison to their dimensions in a

V. Tonkonogyi et al. (Eds.): InterPartner 2020, LNME, pp. 129–138, 2021.
https://doi.org/10.1007/978-3-030-68014-5_13

plan, the use of standard 3D finite elements (FEs) in these cases usually results in numerical instability of solutions. The use of shell and three-dimensional multilayer FE does not allow to model the heterogeneity of the distribution of deformations of transverse shear and extension along with the thickness of the multilayer package due to the hypothetically imposed restrictions on the approximating functions. In this regard, the analysis of the strength of multilayer structures should be made based on refined theories of multilayer systems [1, 2], as well as with the use of particular multilayer FE, allowing considering in the process of deformation the change of boundary conditions between layers due to the emergence of zones of delamination [3].

2 Literature Review

One of the most common types of destruction of objects is destruction due to the formation and development of delamination cracks in them. The delamination along the boundary of contact between layers of multilayered mechanical engineering structures should be distinguished into one of the main mechanisms of cracking. The solution to this evolutionary problem is reduced to the correct modeling of the contact conditions on the formed disconnection and delamination surfaces.

The available publications on this topic can be reduced to three classical main areas, i.e., analytical, numerical, and experimental. Among the analytical methods, it is worth noting the works devoted to refined models of the deformation of multilayer systems to consider the factor of the inhomogeneous distribution of transverse shear strains over the thickness of the layer package [1–7]. An analysis of the criteria for breaking the adhesion layer is given in [8–13].

Numerical methods are used to study physically and geometrically nonlinear problems of defect propagation in multilayer systems. Widespread for solving these problems was the finite element method (FEM) [14–16]. Various approaches to modeling delamination have been developed, based on the development of special contact FE [3, 16, 17], modification of standard FE [18–22] and the use of cohesive models [23–26].

A large number of applied problems are solved based on FEM. In [27–31], the results of studies of the effect of delamination upon deformation of multilayer structures are presented. The influence of joint modeling of crack propagation considering delamination processes is given in [32, 33].

Studies in the field of delamination of metal composites were reflected in [34–38]. Of particular relevance is the use of refined deformation models in the numerical simulation of engineering products from structural polymer composites with thin nanomodified coatings [39, 40].

The above analysis shows the relevance of developing the methods for numerically modeling the evolution of fracture of both individual elements of a multilayer package of layers and the entire structure as a whole.

3 Research Methodology

The theoretical provisions of the present work are based on an iterative-analytical theory of spatial systems [1, 2], the essence of which is as follows. The components of the displacement vector of the multilayer structure are represented as:

$$u_i = \sum_{s=1}^{S} F_s^i(x^3, t) v_i^s(x^1, x^2, t), \tag{1}$$

where v_i^s are the components of the generalized displacement vector of the middle surface of the structure x^1x^2; F_s^i – functions to be determined ($F_1^i = 1$; $F_2^i = x^3$); t - time.

For the vector $\overrightarrow{u}$, for the description of the actual deformed state of the structure, it is necessary to fulfill the Hamilton–Ostrogradsky variational principle, namely:

$$\delta \int_{t_0}^{t_1} (W - K - A)dt = \delta_{v_i^s} \int_{t_0}^{t_1} (W - K - A)dt + \delta_{F_i^s} \int_{t_0}^{t_1} (W - K - A)dt = 0. \tag{2}$$

Here, the notation of the type $\delta_{v_i^s}$ and $\delta_{F_i^s}$ means that the variation of the functional is made by v_i^s or F_s^i, respectively; W is the internal energy of the system; K is kinetic energy; A is the work of external forces.

Given (1) and the independence of the variations and, we obtain the resolving system of equations:

$$\begin{cases} \int_{t_0}^{1} \left\{ \int_V A_j^r \left(v_i^s; \frac{\partial v_i^s}{\partial x^\alpha}; \frac{\partial^2 v_i^s}{\partial x^\alpha \partial x^\beta}; \ddot{v}_i^s; F_s^i; \frac{dF_s^i}{dx^3}; \frac{d^2 F_s^i}{(dx^3)^2}; \ddot{F}_s^i; p_j^r \right) \delta v_j^r dV \right\} dt = 0, \\ \int_{t_0}^{t_1} \left\{ \int_V B_j^r \left(v_i^s; \frac{\partial v_i^s}{\partial x^\alpha}; \frac{\partial^2 v_i^s}{\partial x^\alpha \partial x^\beta}; \ddot{v}_i^s; F_s^i; \frac{dF_s^i}{dx^3}; \frac{d^2 F_s^i}{(dx^3)^2}; \ddot{F}_s^i; p_j^r \right) \delta F_r^j dV \right\} dt = 0 \end{cases} \tag{3}$$

and the corresponding boundary conditions

$$\begin{cases} \int_{t_0}^{t_1} \left\{ \int_S a_j^r \left(v_i^s; \frac{\partial v_i^s}{\partial x^\alpha}; F_s^i; \frac{dF_s^i}{dx^3}; q_j^r \right) \delta v_j^r d\theta \right\} dt = 0, \\ \int_{t_0}^{t_1} \left\{ \int_S b_j^r \left(v_i^s; \frac{\partial v_i^s}{\partial x^\alpha}; F_s^i; \frac{dF_s^i}{dx^3}; q_j^r \right) \delta F_r^j d\theta \right\} dt = 0. \end{cases} \tag{4}$$

Here V is the volume of the multilayer system; θ is the load application surface area; p_j^r – components of the vector of volumetric forces; q_j^r are the components of the external load vector.

A specific feature of the system of Eqs. (3) is that its solution defines not only the vector of generalized unknowns v_i^s, but also the determination of physically justified functions F_s^i, regardless of the physical processes that occur in the material of the structure at any time (plastic deformation, delamination, accumulation of defects, breaking of layers).

Solving the system of Eqs. (3) by direct methods can be difficult. The use of iterative methods consisting of the sequential refinement of the components v_i^s or functions F_s^i, based on the condition of minimization of the residual vector of the three-dimensional operator of the theory of elasticity, makes it much easier to obtain the desired solution. In this case, the second subsystem of Eqs. (3) is represented as a system of ordinary differential equations for functions F_s^i in the sense of the Krylov–Kantorovich method, i.e.:

$$\mathrm{D}_j^{r(n-1)}\left(F_s^{i(n)};\frac{dF_s^{i(n)}}{dx^3};\frac{d^2F_s^{i(n)}}{(dx^3)^2};p_j^r\right)=0 \tag{5}$$

with the appropriate boundary conditions.

$$\mathrm{d}_j^{r(n-1)}\left(F_s^{i(n)};\frac{dF_s^{i(n)}}{dx^3};p_j^r\right)=0. \tag{6}$$

The coefficients in Eqs. (5), (6) are integral characteristics that depend not only on the components of the tensor of the physicomechanical deformation constants of the material of the layer package but also on the components of the deformed state $v_j^{r(n-1)}$ obtained at iteration n – 1, namely:

$$\mathrm{D}_j^{r(n-1)} = \mathrm{D}_j^{r(n-1)}\left\{\int_S\left(v_j^{r(n-1)};p_j^{r(n-1)}\right)dS\right\}. \tag{7}$$

Such a technique for solving the basic system of Eqs. (3) significantly simplifies the solution of the problem since relations (5) are a system of ordinary differential equations, which, with the corresponding grouping of unknowns, also permits obtaining an analytic solution in the exponential form, namely:

$$\begin{cases} u_\alpha^m = v_\alpha^m + \left\{\sum_{i=1}^4 C_i^{\alpha m} e^{r_\alpha^{im}x^3} + G_\alpha^m/\psi_\alpha^m\right\}\xi_\alpha^m \\ u_3^m = \left\{\sum_{\alpha=1}^2\left[\sum_{i=1}^4 \varphi_{\alpha i}^{3m} C_i^{\alpha m} e^{r_\alpha^{im}x^3} + C_i^{3m} e^{r_\alpha^{im}x^3}\right] + \Upsilon_3^m\right\}\omega^1 \end{cases}, \tag{8}$$

where v_α^m, ξ_α^m, ω^1 are the components of the tangential displacement vector, the generalized shear, and the deflection of the middle layer, respectively; $C_i^{\alpha m}$ – integration constants, determined from the conditions of contact between the layers and on the front surfaces of the layer package; r_α^{im} – the integral deformation characteristics of the package, which depends not only on the physical and mechanical characteristics of its layers but also on the components of the deformed state of the layer package, determined at iteration $(n - 1)$; G_α^m/ψ_α^m, Υ_3^m – particular solutions of characteristic equations.

The peculiarity of the obtained solutions is that on their basis, it is possible to simulate not only the effect of slipping of layers but also the effect of compressing (tearing) the layers from each other. The simulation of the delamination process is performed by analogy with [3]. A characteristic feature of this technique is the physically substantiated correction of the tensor of the physicomechanical constants of the fractured region based on the hypotheses taken depending on the type of fracture (Fig. 1).

Numerical modeling of the processes of plastic deformation of the material and fracture is carried out based on a step-iterative algorithm of sequential loads based on the modified Newton–Kantorovich method. The analysis of plastic deformation of the structural layers is carried out based on the basic provisions of the theory of plastic flow with isotropic translational hardening [41–50].

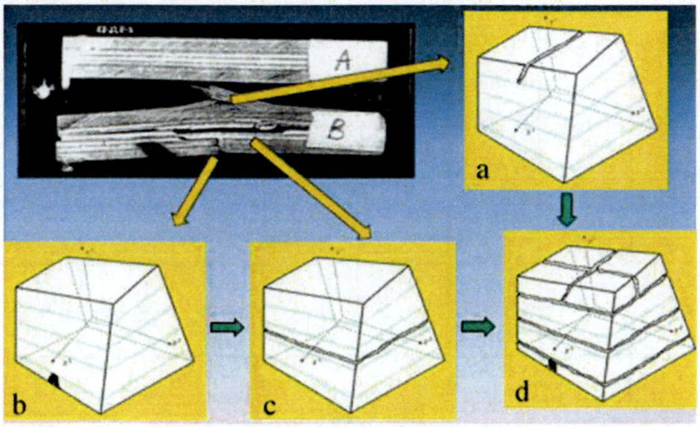

Fig. 1. Schemes of the destruction of multilayer FE: *A* – original multilayer sample; *B* – destroyed sample; *a* is the transverse rupture of the layer; *b* – layer crushing; *c* – delamination; *d* is a general picture of destruction.

The redefinition of the approximating functions of the displacement vector over the thickness of the packet of the FE layers, taking into account the formed delamination zones, is carried out based on an iterative-analytical method of variable approximations [1, 2]. This FE was developed in the APROKS system [1, 2, 16, 27].

APROKS was developed at Igor Sikorsky Kyiv Polytechnic Institute (Ukraine) for solving nonlinear problems of mechanics of a deformed solid based on the FEM.

The practice of using this FE in solving a large number of test and application problems has shown its effectiveness in the study of processes of nonlinear deformation and destruction of multilayer spatial structures [16, 27, 28, 33, 38].

4 Results

4.1 Three-Layer Beam Delamination Test

The possibility of modeling the process of delamination of multilayer structures based on the developed refined FE was a performance at the example of a test problem of the study of the deformation of a three-layer freely supported beam with a thin middle layer (Fig. 2). The modulus of elasticity of the middle layer varies in the range from $E = 1{\cdot}10^{6}$ MPa to $E = 1{\cdot}10^{-4}$ MPa. Beam calculations were performed in parallel using the standard multilayer C3D8I and multilayer shell S4R elements (Tymoshenko's hypothesis) from the ABAQUS FE system library.

The results of the studies (Fig. 2) reflect the dependence of the change in the deflection of the central point of the beam on the value of degree n of the modulus of elasticity of the middle layer E2 = 1•10n MPa (L (beam span) = 20 mm; B (beam width) = 1 mm; thickness t1 = t3 = 0.499 mm; t2 = 0.002 mm; E1 = 1•105 MPa (modulus of elasticity of the upper and lower layers); Poisson's ratio v = 0; P = 10 N.

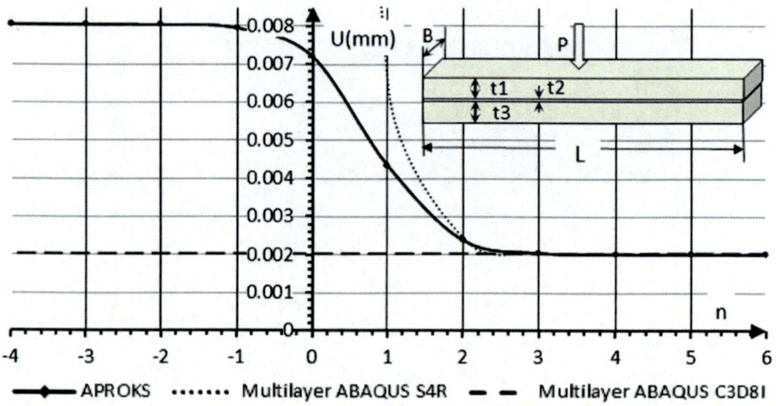

Fig. 2. The deflection of the central point of a beam depending on the exponent $n = \log_{10}(E2)$ of the modulus of elasticity for the middle layer $E2 = 1 \cdot 10^n$ MPa.; P – force, N.

As a result of numerical experiments, it is found that when the middle layer stiffness changes by six orders of magnitude and the outer layers begin to slide independently of each other, and the deflection in the middle of the span of the beam increases four times. The results obtained are in full agreement with the solution of this problem known for the deflection of springs, which is known from the strength of materials, which indicates that a simulation of the bundling process is possible based on the developed FE.

4.2 Plate Coupling with Multilayer Package of Elastic Disks

When transmitting torque and bending moment, the most stressed elements of plate couplings are elastic discs. The package of elastic disks of the considered coupling consists of 28 thin metal layers 0.25 mm thick. The main factor in the operation of plate couplings is its flexibility, which substantially depends on the friction coefficients in the package of elastic disks. Therefore, the lifecycle of the coupling was simulated considering changes in the friction coefficient in the range of C_{fr} = 0–0.5. In the case of angular displacement of the shaft of the coupling by 0.5°, the results are shown in Fig. 3 and Table 1.

At the initial stage of the life cycle, when the friction coefficient changes from C_{fr} = 0.0 to C_{fr} = 0.3, the formation of shear zones is observed in the package of elastic coupling elements. When the coefficient of friction changes from C_{fr} = 0.3 to C_{fr} = 0.4, the disconnection zones are formed in the package of elastic elements. In this case, the package of elastic coupling elements is divided into several independently working packages. Further increase in the coefficient of friction C_{fr} leads to the effect of structural strengthening and accumulation of the internal energy of deformation of the coupling. The last section of the diagram C_{fr} = 0.4–0.5 shows a dramatic change in the shape of the deformation of the disk package, additional disconnection zones appear, and the package of elastic elements of the coupling ceases to fulfill its functional purpose.

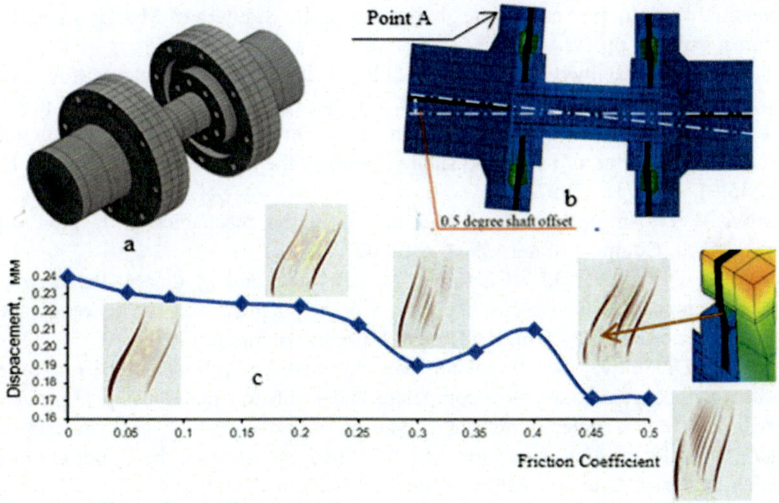

Fig. 3. The process of deformation of the coupling depending on the change of the friction coefficient in the disc package: *a* – FE model of the coupling; *b* – areas of concentration of maximum stresses at angular displacements of shafts; *c* is a diagram of the change in the displacement of point *A* from the friction coefficient of a pack of elastic disks.

Table 1. Horizontal displacements of point *A* at different values of the coefficient of friction.

Friction coefficient	0.00	0.10	0.20	0.30	0.35	0.40	0.50
Displacement, mm	0.235	0.198	0.180	0.092	0.113	0.145	0.034

5 Conclusions

A particular eight-node isoparametric multilayer finite element was developed and numerically implemented to the model of the multilayer structures delamination processes for machine-building products, based on the iterative-analytical theory of multilayer spatial systems. It is established that based on the developed FE, a detailed study of the delamination processes is possible. The analysis of the results showed that the consideration of the factor of heterogeneity of changes in the deformations of the transverse shear along the thickness of the layer package is a prerequisite for the accurate description of the deformation processes of multilayer mechanical engineering structures with changing zones of contact between the layers over time. In this case, the use of multilayer FE based on Tymoshenko's theory can lead to both quantitatively, and qualitatively incorrect results.

References

1. Sakharov, A.S., Kozak, A.L., Gondliakh, A.V., Melnikov, S.L.: Mathematical model of the deformation of multilayer composite shell systems. In: Strength of Materials and Theory of Structures, vol. 44, pp. 13–16 (1984)
2. Gondliakh, A.V.: Refined model of multilayer structures deformation for progressive destruction processes study. Eastern-Eur. J. Enterp. Technol. **2**(7(56)), 52–57 (2012)
3. Sakharov, A.S., Gondlyakh, A.V., Strizhalo, A.V.: On features of numerical integration for the equations of motion of laminated shell systems in the iterative analytic theory. Int. Appl. Mech. **33**(9), 713–718 (1997)
4. Piskunov, V.G.: An iterative analytical theory in the mechanics of layered composite systems. Mech. Compos. Mater. **39**, 1–6 (2003)
5. Tessler, A., Di Sciuva, M., Gherlone, M.: Refined zigzag theory for homogeneous, laminated composite, and sandwich plates: a homogeneous limit methodology for zigzag function selection. NASA, Langley Research Center Hampton (2010)
6. Tessler, A., Di Sciuva, M., Gherlone, M.: A consistent refinement of first-order shear deformation theory for laminated composite and sandwich plates using improved zigzag kinematics. J. Mech. Mater. Struct. **5**(2), 341–367 (2010)
7. Tessler, A., Di Sciuva, M., Gherlone, M.: A refined zigzag beam theory for composite and sandwich beams. J. Compos. Mater. **43**(9), 1051–1081 (2009)
8. Objois, A., Assih, J., Troalen, J.P.: Theoretical method to predict the first microcracks in a scarf joint. J. Adhes. **81**(9), 893–909 (2005)
9. Banks-Sills, L.: Interface Fracture and Delaminations in Composite Materials. Springer, Cham (2018)
10. Pandurangan, P., Buckner, G.D.: Vibration analysis for damage detection in metal-to-metal adhesive joints. Exp. Mech. **46**(5), 601–607 (2006)
11. Martiny, P., Lani, F., Kinloch, A.J., Pardoen, T.: A multiscale parametric study of mode I fracture in metal-to-metal low-toughness adhesive joints. Int. J. Fract. **173**, 105–133 (2012)
12. Guz, A.N.: Establishing the foundations of the mechanics of fracture of materials compressed along cracks (review). Int. Appl. Mech. **50**, 1–57 (2014)
13. Stevanovic, D., Kalyanasundaram, S., Lowe, A., Jarc, P.-Y.B.: FEA of crack–particle interactions during delamination in interlayer toughened polymer composites. Eng. Fract. Mech. **72**(11), 1738–1769 (2005)
14. Zienkiewicz, O.C., Taylor, R.L.: The Finite Element Method for Solid and Structural Mechanics. 6th edn. Butterworth-Heinemann (2005)
15. Ellobody, E., Feng, R., Young, B.: Finite Element Analysis and Design of Metal Structures, 1st edn. Imprint: Butterworth-Heinemann (2013)
16. Gondliakh, A.V.: Adaptation in ABAQUS of the iterated-analytical multilayer user finite element. Eastern-Eur. J. Enterp. Technol. **37**(57), 62–68 (2012)

17. Li, W.D., Ma, M., Han, X., Tang, L.P., Zhao, J.N., Gao, E.P.: Strength prediction of adhesively bonded single lap joints under salt spray environment using a cohesive zone model. J. Adhes. **92**(11), 916–937 (2016)
18. Kafkalidis, M.S., Thouless, M.D.: The effects of geometry and material properties on the fracture of single lap-shear joints. Int. J. Solids Struct. **39**(17), 4367–4383 (2002)
19. Khokhar, Z.R., Ashcroft, I.A., Silberschmidt, V.V.: Interaction of matrix cracking and delamination in cross-ply laminates: simulations with stochastic cohesive zone elements. Appl. Compos. Mater. **18**, 3–16 (2011)
20. Maleki, S., Rafiee, R., Hasannia, A., Habibagahi, M.R.: Investigating the influence of delamination on the stiffness of composite pipes under compressive transverse loading using cohesive zone method. Front. Struct. Civ. Eng. **13**, 1316–1323 (2019)
21. Kesava Rao, B., Balu, A.S.: Assessment of cohesive parameters using high dimensional model representation for mixed mode cohesive zone model. Structures **19**, 156–160 (2019)
22. Silva, D.F.O., Campilho, R.D.S.G., Silva, F.J.G., Carvalho, U.T.F.: Application a direct/cohesive zone method for the evaluation of scarf adhesive joints. Appl. Adhes. Sci. **6**(13) (2018)
23. Mubashar, A., Ashcroft, I.A., Crocombe, A.D.: Modelling damage and failure in adhesive joints using a combined XFEM-cohesive element methodology. J. Adhes. **90**(8), 682–697 (2014)
24. Pascoe, J.A., Alderliesten, R.C., Benedictus, R.: Methods for the prediction of fatigue delamination growth in composites and adhesive bonds - a critical review. Eng. Fract. Mech. **112–113**, 72–96 (2013)
25. Anyfantis, K.N., Tsouvalis, N.G.: A novel traction–separation law for the prediction of the mixed mode response of ductile adhesive joints. Int. J. Solids Struct. **49**(1), 213–226 (2012)
26. Turon, A., Davila, C.G., Camanho, P.P., Costa, J.: An engineering solution for mesh size effects in the simulation of delamination using cohesive zone models. Eng. Fract. Mech. **74** (10), 1665–1682 (2007)
27. Gondliakh, O., Onoprienko, V., Chemerys, A.: Numerical analysis of jerry can strength under static and dynamic loads. Eastern-Eur. J. Enterp. Technol. **3**(7), 23–29 (2015). https://doi.org/10.15587/1729-4061.2015.4438323-29
28. Gondliakh, O., Krytskyi, V., Onopriienko, V., Chemerys, A., Krytska, N.: Computer analysis of thermomechanical state of sealing steel lining for containment of NPPs with VVER-1000/V-320 in emergencies. Nucl. Radiat. Saf. **4**(76), 28–39 (2017)
29. Piskunov, V.G., Grynevyts'kyi, R.V.: Solution of the problem on the vibrations of beams with a variable cross section using the finite-difference method. Strength Mater. **44**, 53–58 (2012)
30. Geleta, T.N., Woo, K., Lee, B.: Delamination behavior of L-shaped laminated composites. Int. J. Aeronaut. Space Sci. **19**, 363–374 (2018)
31. Abdellah, M.Y.: Delamination modeling of double cantilever beam of unidirectional composite laminates. J. Fail. Anal. Prev. **17**, 1011–1018 (2017)
32. Huchette, C., Vandellos, T., Laurin, F.: Influence of intralaminar damage on the delamination crack evolution. In: Riccio, A. (eds). Damage Growth in Aerospace Composites. Springer Aerospace Technology. Springer, Cham (2015)
33. Sakharov, A.S., Gondlyakh, A.V., Mel'nikov, S.L., Snitko, A.N.: Numerical modeling of processes of failure of multilayered composite shells. Mech. Compos. Mater. **25**(3), 337–343 (1989)
34. Van der Sluis, O., et al.: Advances in delamination modeling of metal/polymer systems: atomistic aspects. In: Morris, J. (ed.) Nanopackaging, pp. 129–183. Springer, Cham (2018)

35. Beklemysheva, K.A., Vasyukov, A.V., Petrov, I.B.: Numerical modeling of delamination in fiber-metal laminates caused by low-velocity impact. In: Petrov, I., Favorskaya, A., Favorskaya, M., Simakov, S., Jain, L. (eds.) Smart Modeling for Engineering Systems. GCM50 2018. Smart Innovation, Systems and Technologies, vol. 133, pp. 132–142. Springer, Cham (2019)
36. Jia, Y., Ghazali, S., Bai, Y.: Application of eMMC model to fracture of metal sheets. In: Zehnder, A., et al. (eds). Fracture, Fatigue, Failure and Damage Evolution, vol. 8, pp. 49–55 (2017)
37. Lepikhin, A.M., Moskvichev, V.V., Chernyaev, A.P.: Acoustic-emission monitoring of the deformation and fracture of metal-composite pressure vessels. J. Appl. Mech. Tech. Phys. **59** (3), 511–518 (2018)
38. Gondliakh, O., Onoprienko, V., Nikitin, R.: Numerical Simulation of crack propagation in bimetallic spatial structures. Technol. Audit Prod. Reserves **3**(1(17)), 23–27 (2014)
39. Kolosov, A.E., Sivetskii, V.I., Kolosova, E.P., Vanin, V.V., Gondlyakh, A.V., Sidorov, D. E., Ivitskiy, I.I.: Creation of structural polymer composite materials for functional application using physicochemical modification. Adv. Polym. Technol. **12** (2019). https://doi.org/10.1155/2019/3501456
40. Kolosov, A.E., Sivetskii, V.I., Kolosova, E.P., Vanin, V.V., Gondlyakh, A.V., Sidorov, D. E., Ivitskiy, I.I., Symoniuk, V.P.: Use of physico-chemical modification methods for producing of traditional and nanomodified polymeric composites with improved operational properties. Int. J. Polym. Sci. **18** (2019). https://doi.org/10.1155/2019/1258727
41. Malinin, N.N.: Creep theories in metal forming. In: Sih, G.C., Ishlinsky, A.J., Mileiko, S.T. (eds.) Plasticity and Failure Behavior of Solids. Fatigue and Fracture, vol. 3, pp. 31–59. Springer, Dordrecht (1990)
42. Rösler, J., Bäker, M., Harders, H.: Plasticity and failure. In: Mechanical Behaviour of Engineering Materials, pp. 63–118. Springer, Heidelberg (2007)
43. Sidney, C., Charles, E.A., Borja, E.: A new plasticity and failure model for ballistic application. Int. J. Impact Eng. **38**(8–9), 755–764 (2011).
44. Erice, B., Gálvez, F.: A coupled elastoplastic-failure constitutive model with Lode angle dependent failure criterion. Int. J. Solids Struct. **51**, 93–110 (2014)
45. Kanters, MJW., Kurokawa, T., Govaert, L.E.: Competition between plasticity-controlled and crack-growth controlled failure in static and cyclic fatigue of thermoplastic polymer systems. Polym. Test. **50**, 101–110 (2016)
46. Abe, T., Tsuruta, T.: Advances in Engineering Plasticity and its Applications (AEPA '96). Elsevier (2012)
47. Pineau, A., Benzerga, A.A., Pardoen, T.: Failure of metals I - brittle and ductile fracture. Acta Mater. **107**, 424–483 (2016)
48. Buehler, M.: Atomistic Modeling of Materials Failure. Springer, Boston (2008)
49. Tang, Yi., Bringa, E.M., Meyers, M.A.: Ductile tensile failure in metals through initiation and growth of nanosized voids. Acta Materialia **60**, 4856–4865 (2012)
50. Matta, A.K., Kodali, S.P., Ivvala, J., Kumar, P.J.: Metal prototyping the future of automobile industry: a review. Mater. Today Proc. **5**(9), 17597–17601 (2018)

The Dynamic Model of Conveyor Belt Stresses

Valery Khodusov[1], Oleh Pihnastyi[2], and Georgii Kozhevnikov[2](✉)

[1] V.N. Karazin Kharkiv National University, 4, Svobody Sq., Kharkiv 61022, Ukraine
[2] National Technical University "Kharkiv Polytechnic Institute", 2, Kyrpychova Street, Kharkiv 61000, Ukraine
kozhevnikov.gk@gmail.com

Abstract. The article considers the causes of dynamic stress when moving a conveyor belt with a material. The main factors of the occurrence of the dynamic resistance in the conveyor system are defined. It is demonstrated what the propagation speed of perturbations substantially depends on the loading level of the conveyor system with the material. When constructing a model of dynamic stresses, Hooke's law was used, which is an accurate approximation for most solid bodies, as long as the forces and deformations are small enough. In the article, the main attention is given to the causes of dynamic stresses at the start of the conveyor system. When calculating the deformations, the inertia forces of the moving material and tape are considered. The analysis of the appearance of dynamic stresses at a constant value of the conveyor belt acceleration and the linear nature of the change in the value of the conveyor belt acceleration was carried out.

Keywords: Propagation speed of disturbances · Conveyor belt · Acceleration · Tension force · Oscillatory process

1 Introduction

The transport conveyor system is an integral part of the technological process in many industries at enterprises with a continuous production cycle. The use of conveyor transport is especially common in the mining industry, where the growth of cargo flows and the length of rock transportation has led to the need for the creation of high-performance belt conveyors with considerable length and capacity. Particularly, the length of modern conveyors reaches 100 km and permanently increases. For an increase in the reliability of the transport system, the conveyor is divided into sections. Each section has a length of 1–20 km [1]. This fact allows choosing a separate operating mode for each section, including stopping the conveyor section for repair. A feature of transport systems used in the mining industry is that the flow of rock entering the transport system is non-stationary. This leads to an uneven distribution of material on the belt conveyor along the transport route [2]. To ensure a uniform distribution of material on the conveyor belt, the speed control of the conveyor belt is applied inside a separate section [3, 4]. The change in speed leads to the fact that the

V. Tonkonogyi et al. (Eds.): InterPartner 2020, LNME, pp. 139–148, 2021.
https://doi.org/10.1007/978-3-030-68014-5_14

belt begins to move with acceleration or deceleration. As a result, stresses arise along the belt, the value of which is higher than the maximum allowable ones, which leads to the destruction of the belt [5, 6]. For the prevention of this scenario, additional restrictions should be imposed on the algorithm for controlling pipeline flow parameters [7]. However, these limitations are usually not taken into account when designing tape speed control systems [8, 9], although they are essential in case of uneven loading of material along the route [10]. For solving this problem, the article offers an analytical model of the conveyor, considering the uneven distribution of material.

2 Literature Review

There are a fairly large number of papers devoted to the modeling of conveyor systems. The proposed models should be divided into several large groups. The most common are the models of the conveyor, which use the finite difference method (FDM) [11–13]; finite element method (FEM) [14, 15]; method using the Lagrange equations [16]; the aggregated method [9]; the method using neural network [15, 18]; the system dynamic method [19]; multi-regression analysis [20, 21]. A comparative analysis of several groups of models was performed in [22]. The first group of models (FEM, FDM, and Lagrange equations) use numerical methods to the section flow parameters. It makes it possible to bear in mind the uneven material distribution along the section, but require significant computational resources. The second group of models allows us to give average estimates of the value transport system parameters. Still, they are not used to calculate the distribution of material along the route.

In papers [11, 12, 14], the modes of acceleration and deceleration of the belt are studied. The stresses that occur in the belt are evaluated, and recommendations for controlling the speed in transient conditions are given. An estimate of the propagation speed of the perturbation is presented in [13]. In paper [15], a FEM model of a conveyor with a PID controller, which is based on a neural network model, is presented. A comparative analysis of the values of the maximum belt speed, acceleration, and stresses in the conveyor belt obtained using the simulation model and obtained as a result of the experiment is performed.

3 Research Methodology

Consider the conveyor with a length S_d with material located on a conveyor belt and distributed along a transportation route with linear density $[\chi]_0(t,S)$ [23]. The linear density of the conveyor belt is $[\chi]_{0C}$ known. The equation for the movement of an elementary section of a conveyor belt has the form (Fig. 1):

$$dm\frac{d\langle\mu\rangle}{dt} = R(t,S+dS) - R(t,S) - dF_W,\ dm = \left([\chi]_0(t,S) + [\chi]_{0C}\right)dS. \tag{1}$$

$$R(t,S+dS) = \sigma(t,S+dS)Bh,\ R(t,S) = \sigma(t,S)Bh,\ 0 \le S \le S_d, \tag{2}$$

$$R(t, S + dS) - R(t, S) \approx \nabla R(t, S)dS, \tag{3}$$

where *dm* is the mass of the belt and the material on the segment *dS*; *B*, *h* is width and thickness of belt;$\sigma(t, S)$ is the stresses occurring in the belt; $\langle \mu \rangle = \langle \mu \rangle (t, S)$ is the segment speed of the conveyor belt *dS*; *S* is the coordinate; *t* is the time; F_W is the total resistance to movement [24]:

$$\mathrm{F_W} = \mathrm{F_H} + \mathrm{F_N} + \mathrm{F_{St}} + \mathrm{F_S}. \tag{4}$$

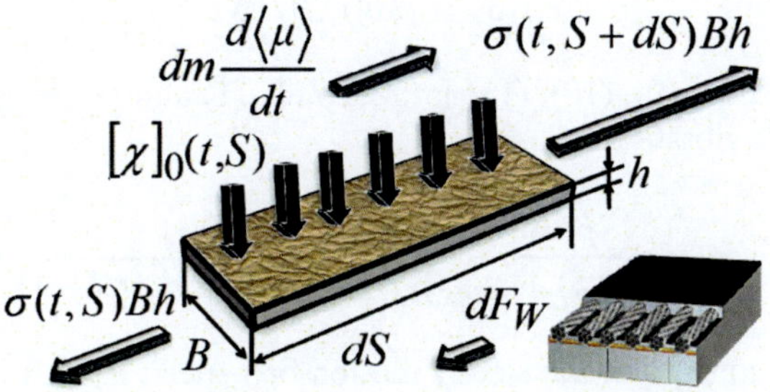

Fig. 1. Scheme of forces acting on the elementary segment.

Primary resistances $\mathrm{F_H}$ are given by the formula [24]:

$$d\mathrm{F_H} = dS \cdot f_C \cdot g_m\big([\chi]_{0R} + \big([\chi]_0(t, S) + [\chi]_{0C}\big)\cos\delta_C\big), \tag{5}$$

f_C is the resistance coefficient to movement [24]; $g_m = 9.81(m/\sec)$; δ_C is the inclination angle of the section of the conveyor [24]; $[\chi]_{0R}$ is linear loading from rotating parts [24], determined by the ratio of the mass roller to the distance between them [25]:

$$[\chi]_{0R} = m_R/S_R. \tag{6}$$

The secondary resistances force $\mathrm{F_N}$ [24], p.17]:

$$\mathrm{F_N} = (C - 1)\mathrm{F_H}. \tag{7}$$

For a long conveyor, $S_d > 2.0(km)$ km C is 1,05. The gradient resistance force $\mathrm{F_{St}}$ is as follows [24]:

$$dF_{\mathrm{St}} = dS \cdot \sin\delta_C \cdot g_m\big([\chi]_0(t, S) + [\chi]_{0C}\big). \tag{8}$$

If we divide (8) by *dm*, we get the following expression:

$$\frac{d\langle\mu\rangle}{dt} = \frac{1}{\big([\chi]_0(t, S) + [\chi]_{0C}\big)} \frac{\partial R(t, S)}{\partial S} - \frac{1}{\big([\chi]_0(t, S) + [\chi]_{0C}\big)} \frac{\partial F_W}{\partial S}. \tag{9}$$

For the model under consideration, it is believed that a linear relationship relates stresses and strain:For the model under consideration, it is believed that a linear relationship relates stresses and strain:

$$\sigma(t,S) = E\varepsilon(t,S),\ \varepsilon(t,S) \approx 10^{-2}, \tag{10}$$

where E is the modulus of elasticity; ε is a fractional extension of the belt element.

The relative element deformation $\varepsilon(t,S)$ is the ratio the element elongation $d\omega(t,S)$ to the segment length dS

$$\varepsilon(t,S) = \partial\omega(t,S)/\partial S. \tag{11}$$

The substitution of Eqs. (9), (10), (11) into formula (1) allows obtaining the following equation of the belt vibrations:

$$\frac{\partial\langle\mu\rangle}{\partial t} + \langle\mu\rangle\frac{\partial\langle\mu\rangle}{\partial S} = \frac{BhE}{([\chi]_0(t,S) + [\chi]_{0C})}\frac{\partial^2\omega(t,S)}{\partial S^2} - \frac{1}{([\chi]_0(t,S) + [\chi]_{0C})}\frac{\partial F_W}{\partial S}. \tag{12}$$

The belt speed $\langle\mu\rangle$ consists of the steady motion belt speed μ_ψ and the vibrational part of the speed $d\omega(t,S)/dt$:

$$\langle\mu\rangle = \mu_\psi + d\omega(t,S)/dt,\ d\omega(t,S)/dt = \partial\omega(t,S)/\partial t + \langle\mu\rangle\partial\omega(t,S)/\partial S. \tag{13}$$

As the relative element deformation $\varepsilon = \varepsilon(t,S)$ is small, then

$$\frac{d\omega(t,S)}{dt} = \frac{\partial\omega(t,S)}{\partial t} + \langle\mu\rangle\varepsilon \approx \frac{\partial\omega(t,S)}{\partial t},\ \langle\mu\rangle\varepsilon << \frac{\partial\omega(t,S)}{\partial t},\ \frac{\partial\mu_\psi}{\partial S} \approx \frac{\partial\omega(t,S)}{\partial S}. \tag{14}$$

$$\begin{aligned}\langle\mu\rangle\frac{\partial^2\omega(t,S)}{\partial t\partial S} &\approx \langle\mu\rangle\frac{\varepsilon(t+\Delta t,S) - \varepsilon(t,S)}{\Delta t} << \frac{1}{\Delta t}\left(\frac{\partial\omega(t+\Delta t,S)}{\partial t} - \frac{\partial\omega(t,S)}{\partial t}\right)\\ &\approx \frac{\partial^2\omega(t,S)}{\partial t^2}.\end{aligned} \tag{15}$$

We substitute the expression (13), (14) into the oscillation Eq. (12), we obtain

$$\frac{\partial^2\omega(t,S)}{\partial t^2} = C_\chi^2(t,S)\frac{\partial^2\omega(t,S)}{\partial S^2} - \frac{1}{([\chi]_0(t,S) + [\chi]_{0C})}\frac{\partial F_W}{\partial S} - f_\psi(t), \tag{16}$$

$$C_\chi^2(t,S) = \frac{BhE}{([\chi]_0(t,S) + [\chi]_{0C})},\ f_\psi(t,S) = \frac{\partial\mu_\psi(t,S)}{\partial t}. \tag{17}$$

When the segment length dS changes, the material density also changes $([\chi]_0(t,S) + [\chi]_{0C})$. If the segment length dS changes and becomes $(dS + d\omega(t,S))$ $dS > > d\omega(t,S)$, the density changes and becomes $([\chi]_0(t,S) + [\chi]_{0C}) + (\Delta[\chi]_0(t,S) + \Delta[\chi]_{0C})$. This allows you to write the balanced equality

$$dS\big([\chi]_0(t,S)+[\chi]_{0C}\big)=(dS+d\omega(t,S))\big([\chi]_0(t,S)+[\chi]_{0C}\big)+\big(\Delta[\chi]_0(t,S)+\Delta[\chi]_{0C}\big), \tag{18}$$

and we obtain

$$\frac{d\omega(t,S)}{dS}\approx-\frac{\Delta[\chi]_0(t,S)+\Delta[\chi]_{0C}}{[\chi]_0(t,S)+[\chi]_{0C}}\approx\varepsilon. \tag{19}$$

This makes it possible to write the function $C_\chi^2(t,S)$ as follows

$$C_\chi^2(t,S)\approx C_\psi^2(t,S)(1-\varepsilon), \tag{20}$$

$$C_\psi^2(t,S)=BhE\Big/\big([\chi]_{0\psi}(t,S)+[\chi]_{0C}\big). \tag{21}$$

where $[\chi]_{0\psi}(t,S)$ is the material density in the absence of the disturbances. The function $C_\psi^2(t,S)$ is the disturbances propagation speed into the belt [13, 26]. Assuming with $\varepsilon(t,S)\ll 1$ (10), the system of equations that will allow calculating the dynamic stresses in the belt takes the form:

$$\frac{\partial^2\omega(t,S)}{\partial t^2}=C_\psi^2(t,S)\frac{\partial^2\omega(t,S)}{\partial S^2}-\frac{1}{\big([\chi]_0(t,S)+[\chi]_{0C}\big)}\frac{\partial F_W}{\partial S}-f_\psi(t). \tag{22}$$

Let supplement Eq. (22) with the boundary conditions for the equation describing the oscillatory processes in the transport system. The stresses at the beginning $\sigma(t,0)$ and the end $\sigma(t,S_d)$ of the conveyor section are defined by the belt tension force T_1 and T_4 (Fig. 2).

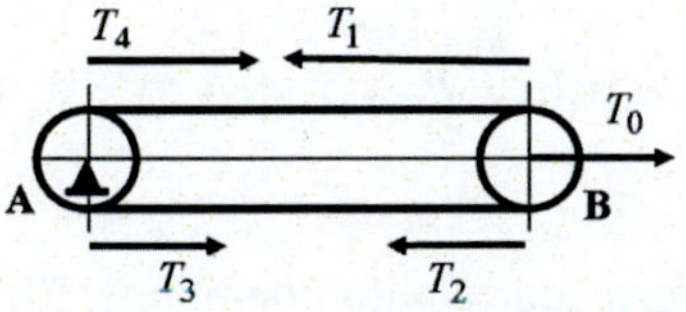

Fig. 2. Scheme of forces of the conveyor belt.

To determine the tensile force T_i at characteristic points of the conveyor section, let use the following system of equations:

$$F_{H(2-3)}=S_d f c g_m\big([\chi]_{0R}+[\chi]_{0C}\big);\ F_{H(4-1)}=F_{H(2-3)}+f c g_m M(t);\ \delta_C=0; \tag{23}$$

$$F_{N(2-3)} = (C-1)F_{H(2-3)};\ F_{N(4-1)} = (C-1)F_{H(4-1)};\ M(t) = \int_0^{S_d} [\chi]_0(t,S)dS; \quad (24)$$

$$F_{St(2-3)} = S_d \cdot \sin\delta_C \cdot g_m[\chi]_{0C} = 0;\ F_{St(4-1)} = F_{St(2-3)} + \sin\delta_C \cdot g_m M(t) = 0; \quad (25)$$

$$F_{\psi(2-3)} = f_\psi(t)[\chi]_{0C}S_d;\ F_{\psi(4-1)} = f_\psi(t)[\chi]_{0C}S_d + f_\psi(t)M(t); \quad (26)$$

$$T_1/T_2 = \exp(k_b\alpha);\ T_3 = T_2 + F_{W(2-3)} + F_{\psi(2-3)};\ T_4/T_3 = k_s;\ T_1 = T_4 + F_{W(4-1)} + F_{\psi(4-1)}; \quad (27)$$

where $F_{\psi(2-3)}$,$F_{\psi(4-1)}$ are inertia forces; k_b is a coefficient of adhesion; k_s is a cylinder loss coefficient; α is the belt angle. For a metal cylinder: $k_b \approx 0.3$, $k_s \approx 1.03$ and $\alpha = \pi$, whence follows $\exp(k_b\alpha) = 2.56$. Keep in mind formulas (23)–(27). The boundary conditions can be defined as:

$$\sigma(t,S_d) = \frac{T_1}{Bh} = \left(\frac{k_sF_{W(2-3)} + k_sF_{\psi(2-3)} + F_{W(4-1)} + F_{\psi(4-1)}}{Bh}\right)\frac{\exp(k_b\alpha)}{\exp(k_b\alpha) - k_s}; \quad (28)$$

$$\sigma(t,0) = \frac{T_4}{Bh} = E\frac{\partial\omega(t,S)}{\partial S}\bigg|_{S=S_d} - \frac{F_{W(4-1)} + F_{\psi(4-1)}}{Bh}. \quad (29)$$

Consider the case of initial conditions when at the moment $\tau = 0$, there are no fluctuations in the conveyor belt

$$\partial\omega(t,S)/\partial t|_{t=0} = 0. \quad (30)$$

The belt stresses at the initial time are determined by the initial material density value $\Psi(S)$ and acceleration of the belt $f_\psi(t)$:

$$\sigma(0,S) = E\omega(0,S/\partial S = \sigma(0,0) + F_{W\Psi(4-1)}(0,S) + F_{\psi\Psi(4-1)}(0,S),\ \sigma(0,0) = E\partial\omega(0,S)/\partial S|_{S=0}. \quad (31)$$

The system of Eqs. (22), together with conditions (28)–(31), can be used to calculate the dynamic belt stresses. The belt stresses depend on the value of the flow parameters $[\chi]_0(t,S)$ and $[\chi]_1(t,S)$ of the section.

4 Results

Consider the solution of Eq. (22). Let's focus on the case in which the conveyor line is underloaded. In this case, the weight of the material on the section of the belt is significantly less than the weight of the section of the conveyor belt itself.

$$[\chi]_0(t,S) << [\chi]_{0C}. \quad (32)$$

Let us use the notation:

$$\tau = t/T_d,\ \xi = S/S_d,\ \omega_0(\tau,\xi) = \omega(t,S)/\omega_{\max},\ \omega_{\max} = \sigma_b S_d/E, \quad (33)$$

$$\theta_R = [\chi]_{0R}/[\chi]_{0\max},\ \theta_C = [\chi]_{0C}/[\chi]_{0\max},\ v_b = Cf_C g_m([\chi]_{0\max} + [\chi]_{0C})S_d/(\sigma_b Bh), \quad (34)$$

$$v_1^2 = v_g^2(1+\theta_C)/\theta_C,\ v_2^2 = v_g^2(v_f + v_b(1+\theta_R/\theta_C)), \quad (35)$$

$$v_f = \frac{([\chi]_{0\max} + [\chi]_{0C})f_\psi(t)S_d}{\sigma_b Bh},\ v_g^2 = \frac{BhE}{([\chi]_{0\max} + [\chi]_{0C})}\left(\frac{T_d}{S_d}\right)^2. \quad (36)$$

where v_b determines the ratio of resistance $Cf_C g_m([\chi]_{0\max} + [\chi]_{0C})S_d$ at maximum load $[\chi]_{0\max}$ to maximum allowable force $\sigma_b Bh$; v_f is the ratio of inertia force $([\chi]_{0\max} + [\chi]_{0C})f_\psi(t)S_d$ at maximum load $[\chi]_{0\max}$ to the maximum permissible stresses force $\sigma_b Bh$. Then Eq. (22), which describes oscillatory processes in a transport system, takes a dimensionless form

$$\ddot{\omega}_0(\tau,\xi) = v_1^2\Delta\omega_0(\tau,\xi) - v_2^2, \quad (37)$$

with boundary conditions (28), (29)

$$\left.\frac{\partial\omega_0(\tau,\xi)}{\partial\xi}\right|_{\xi=1} = \alpha_1,\ \left.\frac{\partial\omega_0(t,\xi)}{\partial\xi}\right|_{\xi=0} = \alpha_1 - \alpha_{12},\ \alpha_1 = \frac{(k_s+1)\theta_C}{1+\theta_C}\frac{v_2^2}{v_g^2}\left(\frac{\exp(k_b\alpha)}{\exp(k_b\alpha) - k_s}\right), \quad (38)$$

and conditions (30), (31) for $\tau = 0$:

$$\partial\omega_0(\tau,\xi)/\partial\tau|_{\tau=0} = 0,\ \partial\omega_0(0,\xi)/\partial\xi = \alpha_1 - \alpha_{12}(1-\xi),\ \alpha_{12} = v_2^2/v_1^2. \quad (39)$$

When the conveyor starts, three characteristic phases should be distinguished: a) the period when the belt goes from rest to moving; b) the formation of a static force; c) acceleration of the belt. Consider the last phase. In this case, a value of traction torque and the acceleration of the elements of the belt is constant. The solution of Eq. (37) is found in the form:

$$\omega_0(\tau,\xi) = \omega_{00}(\tau,\xi) + \alpha_{12}\xi^2/2 + (\alpha_1 - \alpha_{21})\xi + C_0, \quad (40)$$

which allows us to ensure the boundary conditions for $\omega_{00}(\tau,\xi)$ in the form

$$\ddot{\omega}_{00}(\tau,\xi) = v_1^2\Delta\omega_{00}(\tau,\xi), \quad (41)$$

$$\left.\frac{\partial \omega_{00}}{\partial \xi}\right|_{\xi=1} = \left.\frac{\partial \omega_{00}}{\partial \xi}\right|_{\xi=0} = \left.\frac{\partial \omega_{00}}{\partial \xi}\right|_{\tau=0} = 0, \left.\frac{\partial \omega_{00}}{\partial \tau}\right|_{\tau=0} = -\left.\frac{\partial \alpha_{12}}{\partial \tau}\left(\frac{\xi^2}{2} - \xi\right)\right|_{\tau=0} = Z(\xi). \quad (42)$$

The solution of Eq. (41), (42) has the view

$$\omega_{00}(\tau, \xi) = \frac{1}{2v_1} \int_{\xi - v_1\tau}^{\xi + v_1\tau} Z(\alpha) d\alpha. \quad (43)$$

In the case when belt acceleration does not depend on time, follows $\partial\alpha_{12}/\partial\tau = 0$, and $Z(\xi) = 0$. If the belt acceleration (deceleration) linearly depends on time $v_f = v_{f0} + v_{f1} \cdot \tau$, then the function $Z(\xi)$ takes the form:

$$Z(\xi) = -v_g^2 v_{f1}\left(\xi^2/2 - \xi\right)/v_1^2. \quad (44)$$

The constant C_0 is determined from the condition of the min allowable tension, ensuring the adhesion of the belt with the cylinder to create traction in the system $\omega_0(0, \xi) \geq \omega_{\min 0}$. Let us calculate the values of dynamic stresses arising in the conveyor belt (Fig. 3).

$$\partial\omega_{00}(\tau, \xi)/\partial\xi \frac{\partial \omega_{00}(\tau, \xi)}{\partial \xi} = (Z(\xi + v_1\tau) - Z(\xi - v_1\tau))/(2v_1), \quad (45)$$

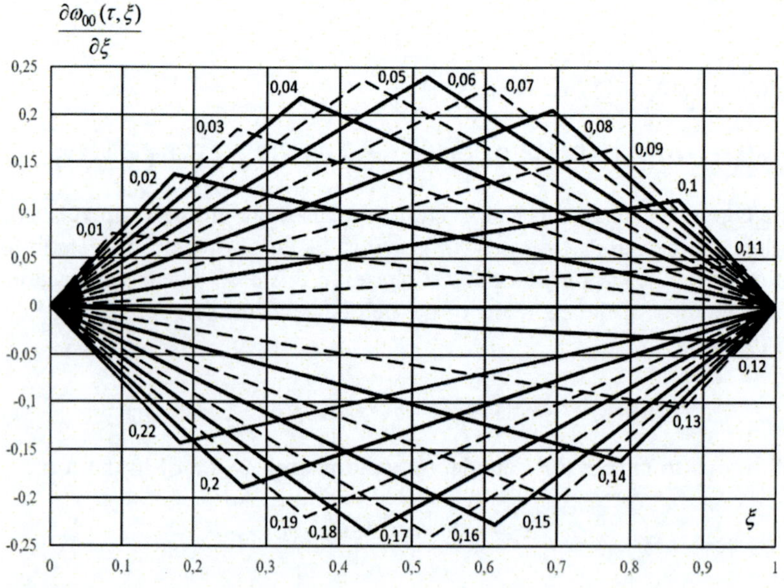

Fig. 3. Conveyor belt dynamic stresses for times $\tau = 0.01 \cdot m$, $m = 0, 1, 2...$

The condition under which the magnitude of the dynamic stress is an insignificant part of the magnitude of the static stress is

$$\frac{v_2^2}{v_1^2} > > |v_d \tau_n|, \tag{46}$$

where $v_{f1} < < (v_{f0} + v_b)/\tau_n$. When the condition (46) is satisfied, the dynamic stresses are small.

5 Conclusions

The model of dynamic stresses arising during the movement of a conveyor belt with a load, proposed in the article, was developed using Hooke's law when the deformation of an elastic object or material is proportional to the stress applied to it. The peculiarity of this model is that it allows you to take into account the impact of the uneven material distribution lengthwise the transport path on the spread of the emerging belt stresses. The mechanism of the occurrence of dynamic perturbations during the belt acceleration is considered in detail. It is shown that when starting a conveyor belt with a linear acceleration character, dynamic stresses occur that can destroy the belt. The prospect of further research is the development of optimal algorithms for controlling the acceleration and deceleration of the conveyor belt.

References

1. Pihnastyi, O., Khodusov, V.: Model of a composite magistral conveyor line. In: IEEE First International Conference on System Analysis & Intelligent Computing (SAIC), Kyiv, Ukraine, pp. 68–72. IEEE (2018). https://doi.org/10.1109/saic.2018.8516739
2. Semenchenko, A., Stadnik, M., Belitsky, P., Semenchenko, D., Stepanenko, O.: The impact of an uneven loading of a belt conveyor on the loading of drive motors and energy consumption in transportation. Eastern Eur. J. Enterp. Technol. **4**/1(82), 42–51 (2016). https://doi.org/10.15587/1729-4061.2016.75936
3. Lauhoff, H.: Speed control on belt conveyors – does it really save energy? Bulk Solids Handling Publ. **25**(6), 368–377 (2005)
4. Pihnastyi, O., Khodusov, V.: Model of conveyer with the regulable speed. Bull. South Ural State Univ. Ser. Math. Model. Program. Comput. Softw. **4**(10), 64–77 (2017). https://doi.org/10.14529/mmp170407
5. Nordell, L.K., Ciozda, Z.P.: Transient belt stresses during starting and stopping: elastic response simulated by finite element methods. Bulk Solids Handling **4**(1), 99–104 (1984)
6. Alspaugh, M.: The evolution of intermediate driven belt conveyor technology. Bulk Solids Handling **3**(23), 168–173 (2003)
7. Lodewijks, G.: Determination of rolling friction of belt conveyors using rubber data: fact or fiction? Bulk Solids Handling **24**, 16–22 (2004)
8. Pihnastyi, O., Khodusov, V.: Optimal control problem for a conveyor-type production line. Cybern. Syst. Anal. **54**(5), 744–753 (2018). https://doi.org/10.1007/s10559-018-0076-2

9. Reutov, A.: Simulation of load traffic and steeped speed control of conveyor. In: IOP Conference Series: Earth and Environmental, vol. 87, pp. 1–7. Saint-Petersburg, Russian Federation (2017). https://doi.org/10.1088/1755-1315/87/8/082041
10. Halepoto, I.A., Khaskheli, S.: Modeling of an integrated energy efficient conveyor system model using belt loading dynamics. Indian J. Sci. Technol. **9**(47), 1–6 (2016)
11. He, D., Pang, Y., Lodewijks, G., Liu, X.: Determination of acceleration for belt conveyor speed control in transient operation. Int. J. Eng. Technol. **8**(3), 206–211 (2016)
12. He, D., Pang, Y.: Lodewijks: belt conveyor dynamics in transient operation for speed control. Int. J. Civ. Environ. Struct. Constr. Archit. Eng. **10**(7), 865–870 (2016)
13. Lodewijks, G.: Two decades' dynamics of belt conveyor systems. Bulk Solids Handling **22** (2), 124–132 (2002)
14. He, D., Pang, Y., Lodewijks, G., Liu, X.: Healthy speed control of belt conveyors on conveying bulk materials. Powder Technol. **327**, 408–419 (2018)
15. Pingyuan, X., Yandong, S.: Application research on BP neural network PID control of the belt conveyor. J. Data Inf. Manag. **9**(6), 266–270 (2011)
16. Mathaba, T., Xia, X.: A parametric energy model for energy management of long belt conveyors. Energies **8**(12), 13590–13608 (2015). https://doi.org/10.3390/en81212375
17. Yang, G.: Dynamics analysis and modeling of rubber belt in large mine belt conveyors. Sens. Transducers **181**(10), 210–218 (2014)
18. Xinglei, L., Hongbin, Y.: The design and application of control system based on the BP neural network. In: Proceedings of the 3rd International Conference on Mechanical Engineering and Intelligent Systems, Yinchuan, China, pp. 789–793 (2015)
19. Wolstenholm, E.: Designing and assessing the benefits of control policies for conveyor belt systems in underground mines. Dynamica **6**(2), 25–35 (1980)
20. Andrejiova, M., Marasova, D.: Using the classical linear regression model in analysis of the dependences of conveyor belt life. Acta Montanistica Slovaca **18**(2), 77–84 (2013)
21. Lu, Y., Li, Q.: A regression model for prediction of idler rotational resistance on belt conveyor. Meas. Control **52**(5–6), 441–448 (2019). https://doi.org/10.1177/0020294019840723
22. Pihnastyi, O.: Control of the belt speed at unbalanced loading of the conveyor. Scientific bulletin of National Mining University. – Dnipro: State Higher Educational Institution "National Mining University" 6, 122–129 (2019). https://doi.org/10.29202/nvngu/2019-6/18
23. Pihnastyi, O., Kozhevnikov, G., Bondarenko, T.: The information controlling model transport system during transient conditions. In: IEEE International Scientific-Practical Conference Problems of Infocommunications, Science and Technology, Kyiv, Ukraine, pp. 197–201 (2019)
24. DIN 22101:2002–08. Continous conveyors. Belt conveyors for loose bulk materials. Basics for calculation and dimensioning. [Normenausschuss Bergbau (FABERG), DIN Deutsches Institut für Normung e.v. Normenausschuss Maschinenbau (NAM)], p. 51 (2002).
25. Budishevsky, V., Sulima, A.: Theoretical Foundations and Calculations of Transport of Energy-Intensive Industries. DonGTU, Donetsk, Ukraine (1999)
26. Pascual, R., Meruane, V., Barrientos, G.: The XXVI Iberian Latin-American Congress on Computational Methods in Engineering, Santo, Brazil, pp. 1–15 (2005)

Modeling of Failure in Laminated Plates by Delamination Buckling

Andrii Kondratiev[1], Viktor Kovalenko[2], Anton Tsaritsynskyi[3(✉)], and Tetyana Nabokina[3]

[1] O.M. Beketov National University of Urban Economy in Kharkiv, 17, Marshal Bazhanov Street, Kharkiv 61002, Ukraine
[2] Yuzhnoye Design Office, 3, Krivorozhskaya Street, Dnipro 49008, Ukraine
[3] National Aerospace University "Kharkiv Aviation Institute", 17, Chkalova Street, Kharkiv 61070, Ukraine
a.tsaritsynskyy@khai.edu

Abstract. The paper deals with the analysis of defects such as composite components' discontinuity and determination of tolerance range for the types of defects considered. The patterns of buckling of disbonds in a thin-walled strip beam made of polymeric composite materials under compression are studied. As a result of calculations, we determined the maximum stress with the average value for the pack, applied to the pack end, at which the delaminated part loses its stability. The geometrical representation of a one-dimensional mathematical model of the discontinuity allowed establishing the limit length of the disbond, located at a known depth at fixed ranges of tolerances on the input parameters. A two-dimensional disbond model, which is approximated by the rectangular plate, allowed obtaining formulas for the conditions of admissibility of the delamination defect of specified dimensions at one-way compression of the defective area for the fixed ranges of tolerances on the input parameters. A numerical example was considered for the composite component with the given structure and pattern of delamination. A comparison with the results of the finite element modeling showed a satisfactory agreement of results.

Keywords: Composite · Tolerance on defect · Local discontinuity · Disbonds · Stability

1 Introduction

It is known that various classes of defects in the structures made of polymeric composite materials (PCM) have a negative impact on their functional properties, in particular, when they are used in the critical engineering products [1, 2]. The specific feature of the above structures is the inevitable presence of the technological defects such as disbonds and initial delaminations, causing the reduction of strength and stiffness [3, 4]. PCM, possessing high specific strength and stiffness, are sensitive to such local discontinuity defects [5, 6]. Disbonds and initial delaminations are the most common defects of the composite structural components, often considered as the determining factor in solving the issue on PCM used in the structures [7]. A local

V. Tonkonogyi et al. (Eds.): InterPartner 2020, LNME, pp. 149–158, 2021.
https://doi.org/10.1007/978-3-030-68014-5_15

discontinuity may occur in the areas of stress concentration and stepwise variation of the material thickness as a result of imperfections in the production technology or effect of the operating loads [8]. The peculiarities of the delamination process are the local loss of stability and buckling of the delaminated layer with subsequent growth of the defect area [9]. It can be a cause of the total failure of the structural component under the load, which is substantially lower than the design load [10]. However, the presence of defects such as discontinuities in structural components made of PCM is not always an obstacle to the structure's trouble-free operation. The degree of risk of such a class of defects should be established by integrated studies of their effect on the bearing capacity and functional properties at design loads [11].

The objective of this paper is to analyze the defects of discontinuities in the structural elements made of PCM and to specify the tolerance limits for the defects under study.

2 Literature Review

The adequate assessment of the bearing capacity of structures with the technological defects under consideration can be made by experimental studies [12, 13]. At the moment, experimental studies of the behavior of composite structure elements with delaminations belong to an underexplored area [14]. It should be noted that for assessment of the strength and stiffness of PCM structural elements, they are to be subjected to direct mechanical tests. With rare exceptions, nondestructive testing of structural elements made of PCM does not allow reliably predicting the moment of failure [15]. Using this method, one of the weakest points in the structure can be identified only. The exact mathematical formulation of the problem of analysis of discontinuities such as disbonds in PCM structural elements is an extremely complex task. It does not have a correct solution in the framework of existing analytical models [16]. In a number of works [17, 18], the laminated composite is considered as a bonded structure. The concept that interlayer stresses in the laminated composite are similar to stresses in the adhesive joint, as presented in these works, undoubtedly deserves our attention. However, there are no reliable mathematical models of PCM discontinuities constructed in this way. Localization of damages in the laminated composites with the use of discrete damage modeling is studied in [19]. In the paper [20], authors proposed to use a multiscale model for calculation of the strength of PCM structures with defects. This model worked well in the analysis of the composite structure strength in the presence of defects on a microscopic scale. The paper proposes the development of this model, considering the combined factors of macroscopic defects of PCM layers, such as disbonds. The most important direction in modeling PCM with interlayer defects is in the development and use of various mathematical models, which allow analyzing the behavior of the defect from the moment of buckling to the loss of bearing capacity. Such models consider the geometry of the defect zone, the nature of the deformation of elements, and the structure of the composite [21]. The paper [22] deals with the method for assessment of the strength of structural elements made of composite materials with the thermosetting matrix in the presence of defects such as delaminations and process-induced deformations. The essential role of stability calculation in the general cycle of calculations for this type of defect is shown in the paper [23]. It includes the theoretical and experimental evidence of the fact that PCM failure can be associated with the loss of its overall stability or local stability of the defective areas. Only a few papers deal with a two-dimensional model of discontinuity of structural elements made of PCM. The publications [24, 25] are among the recent works dealing with this problem; they give the basis for calculating the stability of thin-walled structural elements with defects such as delaminations under the action of static compressive loads. The papers [26, 27] propose solutions related to stability and process of failure (including the analytical and numerical calculations), as illustrated by a thin-walled compressed plate of laminated composite materials with defects between layers. The critical load of compressed plate stability loss is determined. Modeling of cylindrical composite panels with delaminations was carried out in [28, 29]. As shown by the results of studies, the usage of numerical models is advisable to find an adequate solution to the problem under consideration. The finite element method to simulate the delaminations in PCM because of the complexity of the geometry and loading conditions is proposed in these papers. The monographs [30] also contain the results of extensive studies carried out with the use of application program packages of the finite element analysis and additionally developed modules.

In all papers cited above, the development of so-called “allowable” delamination is studied. It is understood as a defect of the shape and dimensions, which allow using the product under compressive force for a specific time. During this time, the defect develops, increasing in size up to the product failure. In our case, the start of local delamination at a regulated (operational) compressive load is taken as the exhaustion of the bearing capacity of the structural element. Under these conditions, tolerances on the input parameters of the resolving inequality are specified.

3 Research Methodology

The local discontinuity zone in the plan view in the general case can take the form of the irregular ellipse with conventional semi-axes a and b. Similarly to the papers [26, 30], we considered the simplest one-dimensional mathematical model of the strip beam under the action of the uniform compressive load (Fig. 1) containing the discontinuity of width l located at the depth $(H - h)$ of the pack, with the delaminating part of layers of h thickness buckled when the load N compressing the pack reaches the critical value N_{cr} [4, 30]. The length of discontinuity l is taken equal to the double semi-axis $2a$ of the actual disbond reduced to the elliptical shape in the plan view, with its larger axis directed along with the effective compressive load (Fig. 1).

The effect of the final value $2b$ of the elliptical discontinuity in the adopted one-dimensional model, where the width of disbond $2b \to \infty$ is approximately taken into account by the averaged coefficient of stability increase, is

$$\xi = \chi(1 + exp(-b/l)), \tag{1}$$

where χ is determined by the numerical experiments based on a two-dimensional model of discontinuity of the elliptical shape taking into account physical-mechanical properties (PMP) of the delaminated part of monolayers and PCM pack as a whole and boundary conditions of bearing.

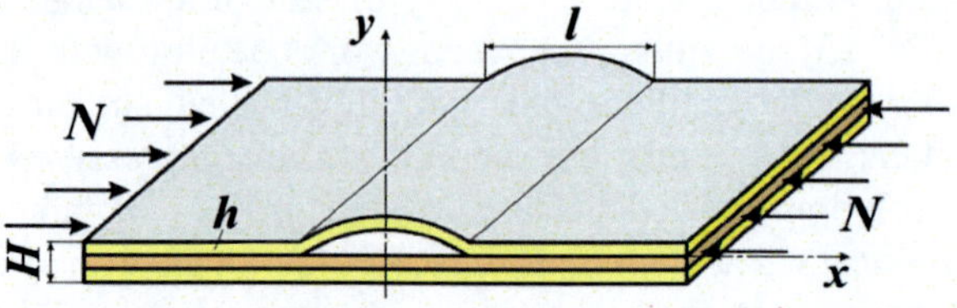

Fig. 1. Geometrical representation of the one-dimensional model of PCM discontinuity

The one-dimensional model assumes that PCM monolayers have equal thickness δ, with the reduced (averaged) elasticity modulus of the pack E_1, elasticity modulus of the delaminating part E_1^0, and of the continuous part under delaminated one – E_2. Accordingly, the effective stresses in the delaminated part of the pack are σ_1^0, and in the continuous part under delaminated one σ_2. Relative deformations are equal to ε_1^0, ε_2, and ε, respectively, for PCM pack as a whole.

The condition of the pack balance is

$$\sigma n\delta = \sigma_1 n_1 \delta + \sigma_2 (n - n_1)\delta, \tag{2}$$

where n, n_1 – number of monolayers of the continuous part and pack as a whole; σ – average (reduced) stresses in the pack from the effective load N.

From the condition of compatibility of deformations $\varepsilon_1^0 = \varepsilon_2$ or $\sigma_1^0/E_1^0 = \sigma_2/E_2$, we obtain the effective stresses in the delaminated part of the pack and its continuous part:

$$\sigma_1^0 = \sigma \frac{n}{n_1 + (n - n_1)E_2/E_1^0}; \quad \sigma_2 = \sigma \frac{n}{n - n_1}\left(1 - \frac{n_1}{n_1 + (n - n_1)E_2/E_1^0}\right); \tag{3}$$

Assuming the conditions of fixing the delaminating part of the PCM pack along with the effective compressive forces, critical stresses of the loss of stability of delaminated part of PCM pack corresponding to restraint may be written as

$$\sigma_{1\,cr}^0 = \frac{4\pi^2 E_1^0}{12\left(1 - \nu_{12}^0 \nu_{21}^0\right)} \frac{n_1^2 \delta^2}{l^2} \xi. \tag{4}$$

Effective stresses in the defective area σ_1^0 should not exceed the critical ones $\sigma_{1\,cr}^0$, that is

$$\sigma \frac{n}{n_1 + (n - n_1)E_2/E_1^0} \le \frac{4\pi^2 E_1^0}{12\left(1 - \nu_{12}^0 \nu_{21}^0\right)} \frac{n_1^2 \delta^2}{l^2} \xi. \tag{5}$$

Regulating the average stresses in the pack from the effective load N, introducing in (5) the tolerances on the deviation of elasticity moduli $\pm\Delta E_1^0$, $\pm\Delta E_2$, Poisson ratios $\pm\Delta\nu$ and neglecting the tolerance on the coefficient of stability increase $\mp\Delta\xi$ from (5), we obtain after transformations:

$$l \leq \pi n_1 \delta \sqrt{\frac{\left(E_1^0 \pm \Delta E_1^0\right)\left[n_1 + (n - n_1)\frac{(E_2 \pm \Delta E_2)}{\left(E_1^0 \pm \Delta E_1^0\right)}\right]\xi}{3n(\sigma \pm \Delta\sigma)\left[1 - \left(\nu_{12}^0 \pm \Delta\nu\right)\left(\nu_{21}^0 \pm \Delta\nu\right)\right]}}. \tag{6}$$

Here approximations of parameters ΔE_1^0, ΔE_2, $\Delta\nu$ and $\Delta\sigma$ are determined according to our data in [26, 30], taking into account the structure (reinforcement pattern) of the delaminating part and below located part of the PCM pack.

As it follows from the formula (6), a one-dimensional model for determination of the tolerance range on the allowable dimensions of the discontinuity (disbond), corrected by the introduction of the coefficient ξ, allows solving the task quite easily. However, the degree of approximation of this solution for the number of high-importance products can be unsatisfactory. Therefore, below we use a two-dimensional mathematical model of the rectangular plate compressed in one direction under boundary conditions of free bearing on the loaded sides and restriction on the other two. Arbitrary, in the general case, disbond shape other than the ellipse is approximated by the rectangular plate of the equivalent (equal) area (Fig. 2).

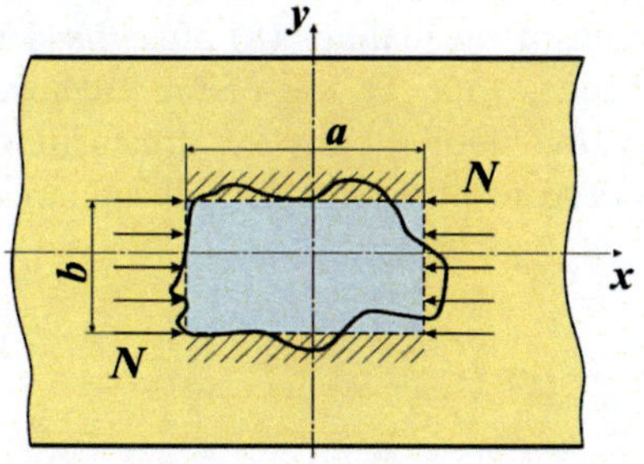

Fig. 2. Approximation of disbond area by the rectangular plate in the plan view

Effective stresses σ_1^0 in the delaminating plate are determined by the formula (3); critical stresses at which it loses its stability are equal, respectively, to:

$$\sigma_{cr}^0 = \frac{\pi^2\sqrt{D_1^0 D_2^0}}{n_1 \delta b^2}\left[\sqrt{\frac{D_1^0}{D_2^0}}\left(\frac{m}{c}\right)^2 + \frac{8}{3}\frac{D_3^0}{\sqrt{D_1^0 D_2^0}} + \frac{16}{3}\sqrt{\frac{D_2^0}{D_1^0}}\left(\frac{c}{m}\right)^2\right], \tag{7}$$

where $c = \frac{a}{b}$; $D_1^0 = \frac{E_1^0 n_1^3 \delta^3}{12\left(1 - \nu_{12}^0 \nu_{21}^0\right)}$; $D_2^0 = \frac{E_2^0 (n - n_1)^3 \delta^3}{12\left(1 - \nu_{12}^0 \nu_{21}^0\right)}$; $D_3^0 = D_1^0 \nu_{21}^0 + 2\frac{G_{12}^0 n_1^3 \delta^3}{12}$; m – number of half-waves in the direction of the x-axis. Here D_1^0 D_2^0, D_3^0 – values of cylindrical stiffness of the delaminating part of PCM pack.

As for the one-dimensional model, effective stresses in the defective area should not exceed the critical value. Introducing the relevant ranges of tolerances, we obtain:

$$\frac{(\sigma\pm\Delta\sigma)n}{n_1+(n-n_1)\dfrac{(E_2\pm\Delta E_2)}{\left(E_1^0\pm\Delta E_1^0\right)}}\le\frac{\pi^2\sqrt{\left(E_1^0\pm\Delta E_1^0\right)\left(E_2^0\pm\Delta E_2^0\right)n_1^3(n-n_1)^3}\,\delta^2}{n_1b^2}\times$$
$$\times\left\{\sqrt{\frac{\left(E_1^0\pm\Delta E_1^0\right)n_1^3}{\left(E_2^0\pm\Delta E_2^0\right)(n-n_1)^3}}\left(\frac{m}{c}\right)^2+\frac{8}{3}\frac{\left[\left(E_1^0\pm\Delta E_1^0\right)\left(\nu_{21}^0\pm\Delta\nu\right)\right]n_1^3}{\sqrt{\left(E_1^0\pm\Delta E_1^0\right)n_1^3\left(E_2^0\pm\Delta E_2^0\right)(n-n_1)^3}}+\right.$$
$$+\frac{8}{3}\frac{\left[2\left(G_{12}^0\pm\Delta G_{12}^0\right)\left[1-\left(\nu_{12}^0\pm\nu\right)\left(\nu_{21}^0\pm\Delta\nu\right)\right]\right]n_1^3}{\sqrt{\left(E_1^0\pm\Delta E_1^0\right)n_1^3\left(E_2^0\pm\Delta E_2^0\right)(n-n_1)^3}}$$
$$\left.+\frac{16}{3}\sqrt{\frac{\left(E_2^0\pm\Delta E_2^0\right)(n-n_1)^3}{\left(E_1^0\pm\Delta E_1^0\right)n_1^3}}\left(\frac{c}{m}\right)^2\right\}; \quad (8)$$

Tolerances included in (8) are determined as before, in accordance with our paper [30] considering the specific reinforcement pattern of the delaminating part and below located part of the PCM pack. The admissibility of the defect in the form of disbond is established by the fulfillment of inequalities (8) after the determination of its dimensions and depth of occurrence, with the use of the advanced nondestructive testing methods. When analyzing the admissibility of dimensions of the defect, such as delamination, it is necessary to use the average reduced stresses σ under the effective base compressive load N.

4 Results

Let's consider the numerical example. The delamination is studied in the structure shown in Fig. 3a. It is assumed that delamination is located under two upper monolayers of the structure (±45). PMP of carbon-fiber-reinforced plastic correspond to properties: elasticity modulus along and across the grain E_1 = 100 GPa, E_2 = 10 GPa, shear modulus G_{12} = 6 GPa, Poisson's ratio ν_{12} = 0.35, the thickness of the monolayer δ = 0.1 mm. PMP of two layers of the delaminating part of the pack (±45) is equal to $E_1^0 = E_2^0$ = 19.96 GPa; G_{12}^0 = 26.07 GPa; $\nu_{12}^0 = \nu_{21}^0$ = 0.663. PMP of four layers of the continuous part of the pack (0, 90, $\mp$ 45), according to formulas of the reinforcement theory [8], are as follows: $E_1 = E_2$ = 83.22 GPa; G_{12} = 32.07 GPa; ν_{12} = 0.3.

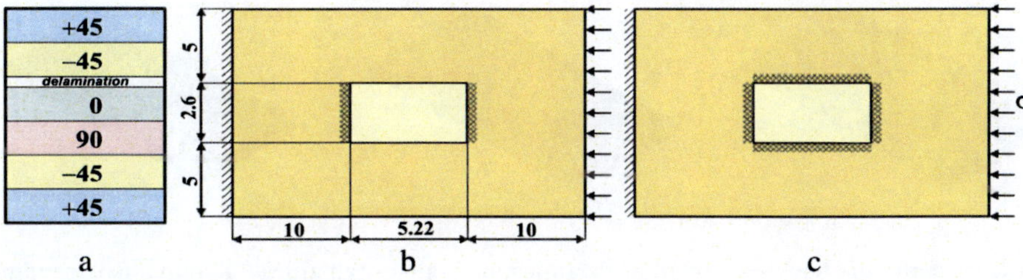

Fig. 3. Studying of delamination: a – the structure of PCM pack under consideration; b, c – considered possible forms of delamination.

Using a one-dimensional model, effective stresses in the defective area are equal to $\sigma_1^0 = 0.321\sigma$. The critical length of the discontinuity at $\xi = 1$ $l_{cr} = 116.7/\sqrt{\sigma}$, corresponding to the stress acting on the thickness of the pack in the product area distant from the discontinuity, is $\sigma = 500$ MPa $l_{cr} = 5.22$ mm.

During consideration of the two-dimensional model, values of cylindrical stiffness of the structure of two monolayers of the delaminated part of the pack, modeled using the plate with the boundary conditions, as shown in Fig. 3, turned to be equal to $D_1^0 = D_2^0 = 23.72$ Hmm; $D_3^0 = 50.49$ Hmm. At $c = a/b = 2$, at which $m = 3$, $a = 5.22$ mm; $b = 2.6$ mm. Critical stresses of the loss of stability are equal, accordingly, to $\sigma_{cr}^0 = 1796$ MPa, i.e., 3.6 times exceeding the effective ones. Calculations correspond to the foreseeable result that the one-dimensional model implemented with free faces of the strip beam gives a very large margin of stability. The same example is calculated in the software package implementing the finite element method. The delaminated part is modeled by a multilayer plate finite element with the reinforcement pattern of ± 45 type. 3D multilayer elements model the remaining part of PCM pack. Two possible patterns of delamination (Fig. 3) are considered. In the first case (Fig. 3b), the delaminated part is connected with the basic part on two short edges only. In the second case (Fig. 3c), the delaminated part is connected with the basic one on all four edges. The pattern of loading and fixing for the area under consideration is shown in the same figure, as distinct from the classical conditions of fixing, here the actual conditions are assumed. As a result of calculations, we determined the maximum stress with the average value for the pack, applied to the pack end, at which the delaminated part loses its stability (Fig. 4). As can be seen from the above results, finite element models show that the one-dimensional model predicts the loss of stability of the delamination at average stresses in the pack being almost three times higher compared to that determined by the finite element method. At the same time, results obtained with two-dimensional and finite element models, differ by 1.5 times only. Therefore, when specifying the tolerances on defects such as local delaminations, it is necessary to proceed from the two-dimensional analytical model.

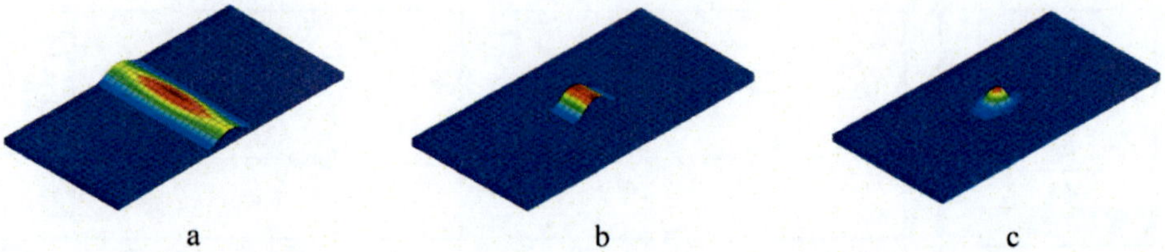

Fig. 4. Pattern of buckling of the delamination: a – cylindrical longitudinal bending (σ_{max} = 140 MPa); b – implementation of the form of Fig. 3b (σ_{max} = 155 MPa); c – implementation of the form of Fig. 3c (σ_{max} = 1167 MPa)

5 Conclusions

Therefore, we analyzed the defects such as discontinuities in the structural elements made of PCM and specified the tolerance limits for the defects under study. The approach proposed in this paper is distinguished by the fact that we obtain the analytical expressions for calculating the critical forces during compression of a laminated plate with interlayer defect, taking into account the difference in the elastic properties of the orthotropic material of specific areas. The start of local delamination at a regulated (operational) compressive load is taken as the exhaustion of the bearing capacity of the structural element.

The geometrical representation of a one-dimensional mathematical model of the discontinuity allowed determining the limit length of the disbond, located at a known depth at fixed ranges of tolerances on the input parameters.

The two-dimensional disbond model approximated by the rectangular plate allowed obtaining formulas, which defined the conditions of the delamination defect admissibility with specified dimensions at one-way compression of the defective area for the fixed ranges of tolerances on the input parameters.

The stability of delaminations for the plate with the delamination was assessed based on the finite element method. Results were compared with the results of the analytical study, and satisfactory agreement of results was obtained. For example, finite element models showed that the one-dimensional model predicted the loss of stability of the delamination at the average stresses in the pack being almost three times higher compared to that determined by the finite element method. At the same time, results obtained for two-dimensional and finite element models are only slightly different.

References

1. Fomin, O., Lovska, A., Kulbovskyi, I., Holub, H., Kozarchuk, I., Kharuta, V.: Determining the dynamic loading on a semi-wagon when fixing it with a viscous coupling to a ferry deck. East.-Euro. J. Enterp. Technologies **2**(7(98)), 6–12 (2019). https://doi.org/10.15587/1729-4061.2019.160456

2. Kondratiev, A.V., Prontsevych, O.O.: Stabilization of physical-mechanical characteristics of honeycomb filler based on the adjustment of technological techniques for its fabrication. East.-Euro. J. Enterp. Tech. **5/1**(95), 71–77 (2018). https://doi.org/10.15587/1729-4061.2018.143674
3. Fomin, O., Gerlici, J., Lovskaya, A., Kravchenko, K., Prokopenko, P., Fomina, A., Hauser, V.: Research of the strength of the bearing structure of the flat wagon body from round pipes during transportation on the railway ferry. In: MATEC Web of Conferences, vol. 235, p. 00003 (2018). https://doi.org/10.1051/matecconf/201823500003
4. Slyvyns'kyy, V., Gajdachuk, V., Kirichenko, V., Kondratiev, A.: Basic parameters' optimization concept for composite nose fairings of launchers. In: 62nd International Astronautical Congress, Cape Town, Red Hook, vol. 9, pp. 5701–5710 (2012). Curran, NY
5. Shupikov, A.N., Ugrimov, S.V., Kolodiazhny, A.V., Yareschenko, V.G.: High-order theory of multilayer plates The impact problem. Int. J. Solids Struct. **35**(25), 3391–3404 (1998). https://doi.org/10.1016/S0020-7683(98)00020-1
6. Kondratiev, A., Gaidachuk, V.: Weight-based optimization of sandwich shelled composite structures with a honeycomb filler. Eastern-European Journal of Enterprise Technologies **1**(1 (97)), 24–33 (2019). https://doi.org/10.15587/1729-4061.2019.154928
7. Rodichev, Y.M., Smetankina, N.V., Shupikov, O.M., Ugrimov, S.V.: Stress-strain assessment for laminated aircraft cockpit windows at static and dynamic loads. Strength Mater. **50**(6), 868–873 (2019). https://doi.org/10.1007/s11223-019-00033-4
8. Gaidachuk, V.E., Kondratiev, A.V., Chesnokov, A.V.: Changes in the thermal and dimensional stability of the structure of a polymer composite after carbonization. Mech. Compos. Mater. **52**(6), 799–806 (2017). https://doi.org/10.1007/s11029-017-9631-6
9. Potter, K.D., Campbell, M., Langer, C., Wisnom, M.R.: The generation of geometrical deformations due to tool/part interaction in the manufacture of composite components. Compos. A **36**(2), 301–308 (2005). https://doi.org/10.1016/S1359-835X(04)00150-2
10. Baker, A., Dutton, S., Kelly, D.: Composite Materials for Aircraft Structures. American Institute of Aeronautics and Astronautics Inc., Virginia, Reston, p. 599 (2004)
11. Kondratiev, A.V., Gaidachuk, V.E., Kharchenko, M.E.: Relationships between the ultimate strengths of polymer composites in static bending, compression, and tension. Mech. Compos. Mater. **55**(2), 259–266 (2019). https://doi.org/10.1007/s11029-019-09808-x
12. Pernicea, M.F., De Carvalhob, N.V., Ratcliffe, J.G., Halletta, S.R.: Experimental study on delamination migration in composite laminates. Compos. Part A **73**, 20–34 (2015). https://doi.org/10.1016/j.compositesa.2015.02.018
13. Forabosehi, P.: Layered plate with discontinuous connection. Exact mathematical model. Compos. Part B **47**, 365–378 (2013). https://doi.org/10.1016/j.compositesb.2012.11.004
14. Ascione, L., Berardi, V.P., Giordano, A., Spadea, S.: Local buckling behavior of FRP thin-walled beams: a mechanical model. Compos. Struct. **98**, 111–120 (2013). https://doi.org/10.1016/j.compstruct.2012.10.049
15. Cárdenas, D., Elizalde, H., Marzocca, P., Abdi, F., Minnetyan, L., Probst, O.: Progressive failure analysis of thin-walled composite structures. Compos. Struct. **95**, 53–62 (2013). https://doi.org/10.1016/j.compstruct.2012.05.022
16. Lemanski, S.L., Wang, J., Sutcliffe, M.P.F., Potter, K.D., Wisnom, M.R.: Modelling failure of composite specimens with defects under compression loading. Compos. A **48**, 26–36 (2013). https://doi.org/10.1016/j.compositesa.2012.12.007
17. Kurennov, S.S., Koshevoi, A.G., Polyakov, A.G.: Through-thickness stress distribution in the adhesive joint for the multilayer composite material. Russ. Aeronaut. (Iz VUZ) **58**(2), 145–151 (2015). https://doi.org/10.3103/S1068799815020026

18. Goodsell, J., Pagano, N.J., Kravchenko, O., Pipes, B.: Interlaminar stresses in composite laminates subjected to anticlastic bending deformation. J. Appl. Mech. **80**(4), 1020–1026 (2013). https://doi.org/10.1115/1.4007969
19. Moure, M.M., Sanchez-Saez, S., Barbero, E.: Analysis of damage localization in composite laminates using a discrete damage model. Compos. B **66**, 224–232 (2014). https://doi.org/10.1016/j.compositesb.2014.05.015
20. Dimitrienko, Yu.I., Dimitrienko, I.D., Sborschikov, S.V.: Multiscale hierarchical modeling of fiber reinforced composites by asymptotic homogenization method. Appl. Math. Sci. **9** (145), 7211–7220 (2015). https://doi.org/10.12988/ams.2015.510641
21. Timonin, A.M.: Finite-layer method: a unified approach to a numerical analysis of interlaminar stresses, large deflections, and delamination stability of composites. Part 1. Linear behavior. Mech. Compos. Mater. **49**(3), 231–244 (2013). https://doi.org/10.1007/s11029-013-9339-1
22. Parrinello, F., Marannano, G., Borino, G.A.: Thermodynamically consistent cohesive-frictional interface model for mixed mode delamination. Eng. Fract. Mech. **153**, 61–79 (2016). https://doi.org/10.1016/j.engfracmech.2015.12.001
23. Zhou, R., Guan, Z., Jia, Y., Wang, K., Zhang, C.: Effects of delamination on compressional properties of composite laminate. J. Beijing Univ. Aeronaut. Astronaut. **41**(2), 311–317 (2015). https://doi.org/10.13700/j.bh.1001-5965.2014.0116
24. Muthurajan, K.G., Sankaranarayanasamy, K., Tiwari, S.B., Nageswara, B.: Rao: post-buckling of a thin film strip delamination in a composite laminate. Trends Appl. Sci. Res. **1**, 48–60 (2006). https://doi.org/10.3923/tasr.2006.48.60
25. Kim, S., Cho, M.: Effcient higher-order shell theory for laminated composites with multiple delaminations. AIAA J. **41**(5), 941–950 (2003). https://doi.org/10.2514/2.2031
26. Bokhoeva, L.A., Rogov, V.E., Chermoshentseva, A.S., Lobanov, D.V.: Stability and process of destruction of compressed plate of layered composite materials with defects. IOP Conf. Ser. Mater. Sci. Eng. **142**, 012077 (2016). https://doi.org/10.1088/1757-899X/142/1/012077
27. Chermoshentseva, A., Pokrovskiy, A., Bokhoeva, L.: The behavior of delaminations in composite materials – Experimental results. IOP Conf. Ser. Mater. Sci. Eng. **116**, 012005 (2016). https://doi.org/10.1088/1757-899X/116/1/012005
28. Liu, P.F., Xing, L.J., Zheng, J.Y.: Failure analysis of carbon fiber/epoxy composite cylindrical laminates using explicit finite element method. Compos. B **56**, 54–61 (2014). https://doi.org/10.1016/j.compositesb.2013.08.017
29. Mue, A., Stawiarski, A.: Identification of damage in composite multilayered cylindrical panels with delaminations. Compos. Struct. **94**, 1871–1879 (2012). https://doi.org/10.1016/j.compstruct.2011.11.026
30. Gajdachuk, A.V., Gajdachuk, V.E., Kondratiev, A.V., Kovalenko, V.A., Kirichenko, V.V., Potapov, A.M.: Methodology for the development of effective structural and technological solutions for composite units of rocket and space technology, vol. 1, p. 263. National Aerospace University Kharkiv Aviation Institute Publ., Kharkiv (2016)

Device for Electric Drives Torque Diagnostics and Its Characteristics Study

Aleksandr Koroliov[1], Igor Kozlov[1](✉), Pavlo Pavlyshyn[2], Raul Turmanidze[3], and Yurii Yeputatov[1]

[1] Odessa National Polytechnic University, 1, Shevchenko Avenue, Odessa 65044, Ukraine
kozlov_i.1_@ukr.net
[2] National Nuclear Energy Generating Company of Ukraine "Energoatom", Kyiv, Ukraine
[3] Georgian Technical University (GTU), 77, Kostava Street, 0175 Tbilisi, Georgia

Abstract. The article is purposed to the development of a torque diagnostical device (TDD) designed to determine and rectify the configuration of power valves' electric drive limit coupling setting for ultimate force. The possibility of promptly adjusting the coupling maximum torque basic settings to limit the electric drive torque according to the technical (passport) data in the repair shops of nuclear power plants is considered; such adjustment will improve the accuracy of the equipment settings and reduce the time of scheduled repair work. Settings refinement experimentally with a step-by-step increase in torque within the electric drive control showed overestimated values of the torques arising in the power valves from the side of electric drive, in contrast to the limit torque coupling manual setting with a torque wrench (static method). With an increase in electric drives power, the overestimated values of torques arising in power valves can involve the working elements damage. Such devices use during the post-repair adjustment of electric drives' power valves allow the subsequent operation of the systems reducing the undesirable loads that occur in power fittings on the drive side, ensuring trouble-free and reliable operation of the pipeline systems of power plants of nuclear power plants.

Keywords: Power fittings · Limit torque coupling · Pipe mains systems · Torque wrench · Manual setting

1 Introduction

Maintaining the design level of power plants' technological equipment operational reliability, including those at nuclear power plants, is largely determined by the reliability of electric drive valves operation [1], so only one 1000 MW nuclear power plant (NPP) unit includes valves of various designs and purposes whose number can range from 8 to 9 thousand items. Therefore, a significant share in complex preventive maintenance works at any nuclear power plant relates to the repair of power valves' electric drives. The electric drives' main parameters post-repair adjustment, in

V. Tonkonogyi et al. (Eds.): InterPartner 2020, LNME, pp. 159–165, 2021.
https://doi.org/10.1007/978-3-030-68014-5_16

particular, torque ratings setting according to passport data, represents an important problem [2]. This procedure, among others, is provided for by the manufacturer's instruction set out in the electric drive's passport. So, the Tula electric drives factory recommends performing the electric drive torque values post-repair adjustment using this factory test benches [3] that is unacceptable for Ukrainian NPPs.

The electric drive's inaccurate settings can lead to excessive or insufficient pressure on the shut-off element of the electric actuator valves, that in turn can cause damage to the valve stem or plate, so at several nuclear power plants, there have been cases of breakdown of the bellows seals, or insufficiently tight valve closure with an unacceptable medium pass [4]. Thus, timely detection of valve damage helps to avoid unplanned shutdowns and accidents of the powerplant's thermal power equipment. And conversely, failure of critical fittings can lead to serious events. Such was the failure of the pressure compensator pulse valve at the Three-mile Island nuclear power plant, which ended in the first serious accident at the nuclear power plants worldwide [5, 6]. A similar failure of pilot operated pressure relief valve has been registered at the Rivne NPP. However, it was discovered and eliminated within due time delay.

A preliminary analysis showed that Ukraine does not produce special compact devices that allow adjusting the valves' drives torque immediately at their regular operation point.

This article is purposed to develop a torque diagnostic device (TDD) designed to determine and clarify the settings of power valves' A and B type [3] electric drives limit coupling in the torque values range – 0–250 N·m with an accuracy of – 0.1 N·m, the measurement results being displayed on a LCD screen.

Such device development shall allow a quick, in-situ, configuration of electric drive limit coupling basic settings according to the passport data, thus reducing the scheduled repair work time and eliminating the from supervisory institutions questions about the correctness of settings made.

2 Literature Review

Mostly, the electric drives currently used in the Ukrainian NPPs, are not equipped with built-in torque meters. Firstly, such design complicates the valves' structure, and accordingly increases the number of their possible failures, and secondly, a fairly simple NPP element becoming more sophisticated this introduces special features into the system operation, and most important, repair of power valves. Therefore, it is advisable to have a compact portable device allowing to quickly configure each electric drive at the NPP's repair shop.

Currently, in the repair shops of nuclear power plants, the electric drive torque is determined by current strength and voltage [7]. The currents at the time of system closing can exceed the nominal value by 5–10 times, respectively, and the torques at closing for a short period of time increase many times. [8, 9]. As a result of torque increment in the shut-off valves from the electric drive side, significant there arises a significant load that reduces the systems reliability during operation [10] besides, this method does not allow to evaluate the torque on the valve stem.

The methods of shut-off valves static (manual) adjustment produced by various domestic manufacturers are based on uniform principles, combined by the standards of the CDB (Central Design Bureau of Valve Engineering, St. Petersburg). In accordance with the standards for electric actuator valves, the closing torque ($M_{\text{closing.ED}}$) is taken equal to the torque calculated for manual adjustment of valves ($M_{\text{man.adjust}}$·), increased by the safety factor [8]:

$$M_{\text{closing.ED}} = K_{\text{res}} \cdot M_{\text{man.adjust}} \cdot \tag{1}$$

where K_{res} - safety factor for the required torque on the drive, for electric drive Kres = 1.10–1.25.

The calculation of efforts on the valve stem proposed by CDB [11], is intended mainly for design organizations, for NPP workers required is a method for those forces rapid determination. The specified approach, in our opinion, is not accurate enough, it leads to a decrease in the reliability of electric valves, which affects the safety of energy enterprises [12].

Currently, there are a large number of torque measurement sensors operating on the basis of indirect measurement methods [13–16].

These methods disadvantage refers to a decrease in accuracy of measuring torque when electric drive operated at dynamic modes, especially when it stops or starts.

3 Research Methodology

For testing the developed methods for monitoring settings of the torque limiting clutch maximum effort (coupling ultimate torque clutch) at power valves' electric drive, an experimental study was conducted using the developed TDD.

The developed TDD is installed on the valve, and on its top the tested electric drive is mounted. That is, it is installed between the armature and the electric drive.

That general device appearance is shown in Fig. 1.

The measured torque, through a square slot 1 (Fig. 1), is transmitted from the electric drive to IM 200 torque sensor installed in housing 2. The signal from the sensor is transmitted to the secondary reader 3 and displayed on the liquid crystal screen of the secondary device. The TDD is mounted on special fasteners (paws) 4, which allow it to be installed on the pipeline next to the tested fittings. If necessary, the legs are removed, and TDD can be mounted on the valve flange or on a special stand.

A schematic representation of torque sensor IM 200 components is shown in Fig. 2.

The TDD experimental study methodology was as follows: TDD attached to the repaired power valves, on which top the tested electric drive was mounted. In this case, the TDD square slot coincided with the landing hole in the electric drive clutch. The incremental increase in torque was regulated on the electric drive using a switch, limiting the torque in the range of 5 serial values, from 1st to 5th positions. In each position, the electric drive operated until it was turned off by a switch in a given position. Data read by the TDD indicator represented the torque value at which the driver stopped.

Fig. 1. TDD exterior: general view 1 – torque transmission slot, 2 – TDD casing, 3 – electronic unit (secondary device); 4 – attachment pieces.

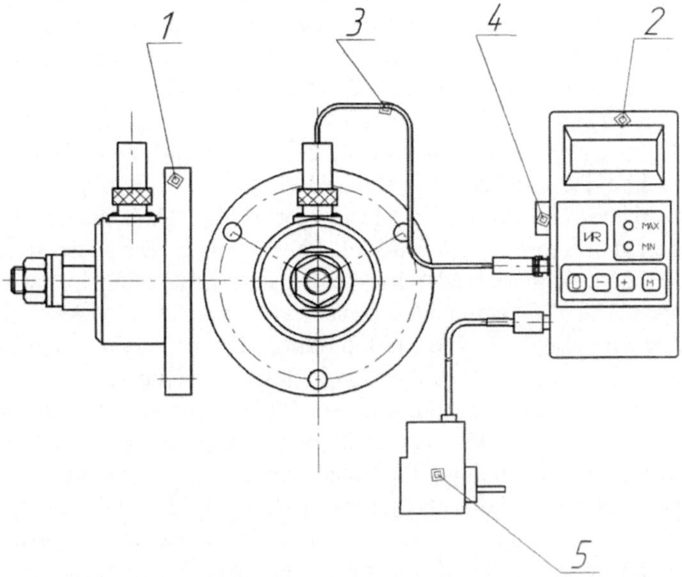

Fig. 2. Schematic view of the electronic torque meter assembly with a secondary device (IM 200): 1 – torque sensor; 2 – electronic unit (secondary device); 3 – connecting harness; 4 – port to connect a computer; 5 – power supply.

4 Results

In the absence of built-in valve closure degree sensors, the NPP repair shop's personnel use torque wrenches, manually closing the valve and adjusting the electric drive's limit coupling as by methodology [17].

In the given experiment, the KD 230 torque wrench was used, which allows measuring torque from 0 to 230 N·m, under basic error of ±2.0% [18].

According to the electric drive's passport data [3], the adjustment curve has the form of a linear graph (Fig. 3).

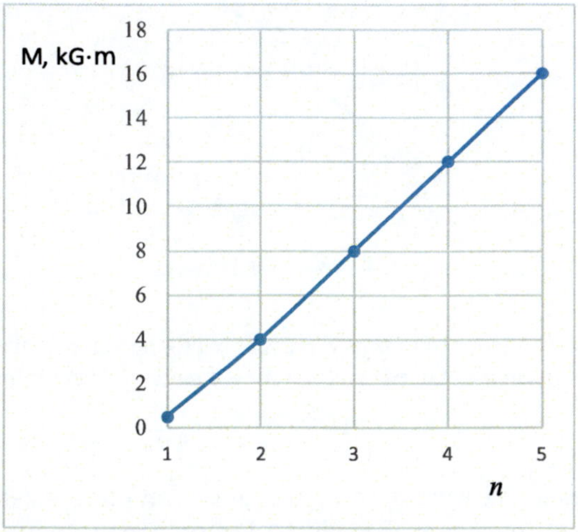

Fig. 3. Dependence of the maximum torque *M* - transmitted to the valve on the values *n*- of the limit coupling torque-limiting switch.

When repair work, these settings are often confusing, which necessitates additional verification of the electric drive maximum torque coupling settings according to the factory schedule (Fig. 3). If the settings do not match, they must be restored. In the absence of special devices, the repair shops of nuclear power plants use the static tuning method with a torque wrench, stepwise setting the torque according to the schedule (Fig. 3) and adjusting the readings of the limit coupling's torque limiter switch.

An underestimation of the torque at which the actuator shall be switched off leads to the valve's incomplete closure and to unacceptable medium leakage. An overestimation of these torque values can lead to damage to the valve stem. Thus, the post-repair limit torque clutch setting operation is necessary and relevant.

Tests of the developed TDD showed that when receiving torque directly from a working electric drive, they give overestimated values of torques in contrast to the static method of manual tightening with a torque wrench. These test results are presented in Fig. 4.

To comply with the factory schedule in Fig. 2, the TDD readings and the torque wrench values obtained in N·m have been converted to kg·m.

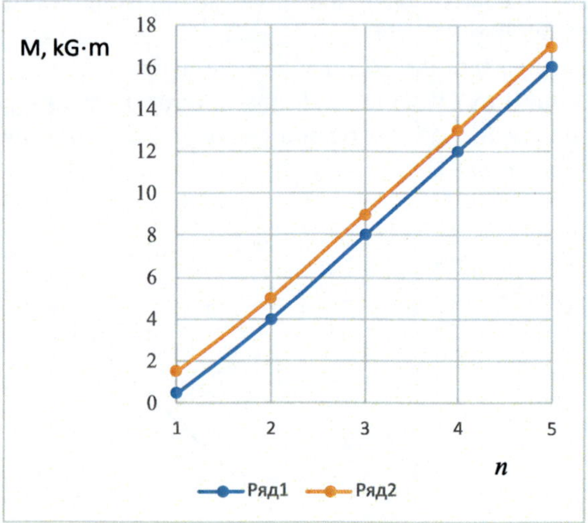

Fig. 4. Comparative dependence of torque M transmitted to the valve on the values *n* of the limit coupling torque limiter switch obtained with a torque wrench – 1 and with TDD on a working electric drive – 2.

From direct torque measurement at the limit, coupling established is a stable deviation of the values obtained "manually". The only explanation for that fact is the limit coupling extra-tightening by an electric motor, which does not stop immediately when the power is turned off, and, due to the inertial rotation, turns the fixture further. It can be assumed that on more powerful electric drives, such valve "after-twisting" can be significantly larger and can lead to the valve damage that should be considered when using torque wrenches at limit coupling manually setting.

5 Conclusions

Represented is the design of a torque diagnostic device for adjusting the power equipment electric drives' limit coupling settings with its test series results.

It is established that the electric drive manual adjustment leads to an underestimation of the real values of torques created by the electric drives.

It seems appropriate to use similar torque diagnostics devices for after-repair adjustments of electric drive valves at nuclear power plants and thermal power plants power units.

References

1. Plahotnikova, E., Eliseeva, T.: Analysis of malfunction motorized stop. Valves Izvestiya Tula State University Engineering. Sci. J. **11**, 352–361 (2013)
2. Podrezova, I., Shutova, L., Ulyanova, Y., Pugacheva, O., Elzhov, Y.: Analysis of the reasons of jamming and rods breaking of pipeline motorized fittings. Global Nucl. Saf. **4**(13), 32–37 (2014)
3. TE099.190M TO: Electric drives with double-sided clutch types M, A, Б, B, Г, Д. Technical description and instruction manual. Tula (1989).
4. Korolev, A., Pavlyshyn, P., Bandurko, I.: Experimental research of valve tightness at different closure forces. Nucl. Radiat. Saf. **4**(80), 21–24 (2018)
5. TMI-2 Analysis Exercise task Group: TMI-2 analysis exercise final report, vol. 91. NEA/CSNI/R (1992).
6. Kozlov, I.: Overview of directions of improving the simulation of severe accidents based on the experience of the Tree Mile Island NPP. East.-Euro. J. Enterp. Tech. **5**(10), 54–59 (2015)
7. Sinelshchikov, P., Babenko, R.: The calculation of the motorized fittings torque by current and voltage signals. Global Nucl. Saf. **4**(13), 28–31 (2014)
8. Vasin, S., Plakhotnikova, E.: Electric drive setting torque calculation method in electric drive shut-off valve system with rectilinea displacement of the valve closure. J. Min. Inst. **232**, 407–412 (2018)
9. Shpakov, O.: On the calculation of loads in electric valves. Armature Eng. **5**(37), 48–51 (2005)
10. Plakhotnikova, E.: Quality and competitiveness improvement of npp safety systems. In: IEEE Conference on Quality Management, Transport and Information Security, Information Technologies (IT&MQ&IS), Nalchik, Russia, pp. 41–42. (2016). https://doi.org/10.1109/ITMQIS.2016.7751946
11. Pipe Fittings. Gate Valves. Force calculation technique: Standard CDBVI 002–2003 (2003).
12. Abramovich, B., Sychev, Y.: Problems of ensuring energy security for enterprises from the mineral resources sector. J. Min. Inst. **195**, 227–230 (2012)
13. Dementev, Y., Umurzakova, A., Udut, L.: Modeling of the asynchronous motor with a device for indirect measurement of time. Fundam. Res. **12**(5), 931–935 (2014)
14. Goryachev, V., Kislyakov, S., Nefediev, D., Varushin, E.: Information-measuring system for measuring torque moments. Measur. Monitor. Manage. Control **3**(25), 5–13 (2018)
15. Altaf, S., Mehmood, M.S., Soomro, M.W.: Advancement of fault diagnosis and detection process in the industrial machine environment. J. Eng. Sci. **6**(2), D1–D8 (2019). https://doi.org/10.21272/jes.2019.6(2).d1
16. Oyegoke, T., Akanji, I.I., Ajayi, O.O., Obajulu, E.A., Abemi, A.O.: Thermodynamic and economic evaluation of gas turbine power plants. J. Eng. Sci. **7**(1), G1–G8 (2020). https://doi.org/10.21272/jes.2020.7(1).g1
17. GOST R 55511–2013: Pipeline valves. Electric actuators. General specifications. Standartinform, Moscow (2014).
18. Electronic torque control devices company INSTRUM-RAND. https://irand.ru/index.php/ru/o-kompanii/novosti-kompanii/188-klyuchi-dlya-izmereniya-krutyashchego-momenta-kd-vklyucheny-v-gosreestr. Accessed 20 May 2020

Synthesis of Elastic Characteristics Based on Nonlinear Elastic Coupling

Victor Kurgan(✉), Ihor Sydorenko, Ihor Prokopovich, Yurii Yeputatov, and Oleksandr Levynskyi

Odessa National Polytechnic University,
1, Shevchenko Avenue, Odessa 65044, Ukraine
kurgan@opu.ua

Abstract. Modern equipment and units operate in a wide range of loads, amplitudes, and vibrational frequencies caused by harmonic, shock, or external influences. On this basis, it is urgent to develop new designs with advanced capabilities to counteract torsional vibrations in the technical system. The design of elastic coupling with nonlinear mechanical feedback is considered in the paper. Developed its design scheme and the main geometrical parameters that determine the elastic characteristic. Expressions were obtained which describing the elastic characteristic of the coupling. Calculations have been made to evaluate the capabilities of the considered device to reproduce nonlinear elastic characteristics. The control function as a step polynomial was used to determine the relationship between the curvature of the control function sections and the relevant sections of the elastic characteristic. A graphical interpretation of the calculations is made. The initial parameters are determined, and the target elastic characteristic with two sections of temporary load loss is calculated. The graphical interpretation showed that the elastic coupling realizes the calculated elastic characteristic. A correction parameter is set to prevent the mechanism from being jammed when using complex control link profiles.

Keywords: Mechanical feedback · Metal elastic elements · Torque

1 Introduction

In modern engineering, elastic couplings with metal elastic elements have become widespread. This is facilitated by the ability of these devices not only to transmit torque but also to prevent negative manifestations of oscillations in the technical system. In studies using mathematical models [1], it has been [2] found that the most favorable results [3] are shown by elastic couplings [4] with nonlinear elastic characteristics [5]. However, elastic couplings that have [6] already been created do not fully [7] meet the stated requirements [8], due to their narrow operating range [9]. But the perspective designs of elastic couplings [10] that realize nonlinear elastic characteristics [11] are not widely used because of the small number of their real mechanical structures.

On the basis of this urgent task is the development and research of new designs of elastic couplings that realize the target nonlinear elastic characteristic.

V. Tonkonogyi et al. (Eds.): InterPartner 2020, LNME, pp. 166–175, 2021.
https://doi.org/10.1007/978-3-030-68014-5_17

2 Literature Review

Previous studies [12] aimed at creating new passive elastic couplings with nonlinear mechanical feedback [13] were conducted using modified kinematic graph theory [14]. The results of these studies are a synthesized graph model of a device with an additional mechanical circuit. The development of the obtained model made it possible to reproduce the kinematic scheme according to which the real design of elastic couplings with mechanical feedback was created in the form of a cam mechanism with kinematic closure. Depending on the design layout, the device implements nonlinear elastic characteristics of the "soft" and "hard" Duffing types, which provides the basis for the reproduction of the design of the combined elastic characteristics. This kind of elastic characteristic enables the resonance to pass through by "jump" without reaching the resonant amplitudes. But this is only possible if there are sections with quasi-zero stiffness in the working range of the displacements.

3 Research Methodology

Consider a perspective design of a nonlinear mechanical elastic coupling. For drawing up of the calculation scheme, we analyze the functional interaction between the coupling elements in the process of operation (Fig. 1).

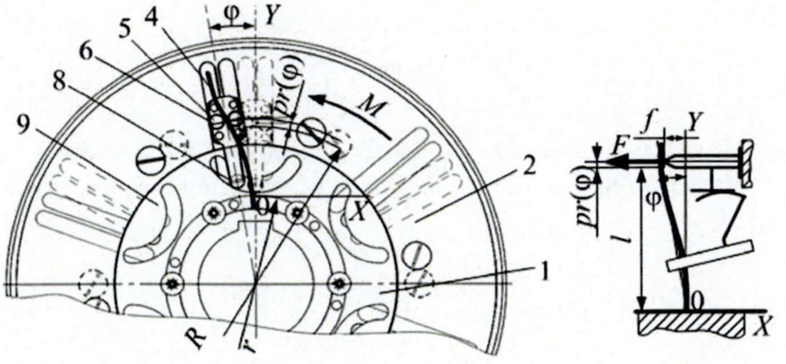

a) b)

Fig. 1. Calculation scheme of elastic coupling with nonlinear mechanical feedback: general (a); simplified (b).

The action of torque on the driving half-coupling 1, causes it to rotate relative to it at some angle φ driven coupling half 2, which is caused by deformation (bending) of the flat spring 4. Deformation of the flat spring 4 occurs since the load rollers 5 of the pusher 6 contacts with free end of the flat spring 4, the other end of which is cantileverly fixed on the coupling 1, and the guide rollers 7 of the pusher 6 in contact with the radial grooves of the coupling half 2. The roller contact 8 of the support pusher 6 with the surface of the curved groove of the disk 9 causes radial displacement of the pusher 6 × (φ) along the radial grooves of the coupling half 2. The magnitude and direction of the radial movement of the pusher 6 due to the shape of the curved groove of the disk 9, and reflect the movement of the contact line of attached load from the load rollers 5 to the cantilever flat spring 4.

Considering the presence of holonomic connections between elements of the device, it is advisable to introduce the coordinate system $X0Y$ in the calculation scheme, placing it so that the point 0 lies on the point of the cantilever attachment of the elastic element at the coupling half 1 (Fig. 1a). In the selected coordinate system, the basic geometrical parameters of the device at the initial time (the radius R of the contact of the rollers 5 and the elastic element 4; the radius r, which determines the cantilever attachment of the elastic element 4 at the coupling half 1) determine the main geometric size (parameter), which determines the elastic coupling characteristic. This main parameter is the length of the elastic element which is subject to bending.

$$\mathrm{l} = \mathrm{R} - \mathrm{r} \tag{1}$$

For determining the elastic characteristics of the device shown in the calculation scheme can be somewhat simplified (Fig. 1b).

The tangent of the twisting angle φ of the coupling half of one relative to the other depends primarily from the amount of deflection f (or the bending angle of the elastic element φ_e) of the elastic elements 4 relatives to the distance between the contact line of the elastic element with the roller 5, and the rotation axle of elastic coupling R. It is with some assumption that the projection of displacement Δl_y is insignificant, from which the assumptions are accepted

$$\varphi_e \approx f, \ \mathrm{tg}\,\varphi = \frac{\varphi_e}{\mathrm{R}} \approx \varphi. \tag{2}$$

Considering the expression that causes the deflection of the cantileverly fixed elastic element under the action of a force whose line is perpendicular to the axis of the elastic element obtained

$$M(\varphi) = \frac{3nEJ_x(R \pm pr(\varphi))^2}{(l \pm pr(\varphi))^3}\varphi. \tag{3}$$

where $F = M/R$ – force that determines the action of the torque M and causes the elastic element to bend; l – load application arm to the elastic member, by expression (1); n – number of elastic elements in the coupling; E – the modulus of elasticity of the material of elastic elements, provided that all elements are made of the same material; J_x – moment of inertia of a section of an elastic element, provided that all sections of the elastic elements have the same shape and size.

Based on the determined calculation scheme and taking into account the presence of additional mechanical structure, the expression that determines the adjustable elastic characteristic of the proposed coupling can be obtained on the basis of expression (5) and has the form

$$M(\varphi) = \frac{3nEJ_x(R \pm pr(\varphi))^2}{(l \pm pr(\varphi))^3}\varphi \quad (4)$$

4 Results

For the evaluation of the capabilities of the device concerning the reproduction of elastic characteristics, a series of calculations were performed using the expression (5). For simplicity, the device was considered to have one elastic element made of steel 65 (GOST 14959–79) with the modulus of elasticity $E = 2.15 \cdot 10^{-5}$ MPa; parameter $l = 50$ mm; parameter $R = 80$ mm; the intersection from of the elastic element is a rectangle $h \times b = 10 \times 1$ (mm); maximum twist angle of the coupling in design parameters $\varphi = 0.5$ rad (30°).

To determine the relationship between the curvature of the sections of the control function and the corresponding sections of the elastic characteristic, the control function, which was given in the form of a step polynomial, was applied

$$pr(\varphi) = 241144\varphi^6 - 394025\varphi^5 + 236937\varphi^4 - 62701\varphi^3 + 6639.4\varphi^2 - 154\varphi + 0.33 \quad (5)$$

In the first design case, the control function determined the change of the shoulder application of the load to the elastic element during the rotation of the coupling half relative to each other by the positive (increasing the denominator and numerator by $+pr(\varphi)$ in expression 4) the value of the control function (Fig. 2a). The control function causes three extremes, which determine $l_1 = 60$ mm at $\varphi = 0.125$ rad, $l_2 = 47$ mm at $\varphi = 0.25$ rad, $l_3 = 50$ mm at $\varphi = 0.45$ rad.

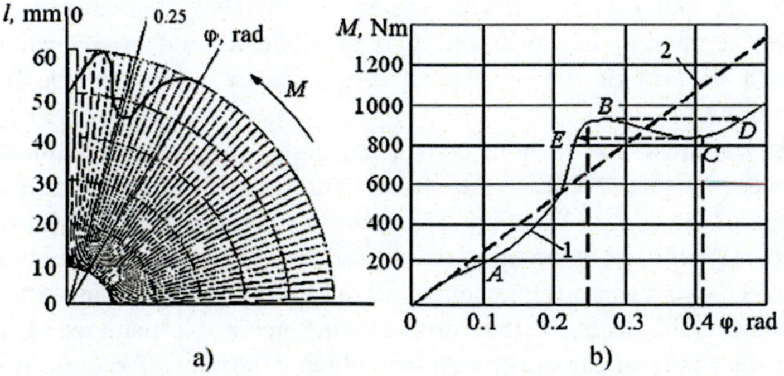

Fig. 2. Analysis of the elastic characteristics of the coupling with the nonlinear feedback: radial graph of the change of the shoulder application of the load $l + pr(\varphi)$ of the elastic element (a); elastic characteristic (b).

In the second calculation case, the control function determined the change of the shoulder application of the load to the elastic element when rotating the half coupling relative to each other by the negative (reducing denominator and numerator by $-pr(\varphi)$ in expression 4) the value of the control function (Fig. 3a). The control function causes three extremes, which determine $l_1 = 70$ mm at $\varphi = 0.125$ rad, $l_2 = 55$ mm at $\varphi = 0.25$ rad, $l_3 = 70$ mm at $\varphi = 0.45$ rad.

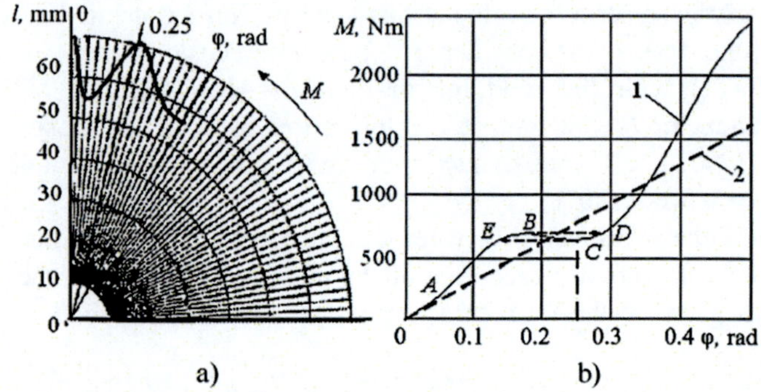

Fig. 3. Analysis of the elastic characteristics of the coupling with the nonlinear feedback: radial graph of the change of the shoulder application of the load $l - pr(\varphi)$ of the elastic element (a); elastic characteristic (b).

Graphical interpretation of the calculations shows that the characteristics in both cases (Fig. 2b, and Fig. 3b) over the entire half-angle range of the twist angle are nonlinear. Both elastic characteristics consist of sections of different nonlinearity, so they can be considered combined. In these areas, a significant increase in deformation is identified with a slight increase or decrease in load (respectively, sections *BD* and *CE*, see Fig. 2b, and Fig. 3b). Elastic systems with these characteristics are called systems with "temporary loss of load". It is theoretically and experimentally proved that such systems provide the passage of resonance by a jump in the forward and reverse directions.

It should be noted that in both considered cases, the points of inflection of the elastic characteristic that determine the corresponding sections on it, exactly correspond to the extrema of the control function. This indicates that the proposed elastic coupling with mechanical feedback is sensitive to the feedback settings since each section of its elastic characteristics can be implemented separately by strengthening or, conversely, weakening the manifestation of the corresponding geometric parameters, namely the shape of the curvature of the curved groove which determines the control algorithm.

Based on the results obtained, it is advisable to calculate the reproducibility, proposed by PM, of the elastic target elastic characteristic.

When choosing the target elastic characteristic, take into account the fact that in the case of the implementation of a combined elastic system with a section of "loss of load capacity" ("LLC") the passage of resonance in the forward and reverse direction is a jump, without the development of critical amplitudes (Fig. 4). The interest is the implementation of the proposed design of a combined elastic characteristic with two such sections.

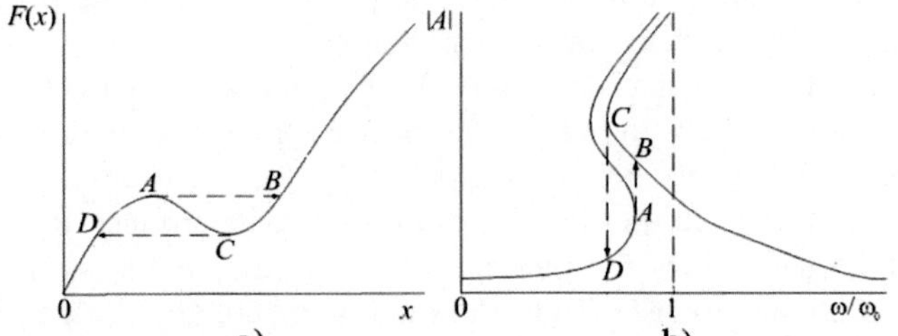

Fig. 4. Direct and inverse overcoming of the "jump" resonance in a nonlinear elastic system: one-section elastic characteristic of load loss (a); the amplitude-frequency response of the system (b).

For the formation of the target elastic characteristics it is necessary to know in advance some starting data, namely: the number of sections LLC $d_{il} = 2$; the parameters of the start points of these sections in $M_{di}(\varphi_i)$; the length of each section in the φ_{di} format; the minimum torque value of M_{dimin} at load loss for each section and the corresponding φ_{dimin} angle. Besides, the target elastic characteristic may have specific points defined under certain conditions, which parameters must also be provided in M_{Qi} (φ_{Qi}) format.

Table 1. The initial settings of the synthesis of the target elastic characteristic.

Parameter, dimension	Marking	Value
Working range torque, N·m	M	50
Twisting angle of the half coupling which determines the operating range, rad	φ_{max}	0.45
Number of sites "loss of load capacity"	*dil*	2
The moment that causes the beginning of the 1st section, N·m	M_{d1}	16
The angle that determines the beginning of the 1st, rad	φ_1	0.1
The minimum moment of the 1st section, N·m	M_{d1min}	16
The angle that determines the minimum moment of the 1st, rad	φ_{1min}	0.1
The length of the 1st section, rad	φ_{d1}	0.1
The moment that causes the beginning of the 2nd section, N·m	M_{d2}	26.5
The angle that causes the beginning of the 2nd section, rad	φ_2	0.3
The moment that determines the characteristic point Q, N·m	M_Q	10
The angle that determines the characteristic point Q, rad	φ_Q	0.05
The moment that determines the characteristic point Q_1, N·m	M_{Q1}	37
The angle that determines the characteristic point Q_1, rad	φ_{Q1}	0.44

As an example, for the synthesis of the target elastic characteristic, the input data entered in Table 1 is used.

In the synthesis of the target elastic characteristic, some simplifications are applied, namely: the basic geometric dimensions of the passive elastic coupling correspond to the basic geometrical parameters of the coupling discussed above.

The presence of points that determine the beginning of the site of loss of load capacity and the minimum value of the moment on them allows one to obtain additional points of the target elastic characteristic (Fig. 5). For example, the start of the LLC section point A upon further loading of the device determines an additional point B. Point C, which determines the minimum moment of the 1st section of the LLC, determines an additional point D. Thus, the set of points 0, Q_1, D, A, C, B, M, E, K, L, Q_2 sufficiently describes the desired target characteristic (graph 1 Fig. 5). The set of coordinates of these points in the format $M_i(\varphi_i)$ are the input data for mathematical processing to obtain the target elastic characteristic in the form of a step polynomial.

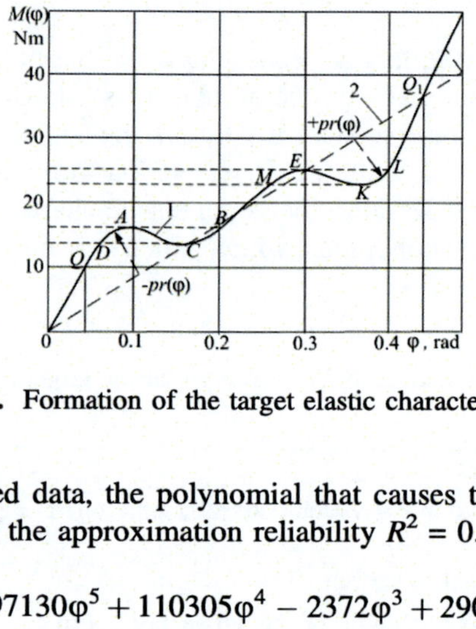

Fig. 5. Formation of the target elastic characteristic.

Based on the received data, the polynomial that causes the target elastic characteristic with the value of the approximation reliability R^2 = 0.998, has the form

$$M_t(\varphi) = 424708\varphi^6 - 397130\varphi^5 + 110305\varphi^4 - 2372\varphi^3 + 2909\varphi^2 + 378\varphi - 2 \cdot 10^{-5}. \quad (6)$$

Further calculations boil down to the definition of the function that determines the curvature of the control groove and accordingly the parameter $pr(\varphi)$. At the beginning of the calculation, you must define the basic design parameters. For this purpose, the polynomial by expression (6) is determined by the averaged value (graph 2, Fig. 5).

$$M_{av}(\varphi) = 110305\varphi. \quad (7)$$

Considering the expressions (1), (4) and (7), an expression is obtained that determines the relationship between the basic geometrical parameters of the coupling (r (on the shaft diameter) and R (free), as well as the initial length l_i, the number n and the material (*E*) the elastic element applied and allows to determine the necessary geometrical parameters of its intersection.

$$J_x = \frac{M_{av}(\varphi_{\max})3nEl_{\Pi}^3}{\varphi_{\max}(l_i + r)^2} \tag{8}$$

According to the results of the calculations according to expression (8), as well as taking into account the initial parameters l_i = 50 mm, $\varphi_{\max}$ = 0.5 rad and r = 35 mm, it is assigned: n = 4 – the number of elastic elements; the intersection shape of the elastic element is a rectangle, $h \times b = 8 \times 0.6$ mm; material of elastic element steel 65 (GOST 14959–79) with the modulus of elasticity $E = 2.15{\cdot}10^{-5}$ MPa.

Comparison of the right-hand side of expression (4) with the polynomial by expression (6) gets an equation whose solution is relative to the parameter $pr(\varphi)$ and will determine the desired unction of the changing of the shoulder application of the load on the elastic element

$$\frac{3 \cdot 4 \cdot 2,15 \cdot 10^5 \cdot 0,054(80 \pm pr(\varphi))^2}{(50 \pm pr(\varphi))^3}$$

$$\varphi = 424708\varphi^6 - 397130\varphi^5 + 110305\varphi^4 - 2372\varphi^3 + 2909\varphi^2 + 378\varphi - 2 \cdot 10^{-5}. \tag{9}$$

The solution of Eq. (9) is obtained in the form of a step polynomial

$$pr(\varphi) = 54708\varphi^6 - 37230\varphi^5 + 94305\varphi^4 - 1332\varphi^3 + 1609\varphi^2 + 378\varphi - 2 \cdot 10^{-5}, \tag{10}$$

that allows it to be used with the smallest transformations for programming the CNC machine tool in the manufacture of a curved groove disc.

Graphical interpretation of the solution of Eq. (9), allows determining the free parameter *R* (Fig. 6). However, given that the additional mechanical structure used is a cam mechanism with kinematic closure, it is necessary to check the function obtained by the pressure angles between the pusher and the cam.

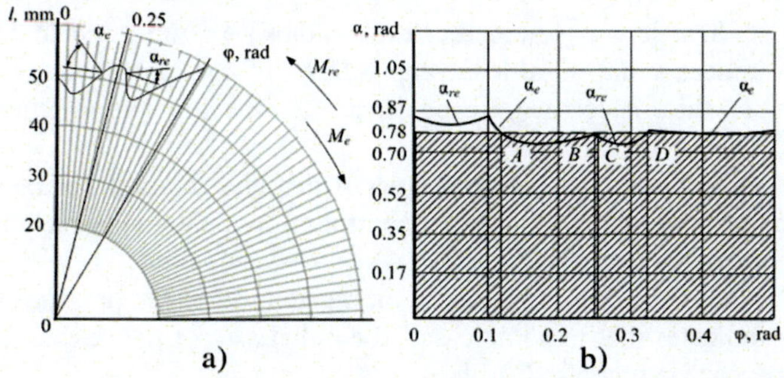

Fig. 6. Synthesis of the elastic characteristic of the coupling with the nonlinear feedback: radial graph of the change of load arm l + pr (φ) of the elastic element (a); control angles of curvature of the control groove without adjustment (b).

It should be noted that the constructive implementation of the proposed device used Autodesk Inventor. This CAD package allows directly in the modeling environment to investigate kinematic schemes of the projected mechanisms at the level of geometric dependencies between their constituent elements. Thus the system is capable of working out not only dependencies of assembly but also contacts of details, thus modeling different types of reactions between them. The validation of pressure angles (as angles between the velocity vector and normal to curve surface at characteristic points) using the built-in CAD package of the Autodesk Inventor Series kinematics module showed that in some sections of *AB*, *CD*, the angle of pressure exceeds or is equal to 45°. Consequently, jamming is possible in these sections (Fig. 6).

For the prevention of this phenomenon, the geometric indices of the curvilinear groove were optimized (increasing the radius of the middle arc). At the same time, a correction parameter was set, which, taking into account the input data, is equal to *Kor* = 20 mm (Fig. 7).

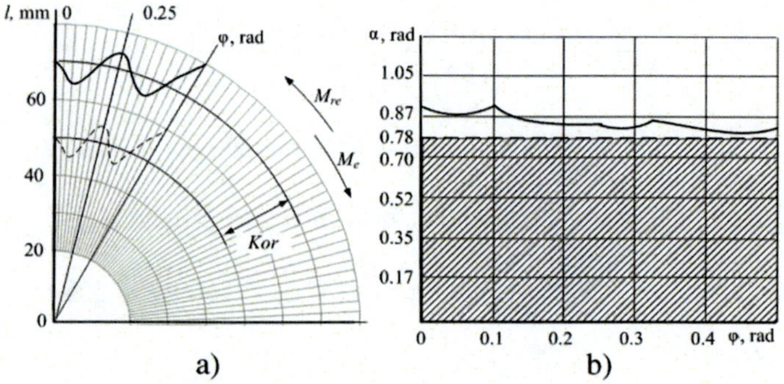

Fig. 7. Synthesis of the elastic characteristic of the coupling with the nonlinear feedback: radial graph of the change of load arm 1 + *pr* (φ) of the elastic element (a); control angles of curvature of the control groove with adjustment (b).

5 Conclusions

Based on the studies, it was found that the synthesis of the adopted target elastic characteristics indicates the significant capabilities of the proposed device for the reproduction of the necessary, according to specific requirements, elastic characteristics.

In the synthesis of complex elastic characteristics, the geometric parameters of the additional mechanical structure should be controlled by contact interaction to prevent jamming phenomena.

The calculation of the correction using the magnitude of the pressure angle of the parameter *Kor* indicates that in some cases, the radial size of the device, to reproduce its functionality, can be increased fully.

The synthesis results and the constructive implementation of a nonlinear mechanical coupling elastic coupling can be used to create rotating elastic support.

References

1. Andrukhiv, A., Sokil, B., Sokil, M.: Asymptotic method in investigation of complex nonlinear oscillations of elastic bodies. Ukrainian J. Mech. Eng. Mater. Sci. **4**(2), 58–67 (2018)
2. Eliseev, S., Eliseev, A.: Construction of mathematical models of mechanical vibrational systems. additional couplings and equivalent transformations. Theory of oscillations. In: Studies in Systems, Decision and Control, vol 252. Springer, Cham (2020).
3. Poparad, H. Methods for modeling an elastic system with permanent contour coupling deformation. In: CONAT 2016 International Congress of Automotive and Transport Engineering. Springer, Cham (2016).
4. Younesian, D., Hosseinkhani, A., Askari, H.: Elastic and viscoelastic foundations: a review on linear and nonlinear vibration modeling and applications. Nonlinear Dyn. **97**, 853–895 (2019). https://doi.org/10.1007/s11071-019-04977-9
5. Bazhenov, E., Buynachev, S., Chernyshev, D.: Mathematical modeling of oscillations of the associated transport and technological complex with the use of the graph theory. In: Proceedings of the 4th International Conference on Industrial Engineering. ICIE 2018. Lecture Notes in Mechanical Engineering. Springer, Cham (2019)
6. Andrukhiv, A., Sokil, B., Sokil, M.: Resonant phenomena of elastic bodies that perform bending and torsion vibrations. Ukrainian J. Mech. Eng. Mater. Sci. **4**(1), 65–73 (2018)
7. Shatokhin, V., Granko, B., Sobol, V.: Synthesis of parameters nonlinear elastic couplings as an extinguisher of oscillations in double transport diesel. Vibr. Eng. Tech. **1**(81), 113–121 (2016)
8. Zyablikov, V., Smirnov, V.: Algorithm for calculating parameters and constructive dimensions of clutch drives with elastic elements as a steel rod round. Eng. J. Appl. **10**, 19–25 (2017)
9. Komarov, C., Pasika, B., Romanchukm, I., Gembara, N.: Investigation of the dynamic characteristics of a coupling with a section of quasinull stiffness taking into account the characteristics of an electric motor. Sci. Bull. NFUU Ukraine **7**(26), 55–64 (2016)
10. Andrianov, I., Awrejcewicz, J., Danishevskyy, V.: Nonlinear elastic problems. In: Asymptotical Mechanics of Composites. Advanced Structured Materials, vol. 77. Springer, Cham (2018)
11. Bociu, L., Guidoboni, G., Sacco, R.: Analysis of nonlinear poro-elastic and poro-visco-elastic models. Arch. Rational Mech. Anal. **222**, 1445–1519 (2016). https://doi.org/10.1007/s00205-016-1024-9
12. Sydorenko, I., Kurgan, V.: Synthesis of nonlinear elastic couplings on the basis of modified kinematic graphs. Proc. Odessa Polytech. Univ. **3**(53), 5–11 (2017)
13. Bauer, F., Römer, U., Fidlin, A.: Optimization of energy efficiency of walking bipedal robots by use of elastic couplings in the form of mechanical springs. Nonlinear Dyn. **83**, 1275–1301 (2016). https://doi.org/10.1007/s11071-015-2402-9
14. Sydorenko, I.: Analysis of the structure of the relaxation shock absorbers and modified the traditional kinematic graphs. ASB Mech. Machines Tools Mater.**1**(26), 88–91 (2014)

Modeling of Pseudoharmonic Oscillations of Vibration Container with Working Mixture

Volodymyr Symoniuk[1], Viktor Denysiuk[1], Yurii Lapchenko[1(✉)], Vasilij Strutinsky[2], and Alexandr Permyakov[3]

[1] Lutsk National Technical University, 75, Lvivska Street, Lutsk 43018, Ukraine
yuralap@gmail.com
[2] National Technical University of Ukraine "Igor Sikorsky Kyiv Polytechnic Institute", 37, Peremohy Avenue, Kyiv 03056, Ukraine
[3] National Technical University "Kharkiv Polytechnic Institute", 2, Kyrpychova Street, Kharkiv 61002, Ukraine

Abstract. The paper presents the technique of calculating the fluctuations of the container with the working mixture. Available methods for investigation of the vibroprocessing are insufficient for the practical use. Additionally, for further use of research material for the grounded design of the technological process of vibrational processing in industrial production, it is expedient to develop a mathematical model of vibroprocessing. This proposed model enables getting detailed information about the course of the process of vibroprocessing without directly experimenting. An elementary and refined model of pseudoharmonic vibrational motion of the container and working mixture, which gives the opportunity to study the oscillatory process is developed. Dependencies output analytically. Based on the developed theory, the influence of individual factors on the process of oscillon is analyzed. According to the results of calculations, the models of vibrating installations, on which the experiments with the use of a partially transparent vibration container, are planned, for confirmation of the most effective use mechanical energy fluctuations.

Keywords: Mixing · Model · Oscillational process · Vibroabrasive · Movement of parts · Vibroabrasive treatment · Acceleration · Fluctuations · Working mixture

1 Introduction

At the present stage of the production development, it is crucial to choose the optimal technological processes, which achieves sufficient quality of machining of parts at a minimum cost of its conduction [1].

Finishing operations, to a large extent, determine the quality of the components obtained [2]. Parts with qualitatively processed surfaces have a longer lifespan, slowly wear out and do not accelerate wear of adjacent parts.

V. Tonkonogyi et al. (Eds.): InterPartner 2020, LNME, pp. 176–185, 2021.
https://doi.org/10.1007/978-3-030-68014-5_18

At present, in machinery and instrument manufacture, the percentage of details requiring a high purity surface is significant. This causes a substantial cost of manufacturing the right assemblies and aggregates [3].

Vibroabrasive method of processing at the present stage is virtually the only thing that allows you to process small parts (they most affect the quality of the product) in batches or continuous flow, rather than individually every detail [4, 5].

Vibroabrasive treatment, in contrast to grinding and polishing with abrasive and felt circles, can easily be mechanized and automated [4, 6]. This allows you to abandon the non-productive manual work, reduce the cost of finishing, and reduce the cost of final production.

For the effective use of the method of vibroabrasive processing in industrial production, it is necessary to have an idea about the processes occurring during processing. This is accomplished by their mathematical modeling [8]. Modeling makes it possible to reduce the complexity of the design of technological processes of vibration processing, to reduce the cost of time and materials, and as a result to ease the start-the overload of the vibromachine and reduce the cost manufactured products [6].

To better understand the processes in mixing, identifying the optimum regimes, spreading the possibilities of progressive application processing using useful vibrations, we had to predict theoretically and practically confirm the basic principles of the effect of the selection of vibration parameters on the quality of processing.

2 Literature Review

Practice has shown that the results of the vibration abrasive process have an impact of about 45–50 different factors [2, 3, 7, 10], of which it is necessary to select the optimum connection. The complexity of this task requires a number of experiences for each case of the new construction of parts of the design vibration processing methods, especially if the technological requirements required is the use of vibrating grinding and polishing. In the study of the process of vibroprocessing main bet make an experiment. There are many research methods [4, 6, 8]. Characterize those of them, which are most typical for vibration treatment.

The easiest is to measure the amplitude of the vibration system. This experiment can be using a wide variety of equipment, from the simplest measuring devices to quite complex measuring systems [3, 7, 9].

A complete process of vibration processing from a dynamic point of view can be investigated with the help of S-Bahn filming. The essence of this method is as follows. The tank in which the vibroprocessing passes is performed with a transparent sidebar, on which the coordinate axes are applied (the moving coordinate system). It is also desirable before the reservoir to establish a stationary coordinate system (done on a transparent basis or simply as several stretched wires). Shooting is carried out through a transparent tank wall with a frequency of at least 1500–2000 frames per second. Based on the footage, build fluctuations graphs of the system within two or three periods of vibration (less time does not allow enough complete understanding of the process progress). The resulting graphs are broken lines as they are built on separate points. These graphs flatten curves (usually not lower than the third-order ones). The disadvantage of this method is its enormous complexity combined with the practical impossibility of automation. Along with this, such researches are conducted by trial and error, i.e., analyzed quite a wide range of possible options, from which it is selected that best satisfies the set goal.

For the measurement of the force of strikes, individual abrasive particles used sensor-detail [8, 10]. This device is a massive body (massive it can be called quite conditionally, because its weight is measured only in grams, simply it makes up most of the weight of the sensor parts) with a sensitive element inside, which perceives strikes of individual particles and converts them into an information signal. Depending on the task, a lot of sensor-details can be changed. Also, using the same device can determine the weight of the polished metal with a unit surface area for a certain period. To realize this, it is necessary to weigh exactly the overlay attached to the sensor-the details, before and after the vibration treatment.

If you want to detect only the grinding metal, use trial parts that weigh before and after processing and determine the loss of their mass [11]. In more detailed studies, the statistical distribution is found.

When experimenting with chemical solutions, be sure to experiment with the basic solution, the results of which are the basis for comparing the effectiveness of other reagents. After dropouts of the least effective additives, conduct experiments with different concentrations of solutions [12].

A significant drawback of experiments is that they are conducted only based on previous experience, without theoretical forecasts of their results [9, 11, 13]. These experiences require high costs of time, intellectual work, and material resources.

3 Research Methodology

Given the insufficiency for the practical use of methods of research of the process of vibroprocessing and considering the purpose of further use of research material for the justified design of the technological process of vibroprocessing in industrial production, it is expedient to develop a mathematical model of vibroprocessing. This model allows getting the complete information about the course of the process of vibroprocessing without direct conducting of an experiment, or at least such that reduces the number of necessary The experiments are minimized [14].

In principle, the scientific approach makes it possible to find a reasonable explanation of any material phenomenon with a satisfactory level of qualitative and quantitative evaluation.

According to the general scheme, in the scientific approach, the researcher tries to disband the investigated object, the process, or the phenomenon on the elementary components, as far as possible to detect in the studied already known and what, what has developed a good mathematical apparatus. Only then, to collect the explained and described (logically or mathematically) components and to trace the theoretical level of their relations and interaction. Only after that, it is advisable to ask an experiment that should confirm, clarify, or refute the proposed assumption. In case of insufficiency of the mathematical apparatus for explanation or insufficient learning, the explanations should be put necessary experiments during the development of the theory.

In this case, considering the sufficient level of study of mechanical processes that occur during vibroprocessing, the ease of dismemberment of the process on the components, it will be advisable to conduct the first theoretical research, and then plan and implement the experiment, which would confirm the correctness of the theoretical conclusions.

When operating the vibration device is extremely important whether the mixture of abrasive with machined parts moves together with the hopper, or at a certain time, there is a separation of it from the bottom of the container [15].

In the first case, the relative movement of parts and abrasive particles is not observed. There is typically a stagnant process. The only thing that changes is acceleration. The force of interaction of abrasive particles with the surface of the parts is proportional to the acceleration, but to carry out the work also requires the movement of cutting protrusions abrasive on the surface parts. In theory, with the periodic variable acceleration medium (a mixture of abrasive and machined parts) should be as periodically flexibly compressed. Since the environment is heterogeneous, the degree of compression of its different parts is different, so there must be relative displacement. The scratching of the surface parts can, in principle, be carried out, but it is necessary to add the fillings with good elastic properties [16] to the mixture. In practice, this process is not applied.

The second case has those useful features, due to which the process of vibroabrasive treatment found widespread. There are as follows: transitional processes in the loss and renewal of contact mixture with container create extreme conditions of abrasive interaction with cultivated surface parts; the mixture is stirring, which results in an all-sided processing of parts; the cut off from the details of shavings and worn abrasive.

The second case is analyzed and taken as a basis for the creation of the mathematical model of processes occurring during vibroprocessing.

4 Results

For a detailed analysis of the description of the movement process of the container and the working mixture is done analytically. For better mathematical reproduction, it is assumed that the mixture moves like a single unit. Without going into the law of change of forced force is believed that the movement of the container is described by a sinusoid. Friction between the working mixture and walls the container is neglected. The axis of the positive direction x is selected.

Container position at a specific time:

$$x = Asin(\omega t), \tag{1}$$

where A – oscillation amplitude, ωt – phase (ω – cyclic frequency, t – time moment).

Container movement speed:

$$\dot{x} = \mathrm{A\omega cos(\omega t)}. \tag{2}$$

Speeding up the container while moving:

$$\ddot{x} = -A\omega^2 sin(\omega t). \tag{3}$$

The working mixture movement is imposed an additional condition. Its acceleration down cannot be greater from the acceleration of free-fall g because, in addition to the gravity, there are no other forces acting on the mixture and aiming down. At a time when the acceleration of the container down becomes greater from the acceleration of free pa-ting, the mixture breaks off and moves separately, accelerating only by the force of gravity.

Thus, to the mixture periodically come off the container, it is necessary that the maximum for the period of acceleration of the container down was greater than the acceleration of free fall, that is necessary to fulfill the conditions:

$$A\omega^2 > g. \tag{4}$$

This inequality shows that increasing the frequency of vibrations more effectively affects the onset of separation than the increase in amplitude. At the same time, at the Choo frequency of operation of vibrodevice, vibration amplitude should be sufficient for the onset of separation. For effective work, it is also necessary that the separation was large enough, that is, to the right and left parts of the irregularities significantly cut-off; However, the over difference reduces the processing efficiency, as the mixture is simply thrown from the working area. The upper limit of the $\mathrm{A\omega^2}$ is dictated by the constructive peculiarities of a particular vibration machine and the level of its workload and cannot be determined based on the assumption of a harmonious (sinusoidal) movement of the container.

For an analytical description of the movement of the mixture defined moments, in which there is a disconnection and connection of the working mixture with container, as well as the coordinate and speed of the mixture at the time of separation it from the container.

Moment t_1 separation mixture from the container is determined by the conditions:

$$\ddot{x} = g, \tag{5}$$

that is

$$sin(\omega t_1) = \frac{g}{-A\omega^2}. \tag{6}$$

At the same time, the formula (1) determines the initial coordinate of the detached mixture x_0' and its initial speed of $\dot{x}_0'$ by the formula (2).

The mixture continues to move independently by law:

$$x' = x_0' + \dot{x}_0'(t - t_1) + 0,5g(t - t_1)^2. \quad (7)$$

After a while, the mixture is faced again with the container. The contact occurs, when the coordinates are as follows:

$$x' = x, \quad (8)$$

that is

$$x_0' + \dot{x}_0'(t_2 - t_1) + 0,5g(t_2 - t_1)^2 = Asin(\omega t_2). \quad (9)$$

Analytical methods do not solve Eq. (9). It can be solved by a graphical method or by successive approximations. Creating a computer program based on the mathematical model graphic method is not suitable. Therefore, for the solution of Eq. (9). The method of successive approximations is used.

In order to find the solution of the equation by successive approximations, you must first determine the range in which it should be found.

Assuming that the mixture is torn away from a container in every period of fluctuations, the

$$t_1 < t_2 < t_1 + T, \quad (10)$$

where $T = \frac{2\pi}{\omega}$ – oscillation period.

Moreover, if it is not the first separation of the mixture, then the most appropriate first nabla-faction search in the area

$$t_2 = t_{2(i-1)} + T, \quad (11)$$

where $t_{2(i-1)}$ – a moment of impact of a mixture on a container in the previous period.

With a consistent substitution of t_2 in the formula (9), its right and left parts to acquire specific values. If the left side is greater than the right, the t_2 in the offensive-the approximation should be reduced, and vice versa. At the end of this calculation are received two approximations of t_2, onc of which is smaller, and the second is truer. The difference between them does not exceed the allowable calculation error. Instead of true, it takes that approximation, which will be less modulo the difference between the right and left parts of the Eq. (9).

If in the interval (10) there was no stroke of the mixture into a container, the time of its free flight is greater than the period of fluctuations in the container. In this case, the search range of the contact time is transferred to the time of entering the mixture into the area of the fluctuations for vibration container. Reaching the t_r in this zone is determined by the terms:

$$x_0^{'} + \dot{x}_0^{'}(t_r - t_1) + 0,5g(t_r - t_1)^2 = -A. \quad (12)$$

Solution Eq. (12) is

$$t_r = t_1 + \frac{-\dot{x}_0^{'} \pm \sqrt{\dot{x}_0^{'2} - 2g(x_0^{'} + A)}}{g}. \quad (13)$$

Solution (13) provides two values of t_r, but the physical content is deprived of a complex number and value, less t_1. Thus, one solution remains.

Stroke time of the mixture on the container is searched in the range:

$$t_r < t_2 < t_r + T. \quad (14)$$

The method of searching for approximation is the same.

It is assumed that the blow is absolutely inelastic. That is, the next system will move as a single unit.

Relative free traffic time:

$$t_{fm} = \frac{t_2 - t_1}{T}. \quad (15)$$

The analysis of the dynamics of the mixture can set the range of possibilities on-line machine by adjusting the amplitude and oscillation frequency of the container.

Along with this, you need to define the scope of fluctuations in the working mixture.

If the mixture manages to fall on the bottom of the container before it starts moving upwards, the extreme lower position of the mixture will be matched with the amplitude of the vibrations of the container:

$$x_d^{'} = A. \quad (16)$$

If the container is already moved away from the bottom point, it captures the mixture. Thus, it limits the movement of the mixture down. In this case, there is a coordinate of the $x_H^{'}$ meeting mixture with the container by the formula (1) at the time of t_2.

The wingspan of the mixture upwards can also be calculated.

If the mixture comes off from a container before it will enter into the extreme upper position, it is possible to increase the scope of fluctuations in the mixture due to its tossing. In the extreme upper position, mix its speed equal to zero. (It is located as the first derivative by time from the coordinates of the mixture).

$$\dot{x}_0^{'} + g(t_u - t_1) = 0, \quad (17)$$

where t_u – achieving the extremely top position mixture.

It is

$$t_u = t_1 - \frac{\dot{x}_0'}{g}. \tag{18}$$

Since $\dot{x}_0'$ negative when you detach the mixture before passing the container extreme to the top position, then the logical relationship $t_u > t_1$ persists.

The coordinate of the mixture at this point is from (6). Substituting tu and turning, we get:

$$x_u' = x_0' - \frac{(\dot{x}_0')^2}{2g}. \tag{19}$$

The scope of fluctuations in the mixture:

$$P_c = x_d' - x_u'. \tag{20}$$

Substituting (7) and (19) in (20), we get

$$P_c = A\sin(\omega t_2) - \left(x_0' - \frac{(\dot{x}_0')^2}{2g}\right). \tag{21}$$

According to the results of calculations to conducting the experiment developed model proposed installation (Fig. 1), which are protected by patents [10]. For better visual observation and photo-and video-recording processes that occur during the operation, the vibrating container is planned to be performed in a transparent.

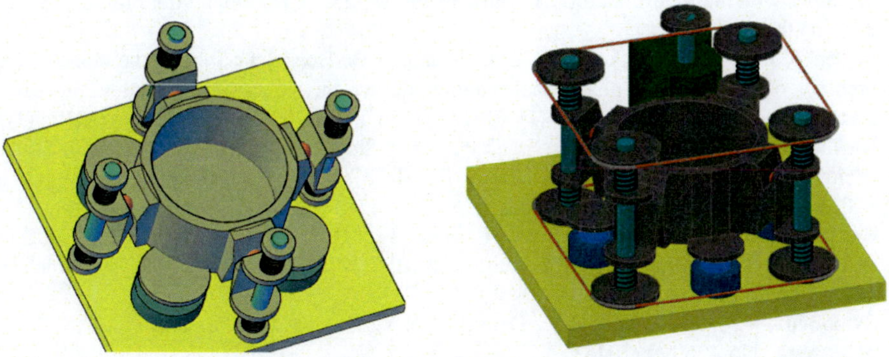

Fig. 1. Vibration units.

5 Conclusions

The following elementary model of mathematical analysis makes it possible to approximate the progress of processes during vibroabrasive processing on sustainable modes. Using this model is desirable with a previous analysis of the vibration device settings. The factor limiting its use is the preliminary assumption about the sinusoidal movement of the container, and as a consequence – the inaccuracy of reproduction of the container and the mixture, as well as that it is not intended to describe transient processes when switching-off device and switch it to another mode of operation, regulating certain parameters directly during operation of the device.

References

1. Blekhman, I.I., Kremer, E.B.: The dynamics of a complex machine assembly: vibration-induced drag on the rotation. J. Mach. Manuf. Reliab. **46**(4), 330–335 (2017). https://doi.org/10.3103/s1052618817030037
2. Filimonihin, G.B., Yacun, V.V.: Investigation of the process of excitation of dual-frequency vibrations by ball auto-balancer of GIL 42 screen. East. Eur. J. Enterp. Technol. **1**(7(79)), 17–23 (2016). https://doi.org/10.15587/1729-4061.2016.59881
3. Strutinski, V.B., Symonyuk, V.P., Denysiuk, V.Y.: Improvement of equipment and process of shock-pulse processing of parts in vibrobunker: monograph. SPD Gadjak Zhanna Volodymurivna, Lutsk (2016)
4. Filimonikhin, G., Yatsun, V., Dumenko, K.: Research into excitation of dual frequency vibrational-rotational vibrations of screen duct by ball-type auto-balancer. East. Eur. J. Enterp. Technol. **3**(7(81)), 47–52 (2016). https://doi.org/10.15587/1729-4061.2016.72052
5. Symonyuk, V., Denysiuk, V., Lapchenko, Y.: Experimental study of circulating vortex movement working environment in vibrobunker. Mech. Mater. Sci. Eng. J. **12** (2017). https://doi.org/10.2412/mmse.58.82.178
6. Blekhman, I.I., Blekhman, L.I., Yaroshevich, N.P.: Upon drive dynamics of vibratory machines with inertia excitation. Ore Enrichment, 49–53 (2017). https://doi.org/10.17580/or.2017.04.09.
7. Pavlenko, I.V., Yukhymenko, M.P., Lytvynenko, A.V., Bocko, J.: Solving the nonstationary problem of the disperse phase concentration during the pneumoclassification process of mechanical mixtures. J. Eng. Sci. **6**(1), F1–F5 (2019). https://doi.org/10.21272/jes.2019.6(1).f1
8. Blekhman, I.I., Indeitsev, D.A., Fradkov, A.L.: Slow motions in systems with inertial excitation of vibrations. J. Mach. Manuf. Reliab. **37**(1), 21–27 (2008). https://doi.org/10.1007/s12001-008-1006-z
9. Balthazar, J.M., Mook, D.T., Weber, H.I., Brasil, M.L.R.F., Fenili, A., Belato, D., Felix, J.L. P.: An overview on non-ideal vibrations. Meccanica **38**(6), 613–621 (2003). https://doi.org/10.1023/a:1025877308510
10. Symonyuk, V.P., Fedorchuk, O.V., Denysiuk, V.Y., Kaidyk, O.L., Lapchenko, Y.S.: Patent of Ukraine, No 126090 (2018)
11. Blekhman, I.I., Vasil'kov, V.B., Yaroshevich, N.P.: On some opportunities for improving vibration machines with self-synchronizing inert vibration exciters. J. Mach. Manuf. Reliab. **42**(3), 192–195 (2013). https://doi.org/10.3103/s1052618813030023.

12. Artyunin, A.I., Eliseyev, S.V.: Effect of "crawling" and peculiarities of motion of a rotor with pendular self-balancers. Appl. Mech. Mater. **373–375**, 38–42 (2013). https://doi.org/10.4028/www.scientific.net/amm.373-375.38
13. Yatsun, V., Filimonikhin, G., Dumenko, K., Nevdakha, A.: Equations of motion of vibration machines with a translational motion of platforms and a vibration exciter in the form of a passive auto-balancer. East. Eur. J. Enterp. Technol. **5**(1(89)), 19–25 (2017). https://doi.org/10.15587/1729-4061.2017.111216.
14. Yatsun, V., Filimonikhin, G., Dumenko, K., Nevdakha, A.: Search for two-frequency motion modes of single-mass vibratory machine with vibration exciter in the form of passive auto-balancer. East. Eur. J. Enterp. Technol. **6**(7(90)), 58–66 (2017). https://doi.org/10.15587/1729-4061.2017.117683
15. Filimonihin, G.B., Yacun, V.V.: Method of excitation of dual frequency vibrations by passive autobalancers. East. Eur. J. Enterp. Technol. **4**(7(76)), 9–14 (2015). https://doi.org/10.15587/1729-4061.2015.47116.
16. Yatsun, V., Filimonikhin, G., Haleeva, A., Nevdakha, A.: On stability of the dual-frequency motion modes of a single-mass vibratory machine with a vibration exciter in the form of a passive auto-balancer. East. Eur. J. Enterp. Technol. **2**(7(92)), 59–67 (2018). https://doi.org/10.15587/1729-4061.2018.128265.

Analysis of Dynamic Mechanic Belt Stresses of the Magistral Conveyor

Oleh Pihnastyi[1], Valery Khodusov[2], Georgii Kozhevnikov[1(✉)], and Tetiana Bondarenko[3]

[1] National Technical University "Kharkiv Polytechnic Institute", 2, Kyrpychova Street, Kharkiv 61002, Ukraine
kozhevnikov.gk@gmail.com

[2] Kharkiv National University named after V.N. Karazin, 4, Svobody Sq., Kharkiv 61022, Ukraine

[3] Ukrainian Engineering-Pedagogics Academy, 16, Universitetskaya Street, Kharkiv 61003, Ukraine

Abstract. Controlling the amount of material flow coming from the accumulated bunker to the input of the conveyor line and controlling the speed of the conveyor belt are common ways to reduce the energy consumption of the transport system. However, most works do not take into account transients associated with a change in the speed of the conveyor belt or a change in the input material flow. As a result of transient processes acceleration or deceleration of the conveyor belt occurs and dynamic stresses arise. These dynamic stresses can exceed the permissible values. In this study, the task of analyzing the causes of dynamic stresses in the transition process is considered. When constructing a model of dynamic stresses, the Hooke's law was used. The model of resistance to the movement of the conveyor belt is adopted in accordance with DIN 22101: 2002–08. The analysis of the propagation of waves of dynamic stresses in the conveyor belt has done. The formation of dynamic stresses as a result of the addition of the forward and backward waves is considered.

Keywords: Dynamic tension · Conveyor belt · Acceleration · Propagation speed of disturbances · Tension force · Oscillatory process

1 Introduction

The introduction of conveyor transport in the mining industry has radically changed the technology of quarrying [1, 2]. The use of this type of transport has reduced the energy consumption and complexity of quarrying [3, 4]. In addition, the use of conveyor systems has significantly improved the efficiency and carrying capacity of the transport system. The growth of cargo traffic and lengths of rock disposal has caused the need to create high-performance belt conveyors of big length and capacity. The characteristics of modern conveyor systems are provided in [5, 6] (Table 1). One of the significant problems of using high-performance belt conveyors of big length is a non-steady value of the ingoing rock flow to the input section of the conveyor. This leads to an uneven distribution of material on the conveyor belt along the transport route [7, 8] and to an

V. Tonkonogyi et al. (Eds.): InterPartner 2020, LNME, pp. 186–195, 2021.
https://doi.org/10.1007/978-3-030-68014-5_19

increase of the specific energy consumption for transporting one tonne of material [9]. For the reduction of the specific energy consumption, it is necessary to regulate the conveyor belt speed [10–12] or the amount of material flow coming from the storage bin [13–15]. This regulation results in acceleration or deceleration of individual conveyor belt sections [5, 16].

Table 1. Characteristics of conveyor transport systems.

	Length (km)	Sections	Power (kW)	Speed (m/sec)	Capacity (t/h)
The Channar Overland, Western Australia (1989) [1]	20.4	2 (10.4; 10.1)	4200	4.1	2200
Neyveli Lignite Corp., India (2007), [2]	14	8	2520	5.4	
Çöllolar Lignite Open Pit Mine, Turkey (2011), [2]	17.4	26	46300		9350
Coarse ore conveyor system Minera Los Pelambres, Chile (1998), [2]	12.7	3	25000		8700
Open Cast Mine Reichwalde, Germany (2010), [2]	13.5	6	19350	5.5	6000
From a mine in India to a cement plant in Bangladesh (2005), [3]	16.5	1		6.5	
Sasol's Impumelelo project in South Africa (2015), [3]	27.5	1		6.5	2400
From the Bu Craa mine to the coast at El Aaiún, Western Sahara, [3]	128.7	11			2000
The Henderson Coarse Ore Conveying System, the North American Continental Divide (2000), [4, 5]	24	3	12700	4.5	2270
Tianjin China Port Authority, China (2005) [6]	8.98	1	4x1500	5.6	6000
Barcelona Tunnel (the Metro (Train) Extension Project), Spain (2005) [6]	4.71	1		3.5	1500
Baumgartner Tunnel, the Metropolitan St. Louis (Missouri), USA [6]	6.18			2.54	200

2 Literature Review

Due to acceleration and deceleration of the conveyor belt sections, dynamic stresses occur, the value of which may exceed the maximum permissible value, which causes damage to the conveyor belt, electric motor and conveyor structure [17, 18]. Despite the existing risks of damage to the transport system [19], restrictions on the value of dynamic stresses are usually not taken into account during the design of belt speed control systems. This assumption does not affect the total energy consumption of the transport system if the total transient time is significantly less than the total time of the main process. The example is a relay system for controlling the conveyor belt speed [12, 20], which involves the adiabatic transition process. In this case, research is reduced to two separate tasks: a) the task of developing an optimal algorithm for controlling parameters of the transport system flow; b) the task of analyzing causes of dynamic stresses in the transition process [21, 22]. In this study, a solution of the second problem is considered.

3 Research Methodology

To analyze dynamic stresses which occur in the conveyor belt under transient modes, we will use the equation with dimensionless parameters:

$$\frac{\partial^2 \omega_0(\tau,\xi)}{\partial \tau^2} = {v_1}^2 \frac{\partial^2 \omega_0(\tau,\xi)}{\partial \xi^2} - {v_2}^2, \tag{1}$$

which is written considering dimensionless variables:

$$\tau = \frac{t}{T_d}, \quad \xi = \frac{S}{S_d}, \quad \omega_0(\tau,\xi) = \frac{\omega(t,S)}{\omega_{\max}}, \quad \omega_{\max} = \frac{\sigma_b S_d}{E}, \quad \theta_C = \frac{[\chi]_{0C}}{[\chi]_{0\max}}, \tag{2}$$

$$\theta_R = \frac{[\chi]_{0R}}{[\chi]_{0\max}}, \quad v_b = \frac{C f_C g_m ([\chi]_{0\max} + [\chi]_{0C}) S_d}{\sigma_b B h},$$

$$v_f = \frac{([\chi]_{0\max} + [\chi]_{0C}) f_\psi(t) S_d}{\sigma_b B h}, \quad {v_g}^2 = \frac{BhE}{([\chi]_{0\max} + [\chi]_{0C})} \left(\frac{T_d}{S_d}\right)^2,$$

$${v_1}^2 = {v_g}^2 \frac{1+\theta_C}{\theta_C}, \quad {v_2}^2 = {v_g}^2 \left(v_f + v_b \left(1 + \frac{\theta_R}{\theta_C}\right)\right),$$

elongation $\omega(t,S)$ is the absolute elongation of the conveyor belt in a moment of time t for a technological position S; S_d is the length of the conveyor; T_d is the characteristic time of the process; $\omega_{\max}$ is the maximum allowable elongation of the conveyor belt; σ_b is the maximum allowable strain of the conveyor belt; E is Young's modulus; is the specific density of the conveyor belt in width B and thickness h; $[\chi]_{0\max}$ is the maximum allowable linear density of the transported material; $[\chi]_{0R}$ is the linear load from the rotating parts [9, p.8], is determined by the ratio of the mass of the rotating element (roller) m_R to the distance S_R between the symmetry axes of the rotating elements [21, p.153]; f_C is coefficient of resistance to movement, includes rolling resistance of driving rollers and belt indentation resistance [9, p.13]; $g_m = 9.81$ (m/sec^2); C – is secondary loss coefficient; $f_\psi(t)$ – conveyor belt acceleration. The coefficient v_b determines the ratio of the resistance force $Cf_C g_m([\chi]_{0\max} + [\chi]_{0C})S_d$ at the maximum load $[\chi]_{0\max}$ to the maximum allowable stress force $\sigma_b Bh$. The coefficient v_f characterizes the ratio of the inertia force $([\chi]_{0\max} + [\chi]_{0C})f_\psi(t)S_d$ at maximum load $[\chi]_{0\max}$ to the maximum allowable stress force $\sigma_b Bh$. The coefficient v_g is the dimensionless speed of disturbance propagation along the conveyor belt.

Let us supplement Eq. (1) with the boundary conditions:

$$\left.\frac{\partial\omega_0(\tau,\xi)}{\partial\xi}\right|_{\xi=1} = \alpha_1, \quad \left.\frac{\partial\omega_0(t,\xi)}{\partial\xi}\right|_{\xi=0} = \alpha_1 - \alpha_{12}, \quad \alpha_1 = \frac{(k_s+1)\theta}{1+\theta}\frac{v_2{}^2}{v_g{}^2}\left(\frac{\exp(k_b\alpha)}{\exp(k_b\alpha)-k_s}\right), \tag{3}$$

and initial conditions:

$$\left.\frac{\partial\omega_0(\tau,\xi)}{\partial\tau}\right|_{\tau=0} = 0, \quad \frac{\partial\omega_0(0,\xi)}{\partial\xi} = \alpha_1 - \alpha_{12}(1-\xi), \quad \alpha_{12} = \frac{v_2{}^2}{v_1{}^2}, \tag{4}$$

were k_b – coefficient of adhesion between the cylinder and the belt; k_s – cylinder loss coefficient “A”; α – the angle of belt drive belts. Three phases are characteristic of transitional modes: a) the period of time of the initial start-up when the conveyor belt moves from rest to moving along the route; b) the phase of formation of the static force along the conveyor belt; c) the phase of acceleration of the conveyor belt to the rated speed. In this paper, we will focus on the analysis of dynamic stresses arising during the phase of acceleration of the conveyor belt to the rated speed.

The solution to Eq. (1) is sought in the form

$$\omega_0(\tau,\xi) = \omega_{00}(\tau,\xi) + \frac{\alpha_{12}}{2}\xi^2 + (\alpha_1 - \alpha_{21})\xi + C_0. \tag{5}$$

The constant C_0 is determined from the condition of the minimum tension of the conveyor belt $\omega_{\min}$, $\omega_0(0,\xi) \geq \omega_{\min 0}$.

Solution (5) allows for a function $\omega_{00}(\tau,\xi)$, satisfying the equation

$$\frac{\partial^2\omega_{00}(\tau,\xi)}{\partial\tau^2} = v_1{}^2\frac{\partial^2\omega_{00}(\tau,\xi)}{\partial\xi^2}, \tag{6}$$

to write the boundary and initial conditions in the following form:

$$\left.\frac{\partial\omega_{00}}{\partial\xi}\right|_{\xi=1} = 0; \quad \left.\frac{\partial\omega_{00}}{\partial\xi}\right|_{\xi=0} = 0 \tag{7}$$

$$\left.\frac{\partial \omega_{00}}{\partial \xi}\right|_{\tau=0} = 0; \left.\frac{\partial \omega_{00}}{\partial \tau}\right|_{\tau=0} = -\left.\frac{\partial \alpha_{12}}{\partial \tau}\left(\frac{\xi^2}{2} - \xi\right)\right|_{\tau=0} = Z(\xi) \tag{8}$$

The solution of Eq. (6) can be represented as follows:

$$\omega_{00}(\tau, \xi) = \frac{1}{2v_1} \int_{\xi - v_1\tau}^{\xi + v_1\tau} Z(\alpha) d\alpha. \tag{9}$$

For the mode with a constant acceleration of the conveyor belt is $\partial\alpha_{12}/\partial\tau = 0$ and, therefore $Z(\xi) = 0$. For the mode in which the acceleration of the conveyor belt linearly depends on time $v_f = v_{f0} + v_{f1} \cdot \tau$, the function $Z(\xi)$ takes the form:

$$Z(\xi) = -\frac{{v_g}^2 v_{f1}}{{v_1}^2}\left(\frac{\xi^2}{2} - \xi\right). \tag{10}$$

The following ratio determines the value of the dynamic stresses of the conveyor belt:

$$\frac{\partial \omega_{00}(\tau, \xi)}{\partial \xi} = \frac{Z(\xi + v_1\tau) - Z(\xi - v_1\tau)}{2v_1}. \tag{11}$$

The properties of the function $Z(\alpha)$ follow from the boundary conditions (7)

$$\left.\frac{\partial \omega_{00}(\tau, \xi)}{\partial \xi}\right|_{\xi=0} = \frac{Z(v_1\tau) - Z(-v_1\tau)}{2v_1} = 0, \ Z(\beta) = Z(-\beta), \tag{12}$$

$$\left.\frac{\partial \omega_{00}(\tau, \xi)}{\partial \xi}\right|_{\xi=1} = \frac{Z(1 + v_1\tau) - Z(1 - v_1\tau)}{2v_1} = 0, \ Z(\beta_{11}) = Z(2 - \beta_{11}), \ \beta_{11} = (1 + v_1\tau).$$

Since the function $Z(\beta)$ (12) is even, and the value τ is arbitrary, it follows

$$Z(2 - \beta) = Z(\beta) = Z(-\beta), \ Z(2 + \beta) = Z(\beta). \tag{13}$$

The function $Z(\beta)$ is an even function of period 2. The general solution of the problem takes the form

$$\omega_0(\tau, \xi) = \frac{1}{2}\frac{{v_2}^2}{{v_1}^2}\xi^2 + \left(\alpha_1 - \frac{{v_2}^2}{{v_1}^2}\right)\xi + \omega_{\min 0} + \frac{1}{2v_1}\int_{\xi - v_1\tau}^{\xi + v_1\tau} Z(\alpha) d\alpha,$$

$$\frac{\partial \omega_0(\tau, \xi)}{\partial \xi} = \frac{{v_2}^2}{{v_1}^2}\xi + \left(\alpha_1 - \frac{{v_2}^2}{{v_1}^2}\right) + \frac{Z(\xi + v_1\tau) - Z(\xi - v_1\tau)}{2v_1}. \tag{14}$$

Let's introduce the variable τ_n so that

$$0 \leq v_1\tau_n - 2n < 2, \quad \tau_n = \tau - \frac{2n}{v_1}.$$

Then, due to the periodicity of function (13), it follows

$$Z(\xi + v_1\tau) = Z(\xi + v_1\tau - 2n) = Z\left(\xi + v_1\left(\tau - \frac{2n}{v_1}\right)\right) = Z(\xi + v_1\tau_n) = Z(\beta_1),$$

$$Z(\xi - v_1\tau) = Z(\xi - v_1\tau + 2n) = Z\left(\xi - v_1\left(\tau - \frac{2n}{v_1}\right)\right) = Z(\xi - v_1\tau_n) = Z(\beta_2)$$
$$\beta_1 = \xi + v_1\tau_n, \quad \beta_2 = \xi - v_1\tau_n$$

Let us calculate the value of dynamic stresses (11) arising in the conveyor belt

$$\frac{\partial\omega_{00}(\tau,\xi)}{\partial\xi} = \frac{Z(\beta_1) - Z(\beta_2)}{2v_1} = Z_0(\beta_1,\beta_2), \quad Z(\beta) = v_d\left(\frac{\beta^2}{2} - \beta\right), \quad v_d = -\frac{{v_g}^2 v_{f1}}{{v_1}^2}.$$

4 Results

Consider the form of the function $Z_0(\beta_1,\beta_2)$ for different values β_1, β_2:

1. $0 \leq \beta_1 < 1, \quad 0 \leq \beta_2 < 1;$

$$Z_0(\beta_1,\beta_2) = \frac{Z(\beta_1) - Z(\beta_2)}{2v_1} = \frac{v_d}{2v_1}\left[\left(\frac{\beta_1^2}{2} - \beta_1\right) - \left(\frac{\beta_2^2}{2} - \beta_2\right)\right]$$
$$= \frac{v_d}{2v_1}\left[\frac{(\beta_1 - \beta_2)(\beta_1 + \beta_2)}{2} - (\beta_1 - \beta_2)\right] = \frac{v_d}{2v_1}[2v_1\tau_n\xi - 2v_1\tau_n] = v_d\tau_n(\xi - 1);$$

2. $0 \leq \beta_1 < 1, \quad -1 \leq \beta_2 < 0;$

$$Z_0(\beta_1,\beta_2) = \frac{Z(\beta_1) - Z(\beta_2)}{2v_1} = \frac{Z(\beta_1) - Z(-\beta_2)}{2v_1} = \frac{v_d}{2v_1}\left[\left(\frac{\beta_1^2}{2} - \beta_1\right) - \left(\frac{\beta_2^2}{2} + \beta_2\right)\right]$$
$$= \frac{v_d}{2v_1}\left[\frac{(\beta_1 - \beta_2)(\beta_1 + \beta_2)}{2} - (\beta_1 + \beta_2)\right] = \frac{v_d}{2v_1}[2v_1\tau_n\xi - 2\xi] = v_d\xi(\tau_n - 1/v_1);$$

3. $1 \leq \beta_1 < 2, \quad -1 \leq \beta_2 < 0;$

$$Z_0(\beta_1,\beta_2) = \frac{Z(\beta_1) - Z(\beta_2)}{2v_1} = \frac{Z(\beta_1 - 2) - Z(-\beta_2)}{2v_1}$$
$$= \frac{v_d}{2v_1}[(2v_1\tau_n - 2)(\xi - 1) + (2v_1\tau_n - 2)] = \frac{v_d}{2v_1}(2v_1\tau_n - 2)\xi = v_d\xi(\tau_n - 1/v_1);$$

4. $1 \le \beta_1 < 2, \quad -2 \le \beta_2 < -1;$

$$Z_0(\beta_1, \beta_2) = \frac{Z(\beta_1) - Z(\beta_2)}{2v_1} = \frac{Z(\beta_1 - 2) - Z(\beta_2 + 2)}{2v_1}$$
$$= \frac{v_d}{2v_1}[(2v_1\tau_n - 4)\xi + 2\xi] = \frac{v_d}{2v_1}(2v_1\tau_n - 2)\xi = \frac{v_d}{v_1}\xi(v_1\tau_n - 1) = v_d\xi(\tau_n - 1/v_1);$$

5. $2 \le \beta_1 < 3, \quad -2 \le \beta_2 < -1;$

$$Z_0(\beta_1, \beta_2) = \frac{Z(\beta_1) - Z(\beta_2)}{2v_1} = \frac{Z(\beta_1 - 2) - Z(\beta_2 + 2)}{2v_1}$$
$$= \frac{v_d}{2v_1}[(2v_1\tau_n - 4)\xi - 2v_1\tau_n + 4] = \frac{v_d}{v_1}\xi(v_1\tau_n - 2)(\xi - 1) = v_d(\tau_n - 2/v_1)(\xi - 1);$$

6. $2 \le \beta_1 < 3, \quad -1 \le \beta_2 < 0;$

$$Z_0(\beta_1, \beta_2) = \frac{Z(\beta_1) - Z(\beta_2)}{2v_1} = \frac{Z(\beta_1 - 2) - Z(\beta_2)}{2v_1}$$
$$= \frac{v_d}{2v_1}[(v_1\tau_n - 1)(2\xi - 2) - (2\xi - 2)] = \frac{v_d}{v_1}(v_1\tau_n - 2)(\xi - 1) = v_d(\tau_n - 2/v_1)(\xi - 1);$$

7. $1 \le \beta_1 < 2, \quad 0 \le \beta_2 < 1;$

$$Z_0(\beta_1, \beta_2) = \frac{Z(\beta_1) - Z(\beta_2)}{2v_1} = \frac{Z(\beta_1 - 2) - Z(\beta_2)}{2v_1}$$
$$= \frac{v_d}{2v_1}[(v_1\tau_n - 1)(2\xi - 2) + (2\xi - 2)] = \frac{v_d}{v_1}v_1\tau_n(\xi - 1) = v_d\tau_n(\xi - 1)$$

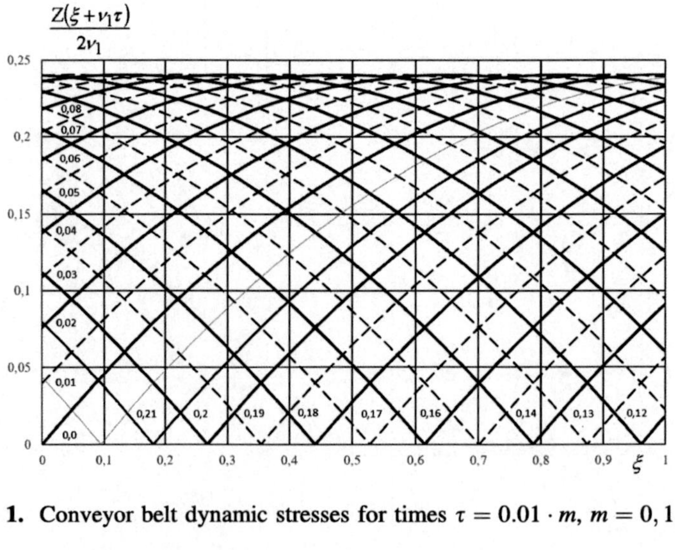

Fig. 1. Conveyor belt dynamic stresses for times $\tau = 0.01 \cdot m$, $m = 0, 1, 2...$

The function $Z_0(\beta_1, \beta_2)$ is linear to the variable ξ. It has an increasing and a decreasing section. The inflection point of the function determines the extreme value of the dynamic stress. Dynamic stresses arising in the forward $Z(\xi + \nu_1\tau)$ and backward perturbation $Z(\xi - \nu_1\tau)$ waves are shown in Figs. 1 and 2. Summary dynamic stresses are presented in Fig. 3.

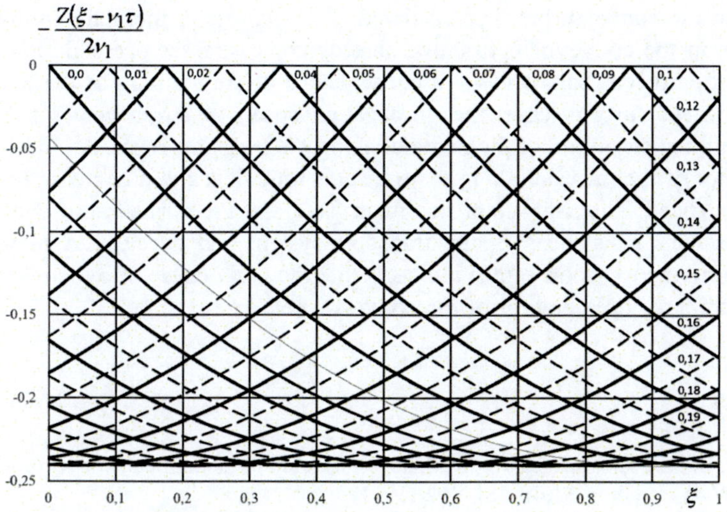

Fig. 2. Conveyor belt dynamic stresses for times $\tau = 0.01 \cdot m$, $m = 0, 1, 2...$

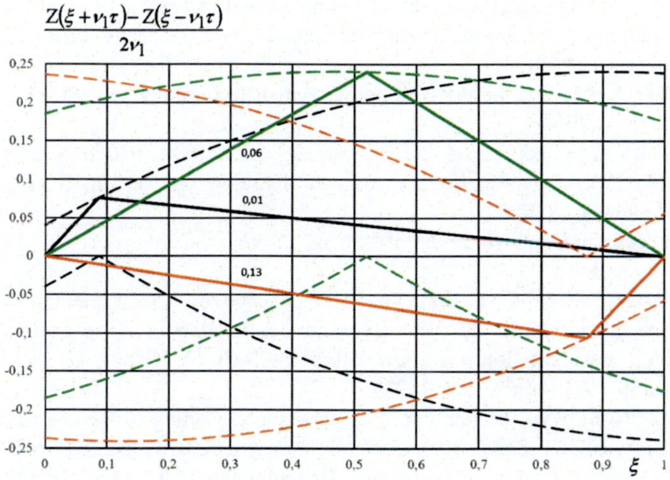

Fig. 3. Conveyor belt dynamic tensions for times $\tau = 0.01$, $\tau = 0.06$, and $\tau = 0.13$.

5 Conclusions

The mechanism of formation of dynamic stresses in transient operation modes of the conveyor belt was considered. The analysis of two main factors of the formation of conveyor belt stresses shows that the main cause of dynamic stresses is the acceleration or deceleration of the conveyor belt.

The mechanism of the formation of dynamic stresses during transient modes of operation of the conveyor belt is considered. The analysis of the main components of the stresses in the conveyor belt shows that dynamic stresses occur during unsteady operation of conveyor belts and are a consequence of the addition of the forward and backward waves. In this case, the propagation speed of stress waves substantially depends on the distribution of the material along the transport route. An increase in the linear density of the material along the transport route leads to a decrease in the wave propagation speed. The analysis of two main factors of the formation of conveyor belt stresses (primary resistances of the transport system and accelerated movement of material along the transport route) indicates that the main cause of dynamic stresses is the acceleration or deceleration of the conveyor belt.

References

1. Nordell, L.: The Channar 20 km overland – a flagship of modern belt conveyor technology. Bulk Solids Handling **11**(4), 781–792 (1991)
2. Siemens. Innovative solutions for the mining industry. www.siemens.com/mining. Accessed 21 Jan 2020
3. Conveyorbeltguide Engineering: Conveyor components. https://conveyorbeltguide.com/engineering.html. Accessed 09 Sept 2019
4. Kung, W.: The henderson coarse ore conveying system. A review of commissioning, start-up, and operation, bulk material handling by belt conveyor 5, society for mining, metallurgy and exploration, Inc. (2004)
5. Alspaugh, M.: Latest developments in belt conveyor technology. In: MINExpo 2004, New York, Las Vegas (2004)
6. https://citeseerx.ist.psu.edu/viewdoc/download?doi=10.1.1.620.5070&rep=rep1&type=pdf
7. Alspaugh, M.: Longer overland conveyors with distributed power. In: Overvand Conveyor Company, Lakewood (2005)
8. https://www.overlandconveyor.com/pdf/Longer_Overland_Conveyors_with_Distributed_Power.pdf
9. Semenchenko, A., Stadnik, M., Belitsky, P., Semenchenko, D., Stepanenko, O.: The impact of an uneven loading of a belt conveyor on the loading of drive motors and energy consumption in transportation. East. Eur. J. Enterp. Technol. **4/1**(82), 42–51 (2016). https://doi.org/10.15587/1729-4061.2016.75936
10. Pihnastyi, O.: Control of the belt speed at unbalanced loading of the conveyor. Sci. Bull. Nat. Min. Univ. **6**, 122–129 (2019). https://doi.org/10.29202/nvngu/2019-6/18.
11. DIN 22101:2002–08. Continous conveyors. Belt conveyors for loose bulk materials. Basics for calculation and dimensioning. [Normenausschuss Bergbau (FABERG), DIN Deutsches Institut für Normung e.v. Normenausschuss Maschinenbau (NAM)], p. 51 (2002)
12. Lauhoff, H.: Speed control on belt conveyors – does it really save energy? Bulk Solids Handling Publ. **25**(6), 368–377 (2005)

13. Pihnastyi, O., Khodusov, V.: Model of conveyer with the regulable speed. Bull. S. Ural State Univ. Ser. Math. Model. Program. Comput. Softw. **10**(4), 64–77 (2017). https://doi.org/10.14529/mmp170407
14. Reutov, A.: Simulation of load traffic and steeped speed control of conveyor. In: IOP Conference Series: Earth and Environmental, vol. 87 (2017). https://doi.org/10.1088/1755-1315/87/8/082041.
15. Pihnastyi, O., Khodusov, V.: The optimal control problem for output material flow on conveyor belt with input accumulating bunker. Bull. S. Ural State Univ. Ser. Math. Model. Program. Comput. Softw. (Bull. SUSUMMCS) **12**(2), 67–81 (2019). https://doi.org/10.14529/mmp190206
16. Wolstenholm, E.: Designing and assessing the benefits of control policies for conveyor belt systems in underground mines. Dynamica **6**(2), 25–35 (1980)
17. Marais, J., Mathews, E., Pelzer, R.: Analysing DSM opportunities on mine conveyor systems. In: Proceedings ICUE, 5th Industrial and Commercial Use of Energy Conference, pp. 5–10. Cape Town, South Africa (2008)
18. https://citeseerx.ist.psu.edu/viewdoc/download?doi=10.1.1.469.574&rep=rep1&type=pdf
19. Nordell, L., Ciozda, Z.: Transient belt stresses during starting and stopping: elastic response simulated by finite element methods. Bulk Solids Handling **4**(1), 99–104 (1984)
20. Alspaugh, M.: The evolution of intermediate driven belt conveyor technology. Bulk Solids Handling **23**(3), 168–173 (2003)
21. Lodewijks, G.: Determination of rolling friction of belt conveyors using rubber data: fact or fiction? Bulk Solids Handling **24**, 16–22 (2004)
22. Halepoto, I., Khaskheli, S.: Modeling of an integrated energy efficient conveyor system model using belt loading dynamics. Indian J. Sci. Technol. **9**(47), 1–6 (2016)
23. Pihnastyi, O., Khodusov, V.: Optimal control problem for a conveyor-type production line. Cybern.Syst.Anal. **54**(5), 744–753 (2018). https://doi.org/10.1007/s10559-018-0076-2. Springer. US [Springer Science+Business Media, LLC, 1060–0396/18/5405–0744]
24. Lodewijks, G.: Two decades dynamics of belt conveyor systems. Bulk Solids Handling **22** (2), 124–132 (2002)
25. Yang, G.: Dynamics analysis and modeling of rubber belt in large mine belt conveyors. Sens. Transducers **181**(10), 210–218 (2014)
26. Budishevsky, V., Sulima, A.: Theoretical Foundations and Calculations of Transport of Energy-Intensive Industries. Advertising and Publishing Agency DonSTU, Doneck (1999)

Dynamics of a Particle on a Movable Wavy Surface

Sergiy Pylypaka[1], Tatiana Volina[2(✉)], Iryna Hryshchenko[1], Iryna Rybenko[2], and Nataliia Sydorenko[3]

[1] National University of Life and Environmental Sciences of Ukraine, 15, Heroyiv Oborony Street, Kyiv 03041, Ukraine

[2] Sumy National Agrarian University, 160, Kondratieva Street, Sumy 40021, Ukraine
t.n.zaharova@ukr.net

[3] Sumy Regional Institute of Postgraduate Pedagogical Education, 5, Rymskogo-Korsakova Street, Sumy 40007, Ukraine

Abstract. The relative motion of a particle on a linear wavy surface with a sinusoid cross-section is considered in the article. The rectilinear generatrices of the surface are parallel to each other and parallel to the horizontal plane. The surface oscillates in such a way that all its points describe circles in horizontal planes. Numerical methods solve the differential equations of the relative displacement of a particle. The trajectories of sliding of a particle on a surface and graphs of its reaction were obtained. The relative trajectory of the particle after movement stabilization can be closed or periodic spatial curves. When the diameter of the circle during oscillations is equal to the period of the sinusoid, the trajectory of the relative motion of the particle can be a periodic curve. In this case, the particle moves in a direction, close to the transverse, overcoming depressions and ridge surfaces. In other cases, the trajectory of sliding is a closed spatial curve whose horizontal projection is close to the circle. A partial example of a surface is a plane, and the trajectory of relative motion (the sliding of a particle) is a circle. An analytical expression of its radius was found.

Keywords: Relative motion · Sinusoid · Circular vibrations · Differential equations · Kinematic parameters

1 Introduction

It is common knowledge that the rough plane is a universal structural element of many machines and devices for sifting and separating particles of technological material. Generally, the most studied is the movement of particles on a horizontal plane, which performs oscillating rectilinear or circular motions. A wavy surface with a sinusoid cross-section as a working surface will significantly change the trajectory of particle sliding. The mathematical description of such a movement will adjust accordingly. Besides, the sliding of a particle over a plane will be a partial case of sliding along a wavy surface when the amplitude of the sinusoid is zero. Doubtless, with vibrations of

V. Tonkonogyi et al. (Eds.): InterPartner 2020, LNME, pp. 196–206, 2021.
https://doi.org/10.1007/978-3-030-68014-5_20

a wavy surface, when all its points describe circles, the movement of the technological material changes significantly.

2 Literature Review

A great deal is being written and said about particles in contact with the surface and moving along it. They may have a different origin: mechanical particles [1], particles in the stream [2], particles of gas [3], or liquid [4]. Indeed, it is easier to study the motion of particles in comparison with the motion of bodies [5], because in the second case, it is necessary to take into account the forces of inertia from the rotation of the body. In some cases, these forces can be neglected or ignored at low angular velocities, for example, in the solving the problem of elimination of load oscillations on a flexible suspension during the swing of a boom crane with simultaneous minimization of dynamic forces in the transmission mechanism [6], in the process of developing of the physical model of the pneumatic classification process for detecting the rotating suspended layer and ensuring the frequency of loading and unloading of a pneumatic classifier [7] or in the process of designing the multistage centrifugal pumps [8]. A great deal is being written and said about the contact of gaseous substances with the working surfaces of the separation equipment, for example, during the modeling of the dynamic deflection elements of separation devices [9] or the gas-cleaning equipment [10], in the fibrous filter layer in the process of inertial filtering separation [11], curved channels of compressor equipment [12], modular separation devices [13], tubes [14], gear clutches [15], complex parts manufacturing [16]. Based on the foregoing, the purpose of our research is to investigate the regularities of the motion of material particles on a wavy surface, which performs circular translational oscillations in horizontal planes. The wavy surface is a cylindrical surface with a horizontal arrangement of rectilinear generatrices and a sinusoid cross-section.

3 Research Methodology

The parametrical equations of a cylindrical surface with the sinusoid cross-section and rectilinear generatrices, which are parallel to the axis OX, are as follows:

$$X = u; \qquad Y = v; \qquad Z = c \sin av, \tag{1}$$

where c is the amplitude constant, a is the frequency constant;

u, v are the independent variables of the surface, where u is the length of the straight line, v is the distance along the axis OY.

Our research explores the case when the cylindrical wavy surface performs translational oscillations in such a way that all its points describe circles (Fig. 1 shows the trajectories of movement of four points of the surface). Therefore, the absolute motion of the particle will be considered regarding the fixed OXYZ coordinate system.

Accordingly, if you attach the moving coordinate system to the surface, the oscillations of the axis of the moving and fixed systems will be parallel all the time.

This means that the absolute trajectory of the particle can be written as the sum of the transportable motion of the surface, points of which describe the circles, the relative motion of the point on the wavy surface is:

$$x = x_e + x_r; \qquad y = y_e + y_r; \qquad z = z_e + z_r, \tag{2}$$

where $x_e = x_e(t); y_e = y_e(t); z_e = z_e(t)$ is the trajectory of transportable surface motion as a function of time t;

$x_r = x_r(t); y_r = y_r(t); z_r = z_r(t)$ is the trajectory of the relative particle motion on the surface as a function of time t.

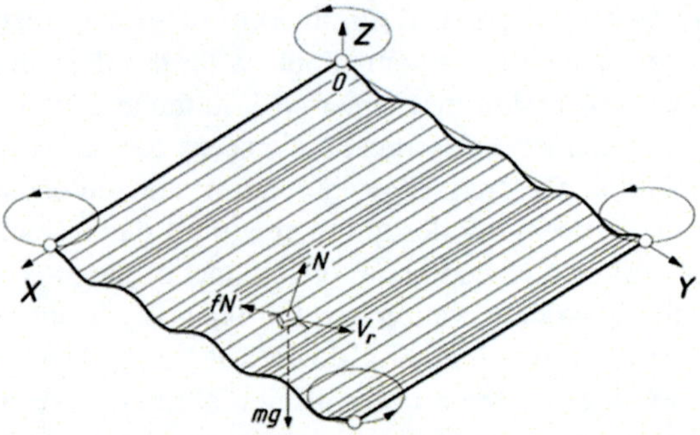

Fig. 1. The scheme circular wavy surface vibrations and forces applied to the particles.

For this particular case, we denote the radius of the circles along which the points of the cylinder move by r. As a consequence, the transportable motion of the points of the wavy surface will be described by the equations:

$$x_e = r \cos \omega t; \qquad y_e = r \sin \omega t; \qquad z_e = 0, \tag{3}$$

where ω is the angular rotation velocity of the points of the surface on the circles.

The point slides on the surface along a specific trajectory. The equation of the trajectory can be obtained by correlating independent variables v and u of the surface (1). Let write this relation in time t. That is, the coordinates of the particle on the wavy surface will be functions of time: v = v (t) and u = u (t). In this case, the relative motion of the particle will be described by the equations:

$$x_r = u; \qquad y_r = v; \qquad z_r = c \sin av. \tag{4}$$

Thus, summing the transportable (3) and relative (4) motions by the formula (2), one can obtain the equation of the absolute trajectory of the particle:

$$x = u + r \cos \omega t; \quad y = v + r \sin \omega t; \quad z = c \sin av. \tag{5}$$

In this regard, the dependencies v = v (t) and u = u (t), which describe the trajectory of relative motion (sliding of a particle along a wavy surface), are unknown functions, which should be found. After differentiating Eqs. (5) by time t, the projections of the absolute velocity of the particle can be found:

$$\dot{x} = \dot{u} - r\omega \sin \omega t\,; \quad \dot{y} = \dot{v} + r\omega \cos \omega t\,; \quad \dot{z} = ac\dot{v} \cos av. \tag{6}$$

Furthermore, by differentiating of expressions (6), we obtain the projections of absolute acceleration:

$$\ddot{x} = \ddot{u} - r\omega^2 \cos \omega t\,; \quad \ddot{y} = \ddot{v} - r\omega^2 \sin \omega t\,; \quad \ddot{z} = ac\left(\ddot{v} \cos av - a\dot{v}^2 \sin av\right). \tag{7}$$

At the same time, let us make the equation of motion in the form $m\overline{w} = \overline{F}$, where m is the mass of the particle, $\overline{w}$ is the vector of absolute acceleration, $\overline{F}$ is the resultant vector of the forces applied to the particle. Such forces (see Fig. 1) are the force of gravity mg (g = 9.81 m/s^2), the reaction N of the surface, and the friction force fN, which arises when the particle slides on the surface (f is the coefficient of friction). It is essential to project all forces on the axis of the OXYZ coordinate system.

One should not forget that the force of gravity is directed downwards, so its projections are written down:

$$\{0; \quad 0; \quad -mg\}. \tag{8}$$

Moreover, the surface reaction N is directed along the normal and is determined from the vector product of two vectors tangent to the coordinate lines of the surface. The projections of these vectors are the partial derivatives of Eqs. (1):

$$\begin{array}{lll} \frac{\partial X}{\partial v} = 0; & \frac{\partial Y}{\partial v} = 1; & \frac{\partial Z}{\partial v} = ac \cos av; \\ \frac{\partial X}{\partial u} = 1; & \frac{\partial Y}{\partial u} = 0; & \frac{\partial Z}{\partial u} = 0. \end{array} \tag{9}$$

Actually, after vector multiplication of vectors (9) and reducing the resulting vector to a single one projections of the normal vector to the surface will be written:

$$\left\{0; \quad -\frac{ac \cos av}{\sqrt{1 + a^2c^2 \cos^2 av}}; \quad \frac{1}{\sqrt{1 + a^2c^2 \cos^2 av}}\right\}. \tag{10}$$

Finally, the friction force is directed tangentially to the trajectory of the relative motion of the particle in the opposite direction of the velocity vector of the particle sliding. The direction of the vector is determined by the first derivatives of Eqs. (4):

$$\dot{x}_r = \dot{u}; \quad \dot{y}_r = \dot{v}; \quad \dot{z}_r = ac\dot{v} \cos av. \tag{11}$$

Thus, the geometric sum of the components (11) gives the value of the sliding speed of the particle on the wavy surface in relative motion:

$$V_r = \sqrt{\dot{x}_r^2 + \dot{y}_r^2 + \dot{z}_r^2} = \sqrt{\dot{u}^2 + \dot{v}^2(1 + a^2c^2 \cos^2 av)}. \quad (12)$$

As a result, a single vector of the tangent in projections on the OXYZ system axis is obtained by dividing the projections (11) by the value of the vector (12):

$$\left\{ \frac{\dot{u}}{V_r}; \quad \frac{\dot{v}}{V_r}; \quad \frac{ac\dot{v} \cos av}{V_r} \right\}. \quad (13)$$

Let us draw a vector equation $m\overline{w} = \overline{F}$ in the projections on the axis of the coordinate system, taking into account, that the friction force fN is directed along a single vector (13) in the opposite direction:

$$m(\ddot{u} - r\omega^2 \cos \omega t) = -fN\frac{\dot{u}}{V_r}; \quad m(\ddot{v} - r\omega^2 \sin \omega t) = -\frac{Nac \cos v}{\sqrt{1 + a^2c^2 \cos^2 av}} - fN\frac{\dot{v}}{V_r};$$
$$mac(\ddot{v} \cos av - a\dot{v}^2 \sin av) = -mg + \frac{N}{\sqrt{1 + a^2c^2 \cos^2 av}} - fN\frac{ac\,\dot{v} \cos av}{V_r}. \quad (14)$$

It can be seen, that system (14) includes three unknown functions: N = N (t), u = u (t) and v = v (t). Solving it regarding N, $\ddot{u}$ and $\ddot{v}$, the following expressions can be complicated:

$$\ddot{v} = \frac{1}{A^2}\left[r\omega^2 \sin \omega t + ac\left(a^2c\dot{v}^2 \sin av - g\right) \cos av\right] - \frac{f\dot{v}}{AV}B;$$
$$\ddot{u} = r\omega^2 \cos \omega t - \frac{f\dot{v}}{AV}B; \qquad N = \frac{m}{A}B, \quad (15)$$
where $A = \sqrt{1 + a^2c^2 \cos^2 av}$;
$$B = \left[g + ac\left(r\omega^2 \cos av \sin \omega t - a\dot{v}^2 \sin av\right)\right].$$

It should be noted that the system (15) cannot be integrated analytically. It is necessary to solve it through numerous methods. It is sufficient to substitute the found dependences v = v (t) and u = u (t) into Eq. (4) to obtain the relative trajectory of particle motion along the wavy surface, that is, the sliding trajectory. Besides, if the amplitude c = 0, the parametrical equations of the surface (1) are converted into the equations of the horizontal plane. In such a case system (15) is greatly simplified:

$$\ddot{v} = r\omega^2 \sin \omega t - fg\frac{\dot{v}}{\sqrt{\dot{u}^2 + \dot{v}^2}}; \ddot{u} = r\omega^2 \cos \omega t - fg\frac{\dot{u}}{\sqrt{\dot{u}^2 + \dot{v}^2}}; N = mg. \quad (16)$$

In Fig. 2,a the trajectory of the particle sliding on a plane at the following parameters is constructed by numerical methods: r = 0.05 m, ω = 10 s^{-1}, f = 0.3. After the transition process, the trajectory of the particle sliding is a circle which radius is smaller than the circle, which are described by all points of the plane.

In Fig. 2b, the graph of sliding velocity change V_r for a period of 3 s is constructed. It shows that after the transition process, when the trajectory of the particle relative motion becomes a circle, the sliding speed becomes constant. Subsequently, this fact can be used for the determination of an analytical description of the particle sliding after movement stabilization. Importantly, when $\sqrt{\dot{u}^2 + \dot{v}^2} = V_r = \text{const}$, system (16) decomposes into two independent differential equations:

$$\ddot{v} = r\omega^2 \sin \omega t - fg\frac{\dot{v}}{V_r}; \qquad \ddot{u} = r\omega^2 \cos \omega t - fg\frac{\dot{u}}{V_r}. \tag{17}$$

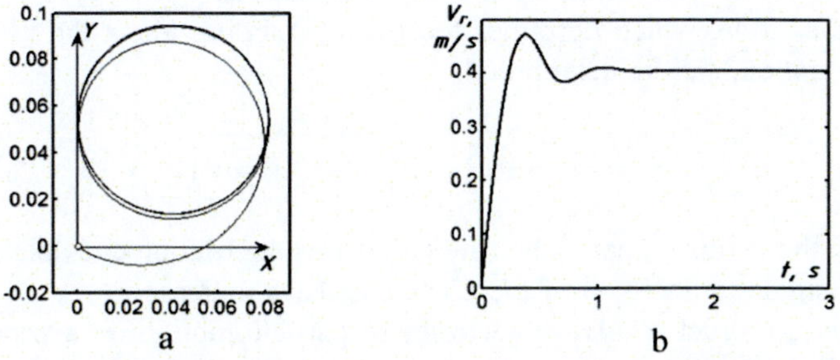

Fig. 2. a) the trajectory of the particle sliding; b) the graph of change of particle velocity.

In the particular case, when the integration constants are zero, the solutions are composed in the form:

$$\begin{aligned} v &= -\frac{rV_r\omega}{f^2g^2 + V_r^2\omega^2}(fg \cos \omega t + V_r\omega \sin \omega t); \\ u &= \frac{rV_r\omega}{f^2g^2 + V_r^2\omega^2}(fg \sin \omega t - V_r\omega \cos \omega t). \end{aligned} \tag{18}$$

By differentiating of Eq. (18), the expression for constant V_r by the given parameters of oscillatory motion can be found:

$$V_r = \sqrt{\dot{u}^2 + \dot{v}^2} = \frac{rV_r\omega^2}{\sqrt{f^2g^2 + V_r^2\omega^2}} \tag{19}$$

The following solution is obtained by solving (19) regarding V_r:

$$V_r = \frac{\sqrt{r^2\omega^4 - f^2g^2}}{\omega} \tag{20}$$

According to it, substitute (20) into (18) and finally receive the parametrical equations that describe the trajectory of the particle sliding on a plane in relative motion after its stabilization:

$$\begin{aligned} v &= -r\sin\omega t + \frac{fg}{r\omega^4}\left(fg\sin\omega t - \sqrt{r^2\omega^4 - f^2g^2}\cos\omega t\right); \\ u &= -r\cos\omega t + \frac{fg}{r\omega^4}\left(fg\cos\omega t + \sqrt{r^2\omega^4 - f^2g^2}\sin\omega t\right). \end{aligned} \quad (21)$$

Hence, if the plane is completely smooth (f = 0), the equations of relative motion (21) have the opposite sign of the equations of transportable motion (3), whereby the particle in absolute motion according to Eqs. (5) remains stationary. From Eqs. (21) the radius r_r of the circle, which describes the particle, sliding along the plane after stabilization of motion, can be determined:

$$r_r = \sqrt{u^2 + v^2} = r\sqrt{1 - \left(fg/r\omega^2\right)^2} \quad (22)$$

Undoubtedly, the radius r_r approaches the radius r as the friction coefficient f decreases or when the angular velocity ω of the plane oscillates increases.

Finally, let construct relative trajectories of particle motion on a wavy surface. It should be noted that the shape of the surface must be limited in some way by the parameters of its cross-section. For instance, a sinusoid with constant c = 0.005 (amplitude) and a = 62.8 (period T = 0.1) was taken, so the ratio of amplitude to period is 1:20. It was found out that at smaller ratios, a separation of the particle from the surface takes place. It is important to emphasize that this ratio is not fixed, it depends on the radius r of the circles on which the surface oscillates, the angular velocity ω, the coefficient of friction f.

4 Results

Let begin by taking a closer look at an example.

In the first case, the diameter of the circles of the transportable surface motion is equal to the period: 2r = T. In fact, this case is typical because the trajectory of a particle sliding along the surface is not a closed line (see Fig. 3). The sliding trajectory is a periodic spatial curve. Besides, the curved sliding of the particle extending in a direction close to the direction of the OY axis, and the particle in its motion overcomes depressions and ridges on the surface. The direction of the spreading of such motion can occur both in the direction of the OY axis and in the opposite direction. Clearly, the behavior of the particle depends on the point of contact with the surface.

In Fig. 3a, the trajectories of the relative motion of the particle on a limited surface area where the transition process occurs, with indicating the point of contact with the surface, are constructed. Figure 3b shows a horizontal projection of the trajectories of the particle relative motion, which moved for 8 s period.

When a particle slides on a wavy surface, its reaction N changes unlike the plane, where N = mg. In Fig. 4a, the graph of the surface reaction for a particle of mass m = 0.01 kg. It is constructed for one of the trajectories shown in Fig. 3b, and is characteristic for another trajectory. As the angular velocity of ω increases, the nature of the spreading of oscillations does not change, but at certain times the reaction of the surface becomes negative. That is, the direction of its vector changes to the opposite. In this case, the trajectory of particle motion would be real if it moved between two parallel surfaces. Since we have only one surface, at the moment when the reaction becomes zero, there is a separation of the particle from the surface, that is, its further trajectory becomes unknown. Thus, we further consider only such cases, when the reaction is positive.

In the second case, special case for the diameter of the circles of the transportable motion of the surface is smaller than the period: 2r < T is considered. In this case, after the transition period, the particle sliding trajectory becomes closed. Figure 5 shows the trajectories of the relative motion of a particle on a wavy surface at all previous parameters, including the incident points, with the change of only one parameter: r = 0.04 m.

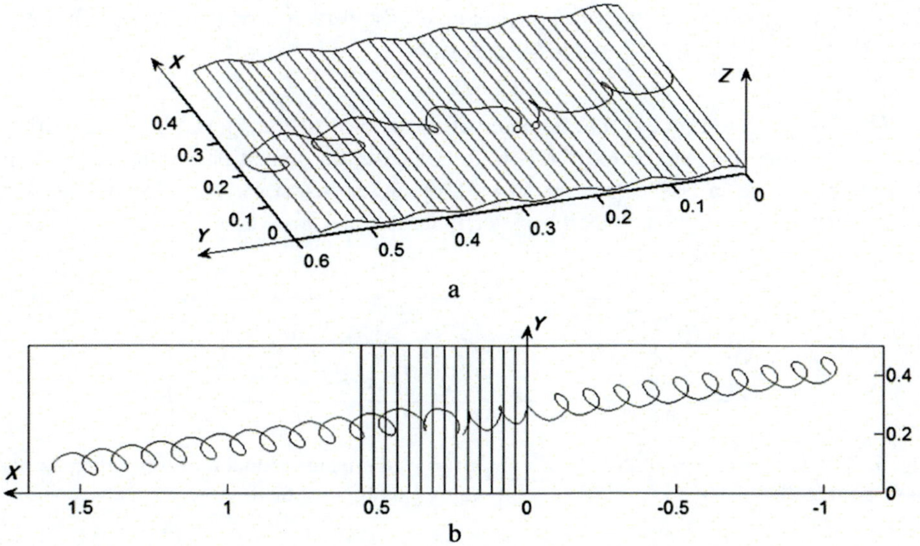

Fig. 3. The trajectories of the relative motion of the particle along the wavy surface at r = 0.05 m, ω = 10 s^{-1}, f = 0.3, s = 0.005, a = 62.8: a) the trajectory of particle motion on a limited area in axonometry; b) the trajectory of particle motion on a horizontal projection for 8 s period.

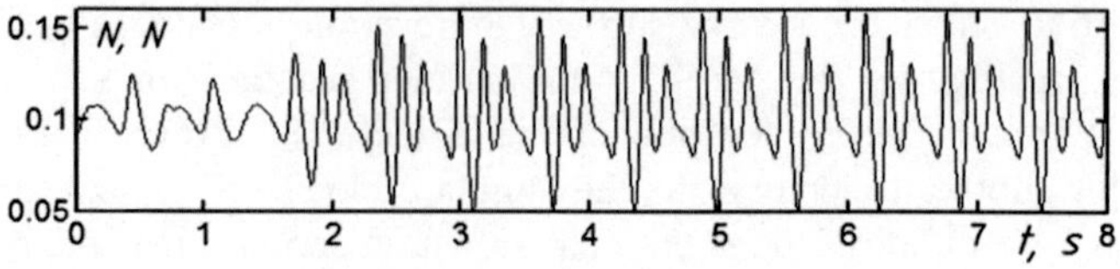

Fig. 4. Schedule of change of surface reaction on a particle sliding on a wavy surface.

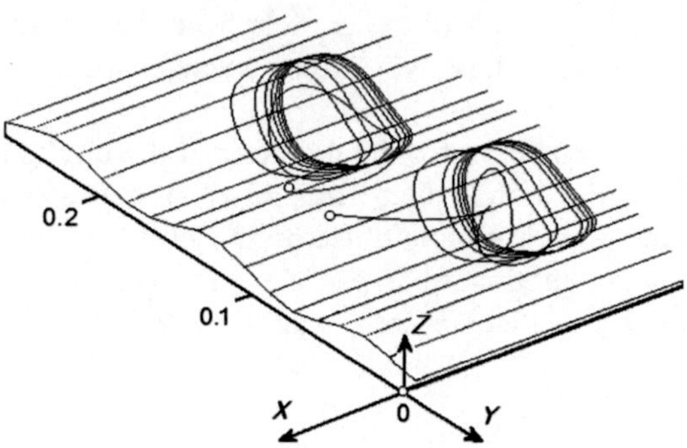

Fig. 5. The trajectories of the relative motion of the particle along the wavy surface at r = 0.04 m, $\omega = 10\ s^{-1}$, f = 0.3, c = 0,005, a = 62.8.

The Third Case. Finally, the diameter of the circles of transportable motion of the surface is higher than the period: 2r > T. In this case, the transition period can be long enough, as shown in Fig. 6 (at $\omega = 10\ s^{-1}$), or rather overcome to a closed trajectory with an increasing angular velocity of oscillations ω (at $\omega = 15\ s^{-1}$).

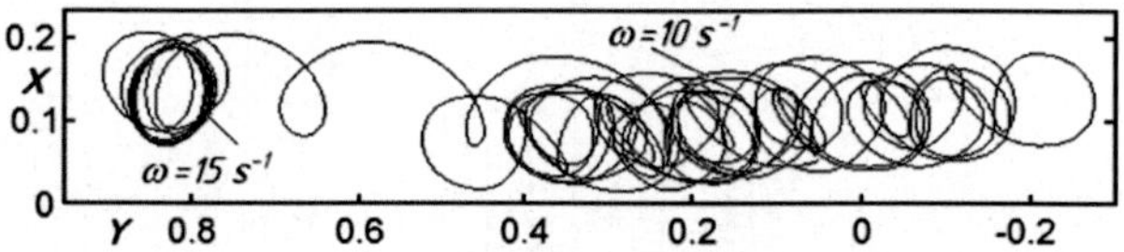

Fig. 6. The horizontal projection of the trajectories of the relative motion of the particle on the wavy surface at r = 0.06 m, $\omega = 10\ s^{-1}$, f = 0.3, c = 0.005, a = 62.8.

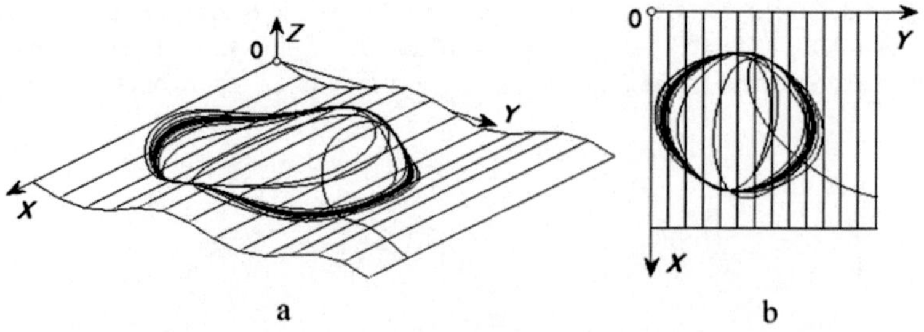

Fig. 7. The trajectory of the relative motion of the particle along the wavy surface at r = 0.07 m, $\omega = 10\ s^{-1}$, f = 0.3, s = 0.005, a = 62.8: a) axonometric image; b) horizontal projection.

As the radius r increases further, the particle moves along a closed curve after the transition period, and this curve increases in size according to the radius r of circular oscillations. In Fig. 7, the trajectory of the particle sliding with increasing radius r of oscillation circles to 0.07 m is constructed. With reducing the amplitude of the sinusoid, the wavy surface will approach the plane, and the closed sliding curve will approach the circle, which confirms the new theory.

To summarize, the considered cases fully reflect the possible trajectories of the relative motion of a particle on a wavy surface during its circular oscillations. The influence of the friction coefficient on the particle trajectory is not shown in the article. However, its value does not make any significant changes to the character of the particle sliding curves and is limited by the considered cases.

5 Conclusions

In circular oscillations of a wavy linear surface with a sinusoid cross-section, the relative trajectory of the particle after movement stabilization can be closed or periodic spatial curves. To avoid detachment of the particle from the surface, the set oscillation mode should take into account the shape of the surface and kinematic oscillation parameters. When the diameter of the circle, which describes all points of the surface during its oscillation equal to the period of the sinusoid, the trajectory of the relative motion of the particle can be a periodic curve. In this case, the particle moves in a direction close to the transverse, overcoming depressions and ridge surfaces. In other cases, the trajectory of sliding is a closed spatial curve, in which horizontal projection is close to the circle.

References

1. Isaev, Y., Semashkin, N.M., Nazarova, N.N.: Justification of the moving process of the seeds by the spiral-spiral working part. Bull. Ulyanovsk State Agric. Acad. **1**, 97–99 (2011)
2. Tauroa, F., Piscopiab, R., Grimaldi, S.: PTV-stream: a simplified particle tracking velocimetry framework for stream surface flow monitoring. CATENA **172**, 378–386 (2019). https://doi.org/10.1016/j.catena.2018.09.009
3. Batluk, V., Basov, M., Klymets, V.: Mathematical model for motion of weighted parts in curled flow. Econtechmod Int. Q. J. **2**(3), 17–24 (2013)
4. Liaposchenko, O., Pavlenko, I., Nastenko, O.: The model of crossed movement and gas-liquid flow interaction with captured liquid film in the inertial-filtering separation channels. Sep. Purif. Technol. **173**, 240–243 (2017)
5. Pylypaka, S., Klendiy, M., Zaharova, T.: Movement of the particle on the external surface of the cylinder, which makes the translational oscillations in horizontal planes. In: Ivanov, V., et al. (eds.) Advances in Design, Simulation and Manufacturing II. DSMIE-2019. Lecture Notes in Mechanical Engineering, pp. 336–345. Springer, Cham (2020). https://doi.org/10.1007/978-3-319-93587-4_35
6. Loveikin, V.S., Romerevych, Y.: Optimization of the swinging mode of the boom crane upon a complex integral criterion. J. Theoret. Appl. Mech. Sofia **49**, 285–296 (2019)

7. Lytvynenko, A., Yukhymenko, M., Pavlenko, I., Pitel, J., Mizakova, J., Lytvynenko, O., Ostroha, R., Bocko, J.: Ensuring the reliability of pneumatic classification process for granular material in a rhomb-shaped apparatus. Appl. Sci. (Switz.) **9**(8), 1604 (2019). https://doi.org/10.3390/app9081604
8. Pavlenko, I.: Static and dynamic analysis of the closing rotor balancing device of the multistage centrifugal pump. Appl. Mech. Mater. **630**, 248–254 (2014). https://doi.org/10.4028/www.scientific.net/amm.630.248
9. Pavlenko, I., Liaposhchenko, A., Ochowiak, M., Demyanenko, M.: Solving the stationary hydroaeroelasticity problem for dynamic deflection elements of separation devices. Vibr. Phys. Syst. **29**, 2018026 (2018)
10. Plyatsuk, L., Ablieieva, I., Vaskin, R., Yeskendirov, M., Hurets, L.: Mathematical modeling of gas-cleaning equipment with a highly developed phase contact surface. J. Eng. Sci. **5**(2), F19–F24 (2018). https://doi.org/10.21272/jes.2018.5(2).f4
11. Sklabinskyi V., Liaposhchenko O., Pavlenko I., Lytvynenko O., Demianenko M.: Modelling of liquid's distribution and migration in the fibrous filter layer in the process of inertial filtering separation. In: Ivanov, V. et al. (eds.) Advances in Design, Simulation and Manufacturing. DSMIE-2018. Lecture Notes in Mechanical Engineering, pp. 489–497. Springer, Cham (2019). https://doi.org/10.1007/978-3-319-93587-4_51
12. Liaposhchenko, O., Sklabinskyi, V., Zavialov, V., Pavlenko, I., Nastenko, O., Demianenko, M.: Appliance of inertial gas-dynamic separation of gas dispersion flaws in the curvilinear convergent-divergent channels for compressor equipment reliability improvement. In: IOP Conference Series: Materials Science and Engineering, vol. 233, no. 1, p. 012025 (2017). https://doi.org/10.1088/1757-899X/233/1/012025
13. Liaposhchenko, O., Pavlenko, I., Monkova, K., Demianenko, M., Starynskyi, O.: Numerical simulation of aeroelastic interaction between gas-liquid flow and deformable elements in modular separation devices. In: Ivanov, V. et al. (eds.) Advances in Design, Simulation and Manufacturing II. DSMIE-2019. Lecture Notes in Mechanical Engineering, pp. 765–774. Springer, Cham (2020). https://doi.org/10.1007/978-3-030-22365-6_76
14. Petinrin, M., Towoju, O., Ajiboye, S., Zebulun, O.: Numerical study of the effect of changing tube pitches on heat and flow characteristics from tube bundles in cross flow. J. Eng. Sci. **6**(2), E1–E10 (2019). https://doi.org/10.21272/jes.2019.6(2).e1
15. Shevchenko, S., Mukhovaty, A., Krol, O.: Gear clutch with modified tooth profiles. Procedia Eng. **206**, 979–984 (2017). https://doi.org/10.1016/j.proeng.2017.10.581
16. Ivanov, V., Dehtiarov, I., Pavlenko, I., Kosov, I., Kosov, M.: Technology for complex parts machining in multiproduct manufacturing. Manage. Prod. Eng. Rev. **10**(2), 25–36 (2019). https://doi.org/10.24425/mper.2019.129566

Synthesis of a Passive Pressure Reducing Valve Using Modified Kinematic Graphs

Ihor Sydorenko(✉), Vladimir Semenyuk, Valeriy Lingur, Liubov Bovnegra, and Olena Pavlyshko

Odessa National Polytechnic University, 1, Shevchenko Avenue, Odessa 65044, Ukraine
igs.ods@gmal.com

Abstract. Pressure reducing valves are widely used in the industry. The growing requirements for precise pressure control in industrial hydraulic or pneumatic systems, as well as automatic pressure control with maximum energy independence and autonomy of these operations, require the introduction of new design solutions. One of the effective tools for solving the synthesis of new devices is graph theory, which is a convenient tool for modeling the structural properties of technical systems and relationships between objects of various nature. The article presents the result of the synthesis of a fundamentally new passive pressure reducing valve based on its model in the form of a modified kinematic graph. The analysis of the obtained solution is carried out, and possible options for its optimization are shown. It is established that the presence of a mechanical control system in the synthesized pressure-reducing valve makes it possible to realize the target characteristics of the reduction of various types of nonlinearity. The results obtained indicate the possibility of widespread use of the proposed device.

Keywords: Graph theory · Kinematic graphs · Device structure · Pressure reducing valve · Mechanical control system

1 Introduction

Modern methods of synthesis of technical systems are based on a morphological approach, the main features of which can be considered: a genetic approach to the classification, description, evolution of development and prediction of various types of mechanical products as complex technical systems of natural and anthropogenic origin; multilevel morphological approach to the synthesis of complex technical systems; the aggregate-modular principle of the layout of technical systems; use of advanced information technologies and intelligent computer systems.

By analyzing the structures of controlled technical devices to identify elements of the same purpose and, as a result, the same properties, we can distinguish elements whose functional purpose is determined only by the specifics of the design. In this case, an additional study of these elements is required in terms of their functional expediency and possible substitution or association with existing structural elements. The analysis results create the basis for a scientifically based synthesis of fundamentally new

V. Tonkonogyi et al. (Eds.): InterPartner 2020, LNME, pp. 207–216, 2021.
https://doi.org/10.1007/978-3-030-68014-5_21

devices, the functionality of which can be significantly expanded. One of the effective tools for solving synthesis problems is graph theory, which provides a convenient tool for modeling the structural properties of systems and relationships between objects of different nature.

Graphs are widely used in the synthesis of technical systems as structural models, which represent their idealized representation in the form of circuits with lumped components. However, the specificity of the tasks solved with their help in most cases requires a corresponding change in the model itself in the form of a graph using a modification.

Proceeding from this, the development of graph theory, methods of its application for analysis and synthesis of structures of controlled technical devices is of scientific interest, and research conducted in this direction, in which the structure of a new real device is synthesized, should be considered relevant.

2 Literature Review

Structural synthesis is the main stage in the design of any mechanical system that ultimately determines the effectiveness of its application in various fields of technology [1].

One of the effective tools for solving the synthesis problem in graph theory, which provides a convenient tool for modeling structural properties of systems [2]. Graph theory is especially useful in modeling relationships between objects of diverse nature [3].

Graphs are widely used as structural models of physical systems that admit an idealized representation in the form of circuits with lumped components [4]. In the case of mechanical systems, their idealized representation in the form of graphs is widespread [5, 6]. Models of mechanical systems in the form of graphs are often used in a structural analysis of mechanisms [7, 8]. The results of the study, with such modeling, very often are a fairly solid basis for the synthesis of new mechanical systems [9, 10].

Considering that the graph-based model is a very flexible tool, its development for solving the problems of analysis and synthesis of mechanical systems taking into account their specificity is quite obvious [11, 12]. There is an approach to the analysis and synthesis of a self-governing mechanical system based on a modification of the kinematic graph of such a system [13].

The modification of the kinematic graph is that non-kinematic objects are added to the system of relations between objects of kinematic nature [14]. In this case, an idealized representation of the structure of a planar passive device, with one or more parameters that are subject to controlled change, is a modified kinematic graph (not bipolar), which characterizes the negative mobility of the device in question by the following expression.

$$W = 3(p_{\Sigma} - 1) - 2q_5 - q_4 - q_c \leq 0, \quad (1)$$

where p_{Σ} is the total number of graph vertices corresponding to the number of hard parts of the device; q_i is the number of ribs of the graph equal to the number of kinematic pairs of the i-th class; q_c is the number of ribs of the graph, which is equal to the number of non-kinematic bonds (elastic, dissipative, etc.) between the hard parts of the device, while this component does not affect the kinematic characteristics of the device and can be considered as a virtual kinematic pair of the 4-th class.

The possibility of kinematic control of one or another parameter of the device, which is subject to controlled change, can be judged by using the indicator for the modified kinematic graph used in the form of a cyclomatic number

$$\sigma = q - p_\Sigma + 1 \geq 2 \quad (2)$$

where q is the number of ribs of the graph that determine both kinematic and non-kinematic interactions; p_Σ is the total number of vertices of the graph.

With the use of this approach, more detailed information about the mechanism with mechanical control to be analyzed or synthesized is provided by the assembly matrix, which is a more detailed analog of the adjacency matrix [15].

3 Research Methodology

Using the presented approach, based on the modification of the kinematic graph, the analysis of the existing direct pressure reducing valve has been carried out [16]. A model in the form of a modified kinematic graph was obtained by a logical transition from a real construction to its kinematic diagram and then to a model in the form of a modified kinematic graph (Fig. 1).

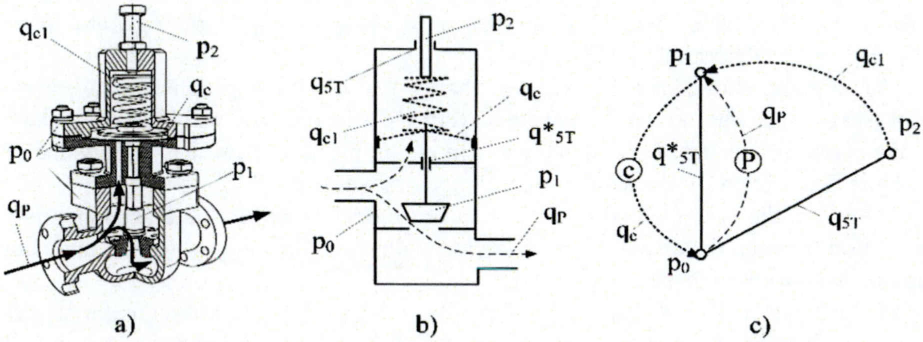

Fig. 1. Simulation of a direct pressure reducing valve: design (a); kinematic scheme (b); model in the form of a modified kinematic graph (c).

The model is a bipartite directed modified kinematic graph that determines the degree of mobility according to expression (1)

$$W = 3(3 - 1) - 2 \cdot 2 - 0 - 3 = -1 \leq 0.$$

The obtained result indicates that the presented design of the pressure reducing valve can be considered as the design of a flat passive device with one or more parameters, which is subject to controlled change.

The nature of the kinematic control for the properties of the device can be judged with the use of the indicator as a cyclomatic number. According to expression (2), it can be obtained the following one:

$$\sigma = 5 - 3 + 1 = 3 \geq 2$$

The cyclomatic number indicates that three cycles with one typical rib can be distinguished in the structure under consideration, namely: $p_0 - q^*_{5T} - p_1 - q_c$, $p_0 - q^*_{5T} - p_1 - q_P$ and $p_0 - q^*_{5T} - p_1 - q_{c1} - p_2 - q_{5T}$.

The selected cycles correspond to specific functional interactions of the device elements and compare to the equilibrium equation, which determines the operability of the device in question. The equilibrium equation of the device has the form

$$P \cdot S - c \cdot x = c_1(x_2 + x_1) \tag{3}$$

where P is the pressure acting on the membrane; S is the membrane area; c - rigidity of the membrane; c1 - spring stiffness; x is the deformation of the membrane under the action of the pressure force of the medium; x2 - pre-compression of the spring when installing in the device; x1 - adjusting the compression of the spring when adjusting the device.

Considering that the order of traversal of vertices for ribs q_P and q_c is different, the existing pair of cycles with a common rib q^*_{5T} ($p_0 - q^*_{5T} - p_1 - q_c$ and $p_0 - q^*_{5T} - p_1 - q_P$) determine the digraph.

This digraph shows that the pressure change (rib q_P) causing translational movement (rib q^*_{5T}) of the movable valve stem (vertex p_1) on one side will be compensated by the elastic resistance of the membrane (rib p_c) on the stem (vertex p_1), the value of which is determined by the same by continuous movement (rib q^*_{5T}).

Therefore, the presented part of the obtained model describes the main functional interaction between the elements of the device under consideration, which are the main and adequate to the left side of Eq. (3).

In this case, the cycle $p_0 - q_{5T} - p_2 - q_{c1} - p_1 - q^*_{5T}$ shows the functional interaction between the primary (vertices p_0, p_1) and additional (vertex p_2) elements of the device, which determines the change in its properties and is adequate to the right side of the Eq. (3).

Additional characteristics of the model used in the analysis are its adjacency matrix

$$M_R = \begin{pmatrix} 0 & 1 & 1 \\ 1 & 0 & 1 \\ 1 & 1 & 0 \end{pmatrix} \tag{4}$$

and an assembly matrix defining the type of interaction between the vertexes of the model

$$M = \begin{pmatrix} 0 & q_{5T}^{*}, q_P, q_c & q_{5T} \\ q_{5T}^{*}, q_P, q_c & 0 & q_{c1} \\ q_{5T} & q_{c1} & 0 \end{pmatrix}. \tag{5}$$

Based on the results of the analysis of the existing direct passive pressure-reducing valve, the study aims to synthesize a new direct pressure-reducing valve with a mechanical control system on its basis, to analyze the resulting solution and possible options for its optimization.

4 Results

A modified kinematic graph of the existing direct passive pressure reducing valve (Fig. 1c) and its assembly matrix (5) were taken as initial data for solving the synthesis of a new direct pressure reducing valve with a mechanical control system.

Fo the minimizayion of possible solutions in the synthesis, the following restrictions were adopted: the addition of new vertexes and the organization of new cycles in the model are not allowed, and the fulfillment of the conditions determining the functioning of the device in the form of expressions (1) and (2) is mandatory.

When solving the synthesis problem of new construction with the indicated limitations, it is quite natural to obtain a solution by replacing the virtual kinematic pairs of the 4th class with real kinematic pairs of the same class (Fig. 2a).

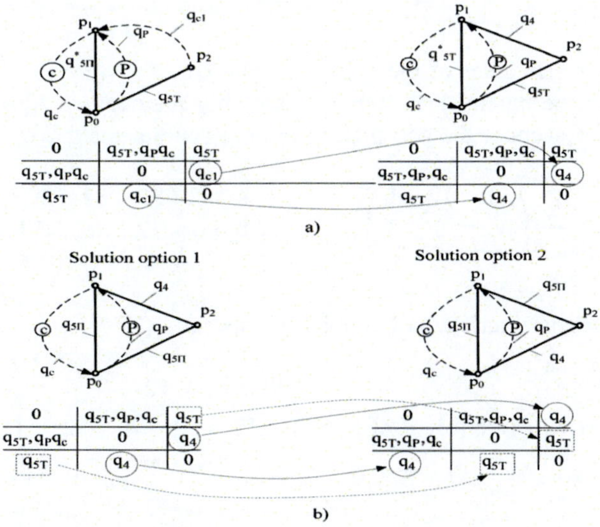

Fig. 2. Modified kinematic graphs and assembly matrices: search for a solution (a); identification of several solutions (b).

However, this, in turn, when conditions (1) and (2) are fully satisfied, determine several solutions (Fig. 2b). This, in turn, requires the selection of one optimal solution.

There are several methods for choosing the optimal solution in the case of several solutions obtained using graph models [2, 4, 6]. One of them is based on the principle of minimum energy of the optimal mechanical system, which is determined by the graph [6]. It is this approach that is proposed to be used to solve the synthesis problem under consideration.

For the obtained solutions I and II, it was proposed to assign to the ribs of the corresponding modified kinematic graphs weight coefficients that take into account the priority of the functional interaction, as well as structural or technological considerations (Table 1).

Table 1. The weights of the edges of the modified kinematic graphs in the synthesis of the device.

Name of indicator (Designation)	Weight coefficient	Note
Pressure (p)	4	Main parameter
Spring stiffness (c)	3	Auxiliary parameter
Kinematic pair of 5 classes, translational	2	Technologically simpler
Kinematic pair 4 classes	1	Technologically harder

The energy of the graph is defined as the sum of the absolute values of the adjacency matrix

$$E(M) = \sum_{i=1}^{n} \lambda_i, \tag{6}$$

where λ_i are the eigenvalues of the adjacency matrix.

To determine the eigenvalues, we use assembly matrices as adjacency matrices, which, taking into account the accepted weighting coefficients, have the form

$$M_I = \begin{pmatrix} 0 & 12 & 1 \\ 12 & 0 & 2 \\ 1 & 2 & 0 \end{pmatrix}, \qquad M_{II} = \begin{pmatrix} 0 & 12 & 2 \\ 12 & 0 & 1 \\ 2 & 1 & 0 \end{pmatrix}. \tag{7}$$

We introduce the characteristic polynomial for matrices (7)

$$\begin{aligned} P_I(\lambda) = Det(M_I - \lambda E) &= \begin{pmatrix} 0-\lambda & 12 & 1 \\ 12 & 0-\lambda & 2 \\ 1 & 2 & 0-\lambda \end{pmatrix}, \\ P_{II}(\lambda) = Det(M_{II} - \lambda E) &= \begin{pmatrix} 0-\lambda & 12 & 2 \\ 12 & 0-\lambda & 1 \\ 2 & 1 & 0-\lambda \end{pmatrix}. \end{aligned} \tag{8}$$

The characteristic polynomial for both matrices (8) can be determined according to the expression

$$P_{I(II)}(\lambda) = \lambda^4 - 4\lambda^2 = 0 \quad (9)$$

The roots of the characteristic polynomial (9) for each of the matrices (8): $\lambda_{I1} = -1{,}41$; $\lambda_{I2} = -1{,}41$; $\lambda_{I3} = 1{,}41$; $\lambda_{I4} = 1{,}41$. $\lambda_{II1} = -2$; $\lambda_{II2} = -0{,}00018$; $\lambda_{II3} = 0{,}00018$; $\lambda_{II4} = 2$. Corresponding graph energies by expression (6): $E(M_I) = 5{,}64$, $E(M_{II}) = 4{,}00036$.

Thus, according to the accepted selection conditions, synthesis option II is optimal, since the energy of its graph is the smallest.

Further work with the optimal modified kinematic graph allowed us to obtain a kinematic diagram of the device for its further design implementation (Fig. 3).

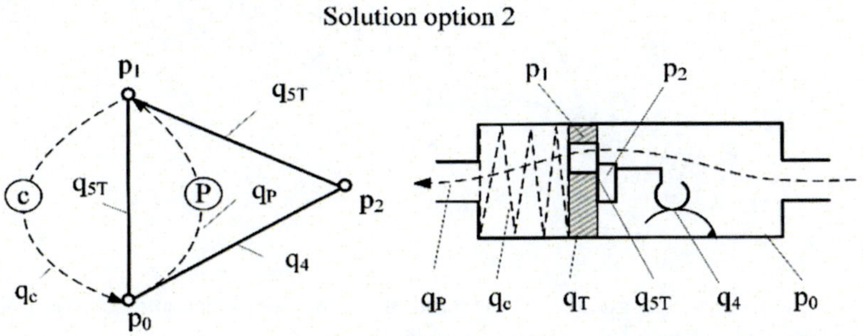

Fig. 3. Device synthesis: modified kinematic graph (a); kinematic scheme (b).

The design implementation of the obtained kinematic scheme in the form of a 3D model of the device (the designation of the device elements on the kinematic scheme and the 3D model coincide) was performed using the CAD part of the Autodesk Inventor Series package (Fig. 4).

The synthesized pressure reducing valve operates as follows. The housing 1 is oriented in the pipeline concerning the direction of movement of the medium using inlet 2 and outlet 3 nozzles. The pressure of the medium at the inlet pipe 2 leads to the appearance of a pressure force on the composite piston 4, causing it to move along the housing by an amount x, depending on the stiffness of the spring 12. When moving the piston 4 along the housing 1, the shutter 7 located inside it moves in the radial direction by ±y (x) due to the contact of its axis 8 with the curved groove of the guide plate 9. The magnitude of the shift of the shutter 7 in the radial direction ±y (x) depends from the shape of the curved groove of the guide plate 9 and determines the total or partial overlap of the throttle holes (change in their cross-sectional area) and thereby the pressure on the outlet pipe 3.

The change in pressure at the inlet pipe 2 leads to a corresponding displacement of ±x of the piston 4 along the housing 1, thereby causing an autonomous, determined by the shape of the curved groove of the guide plate 9, displacement of ±y (x) of the valve

7 in the radial direction, which determines - the corresponding change in the area of the throttle holes of the piston 4 and the pressure on the outlet pipe 3.

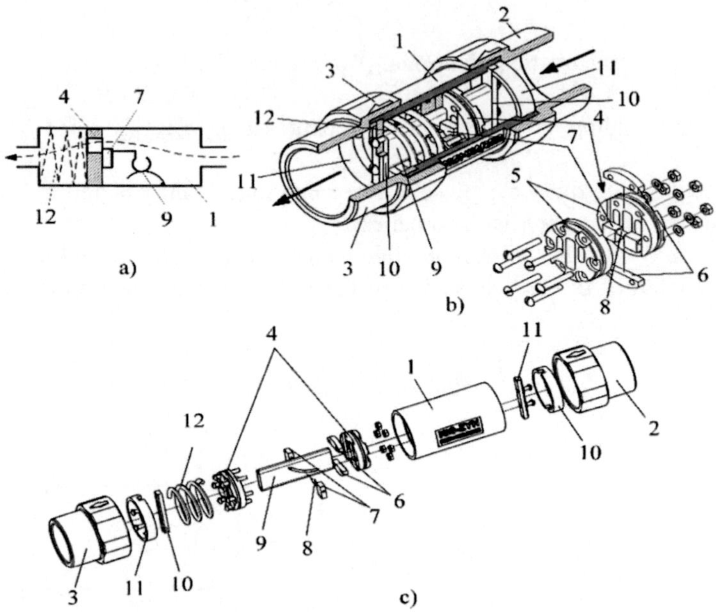

Fig. 4. Constructive implementation of the synthesized device: kinematic scheme (a); design implementation of the device (b); device assembly diagram (c).

The presence of 3-D models of the constituent elements of the synthesized device made it possible to pre-select the material for each of them, clarify their shape, and also indicate the nomenclature of standard products that can be used to develop the design of this device.

Checking the kinematics of the developed device using the Autodesk Inventor Series package showed that the change in the throttle opening area is closely related to the geometric parameters of the elements that determine the class 4 kinematic pair. Implementation of the control system in the form of a cam mechanism with a kinematic closure suggests the possibility of synthesizing the device's operating characteristics various forms of nonlinearity. Moreover, the form of nonlinearity will be closely related to the nonlinearity of the cam, which in this case, defines an algorithm for mechanically controlling the properties of the device.

5 Conclusions

The research results show that the adopted approach, based on the modification of kinematic graphs, as applied to the synthesis of hydraulic equipment elements, should be considered successful since the results obtained make it possible to obtain a prototype device with the required functional interaction between its constituent elements.

It was found the following results. The synthesis of devices based on modified kinematic graphs allows several options for possible solutions. Thus, in the synthesis of a new device, two possible solutions were obtained, although strict restrictions were imposed on its graph model for adding new poles and organizing new cycles.

If the restrictions on the number of elements of the graph model are removed, then the number of possible solutions can increase significantly, which will require the development of a methodology for choosing the optimal solution. In this case, it is advisable to build the choice of the optimal solution on a technique that is based on the principle of the minimum energy of the optimal mechanical system, which is determined by the kinematic graph.

To select the optimal solution, the principle of minimum energy of the optimal mechanical system determines the kinematic graph. It is advisable to use the weighting coefficients of its edges based on priority in descending order: the main variable parameter is an auxiliary parameter - technological complexity.

It should be noted that the original device synthesized as an example, in the form of a passive direct pressure reducing valve with a mechanical control system, is self-governing. Further studies related to the determination of the optimal geometric parameters of the synthesized device and the kinematic parameters of its mechanical control system will determine the area of its application and the range of the technological problems that can be solved with its help.

References

1. Kraynev, A.F.: Mechanics (the art of building) machines. Fundamental Dictionary. Engineering, Moscow (2014)
2. Schlick, T.: Adventures with RNA graphs. Methods **143**, 16–33 (2018)
3. Soni, A.: Structural analysis of two general constraint kinematic chains and their practical application. J. Eng. Ind. **93**, 231–238 (2010)
4. Periyasamy, B.: Solution of linear differential equations in chemical kinetics through flow graph theory approach. J. Taiwan Inst. Chem. Eng. **57**, 9–17 (2015)
5. Dao, T.-S., McPheeю, J.: Dynamic modeling of electrochemical systems using linear graph theory. J. Power Sources **196**, 10442–10454 (2011)
6. Yang, F., Deng, Z., Tao, J., Li, L.: A new method for isomorphism identification in topological graphs using incident matrices. Mech. Mach. Theory **49**, 298–307 (2012)
7. Bendsoe, M.P., Sigmund, O.: Topology Optimization: Theory, Methods, and Applications. Springer, Berlin (2013)
8. Yan, H.-S., Hung, C.-C.: Identifying and counting the number of mechanisms from kinematic chains subject to design constraints. ASME J. Mech. Des. **128**(5), 1177–1182 (2016)
9. Chiu, T.: An improved algorithm for the construction of generalized kinematic chains. Mech. Mach. Theory **78**, 229–247 (2014)
10. Ding, H., Zi, B., Huang, P., Kecskeméthy, A.: The whole family of kinematic structures for planar 2- and 3-DOF fractionated kinematic chains. Mech. Mach. Theory **70**, 74–90 (2013)
11. Simoni, R., Carboni, A.P., Martins, D.: Enumeration of kinematic chains and mechanisms. Proc. Inst. Mech. Eng. Part C J. Mech. Eng. Sci. **223**(4), 1017–1024 (2009)

12. Tian, C., Fang, Y., Ge Jeffrey, Q.: Structural synthesis of parallel manipulators with coupling sub-chains. Mech. Mach. Theory **118**, 84–99 (2017)
13. Mirafzal, S.M.: More odd graph theory from another point of view. Discrete Math. **341**, 217–220 (2018)
14. Martins, D., Carboni, A.P.: Variety and connectivity in kinematic chains. Mech. Mach. Theory **43**(10), 1236–1252 (2008)
15. Sydorenko, I.: Passive Vibration Isolation Device with Elements of Active Systems. Palmarium Academic Publishing, Saarbrücken (2011)
16. Sydorenko, I.: Analysis of the structure of the relaxation shock absorbers and modified the traditional kinematic graphs. ASB Mech. Mach. Tools Mater. **1**(26), 88–91 (2014)

Reverse Engineering Based on Information Model

Volodymyr Tigariev(✉), Vira Salii, Olga Rybak, Yuliia Barchanova, and Oleksii Lopakov

Odessa National Polytechnic University, 1, Shevchenko Avenue, Odessa 65044, Ukraine
volodymyr_t@ukr.net

Abstract. This article is focused on the application of the information model during the process of reverse engineering to advance it both as part of theoretical research and for practical purposes. Reverse engineering technology is presented in detail. A complete solid 3D model of a workpiece is created based on a mathematical model consisting of a 3D point cloud produced using the results of 3D scanning. Reverse engineering of the workpiece is carried out in the Siemens Solid Edge CAD system. The quality of the final 3D model depends on the number and density of facets in an obtained mesh model of an object. The mesh model is adjusted and optimized using the tools of the Siemens Solid Edge CAD system. The mesh model is used as a fundament for making a virtual prototype of the workpiece. After the steps for creating a suggested information model are analyzed and examined in detail, they are applied to the process of reverse engineering of an implant for thigh bone reinforcement. As a result, the information model undergoes significant changes, particularly for the shape analysis and stress simulation stages. Along with creating a virtual prototype of the real object on the example of the thigh bone, further designing of an anatomically shaped implant is also considered. For the improvement of the mesh model and the design of a solid model with sufficient quality, MeshLab and Fusion 360 software was applied. Required stress simulations are carried out in ANSYS applications. Thus, applying an information model for creating an object provides advantages in the reverse engineering process.

Keywords: 3D scanning · Mesh model · Solid model

1 Introduction

An information model includes all the information necessary for each step of creating an object with the use of reverse engineering technology [1–3]. Reverse engineering concept can be applied in mechanical engineering for producing specific workpieces, in medicine for designing bones of a skeleton, in architecture to restore parts of the buildings and decoration, for cultural heritage preservation, etc. There are frequent situations that require applying reverse engineering as it is optimal or even the only way to solve challenging problems in the technological progress of the modern world [3]. As part of mechanical engineering, reverse engineering is often used for the

V. Tonkonogyi et al. (Eds.): InterPartner 2020, LNME, pp. 217–226, 2021.
https://doi.org/10.1007/978-3-030-68014-5_22

generation of digital copies of different models in the process of searching the form of completely new objects (e.g., in the design, modernization, and repair of equipment, as well as in cases when the required workpiece is very difficult or impossible to obtain). The reverse engineering method is applied to create a computer 3D model, according to which a new object can be easily designed with the use of additive technologies [1, 2, 4].

2 Literature Review

The reverse engineering process starts with 3D scanning, which enables the generation of a mathematical image consisting of a 3D point cloud. This point cloud is processed in the CAD system and restructured with NURBS into sets of polygons [5–7]. The result of 3D scanning is a polygonal model that is never perfect (see Fig. 1) [5, 7], regardless of the ideal shape of the scanned object and advanced technological opportunities of a 3D scanner. That happens because the obtained model consists of a large number of single surfaces that are not parallel or perpendicular to one another. Besides, there are problems with inner "pockets" scanning [5, 6] and surface closure [7]. Thus, the final objective of 3D scanning is to get a high-level model of the general form of an object.

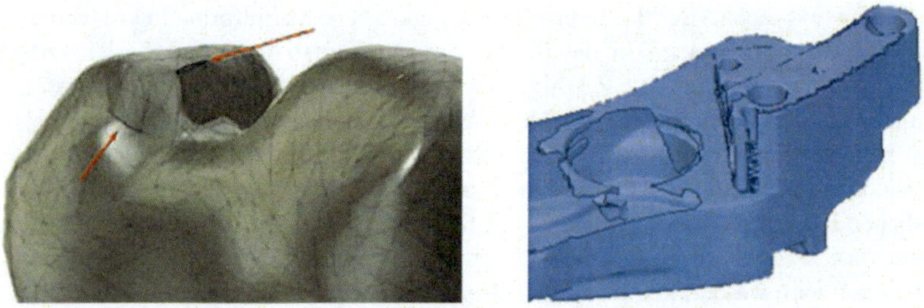

a b

Fig. 1. Examples of disadvantages of the polygonal model in the scanning process: a – open polygons; b – some surfaces are fully or partially missing.

The presented research aims at the development of an information model and its application into the process of reverse engineering of a real object. Such an approach is expected to reduce the time required to analyze, design, and carry out stress testings of the model.

Information model development requires gathering data, which describes essential properties and parameters of an object, correlations between them, and allows simulating possible changes of state of an object [8–10]. In this study information model of the reverse engineering process is presented. The practical application of the suggested technology is considered for a typical workpiece and a skeleton bone that will be used in the further development of adaptive parametric models.

3 Research Methodology

In many scientific works, the research is aimed at the development of an information model of an object. However, these studies often lack sufficiently accurate analytical solutions to the problems of reducing the time spent on designing, testing, and analyzing stress distribution [9, 10]. As a result of studying this question, the information model of a virtual object creating (see Fig. 2) was developed and suggested not only as part of theoretical research but also for practical use.

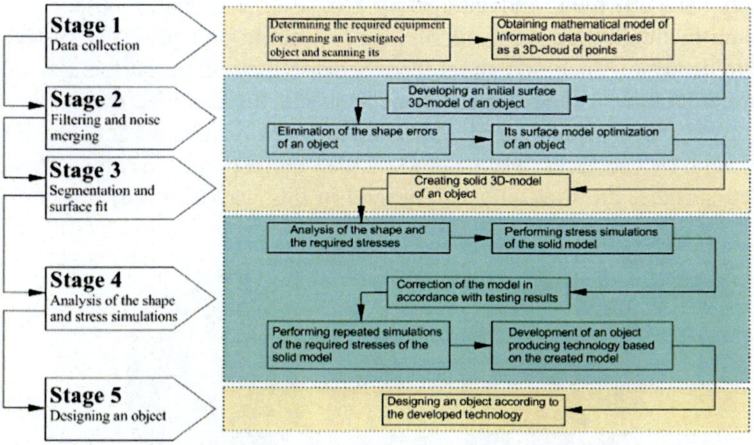

Fig. 2. The information model of designing an object using reverse engineering.

Designing an object is thoroughly considered using the modern technology of reverse engineering at Siemens Solid Edge. For obtaining the input data at the first stage of information model development, a real workpiece is scanned (Fig. 3a) using a modern 3D scanner. As a result, we obtain a sufficiently high-quality mathematical model consisting of a 3D point cloud. At the second stage of information model development, the received geometrical dimensions are imported into the Solid Edge software application as a mesh model (Fig. 3b). The mesh model consists of triangular polygons called facets (Fig. 3 c).

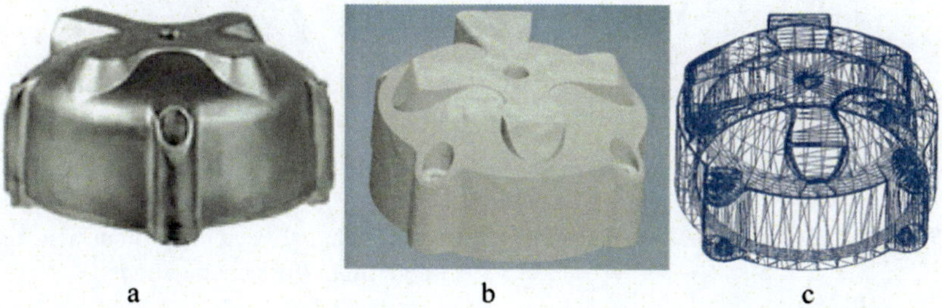

Fig. 3. Photo of the original workpiece (a) and the imported mesh model, which were obtained by 3D scanning (b, c).

The quality of the 3D model depends on the quantity and density of facets [3–6]. However, there are still disadvantages on the surface of the obtained model, such as open polygons and errors during inner pockets scanning [5]. Thus, the surface should be optimized to eliminate these disadvantages and reduce the number of polygons in homogeneous areas. For making the surfaces completely cylindrical and concentric as required by the construction [3–6], all the holes on the surface are also removed.

At the stage of creating a solid model of a workpiece, the mesh model is firstly investigated to identify areas that appear to be planar, cylindrical, conical, spherical. In Solid Edge, flat polygon collections are colored in yellow, cylindrical in pale blue, spherical in the dark blue, and conical in red. Both automatic surface recognition (Fig. 4a) and manual mode are available. As for purple polygons, the system has not been able to determine their analytical form because they often represent a "boundary" between different types of the model. These areas can be identified manually as well as the areas of the model which are identified incorrectly. So, the borders are highlighted in purple, and the surfaces are painted in the corresponding color. Thus, all the unidentified surfaces are manually determined in this way (Fig. 4b) [3–6].

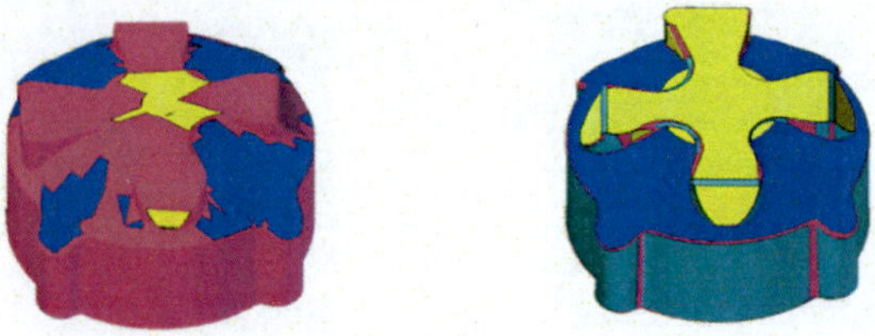

a b

Fig. 4. Mesh model after automatic surfaces recognition (a) and identification of all areas on the body of the model with the applied boundaries between them (b).

The next step of designing a solid model of the workpiece is stretching the required surfaces from the mesh model and extending them to a complete intersection. It is possible to stretch surfaces both in automatic (Fig. 5) and manual mode (Fig. 6).

Fig. 5. Surface stretching in automatic mode.

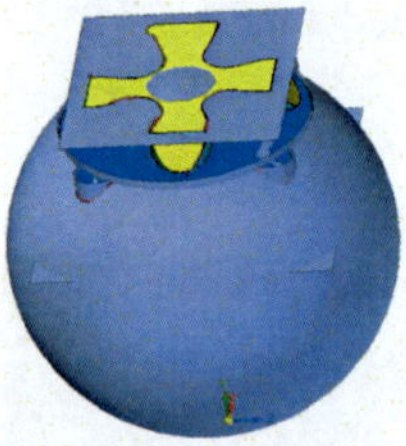

Fig. 6. Manual stretching of surfaces required to determine the general shape.

The applied software allows us to create a solid model that is formed by the nested set of selected surfaces (Fig. 7). When the mesh model is imposed on the solid body of a workpiece, it is possible to verify how closely these two objects correspond to each other (Fig. 8). All other surfaces in the mesh model are stretched and expanded the same way. After solid model components are formed, all the structural elements, such as holes, threaded holes, chamfers, and joints, are completed (Fig. 9).

At the stage of the shape analysis and required stress simulations, material properties are attributed to the created solid model, and a virtual workpiece is fixed at the mounting points. The stresses are applied according to the subassembly documentation in order to determine their effect on the model. The obtained results of simulations are analyzed to indicate the sufficient and necessary strength of the designed product. Then the model can be transmitted to a 3D printer, or it can be machined for further application.

Fig. 7. The solid model formed of selected mesh surfaces.

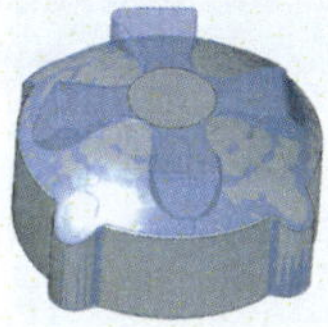

Fig. 8. Mesh model imposed on the solid body of a workpiece.

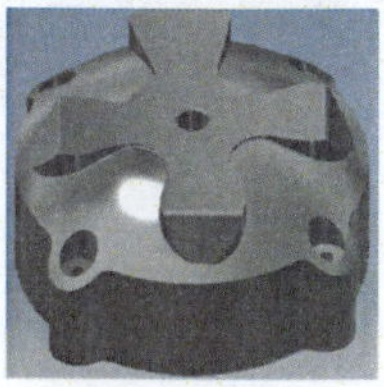

Fig. 9. The result of reverse engineering.

4 Results

Along with the available and verified information model of the workpiece, the presented method might be used for other practical applications. As an example, reverse engineering technology is considered in the process of creating an implant for thigh-bone reinforcement [7, 11, 12]. Thus, stages of information model development (Fig. 10) are adapted according to the specificity of the process of implant design. They are based on computed tomography scanning of a limb and imply other operations compared to mechanical engineering.

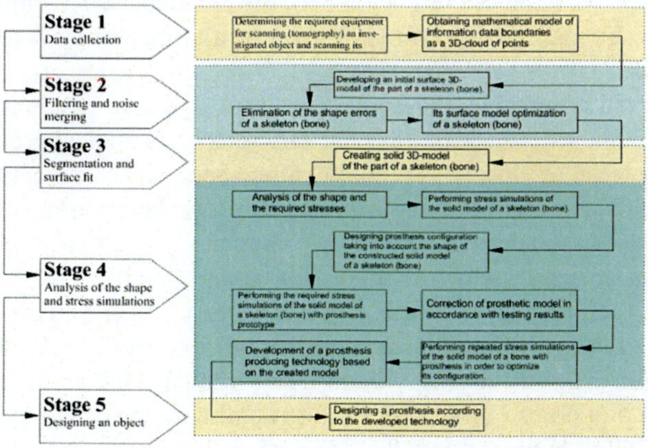

Fig. 10. Information model for creating a part of a skeleton using a reverse engineering approach.

The first stage of information model development is data collection that consists of several steps. Images obtained using computed tomography scan methods provide a basis for a femur design. After being analyzed with In Vesalius 3.1 computer software, these images form a 3D point cloud (Fig. 11) [12], which describes edges of the femur. The point cloud is a basis for the virtual model development of this bone in the computer simulation system. Since the mathematical model of a thighbone is created as a virtual object, it is imported into Fusion 360 software for further editing [7, 11, 12].

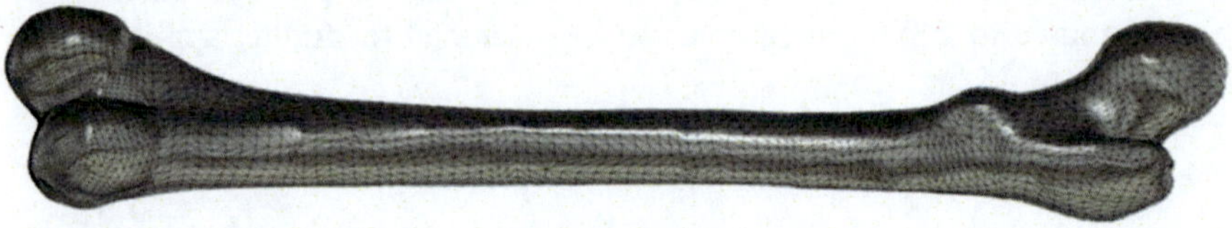

Fig. 11. Model of a bone obtained using a 3D scan.

The next step in the process of 3D editing is a point processing. During this stage, it is necessary to reduce noise to decrease the number of points of the scanned data in uniform areas. Model simplification allows us to significantly reduce the time of the model processing as well as make it possible to control the accuracy of geometrical constructions. Simplification was accomplished in MeshLab software, which enables us to carry out different types of editing of three-dimensional models. The model consists of about $6.5 \cdot 10^3$ polygons. This number of polygons makes a calculation of the model significantly more complicated and slows down further research, especially in the reinforced state. The number of polygons should be reduced by six times to prevent holes forming in the created model [7, 11–17].

At the stage of segmentation, the mesh model is built of triangles controlled according to the requirements of the program. For obtaining the model of acceptable quality, the density of the mesh model should be adjusted. The reason for that is inaccuracies in the form of open polygons that occur when the object is being created using 3D scanning. With the use of Fusion 360, the deficiencies are eliminated, and the density of the mesh model is changed [12]. As a result, the complete mesh model of a bone is finalized.

For receiving an acceptable and correct result of stress simulations, the computer model of a bone should consider physical characteristics that approximate the real features of the material as closely as possible. Since bone tissue is a complicated multi-layered bio-composite material, it is simplified to a two-layer computable model. An outer layer is a hard tissue determined by its homogeneity and density. The inner layer is cancellous bone material, which is more porous and flexible.

Designing of the outer layer of a bone requires to transform mesh-body into a B-Rep model (Fig. 12) [12]. Solid models in Fusion 360 are presented with a set of surfaces that form a tightly closed volume. This model enables us to calculate the volume, physical properties of the object and to determine the center of mass. Creating an inner layer of thigh bone requires the offset of each polygon in the model by about 10%. For the determination of the center of the model, Scale tool with appropriate settings and the center of mass search instrument is applied. After that, the inner layer is separated from the outer one with the help of a relevant function. This function uses settings that allow selecting the main and instrument bodies (Fig. 13) [11, 12].

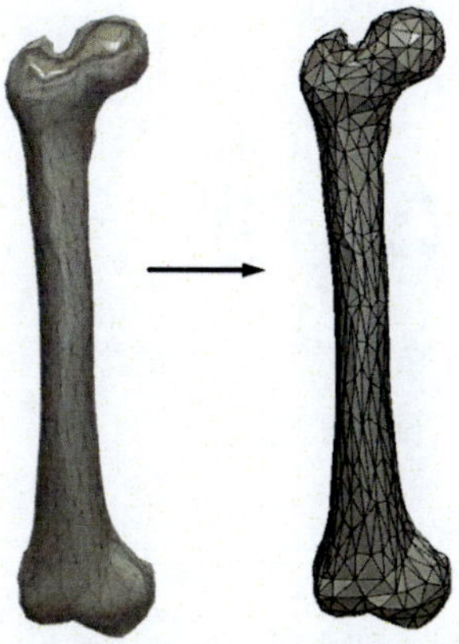

Fig. 12. Demonstration of applying transformation of the Mesh model into B-Rep.

Fig. 13. Result of applying the function of combining bodies.

The last stage of 3D editing is a phase of designing. At this stage, the model is transformed into NURBS surfaces that represent shape segmentation. All of the segments fit closely with one another, which provides shape smoothness. After this stage is complete, the NURBS surface (Fig. 14) [11] can be exported into STL, IGES, or STEP file depending on the type of the program [13–17].

Fig. 14. Created a solid model of the thighbone.

Since implants and their effect on stress distribution are considered in motion, analysis of shape and required stresses for the created solid model of a bone is provided for the whole proximal, including greater trochanter and the neck of the femur. Calculation of stress distribution is performed in Fusion 360 software, but ANSYS application might also be used for its finalization and correction. Required stress simulations of the solid model of a bone are carried out in three directions (for horizontal, vertical, and rotational force application). Results received during these tests are shown in Fig. 15 [7, 11, 12].

Proximal femur reinforcement is performed to strengthen bio-composite material and prevent hip fractures of the femur for patients belonging to the high-risk group. The shape of the implant is designed in Fusion 360 software based on the constructed solid model of the thigh bone. The device consists of bent flexible trihedral spokes, inner guide bushes, diaphyseal plate, and monocortical screws. An optimal model of an implant that meets all operational requirements is designed applying geometric and polygonal modeling. The process of the plate designing and positioning of created components of an implant is presented in Fig. 16 [7, 11, 12].

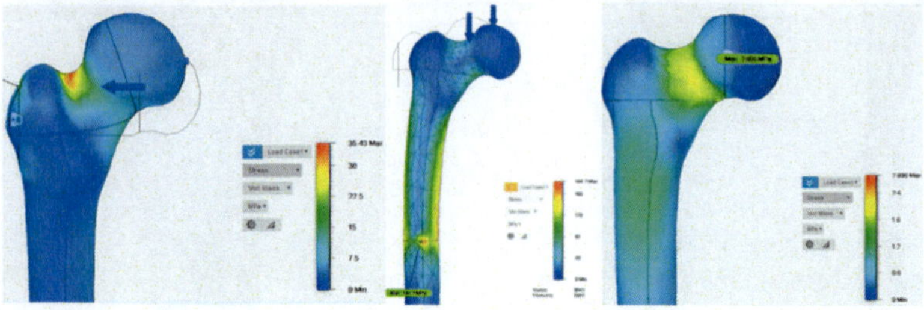

a b c

Fig. 15. Results of simulation of stress distribution in the model of a bone for the cases of horizontal (a), vertical (b), and rotational (c) force application.

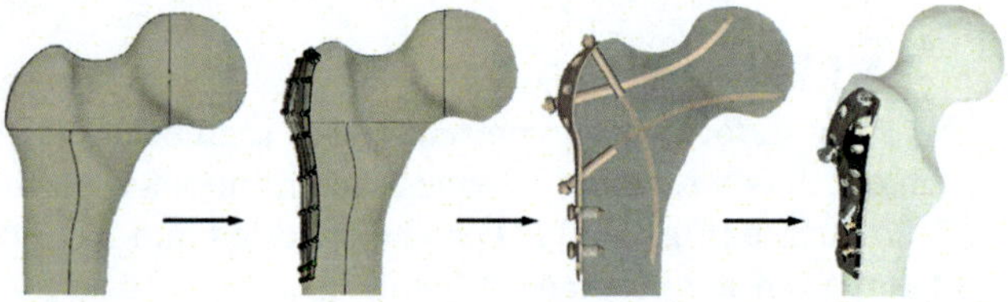

Fig. 16. Process of the plate designing and positioning of created components of an implant.

At the stage of required stress simulations, material properties attributing to the solid model of a bone with an implant is fundamentally essential. After that, either knee joint or part of the greater trochanter and one-third of diaphysis are fixed. In this case, stress distribution also takes place in three directions (Fig. 17) [7, 11, 12].

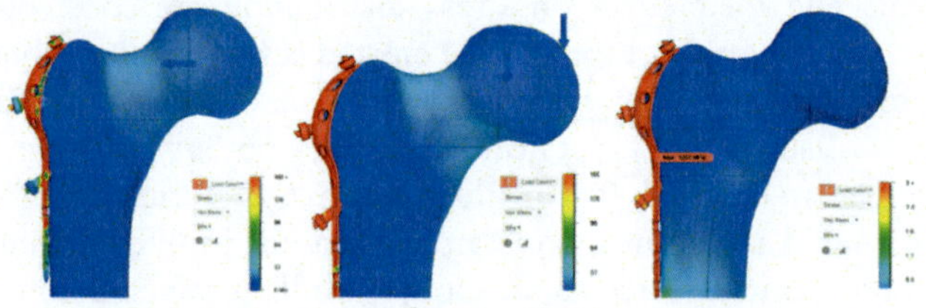

a b c

Fig. 17. Results of simulation of stress distribution in the model of a bone with an implant for the cases of horizontal (a), vertical (b), and rotational (c) force application.

As can be seen from the comparison of stresses, applying an implant significantly reduces the level of strain in a bone as it accepts maximum general stress in all cases. The bone remains in its natural state and does not deviate due to the implant installation. Figure 17 shows areas that experience maximum stresses in a bone and in an implant. Correction of the implant design is carried out repeatedly according to the testing results to obtain a reliable but still anatomically correct construction [11–17].

At the final stage, a real implant for femur reinforcement is created using additive or other modern technologies. Therefore, reverse engineering makes it possible to design an implant using a computed tomography scan method to fit individual customer demands. If the suggested implant does not show a relevant result, another type of implant construction can be used for this purpose (Fig. 18) [12]. In case of difficult injuries of the proximal or missing parts of a bone, spherical metallic the implant might also be applied (Fig. 19) [7, 11, 12].

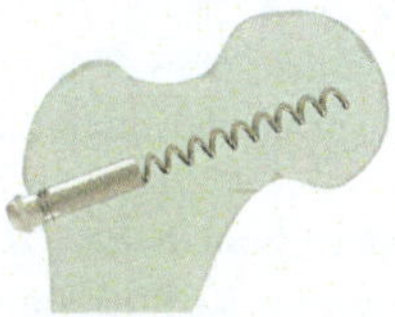

Fig. 18. Alternate version of proximal femur reinforcement.

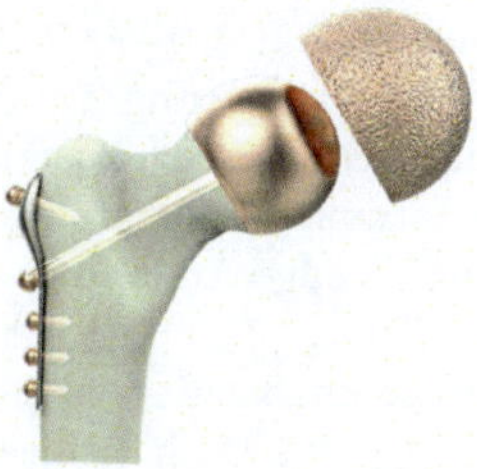

Fig. 19. Spherical metallic implant installation.

5 Conclusions

This study is focused on a new approach to development, analysis, and application of the information model, which includes the entire data for each step of creating an object using reverse engineering technology. The reverse engineering concept is presented in detail and applied to a typical workpiece. The mathematical model obtained as a result of 3D scanning provides the basis for creating a solid model of an object. The solid model is subsequently improved into a 3D model of sufficient quality.

Applying an available and verified information model, this approach is suggested to be used in other areas. It can be adopted, for instance, to a process of designing an implant for femur reinforcement based on reverse engineering technology. In this case, the information model is transformed in order to solve a specific problem. Along with creating a virtual prototype of the real object, further design of an anatomically shaped implant is also considered to make an implant fitting closely to the complicated surface of a bone. The most probable variants of applied force direction (horizontal, vertical, and rotational) have been taken into account during stress simulations. Analysis of the results received in Fusion 360 software proves that applying an implant significantly reduces the level of strain in a bone as an implant accepts maximum general stress. ANSYS software tools may be used for the model development to get more detailed results of the simulations.

Applying information model of an object is proved to be feasible in reverse engineering process. It reduces the amount of time needed to design, analyze, and carry out stress testings of the model both as part of theoretical research and for practical purposes. The presented information model might be developed to design objects which have missing parts of the shape. A case in point is prosthetics of different parts of a skeleton, including skull bones, as a result of injuries or surgery. The proposed method of design allows us to advance the process of prosthetics and patients' recovery.

References

1. Wagner, G.: Information and process modeling for simulation. J. Simul. Eng. **1**(1), 1–25 (2018)
2. Levenchuk, A.: How to describe information models of large engineering objects. Livejournal **2**(5), 1–7 (2016)
3. Reverse Engineering Tutorial Part 1 – Get Your Feet Wet. https://blogs.sw.siemens.com/solidedge/reverse-engineering-tutorial-part-1-get-your-feet-wet/. Accessed 06 Nov 2019
4. An example of literal "Reverse Engineering" using Solid Edge. https://blogs.sw.siemens.com/solidedge/An-example-of-literal-Reverse-Engineering-using-Solid-Edge/. Accessed 10 Nov 2019
5. Reverse Engineering. https://sapr.ru/article/25559. Accessed 10 Nov 2019
6. Dealing with Mesh Bodies in Solid Edge. https://blogs.sw.siemens.com/solidedge/dealing-with-mesh-bodies-in-solid-edge/. Accessed 06 Nov 2019
7. Savelyeva, O., Pavlyshko, A.: Creation of an imitation implant model for the treatment of femoral neck fractures using the system Delcam PowerSHAPE. CAD J. Graph. **1**(1), 79–81 (2014)
8. Ghafoor, H.: Reverse engineering in orthodontics. Turk. J. Orthod. **31**, 139–144 (2018)
9. Tigariev, V., Salii, V.: Development of the information model support for children of patients with calculations. Inf. Math. Methods Model. **2**(9), 96–105 (2019)
10. Tigariev, V., Salii, V.: Designing of engineering objects using the information model. Proc. Odessa Polytech. Univ. **2**(58), 91–96 (2019)
11. Savelyeva, O., Prokopovich, I.: Computer modeling of implant for femur reinforcement. Proc. Odessa Polytech. Univ. **1**(54), 51–61 (2018)
12. Savelyeva, O., Starushkevitch, T.: Mathematical analysis and computer simulation of reinforced femur bone stressed states. Geom. Model. Inf. Technol. **2**(6), 65–71 (2018)
13. Kaleev, A., Kashapov, L.: Application of reverse engineering in the medical industry. In: IOP Conference Series: Materials Science and Engineering (ISTC-IETEM-2016), Kazan, Russian Federation, vol. 1, pp. 1–4 (2016)
14. Karbowski, K, Sujka, W.: Reverse engineering in medicine. In: 1th National Scientific Conference "Rapid prototyping Modeling - Production – Measurements", Warsaw, Poland, vol. 12, pp. 111–113 (2015)
15. Jamrozik, K., Szozda, D.: Application of computer modeling for planning plastic surgeries. Manage. Prod. Eng. Rev. **5**(4), 18–25 (2018)
16. Chrzan, R., Urbanik, A.: Cranioplasty prosthesis manufacturing based on reverse engineering technology. Med. Sci. Monit **1**(18), 1–6 (2012)
17. Karbowski, K., Chrzan, R.: Reverse engineering in cranioplasty. Innovative Technol. Biomed. **1**(1), 4–15 (2015)

Mathematical Model of the Hydraulic Drive for Special Machines Tools

Valentin Tikhenko(✉) and Aleksandr Volkov

Odessa National Polytechnical University, 1, Shevchenko Avenue, Odessa 65044, Ukraine
vntikhenko@gmail.com

Abstract. The article considers the urgent problem, which is to ensure the stability of the rectilinear movements of the working parts during finishing metalworking operations on special metalcutting machines. The use of hydraulic drives for these purposes (especially in heavy machines) is due to a number of their advantages compared to electromechanical drives. The research goal of the article is to create a mathematical model of the hydraulic drive for subsequent analysis of the stability of movement of the working part of the machine. The mathematical drive model is developed taking into account the characteristics of the pressure regulating valve, flow regulator, working fluid, as well as the dependence of the friction force on the speed of movement of the working parts. Based on the mathematical model, a block diagram is compiled with the help of which the dynamic characteristics of the hydraulic drive are carried out under different operating modes. As an example, the logarithmic amplitude-frequency and logarithmic phase-frequency characteristics of the hydraulic drive are given.

Keywords: Mathematical modeling · Machine tool · Hydraulic drive · Rate of feed · Stability · Modeling of working process

1 Introduction

In the modern industry, metalcutting machines play an irreplaceable role. Productivity, quality, and cost of machining in many respects depend on the quality of the machine tools. The functionality, technical excellence, and competitiveness of machine tools are largely determined by the technical characteristics of the feed drives used in them [1]. These drives provide relative movements of the metalcutting tools and the workpieces in the mode of forming specified surfaces, and their number corresponds to the number of controlled coordinates. Electromechanical drives with a gearbox and a rolling screw-nut transmission are mainly used. However, for some machines it is advisable to use hydraulic or electrohydraulic feed drives. The most effective is the use of hydraulic drives in machines with reciprocating movements of the working parts (obtaining a linear motion is carried out using a hydraulic cylinder without kinematic transformations as in electromechanical drives). In these cases, a number of advantages of hydraulic drives are manifested: the ability to obtain large forces with limited sizes of the hydraulic cylinder, good performance with the required quality of transients, high reliability in operation [2]. They are used for heavy machines when high power density

V. Tonkonogyi et al. (Eds.): InterPartner 2020, LNME, pp. 227–236, 2021.
https://doi.org/10.1007/978-3-030-68014-5_23

is required. Hydraulic feed drives with throttle speed control, which are made in the form of an autonomous unit of an aggregate-modular layout (power head or power table), are widely used in special, modular machines and automatic lines [3]. The work cycle of the aggregate power head includes its accelerated approach to the workpiece, the working feed, the stand-up stop, and the accelerated retraction to the initial position. The purpose of the article is to create a mathematical model of hydraulic feed drives for the subsequent analysis of the stability of the movement of the working part of the machine tools, the choice of the required transient parameters.

2 Literature Review

Constantly growing requirements for workloads, positioning accuracy, speeds, and accelerations of the movement of the working parts of machine tools necessitate further improvement of hydraulic drives, which is associated with an increase in their dynamic stability, efficiency, reliability, and other quality indicators.

From the point of view of dynamics, the hydrodrive should be considered as an interconnected part of the general hydromechanical system, which contains circuits that display the processes of cutting, friction, and in the carrier system of the machine [1].

The following notation is shown in Fig. 1: F_{cut} – perturbation on the cutting process; F_0 – efforts that develop the hydraulic cylinder; F_{fr} – a force of friction in guides; $\overline{f}_{cut}$ – vector of cutting forces; $\overline{\Delta}_H$ – vector displacement in the moving joint of the guides; $\overline{y}$ – displacement vector in the cutting zone; z – displacement of working parts; V_z – velocity; U – control signal voltage.

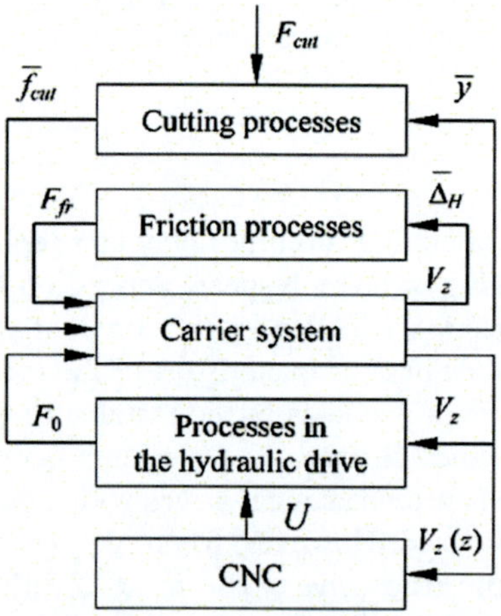

Fig. 1. Hydraulic drive as an interconnected part of the general hydromechanical system [1].

In this case, both linear and nonlinear factors are manifested due to the quadratic dependence of the flow rate on pressure, a change in the elastic modulus of the fluid, and a nonlinear friction characteristic in the hydraulic cylinder and the working parts of the machine [4, 6, 16, 17, 20].

The most effective means of analyzing complex dynamic systems is a simulation using software products [8, 16, 17]. However, this requires qualified personnel and significant machine time. Therefore, methods for representing dynamic systems in the form of structural schemes are widely used in engineering practice. These schemes are characterized by clarity and simplicity of stability analysis and quick results in the form of frequency characteristics [6, 7, 9, 12, 13].

In machines designed for the final processing of surfaces, for example, grinding, finishing, boring, honing, where the movement of the table or tool is carried out by hydraulic drives, it is often necessary to obtain small and uniform feeds. The phenomenon of «stick-slip» oscillations of the feed drives of machine tools is an essential problem in the design of certain types of machine tools [5]. An analysis of the literature [5–20] indicates that the problem of complex modeling, which determines the influence of the parameters of the hydromechanical system of the machine on the stability of movement at low speeds, has not yet been solved.

3 Research Methodology

The initial data for the design scheme are the hydraulic circuit, the parameters of hydraulic units, hydraulic cylinders, pipelines, the mass of the moving parts, as well as the resistance forces (cutting, friction). Directional control valves which switch the working fluid flow in the cavity of the hydraulic cylinder is not involved in the established processes and are not included in the calculations.

The design scheme of the hydromechanical feed drive system (Fig. 2) includes a pump 1, a pressure regulating valve 2, a working part of the machine 3, a hydraulic cylinder 4, and a flow regulator 5.

According to the scheme (Fig. 2), a mathematical model is compiled with the following:

1. The pump provides constant performance.
2. The connection of the output link of the hydraulic cylinder with the working body is absolutely rigid.
3. Inertia and cutting forces are small compared with the friction force in the guides of the working body and the seals of the hydraulic cylinder.
4. Wave processes in hydraulic lines due to their short length do not affect drive dynamics.
5. The temperature of the working fluid does not change during the considered dynamic process.
6. The coefficients of flow through the throttling and spool elements are constant.

To describe the dynamics, the following notation of variables and parameters was adopted:

m – a mass of motion parts reduced to the cylinder rod;
v – the speed of movement of moving parts;

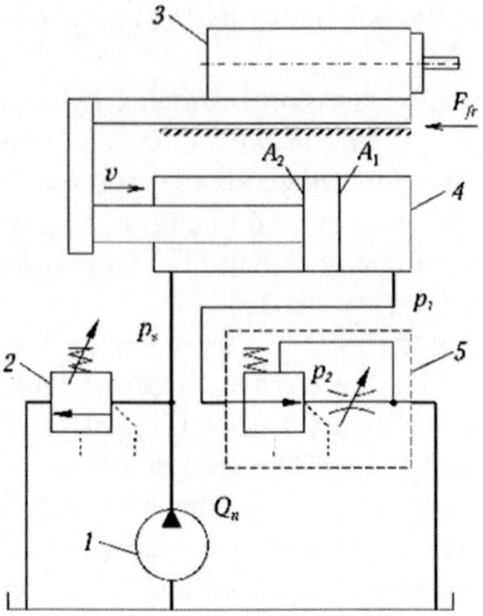

Fig. 2. The design scheme of the hydromechanical feed drive system.

F_{fr} – frictional force;
F_{fr0} – a force of static friction;
α – the coefficient for falling friction characteristic;
p_s – supply pressure;
p_1 –pressure in the piston chamber;
e_1 – total piston chamber compliance factor;
e_2 – total rod chamber compliance factor;
A_1 – effective area of the piston chamber of the hydraulic cylinder;
A_2 – effective area of the rod chamber of the hydraulic cylinder;
Q_n – nominal flow rate;
f_r – pressure regulating valve spool control area;
f_f – flow regulator spool control area;
r_r – hydraulic resistance of the pressure regulating valve damper;
r_f – hydraulic resistance of the flow regulator damper;
h_r – valve opening of the pressure regulating valve;
h_f – valve opening of the flow regulator damper;
L_r – coefficient of inertial pressure in the channels of the pressure regulating valve dampers;
L_f – coefficient of inertial pressure in the channels of the flow regulator dampers;
m_r – a mass of pressure regulating valve spool;
m_f – a mass of flow regulator spool;
c_r – spring stiffness of pressure regulating valve;
c_f – spring stiffness of flow regulator valve;
k_{hr} – hydrodynamic force of a pressure regulating valve;
k_{hf} – hydrodynamic force of the flow regulator;
b_r – length of the slit valve of a pressure regulating valve;
b_f – length of the slit of the valve flow regulator;

p_2 – pressure in the rod chamber of the hydraulic cylinder;
R_r – flow resistance of the pressure valve port;
R_f – flow resistance of the flow regulator port;
R_t – flow resistance of the throttle port;
μ – turbulent fluid flow coefficient;
γ – specific gravity;
g – free-fall acceleration;
V_1 – a volume of the working fluid in the piston chamber of the hydraulic cylinder;
V_2 – a volume of the working fluid in the rod chamber of the hydraulic cylinder;
β – effective bulk modulus;
k_{e1} and k_{e2} – coefficients that take into account the air content in the working fluid and the pressure in the piston and rod cavity of the hydraulic cylinder.

Fine boring operations are characterized by low feed rates with low small cutting forces. In this case, the friction force in the moving joints is approximated by the expression $F_{fr} = (F_{fr0} - \alpha \cdot v) \cdot \mathrm{sign}(v)$.

The total compliance factors for the piston and rod chambers of the hydraulic cylinder are determined from the expressions $e_1 = V_1/k_{e1}\beta_{01}$ and $e_2 = V_2/k_{e2}\,\beta_{02}$.

The basic equation of dynamics in relation to a loaded hydraulic cylinder

$$m\frac{dv}{dt} + \left(F_{fr0} - \alpha v\right) = A_2 p_s - A_1 p_1. \tag{1}$$

Flow equations that take into account the continuity of the fluid flow for the piston and rod chambers of the hydraulic cylinder

$$Q_n = A_2 v + e_2 \frac{dp_s}{dt} + \frac{1}{R_r}\sqrt{p_s}, \tag{2}$$

$$A_1 v = e_1 \frac{dp_1}{dt} + \frac{1}{R_f}\sqrt{p_s - p_2}. \tag{3}$$

Pressure-regulating valve spool equation

$$f_r\left(p_s - r_r f_k \frac{dh_r}{dt} - L_r f_r \frac{d^2 h_r}{dt^2}\right) = m_r \frac{d^2 h_r}{dt^2} + c_r(h_{r0} + h_r) + k_h b_r h_r p_s. \tag{4}$$

Flow regulator spool equation

$$f_f\left(p_2 - r_f f_f \frac{dh_f}{dt} - L_f f_f \frac{d^2 h_f}{dt^2}\right) = m_f \frac{d^2 h_f}{dt^2} + c_f(h_{f0} + h_f) - k_h b_f h_f (p_1 - p_2). \tag{5}$$

The equation of continuity of fluid flow in a flow regulator

$$\frac{1}{R_f}\sqrt{p_1 - p_2} = \frac{1}{R_t}\sqrt{p_2} - f_f \frac{dh_f}{dt}. \tag{6}$$

Relations equations between movements of valve spools and their hydraulic resistances

$$\frac{1}{R_r} = \mu b_r h_r \sqrt{\frac{2g}{\gamma}}, \tag{7}$$

$$\frac{1}{R_f} = \mu b_f h_f \sqrt{\frac{2g}{\gamma}}. \tag{8}$$

Equations can be reduced, taking into account basic values to dimensionless form.

And then, applying the Laplace transform for relative coordinates under zero initial conditions, we obtain them in operator form ($S = d/dt$ is the Laplace operator). To simplify the presentation, coordinate images have the same designation as the originals.

$$A_2 \Delta p_s - A_1 \Delta p_1 - \alpha \left(\frac{m}{\alpha} S - 1 \right) \Delta v = 0, \tag{9}$$

$$e_2 S \Delta p_s + A_2 \Delta v + \left(\frac{1}{R_r} \sqrt{p_s} - \frac{1}{R_{r0}} \sqrt{p_{s0}} \right) = 0, \tag{10}$$

$$A_1 \Delta v - e_1 S \Delta p_1 - \left(\frac{1}{R_f} \sqrt{p_s - p_1} - \frac{1}{R_{f0}} \sqrt{p_{10} - p_{20}} \right) = 0, \tag{11}$$

$$(f_r + k_{hr} b_r h_{r0}) \Delta p_s - c_r \left(\frac{m_r + L_r f_r^2}{c_r} S^2 + \frac{r_r f_r^2}{c_r} S + 1 \right) \Delta h_r - k_{hr} b_r p_{s0} \Delta h_r = 0, \tag{12}$$

$$(f_f - k_{hf} b_f h_{f0}) \Delta p_2 - c_f \quad \frac{m_f + L_f f_f^2}{c_f} S^2 + \frac{r_f f_f^2}{c_f} S + 1 \Bigg) \Delta h_f - k_{hf} b_f (p_{10} - p_{20}) \Delta h_f$$
$$+ k_{hf} b_f h_{f0} \Delta p_1 \tag{13}$$

$$\frac{1}{R_f} = \mu b_f (h_{f0} - \Delta h_f) \sqrt{\frac{2g}{\gamma}}, \tag{14}$$

$$\frac{1}{R_r} = \mu b_r (h_{r0} + \Delta h_r) \sqrt{\frac{2g}{\gamma}}, \tag{15}$$

$$\frac{1}{R_f} \sqrt{p_1 - p_2} - \frac{1}{R_t} \sqrt{p_2} - f_f S \Delta h_f = 0. \tag{16}$$

In order to construct a structural diagram, the resulting equations are better represented as follows

$$\Delta v = \frac{1}{\alpha(T_m S - 1)}(A_2 \Delta p_s - A_1 p_1), \tag{17}$$

$$\Delta p_s = \frac{1}{e_2 s}\left[-A_2 \Delta v - \left(\frac{1}{R_r}\sqrt{p_s} - \frac{1}{R_{r0}}\sqrt{p_{s0}}\right)\right], \tag{18}$$

$$\Delta p_1 = \frac{1}{e_1 S}\left[A_1 \Delta v - \left(\frac{1}{R_f}\sqrt{p_1 - p_2} - \frac{1}{R_{f0}}\sqrt{p_{10} - p_{20}}\right)\right], \tag{19}$$

$$\Delta h_r = \frac{1}{c_r(T_r^2 S^2 + 2\zeta_r T_r S + 1)}[(f_r + k_{hr} b_r h_{r0})\Delta p_s - k_{hr} b_r p_{s0} \Delta h_r], \tag{20}$$

$$\Delta h_f = \frac{1}{c_f(T_f^2 S^2 + 2\zeta_f T_f S + 1)}[(f_f - k_{hf} b_f h_{f0})\Delta p_2 - k_{hf} b_f (p_{10} - p_{20})\Delta h_f], \tag{21}$$

$$\frac{1}{R_r} = \mu b_r (h_{r0} + \Delta h_r)\sqrt{\frac{2g}{\gamma}}, \tag{22}$$

$$\frac{1}{R_f} = \mu b_f (h_{f0} + \Delta h_f)\sqrt{\frac{2g}{\gamma}}, \tag{23}$$

$$p_2 = R_t^2 \left(\frac{1}{R_r}\sqrt{p_1 - p_2} + f_r S \Delta h_r\right)^2. \tag{24}$$

4 Results

Based on Eqs. (17)–(24), a block diagram of the dynamic model of the hydraulic drive (Fig. 3) is obtained (T_r and T_f – time constants, ξ_r, and ξ_f – damping coefficients). In the diagram, blocks are identified that correspond to the pressure regulating valve (1), flow regulator (2), piston (3) and rod (5) chamber of the hydraulic cylinder, as well as the working parts (4) of the machine. The mathematical model of the hydraulic drive obtained as a result of this description allows us to perform calculations and studies using software products of various working processes that occur during the cycle of the machine. At the same time, performing engineering calculations using such a complex model causes a number of difficulties.

However, under certain conditions, when the essential frequencies (the natural frequencies of the system and the frequency of periodic effects) significantly differ from the natural frequencies of the circuits, consisting of hydraulic units, the problem can be simplified. Using the superposition method and algebra for the structural scheme under the theory of automatic control, transfer functions, and frequency characteristics of the hydromechanical drive system can be obtained. To verify the results of theoretical dependencies, studies on the hydraulic drive of a special boring machine with a horizontal table were conducted.

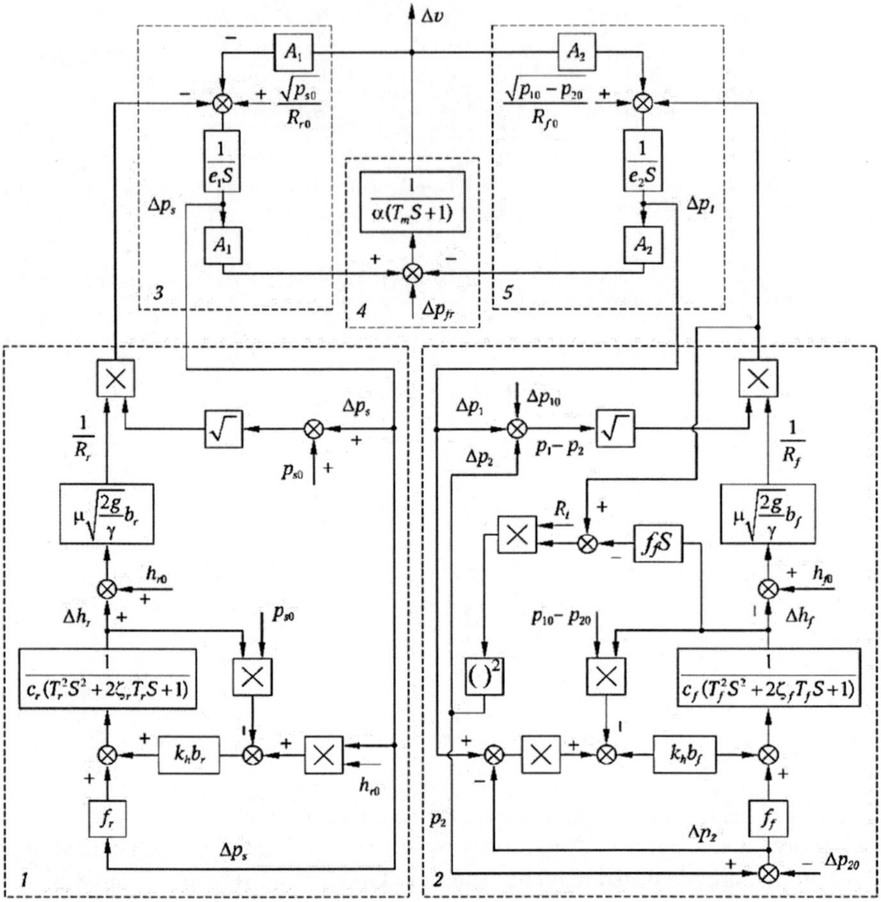

Fig. 3. Block diagram of the dynamic model.

The lubrication of the sliding guides of the table was carried out periodically by water. The value of the friction force in the guide was twice less than in the hydraulic cylinder. In the speed range from 22 to 110 mm/min, an intensive decrease in the friction force was noted with an increase in the speed of movement. The natural frequency of the hydromechanical system was $f = 29.6$ Hz. Calculations of the natural frequencies of the pressure regulating valve and the flow regulator valve showed that their values were several times different from f. Studies of hydraulic drive dynamics in a speed range from 47 to 100 mm/min showed table vibrations with a frequency of 32 Hz, which is close to f. On this basis, it is possible, by analyzing the frequency characteristics of the hydraulic system, to select the parameters of the hydraulic drive elements according to the criterion of ensuring the necessary stability of movement.

A comparison of the obtained values with the characteristics of the mechanical system, as well as with the results of experimental records of fluctuations in the speed of the working part, can be the basis for a simplified determination of dynamic stability. For example, the dynamic characteristic of a pressure regulating valve is replaced by its static characteristic. In turn, instead of the dynamic characteristics of the flow controller, its static linearized characteristic is used. This allows to build calculated by logarithmic amplitude and phase-frequency characteristics of the hydraulic drive, an example of which is shown in Fig. 4. In this case, the hydromechanical system is stable and has a phase margin. Experimental studies, which were carried out on a boring finishing machine, showed the stable operation of the hydraulic table feed at a speed from 2 to 5 mm/min [5]. To increase the stability margin, it is advisable to use external damping, for example, using a controlled hydraulic damper with magnetic fluid.

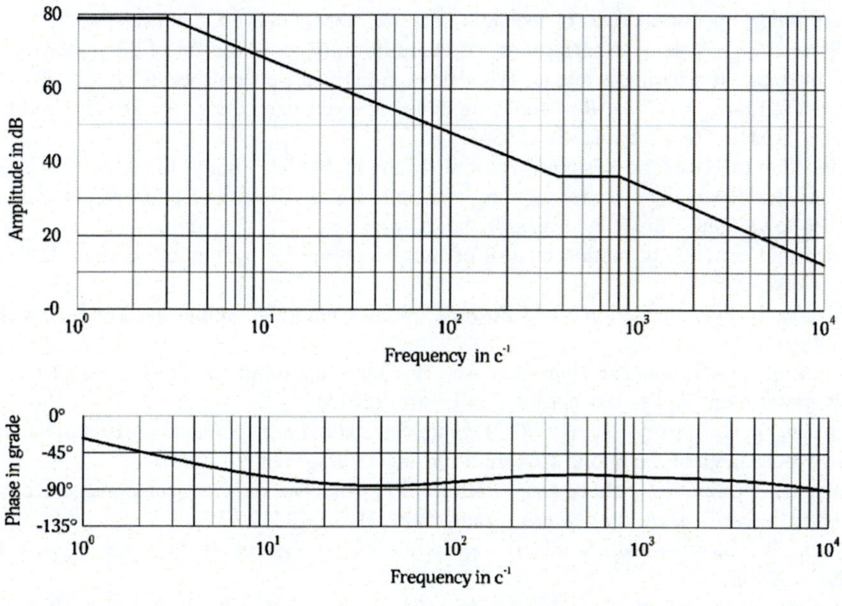

Fig. 4. Logarithmic amplitude and phase-frequency characteristics of the hydraulic drive.

5 Conclusions

As a result of studies of the hydraulic drive of special machines, a mathematical model of the drive was compiled, which takes into account the characteristics of the pressure control valve, flow regulator, working fluid, as well as the dependence of the friction force on the speed of the working parts. The possibility of using a mathematical model to obtain the transmission functions and frequency characteristics of the hydromechanical system for a reasonable choice of structural and tuning parameters of the drive elements is shown. The block diagram can be used both at the preliminary stages of design and for advanced research of existing hydraulic drives. Experimental studies on the boring machine support the conclusion of the strong influence of the friction in the hydraulic cylinder on the total frictional force and dynamics of the hydraulic drive. The occurrence of velocity fluctuations in the region of the incident friction characteristic at a frequency close to the natural frequency of the hydromechanical system confirms the adequacy of the mathematical model of the hydraulic drive.

References

1. Mechanical Engineering. Encyclopedia in 40 volumes. Volume 4–7. Metalcutting machines and woodworking equipment. Mechanical Engineering, Moscow (2002)
2. Sveshnikov, V.: Machine hydraulic drives. Mechanical Engineering, Moscow (2008)
3. Aggregate-modular technological equipment in 3 parts. Under the editorship of Yu. Kuzetsov. Imex LTD, Moscow (2014).
4. Pinchuk, V., Sheleg, V.: Calculation and design of aggregate-modular hydroblocks for controlling hydraulic drives of technological machines. GSTU, Gomel (2010)
5. Tikhenko, V., Volkov, A.: The study of the uniformity of movement of the hydraulic drive of the table of a boring machine. Promislova gidravlika i pneumatics **1**(35), 55–58 (2012)
6. Xie, Q., Liang, S., Chen, R.: Modelling of linear motor feed drives for grinding machines. Int. J. Manuf. Res. **1**(1), 41–58 (2006)
7. Modeling of Hydraulic Systems. Tutorial for the hydraulics library, Modelon (2013)
8. Kline, K.: Hydraulic system modeling and optimization to achieve performance characteristics Iowa State University Ames, Iowa (2016)
9. Hochwallner, M.: On motion control of linear incremental hydraulic actuators. Linkoping University (2017)
10. Watton, J.: Fundamental of Fluid Power Control. Cambridge University Press, Cambridge (2009)
11. Landerg, M.: New linear Hydraulic Actuator In: Proceedings of the FSME/BATH 2015 symposium on fluid power and motion control (2015).
12. Licsko, G., Champneys, A., Hos, C.: Dynamical analysis of a hydraulic pressure relief valve. In: Proceedings of the World Congress on Engineering, vol. 2 (2009)
13. Wu, D., Burton, R., Schoenau, G., Bitner, D.: Analysis of a pressurecompensated flow control valve. J. Dyn. Syst. Meas. Control **129**(2), 203–211 (2007)
14. Akers, A., Gassman, M., Smith, R.: Hydraulic Power System Analysis. CRC Press, Boca Raton (2006)
15. Koster, M.: On modeling, analysis and nonlinear control of hydraulic systems. Dissertation von Dipl.Wirtsch.-Ing. Karlsruher Institut fur Technologie, p. 227 (2017)
16. Mickoski, H., Djidrov, M.: Modeling of mechanism with linear hydraulic drive system. Math. Models Eng. **5**(3), 73–78 (2019)
17. Orošnjak, M., Jocanović, M., Karanović, V.: Simulation and modeling of a hydraulic system in FluidSim. In: XVII International Scientific Conference on Industrial Systems (IS 2017), Novi Sad, Serbia (2017).
18. Bureček, A., Hružík, L., Vašina, M.: Simulation of dynamics of system with hydraulic lines and linear hydraulic motor with mass load. In: EPJ Web of Conferences (2013)
19. Feng, H., Du, Q., Huang, Y., Chi1, Y.: Modelling study on stiffness characteristics of hydraulic cylinder under multi-factors. J. Mech. Eng. **63**, 447–456 (2017)
20. Yanada, H., Khaing, W., Tran, X.: Effect of friction model on simulation of hydraulic actuator. In: The 3rd International Conference on Design Engineering and Science, ICDES 2014, Pilsen, Czech Republic (2014)

Particle Movement on Concave Coulter of the Centrifugal Distributor with Radially Installed Vertical Blades

Tatiana Volina[1(✉)], Sergiy Pylypaka[2], Alla Rebrii[1], Olexandr Pavlenko[3], and Yaroslav Kremets[2]

[1] Sumy National Agrarian University, 160, Kondratieva St., Sumy 40021, Ukraine
t.n.zaharova@ukr.net
[2] National University of Life and Environmental Sciences of Ukraine, 15, Heroyiv Oborony St., Kyiv 03041, Ukraine
[3] Bohdan Khmelnytsky Melitopol State Pedagogical University, 20, Hetmanska St., Melitopol 72300, Ukraine

Abstract. The relative motion of a particle along the inner surface of a horizontal concave coulter along a vertical blade installed in the radial direction is considered in the article. The disk rotates around a vertical axis with a given angular velocity. A system of differential equations of particle motion, which is solved by numerical methods, is complicated. Kinematic characteristics were found, the regularities of the relative motion of a particle on the spherical surface were determined. The spherical working surface of the disk allows increasing the dispersion zone of technological material. Analytical description of the particle motion makes it possible to investigate the acceleration during the motion on the disk along with the blades, as well as to find the relative and absolute velocities at the moment of the ascent particles from the disk. The obtained analytical dependencies allow determining the influence of structural and technological parameters on the particle acceleration process.

Keywords: Working surface · Spherical disk · Fertilizer particle · Coordinate system · Differential equations · Absolute velocity

1 Introduction

It is a well-known fact that the efficiency of production of various crops depends on the use of mineral fertilizers, which are most often introduced into the fields by surface method, spreading, using machines equipped with centrifugal scattering working organs [1]. These working bodies have undoubted advantages over other types of working bodies. The operating parts of the centrifugal type devices may have different constructions. The principle of operation of such devices is that the material particle (the fertilizer particle) slides on the working surface, which rotates around the vertical axis, accelerates by the centrifugal force, and moves to the periphery from the axis of rotation. For the increase in the dispersal speed, such devices are additionally equipped with vertical blades, along which the particle is forced to move in the radial direction.

V. Tonkonogyi et al. (Eds.): InterPartner 2020, LNME, pp. 237–246, 2021.
https://doi.org/10.1007/978-3-030-68014-5_24

The kinematic characteristics of the movement of technological material depend on the form and construction parameters of the operating parts of the device. The research of the laws of particle motion depending on the shape and construction features of the working parts allows improving the device design.

2 Literature Review

A great deal is being written and said about particles in contact with the surface and moving along it. They may have a different origin: mechanical particles [2], particles in the stream [3], particles of gas [4], or liquid [5]. The movement of particles that are in contact with moving spiral working parts is considered in the article [6]. The motion of a single particle cannot be identified with the motion of a body or technological material, which consists of individual particles. Indeed, it is easier to study the motion of particles in comparison with the motion of bodies [7], because in the second case, it is necessary to consider the forces of inertia from the rotation of the body.

In some cases, these forces can be neglected or ignored at low angular velocities, for example, in the solving the problem of elimination of load oscillations on a flexible suspension during the swing of a boom crane with simultaneous minimization of dynamic forces in the transmission mechanism [8], in the process of developing of the physical model of the pneumatic classification process for detecting the rotating suspended layer and ensuring the frequency of loading and unloading of a pneumatic classifier [9] or in the process of designing the multistage centrifugal pumps [10]. However, based on the motion of a single particle, it is possible to identify dependencies that can be applied to a body or material or to identify the direction in which further research should be continued. The analysis of particle motion in combination with other studies is a prerequisite for designing the fibrous filter layer in the process of inertial-filtering separation [11], for simulation of aeroelastic interaction between gas-liquid flow and deformable elements [12], as well as for the analysis of the stationary hydroaeroelasticity problem for dynamic deflection elements of separation devices [13]. Moreover, the abovementioned methodology can be applied in the field of manufacturing engineering for designing profiles of gears [14] and clutches [15], as well as for the parametric optimization of fixtures for multiaxis machining of parts in multiproduct manufacturing [16]. Against the background, for studying the movement of a particle depending on the design parameters of the surface, it is essential to have analytical dependencies that describe this movement. That is why the central question that motivates this paper is to investigate the regularities of motion of material particles in the centrifugal distributor in the form of the surface of a spherical segment with radially installed vertical blades.

3 Research Methodology

It is commonly known that the parametric equations of a sphere with the origin coordinates in its lower pole are:

$$X = R\sin\varepsilon\cos\alpha;\, Y = R\sin\varepsilon\sin\alpha;\, Z = R(1-\cos\varepsilon), \tag{1}$$

where R is the radius of the sphere; ε, α – independent variables of the sphere (angles that define the position of a point on the sphere surface in the direction of the meridian and the parallel).

We then review the factor, that two following constructive parameters give the spherical segment: the angle of particle lead β at the moment of its ascent from the surface of the segment along the vertical blade and sphere diameter d (Fig. 1). Based on these conditions, the radius R of the sphere can be found:

$$R = d/2\sin\beta. \tag{2}$$

It is important to emphasize, that to construct by Eqs. (1) according to formula (2) a spherical segment with given construction parameters d, β and R, the angles ε and α must vary within the following limits: ε = [0, β], α = 0 – 2π.

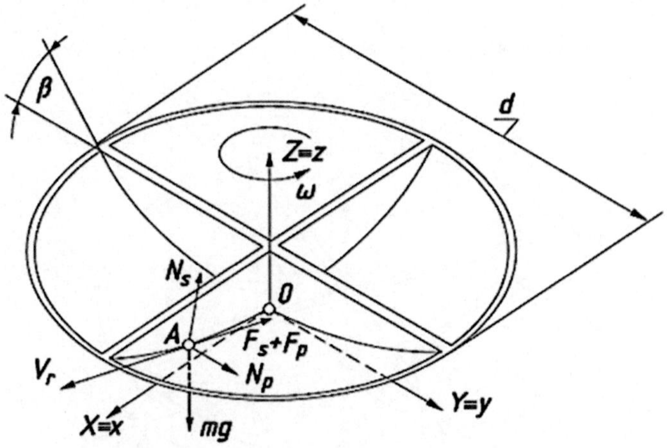

Fig. 1. Operating part of the centrifugal distributor with vertical blades and the position of the material particle at point A with the applied forces.

Undoubtedly, the material particle during of the disk rotation with an angular velocity ω will have contact with the surfaces of the segment and the blade and slide along their common line – a circle of radius R. Its equation is obtained from (1) at α = 0:

$$\{0;\quad 0;\quad -1\} \tag{3}$$

The line (3) is the trajectory of the relative motion of the particle, that is, the trajectory of its sliding. During the disk rotation for some time t, the curve together with the disk will rotate to the angle $\theta = \omega t$. During this time, the particle will move along the curve (3) towards the periphery of the disk (Fig. 2). The process of particle motion will be described relatively to two coordinate systems: inertial *OXYZ* and non-inertial *Oxyz*. The movable coordinate system is rigidly connected with the disk and is rotated with it. Initially, these two systems coincide (Fig. 1).

Generally, the relative trajectory of particle motion along the curve (3) is given in the form of the dependence $\varepsilon = \varepsilon\ (t)$, which must eventually be found.

It is necessary to find the absolute trajectory of particle motion. During time *t*, it made a relative displacement along the curve by Eqs. (3), where the value of the angle ε is a function of time t and a rotation to the angle $\theta = \omega t$ (Fig. 2). The vertical coordinate for both systems is the same: $Oz \equiv OZ$. Therefore, the absolute trajectory of particle movement is found by rotating the movable system with the coordinates (3) of the particle relative to the fixed coordinate system around *OZ* axis by the known formulas:

$$x_a = x\cos\theta - y\sin\theta; y_a = x\sin\theta + y\cos\theta; z_a = z. \tag{4}$$

One can substitute in (4) Eq. (3) and the expression of the angle $\theta = \omega t$ and as a result, obtain the parametric equations of the absolute trajectory of particle motion:

$$x_a = R\sin\varepsilon\ \cos(\omega t); y_a = R\sin\varepsilon\ \sin(\omega t); z_a = R(1 - \cos\varepsilon)\,. \tag{5}$$

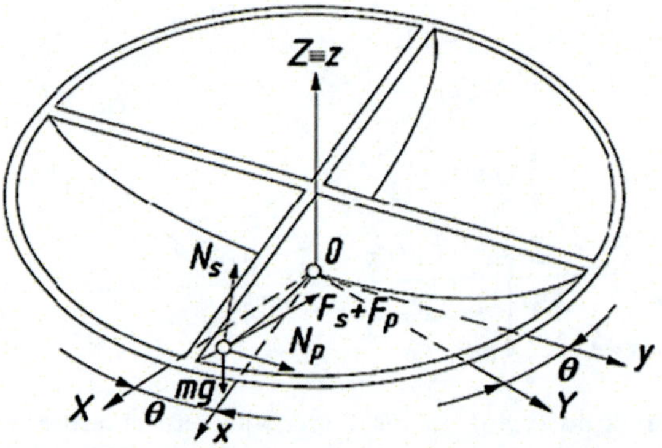

Fig. 2. Operating part of the centrifugal distributor with vertical blades and the position of the material particle with the applied forces after rotating to the angle $\theta = \omega t$.

The differential equation of the absolute particle motion in a vector notation has the form: $m\overline{w} = \overline{F}$, where *m* is the mass of the particle, $\overline{w}$ is the vector of absolute acceleration, $\overline{F}$ is the resultant vector of the applied to the particle forces. Such forces are as follows: the weight force *mg* ($g = 9.81$ m/s^2); the reaction force N_s of the surface of the spherical disk; the reaction force N_p of the surface of the blade; the friction force $F_s = f_s \cdot N_s$ during the particle slides on the disk surface; the friction force $F_p = f_p \cdot N_p$ during the particle slides on the blade surface (f_s and f_p are the coefficients of friction of the particle along with the disk and the surfaces of the blades). All these forces have a strictly defined direction of action. For the initial moment of motion (when both coordinate systems coincide), it is shown in Fig. 1. We begin by taking a closer look at applied forces.

Firstly, the force of the weight is directed downwards. Therefore, its projections will be recorded:

$$\{0;\quad 0;\quad -mg\}. \tag{6}$$

Besides, the reaction N_p of the blade is directed perpendicularly to it, that is, it coincides with the direction of the axis Oy. The projections of a unit reaction vector N_p are:

$$\{0;\quad 1;\quad 0\}. \tag{7}$$

Moreover, the reaction N_s of a spherical disk (segment) is directed perpendicularly to its surface. The radius vector of the circle (3), along which the directing vector of the reaction N_s is determined, is directed from the sphere center to the point on the curve. Thus, the directing vector will be directed in the opposite direction. Its projections are as follows:

$$\{-\sin\varepsilon;\qquad 0;\qquad \cos\varepsilon\}. \tag{8}$$

Furthermore, both friction forces F_s and F_p are directed tangentially to the trajectory of the relative motion of the particle in the opposite direction of the velocity V_r of its sliding (Fig. 1). It is essential to find a unit vector of the direction of relative velocity V_r along which the force of the total friction force ($F_s + F_p$) acts. For this, it is necessary to differentiate the relative trajectory (3), implying that $\varepsilon = \varepsilon(t)$:

$$\dot{x} = R\dot{\varepsilon}\cos\varepsilon;\, \dot{y} = 0;\, \dot{z} = R\dot{\varepsilon}\sin\varepsilon. \tag{9}$$

Hence, the geometric sum of the components (9) will give the expression of the relative velocity of the particle:

$$V_r = \sqrt{\dot{x}^2 + \dot{y}^2 + \dot{z}^2} = R\dot{\varepsilon}^2. \tag{10}$$

By dividing the projections of the relative velocity (9) to its value (10), one obtains the expressions of a unit vector of the relative velocity V_r in the projections on the axis of the movable system $Oxyz$:

$$\{\cos\varepsilon;\quad 0;\quad \sin\varepsilon\}. \tag{11}$$

Obviously, after the disk rotating together with the movable $Oxyz$ coordinate system to the angle $\theta = \omega t$, the particle will occupy a new position on the curve, and the curve itself will also rotate to this angle. The direction of action of all vectors of applied forces is determined. As the disc is rotated, the particle and the vectors of forces applied to it, which do not change direction in the moving system, rotates too. So, after the vectors (7), (8), (11) are rotated to the angle $\theta = \omega t$ around the OZ axis, the following expressions describe their projections to the fixed system.

A unit reaction vector N_p of the blade:

$$\{-\sin\omega t;\quad \cos\omega t;\quad 0\}. \tag{12}$$

A unit reaction vector N_s of a spherical disk:

$$\{-\sin\varepsilon\cos\omega t;\quad -\sin\varepsilon\sin\omega t;\quad \cos\varepsilon\}. \tag{13}$$

A unit relative velocity vector V_r:

$$\{\cos\varepsilon\cos\omega t;\quad \cos\varepsilon\sin\omega t;\quad \sin\varepsilon\}. \tag{14}$$

Importantly, the direction of action of the particle weight force (6) within rotating the disk does not change.

The projections of absolute velocity and absolute acceleration of the particle on the axis of the immovable $OXYZ$ coordinate system are found by the sequential differentiation of Eqs. (5). After differentiating of Eqs. (5), the projections of absolute velocity are obtained:

$$\begin{aligned} \dot{x}_a &= R\dot{\varepsilon}\cos\varepsilon\cos\omega t - R\omega\sin\varepsilon\sin\omega t; \\ \dot{y}_a &= R\dot{\varepsilon}\cos\varepsilon\sin\omega t + R\omega\sin\varepsilon\cos\omega t; \\ \dot{z}_a &= R\dot{\varepsilon}\sin\varepsilon. \end{aligned} \tag{15}$$

Thus, the value of absolute velocity is obtained by geometric addition of the components (15):

$$V_a = \sqrt{\dot{x}_a^2 + \dot{y}_a^2 + \dot{z}_a^2} = R\sqrt{\dot{\varepsilon}^2 + \omega^2\sin^2\varepsilon}. \tag{16}$$

After differentiating expressions (15), the projections of the vector of absolute acceleration are:

$$\begin{aligned} \ddot{x}_a &= R(\ddot{\varepsilon}\cos\omega t - 2\omega\dot{\varepsilon}\sin\omega t)\cos\varepsilon - R(\dot{\varepsilon}^2 + \omega^2)\sin\varepsilon\cos\omega t; \\ \ddot{y}_a &= R(\ddot{\varepsilon}\sin\omega t + 2\omega\dot{\varepsilon}\cos\omega t)\cos\varepsilon - R(\dot{\varepsilon}^2 + \omega^2)\sin\varepsilon\sin\omega t; \\ \ddot{z}_a &= R\ddot{\varepsilon}\sin\varepsilon + R\dot{\varepsilon}^2\cos\varepsilon. \end{aligned} \tag{17}$$

It is necessary to decompose the vector equation $m\overline{w} = \overline{F}$ in the projections on the axis of the fixed coordinate system $OXYZ$. The directions of actions of reaction N_p of blades and N_s of the spherical disk are given by unit vectors (12) and (13). The direction of the total friction force $(F_s + F_r) = (f_s N_s + f_p N_p)$ is given by the unit vector (14), which should be taken with the opposite sign. As a consequence, three equations are obtained:

$$\begin{aligned}\ddot{x}_a &= -N_p \sin \omega t - N_s \sin \varepsilon \cos \omega t - (f_s N_s + f_p N_p) \cos \varepsilon \cos \omega t;\\ \ddot{y}_a &= N_p \cos \omega t - N_s \sin \varepsilon \sin \omega t - (f_s N_s + f_p N_p) \cos \varepsilon \sin \omega t;\\ \ddot{z}_a &= N_s \cos \varepsilon - (f_s N_s + f_p N_p) \sin \varepsilon - mg.\end{aligned} \tag{18}$$

Substitute in (18) the expressions of absolute acceleration from (17) and obtain a system of three differential equations with three unknown functions: $\varepsilon = \varepsilon(t)$, $N_s = N_s(t)$, and $N_p = N_p(t)$. The system should be solved relatively to $\ddot{\varepsilon}$, N_s, N_p:

$$\ddot{\varepsilon} = \left(\omega^2 \sin \varepsilon - 2\dot{\varepsilon}\omega f_p - \frac{g}{R} f_s\right) \cos \varepsilon - \left(\frac{g}{R} + f_s \omega^2 \sin \varepsilon\right) \sin \varepsilon - \dot{\varepsilon}^2 f_s. \tag{19}$$

$$N_s = m(R\dot{\varepsilon}^2 + g \cos \varepsilon + R\omega^2 \sin^2 \varepsilon)\ . \tag{20}$$

$$N_p = 2mR\omega\dot{\varepsilon} \cos \varepsilon. \tag{21}$$

To summarize, differential Eq. (19) is independent of (20) and (21), so it can be solved separately. The reactions of surfaces (20) and (21) can be found after solving the Eq. (19). Numerical methods are required to solve it.

4 Results

The following example is considered. For instance, the diameter of the spherical disk is $d = 0.4$ m; the angle is $\beta = 20°$. By the formula (2), the third constructive parameter $R = 0.58$ m can be found. Within numerical integration of the differential Eq. (19), the time t must be given in limits, which provide the angle ε is smaller than the angle β. When $\varepsilon = \beta$, the particle is at the outer edge of the disk.

In Fig. 3, the dependence $\varepsilon = \varepsilon(t)$ is constructed for the given constructive parameters and for the following values: $f_s = f_p = 0.3$, $\omega = 25$ rad/s. The initial value of the angle ε is not zero because the particle cannot be placed on the axis of rotation. In this case, there is no centrifugal force, and the particle cannot start moving. As Fig. 3a shows, the particle accelerates during 0.32 s to reach the periphery of the disk.

Graphs of change of relative and absolute speeds are shown in Fig. 3b. It can be seen that at the moment of convergence from the disk, the relative and absolute velocities have the following values: $V_r = 3.2$ m/s, $V_a = 5.8$ m/s. Figure 4 shows graphs of the change in the reaction force N_s of a spherical disk and N_p of the blade for a particle with mass $m = 0.01$ kg by formulas (20) and (21).

If the angle β did not bound the spherical segment, the particle would rise along the inner surface of the sphere to a certain height. Numerical integration of the differential Eq. (19) with the duration of motion $t = 2.5$ s shows that the angle ε reaches the maximum value for half a second ($\varepsilon = 90°$). Then its decaying oscillations begin (Fig. 5). It means that the particle rises to the height of the hemisphere and then begins to descend, and the lowering is alternating with the raising until the full stop of the particle at a certain height. For the case in Fig. 5, the particle stop corresponds to an angle ε = 70°, that is, the particle "sticks" at a height close to the hemisphere and then rotates with it.

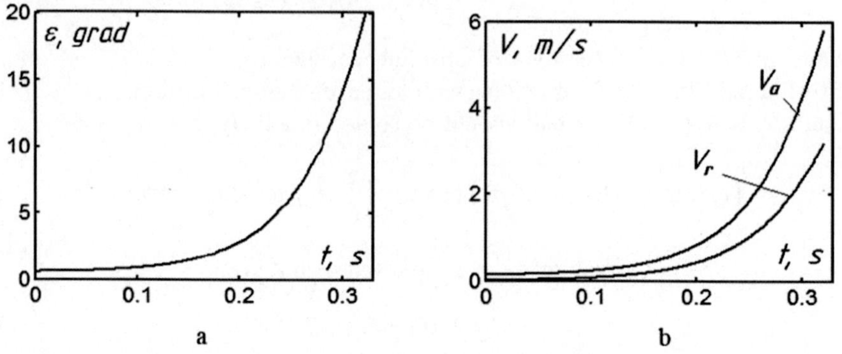

Fig. 3. Graphs characterizing the particle motion at $f_s = f_p = 0.3$, $\omega = 25$ rad/s:

a – the chart of angle ε change; b – graphs of changes in relative V_a and absolute V_r velocities.

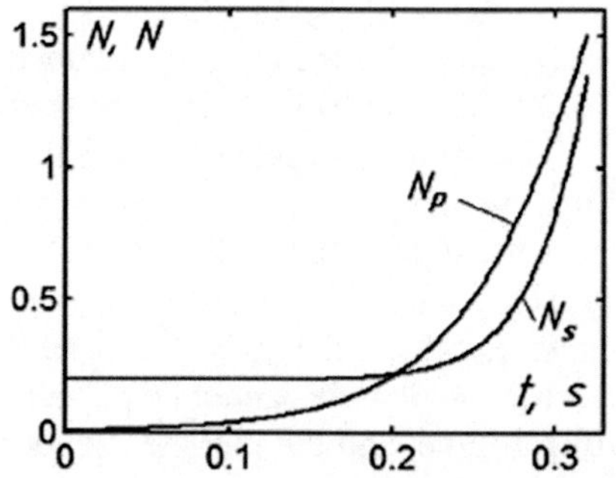

Fig. 4. Graphs of change of reaction force N_s of a spherical disk and N_p of a blade.

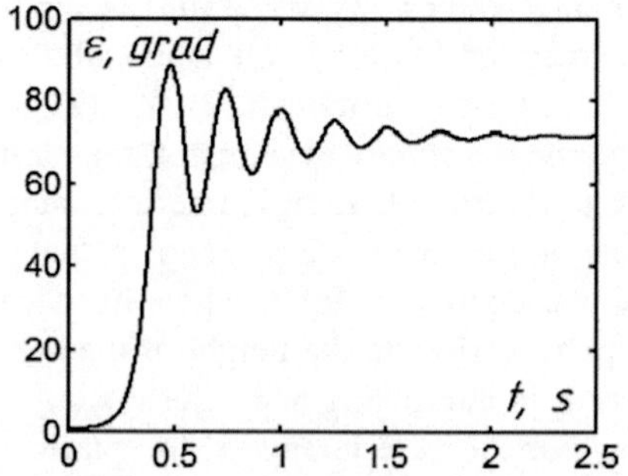

Fig. 5. Charts of angle change.

The sticking angle can be found analytically. When it happened, $\dot{\varepsilon} = \ddot{\varepsilon} = 0$. Substituting these values into differential Eq. (19) leads to a trigonometric equation relatively to the unknown angle ε:

$$(R\omega^2 \sin\varepsilon - gf_s)\cos\varepsilon - (g + Rf_s\omega^2 \sin\varepsilon)\sin\varepsilon = 0\,. \tag{22}$$

Numerical methods must be used to solve Eq. (22). However, at $f_s = 0$, that is, at a completely smooth surface of the disk, it is significantly simplified and has a solution:

$$\varepsilon = \mathrm{Arccos}\frac{g}{R\omega^2}\,. \tag{23}$$

Numerical calculation methods have shown that at $f_s = 0$ and $f_p \neq 0$, the decaying oscillations of the angle ε approach to the value that can be found by formula (23). According to it, with an unlimited increase of the angular velocity of disk rotation, the angle ε tends to 90°, that is, the particle “sticks” at the height of the sphere center. If both surfaces are smooth, that is, $f_s = f_p = 0$, then the oscillations take place infinitely long.

5 Conclusions

Manufacturing of the working surface of the disk of the centrifugal distributor in the form of a spherical segment provides the beginning of the flight of the particle at the time of ascent from the disk at a given angle to the horizontal plane. Consequently, this increases the dispersion zone of technological material. Analytical description of the particle motion makes it possible to investigate the acceleration on the disk along with the blades, as well as to find the relative and absolute velocities at the moment of the ascent of the particle from the disk. The obtained analytical dependencies allow determining the influence of structural and technological parameters on the process of particle acceleration. Prospects of further researches are to study the impact of individual parameters upon the particle acceleration process.

References

1. Cunha, J.O.P., Filno, R.S.: Broadcast distribution uniformity of fertilizer with centrifugal spreaders used in variable rate application. Engenharia Agrícola Jaboticabal **36**(5), 928–937 (2016). https://doi.org/10.1590/1809-4430-Eng.Agric.v36n5p928-937/2016
2. Isaev, Yu.M., Semashkin, N.M., Nazarova, N.N.: Justification of the moving process of the seeds by the spiral-spiral working part. Bull. Ulyanovsk State Agric. Acad. **1**, 97–99 (2011)
3. Tauroa, F., Piscopiab, R., Grimaldi, S.: PTV-Stream: a simplified particle tracking velocimetry framework for stream surface flow monitoring. CATENA **172**, 378–386 (2019). https://doi.org/10.1016/j.catena.2018.09.009
4. Batluk, V., Basov, M., Klymets, V.: Mathematical model for motion of weighted parts in curled flow. Econtechmod Int. Q. J. **2**(3), 17–24 (2013)

5. Liaposchenko, O., Pavlenko, I., Nastenko, O.: The model of crossed movement and gas-liquid flow interaction with captured liquid film in the inertial-filtering separation channels. Sep. Purif. Technol. **173**, 240–243 (2017). https://doi.org/10.1016/j.seppur.2016.08.042
6. Matveev, A.I., Lebedev, I.F., Nikiforova, L.V., Yakovlev, B.V.: Modeling of the particles movement in a screw pneumatic separator. Min. Inf. Anal. Bull. **10**, 172–178 (2014)
7. Pylypaka, S., Klendiy, M., Zaharova, T.: Movement of the particle on the external surface of the cylinder, which makes the translational oscillations in horizontal planes. In: Ivanov, V., et al. (eds.) Advances in Design, Simulation and Manufacturing. DSMIE 2019. Lecture Notes in Mechanical Engineering, pp. 336–345. Springer, Cham (2019). https://doi.org/10.1007/978-3-319-93587-4_35
8. Loveikin, V.S., Romerevych, Y.: Optimization of the swinging mode of the boom crane upon a complex integral criterion. J. Theoret. Appl. Mech. Sofia **49**, 285–296 (2019)
9. Lytvynenko, A., Yukhymenko, M., Pavlenko, I., Pitel, J., Mizakova, J., Lytvynenko, O., Ostroha, R., Bocko, J.: Ensuring the reliability of pneumatic classification process for granular material in a rhomb-shaped apparatus. Appl. Sci. (Switzerland) **9**(8), 1604 (2019). https://doi.org/10.3390/app9081604
10. Pavlenko, I.: Static and dynamic analysis of the closing rotor balancing device of the multistage centrifugal pump. Appl. Mech. Mater. **630**, 248–254 (2014). https://doi.org/10.4028/www.scientific.net/amm.630.248
11. Sklabinskyi, V., Liaposhchenko, O., Pavlenko, I., Lytvynenko, O., Demianenko, M.: Modelling of liquid's distribution and migration in the fibrous filter layer in the process of inertial-filtering separation. In: Ivanov, V., et al. (eds.) Advances in Design, Simulation and Manufacturing. DSMIE-2018. Lecture Notes in Mechanical Engineering, pp. 489–497. Springer, Cham (2019). https://doi.org/10.1007/978-3-319-93587-4_51
12. Liaposhchenko, O., Pavlenko, I., Monkova, K., Demianenko, M., Starynskyi, O.: Numerical simulation of aeroelastic interaction between gas-liquid flow and deformable elements in modular separation devices. In: Ivanov, V., et al. (eds.) Advances in Design, Simulation and Manufacturing II. DSMIE-2019. Lecture Notes in Mechanical Engineering, pp. 765–774. Springer, Cham (2020). https://doi.org/10.1007/978-3-030-22365-6_76
13. Pavlenko, I., Liaposhchenko, A., Ochowiak, M., Demyanenko, M.: Solving the stationary hydroaeroelasticity problem for dynamic deflection elements of separation devices. Vibr. Phys. Syst. 2018026 (2018)
14. Lishchenko, N.V., Larshin, V.P.: Profile gear grinding temperature reduction and equalization. J. Eng. Sci. **5**(1), A1–A7. https://doi.org/10.21272/jes.2018.5(1).a1
15. Shevchenko, S., Mukhovaty, A., Krol, O.: Gear clutch with modified tooth profiles. Procedia Eng. **206**, 979–984 (2017). https://doi.org/10.1016/j.proeng.2017.10.581
16. Ivanov, V., Dehtiarov, I., Pavlenko, I., Liaposhchenko, O., Zaloga, V.: Parametric optimization of fixtures for multiaxis machining of parts. In: Hamrol, A., Kujawińska, A., Barraza, M. (eds.) Advances in Manufacturing II, MANUFACTURING 2019. Lecture Notes in Mechanical Engineering, pp. 335–347 (2019). https://doi.org/10.1007/978-3-030-18789-7_28

Strength Calculation Method for Steel Wire Rope Considering Broken Wires

Ivan Chaiun and Pavlo Vovk(✉)

Odessa National Polytechnic University,
1, Shevchenko Avenue, Odessa 65044, Ukraine
vovk.pavel.1995@gmail.com

Abstract. Relying upon a study of elastoplastic deformation, the authors define the load capacity of wire ropes subject to pure tension and allowing for the resistance of broken wires. Reaching by the most loaded wire under the tension of its ultimate balanced strain, whose value is determined based on a diagram σ-ε of tensile testing, is used as a criterion for the exhaustion of load capacity. For determining the resistance force for broken wires, two problems have been solved. The first one regards friction of a helically-formed thread (wire) against a cylinder having a straight element. The second problem (concerning double-lay ropes) focuses on the friction of wire against the surface of a cylinder having a helically-formed element. The parameters determining the resistance of broken wires are: a friction coefficient, a distance of breakage from the reference section of a rope, angles of wire twist in strands (single-lay ropes), and strand lay angles in a double-lay rope. Being illustrated by the particular rope structures, a dependence of their load capacity from a friction coefficient, number of broken wires. The divergence between theoretical values of the load capacity of ropes and the empirical data does not exceed 6%.

Keywords: Load capacity of wire ropes · Deformation of wire ropes · Friction in a wire rope

1 Introduction

In the course of the use of a rope, some wires become broken, and that is their natural operational condition. Safety of their use is achieved subject to compliance with rejection regulations, given a number of broken wires within the limits of certain lengths of rope segments. The rejection regulations were set forth mainly on the basis of empirical results of the use of wire ropes of certain types and some particular conditions of their use, without any sufficient analytic study of the mechanics of involvement of broken wires in the wire ropes' strength performance. As is known, broken wires contribute to the load capacity of wire ropes to a certain extent, due to friction, i.e., are reflected in their strength performance.

V. Tonkonogyi et al. (Eds.): InterPartner 2020, LNME, pp. 247–257, 2021.
https://doi.org/10.1007/978-3-030-68014-5_25

2 Literature Review

Optimization of a solution concerning exhaustion of the load capacity of a wire rope under some particular conditions of use is a crucial improvement of their strength predictions [1–11]. Avast information as to the main criterion for safe use, being an admissible number of breakages in a lay length, is given in the mono-graph [12]. Thus, for double-lay wire ropes used as the main rope in shaft sinking, the criterion numeral is equal to 5% from a whole number of wires in a wire rope. As regards this one and other parameters, the criteria of the limit state differ from country to country. For example, in the Republic of South Africa for cable-laid ropes, a sectional area of broken wires with respect to their aggregate area shall not exceed [13]:

- 4% if there are breakages are on one strand or two adjacent strands;
- 7% if at least half a total number of all the breakages are on more than two adjacent strands.

Pursuant to the Rules [14], an analysis for static load strength of ropes shall be performed according to a formula as follows:

$$P_{az} = K \cdot P_c \geq S \cdot [n], \tag{1}$$

where: P_{az} is a strength property (load capacity of a rope) referred to in [13] as an actual aggregate breaking strength; $K = 0.83$ – reduction factor P_{az} in comparison with P_c; P_c; $P_c = \sigma_в \cdot A$ is an actual aggregate breaking strength of all the wires of a rope ($\sigma_в$ is an ultimate stress limit of wires according to their marking group; A is an aggregate area of wires); S is the maximum tension among the parts of a rope; $[n]$ – standard safety margin.

In the Codes and Standards of DIN [15] a similar characteristic P_{az} is called “Mindest bruchkraft”, which is a minimum breaking tension of a rope, and also $K = 0.75{-}0.90$: a less value pertains to a three-layer double-lay rope. A more value pertains to a one-layer rope.

The thesis [16] also offers a theoretical definition of a coefficient K based on a study of elastoplastic deformation of a rope, with the use of a diagram representing wire extension according to an elastoplastic body model with linear hardening. The study contains formulas calculating coefficients K for main types of ropes, but not allowing for wire breakages. Now, it seems expedient to develop formulas allowing for wire breakages.

The authors [17] offer a study of resistance of broken wires due to friction and their contribution to the load capacity of ropes. The calculation formulas, however, contain some errors and may not be deemed universal for revealing a connection among wire breakages, their resisting forces caused by friction, and any change in ropes’ load capacity. The essence of the said errors is described in detail in Sect. 3 of this thesis, with a scholarly discussion thereof.

3 Research Methodology

The purpose of this work lies in the analytical study of frictional force caused by extension of wires in a rope, with respect to their helical shape, and, on its basis, determination of the contribution of broken wires to a rope's load capacity (aggregate breaking strength), as well as a discussion of the supposed errors [17].

3.1 The Friction of a Helically Formed Wire upon Extension on a Cylinder with Straight Generatrix

This study is similar to an Euler problem dedicated to the friction of a thread on a circular cylinder. The difference is that a thread (wire) is arranged on the cylindrical surface not concentrically but in a helical curve [18]. The applied relevance of the problem pertains to a study of the deformed state of tension of wires in single-lay ropes.

The main parameter of a helical line is the radius of curvature (Fig. 1).

$$\rho = r / \sin^2 \alpha, \tag{2}$$

where: r is the radius of the base cylinder; α is an angle of obliquity of the helical line.

Let us consider an equilibrium of the helical element ds with the x, y, z coordinate axes.

$$\sum z = (T + dT)\cos\frac{d\varphi}{2} - T\cos\frac{d\varphi}{2} - dF = 0; \tag{3}$$

$$\sum z = dN - T\sin\frac{d\varphi}{2} - (T + dT)\frac{\sin d\varphi}{2} = 0, \tag{4}$$

where: dF is the frictional force within the limits of the element ds; dN is a reaction to the effect of pressing force caused by the tension of a wire.

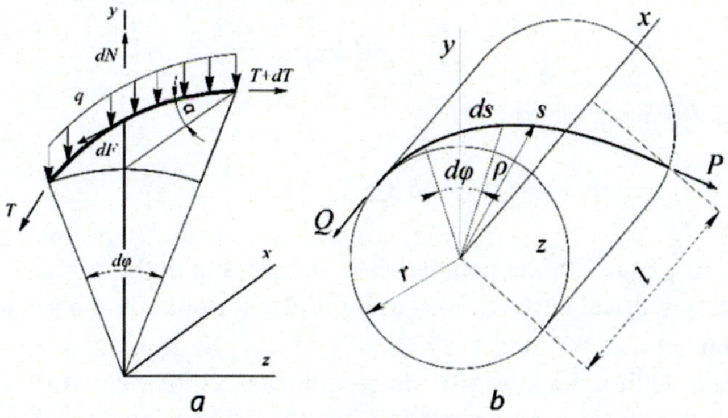

Fig. 1. Analytical model of friction of a thread (wire).

The frictional force within the limits of *ds:*

$$dF = dNf, \tag{5}$$

where f is a friction coefficient.

Thus,

$$dT = Tfd\varphi. \tag{6}$$

Let us also present the parameters of the helical line through an arch

$$d\varphi = \frac{2\pi \sin \alpha}{h} dx, \tag{7}$$

where h is a lead of helix.

Then, the presentation of (6, 10) with due regard for (7, 11)

$$dT = Tfd\varphi = Tf\frac{2\pi \sin \alpha}{h} dx. \tag{8}$$

Separation of variables results in (see Fig. 1, b)

$$\int_Q^P \frac{dT}{T} = f\frac{2\pi \sin \alpha}{h} \int_0^l dx. \tag{9}$$

Upon integrating, we get

$$ln\frac{P}{Q} = f2\pi\bar{l}\sin \alpha \Rightarrow \frac{P}{Q}e^{f\gamma}, \tag{10}$$

where $\bar{l} = l/h$ in leads of helix the length of its segment, which is in contact with the base cylinder;

$$\gamma = 2\pi\bar{l}\sin \alpha. \tag{11}$$

Given (10), the frictional force

$$F = Q\left(e^{f\gamma} - 1\right) \Leftrightarrow F = Q\left(1 - e^{-f\gamma}\right). \tag{12}$$

In the resultant formula (12), the parameter γ is equivalent to the angle of contact of the thread with the cylindrical surface from to the Euler formula. In a particular case, with the angle of contact $\alpha = 90°$ and the parameter $\bar{l} = 1$ the formula (12) coincides with the Euler formula with the angle of contact with the cylinder equal to 2π.

Figure 2 is, therefore, a representation, in dependency graphs, of frictional force F of a thread in the helical form under tension on the cylindrical surface of a circular cylinder with straight generatrix. The angle of α in degrees of a helical line to the axis of the cylinder, a number of wreaths from $\bar{l} = 5$ to 0.1 of contact with the cylindrical surface and friction coefficient f are predictor variables.

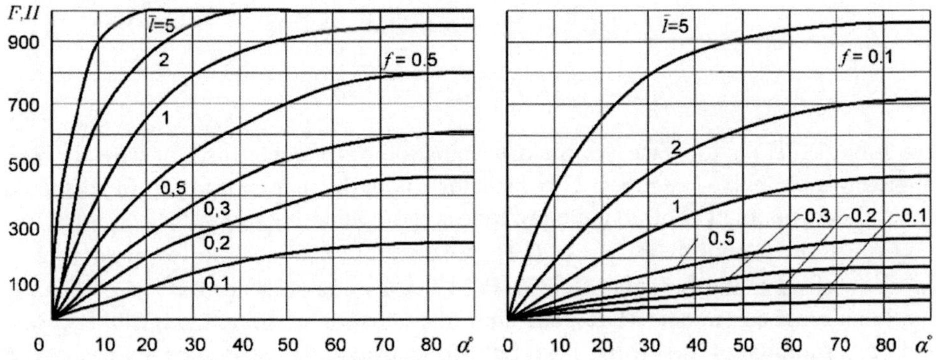

Fig. 2. Frictional force of a thread in helical form under tension on a cylindrical surface.

3.2 The Friction of a Helically Formed Wire Under Tension on a Cylinder with Helical Generatrix

This problem has applied relevance for double-lay ropes. For the central wire of a strand, the angle of its lay to the axis of a strand is $\alpha = 0$, and to the axis of a wire rope itself is equal to β. It follows from here that the central wire of a strand is the base cylinder with a helical generatrix at the angle β to its axis. On the surface of such a cylinder, in helical curve at the angle α there are wires of the strands under tensioning. Thus, we get an analytical model of friction of a wire laid in the helical curve, under tension, on the cylindrical surface with a helical generatrix. With that, the differential of a length of the central wire of a strand is the differential of a helical generatrix of the base cylinder.

$$\tilde{d}s = dx/\cos\beta, \tag{13}$$

where β is the angle of obliquity of the base cylinder to its axis, i.e., to the axis of the wire rope.

The wires in the strand's layers are wound around the central wire at the angle of α. With that, their differential of length

$$ds = \tilde{d}s/\cos\beta = dx/\cos\beta\cos\alpha.$$

The angle corresponding to the arch ds

$$d\varphi = \frac{2\pi\sin\alpha}{h\cos\beta}dx. \tag{14}$$

Equilibrium of the helical element ds in the system of coordinates x_n, y_n, z_n of the sectional view of a strand is represented with equations analogous to (3) and (4), and the representations of frictional forces are similar, too:

$$F = Q\left(e^{fb} - 1\right) \Leftrightarrow F = P\left(1 - e^{-fb}\right), \tag{15}$$

$$b = 2\pi\bar{l}\sin\alpha/\cos\beta = \gamma/\cos\beta \tag{16}$$

The formula (15) is universal for the determination of frictional force of wires in both single-lay and double-lay ropes. The distinction is in the index of power: there should be used the formula (11) for single-lay ropes, and double-lay ropes, the formula (16).

Errors in formulas at work [17] are next. First, in the expression $b = 2\pi\bar{l}\sin\alpha/\cos\beta$ of the formula of the work [17] $tg\alpha$ is stated instead of $\sin\alpha$, which is due to an error committed while generating the equation of strain compatibility. Second, the formula of the work [17] fails to consider the resistance of each wire depending on the location of its breakage because all the breakages are deemed as located in one and the same sectional view at a distance of l equal to a length of a segment with some concentration of wire breakages in a wire rope. That reduces the accuracy of the calculations.

3.3 Contribution of Broken Wires to Load Capacity of a Wire Rope

Given a certain value of the length l_0. of a wire from the point of breakage, resistance force almost reaches its ultimate tension. Such wire contributes fully to the load capacity of a wire rope. With distances of $l < l_0$ resistance force becomes reduced: a wire only contributes in part to the load capacity of a wire rope. With a distance of $l = 0$ resistance force becomes reduced: a wire only contributes in part to the load capacity of a wire rope. Hence, reasoning from (15), the calculation formula of resistance force caused by friction of a broken wire, in connection with extension in a wire rope, given its helical form, is as follows:

$$F = P(1 - e^{-fb}) = \frac{\pi\delta^2}{4}\sigma_в(1 - e^{-fb}); \;\; b = 2\pi\bar{l}\sin\alpha/\cos\beta\,; \;\; \bar{l} = l/h\,, \tag{17}$$

where: $P = \frac{\pi\delta^2}{4}\sigma_в$ is breaking tension of a wire; δ and $\sigma_в$ is the diameter and ultimate stress limit of a wire; l is a distance from a reference section of a wire rope to the point of breakage.

Contribution of broken wires to the load capacity of a wire rope may be represented through the loss of its load capacity (Fig. 3):

$$\Delta P = \frac{\pi\delta^2}{4}\sigma_в(n - \Sigma_{i=1}^{n}(1 - e^{-fb}))\cos\alpha\cos\beta\,, \;\; b = 2\pi\bar{l}\sin\alpha/\cos\beta, \tag{18}$$

where n – is a number of wires with breakages in a layer of a single-lay rope or in a layer of a strand in a double-lay rope.

For a whole wire rope, the loss of its load capacity conditioned by the factor analyzed is represented by a sum as follows:

$$\Delta P_c = \sum\nolimits_1^{m_1} \Delta P_1 + k_2 \sum\nolimits_1^{m_2} \Delta P_2 + k_3 \sum\nolimits_1^{m_3} \Delta P_3 + k_4 \sum\nolimits_1^{m_4} \Delta P_4, \quad (19)$$

where: m_1 is a number of wire layers in a single-lay rope or in a metal core of a double-lay rope; m_2, m_3, m_4 are a number of wire layers in a strand of the 2nd; 3rd; 4th layer in a double-lay rope; k_2, k_3, k_4 are a number of strands in the 2nd; 3rd; 4th layer of a double-lay rope.

The load capacity of a wire rope with due regard for wire breakages can be represented as follows:

$$P_{об} = (P_{аг} - \Delta P_c). \quad (20)$$

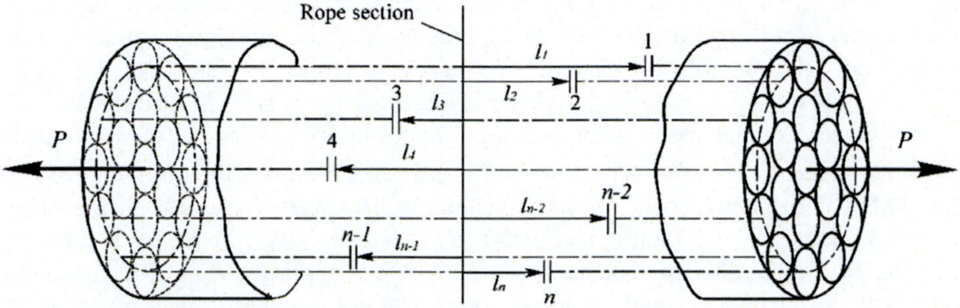

Fig. 3. The analytical model of broken wires' resistance: 1, 2, 3, …, n – points of wires breakage; $l_1, l_2, \ldots, l_n$ is a distance in modulus from a reference section of a wire rope to the point of breakage of wires; P is tension strain.

Given the analytical model [16] and allowing for broken wires, the coefficients K can be represented in a manner as follows:
for single-lay ropes

$$K = \frac{\sum_{i=1}^{s} \left(1 + \left(\bar{\varepsilon}_{np}^{np} \cos^2 \alpha_i - 1\right) \bar{E}_T\right) \left(1 - e^{-f\gamma_i}\right) \delta_i^2 \cos \alpha_i}{\left(1 + \left(\bar{\varepsilon}_{np}^{np} - 1\right) \bar{E}_T\right) \sum_{i=1}^{s} \delta_i^2} v_1; \quad (21)$$

for metal-core double-lay ropes

$$K = \frac{\sum_{i=1}^{s} \left(1 + \left(\bar{\varepsilon}_{np}^{np} \cos^2 \alpha_i \cos^2 \beta_i - 1\right) \bar{E}_T\right) \left(1 - e^{-f\gamma_i}\right) \delta_i^2 \cos \alpha_i \cos \beta}{\left(1 + \left(\bar{\varepsilon}_{np}^{np} - 1\right) \bar{E}_T\right) \sum_{i=1}^{s} \delta_i^2} v_1 v_2; \quad (22)$$

for fiber-core double-lay ropes

$$K = \frac{\sum_{i=1}^{s}\left(1+\left(\overline{\varepsilon}_{np}^{np}\lambda^{-1}\cos^2\alpha_i\cos^2\beta_i-1\right)\overline{E}_T\right)\left(1-e^{-f\gamma_i}\right)\delta_i^2\cos\alpha_i\cos\beta_i}{\left(1+\left(\overline{\varepsilon}_{np}^{np}-1\right)\overline{E}_T\right)\sum_{i=1}^{s}\delta_i^2}\nu_1\nu_2. \quad (23)$$

In the formulae (21)–(23), allowing for the resistance of broken wires is represented by expressions $\left(1-e^{-f\gamma_i}\right)$; $\gamma = 2\pi\bar{l}\sin\alpha$. Concerning the very degree of contribution of every *Nth* broken wire in a strand's layer, it is determined by a distance $\overline{l_i} = l_i/h$ from a reference section of a wire rope to the point of breakage. In calculating by (21)–(23) for broken wires, distances shall be taken as $\overline{l_i} \geq 10$.

For illustrative purposes, we would like to mention a calculation of the coefficient K of a wire rope designed as $1/1.15+6/1+12/1$ with a diameter of 5.15 mm; with an angle of twist being $\alpha_1 = 0°; \alpha_2 = 16.75°; \alpha_3 = 16.16°$. A wire pitch in the outer layer of wire is $h = 44.44$. Let us imagine that within one wire pitch, there are arranged symmetrically four broken wires. Breakages are equally spaced within a pitch (a distance from the reference section of the wire rope to the 1st breakage $l_1 = h/4 = 44.44/4 = 11.11\,\text{mm}$; to the last, 4th breakage is $l_4 = 44.44\,\text{mm}$). In case of an ultimate homogeneous deformation of the wires $\overline{\varepsilon}_{np}^{np} = \varepsilon/\varepsilon_T = 2$ and homogeneous lay tension, i.e., for $\nu_1 = 1$ a coefficient calculated by the formula (21) is $K = 0.869$. If the length of a segment containing breakages increases, all the other conditions remaining the same, the coefficient K will increase. Thus, for a segment equal to 2h, the coefficient will be $K = 0.797$. For a segment being 20h long, the coefficient is practically equal to a state of a wire rope without any broken wires $K = 0.951$. In case of a breakage of all the 4 wires in one sectional view of a wire rope (according to (25) $\Delta P = 0$) $K = 0.761$.

The coefficients K depend considerably on plastic properties of a rope wire, i.e. on $\overline{\varepsilon}_{\text{np}}^{\text{np}}$. For a high-tensile wire, $\overline{\varepsilon}_{\text{np}}^{\text{np}} < 2$. Thus, for $\overline{\varepsilon}_{\text{np}}^{\text{np}} = 1.5$ we get $K = 0.774$.

For pilot testing of the resultant formulae (18) ÷ (20), let us use the results of the tests with 20 samples of a wire rope in work [7], which contains data concerning rupture of samples being 3.0–3.5 m long with broken wires on segments l_k divisible by 0.5 rope lay. The breakages in the samples were evenly distant from one another within the limits of the segments l_k, the number of wire breakages differing for different samples. The breakages were performed by cutting the wires through with pliers. The samples were made of a wire rope type LK-RO 6 × 36(1 + 7 + 7/7 + 14) + o.c. with a diameter of 39 mm, with an aggregate breaking strength $P_{a.o.} = 870$ kN, with an aggregate area of wires being $F_c = 615.95\,\text{mm}^2$, with a rope lay $h_k = 256\,\text{mm}$, and a strand pitch $h = 112\,\text{mm}$.

Table 1 shows the results of a calculation of the load capacity of 20 samples of the said wire rope with a different number of wire breakages in the outer layer of strands. The calculations were performed by the resultant formulae (18)–(20), the friction coefficient being $f = 0.14$, according to [7] for all the samples of a wire rope.

Table 1. Results of a calculation of the load capacity of wire rope.

Wire breakages		Theoretical breaking tension of a wire rope, *kN*							
		Length of a segment with breakages in pitches h_k							
Quantity	$\overline{F}$, %	0	0.5	1	1.5	2	3	4	5
24	14.8	$\frac{825}{–}$	$\frac{836}{5.8}$	$\frac{855}{4.2}$	$\frac{870}{–}$	$\frac{882}{4.5}$	$\frac{901}{–}$	$\frac{914}{–}$	$\frac{923}{–}$
42	25.8	$\frac{729}{–}$	$\frac{744}{1.9}$	$\frac{775}{0.5}$	$\frac{800}{-1.2}$	$\frac{820}{-1.1}$	$\frac{852}{–}$	$\frac{874}{–}$	$\frac{890}{–}$
60	37.1	$\frac{632}{–}$	$\frac{652}{-1.1}$	$\frac{694}{-4.8}$	$\frac{729}{-5.2}$	$\frac{758}{-6.3}$	$\frac{802}{–}$	$\frac{833}{–}$	$\frac{856}{–}$
84	51.8	$\frac{503}{–}$	$\frac{530}{-8.2}$	$\frac{588}{-12.8}$	$\frac{635}{-12.8}$	$\frac{675}{-12.3}$	$\frac{736}{–}$	$\frac{779}{–}$	$\frac{811}{–}$

Note: the numerators show theoretical values of the load capacity of the wire ropes, and the denominators show discrepancies between the theoretical and empirical values in percentage [17].

The error increased on the samples of a wire rope with 84 breakages Δ = –8.2–12.8% may be explained as follows. Breakages in the samples, according to [17] were performed by cutting the wires through with pliers. Break-off pips increase the surface roughness, which means an enhancement of the friction coefficient *f*. The most value of *f* will be reached with 84 wire breakages. But the theoretical values of the load capacity of all the samples were calculated at *f* = 0.14. The calculations made at *f* = 0.20–0.25 (instead of 0.14) show discrepancies also for the samples with 84 breakages as equal Δ = –5.4–0.9%. The features of the experimental determination of the coefficient of friction are considered in [19, 20]. Thus, the suggested analytical model for taking account of broken wires in the determination of the load capacity of steel wire ropes under pure tension is quite correct.

4 Results

- In the case of wire breakages in the same sectional view ($l_k = 0$) and their symmetrical location, the load capacity of a wire rope becomes reduced by an aggregate load capacity of wires so broken.
- As a distance between a wire breakage and a reference section of a wire rope increases, the effect on the reduction of its load capacity decreases. In case of a distance of (4–5) h_k rope lays, wire breakages have almost no effect on the load capacity of a wire rope.
- Dependence of the load capacity of a wire rope on the length of the segments with breakages may be deemed linear.
- The load capacity of a wire rope depends considerably from the ultimate balanced strain ε_{np}^{np} of a wire, which is one of the parameters determining its flexibility.

5 Conclusions

The method suggested is deemed to be a basis for considering wire breakages under some other loading conditions: extension with torsion, extension with torsion, and bent, including with different locations of wire breakages in the sectional views of wire ropes. Such research will lay the groundwork for optimization of the calculations of strength of wire ropes with all ensuing positive values of cost efficiency while maintaining the reliability.

References

1. Marais, J., Bester, N.: A holistic approach to continuous rope monitoring/safe use of ropes. In: Proceedings of the OIPEEC Conference, pp. 85–101. College Station, Texas (2011)
2. Ernst, B., Wehking, K.-H.: The influence of twist on tensile fatigue of wire ropes. In: OIPEEC/MTS Conference Proceedings, pp. 121–137. College Station, TX/USA (2011)
3. Ernst, B., Wehking, K.-H.: The influence of twist on tensile fatigue of wire ropes. OIPEEC Bull. **102**, 5–32 (2012)
4. Ivan, A.: Response of a wire rope strand to axial and torsional loads: asymptotic modeling of the effect of inter wire contact deformations. Int. J. Solids Struct. **48**, 1413–1423 (2011)
5. Kalentev, E., Václav, Š., Božek, P., Tarasov, V., Korshunov, A.: Numerical analysis of the stress-strain state of a rope strand with linear contact under tension and torsion loading conditions. Res. J. Adv. Sci. Technol. **11**, 231–239 (2017)
6. Xiang, L., Wang, H.Y., Chen, Y., Guan, Y.J., Dai, L.H.: Elastic-plastic modeling of metallic strands and wire ropes under axial tension and torsion loads. Int. J. Solids Struct. **129**, 103–118 (2017)
7. Vîlceanu, L., Ghita, E., Puţan, V.: Research on the working life at steel wire ropes in contact. Int. J. Mech. **2**, 140–147 (2012)
8. Vukelic, G., Vizentin, G.: Damage-induced stresses and remaining service life predictions of wire ropes. Int. J. Appl. Sci. **7**, 50–57 (2017)
9. Chaiun, I., Vovk, P.: The formation and development of structural mechanics of the rope. In: Surjaninov, N. (eds.) Abstracts of the 5-th International Scientific-practical Conference "Actual problems of engineering mechanics" 2018, pp. 237–240. Odessa State Academy of Civil Engineering and Architecture, Odessa, Ukraine (2018)
10. Chaiun, I., Vovk, P.: Tensile strength of ropes taking into account wire breaks. Sci.-Tech. Prod. J. Lifting Equip. Spec. Devices **203**, 17–19 (2019)
11. Malinovskiy, V.: Steel Ropes. Astroprint, Odessa, Ukraine (2016)
12. Berezhinskiy, V., Shatilo, A.: Mine hoist ropes. University book, Moscow, Russia (2015)
13. Condition assessment of steel wire rope on mine winders. Code of practice SABS 0293: 1996. South Africa by the South African Bureau of Standards, Republic of South (1996)
14. Rules for the construction and safe operation of cranes. Fort, Kharkiv, Ukraine (2007)
15. DIN-Taschenbuch. Drahtseile: Normen. Berlin-Koln, Beuth (1990)
16. Chauin, I.: The relationship between the accuracy of the design scheme and the safety margins of the ropes. Steel ropes, no. 9, pp. 126–141. Astroprint, Odessa, Ukraine (2013)
17. Beliy, V., Lesin, K., Samarskiy, A.: Selection, linkage, exploitation and monitoring of the condition of mine ropes. Nedra, Moscow, Russia (1967)
18. Chaiun, I., Vovk, P.: Friction of a thread located on a screw line during a cylindrical surface. J. Handling Equip. **2**, 75–81 (2019)

19. Chang, X.-D., Peng, Y.-X., Zhu, Z.-C., Gong, X.-S., Yu, Z.-F., Mi, Z.-T., Xu, C.-M.: Effects of strand lay direction and crossing angle on tribological behavior of winding hoist rope. Materials **10**, 1–20 (2017)
20. Chang, X.-D., Peng, Y.-X., Zhu, Z.-C., Zou, S.-Y., Gong, X.-S., Xu, C.-M.: Evolution properties of tribological parameters for steel wire rope under sliding contact conditions. Metals **10**, 1–6 (2018)

Engineering Methodology for Determining Elastic Displacements of the Joint «Spindle Assembly-Face Milling Cutter» While Machining Planes

Heorhii Vyhovskyi, Mykola Plysak, Nataliia Balytska, Oleksandr Melnyk(✉), and Larysa Hlembotska

Zhytomyr Polytechnic State University,
103, Chudnivska Street, Zhytomyr 10005, Ukraine
o.l.melnyk@ukr.net

Abstract. The work is devoted to the development of a methodology for determining elastic displacements joint of the vertical milling machine spindle assembly with a milling cutter by using the software SolidWorks. It has been proposed to divide simulation modeling into two stages in order to achieve detail of task and simultaneously to limit the tack dimension increase. In the first stage, the calculation of the bearing's stiffness and displacement of the inner ring in the radial and axial directions based on the known values of their radial has been performed. In the second stage, the geometric model of the bearing is replaced by a corresponding virtual connector, which is obtained in the first stage where the stiffness indicators and simulation process carries out. The performed research made it possible to determine the elastic displacements of the spindle connection that occurs during the cutting process. The determined magnitudes of displacement of the spindle joint allowed to determine the angular misalignment of the mill axis relative to the machined surface. Thus, the developed methodology makes it possible to predict the flatness deviation in face milling of flat surfaces, as well as to the acceleration of designing the cutting tool and machine components.

Keywords: Bearing stiffness · Flatness deviation · SolidWorks Simulation · Static analysis · Angular misalignment

1 Introduction

The machines' functionality depends to a large extent on the parts' quality characteristics. The parts exploitation process is often accompanied by alternating power loads, the action of heat sources, corrosive environments, etc. Parts must be able to resist external influences over the plan period – it will ensure the reliable operation of the mechanisms and machines in general. Operational properties such as fatigue resistance, wear resistance, corrosion resistance, etc. depends on the surface layer performance of parts. Achieving the required parameters of the machined surfaces requires an in-depth study of the machining processes and the influence of individual components

V. Tonkonogyi et al. (Eds.): InterPartner 2020, LNME, pp. 258–268, 2021.
https://doi.org/10.1007/978-3-030-68014-5_26

characteristics of the technological system on the final parameters of the quality of machining.

2 Literature Review

One of the most widespread methods of flat surfaces producing is the clean face milling machining. It is known that the quality of machining of flat surfaces to a large extent considerably determined by deformations and displacement of the technological system [1], and at face milling machining of error of machining may constitute 90% of the total error of machining.

One face milling characteristic feature is the variability of the thickness of the cut along the arc of contact of the mill cutting elements with the workpiece. Cutting forces are modifiable while machining, and thus that changes their magnitude and direction of action, resulting in machining errors in the transverse direction due to dynamic deformations of the technological system.

In the longitudinal direction (in the direction of the feed), the errors of machining are related primarily due to the variability of the milling width in the drilled sections of the cutter into the workpiece and the output. Variable milling width results in fluctuations of cutting forces while drilled and at the exit of the mill from contact with a workpiece and variable deformations of the technological system and errors of the longitudinal profile.

Authors [2, 3] using mathematical and computer simulations investigated the impact of structural and geometric parameters of a stepped face mill on characteristics of the cut layer for all the teeth, which cause the load of the cutting edges and, therefore, the fluctuations of the cutting forces.

The questions of dynamic stability research and mathematical modeling of dynamic characteristics of face mills of different designs are considered in [4–7]. The obtained results might be useful in the design stages of new face mills designs with high dynamic stability.

Researchers pay great attention to the study of the influence of machining conditions on the quality characteristics of machined surfaces of parts made of different materials [8–11], this is especially important for finishing and final milling. In [12, 13], models have been developed to predict the quality of the flat surfaces of the workpiece's compounds after milling, which take into account the kinematic and dynamic characteristics of the machining process.

Each element of the technological system contributes to the deformation of the system and thus influences the total processing error. Thus, work [14] is devoted to the determination of elastic deformations on a tool. The authors investigated the stress-strain state of the face oblique mills and its influence on the formation of the macro-profile of the treated surfaces. Studies [15–20] show that the characteristics of the spindle unit significantly influence the formation of relative position accuracy indexes and machined parts shape. The authors of these works, after a large number of calculations and experiments, agree that the influence of the static axis inclination angle of the spindle on the surface flatness error is very significant.

The spindle units have high requirements on rotation uniformity, vibration resistance, speed, and also high elastic characteristics. The accuracy of the spindle unit is evaluated by taking into account the amount of error it contributes to the total processing error.

In [21, 22], the problems with determining the contact stiffness of a taper connection between the face milling cutter and the machine spindle were investigated. The authors of [21] proposed to increase the radial stiffness of a conical connection 7:24 of a cutter with a machine spindle by performing a cutter shank with two centering faces.

The impact of the spindle units on forming indicators of the relative positioning accuracy of the surfaces and the accuracy of their shape is from 50% to 90% in the overall accuracy of machine distribution. However, the known methodologies of both analytical and simulation studies of stiffening characteristics are nowadays characterized by high complexity, time-consuming, and computational power.

In the paper [23], the author proposes to use a preliminary numerical estimation of the bearings stiffening characteristics with the help of SolidWorks Simulation to simplify and accelerate studies of the antenna resonance characteristics. The approaches proposed in [23] can form the basis of a methodology for investigating the spindle units stiffening characteristics of the metal cutting machines using SolidWorks Simulation.

Thus, the engineering methodology development for the simplified determination of elastic displacements of the connection "Spindle Unit and cutting tool" when milling planes is an important scientific and technical task. This methodology can be used to accelerate the parametric design process of the cutting tool and machine units, as well as to study the influence of both the individual component's design parameters of the technological system and the cutting conditions on the final quality of machining.

3 Research Methodology

The 3D model of the vertical milling machine spindle unit, which is shown in Fig. 1, was made in SolidWorks. In the spindle unit back support there is the double row cylindrical roller bearing NN3022 (Fig. 1, position 4), and in the front support - the double row cylindrical roller bearing NN3024 (Fig. 1, position 7) and the double row angular contact ball bearing 234424 (Fig. 1, position 6), which fixes the spindle in relativity to the housing in the radial and axial directions [5].

The stress-strain state comprehensive engineering analysis of the vertical milling machine spindle was performed using the Simulation module.

The calculation was performed for a vertical milling machine with a fixed face mill (diameter D = 440 mm, number of inserts z = 24, milling width B = 350 mm) for two cases:

1. Finishing of cast iron with a hardness of 160…250 HB (cutting depth 0,75 mm, feed per tooth 0,1 mm/t, spindle speed 380 min-1, tangential cutting force 367 N, radial cutting force 162,350 N, axial cutting force 91,861 N, spindle torque 800 Nm;

2. Roughing of high-temperature corrosion resistance steel with a tensile strength 610 MPa (cutting depth 6 mm, feed per tooth 0,25 mm/t, spindle speed 100 min-1, tangential cutting force 9728 N, radial cutting force 4377,6 N, axial cutting force 2432 N, spindle torque 21400 Nm.

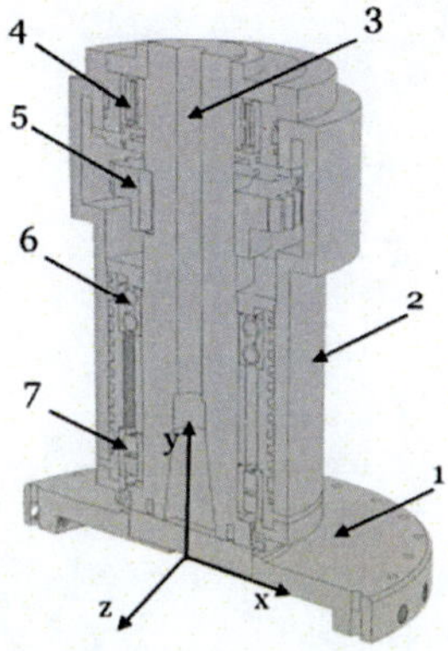

Fig. 1. The spindle unit 3D model in section: 1– face mill; 2 – spindle housing; 3 – spindle; 4 – double row cylindrical roller bearing NN3022; 5 – gear; 6 – double row angular contact ball bearing 234424; 7 – double row cylindrical roller bearing NN3024

It is necessary to know the stiffening characteristics of the spindle bearings under radial load (perpendicular to the axis) and at an axial load (along with it) to calculate the spindle unit displacement. For determination stiffness characteristics of bearings, analytical assessments exist, but the corresponding formulas are quite complex and approximate [23]. In this regard, it was decided to obtain a numerical assessment of the stiffness characteristics of bearings using SolidWorks Simulation.

The geometric 3D models of bearings were formed in order to obtain the stiffness characteristics perpendicularly (X, Z axes) and along the axis (Y-axis). A model, for example, angular contact ball double-row bearing, is shown in Fig. 2.

Preparation of a Geometric Model. Using the SplitLine command, on the balls and rings (outer/inner) were allocated faces for localization contact (Fig. 3), which should be completely covering the contact area of the balls with rings (outer/inner) [5]. For reproducing the bearing to fit in a rigid bushing, fixation of the outer rings in all directions, the “Fixed Geometry” command was used, and radial force was applied to the inner rings from Table 1. The load on the inner rings was applied using the option “Remote Loads” command in the context “Rigid connection”. The calculated values of the radial forces are given in Table 1.

Contact boundary conditions between balls, rollers, and bearing rings were appointed using auxiliary faces on the rings, which were created to localize contact and to simplify the calculation (Fig. 4). The “No Penetration” contact was used option in the context “Surface to surface”. In order to accelerate the analysis, a virtual connector

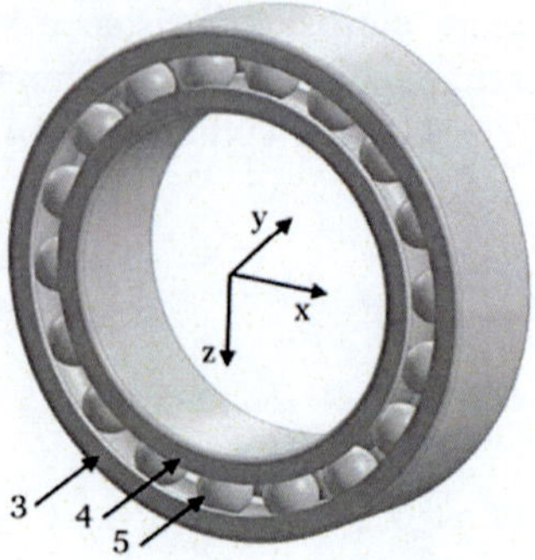

Fig. 2. Angular contact bearing 234424: 3 – outer ring; 4 – inner ring; 5 – balls.

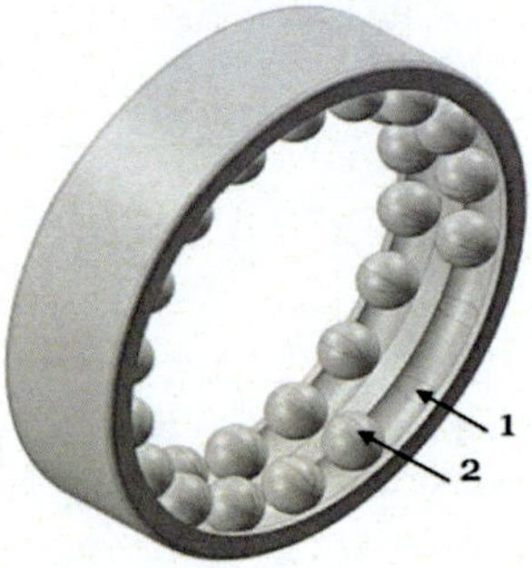

Fig. 3. The geometric model of the bearing for calculated 234424 with allocated faces for localization contact: 1 – on the bearing ring; 2 – on the bearing balls.

Table 1. The radial force values P, which are an action on the bearings' inner rings.

Bearings:	Radial force *P*, N	
	Processed material – cast iron	Processed material – high-temperature corrosion resistance steel
NN3022	115	3179
234424	181	4877
NN3024	550	14600

"Elastic support" was used instead of a separator [23] with properties, as shown in Fig. 5.

The finite element meshes were a generation in the following way: to faces outer ring and inner rings, balls, and areas of allocated contact was used «Mesh Control» parameter. A density mesh (size final element is 0.5 mm) in the areas increases the accuracy results, and also reduces the discretization errors [24, 25]. Static analysis parameters were set up with the following parameters: "Improve accuracy for no

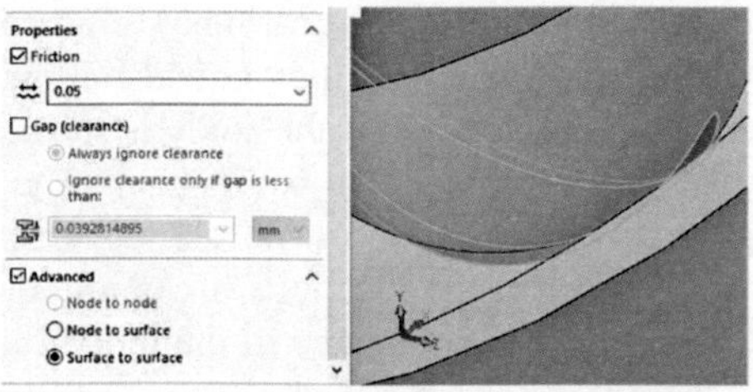

Fig. 4. Appointing contact conditions between bearing's rings and balls.

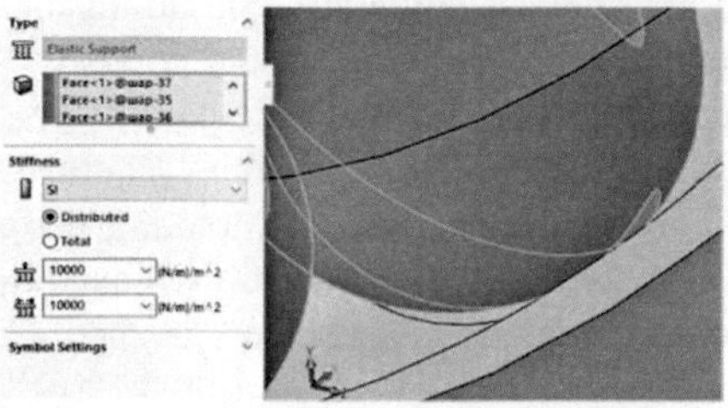

Fig. 5. Appointing virtual connector to stabilize a movement of the balls.

penetration contacting surface", "Use soft spring to stabilize model," and "Direct Spears" program was used for calculation.

Static studies have been determined values of the radial and axial displacement inner rings bearings. On this basis, bearings radial and axial stiffness were calculated by the following dependence:

$$J = P/S, \quad (1)$$

where P is a radial load on the bearing, N; S – displacements inner ring in the radial and axial direction, mm.

The calculated values radial J_X and axial J_Y stiffness are given in Table 2.

Table 2. The calculated values radial and axial bearings stiffness.

Bearings:	The calculated values radial and axial bearings stiffness, kN/mm			
	Cast iron		High-temperature corrosion resistance steel	
	J_X	J_Y	J_X	J_Y
NN3022	1150	575	1059,7	635,8
234424	362	301,7	487,7	243,85
NN3024	1375	785,7	1622,2	730

The calculated values of radial and axial bearings stiffnesses make it possible to determine the stiffness of the spindle unit of the vertical milling machine.

The simulation spindle assembly model for the study is shown in Fig. 6. 3D models have been left housing parts which contact the bearings and the spindle, and also face mill, which the cutting forces were applied. The spindle displacements have been formed three displacement sensors at the locations in spindle unit supports (Fig. 6).

For correctly appointment bearing parameters in the model, it was decided to make the following changes, as follows: were selected faces on the spindle surface and housings in the locations bearings using the Split Line command. The created faces should cover the width interaction area outer and inner rings bearings with spindle and housing. It was applied to contact boundary conditions between the spindle and housing parts using contact "No Penetration" in the context "Surface to surface".

Specificities bearing parameters set up, with the parameters "Allow self-alignment" and "Flexible", are shown in Fig. 7. The "Flexible" parameter makes it possible to deform or move the shaft along the axis without making it rigid [23]. The parameter "Stabilize shaft rotation" was used to prevent rotational instability (caused by rotation), which can lead to erroneous calculations and errors in the calculation process.

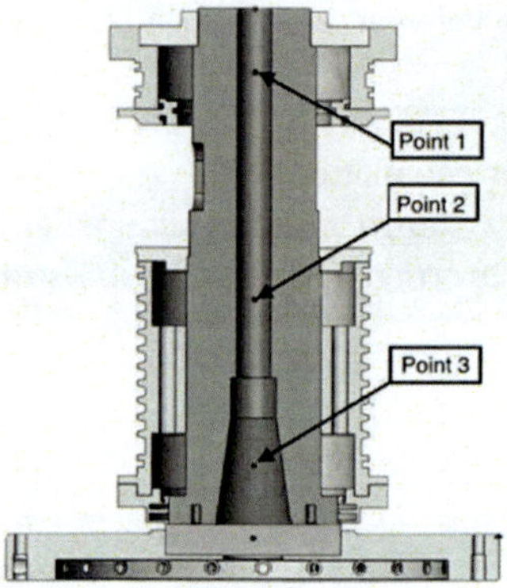

Fig. 6. Geometric 3D model of spindle connection and location sensors for determining the elastic displacements of the spindle.

The parameters of static analysis were adjusted with the following parameters: "Improve accuracy for no penetration contacting surface", "More accurate", "Compute free body forces", "Use soft spring to stabilize model", and for the calculation of the program "FFEPlus" was used.

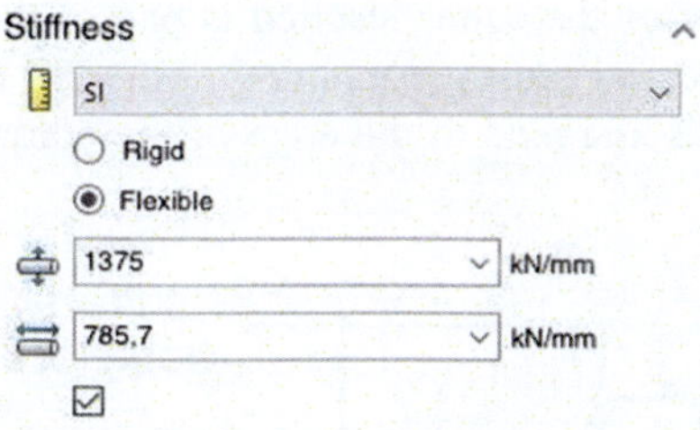

Fig. 7. Assigning connector-bearing parameters.

4 Results

The results of determining radial spindle displacements are presented in Fig. 8. It demonstrates the variability of displacement values for different cutting conditions (finishing/roughing, material to be processed), which determine the magnitude and ratio of the components of the cutting forces.

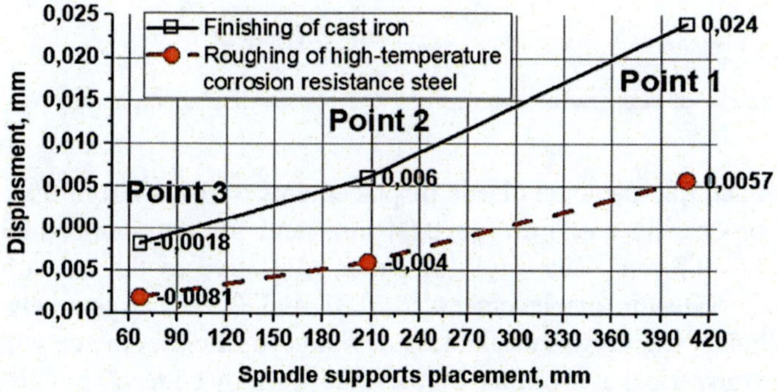

Fig. 8. Graphs of displacements of the spindle relative to its axis in the radial direction.

Deviation of the axis of the spindle due to elastic deformation leads to the warp of the milling cutter in the machining process. The equation allows calculating the maximum flatness deviation (Fig. 9) of the machined surface of the workpiece:

$$f = \left(D_m - \sqrt{D_m^2 - B^2}\right) \cdot \frac{sin\alpha}{2}, \quad (2)$$

where f – is the maximum deviation from the plane, mm; D_m – diameter of a mill, mm; B – milling width, mm; α – is the axis inclination angle of the mill, deg.

In Fig. 9, parameter C is the offset of the milling axis relative to the axis of symmetry of the workpiece.

The axial displacement of the cutter inserted is one of the reasons for the flatness deviation of the workpiece and causes machining errors. To determine the position of the cutter insert sensor was assigned to the edge of the cutting tool Fig. 10 (point 4).

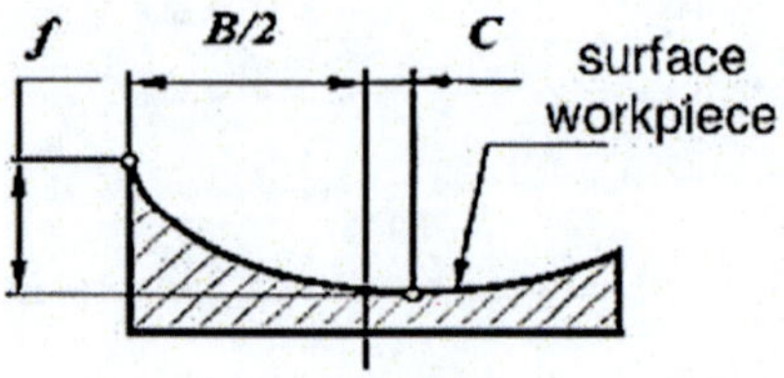

Fig. 9. The profile of the machined surface of the part when inclination the face mill.

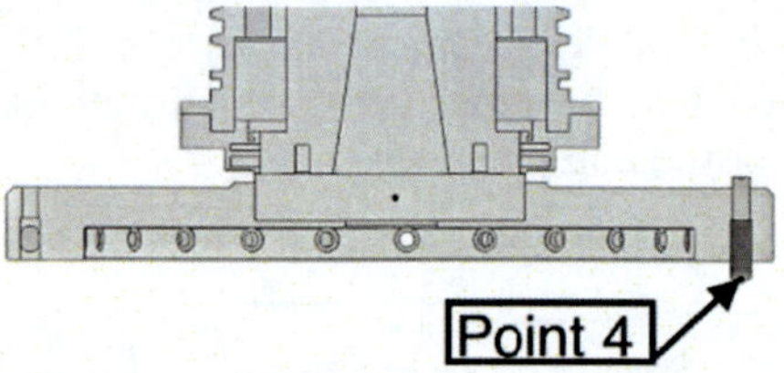

Fig. 10. Placement of sensors for determining the elastic displacements of the cutter insert.

It is determined that the axial elastic displacements of the insert of the cutter for the conditions of processing cast iron are 0,04 mm, and for high-temperature corrosion resistance steel – 0,3 mm. The angle of cutter α relative to the axis of the spindle connection due to elastic displacements is 0,01 and 0,09°, respectively. It made it possible to calculate the maximum flatness deviation of the workpiece with a width of 350 mm for symmetrical machining of the face milling cutter $D = 440$ mm for the above conditions. Maximum flatness deviations are 0,015 mm in the processing of cast iron and 0,14 mm in the processing of high-temperature corrosion resistance steel.

5 Conclusions

The research presents the engineering methodology for a static calculation of the joint of the vertical milling machine spindle assembly with a milling cutter in the Solid-Works Simulation environment to determine the elements' elastic displacements. A feature of the proposed methodology is two-stage modeling to detail the task and limit its dimension, which speeds up the calculation.

At the first stage, the stiffness of the bearings is calculated according to the known values of their radial load and displacements of the inner ring in the radial and axial directions. In the second stage, the geometric model of the bearing is replaced by a suitable virtual connector with the stiffness that was set in the first stage, and simulation is performed.

The stiffness of the bearings is calculated according to the known values of the radial load and the displacements of the inner ring in the radial and axial directions.

In order to obtain the values of elastic displacements of the spindle joint in the machining process, sensors were formed at the locations of the supports of the spindle assembly and the milling cutter insert.

The maximum flatness deviation of the machined surface of the workpiece with a width of 350 mm was determined for symmetrical machining of the face milling cutter D = 440 mm, namely 0,015 mm at the finishing processing of cast iron and 0,14 mm at the rough processing of high-temperature corrosion resistance steel.

The developed methodology can be used to analyze and predict the influence of the connection design between machine spindle assemblies and tools on the geometric accuracy of machined flat surfaces. It will reduce the number of costly full-scale tests.

References

1. Schmitz, T.L., Ziegert, J.C., Canning, J.S., Zapata, R.: A comparison of error sources in high-speed milling. Precis. Eng. **32**(2), 26–133 (2008)
2. Hlembotska, L., Melnychuk, P., Balytska, N., Melnyk, O.: Modelling the loading of the nose-free cutting edges of face mill with a spiral-stepped arrangement of inserts. East.-Eur. J. Enterp. Technol. **1**(91), 46–54 (2018)
3. Hlembotska, L., Balytska, N., Melnychuk, P., Melnyk, O.: Computer modelling power load of face mills with cylindrical rake face of inserts in machining difficult-to-cut materials. Sci. J. TNTU **93**(1), 70–80 (2019)
4. Stepchin, Ya.: Comparative characteristic of dynamics of processes of face milling by mills of standard and special designs. Bull. ZhSTU Tech. Sci. **1**(72), 51–56 (2015)
5. Stepchin, Ya.: Mathematical model of transverse oscillations of a face mill at a finish milling. Bull. ZhSTU Tech. Sci. **1**(64), 11–15 (2013)
6. Stepchin, Ya.: Mathematical model of dynamics of face milling of different cutting schemes. Bull. ZhSTU Tech. Sci. **3**(66), 37–41 (2013)
7. Zhenyu, S., Luning, L., Zhanqiang, L.: Influence of dynamic effects on surface roughness for face milling process. Int. J. Adv. Manuf. Technol. **80**(9–12), 1823–1831 (2015)
8. Sheth, S., George, P.M.: Experimental investigation and prediction of flatness and surface roughness during face milling operation of WCB material. Procedia Technol. **23**, 344–351 (2016)
9. Bajić, D., Lela, B., Živković, D.: Modeling of machined surface roughness and optimization of cutting parameters in face milling. Metalurgija **47**(4), 331–334 (2008)
10. Sheth, S., George, P.: Experimental investigation and optimization of flatness and surface roughness using grey relational analysis for WCB material during face milling operation. Advances in Intelligent Systems Research, vol. 137, pp. 65–70 (2017)
11. Moskvin, P., Balytska, N., Melnychuk, P., Rudnitskyi, V., Kyrylovych, V.: Special features in the application of fractal analysis for examining the surface microrelief formed at face milling. East. Eur. J. Enterp. Technol. **2**(1(86)), 9–15 (2017)
12. Denkena, B., Köhler, J., Hasselberg, E.: Modeling of workpiece shape deviations in face milling of parallel work piece compounds. In: 2013 14th CIRP Conference on Modeling of Machining Operations, Procedia CIRP, vol. 8, pp. 176–181. Elsevier B.V. (2013)

13. Denkena, B., Hasselberg, E.: Influence of the cutting tool compliance on the workpiece surface shape in face milling of workpiece compounds. In: 2015 15th CIRP Conference on Modelling of Machining Operations, Procedia CIRP, vol. 31, pp. 7–12. Elsevier B.V. (2015)
14. Vyhovskyi, H., Plysak, M.: Investigation of the stress-strain state of the face mill for finishing planes. Bull. ZhSTU Tech. Sci. **1**(83), 53–58 (2019)
15. Gu, F., Melkote, S.N., Kapoor, S.G., DeVor, R.E.: Model for the prediction of surface flatness in face milling. J. Manuf. Sci. Eng. Trans. ASME **119**(4), 9 (1997)
16. Nguyen, H.T., Wang, H., Hu, S.J.: Modeling cutter tilt and cutter-spindle stiffness for machine condition monitoring in face milling using high-definition surface metrology. Int. J. Adv. Manuf. Technol. **5**(8), 1323–1335 (2014)
17. Chauhan, P., Patel, S., Patel, K.: Experimenting and predicting flatness and surface roughness during face milling operation of CF8M grade A-351. In: International Conference on Research and Innovations in Science, Engineering &Technology, vol. 1, no. 1, pp. 537–547 (2017)
18. Pimenov, D.Y., Guzeev, V.I., Krolczyk, G., Mia, M., Wojciechowski, S.: Modeling flatness deviation in face milling considering angular movement of the machine tool system components and tool flank wear. Precis. Eng. **54**, 327–337 (2018)
19. van Luttervelt, C.A., Willemse, H.R., Pekelharing, A.J.: Stifness measurements as a check of machine-tool spindles. CIRP Ann. Manuf. Technol. **31**(1), 251–254 (1982)
20. Pimenov, D.Y., Guzeev, V.I., Mikolajczyk, T., et al.: A study of the influence of processing parameters and tool wear on elastic displacements of the technological system under face milling. Int. J. Adv. Manuf. Technol. **92**, 4473–4486 (2017)
21. Namazi, M., Altintas, Yu., Abe, T., Rajapakse, N.: Modeling and identification of tool holder–spindle interface dynamics. Int. J. Mach. Tools Manuf. **47**(9), 1333–1341 (2007)
22. Melnyk, O., Hlembotska, L., Balytska, N., Holovnia, V., Plysak, M.: The imitation study of taper connections stiffness of face milling cutter shank using machine spindle in the SolidWorks Simulation Environment. In: Advances in Design, Simulation and Manufacturing II. DSMIE 2019. Lecture Notes in Mechanical Engineering, pp. 602–612. Springer, Cham (2020)
23. Aliamovskyi, A.A.: SolidWorks Simulation. Engineering analysis for professionals: tasks, methods, recommendations. DMK, Moscow (2015)
24. Aliamovskyi, A.A.: SolidWorks/COSMOSWorks 2006–2007. Engineering analysis by finite element method. DMK, Moscow (2007)
25. Aliamovskyi, A.A.: Engineering calculations in SolidWorks Simulation. DMC Press, Moscow (2015)

Advanced Materials

Effect of Barocryodeformation Degree at 77 K on the Precipitation Structure in CuCrZr Alloy

Alla Belyaeva[1], Ivan Kolenov[2,3](✉), Pavel Khaimovich[4], Alexey Galuza[1], and Alla Savchenko[1]

[1] National Technical University "Kharkiv Polytechnic Institute", 2, Kyrpychova Street, Kharkiv 61002, Ukraine
alla.iv.belyaeva@gmail.com

[2] Institute of Electrophysics & Radiation Technologies of NAS of Ukraine, 28, Chernyshevsky Street, Kharkiv 61002, Ukraine

[3] O.Ya. Usikov Institute for Radiophysics and Electronics of NAS of Ukraine, 12, Ac. Proskura Street, Kharkiv 61085, Ukraine

[4] National Scientific Center "Kharkiv Institute of Physics and Technology", 1, Akademicheskaya Street, Kharkiv 61108, Ukraine

Abstract. The effect of severe plastic deformation by barocryodeformation at liquid nitrogen temperature on the structure of precipitation-hardened commercial CuCrZr alloy is investigated. In particular, the dependence of the microstructure, the preferential orientation of the crystallites, and the lattice period of the alloy on the deformation degree were studied. The samples were preliminarily water quenched from a premelting temperature to create a supersaturated solid solution of dopants in the copper matrix. It was shown that the evolution of the average size and distribution density of the precipitates with the degree of deformation increase indicates the occurrence of competing processes in the alloy. It was shown that the decomposition of the solid solution due to the strain-induced acceleration of diffusion processes and strain-induced dissolution of the precipitates definitely contribute to the evolution of the structure. The conclusion about the simultaneous occurrence of kinetically opposite phase transformations was confirmed by the observation of a nonmonotonic change in the lattice constant of the matrix and the precipitates density with increasing accumulated deformation.

Keywords: Severe plastic deformation · Deformation-induced precipitation · Deformation-induced dissolution · Copper alloy

1 Introduction

Low-alloyed precipitation-hardened (PH) coppers, in particular, CuCrZr ternary systems, have high electrical and thermal conductivity and, at the same time, can stand large mechanical loads under high-temperature conditions. Due to the high stability of their properties under radiation conditions [1], these systems are considered promising materials for structural elements of nuclear reactors [2–5]. CuCrZr alloys are in the focus of many investigations for several decades, and to date, the possibilities of traditional approaches to improving their properties have been exhausted. Recently, it

V. Tonkonogyi et al. (Eds.): InterPartner 2020, LNME, pp. 271–281, 2021.
https://doi.org/10.1007/978-3-030-68014-5_27

has been demonstrated that the physical and mechanical properties of PH copper alloys can be significantly improved by nanostructuring using severe plastic deformation (SPD) methods [6–10]. SPD allows one to accumulate large deformations in a material. For example, during equal-channel angular pressing (ECAP), the plastic strain value is about 8–12. Various methods of SPD can be used both individually and in combination with other SPD methods, mutually complementing each other to obtain optimal physical and mechanical properties. Below several studies devoted to the effect of different SPD on the structure and properties of CuCrZr alloy are discussed.

2 Literature Review

The results of experimental studies of the CuCrZr alloy after heat treatment and ECAP with subsequent aging were presented in [4, 11]. It was shown that ECAP allows controlling the structure and properties of this material, causing the transformation of the precipitation-hardened composite into a mixed type composite, namely, into a dispersion-fiber metal composite. ECAP may lead not only to significant refinement of the structure fragments (up to 200–300 nm) [4] and an increase in the defects density but also to the formation of an anisotropic structure oriented along the deformation axis. Since the precipitates are much larger than the Cu matrix structure elements, their distribution is not related to boundaries between the matrix elements. It contributes to the homogenization of the structure and the improvement of the alloy properties. Below several studies of various SPD methods, combinations are discussed.

Kulczyk et al. (2007) [12] have shown that the combined hydrostatic extrusion (HE) and ECAP deformation of copper greatly increases the homogeneity of the microstructure and improves its mechanical properties considerably as compared to a single technique (either HE or ECAP). In [13, 14] showed that a combination of ECAP and barocryodeformation at 77 K allows improving the CuCrZr alloy properties.

P.A. Khaimovich et al. in their studies [15], laid the foundations of low-temperature quasi-hydroextrusion (QHE) of metals and alloys. Currently, QHE-based methods are more generally referred to as barocryodeformation (BCD). This name emphasizes two main points: (i) the billet is subjected to isotropic three-axial compression, which forces the billet to squeeze through the container and the die, and (ii) the processing is performed under deep cooling conditions. Modern BCD allows plastic deformation of various metals and alloys, including those with low ductility. During deformation, quasihydrostatic conditions take place in the extrusion unit channel. As a result, the workpiece is completely squeezed out of the container (only the pressure transfer medium remains in the butt), and the constant diameter of the workpiece in the container during it is being extruded through the die is ensured. But the most important is that BCD provides high homogeneity of the structure and properties in the extruded rod cross-section. The cryogenic temperature during BCD makes it impossible to use a liquid as a pressure transfer medium. Therefore, a solid which is more ductile at operating temperatures than the workpiece material is used. Usually, indium is used as such a solid [15].

Processing by pressure before thermal aging is known to affect the kinetics of a supersaturated solid solution decomposition and its final physicomechanical properties significantly [16–23]. For obtaining the optimal properties of a material, it is necessary to understand the mechanisms of its structure formation during an SPD (in our case, BCD) processing. So, according to literature data [6–10], the following mechanisms of the microstructure modification may take place under SPD: the matrix grain refinement; dynamic aging of the supersaturated solid solution; the matrix recrystallization; dissolution of the precipitates in the matrix, etc. Some of these processes are competing. For instance, an experimental study of the simultaneous decomposition of supersaturated solid solution and dissolution of precipitates in the matrix during ECAP at room temperature was carried out in [24].

This work is devoted to the study of the influence of BCD at 77 K with various deformation degree on the CuCrZr alloy structure. In particular, it is aimed at the search for experimental evidence of strain-induced precipitation-dissolution of precipitates in low-alloyed CuCrZr alloys during BCD at 77 K temperature and the study of the role of these processes in the complex structural and phase transitions occurring in this alloy.

3 Research Methodology

In this work, the commercial Cu-(0.4–1)Cr-(0.03–0.08)Zr (% wt.) alloy was studied. The standard industrial processing of such alloys consists of annealing, quenching, and aging. The material was supplied in the form of a 25 mm diameter rod. Cylindrical billets of 10 mm diameter and 25 mm length were cut from the initial rod, thermally treated in liquid $BaCl_2$ at 1000 °C for 1 h, and water quenched to obtain a supersaturated solid solution of dopants in the copper matrix. Images of the microstructure of the initial material and the billets after thermal treatment are shown in Fig. 1.

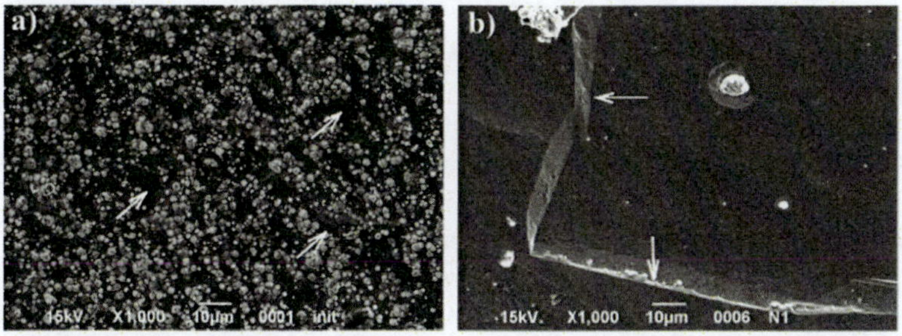

Fig. 1. The alloy SEM images: the initial state (a) and after the thermal treatment (b). The arrows indicate the precipitates-free areas of the matrix (a) and the matrix grains boundaries (b).

One can see that high-temperature annealing and water quenching resulted in the formation of the coarse-grain structure of the copper matrix and almost complete dissolving of the dopants in it. After the above heat treatment, the billets were subjected to BCD at 77 K up to the following deformation degrees: 25, 32, 40, 48, 60, and 74%. Deformation degree ε is determined as follows:

$$\varepsilon = (S_o - S)/S_o \cdot 100\%, \quad (1)$$

where S_o and S are the cross-sectional areas of the billet and the extruded rod.

The schematic of the BCD is presented in Fig. 2 a. Since the structure of the material after BCD is anisotropic, each extruded rod was cut into two samples (see Fig. 2 b): a disk 1, which surface is perpendicular to the BCD axis (transverse section), and a plate 2, which surface is parallel to the BCD axis (longitudinal section). Below, the structure of the longitudinal and transverse samples is considered.

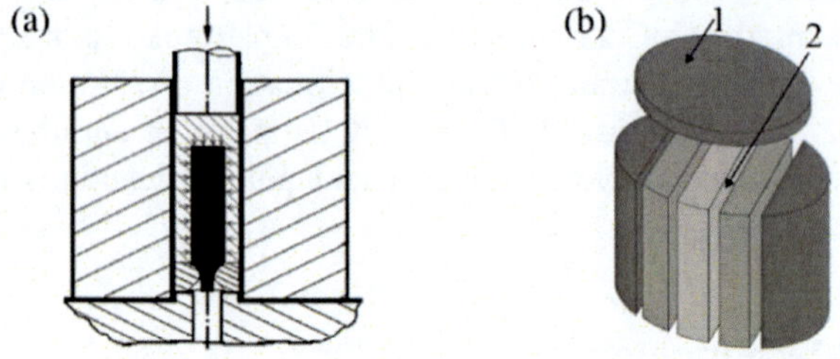

Fig. 2. BCD scheme (a) and the extruded rod cut scheme (b).

The following set of experimental techniques was used in work for studying the structure of the samples. Fine structure parameters were determined using X-ray diffractometry with the Bragg-Brentano focusing geometry with a Mo anode (DRON-4) [21]. The surface microstructure was studied using scanning electron microscopy (JSM-6390LV). The surface microstructure was preliminary developed using the sputtering by argon ions in the DSM-2 installation [14]. The diffraction patterns were processed using standard methods [25] using special software.

4 Results

SEM images of the longitudinal and transverse sections of the samples after BCD for different deformation degrees (32, 40, 48, 60, and 74%) are shown in Fig. 3.

One can see that deformation-induced precipitation takes place, the microstructure of the copper matrix, and the number, distribution, and shape of the precipitates change with the deformation degree increase. For estimating the relative intensity of the precipitation, the precipitates density was calculated from the area of contrast (light) regions on the SEM-images of both sections (Fig. 4). The relative density was calculated as the ratio of the precipitates area to the whole image area. The obtained values were normalized to the density of precipitates in the sample after thermal treatment, taken for 1.

Figure 5 shows diffraction patterns of the transversal section of the CuCrZr samples with different deformation degree. The patterns show peaks corresponding to Cu (fcc lattice), and there are no peaks, corresponding to Cr or other phases. The patterns in

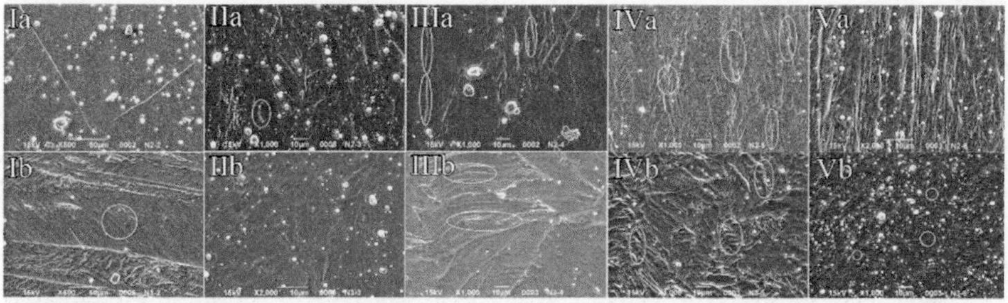

Fig. 3. SEM images of the longitudinal (a) and transverse (b) sections of the samples after BCD for different deformation degrees (I – 32%, II – 40%, III – 48%, IV – 60%, V – 74%).

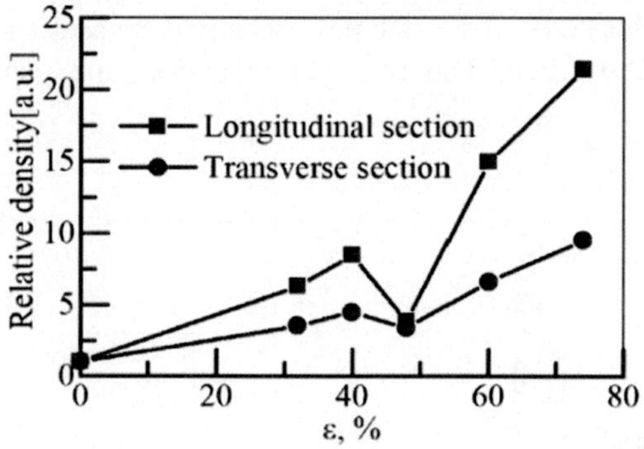

Fig. 4. Dependence of the relative density of precipitates upon the deformation degree for longitudinal and transverse sections.

Fig. 5 were used to determine the degree of preferred crystallographic orientation of the copper crystallites and to calculate the lattice parameter of the copper matrix.

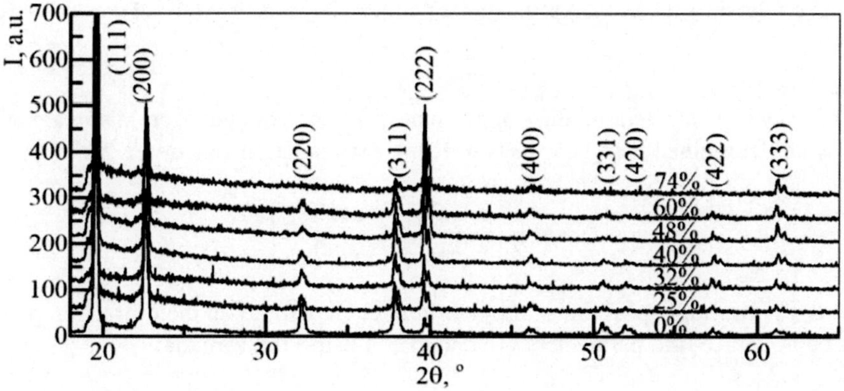

Fig. 5. X-ray diffraction patterns of the samples obtained using a DRON-4 diffractometer.

For determining the characteristics of the texture of the samples, the pole densities (p_{hkl}) were calculated as follows:

$$p_{hkl} = \left(\frac{I_{hkl}}{I'_{hkl}}\right) \Big/ \frac{1}{n}\sum_{n}\left(\frac{I_{hkl}}{I'_{hkl}}\right), \quad (2)$$

where I_{hkl} and I'_{hkl} are the integral intensities of the interference peaks for a textured and textureless samples, correspondingly, n is the number of the peaks.

Figure 6 shows the dependencies of the pole densities (p) on the ε for different crystallographic planes. One can see that BCD leads to the formation of an axial texture with the preferred crystallographic direction [111]. Wherein, the pole density for (111) planes growth nonlinearly with the ε, and the maximum is observed at 74%. The pole density for (200) planes practically does not change with the increase in the deformation degree. For (220) planes the pole density decreases and practically reaches zero at 74% deformation. Pole density (311) decreases with increasing ε to a value of ~0.03.

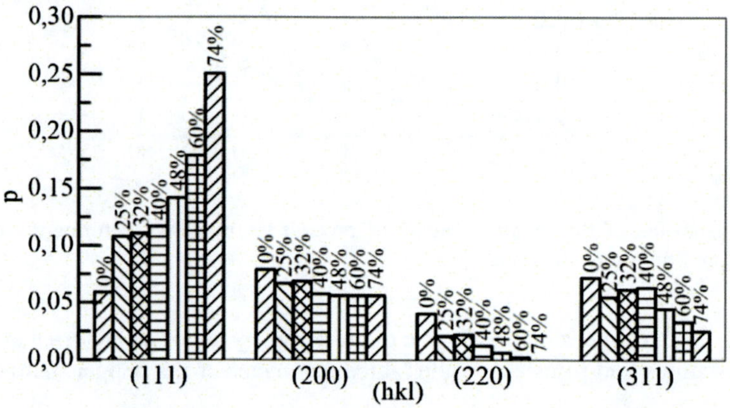

Fig. 6. Pole densities (p) for (hkl) planes in the CuCrZr alloy samples with different deformation degree.

The lattice parameters of the copper matrix of the samples were also calculated. They were determined using the Nelson-Riley extrapolation function [25]:

$$\xi_{hkl} = \frac{1}{2}\left[\frac{\cos^2\theta_{hkl}}{\theta_{hkl}} + \frac{\cos^2\theta_{hkl}}{\sin\theta_{hkl}}\right], \quad (3)$$

where θ_{hkl} is the angle position of the interference maximum on the diffraction pattern. For a cubic lattice, the parameters are calculated using the formula:

$$a_{hkl} = \lambda\left(h^2 + k^2 + l^2\right)/2sin\theta_{hkl}, \tag{4}$$

where λ is the k_α wavelength of the anode radiation, h, k and l are the Miller indices. The actual value of the lattice parameter is obtained by extrapolating the $a_{hkl}(\xi_{hkl})$ dependence by a straight line to the value, corresponding to $\xi_{hkl} = 0$. The dependence of the lattice parameter upon the deformation degree is shown in Fig. 7. One can see that generally, the lattice parameter tends to decrease with the deformation degree increase. However, in the neighborhood of 40%, there is a region where the lattice parameter increases and then continues to decrease.

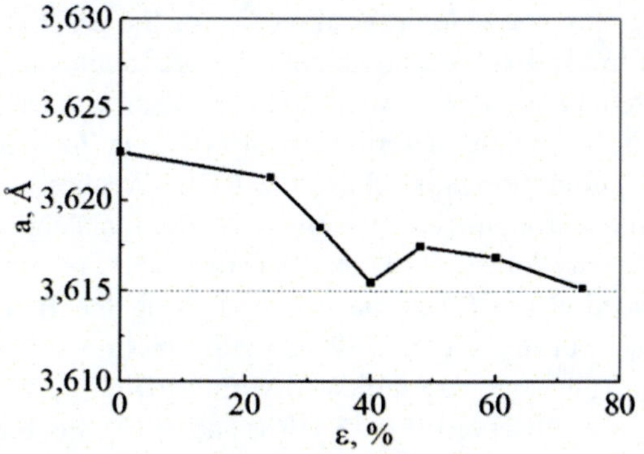

Fig. 7. The dependence of the lattice parameter of the Cu matrix upon the deformation degree. The horizontal dotted line corresponds to the lattice parameter of the pure annealed Cu [26].

In the initial state, the CuCrZr alloy matrix is rather homogeneously filled with Cr precipitates, mostly spherical (Fig. 1(a)). The average size of the precipitates is ~1 μm. There are also small precipitates-free regions in the copper matrix (indicated with arrows in Fig. 1 a). After water quenching and annealing, the sample has a coarse-grained (50–100 μm or more) microstructure; most of the dopants are dissolved in the matrix (Fig. 1 b), but there is a small amount of large (~10 μm) and small (<1 μm) precipitates. This state of the material is characterized by the largest lattice parameter (Fig. 7) due to the dissolved dopants and the absence of a texture (Fig. 6).

Subsequent BCD 77K leads to significant changes in the structure. In SEM images (Fig. 3), one can see that when the $\varepsilon \leq 32\%$, the precipitates density significantly increases due to deformation-induced precipitation, which is especially noticeable in the longitudinal section. It is also reflected in the reduction of the lattice period. The average size of the precipitates is several microns. After deformation, the precipitates density in the longitudinal section is higher, than in the transversal one, except for the 48% deformation (Fig. 3). In the transverse section, the formation of a fibrous microstructure occurs (circled with white in Fig. 3 I b, III b, IV b, V b). Also, the formation of an axial texture with a preferred [111] orientation along the extrusion direction takes place.

The maximum of the precipitations density, both in the longitudinal and transverse sections, is observed at a 40% deformation. This correlates with the dependence of the lattice parameter on the deformation degree, where there is a minimum.

First of all, the density of precipitates smaller than 1 μm grows. For the longitudinal sections, chains of spherical precipitates, oriented along the direction of the BCD, are observed. At 48% deformation, the precipitates density decreases significantly, along with the density anisotropy in the considered sections (Fig. 4). Such behavior is due to the fact that starting from deformation degrees close to 50%, the dissolution of the secondary phases (deformation-induced dissolution) becomes significant. An increase in the lattice parameter also confirms this effect. It is worth noting the presence of the fibrous structure in the matrix, at that in the longitudinal section, the fiber orientation is more pronounced along the direction of the BCD (Fig. 2 III a, b).

With an increase in the ε (60% and above), the predominance of the deformation-induced precipitation and a more pronounced anisotropic microstructure formation processes is observed. So, for the deformation of 60% and 74%, a sharp grows of the precipitate's density (often elongated along the BCD direction) and the fibers in the matrix. There is a more pronounced anisotropy of the inclusion density and a sharp increase in the pole density for the (111) planes (Fig. 3 and Fig. 6). At ε 60% and 74%, in the longitudinal section the fibers are oriented along the BCD axis (encircled in Fig. 3 IV(a) and V(a)), but in the case of 74% their direction is more pronounced, and their density is higher (the fibers are indicated with arrows). In the transversal section, fibrous structures are also observed (Fig. 3 IV(b) and V(b)), but their orientation is not pronounced. At 74% deformation, the dispersity of the fibers and the contrast in the transversal section is reduced, which is more likely due to a decrease in the dopants concentration (see Fig. 3 V b, circles). At the same time, the number of spherical precipitates increases, which is associated with the heating of the sample during extrusion. It follows from this study that the BCD of the CuCrZr alloy leads to the formation in the copper matrix of fibrous structures with elongated Cr precipitates. The quantity, shape, and distribution of the precipitates also change, which is associated with the deformation-induced precipitation and deformation-induced dissolution processes. Apparently, the process of precipitation proceeds mostly by a heterogeneous mechanism, i.e., at the boundaries of grains or subgrains. So, we can assume that the orientation Kurdyumov–Sachs ratio $\{111\}_{fcc}||\{110\}_{bcc}$ and $<110>_{fcc}||<100>_{bcc}$ [27] occurs. This explains the absence of chromium interference maxima on the diffraction patterns. Namely, interplanar spacings $\{111\}_{Cu}$ and $\{110\}_{Cr}$ are nearly equal, and the $(110)_{Cr}$ peak is rather difficult to isolate, taking into account the low concentration of chromium. The absence of other chromium peaks is most likely because of their very low intensity due to the strongly pronounced texture of the chromium phase (110) along the BCD direction.

It should be underlined that the lattice parameters at 74% and 40% degrees of deformation are very close (Fig. 7), but the precipitates density at 74% is times higher (Fig. 3). This difference is caused by the fact that at the deformation of 40%, there is a large number of precipitates smaller than the SEM resolution. The techniques used in this study did not detect Zr based precipitates. However, according to published data, Zr can precipitate into a separate phase in the form of intermetallic compounds [28–30].

5 Conclusions

The primary results can be summarized as follows. BCD at 77K significantly affects the structure of the CuCrZr alloy. In particular, it stimulates the decomposition of the supersaturated solid solution. The deformation degree influences the density, shape, and distribution of the particles of the secondary phases, which in turn determines the strength properties of the material.

With the deformation degree increase, two kinetically opposite processes occur dynamic aging (precipitation of secondary phase particles from the matrix) and dynamic doping (precipitates dissolution). Moreover, the influence of both mechanisms on the microstructure depends not monotonically on the deformation degree.

In the neighborhood of the 48% deformation degree, a decrease in the effect of dynamic aging is observed, which is confirmed by an increase in the lattice period and a decrease in the precipitates density. XRC shows the increase of the number of grains of the copper matrix with predominant orientation along the [111] direction, which may indicate the fibrous nature of its structure. At deformations 60% and more, fibrous precipitates are observed, which may lead to the anisotropy of the mechanical properties of the material.

Properties of precipitation-hardened alloys are mainly determined by the ensemble of precipitates. Therefore, the results of this study can be applied for the development of scientifically based approaches to the use of SPD to achieve enhanced functional properties of alloys of the CuCrZr system. These results are also of interest from a fundamental point of view, since the straightforward application of quasi-equilibrium thermodynamics cannot explain the process of deformation-induced dissolution and, therefore, includes non-trivial mechanisms.

Thus, electromechanical properties of the alloy after BCD and aging, as well as the influence of other BCD modes on the structure and properties of the CuCrZr alloy, will be the subject of further studies.

References

1. Pokrovsky, A., Fabritsiev, S., Edvards, D., Zinkle, S., Roweliffe, A.: Effect of neutron dose and irradiation temperature on the mechanical properties and structure of dispersion strengthened copper alloys. J. Nucl. Mater. **283**, 404–408 (2000)
2. Lipa, M., Durocher, A., Tivey, R., Huber, T., Schedler, B., Weigert, J.: The use of cooper alloy CuCrZr as a structural material for actively cooled plasma facing and in vessel components. Fusion Eng. Des. **75**, 469–473 (2005)
3. Doceul, L., Bucalossi, J., Larroque, S., Lipa, M., Portafaix, C., Saille, A., Samaille, F., Seler, B., Ferlay, F., Verger, J.: Design, integration and feasibility studies of the Tore-Supra West divertor structure. Fusion Eng. Des. **88**, 814–817 (2013)
4. Belyaeva, A.I., Galuza, A.A., Savchenko, A.A., Faizova, S.N., Raab, G.N., Aksenov, D.A.: Effect of microrelief on the optical characteristics of light CuCrZr copper alloys bombarded by ions of deuterium plasma. Bull. Russ. Acad. Sci. Phys. **76**(7), 764–767 (2012)
5. Voitsenya, V.S., Bardamid, A.F., Belyaeva, A.I., Bondarenko, V.N., De Temmerman, G., Konovalov, V.G., Lipa, M., Litnovsky, A., Ryzhkov, I.V., Schunke, B.: Interpretation of tore supra in-vessel mirror experiments. Plasma Devices Oper. **16**(1), 1–10 (2008)

6. Valiev, R.Z., Langdon, T.G.: Principles of equal-channel angular pressing as processing tool for grain refinement. Prog. Mater. Sci. **51**, 881–981 (2006)
7. Shakhova, I., Yanushkevich, Z., Fedorova, I., Belyakov, A., Kaibyshev, R.: Grain refinement in a Cu-Cr-Zr alloy during multidirectional forging. Mater. Sci. Eng. A. **606**, 380–389 (2014)
8. Zhilyaev, A.P., Shakhova, I., Morozova, A., Belyakov, A., Kaibyshev, R.: Grain refinement kinetics and strengthening mechanisms in Cu-0.3Cr-0.5Zr alloy subjected to intense plastic deformation. Mater. Sci. Eng. A. **654**, 131–142 (2016)
9. Purcek, G., Yanar, H., Demirtas, M., Shangina, D.V., Bochvar, N.R., Dobatkin, S.V.: Microstructural, mechanical and tribological properties of ultrafine-grained Cu–Cr–Zr alloy processed by high pressure torsion. J. Alloy. Compd. **816**, 152675 (2020)
10. Huang, A.H., Wang, Y.F., Wang, M.S., Song, L.Y., Li, Y.S., Gao, L., Huang, C.X., Zhu, Y.T.: Optimizing the strength, ductility and electrical conductivity of a Cu-Cr-Zr alloy by rotary swaging and aging treatment. Mater. Sci. Eng. A **746**, 211–216 (2019)
11. Belyaeva, A.I., Galuza, A.A., Kolenov, I.V., Faizova, S.N., Raab, G.I., Faizov, I.A.: Anisotropy of structure and strength properties of high-temperature Cu-Cr-Zr composite, induced by equal-channel angular pressing. J. Nano-Electron. Phys. **8**(4), 04082 (2016)
12. Kulczyk, M., Pachla, W., Mazur, A., Sus-Ryszkowska, M., Krasilnikov, N., Kurzydłowski, K.J.: Producing bulk nanocrystalline materials by combined hydrostatic extrusion and equal channel angular pressing. Mater. Sci. Pol. **25**, 991–999 (2007)
13. Belyaeva, A.I., Galuza, A.A., Khaimovich, P.A., Kolenov, I.V., Savchenko, A.A., Ryzhkov, I.V., Shtan', A.F., Solodovchenko, S.I., Shul'gin, N.A.: Effect of quasi-hydrostatic extrusion on microhardness in CuCrZr alloy. Probl. At. Sci. Technol. **1**(95), 170–173 (2015)
14. Belyaeva, A.I., Galuza, A.A., Khaimovich, P.A., Kolenov, I.V., Savchenko, A.A., Solodovchenko, S.I., Shul'gin, N.A.: Effect of various kinds of severe plastic deformation on the structure and electromechanical properties of precipitation strengthened CuCrZr alloy. Phys. Met. Metall. **117**(11), 1170–1178 (2016)
15. Khaimovich, P.A.: Cryodeformation of metals under isotropic compression (review). Low Temp. Phys. **44**, 349 (2018)
16. Vinogradov, A., Patlan, V., Suzuki, Y., Kitagawa, K., Kopylov, V.I.: Structure and properties of ultra-fine grain Cu–Cr–Zr alloy produced by equal-channel angular pressing. Acta Mater. **50**, 1639–1651 (2002)
17. Nagai, T., Henmi, Z., Sakamoto, T., Koda, S.: Effect of precipitates on recrystallisation temperature in Cu-Cr Cu-Zr and Cu-Zr-Cr alloys. Trans. JIM **14**, 462–469 (1973)
18. Henmi, Z., Nagai, T.: Transmission electron microscopy of Cu-Cr-Zr alloy. Trans. JIM **10**, 305–313 (1969)
19. Luo, C.P., Dahmen, U., Westmacott, K.H.: Morphology and crystallography of Cr precipitates in a Cu-0.33 wt% Cr alloy. Acta Metall. Mater. **42**, 1923–1932 (1994)
20. Dahmen, U., Witcomb, M.J., Westmacott, K.H.: Morphology of Cr precipitates in an overaged Cu-0.3% Cr alloy. Scripta Metall. **22**, 1867–1872 (1988)
21. Rdzawski, Z., Stobrawa, J.: Structure of coherent chromium precipitates in aged copper alloys. Scripta Metall. **20**, 341–344 (1986)
22. Stobrawa, J., Rdzawski, Z.: Inhomogeneous precipitation in aged copper-chromium alloy. Scripta Metall. **21**, 1269–1271 (1987)
23. Kamijo, T., Furukawa, T., Watanabe, M.: Homogeneous nucleation of coherent precipitation in copper-chromium alloys. Acta Metall. **36**, 1763–1769 (1988)
24. Faizov, I.A., Mulyukov, R.R., Aksenov, D.A., Faizova, S.N., Zemlyakova, N.V., Cardoso, K.R., Zeng, Y.: Dissolution of the second phase particles in the course of the equal channel angular pressing of diluted Cu-Cr-Zr alloy. Lett. Mater. **8**(1), 110–114 (2018)

25. Ladd, M.F.C., Palmer, R.A.: Structure Determination by X-ray Crystallography, 3rd edn. Plenum Press, New York (1993)
26. West, A.R.: Solid State Chemistry and Its Applications. Wiley, New York (2014)
27. Batra, I.S., Dey, G.K., Kulkarni, U.D., Banerjee, S.: Precipitation in a Cu–Cr–Zr Alloy. Mater. Sci. Eng. A **356**, 32–36 (2002)
28. Zeng, K.J., Hamalainen, M., Lilius, K.: Phase relationship in Cu-rich corner of the Cu-Cr-Zr phase diagram. Scripta Metall. Mater. **32**(12), 2009–2014 (1995)
29. Chbihi, A., Sauvage, X., Blavette, D.: Atomic scale investigation of Cr precipitation in copper. Acta Mater. **60**(11), 4575–4585 (2012)
30. Zeng, K.J., Hamalainen, M.: A theoretical study of the phase equilibrium in the Cu-Cr-Zr alloy. J. Alloy. Compd. **220**, 53–61 (1995)

Determination of Reactant Gases Mass Balance in the Process of Deposition of Ionic-Plasma Coatings

Katerina Diadiun(✉)

Kherson Polytechnic College of Odessa National Polytechnic University, 23, Nebesna Sotnya, Kherson 73000, Ukraine
homutovakaterina@gmail.com

Abstract. One of the ways to increase the service characteristics of the tools used is to apply wear-resistant coatings with special properties to their working surfaces. Such methods include the method of condensation in a vacuum on the product surface of a substance from the plasma phase with ion bombardment – the CIB method. However, this method has many disadvantages, one of which is the lack of quality control of the coating deposition - the supply of several reagent gases to the chamber without preliminary mixing. In such a situation, the operator will not be able to manage the coating process at all, while alternative methods of controlling this process have not yet been developed. Today, an urgent problem is to obtain high-quality coating on the surface of an object using the CIB method, namely, to solve the problem of controlling the process of supplying two or more gases - reagents to a vacuum chamber. The article is devoted to the urgent problem of automation control of equipment for applying ion-plasma coatings for cutting tools. Issues related to the control of the coating process are considered when several gaseous reactants are simultaneously fed into a vacuum chamber with the formation of a multicomponent coating. Also, the article proposed adequate mathematical models of the mass balance of gaseous reagents and, on this basis, performed the structural synthesis of an automatic gas supply control system when applying a multicomponent coating.

Keywords: CIB method · Wear-resistant coating · Mathematical modeling of processes · Ion bombardment · Ion plasma technology · Multicomponent coating

1 Introduction

The requirements in manufacturing engineering for the wear resistance of a cutting tool and an increase in its cutting speed are very high. One of the most important indicators of the use of a cutting tool is its ability to maintain its functional parameters for a long time. The development of engineering requires the use of modern technology. There are many technologies for machining the working surfaces of cutting tools, which gives them additional hardening. One of these technologies is the use of ion bombardment. Modern installations for applying ion-plasma coatings by the CIB method do not provide control and monitoring systems when two or more gases are supplied - a

V. Tonkonogyi et al. (Eds.): InterPartner 2020, LNME, pp. 282–289, 2021.
https://doi.org/10.1007/978-3-030-68014-5_28

reagent in a vacuum chamber. From the above, it follows that the solution of automation and control of gas supply to the vacuum chamber is an urgent task.

2 Literature Review

Improving the cutting tool performance can be achieved by improving the properties of the surface layer of the tool material, by which the working surface of the tool most effectively resists the specific types of wear [1]. Such material must have a significant safety margin in bending, compression, and withstand shock loads [2]. One of the ways to increase the service characteristics of the tools used is to deposit wear-resistant coatings with special properties on their working surfaces [3]. Such methods include the method of condensation in a vacuum on the product surface of a substance from a plasma phase with ion bombardment - the CIB method developed in Ukraine [4].

However, this method has many disadvantages, one of them is a decrease in the surface quality of the product [5], as a result of ineffective control of the process [6] of deposition of ion-plasma coatings [7]. It can lead to waste [8]. The supply of several reagent gases can complicate the problem of quality control of the deposition of coatings to the chamber without preliminary blending [9]. In such a situation [10, 11], the operator will not be able to manage the deposition of coatings process at all [12], while alternative methods for controlling this process have not been developed yet [13].

The elimination of mentioned correcting actions requires the development of a fundamentally new model for the supply of several gases and [14], as a result [15], an automated control system for the balance of the supply of reactant gases [16] to the vacuum chamber without the participation of the human factor and as a result, the creation of a high-quality multicomponent coating [17].

3 Research Methodology

Dynamic mathematical model building of the deposition of coatings process based on the description of the mass balance for n_i in a vacuum chamber is presented in the form of a system of Eqs. (1):

$$\begin{cases} \frac{dM_1}{d\tau} = m_{supply(1)} + m_{airltak(1)} - m_{vent(1)} - m_{chemicalreaction(1)} \\ \frac{dM_2}{d\tau} = m_{supply(2)} + m_{airltak(2)} - m_{vent(2)} - m_{chemicalreaction(2)} \\ \dots\dots\dots\dots\dots\dots\dots\dots\dots\dots\dots\dots \\ \frac{dM_{n_i}}{d\tau} = m_{supply(n_i)} + m_{airltak(n_i)} - m_{vent(n_i)} - m_{chemicalreaction(n_i)} \end{cases} \quad (1)$$

where: M gases (1), (2),.. (n_i) – mass ni of gases in the vacuum chamber, kg; m supply (1), (2),.. (ni) - gases input due to its forced supply to the chamber, kg/c; m air leak (1), (2),.. (ni) - gases input due to air leak from the atmosphere because of chamber non - hermeticity, kg/c; m vent (1), (2),.. (ni) - gases consumption due to the operation of the vacuum pump, kg/c; m chemical reactions (1), (2),.. (ni) - gases consumption for chemical reactions in the chamber, kg/c.

At the stage of the deposition of coating, the volume of the vacuum chamber and the reactant gas temperature in it are constant. Referring to this fact, we will describe the system of equations based on Mendeleyev's - Clapeyron Law (2):

$$\begin{cases} M_{(1)} = \frac{\mu_{(1)} V_1}{RT_1} \\ M_{(2)} = \frac{\mu_{(2)} V_1}{RT_1} \\ M_{(n_1)} = \frac{\mu_{(n_1)} V_1}{RT_1} \end{cases} \tag{2}$$

where: μ - molecular weight of the gases coming into the chamber; V chamber – the volume of the vacuum chamber, m^3; R – absolute gas constant (R = 8.314 Jg-mol / K), m^3; T chamber – the temperature of the gas mixture in the vacuum chamber, K; P gases (1), (2),.. (ni) - partial pressure P.

In differential form, the system of equations will take the form (3):

$$\begin{cases} \frac{dM_{(1)}}{dt} = \frac{\mu_{(1)} V_k}{RT_k} \cdot \frac{dP_{(1)}}{dt} \\ \frac{dM_{(2)}}{dt} = \frac{\mu_{(2)} V_k}{RT_k} \cdot \frac{dP_{(1)}}{dt} \\ \frac{dM_{(n_j)}}{dt} = \frac{\mu_{(n_j)} V_k}{RT_k} \cdot \frac{dP_{(ni)} V_k}{RT_k} \end{cases} \tag{3}$$

The solution of this system of equations will make it possible to obtain a system of differential equations of the dynamics of gas pressure changes in a vacuum chamber during the deposition of the coating process. Further transformation of this system of equations, according to Laplace, will allow performing structural synthesis of the automated control system for the balance of several reactant gases in the deposition of ion-plasma coatings technologies.

The differential equation of dynamics of changes in volumes for two nitrogen and carbon gases coming into the vacuum chamber during the deposition of the coating process in the form of:

$$\begin{cases} \frac{\mu_{(C_2)} \cdot V_k}{R \cdot T_k} \cdot \frac{dV_{(C_2)}(\tau)}{d\tau} = \frac{\mu_{(C_2)} \cdot V_k}{R \cdot T_k} P_{(C_2)}(\tau) + m_{supply(C_2)}(\tau) - \frac{\mu_{(C_2)}}{\mu} \cdot m(\tau) + \frac{\mu_{(C_2)} \cdot H}{RT_k} \\ \frac{\mu_{(N_2)} \cdot V_k}{R \cdot T_k} \cdot \frac{dV_{(N_2)}(\tau)}{d\tau} = \frac{\mu_{(N_2)} \cdot V_k}{R \cdot T_k} P_{(N_2)}(\tau) + m_{supply(N_2)}(\tau) - \frac{\mu_{(N_2)}}{\mu} \cdot m(\tau) + \frac{\mu_{(N_2)} \cdot H}{RT_k} \end{cases}$$

In static mode, $V_{N_2} = V_{N_{2,0}}, V_{C_2} = V_{C_{2,0}}$. When N_2 is supplied to the vacuum chamber, the coefficients necessary for building the mass model will be equal to:

$$a_{1_{N_1}} = \frac{\mu_{N_1} V_K}{RT_K}; \tag{4}$$

$$a_{2_{N_2}} = \frac{\mu_{N_2} V_K}{RT_K}; \tag{5}$$

$$a_{3_{N_2}} = \frac{0,5\mu_{N_2}}{\mu_{\mathrm{Ti}}}; \quad (6)$$

$$a_{4N_2} = 0,79 \cdot \left(\frac{28 \cdot 16,66 \cdot 10^{-6}}{8,314 \cdot 300} \right) = 0,24 \cdot 10^{-6} \Gamma/c \quad (7)$$

When C_2 is supplied to the vacuum chamber, the coefficients necessary for building the mass model will be equal to:

$$a_{1_{C_2}} = \frac{\mu_{C_2} V_K}{RT_K}; \quad (8)$$

$$a_{2_{C_2}} = \frac{\mu_{C_2} V_K}{RT_K}; \quad (9)$$

$$a_{3_{C_2}} = \frac{0,37\mu_{C_2}}{\mu_{Ti}}; \quad (10)$$

$$a_{4_{C_2}} = 0,195 \cdot 10^{-6} \Gamma/c. \quad (11)$$

The transition function of the dependence of the output parameter V on the control m supply C_2 and the perturbation m_{Ti} will have the form:

$$\begin{cases} a_{1(c_2)} \cdot \frac{dP_{C_2}(\tau)}{d\tau} = -a_{2(C_2)} P_{(C_2)}(\tau) + m_{air\,leak}(\tau) - a_{3(C_2)} \cdot m_{supply}(\tau) + a_{4(C_2)} \\ a_{1(N_2)} \cdot \frac{dP_{C_2}(\tau)}{d\tau} = -a_{2(N_2)} P_{(N_2)}(\tau) + m_{air\,leak}(\tau) - a_{3(N_2)} \cdot m_{supply}(\tau) + a_{4(N_2)} \end{cases} \quad (12)$$

Supposing in $m_{\text{подC}_2}(\tau) = \text{const}$, $m_{\text{подN}_2}(\tau) = \text{const}$ and $m_{\mathrm{Ti}}(\tau) = \text{const}$, as well as:

$$A_{N_2} = \frac{a_{2N_2}}{a_{1N_2}}; B_{N_2} = \frac{m_{supplyN_2} - a_{3N_2} m_{Ti} + a_{4N_2}}{a_{1N_2}} \quad (13)$$

$$A_{C_2} = \frac{a_{2C_2}}{a_{1C_2}}; B_{C_2} = \frac{m_{supplyC_2} - a_{3C_2} m_{Ti} + a_{4C_2}}{a_{1C_2}} \quad (14)$$

Let us transform the system of linear differential Eqs. (13) to the form:

$$\begin{cases} \frac{dV_{(C_2)}(\tau)}{d\tau} + A_1 V_{(C_2)}(\tau) - B_1 = 0 \\ \frac{dV_{(N_2)}(\tau)}{d\tau} + A_2 V_{(N_2)}(\tau) - B_2 = 0 \end{cases} \quad (15)$$

$$V_{N_2}(0) = V_{N_2}, 0;$$

$$V_{C_2}(0) = V_{C_2}, 0.$$

Equation (16) has the following solution [5]:

$$\begin{cases} V_{(C_2)} = \left(V_{(C_2)_0} - \frac{B_{(C_2)}}{A_{(C_2)}}\right)e^{-A\tau} + \frac{B_{(C_2)}}{A_{(C_2)}} \\ V_{(N_2)} = \left(V_{(N_2)_0} - \frac{B_{(N_2)}}{A_{(N_2)}}\right)e^{-A\tau} + \frac{B_{(N_2)}}{A_{(N_2)}} \end{cases} \tag{16}$$

The program decision for this stage of the process is a constant value of the volumes of the gases coming into the vacuum chamber, and the program control is a certain value m supply N_2', m supply C_2', which provides the specified pressure value in the static mode.

Let us turn to the deviations:

$$y_{C_2} = V_{C_2} - V_{(C_2),0}; \tag{17}$$

$$V_{C2} = m_{supply\,C2} - m_{supply\,C2'}; \tag{18}$$

$$y_{N_2} = V_{N_2} - V_{(N_2),0}; \tag{19}$$

$$V_{N2} = m_{supply\,N2} - m_{supply\,N2'} \tag{20}$$

In static mode, the ratio is:

$$\begin{cases} 0 = -a_{2(C_2)}V_{(C_2)}(\tau) + m_{supply(C_2)}(\tau) + a_{4(C_2)} \\ 0 = -a_{2(N_2)}V_{(N_2)}(\tau) + m_{supply(N_2)}(\tau) + a_{4(N_2)} \end{cases} \tag{21}$$

Substituting (18), (19), (20), and (21) into (22), we obtain:

$$\begin{cases} a_{1(C_1)}\frac{dy_{(C_2)}}{d\tau} = -a_{2(C_2)}y_{(C_2)} - a_{2(C_2)}P_{(C_2),0} + V_{(C_2)} + m_{supply(C_2),0} - a_{3(C_2)}m_{Ti}(\tau) + a_{4(C_2)} \\ a_{1(N_1)}\frac{dy_{(N_2)}}{d\tau} = -a_{2(N_2)}y_{(N_2)} - a_{2(N_2)}P_{(N_2),0} + V_{(N_2)} + m_{supply(N_2),0} - a_{3(N_2)}m_{Ti}(\tau) + a_{4(N_2)} \end{cases} \tag{22}$$

The equation in the deviations becomes the form:

$$\begin{cases} a_{1(C_2)}\frac{dy_{(C_2)}}{d\tau} = -a_{2(C_2)}y_{(C_2)} + v_{(C_2)} - a_{3(C_2)}m_{Ti}(\tau) \\ a_{1(N_2)}\frac{dy_{(N_2)}}{d\tau} = -a_{2(N_2)}y_{(N_2)} + v_{(N_2)} - a_{3(N_2)}m_{Ti}(\tau) \end{cases} \tag{23}$$

Let us construct the transition function of the dependence of the output parameters of V_{C2} and V_{N2} on the control m supply N_2 and m supply C_2 and the perturbation m_{Ti}. After the Laplace transform, Eq. (24) becomes the form:

$$\begin{cases} Y(s)_{(C_2)} = U(s)_{(C_2)}\frac{1}{\left(a_{1(C_2)}S + a_{2(C_2)}\right)} - Q(s)_{(C_2)}\frac{a_{3(C_2)}}{(a_{1(C_2)}S + a_{2(C_2)})} \\ Y(s)_{(N_2)} = U(s)_{(N_2)}\frac{1}{(a_{1(N_2)}S + a_{2(N_2)})} - Q(s)_{(N_2)}\frac{a_{3(C_2)}}{(a_{1(N_2)}S + a_{2(N_2)})} \end{cases} \tag{24}$$

So, the transfer function of control for two gases will have the form:

$$\frac{Y(S)_{N_2}}{U(S)_{N_2}} = \frac{1}{(a_{1N_2}s + a_{2N_2})} \tag{25}$$

$$\frac{Y(S)_{C_2}}{U(S)_{C_2}} = \frac{1}{(a_{1C_2}s + a_{2C_2})} \tag{26}$$

Transfer function of perturbation for two gases will have the form:

$$\frac{Y(s)_{C_2}}{Q(s)C_2} = \frac{a_{3C_2}}{(a_{1C_2}S + a_{2C_2})}; \tag{27}$$

$$\frac{Y(s)_{N_2}}{Q(s)N_2} = \frac{a_{3N_2}}{(a_{1N_2}S + a_{2N_2})} \tag{28}$$

4 Results

The following problems should be solved in order to increase the efficiency of ion-plasma technologies by developing and implementing an automated system for analyzing and controlling the mass balance of reagent gases under the conditions of supplying several gases for this purpose:

1. Selected and justified criteria for evaluating the parameters of gas supply in the CIB technology;
2. A mathematical model of the coating process has been developed based on the description of the mass balance for supplying several gases to the vacuum chamber;
3. The structural synthesis of the automated control system for the balance of several reactant gases supplied to the vacuum chamber in the technology of applying ion-plasma coatings has been performed.
4. After the mathematical modeling of the coating process is carried out based on the description of the mass balance for supplying several gases to the vacuum chamber, the block diagram of the control of the process of letting gas reagents into the vacuum chamber will look like (Fig. 1).

The following main parameters have a direct effect on the structure and physical properties of coatings obtained by ionic - plasma deposition: arc current, reaction gas pressure in the working chamber, and substrate temperature.

The influence of the program parameters for deposition of multi-element ionic-plasma coatings on their quality, that is, the ability to control the balance of several reactant gases supplied to the vacuum chamber, is also little studied. So, the mobility of the use of ionic-plasma technologies significantly increases. The block diagram of the control of the gas puffing process of reactant gases into the vacuum chamber describes the transition function of the dependence of the output parameters VC2 and VN2 on the control of m supplyN2 and m supplyC2 and perturbation mTi.

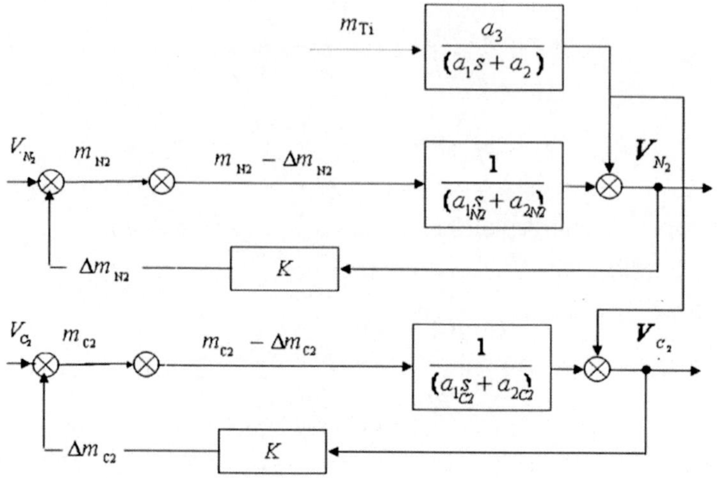

Fig. 1. Structural chart of reactant gases puffing process into the vacuum chamber.

5 Conclusions

The commonality of the investigated method for controlling the reactant gas supply in the process of ionic-plasma coatings deposition is its main advantage.

As a result of the work, a mathematical model of the coating process is presented, based on the description of the mass balance for supplying several gases to the vacuum chamber using the example of the supply of two gases to the vacuum chamber - reagents. For this, criteria for evaluating the parameters of the supply of gases - reagents using CIB technology were selected and justified, as well as the structural synthesis of an automated balance control system for several gases - reagents supplied to the vacuum chamber by technology deposition of ion-plasma coatings.

Further work aims to develop a simulation model for controlling the simultaneous supply of two or more gases when applying composite ion-plasma coatings to a cutting tool, the hardware for such control, as well as the practical implementation and implementation of the proposed solutions.

References

1. Vereshchaka, A.S., Tretiakov, I.P.: Cutting tool with wear resistant coatings. Engineering, Moscow (1986). (in Russian)
2. Dytnerskiy, Yu.I.: Processes and equipment of chemical technology. Chemistry, Moscow (1995). (in Russian)
3. Boxman, R.L.: Vacuum arc deposition: early history and resent developments. In: Proceedings of the XIXth ISDEIV, Xi'an. Journal 9, pp. 57–82 (2002)
4. Mattox, D.M.: The History of Vacuum Coating Technology. Albuquerque, NM, USA (2002)

5. Tonkonogiy, V.M.: Reactant gas supply control while depositing of ion-plasma coatings with forecasting of the vacuum units non – hermeticity. Refrig. Technol. **3**(89), 70–73 (2004)
6. Yeh, W., Chen, S.K., Lin, S.J.: Nanostructured high-entropy alloys with multiple principal elements. Adv. Eng. Mater. **2**(6), 67–78 (2004)
7. Aksenov, I.I.: Vacuum arc in erosive plasma sources. NNTs KhFTI, Kharkov (2005)
8. Otto, F., Yang, Y., Bei, H.: Relative effects of enthalpy and entropy on the phase stability of equiatomic high-entropy alloys. Acta Materialia (2013)
9. Aksenov, I.I., Aksenov, D.S., Belous, V.A.: Deposition of vacuum – arc coatings technique. NNTs KhFTI, Kharkov (2014)
10. Tsai, M.H., Yeh, J.W.: High-entropy alloys: a critical review. Mater. Res. Lett. **2**, 107–123 (2014)
11. Dunstan, D.J., Bushby, A.J.: Grain size dependence of the strength of metals: the Hall-Petch effect does not scale as the inverse square root of grain size. Int. J. Plast. **53**, 56–65 (2014)
12. Schuh, B., Mendez-Martin, F., Vulker, B.: Mechanical properties, microstructure and thermal stability of a nanocrystalline CoCrFeMnNi high-entropy alloy after severe plastic deformation. Acta Mater. **96**, 258–268 (2015)
13. Aksenov, I.I., Aksenov, D.S., Andreyev, A.A., Belous, V.A., Sobol, V.A.: Vacuum – arc coatings: technologies, materials, structure, properties. NNTs KhFTI, Kharkov (2015)
14. Kirkopulo, K., Tonkonogyi, V., Stopakevych, O., Stopakevych, A.: Design of a set of nonlinear control systems of the ARC PVD ion-plasma installation. Eastern-Eur. J. Enterp. Technol. **2**(12), 65–74 (2018)
15. Eremin, E.N., Yurov, V.M., Laurynas, V.C.: Method for determining the surface energy of nitrides, carbides and borides. IOP Conf. Ser. **10**(12), 35–41 (2019)
16. Usov, A., Tonkonogyi, V., Dašic, P., Rybak, O.: Modelling of temperature field and stress-strain state of the workpiece with plasma coatings during surface grinding. Mchines Swiland **1**(7), 20–26 (2019)
17. Tonkonogyi, V., Dašić, P., Rybak, O., Lysenko, T.: Application of the modified genetic algorithm for optimization of plasma coatings grinding process. In: New Technologies, Development and Application. NT-2019. Lecture Notes in Networks and Systems, vol. 6, no. 10, pp. 199–211 (2020)

Check for updates

Energy-Efficient Technology of Epoxy Polymers Producing by Using Ultrasonic Treatment

Aleksandr Kolosov[1(✉)], Aleksandr Gondlyakh[1], Elena Kolosova[1], Dmitro Sidorov[1], and Anish Khan[2]

[1] National Technical University of Ukraine "Igor Sikorsky Kyiv Polytechnic Institute", 37, Prospect Peremohy, Kyiv 03056, Ukraine
a-kolosov@ukr.net

[2] King Abdulaziz University, Jeddah 21589, Saudi Arabia

Abstract. Various aspects of the developed technological foundations of energy-efficient production of epoxy polymers with improved performance properties by the combined use of ultrasonic treatment of liquid polymer media are considered. Two compounds of hot-cured epoxy compositions were investigated: for the preparation of epoxy polymers with shape memory effect and for the preparation of epoxy adhesive compositions. The first of the investigated epoxy compositions were used to form products that, in the hardened state, have a glass transition temperature in the range 50–100 °C. The second investigated epoxy composition was plasticized with a carboxyl rubber. As the controlled parameters of ultrasonic vibrations, frequency in the low-frequency and mid-frequency ranges, intensity and amplitude of elastic vibrations were chosen. As the operational characteristics of the hardened epoxy adhesive composition, the tensile strength, the adhesive pull strength, and adhesive shear strength were measured, and for the hardened heat-shrinkable epoxy polymer, the tensile strength, failure deformation, and glass transition temperature were measured. As the main parameter characterizing energy efficient or energy saving, we selected the total curing time of the thermoplastic polymers by constructing experimental hardening cyclograms. It has been established that the use of combined ultrasonic treatment leads to both improvements of the quality for homogenization of epoxy compositions and reduction in the total time of their hardening by at least twice while increasing the physicomechanical properties of the obtained hardened epoxy polymers and epoxy adhesive compositions.

Keywords: Energy efficient · Epoxy composition · Hardening cyclogram · Ultrasound · Cavitation · Frequency · Excess static pressure

1 Introduction

Traditional polymer composite materials (PCMs) in the form of thermoplastic and thermosetting unfilled compositions, as well as PCMs, reinforced with various types of fibrous fillers, are currently mass-produced by the industry on an ever-increasing scale [1–4]. Fields of application of such PCMs are constantly expanding, for example,

V. Tonkonogyi et al. (Eds.): InterPartner 2020, LNME, pp. 290–299, 2021.
https://doi.org/10.1007/978-3-030-68014-5_29

military affairs [5], biomedicine [6], and other industries can be attributed here. New types of PCMs also appear, such as nanomodified ones [7–11], as well as "intelligent" ones [11–15]. The latter type of innovative PCMs is capable of optimally "monitoring" the stress-strain state of the structure under difficult operating conditions (f.e., temperature + pressure) and reacting accordingly (depending on the degree of "intellectualization" due to the presence of intelligent sensors in the structure of such PCMs). The above emphasizes the importance of considering design issues for the production technology of both traditional PCMs [16–19] and innovative PCMs [11–15].

PCMs based on epoxy polymers (EPs) from an epoxy matrix in the form of epoxy oligomers (EOs), epoxy compositions (ECs), and epoxy adhesive compositions (EACs) are especially widely used [20]. In the production of the above PCMs, a whole range of chemical and physical methods are used to modify the components of PCMs that change the structure and properties of the components, in which ultrasound (US) plays an important role. It is due to the resulting complex of valuable technological and operational properties of both individual PCMs components processed by US and cured PCMs based on them.

The development of the scientific basis for the use of US modification concerning PCMs molding and adjust the properties of the final PCMs obtained, based on the initial compositions and technological properties of the PCMs components [16–19], as well as parameters of US processing (sonicating). At the same time, the achievement of energy efficiency (saving) in the formation of such PCMs plays a particularly important role. This is since curing of thermosets, especially "hot" curing, as a rule, occurs over a long period (several hours or even days).

Therefore, the study of US modification methods to the acceleration of curing of such polymers, that is, energy saving during their formation, while improving the properties of cured PCMs, is an important area of polymer materials engineering.

2 Literature Review

During US treatment of liquid media (polymer compositions), the energy of the excited mechanical vibrations actively influences the processed liquid polymer medium [21–26]. At the optimum, this helps to improve the adhesion of ECs [27], which determines the operational properties of cured EPs and PCMs reinforced on their basis with fillers [28–31].

To implement the effective regime of US treatment of liquid media, it is necessary first of all to determine the effective values of the intensity I and frequency f of US treatment. Since most US of liquid technological processes is associated with cavitation [22, 24] and the sound-capillary effect [32], changing the conditions of the course of cavitation, it is possible to enhance or weaken various cavitation effects. Besides, the physicochemical effect of elastic US treatment in liquid media, as a rule, manifests itself in medium and high-intensity US fields.

In this case, other than US cavitation, effects such as radiation pressure and sound wind become significant. Another concomitant US effect (namely sonochemistry) studies the chemical reactions occurring in substances when they are sonicated [22, 23, 33]. A derivative effect in sonochemistry is luminescence is a phenomenon of the appearance of a flash of light upon the collapse of cavitation bubbles formed in a liquid under the action of a powerful US wave [32]. The use of US in the synthesis of new nanomodified thermoplastic PCMs is also promising [34, 35].

According to a number of studies, in some cases, it is promising to carry out US modification of liquid media in the different frequency US range [36], as well as use excessive static pressure P_o [37]. In this case, according to calculations [24, 25], an increase in intensity and a reduction in the time of the sonicating of liquid media are possible. An important aspect of the production of high-strength reinforced PCMs is the design of the technological operation of the US impregnation of reinforcing fillers with polymer binders. Such design is carried out from the standpoint of the theory of capillary impregnation [38, 39] and sound capillary effect.

As experiments show [40, 41], when using US, the wettability of EC to the surface of reinforcing fillers improves, and the homogenization of EC and capillary impregnation are intensified. The use of US is promising for improving the operational properties of both cured EPs [42, 43] and PCMs reinforced on their basis [44], including for intensifying the process of their application [45]. It should also be noted the relevance of designing technical means for molding thermoplastics [46, 47], including the use of low-frequency US [48, 49].

A number of issues related to modeling of technology for producing high-strength PCMs, e.g., using calculation and analytical methods, are considered in studies [17, 50–52]. The issues of designing US treatment tools in the form of separate dosing devices were investigated in the article [53], and approaches to the design of the technological cycle as a whole with the use of US treatment were considered in work [54], which also describes directions for further research.

The above aspects served as the basis for the development of the scientific foundations of energy-efficient (energy saving) technology for the production of EPs by using effective US treatment regimes.

3 Research Methodology

We studied two compositions of ECs of "hot" curing: for the preparation of EPs and EACs. First investigated EC was used to form products (in particular, couplings), which in the hardened state have a glass transition temperature T_g in the range of 50–100 °C. This EC consisted of a rigid and elastic component. The aromatic complex of diglycidyl ether in the form of EO of ED-20 and UP-640 brands was used as a rigid component. As an elastic component, a block oligomer of aliphatic EO and acid oligoester of the UP-599 brand was used. The mass ratio of the rigid and elastic components in EC was 3:2, respectively.

A hardener mixture of iso-methyl-tetrahydro phthalic anhydride (iso-MTHFA) with accelerator UP-606/2, taken in stoichiometric ratio to the resin part, was used as a hardener for the specified two-component EC. The curing mode of the two-component EC was: 70 °C/8 h + 100 °C/4 h + 120 °C/2 h.

To prepare an EAC of "hot" curing, plasticized with carboxyl-containing rubber of the SKN-30 brand, a halogen-containing EO of the UP-631 brand was used. Iso-MTHFA was used as a hardener for this EAC. The composition of the EAC was as follows: 100 wt.p. of UP-631 + 10 wt.p. SKN-30 + 100 wt.p. of iso-MTHFA.

EAC curing mode was: 80 °C/3 h + 130 °C/4 h + 160 °C/5 h.

As a source of low-frequency US treatment, a US generator of the UZG 3–4 brand was used, which fed the magnetostrictive transducers of PMS 15A-18 brand with an output power of 4 kW. Alternatively, piezoceramic transducers were used. The frequency f kHz, the intensity I, the amplitude A of the US vibrations (USVs), and sonicating time τ were chosen as the controlled parameters of US treatment.

To generate mid-frequency US in the megahertz range $f = 1.0–1.5$ MHz in the experimental setup, together with the transducer, focusing piezoceramic transducers were used, which were mounted on the sidewalls of the sonicating cell, to the bottom of which piezoceramic (or magnetostrictive) transducers of low-frequency US were also attached. They also used a compressor to create excess static pressure P_o. The effective value of the excess static pressure P_o during the sonicating of liquid ECs was determined by measuring the operational properties of the cured EACs, as well as heat-shrinkable EPs.

As the operational characteristics of EPs (hardened ECs and EACs), tensile strength σ_{ts}, adhesive pull strength σ_{ps}, and adhesive shear strength σ_{ss} were measured. Additionally, failure deformation ε_f and glass transition temperature T_g were measured for the last two characteristics and only for two-component ECs. The number of tests per each variable parameter under investigation was $N^o = 7–10$ with the reliability of $P = 95\%$. As the main parameter of energy efficiency (energy saving), the total curing time of the thermoplastic polymers was chosen by constructing experimental hardening cyclograms.

4 Results

4.1 Operational Properties of Sonicated Epoxy Polymers

Upon receipt of the EAC, the resinous part of the initial EC before its hardening was subjected to the simultaneous volumetric effect of low-frequency US treatment of $f = 14–18$ kHz at the temperature of $T = 60–80$ °C and constant static pressure $P_o = 0.4–0.5$ MPa during $\tau = 15–20$ min.

Figure 1 shows the effect of the sonicating frequency f upon variation of the static pressure P_o on the strength properties of the hardened EAC. The maximum value of hardening for EAC, modified with carboxyl rubber, at excess pressure $P_o = 0.4$ MPa was: for adhesive shear strength σ_{ss} on 135% and for adhesive pull strength σ_{ps} on 170%.

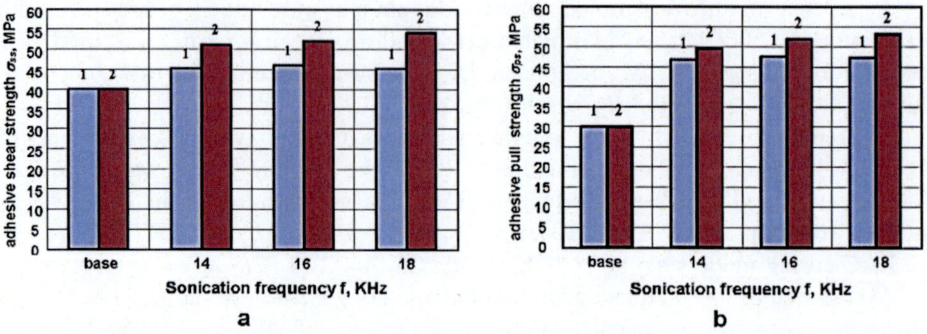

Fig. 1. The effect of the US treatment frequency f at normal pressure (1) and excess static pressure P_o = 0.4 MPa (2) on the adhesive shear strength σ_{ss}, MPa (*a*) and adhesive pullstrength σ_{ps}, MPa (*b*) hardened EAC modified with carboxyl rubber.

Upon receipt of an EP with a shape memory effect, the resinous part of the initial EC before its hardening is subjected, in addition to the above modes of obtaining EAC, to the influence of different frequency USVs. Figure 2 shows the average values of the contribution of six options for sonicating liquid EC in different frequency ranges with varying working pressures to increase the strength and productivity characteristics of heat-shrinkable EPs.

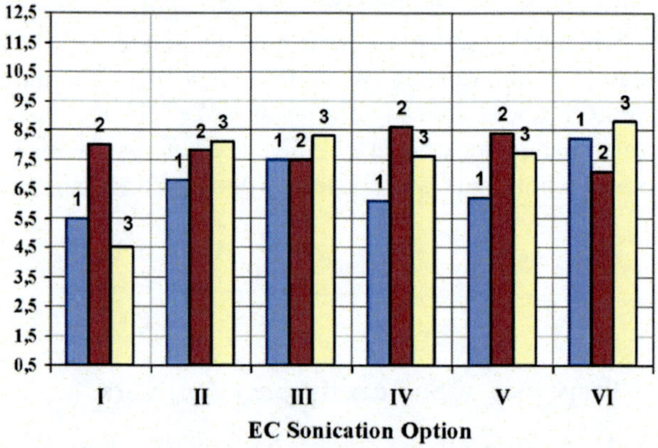

Fig. 2. Average values of the contribution according to options (*I – VI*) for sonicating liquid EC in the low-frequency (f = 18 kHz) and mid-frequency (f = 1 MHz) ranges at normal and excess (P_o = 0.4 MPa) static pressure to increase the strength and operational characteristics of heat-shrinkable EP based on EC: *I* – basic EC; *II* – basic EC, sonicated at atmospheric pressure; *III* – basic EC, sonicated at overpressure; *IV* – basic EC, sonicated by mid-frequency US at atmospheric pressure; *V* – basic EC, sonicated by mid-frequency US at overpressure; *VI* – basic EC, sonicated simultaneously by low-frequency and mid-frequency US at overpressure; 1 – σ_{ts} = 0.1 MPa; 2 – ε_f, %; 3 – $T_{g'}$ = 0.1 °C.

In the optimal (combined) version of sonicating (*VI*), the operational properties of heat-shrinkable EPs are maximally improved. Namely, tensile strength σ_{ts} on 41%; glass transition temperature T_g on 190%. At the same time, failure deformation ε_f decreased by 13%, the absolute value of which, however, remained within the technological tolerances. This yaayce occurs with simultaneous sonicating in the low-frequency and mid-frequency ranges, as well as at overpressure. The first is carried out at a frequency *f* from 15 kHz to 18 kHz with a sonicating amplitude *A* from 3 μm to 6 μm and intensity *I* = 4–8 W/cm^2, the second - at a frequency *f* = 1.0–1.5 MHz, amplitude *A* from 0.1 to 0.2 μm and intensity *I* from 20 to 30 W/cm^2.

It was found that a deviation from the specified parameters of US modification leads to a decrease in the strength of heat-shrinkable EPs.

4.2 Energy Saving on Obtained of Sounded Epoxy Polymers

The hardening cyclograms of the EACs are shown in Fig. 3, and the hardening cyclograms upon receipt of EPs with shape memory effect are shown in Fig. 4.

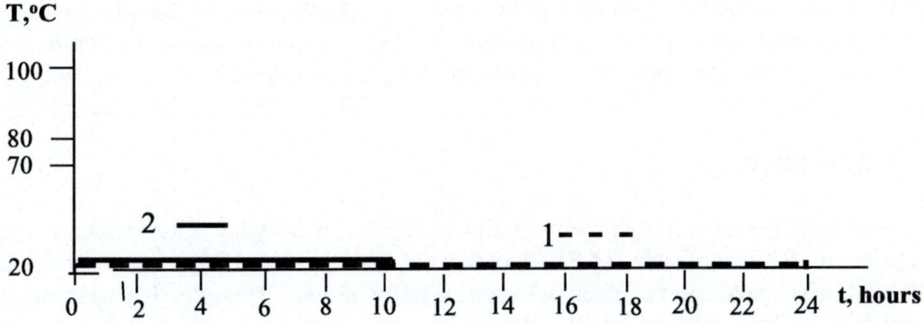

Fig. 3. Hardening cyclograms and the total time *t* of obtaining EACs: *1* - basic mode of obtaining EACs; *2* - the mode of obtaining EACs using US treatment at an excess pressure of P_o = 0.4 MPa.

It was experimentally established that the total curing time of the sonicated EAC is less than 24: 10.5 = 2.3 times compared with the original (unsonicated) EAC. And when forming EP with the shape memory effect, it was found that, in contrast to the traditional hardening mode (70 °C/8 h + 100 °C/4 h + 120 °C/2 h), which lasts *t* = 14 h, after sonicating the EC at normal pressure, an accelerated hardening mode for 7 h is quite acceptable (Fig. 4).

In turn, the joint conduct of US treatment at overpressure allows reducing this time by 0.5 h under the optimal mode, namely as follows: 0.5 h (US treatment) + 70 °C/2.5 h + 100 °C/2 h + 120 °C/1.5 h.

It was also experimentally established that the performance of US mixing (homogenization) of the components of the investigated ECs increases by at least three times, compared with the traditional technology of mechanical mixing.

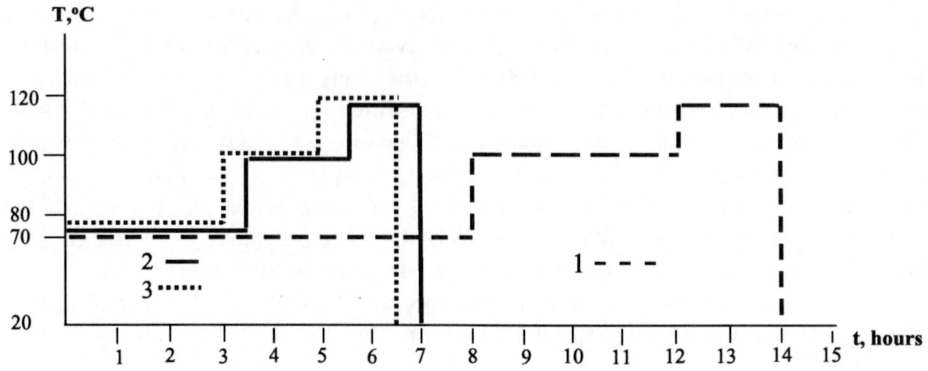

Fig. 4. The hardening cycle and the total time *t* of obtaining the EPs with the shape memory effect: *1* – the basic mode of obtaining EPs; *2* – US processing of ECs at normal pressure; *3* – US processing of ECs simultaneously in the low-frequency (f = 18 kHz) and mid-frequency (f = 1 MHz) ranges at overpressure P_o = 0.4 MPa.

Thus, the conducted studies can be used as a justification of the principles of energy-efficient technology (achieving energy saving) in the development of processes for obtaining products from EP compositions with US treatment.

5 Conclusions

The effectiveness of US treatment of liquid ECs with varying static pressure and frequency in the manufacture of EACs based on plasticizers and ECs used to fabricate EPs with a shape memory effect is experimentally shown. The effective operational parameters of the combined US modification are found. It was established that the use of US treatment leads to an improvement in the quality of homogenization of EC, to a reduction in the total time of their hardening, i.e. energy saving not less than 2 times, while increasing the physicomechanical properties of the hardened EPs and EACs obtained. This fact allows one to select the effective parameters of the energy-saving technological foundations of their production directionally even at the stage of designing effective modes based on the combined use of US.

References

1. Kar, K.K.: Composite Materials. Processing, Applications, Characterizations. Springer, Berlin (2017)
2. Krishan, K.: Chawla: Composite Materials Science and Engineering. Springer, New York (2012)
3. Yi, X.S., Du, S., Zhang, L.: Composite Materials Engineering. Vol. 1. Fundamentals of Composite Materials. Springer, Singapore (2018)
4. Yi, X.S., Du, S., Zhang, L.: Composite Materials Engineering. Vol. 2. Different Types of Composite Materials. Springer, Singapore (2018)

5. Benzait, Z., Trabzon, L.: A review of recent research on materials used in polymer-matrix composites for body armor application. J. Compos. Mater. (2018). https://doi.org/10.1177/0021998318764002
6. Grumezescu, V., Grumezescu, A.: Materials for Biomedical Engineering: Thermoset and Thermoplastic Polymers. Elsevier, Amsterdam (2019)
7. Wang, X., Yong, Z.Z., Li, Q.W., et al.: Ultrastrong, stiff and multifunctional carbon nanotube composites. Mater. Res. Lett. **1**(1), 19–25 (2013)
8. Kondrashov, S.V., Shashkeev, K.A., Popkov, O.V., Solovyanchik, L.V.: Perspective technologies for obtaining of functional materials of structural purpose on the basis of nanocomposites with CNTs (review). Electron. Sci. J. «Proc. VIAM» **3**(39), 54–64 (2016). https://doi.org/10.18577/2307-6046-2016-0-3-7-7
9. Garcia, E.J., Saito, D.S., Megalini, L., et al.: Fabrication and multifunctional properties of high volume fraction aligned carbon nanotube thermoset composites. J. Nano Syst. Technol. **1**(1), 1–11 (2009)
10. Cheng, Q., Wang, J., Jiang, K., Li, Q., Fan, S.: Fabrication and properties of aligned multiwalled carbon nanotube-reinforced epoxy composites. J. Mater. Res. **23**(11), 2975–2983 (2008)
11. Tiwari, A., Mishra, Y.K., Kobayashi, H., Turner, A.P.F.: Intelligent Nanomaterials, 2nd ed. Scrivener Publishing LLC (2017). https://doi.org/10.1002/9781119242628
12. Mihajlin, Ju.A.: Special polymer composite materials, Scientific fundamentals and technologies. St. Petersburg (2008). (in Russian)
13. Ivitskiy, I., Sivetskiy, V., Bazhenov, V., Ivitska, D.: Modeling the electrostatic control over depth of the introduction of intelligent sensors into a polymer composite material. Eastern-Eur. J. Enterp. Technol. **1**(5(85)), 4–9 (2017). https://doi.org/10.15587/1729-4061.2017.91659
14. Bazhenov, V., Protasov, A., Ivitskiy, I., Ivitska, D.: Simulation of nanomodified polymers testing by the electric capacitive method. Eastern-Eur. J. Enterp. Technol. **4**(5(88)), 4–9 (2017). https://doi.org/10.15587/1729-4061.2017.108460
15. Sivetskyy, V.I., Khalimovskyy, O.M., Sokolskyyy, O.L., Ivitskiy, I.I.: Automation of intelligent sensor injection inlet in polymer moldings by using vector controlled electric drive. In: 2017 IEEE First Ukraine Conference on Electrical and Computer Engineering (UKRCON), Kiev, pp. 534–537 (2017). https://doi.org/10.1109/UKRCON.2017.8100297
16. Muralisrinivasan, N.S.: Basics of Troubleshooting in Plastics Processing: An Introductory Practical Guide. Handbook of Plastics Testing and Failure Analysis. 3rd edn. Wiley (2011).
17. Bratukhin, A.G., Bogolyubov, V.S.: Composite Manufacturing Technology. Springer, Heidelberg (2012)
18. Shalin, R.E.: Polymer Matrix Composites. Springer, Heidelberg (2012)
19. Mallick, P.K.: Processing of Polymer Matrix Composites, 1st edn. CRC Press, Boca Raton (2017)
20. Pascault, J.-P., Williams, R.J.J.: Epoxy Polymers: New Materials and Innovations. Wiley VCH, Weinheim (2009)
21. Gallego-Juarez, J.A., Graff, K.F.: Power Ultrasonics: Applications of High-Intensity Ultrasound. Woodhead Publishing, Cambridge (2014)
22. Margulis, M.A.: Sonochemistry and Cavitation. Gordon & Breach, London (1996)
23. Suslick, K.S.: Sonochemistry. In: Kirk-Othmer Encyclopedia of Chemical Technology, 4th edn., vol. 26, pp. 516–541. Wiley, New York (1998)
24. Fedotkin, I.M., Gulyi, I.S.: Cavitation, Cavitation Engineering and Technology, Their Use in Industry. Part II. OKO, Kiev (2000). (in Russian)
25. Leighton, T.G.: The Acoustic Bubble. Academic Press, London (1994)
26. HIELSCHER. https://www.hielscher.com/sonochem. Accessed 02 Sept 2018

27. Baronin, G.S., Zavrazhin, D.G., Kobzev, D.E.: The intensification of solid-phase forming process of polymers and composites with electrophysical effects of microwave and ultrasonic fields. In: Scientific Proceedings IX International Congress "Machines, Technologies, Materials", Varna, Bulgaria, , pp. 16–18 (2012)
28. Mozetič, M.: Surface modification to improve properties of materials. Materials **12**(3), 441 (2019). https://doi.org/10.3390/ma12030441
29. Crutchley, E.B.: Innovation Trends in Plastics Decoration and Surface Treatment. Smithers Rapra (2014)
30. Pestov, A.V., Kuznetsov, V.A., Mekhaev, A.V., Gorbunova, T.I., Saloutin, V.I., Smirnov, S. V., Vichuzhanin, D.I., Matafono, P.P.: Designing new adhesive materials based on epoxy oligomers filled with organic compounds. Polym. Sci. Ser. D **8**(2), 149–152 (2015). https://doi.org/10.1134/S1995421215020136
31. Banea, M.D., daSilva, L.F.M.: Adhesively bonded joints in composite materials: an overview. Review paper 1. Proc. Inst. Mech. Eng. Part L. J. Mater. Des. Appl. **223**(1), 1–18 (2008). https://doi.org/10.1243/14644207JMDA219
32. Prokhorenko, P.P., Dezhkunov, N.V., Konovalov, G.E.: Ultrasonic capillary effect. Tekhnika, Minsk, Belorussia (1981). (in Russian)
33. Lanin, V.L.: Physical effects of ultrasound in liquid media and their use in engineering. Technol. Electron. Ind. **2**, 11–15 (2013)
34. Bang, J.H., Suslick, K.S.: Applications of ultrasound to the synthesis of nanostructured materials. Adv. Mater. **22**, 1039–1059 (2010). https://doi.org/10.1002/adma.200904093
35. Hielscher, K.: Ultrasound for the preparation of graphene. In: Catalogue of Hielscher Ultrasonics (2018). https://pdf.directindustry.com/pdf/hielscher/ultrasound-preparation-graphene-hielscher-ultrasonics/24814-379925.html
36. Kolosov, A.E.: Efficiency of liquid reactoplastic composite heterofrequency ultrasonic treatment. Chem. Pet. Eng. **50**(3–4), 268–272 (2014). https://doi.org/10.1007/s10556-014-9893-y
37. Kolosov, A.E.: Low-frequency ultrasonic treatment of liquid reactoplastic media with pressure variation. Chem. Pet. Eng. **50**(5–6), 339–342 (2014). https://doi.org/10.1007/s10556-014-9904-z
38. Kolosov, A.E.: Impregnation of fibrous fillers with polymer binders. 1. Kinetic equations of longitudinal and transverse impregnation. Mech. Compos. Mater. **23**, 625–633 (1988). https://doi.org/10.1007/bf00605688
39. Konovalov, V.I., Kudra, T., Gatapova, N.Ts.: Modern issues of transport theory during drying. Bull. Tambov State Tech. Univ. **14**(3), 538–559 (2008)
40. Mouritz, A.P.: Ultrasonic and interlaminar properties of highly porous composites. J. Compos. Mater. **3**(34), 218–239 (2000)
41. Liu, L., Huang, Y.D., Zhang, Z.Q., Yang, X.B.: Effect of ultrasound on wettability between aramid fibers and epoxy resin. J. Appl. Polym. Sci. **99**(6), 3172–3177 (2006)
42. Kolosov, A.E.: Low-frequency ultrasonic treatment as an effective method for modifying liquid reactoplastic media. Chem. Pet. Eng. **50**(1–2), 79–83 (2014). https://doi.org/10.1007/s10556-014-9859-0
43. Kolosov, A.E.: Effect of low-frequency ultrasonic treatment regimes on reactoplastic polymer composite material operating properties. Chem. Pet. Eng. **50**(3–4), 150–155 (2014). https://doi.org/10.1007/s10556-014-9871-4
44. Kolosov, A.E., Sakharov, A.S., Sivetskii, V.I., Sidorov, D.E., Sokolskii, A.L.: Substantiation of the efficiency of using ultrasonic modification as a basis of a production cycle for preparing reinforced objects of epoxy polymer composition. Chem. Pet. Eng. **48**(5–6), 391–397 (2012). https://doi.org/10.1007/s10556-012-9629-9

45. Kolosov, A.E.: Prerequisites for using ultrasonic treatment for intensifying production of polymer composite materials. Chem. Pet. Eng. **50**(1–2), 11–17 (2014). https://doi.org/10.1007/s10556-014-9846-5
46. Sakharov, A.S., Kolosov, A.E., Sivetskii, V.I., Sokolskii, A.L.: Modeling of polymer melting processes in screw extruder channels. Chem. Pet. Eng. **49**(5–6), 357–363 (2013). https://doi.org/10.1007/s10556-013-9755-z
47. Sidorov, D.É., Kolosov, A.E., Pogorelyi, O.V., Gur'eva, A.A.: Engineering approach to the determination of the radiation field of a Polyethyleneterephthalate (PET) medium under radiant heating. J. Eng. Phys. Thermophys. **88**(6), 1409–1415 (2015). https://doi.org/10.1007/s10891-015-1325-0
48. Ávila-Orta, C., Espinoza-González, C., Martínez-Colunga, G., Bueno-Baqués, D., Maffezzoli, A., Lionetto, F.: An overview of progress and current challenges in ultrasonic treatment of polymer melts. Adv. Polym. Technol. **32**(S1), E582–E602 (2013). https://doi.org/10.1002/adv.21303
49. Liu, J., Chen, X., Diao, S.: Ultrasonic technology for the injection molding of polymer microstructure. Microsyst. Technol. **24**(5), 2253–2264 (2018)
50. Wagner, J.R.: Handbook of Troubleshooting Plastics Processes: A Practical Guide. Wiley, Hoboken (2012)
51. 51.Shalun, G.B., Surzjenko, E.M.: Laminated Plastics. Chemistry, Leningrad (1978)
52. Vasiliev, V.V., Morozov, E.V.: Advanced Mechanics of Composite Materials and Structures, 4th edn. Elsevier, Amsterdam (2018)
53. Kolosov, A.E., Sivetskii, V.I., Kolosova, E.P., Lugovskaya, E.A.: Procedure for analysis of ultrasonic cavitator with radiative plate. Chem. Pet. Eng. **48**(11–12), 662–672 (2013). https://doi.org/10.1007/s10556-013-9677-9
54. Kolosov, A.E., Virchenko, G.A., Kolosova, E.P., Virchenko, G.I.: Structural and technological design of ways for preparing reactoplastic composite fiber materials based on structural parametric modeling. Chem. Pet. Eng. **51**(7–8), 493–500 (2015). https://doi.org/10.1007/s10556-015-0075-3

Research of the Influence of Conditions of D-gun Spraying on Properties of Tungsten and Chromium Carbides Coatings

Yuriy Kharlamov, Volodymyr Sokolov, Oleg Krol(✉), and Oleksiy Romanchenko

Volodymyr Dahl East Ukrainian National University, 59-a, Central Pr., Severodonetsk 93400, Ukraine
krolos@i.ua

Abstract. The influence of conditions of D – gun spraying on properties of coatings from tungsten and chromium carbides are investigated. It is proved that the formation of structure and properties of D-gun spraying coatings is determined by character and intensity of physical and chemical transformations in powder particles and their interaction with the environment at all stages of technological process, including manufacture and preparation of powder, as well as the formation of interparticle connections. It is shown that in the development of technological processes, it is necessary to consider and provide optimal conditions for conductive transformations and phenomena, which have a decisive influence on the structure and properties of applied coatings. For this can be used regulation of combustible mixture composition, geometry, and dimensions of a barrel, conditions of powder introduction into the barrel, a single dose of powder, spraying distance, cyclogram of the working cycle, and other technological parameters. In the design of technologies and equipment for D-gun spraying of coatings from tungsten and chromium carbides, special attention should be paid to the formation of a local powder cloud in combustion detonation chamber, as well as its location in chamber taking into account the spraying distance. D-gun spraying can be used to obtain high-quality coatings from fine powders of tungsten carbide-cobalt. Still, it is necessary to design specialized devices for dosing and local introduction of fine powders into the combustion detonation chamber.

Keywords: D-gun sprayed coatings · Adhesion strength · Spraying distance · Tungsten carbide · Chromium carbide · Hardness

1 Introduction

Thermal spraying technologies (TST) are science-intensive and are used primarily in high-tech industries, e.g., aerospace, and precision engineering, which encourages further intensive research [1–5]. More recently, the theory and practice of TST have evolved toward finding and upgrading methods based on thermal mechanisms of coating formation. Currently, a gradual transition to high – velocity spraying methods are implemented, in which coating formation is based on the use of mechanical

V. Tonkonogyi et al. (Eds.): InterPartner 2020, LNME, pp. 300–310, 2021.
https://doi.org/10.1007/978-3-030-68014-5_30

activation of interacting materials [6–10]. D-gun spraying technologies have been supplemented by such new methods as warm spraying, high-velocity oxy-fuel spraying, and at the end of the last century – cold spraying method [11–15]. For D-gun spraying, powders specially adapted for these processes are used, and in the nomenclature of such materials, powders based on tungsten and chromium carbides occupy a considerable part.

However, there are conflicting data in scientific and technical literature by the comparative properties of D – gun spraying coatings, which contain refractory carbides. Additional research needs to be conducted in which comparison of coating properties would be carried out for the same material composition and taking into account whether the spraying mode is optimal for material structure, is it provides the best performance properties of the coating. It makes sense to compare coatings of similar composition to provide the best performance that can be obtained by varying spraying modes.

The objective of this article is to analyze the results of experimental research on the influence of conditions of D-gun spraying on properties of coatings based on tungsten and chromium carbides. Modern science and technique are characterized by extensive application of various protective and functional coatings [1, 2]. In the relevant field – surface engineering [6] – numerous methods of obtaining coatings and films are widely used. Among them, the technique and technology of thermal spraying [7–11] are intensively developing. If at first the methods of thermal spraying were based on the mechanism of thermal activation of interacting materials, then at the present stage of development of science and technology, high – velocity methods in which mechanical mechanism of activation plays a predominant role are actively developing. Such practices include D-Gun spraying [1, 12–16].

Among the main advantages of D-Gun sprayed coatings are following: the possibility of obtaining durable coatings during spraying; extensive opportunities of regulation of the thermal cycle of formed coating and product; a high growth rate of coating thickness; reduced requirements for quality of preparation of sprayed surface; relative simplicity of designs and high reliability of technological equipment; high density of resulting coatings.

2 Literature Review

Tungsten and chromium carbides have a high melting point, high hardness over a wide temperature range, have high heat resistance and resistance to aggressive environments, and others. They have been widely used as the main component of sintered hard alloys for various functional coatings, including D-gun spraying, and others [1, 6, 7, 10–12, 15, 17]. Tungsten carbide has a hexagonal crystal lattice. Tungsten carbide has a high hardness (9 by Mohs scale) and wear resistance. Tungsten carbide crystals have anisotropy of hardness in different crystallographic planes. So, depending on the orientation, the minimum value of microhardness is 13 GPa, and the maximum is 22 GPa [16, 18]. Other properties are as follows: Rockwell hardness 92–94 HRA; modulus of elasticity 710 GPa; linear thermal expansion coefficient $(3.84–3.90)\cdot 10^{-6}\ K^{-1}$. There are two tungsten carbides – WC and W_2C. WC monocarbide contains 6.12% of C (by mass) and decomposes at 1600 °C in the peritectic reaction. During it produces a liquid phase that contains 5.5% of C and graphite. Monocarbide or γ – phase (solid solution on the base of WC) has a very low homogeneity area. Minor carbon losses cause the transition of WC to a lower W_2C carbide (or β – phase – solid solution on the base of W_2C) and graphite: $2WC \rightarrow W_2C + C$. Carbide W_2C contains 3.2% of C (by mass). However, the solid solution based on it (β – phase) has a carbon concentration from 2.5 to 3.6% (by mass). Therefore, there is a wide area of phase homogeneity based on W_2C. The melting temperature of W_2C is 2760 °C, the density is 17.2 g/cm^3, and the microhardness (28–31) GPa. Alloys of WC – Co are the most durable of all known hard alloys.

Chromium carbide (Cr_3C_2) is an ordered phase with a very narrow homogeneity region. Carbon content is 13.34% by mass [16, 18]. It has a rhombic crystal lattice in which each of 8 carbon atoms is located in the center of a triangular prism, in corners of which are the chromium atoms. The coefficient of linear thermal expansion is $11.7 \cdot 10^{-6}$ K^{-1} at 20–100 °C. The microhardness of chromium carbide is about 10.4–20.2 GPa, which is related to the anisotropy of crystal lattice. The modulus of elasticity is 372 GPa. Among carbides of a transition metal of IV, V, VI groups, chromium carbide is the most stable at high-temperature oxidation. Chromium carbide can be a part of various cermets and coatings that work in conditions of wear, high temperatures, and harsh environments. It is the main component of KHN hard alloys of (chromium carbide-nickel). The proximity of the coefficient of thermal expansion of Cr_3C_2 to CTE of steel makes it possible to apply carbide – chromium coatings in which there are no significant residual stresses.

Carbide powders for D – gun spraying are obtained by sintering, conglomerating and sintering, cladding, and spheroidization. They are used mainly in such spraying methods as flame spraying, high – velocity oxy-fuel spraying, D-gun, and less limited by other ways with the use of special technological equipment [19–23]. The characteristics of powders and methods of their production are described in detail in the literature [5, 16, 17, 24–27], more information can be found in the prospectuses of according manufacturers. For example, FST (FLAME SPRAY TECHNOLOGIES), Netherlands, produces around 60 grades of tungsten carbide powder for thermal spraying. For high – velocity oxy-fuel spraying, agglomerated and sintered low – carbon (3.6–4.1% C), coarse-grained spherical powders WC – Co – 88/12 of brands AMPERIT 512.059 (particle size 30/5 μm) and AMPERIT 512.074 (45/15 μm) are intended. For plasma spraying sintered and grounded with high large content of WC (carbon: 3.9–4.3%) powders of brand WC – Co – 88/12 of marks AMPERIT 515.001 (45/22 μm), AMPERIT 515.002 (90/45 μm), and AMPERIT 515.074 (45/15 μm) are intended.

Also, get solid, dense coatings with good abrasion, erosion, and adhesion resistance of machine parts. In the same group are also manufactured powders of AMPERIT of brand: 515.203; 515.400; 515.401; 515.830; 515.851; 615.949. The line of powders of AMPERIT provides unique flexibility in the choice of powders with required sizes of carbide particles and metal matrix. Powders for thermal spraying with tungsten carbide, for example, such as AMPERIT 554 – 558 WC – Co – Cr, AMPERIT 512–528 WC – Co and AMPERIT 547 WC – Ni can be used at temperatures below 500 °C. Tungsten carbide coatings are characterized by extremely high wear resistance, and the addition of chromium to metal matrix greatly improves corrosion resistance.

At temperatures above 500 °C, tungsten carbide coatings do not provide the necessary resistance to oxidation. In this case, it is advisable to use chromium carbide coatings (for example, powders AMPERIT 584–588 with nickel chromium matrix (CrC – NiCr)), which provide effective protection against hot gas corrosion in parts of turbine or against oxidation and wear of pump\ bodies, shafts, and machine parts – even at temperatures up to 870 °C. By changing the content of the matrix from 10 to 50 percent, it is possible to find the right combination of hardness and plasticity of coating for a specific application. The application and development of D-gun sprayed coatings are associated with the requirements of strong and dense coatings. In most publications on this issue, much attention is paid to such an essential property of coatings as adhesion strength [12, 16–19]. However, to date, the mechanisms of formation and formation of strong coatings for D – Gun spraying have not yet been sufficiently researched, which complicates the development of technological processes and designs of coated products, as well as corresponding technological equipment [20–25].

The quality and performance characteristics of thermal sprayed coatings, regardless of spraying method, strongly depend on adhesion strength of the connection between the coating and basic material of the product, since the failure of coating leads to destruction of “coating-basic material system”. As a rule, this is an unacceptable failure. However, the prediction and control of adhesion strength of coating is difficult because it depends on many factors [12, 20, 26, 27]: method of thermal spraying and technological modes of coating and their subsequent processing; operating conditions of coated product and methods of adhesion strength determination; composition and type of starting material of coating, distribution of powder particles by size and morphology; primary material of product for spraying and the condition of sprayed surface, including presence and condition of oxide films (composition and thickness of oxide); roughness (the height of projections relative to average size of splats, and distance between projections of base roughness, which is characterized by root mean square roughness $R\Delta q$, that is, root mean square of ordinates of roughness profile); cleanliness (removal of surface impurities and abrasive residues); temperature and preheating method before spraying at temperature sufficient to remove adsorbents and condensates, etc.; residual stresses formed in «coating-basic material» system both during spraying process and during further processing and operation. These stresses also depend on many factors, including a combination of materials in the “coating-basic material” system, thermal spraying conditions and average surface temperature of substrate, and so on.; environmental conditions, such as temperature and humidity, ambient vibration transmitted to sample during test, and so on.

Besides, the structure of thermal sprayed coatings differs from the structure of compact materials, as they are usually composed of single particles (splats), the real contact surface of which is from about 15 to 60% of their surface: non-molten particles, pores of different shape, cracks and so on. The main mechanisms of adhesion between thermal spraying coating and lining are highly dependent on actual contacts between splats and between splats and substrate. They are divided into three main groups:

(1) mechanical (or anchor) adhesion; (2) metal-metal connection (diffusion phenomenon); (3) chemical compound (formation of an intermetallic compound with substrate).

The character of coating failure from the substrate may be adhesive, cohesive, or mixed. The adhesion strength can be determined by fracture mechanics [15, 20], which takes into account energy required for origin or propagation of cracks and evaluates adhesion in the "coating-basic material" system in terms of fracture toughness. It is necessary to experimentally establish conditions of equilibrium under which the propagation of a stable crack balances elastic energy created by an external force (determined by the geometry of sample and applied load). At the critical value of the rate of deformation energy release G_c, J/m^2, the crack propagates and, consequently, adhesion of the "coating-basic material" system breaks down. However, this requires the application of specialized laboratory equipment and samples of coated materials, as well as appropriate methods of analysis [20, 28–31].

The objective of this article is to research adhesion strength of D-gun sprayed coatings based on tungsten and chromium carbides, as well as to analyze the existing ideas about formation mechanisms of strong adhesion of D-gun sprayed coatings.

3 Research Methodology

In this work, the structure and properties of composite coatings of tungsten carbides and chromium were studied using unique research methods [11, 28–31]. As a starting point, the following powders were used: standard mechanical mixtures of powders WC – 8% of Co (VK8), and WC – 15% of Co (VK15); powders of alloys WC – 4% of Co (VK4), and WC – 8% of Co (VK8) obtained by crushing of sintered billets; conglomerated powders of alloy WC – 15% – NiCrSiB (VSNGN – 85); conglomerated powders of alloys WC – 25% of Co (VK25), and WC – 18% of Co (VK18C); mechanical mixtures of powders of WC – Ni; conglomerated powders of CrxCy – Ni alloy (KHN); mechanical mixtures of powders of chromium carbides and nickel type KHN.

Tungsten and chromium carbides are compounds of metalloid type. During the application of this coating by the method of TSP, it is necessary to consider selective oxidation and evaporation of metalloid elements during heating, dissociation, and sublimation. The formation of D-gun coatings can be carried out from molten powder particles and in viscous – plastic state at a sufficiently high velocity of their impact with part surface. The degree of melting of powder particles in spray jet can be estimated using the parameter of difficulty of material melting [7, 16, 18] $D = (i_{\mathrm{ml}} - n_Q Q_{\mathrm{ex}})^2 \rho_p (1 + 0,2Bi)^2$, where i_{ml} – weight enthalpy of material of molten particles; Q_{ex} – specific exothermic effect in volume of particle; n_Q – degree of use of exothermic effect when particle is heated; ρ_p – the density of particle material; Bi – Bio criterion, that takes into account the thermophysical properties of particle material and heat sources in process of heat exchange, and in the absence or smallness of exothermic thermal effect and values of Bi < 0,2 $D = i_{ml}^2 \rho_p$.

The parameter of melting difficulty expresses the ratio of heat amount required to melt a unit volume of a particle to the residence time of the particle in the heating zone, which is determined by the material density of the particle. During the impact of particles with substrate and formation of a single coating layer, an important role has an accumulation factor of heat by particle, $b = \sqrt{C_p p_p \lambda_p}$ where C_p, ρ_p, λ_p – heat capacity, density, and thermal conductivity of particle material. Tungsten and chromium carbides have relatively low values of melting difficulty parameters: for WC $D = 2{,}1 \cdot 10^{10}$ $kJ^2 \cdot kg^{-1} \cdot m^{-3}$, for Cr_3C_3 $D = 2{,}28 \cdot 10^{10}$ $kJ^2 \cdot kg^{-1} \cdot m^{-3}$, which is favorable for thermal spraying of coatings. The coefficients of heat accumulation of these materials are lower or same order with pure metals, its value for WC is 10^5, for Cr_3C_3 – 10^6 $W \cdot m^{-2} \cdot K \cdot s^{-0.5}$ [7, 16, 18]. The density of Cr_3C_3 carbide is 6600 kg/m^3, WC – 15600 kg/m^3. The chromium carbide particles have a lower value of time of velocity relaxation and, accordingly, stay time in the high – temperature jet. When spraying tungsten carbide coatings, its high tendency for decarburization should be taken into account, and after oxidation, the monocarbide decomposes. The process of oxidation of chromium carbide particles is terminated by the creation of Cr_2O_3 oxide film on their surface. Carbon losses during thermal spraying of chromium carbide are small (0.1–0.3%).

4 Results

One of the essential tasks in the development of technology of D – gun spraying is the organization of proper injection of sprayed powder into the detonation combustion chamber, which affects the interaction of the particles with the substrate.

This includes the mass of single doses of powder, its location in the barrel, and spraying distance. In the first series of experiments, the weighed portions of powder were introduced into the barrel at a certain distance from the barrel exit section S. Barrel length l = 1.8 m, and the inner diameter was 20 mm, which was filled with a mixture of acetylene and oxygen in a ratio of 1:1.1. Coatings were also applied to fixed steel specimens. The spraying distance was also changed.

The results of microhardness measuring for different powders at different positions of injection point of powder and spraying distance are shown in Fig. 1. From various ways of injection of powder (100 mg) into the barrel, the highest values of microhardness of obtained coatings are achieved at the initial placement of the weighed portions in a center of the barrel (Fig. 1).

There are also changes in conditions of formation of a powder particle cloud in the barrel and its interaction with the impulse flow of hot products of gas detonation. Under different spraying conditions, the temperature and velocity of powder particles upon contact with the part surface, as well as the time of their preheating and acceleration, are changed. Particles undergo varying degrees of transformations inflow and at collisions during the process of coating formation. The cohesive strength of coatings is also changing.

The boundaries of separation between individual particles affect the mechanical properties of coatings, including their hardness. Thus, the conditions of formation of a pulsed heterogeneous flow and the change of residence time of particles in the flow in this series of experiments also affect the properties of obtained coatings.

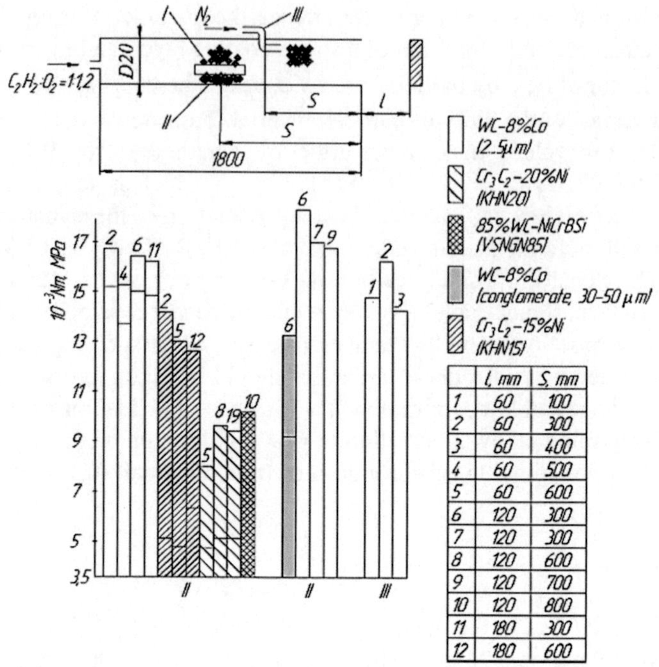

Fig. 1. The microhardness of coatings at various ways of powder injection into a barrel.

When the powder is injected into the middle part of the barrel automatically and it dissipation by nitrogen in the direction of open-end with increasing spraying distance, the microhardness of coatings from powder VK8 decreases almost linearly (Fig. 2).

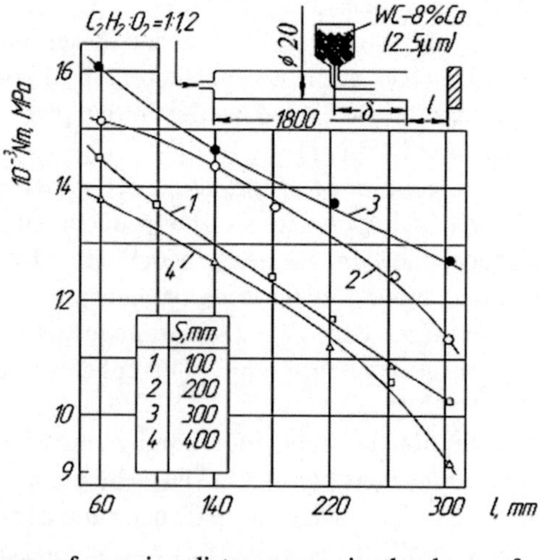

Fig. 2. Influence of spraying distance on microhardness of coatings VK8.

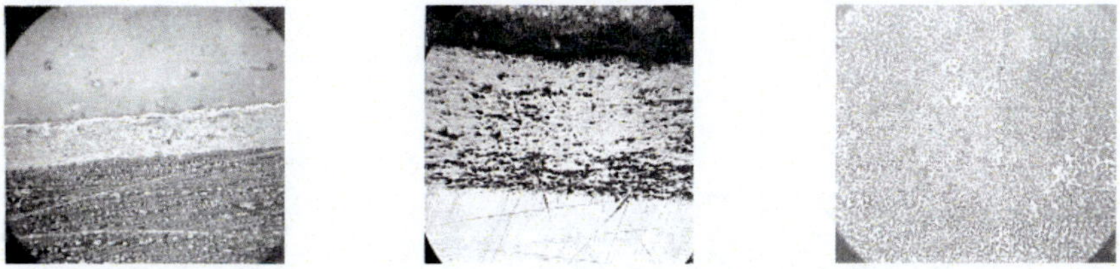

a b c

Fig. 3. The microstructure of coatings from powderVK8 on steel substrate: a – × 130; b – × 500; c – × 500 (polished lengthwise specimens).

The coatings have a high density (Fig. 3). The carbon content in coatings from fine powder VK8 is reduced to 60–80% of stoichiometric composition. The cobalt content in coating corresponds to the original powder. On the micrograph of the transverse thin section of coating, inclusions resembling the η1 – phase (Co3W3C), with irregular shape and distributed one-dimensionally by coating are visible. In coating layer grains of phase, WC are stored (Fig. 4b). With prolonged etching, this phase is completely etched, and characteristic distribution of phase η1 in form of laces is revealed (Fig. 4c). There is a significant variation of microhardness in the coating layer from 8 to 16 GPa is observed (Fig. 5). This is due to the heterogeneous nature of the coating structure and uneven distribution of phase components.

a b c

Fig. 4. The microstructure of coatings VK8 on titanium alloy: a – ×500; b – ×800; c – ×1500.

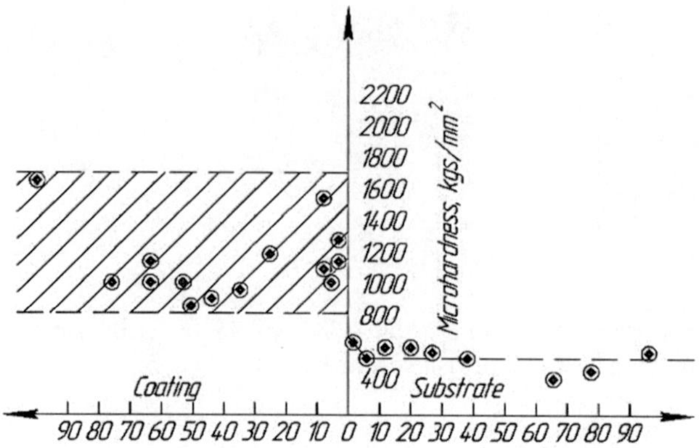

Fig. 5. Microhardness in cross-section of a sample from titanium alloy VT3–1 with coating from powder VK8.

5 Conclusions

The formation of the structure and properties of D-gun spraying coatings is determined by nature and intensity of the processes of physicochemical transformations in powder particles and their interaction with the environment at all stages of technological process, including the manufacture and preparation of powder, as well as the formation of interparticle bonds. When developing the technological processes, it is necessary to consider and ensure optimal conditions for the occurrence of leading transformations and phenomena that have a decisive influence on the structure and properties of applied coatings. For this, regulation of combustible mixture composition, geometry and dimensions of the barrel, conditions for injection of powder into the barrel, a unit dose of powder, spraying distance, the cyclogram of the working cycle, and other technological methods can be used.

In developing technologies and equipment for D-gun spraying of coatings from tungsten and chromium carbides, special attention should be paid to the formation of a local powder cloud in the detonation combustion chamber, as well as its location in the chamber, taking into account the spraying distance.

High quality coatings from powders of tungsten carbide-cobalt can be obtained by D-gun spraying, but this requires the development of specialized devices for dosing and local introduction of fine powders in the detonation combustion chamber.

When spraying powders with a particle size of tens of micrometers, their predispersion in a fresh combustible mixture should be ensured.

References

1. Kharlamov, Y.: The development of D-Gun spraying technologies. Bulletin of V. Dahl EUNU **7**(237), 114–132 (2017)
2. Yushchenko, K., Borisov, Y., Kuznetsov, V., Korzh, V.: Surface Engineering. Nauk. Dumka, Kyiv (2007)

3. Fesenko, A., Basova, Y., Ivanov, V., Ivanova, M., Yevsiukova, F., Gasanov, M.: Increasing of Equipment Efficiency by Intensification of Technological Processes. Periodica Polytechnica Mech. Eng. **63**(1), 67–73 (2019). https://doi.org/10.1007/978-3-030-04792-4_37
4. Krol, O., Sokolov, V.: Parametric modeling of gear cutting tools. In: Gapinski et al. (eds) Advances in Manufacturing II 2019, LNME, vol. 4, pp. 3–11. Springer, Cham (2019), doi:https://doi.org/10.1007/978–3–030–16943–5_1
5. Zablotskyi, V., Moroz, S., Tkachuk, A., Prystupa, S., Zabolotnyi, O.: Influence of diamond smoothering treatment power parameters on microgeometry of working surfaces of conjugated parts. In: Tonkonogyi V. et al. (eds) Advanced Manufacturing Processes. InterPartner 2019. LNME, pp. 372–381. Springer, Cham (2020), doi:https://doi.org/10.1007/978-3-030-40724-7_38
6. Ilyushchenko, A.F., Okovityj, V.A., Kundas, S.P., Formanek, B.: Formation of Thermal Sprauing Coatings: Theory and Practice. Bestprint, Minsk (2002)
7. Budag'janc, N.: D-Gun Processes in Industry. Publishing house of VUGU, Lugansk (1998).
8. Sokolov, V.: Diffusion of circular source in the channels of ventilation systems. In: Fujita, H. et al. (eds.) Advances in Engineering Research and Application. ICERA 2018. LNNS, vol. 63, pp. 278–283. Springer, Cham (2019), doi.org/https://doi.org/10.1007/978-3-030-04792-4_37
9. Rogovyi, A.: Energy performances of the vortex chamber supercharger. Energy **163**, 52–60 (2018). https://doi.org/10.1016/j.energy.2018.08.075
10. Pierre, L., Fauchais, J., Heberlein, V., Maher, R., Boulos, I.: Thermal Spray Fundamentals: From Powder to Part. Springer, New York (2014)
11. Anciferov, V.N., Bobrov, G.V., Druzhinin, L.K.: Powder Metallurgy and Spray Coatings. Metallurgy, Moscow (1987)
12. Davis, J.R. (eds.): Handbook of Thermal Spray Technology. ASM International, Ohio (2004).
13. Pavlenko, I., Trojanowska, J., Ivanov, V., Liaposhchenko, O.: Parameter identification of hydro-mechanical processes using artificial intelligence systems. Int. J. Mech. Appl. Mech. **2019**(5), 19–26 (2019)
14. Pavlenko, I., Liaposhchenko, A., Ochowiak, M., Demyanenko, M.: Solving the stationary hydroaeroelasticity problem for dynamic deflection elements of separation devices. Vibr. Phys. Syst. 2018026 (2018) .
15. Pawlowski, L.: The Science and Engineering of Thermal Spray Coatings, 2nd edn. Wiley , Chichester (2008)
16. Tret'jakov, V.: Fundamentals of metal science and production technology of sintered hard alloys. Metallurgy, Moscow (1976).
17. Javanbakht, T., Hadian, H., Wilkinson, K.J.: Comparative study of physicochemical properties and antibiofilm activity of graphene oxide nanoribbons. J. Eng. Sci. **7**(1), C1–C8 (2020). https://doi.org/10.21272/jes.2020.7(1).c1
18. Samsonov, G., Vinickij, I.: Refractory Compounds (Reference Book). Metallurgy, Moscow (1976)
19. Sokolov, V., Krol, O., Stepanova, O.: Automatic control system for electrohydraulic drive of production equipment. In: International Russian Automation Conference (RusAutoCon) 2018. IEEE (2018). https://doi.org/10.1109/RUSAUTOCON.2018.8501609
20. Rogovyi, A., Khovanskyy, S., Grechka, I., Pitel, J.: The wall erosion in a vortex chamber supercharger due to pumping abrasive mediums. In: Advances in Design, Simulation and Manufacturing II. DSMIE 2019. Lecture Notes in Mechanical Engineering, pp. 682–691. Springer, Cham (2020).

21. Sokolov, V., Krol, O., Stepanova, O.: Choice of correcting link for electrohydraulic servo drive of technological equipment. In: Ivanov, V. et al. (eds.) Advances in Design, Simulation and Manufacturing II. DSMIE 2019. LNME, pp. 702–710. Springer, Cham (2020). https://doi.org/10.1007/978-3-030-22365-6_70
22. Fedorovich, V., Mitsyk, A.: Mathematical simulation of kinematics of vibrating boiling granular medium at treatment in the oscillating reservoir. Key Eng. Mater. **581**, 456–461 (2014). https://doi.org/10.4028/www.scientific.net/KEM.581.456
23. Sokolov, V., Krol, O., Stepanova, O.: Nonlinear simulation of electrohydraulic drive for technological equipment. J. Phys: Conf. Ser. **1278**, 012003 (2019). https://doi.org/10.1088/1742-6596/1278/1/012003
24. Heintze, G.N., McPherson, R.: Fracture toughness of plasma sprayed zirconia coatings. Surf. Coat. Tech. **134**, 15–23 (1988)
25. Movshovich, A.Ya., Chernaja, Yu.A., Ishchenko, G.I.: Investigation of influence of physico-mechanical characteristics of D-gun spraying coatings on wear resistance of cutting elements of reworkable dies. Bull. NTU "KhPI" Collect. Sci. Pap. Thematic Rel. New Sol. Modern Technol. **45**, 22–28 (2011).
26. Faradzhallah, M.A., Panarin, V.E., Bys,' S.S.: The problem of residual technological stresses in improving the quality of D-gun spraying coatings. Bull. Khmelnytskyi Natl. Univ. **2**, 14–18 (2011).
27. Rjahovskij, A.V., Kosenko, V.V., Vlasenko, V.N.: Features of estimation of adhesion strength of D-Gun spraying coatings. Weapons Syst. Military Equip. **3**(31), 215–217 (2012)
28. Lin, C.K., Berndt, C.C.: Measurement and analysis of adhesion strength for thermally sprayed coatings. J. Therm. Spray Technol. **3**(1), 75–104 (1994)
29. Maksimovich, G.G., Shatinskij, V.F., Kopylov, V.I.: Physicochemical processes during plasma spraying and destruction of coated materials. Nauk. , Dumka Kyiv (1983)
30. Fedorov, V.V.: Kinetics of Damage and Destruction of Solids. Fan, Tashkent (1985)
31. Korobov, Yu.S.: Analysis of Properties of Thermal Coatings. Part 2. Evaluation of Coating Parameters. Publishing House of Ural University, Yekaterinburg (2016).

Check for updates

Improvement of the Polyurethane Foam Molding by the DoE Method: Case Study

Agnieszka Kujawińska(✉), Radosław Kowalski, and Adam Hamrol

Poznan University of Technology, 5, Pl. M. Skłodkowskiej-Curie, 60-965 Poznań, Poland
agnieszka.kujawinska@put.poznan.pl

Abstract. Design of Experiments is one of the tools for improvement and optimizing of products and manufacturing processes. Experiments conducted in industrial conditions should marginally interfere with the course of a studied process. The subject of the paper is the analysis of the impact of polyurethane foam-forming parameters on its strength properties, i.e., tensile strength and elongation at break. The tests were carried out on flexible foam molded for car seats. The measured parameters were: mixing pressure in the head, the volume of the mixing chamber, and time to prepare the pressure. The purpose of experiments was also to analyze the interactions of parameters on foam properties. A full factorial experiment was used for this. Eight combinations of parameters were developed, and on their basis, foams were prepared from which test samples were cut out. The foams were made using a one-step method. The results showed that both the elongation and strength have the greatest influence on the mixing pressure.

Keywords: Design of experiments · Moulding · Optimisation

1 Introduction

Manufacturing optimization problems are a major concern of all industrial companies, and continue to remain improperly solved, it continues to be of utmost importance and necessity, being even more crucial nowadays, in the context of the currently rapidly upcoming 4th Industrial revolution [1]. Thus, there are being put forward many different solutions, including ontologies [2], models [3–5], along with platforms and tools, for improving the manufacturing management process [6–9], and a specific and also very important issue arises in the context of the improvement of the polyurethane foam molding process, for which a contribution is put forward in this paper.

Polyurethane foams are materials which, due to their extensive application, currently play an important functional role in everyday life. Due to their widespread use in the automotive industry, they must comply with high requirements for safety and comfort of use. Polyurethane foams constitute 67% of products made of PUR. Flexible foams are most often used as products for the furniture industry (upholstery foams, mattresses) as well as backrests and seat pans in car seats. Rigid foams can be found in the construction industry as an insulating material. Other applications include mainly

V. Tonkonogyi et al. (Eds.): InterPartner 2020, LNME, pp. 311–321, 2021.
https://doi.org/10.1007/978-3-030-68014-5_31

fibers, footwear, and clothing parts (shoe soles, outer layers of jackets and vests). They constitute 18% of products made of polyurethane. Various methods can provide making foams from polyurethane (PUR). Depending on the technology used, products of different structures are obtained [10].

2 Literature Review

The properties of the ready-made foam and its structure also depend on raw materials used for its production. The basic components used in the production of PUR foams include polyols, isocyanates, and cross-linking and chain-extending agents. Other materials that support the manufacturing process include blowing agents, catalysts, and surfactants. Their presence affects not only the process but also the final foam structure.

The most important properties of PUR foams, distinguishing it as a material in the manufacturing process, include: high flexibility, a wide range of hardness, high tensile strength and tear resistance, low thermal conductivity and resistance to extreme weather conditions (e.g., high and low temperatures), high elasticity - after removing the load, it returns to its original shape, good electrical properties - thanks to this property they can be used as, e.g., insulation.

The proportion of components used in the process has a great impact on the properties listed above. The most important are the main substrates: isocyanate and polyol. The first of these affects the number of rigid segments in the polymer structure. They consist mainly of isocyanate groups but also urethane groups and chain extenders. A large number of them increase the mechanical properties of polyurethane, e.g., tensile strength, hardness, flexibility. The type of isocyanate used is also important. In the case of symmetrical construction of its chains, it has a greater impact and allows obtaining better properties. The second most important component of the reaction mixture is a polyol. The existence of elastic segments is associated with it. An increase in their number lowers tensile strength but improves elongation at break and flexibility [10].

Polyurethane foams are one of the most developed product groups in the PUR market. Depending on specifically required properties, they are produced in various forms. The most important of them include flexible polyurethane foams (block foams, molded foams), viscoelastic, rigid, and semi-rigid foams [11, 12].

Block foams are usually intended for the production of large products. The produced foam block is cut into smaller pieces to be processed into ready-made products (e.g., mattresses). These foams can be obtained by periodic and continuous methods (used with very high demand for the product, capacity up to 50000 t/y).

A separate group of foams is made up of molded foams – the reaction mixture is poured into a mold that reflects the shape of the finished product. Foams produced by the blowing method in the mold are characterized by accurate reproduction of the assumed shape. They are used, among others, as backrests and seat pans in the automotive industry. In order to provide user comfort, they must meet many requirements. One of them is to ensure the ergonomic shape of the foam. Also, car seat foams must comply with environmental protection regulations [13].

In addition to meeting the above requirements, car seat foams must, above all, guarantee safety. It is related to properties such as foam strength, foam hardness, flammability, and many others. Failure to meet any of the requirements results in the decision to destroy the intermediate product in the manufacturing process [14, 15].

3 Research Methodology

The enterprise in which the tests were conducted specializes in the production of car seat pans and backrests. The manufactured products are designed for two types of cars. Hence the foams differ in size and are divided into a group of small and large seat pans (Fig. 1).

The foam production process begins with the supply of raw materials to the plant premises in barrels placed on pallets. Special care is taken both during transport and use of the substance due to irritating and toxic effects on the skin and possible allergic reactions. The components are pumped into storage tanks. Each of them is equipped with a component level indicator and a pump feeding the components. Also, the polyol tank is equipped with mixers and a raw material recirculation circuit.

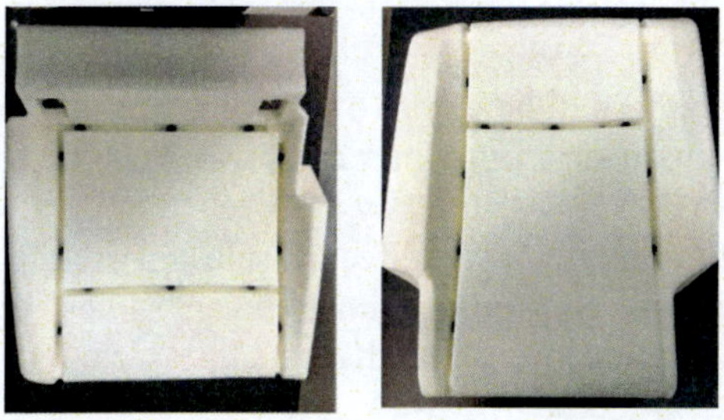

Fig. 1. Example of foams for a small seat pan.

Depending on the demand, raw materials are pumped automatically from storage tanks to day tanks. Their task is the maintenance of the material at the temperature range of 20 ÷ 30 °C. Temperature stabilization takes place thanks to two power generators connected to two heat exchangers and a water jacket. The cooling system uses water at the temperature not exceeding 20 °C.

The process of molding polyurethane foam requires compliance with many strictly defined parameters related to the mold. These include mold temperature and cycle length, i.e., the time during which the raw material is molded. The mold is made of aluminium. It is a closed unit with an internally grooved shape of the produced seat pan.

The preparation of the mold requires the application of a release agent on its surface. An employee applies the separator using a spray gun. The main component of the continuous phase of the release agent is water, which must evaporate completely before the foam is poured. A water heating circuit is used to maintain a constant mold temperature. The water temperature is in the range of 35 ÷ 75 °C. In the case of a seat pan foam, before pouring the raw material, an employee provides the mold with all integral foam elements, i.e., clips, hook-and-loop fasteners. The presence of these parts is necessary for the further processes of providing seats with outer coverings. The foam production process is carried out using a one-stage method using pressure heads. The components in the head are mixed, and then, after pouring, blowing in the mold and cross-linking take place. High-pressure pumps are responsible for delivering raw materials to the head. Their task is also to maintain constant pressure when dosing the raw material. The mold is filled automatically using heads hanging on robotic arms.

After the foam leaves the mold, an employee places the product on a conveyor belt, which then directs the foam under the cell-opener. The lid is lowered, and the foam closes, followed by a gas exchange process. The outlet of the cell-opener is connected to the ventilation system of the production hall. After the process is completed, the lid is lifted, and the foam is removed from the conveyor station. This station allows for gas exchange between the inside of the cells and the surroundings. The finished product, after having been released, contains mainly reactant carbon dioxide in the cells. Its removal ensures the dimensional stability of the product. Failure to do so results in the loss of the required foam shape - the form shrinks, and its upper surface becomes concave.

After the foam molding process has been completed, it must be checked visually. Foams are checked for inadequate pouring, tears, and deformations. Foams meeting the visual criteria end up on the mobile conveyor belt. Damaged foams with inadequate pouring and tears, depending on the extent of damage, are then repaired or destroyed. The produced foams remain on the mobile conveyor belt for another 24h to further cross-link the polymer. After this time, they are packed collectively.

The research aimed to determine the effect of selected molding parameters (mixing pressure in the head, mixing chamber volume, and pressure preparation time) on the tensile strength and elongation of polyurethane foam as well as to analyze the interaction between the studied factors.

3.1 Plan of the Experiment

A full factor experiment was conducted to analyze the effects of factors and their interactions. It involved determining two levels of factors and then conducting experiments for all level combinations. Table 1 lists the adopted factor levels (the so-called operating window in the experiment).

Table 1. Determination of factor levels in the experiment.

Code	Factor	Factors at level (+)	Factors at level (−)
X_1	Mixing pressure in the head	High (polyol - 145 bar, isocyanate - 144 bar)	Low (polyol - 20 bar, isocyanate - 15 bar)
X_2	Volume of the mixing chamber	Large (4mm)	Small (10 mm)
X_3	Pressure preparation time	Long (5 s)	Short (0.5 s)

Experimenting with the three examined factors required the performance of eight experiments ($N = 2^n$ where n – number of factors) is realized (Table 2).

Table 2. Plan of factor levels for particular experiments.

Experiment no.	Main factors		
	X_1	X_2	X_3
I	+	+	+
II	+	−	+
III	+	+	−
IV	+	−	−
V	−	+	+
VI	−	−	+
VII	−	+	−
VIII	−	−	−

3.2 Organization of the Research Site and Samples

In the experiment, the PUR mixture was poured into cuboidal cardboard packaging. After pouring, the packaging was transferred to a stacking table to allow the foam to rise freely and to cross-link the polymer. After 24 h, the molded foams were removed from the packaging and then cut into strips of assumed thickness. A sample was cut out from each strip with a paddle-shaped cutter (Fig. 2).

Fig. 2. Cutter for samples and correctly (top) and incorrectly (bottom) broken sample during the test.

The paddles that took part in the experiments did not have visual defects. The samples had a thickness of 7.9 mm to 8.4 mm with a tolerance of 0.5 mm. It was assumed that its correct breaking occurs if it happens in the middle of the paddle. Figure 2 shows a correctly and incorrectly broken sample. The cut-out test samples were conditioned in the laboratory for 24 h at a temperature of 23 °C ± 2 °C and relative humidity (50 ± 5)%. For each experiment, 3 cuboidal foams were obtained, which were cut into strips, and then 5 paddle-shaped samples were cut out.

The tensile strength test was carried out following the recommendations of the PN-EN ISO 1798:2008 standard – *"Flexible cellular polymeric materials. Determination of tensile strength and elongation at break"*. Tensile strength and elongation measurements were taken on the ZWICK 1447 SKP-248 testing machine.

4 Results

The initial visual assessment of the foam structure showed that the foams differed from one experiment to another. The difference was particularly evident between samples from an experiment I to IV (pressure set at a high level) and samples from experiment V–VIII (pressure set at a low level). The previous ones had a uniform structure. The foams were very soft and delicate to the touch. No visible bubbles indicated that the components had been thoroughly mixed. The opposite situation was observed in the latter ones (experiments V–VIII). Their structure resembled a sponge. Also, the transition from the soft to hard phase was observed in the sample from experiments VI and VIII. The structure of the material was not uniform. It is related to the second parameter - the volume of the mixing chamber. At low pressure and a small mixing chamber, the reaction substrates had not been mixed well. Hence, different foam composition was noted. For samples prepared at high pressure, the volume of the chamber had no significant effect.

4.1 Impact of Tested Factors on Tensile Strength

The greatest impact on the foam strength is exerted by the X_1 factor - mixing pressure in the head (Fig. 3). The effect of its impact is −19.66. It means that the difference between the two levels (+) and (−) of the examined factor is large. The negative effect value indicates that at (−) level, the strength values are higher.

The second factor having a significant impact is the volume of the chamber (X_2). The effect of its action is 5.27. The positive value indicates that with a larger volume of the mixing chamber, the foam strength is greater.

The last of the main factors is the time of pressure preparation (X_3). Its influence is smaller than the others. It takes the value of −3.20. With shorter pressure preparation time, the obtained tensile strength values were slightly higher.

Analyzing the mutual influence of factors on tensile strength, it can be concluded that the interaction of factors X_1X_2 for which the value of the effect is that −0.10 has the lowest impact (Fig. 3). It means that changing the volume of the chamber at different pressure levels has no major effect on strength. The slope of the impact line of these factors is similar. Changing the time level at high pressure does not affect strength. Low pressure significantly increases strength (stress). Analyzing the interaction of volume and time, it can be concluded that with a long time, foam strength values for the chamber volume at (+) will be higher and for (−) lower. With a short time, the volume of the chamber does not matter much.

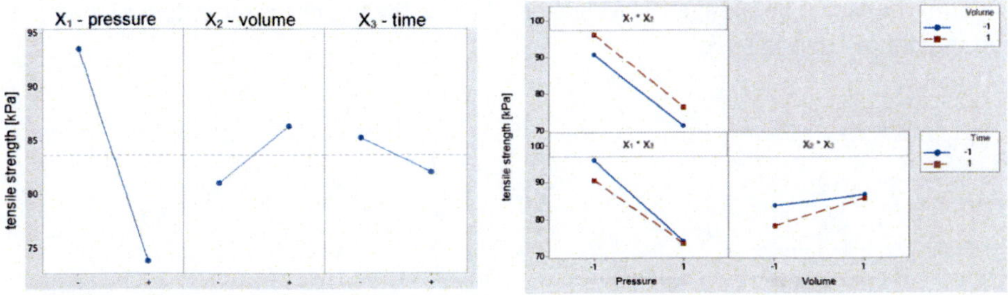

Fig. 3. Graphic representation of the impact of the main effects and interactions.

The received impact effects were subjected to statistical verification in order to examine their significance. A Pareto diagram was created (Fig. 4) illustrating the standardized effects of main factors and interactions. The mixing pressure in the head and the volume of the mixing chamber have a statistically significant impact on tensile strength. Other factors are not statistically significant.

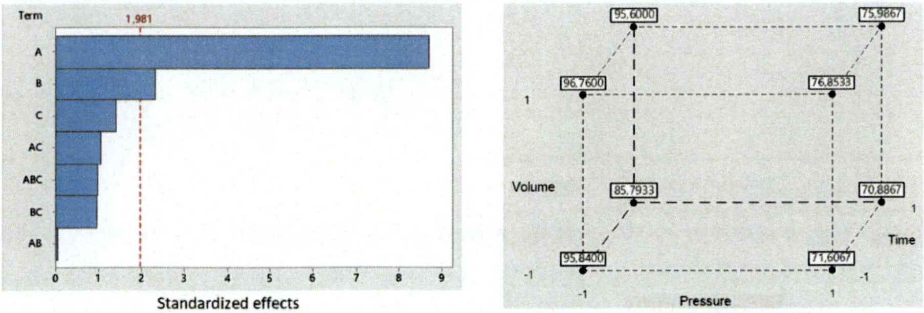

Fig. 4. Pareto diagram ($\alpha = 0{,}05$) and the average value of tensile.

The results for individual combinations of the experiment are shown in a cubic graph (Fig. 4). The highest value of tensile strength was obtained for the set of parameters: pressure (−), volume (+), time (−). It is worth noticing that at low pressure, the average tensile strength values are much higher than at high pressure. It is related to the physical structure of the produced foam. At low pressure, the reaction mixture is not well mixed in the head. The produced foam has much larger cells, and there are fewer of them, compared to foams prepared with high pressure. It is hard, and its polymer chains are stronger (greater cell wall thickness).

4.2 Impact of Main Factors on Elongation at Break

As with tensile strength, pressure also has the greatest effect on elongation. The effect of its action is 21.69. It is many times larger than the effect of other factors (Fig. 5). At high pressure, the elongation values are significantly higher. The X_2 volume factor also has a big impact. The elongation values are higher for the chamber volume set at (+). The impact of time is small.

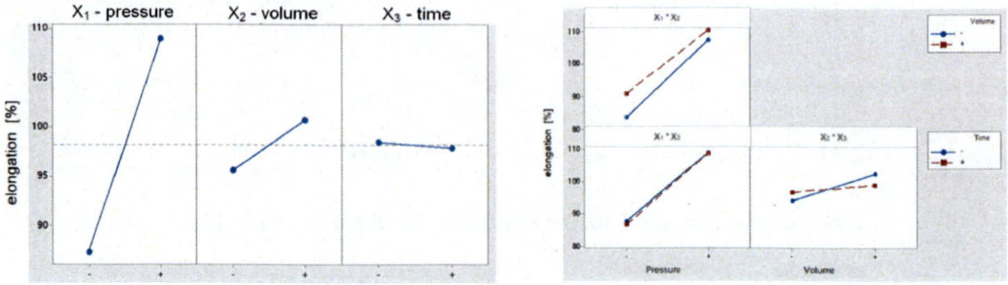

Fig. 5. Impact of main factors on elongation at break and interaction of factors.

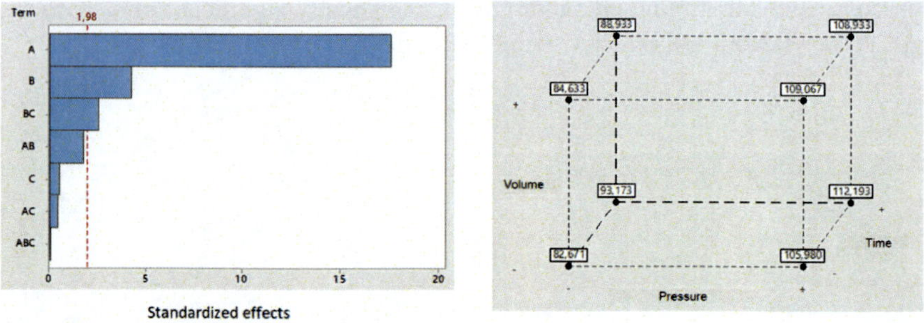

Fig. 6. Pareto diagram (α = 0,05) and the average value of elongation at break for individual experiments.

The analysis of the interaction graphs of factors for elongation shows that it only occurs between the volume of the chamber and the time of reaching the pressure (Fig. 6). In order to check the significance of the effects of factors, they are summarized on a Pareto diagram (Fig. 6). The X_1 factor statistically significantly affects the tensile strength. The high mixing pressure in the head allows obtaining a much better foam structure. Figure 6 also shows the average elongation values obtained from the results in each experiment. Whereas elongation should be as large as possible, the process should be carried out at a high level of mixing pressure. It is a factor that plays a decisive role in the results of this property and allows to obtain good-quality foam. For different levels of volume and time, the obtained differences in elongation values are small.

4.3 Conclusions

Produced foams are mainly assessed in terms of elongation at break, which should be as high as possible. Foam is then characterized by high flexibility, which affects the comfort of its use. Tensile strength should not be less than 80 [kPa].

The results obtained in the experiment allowed to state that with high elongation values of the tested samples, their strength is low and vice versa: with low elongation, higher tensile strength was obtained. For mixing pressure in the head at a low level (−), elongation values are much lower. The smallest elongation was obtained with a small chamber volume. Regardless of the set time level, the obtained average results of the tested properties for different levels of other factors are similar. It means that this factor has an insignificant effect on elongation at break.

To sum up, in the case of elongation, two main factors have the greatest impact: mixing pressure in the head (X_1) and volume of the mixing chamber (X_2). These parameters should be set at a high level to obtain the highest elongation value.

5 Summary

The article presents the results of an ongoing experiment that allows for the optimization of the PUR foam production process with a focus on its strength and elongation.

In order to analyze the influence of mixing pressure in the head, the volume of the mixing chamber and time of pressure preparation on the features as mentioned above, a full factor experiment was planned and carried out. The elaboration of the results allowed to indicate a favorable setting of the PUR foam molding process in the production of car seats.

Pressure has the greatest effect on both elongation and tensile strength. The structure and properties of produced foams depend mainly on the level of this factor. With pressure going up, elongation increases, and the foam structure improves. However, tensile strength decreases. The obtained results showed the influence of volume. However, it is smaller than for pressure. The same interdependence exists for both properties: with a large volume of the mixing chamber, elongation and strength values are higher. Time has no significant impact on the properties tested.

The interaction of the factors turned out not to be of great importance for tensile strength. None of them exceeded the limit of the significance of the effect of their action. In the case of elongation, volume and time interact. In order to increase elongation, the process should be carried out with a long time and a large chamber volume.

Foams made at high pressure, and low volume have the best properties. Their elongation is the highest, and the results obtained did not differ (low standard deviation). It means that the process is repeatable. The worst solution is to set all parameters at (−). Foams produced in this way are not very flexible, and their structure does not meet the requirements. Streaks of components that have not properly reacted are visible.

The research results allowed learning more about the process of molding polyurethane foams and to gain knowledge about the interdependence of parameters and their impact on the finished product.

Acknowledgments. The paper is prepared and financed by scientific statutory research conducted by the Chair of Management and Production Engineering, Faculty of Mechanical Engineering and Management, Poznan University of Technology, Poland, supported by the Polish Ministry of Science and Higher Education from the financial means in 2019 (02/23/DSPB/7716).

References

1. Jasiulewicz-Kaczmarek, M., Gola, A.: Maintenance 4.0 technologies for sustainable manufacturing – an Overview. IFAC PapersOnLine **52**(10), 91–96 (2019). https://doi.org/10.1016/j.ifacol.2019.10.005
2. Varela, M.L.R., Silva, S.C.: An ontology for a model of manufacturing scheduling problems to be solved on the web. In: Innovation in Manufacturing Networks (Pro-VE 2008), IFIP International Federation for Information Processing, vol. 266, pp. 197–204. Springer (2008). https://doi.org/10.1007/978-0-387-09492-2_21.
3. Antosz, K., Ratnayake, R.M.C.: Machinery classification and prioritization: empirical models and AHP based approach for effective preventive maintenance. In: IEEE International Conference on Industrial Engineering and Engineering Management, Bali, pp. 1380–1386 (2016). https://doi.org/10.1109/IEEM.2016.7798104.
4. Antosz, K., Ratnayake, R.M.C.: Classification of spare parts as the element of a proper realization of the machine maintenance process and logistics - case study. IFAC-PapersOnline **4**(12), 1389–1393 (2016). https://doi.org/10.1016/j.ifacol.2016.07.760
5. Campanella, G., Ribeiro, R.A., Varela, M.L.R.: A model for B2B supplier selection. In: Melo Pinto, P., Couto, P., Serodio, C., et al. (eds.) EUROFUSE 2011: Workshop on Fuzzy Methods for Knowledge-Based Systems. Advances in Intelligent and Soft Computing, vol. 107, pp. 221–228 (2011). https://doi.org/10.1007/978-3-642-24001-0_21.
6. Varela, M.L.R., Putnik, G.D., Cruz-Cunha, M.M.: Web-based technologies integration for distributed manufacturing scheduling in a virtual enterprise. Int. J. Web Portals **4**(2), 19–34 (2012). https://doi.org/10.4018/jwp.2012040102
7. Varela, M.L.R., Ribeiro, R.A.: Distributed manufacturing scheduling based on a dynamic multi-criteria decision model. Recent developments and new directions in soft computing. In: Studies in Fuzziness and Soft Computing, vol. 317, pp. 81–93, Springer, Cham (2014). https://doi.org/10.1007/978-3-319-06323-2_6.
8. Arrais-Castro, A., Varela, M.L.R., Putnik, G., Ribeiro, R., Dargam, F.: Collaborative negotiation platform using a dynamic multi-criteria decision model. Int. J. Decis. Support Syst. Technol. **7**(1), 14 (2015). https://doi.org/10.4018/ijdsst.2015010101
9. Jasiulewicz-Kaczmarek, M., Bartkowiak, T.: Improving the performance of a filling line based on simulation. In: ModTech International Conference - Modern Technologies in Industrial Engineering IV, IOP Conference Series: Materials Science and Engineering, Iasi, Romania, 15–18 June 2016, vol. 145, p. 042024 (2016). https://doi.org/10.1088/1757-899X/145/4/042024.
10. Prociak, A., Rokicki, G., Ryszkowska, J.: Poliuretan materials (in Polish: Materiały poliuretanowe). PWN, Warszawa (2014)
11. Fesenko, A., Yevsiukova, F., Basova, Y., Ivanova, M., Ivanov, V.: Prospects of using hydrodynamic cavitation for enhancement of efficiency of fluid working medium preparation technologies. Periodica Polytech. Mech. Eng. **62**(4), 269–276 (2018). https://doi.org/10.3311/PPme.11877
12. Liaposhchenko, O., Pavlenko, I., Monkova, K., Demianenko, M., Starynskyi, O.: Numerical simulation of aeroelastic interaction between gas-liquid flow and deformable elements in modular separation devices. In: Ivanov, V. et al. (eds.) Advances in Design, Simulation and Manufacturing II. DSMIE-2019. Lecture Notes in Mechanical Engineering, pp. 765–774. Springer, Cham (2020). https://doi.org/10.1007/978-3-030-22365-6_76.

13. Kujawińska, A., Diering, M., Rogalewicz, M., Żywicki, K., Hetman, Ł.: Soft modelling-based methodology of raw material waste estimation. In: Intelligent Systems in Production Engineering and Maintenance – ISPEM 2017, Advances in Intelligent Systems and Computing, vol. 637, pp. 407–417 (2017). https://doi.org/10.1007/978-3-319-64465-3_39.
14. Kujawinska, A., Rogalewicz, M., Diering, M., Hamrol, A., Żywicki, K., Hoffmann, P., Konstanczak, M.: Improvement of surface layer grinding process with the use of design of experiments methodology. In: Proceedings of the 24th International Conference on Production Research (ICPR), DEStech Transactions on Engineering and Technology Research, pp. 519–524 (2017). https://doi.org/10.12783/dtetr/icpr2017/17664.
15. Rogalewicz, M., Smuskiewicz, P., Hamrol, A., Kujawińska, A., Luis, P.R.: Possibilities and limitations of passive experiments conducted in industrial conditions. In: Advances in Manufacturing, Part of the Lecture Notes in Mechanical Engineering, pp. 837–846. Springer (2018). https://doi.org/10.1007/978-3-319-68619-6_84.

Comparison of Technologies for the Magnesium Alloys Protection Using the Quality Assessment and Quantitative Metallography

Tatiana Lysenko(✉), Kyryll Kreitser, Oleksandr Dereviančhenko, Evgeny Kozishkurt, and Dmytro Vasylyev

Odessa National Polytechnic University, 1, Shevchenko Avenue, Odessa 65044, Ukraine
tvl12odessa@gmail.com

Abstract. Trends in the development of industry and engineering require the development of technologies for casting light alloys. In modern technology, the development of magnesium production is most relevant. Technological processes of melting of magnesium alloys in a protective gas atmosphere with their continuous or pulsed supply to the melting or distribution unit, as well as an automated system of control and management of the technological process of gas protection of magnesium melt in melting and distribution units of complexes. Two technologies for the protection of magnesium alloys during melting, flux, and flux-free pulsed, were investigated. A large number of non-metallic inclusions are associated with the use of flux and flux-free protection. To reduce non-metallic inclusions in magnesium castings, the authors propose using a pulsed supply of protective gases. For the analysis of the casting results, a special software package was used. The study aims at a quantitative analysis of the microstructures of alloys using the computer metallography. The paper presents the positive results of the introduction of a new technological process. The number of non-metallic inclusions in castings from magnesium alloy AZ91D obtained by the new technology has significantly decreased.

Keywords: Magnesium alloys · Metallographic analysis · Non-metallic inclusions · Castings

1 Introduction

In modern technology, the development of magnesium production is very relevant. Magnesium, a rare material for the space industry, over the past decade, has become the third most commonly used metal after steel and aluminum.

Magnesium is the lightest structural material in the industry; its density is 1800–1900 kg/m^3. It is six times lighter than steel and 1.5 times lighter than aluminum. It allows for reducing the weight of the product up to 30%. At such a low density, magnesium has high specific strength (second only to titanium in this indicator), specific vibration resistance, and rigidity [1].

V. Tonkonogyi et al. (Eds.): InterPartner 2020, LNME, pp. 322–329, 2021.
https://doi.org/10.1007/978-3-030-68014-5_32

Despite all the advantages of magnesium, it has a number of disadvantages that hinder its introduction. One of the main problems is its fire when interacting with the atmosphere of the workshop. A significant problem in the manufacture of castings from magnesium alloys is the presence of non-metallic inclusions, leading to a decrease in the operational properties of products.

Research purpose. The article aims to present a new technology for the protection of magnesium floats, supplemented by elements for assessing their quality using computer metallography.

The analysis was carried out using computer metallography to estimate the amount of non-metallic inclusions in castings from magnesium alloy AZ91D. Old and new technologies allows obtaining castings.

2 Literature Review

There are various technologies for producing magnesium alloys. Old flux smelting technology is widespread based on $MgCl_2$, Cl, and F. The use of flux protection reduces the quality of castings due to the large number of non-metallic inclusions. Therefore, research is being conducted, and the development of new smelting methods is being carried out [2].

The gaseous medium should delay the evaporation of magnesium vapor during melting and prevent further interaction of the gaseous medium with the magnesium melt after the formation of the film. For these purposes, a continuous supply of a mixture of active and inert gases to the alloy mirror is used. Constant gas supply increases the thickness of the protective film, which creates a large temperature difference. In this case, temperature stresses in the entire layer begin to grow. As a result of this, cracks (loosening) appear in the film through which magnesium vapor passes from the side of the liquid metal, and oxygen diffuses into the liquid metal. The most promising protective composition of gases is a mixture of dried air and SO_2. Indirect disadvantages of this process include high consumption of active gases, loss of alloy, reduced quality of magnesium alloy, clogging with non-metallic inclusions, and, as a result, reduced casting quality [3].

To eliminate these shortcomings, we suggest using a pulsed system of protective gas supply to the molten metal mirror.

A dynamic control system provides the effectiveness of this system with feedback.

Pressure die casting of magnesium alloys, with a flux-free protection option, is mainly implemented on the basis of pressure die casting machines with a hot pressing chamber [4].

For the operation of such a system, it is necessary to experimentally obtain a mathematical control model, as a function of the inertial factors and the temperature of the alloy surface alloy and the components of the gas mixture – inertia-free factors. Use a tact time without ignition as a generalized response.

Upon receipt of castings from magnesium alloys using various technologies, the question of quality control of materials arises [5–7].

Currently, the most advanced method of protecting magnesium alloys is flux-free melting under a layer of protective gases. The new technology of flux-free melting of magnesium alloys is based on the use of gas protection, which allows us to isolate the alloy from contact with air better than flux coating. A gaseous medium is capable of fulfilling its protective function if it, chemically inter-acting with a liquid magnesium alloy, forms a thin dense film [8–10].

Let's consider some of the modern publications on the issues of quality control of new materials.

As stated in [11], digital technology is the key to competitive, sustainable development in the long term. Intelligent (digital, computer) technologies are developing rapidly in all areas of modern engineering and, in particular, in modern foundry and materials science.

Modern metallographic makes extensive use of computer microscopes and digital imaging techniques [12]. In the publication [13], the changes and prospects of quantitative evaluation of microstructure morphology were considered. The changes in microstructures evaluation are reviewed from the viewpoint of fusion of material science and science of shape.

In the article [14], the methods of computer vision and machine learning were applied to the challenge of automatic microstructure recognition.

The material characterization in sheet metal forming is significantly improved with the introduction of optical measurement systems reviewed in publication [15].

A review of the above articles indicates the promise of applying methods of computer quantitative and qualitative metallographic analysis to assess the quality of new materials.

3 Research Methodology

For the study, the processing of melting and distributing electric furnaces with a removable crucible was carried out. Thermocouples, a movable measuring head for determining the surface tension of the liquid alloy, are mounted through a cover with sealed units in the zone of the magnesium alloy. Glasses are inserted into the lid for the operation of laser sensors for detecting the appearance of white smoke (magnesium vapor). Metal wares have been added for supplying heated ingots and pumping a liquid alloy under the mirror level of the metal, as well as metalworkers for mixing the alloy.

To regulate the flows of inert and active gases, proportional devices for the consumption of dry air, sulfur hexafluoride, and sulfur dioxide were used.

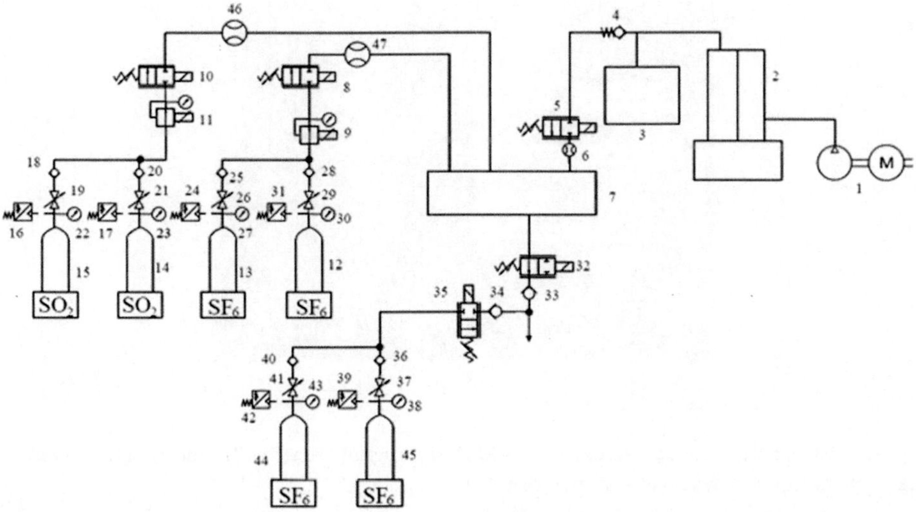

Fig. 1. Pulse flux-free protection circuit: 1 – a compressor; 2 – a dehumidifier; 3 – a device for reducing oxygen; 4 – a non-return valve with back pressure; 5 – a proportional valve for supplying dried air with oxygen content up to 6%; 6 – a flow meter; 7 – an air mixer, SO_2, and SF_6; 8 – proportional valve SF_6; 9 – pressure reducing valve SF_6; 10 – proportional valve SO_2; 11 – pressure reducing valve SO_2; 12, 13 – cylinders, SF_6; 14, 15 – cylinders, SO_2; 16, 17, 24, 31, 42, 39 – electric pressure switches; 19, 21, 24, 29 – shut-off valves; 22, 23, 27, 30, 38, 43 – pressure gauges; 44, 45 – SF_6 cylinders of the fire circuit; 47, 47 – flow meters of active gases.

Figure 1 shows a new diagram of the system for the pulsed supply of a mixture of protective gases. Air from a compressor with operating pressure up to 16 atm. Was supplied to the inlet of an adsorption dryer in the system of which air was cleaned from moisture to the level of dew point (–70 °C), oil, and mechanical impurities. Dried air enters the oxygen reduction system up to 6%. The mixer (7) made it possible, by reducing oxygen and nitrogen compensation, to supply the necessary gas level to the dehumidifier. Into the mixer, through a proportional valve (5, 8, 10) with control over a given content, a dynamic flow was provided in a pulsed manner, taking into account calculations of a mathematical model of the content of active gases. The system provides an independent fire extinguishing circuit for ignition languages.

Metallographic studies of micro sections were carried out using a computer microscope - the result of equipping the MIM-7 microscope with a vision system (Fig. 2).

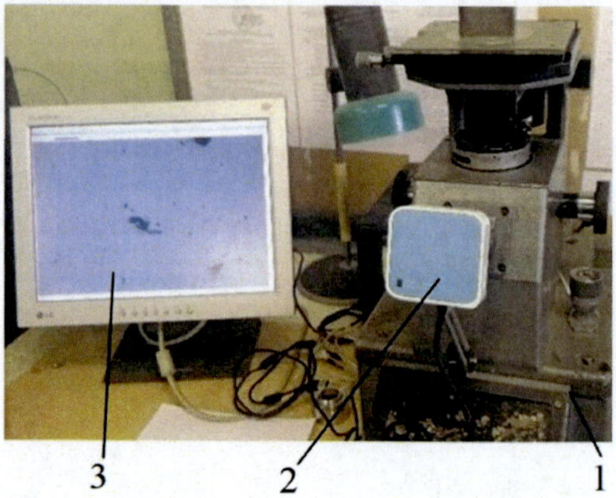

Fig. 2. The metallographic microscope MIM-7 (1), digital camera (2), and monitor with the image of the zone of non-metallic inclusion (3).

The article aims to present a new technology for the protection of magnesium floats, supplemented by elements for assessing their quality using computer metallography [13].

4 Results

Magnifications in the range of 100–300 was used. The microstructure of the alloy, made according to the old technology, consisted of small light angular grains of a solid solution of aluminum in magnesium and dark inclusions of the g-phase. With an increase of 100, a large number of dark brown inclusions up to 500–600 μm in size were found. Inclusions are not evenly distributed.

With a large increase in the immediate vicinity, a large number of small satellite inclusions with the same composition of the physical state are detected.

Separate sections of the thin section contained inclusions along grain boundaries. It has an acute-angled or elongated shape.

Preliminary observations showed, that the microstructure of the alloy smelted according to the new technology, contained light round grains, surrounded by less contrasting small non-metallic inclusions.

On the surface of the alloy smelted according to the new technology, with the indicated increases in non-metallic inclusions larger than 0.5 μm in size, they were not found.

A general view of one of the working panels for the contours of microstructures selection, and determining their primary features is shown in Fig. 3.

For processing digital images of microstructures and their quantitative metallographic analysis, the software was developed using $C^{\#}$.

In the upper left window, there is an image “in shades of gray” (3), in the upper right – a binary image of the microstructure (4).

Let us consider the main stages of image processing of microstructure of alloys and quantitative analysis of microdefects: input of the microstructure image into the software package; preliminary processing; formation of a binary image; selection of a set of contours of the microstructures components; the quantitative parameters of the component's determination.

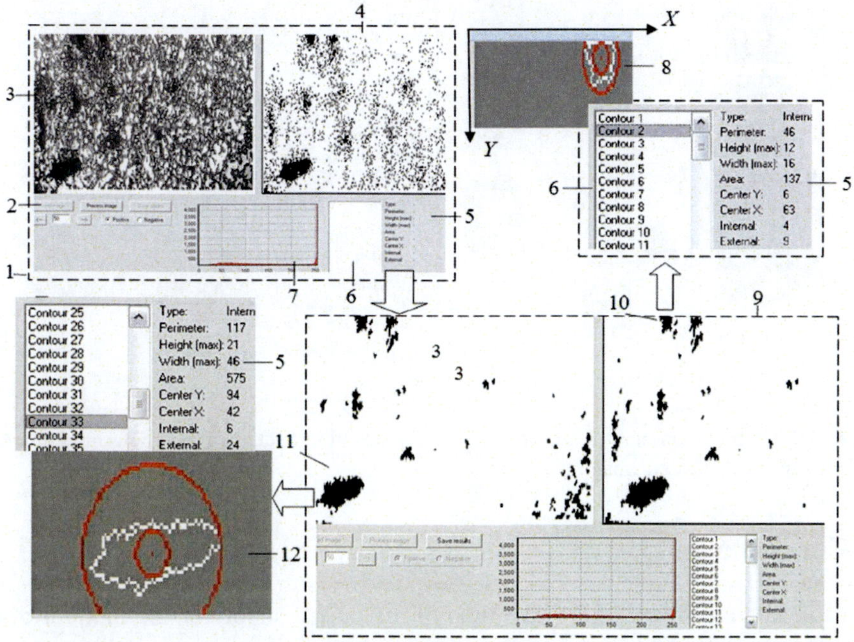

Fig. 3. Images of the working panels of a software package for the quantitative analysis of microstructures of an alloy made according to old technology (Sect. 1): 1 – panel loading and image type conversion; 2 – zone of operation buttons for controlling the conversion process; 3 – image of the microsection "in shades of gray"; 4 – a binary image of a microsection; 5 – zone for the allocation of parameters of microstructures elements (here, non-metallic inclusions); 6 – zone indicating the numbers of the selected contours of the elements of the microstructures; 7 – a histogram; 8 – selected contour non-metallic inclusion 10; 9 – panel of the selected components of the microstructures; 11 – non-metallic inclusions; 12 – its selected contour XY – image coordinate system (its beginning is in the upper left corner of the screen).

The purpose of quantitative analysis is to determine the total area of defects and their comparison with the image area. This requires the operation of selecting the contours of all defects and determining the set of their parameters.

For the quantitative analysis, the geometric features indicated in the data tables were used as the primary features of the components of the structures (Fig. 3, pos. 5).

For each defect in this system, the overall dimensions, perimeter, area, coordinates of the center of gravity are determined (Center Y, Center X), a radius of the inscribed and circumscribed circles (Internal, External).

The ratio of the total area of the selected non-metallic inclusions (Fig. 4a) to the image area (Fig. 4b) is a preliminary quantitative characteristic of the studied microstructure.

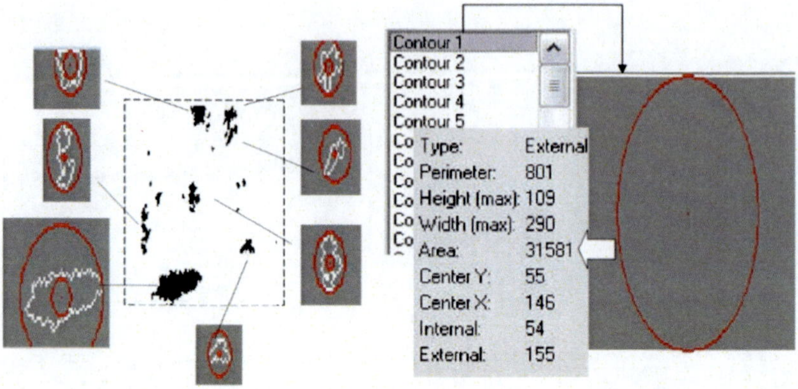

a b

Fig. 4. Highlighted contours of non-metallic inclusions: thin Sect. 1 (a) and determination of the total area of the analyzed image (b).

5 Conclusion

After the studies, the optimal cycle time of the pulsed supply of the mixture of protective gases was established, which is 8 min; characteristics of the active gas supply pulse is determined as a function of the temperature of the alloy according to surface tension and is calculated by a mathematical model. Using a pulsed system allowed to reduce SO_2 consumption by 12 times, and SF_6 consumption by 19 times.

When comparing the results of the study of non-metallic inclusions according to the new technology with magnifications of 100 and 300 times in samples of non-metallic inclusions larger than 0.5 μm, it was not found.

The overall result of the preliminary quantitative analysis: it is showed a significantly higher quality of the alloy obtained by the new technology.

References

1. Lysenko, T.V., Zamytin, N.I., Khudenko, N.P., Tur, M.P.: Spline interpolation for data processing at determining heat conduction coefficient of antistich coating of frozen mold. Metall. Min. Ind. **4**, 37–41 (2014)
2. Shinsky, O., Shalevska, I., Kaliuzhnyi, P., Shinsky, V., Lysenko, T., Shevchuk, T., Sliusarev, V., Pohrebach, I., Kolomiitsev, S.: Principles of construction and identification of a multi-level system for monitoring parameters of technological cycle of casting. East.-Eur. J. Enterp. Technol. **5**, 25–32 (2018)

3. Czerwinski, F.: Magnesium Alloys-Design Processing and Properties. INTECH, Rijeka (2011)
4. Xie, X., Shen, J., Cheng, L., Li, Y.: Effects of nano-particles strengthening activating flux on the microstructures and mechanical properties of TIG welded AZ31 magnesium alloy joints. Mater. Des. **81**, 31–38 (2015)
5. Mukhina, I.Y., Duyunova, V.A., Uridiya, Z.P.: Promising foundry magnesium alloys. Foundry **5**, 2–5 (2013)
6. Uridia, Z.P., Mukhina, I.Y.: Investigation of the microstructure of magnesium-zirconium ligature and heat-resistant foundry magnesium alloy ML10. VIAM 10 (2015).
7. Hwang, I.S., Kim, D.C., Kang, M.J.: Inverter DC resistance spot welding of magnesium alloy AZ31. Mater. Manufact. Process. **48**(2), 112–117 (2011)
8. Lysenko, T.V., Tonkonogyi, V.M., Bovnegra, L.V., Kreutzer, K.O.: Magnezium alloys casts quality control at lowpressure casting technology. In: 5th International Conference Economics and Management-Based on New Technologies 2015, pp. 307–311. Vrnjacka Banja, Serbia (2015)
9. Duyunova, V.A., Uridia, Z.P.: The flammability study of cast magnesium alloys of the Mg–Zn– Zr system. Foundry Work. Russ. **11**, 21–23 (2012)
10. Luo, A.A., Sadayappan, K.: Technology for Magnesium Castings. American Foundry Society, Schaumburg (2011)
11. Tokody, D.: Digitizing the European industry - holonic systems approach. Procedia Manufact. **22**, 1015–1022 (2018). https://doi.org/10.1016/j.promfg.2018.03.144
12. Decost, B.L., Holm, E.A.: A computer vision approach for automated analysis and classification of micro-structural image data. Comput. Mater. Sci. **110**(29), 126–133 (2015). https://doi.org/10.1016/j.commatsci.2015.08.011
13. Sato, N.A., Sadamatsu, S.: Change and prospect of quantitative evaluation of micro-structure Morphology. J. Iron Steel Inst. Jpn. **100**(10), 1182–1190 (2014). https://doi.org/10.2355/tetsutohagane.100.1182
14. Chowdhury, A., Kautz, E., Yener, B., Lewis, D.: Image driven machine learning methods for micro-structure recognition. Comput. Mater. Sci. **123**(1), 176–187 (2016). https://doi.org/10.1016/j.commatsci.2016.05.034
15. Emanuela, A., Marion, M.: Metallographic analysis of Nakajima tests for the evaluation of the failure developments. Procedia Eng. **183**, 83–88 (2017)

Thermo-Mechanical Properties of Perlite Composite

Mykola Melnychuk[1](✉), Mykhailo Poteichuk[1], Vitalii Kashytskyi[1], Marcin Sosnowski[2], and Serhii Kutsyk[1]

[1] Lutsk National Technical University, 75, Lvivska Street, Lutsk 43018, Ukraine
m.melnychuk@lntu.edu.ua

[2] Jan Dlugosz University in Czestochowa, 13/15, Aleja Armii Krajowej, 42-200 Czestochowa, Poland

Abstract. Currently, perlite is used in the insulation of facades, roofs, ceilings, based on perlite, heat-insulating plasters, masonry mortars, and even adhesive compositions are produced. However, practically not used as a monolithic-structural material. Expanding the production of heat-insulating perlite materials for various purposes will inevitably be allowed to reduce heat loss and the high cost of fuel. Perlite materials can improve the physical properties of existing structures. A significant advantage of the use of perlite as a heater is a reduction in a fire hazard and an increase in the fire resistance of building structures. The article presents the results of a study of the monolithic heat-insulating material based on expanded perlite of deposit "Fogosh" (Zakarpatskyi region, Ukraine) and liquid silicate sodium glass. The basic thermal, mechanical properties, and structures of this material and compliance with the standard are investigated. In this article, describe a complex of properties of new thermal insulation material that has: ρ = 219 kg/m^3, λ = 0.087 W/(m K), σ_c = 9.1 MPa. It is established that the investigated perlite composite according to its strength and thermal conductivity characteristics can be recommended as thermal insulation and structural material for thermal power equipment isolation.

Keywords: Expanded perlite · Composite of perlite · Liquid glass · Thermal insulation · Water absorption · Thermal conductivity

1 Introduction

Perlite is widely used as loose-fill insulation, especially in masonry construction, because of the following qualities that make it desirable: low toxicity. According to the Perlite Institute, "No test result or information indicates that perlite poses any health risk » Other insulators, such as asbestos, vermiculite (which may contain asbestos), and fiberglass are more hazardous; chemical inertness, meaning it will not corrode piping, electrical or communications conduits. Perlite has a pH of around 7, which is similar to freshwater; pliability.

It was rapidly heating perlite ore (Fig. 1) to temperatures of about 900 °C (1,700 °F), so ends the volcanic glass, causing entrapped water molecules in the rock to turn to steam and expand the particles like popcorn.

V. Tonkonogyi et al. (Eds.): InterPartner 2020, LNME, pp. 330–338, 2021.
https://doi.org/10.1007/978-3-030-68014-5_33

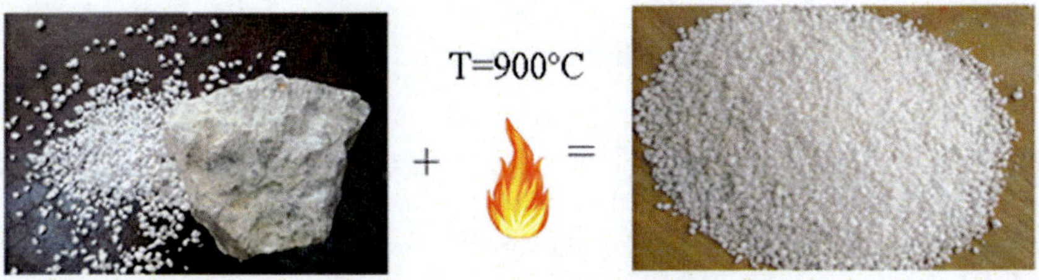

Fig. 1. Perlite ore and expanded perlite. Perlite has a bulk density of around 1100 kg/m^3, while typically expanded perlite has a bulk density of about 30–150 kg/m^3.

The leader in the production of expanded perlite and its products is the USA, where about 7 million m^3 of this product is produced every year. Until 2005, the leadership in this area belonged to Greece, and major manufacturers in the market are Japan and China too.

Since 1960, the Fogosh perlite deposit (Transcarpathian region) has been operating in Ukraine, with reserves of 13,4 million tonnes. The Transcarpathian deposit is known not only in Ukraine but also far beyond its borders. In Ukraine, perlite is mainly used as backfill in pure form and about 30% in the production of dry cement mortar or gypsum binder by “Knauf Gypsum”, “SMS-Knauf” (Moldova), “Henkel”, “Artel”. The analysis of international and domestic experience in the use of expanded perlite shows that in addition to the traditional spheres of application of this material, known both in Ukraine and abroad, new directions have emerged and are under development [1–5].

In particular, syntactic foams filled with metal matrix perlite (MMSF) are widely explored as new materials. They have great potential in energy absorption products [1], and since MMSFs are durable, lightweight, and flame retardant, they can be used as structural materials, such as lightweight cores for panels and pipes.

These foams are a good substitute for conventional metal foams because they have a low density (0.75–1.10 g/cm^3) and high energy absorption efficiency [6–8].

2 Literature Review

In the course of the research, scientific articles of such scientists as G.M. Samadova, U. Usmanova, S.V. Bukharov [1–3], whose works investigated the thermal insulation properties of materials based on perlite. In the last decade in the world have been developed and passed all the necessary environmental and fire tests such thermal insulation materials based on perlite, such as lignoperlite, epsoperlite, thermoperlite and perlitodiatomite [2–5, 10]. The vast majority of patented monolithic composites are filled with perlite based on powdered mineral binders [12–16]. There is a rather limited volume of published works devoted to the study of monolithic thermal insulation materials based on perlite [17] with other types of binders, and there is practically no information on the study of perlite from Ukrainian deposits, which have their unique chemical composition and properties.

Based on the above, it is possible to formulate the task of the study, which is to determine the optimal composition, the method of formation of monolithic materials filled with perlite, and to establish the relationship between composition, method of manufacture and physical and mechanical properties.

3 Research Methodology

Experiments on the development of the optimal composition of the material were carried out using a second-order simplex lattice plan (Shaffe). The standard physical and mechanical parameters for thermal insulation materials according to the Ukrainian Standard 17177–94 were determined as initial values: Y1 – material density, ρ, kg/m^3; Y2 – strain compression strength [σ_{st}], MPa; Y3 – the coefficient of thermal conductivity. Variation factors in the experiment were (in terms of 1 m^3 of finished material): X1 – the amount of filler in a completely dry state, kg; X2 – the amount of binder, % of X1; X3 – the amount of plasticizer, % of X1. The optimal composition was considered that, in which the minimum density and thermal conductivity were reached with satisfactory values of physical and mechanical properties. The optimization problem can be formulated as follows: it is necessary to find in the range of permissible values of factors Ω those values for which the initial parameters take the minimum or maximum values.

Studies of compressive strength and bending were performed by known methods of study of building materials. According to the experimental data of the destructive force, calculated the ultimate compressive strength of formula (1) and the flexural strength of formula (2):

$$\sigma_c = \frac{P_{max}}{S}, \quad (1)$$

where *P* – is the destructive force, kg; S – is the sample area, cm^2.

$$\sigma_A = \frac{3P_{max} * l}{2b * h^2}, \quad (2)$$

where *P* – is the destructive force, Pa; *l* – is the distance between the supports, cm; *b*, *h* – the width and thickness of the specimen, respectively, cm.

Surveys of the state of the surface were carried out using an optical digital microscope LEICA DMS300. The microstructure of thermoperlite specimens was examined using an SEM Hitachi TM3000 scanning electron microscope.

4 Results

For the preparation of the compositions were used: perlite fraction of 1.8–2.2 mm, which was obtained using laboratory sieves, liquid silicate sodium glass (Ukrainian Standard 13078–81) and plasticizer. The components were added in the following order: firstly, the plasticizer was mixed, and then the binder was gradually mixed mechanically with perlite. The samples were formed in the metal forms (they are shown in Fig. 2). Further, the compositions were compressed and sintered in the temperature range from 600 to 800 °C for 1–2 h.

Fig. 2. General view of molds (a) and specimens of sintered thermal perlite (b).

The samples were examined for water absorption of the material (according to the Ukrainian Standard 14457) is to measure the mass of water absorbed by the sample of dry material, partially immersed in water for a given time [10].

The sample was placed in a tub on a mesh stand and fixed with a mesh load. Then it was poured into the bathwater at a temperature of 22 °C so that the water level was higher than the loader by 20–40 mm. After 24 h after pouring water, the sample was transferred to a stand and weighed on a dry pallet after 30 s. The mass of water leaking from the sample during weighing into the pallet was included in the mass of the water-saturated sample.

As a result of mathematical planning, optimal composition ratios were obtained for five compositions whose properties were investigated empirically. The average values of the specific density of the different compositions are shown in Table 1.

Table 1. The average values of the specific density of the different compositions.

Compositions	1	2	3	4	5
ρ, kg/m^3	188	204	219	326	458

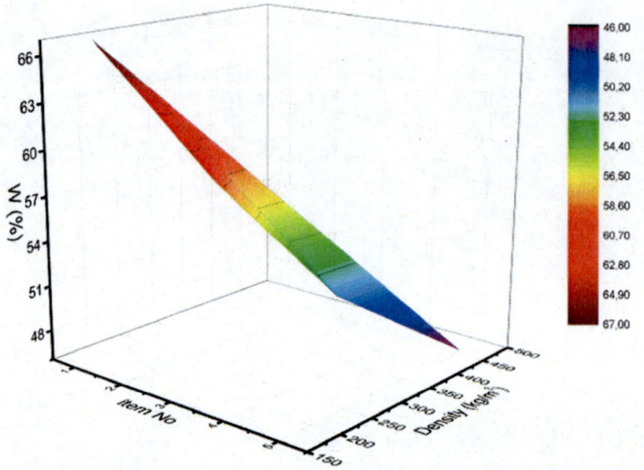

Fig. 3. Dependence of change of water absorption coefficient for different compositions on specific density.

Figure 3 shows the results of measuring water absorption by weight. Comparing the results that were reported in [10], it can be observed that water absorption decreased on average by 20%. This is due to the decrease in the number of open pores, possibly due to pressing and sintering at temperatures up to 800 °C, but the adsorption of water by perlite is still quite high, and the material absorbs a lot of water up to 68%. Therefore, for products where possible contact with water, the material requires hydrophobization of the surface layer, which will significantly reduce the water absorption.

So in Fig. 4, it can be seen visually that the number of surface pores increases with the decrease in the amount of binder from composition No. 5 to No. 1. Inclusions of dark color are particles of graphite powder, which was used as a non-stick layer in metallic forms.

The results of electron microscopy are shown in Fig. 5, it can be seen that all the composite samples have a uniform structure and evenly distributed perlite particles in the matrix.

However, cavities are observed in compositions No. 1 and No. 3, as shown in Fig. 5 b, c. Obviously, this was due to insufficient binder being insufficient to completely wet the filler, as a result of cavities of up to 400 μm in size. Accordingly, such

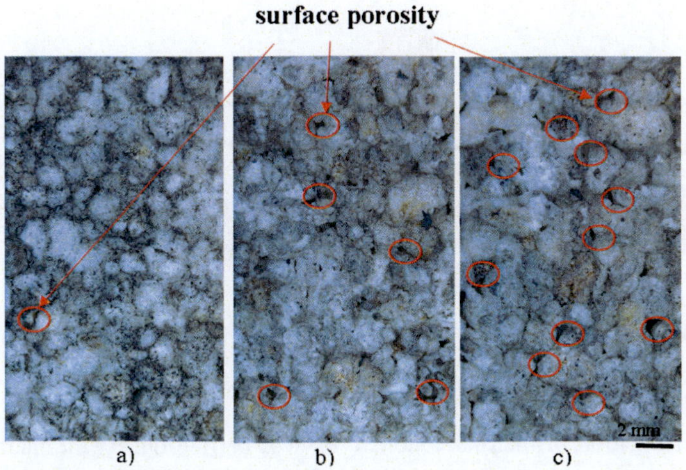

Fig. 4. Surface porosity of compositions: a - №5; b - №3; c - №1, x 3.2.

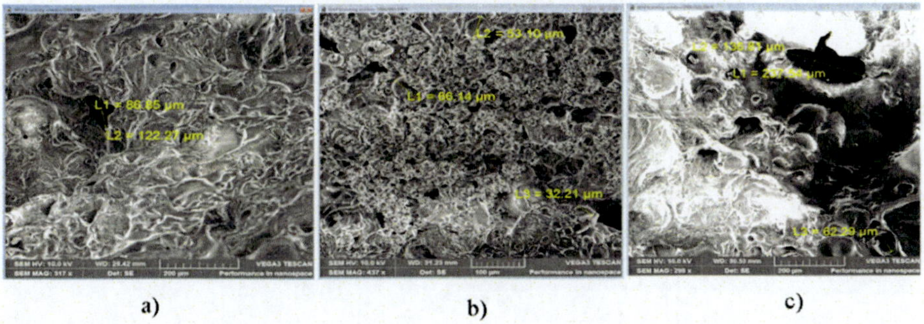

Fig. 5. SEM microstructure of the fracture surface of thermoperlite compositions: a – No. 5; b – No. 3; c – No. 1.

cavity dimensions are stress concentrators and significantly weaken the matrix, which correlates well with the results of the study of the mechanical characteristic [11]. In sample № 3, the pores on the border of the filler matrix are smaller and more evenly distributed across the cross-section of the sample. Sample No. 5 contains the least amount of defects due to its high liquid glass content and extrusion. Further studies are needed to establish a relationship between seal pressure and pore number.

According to the results of the strength study (Fig. 6), with increasing binder content, the composite strength is almost linearly increased, which provides the matrix and the formation of strong interphrase bonds between the matrix and the filler. All compositions have sufficient strength not only as thermal insulation lining material but also as a structural material. In particular, the composition No. 5 withstands the load corresponding to the strength of the brick brand M-150.

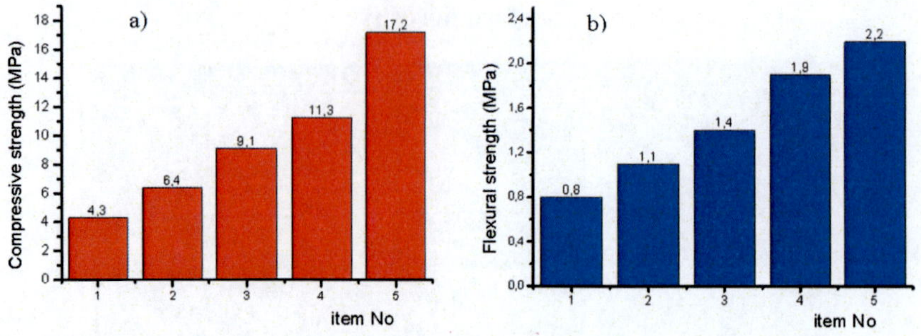

Fig. 6. Dependence of change of compressive strength (a) and flexural strength (b) for different compositions.

The determination of thermal conductivity was carried out by the method of simulation. The sample is placed between the heater and the cooling medium, as shown in Fig. 7.

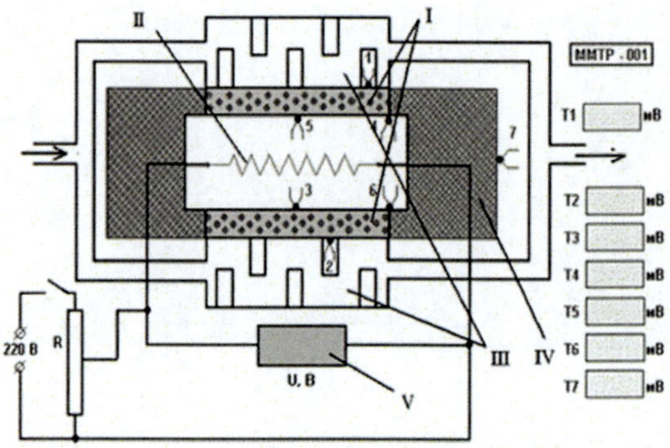

Fig. 7. Experimental setup for determining thermal conductivity: 1–7 - thermocouple junctions.

The heat flow from the heater II through the test samples and the discharge flows through the refrigerator III with water. The fridge is a container with spiral grooves, which creates the circulation of cooling water, which provides the same temperature on the cooled surfaces of the samples. To reduce heat loss through the face surfaces of the specimens into the environment, a heat-insulating casing I extruded with expanded polystyrene is provided.

Hot junctions 1 and 2 thermocouples are located on the outer surface (cooled), hot junctions 3 to 6 thermocouples are located on the inner surface (heated) of the sample, hot junction thermocouple 7 is installed on the outer surface of the insulating casing and is used to determine thermal losses through the torque the surface of the samples.

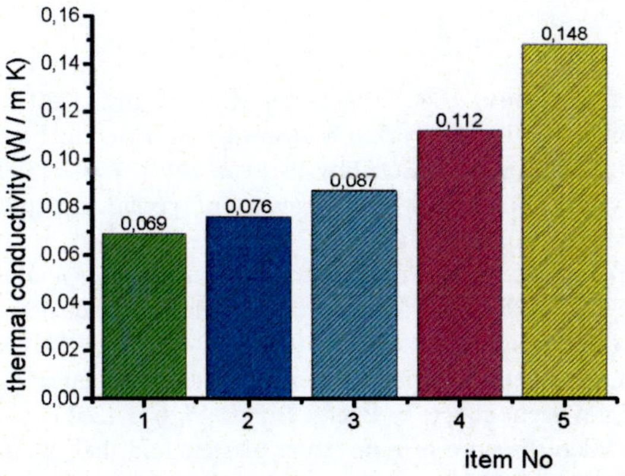

Fig. 8. The dependence of the change in the thermal conductivity coefficient for different compositions.

According to research results, the coefficient of thermal conductivity increases with the increase in the composition of the proportion of liquid glass, which is logical because not only is the decrease in the amount of perlite in the composition which provides low thermal conductivity (Fig. 8), but also the decrease in the number of internal air pores, which is consistent with the increase in the specific density of the samples.

The obtained thermophysical properties of the compositions are provided with the appropriate composition of the components, the method of molding, and the sintering mode.

5 Conclusions

The article focuses on the development of the optimum composition of new composite material based on perlite for the manufacture of thermal insulation structures. According to the results of mathematical planning, the optimal ratios were determined for five compositions based on perlite and binder liquid glass. However, not all samples have been tested for water absorption and therefore require additional hydrophobization measures for use in industrial and residential construction. Specifically, specimens 4 and 5 have a density 326–458 kg/m^3 and relatively thermal conductivity 0.112–0.148 W/(m K); specimens 1 and 2 have water absorption more than 50% and not very high compressive and flexural strength. Monolithic material number 3 has the most optimal complex of properties: $\rho = 219$ kg/m^3, $\lambda = 0.087$ W/(m K), $\sigma_c = 9.1$ MPa. Consequently, that material can be recommended as insulation material for the isolation of power and heating technological equipment with temperatures up to plus 970 K. Further studies of other characteristics of the material, namely vapor permeability, durability, frost resistance, porosity, porosity, porosity, are planned strengths and others.

References

1. Samadova, G.M., Usmanov, U.R., Usmanov, R., Nazarov, H.M.: Investigation of the possibilities of obtaining thermal perlite heat-insulating materials based on new promising rocks. Rep. Acad. Sci. Tajikistan, Ser. Phys. Chem. **56**(9), 708–713 (2013)
2. Celik, S., Family, R., Mengus, M.P.: Analysis of perlite and pumice based building insulation materials. J. Build. Eng. **6**, 105 (2016)
3. Pokorný, J., Pavlíková, M., Pavlík, Z.: Properties of cement-lime render containing perlite as lightweight aggregate. In: IOP Conferences Series: Materials Science and Engineering, vol. 596, pp. 012015 (2019). https://doi.org/10.1088/1757-899x/596/1/012015
4. Sengul, O., Azizi, S.: Effect of expanded perlite on the mechanical properties and thermal conductivity of lightweight concrete. Energ. Build. **43**, 671 (2011)
5. Brachaczek, W.: Microstructure of renovation plasters and their resistance to salt. Constr. Build. Mater. **182**, 418 (2018)
6. Nacievskij, S.J.: Perlite in modern concrete, dry mortar, and thermal insulation products. Build. Mater. **6**, 78–82 (2006)
7. Kadar, C., Chmelik, F., Ugi, D., Mathis, K., Knapek, M.: Damage characterization during compression in a perlite-aluminum syntactic foam. Materials **12**, 3342 (2019)
8. Zhao, Y., Tao, X., Xue, X.: Manufacture and mechanical properties of metal matrix syntactic foams. Mater. Sci. Technol. **4**, 2607–2615 (2008)
9. Ovcharenko, E.G.: Expanded perlite insulation. J. Build. Mater. XXI Century. Technol. Equipment **2**, 18–21 (2003)
10. Melnychuk, M.D., Skuba, V.M., Gusachuk, D.A., Lysyuk, P.O.: Development of monolithic heat-insulating material on the basis of expanded perlite. Interuniversity Collect. Sci. Notes **54,** 214–219 (2016). [in Russian]
11. Savchuk, P.P., Kashytskyi, V.P., Melnychuk, M.D., Sadova, O.L.: The structuring of tribotechnical epoxy composite materials in the electromagnetic field. Funct. Mater. **26**(3), 621–628 (2019)
12. Cui, H., Memon, S.A., Liu, R.: Development, mechanical properties and numerical simulation of macro encapsulated thermal energy storage concrete. Energ. Build. **96**, 162–74 (2015)
13. Meshgin, P., Xi, Y., Li, Y.: Utilization of phase change materials and rubber particles to improve thermal and mechanical properties of mortar. Constr. Build. Mater. **28**(1), 713–21 (2012)
14. Memon, S.A.: Phase change materials integrated in building walls: a state of the art review. Renew. Sustain. Energ. Rev. **31**, 870–906 (2014)
15. Jedidi, M., Benjeddou, O., Soussi, C.: Effect of expanded perlite aggregate dosage on properties of lightweight concrete. Jordan J. Civil Eng. **9**(3), 278–291 (2015)
16. Vaou, V., Panias, D.: Thermal insulating foamy geopolymers from perlite. Minerals Eng. **23** (14), 1146–1151 (2010)
17. Akimov, O.V., Marchenko, A.P., Alyokhin, V.I., Soloshenko, V., Shinsky, O.Y., Klymenko, S.I., Kostyk, K.O.: Computer engineering and design of cast parts for internal combustion engine crankcase. J. Eng. Sci. **6**(2), E24–E30 (2019). https://doi.org/10.21272/jes.2019.6(2).e4

Simulation Permeable Porous Materials of the Complex Shape During Radial-Isostatic Compression

Oleksandr Povstyanoy[1(✉)], Anatoliy Mikhailov[2], Nataliya Imbirovich[1], Oksana Dziubynska[1], and Halyna Herasymchuk[1]

[1] Lutsk National Technical University, 75, Lvivska Street, Lusk 43018, Ukraine
povstjanoj@ukr.net
[2] National University of Food Technology, 68, Volodymyrska Street, Kiev 01033, Ukraine

Abstract. Recent years have been characterized by a significant increase in the use of compaction processes for permeable porous materials. It is at this moment that the traditional schemes and technologies for producing products are continually being improved, and progressive methods of pressing, in particular, radial-isostatic pressing, are being used. Therefore, along with traditional research methods, the method of preliminary computer modelling and prediction of the behaviour of powder materials in the process of compaction, and the creation of appropriate mathematical models, are increasingly used. In this article, the process of compaction of permeable porous materials of complex shape made of BBS15 steel powder by radial-isostatic pressing is studied using computer simulation. The regularities of compaction of products-filters of complex shapes in the form of a flask are considered. It was found that when compressing permeable porous materials of complex shape, the distribution of the porosity value is uneven. As the radius increases, the porosity increases. It is shown that during the manufacture of filters of complex shape as a flask, the porosity distribution depends on the sealing scheme. Particularly, the wall material is compacted more intensively during radial pressing. Additionally, the filter bottom material is compacted more intensively during axial pressing. The simulation is based on a continuum approach. The ratio of the porous body's plasticity theory was used as the determining relations. The determination of the workpiece shape, compaction, and density, stress, and strain fields is based on the finite element method.

Keywords: Modeling · Forecasting · Compaction · Porosity distribution · Deformation · Stress

1 Introduction

Powder metallurgy with each newly developed technological process demonstrates the advantages that allow you to obtain materials with the best or with all new properties, to produce products in the most cost-effective way. These products include permeable porous materials (PPM).

V. Tonkonogyi et al. (Eds.): InterPartner 2020, LNME, pp. 339–348, 2021.
https://doi.org/10.1007/978-3-030-68014-5_34

The choice of optimal parameters for the process of creating permeable porous materials is a complex and important problem of efficient up-to-date production. The success of solving the problems that arise in this process is measured by the quality and degree of forecasting the processes and phenomena that accompany these technologies, namely, the compaction of powder material. It is possible to increase the efficiency of traditional technologies, as well as to introduce waste-free production of products for a wide range of purposes, save energy, reduce labor costs, and control the parameters of the structure of powder permeable materials in the process of their production by means of forecasting using modern modelling tools.

PPMs are widely used in various branches of modern industry [1–3]. In particular, they are used as filters for cleaning liquids and gases.

A promising method for producing filter elements is the radial isostatic pressing of powders [4–7]. The method allows you to produce long products of simple and complex shapes with a uniform distribution of porosity. Compaction of the powder occurs under the influence of an elastic element, which makes it possible to effectively mechanize and automate the process, increase production culture, reduce energy consumption, and so on. Small volumes of intermediate media can reduce the metal content, size, and cost of equipment compared to other types of pressing.

2 Literature Review

At the moment, there are a number of publications that have conducted experimental [8–10] and theoretical studies [11, 12] to determine the porosity distribution in films whose shape is a hollow cylinder. At the same time, the regularity of compaction of powder filter elements of complex shape is not observed.

Analysis of the literature data [13, 14] shows that promising methods for obtaining PM are methods that are aimed at creating a PM with an anisotropic structure, in which the size and number of pores change in the direction of filtration. In such PPMS, a layer with a minimum pore size will determine the filtration fineness. The permeability is an integral value that is determined by the porous structure of the entire material, and the contaminant in the filtration process is distributed over the whole volume of the filter element, which allows increasing the filter life.

Filter PPMs in the form of rotation bodies (pipes, disks, flasks, etc.) are becoming more common in various branches of technology, as they have a high technological design. The main requirements for the geometric of such products are to ensure the accuracy of external and internal dimensions.

The most common method of modeling the deformation treatment of powdered porous materials is the finite element method [15, 16].

It is essential to predict the sealing behavior of permeable porous materials of complex shape during radial-isostatic pressing. A well-modeled model of compacting the powder into a suitable shape will be able to predict the direction of movement and changes in the concentration of the powder over time in a particular volume. With the help of the developed model of consolidation, it is possible to eliminate or control the movement of powder [17].

Developed models that allows developing mechanical properties and simulating partial relevant processes are widely used and applied in the research works [18, 19].

But at present, the issue of modeling powder compaction in complex forms, namely in radial-isostatic pressing, is not thoroughly investigated and studied.

The purpose of this work is to study the influence of the radial-isostatic compression scheme on the regularities of compaction and form change of the filter element, which has a complex shape in the form of a flask, using computer modelling.

3 Researches Methodology

Technological parameters of the PPM pressing process determine the size, shape, and distribution of properties by volume of products, which, in turn, affects the performance properties of filters.

Choosing the optimal parameters for the pressing process is a difficult task. That is, along with traditional research methods, the method of preliminary computer modelling is now being used more and more widely. This was made possible due to significant progress in understanding the main features of the behaviour of powder materials in the process of compaction, and the creation of appropriate mathematical models.

Filter PPMs in the form of rotation bodies (pipes, disks, flasks, etc.) are becoming more common in various branches of technology, as they have a high technological design. The main requirements for the geometric of such products are to ensure the accuracy of external and internal dimensions.

Widely used such PPMs are characterized by a high production efficiency. The dimensions of the devices are part of determining the size and shape of such products. Increasing the performance of these devices while maintaining their size allows you to increase the efficiency of their use significantly.

The simulation is based on a continuum approach. The relations of the theory of plasticity of a porous body were used as determining relations [20, 21].

Elastic shells made of polyurethane, which can withstand multiple deformations at pressures up to 600 MPa, are the most suitable for pressing. These shells retain their elastic properties well, are technologically efficient when manufactured, and have good adhesion properties to metals and ceramics [22].

The accuracy of the compression obtained by radial-isostatic compression depends on the accuracy of manufacturing the inner surface of the flexible tool (Fig. 1).

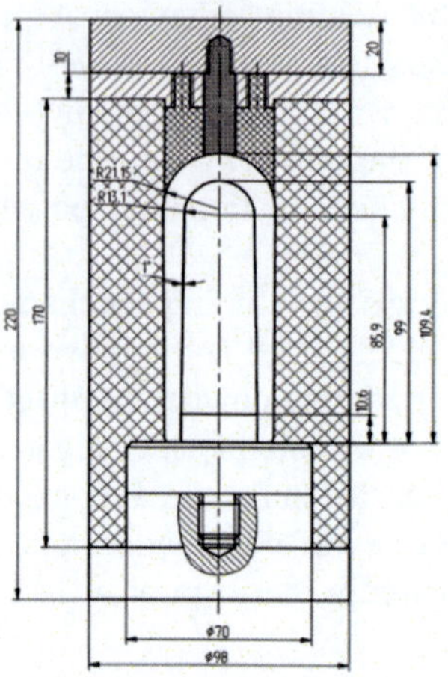

Fig. 1. Form for pressing filters with the spherical bottom.

For comparison and analysis of the finished product, two sealing schemes were considered: radial (Fig. 2a) and axial (Fig. 2b). The powder material, which is compacted – powder of steel BBS15. The initial porosity of the filter element was 0.7. The compaction occurred on the mandrel under the influence of the elastic environment, the material of which is polyurethane.

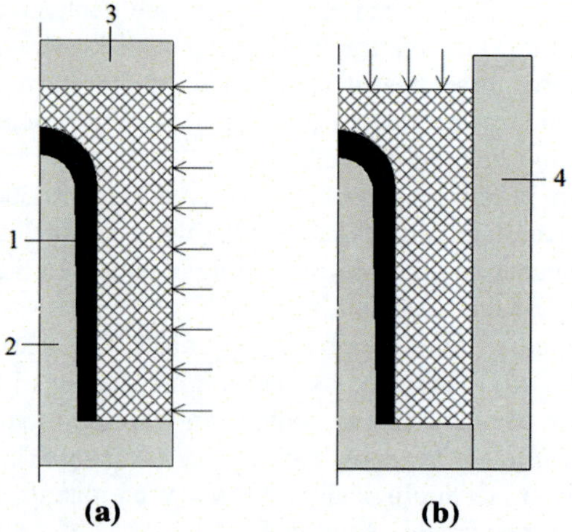

Fig. 2. Schemes of radial (a) and axial (b) sealing: 1 – powder compacted; 2 – mandrel; 3 – cover; 4 – wall.

Due to the symmetry, half of the axial section was considered in the simulation. It was believed that the mandrel was stationary. It was also considered that the radial seal has a fixed cover (3 in Fig. 2), and for axial compression, the fixed wall (Fig. 2, pos. 4). The surface of the polyurethane, on which the force load was applied, moved at a constant speed in the radial (radial seal), or the axial (axial seal) direction.

4 Results

At the initial moment, the wall of the filter element is sealed. The distribution of porosity along the radius of the filter wall is uneven. As the radius increases, the porosity increases. The distribution of the amount of accumulated plastic deformation is also uneven. The maximum value of accumulated deformation is in the inner surface of the filter wall, and the minimum value is in the outer surface.

Figures 3, 4 shows the distribution of porosity and accumulated plastic deformation under axial compression. The powder is most intensively compacted in the area of the bottom of the filter. The distribution of porosity along the bottom radius is uneven (Fig. 3a). As the radius increases, the porosity increases.

During further pressing, the relative density of the wall increases, and the process spreads, compacting to the bottom of the filter element. However, as can be seen in Fig. 3, at the end of the pressing process, the porosity in the filter bottom region is significantly higher than in the wall region. The nature of the change in porosity along with the radius of the wall remains the same as for the beginning of pressing. The distribution of porosity along the wall height is uneven: in the lower and upper parts (Fig. 3a), the porosity is higher.

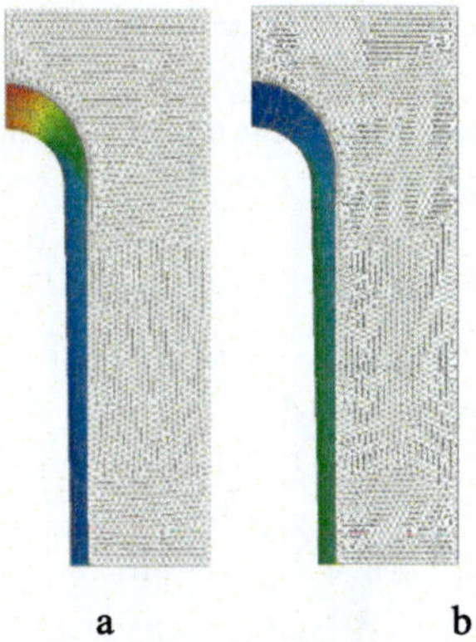

Fig. 3. Schematic representation of the porosity distribution (a) and the amount of accumulated plastic deformation (b) under axial compression.

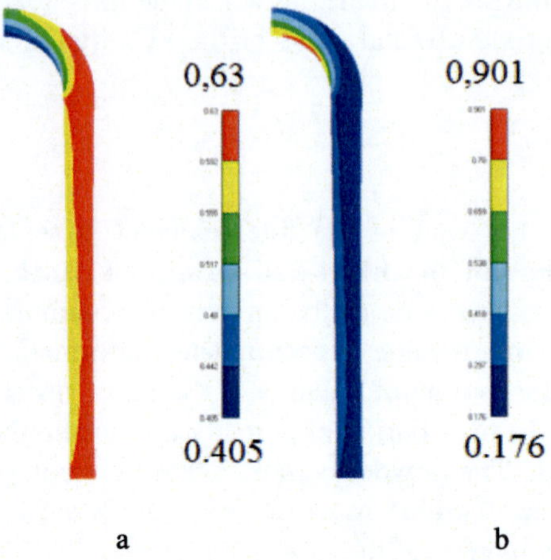

Fig. 4. Results of porosity distribution (a) and accumulated plastic deformation (b) under axial compression.

The amount of accumulated plastic deformation is distributed over the filter volume in the same way as the relative density distribution. In the bottom area of the filter element, it is lower.

The compaction of the powder in the wall area occurs to a lesser extent. The porosity is lower in the inner surface of the wall and higher in the outer surface. The amount of accumulated plastic deformation is higher in the area of the inner wall surface.

In this regard, a radial compaction scheme was considered, in which the powder was compacted in the radial direction (Fig. 5). The simulation results are shown in Figs. 6–8.

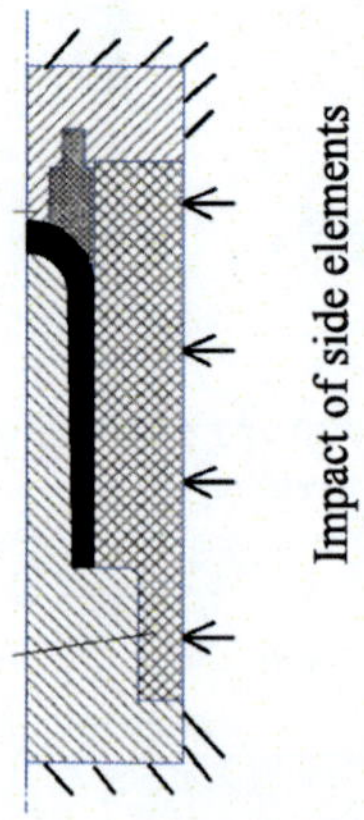

Fig. 5. Scheme of radial-isostatic pressing for obtaining PPM in the form of a flask.

At the first stage, the wall is compacted more intensively, and at the second stage - the bottom of the filter element. As a result, the porosity and accumulated plastic deformation are distributed more evenly.

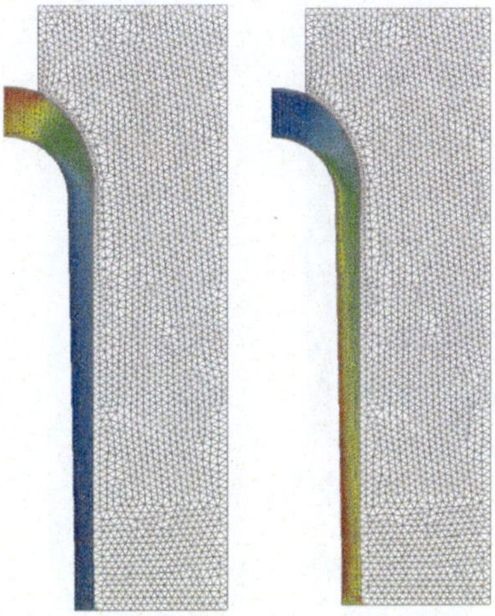

a b

Fig. 6. Schematic representation of the porosity distribution (a) and the amount of accumulated plastic deformation (b) during radial compaction.

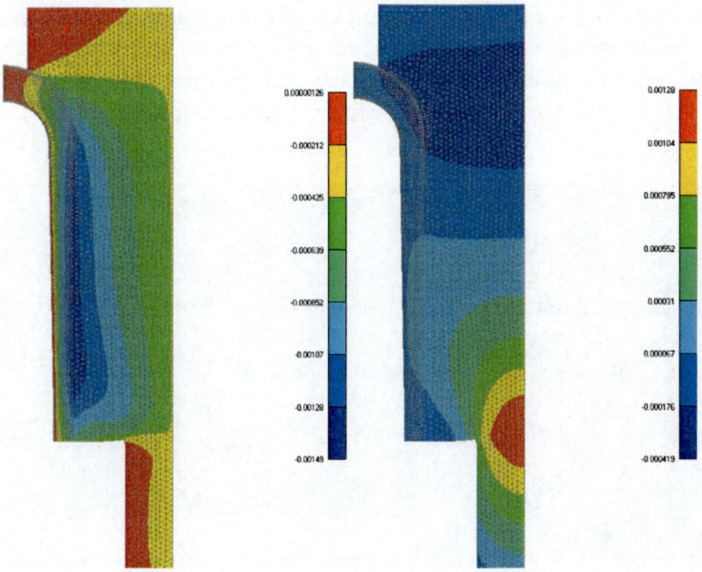

a b

Fig. 7. Movement of the powder in the PPM in horizontal (a) and vertical (b) with radial compaction.

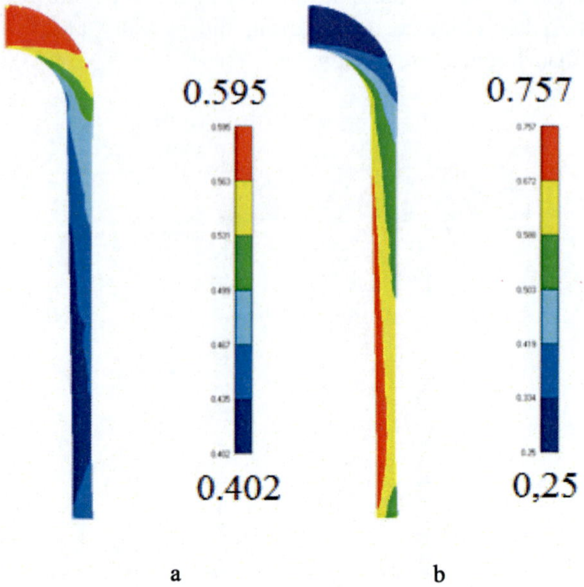

Fig. 8. Results of porosity distribution (a) and accumulated plastic deformation (b) in radial compaction.

When using the radially-isostatic pressing of a powder filter element as a flask, the distribution of porosity values and accumulated plastic deformation over the volume of the product is uneven and depends on the sealing scheme.

When the load is applied radially, the material is compacted more intensively in the area of the filter wall. The bottom of the filter is less compacted (Fig. 9).

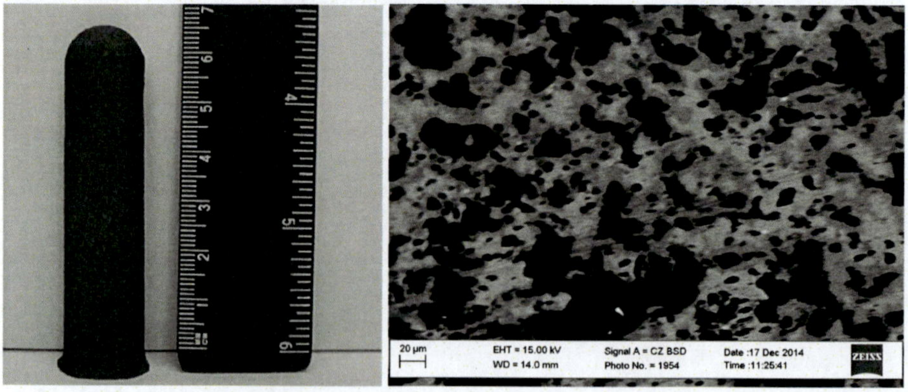

Fig. 9. Filter and structure PPM in the form of flasks, which is obtained from a powder BBS15.

5 Conclusions

When radially isostatic pressing of a cylindrical powder filter element, the porosity distribution is uneven. As the radius increases, the porosity increases.

When pressing a powder filter in the form of a flask, the distribution of the values of porosity and accumulated plastic deformation over the volume of the product is uneven and depends on the sealing scheme.

When the load is applied radially, the material is compacted more intensively in the area of the filter wall. The bottom of the filter is less compacted.

In the case of using the axial load scheme, there is a more intensive increase in the sealing of the filter bottom. The filter wall is also compacted but less intensive.

The use of a scheme in which the powder is first compacted in the radial and then in the axial direction makes it possible to obtain a more uniform distribution of the cost and accumulated plastic deformation.

As the radius increases, the porosity of the filter wall and bottom increases, and the amount of accumulated plastic deformation decreases.

Acknowledgment. The main results of this work are implemented in production at the Lutsk place of activity of state-owned enterprise "Ukrspyrt" (Lutsk, Ukraine) and LLC "WOG TRADE" (Kyiv, Ukraine) for cleaning technical liquids and fuel from mechanical contamination [23]. The use of developed single-layer filter PPMs made of BBS15 steel powder, obtained by means of radially static compression, increases the uniformity of porosity separation of filter materials by 20–30% and increases the penetration by 15–20% in comparison with similar traditional filter materials.

References

1. Belov, S.: Porous permeable materials. Reference book. Metallurgy (1987)
2. Osipov, S.N.: Energy-efficient small-sized heat exchangers made of porous heat-conducting materials. Energ. News High. Educ. Institutions Energ. Associations CIS **61**(4), 346–358 (2018)
3. Baklanov, A.E., Kanapinov, M.S., Malashina, S.A., Novoselova, T.V., Sitnikov, A.A., Tubalov, N.P.: Production of porous permeable metal-ceramic materials using ores in the form of limestones instead of rare earth elements. Polzunovskii Vestn. **2**, 205–212 (2016)
4. Yu, L., Yupeng, R., Dejian, P., Hongjian, S., Yaodong, Y., Daqiang, C.: Mechanism of pore formation in novel porous permeable ceramics prepared from steel slag and bauxite tailings. ISIJ Int. **59**, 1723–1731 (2019)
5. Rud, V.D., Povstianoi, O.Y., Zabolotnyi, O.V., Bohinskyi, L.S.: Technologies, structure, properties of porous permeable materials. Lutsk (2016)
6. Orlov, M.P.: Formation of pores in single crystal cooled turbine vanes in operation. Zh. Fiz. Metall. **8**, 306–312 (2007)
7. Van Nguyen, C., Bezold, A., Broeckmann, C.: Inclusion of initial powder distribution in FEM modelling of near net shape PM hot isostatic pressed components. Powder Metall. **57**, 295–303 (2014)

8. Povstyanoy, O., Zabolotnyi, O., Rud, V., Kuzmov, A., Herasymchuk, H.: Modeling of processes for creation new porous permeable materials with adjustable properties. In: Ivanov, V., et al. (eds) Advances in Design, Simulation and Manufacturing II. DSMIE-2019. Lecture Notes in Mechanical Engineering. Springer, Cham (2020). https://doi.org/10.1007/978-3-030-22365-6_46
9. ElRakayby, H., Kim, K.: Deformation and densification behaviours of nickel-based superalloy during hot isostatic pressing. Powder Metall. **11**, 1–8 (2017)
10. Flodin, A., Andersson, M., Miedzinski, A.: Full density powder metal components through hot isostatic pressing. Met. Powder Rep. **72**, 2–5 (2016)
11. Liu, M., Cui, Z., Li, Y.: Modeling and simulation of porosity in spray deposition. Metall. Mater. Trans. **B50**, 1908–1920 (2019)
12. Yang, J., Huang, Y.: Generation, development, inheritance, and control of the defects in the transformation from suspension to solid. In: Novel Colloidal Forming of Ceramics. Springer, Singapore (2020)
13. Povstyanoy, O., Sychuk, V., Makmyllan, A., Rud, V., Zabolotnyy, O.: Metallographic analysis and processing of images of microstructure of nozzles for sandblasting which are made by powder metallurgy. Powder Metall. **3**(4), 234–240 (2015)
14. Reut, O., Boginskyi, L., Petiushik, Y.: Dry isostatic pressing of compactable materials. Minsk (1998)
15. Bruno, G., Efremov, A.M., Levandovskyi, A.N.: Connecting the macro- and microstrain responses in technical porous ceramics: modeling and experimental validations. J. Mater. Sci. **46**(1), 161–173 (2010)
16. Häffelin, A., Niedrig, C., Wagner, S.F., Baumann, S., Meulenber, W.A., Ivers-Tiffée, E.: Three-dimensional performance model for oxygen transport membranes. J. Electrochem. Soc. **161**(14), 1409–1415 (2014)
17. Shtern, M.B., Mikhailov, O.V.: Numerical modeling of the compaction of powder articles of complex shape in rigid dies: effect of pressing method on density distribution. 1. Mechanical model of powder densification. Powder Metall. Metal Ceram. **41**, 581–587 (2002)
18. Rud, V., Saviuk, I., Samchuk, L., Povstyana, Y.: Research of mechanical properties of thermite material on the basis of steel dross. J. Eng. Sci. **5**(1), C6–C10 (2018). https://doi.org/10.21272/jes.2018.5(1).c2
19. Coube, O.: Modeling and numerical simulation of powder die compaction with consideration of cracking. PhD. Thesis. University Pierre et Marie Curie, Paris (1998)
20. Shtern, M.B.: Modified models of deformation of powder materials based on plastic and hard-deformed powders. Powder Metall. **3**, 13–19 (2011)
21. Liu, X., Gui, N., Wu, H., et al.: Numerical simulation of flow past stationary and oscillating deformable circles with fluid-structure interaction. Exp. Comput. Multiph. Flow **2**, 151–161 (2020)
22. Vitiaz, P.A.: Porous powder materials and products thereof. Minsk (1987)
23. Tkachuk, V., Rechun, O., Merezhko, N., Bozhydarnik, T., Karavaiev, T.: Assessment of the quality of alternative fuels for gasoline engines. In: Ivanov, V., et al. Advances in Design, Simulation and Manufacturing II. DSMIE-2019. Lecture Notes in Mechanical Engineering. Springer, Cham, pp. 871–881 (2020)

Study of the Porosity Based on Structurally Inhomogeneous Materials Al-Ti

Oleg Zabolotnyi[1(✉)], Viktoriya Pasternak[1], Nataliia Ilchuk[1], Dagmar Cagáňová[2], and Yurii Hulchuk[1]

[1] Lutsk National Technical University, 75, Lvivska Street, Lutsk 43018, Ukraine
volynasi@gmail.com
[2] Slovak University of Technology, Bratislava, Slovakia

Abstract. A technology for forming structurally inhomogeneous materials of Al-Ti samples has been developed. The components of the initial mixture were calculated based on the bulk of real powders, taking into account the mass, stoichiometric coefficients of the original components, the purity, and the bulk density. The microstructure of structurally inhomogeneous materials samples was studied using application programs. Besides, image recognition is performed at various magnifications before etching and after etching. The main properties of Al-Ti materials at different temperatures are obtained. An interpolation of the porosity dependence is constructed based on theoretical and experimental data. It is proved that the granulometric composition of structurally inhomogeneous materials has been improved based on the correct percentage of the initial mixture. The dependences of porosity on pressure and melting temperature for manufacturing structural parts are substantiated and constructed. Thus, there is reason to argue that improving the granulometric composition of Al-Ti based on a correctly selected percentage of the initial mixture, taking into account the further process of forming structurally inhomogeneous materials, allows to control such a parameter as porosity, as well as to study the physical and mechanical properties in the manufacture of various structural purposes parts to predict their structures.

Keywords: Technology of sample formation · Microstructure · Components of the initial mixture · Percentage ratio · Powder filling · Temperature · Pressing pressure · Physic-Mechanical properties · Particle size distribution

1 Introduction

One of the areas of materials science is the development of new materials using powder metallurgy technology. The development objective needs are caused by the need to create fundamentally new structural and functional materials that have a sufficient level of mechanical strength at high loads, increased wear resistance, heat resistance, and low density. Currently, when creating a number of parts and assemblies in mechanical engineering, shipbuilding, aviation and rocket and space technology, high-strength, lightweight composite materials are increasingly used, in particular, composites that combine components with a high young's modulus, and elements with significantly

V. Tonkonogyi et al. (Eds.): InterPartner 2020, LNME, pp. 349–359, 2021.
https://doi.org/10.1007/978-3-030-68014-5_35

lower values of the elastic modulus. By combining the volume content of the components, it is possible to obtain structurally inhomogeneous materials with the required values of the main physical, mechanical, and functional properties.

In recent years, significant advances have been made in the design of such materials for various applications. However, the specific structure of such composites significantly limits the possibility of using traditional research methods to obtain them. During manufacturing, such materials with the necessary set of properties, it is necessary to control the parameters of their structure during the implementation of the technology at all its stages. This applies to the operation of pouring into the mould, which is accompanied by the correct percentage of the original mixture. It is well-known that a wide range of chemical, physical, and technological components leads to heterogeneity in the properties of materials inside, and does not allow obtaining structural characteristics at a qualitative level. Therefore, progress in creating new materials requires a broader study and prediction of the structural characteristics of these materials, as well as the traditional methods of studying the structure of labor intensity. Thus, an urgent task is to improve known and to develop new methods for predicting the structural and physical-mechanical characteristics of materials.

2 Literature Review

As described in the paper [1], powders from aluminium alloys were studied using the casting method. The initial mixture was diluted with various chemical compositions of the pore-forming agents. During the research, the materials of different porosity were obtained in different ways. Also, using various compositions of amorphous Ni-Ti alloys, the heterogeneous properties of the source materials were significantly enhanced. A considerable number of researchers were engaged in the study of the mechanical properties of powders at the time, among which it is important to note [2, 3]. The results obtained in these studies are provided by cleaning the starting materials in liquid, gas, and other states. In turn, these results put increased demands on the microstructure properties of structurally inhomogeneous materials. Considerable attention is paid to maximum insight, dirt and other operational properties. As discussed in [4], the Ti-6Al-4 V powder is investigated. The authors of the paper have applied heat treatment at a range of certain tensile temperatures of Ti-6Al-4 V samples produced by SLM. The obtained data show a significant loss of force due to an increase in the annealing temperature due to grain growth, while there was no noticeable tendency to deformation of the samples. According to [5], the authors of the paper have investigated a general edge approach for analyzing thermo-electroelastic structurally inhomogeneous materials containing a shell that allows permissive inclusions with specific boundary conditions. It should be noted that the results obtained are reduced to a system of singular boundary integral equations, which is solved numerically by the boundary element method. Besides, special attention is paid to modelling the functions of various forms of materials that explain this feature and allow to determine the age of the intensity coefficients accurately.

According to the paper [6], individual and combined addition of Ti to the processed Al-Si-Cu-Fe-Mn alloy was performed. The peculiarity of these works is that the microstructure and phases of these alloys were studied using an optical microscope and x-ray testing in combination with EDS. The morphology of these alloys was quantified using SDAS. In addition, the strengthening of these manufactured materials was due to grain purification for α-Al and modification for the coarse, secondary phase of the substance. However, such methods do not allow to fully realize the possibilities of dispersed hardening due to the unsatisfactory wettability of material particles due to the presence of oxide films on their surface. It is known that the process of forming blanks consists of compacting the powder under the action of a certain pressure to obtain blanks of a certain shape from it [7, 9]. For an increase in their strength, blanks formed from powders are sintered [8]. At the same time, the melting temperature parameters, pressure, and physical and chemical properties of powder materials do not reflect the results of real parameters that should be taken into account when obtaining laboratory (model) samples, more real industrial products [10–12]. However, despite the efforts of scientists, a number of problems in the field of materials science remain open. Therefore, it is essential to further study and improve them by strengthening the particles of structurally inhomogeneous materials that need to be introduced either by mechanical mixing with aluminium powder (using powder metallurgy methods) or by direct introduction into the aluminium melt. It should be noted that the solution to such problems would allow more control of such parameters like porosity, as well as to study the physical and mechanical properties in the manufacture of parts for various structural purposes to predict their structures.

3 Researches Methodology

3.1 Calculation of the Initial Components Mixture

The purpose of the research is to investigate the physical and mechanical properties in the manufacture of structural parts, and to construct interpolation dependences of porosity on the pressure and melting temperature, as well as on the distance to the walls of the hopper.

The calculation of the initial mixture components was carried out considering the mass and stoichiometric coefficients of the initial components, the purity, and bulk density of the initial charge. The data required for calculating the initial components are shown in Table 1.

The total mass of the initial components forms an aluminium alloy, which must be manufactured by using individual technological processes of powder and lapidary metallurgy. The mass of this mixture was calculated using the formula:

$$M_{comp} = \sum_{i=1}^{m} z_i M_i \quad (1)$$

where m – number of source components; M_i – the atomic or molecular weight of the original component (*mol*).

Table 1. Components of the initial Al-Ti mixture and their properties.

Component	Number of components, mol, z_i	Atomic or molecular weight, M_i	The density of matter, ρ, kg/m^3 (g/cm^3)
Al	1	26,9 g/mol	2,7 (g/cm^3)
Ti	2	10,6 g/mol	4,54 (g/cm^3)

The volume containing the components according to their theoretical (single-crystal) density is:

$$V_{comp} = \sum_{i=1}^{n} z_i \frac{M_i}{\rho_i} \tag{2}$$

where V_{comp} – the volume that is filled with the original components ($cм^3$); ρ_i – the density of the corresponding component ($g/cм^3$); z_i – a number of initial components (*mol*).

The theoretical density of the initial mixture is G_{theor}:

$$G_{theor} = \frac{M_{comp}}{V_{comp}} \tag{3}$$

The bulk density of the initial mixture G_{bd} is found from the expression:

$$G_{bd} = b \times G_{theor}, \; M_{cm} = G_{bd} \times V_{cm} \tag{4}$$

Knowing the bulk density of the initial mixture and the volume, that is, the volume of the sample that was prepared for pressing and subsequent sintering, the authors of the paper find the theoretical mass of the initial mixture. In this case, the authors of the paper have used a cylindrical mould with dimensions: Ø = *30* mm, h = *60* mm, after which we can find the volume according to the following formula:

$$V_{cm} = \pi r_{sample}^2 \cdot h_{sample} \tag{5}$$

where r_{sample} – the radius of a cylindrical sample; h_{sample} – height of the cylindrical sample.

The calculated values of the mixture mass and the components are presented for the case of complete interaction of the components with the formation of a stable Al-Ti bond. However, in practical applications, it is necessary to ensure that the desired porosity, wear resistance, and density of parts are also obtained. This can be achieved by changing the percentage of components, as well as by adding pore-forming components. In this case, titanium and aluminium can interact, thus not fully forming a solid mixture (alloy).

3.2 Typical Fillings and Basic Properties of Structurally Inhomogeneous Materials

Thermal synthesis of powder mixtures Al-Ti was carried out according to the following technological scheme. The initial powder charge was mixed in a drum mixer for 15 min. The powder mixture was weighed (based on the volume of the working cavity of the mold and the bulk density of the powder), and weighed on a Radwag WLC 0.6/C/1 precision scale. The particle dispersion of the prepared mixture was controlled using sieve analysis. The criterion for evaluating the degree of grinding of a aluminium and titanium powders mixture was the presence of a large number of particles with an average size that varied from 0.5 μ to 1 μ. This powder mixture was etched with acetone and dried in the air, to obtain a clearer image of the grain boundaries of the micro-grinders. The mass of the initial components was pressed under pressure into cylindrical blanks with a diameter of Ø = 30 mm, h = 60 mm. The pressing pressure varied from 50 MPa to 90 MPa. Sintering was performed at temperatures of 200, 500, and 1000 °C. The cooling rate after sintering was about 0.2 deg/sec. The porosity of the obtained samples ranged from 14% to 25%. The main physical and mechanical properties of structurally inhomogeneous Al-Ti materials are presented in Table 2.

Table 2. Basic properties of Al-Ti materials at different temperatures.

Property	Temperature, 200 °C	Temperature, 500 °C	Temperature, 1000 °C
Volume content, %	99.7	98.9	98.1
Outer diameter, mm	24.10	24.00	24.00
Internal diameter, mm	17.71	17.87	17.84
Weight, g	3.870	3.439	3.042
Density, g/cm^3	6.78	6.57	6.40
Hardness, GPa	25	10	4.4
Porosity, %	14	18	25
The tensile strength in bending, MPa	240	200	137
The limit of compressive strength, GPA	1,7	–	–
Modulus of elasticity, GPA	81	79	71
Thermal conductivity, W/(m · K)	25.061	51	79.2

It should be noted that these raw materials Al-Ti have a low density, which reduces the weight of structural parts (piston) and, consequently, reduces the inertial load on the elements of the cylinder-piston group of materials. This also simplifies the problem of reducing the thermal resistance of parts elements, which in combination with good thermal conductivity, reduces the heat stress of the final products. It should also be noted that when titanium and aluminium components interact, the formation of four intermetallics is possible, particularly TiAl, $TiAl_2$, and $TiAl_3$ with a tetragonal structure, and Ti_3Al – with a hexagonal structure. Therefore, the main advantages of this selected percentage of raw materials include: small mass of the original components (at least 30%); high thermal conductivity (3–4 times higher); good anti-friction properties.

4 Results

4.1 Dependence of Porosity on Pressure and Melting Temperature for Manufacturing Structural Parts

In this section, the authors of the paper have observed the process of obtaining Al-Ti materials using a reaction between molten aluminium and a pre-made mixture of Ti powders. Particular attention was paid to the reaction temperature, which significantly affects the final processing of the product microstructure. It should be noted that the introduction of aluminium powder into the initial charge leads to the formation of thermal energy. As a result, the reaction between Al and Ti takes place, which further initiates the Ti reaction. After that, the aluminium melts, the melt envelops the titanium particles, so that a structurally inhomogeneous Al-Ti material is formed in the boundary zones. It was found that when the temperature reaches 1000 °C, the sample begins to glow, after which it self-ignites after 3–5 s. The use of a thermocouple installed inside the container, the self-ignition temperature of the samples was recorded, which is (1000 ± 10) °C and depends on the percentage of aluminium, the more aluminium, the lower the self-ignition temperature. The porosity is relatively small.

In Fig. 1, the dependencies of porosity on pressure and melting temperature for manufacturing structural parts are presented.

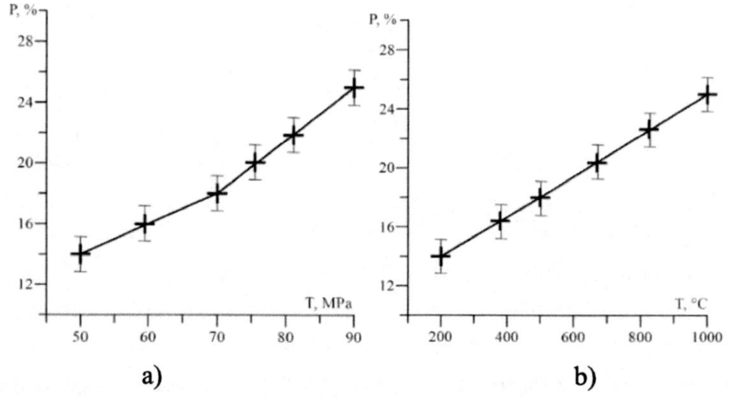

Fig. 1. The dependence of porosity on pressure and melting temperature: a – dependence of porosity on pressure; b – dependence of porosity on melting temperature

It should be noted that any method is chosen for obtaining structurally inhomogeneous materials (Al-Ti) provided a uniform distribution of components throughout the matrix volume, the formation of the maximum strong bond between the matrix and the reinforcing phase, the invariance of the structural-phase composition, and was also cost-effective and environmentally friendly. In addition, the advantage is that the pressing method allows us to quickly compact the powder body and get the workpiece with minimal residual porosity when sufficient low pressure (50–90 MPa).

4.2 The Microstructure of Structurally Inhomogeneous Al-Ti Materials Samples

For a complete and qualitative assessment of Al-Ti samples, it is necessary to determine and investigate the main morphological parameters of their microstructure, namely: 1 – determination the number of different sizes and shapes particles; 2 – determination of the sample structural defects; 3 – determination of the pore shape and particle shape; 4 – determination of the total pore distribution in the section and throughout the volume; 5 – determination the total distribution of certain particle shapes along the perimeter and volume.

In Fig. 2, the structure of Al-Ti samples at an increase of 800 μm (before and after etching) is presented.

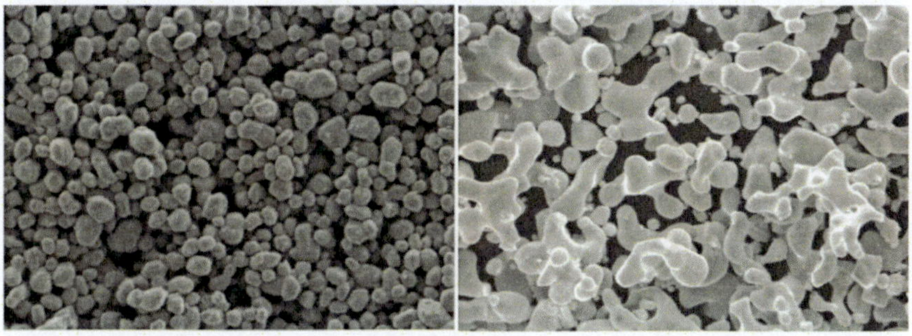

a) b)

Fig. 2. The microstructure of samples of structurally inhomogeneous Al-Ti materials at an increase of × 800 μ, where: a - before etching; b - after etching.

There is a correlation between the original Al-Ti components and their properties. A structure that has dissolved to a greater depth has a dark color, and a structure that has dissolved less has a light color. This allows us to claim that the percentage of source materials is correctly selected. From the obtained graphic dependencies, it was found that the surfaces of structurally inhomogeneous materials micro shifts are inhomogeneous, and their morphology depends on digestion. After etching the surface of the cuts, their relief increased due to stronger etching of the edges around the pores that were worn during the grinding and polishing of the samples. At the same time, the pore sizes increased, which made it possible to record clearly the difference in the porosity of structurally inhomogeneous materials of different compositions. The etching duration was the same to ensure the same effect on the relief of the cuts.

4.3 Interpolation Dependence of the Porosity Based on Theoretical and Experimental Data

Based on the results obtained and the proposed numerical methods for calculating the components of the initial mixture, which was carried out taking into account the mass and stoichiometric coefficients of the initial components, the purity and bulk density of the initial charge, the porosity dependence on the distance to the mould wall was calculated using computer simulation models developed and based on experimental studies. The data obtained are shown in Table 3. The authors of the paper have also built an interpolation of the relationship (Fig. 3) porosity on the distance to the hopper walls, where 1 is at a distance of L_1 = 50 μ; 2 – at a distance of L_2 = 100 μ; 3 – at a distance of L_3 = 150 μ; 4 – at a distance of L_4 = 200 μ; 5 – at a distance of L_5 = 250 μ (at the bunker wall located at a distance of 280 μ). Solid line – theoretical data, the dash-dotted curve corresponds to the experiment.

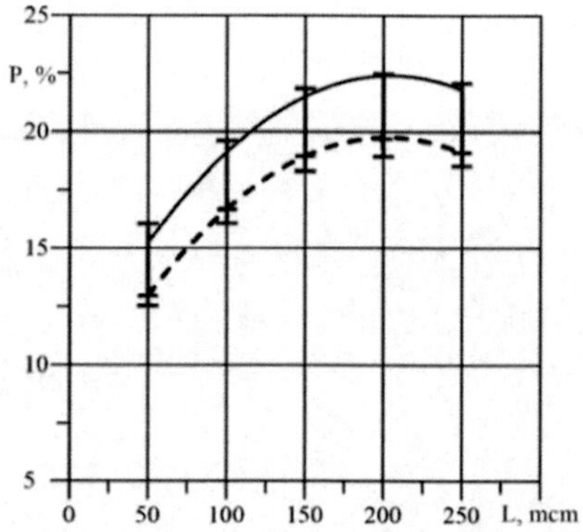

Fig. 3. Interpolation dependence of porosity on the distance to the hopper walls.

Table 3. Dependence of the porosity on the distance to walls of the silo.

L, mcm	Porosity, %
L_1 = 50	14
L_2 = 100	16
L_3 = 150	18
L_4 = 200	19
L_5 = 250	24

The interpolation dependence of porosity on the distance to the hopper walls provides a visual representation of the particle size distribution, provided that the radius intervals in the fractions are the same. When calculating the n-number of particles in the range of radii from r_i to r_j belong to the average r_i, that is, it satisfies the condition $Q_n = \bar{h}(r)$, which characterizes the differential distribution of particles over the volume of the hopper.

4.4 Experimental and Industrial Justification of the Results Obtained

Structurally inhomogeneous materials based on aluminium alloy Al-Ti are characterized by a set of properties that open up wide opportunities for their application in various industries.

For Fig. 4, a general view of the structural part, in particular, the piston, is presented.

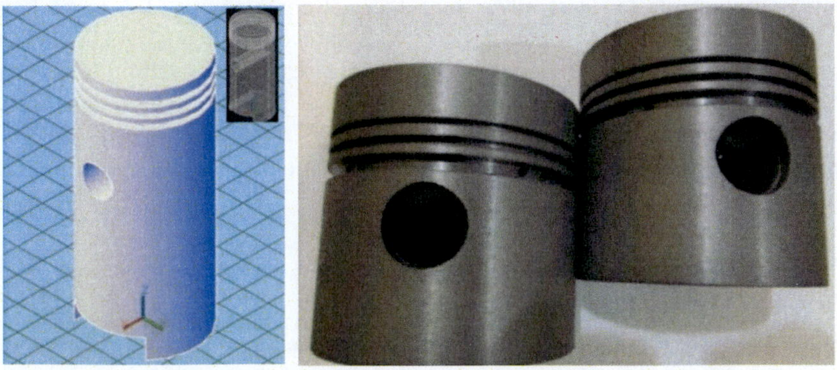

a) b)

Fig. 4. General view of the piston, where: a – in the COMPASS software window; b – photos of real cylindrical pistons

Based on the results of porosity calculations, as well as heat treatment modes, the economic effect of Al-Ti was calculated and determines the reliability and durability of the piston, in particular, power, mechanical and technological properties, its cost and scarcity. The results obtained make it possible to optimize the version of the charge composition, which provides a small mass of the initial components (at least 30%); high thermal conductivity (3–4 times higher); good anti-friction properties, which fully satisfies the technical conditions of operation of structural parts.

5 Conclusions

A method for evaluating the main parameters of the full-scale powder filling process has been developed. By using the metallographic analysis, the authors of the paper have investigated the regularities of structure formation and established the dependencies of the structure's influence on mechanical, physical, and mechanical properties, as well as porosity from the distance to the mould wall. It was found out that the results obtained provide a visual representation of the particle size distribution, provided that the radius intervals in the fractions are the same.

It should be noted that the results obtained allow to optimize the porosity of the resulting material for specific functional properties and technological requirements of specific products, as well as to predict the physical and mechanical properties of their structures. The calculations indicate that the porosity of the obtained samples of structurally inhomogeneous materials decreases with an increase in the width of the hopper.

Experimental and industrial justification of the results obtained showed that the proposed percentage ratio of the charge composition variant provides a small mass of the initial components (at least 30%); high thermal conductivity (3–4 times higher); good anti-friction properties, which fully satisfies the technical conditions of structural parts operation. Moreover, conduct the bulk of research without setting up expensive and time-consuming field experiments. This makes it possible to introduce virtually waste-free production of products for a wide range of purposes, save energy and materials, and reduce labor costs by reducing the number of technological operations and automating processes.

Acknowledgments. The paper has been written with the support of H2020 project titled "Directional Composites through Manufacturing Innovation" (acronym "DiCoMi") and H2020 project titled "A Policy Tool Kit for the Promotion of Intercultural Competence and Diversity Beliefs, Reduction a Discrimination and Integration of Migrants into the Labour Market" (acronym "Fair Future").

References

1. Lin, Z., Tian-Shu, L., Tao-Tao, D., Tao-Tao, L., Feng, Q., Hong-Yu, Y.: Design of a new Al-Cu alloy manipulated by in-situ nanocrystals withsuperior high temperature tensile properties and its constitutive equation. Mater. Des. **181**(2), 1–12 (2019)
2. Vijay Ponraj, N., Azhagurajan, A., Vettivel, S.: Microstructure, consolidation and mechanical behaviour of Mg/n-TiC composite. Alexandria Eng. J. **55**(2), 2077–2086 (2016)
3. Qijun, L., Lin, Zh., Dongbin, W., Shubin, R., Xuanhui, Qu: Porous Nb-Ti based alloy produced from plasma spheroidized powder. Results Phys. **7**(3), 1289–1298 (2017)
4. Haar, G., Becker, T., Blaine, D.: Influence of heat treatments on the microstructure and tensile behaviour of selective laser melting-produced Ti-6Al-4 V parts. S. Afr. J. Ind. Eng. **27**(3), 174–183 (2016)
5. Sulym, H., Pasternak, I., Pasternak, V.: Boundary element modeling of pyroelectric solids with shell inclusions. Mech. Mech. Eng. **22**(3), 727–737 (2018)
6. Kang, W., Peng, T., Yi, H., Yanjun, Z., Wenfang, Li., Tian, J.: Characterization of microstructures and tensile properties of recycled Al-Si-Cu-Fe-Mn alloys with individual and combined addition of titanium and cerium. Hindawi. Scan. **34**(7), 1–14 (2018)
7. Povstyanoy, O., Zabolotnyi, O., Polinkevich, R., Somov, D., Redko, O.: Modeling the structural characteristics of porous powder materials with application models of casual two-dimensional packaging. In: Beltran, J.A., Lontoc, Z., Conde, B., Serfa Juan, R., Dizon, J. (eds) World Congress on Engineering and Technology; Innovation and its Sustainability 2018. WCETIS 2018. EAI/Springer Innovations in Communication and Computing, pp 15–25. Springer, Cham (2019)
8. Wei-min, M., Wen-zhi, Z.: Tensile properties and microstructure of rheo-diecast 7075 alloy prepared by serpentine channel process. Res. Dev. **16**(3), 161–167 (2019)

9. Povstyanoy, O., Zabolotnyi, O., Rud, V., Kuzmov, A., Herasymchuk, H.: Modeling of processes for creation new porous permeable materials with adjustable properties. In: Ivanov, V., et al. (eds) Advances in Design, Simulation and Manufacturing II. DSMIE 2019. Lecture Notes in Mechanical Engineering, pp. 456–465. Springer, Cham (2019)
10. Fathy, N., Ramadan, M., Hafez, K., Alghamdi, A., Halim, A.: Microstructure and induced defects of 6061 Al alloy after short times cyclic semi-solid heat treatment. MATEC Web Conf. **67**(5), 1–6 (2016)
11. Samar Reda, A., Hamid, A., Menam, A., Salah Elden, I., Haytham Abdelrafea, E., Hassan Abdel, S.: Laser powder cladding of Ti-6Al-4 V α/β alloy. Materials **10**(11), 2–16 (2017)
12. Berladir, K., Hovorun, T., Bondarenko, M., Shvetsov, D., Vorobiov, S.: Application of reinforcing thermocycling treatment for materials of stamps hot deformation. J. Eng. Sci. **6** (2), C6–C10 (2019). https://doi.org/10.21272/jes.2019.6(2).c2

Manufacturing Technology

The Manufacture of Cylindrical Parts by Drawing Using a Telescopic Punch

Roman Arhat[1(✉)], Ruslan Puzyr[2], Viktor Shchetynin[1], and Mykola Moroz[1]

[1] Kremenchuk Mykhailo Ostrohradskyi National University, 20, Pershotravneva Street, Kremenchuk 39600, Ukraine
argat.rg@gmail.com

[2] Kremenchuk Mykhailo Ostrohradskyi National University College, 7, Chumatskyi Shliakh Street, Kremenchuk 39621, Ukraine

Abstract. The paper deals with the analysis of the possibility of intensifying the process of drawing sheet workpieces by changing the design of equipment. It is shown that this method of sheet stamping is characterized by the introduction of non-waste technologies and the production of parts that do not require final processing. However, the task of developing rational technological processes of drawing, requiring the minimum labor input and the cost of manufacturing specified parts with the best quality, is still particularly urgent. Modern enterprises operating in conditions of frequent changes in the range of manufactured parts seek to minimize the cost of tools and equipment to reduce the cost of the process. Therefore, the unification of the tool, simplification of the design of die tooling in the conditions of flexible serial production will make it possible, with minimal loss of time, to switch to the release of new products. A method of pressure-free stamping using a telescopic punch is proposed. Also, the fundamental possibility of drawing by the proposed method is confirmed based on the performed research. A method for calculating the punch lock springs is developed. The demonstrated technological calculations make it possible to proceed to the design and calculation of die tooling and punch power springs, which contributes to the manufacture of high-quality parts, reducing labor intensity and eliminating the use of a double-acting press. As a result, the proposed technical measures will reduce the technological cost of the product.

Keywords: Sheet workpiece · Flexible production · Stamping equipment

1 Introduction

Manufacturing of cylindrical parts from a flat round workpiece is based on the operation of cold sheet stamping– drawing. This operation was widely used in production in the manufacture of a wide range of engineering products, aircraft, automotive, shipbuilding, instrument making, and other industries [1, 2]. It is also popular due to its high productivity and economic efficiency. It is characterized by the introduction of non-waste technologies and obtaining parts that do not need finishing. However, the task of developing rational technological processes of drawing, which require the

V. Tonkonogyi et al. (Eds.): InterPartner 2020, LNME, pp. 363–372, 2021.
https://doi.org/10.1007/978-3-030-68014-5_36

minimum labor input and the cost of manufacturing specified parts with the best quality, is still of particular relevance. So, for example, this method is used for making up to 70% of parts of electrical appliances, 50% of car parts, and 80% of consumer goods. And saving material or energy resources at the manufacturing stage of each part will result in a significant reduction in the cost of the product [3]. This will increase the competitiveness of enterprise products in the markets. Therefore, the development and implementation of new schemes and methods of extraction will allow intensifying the process as a whole and in each particular case of production to achieve optimal performance in terms of price/quality.

2 Literature Review

The existing drawing methods and equipment designs are aimed at reducing the deformation energy, increasing the thickness of the semi-finished product in a dangerous section, the degree of deformation in one step, and eliminating folding, reducing the effect of tensile stresses in a critical section, which results in the improvement of the efficiency of the forming process.

So, for example, to increase the degree of deformation during one drawing step, and end support of the workpiece flange is repeatedly used. In this case, elastic insert elements or liquid are used as a backing medium [4, 5]. Creating high hydrostatic compression requires additional energy costs, expressed in an increase in the deformation force, and taking measures to prevent the workpiece material from flowing into the small gaps between the bushings, the punch, and the clips (otherwise, tool jamming is possible) (Fig. 1). The use of a transmission medium as a liquid introduces significant complications in the design of the equipment, greatly complicating it and thereby contributing to an increase in the cost of the finished part.

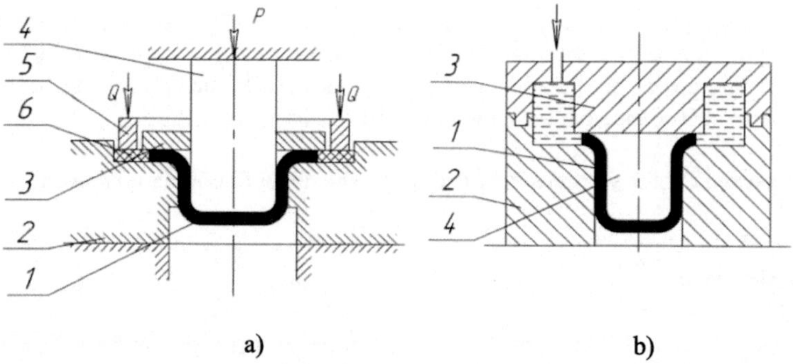

Fig. 1. The tool flange support at drawing a cylindrical barrel: a – flange support using an elastic element; b – workpiece flange support using pressurized process liquid; 1 – semi-finished product; 2 – die; 3 – blank holder; 4 – punch; 5 – fit ring; 6 – polyurethane ring.

Changing the shape of the entry part of the die allowed the authors [6] to obtain a cylindrical part in one drawing step in place of three (Fig. 2). Such a number of stages allows the product to be shaped without exhausting the resource of the ductility of steel and eliminating the loss of stability of the workpiece geometry [7, 8]. This die design implies an increased stroke of the press beam and time for the operation. The elimination of stability loss claimed by the authors is also controversial since most of the workpiece does not come into contact with the working surfaces of the stamp for a long time, which can lead to folding.

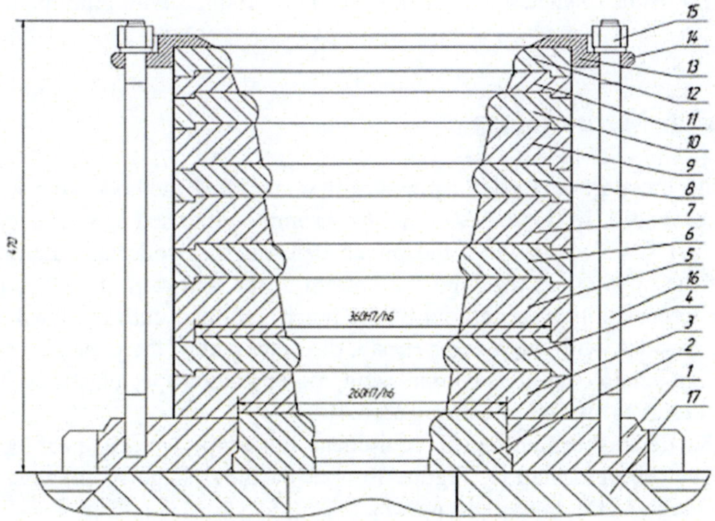

Fig. 2. The design of a special profile die: 1 – plate, 2 – die, 3 – 12 – fixtures, 13 – fixing ring, 14 – washer, 15 – nuts, 16 – pins, 17 – press table.

According to the method [9], at the stage of preliminary drawing, the flat workpiece from the sheet is formed by conical pressing into a cup. In this case, the diameter of the smaller base of which is determined by the dependence $d = D_b - (35\text{–}40)\ S$, where S – the thickness of the initial workpiece, D_b – the diameter of the initial sheet workpiece. The technical result of the invention is the increase of the drawing depth in one stamp. However, it is known that conical dies are difficult to manufacture, and their height is greater compared with the dies with a radial inlet, and hence there is increased consumption of expensive tool steel, which should be minimized [10, 11]. But the undoubted advantage of this method of drawing is that folding is eliminated by pre-forming the workpiece, and there is no need to use a blank holder.

A large number of ways to improve the drawing without pressing the flange of the workpiece are aimed at the preliminary preparation of workpieces. This preparation includes the use of square or hexagonal in plan workpieces instead of round ones [12], the creation of thinning on the flange in the form of cylindrical grooves [13], and the formation on the workpiece of a flat central part with an adjacent flange zone, which is made corrugated with a sinusoidal section profile [14]. It is claimed to increase the degree of deformation and eliminate corrugation. The analyzed inventions allow saving metal and increasing the degree of deformation during the stamping step. However, they lengthen the production cycle of the part, which is not always advantageous [15, 16]. Creating gutters or a periodic profile requires the introduction of an additional step and equipment, so it is more advisable to use these solutions in small-scale production [17, 18].

Most metal processing enterprises operate in conditions of frequent changes in the range of manufactured parts, which leads to minimizing the cost of tools and equipment to reduce the cost of the process [19, 20]. Therefore, the unification of the tool, simplification of the design of die tooling in the conditions of flexible serial production will make it possible to switch to the release of new products with minimal loss of time [21, 22].

3 Research Methodology

Based on the literature analysis, a pressure-free stamping method using a telescopic punch was proposed. In this case, the above example was used as a prototype [6].

At the first stage of drawing, the punch deforms the workpiece along the outer diameter of the punch, and reaching the bottom of the first step, the outer part stops, and further deformation is carried out by the inner main part until the finished product is obtained. Besides, with the reverse stroke, the outer part of the punch takes its initial position with the help of springs compression, moving along special guides. The design of a punch deserves special attention [23, 24].

Figure 3a depicts technological equipment (telescopic punch, profiled die, and workpiece) before deformation. Figure 3b – at the final stage of drawing. Figure 4 shows the design of the telescopic punch.

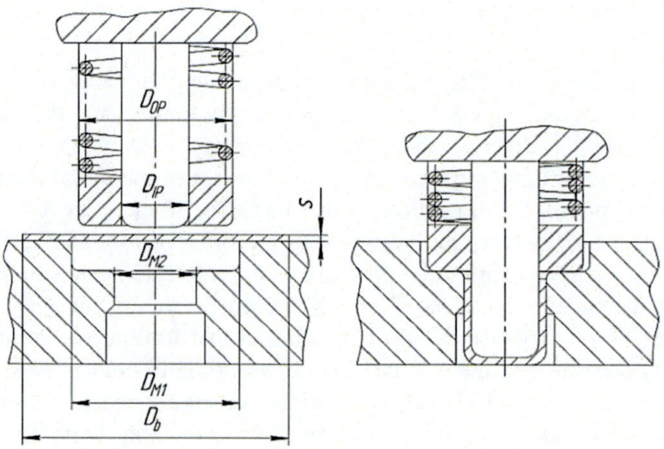

a) b)

Fig. 3. The operation of the telescopic punch: a – stamping equipment before the deformation; b – the punch at the end of the operation stroke.

The design of the telescopic punch contains the following parts: 1 – main internal shaping punch, 2 – moving external shaping punch 3 – thrust ball, 4 – internal spring, 5 – wedge block, 6 – wedge screw, 7 – external spring, 8 – recess, 9 – keyway.

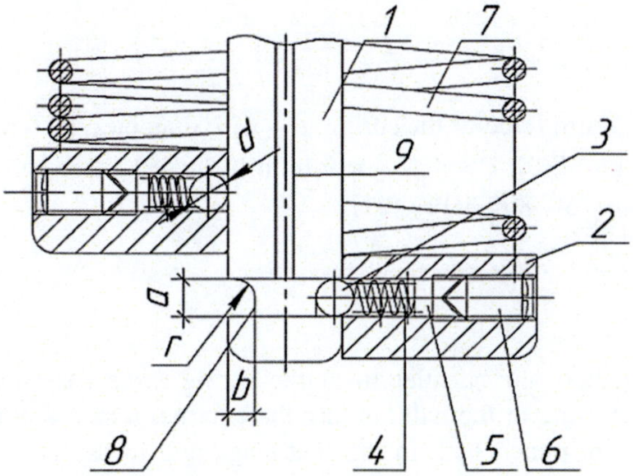

Fig. 4. The design of a telescopic punch.

Internal punch 1 has a circular recess with dimensions b = d/2, a = d + 2 (mm), were d – the diameter of the thrust ball fixing the external punch at the beginning of the operation stroke. When the external punch reaches the first stage of the die and presses the workpiece against it, it stops. Further lowering of the beam will cause the thrust ball to disengage from the internal punch, which is achieved by pressure of spring 4. After this, the shape is changed only by the internal punch, since the outer part will stop and slide along the surface of the internal punch up the key (not shown in Fig. 4), which is installed in keyway 9 on the internal punch. During the reverse stroke, the external punch returns to its original position under the action of the force of spring 7. Recess 8 has the radius of curvature r = d/2 in the upper part to release the thrust ball out of engagement and forms a step from below – to fix the position of the external punch. The force of the pressure of the thrust ball to the internal punch is regulated using block 5 and screw 6.

The pressure force of the thrust ball is calculated depending on the effort of the first stage of drawing and should be $P \geq 1.2 \cdot P_B$ [25, 26], where P – the pressure force of the thrust ball, P_B – the force of the first stage of drawing, k = 1.2 – reserve coefficient taking into account the large spring compression.

The drawing force is calculated by formula [27, 28]:

$$P_B = \pi d_1 s \sigma_B k_1 \quad (1)$$

where d_1 – the diameter of the cylindrical semi-finished product at the first drawing step; s – material thickness; $\sigma_в$ – ultimate stress; k_1 –coefficient depending on relative thickness s/D relative diameter and drawing coefficient (it is found by the reference data).

The choice of spring is carried out according to the tables and normals, or by calculation [29]:

$$P = \frac{\pi d^2}{8D}[\tau] \quad (2)$$

where P – the maximum force of the spring; d – wire diameter; $[\tau]$ – admissible torsion stress ($[\tau]$ = 5–6 MPa) [30]; $F = \frac{n\pi d^2}{dG}$ – maximum admissible spring compression; G – shear modulus (G = 750–800 MPa) [31].

4 Results

In this drawing method, the calculation of the spring force has decisive importance since insufficient pressing of the ball holding the external punch will result in drawing only with the internal punch both in the first stage and in the second. An excessive spring force will clamp both punches after the first stage and increase the pressing force the external punch of the formed semi-finished product and can lead to rejecting in the form of a separation of the bottom.

Therefore, it is proposed based on research to refine formula (1). The flange resistance to deformation and, accordingly, the highest tensile meridional stress during drawing without pressing the flange will depend on the following summands: stresses from the ratio of the diameters of the workpiece and the semi-finished product stresses from bending of the workpiece at the radius of the die, friction stresses on the radius of the die [32, 33].

Then the drawing force will be determined by expression:

$$P_B = 2\pi R_d s \sigma_s \left(\ln \frac{R_b}{R_d + r_m} - \frac{r_m}{R_d + r_m} + 0,1 \frac{R_b}{R_b - R_d} \left(\frac{s}{r_m} \right)^2 + \ln \frac{R_b}{R_d} + \mu \frac{s}{r_m} \right). \quad (3)$$

This formula allows taking into account all the technological and design factors of the drawing process without a folding holder but does not consider the hardening of the metal during plastic deformation. This can be done by substituting the yield point by the ultimate stress of the drawn metal by analogy with the formula (1) or by approximating this dependence according to [34, 35].

For the verification of the operation of the designed punch, experimental research was conducted. It aimed at measuring the drawing force of the proposed method and the operability of dependences (1)–(3) [36].

With this purpose in view, the limits of stamping without pressure were determined according to V.P. Romanovskii: D – d_1 > 18·s – 410 – d_1 > 72. So, d_1 = 330 mm.

The semi-product diameter was calculated according to the V.P. Romanovskii coefficient D – d_1 > 22–410 – d_1 > 88; then d_1 = 313 mm [12]. Considering the positive tolerance for rolling, it was finally assumed that d_1 = 310 mm. I.e., the workpiece diameter made D = 410 mm, the material – Steel DIN DD11. Then the drawing coefficients by the steps were – $m_1 = \frac{310}{401} = 0.77$ and $m_2 = \frac{210}{310} = 0.68$. The value of the second drawing coefficient is somewhat underestimated in comparison with the recommendations of the reference tables, however, due to the design features of the recommended method of drawing in a profiled die, where the second step is carried out with a fixture, we accept it without changes [37, 38]. Figure 5 contains an experimental curve representing the measurement of the force of drawing by means of a telescopic punch. The experiment was performed on the UME-100 TM tension testing machine, applying a force of 1 about MN.

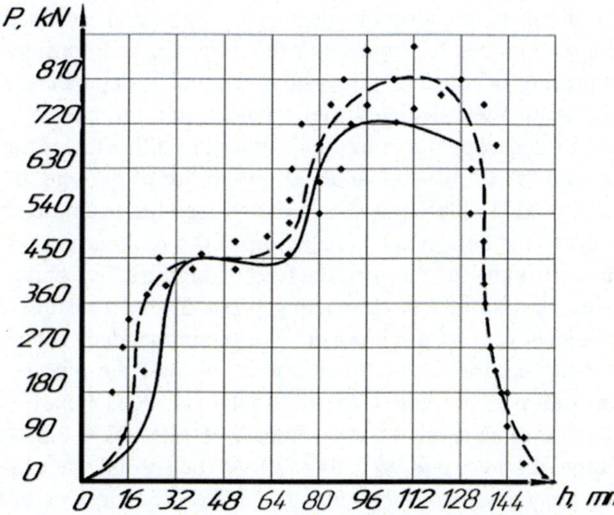

Fig. 5. The drawing force: P – the drawing; h – punch stroke; the drawing, calculated according to (4) "—" the drawing, approximated by experimental measurements "———".

The stage-by-stage height of the drawn semi-finished product made: drawing by the external punch – $h_1 = 0.25\left(\frac{D}{m_1} - d_1\right) = 0.25\left(\frac{401}{0.77} - 310\right) = 53\ mm$; drawing by the internal punch – $h_2 = 0.25\left(\frac{D}{m_1 m_2} - d_2\right) = 0.25\left(\frac{401}{0.77 \cdot 0.68} - 210\right) = 139\ mm$ [39].

The radii of curvature of the punch and the die were chosen according to the reference tables [12, 40]. At the first stage, when the drawing process is carried out without pressing the workpiece flange, the radius of the die is taken $r_m = 4$ mm; the punch curvature radius $r_p = 4$ mm. Reduced rounding radii compared with the recommended ones increase the meridional tensile stresses, which prevents the formation of corrugations. At the second stage, when drawing is performed with pressing the flange – $r_m = 35$ mm; $r_p = 25$ mm. The one-sided gap between the drawing punches and the die was chosen the same for both steps; it was z = 4 + 0.2 + 0.35 = 4.55 mm.

The performed research made it possible to recommend this drawing method to production. Due to it, the use of a double-action press, whose price is many times higher than the price of a single-action press, is eliminated. There is no need for a buffer device of a single-action press. The cost of production of telescopic equipment is commensurable with the cost of two common drawing steps. However, the possibilities of the method are restricted due to the use of compressing springs. The recommended intervals of drawing are the relative thickness of the workpieces s/D·100 = 0.8–1.2 at the overall coefficient of drawing m_g = 0.50–0.52.

5 Conclusions

The presented method of drawing a cylindrical semi-finished product without a fixing ring contributes to the intensification of this shaping operation, is based on a detailed analysis of existing techniques for improving this process, and considering almost all technological parameters of shaping. The above graphical dependencies allow us to confirm the adequacy of the calculated dependencies for the choice of punch compression springs and the accepted assumptions about the influence of the main process factors on the drawing force. The fundamental possibility of drawing by the proposed method is also confirmed. Technological calculation (see the example above) makes it possible to move on to the design and calculation of die tooling and power springs of the punch, which contributes to the manufacture of high-quality parts, reduces labor costs, and eliminates the use of a double-acting press. These measures will ultimately reduce the technological cost of the product. The recommended materials (steels) for the manufacture of the proposed equipment are as follows: the internal punch – DIN X210Cr12; the external punch – DIN C45; die – DIN X210Cr12; balls – DIN 100Cr6; springs – 66Mn4. The dimensions of the workpieces s/D·100 = 0.8–1.2. The coefficient of the first stage of drawing m_1 = 0.70–0.72. At the second stage m_2 = 0.70–0.72. The tolerances for the manufacture by punch and the dies meet class 8 of accuracy quality, the working surfaces roughness Ra = 0.16 – 0.125.

References

1. Bedford, A., Hill, C.: Numerical and experimental investigations of multistage sheet-bulk metal forming process with compound press tools. Key Eng. Mater. **651–653**, 1153–1158 (2015). https://doi.org/10.4028/www.scientific.net/KEM.651-653.1153
2. Haikova, T., Puzyr, R., Dragobetsky, V., Symonova, A., Vakylenko, R.: Finite-element model of bimetal billet strain obtaining box-shaped parts by means of drawing. In: Ivanov, V. et al. (eds) CONFERENCE 2019, DSMIE, pp. 85–94. Springer, Lutsk, Ukraine (2019). https://doi.org/10.1007/978-3-030-22365-6_9
3. Abe, Y., Mori, K., Ito, T.: Multi-stage stamping including thickening of corners of drawn cup. Procedia Eng. **55**(637), 148–152 (2014). https://doi.org/10.1016/j.proeng.2014.10.083
4. Wang, X., Luo, W., Xia, J.: Investigation of warm stamping-forging process for car flywheel panel. Forging Stamping Technol. **34**(5), 43–46 (2009). https://doi.org/10.3969/j.issn.1000-3940.2009.05.013
5. Luo, W.: An Investigation of Warm Stamping-Forging Process for Flywheel Panel of Automatic Transmission. Huazhong University of Science & Technology, Wuhan (2009)

6. Kalyuzhnyi, O.: Reduction of the number of transitions of stretching of axisymmetric products using a special profile matrix. Press. Process. Collect. Sci. Pap. Kramatorsk DSEA **4**(37), 93–96 (2013)
7. Sosenushkin, E., Yanovskaya, E., Sosenushkin, A., Emel'yanov, V.: Mechanics of nonmonotonic plastic deformation. Russ. Eng. Res. **35**(12), 902–906 (2015)
8. Ogorodnikov, V., Derevenko, I., Sivak, R.: On the influence of curvature of the trajectories of deformation of a volume of the material by pressing on its plasticity under the conditions of complex loading. Mater. Sci. **54**(3), 326–332 (2018)
9. The method of stamping hollow parts: US Pat. 2157287 of the Russian Federation: IPC B21D22/ 20 No. 97119363/02; declared 11/20/97; publ. 10.10.00
10. Puzyr, R., Savelov, D., Shchetynin, V., Levchenko, R., Haikova, T., Kravchenko, S., Yasko, S., Argat, R., Sira, Y., Shchipkovakyi, Y.: Development of a method to determine deformations in the manufacture of a vehicle wheel rim. Eastern-Eur. J. Enterp. Technol. **4**(1 (94)), 55–60 (2018). https://doi.org/10.15587/1729-4061.2018.139534
11. Puzyr, R., Haikova, T., Trotsko, O., Argat, R.: Determining experimentally the stress-strained state in the radial rotary method of obtaining wheels rims. Eastern-Eur. J. Enterp. Technol. **4**(1(82)), 52–60 (2016). https://doi.org/10.15587/1729-4061.2016.76225
12. Romanovsky, V.: Handbook of cold stamping. Engineering (1976)
13. Billet for the extraction of a cylindrical semi-finished product: US Pat. 2056199 RF: IPC 7 B21D22/ 20 No. 93006727/08; declared 02/03/93; publ. 03/20/96
14. Blank for hoods: US Pat. 1542665 USSR: IPC 7 B21D22/ 30 No. 4405006/ 31-27; declared 04/08/88; publ. 02.15.90, Bull. No. 6. 3 p
15. Altinbalik, T., Tonka, A.: Numerical and experimental study of sheet thickness variation in deep drawing processes. Int. J. Mod. Manuf. Technol. **4**(2), 9–16 (2012)
16. Markov, O., Gerasimenko, O., Khvashchynskyi, A., Zhytnikov, R., Puzyr, R.: Modeling the technological process of pipe forging without a mandrel. Eastern-Eur. J. Enterp. Technol. **3**(1(99)), 42–48 (2019). https://doi.org/10.15587/1729-4061.2019.167077
17. Kim, S., Huh, H., Bok, H., Moon, M.: Forming limit diagram of auto-body steel sheets for highspeed sheet metal forming. J. Mat. Pro. Tech. **211**(5), 851–862 (2011)
18. Maslov, A., Batsaikhan, J., Puzyr, R., Salenko, Y.: The determination of the parameters of a vibration machinef the internal compaction of concrete mixtures. Int. J. Eng. Technol. **7**(4.3), 12–19 (2018). https://doi.org/10.14419/ijet.v7i4.3.19545
19. Zagirnyak, M., Drahobetskyi, V.: New methods of obtaining materials and structures for light armor protection. In: Military Technologies (ICMT), International Conference (19–21 May 2015). Brno, Czech Republic, vol. 1, pp. 705–710 (2015)
20. Demirci, H., Esner, C., Yasar, M.: Effect of the blank holder force on drawing of aluminum alloy square cup: theoretical and experimental investigation. J. Mat. Pro. Tech. **206**, 152–160 (2008)
21. Pottier, T., Vacher, P., Toussaint, F.: Out-of-plane testing procedure for inverse identification purpose: application in sheet metal plasticity. Exp. Mech. **52**, 951–963 (2012). https://doi.org/10.1007/s11340-011-9555-3
22. Savelov, D., Dragobetsky, V., Puzyr, R., Markevych, A.: Peculiarities of vibrational press dynamics with hard-elastic restraints in the working regime of metal powders molding. Metall. Min. Ind. **2**, 67–74 (2015)
23. A method of extracting cylindrical parts in a special profiled punch matrix: US Pat. 107394 Ukraine: IPC (2016.01) B26F 1/00. № 201508823; claimed 09/14/15; publ. 06/10/16, Bul. № 11. P. 4

24. Puzyr, R., Savelov, D., Argat, R., Chernish, A.: Distribution analysis of stresses across the stretching edge of die body and bending radius of deforming roll during profiling and drawing of cylindrical workpiece. Metall. Min. Ind. **1**, 27–32 (2015)
25. Wang, X., Ouyang, K., Xia, J.: FEM analysis of drawing-thickening technology in stamping-forging hybrid process. Forging Stamping Technol. **34**(4), 73–78 (2009). https://doi.org/10.3969/j.issn.1000-3940.2009.04.017
26. Mori, K., Nakano, T.: State-of-the-art of plate forging in Japan. Prod. Eng. **10**(1), 81–91 (2016). https://doi.org/10.1007/s11740-015-0648-1
27. Yan, G., Wang, X., Deng, L.: A study of hole flanging-upsetting process. Adv. Mater. Res. **939**, 291–298 (2014). https://doi.org/10.4028/www.scientific.net/AMR.939.291
28. Cai, L., Sánchez-Dehesa, J.: Analysis of equivalent anisotropy arising from dual isotropic layers of acoustic media. J. Acoust. Soc. Am. **132**(4), 15–22 (2012)
29. Dell, H., Gese, H., Kepler, L., Werner, H., Hooputra, H.: Continuos Failure Prediction Model for Nonlinear Load Paths in Successive Stamping and Crash Processes, SAE – Paper 2001–01–1131. New Sheet Steel Products and Steel Metal Stamping (SP–1614). SAE 2001 world Congress, (Michigan, March 5–8), pp. 113–122 (2001)
30. Leyu, W., Daxin, E.: Numerical simulation analysis of variable BHF drawing of rectangular cup on curve blank-holder. Mod. Manuf. Eng. **2**, 73–74 (2006)
31. Yang, C., Li, P., Fan, L.: Blank shape design for sheet metal forming based on geometrical resemblance. Procedia Eng. **81**, 1487–1492 (2014)
32. Roberto, B., Terence, G.: Using severe plastic deformation for the processing of advanced engineering materials. Mater. Trans. **50**(7), 1613–1619 (2009). https://doi.org/10.2320/matertrans.MF200913
33. Puzyr, R., Haikova, T., Majerník, J., Karkova, M., Kmec, J.: Experimental study of the process of radial rotation profiling of wheel rims resulting in formation and technological flattening of the corrugations. Manuf. Technol. **18**(1), 106–111 (2018). https://doi.org/10.21062/ujep/61.2018/a/1213-2489/mt/18/1/106
34. Akbari-Mousavi, S., Barrett, L., Al-Hassani, S.: Explosive welding of metal plates. J. Mater. Process. Technol. **202**(1–3), 224–239 (2008)
35. Gavas, M., Izciler, M.: Effect of blank holder gap on deep drawing of square cups. Mater. Des. **28**, 1641–1646 (2007)
36. Grushko, A., Kukhar, V., Slobodyanyuk, Y.: Phenomenological model of low-carbon steels hardening during multistage drawing. Solid State Phenom. **265**, 114–123 (2017). https://doi.org/10.4028/www.scientific.net/SSP.265.114
37. Ferran, G., Barros, C., di Pasquale, E., Yamashita, E.: Simplified numerical simulation of forming a car wheel disk using the SIME Xexplicit FEM code, and comparisons with press-shop results. J. Mater. Process. Technol. **60**, 523–528 (1996)
38. Sotiria, H., Spyros, A.: Stability of long transversely-isotropic elastic cylindrical shells under bending. Int. J. Solids Struct. **47**, 10–24 (2010)
39. Logesh, K., Raja, V., Krishnan, D., Azeemudeen, M., Andrew, A., Hariprasath, K.: Experimental investigation and analysis of tensile and forming behaviour of cold rolled sheet metal. Int. J. Mech. Eng. Technol. (IJMET) **8**(11), 252–264 (2017)
40. Puzyr, R., Kukhar, V., Maslov, A., Shchipkovskyi, Y.: The development of the method for the calculation of the shaping force in the production of vehicle wheel rims. Int. J. Eng. Technol. **7**(4.3), 30–34 (2018). https://doi.org/10.14419/ijet.v7i4.3.20128

Volumetric Vibration Treatment of Machine Parts Fixed in Rotary Devices

Volodymyr Borovets, Oleksiy Lanets, Vitaliy Korendiy(✉), and Petro Dmyterko

Lviv Polytechnic National University, 12, S. Bandera St., Lviv 79013, Ukraine
UJofME@gmail.com

Abstract. The research aims to increase the efficiency of machines for vibration treatment of parts by substantiating their design parameters. The design diagram of the machine with a rotary device for volumetric vibration treatment of parts is considered. The mathematical model of a vibration machine with inertial vibration exciter is developed. The working medium motion and the oscillations of the machine working body and of the rotary device with the fixed parts being treated are taken into account. Using the Lagrange equations of the second order, the dynamic model of machine operation is formed. Using the MathCad software, the mathematical analysis of the model has been carried out for various values of the system parameters, in particular the filling coefficient and the factor of regulation of the unbalanced units. The simulation results of the system's operation, in particular the trajectories of vibrations of the working chamber and the working medium, as well as the motion speeds of the medium under different operational conditions are analyzed.

Keywords: Mathematical model · Vibration machine · Operational efficiency · Inertial vibration exciter · Working medium · Motion trajectory

1 Introduction

Modern manufacture and requirements set to finished products demand the creation of new models of vibration machines with high technical and economic indexes. Therefore, the improvement of the operational efficiency of existing equipment for vibration treatment of machine parts and the development of new designs of machines are important problems set for designers and manufacturers since the minimal improvement of these indexes can lead to a significant economic effect. At the same time, based on rational principles of the working process modeling, it is particularly important to clarify the laws of motion of the vibration machine elements taking into account the motion of the working medium.

The volumetric vibration treatment of machine parts is a promising direction for the finishing treatment technology. Various methods can implement it, and its efficiency is directly related to the reciprocal motion of the working medium and the parts being treated. That is why the investigation of the working medium motion is one of the most important problems while analyzing the treatment process.

V. Tonkonogyi et al. (Eds.): InterPartner 2020, LNME, pp. 373–383, 2021.
https://doi.org/10.1007/978-3-030-68014-5_37

2 Literature Review

The models and techniques of vibration treatment are studied in numerous publications. One of the most widely used techniques is considered in papers [1–3], in which the authors presented the results of investigations of vibration machining using a loose abrasive medium and liquid lubricant in a closed vibrating tumbler. In [4–6], the authors proposed to use the effect of a shock wave of the working medium impacting the parts being treated. The comprehensive study of various parameters influencing the process of finishing treatment is presented in [7–9]. The papers [10–12], and [13] analyze the kinematic parameters and the dynamic behavior of vibratory finishing machines under different operational conditions.

The treated parts are distributed within the working medium, and their motion is provided only by the interaction with this medium. One of the problems that have not yet been thoroughly investigated consists of the following. The intensification of vibration treatment can be ensured by increasing the relative speed of parts being treated using a special rotary device, which rotates in the opposite direction to the motion of the working medium.

In order to study the working medium motion, in some cases, it is possible to adopt a single-mass model [14]. It is based on the fact that vibration treatment in the machine with parts fixed in special rotary devices is mainly related to the motion of the certain "active" layer since the rest of the volume of the medium is relatively less significant [15, 16]. While carrying out the following investigations, let us use the design diagram of the machine with a rotary device for vibration treatment of parts, which is developed by V. Borovets and considered in detail in the paper [14].

The purpose of this research paper is to increase the efficiency of machines for vibration treatment of parts by substantiating their design parameters based on the development and usage of special devices for intensifying the working process of treatment and for ensuring the possibilities of automation of auxiliary operations.

3 Research Methodology

The calculation diagram of the machine for vibration treatment of parts can be presented in a fixed coordinate system, the center of which coincides with the position of the projection of the working chamber longitudinal axis in a state of rest (Fig. 1 a). We denote x_k, y_k, x_3, y_3 as the coordinates of the mass centers of the working chamber and the medium.

In the state of rest, the working medium is in the bottom part of the container, and the values of corresponding coordinates are as follows $x_3 = x_{30} = 0$, $y_3 = y_{30} \neq 0$, $y_{30} \approx r_k \cdot (\eta_3 - 1)$, where r_k is the radius of the working chamber; η_3 is the filling coefficient ($\eta_3 = 0.6...0.75$) [14]. The treatment of parts is carried out in the outer layer of the working medium (Fig. 1 b), which fits the wall of the container. In order to analyze the motion of the working medium, we introduce into consideration the angular velocity of its surface layer $\dot{\psi}_3^n$.

In order to evaluate the kinetic energy of the working medium associated with its rotation, we assume the following facts: there is no rotation in the region adjacent to the mass center of the working medium, and in the direction from the mass center to the outer layer, the angular velocity changes according to the linear law:

$$\dot{\psi}_{(r)} = \dot{\psi}_3^n \frac{r_i}{r_3}, \tag{1}$$

where $r_3 = \sqrt{\eta_3}\, r_k$ is the radius of the cylinder, by which we conditionally replace the medium; r_i is the radius of the i-th layer; r_3 is the radius of the machine's container.

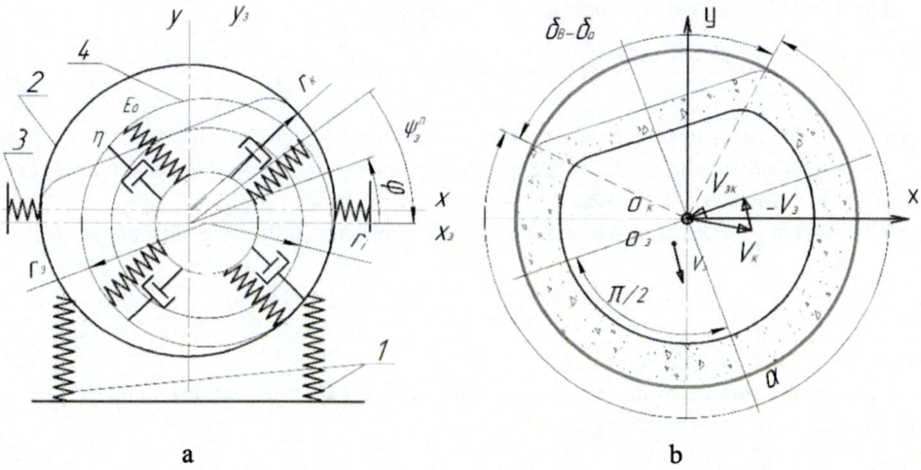

Fig. 1. Calculation diagrams of the vibration machine (a) and of the contact surface between the working medium and the container (b): 1, 3 – elastic supports; 2 – working chamber; 4 – working medium.

For friction forces between the outer layer of the working medium and the container wall, taking into account the fact that the medium has properties of viscous fluid and certain properties of the bulk material, we assume the following:

$$F_m \approx \eta_1 S_k \Delta V, \tag{2}$$

where η_1 is the coefficient of internal friction; S_k is the area of the contact surface between the working medium and the container wall; ΔV is the difference between the tangential components of the linear velocities of the container surface and the adjacent layer of the working medium.

The velocity of the working medium layer located at a distance r_i from the axis of rotation of the working medium "cylinder" is as follows [14]:

$$V_{(r)} = r_i \cdot \dot{\psi}(r_i) = \frac{r_i^2}{r_3} \dot{\psi}_3^n. \tag{3}$$

The formula for the kinetic energy of rotation of the working medium is the following:

$$T_{wm}^{rot} = \frac{\rho_3 L}{2} \int_0^{r_3} \int_0^{2\pi} V^2(r_i)\, r_i\, dr_i\, d\alpha = \frac{I_3 \left(\dot{\psi}_3^n\right)^2}{3} \approx 0.333 \cdot I_3 \left(\dot{\psi}_3^n\right)^2, \tag{4}$$

where $I_3 = 0.5 \cdot m_3 r_3^2$ is the moment of inertia of the working medium "cylinder"; ρ_3 is the volumetric density of the medium; m_3 is the mass of the charged medium.

The total kinetic energy of the working medium is equal to:

$$T_{wm} = 0.5 \cdot m_3(\dot{x}_3^2 + \dot{y}_3^2) + 0.333 \cdot I_3(\dot{\psi}_3^n)^2. \tag{5}$$

The kinetic energy of the container is as follows:

$$T_k = 0.5 \cdot m_k(\dot{x}_k^2 + \dot{y}_k^2) + 0.5 \cdot I_k \varphi^2, \tag{6}$$

where m_k is the mass of the container; I_k is the moment of inertia of the container; $\dot{x}_k$, $\dot{y}_k$ are the projections of the speed of the working chamber center on the x and y axes; $\dot{\varphi}$ is the angular velocity of the center of the working chamber.

The working medium moves relative to the container in the direction of the vector:

$$\overline{V}_{3k} = (\dot{x}_3 - \dot{x}_k\ ,\ \dot{y}_3 - \dot{y}_k) = \overline{V}_3 - \overline{V}_k\ , \tag{7}$$

where $\overline{V}_3 = (\dot{x}_3,\ \dot{y}_3)$, $\overline{V}_k = (\dot{x}_k,\ \dot{y}_k)$.

The contact surface of the working medium is limited by the angles $\delta_a.\delta_B$ (Fig. 1 b) and for the considered case we have:

$$\delta_B = \min(\gamma_{3k} + 0.5\pi\ ,\ \gamma_3), \tag{8}$$

$$\delta_a = \max(\pi\eta_3, \gamma_{3k} - 0.5\pi), \tag{9}$$

where $\gamma_3 = \text{arctg}\left[(y_3 - y_3^o)/(x_3 - x_3^o)\right], \gamma_{3k} = \text{arctg}[(\dot{y}_3 - \dot{y}_k)/(\dot{x}_3 - \dot{x}_k)]$.

The speed of the container's surface within angle $\Delta\alpha$ consists of the speed of its rotation and the speed of plane motion $\overline{V}_k$.

The projection of the total speed of the mentioned area of the container on the tangent vector drawn through the point that corresponds to the angle α is equal to:

$$V_{k\alpha}^m = r_k\dot{\varphi} - \dot{x}_k \sin\alpha + \dot{y}_k \cos\alpha. \tag{10}$$

For the certain value of the angle α, the difference between the tangential components of speeds of the working medium and the container is equal to:

$$r_k\,\dot{\varphi} - r_3\,\dot{\psi}_3^n = (\dot{x}_k - \dot{x}_3)\ \sin\alpha + (\dot{y}_k - \dot{y}_3)\ \cos\alpha. \tag{11}$$

The projections of the resulting friction force, which can be considered as the force applied in the point that corresponds to the angle $\delta_3 = 0.5 \cdot (\delta_a + \delta_B)$, can be determined by integrating this force within the limits of the contact surface:

$$F_{mx}^{k} = -\eta_1 L r_k \left\{ \begin{array}{l} 2(r_k\dot{\varphi} - r_3\dot{\psi}_3^n)\sin\delta_3\sin(0.5\delta) + 0.5(\dot{x}_k - \dot{x}_3)\delta \\ +0.5\sin\delta[(\dot{x}_k - \dot{x}_3)\cos 2\delta_3 + (\dot{y}_k - \dot{y}_3)\sin 2\delta_3] \end{array} \right\}, \quad (12)$$

$$F_{mx}^{k} = -\eta_1 L r_k \left\{ \begin{array}{l} 2(r_k\dot{\varphi} - r_3\dot{\psi}_3^n)\cos\delta_3\sin(0.5\delta) + 0.5(\dot{y}_k - \dot{y}_3)\delta \\ +0.5\sin\delta[(\dot{y}_k - \dot{y}_3)\cos 2\delta_3 + (\dot{x}_k - \dot{x}_3)\sin 2\delta_3] \end{array} \right\}, \quad (13)$$

where $\delta = \delta_B - \delta_a$.

The tangential component of the friction force acting upon the container within the angles (δ_a, δ_B) is equal to:

$$M_m^k = -\eta_1 L r_k^2 \left\{ \left(r_k\dot{\varphi} - r_3\dot{\psi}_3^n \right)\delta + 2\sin(0.5\delta)[(\dot{x}_k - \dot{x}_3)\sin\delta_3 + (\dot{y}_k - \dot{y}_3)\cos\delta_3] \right\} \quad (14)$$

The projections of friction forces acting upon the working medium are equal in magnitude and have the opposite direction to the corresponding projections of friction forces acting upon the container. The moment of friction forces applied to the medium is equal to:

$$M_m^3 = -\frac{r_3}{r_k} M_m^k. \quad (15)$$

Assuming that the filling coefficient η_3 is closer to 0.5 than to 1, we introduce an additional condition for the presence of the contact. It means that we neglect the contact interaction if the working medium "hangs up" in the container space that occurs in the case when the distance from the center of mass of the medium to the center of mass of the container is less than its value in a state of rest. The dependence can determine a function considering the "hanging up" of the medium:

$$B = B(x_3, y_3, x_k, y_k) = \frac{1}{2}\left\{ \mathrm{sign}\left[\sqrt{(x_k - x_3)^2 - (y_k - y_3)^2} - \left| y_3^o \right| \right] + 1 \right\}. \quad (16)$$

While calculating the virtual work, taking into account the fact that the working medium has elastic properties, the work of the resistance forces can be defined as the virtual work of deformation:

$$\tilde{\delta} A_q = E_c S_k \tilde{\delta} q, \quad (17)$$

where E_c is the value that characterizes the elastic properties of the working medium; $\tilde{\delta} q$ is the value of deformation.

We also take into account the fact that during the separation process of the medium from the surface of the working chamber, there is a loosening of material (within a certain depth d_o). While calculating the deformation work, let us determine the value E_c by the following formula:

$$E_c = E_c(d) = E_o\left(1 - e^{-\frac{d}{d_o}}\right), \tag{18}$$

where $d = \sqrt{(x_k - x_3)^2 + (y_k - y_3)^2} - |y_3^o|$ is the value of deformation of the working medium; E_o is the elasticity modulus of the compacted working medium. The case when $d < 0$ is taken into account with the help of the previously applied function B.

For the internal friction between the layers of the working medium, taking into account the presence of the liquid properties, we assume:

$$F_m \approx \mu \cdot (dV/dr), \tag{19}$$

where μ is the coefficient of friction.

To define the moment of forces of internal friction, the following formula is used:

$$L_3 = \left(2\mu \cdot M_3 \cdot \dot{\psi}_3^n\right) / d_3. \tag{20}$$

The diagram of a vibration machine is presented in Fig. 2, considering the working medium is placed in a container, and the workpieces are fixed on the rotary device. The fixing of parts ensures their location in the active zone of the working medium. The angular velocity of the medium is considered to be constant and equal to the speed of its surface layer $\dot{\psi}_3^n$.

It is assumed that the parts are of the same type, and the center of inertia of the mentioned projection is at a distance r_s from the axis of the rotary device and is equal to the radius of the active layer of the medium (r_3).

The total area of projections for all parts being treated is as follows:

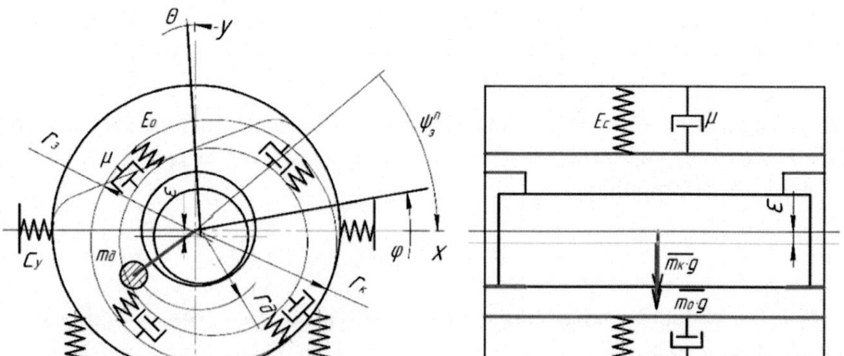

Fig. 2. The generalized calculation diagram of the vibration machine with a rotary device filled with the working medium.

$$S_d = \sum_i S_i, \tag{21}$$

where S_i is the area of the projection of the part placed on the spinning device.

Taking into account the liquid properties of the medium, the resistance force between the working medium and the rotary device with parts is equal to:

$$\left|\overline{F_s}\right| = 0.5\rho_3\, u^2\, S_d\, k_3, \tag{22}$$

where u is the average speed of parts motion related to the motion of the working medium; k_3 is the coefficient that takes into account the properties of the medium.

Thus, the system of differential equations describing the motion of the working elements of the vibration machine, of the rotary device, and the working medium can be formed using the Lagrange equations of the second order as follows:

$$\begin{cases} (m_k + m_o)\ddot{x}_k - m_o\,\varepsilon\left(\ddot{\psi}\ \sin\psi + \dot{\psi}^2\ \cos\psi\right) \\ \quad = -c_x x_k - k_x\dot{x}_k - BS_kE_c\cos\gamma_k + F_m\cos\omega t; \\ (m_k + m_o)\ddot{y}_k - m_o\,\varepsilon\left(\ddot{\psi}\ \cos\psi + \dot{\psi}^2\ \sin\psi\right) \\ \quad = -c_y y_k - k_y\dot{y}_k + BS_kE_c\sin\gamma_k + F_m\sin\omega t; \\ m_3\ddot{x}_3 = -BS_kE_c\cos\gamma_{31} - 0.125\cdot(\dot{x}_3 - \dot{x}_k)\,|\dot{x}_3 - \dot{x}_k|\rho_3S_d\eta_3k_3; \\ m_3\ddot{y}_3 = -B\,S_kE_c\sin\gamma_{31} - m_3\,g - 0.125\cdot(\dot{y}_3 - \dot{y}_k)|\dot{y}_3 - \dot{y}_k|\rho_3S_d\eta_3k_3; \\ I_k\ddot{\varphi} = -c_\varphi\varphi - k_\phi\dot{\varphi} - \lambda\left(r\dot{\varphi} - r_o\dot{\theta}\right)r + M_m^k; \\ 0.667\cdot I_3\ddot{\psi}_3^n = M_m^3 - \left(\dot{\psi} - \dot{\theta}\right)\left(\dot{\psi}_3^n - \dot{\theta}\right)|\dot{\psi}_3^n - \dot{\theta}|\rho_3\left(\dot{\psi}_3^n r_3\right)^2 S_d\eta_3r_3k_3; \\ m_o\varepsilon^2\ddot{\psi} - m_o\varepsilon\left(\dot{y}_k\sin\psi + x_k\cos\psi\right)\dot{\psi} = -m_o\,g\ \cos\psi\,\varepsilon - \lambda\left(r\dot{\psi} - r_o\dot{\theta}\right)r\,; \\ I_o\ddot{\theta} = \lambda\left(r\dot{\psi} - r_o\dot{\theta}\right)r_o + 0.5\cdot\rho_3\left(\dot{\psi}_3^n\,r_3\right)^2\ S_d\,sign\left(\dot{\theta} - \dot{\psi}_3^n\right). \end{cases} \tag{23}$$

4 Results

Let us carry out the solving of the obtained systems of Eqs. (23) using the numerical methods with the help of MathCad software.

In order to check the mathematical model of the vibration machine operation with the simultaneous motions of the container, of the rotary device with parts and of the working medium, the numerical investigation of the differential Eqs. (23) was carried out for the following values of parameters: container mass $m_k = 96$ kg; container moment of inertia $I_k = 13.2$ kg $\cdot$ m^2; unbalanced unit mass $m_u = 2$ kg; unbalanced unit radius $r_u = 0.125$ m; container length $L = 0.7$ m; linear and angular stiffness of the container spring system $c_x = 10^5\ \mathrm{N \cdot m^{-1}}$, $c_y = 5 \cdot 10^5\ \mathrm{N \cdot m^{-1}}$, $c_\varphi = 1000^5\ \mathrm{N \cdot rad^{-1}}$; damping coefficients $k_x = k_y = 80\ \mathrm{N \cdot s \cdot m^{-1}}$, $k_\varphi = 20\ \mathrm{N \cdot s \cdot rad^{-1}}$; filling coefficient $\eta_3 = 0.55...0.75$; the maximal mass of the charged working medium $m_3 = 300$ kg; volumetric density of the working medium $\rho_3 \approx (2.2...2.4) \cdot 10^3\ \mathrm{kg \cdot m^{-3}}$; rotary device mass $m_o = 18$ kg; rotary device moment of inertia $I_o = 0.8\ \mathrm{kg \cdot m^2}$; the radius of the rotary device casing $r_o = 0.11$ m; rotary device eccentricity $\varepsilon = 0.003$ m; sliding coefficient $\lambda = 1.35\ \mathrm{N \cdot s \cdot m^{-1}}$. The results of the calculations are presented in Figs. 3, 4 and 5.

Analyzing the presented plots (Fig. 3), it can be concluded that the mass centers of the container and the medium oscillate along the trajectories other than elliptic ones. The maximal deviations are approximately 5 mm and 3 mm, respectively.

The velocity V_{sl} of the outer surface layer of the working medium is 1.0 – 2.5 dm/s (Fig. 4). The obtained simulation results correspond to the experimental data [14] and approximately linearly depends on the amplitude of the container vibrations. The outer surface layer reaches the stable operation mode in 25 s after the machine turning on.

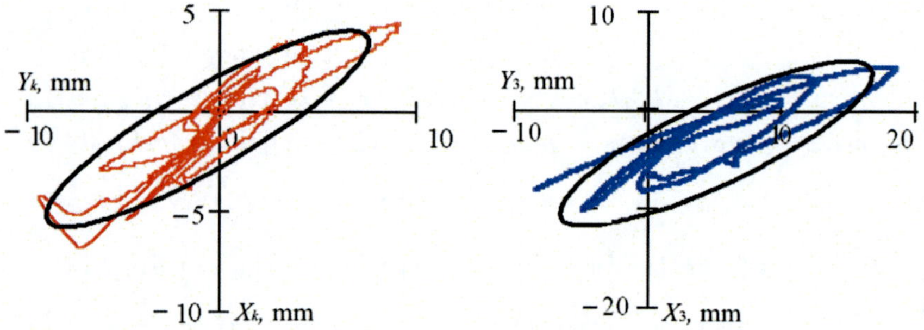

Fig. 3. Motion trajectories of the mass centers of the container (X_k, Y_k) and the working medium (X_3, Y_3).

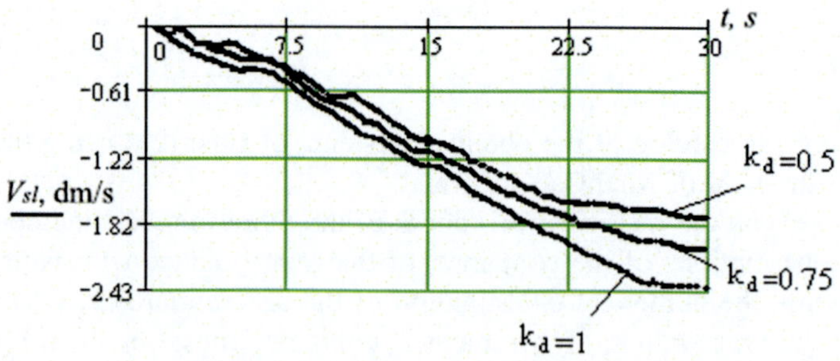

Fig. 4. The time dependence plot of the working medium motion speed V_{sl} for different values of the factor k_d of regulation of the unbalanced units.

The change of the operation amplitude is characterized by a factor of regulation of the unbalanced units ($k_d = 0...1$).

Analyzing the obtained values of x_3 y_3 (Fig. 3), it can be concluded that the working medium occupies the position at the bottom of the container during the operation of the machine. This fact substantiates the validity of the assumptions made while modeling the deformation surface of the working medium when it contacts the chamber's wall.

Based on the results of the carried out investigations, we may conclude that the derived Eqs. (23) adequately simulate the motion of the working chamber and the working medium within the filling coefficient of $\eta_3 = 0.5...1$ (Fig. 5).

5 Conclusions

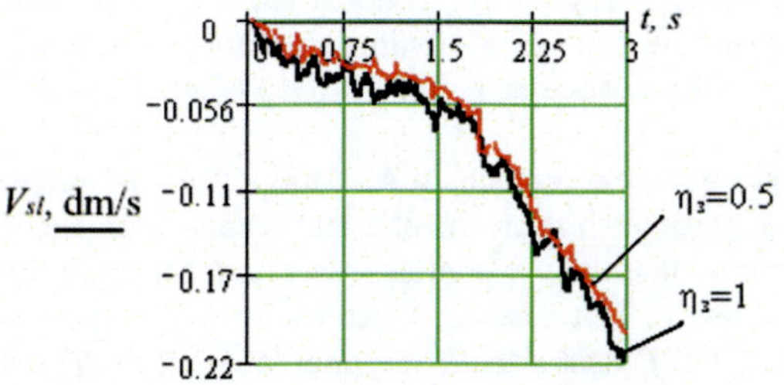

Fig. 5. The time dependence plot of the working medium motion speed for different values of the filling coefficient η_3.

Unlike most of the existent investigations related with vibration treatment and describing the kinematic parameters and dynamic behavior of chambers of vibratory machines, working media, and parts freely placed inside the chambers, the present paper analyses the process of volumetric vibration treatment of parts fixed on a special rotary device, which rotates in the opposite direction to the motion of the working medium. The calculation diagram of the machine with a rotary device for volumetric vibration treatment of parts is considered. The mathematical model describing the operation of the vibration machine is developed. The derived differential equations take into account the working medium motion, as well as the vibrations of the machine body and of the rotary device with the fixed parts being treated. The simulation of the working medium motion under different operational conditions, in particular for different values of the filling coefficient and of the factor of regulation of the unbalanced units, is carried out using the MathCad software. The results of computer modeling of the system's operation, in particular the trajectories of vibrations of the working chamber and the working medium, as well as the motion speeds of the working medium under different operational conditions are analyzed. Reviewing numerous information sources and comparing the presented results with the ones obtained experimentally, we can conclude about the adequacy of the proposed mathematical model of the vibration machine for volumetric treatment of parts, as well as about the increased relative velocities between the parts and the working medium. This effect will be analyzed in further investigations in order to substantiate the increased intensity and efficiency of the vibration treatment process.

References

1. Bańkowski, D., Spadło, S.: The application of vibro-abrasive machining for smoothing of castings. Arch. Foundry Eng. **17**(1), 169–173 (2017)
2. Bańkowski, D., Spadło, S.: Vibratory machining effect on the properties of the aluminum alloys surface. Arch. Foundry Eng. **17**(4), 19–24 (2017)
3. Bańkowski, D., Spadło, S.: The influence of abrasive paste on the effects of vibratory machining of brass. Arch. Foundry Eng. **19**(3), 5–10 (2019)
4. Kundrák, J., Morgan, M., Mitsyk, A.V., Fedorovich, V.A.: The effect of the shock wave of the oscillating working medium in a vibrating machine's reservoir during a multi-energy finishing-grinding vibration processing. Int. J. Adv. Manuf. Technol. **106**(9–10), 4339–4353 (2020)
5. Kundrák, J., Mitsyk, A.V., Fedorovich, V.A., Morgan, M., Markopoulos, A.P.: The use of the kinetic theory of gases to simulate the physical situations on the surface of autonomously moving parts during multi-energy vibration processing. Materials **12**(19), 3054 (2019)
6. Schulze, V., Gibmeier, J., Kacaras, A.: Qualification of the stream finishing process for surface modification. CIRP Ann. Manuf. Technol. **66**(1), 523–526 (2017)
7. Vijayaraghavan, V., Castagne, S.: Sustainable manufacturing models for mass finishing process. Int. J. Adv. Manuf. Technol. **86**(1–4), 49–57 (2016)
8. Makiuchi, Y., Hashimoto, F., Beaucamp, A.: Model of material removal in vibratory finishing, based on Preston's law and discrete element method. CIRP Ann. Manuf. Technol. **68**(1), 365–368 (2019)
9. Da Silva Maciel, L., Spelt, J.K.: Influence of process parameters on average particle speeds in a vibratory finisher. Granular Matter **20**(4), 65 (2018)
10. Zhang, C., Liu, W., Wang, S., Liu, Z., Morgan, M., Liu, X.: Dynamic modeling and trajectory measurement on vibratory finishing. Int. J. Adv. Manuf. Technol. **106**(1–2), 253–263 (2020)
11. Lachenmaier, M., Brocker, R., Trauth, D., Klocke, F.: Analysis of the relative velocity and its influence on the process results in unguided vibratory finishing. J. Manuf. Sci. Eng. Trans. ASME **140**(3), 031012 (2018)
12. Hashimoto, F., Johnson, S.P.: Modeling of vibratory finishing machines. CIRP Ann. Manuf. Technol. **66**(1), 313–316 (2017)
13. Kang, Y.S., Hashimoto, F., Johnson, S.P., Rhodes, J.P.: Discrete element modeling of 3D media motion in vibratory finishing process. CIRP Ann. Manuf. Technol. **64**(1), 345–348 (2015)
14. Borovets, V.M., Borovets, Ya.V.: The influence of the working medium on the kinematics of vibration machines. Vibr. Tech. Technol. **2**(82), 10–15 (2016). (in Ukrainian)

15. Fedorovich, V.A., Mitsyk, A.V.: Mathematical simulation of kinematics of vibrating boiling granular medium at treatment in the oscillating reservoir. In: Proceedings of the 7th International Congress of Precision Machining. ICPM, Miskolc, Hungary, vol. 581, pp. 456–461 (2013)
16. Topilnytskyy, V., Rebot, D., Sokil, M., Velyka, O., Liaskovska, S., Verkhola, I., Kovalchuk, R., Dzyubyk, L.: Modeling the dynamics of vibratory separator of the drum type with concentric arrangement of sieves. Eastern-Eur. J. Enterp. Technol. **86**(2/7), 26–35 (2017)

Check for updates

Environmental Impact of Additive Manufacturing for Individual Supplies

Filip Górski(✉), Filip Osiński, Natalia Wierzbicka, and Magdalena Żukowska

Poznan University of Technology, 3 Piotrowo, St., 61–138 Poznan, Poland
filip.gorski@put.poznan.pl

Abstract. The paper presents a comparison between the traditional and modern process of production of customized ankle-foot orthoses, assuming automation of the latter, and taking into account the environmental impact of both methods. Both processes are presented. The authors' proposal of a new, automated way of designing and producing the orthoses using additive manufacturing (3D printing) is shown in greater detail, on the case of a 5-year old patient, for whom orthoses were made using both processes. The modern, automated process, consisting of 3D scanning, then automated data processing and design, as well as 3D printing of a resulting 3D model, was realized successfully, with promising results. The product carbon footprint (PCF) was calculated and compared with a traditionally produced orthosis. The 3D printing, a modern process using a biodegradable PLA material, was proven to have almost three times lower PCF than a traditional one, hence proving the feasibility of this approach in terms of environmental friendliness.

Keywords: 3D printing · Individualized orthoses · Product carbon footprint

1 Introduction

The orthoses are medical supplies used for keeping a selected part of the patient's body rigid and safe during healing or convalescence. Usually, it is realized by immobilizing and protecting from deformations and physical damage of the body around a selected joint. The orthoses may also be used for enforcement of a specific position and mutual orientation of various body parts [1]. There can be universal, relatively inexpensive orthoses or customized (much better in terms of healing function and comfort) products, made for a specific patient based on their anatomical measurement [2].

Orthopedic supplies are more and more often manufactured using additive manufacturing (3D printing) technologies [3–5]. The technology is invaluable when customized shapes must be produced, as no tooling is required. The most popular process of Fused Deposition Modelling is very cheap both in terms of machines and thermoplastic materials available. Therefore, it is widespread in medical applications [6, 7].

In the modern world, environmental impact is a very important factor, often a decisive one in terms of the decision on a specific technology being implemented. The 3D printing processes are relatively scarcely studied. It is known that the FDM process with ABS material produces potentially harmful effects. However, there are

V. Tonkonogyi et al. (Eds.): InterPartner 2020, LNME, pp. 384–393, 2021.
https://doi.org/10.1007/978-3-030-68014-5_38

biodegradable materials, such as PLA (polylactic acid). The process of additive manufacturing also takes less time and materials than the traditional one, when comparing products such as the orthoses mentioned above. This paper contains a comparison of the two approaches in terms of their efficiency and environmental aspect.

The aim of the presented work was to, first of all, verify if it is possible to produce a fitting, working 3D printed ankle-foot orthosis (AFO) using an automated design process. The second main goal was to verify the carbon footprint of such a product, in comparison with a traditional, manual process of composite orthosis manufacturing.

2 Literature Review

An individually made AFO apparatus covers the shin with the foot and ankle. The AFOs are used in all situations requiring long-term immobilization of the foot and ankle: in inflammations, in arthrosis, post-traumatic conditions, treatment of orthopedic operations as temporary protection against contractures, and in cases of early paralysis. The use of AFOs includes both injuries and chronic ankle instability and dysfunctions within it, caused by various factors.

A typical process of manufacturing of orthopedic supplies (such as AFOs) involves manual activities, such as measuring a patient by making a "negative" using a plaster cast, then making a positive model out of it. It is usually done by a manual layering of resin and fabrics of glass fiber, obtaining typical laminate composites [8]. Many manufacturers use this process (Fig. 1, "A"), often the largest companies in the industry. The traditional process results are of low repeatability, as they are dependent on manual skills, time of delivery is therefore very long. The process is difficult to perform, and it generates much waste in the form of the "negative" and "positive" disposable props. Moreover, the obtained laminate is quite difficult to recycle.

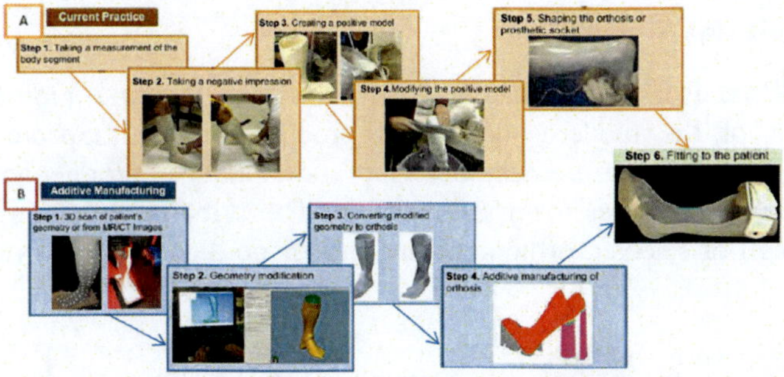

Fig. 1. Traditional (A) and modern (B) process of manufacturing customized orthoses [8].

The modern process (Fig. 1, "B") introduces repeatability, as the patient's anatomy is digitized and stored, using non-contact measurements using 3D scanning. After gathering data, typically, a work of a biomedical engineer is required to design a customized product, maintaining anatomical and technical correctness. Then it can be manufactured, often by industrial 3D printing. It can be made of ABS, using the Selective Laser Sintering technology, having comparable or better properties [9]. They can also be made using the FDM process of bio-degradable material (PLA) [10].

One of the largest problems in 3D printing of customized orthopedic supplies is the requirement of specialized engineering knowledge. The patient anthropometric data must be gathered and processed. It can generate a lot of inaccuracies [11]. Obtaining a shape requires many hours of advanced modeling in CAD systems. Additionally, 3D printing of thermoplastic products with satisfying values of accuracy and strength is difficult [12]. That is why a traditional process of making plaster casts has still not been replaced with 3D printing. There are constant studies on how to make the data gathering, processing, and manufacturing easier and more available in general medical practice. Automation of certain engineering tasks seems a promising direction [13]. As the authors of the paper [14] note, the lack of dedicated, easy-to-use design software significantly limits the possibility of widespread use of additively manufactured orthoses. They offered their technical solution, which is so simple and intuitive that it can also be used by medical personnel with no experience with traditional CAD 3D systems for engineers. The authors of the work [15] suggest that the software for the design of orthoses should use only the simplest modeling operations performed directly on the meshes coming from the scanning process.

The modern process of orthoses generation is more robust, generates less waste, and allows obtaining recyclable products, but it takes time and needs to be performed by a skilled engineer. It makes it much less available for large groups of patients than the traditional process, and the time of delivery can also be very long, although shorter than in the traditional process [13]. Hence, a need for automation arose and was addressed by the authors.

3 Research Methodology

3.1 AutoMedPrint System

The AutoMedPrint is a system in development by the authors, aiming at effective implementation of the modern method of production of orthopedic supplies by automation of patient's measurement, design of individualized product and preparation of its manufacturing process using 3D printers. The principle of the system for the automated design of selected orthopedic and prosthetic devices is shown in Fig. 2.

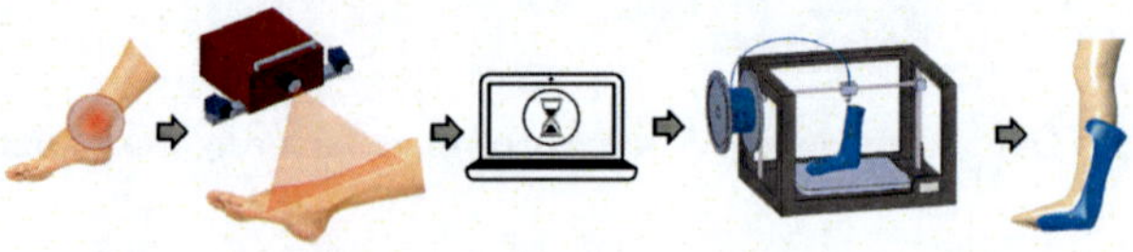

Fig. 2. The AutoMedPrint system – the main idea.

The system allows to implement the following activities:

- defining the basic features of the product (product type and its purpose),
- entering the recipient's anthropometric data by using non-contact measurements, carried out semi-automatically (simple user interface),
- completely automated design of a new customized product for a given recipient using a CAD system, based on collected anthropometric data,
- automatic preparation of the manufacturing program for the additive manufacturing machine based on the developed CAD model,
- manufacturing and assembly process without the participation of the recipient, assisted by technical staff.

3.2 Design of Orthosis

The AFO was designed automatically, according to the process scheme (Fig. 2). The patient was a 5-year old boy, born prematurely and suffering from many conditions, including cerebral palsy. The patient uses AFOs daily. First, the patient was 3D scanned, using a semi-automated scanning workstation, with an assist from a qualified physiotherapist (Fig. 3). The patient's leg was scanned in 8 positions, from different angles using David SLS-3 device. The scanning took approximately 20 min. Then, the patient's data were automatically processed using MeshLab software and a set of macros, prepared earlier by the authors. The scans were joined, trimmed, cleaned, and then the leg's shape was reconstructed using Screened Surface Poisson Reconstruction (Fig. 4a).

Fig. 3. 3D scanning of the patient.

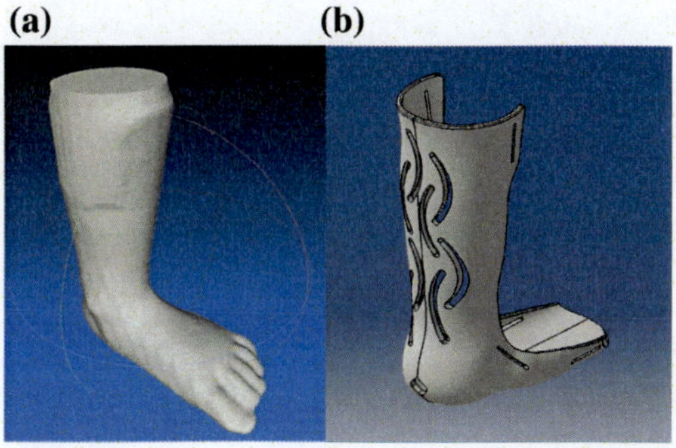

Fig. 4. Result of 3D scanning (a) and automated CAD design (b).

The next step was automatic extraction of point coordinates from the reconstructed mesh (a set of macros in MeshLab), filtering and conversion, to obtain a set of points stored in an Excel spreadsheet. The spreadsheet contains point coordinates used as parameters in an intelligent CAD model created in Autodesk Inventor software. The model changes its shape according to the point data – it is fully customizable with regards to the anatomy of a given patient. The resulting model is shown in Fig. 4b. The automated model processing took approximately 15 min.

3.3 Manufacturing

The product was manufactured additively using Fused Deposition Modelling technology. The machine used was of Delta type – all the movement is realized by the extrusion head located at the end effector of the vertical, inverted Delta robot. The TEVO Little Monster low-cost machine was used for that purpose. PLA (polylactic acid) filament manufactured in Poland was used as a building material. No support structures were used. The layer thickness was 0,3 mm, infill was 30%, and the build orientation was vertical (leg axis approximately parallel to the machine vertical axis). The program was prepared using Simplify3D software. Two pieces of orthosis were made this way, to test the repeatability of the process.

3.4 Calculation of Environmental Impact

The environmental impact measure used was the product carbon footprint (PCF). The 3D printed orthosis was juxtaposed with a traditionally made one, supplied by a professional company. The orthosis was a custom product for the same patient. Both traditional and 3D printed orthoses were analyzed by their environmental impact.

Product Carbon Footprint is the amount of carbon dioxide released to the atmosphere during all Life Cycle Assessment of the product [16] (Fig. 5). Since not all of the emissions during the production process are CO2, other gases have been converted into CO_2eq with the GWP - 100-year Global Warming Potential in comparison to CO2 [17].

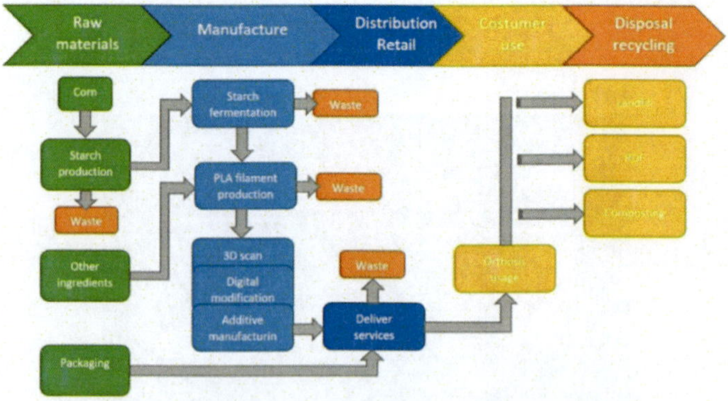

Fig. 5. Life cycle assessment of PLA orthosis shell.

The calculations were carried out based on data available in the literature on GHG emission, energy consumption as well as ready CO_2eq emission indicators for individual production processes and raw materials [18]. In all processes except composting, studies developed within the European Union were used, which allowed for the unification of data in terms of location. In both cases, the weight of raw material for calculations was 250 g. In the case of the laminate orthosis, 10% of the waste was assumed, while FDM production was assumed waste-free.

The following equations were used to calculate PFC at specific stages of the product life cycle [19]:

$$E = \frac{a[Fuel_a \times EF_a] \times M}{L_w}$$

Where: E is the greenhouse gasses emissions converted into CO_2eq in transportation processes, $Fuel_a$ is the total consumption of fuel a for transportation process, EF_a is the emissions factor of fuel a, M is the mass of the transportation part, L_w is the total loading weight of the vehicle.

$$E_e = FC_e \times EF_{eGHG}$$

Where: E_e is the greenhouse gasses (CO_2eq em) emitted during electricity production and distribution, FC_e is the total consumption of electricity (kWh) during work of electrical devices in manufacturing processes, EF_{eGHG} is the emission factor of electricity production. For manufacturing processes, the CO2eq has been was adopted at the level of 0,836 CO_2/kwh –factor calculated for Poland in 2017 [20]. The calculations did not include energy expenditure related to infrastructure maintenance (e.g., heating and lighting of production buildings).

4 Results

4.1 Manufacturing Result

The orthosis was manufactured in a time of 5,5 h. Another 30 min were spent for the manual post-processing, consisting of support structure removal, polishing of sharp edges, and cutting and gluing the EVA foam from a sheet, to create a comfortable, soft surface for the limb (similarly as in the traditional orthoses). A total of 250 grams of PLA material was used in the process, including waste (support at the bottom of the orthosis). The process was fully stable and repeatable, with no direct errors observed. The resulting orthosis is shown in Fig. 6.

Fitting of the product was checked twofold. Firstly, the raw 3D scan was juxtaposed with the orthosis scan – minor collisions were noted, but otherwise, the fitting was achieved. Secondly, the patient was asked to try the orthosis. The product was tested in the presence of a qualified physiotherapist, and generally, it was concluded that it performs its task of keeping the leg rigid and straight properly. However, as the patient was given the orthosis several months after the original scan was performed, slight discomfort was noted, due to changing dimensions of the leg.

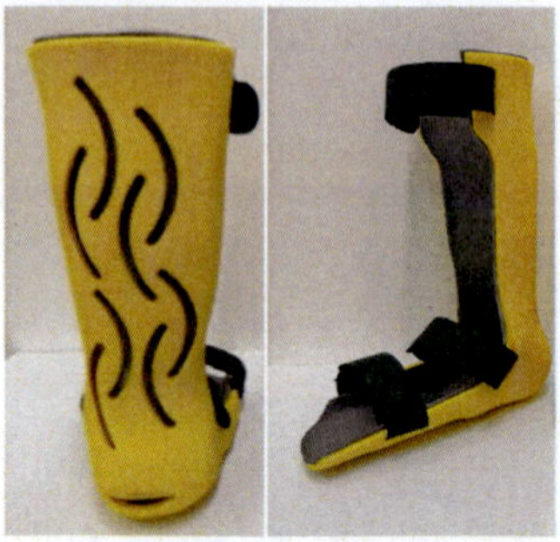

Fig. 6. AFO manufactured additively from PLA material.

No strength tests were made, but during 15 min of use for walking, jumping, and other activities, the orthosis did not break, nor any visible traces of wear were visible. The PLA material used had appropriate properties. As a side note, different materials were tested in other studies – the PLA was the best strength-wise, alongside nylon, as opposed to ABS, which was considerably more prone to break.

After the functional tests, it was assumed that the product fulfills its task, and the modern method based on 3D scanning can be used to replace a traditional method, production-wise.

4.2 Environmental Impact Calculation Results

The results of the calculation are summarized in Table 1. A big difference in stages of completing resources, manufacturing, and end-of-life can be observed in LCA of orthosis shell. The main reason is the origin of the resource. Fiberglass is a product of chemical and physical treatment of silicates. Many of the processes used during production are extremely energy consuming (e.g., batch melting) and produce a lot of gaseous emissions. PLA is the product of polymerization of dextrose extracted mostly from corn and white beets. PLA data considered total carbon balance, including CO2 uptake by plants during their growth.

In the stage of manufacturing, the main part of the carbon footprint for PLA shell took the process of fused deposition modeling – mainly plastic heating. The process took about 5.5 h on a 600 W printer. Because the process of manufacturing a composite shell is mostly manual, most of the CO2eq is the emission of gaseous products of resin curing, e.g., styrene, toluene, and other VOC's [25].

The end-of-life stage refers to the materials used and the possibility of their disposal and recycling. In the case of the AFO orthosis, one of the basic problems associated with the end-of-life cycle is the complexity of the product, and more precisely, the use of many different materials. Each orthosis consists of shells and auxiliary elements, which, however, are the same for both orthoses and can be easily removed from the basic part. The main difference between the two types of production appears in the shells.

In the case of component shells, from the technological and legal (EU) point of view, it is possible only to dispose of landfill sites or recovery at waste incineration plant (adopted in PCF). These methods are the two least desirable processes according to the hierarchy of waste pyramid [26]. The adopted CO_2eq emission factor includes heat and energy recovery in the waste incineration process.

Table 1. CO_2eq in life cycle assessment of orthosis shell, some values from [21–24].

LCA	AutoMedPrint - PLA shell [g CO_2eq/unit]	Traditional process - laminate shell [g CO_2eq/unit]
Resources	PLA filament - 125	Gypsum – 281 Carbon fiber composite - 8250
Transportation	100	140
Manufacturing	FDM -2792 Scanning and design - 253	200
Distribution	70	70
Use stage	0	0
End-of-life stage	Composting - 85 Grinding - 1133	Waste incineration - 400
Total	**3172-4220**	**9341**

In the AutoMedPrint production using PLA, there is a much wider range of development options. A fist is recycling through material regranulation – allow to reuse as a raw material in production processes, possible if good quality filaments and appropriate waste segregation are used. Second is recycling through composting - production of organic compost in industrial composting plants, possible in the case of PLA with a compostability certificate (e.g., OK Compost; EN 13432). It is required due to the use of various dyes, antioxidants, and other additives in different PLA filaments. FDM printed shells can also be used for energy recovery as a Refuse-Derived Fuel (RDF), which can be used as co-incineration in technological processes, e.g. in cement plants. Because of a large variety of PLA waste management processes, less damaging to the environment than disposal of composite, the carbon footprint results are lower.

5 Conclusions

The conducted studies allowed to prove that, first of all, it is possible to utilize the modern process of leg orthosis production, based on 3D scanning and 3D printing, to create a functional product, as well as automate this process to deliver such product in a time of one working day (one hour of 3D scanning and design, 6 h of manufacturing, one hour for final adjustments). That is a huge improvement over the existing methodologies, as well as traditional processes. Moreover, it was proven by approximate calculation of product carbon footprint, that such an orthosis, made of PLA, is far more environmentally friendly than a traditional orthosis, by a factor of three in terms of CO2eq. In other words, the automated process is not only more efficient but also more ecological, which is invaluable in the present day.

There are still problems to solve to enable more widespread use of the process. The 3D scanning itself is a demanding process for the patient and the operator (who does not need to be a qualified engineer), as it requires the patient not to move their lower limb for several minutes, which is difficult for small children (main target group using AFOs). The automated design algorithms have not been tested over a wider population, to prove whether they can generate geometry without errors, and – more important – always fitting to the patient and fulfilling its main purpose of keeping the limb in the right position. These are the main concerns of the authors to be tackled in further studies. However, it has been proven that the assumed approach should be continued.

More studies are needed to validate the findings –further studies will include a larger sample of patients. The project assumes 50 patients will be supplied with orthopedic devices, which will make for a good study sample.

Acknowledgments. The studies were realized with support from the Polish National Center for Research and Development, in the scope of the "LIDER" program (grant agreement no. LIDER/14/0078/L-8/16/NCBR/2017).

References

1. Baronio, G., Volonghi, P., Signoroni, A.: Concept and design of a 3D printed support to assist hand scanning for the realization of customized orthosis. Appl. Bion. Biomech. **2017**, 8171520 (2017). https://doi.org/10.1155/2017/8171520
2. Andringa, A., van de Port, I., Meijer, J.W.: Long-term use of a static hand-wrist orthosis in chronic stroke patients: a pilot study. Stroke Res. Treat. **2013**, 546093 (2013). https://doi.org/10.1155/2013/546093
3. Mavroidis, C., Ranky, R.G., Sivak, M.L., et al.: Patient specific ankle-foot orthoses using rapid prototyping. J. NeuroEng. Rehabil. **8**(1) (2011). https://doi.org/10.1186/1743-0003-8-1
4. Palousek, D., Rosicky, J., Koutny, D., et al.: Pilot study of the wrist orthosis design process. Rapid Prototyping J. **20**(1), 27–32 (2014). https://doi.org/10.1108/RPJ-03-2012-0027
5. Paterson, A., Bibb, J.R., Campbell, R.I., Bingham, G.A.: Comparing additive manufacturing technologies for customised wrist splints. Rapid Prototyping J. **21**(3), 230–243 (2015)
6. Otawa, N., Sumida, T., Kitagaki, H., Sasaki, K., Fujibayashi, S., Takemoto, M., Nakamura, T., Yamada, T., Mori, Y., Matsushita, T.: Custom-made titanium devices as membranes for bone augmentation in implant treatment: modeling accuracy of titanium products constructed with selective laser melting. J. Craniomaxillofac. Surg. **43**(7), 1289–1295 (2015)
7. Banaszewski, J., Pabiszczak, M., Pastusiak, T., Buczkowska, A., Kuczko, W., Wichniarek, R., Górski, F.: 3D printed models in mandibular reconstruction with bony free flaps. J. Mater. Sci. Mater. Med. **29**, 23 (2018)
8. Cha, H.Y., Lee, K.H., Ryu, H.J., Joo, I.W., Seo, A., Kim, D., Kim, S.J.: Ankle-foot orthosis made by 3D printing technique and automated design software. Appl. Bionics Biomech. **2017**, 9610468 (2017). https://doi.org/10.1155/2017/9610468
9. Jina, Y., Plotta, J., Chena, R., Wensman, J., Shih, A.: A review of the traditional and additive manufacturing of custom orthoses and prostheses. Procedia CIRP **36**, 199–204 (2015)

10. http://www.3ders.org/articles/20160614-gonzaga-university-students-develop-affordable-3d-printed-ankle-foot-orthosis-for-children.html. Accessed 30 Oct 2019
11. Huotilainen, E., Jaanimets, R., Valasek, J., Marcian, P., Salmi, M., Tuomi, J., Makitie, A., Wolff, J.: Inaccuracies in additive manufactured medical skull models caused by the DICOM to STL conversion process. J. Craniomaxillofac. Surg. **42**, 259–265 (2014). https://doi.org/10.1016/j.jcms.2013.10.001
12. Górski, F., Wichniarek, R., Zawadzki, P., Hamrol, A.: Computation of mechanical properties of parts manufactured by fused deposition modeling using finite element method. In: Advances in Intelligent Systems and Computing, vol. 368, pp. 403–413. Springer (2015). https://doi.org/10.1007/978-3-319-19719-7_35
13. Górski, F., Wichniarek, R., Zawadzki, P., Wierzbicka, N., Wesołowska, I., Żukowska, M.: Automated design of customized 3D-Printed wrist orthoses on the basis of 3D scanning. In: Okada, H., Satya, N. (eds.) Computational and Experimental Simulations in Engineering: Proceedings of ICCES 2019. Atluri, pp. 1133–1143. Springer International Publishing (2020)
14. Li, J., Tanaka, H.: Feasibility study applying a parametric model as the design generator for 3D–printed orthosis for fracture immobilization. 3D Printing Med. **4**(1) (2018). https://doi.org/10.1186/s41205-017-0024-1
15. Baronio, G., et al.: A critical analysis of a hand orthosis reverse engineering and 3d printing process. Appl. Bion. Biomech. **2016**, 8347478 (2016). https://doi.org/10.1155/2016/8347478
16. ISO, ISO 14067:2018 Greenhouse gases—Carbon footprint of products—Requirements and guidelines for quantification and communication. International Organization for Standardization, Geneva, Switzerland (2018)
17. Pachauri, R.K., Meyer, L.A. (eds.): Climate Change 2014: Synthesis Report. Contribution of Working Groups I, II and III to the Fifth Assessment Report of the Intergovernmental Panel on Climate Change IPCC. Geneva, Switzerland (2014)
18. Cushman-Roisin, B., Tanaka Cremonini, B.: Useful numbers for environmental studies and meaningful comparisons useful numbers for environmental studies and meaningful comparisons. Dartmouth College (2019)
19. Wang, S., Wang, W., Yang, H.: Comparison of product carbon footprint protocols: case study on medium-density fiberboard in China. Environ. Res. Public Health **15**(10), 2060 (2018)
20. Carbon Footprint homepage, https://www.carbonfootprint.com/. Accessed 27 Jan 2020
21. Deesing, B.: Comparing greenhouse gases from composting and landfilling. In: Proceedings of the National Conference on Undergraduate Research, Ashville (2016)
22. Bałazińska, M., Zuwała, J., Tokarski, S.: Carbon foot related to transport of fuels for energy purposes. Rynek Energii **4**, 68–73 (2013)
23. Fořt, J., Černý, R.: Carbon footprint analysis of calcined gypsum production in the Czech Republic. J. Clean. Prod. **177**, 795–802 (2018)
24. Vähk, J.: The impact of Waste-to-Energy incineration on climate. Policy Briefing, Zero Waste Europe (2019)
25. Malik, M., Choudhary, V., Varma, I.K.: Current status of unsaturated polyester resins. J. Macromol. Sci. **40**, 139–165 (2000)
26. Cole, C., Gnanapragasam, A., Coopera, T., Singh, J.: An assessment of achievements of the WEEE Directive in promoting movement up the waste hierarchy: experiences in the UK. Waste Manage. **87**, 417–427 (2019)

Minimizing Surface Roughness and Radius Error in Laser Cutting of EN10346 Steel Plate

A. Mustafa Kangal[1], Alper Uysal[1(✉)], Eshreb Dzhemilov[2], and Ruslan Dzhemalyadinov[2]

[1] Yildiz Technical University, 34349 Besiktas Istanbul, Turkey
auysal@yildiz.edu.tr
[2] Crimean Engineering and Pedagogical University, 8 Uchebnyy side St, Simferopol 29501, Republic of Crimea, Russia
eshrebdzhemilov@gmail.com

Abstract. In this study, the effects of cutting parameters on radius error and surface roughness during laser cutting of EN10346 (DX51D + Z) hot-dip rolled galvanized steel plate were discussed. The process parameters were chosen based on the literature and technical catalog of the laser machine tool. The experiments were planned according to the Taguchi L_9 design of experiments to reduce the number of experiments. After laser cutting operations, the surface roughness values were measured by the Mituyoto Surftest SJ-210 surface roughness measurement device, and the edge radius values were measured by Mshot MD30 microscope. The optimum cutting parameters for minimum surface roughness were calculated as laser power of 4 лW, feed rate of 5000 m/min, frequency of 4 kHz, and focus of −0.5. This combination was confirmed with the minimum surface roughness of 2.241 μm. The optimum cutting parameters for minimum radius error were calculated as laser power of 2 kW, feed rate of 4000 mm/min, frequency of 5 kHz, and focus of −0.8. This result was also confirmed with the minimum radius error of 0.08%.

Keywords: Laser cutting · EN10346 steel plate · Radius error · Surface roughness

1 Introduction

Modern requirements in manufacturing methods demand increasing productivity and part quality, including the stage of blank productions, one of the most common operations of which is the cutting of sheet materials. These requirements can be ensured by using physicomechanical methods of processing materials, which include laser cutting. Nowadays, fiber lasers are widely used due to having some advantages compared to CO_2 lasers, such as processing speed and low maintenance. Unlike traditional machining methods, laser cutting machine tools have a wide range of regulation of cutting parameters, which include feed rate, power, frequency, focus, a diameter of the laser nozzle, and assist gas. In literature, a lot of work has been performed to improve the quality during laser cutting of sheet materials.

V. Tonkonogyi et al. (Eds.): InterPartner 2020, LNME, pp. 394–401, 2021.
https://doi.org/10.1007/978-3-030-68014-5_39

2 Literature Review

Scintilla [1] presented the influences of cutting speed, focus position, and assist gas pressure on the scale height, mortise width, and roughness parameters in laser cutting. Chen et al. [2] investigated the laser processing of the AA2219 aluminum alloy and identified three stratification zones as horizontal striation zone (HSZ), vertical striation zone (VSZ), and oblique striation zone (OSZ) in sequence. After laser cutting, the scale height was also measured to reveal the influence of processing parameters on surface morphology. Shin et al. [3] performed the experimental studies in fiber laser cutting of thick-walled stainless steel plates (thickness of 60 mm) by using a new design of fiber laser cutting head. Olsen et al. [4] presented the first results of proof-of-principle multi-beam fiber laser head in laser cutting with high brightness and short-wavelength lasers. Burr free cuts in steel and aluminum materials having a thickness of 1 mm and in AISI 304 stainless steel having thicknesses of 1 mm and 2 mm were demonstrated over a wide range of cutting rates. Oh et al. [5] investigated the laser cutting of carbon fiber reinforced plastic (CFRP) sheets. The main parameters were selected as the width of the cut, the depth of cut, the width of matrix evaporation, the width of matrix recession, the angle of the cut cone, the damage zone of the matrix, and the morphology of the cut surface. It was found that an increase in cutting speed and the number of passes were the main parameters to minimize thermal damage to polymeric materials. Riveiro et al. [6] focused on the role of auxiliary gas in the process of laser cutting by fusion. The aerodynamic interactions between the auxiliary gas and the workpiece were considered. According to the results, the shock waves, choking, and boundary layer separation caused a reduction in the quality and productivity of the process. In some studies [7, 8], the researchers used an InGaAs photodiode sensor system for quality control during laser cutting in the near-infrared range when measuring thermal radiation in the process zone. In some studies [9–15] the influence of cutting parameters was evaluated to achieve the optimal accuracy and surface roughness in laser cutting of different steel plates.

In this paper, the influences of cutting parameters on surface quality when laser cutting EN10346 (DX51D+Z) hot-dip rolled galvanized steel plate were investigated. The laser power, feed rate, frequency, and focus were selected as the process parameters.

3 Research Methodology

In the experiments, EN10346 (DX51D+Z) hot-dip rolled galvanized steel plate having a tensile strength of 377 N/mm^2, the yield point of 307 N/mm^2 and elongation of 36% was selected as workpiece, and its chemical composition was given in Table 1.

The laser cutting parameters were chosen based on the literature and technical catalog of the laser machine tool and given in Table 2. Apart from these parameters, there are some constant parameters, which are the nozzle (diameter of 2 mm and conical shape), the gas (Nitrogen, N_2), and the gas pressure (15 bar). Besides, the experiments were planned according to the Taguchi L_9 design of experiments, as seen in Table 3.

Table 1. The chemical composition of EN10346 (DX51D+Z) hot-dip rolled galvanized steel plate workpiece.

C (%)	Mn (%)	P (%)	S (%)	Si (%)	Al (%)	Tl (%)	Cr (%)	Cu (%)	Ni (%)
0.045	0.193	0.009	0.009	0.001	0.051	0.001	0.029	0.034	0.047

The objective of Taguchi Methods is to improve product quality and reduce cost-efficiently in real-life industrial applications. The following key elements can summarize Taguchi Methods: (1) uncompromised quality improvement and cost reduction based on social justice; (2) engineering creation to improve the living standards based on social responsibility; and (3) keen observation of experimental results and sufficient mathematical expressions and analyses to drive engineering improvement [16].

Table 2. Laser cutting parameters.

Feed rate (mm/min)	Laser power (W)	Frequency (Hz)	Focus
4000	2000	3000	−0.5
5000	3000	4000	−0.8
6000	4000	5000	−1.0

Table 3. Taguchi L_9 design of experiments.

Experiment number	Process parameters			
	Feed rate (mm/min)	Laser power (kW)	Frequency (kHz)	Focus
1	4000	2	3	−1.0
2	4000	3	4	−0.8
3	4000	4	5	−0.5
4	5000	2	4	−0.5
5	5000	3	5	−1.0
6	5000	4	3	−0.8
7	6000	2	5	−0.8
8	6000	3	3	−0.5
9	6000	4	4	−1.0

The cutting operations were conducted by using Durma HD-F 3015 fiber laser machine tool, as shown in Fig. 1. The dimensions of the cut samples were 30 × 30 mm, the thickness was 2.0 mm, and the edge radii were 2.5 mm.

After laser cutting operations, the surface roughness values were measured by the Mituyoto Surftest SJ-210 surface roughness measurement device, as seen in Fig. 2. Six measurements were performed on each sample to determine average surface roughness values. Mshot MD30 microscope were utilized to measure the edge radius values and then absolute radius errors were calculated.

Fig. 1. The view of laser cutting operation.

Fig. 2. Surface roughness measurement.

4 Results

During Taguchi analysis, the "Smaller is better" option was chosen to obtain optimum process parameters for minimizing the radius error and surface roughness based on Eq. 1. The signal-to-noise (S/N) results for surface roughness were given in Fig. 3.

$$S/N = -10\log\left[\frac{1}{n}\sum\nolimits_{i=1}^{n} y_i^2\right] \tag{1}$$

where y_i denote the n observations of the response variable.

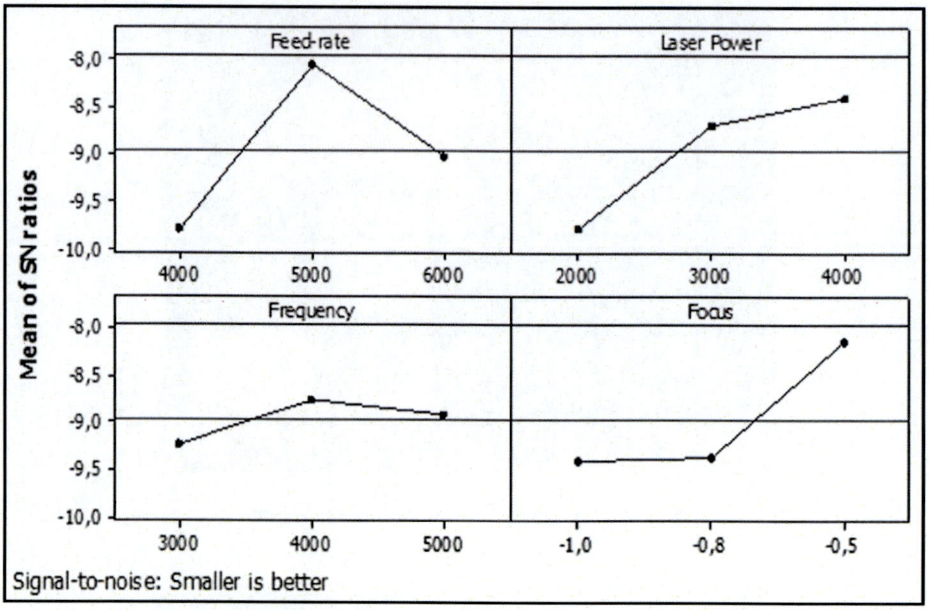

Fig. 3. S/N results for surface roughness.

Depending on the results, the optimum process parameters for the minimum surface roughness are laser power of 4 kW, the feed rate of 5000 mm/min, frequency of 4 kHz, and focus of −0.5. This parameter combination is not included in the design of the experiment. Therefore, a confirmation experiment was performed with this combination. According to the confirmation experiment, the surface roughness was measured as 2.241 µm. This result showed that the suggested combination gave the minimum surface roughness.

In Table 4, there are delta values, and the highest delta value corresponds to the most effective cutting parameter on the surface roughness. Accordingly, the most effective parameter on the surface roughness was found as feed rate, whereas frequency was determined as the least effective parameter.

In addition to analyzing the surface roughness, the radius error was also investigated in the laser cutting process. The S/N results for radius errors were given in Fig. 4.

Table 4. Response table for S/N ratios of surface roughness.

Level	Feed rate (mm/min)	Laser power (W)	Frequency (Hz)	Focus
1	−9.805	−9.807	−9.235	−9.407
2	−8.067	−8.702	−8.753	−9.376
3	−9.041	−8.404	−8.926	−8.130
Delta	1.739	1.403	0.482	1.276

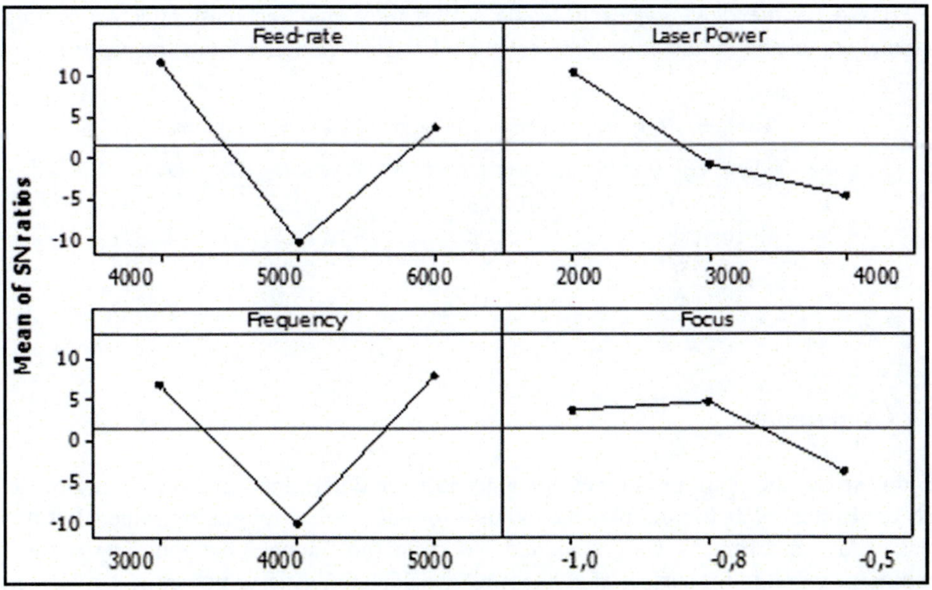

Fig. 4. S/N results for radius error.

Based on the results, the optimum cutting parameters to minimize the radius error are laser power of 2 kW, the feed rate of 4000 mm/min, frequency of 5 kHz, and focus of –0.8. For this parameter combination, a confirmation experiment was conducted, and the radius was measured as 2.502 mm as given in Fig. 5. The radius error was determined as 0.08%. This result showed that the suggested combination gave the minimum radius error.

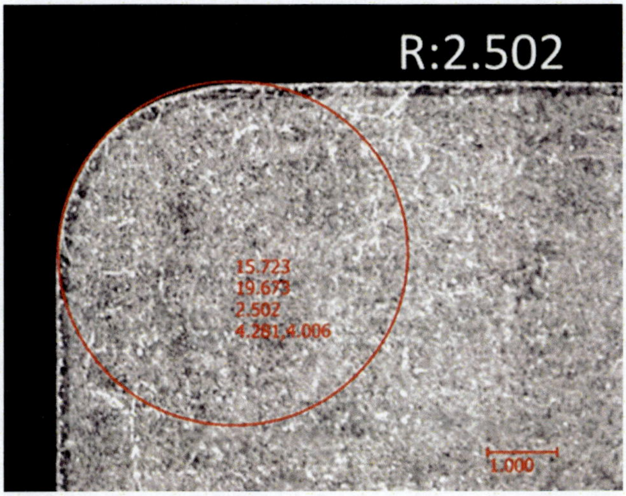

Fig. 5. Radius value for the optimal parameter combination.

Based on the delta values in Table 5, the most and the least effective cutting parameters on the radius error were calculated as feed rate and focus, respectively.

Table 5. Response table for S/N ratios of surface roughness.

Level	Feed rate (mm/min)	Laser power (W)	Frequency (Hz)	Focus
1	11.6861	10.4023	6.9232	3.7600
2	−10.4288	−0.8208	−10.1159	4.7877
3	3.5556	−4.7687	8.0055	−3.7348
Delta	22.1148	15.1711	18.1214	8.5225

5 Conclusions

In this study, the optimum process parameters were determined in laser cutting of the EN10346 steel plate to minimize the radius error and surface roughness values. For this reason, the experiments were designed based on the Taguchi method. The optimal cutting parameters for surface roughness were calculated as laser power of 4 kW, feed rate of 5000 mm/min, frequency of 4 kHz, and focus of −0.5. This combination was confirmed with the minimum surface roughness of 2.241 μm. The optimal cutting parameters for radius error were calculated as laser power of 2 kW, the feed rate of 4000 mm/min, frequency of 5 kHz, and focus of −0.8. This result was also confirmed with the minimum radius error of 0.08%. By selecting the optimum process parameters, the efficiency of the process can be increased while providing economic benefits for industrial applications. In addition, the most and the least effective laser cutting parameters on the radius error and surface roughness values were found. The feed rate was found the most effective parameter for both but the least effective parameter on the surface roughness was determined as frequency whereas it was found as focus for radius error.

References

1. Scintilla, L.D.: Experimental investigation on fiber laser cutting of aluminium thin sheets. Opt. Eng. **53**(6), 066113 (2014)
2. Chen, C., Gao, M., Jiang, M., Zeng, X.: Surface morphological features of fiber laser cutting of AA2219 aluminum alloy. Int. J. Adv. Manuf. Technol. **86**, 1219–1226 (2016)
3. Shin, J.S., Oh, S.H., Park, H., Chung, C., Seon, S., Kim, T., Lee, L., Choi, S., Moon, J.: High-speed fiber laser cutting of thick stainless steel for dismantling tasks. Opt. Laser Technol. **94**, 244–247 (2017)
4. Olsen, F.O., Hansen, K.S., Nielsen, J.S.: Multibeam fiber laser cutting. J. Laser Appl. **21**, 133–138 (2009)
5. Oh, S., Lee, I., Young-Bin Park, Y., Ki, H.: Investigation of cut quality in fiber laser cutting of CFRP. Opt. Laser Technol. **113**, 129–140 (2019)
6. Riveiro, A., Quintero, F., Boutinguiza, M., Val, J., Comesaña, R., Lusquiños, F., Pou, J.: Laser cutting: a review on the influence of assist gas. Materials **12**(1), 1 (2019)

7. Adelmann, B., Schleier, M., Neumeier, B., Hellmann, R.: Photodiode-based cutting interruption sensor for near-infrared lasers. Appl. Opt. **55**, 1772–1778 (2016)
8. Schleier, M., Adelmann, B., Neumeier, B., Hellmann, R.: Burr formation detector for fiber laser cutting based on a photodiode sensor system. Opt. Laser Technol. **96**, 13–17 (2017)
9. Bernát, R., Záležák, Z., Žarnovský, J., Kecskés, N., Peniaško, M., Midor, K.: Monitoring the quality of laser cutting. Multi. Aspects Prod. Eng. **1**(1), 137–143 (2018)
10. Gao, W., Lei, M., Li, B., Li, G., Li, K., Feng, Q., Wang, J.: Investigations on the laser cutting of LiNbO3. Optik **201**, 163508 (2020)
11. Kumar, S., Govindarajalu, J., Thilak, M.: Parametric analysis of laser cutting of mild steel material. J. Chem. Pharm. Sci. **10**(1), 385–388 (2017)
12. Wandera, C., Kujanpaa, V.: Characterization of the melt removal rate in laser cutting of thick section stainless steel. J. Laser Appl. **22**, 62–67 (2010)
13. Sołtysiak, R., Sołtysiak, A., Wasilewski, P.: Development of laser cutting technology with high quality of the cut surface. In: 5th International Scientific and Business Conference — Future Engineering 2019, pp. 111–119. DEStech Transactions, Ołtarzew, Poland (2019)
14. Ahn, D., Byun, K.: Influence of cutting parameters on surface characteristics of cut section in cutting of Inconel 718 sheet using CW Nd: YAG laser. Trans. Nonferrous Metals Soc. China **19**, 32–39 (2009)
15. Runchev, D., Zdraveski, F., Ivanova, I.: Influence of cutting parameters on the quality of the cut surfaces of steel switch a laser beam. Adv. Technol. Mater. **44**, 21–26 (2019)
16. Mori, T.: Taguchi Methods: Benefits, Impacts, Mathematics, Statics, and Applications. Translated by Tsai, S.C. ASME Press, New York (2011)

Solid Lubricants Used in Small Diameter Drilling

Natalia Lishchenko[1], Vasily Larshin[2(✉)], and Irina Marchuk[3]

[1] Odessa National Academy of Food Technologies, 112 Kanatna St., Odessa 65039, Ukraine
[2] Odessa National Polytechnic University, 1 Shevchenko Avenue, Odessa 65044, Ukraine
vasilylarshin@gmail.com
[3] Lutsk National Technical University, 75 Lvivska St., Lutsk 43018, Ukraine

Abstract. Lubricants are either design materials or additional means for metalworking. In the first case, the lubricants should reduce friction and wear of working machines parts. In the second case, they should improve the machinability of the part material when machining this part by cutting or grinding. The paper studies the features of the solid lubricants based on stearic and oleic fatty acids. A comparative experimental study of both solid and liquid lubricants is carried out on a special friction stand. It is established that the solid lubricants provide a lower coefficient of friction and have the technological advantage in drilling small holes without the use of drilling fluid. The effect of molybdenum diselenide, sulfur, and serpentinite additives on drilling torque and axial cutting force, as well as on the increase in the life of drill bits, has been experimentally studied. The use of an electron microscope for chemical analysis made it possible to establish the solid lubricant components diffusion into the surface layer of the drill cutting blade. The solid lubricant formulations experimental studies are performed on modern CNC machine tools using a modern computer data acquisition system NI-DAQ_{mx} (hardware) with NI-LabVIEW (software). Each experiment was repeated the required number of times to reduce the influence of random errors on the measurement results.

Keywords: Metalworking fluid · Friction coefficient · Serpentinite powder

1 Introduction

Technical lubricants (solid, liquid, gaseous) in engineering and technology are used in two main functional areas. The first area is that when the lubricant is design material that ensures functioning machines and mechanisms at the stage of their operations (lubricants to decrease friction and wear). The second area is that when the lubricant is a necessary additional means for improving the machinability of machine parts by cutting and grinding of these parts at the stage of their manufacturing (lubricants for metal cutting and grinding).

The functional requirements for the lubricants are due to these two directions. In the first case, the lubricants should reduce friction and wear of parts during the machine's

V. Tonkonogyi et al. (Eds.): InterPartner 2020, LNME, pp. 402–411, 2021.
https://doi.org/10.1007/978-3-030-68014-5_40

operation, and contribute to an increase in their durability. In the second case, the lubricants must be destructive concerning the material being machined, i.e., they should facilitate the machining of these materials, and concurrently they should be stabilizing and resource-restorative concerning the cutting tool. In both cases, these lubricants must be environmentally safe and comply with sanitary standards for their use.

The paper focuses on such aspects of the performance properties of the lubricants, which are not adequately reflected in the literature, namely on the features of the solid lubricants (hereinafter SL or SLs for single and plural, respectively) based on technical stearic and oleic fatty acids. Such SLs are environmentally friendly; they are cheaper in terms of the cost of their production and operation.

A comparative experimental study of both SLs and metalworking fluids (hereinafter MWF or MWFs) on a special friction stand was carried out. It is established that SL provides a lower coefficient of friction. Also, SLs have the technological advantage when drilling small holes (diameter less than 5 mm) without MWF (i.e., drilling fluid), particularly when drilling holes in parts of stainless steel.

The role of friction modifiers (molybdenum diselenide, sulfur, serpentinite) in increasing the lifecycle of drill bits made from tool steel, as well as in reducing torque and axial cutting force in drilling, under otherwise equal conditions, has been experimentally studied.

The effectiveness of SL for surface grinding workpieces of hardened steels in the combined use of the SL and conventional MWF is shown. In this case, the SL improves the machinability of the workpiece material in the grinding system, while the MWF cools the workpiece being ground to prevent its temperature defects.

Additional features of the research performed are the use of an electron microscope for drill bits chemical analysis and automated research computer system. This system includes the modern computer data acquisition device NI-DAQ$_{mx}$ (hardware) with NI-LabVIEW (software) based on which the cutting monitoring and diagnosing computer subsystems can be developed for relevant CNC machines.

2 Literature Review

The literature analysis allowed establishing two main classes of technical lubricants according to their functional application. The first lubricant class (nondestructive) represents the technical lubricant as a working construction material without which the work of the machine interacting parts is impossible. For example, the interaction of the bearing shell with the engine crankshaft is impossible without the introduction of oil into the contact area of these parts. The second lubricant class (destructive) represents the technical lubricants with the aid of which the machinability of the material by cutting and grinding is greatly improved. Concurrently, the cutting forces and the temperature in the contact zone are reduced, and the cutting tool life is increased.

The first class of lubricants (nondestructive) is discussed in detail in the works [1–4]. In most tribological applications, liquid or grease (semisolid) lubricants are used to reduce friction and wear. Anti-wear additives and friction modifiers are added to them to increase the efficiency of lubricants. These additives are active ingredients that can be added to base oils during the mixing process to enhance their existing characteristics or to add new properties that they lack. There are new trends in this class of lubricants, namely: anti-wear antifriction composition with natural materials of the serpentine group [2].

Lubricants for metal cutting and grinding (liquid, grease, and solid) are working materials that affect the cutting and grinding system performances (operation quality). Besides, lowering the temperature in the cutting and grinding zone is also one of the main tasks [5, 6]. MWF is the conventional choice to act as both lubricant and coolant. There are some trends in this class of the lubricants which is related to limiting the MWFs use. The new technique of minimum quantity lubrication (MQL) is proposed to ensure reduced MWF flow [7–9]. MQL uses the minimal quantity of MWF, which is jetted with high-pressure gas to the cutting or grinding zone.

Adding to MWF, about 2% of the special means named as 'antifriction regenerative composition' (ARC) increases the life of both the cutting tool and the grinding wheel in 2–20 times [10]. The ARC is similar to anti-wear antifriction composition and also includes the serpentine substance. The similar mineral-containing additive reduced friction and wear, showing the opportunity to formulate the so-called energy-saving lubricants [11]. At the same time, the use of MWFs is potentially dangerous (a source of pollution and work-related diseases) and expensively. Hence, there arises a need to identify eco-friendly and user-friendly alternatives to conventional cutting fluids [12]. Modern tribology has facilitated the use of SLs as an alternative to conventional MWFs in machining. For example, the benefits of SLs for cutting are noted in work [13], whose authors investigate the effect of the SLs with graphite and molybdenum disulfide on surface quality, cutting forces, and specific energy in end milling on the experimental setup for SL powder assisted machining. The fine SL powder, with 2 μm average particle size, was loaded into the hopper of the feeder directly without any binder (base).

So, the machining process performance of SLs containing serpentinite has not been studied yet, while studies of liquid lubricants with the serpentine substance have shown positive results with reduced friction and wear. The literature mentioned, as well as total reviews [14, 15], did not reflect the direction associated with the use of the metalworking SLs based on oleic and stearic fatty acids, including the use of serpentinite powder (more pure mineral than serpentine), with about 1 μm average particle size, as one of the SL components both in drilling and grinding.

3 Research Methodology

As it is known from experience, when drilling small holes with a diameter of less than 5 mm, the drill bit does not require additional cooling, since the heat released during cutting is absorbed by the workpiece being drilled. It can be explained as follows: around the rotating drill bit, there is a metal environment of the material being drilled, which takes heat from the drill off. In this case, SLs used in drilling have a significant effect on the lubricating, chemical, and adsorption actions. Most often, SL is applied to the working surface of the tool (drill bit, tap, abrasive wheel, etc.) with a mechanical touch and pressure. For comparative tests of solid and liquid lubricants, an experimental setup in the form of a laboratory stand was created and used with the friction scheme of the 'ring end face-plane' (Fig. 1).

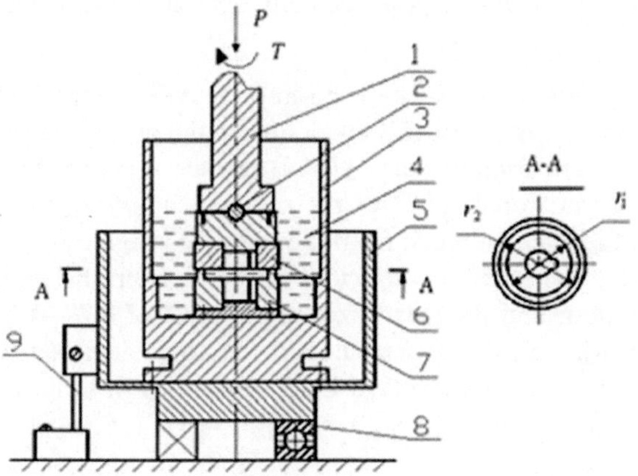

Fig. 1. Scheme of the stand for studying friction and wear: 1 – spindle; 2 – centering ball; 3 – inner cup; 4 – liquid lubricant; 5 – external cup; 6 – moving sample; 7 – fixed sample; 8 – ball thrust bearing; 9 – beam with strain gauges.

The materials from which samples 6 and 7 (Fig. 1) were made corresponding to the pair of 'instrumental material' (R6M5, Germany analog 1.3343, HRC 60…63) – 'processed material' (steel 45, Germany analog C45E, HB 200…212). The formula calculated the coefficient of friction

$$f = \frac{T_f}{P \cdot r}, \tag{1}$$

where T_f is the moment of friction, which is fixed by strain gauges, N·m; P is the axial load, which is given by the corresponding loads on the handle of the vertical feeding of the spindle, N; r is the average radius of friction ring, m. The inner r_1 and outer r_2 radiuses of the ring are 15 and 12 mm, respectively; therefore, $r = 13.5$ mm.

To increase the reliability of experimental data, each test was performed on the stand (Fig. 1) three times when loading and unloading the friction pair with subsequent averaging of the result. The subsequent tests are carried out on the coordinate boring machine 24K40SF4 equipped with 'Flex NC' system when drilling the blind holes (drilling length 10 mm) in a workpiece of steel 35 (Germany analog C35E, HB 190… 210). Drilling parameters: drill speed 2000 rpm; vertical feed 50 mm/min. Drill bits with a diameter of 2.85 mm are made of steel P18 (Germany analog 1.3355). The number of drilled holes for each SL formulation was equal to 30 pcs. Each experiment was also repeated three times, i.e., each SL formulation to be tested was successively tested with three different drills (taken from one batch of the drills) with further averaging of the measured parameters over 90 drilled holes.

4 Results

At the first stage, exploratory, experimental studies of the friction coefficient in friction pairs were performed using both liquid and solid lubricants. The technique of this study is as follows.

1. For one hour in industrial oil I-20, a run-in of samples 6 and 7 (Fig. 1) was made (Fig. 2 a) with the registration of the friction moment and temperature of the oil. The axial load in the friction pair (470 N) creates a pressure of 1.8 MPa in the contact area. The rotational speed of the moving sample 6 is 450 rpm. The axial load is equal to the average value from the possible range of its variation. It can be seen (Fig. 2a) that the friction coefficient increases from the initial value of about 0.14 within one hour and then stabilizes at the level of 0.22–0.23.
2. After the run-in was up, the dependence of the coefficient of friction on the axial load was determined using the following formulations of both the liquid and solid lubricants (Fig. 2b):

- I-20 (industrial oil);
- I-20 + 2% ARC (the antifriction regenerative composition of the company 'Venture –N' in the kind of tribo-polymer-forming additive EF-357);
- SL based on stearin (25% oleic acid, 10% chromium oxide, stearin – the rest).

Let's analyze the results of the experiment on the stand, for example, at a load of 1100 N (Fig. 2b). The highest friction coefficient (minimum value is $f = 0.120$) was obtained by testing the friction pair on I-20 oil, and the lowest (minimum value $f = 0.04$) – on SL. The intermediate result ($f = 0.1$) was obtained on the compound (I-20 + ARC). In this regard, further studies were carried out to optimize the SL formulation.

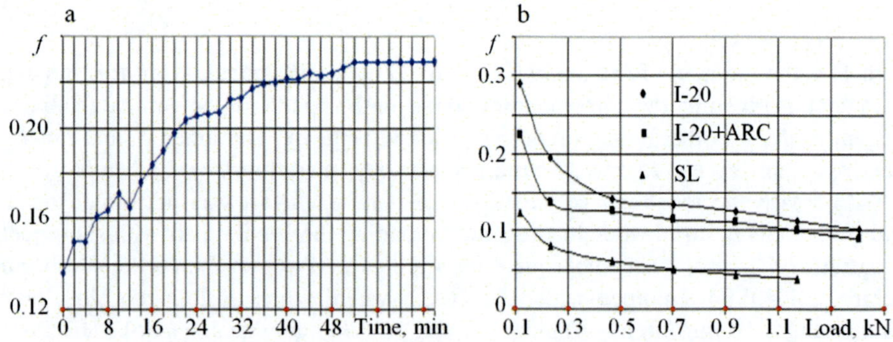

Fig. 2. Friction coefficient f vs. running-in time (a) and vs. the load applied (b).

In the drilling, the problem is to carry out a comparative analysis of various SLs in order to determine the most effective of them, for example, in the drilling small-diameter holes with the diameter up to 5 mm. The initial data for this analysis are the results of measurements of parameters characterizing the cutting process in drilling.

In small holes drilling, these parameters include the torque on the drill *T* and axial cutting force *F*.

To carry out such an analysis, it was used the measurement means, including sensors, data acquisition computer device NI-DAQmx with the appropriate software NI-LabVIEW. Both the sensors that were built into the drilling system and primary information converters had the necessary sensitivity and linearity of their characteristics. The absence of such means or their low quality would not allow obtaining reliable results about the effectiveness of the cutting process with the use of different SL formulations.

In the experiment, the torque *T* and axial cutting force *F* during drilling were measured with a frequency of 200 Hz. As the electrical *T* and *F* signals, there were used the signals which were proportional to the quadrature currents of the respective machine tool electro-drives: main motion electro-drive (for the *T* signal) and vertical feed electro-drive (for the *F* signal). These signals are used in these electro-drives for precision automatic control of the speed of the corresponding asynchronous motors. In the 'Flex NC' CNC system, these signals are made available for diagnostics of the machining system (drilling system diagnosing). When the CNC machine tool operates, the measurement results T_Σ and F_Σ files with the '.ppl' extension are automatically generated as Fig. 3 shows.

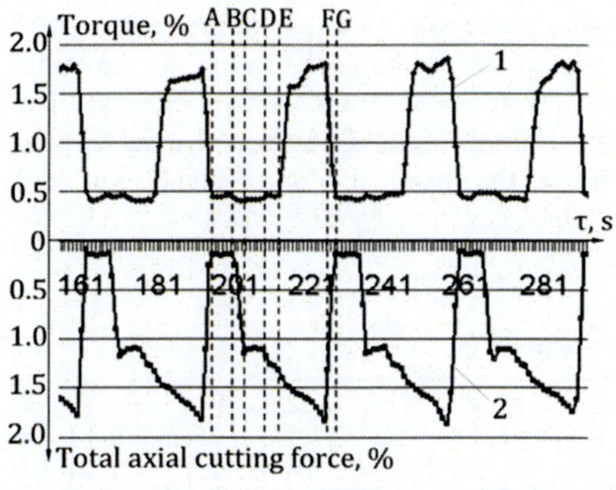

Fig. 3. Total torque T_Σ (curve 1) on the machine tool spindle and total axial cutting force F_Σ (curve 2) signals after 'sliding averaging': A-G is the time interval of the drilling cycle.

Curve 1 (Fig. 3) shows an area of idling (interval A-D), cutting area (E-F), and tool (drill bit) output area (F-G). On curve 2, the idling section has two levels (intervals A-B and C-D). It is due to the switching of the axial feed from the accelerated value (A-B) to the working value (C-D). The time interval (D-E) corresponds to the infeeding time of the drill bit into the workpiece. It is seen that this interval is accompanied by an increase in the total axial cutting force F_Σ, while the total torque T_Σ is at the level of idling (does not increase).

The following steps of processing measurement information were developed.

1. In the Excel software, open the primary measurement data file with the extension '.ppl' and delete the service information.

2. Choose a method of forming an approximating dependence necessary to solve the problem, for example, 'sliding averaging', which allows selecting a trend from current high-frequency measurements. To realize the 'sliding averaging', the calculation algorithm was carried out according to 100 previous and 100 subsequent signals readouts. At the same time, the number of points corresponding to drilling, for example, 30 holes, is reduced 100 times. To form the signal trend, the following formula for the 'sliding averaging' was used

$$y_p = \frac{1}{200} \sum\nolimits_{i=[100(p-1)]+1}^{100p+100} B_i, \tag{2}$$

where B_i is the current signal level in the corresponding units; p is the order number of the average signal values; y_p is the p-th value of the signal average.
3. Since in Excel software, the source data files (.ppl) are limited to the maximum number of measurement points, it is necessary to combine these files into a single one using the file manager, e.g., FAR manager.
4. In MATLAB media create *m*-file (that is a program in MATLAB language) that implements the selected data processing algorithm (the 'sliding averaging'). To process the resulting single file, launch MATLAB program, in the working window of which you must enter the name of the *m*-file and the linked single file (.ppl). The final file obtained after processing (it's read in Excel) has two columns of new data corresponding to two information signals: torque T_Σ and axial cutting force F_Σ.
5. In MATLAB media, create *m*-file that allows selecting available information of the trend in both useful drilling torque T and useful axial drilling force F. Thus, it is possible to watch (trace) the change of T and F signals within the drilling interval of 30 holes (Fig. 4).

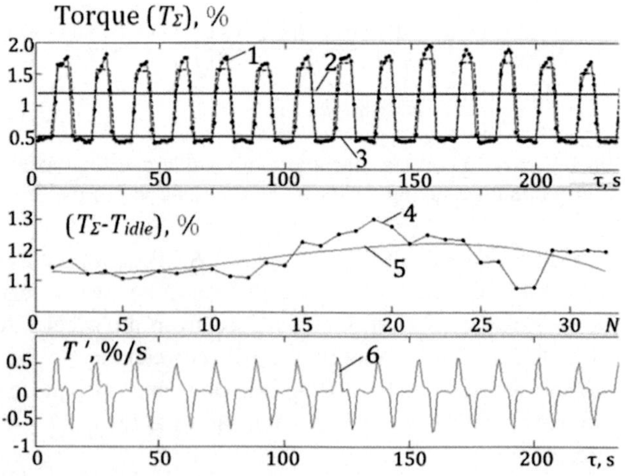

Fig. 4. Mathematical models and levels of the torque signal: 1 – total torque; 2, 3 – lines of the upper and lower thresholds, 4 – useful torque, 5 – useful torque trend; 6 – the derivative of the total or useful torque as a synchronous signal.

The magnitude of the useful signals T and F was determined as the difference between their total and idling levels. For example, the total torque when drilling T_{Σ} and the idle one T_{idle}, which is determined in the absence of drilling. The difference (T_{Σ} - T_{idle}) was determined before the start of cutting and after its completion, i.e. two times when drilling one hole (Fig. 5, a). The level T_{idle} is not constant, for example, due to the warming up of the drive mechanisms when parts of the mechanisms are heating during cutting and cooling during idle, etc. Similar signal transformations were made when determining the useful signal of axial (vertical) drilling force F (Fig. 5, b).

The following seven candidate SL formulations (SL1-SL7) presented below were tested. SL1: oil I-20 – 15%; oleic acid – 20%; stearin – the rest. SL2: oil I-20 – 10%; oleic acid – 20%; ARC (firm 'Venture-N') – 2%; stearin – the rest. SL3: oil I-20 – 30%; technical sulfur – 40%; stearin – the rest. SL4: oil I-20 – 3%; chromium oxide – 20%; stearin – the rest. SL5: oil I-20 – 2…3%; technical sulfur – 30…40%; molybdenum diselenide – 10…20%; polyisobutylene – 0.5…1%; stearin – the rest. SL6: stearin – 60…65%; oleic acid – 20…25%; acetanilide – the rest. SL7: oil I-20 – 10%; oleic acid – 20%; serpentinite – 3%; stearin – the rest.

Figure 5 shows the final results obtained for both T (Fig. 5, a) and F (Fig. 5, b) parameters. On the left part of Fig. 5 (Fig. 5, a) we have: 1 – 'dry' drilling; 2 – SL4; 3 – SL2; 4 – SL1; 5 – SL6; 6 – SL5; 7 – SL7; 8 – SL3. On the right part of Fig. 5 (Fig. 5, b): 1 – 'dry' drilling; 2 – SL2; 3 – SL1; 4 – SL4; 5 – SL6; 6 – SL3; 7 – SL5; 8 – SL7.

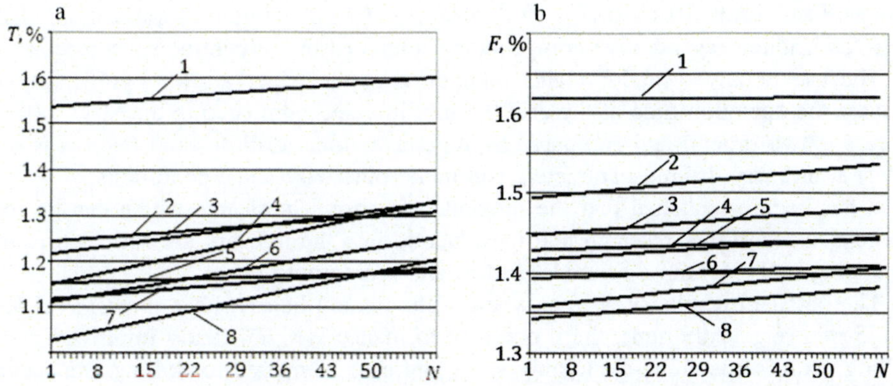

Fig. 5. Changing the values of the useful parameters T(a) and F(b).

Analysis of the results has shown that with an increase in the drill operation time, the useful torque T and useful axial cutting force F about proportionally increase by 4–16% and 1–4%, respectively. The resolution of the measuring system for the T and F parameters is 0.1% and 0.05%. It allows reliably identifying the difference in these parameters for different SL formulations to be tested. It has been established that the largest T and F values take place when drilling is 'dry', i.e., without the use of any lubricants at all, while the smallest ones – when using SL formulations containing serpentinite, sulfur and molybdenum diselenide.

To prove the SL components present in the surface layer of the drill cutting blade, the local chemical analysis was performed using an instrumental method base on the scanning electron microscope JSM-6490LV JEOL with microanalysis system INCA Energy 450. The study object was the back surface of the cutting blade at its certain points. At each of these points, the presence of chemical elements was determined three times with the aid of electron backscatter diffraction (EBSD) analysis (beam diameter 20 μm). All the drill bits working with the SL7 mentioned above were investigated. The presence of the following chemical elements was revealed: V, Cr, Mn, Fe, Ni, and W. Also, the presence of such chemical elements as Mg, Ti, Na, Ca, and an increased Si content indicates that some SL7 components, including Mg and Ca, due to diffusion is in the surface layer of the drill cutting blade.

The surface grinding in the combined use of both the SL7 and conventional MWF (water-based 3–5% emulsion) has also been experimentally studied when grinding workpieces of hardened steels. The positive results in the grinding productivity-increasing without both the grinding burns and temperature distortions were shown.

5 Conclusions

In the load range 300…1100 N, the highest friction coefficient (0.175–0.120) was obtained by testing the friction pair with the liquid lubricant (I-20 oil) and the lowest friction coefficient (0.075–0.040) – with the solid lubricant on the stearin base. The intermediate result (0.13–0.10) was obtained for the liquid composition: I-20 oil + 'antifriction regenerative composition' contained the serpentine substance.

The use of any solid lubricants mentioned in the paper allows significantly to increase the operation time of the drill bit until it breaks compared to the 'dry' drilling (excess is two orders) and compared to the conventional drilling fluid (excess is one order) at ordinary drilling parameters and under otherwise equal conditions.

It has been established that the most effective solid lubricants which can be recommended for further research and introduction into the industry are solid lubricants containing molybdenum diselenide, sulfur, and serpentinite.

The obtained results show that when drilling small holes with a diameter of less than 5 mm (e.g., 2.85 mm), the process performance is significantly improved when drilling using serpentinite-containing solid lubricant compared to drilling using conventional metalworking fluids.

The use of modern electron microscopy with the INCA microanalysis system allowed confirming experimentally the diffusion penetration of components of solid lubricant containing serpentinite into the surface layer of the cutting blade of the drill bit when drilling small holes that open up the opportunity for further research on the mechanism for reducing tool wear.

The effectiveness of the serpentinite-containing solid lubricant for surface grinding workpieces of hardened steels in the combined use of this lubricant and conventional grinding fluid is shown. In this case, the solid lubricant improves the machinability of the workpiece material in the grinding system, while the grinding fluid cools the workpiece being ground to prevent its temperature burns and deformations.

Acknowledgment. This work was carried out under the State (Ukraine) budget theme of the Odessa National Polytechnic University (2018–2021, registration number 0118U004400).

References

1. Levanov, I., Doykin, A., Zadorozhnaya, E., Novikov, R.: Investigation antiwear properties of lubricants with the geo-modifiers of friction. Tribol. Ind. **39**(3), 302–306 (2017)
2. Duradji, V., Kaputkin, D., Duradji, A.: Tribological studies of antiwear antifriction composition and its application. Tribol. Ind. **38**(4), 496–507 (2016)
3. Xu, Y., Gao, F., Zhang, B., Nan, F., Xu, B.-S.: Technology of self-repairing and reinforcement of metal worn surface. Adv. Manuf. **1**, 102–105 (2013)
4. Sharma, S., Anand, A.: Solid lubrication in iron based materials – a review. Tribol. Ind. **38**(3), 318–331 (2016)
5. Larshin, V., Lishchenko, N.: Adaptive profile gear grinding boosts productivity of this operation on the CNC machine tools. In: Ivanov V. et al. (eds). Advances in Design, Simulation and Manufacturing. DSMIE-2018. Lecture Notes in Mechanical Engineering, pp. 79–88. Springer, Cham (2019). https://doi.org/10.1007/978-3-319-93587-4_9
6. Lishchenko, N., Larshin, V.: Temperature field analysis in grinding. In: Ivanov V. et al. (eds). Advances in Design, Simulation and Manufacturing II. DSMIE-2019. Lecture Notes in Mechanical Engineering, pp. 199–208. Springer, Cham (2020). https://doi.org/10.1007/978-3-030-22365-6_20
7. Singh, P., Dureja, J.S., Singh, H., Bhatti, M.S.: Nanofluid-based minimum quantity lubrication (MQL) face milling of inconel 625. Int. J. Automot. Mech. Eng. **16**(34), 6874–6888 (2019)
8. Gupta, M.K., Sood, P.K.: Surface roughness measurements in NFMQL assisted turning of titanium alloys: an optimisation approach. Friction **5**(2), 155–170 (2017)
9. Hebab, H., Darras, B., Kishawy, H.A.: Sustainability assessment of machining with nano-cutting fluids. Procedia Manuf. **26**, 245–254 (2018)
10. Novikov, V., Gostev, Y., Zaslavskiy, R., Skobeltsin, A., Buyanovskiy, I.: Antifriction and antiwear resource restoring lubricant composition (ARRC). In: Proceedings of the scientific-practical conference-exhibition 'Tribotech 2003', 17 (2003). (in Russian)
11. Bai, Z.M., Yang, N., Guo, M., Li, S.: Antigorite: mineralogical characterization and friction performances. Tribol. Int. **101**, 115–121 (2016)
12. Krishna, P.V., Srikant, R.R., Nageswara, R.D.: Solid lubricants in machining. Proc. IMechE Part J: J. Eng. Tribol. **225**, 213–227 (2011)
13. Reddy, N.S.K., Rao, P.V.: Experimental investigation to study the effect of solid lubricants on cutting forces and surface quality in end milling. Int. J. Mach. Tools Manuf. **46**, 189–198 (2006)
14. Anand, A., Vohra, K., Ul, Haq, M., Raina, A., Wani, M.: Tribological considerations of cutting fluids in machining environment: a review. Tribol. Ind. **38**(4), 463–474 (2016)
15. Astakhov, V.P., Joksch, S.: Metalworking Fluids (MWFS) for Cutting and Grinding: Fundamentals and Recent Advances. Woodhead Publishing, Cambridge (2012)

Technological Support of Surface Layer for Optical Metalware

Fedir Novikov[1], Viktor Marchuk[2], Irina Marchuk[2(✉)], Valentin Shkurupiy[1], and Vladimir Polyansky[3]

[1] Simon Kuznets Kharkiv National University of Economics, 9-A, Nauky Avenue, Kharkiv 61166, Ukraine
[2] Lutsk National Technical University, 75, Lvivska Street, Lutsk 43018, Ukraine
MarchukV@bigmir.net
[3] LLC "Empire Metals", 88, Hryhorovske Road, Kharkiv 61020, Ukraine

Abstract. In the article, the justification of the parameters of the polishing regimes in the processing of surfaces of parts to smooth their surface layer. The procedure was developed, calculating the time of the whole process of processing parts made of copper and aluminum, the time of each transition, and the grain size of the abrasive at each transition. The paper considers the issues of technological support for smoothing the surface layer of optical metal products under conditions of abrasive polishing. A significant effect of processing time on the surface roughness parameters has been established. It was proposed that abrasive polishing be carried out in several technological transitions, reducing the granularity of the abrasive at each transition, up to a grain size of 1/0. The minimum number of transitions of the technological cycle is established to obtain the minimum values of the height parameters of surface roughness. It is shown that the stabilization time of the formation of the altitude parameter of surface roughness depends little on the granularity of the abrasive and is determined by the initial surface roughness before processing. A technique has been developed for calculating the time of the entire process of processing parts made of copper and aluminum to obtain a mirror surface, the number of transitions, the time of each transition, and the grain size of the abrasive at each transition. The results can be used for abrasive polishing of surfaces of laser mirrors with high reflectivity.

Keywords: Abrasive polishing · Roughness parameters · Granularity abrasive · Copper · Aluminum · Processing time

1 Introduction

Reduced labor input and production costs and the improvement of their quality is the essential task. Numerous studies have established that the decisive role in providing the state of the surface layer plays the active characteristics, which finally is formed at the finish operations. It is especially true for metal products that work under conditions of light exposure and lose their performance due to the appearance of temperature deformations. The creation of optical surfaces on these metal products makes it possible to reduce the temperature of their heating, temperature deformations, and,

V. Tonkonogyi et al. (Eds.): InterPartner 2020, LNME, pp. 412–421, 2021.
https://doi.org/10.1007/978-3-030-68014-5_41

accordingly, to increase operational characteristics. One of the effective solutions in this direction is the creation of optical surfaces of metal products by reducing their roughness during mechanical and physical-technical processing. The most significant effect is achieved in the process of abrasive polishing. However, at present, issues related to the possibility of a substantial reduction in surface roughness during abrasive polishing and ensuring the required optical properties of metal products have not been sufficiently studied. Therefore, the urgent task is to determine the rational conditions of abrasive polishing, providing the required surface roughness.

2 Literature Review

The work [1, 2] is devoted to the problem of improving the operational characteristics of parts by technological methods, where the main scientific points are formulated. The regularities of the formation of surfaces of parts, considering their optical properties during abrasive polishing, are disclosed in [3–7]. They focus on the issue of choosing rational technological environments (abrasive grains, grain size, and concentration). In [4, 8], the features of magnetically abrasive polishing are disclosed, which provide higher technological parameters of the machined surfaces than with conventional abrasive polishing. The conditions [1, 2, 9] are devoted to determining the conditions for achieving the minimum surface roughness during abrasive treatment and, accordingly, abrasive polishing. In these works, the authors associate a change in the optical properties of surfaces with a difference like the nonmetallic film and do not consider the effect of roughness. In [10–14], attention is drawn to the need to smooth the surface layer of a part during abrasive polishing to increase reflectivity. In the works [1, 2], it was shown that the efficiency and productivity of abrasive polishing depend on the technological environment, which includes a polishing pad, an abrasive and non-abrasive component of the technological composition and the material of the workpiece. At the same time, traditional approaches [1, 2], having a developed apparatus, do not allow explicitly taking into account the specific features of the dynamics of the polishing process concerning grinding [15–17]. All this reduces the effectiveness of technological decisions and makes them unsuitable in practice. In the proposed recommendations for the use of abrasive materials for polishing, there is insufficient information on the processing time necessary to achieve the greatest smoothing of the surface layer. Therefore, the objective of the study is to develop recommendations for reducing the polishing time to achieve a given smoothing of the surface layer.

3 Research Methodology

Abrasion polishing, depending on the nature of the abrasive media used and technological fluids, is a mechanochemical smoothing process surface layer by plastic deformation of microroughness, removal of oxides from the surface being treated. The polishing process is followed by successive application to the surface processed parts of a large number of scratches and traces of plastic deformation when they overlap and intersect. The technological liquid ensures the removal of wear products (metal particles and abrasive particles) from the surface of the processed details, helps to cool the surface layer of the workpiece.

The intensity of processing depends on the dynamic parameters determined by the polishing modes, the duration of polishing, the characteristics and dimensions of the abrasive particles, and the characteristics of the mechanical properties of the material of the part.

The cutting tool is formed directly during processing as abrasive an environment with special properties and certain internal connections. The complex geometric shape of the grains and their cutting parts is one of the essential characteristics of abrasive tools.

The abrasive tool will first contact the projections of the initial roughness of the surface, with gradual rounding of the protrusions. In the polishing process, the height parameters of the profile of the initial roughness of the surface will decrease, and the step parameters will change insignificantly. If the polishing process is stopped after 30 s from the start of polishing, then at the initial roughness after grinding a part of the initial microrelief will remain, while the roughness of the surface of the part will consist of smoothed protrusions.

The reference surface of the machined part at levels of 10, 20, 30, and 40% will be significantly increased in comparison with the original, and the basins of the microrelief will remain unchanged. Continuation of the polishing process will lead to the complete removal of the protrusions of the initial roughness. The ratio of altitude parameters $R_a/R_{\max}$ will decrease in this case. This indicates the occurrence of a large number of scratches on the polished surface, associated with the presence of an enlarged fraction in commercially available abrasive powders. In the future, the polishing process is stabilized. The relief characteristic for the polishing process will be reproduced continuously; its parameters will not change over time but will be determined by processing modes and graininess applied abrasive. Based on this, experimental studies of the parameters of surface roughness, material removal rate should be carried out, and the minimum number of transitions of the technological cycle to establish the minimum values of the height parameters of surface roughness should be established. This will allow a scientifically sound approach to the selection of optimal conditions for abrasive polishing.

4 Results

We studied the influence of abrasive graininess and the duration of preliminary treatment on the values of the height parameters of the surface roughness for samples from PN 30HGSA Steel (Fig. 1). From the graph (Fig. 2), it can be seen that the intensity of the change $R_{\max}$ does not correspond to the rate of change of values R_a. With increasing grain size, the abrasive value $R_{\max}$ is increasing.

With increasing processing time, the intensity of the change $R_{\max}$ sharply increases that can be explained by crushing the grains during processing, hence the processing process is necessary to stop before the intensive destruction of grains. As with the reduction in the grain size (with the force of pressing the polishing pad unchanged), the contact pressure increases and, accordingly, the depth of scratching by a single crushed

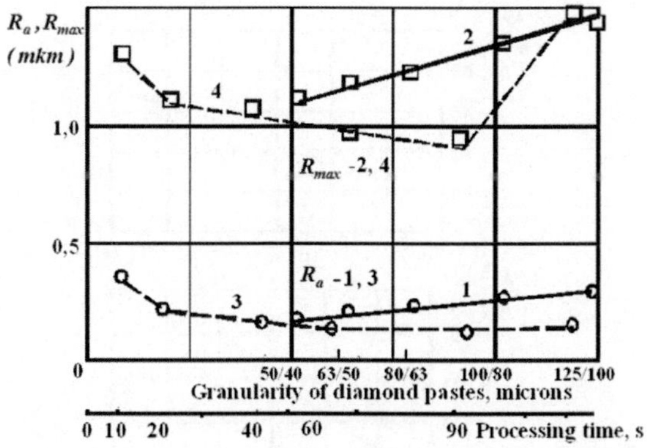

Fig. 1. Effect of the graininess of diamond pastes 1, 2 and treatment time 3 and 4 on the intensity of change in the values of the altitude parameters of the roughness, surfaces for samples from PN 30HGSA Steel: pressure 40 MPa; Circumferential speed 30 m/s; time of processing 20 s for 1 and 2; Granularity of abrasive for 3 and 4 – graininess 50/40

grain should be greater, R_{max} increases while preserving the smoothing effect. Ascending contact pressure contributes to the simultaneous collision of many abrasive grains, increasing the energy of motion of crushed grain. From the graph, we see that the ratio R_a/R_{max} in the initial period of polishing to 90 s decreases slightly, and with an increase in processing time above 90 s, the value of this ratio decreases and is 0.087. This reduces the quality of the processed surface. The value of the ratio of the altitude parameters depending on the graininess of the abrasive tape varies slightly, but when polishing with diamond pastes small granularity 2/1, the values of this ratio are sharply reduced. Therefore, it is necessary to establish optimal values for the processing time and grain size of the abrasive, based on the requirements for surface roughness.

To solve this problem, we use the paper [1], which gives the dependence of the removal rate Q on the parameters of the polishing process:

$$Q = k \cdot V^a \cdot c^{b_1} \cdot e^{b_2 c} \cdot p^{d_1} \cdot e^{d_2 p}, \tag{1}$$

where V – is the cutting speed, m/s; P – pressure, kPa; c – is the density of the abrasive slurry, G/ml; K, a, b, d are the coefficients.

We have obtained dependencies characterizing the change in the removal of the material from the part from one variable for fixed values of two variables.

Cutting speed with changing cutting speed:

$$Q(V) = 2,39 \cdot 10^{-2} \cdot V^{1,15}\,(\text{with c} = 2 \cdot 10^{-2}\text{g/ml; p} = 12 \cdot 10^2\,\text{kPa}).$$

Rate of removal when pressure changes:

$$Q(p) = 0,68 \cdot 10^4 \cdot p^{-3,38} \cdot e^{0,0173p}\,(\text{with V} = 0,2\,\text{m/s; p} = 12 \cdot 10^2\,\text{kPa}) \tag{2}$$

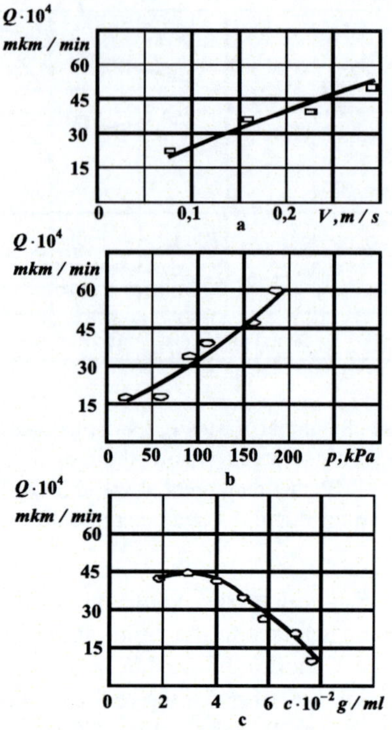

Fig. 2. Dependence of the removal rate Q on the parameters of the abrasive process: a – the linear velocity V; (b) – pressure; c – the density of the suspension c.

The rate of removal when the concentration of abrasive suspension is changed:

$$Q(c) = 1,14 \cdot 10^8 \cdot c^{5,64} \cdot e^{1,46c} \,(\text{with V} = 0,2\,\text{m/s; p} = 12 \cdot 10^2\,\text{kPa})$$

Figure 2 shows the experimental and calculated points obtained by the formulas (2). The discrepancy between the calculated and experimental values of material removal no more than 3%. To smooth the surface layer of the parts, a minimum тumber of technological cycle transitions to obtain minimum values poneycomb parameters of surface roughness. At the end of the first processing cycle, we get surface, the roughness of which R_1, and the altitude parameter of the initial liability R_0. When performing cycles, we obtain N surfaces with intermediate values of high-altitude roughness parameters R_i in accordance with the different transitions. The number of transitions and the intermediate value of the surface roughness depends on the physicochemical properties of the surface to be treated, its shape, processing time, properties, and graininess of abrasive material.

During abrasive polishing, we believe that the maximum depth of grain penetration is equal to the diameter of the abrasive particle or its maximum size. In this case, the height parameters of the surface roughness and the material removal rate at each transition are proportional to the dimensions of the abrasive particles:

$$R_{z_i} = \beta \cdot D_i; \quad Q_i = \alpha \cdot D_i, \tag{3}$$

where D_i – diameter of abrasive particles at the i-th junction (i = 1, 2… N−1); α and β – Coefficient the proportionality factors, determined empirically, are constant for this processing process.

It follows from (3) that

$$Q = \frac{\alpha}{\beta} \cdot R_{z_i}. \tag{4}$$

Table 1 shows the values of the rate of polishing and roughness during processing for one-hour parts from copper by various fractions of diamond micropowders.

Table 1. Results of abrasive polishing.

Surface roughness after milling R_1, mkm	The diamond-micropowders, graininess	Spee Polishing $Q \cdot 10^3$ mkm/min	Roughness after polishing, R_z, mkm
0,32	5/3	5,325	0,032
0,32	3/2	3,225	0,050
0,32	1/0	1,35	0,020

The size of the allowance h, corresponding to the depth of the defective layer, has the form:$h_i = k \cdot R_{z_{i+1}}$, where k is the coefficient of proportionality, which determines the amount of material removed in time t_1

$$t_i = \frac{F_i}{Q_i} = \gamma \cdot \frac{R_{z_{i-1}}}{R_{z_i}}, \tag{5}$$

where $\gamma = \frac{k\beta}{\alpha}$ with i = 1, 2, …, N.

The total processing time for all cycles is:

$$T_i = \sum_{i=1}^{N} t_i = \gamma \cdot \sum_{i=1}^{N} \frac{R_{z_{i-1}}}{R_{z_i}}. \tag{6}$$

To optimize the process by the minimum criterion, the total processing time (5) it is necessary to determine the optimum values of the intermediate surface roughness R_{z_i}, $i = 1, 2, \ldots, N-1$: $R_{z_i} = R_{z_{i-1}} = R_{z_{i+1}}$, hence $\frac{R_{z_{i-1}}}{R_{z_i}} = \frac{R_{z_i}}{R_{z_{i+1}}}$.

Taking formula (6) into account, it follows from Eq. (5) that under the optimal process, the transition time is the same, i.e. $t_1 = t$. This is true for surface treatment details with the same value of the height parameters of the roughness of the initial surface. However, this is not confirmed for samples with different initial surface roughness, since with decreasing initial roughness of the surface, the processing time sharply decreases. This is confirmed by the results of the experiment (Fig. 3).

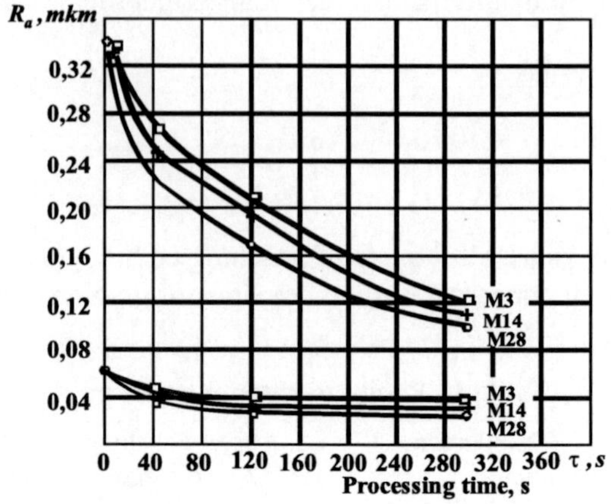

Fig. 3. Dependence of the height parameter of the surface roughness on time no polishing with abrasive materials of different granularity.

Dependency analysis shows that:

- time of stabilization of the process of formation of the altitude parameter of roughness the surface depends little on the grain size of the abrasive (from M3 to M28);
- time of stabilization of the process of formation of the altitude parameter of roughness surface significantly decreases with a decrease in the height parameter of the initial roughness of the surface before processing. When the initial roughness parameter of the surface is decreased R_a in 5.23 times (from 34 microns to 0,065 microns), the stabilization time R_a the treated surface is reduced eight times (from 320 to 40 s).

Thus, with smoothing, the top layer of the part, the cycle time of the subsequent finish operation will decrease more intensively than the reduction of the initial roughness parameter before processing;

- for each grain of abrasive material, there is a limit on stabilization values of the altitude parameter of the surface roughness, and this is very important when assigning a sequence of use of working media for smoothing the surface layer details. It should be noted that this limit will depend on the initial state of the surface of the part before processing. We are interested in the smallest limiting value R_a, achieved by surface polishing under different processing conditions.

The optimal values of the total processing time can be determined from the following expression: $T = \gamma \cdot N \cdot (\frac{R_{z0}}{R_{zN}})^{1/N}$. The optimal number of cycles is obtained by considering the total time T as a function of the number of cycles N. Define its minimum: $N_{\min} = \ln \frac{R_{z0}}{R_{zN}^{1/N}}$.

The optimum value of diameters of abrasive particles at each transition is:

$$D_i = \frac{1}{\beta} \cdot \frac{R_{ZN}^{1/N}}{R_{Z0}^{1/(N-1)}} = \frac{1}{\beta} \cdot R_{z_0} \cdot \left(\frac{R_{ZN}}{R_0}\right)^{1/N} \text{ at i} = 1,\, 2,\, \ldots,\, \text{N}$$

Carrying out similar calculations for the cases of surface treatment with changing physical and chemical properties, the expressions will take the form:

$$T_{opt} = \gamma \cdot N \cdot \left(x \cdot \frac{R_{Z0}}{R_{ZN}}\right) 1/N \text{ at N} = 2, 3 \cdot \ln \left(\text{x} \cdot \frac{\text{R}_{Z0}}{\text{R}_{ZN}}\right).$$

From the foregoing, it can be seen that the coefficients x and γ reflect the dependence of the polishing rate on the microhardness, the density of the material being processed, and the grain size diamond micropowder. It was experimentally established that for diamond micropowders of the corresponding granularity the value x is 0.06, γ is 1.7. Consequently, the optimal variant of the polishing process from the point of view of minimum transitions when processing a metallic mirror surface is determined only by the roughness height surface before and after treatment. It should be noted that the minimum number of transitions depends on the physical and chemical properties of the abrasive, the initial roughness of the surface $R_{z_0} = D_0 \cdot \beta$, Granularity of diamond micropowders $D_N = \beta \cdot R_{z_N}$ used at the last transition. The developed technique was used to optimize the polishing process of a copper mirror surface [18]. For the initial state, samples were taken with the surface treated to $R_a = 0,5$ mkm, at the final stage of processing, the roughness was $R_z = 0.032$–0.025 mkm. Then the optimal number of cycles is 3, and the time (averaged for the upper and lower limits) is $T = 7.87$ min. Studies have shown that the estimated time from the experimental difference is 20% ($T_{econ} = 9.5$ min), which corresponds to an error $\varepsilon = 0.01$. For abrasive compounds, the grain sizes at the respective stages processing: $D_1 = 3.121$–5.000 mkm; $D_2 = 1.154$–2.050 mkm; $D_3 = 0.425$–0.800 mkm, which corresponds to granularity of abrasives of 5/3, 3/2, and 1/0, respectively.

5 Conclusions

In work theoretically and experimentally revealed patterns the formation of surface roughness when polishing with abrasive materials of various grain sizes of copper and aluminum parts. It has been established that the ratio of surface roughness parameters $R_a/R_{\max}$ with an increase in the processing time significantly decreases (to a value of 0.087) and reduces the quality of the surface being treated. This is due to the crushing of abrasive grains. Therefore, for the effective implementation of the process of abrasive polishing, it is necessary to establish the values of the processing time based on the requirements for surface roughness.

The rationale for the optimal parameters of the polishing modes during surface treatment of parts to smooth their surface layer and ensure optical properties is given. The minimum number of transitions of the technological cycle is established to obtain the minimum values of the height parameters of surface roughness. The grit of the abrasive, in this case, should be reduced at each transition.

Based on the studies carried out, the procedure for calculating the time of the entire processing process, the number of transitions, the time of each transition, and the grain size of the abrasive at each transition.

References

1. Ryzhov, E.V., Suslov, A.G., Fedorov, V.P.: Technological support of operational properties of machine parts. In: Mechanical Engineering, Moscow (1979)
2. Yashcheritsyn, P.I., Zaitsev, A.G., Barbatko, A.I.: Subtle finishing processes for machining machine parts and devices. In: Science and technology, Minsk (1976)
3. Zverintsev, V.V., Zakhirintinskaya, Y.S., Tyaguseva, Y.I., Zverintseva, L.V.: The physical essence of the process of abrasive polishing. In: Actual Problems of Aviation and Cosmonautics. Technical Science, pp. 11–12 (2014)
4. Shather, S.K.: Enhancement of surface roughness and metal removal rate by using combined abrasives during magnetic abrasive finishing. Eng. Technol. Int. J. Res. **7**(8), 1–8 (2019)
5. Fu, G., Chandra, A.: A model for wafer scale variation of material removal rate in chemical mechanical polishing based on viscoelastic pad deformation. J. Electron. Mater. **31**(10), 1066–1073 (2002)
6. Wang, Y.G., Zhao, Y.W., Li, X.: Modeling the effects of abrasive size, surface oxidizer and binding energy on chemical mechanical polishing at molecular scale. Tribol. Int. **41**, 202–210 (2008)
7. Kim, J.-S., Lim, E.-S., Jung, Y.-G.: Determination of efficient super-finishing conditions for mirror surface finishing of titanium. J. Cent. South Univ. **19**, 155–162 (2012)
8. Yebing, T., Zenghua, F., Chen, S., Qiang, Z.: Experimental investigations on magnetic abrasive finishing of Ti-6Al-4V using a multiple pole-tip finishing tool. Int. J. Adv. Manuf. Technol. **106**, 3071–3080 (2020)
9. Novoselov, Y.: Dynamics of Surface Shaping in Abrasive Processing. LAP LAMBERT Academic Publishing. Saarbrucken, Deutschland (2017)
10. Nazarov, Y., Melnikov, O.N.: The choice of the optimal route for processing parts taking into account technological heredity. Bull. Mach. Build. **6**, 47–49 (1986)
11. Ruban, V.M., Nazarov, Y., Lurie, G.B., Romanova, V.I.: Investigation of the dependence of polishing speed on technological processing factors. Diamonds Superhard Mater. **10**, 89 (1980)
12. Shkurupiy, V.G., Nazarov, Y.F.: Smoothing the surface layer of copper and aluminum parts during their abrasive polishing. In: Protection of Metallurgical Machines from Breakdowns, vol. 12, pp. 281–286. PSTU, Mariupol (2010).
13. Koroleva, L.F.: Nanoparticulate zirconia-modified solid solutions of aluminum-iron oxides for polishing titanium metal. Diagn. Resour. Mech. Mater. Struct. **1**, 90–102 (2015)
14. Kacalak, W., Shafraniek, F., Tandecka, K.: Analysis of the active abrasive grains in the films abrasive finishing process. MECHANIK NR (2017)
15. Matarneh, M.E., Al Quran, F.M., Novikov, F., Andilakhay, V.: Theoretical corroboration for the temperature reduction conditions in the cutting zone during treatment. Eur. J. Mech. Eng. Res. **5**(3), 1–8 (2018)

16. Novikov, F., Polyansky, V., Shkurupiy, V., Novikov, D., Hutorov, A., Ponomarenko, Y., Yermolenko, O.O., Yermolenko, O.A.: Determining the conditions for decreasing cutting force and temperature during machining. East. Eur. J. Enter. Technol. Ser. Eng. Technol. Syst. **6**(102), 41–50 (2019)
17. Matarneh, M.E.: Improvement of abrasive and edge cutting machining efficiency through theoretical analysis of physical conditions. Int. J. Mech. Prod. Eng. Res. Develop. **8**(2), 249–262 (2018)
18. Tsesnek, L.S., Sorokin, O.V., Zolotukhin, A.A.: Metal mirrors. In: Mechanical Engineering, Moscow (1983)

Geometric Shape of the Projection and the Characteristics of the Cutting Edges of the Grains Synthetic Diamond Grinding Powders of Continuous Series Their Grades and Granularities

Grygorii Petasyuk(✉), Valerii Lavrinenko, Yurii Sirota, and Vladimir Poltoratskyi

V. Bakul Institute for Superhard Materials of the National Academy of Sciences of Ukraine, 2 Avtozavodska Street, Kyiv 04074, Ukraine
petasyuk@ukr.net

Abstract. The analytical processing of a large array of experimental static strength data and morphometric characteristics of the standard brand synthetic diamond grains has been performed. The relationship between the number and average value of the corners of the grain cutting edge sharpening and index of shape similarity their projection with the static strength and granularity were analyzed. The trends and the degree of variation of these characteristics in a wide range of standard grades and grain sizes of diamond grinding powders are determined. An empirical mathematical model for the relationship between the static strength of synthetic diamond grinding powders with the highest he index of shape similarity their projection and the morphometric characteristics was obtained. A review of publications on this topic has confirmed the originality of such studies. The applied significance of the research results was formulated. Possible and important for science and practice directions of continuing research carried out in this work are indicated.

Keywords: Morphometric characteristics · Static strength · Grain projection · Empirical mathematical model · Shape similarity · Particle angularity

1 Introduction

The characteristics of the 3D grain shape of abrasive powders are an important sign of their quality. As the characteristics indices of the 3D form, the most frequently used are various relationships between the spatial geometric parameters of the grain. For example, between the volume and the surface area of the grain, between the surface areas of the real grain and the sphere equal in volume. Important indices of the characteristics of the 3D grain shape are also its maximum and minimum dimensional parameters, their ratio, the magnitude of solid angles at the tops of the protrusions and elevations. The beginnings of the basics of the conceptual apparatus of the characteristics of the dispersed materials grains shape (mainly geological), their identification and quantitative assessment were laid down by works [1–3]. In these works, the names

V. Tonkonogyi et al. (Eds.): InterPartner 2020, LNME, pp. 422–432, 2021.
https://doi.org/10.1007/978-3-030-68014-5_42

of such characteristics were introduced simultaneously: form factor, edges parameters, circularity, elongation, particle angularity, sphericity. A detailed description of the essence of 3D and 2D characteristics of the dispersed materials shape particles, the description of the methods as well as algorithms for determining the 2D characteristic indices, were presented in [4, 5].

2 Literature Review

It should be noted that the sphere of application of dispersed materials in the studies mentioned above relates mainly to the enrichment of minerals and the production of building materials. These researches are not oriented on abrasive treatment, which has their formation history and solves other problems. However, referring to [6–11], we can rightly conclude that for abrasive powders, there is complete methodological continuity for the grain shape characteristics. The same grain characteristics and their determination methods are relevant and suitable also for the area of abrasive treatment. The researchers are unanimous in their opinion that 3D characteristics should be considered the adequate exponent of the geometry and morphology of the grains of real powders as spatial bodies. However, the implementation of such an approach, which is acceptable for practical use, can only be implemented for abrasive powders, the grains of which are a priori in the form of correct polyhedra and the bodies derived from them. For example, in the case of high-strength synthetic diamond (SD) powders, these are such bodies as octahedron, cubooctahedron, truncated octahedron. For other abrasive powders, this list, in some cases, can be supplemented with a cube, a tetrahedron, and other regular polyhedra. This limitation is due to two reasons. First, the 3D shape of the grains of most abrasive powders is different from regular or semi-regular bodies. Secondly, there are methodological and technical difficulties of practical 3D diagnostics and unambiguous generally accepted interpretation of the dimensional, geometrical, and morphological (cumulatively morphometric) characteristics of abrasive powders. At present, it is difficult to perform automated diagnostics of such 3D morphometric characteristics as volume and surface area, many cutting edges and cutting blades, the angles of their sharpening even for one grain, not to mention their large number (about 2–6 thousand grains). This circumstance is the reason for the use of 2D analogs of the spatial (3D) form of abrasive grains that were discussed in [5].

Usually, as 2D morphometric characteristics imply the characteristics of one (separate) projection of grain of powder in from a large number of its possible variants. The implementation of the projection depends on the steady-state of equilibrium that the grain tends to take on the microscope table. The position of such a stable equilibrium is determined by the principle of minimum potential energy and is unknown a priori. It is probably the main reason why the 2D morphometric characteristics of abrasive powders is a very conditional concept. Another significant reason for this conditionality is the lack of direct correspondence between the 3D and 2D characteristics of abrasive powders.

Moreover, when moving from a 3D grain shape to its projection (2D shape), some geometrical parameters are lost. For example, the volume and surface area, as well as the height of the grain. Furthermore, if for 2D analogs of volume and surface area, you can conditionally take the area and perimeter of the projection, then for the minimum size (height) of a 3D grain, there is not even such a conditional 2D analog. This circumstance is especially important for grains of irregular shape. However, this approach has won the right to life. It lies at the basis of mutual understanding and common views on the quality of abrasive powders between their manufacturers, consumers, and researchers of abrasive machining. On this approach, many standards for abrasive powders are based.

Along with this, in publications, one more aspect is considered - applied, which is connected with the practical use of the characteristics of the shape of grains. It concerns the relationship of these characteristics with the effectiveness of the use of dispersed materials in a particular area. For example, in the field of abrasive powders, on which the present work is oriented, this is the interrelation of the characteristics of the shape of the powder grains with their abrasiveness and the treatment indices. As the most informative in this sense, 2D morphometric characteristics of the grain shape are recognized. It is generally accepted as such to consider the form factor (roundness), the parameters of sharp protrusions, elongation, the angularity of particles [8–11]. The set of 2D morphometric characteristics is the most often used the latter, that is, the angularity of the projection of grains.

Along with this very informative characteristic, when solving a similar problem, there may be cutting edges [12]. As indices of the cutting edges, their number and sharpening angles are taken. In [13, 14], a method is proposed for efficient indirect express analytical determination of the number of cutting edges and the sharpening angles. It is oriented on modern automated diagnostics of the morphometric characteristics of abrasive powders.

In this paper, for the first time, an analysis will be made of the trend and essentiality (degree) of a change in the geometric shape of the projection and characteristics of the cutting edges of grains, which have a great influence on the abrasiveness of grinding SD powders. The object of the study is SD grinding powders of a wide range of a continuous range of their grades and granularity. Although abrasive machining is one of the oldest types of machining, it, as before, is important for modern engineering. The quality of the abrasive tool is largely determined by the quality of the abrasive powder used in its cutting layer [15].

SD grinding powders used in abrasive machining operations are characterized by a wide range of static strength indices. In this, less durable grains (grinding powders of grades AC6–AC20) are used in diamond wheels for grinding treatment of engineering products, stronger grains (grinding powders of grades AC32–AC160) - for cutting wheels and wheels used for machining stone and concrete. Grains of the next range of static strength (grinding powders of grades AC100–AC300) are characteristic for rock-destroying (geological exploration) tools and, finally, high-strength grinding powders of grades AC200–AC400 are used in precision dressers. Note that for the sphere of diamond-abrasive treatment, the use of grinding powders of all 4 ranges of their grades indicated above is typical in the tool. Therefore, the work aimed to study the tendency and essentiality of changes in the characteristics of the grain's projection and cutting edges of SD grinding grains during the transition from one range of a continuous series of grades and granularities to another.

3 Research Methodology

Regulated by normative documents [16], the set of SD grinding powders consists of 14 grades, and 13 granularities amount 180 of their samples. The practical implementation of this goal for all 180 samples of SD grinding powders is a long-term and large-scale research project. In consideration of this circumstance, in the present work, the following scheme of solving the problem in question is used. The entire set of according standard to [16] of grinding powders is divided into separate groups. Such groups are formed according to the principle - grinding powders of a single granularity of a continuous series of grades and grinding powders of a single grade of a continuous series of granularities. In both cases, under the continuity refers to a complete list of grinding powders of either the same grade or one the granularity, provided for by the standard [16].

Research carried out in this work covered three groups of standard grinding powders SD. The first group consisted of grinding powders with a granularity of 250/200 (in microns) grades AC6–AC200 (conditionally area A). The second group consists of grinding powders with a granularity of 500/400 (in microns) grades AC100, AC200–AC400 region B). In the third group, AC200 grade grinding powders with a granularity of 250/200 to 800/630 (region C) were combined. As a numerical identifier of the grade, an index of static strength was taken, determined to the standard [16]. The numerical index of granularity was the minimum Fere diameter [17], which is the closest analog of the width of the grain projection [18].

Static strength tests, taking into account a wide range of grades of grinding powders under investigation, were carried out on the DA-2, DA-2M, and DDA-33 devices [19]. Automatized diagnostics of morphometric characteristics of SD grinding powder was carried out using the DiaInspect.OSM instrument [17]. The geometric shape of the projection grains was identified and analyzed; the characteristics of their cutting edges, as well as the static strength of SD grinding powders from these three areas, were studied.

In [13], an original method of indirect analytical determination of the average value of the angle of sharpening (F) and the number of cutting edges (n) of abrasive powders was proposed. According to this method, these characteristics are found from the system of equations obtained in [11]:

$$f_r - \frac{n \cdot Rg^2 \cdot tg(\pi/n)}{\pi[Rg - tg(\pi/n) \cdot \sqrt{Rg^2 - 1}]} = 0, \; \Phi = \frac{\pi(n-2)}{2n} - \arccos\left[\frac{1}{Rg}\right]. \tag{1}$$

The initial data for the application of the developed method are the form factor *fr*, roughness *Rg*, and the projection area A_t of the grains of the abrasive powder. The form factor characterizes the degree of roundness of the projection of grain, the roughness *Rg* - the degree of development of its projection [19].

Identification and quantification of the geometric shape of the projection of grains were carried out by the system-analog method [20]. This method is an improvement of the system-criterial method, first proposed in [20]. It should be noted that the indices of differential shape-similarity were used. They represent the relative share (%) of grains in their sample with a certain form of a projection from the list of accepted basic figure analogs (BFA). The dependence expresses the numerical value of the differential in indeces

$$f_n^{(d)} = \frac{N_n}{N} 100, \tag{2}$$

where n is the BFA number, n = 1, 2,…, M; M is the quantity of BFA; N is the quantity of grains in the control sample of the powder, N_n is the quantity of grains, the projection shape of which was identified as BFA with number n. At the same time, if any figure from the set of BFA among the actual forms of the projection is not identified, then the index of differential shape-similarity to this basic figure is assumed to be zero.

The error of the differential form replacement, which is a quantitative assessment of its adequacy for each of the classes N_m, is determined by the dependence

$$\Delta_m^{(d)} = \frac{1}{n_m} \sum_{n=1}^{n_m} \delta_n, \tag{3}$$

where $\delta_n = \min\limits_{1 \le m \le M} \{q_{mn}\}$ is the minimum value of this error for all identifiable basic analogs, $q_{n,m}$ is the relative error of the shape-substitution of the grain projection with the number n by the basic analog with the number m. The summation in (3) is carried out over all classes N_m, which are not empty.

The subject of the study was the tendency to change the geometric shape of the projection and the characteristics of the sharpening angles of cutting edges of the grains of the SD grinding powders during the transition from one continuous series of their grades and granularities another. From the list of morphometric characteristics that were diagnosed by DiaInspect.OSM, in this study the fj are used the following: Fere-elongation $F_e = F_{max} / F_{min}$, as the ratio of the maximum (F_{max}) and minimum (F_{min}) diameters of Fere grains, compactness or form factor of the actual image of grains (f_r), roughness (Rg) and area (A_t) of the projection of grains, equivalent diameter (d_e), the coefficient of flattening grains k_{fl} and the relative fraction (A_{lg}) of the light part of the grain projection in its total area. A complete description of the above and other modern morphometric characteristics of abrasive powders, their geometric and conceptual meaning is given, for example, in [18].

4 Results

The subject of discussion is the values and trends in the change in the SD grinding powder characteristics in each of the three (A, B, C) regions of their grades and granularities. Analysis of the information received and its graphical interpretation are submitted separately for the geometric shape of the projection of grains and for the characteristics of their cutting edges.

Based on the analysis of the obtained quantitative information of the shape-similarity of projection of grains, the following conclusions can be drawn. In the transition from low-strength grades of grinding powders SD (AC6, AC15,…) to their high-strength grades (AC125, AC200) within the same granularity (region A), as expected, the static strength also increases. Furthermore, this is understandable, given the physical meaning of the ratio of the concepts of static strength and grade. The same tendency is observed for indices of differential shape-similarity to some basic forms of grain projection. In particular, for grinding powders of higher strength grades, there is a steady and growing tendency to increase in their composition of grains, whose projection the shape has of a regular hexagon (Fig. 1). Based on this, it can be concluded that the presence in the grinding powder of a larger number of grains of this shape projection is a sign of high static strength.

Less pronounced in area A is a tendency to change the shape-similarity of projection of grains to the base analog in the shape of a square. Up to grade AC50, an increase in the relative proportion of grains with this form of projection is observed. For higher grade, there is a variable nature of change - an increase alternates with a decrease. At the same time, low-strength (AC6–AC20) grinding powders are characterized by an increased quantity of grains with a projection in the form of a rectangle, an isosceles trapezoid, and a triangle (Fig. 1). On this basis, it can be said that the increased content of grains with such a projection shape in the grinding powder leads to a decrease in its static strength. For the other BFA projections of grains (oval-like figures, rhombus, square, parallelogram), it is impossible to draw unambiguous conclusions regarding this relationship.

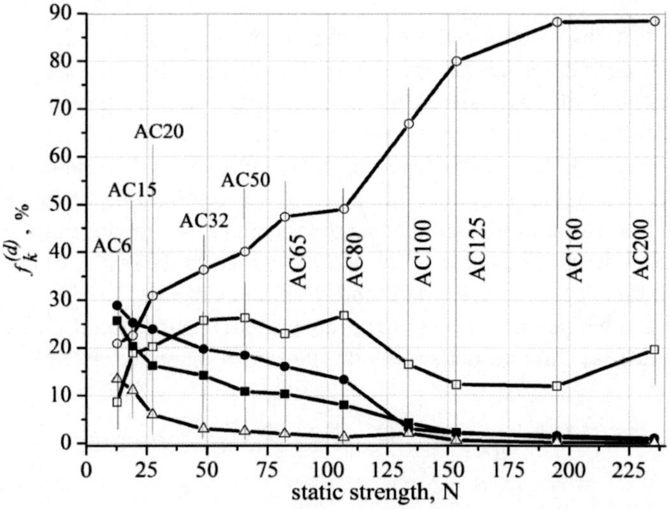

Fig. 1. The relationship between shape-similarity (by differential index $f_m^{(d)}$) to some the BFA (○ - regular hexagon, □ - square, Δ - triangle, ■ - rectangle, ● - isosceles trapezium) and static strength of grinding powders in region **A**

The relationship between the differential shape-similarity of projection of grains with static strength is also observed in region B. There is a further increase in the indices of form similarity to BFA in the form of a regular hexagon against the background of a decrease in these indices for others BFA of the projection form. For region C, i.e., during the successive transition from smaller granularities to large granularities within the same grade, for any of the possible forms of projection of grains, there is no stability like the change in the indices of differential shape-similarity. However, they are lower in level than the similar indices of the shape-similarity to these BFAs in region A. For each granularity (with slight deviations), the same distribution of grains over a reduced amount of basic projection forms and the ratio between the indices of shape-similarity of these basic forms are observed. On this basis, we can conclude that the shape of the grain projections is more related to SD powder strength than to their granularity.

Together with a qualitative analytical analysis, a quantitative correlation analysis was also carried out. In this case, static strength was adopted as the dependent factor and the dependent variable of empirical mathematical models (EMM). All other characteristics had the status of independent factors and, accordingly, independent EMM variables. Among the morphometric characteristics, these were: the coefficient of the grains flattening, the roughness of the grain projection, the relative proportion of the light part of the grain projection in its total area, and the equivalent grain diameter. Our previous studies [14] showed that it is these morphometric characteristics of SD grinding powders that are most closely related to their static strength. As a characteristic of the shape of the grain projection, we have included in this set the differential shape-similarity index to the basic figure of a regular hexagon $f_{F7}^{(d)}$.

The highest correlation coefficient occurs for a pair of factors "static strength - an index of differential shape-similarity to the base figure in the form of a regular hexagon", $r_{1,4} = 0.908$. Among the morphometric characteristics, the highest correlation coefficient corresponds to a pair of factors "static strength - the relative proportion of the light part of the grain projection area", $r_{1.3} = 0.895$. For other morphometric characteristics, the correlation coefficients are less than $r_{1.3}$, but close to it. The multiple correlation coefficient turned out to be 0.979, which indicates a close relationship between a combination of independent and dependent factors.

Empirical mathematical models were constructed using the LrAprox computer software system [14]. In automatized mode, the most adequate EMM was obtained:

$$P = b_0 + \frac{b_1}{\lg(K_{\text{сп}}^2)} + b_2 Rg^2 + b_3 A_{\lg}^2 + b_4 (f_7^d)^2 + b_5 \sqrt{d_{\text{eqv}}}, \tag{4}$$

where $b_0 = -689.27$ N; $b_1 = -17.62$ N; $b_2 = 99.75$ N; $b_3 = 1246.08$ N; $b_4 = 0.02$ N; $b_5 = 3.26$ N/μm$^{0.5}$. The average value of the relative error (Δ, %) is 3.58%, the maximum is 10.26%. The relative value of the average deviation on absolute meaning is 4.77%, the maximum - 10.67%.

The number of cutting edges of the grains with increasing static strength tends to increase. The relative estimate of this increase is as follows. In region A (Fig. 2), in the transition from grade AC6 to grade AC200, the increase is approximately 1.8 times.

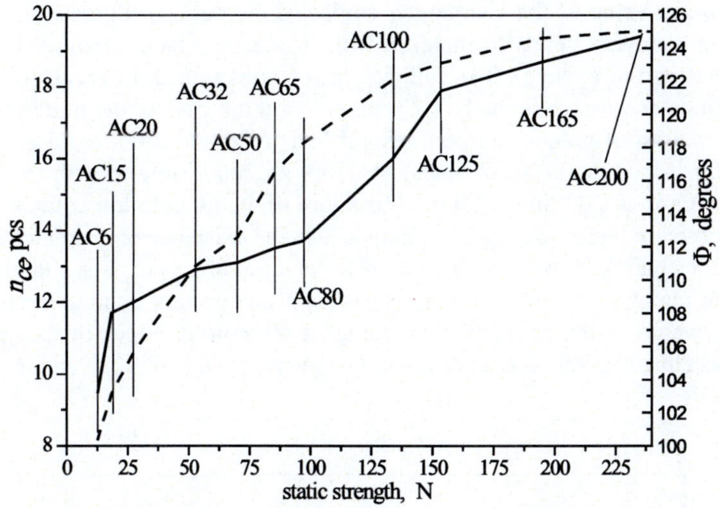

Fig. 2. Relation of the number of cutting edges (Δ, pcs) and the average value of sharpening corner (□, Φ degrees, right scale) of the grinding powders static strength of AC6-AC200 (granularity 250/200, region **A**)

Moreover, the main increase in the number of cutting edges falls on the interval of grades AC6–AC15 (50%). In the range of grades from AC15 to AC200 - half as much (25%). A similar picture of the behavior of the number of cutting edges, but with a lesser tendency to change, also occurs in region B (Fig. 3). Regarding region C (Fig. 4), we can say that there is no pronounced tendency for their change. The index of the number of cutting edges varies around their average value of 18 pcs. Variational amplitude swings - 4.2 pcs.

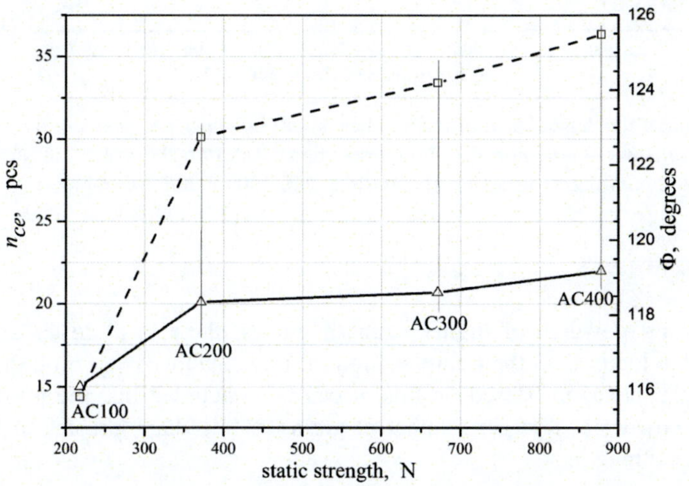

Fig. 3. Graphs of the dependence of the number of cutting edges (Δ, pcs) and the average value of their sharpening angles (□, Φ degrees, right scale) on the static strength of grinding powders of grades AC100, AC200–AC400 (granularity 500/400, region B).

The average value of the sharpening angles of the cutting edges of the grains, as well as their numbers, tends to increase with increasing static strength. In region A (Fig. 2), the trend of changes is as follows. In the transition from grade AC6 to grade AC200, Φ increases approximately 1.23 times. As in the case of the number of cutting edges, two ranges of grades can be distinguished: AC6–AC100 - rapid growth (from 101° to 122°, an increase of 1.2 times); AC100–AC200 - slow growth (from 122° to 125°, an increase of 1.03 times). Similar behavior of Φ, but with lower rates of change, is also observed in region B (Fig. 3). There is also the existence of characteristic grades (points on the charts). However, the ratio of Φ values at these points is slightly different compared to region A. In region C (Fig. 4), there is no constant trend in the behavior of the average value of the angle of sharpening of the cutting edges of the grains. The index of this characteristic fluctuates around an average value of 123° with a variational range of 4.3°.

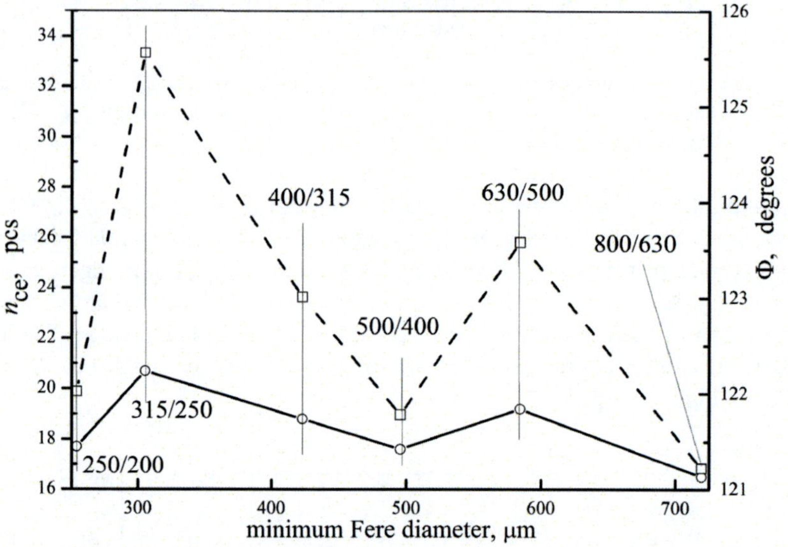

Fig. 4. Graphs of the dependence of the number of cutting edges (Δ, pcs) and the average value of the angles of their sharpening (□, Φ degrees, right scale) on the static strength of grinding powders AC200 continuous series of granularities 250/200 - 800/630, (region C).

5 Conclusions

– For grinding powders of higher strength grades, there is a steady and growing tendency to increase in their composition of grains, whose projection the shape has of a regular hexagon. Based on this, it can be concluded that the presence in the grinding powder of a larger number of grains of this shape projection is a sign of high static strength.

- For higher grade, there is a variable nature of change - an increase alternates with a decrease. At the same time, low-strength (AC6–AC20) grinding powders are characterized by an increased quantity of grains with a projection in the form of a rectangle, an isosceles trapezoid, and a triangle.
- The number of cutting edges of the grains with increasing static strength tends to increase. In the transition from grade AC6 to grade AC200, the increase is approximately 1.8 times. Moreover, the main increase in the number of cutting edges falls on the interval of grades AC6–AC15 (50%). In the range of grades from AC15 to AC200 - half as much (25%).

Subsequently, the studies carried out in this work should be continued in the following directions:

- the study of the dynamics, degrees, and trends Changing of the initial, intermediate and final geometric shapes of the cutting edges of the grains;
- development of a scientific and methodological apparatus for analyzing the characteristics and geometric shape of cutting edges as elements 3D relief of the actual spatial shape of grains.

References

1. Krumbein, W.C.: Measurement and geological significance of shape and roundness of sedimentary particles. J. Sediment. Petrol. **11**(2), 64–72 (1941)
2. Lees, G.: A new method for determining the angularity of particles. Sedimentology **3**, 2–1 (1964)
3. Sukumaran, B., Ashmawy, A.K.: Quantitative characterisation of the geometry of discrete particles. Geotechnique **51**(7), 619–627 (2001)
4. Cavarretta, C., O'Sullivan, M., Coop R.: Applying 2D shape analysis techniques to granular materials with 3D particle geometries. In: AIP Conference Proceedings Citation: AIP Conference Proceedings, vol. 1145, p. 833 (2009). https://doi.org/10.1063/1.3180057
5. Fernlund, J.M.R.: Image analysis method for determining 3-D shape of coarse aggregate. Cem. Concr. Res. **35**(8), 1629–1637 (2005)
6. Powers, M.C.: A new roundness scale for sedimentary particles. J. Sediment. Petrol. **23**(2), 117–119 (1953)
7. Rabinowicz, E.: Friction and Wear of Materials, 2nd edn. Wiley, New York (1995)
8. Hamblin, M.G., Stachowiak, G.W.: Description of abrasive particle shape and its relation to two-body abrasive wear. Tribol. Trans. **39**(4), 803–810 (1996)
9. Pintaude, G., Coseglio, M.: Remarks on the application of two-dimensional shape factors under severe wear conditions. Friction **4**(1), 65–71 (2016)
10. Pintaude, G.: Characteristics of abrasive particles and their implications on wear. In: Taher, G. (ed.) New Tribological Ways, pp. 117–130. Intech, Rijeka (2011)
11. Coseglio, M, Pintaude, G.: Abrasive particle characterization following different measurements of shape factor. In: Proceedings of the 20th International Congress of Mechanical Engineering, Gramado, Brazil, 2009: paper COB09–140 (2009)
12. Nikitin, Y., Uman, S.M., Kobernichenko, L.V., et al.: Synthetic Diamond Powders and Pastes. Naukova Dumka, Kyiv (1992). (in Russian)

13. Petasyuk, G.A.: Interpretative and applied aspects of some morphological characteristics of superabrasive powders. J. Superhard Mater. **32**(2), 128–139 (2010)
14. Petasyuk, G.A., Lavrinenko, V.I., Sirota, Y., Muzichka, D.G.: A study of the tendency and extent of variation of morphometric characteristics of standard synthetic diamond grits in a continuous series of their grades and sizes. J. Superhard Mater. **40**(6), 414–423 (2018)
15. Linke, G.: A review on properties of abrasive grits and grit selection. Int. J. Abrasive Technol. **7**(1), 46–58 (2015)
16. DSTU (Ukrainian State Standard) 3292–95: Synthetic Diamond Powders. General Technical Specifications (1995)
17. List, E., Frenzel, J., Vollstadt, H.: A new system for single particle strength testing of grinding powders. Ind. Diamond Rev. **1**, 42–47 (2006)
18. Novikov, N.V., Nikitin, Y., Petasyuk, G.A.: Computer-based diagnostic sieve for identification of grit size and grit size distribution in microscopic samples of diamond grits. J. Superhard Mater. **25**(3), 68–78 (2003)
19. Loshak, M., Alexandrova, L., Kosenchuk, T.: Density of diamond crystals for static loading. Metrol. Instrum. **2**(40), 11–15 (2013). (in Russian)
20. Petasyuk, G.A.: A system-analogue method of identification of geometric shape of the abrasive grain projection. J. Superhard Mater. **38**(4), 277–287 (2016)

Finite-Element Simulation of the Process of the Tubular Workpiece Expansion in the Manufacture of Automotive Parts

Ruslan Puzyr[1(✉)], Oleg Markov[2], Dmytro Savielov[3], Andrii Chernysh[3], and Yuliia Sira[1]

[1] Kremenchuk Mykhailo Ostrohradskyi National University College, 7 Chumatskyi Shliakh Street, Kremenchuk 39621, Ukraine
puzyruslan@gmail.com
[2] Donbass State Engineering Academy, 39, Mashinostroiteley Bul., Kramatorsk 84313, Ukraine
[3] Kremenchuk Mykhailo Ostrohradskyi National University, 20, Pershotravneva Street, Kremenchuk 39600, Ukraine

Abstract. The paper deals with the results of the finite-element simulation of the process of the tubular workpiece expansion. The operability of the criteria of the stability and ultimate strain during the expansion by conical punches is verified. The expansion coefficient is limited by the destruction of the workpiece in the form of a longitudinal crack along the weld joint in the heat-affected zone or in the solid metal, which mainly starts at the workpiece end, and also by the loss of the stability of the rigid part of the workpiece, causing the occurrence of a transversal circumferential fold. It is demonstrated that the criteria used in production, which are obtained based on various hypotheses and assumptions of experimental information, correlate with numerical models, and their results practically coincide. The expansion of the isotropic workpiece is compared with the workpiece with cylindrical anisotropy at the equal conditions of strain. It enabled the determination of the initial anisotropy influence on the distribution of stresses and strains during the expansion. It is shown that an increase in the anisotropy parameter in the axial direction, as compared with the cylindrical plane, makes it possible to increase the coefficient of the expansion in comparison with the isotropic model of the workpiece. The performed development does not exclude the possibility of using the proposed simulation method to solve engineering problems concerning production optimization.

Keywords: Plastic deformation · Pipe · Anisotropy

1 Introduction

The increase in the diameter of the ends of the cylindrical tubular workpieces using expansion is a theoretically sufficiently studied process of cold sheet-metal stamping. However, the existing formalized dependences do not fully take into account the influence of various factors of the process (the technological, structural ones, the material properties) on the finite change of the form. The desire to increase deformation

V. Tonkonogyi et al. (Eds.): InterPartner 2020, LNME, pp. 433–442, 2021.
https://doi.org/10.1007/978-3-030-68014-5_43

in one expansion step is accompanied by two types of loss of stability or crack formation, originating at the end of the deformed pipe.

Obtaining a complete analytical model of the workpiece strain at certain stages of its production, the use of which will make it possible to implement most research tasks of the improvement of the quality and resource of finished product operation is a rather topical research problem [1]. So, for example, in wheel production, in the manufacture of steel wheel rims, various automobile connecting pipes, the expansion of a cylindrical semi-finished product is the first manufacturing step. The expansion aims to decrease the degree of the strain at the subsequent radial rotation operation. In this case, one tries to obtain a great increase in the end diameter as possible to meet particular strain conditions [2]. The coefficient of the expansion is limited by the destruction of the workpiece in the form of a longitudinal crack along the weld joint, in the heat-affected zone or in solid metal, starting mainly at the end of the workpiece, as well as by the loss of stability of the rigid part of the workpiece in the form of a transverse circumferential fold [3].

These phenomena result in the deterioration of the process efficiency. The use of numerical methods of calculation will allow creating adequate solutions and substantiating new technological methods that eliminate the loss of stability or destruction of the workpiece during expansion.

2 Literature Review

Many papers of both foreign and Ukrainian researchers deal with the theoretical study of pressing and expansion processes [4, 7]. The determination of the stress-strain state reduces to solving simplified balance equations in an axisymmetric formulation together with the plasticity condition of Huber–Mises [5] or Tresca–Saint-Venant [6]. Here the problem is solved in stresses, and the calculation of strains presents certain difficulties. The workpiece's anisotropy influence on the distribution of the stresses and strains during the form changing is considered in papers [8, 9]. It specifies the analytical dependencies but makes them lengthier. It makes it difficult to identify the most significant factors of the straining process that affect expansion-stressing, and the determination of strains, in general, becomes an almost indefinable task [10]. Many research studies deal with taking into account the variation of the workpiece width during the strain [11, 12]. It resulted from the wish to make full-strength parts and mechanism units to prolong their operating life and reliability [13]. In our opinion, Popov E.A. obtained the simplest and most obvious solution [7]. However, there are more accurate solutions based on the theory of plastic flow taking friction into account [14], and on the theory of viscous-plastic flow [15] and the strain theory of plasticity [16, 17].

The formulas are valid for a cone tool but cause some difficulties with calculations. Recently, numerical methods for the analysis of plastic forming processes have become widely used by researchers and engineers [18, 19]. Their implementation using applied computer systems allows one to take into account material hardening, initial and acquired anisotropy, friction between the workpiece and tools according to various laws, fracture criteria and the accumulation of plastic deformation defects, speed effects, various conditions at the boundary, and also take into consideration both plastic and elastic constituents [20, 21]. However, the user needs skills in this software environment and knowledge based on the theory and practice of cold forging processes. In this case, the solution to one specific task of shaping may not cause difficulties [22]. The researcher has visual solutions with the breakdown of the part into zones by a finite-element grid and color grading of stress, strain, and displacement fields. However, obtaining an adequate statistical dependence for a wide range of workpiece sizes, conditions of strain, and optimization of tool parameters already leads to the need for a huge number of numerical experiments and a regression analysis of the obtained results. This does not always give adequate results, the values of which coincide with the theoretical and experimental data.

3 Research Methodology

A comprehensive study of the phenomena that occur during the expansion of tubular workpieces for automobile parts and other equipment still leaves less discussed the problem of the loss of the stability of the cylindrical workpiece in the force transfer zone in the form of a circumferential transversal fold. The opinions of some authors as to this problem coincide, and they propose ways for its solution [23]. Therefore, the purpose of this paper is in the finite-element simulation of the stability loss for tubular workpiece during the expansion, as well as in the comparison of results with the existing theoretical and experimental conditions for the transition of the rigid part of the workpiece to this state.

The formula of Averkiiev Yu. A. is the most well-known and widely used empiric dependence of the loss of the stability of the second type [24].

$$\frac{s_0}{D_0}100 > 2.5\ldots3.0, \tag{1}$$

where s_0 – the initial width of the workpiece;
D_0 – the external diameter of the cylinder.

E.A. Popov's formulas are also used for the relevant radius of the part without occurring cracks at the end.

$$R_{\max} = \frac{r_b}{1-\psi_p}\exp\sqrt{\frac{s}{r_b}\sin\alpha}, \tag{2}$$

where s – the initial width of the workpiece;
r_b – the radius of the workpiece;
ψ_p – the relative decrease of the cross-section area at the time the neck begins to form (characterizes the intensity of the workpiece metal hardening);
α – the angle of the punch taper,

The I.P Renne's empiric formula for the punches with the taper angle of 30–55° and relative wall width $\hat{s} = s_0/D_0 = 0.02–0.05$ [27].

$$k_b = \frac{(2.377\bar{s}+0.0328)}{1.475\bar{s}+0.0277}, \tag{3}$$

where k_b – the degree of the expansion at which the stability is lost,

M.N. Gorbunov's formula in stresses [25].

$$\sigma_{max} \approx \sigma_s, \tag{4}$$

where σ_{max} - the maximal axial stress;

σ_s – the yield point of the workpiece metal, and the formula of Sosenushkin–Yakovlev–Pilipenko considering the material anisotropy [26].

$$\frac{h_0}{s_0} = \frac{2\pi^2 e^{3\varepsilon_x} R_m}{\sqrt{3}} \sqrt{B(R_i) E_k B_1(R_i)} \Big/ \sqrt{\left[\sigma_{s0} + AC^n(R_i)\varepsilon_x^n\right] 4\pi^2 R_m^2 e^{2\varepsilon_x} - B_1(R_i) E_k B_2(R_i) h_0^2}, \tag{5}$$

where h0 – the initial length of the tubular workpiece;
ε_x – the real strain along the workpiece length;
R_m – the average radius of the medium surface of the workpiece;
$B(R_i)$, $B_1(R_i)$, $B_2(R_i)$, $C(R_i)$ – the relations of the anisotropy coefficients in the tangential and meridional direction;
E_k – tangent modulus of metal hardening;
A, n – the coefficients of the metal hardening curve.

The application complex Simullia Abaqus – student edition was used to model the expansion of the cylindrical workpiece. It was provided by the engineering company "TESIS" as a get-to-know one and differed from the professional versions by a limited number of units of the finite-element model.

At the first stage of the simulation the ultimate geometric parameters of the workpiece were chosen concerning stability loss (1). Consequently, in this case, condition (4) is to be met accordingly. The geometric dimensions of the workpiece and the punch were, respectively: D_0 = 80 mm, s_0 = 3 mm, h_0 = 250 mm, the biggest diameter of the punch D_p = 128 mm, the height of the punch H_p = 50 mm, the taper angle of the cone generatrix α = 38°, the expansion coefficient k_d = 1.6.

EN 1.0322 Carbon Steel was modeled as the workpiece metal. Its mechanical characteristics are as follows: modulus of elasticity 210 GPa, the Poisson coefficient 0.28; ultimate strength 320 MPa; yield point 230 MPa [28, 29]. The diagram of real stresses was approximated by dependence $\sigma_{0.2} = 230 + 3.46 \cdot \varepsilon^{0.6}$ [30, 31]. The problem was solved with the following assumptions: the first variant – the workpiece metal is isotropic; the second variant – the workpiece metal has cylindrical anisotropy, the sheet metal anisotropy acquired in the process of plastic deformation is low and does not essentially influence its initial anisotropy; the workpiece material is incompressible; the metal hardening is isotropic; the Baushinger effect is absent.

The punch was assigned as a rigid solid 3D body obtained by rotation, the workpiece – as a strained solid body in a 3D presentation obtained by extrusion.

The "assembly" inlay contains the mounting of the workpiece in alignment with the punch. In this case, half of the workpiece was considered as the problem is axisymmetric [34, 35]. When generating a finite-element grid for the tool, the free method of its construction was determined, and the element type R3D4 was selected, setting the bilinear order of the element that is contained in the standard library from the category of solid quadrangular ones. A hexahedral form was chosen for the workpiece; expansion was taken as the method of its creation, element C3D8R with the linear order from the standard library, category 3D Stress, i.e., operating in all three directions across the whole volume was selected.

Next, we set the type of interaction as contact between the workpiece and the forging tooling. As the mechanical constraints in the corresponding module of the program, we determined the kinematic contact method with the final slip formulation. As the contact properties, we used normal and tangential interaction with the coefficient of friction 0.25 [33]. The cylindrical workpiece was fixed in all degrees of freedom, and to simulate the force impact, the vertical movement was imparted to the punch at a speed of 3 mm/s [36, 37]. The period of calculation was limited to 300 s. Thus, the depth of its introduction made 50 mm. As a result, the semi-finished product received an increase in diameter with an expansion coefficient $k_d = 1.6$.

The first approximation was the solution of the strain problem for the isotropic metal of the workpiece. The second was for the anisotropic one.

The cylindrical anisotropy of the sheet metal was assigned by the following engineering constants [38]: E1 = 168 GPa; E2 = 270 GPa; E3 = 168 GPa – Young's modulus in the orthotropy planes; $\mu_1 = \mu_2 = \mu_3 = 0.28$ – Poisson's coefficients; $G_1 = G_2 = G_3 = 78$ GPa – shift moduluses [39, 40].

4 Results

Figures 1 and Fig. 2 contain the results of the numerical simulation for the isotropic and anisotropic workpiece metal. There is a loss of stability in the form of the formation of two transversal folds. In this case, the stability conditions (1) and (4) are met. These figures contain the distribution of the stress intensity, according to von Mises.

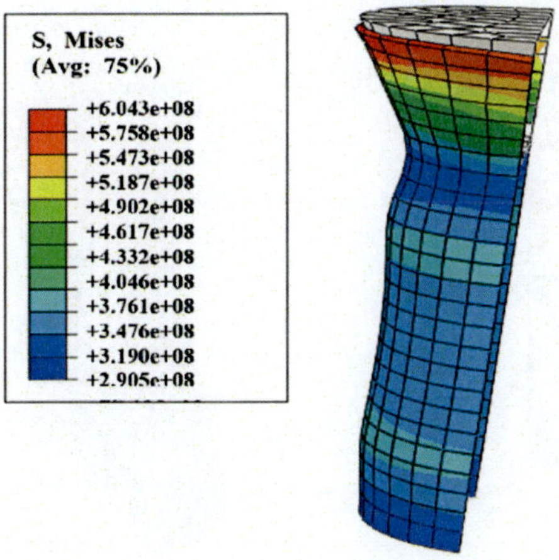

Fig. 1. The stress intensity distribution, according to von Mises for the isotropic workpiece.

It is typical that for an isotropic workpiece, the yield stresses at the end face of the workpiece exceed those for anisotropic metal by 2.2% and make respectively 604 MPa and 591 MPa. It is related to the accepted cylindrical anisotropy in the sheet plane, the indices of which are lower than the elastic modulus for isotropic metal.

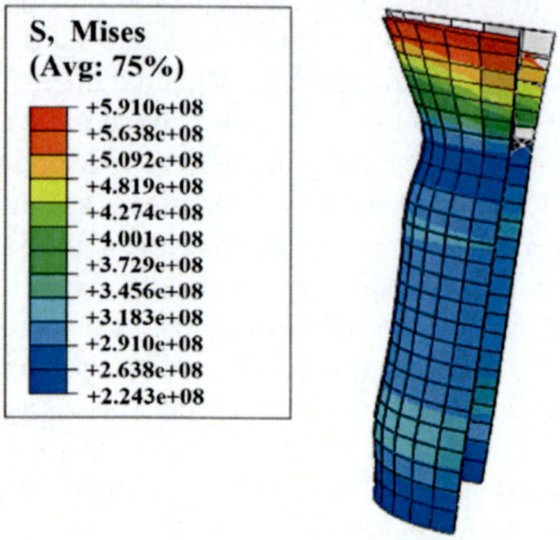

Fig. 2. The stress intensity distribution, according to von Mises for the anisotropic metal of the workpiece.

To verify the ultimate change of shape, a workpiece with the following geometric characteristics was modeled: $D_0 = 80$ mm, $s_0 = 2$ mm, $h_0 = 150$ mm. The tool dimensions, the mechanical characteristics of the workpiece metal, the loading, and contact conditions remained unchanged. The movement of a punch was 43.8 mm, which corresponds to the expansion coefficient $k_d = 1.51$. Figure 3 contains the results of the research.

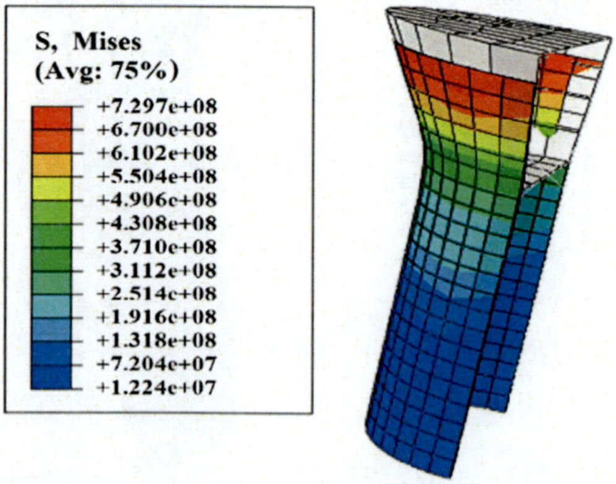

Fig. 3. The strain model for the verification of criteria (2) and (3).

As can be seen in the stress intensity distribution, according to von Mises in Fig. 3, the end part of the workpiece undergoes the highest stresses reaching the ultimate strength according to the adopted hardening curve. In the next loading step, the workpiece loses stability in the form of localization of the strains and the occurrence of a longitudinal crack.

5 Conclusions

The performed research revealed that the results of finite-element simulation of the process of expansion of a cylindrical workpiece are consistent with the considered criteria for the ultimate strain and stability conditions. However, the question of matching the simulation results with criterion (5) remains open, as the volume of this paper prevents it. In our opinion, the outlined trend of folding will continue for the relative sizes of workpieces with $h0/R_m < 6$. This conclusion can be made by relying on the empirical stability criterion (3), which, in its turn, was obtained for low tubular workpieces based on the performed experiments. The ultimate coefficients of the expansion practically coincide by criterion (2) - $k_d = 1.395$ and criterion (3) – $k_d = 1.46$, and also correspond to the obtained simulation data $k_d = 1.5$ (Fig. 3). Small discrepancies are since the model of form change used in the derivation of criterion (2) does not fully take into account all the features of the process. In particular, it is erroneous to assume that the curvature of the zone of the conical part of the workpiece, with its rigid part, equally depends on the degree of the expansion of the pipe at all stages of shape changing.

The finite-element formulation of the problem adequately describes the expansion process. It can be used in this form to optimize this method of form changing in the manufacture of automobile parts. It will make it possible at each stage of the strain to determine the stress-strain state, as well as vary the process parameters to reduce rejects. So, for example, it follows from the above that a decrease in the anisotropy coefficient in the plane of the cylinder, compared with the axial one, will make it possible to increase the ultimate coefficient of expansion, which means to intensify the process. However, this research is limited due to the small number of the considered workpiece sizes. Its further development will be associated with the increase in the size range for the experimental batches, as well as with the formation of an empirical stability criterion. Besides, a more detailed study of the workpiece material anisotropy influence on the stability of the deformation process is of interest.

References

1. Titov, V., Zlochevskaya, N., Kachan, A.: Features of integrated manufacture of blade workpieces from eutectically strengthened titanium alloys. Metallurgist **58**(1–2), 141–148 (2014)

2. Puzyr, R., Savelov, D., Shchetynin, V., Levchenko, R., Haikova, T., Kravchenko, S., Yasko, S., Argat, R., Sira, Y., Shchipkovakyi, Y.: Development of a method to determine deformations in the manufacture of a vehicle wheel rim. East. Eur. J. Enterp. Technol. **4**(1 (94)), 55–60 (2018). https://doi.org/10.15587/1729-4061.2018.139534
3. Sosenushkin, E., Yanovskaya, A., Sosenushkin, A., Emel'yanov, V.: Mechanics of nonmonotonic plastic deformation. Russ. Eng. Res. **35**(12), 902–906 (2015). https://doi.org/10.3103/s1068798x15120199.
4. Yang, C., Li, P., Fan, L.: Blank shape design for sheet metal forming based on geometrical resemblance. Procedia Eng. **81**, 1487–1492 (2014). https://doi.org/10.1016/j.proeng.2014.10.178
5. Wang, X., Ouyang, K., Xia, J.: FEM analysis of drawing-thickening technology in stamping-forging hybrid process. Forging Stamp. Technol. **34**(4), 73–78 (2009). https://doi.org/10.3969/j.issn.1000-3940.2009.04.017
6. Pereira, M., Weiss, M., Rolfe, B., Hilditch, T.: The effect of the die radius profile accuracy on wear in sheet metal stamping. Int. J. Mach. Tools. Manuf. **66**, 44–53 (2012). https://doi.org/10.1016/j.ijmachtools.2012.11.001
7. Popov, E.: Fundamentals of the theory of sheet stamping. In: Mechanical Engineering, Moscow (1977)
8. Dasappa, P., Inal, K., Mishra, R.: The effects of anisotropic yield functions and their material parameters on prediction of forming limit diagrams. Int. J. Solids Struct. **49**, 3528–3550 (2012). https://doi.org/10.1016/j.ijsolstr.2012.04.021
9. Cardosoa, R., Adetorob, O.: A Generalisation of the Hill's quadratic yield function for planar plastic anisotropy to consider loading direction. Int. J. Mech. Sci. **128–129**, 253–268 (2012). https://doi.org/10.1016/j.ijmecsci.2017.04.024
10. Yoon, J., Dick, R., Barlat, F.: A new analytical theory for earing generated from anisotropic plasticity. Int. J. Plast. **27**, 1165–1184 (2011). https://doi.org/10.1016/j.ijplas.2011.01.002
11. Savelov, D., Dragobetsky, V., Puzyr, R., Markevych, A.: Peculiarities of vibrational press dynamics with hard-elastic restraints in the working regime of metal powders molding. Metall. Min. Ind. **2**, 67–74 (2015)
12. Kukhar, V., Yelistratova, N., Burko, V., Nizhelska, Y., Aksionova, O.: Estimation of occupational safety risks at energetic sector of Iron and Steel Works. Int. J. Eng. Technol. **7**(2), 216–220 (2018)
13. Maslov, A., Batsaikhan, J., Puzyr, R., Salenko, Y.: The determination of the parameters of a vibration machinef the internal compaction of concrete mixtures. International Journal of Engineering and Technology 7(4.3), 12–19 (2018).
14. Malinin, N.: Applied Theory of Plasticity and Creep. Engineering, Moscow (1975)
15. Sokolovsky, V.: Drawing of a thin-walled pipe through a conical matrix. Applied Mathematics and Mechanics **24**(5), 27–31 (1960)
16. Neto, D., Oliveira, M., Alves, J., Menezes, L.: Influence of the plastic anisotropy modelling in the reverse deep drawing process simulation. Mater Des **60**, 368–379 (2014). https://doi.org/10.1016/j.matdes.2014.04.008
17. Ilyushin, A.: Plasticity. Publishing House of the USSR Academy of Sciences, Moscow (1963)
18. Harpell, E., Worswickb, M., Finn, M., Jain, M., Martin, P.: Numerical prediction of the limiting draw ratio for aluminum alloy sheet. J. Mater. Process. Technol. **100**, 131–141 (2000)
19. Akbulut, H.: On optimization of a car rim using finite element method. Finite Elem. Anal. Des. **39**, 433–443 (2003)

20. Altinbalik, T., Tonka, A.: Numerical and experimental study of sheet thickness variation in deep drawing processes. International Journal of Modern Manufacturing Technologies **2**, 9–16 (2012)
21. Haikova T., Puzyr R., Dragobetsky V., Symonova A., Vakylenko R. Finite-Element Model of Bimetal Billet Strain Obtaining Box-Shaped Parts by Means of Drawing. In: Ivanov V. et al. (eds) CONFERENCE 2019, DSMIE, p. 85 – 94. Springer, Lutsk, Ukraine (2019), doi: https://doi.org/10.1007/978-3-030-22365-6_9.
22. Wang, G., Zhao, G., Xia, S., Luan, Y.: Numerical and experimental study of new cold precision fording technique of spear gears. Trans. Nonferrous Metal Soc. **13**(14), 798–802 (2003)
23. Kalyughniyi, V., Zaporozchenko, A., Pimanov, V.: Intensification of technologies of making of good is «Bulb 180x184». Treatment of materials pressure: Collection of scientific labour **2**(31), 136–140 (2012)
24. Averkiev, A.: Forming of a pipe billet during distribution and crimping. Forging and stamping production **1**, 6–9 (2000)
25. Gorbunov, M.: Distribution of tubular billets on a conical punch with support at the edge. Forging and stamping **8**, 22–26 (1968)
26. Pilipenko, O., Feofanova, A., Zharkov, A.: Creasing during the distribution of a tube billet from anisotropic material. Ivestia TulSU. Series. Mechanics of a deformable solid and metal processing 2, 140–145 (2006).
27. Renne, I., Kayushin, V.: Distribution of the pipe with a conical punch, limited by the loss of stability of the cylindrical shape of the workpiece. Forging Stamp. **3**, 20–21 (1988)
28. Houliara, S., Karamanos, S.: Stability of long transversely-isotropic elastic cylindrical shells under bending. Int. J. Solids Struct. **47**, 10–24 (2010)
29. Puzyr, R., Savelov, D., Argat, R., Chernish, A.: Distribution analysis of stresses across the stretching edge of die body and bending radius of deforming roll during profiling and drawing of cylindrical workpiece. Metall. Min. Ind. **1**, 27–32 (2015)
30. Kukhar, V., Kurpe, O., Klimov, E., Balalayeva, E., Dragobetskii, V.: Improvement of the method for calculating the metal temperature loss on a coilbox unit at the rolling on hot strip mills. Int. J. Eng. Technol. (UAE) **7**(4), 35–39 (2018)
31. Geselbracht, M.J., Ellis, A.B., Penn, R.L., Lisensky, G.C., Stone, D.S.: Mechanical properties of metals: experiments with Steel, Copper, Tin, Zinc, and Soap Bubbles. J. Chem. Educ. **71**(3), 254 (1994)
32. Wagener, H., Wolf, J.: Coefficient of friction in cold extrusion. J. Mater. Process Technol. **44**, 283–291 (1994)
33. Matsuda, K., Hashimoto, D., Nakamura, K.: Real contact area and friction property of rubber with two-dimensional regular wavy surface. Tribol. Int. **93**(part B), 523–529 (2016)
34. Puzyr, R., Kukhar, V., Maslov, A., Shchipkovskyi, Y.: The Development of the Method for the Calculation of the Shaping Force in the Production of Vehicle Wheel Rims. Int. J. Eng. Technol. **7**(4.3), 30–34 (2018)
35. Jin, J.-S., Deng, L., Wang, X.-Y., Xia, J.-C.: A new rotary forming process for rim thickening of a disc-like sheet metal part. J. Mater. Process. Technol. **212**(11), 2247–2254 (2012). https://doi.org/10.1016/j.jmatprotec.2012.06.013
36. Rout, M., Pal, K., Singh, S.: Prediction of edge profile of plate during hot cross rolling. J. Manuf. Process. **31**, 301–309 (2018). https://doi.org/10.1016/j.jmapro.2017.11.024
37. Puzyr, R., Haikova, T., Majerník, J., Karkova, M., Kmec, J.: Experimental study of the process of radial rotation profiling of wheel rims resulting in formation and technological flattening of the corrugations. Manuf. Technol. **18**(1), 106–111 (2018). https://doi.org/10.21062/ujep/61.2018/a/1213-2489/mt/18/1/106

38. Puzyr, R., Haikova, T., Trotsko, O., Argat, R.: Determining experimentally the stress-strained state in the radial rotary method of obtaining wheels rims. East. Eur. J. Enterp. Technol. **4**(1(82)), 52–60 (2016). https://doi.org/10.15587/1729-4061.2016.76225
39. Wang, Z., Zhang, Q., Liu, Y., Zhang, Z.: Robust and accurate geometric model for automated design of drawbeads in sheet metal forming. Comput. Aided Des. **92**, 42–57 (2017). https://doi.org/10.1016/j.cad.2017.07.004
40. Wang, X., Jin, J., Deng, L.: Review: state-of-the-art of stamping-forging process with sheet metal blank. J. Harbin Inst. Technol. **24**, 1–6 (2017). https://doi.org/10.11916/j.issn.1005-9113.17005

The Influence of Technological Factors on the Reliability Connection for Tungsten Carbide Insert Cutter with Cone in the Roller Cone Bits

Andrii Slipchuk[1(✉)], Roman Jakym[2], Vladimir Lebedev[3], and Emil Kurkchi[4]

[1] Lviv Polytechnic National University, 12, Bandera Street, Lviv 79013, Ukraine
andsl@ukr.net
[2] Drohobych State Pedagogical University by name I. Franko, 24, Ivan Franko Street, Drohobych 82100, Ukraine
[3] The Physical-Technical Institute of the National Academy of Science of Belarus State Scientific Institution, 10, Cuprevich Street, 220141 Minsk, Belarus
[4] Crimean Engineering and Pedagogical University, 8, Uchebnyy Side Street, Simferopol 295015, Republic of Crimea, Russia

Abstract. The scientific work investigates the influence of technological parameters such as the roughness of the holes in the cone for the tungsten carbide insert cutter, speed of pressing, and the value of tension in the connection "cone - tungsten carbide insert cutter" of the roller cone bits. Researches were conducted based on mathematical planning of experiments in the real production of roller cone bits drill bits. Methods for improving the reliability of connection "cone–tungsten carbide insert cutter" are proposed based on the results obtained. To ensure a stable profile and the value of the holes, roughness can be achieved by introducing an additional final reaming operation, which would serve as the final calibration of the hole. For this purpose, it is effective to use monolithic hard-alloyed reamers, which provide higher stability and quality of the machined holes in contradistinction to the reamers with soldered hard-alloyed reamers. According to the obtained results, it was determined the effect of the roughness of the walls of the holes in the cone on the efforts of pressing the inserted tungsten carbide insert cutters of a typical profile of the shank.

Keywords: Templates of cones · Accuracy · Pressing speed · Cone · Stress state · Tension · Joints · Roughness

1 Introduction

The existing technology for the production of roller cone bits involves the cold press connection of tungsten carbide insert cutter with the body of the cone [1–3]. In manufacturing, the system of shrink fit is used for each pattern size of the insert cutter according to the hardness of the cone. To ensure connection accuracy within specified

V. Tonkonogyi et al. (Eds.): InterPartner 2020, LNME, pp. 443–452, 2021.
https://doi.org/10.1007/978-3-030-68014-5_44

limits, the quality of the hole shaping is monitored at all stages: core drilling after carburizing, hardening and high-temperature tempering, reaming after tempering, and low-temperature tempering [4–6].

2 Literature Review

Special types of roller cone bits steel are used for various pattern sizes of the cone, such as: AISI 9315H, AISI 4815H, AISI 8720. Cones are subjected to a complicated technology of hardening, which is based on the process of carburization, which is thoroughly described in [1, 7, 8]. It provides high wear resistance rates and impact viscosity of cones. However, the high indexes of the hardness of gage row insert on the cones, into which the tungsten carbide insert cutters were pressed, did not provide a reliable "cone–tungsten carbide insert cutter" connection because of the creation of cracks and insert cutter loss. The model is derived based on the mechanism of rock fragmentation under a single tooth impact indentation [4]. For a long time in the roller cone bits making industry the technology of milling a plane on the surfaces of the rolling-cutter row or cutting to a certain depth of the cemented layer was used, as it is recommended in [9, 10]. There is a study where made to assist engineers in identifying the critical and sensitive subsystems [7]. During the last decade, such technology has ceased to be used because of the considerable complexity of the process, high consumption of the cone steel and instruments, rapid wear of the tool, etc. A more economical technology is the protection of the gage rows with anti-cement paste just before the process of cones cementation [1, 8]. However, the formation of carbon-saturated sites and the occurrence of unforeseen "breakdown" of the protective layer are the disadvantages of such technology. It requires many advanced requirements for the durability and quality of the tool and for improving the precision of the machining of the holes for the tungsten carbide insert cutters insertion. A reliable "tungsten carbide insert cutter – rolling-cutter teeth row" connection is also provided in the structures with less than 10 mm between adjacent holes edges [11]. However, such design solutions require high rates for steel plasticity due to the risk of cracks even in the rolling-cutter row made without a cemented layer. Some researches have detailed the principles of these three features and their application to rock bit design [12–14]. Moreover, it also defines the rigid requirements for the technology of production of cones, and especially the parameters of technological operations for molding the holes and assembling of the connections [15]. Some scientists have developed a new bearing assembly, specifically developed for rock bits, that enables significant increases in the service life of bits using roller-cones. [16].

3 Research Methodology

In this research, the results of the experiments of one of the key aspects in ensuring the quality of production of cones for the roller cone bits equipped with tungsten carbide insert cutters are shown. The influence of the roughness parameters on the formation of the holes for tungsten carbide insert cutters and the subsequent operation of assembling the joint of "tungsten carbide insert cutter - cone" were researched.

To investigate the geometric data of the holes and tungsten carbide insert cutters made of hard-alloyed materials, measuring tools and laboratory equipment were used, following certain recommendations [17]. The micrometric calipers of 5–30 mm Δ ± 0,01 mm according to GOST 10-88, the caliper depth gauge with an electronic indication of the reference 0–150 mm Δ ± 0,03 mm according to GOST166-90 were used.

To carry out this study, special templates cut off the cones of roller cone bits were specially prepared.

3.1 Effect of the Roughness

Measurements such as roughness were performed using a HOMVEL TESTER T1000 profiler, as well as a Model 296 profiler (Fig. 1).

Express measurements of the holes in the cones were performed using the workshop equipment based on the "Western L-10" pneumatic comparator, which gives a maximum error of up to 2.5 microns.

Fig. 1. Measurements of roughness on the surfaces of shanks of tungsten carbide insert cutters and holes in the templates of cones.

For the final processing of the holes for the insertion of tungsten carbide insert cutters, a cutting tool was used, the cutting parts of which are shown (Fig. 2).

An analysis has found that most of the quality assurance issues occur in the stages of the holes fine reaming. This operation is performed using a reamer with hard-alloyed plates. The documentation provides the roughness of Ra 1,25–2,5 microns.

For a long time, the pressing of tungsten carbide insert cutters was performed manually at a depth of 3 to 5 mm, and then a hydraulic press was used to put them to the stop in the butt end part of the hole bottom [18, 19]. It was found out that the peculiarities of the design and the dimension of the rates of the shanks roughness of tungsten carbide insert cutters, and also the parameters of the technological process of the displacement, influence not only the quality of the connection but also the integrity of a tungsten carbide insert cutter. In particular, with poor quality facet or its wrong parameters in the rear part of a tungsten carbide insert cutter, there appeared distortions while putting it between the axis of the hole symmetry and an insert cutter [20]. Also, sometimes while pressing, some crackling sounds occurred.

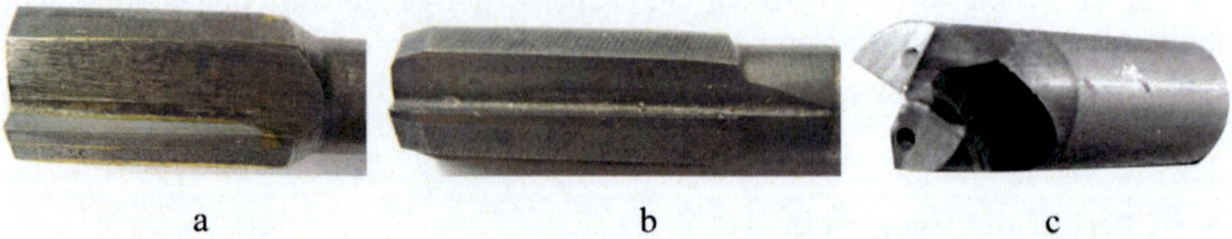

Fig. 2. Cutting parts of the tool for various options of finishing process of the holes for tungsten carbide insert cutters: a – reamer with hard-alloyed plates BK6M, which allows selection of the hole bottom, b – reamer of solid hard-alloy BK6OM scan, c – the drill of solid hard-alloyed bit "Hertel" used for the final fine reaming and holes bottom processing

3.2 Effect of the Interference Fit

The plates (Fig. 3), the parameters of deviations in the cross-sections of the holes were defined, for the tapered shape and deviations in the entrance part of the hole and a certain place on its bottom. Also, the technological parameters at which the tungsten carbide insert cutters were pressed and pressed off were determined.

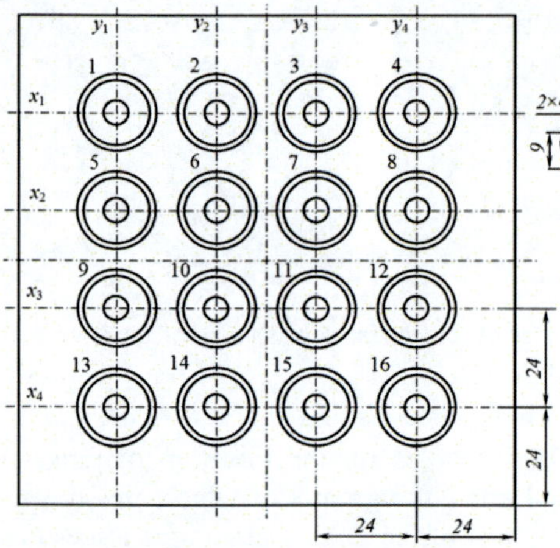

Fig. 3. Plate diagram for investigation of the parameters of the hole for the inserting tungsten carbide insert cutters (as an example, the sizes for testing tungsten carbide insert cutters with a diameter of 16 mm are given).

3.3 Effect of the Pressing Speed

According to the numerous tests on various typical sizes of tungsten carbide insert cutters, the speed of pressing on is of extraordinary importance. In particular, with slow pressing, cracks, and considerable plastic deformation occur (at different stages) in the rolling-cutter row of the cones. During the very fast process, often an impact, with its frequent application, breakdowns of tungsten carbide insert cutters, and no axes symmetry between them, often occurred, which was caused by the inconsistency between the axes of the holes and the pressed tungsten carbide insert cutters. Because of the physical and mechanical features of the melting of different steels, the design options for placement of tungsten carbide insert cutters in the rolling-cutter row of a cone, and the typical sizes of the tungsten carbide insert cutters give a significant dispersal of rates (Fig. 5, 7). Thus the development of rational parameters of the pressing process is relevant today.

To improve the accuracy of the technological operation of pressing, a special device was developed [1], which allows accurately orienting the inserted tungsten carbide insert cutters before pressing (Fig. 4), as well as to reduce the spoilage significantly. Also, such technology makes it possible to increase the speed of the process. In particular, in rapid pressing, as it turned out, the effect of fast slipping of the shank body into the hole of the rolling-cutter row can be applied. Thus the contacting surfaces undergo elastic deformation, and the insert cutter is not damaged.

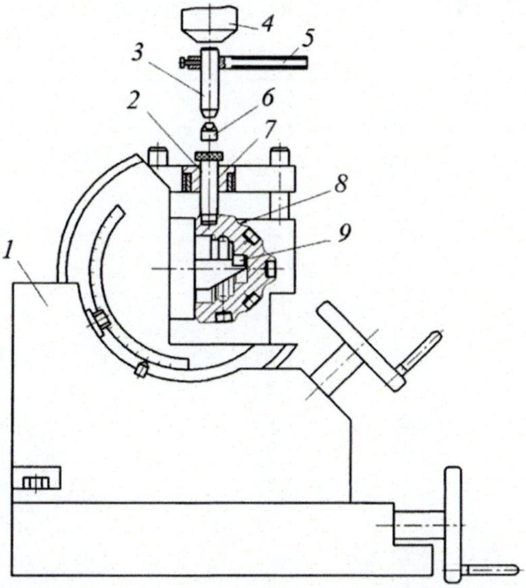

Fig. 4. Scheme of the device for the operation of pressing tungsten carbide insert cutters into the body of the cone on the press 6234: 1 - the base of the device; 2 - the hole catcher; 3 - the pin; 4 - the mandrel for the press pusher; 5 - arms; 6 - tungsten carbide insert cutter; 7 - guide sleeve; 8 - cone; 9 - mandrel installation

4 Results

4.1 Effect of the Roughness

The analysis of cut and taken apart connections showed that on the hole walls, there appeared considerable plastic deformation, microcracks, and, in some places, shavings on the bottom of a hole. On the body of the shank, chippings of different sizes were detected.

It made it possible to obtain a general picture of the roughness distribution over the entire cylindrical surface (Table 1). The study of similar parameters at the holes in the rolling-cutter row of the cones was carried out on pre-prepared templates cut and prepared from the roller cone bits.

Table 1. Results of roughness measurement on shanks of some types of tungsten carbide insert cutters.

№	R_a, µm			
	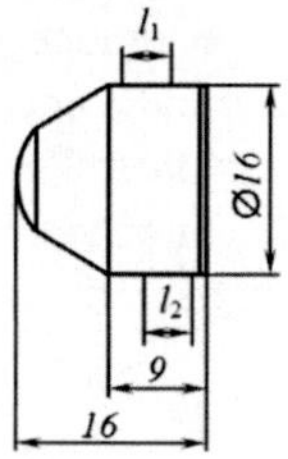		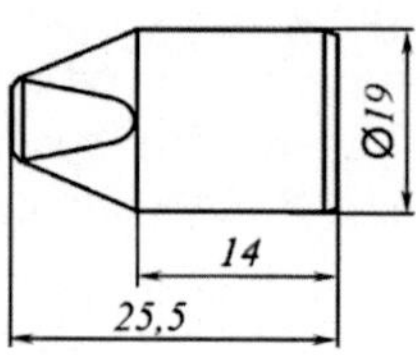	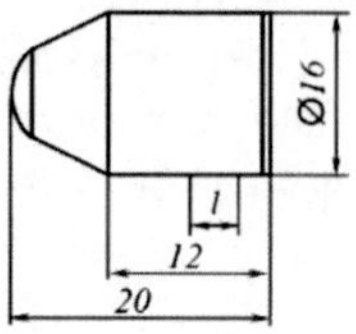
	l_1	l_2		
1	2,11	1,61	1,21	1,35
2	2,29	1,37	1,49	1,25
3	2,15	1,43	1,61	1,13
4	2,02	1,37	1,37	1,23
5	1,92	1,43	1,31	1,35
6	1,86	1,51	1,31	1,29
7	1,61	1,63	1,29	1,07
8	1,55	1,71	1,45	1,11
9	1,35	1,73	1,51	1,23
10	1,39	1,87	1,43	1,13
11	1,39	1,95	1,93	1,37
12	1,57	2,06	1,75	1,09
13	1,67	2,13	1,72	1,25
14	1,83	2,19	1,51	1,23
15	1,57	2,25	1,61	1,17
16	1,41	2,19	1,37	1,45
17	1,53	1,92	1,31	1,41
18	1,37	1,85	1,34	1,62
19	1,43	1,78	1,29	1,75
20	1,53	1,62	1,46	1,53
21	1,64	1,53	1,55	1,35
22	1,76	1,46	1,95	1,26
23	1,83	1,37	1,76	1,12
24	2,05	1,41	1,58	1,18
25	2,12	1,57	1,34	1,26
mean value	**1,718**	**1,718**	**1,498**	**1,287**
Step	~2 mm	~2 mm	~2,4 mm	~2 mm

Using the recommendations [23, 24, 25], it was found out that the pre-immersion of the tungsten carbide insert cutters into oxidized paraffin before pressing gives an almost double reduction of the plastic deformation of the walls in the hole. It makes it possible to increase the value of tension and connection.

The parameters of the roughness deviation of such surfaces were also investigated. To measure the distribution of roughness on the cylindrical surface of the tungsten carbide insert cutters shanks the rotation was carried out at a certain stage around the axes of the tungsten carbide insert cutters (see Fig. 5).

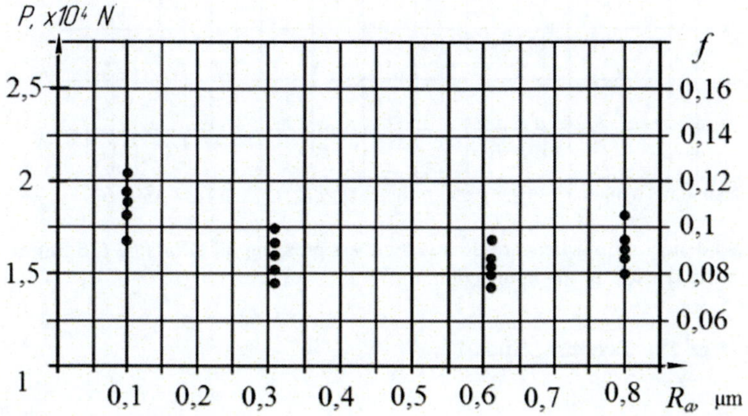

Fig. 5. The effect of the roughness of the hole's walls in the cone and the coefficient of friction on the efforts of pressing the inserted tungsten carbide insert cutters of a typical profile of the shank.

4.2 Effect of the Interference Fit

The analysis of the walls of the opening on AISI 9315H steel revealed that elastic deformation took place at tensions up to 0.068 mm, then plastic deformation prevailed up to 0.148 mm (Fig. 6). The intensive formation of significant plastic deformations and microcracks was further recorded. At tensions above 0.158 mm, there was chipping at the base of the tungsten carbide insert cutters shank.

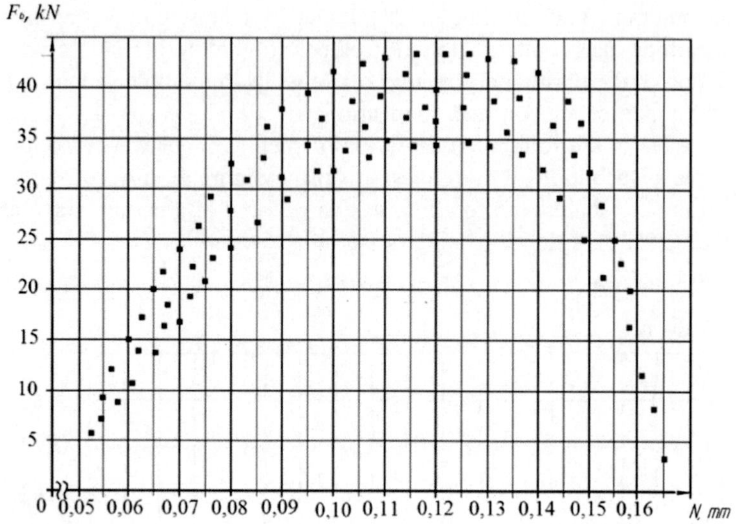

Fig. 6. Dependence of tungsten carbide insert cutters pressing off efforts F_B (16 mm insert cutter diameter) on the tension in the connection.

4.3 Effect of the Pressing Speed

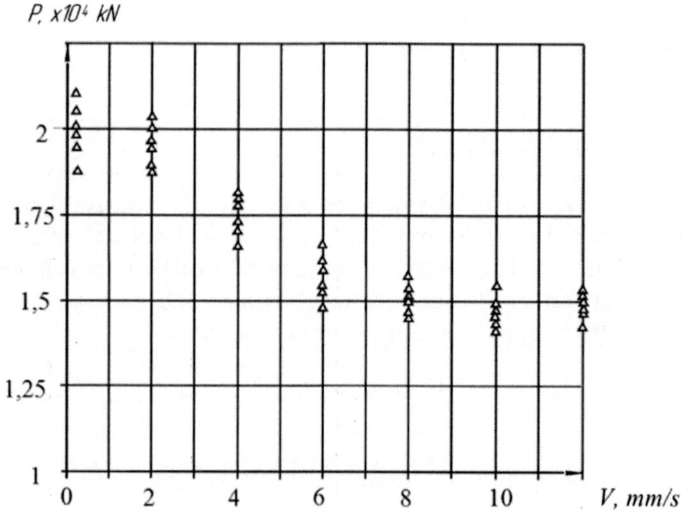

Fig. 7. The effect of the pressing speed process on the efforts of pressing off the inserted tungsten carbide insert cutters.

5 Conclusions

According to the obtained results, it was determined the effect of the roughness of the walls of the holes in the cone on the efforts of pressing the inserted tungsten carbide insert cutters of a typical profile of the shank. As you can see from Fig. 5, the best roughness of the walls of the hole in the cone is Ra 0,1 microns for roughness Ra 1,25–2,5 on shanks of tungsten carbide insert cutters.

To ensure a stable profile and the value of the holes, roughness can be achieved by introducing an additional final reaming operation, which would serve as the final calibration of the hole. For this purpose, it is effective to use monolithic hard-alloyed reamers, which provide higher stability and quality of the machined holes in contradistinction to the reamers with soldered hard-alloyed reamers.

From the graphic (Fig. 7), it is gathered that pressing speed has affected the efforts of pressing off the inserted tungsten carbide insert cutters. Numerous studies have shown that rational pressing speed must have been between 0,1 to 2 m/s. It significantly improves the reliability of the connection "Cone-tungsten carbide insert cutter".

The pressing parameters should be selected according to the strength and plasticity of the conical steel. It is necessary to take into account the nature of stress distribution is not only closely deployed two tungsten carbide insert cutters, but in groups of tungsten carbide insert cutters in the cones rolling-cutter row.

To fit the tungsten carbide insert cutters on the main rolling-cutter row (as the most loaded), it should be pressed under conditions of high tension. To improve the quality of pressing, before pressing, tungsten carbide insert cutters should be to moistened in oxidized paraffin. The tungsten carbide insert cutters must be calibrated to eliminate the uneven roughness on the forming shank.

Further improvement of the selective assembling technology should be used.

References

1. Gupta, A., Chattopadhyaya, S., Hloch, S.: Critical investigation of wear behaviour of WC drill bit buttons. Rock Mech. Rock Eng. **46**(1), 169–177 (2013). https://doi.org/10.1007/s00603-012-0255-9
2. Jakym, R., Slipchuk, A.: Assessment of reliability and criteria for improving the quality of rock cutting equipment of tricone drilling bits for well-boring especially hard rock. Bull. NTU "KhPI" **45**(1321), 77–86 (2018)
3. Yong, D., Chen, M., Jin, Y., Yakun, Z., Daiwu, Z., Lu, Y.: Theoretical and experimental study on the penetration rate for roller cone bits based on the rock dynamic strength and drilling parameters. J. Nat. Gas Sci. Eng. **36**, 117–123 (2016)
4. Nickel, J., Shuaib, A.N., Allam, I.M.: Wear studies of tungsten carbide cutting tools using micro-beam techniques. J. Tribol. **121**(1), 177–184 (1999). https://doi.org/10.1115/1.2833800
5. Rezvanizaniani, S.M., Barabady, J., Valibeigloo, M., Asghari, M., Kumar, U.: Reliability analysis of the rolling stock industry. Int. J. Perform. Eng. **5**(2), 167–175 (2009)
6. Slipchuk, A., Kuk, A.: Evaluation of the permissible moment in a roller cone drill bit providing the prescribed reliability of work. Ukrainian J. Mech. Eng. Mater. Sci. **4**(1), 116–124 (2018)

7. Pandey, P.A., Mukhopadhyay, K., Chattopadhyaya, S.: Reliability analysis and failure rate evaluation for critical subsystems of the dragline. J. Braz. Soc. Mech. Sci. Eng. **40**(2), 50 (2018). https://doi.org/10.1007/s40430-018-1016-9
8. Yakim, R., Petrina, Y.U., Yakim, I.: Scientific and Practical Bases of the Technology for Manufacturing Cone Drill Bits and Improving Their Quality and Efficiency. IFNTUNG Publ, Ivano-Frankovsk (2011)
9. Kershenbauma, V.: Drilling Rock Cutting Tool. International Engineering Encyclopedia, vol. 1. Cone bits. Neft' i Gaz, Moskow (2003). [in Russian]
10. Zhidovtsev, N., Kershenbaum, V., Ginzburg, E.: Durability of Roller Bits. Nedra, Moskow (1992). (in Russian)
11. Slipchuk, A., Yakim, R.: Refinement technology pressing of tungsten carbide insert cutter in roller cone bit. New Sol. Mod. Technol. **7**(1229), 134–143 (2017)
12. Chen, S., Dahlem, J., Rayburn, C.: A study of drilling performance of energy balanced roller cone bit. In: SPE Asia Pacific Oil and Gas Conference and Exhibition. Society of Petroleum Engineers Asia Pacific Oil and Gas Conference and Exhibition, Jakarta (2003). https://doi.org/10.2118/80493-MS
13. Huang, Z., Li, Q., Zhou, Y., Jing, S., Ma, Y., Wengang, H., Fan, Y.: Experimental research on the surface strengthening technology of roller cone bit bearing based on the failure analysis. Eng. Fail. Anal. **29**, 12–26 (2013)
14. Shigemi, N.: Feasibility study on roller-cone bit wear detection from axial bit vibration. J. Petrol. Sci. Eng. **82**, 140–150 (2012)
15. Schroder, J.: Cone retention and tapered bearing preload system for roller cone bit. Patent of US, № 20130105229A1 (2011)
16. Jon, S., Maurizio, D.P., Alun, R., Jesse, Y.: Bearing innovations extend roller-cone bit life. Oil Gas J. **114**(6), 50–55 (2016)
17. Belkin, I.: Means of Linear-Angular Measurements. Directory. In: Mechanical Engineering, Moscow (1987). [in Russian]
18. Gustafson, A., Schunnesson, H., Kumar, U.: Reliability analysis and comparison between automatic and manual load haul dump machines. Qual. Reliab. Eng. Int. **31**(3), 523–531 (2015). https://doi.org/10.1002/qre.1610
19. Prakash, S.: Mukhopadhyay: reliability analysis of tricone roller bits with tungsten carbide insert in blasthole drilling. J. Int. J. Min. Reclam. Environ. **34**(2), 101–118 (2020). https://doi.org/10.1080/17480930.2018.1530055
20. Ratnam, M.: Factors affecting surface roughness in finish turning. Ref. Mod. Mater. Sci. Mater. Eng. **1**(1), 1–25 (2016)
21. Bybee, K.: Drilling performance of an energy-balanced roller-cone bit. J. Petrol. Technol. **12**, 49–50 (2003)
22. Zhou, R.S., Nixon, H.: A contact stress model for predicting rolling contact fatigue. J. Commer. Veh. **101**(2), 556–563 (1992)
23. Meeker, W.Q., Escobar, L.A.: Statistical Methods for Reliability Data. John Wiley & Sons, New York (2014)

Methods of Evaluating the Wear Resistance of the Contact Surfaces of Rolling Bearings

Kostiantyn Svirzhevskyi[1], Oleg Zabolotnyi[1(✉)], Anatolii Tkachuk[1], Valentyn Zablotskyi[1], and Dagmar Cagáňová[2]

[1] Lutsk National Technical University, 75, Lvivska Street, Lutsk 43018, Ukraine
volynasi@gmail.com
[2] Slovak University of Technology, 25, Bottova, 917 24 Trnava, Slovakia

Abstract. In modern conditions of development of mechanical engineering, high priority is given to experimental and material science aspects of assessing the wear resistance of conjugate friction surfaces, which consist of the selection of materials with high wear resistance characteristics. However, this way, you can only reduce the intensity of the wear process, but not control the wear process itself and, most importantly, the changes in the state and performance of the mating surfaces that occur as a result of the wear process. The durability of machines is laid down at the design stage and depends on the design scheme, materials used, lubricants, and other factors. An increase in the durability of mated machine parts is impossible without creating modern engineering methods for calculating wear resistance, which would consider the physical and mechanical characteristics of materials (friction pairs, modes of operation of the load node, and angular velocity), external friction conditions (environment, and lubrication), as well as the design and technological features of the friction interface. During the wear of two contiguous bodies, the unevenness of one surface is exposed to the unevenness of the other surface. In this case, the irregularities of the more durable material act similarly to the abrasive elements, cutting off the thin chip from the micro-irregularities of the surface of the less durable material. At the same time, the irregularities that cut this chip wear themselves out, just as cutting abrasive tools wear out.

Keywords: Tests · Wear rate · Burn-in period · Roller bearing ring · Abbott-firestone curve

1 Introduction

The existing methods of the mathematical description of the wear of coupled machine parts can be divided into two groups. The first one is based on the physical and mechanical laws of wear, considering the influence of various factors on the wear process. The second one is based on the analysis of quantitative changes in the wear process without taking into account physical processes [1].

Using computational methods to find optimal design solutions can significantly facilitate the process of creating durable machines in terms of wear resistance of their

V. Tonkonogyi et al. (Eds.): InterPartner 2020, LNME, pp. 453–463, 2021.
https://doi.org/10.1007/978-3-030-68014-5_45

parts [2, 3]. Using them, you can solve a number of scientific and practical problems. These problems are as follows: to choose and justify the optimal design parameters [4] of parts that provide a minimum wear rate; to set the limits of wear of parts; to select standard sizes of unified elements; to ensure uniform stability [5] of a node or part with several functional surfaces; assign wear-resistant materials and apply technological methods to strengthen them, justify the requirements for physical and mechanical properties; conduct a comparative assessment of the service life of parts (components), namely, several variants of design; predict the service life of parts based on the results of the short-term bench or operational tests [6–8].

2 Literature Review

A number of scientists [9–11] have developed analytical expressions describing the wear process (Table 1). However, they do not fully reflect all aspects of the wear process, due to the inability consider the specific contact of various tribological materials.

Table 1. Basic analytical expressions for determining the intensity of the wear process.

№	Expression	Type of wear
1	$J = \frac{\epsilon h_{max} A_r}{(v+1)\, d\, n_p A_c}$	Fatigue wear [7]
2	$J = \frac{h\, A_r}{(v+1)\, d\, n_p A_c}$	Mechanical wear [7]
3	$J = \frac{\rho \psi 1 P}{P_{cp} K \phi}$	Abrasive wear [7]
4	$J = \frac{tg\theta\, p}{6P}$	Mechanical wear [8]
5	$J = \frac{\pi h^2 (R - \frac{h}{3})\, n_a}{A_a\, d\, n_p}$	Abrasive wear [8]

Note: A_r and A_c – actual and contour contact area; P-load; h_{max}, tgθ, v – surface roughness parameters; H – hardness; ρ – density; P_{cp}, P_a – average and nominal pressure; f – coefficient of friction; I – the path of friction; R-gas constant; T – absolute temperature; ψ, n, r, k, λ, φ – empirical coefficients; ε – relative deformation; d – diameter of the contact zone

For the confirmation of the theoretical assumptions, tribological machines are used for durability tests. For kinematic characteristics, all installations for tribotechnical testing of materials are divided into two classes: unidirectional and alternating relative movement. Within each class, the installations are divided into two groups: face friction machines and friction contact machines, in each group two more subgroups are distinguished: by the coefficient of mutual overlap K_{mo}, so there are two boundary cases: $K_{mo} \rightarrow 1$ and $K_{mo} \rightarrow 0$, in practice, test machines are used, the diagrams of which are presented in Fig. 1 [2, 9, 10, 12].

We introduce the following notation, in the case of $0.5 < K_{mo} < 1$: for a – unidirectional end friction; fib – unidirectional friction on the formation; c – alternating end friction; d – alternating friction on the formation.

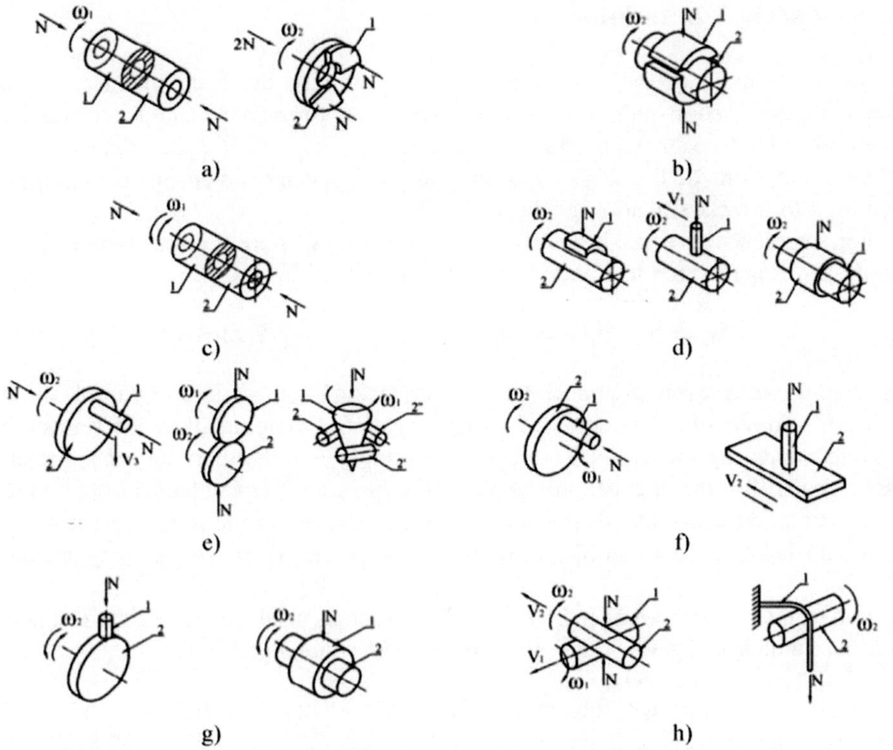

Fig. 1. Kinematic schemes of installations for determination of durability.

In the case of $0 \leq K_{mo} < 0.5$: for d – unidirectional end friction; h – unidirectional friction on the formation; same f – alternating face friction; g – alternating friction on the formation. The main criteria for evaluating tribotechnical characteristics are the wear intensity (wear resistance criterion) and the coefficient of friction (the criterion of mechanical losses during friction – the ratio of the friction force to the value of the normal force. The wear rate is the ratio of the thickness of the destroyed layer of material to the path where friction occurred:

$$I = h / L \quad (1)$$

where h – the value of the destroyed layer, mm, for the friction path L, mm, under specific regulated test conditions (lubricants, presence of abrasive, sample temperature, loads).

3 Research Methodology

For the assessment of the wear resistance of materials, the wear intensity of each sample of the friction pair is determined based on formula 1. But to consider the structural and tribotechnical properties of different types of friction pairs, let us consider different variations of the ways of contact of parts and movement of test equipment according to the classification given in Fig. 1.

For samples with a smaller surface of reciprocating friction pairs (sample 1, e.g., piston, and ring) for the test period with the number of cycles n:

$$I_1 = h / 2nH = \Delta q / 2\rho nHbl_{k1} = \Delta q_1/2\rho_1 nHA_a \tag{2}$$

where h1 – worn layer of sample 1 for n cycles; Δq1 – mass loss of sample 1 for n cycles; H – move rolling sample; 2H – friction path for all points of the friction surface of sample 1 during the cycle; 1 – the size of sample 1 in the direction of the relative displacement; b – the size of the sample in the direction perpendicular to the relative movement; what characterizes the nominal area of contact of the friction pair; $A_a = l_k$ b – nominal contact area pair (the working area of sample 1); ρ is the density of sample 1.

For samples 2 with a larger friction surface, such as a sleeve, during the test period with the number of cycles n, the wear intensity is defined as follows:

$$I_1 = h_2/ 2nl_k = \Delta q_2/2\rho_2 nl_2 bH = \Delta q_2/2\rho_2 nHA_a \tag{3}$$

where $2l_k$ – the greatest way of friction on the surface of sample 2 in one cycle; h_2 is the mean value of worn-out layer of sample 2 over n cycles; Δq2 – mass loss of sample 2 for n cycles; ρ_2 – the density of sample 2.

For friction pairs of rotational motion according to the "pad-roller" scheme, when determining the mass loss of each of the samples, the wear intensity is determined using these formulas.

For pads for the test period with the number of revolutions n:

$$I_1 = h_1/ L_1 = \Delta q_1/2\pi RnF\gamma_1 \tag{4}$$

where h1 – worn layer of the sample for n turns (taken uniformly on the friction surface of the sample); Δq1 – mass loss of the sample in n revs; L 1 = 2πRn is the path of friction of the sample in n of turns; R is the radius of the roller; F = bl – of the nominal contact area pair; l – the sample size-pads in the direction of the relative displacement; b – the sample size-pads in the direction perpendicular to the relative movement; γ_1 is the specific density of the material of the sample pad.

For the roller sample during the test period with the number of revolutions n, the wear intensity is equal to:

$$I_2 = h_2/ L_2 = \Delta q_2/2\pi Rnb\gamma_1 \tag{5}$$

where h_2 – the average thickness of the worn layer of the sample clip for n revolutions; L_2 = l is the maximum path friction points of the sample surface of the roller in one revolution; Δq_2 – mass loss of the sample-roller n speed; γ_2 is the specific density of the material sample clip.

At end friction:

$$I = \Delta h \,/2\pi r_{av} n \tag{6}$$

where Δh – average linear wear for n test cycles; r_{av} – the average radius of the contact area; n – the total number of revolutions of the sample.

In the case of wiping the socket with a cylindrical sample on a flat sample when measuring the total wear, the average wear intensity during the increase in the number of revolutions n_i is calculated using the formula:

$$I_{hi} = hi/30\pi n_i \tag{7}$$

where hi – the average value of the displacements, mm.

In addition to the wear rate, the test results are evaluated using the relative wear resistance, that is, by weighing samples before and after the tests, thereby determining the average arithmetic value of the mass loss g_{st}, reference samples and the average arithmetic value of the mass loss of samples of the test material g_{ts} (8, 9):

$$g_{st} = \frac{\sum_{i=1}^{m} g_{sti}}{m}, \tag{8}$$

$$g_{ts} = \frac{\sum_{i=1}^{m} g_{tsi}}{m}, \tag{9}$$

where g_{sti} , g_{tsi} – mass loss, g, during testing of reference samples and samples of the test material, respectively; m – the number of samples of the test material.

The relative wear resistance of the material under study is expressed as:

$$K_{WR} = (g_{st} \cdot \rho_{ts} \cdot n_{ts}) \,/\, (g_{ts} \cdot \rho_{st} \cdot n_{st}) \tag{10}$$

where ρ_{st} , ρ_{ts} – density of the reference and test materials, respectively, g/cm^3 ; n_{st} , n_{ts} – the number of revolutions of the roller during testing of the reference and test materials.

During the measurement of sample sizes before and after testing, the relative wear resistance is determined using the formula:

$$\epsilon = \Delta h_{st}/\,\Delta h_{ts} \cdot (d_{st}/\, d_{ts})^2 \tag{11}$$

where Δh_{st} – absolute linear wear of the reference sample; Δh_{ts} – absolute linear wear of the test sample; d_{st} – the actual diameter of the reference sample; d_{ts} – the actual diameter of the test sample.

The effect of the longitudinal feed (or a value proportional to the thickness of the removed metal layer) during machining, is expressed as follows:

$$J = H_{max} = H_p + H_y \cdot \left(1 - \frac{s^2}{2R^2}\right) \quad (12)$$

where H_{max} – the real height of the micro-irregularities, in microns; H_p – the calculated height of the irregularities obtained from the geometric dimensions of the tip of the cutter; H_y – elastic recovery of metal after removal of load from the indenter; s – longitudinal flow; R – the radius of the tip of the cutter.

The nature of the dependence H_{max} = f(s) during cutting is, to some extent, similar to the nature of the dependence of wear on the specific load, if the tangent surfaces have a different micro geometric structure, that is, different height and pitch between adjacent irregularities.

Table 2. Values of the parameters of the reference surface curves for transverse and longitudinal roughness.

v_1	1	2	3	2
v_2	1	1	1	2
A_{r2}/A_{r1}	2	3	4	5.87

Studies of the wear process and determination of the burn-in period were performed by the installation VNIPP-542 (Fig. 2).

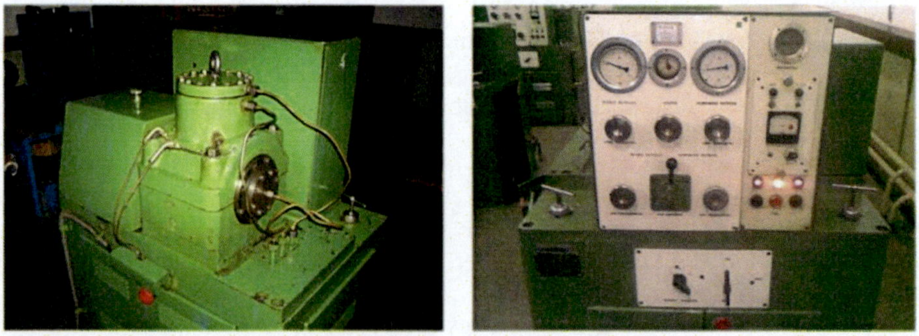

Fig. 2. Installation of VNIPP-542 to test the wear resistance of bearings.

Tests were performed on new roller bearings. The experiment ended when the bearing began to collapse. Thus was determined by the period of stable operation of the roller bearing, depending on the method of finishing.

The samples are used roller bearings of the 700 series (steel 100Cr6, hardness HRC 60–62), which were made according to various finishing technologies with the microstructure of martensite and carbides [13–15]. For testing 5, gradations of purity of treatment were taken. The first sample was made by the method of hardening-smoothing treatment (height of inequalities 0.2 μm). The remaining four samples were prepared by necessary technology as superfinishing with the height of micro-inequalities 0.8; 1.5; 3, and 6 μm. The inner and outer rings had the same roughness.

If the material densities of the reference and test samples are equal, the ratio of absolute linear wear in formula (11) may be replaced by the ratio of absolute wear of the mass.

4 Results

Thus, it would be justified to say that wear is the same chip removal process as other abrasive machining processes [1, 16]. The only difference is that the machining is performed with a cutting tool with a controlled geometry (deterministic approach), and the treated surfaces are regularly located irregularities of the same size, one layer of chips is removed in one pass (stochastic approach). During the wear process, the irregularities of the contact body of the stronger material have a stochastic geometric shape on the destroyed surface after burn-in, the irregularities remain in random order. Besides, during wear, the phenomenon of molecular setting has a determining effect. Therefore, for the characteristic of wear resistance, the dependencies for the abrasive processing processes will be set fair [17, 18]. If you select the prevailing factors that affect the wear process, the effect of these factors will be the same as the effect of similar factors in the process of cutting metals.

The dependence of $H_{max} = f(s)$ is expressed by a logarithmic curve (Fig. 3), which for values $s = R\sqrt{2}$ rises above the curve $H_p = f(s)$, and at a point intersects with this curve and goes down. Thus, it is fair to say that if you feed less than the real height of asperities higher than the calculated height H_p, and after the specified flow value of H_{max} is less than H_p.

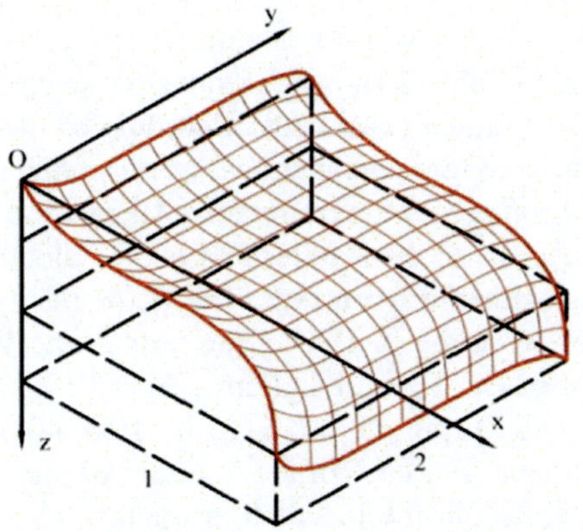

Fig. 3. Scheme for constructing a reference curve of the Abbot-Firestone surface: 1 – transverse profile of the curve (ox); 2 – the longitudinal profile of the curve (oy).

One of the most common methods for evaluating the microgeometry of the surface layer of solids is the construction of the reference surface curve (Abot-Firestone curve). This curve characterizes the filling of zones of the normal cross-section of a solid body with material located between parallel lines drawn along the body profile [16, 19]. The most objective characteristic is the curve of the dependence of the material filling zones of the surface of a solid formed by planes parallel to its surface. In [10], recommendations are given for experimentally obtaining such a dependence. This method involves the construction of transverse and longitudinal curves of the reference surface. The abscesses of the obtained curves corresponding to the same level are multiplied to obtain the values of the cross-section areas. It is well-known that depending on the processing method, there is either a specific orientation in the location of micro-steps and their shape or an isotropic geometric structure for all directions. A clear focus is observed during turning, milling, and grinding. The isotropic distribution of protrusions is typical for polishing and strengthening-smoothing treatment [12]. The curves of the support surface in the case of an isotropic distribution of protrusions are the same for all directions. And the cross-sectional area of the material at this level corresponds to the square of the abscissa of the support surface curve. An example of such a surface is shown in Fig. 3.

The equations of curves of reference surfaces in their initial part can be represented as:

$$\begin{aligned} x &= b_1 z^{v_1}, \\ y &= b_2 z^{v_2}, \end{aligned} \tag{13}$$

where b_1, v_1, b_2, v_2 – parameters of support surface curves for transverse and longitudinal roughness. Given (13), the equation of the cylindrical surface will take the form:

$$z = \frac{1}{b_1^{\frac{1}{v_1}}} x^{\frac{1}{v_1}} + \frac{1}{b_2^{\frac{1}{v_2}}} y^{\frac{1}{v_2}} . \tag{14}$$

The dependence (14) allows you to determine the cross-sectional area of the material at a given level z. In Table 2 shows the value of the cross-sectional areas of the material defined using the formula (14) – Ar_1 according to the given Ar_2 method.

After testing, the volume of removed metal was determined using a research facility based on the model 201 Profiler. Profilograms were taken from the forming functional surfaces of the rings. The probe of the Profiler first moved along the unworn part of the surface layer, then along with the worn one. The particular type of the profilogram determines the volume of worn metal. During the tests at the VNIPP-542 installation, the following modes were set: F_a = 21 kgf/cm^2, F_r = 12 kgf/cm^2 (axial and radial loads, respectively); n = 6000 rpm; T = 0.5–15 h. 10W-40 oil with 10 drops per minute was used as a lubricant. For each of the 5 grades of purity, 10 identical samples were taken, which were tested in the installation for 0.5–15 h. A total of 50 control points were obtained, the location of which is shown in Fig. 4. The nomogram shows the mass of the cut material in grams along the ordinate axis, the test time in hours along the abscissa axis (to the right), and the maximum height of irregularities in the roller bearing rings on the applicant axis (to the left). The nomogram shows that the dependence of wear J during the burn-in period on the surface layer roughness (H_{max}) is expressed for all H_{max} values by a line that intersects the ordinate axis above the origin point.

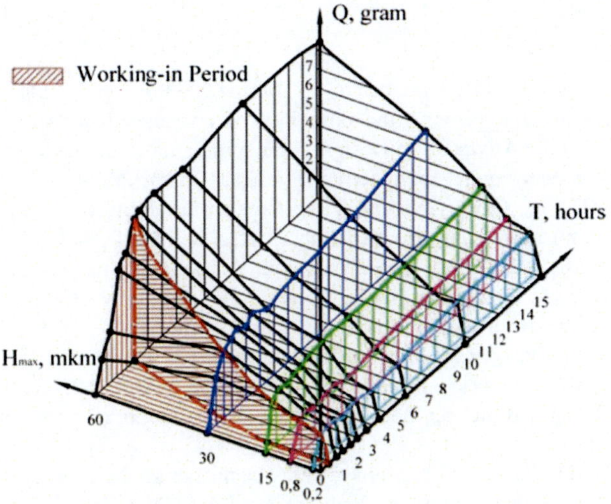

Fig. 4. Nomogram for determining the period of wear of friction surfaces by the amount of material removed due to wear.

The dependence of wear on the duration of tests for all grades of surface layer purity is approximately the same.

5 Conclusions

The proposed dynamic calculation is based on the analysis of the mechanism of the process of destruction of parts under different types and modes of friction and allows obtaining approximate wear rates of various materials and their interfaces. The value of this method also lies in the fact that, based on the calculated data, it is possible to lay the foundations of high wear resistance of friction pairs at the design stage. It was found that the amount of wear increases sharply during the first 0.5–2.5 h, and then growth slows down. The burn-in area, when the micro-roughness of the machining is cut off, is shown in Fig. 4 is marked in red and shaded. In this area, wear largely depends on the burn-in time. If the radial load wear is very high and does not rely on the microgeometry of the contacting surfaces, then a mathematical description of the wear phenomenon is carried out based on the results of processing data obtained during operation. Experimental data confirm the proposed method for analyzing the wear resistance of the mated surfaces of parts to ensure the maximum duration of the period of stable operation by assigning technological modes of mechanical processing. It can be used in the design of technological processes for manufacturing parts.

It is planned to develop a technology for machining the parts of the bearings, which will minimize the working-in period. This will allow the machines to be operated immediately at rated modes bypassing the running-in period.

References

1. Tkachuk A., Zablotskyi V., Kononenko A., Moroz S., Prystupa S. Directed formation of quality, as a way of improving the durability of conjugated parts of friction pairs. In: Ivanov V. et al. (eds) Advances in Design, Simulation and Manufacturing II, DSMIE-2019. Lecture Notes in Mechanical Engineering. Springer, Cham (2020).
2. Kaplun, P.V., Dykha, O.V., Gonchar, V.A.: Contact durability of 40 Kh steel in different media after ion nitriding and nitroquenching. Mater. Sci. **53**(4), 468–474 (2018)
3. Marchuk, V., Kindrachuk, M., Kryzhanovskyi, A.: System analysis of the properties of discrete and oriented structure surfaces. Aviation **18**(4), 161–165 (2014)
4. Ivanov, V., Dehtiarov, I., Pavlenko, I., Liaposhchenko, O., Zaloga, V.: Parametric optimization of fixtures for multiaxis machining of parts. In: Hamrol, A., Kujawińska, A., Barraza, M. (eds.) Advances in Manufacturing II, MANUFACTURING 2019. Lecture Notes in Mechanical Engineering, pp. 335–347 (2019). https://doi.org/10.1007/978-3-030-18789-7_28.
5. Liaposhchenko, O., Pavlenko, I., Monkova, K., Demianenko, M., Starynskyi, O.: Numerical simulation of aeroelastic interaction between gas-liquid flow and deformable elements in modular separation devices. In: Ivanov, V., et al. (eds) Advances in Design, Simulation and Manufacturing II. DSMIE-2019. Lecture Notes in Mechanical Engineering, pp. 765–774. Springer, Cham (2020). https://doi.org/10.1007/978-3-030-22365-6_76.
6. Kukhar, V.V., Vasylevskyi, O.V.: Experimental research of distribution of strains and stresses in work-piece at different modes of stretch forging with rotation in combined dies. Metall. Min. Ind. **3**, 71–78 (2014)
7. Vasyliev, M.O., Mordyuk, B.M., Sidorenko, S.I., Voloshko, S.M., Burmak, A.P., Kindrachuk, M.V.: Synthesis of deformation-induced nanocomposites on aluminium D16 alloy surface by ultrasonic impact treatment. Metallofiz. Noveishie Tekhnol. **4**(38), 545–563 (2016)
8. Wang, Z., Yang, M.: Laser-guided discharge surface texturing. Laser Surf. Eng. 455–467 (2015).
9. Chernets, M.V.: Prediction of the life of a sliding bearing based on a cumulative wear model taking into account the lobing of the shaft contour. J. Frict. Wear **36**(2), 163–169 (2015)
10. Dykha, A., Sorokatyi, R., Makovkin, O., Babak, O.: Calculation-experimental modeling of wear of cylindrical sliding bearings. East.-Eur. J. Enterp. Technol. **5**(1(89)), 51–59 (2017)
11. Teja, P.S., Kumar, M.D., Krishna, R., Sreenivasan, M.: Simulation and optimization studies on the ring rolling process using steel and aluminum alloys. J. Eng. Sci. **6**(2), E36–E40 (2019). https://doi.org/10.21272/jes.2019.6(2).e6
12. Du, J., Liu, Z., Lv, S.: Deformation-phase transformation coupling mechanism of white layer formation in high speed machining of FGH95 Ni-based superalloy. Appl. Surf. Sci. **292**, 197–203 (2014)
13. Jiang, J., Lijue Xue, L., Shaodong Wang, S.: Discrete laser spot transformation hardening of AISI O1 tool steel using pulsed Nd:YAG laser. Surf. Coat. Technol. **205**(21–22), 5156–5164 (2011)

14. Kryshtopa, S., Petryna, D., Bogatchuk, I., Prun'ko, I., Mel'nyk, V.: Surface hardening of 40 Kh steel by electric-spark alloying. Mater. Sci. **53**(3), 351–358 (2017)
15. Umbrello, D., Jawahir, I.S.: Numerical modeling of the influence of process parameters and workpiece hardness on white layer formation in AISI 52100 steel. Int. J. Adv. Manuf. Technol. **44**(9–10), 955–968 (2009)
16. Kindrachuk, M., Radionenko, O., Kryzhanovskyi, A., Marchuk, V.: The friction mechanism between surfaces with regular micro grooves under boundary lubrication. Aviation **8**(2), 64–71 (2014)
17. Etsion, I.: Improving tribological of mechanical components by laser surface texturing. Tribol. Lett. **17**(4), 733–737 (2004)
18. Sychuk V., Zabolotnyi O., Somov D.: Technology of effective abrasive jet machining of parts surfaces. In: Ivanov V. et al. (eds) Advances in Design, Simulation and Manufacturing. DSMIE-2018. Lecture Notes in Mechanical Engineering. Springer, Cham (2019)
19. Zablotskyi, V., Moroz, S., Tkachuk, A., Prystupa, S., Zabolotnyi, O.: Influence of diamond smoothering treatment power parameters on microgeometry of working surfaces of conjugated parts. In: Tonkonogyi, V., et al. (eds.) Advanced Manufacturing Processes. InterPartner-2019. Lecture Notes in Mechanical Engineering, pp. 372–381. Springer, Cham (2020)

Stability of the Quality Parameters for the Surface Layer of Parts During Circular Grinding Operations

Alexey Yakimov[1](✉), Isak Karabegovic[2], Sergey Uminsky[3], Viktor Strelbitskyi[4], and Julia Shichireva[1]

[1] Odessa National Polytechnic University, 1, Shevchenko Avenue, Odesa 65044, Ukraine
yakimov_aa@ukr.net
[2] University of Bihac, Pape Ivana Pavla II 2, Bihac, Bosnia and Herzegovina
[3] Odessa State Agrarian University, 13, Panteleimonovska Street, Odessa 65012, Ukraine
[4] Odessa National Maritime University, 34, Mechnikova Street, Odessa 65029, Ukraine

Abstract. It is established that the regulation of the contact duration between the workpiece and the intermittent abrasive circle by changing the size of protrusions, depressions and their amount allows not only to control the thermal processes occurring in the cutting zone but also prevent the appearance of resonant states in the elastic machine system. Theoretically, the conditions for the occurrence of elastic system resonant states of a flat grinding machine during a periodic cutting interruption process are determined. The conditions for a significant decrease in the intensity of forced oscillations and a probability decrease of parametric excitation in the elastic system of the machine are revealed. It is shown that increasing the number of cutting protrusions on the circle contributes to its operation in the mode of intensive self-sharpening, which makes it possible to reduce the number of the grinding wheel edits and thereby increase the processing performance. It is theoretically proved that it is possible to reduce the penetration depth into the part of critical temperatures that cause structural and phase changes by increasing the heat source speed of movement and increasing the number of cutting protrusions on the working surface of the abrasive tool. It is determined that it is possible to reduce the amplitude of forced oscillations, decrease the temperature in the cutting area, reach high cutting ability of the grinding wheel and, consequently, improve the quality and processing performance by using circles with lots of slots on their working surface.

Keywords: Traverse feeding · Structural transformations · Thermal source · Intermittent grinding · Resonant states

V. Tonkonogyi et al. (Eds.): InterPartner 2020, LNME, pp. 464–476, 2021.
https://doi.org/10.1007/978-3-030-68014-5_46

1 Introduction

Ensuring the stability of the quality parameters of the surface layer of parts during circular grinding operations with an intermittent working surface was discussed in the research studies [1, 2]. Performance indicators of machine parts are closely connected with the state of their working surfaces. In many cases, the required accuracy and roughness of these surfaces can only be achieved by grinding [3]. When grinding in the cutting zone, thermodynamic processes occur that hurt the treated surfaces [4, 5]. Burnouts and internal stresses appear in the surface layer of the parts. Burns reduce the specified surface hardness, and internal stress can lead to cracks [6, 7].

2 Literature Review

Using abrasive wheels with intermittent working surfaces is one of the possible ways to reduce the temperature during grinding and, as a result, prevent defects [8–11].

The disadvantages of these circles were the significant consumption of abrasive when editing and profiling working surfaces, as well as time-consuming spent on forming their cutting macro-relief outside the grinding machine [12, 13]. The discreteness of cutting carried out by intermittent circles causes the periodically acting force, which is a source of additional vibrations in the elastic system of the machine [14, 15].

There are two types of unstable conditions that can negatively affect the quality of the treated surface during intermittent grinding. It is a shock resonance that occurs when one of the natural frequencies of the elastic system of the grinding machine coincides with the contact frequency of cutting fragments of the working surface of the discontinuous circle with the processed material [16, 17]. The risk of parametric resonance occurs when the frequency of changing in cutting rigidity is equal to twice the natural frequency of the elastic system of the grinding machine [18, 19].

The purpose of this work is to create scientific and practical recommendations for stabilizing the quality parameters of the parts' surface layer during intermittent grinding.

3 Research Methodology

The research was conducted based on the theory of metal cutting and the theory of vibrations. It is known [20, 21] that if you increase the speed of longitudinal table feed surface grinding machine and simultaneously reduce the depth of cut, keep maintaining processing performance, then with a slight increase in surface temperature, it is possible to reduce the depth of its penetration into the material significantly and thereby reduce the thickness of the defective (released) layer. Let us check the validity of this statement concerning gear grinding on the MAAG machine with zero adjustments of abrasive disc wheels.

During the period of the grinding wheel running on the wheel tooth, the heat source goes from the head to the leg and back. Each point of the tooth side surface is subjected to multiple thermal effects in the process of tooth grinding. A large number of rolling movements, accompanied by a large number of heating and cooling phases occur during the period of the heat source movement in the longitudinal feed direction (along the axis of the processed wheel).

4 Results

The speed of the heat source moving along the involute tooth profile changes according to the sinusoidal law while grinding the gears on the MAAG machine with two disc wheels set to zero initial cutting rail. Therefore, the temperatures in the contact zone and the penetration depth of the critical T^*_{cr} temperatures that cause structural and phase transformations in the treated material will differ at different points in the involute profile of the tooth. The calculated dependences of the speed of running the abrasive wheel along the side surface of the tooth on the metric characteristics of the processed wheel are presented in Fig. 1. Calculations of V_{obk} speeds were made in the middle part of the profile of the processed tooth (in the area of the dividing circle). The diameter of the gear wheel increases, the V_{obk} running speed also increases.

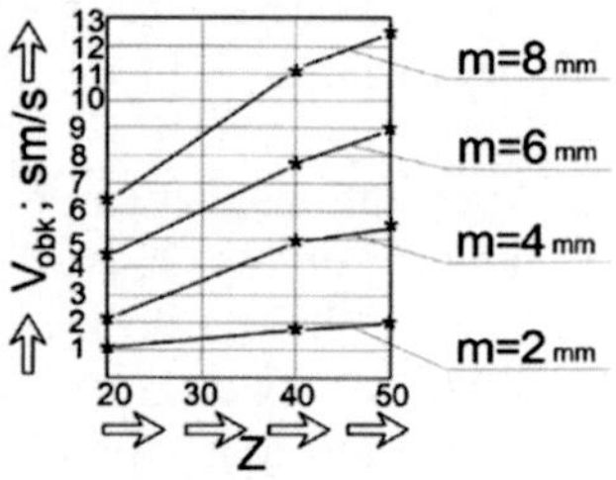

Fig. 1. Dependence of the run-in speed in the area of the dividing circle of the processed gear-wheel on the number of teeth Z for modules m = 2mm, 4mm, 6mm, 8mm, n' = 112 motion in the minute.

The temperature at the point located near the top of the tooth (let's call this point the letter "C") can be calculated using the following formula:

$$T* = \sum_{k'=0}^{k'=n_c-1} \frac{1.3 \cdot q_1 \cdot a}{\pi \cdot \lambda \cdot V_c} \int_{z'-(n_c-k')\cdot L'-H'}^{z'-(n_{c-k'})\cdot L'+H'} e^{-\xi} \cdot K_0'(\xi) \cdot d\xi + \sum_{k'=0}^{k'=n_c-1} \frac{q_2 \cdot a}{\pi \cdot \lambda \cdot V_c} \int_{z'-(n_c-k')\cdot L'-H'}^{z'-(n_{c-k'})\cdot L\prime+H'} e^{-\xi} \cdot K_0'(\xi) \cdot d\xi + \frac{q_2 \cdot a}{\pi \cdot \lambda \cdot V_c} \int_{-2\cdot H'}^{0} e^{-\xi} : K_0'(\xi) \cdot d\xi \quad (1)$$

where: $Z = n_c \cdot L'$

$n_c = \frac{\sqrt{D_{kr} \cdot 2.2 \cdot m + (2.2 \cdot m)^2}}{S}$ – number of heating; q_1 and q_2 – the intensity of heat flows in the direction of departure from the tooth head and, respectively, in the direction of approach to the head, W/m^2; λ – thermal conductivity of the processed material, I/(m•S•°C); S – longitudinal feed (mm) for one double stroke of the table; a – temperature conductivity of the processed material, m^2/S; V_c – the speed of the heat source on the tooth head, m/s; $K_0'(\xi)$ – the Bessel function of the second kind of zero-order; ξ – the variable of integration;

$$L' = \frac{V_{sr} \cdot l}{2 \cdot a}; l = (m + 1.25 \cdot m + 4L_{sr})$$

m – gear module, m; Lsr – average width of the heat source within the height of the tooth, mm; Vsr – average movement speed of the heat source, m/s; H' – average dimensionless half-width of the heat source;

$$H' = \frac{V_{sr} \cdot L_{sr}}{4 \cdot a}$$

The equation describing the temperature in the dividing circle contact zone on around the point "A" has the form:

$$T* = \sum\nolimits_{k'=0}^{k'=n_A-1} \frac{1.3 \cdot q_1 \cdot a}{\pi \cdot \lambda \cdot V_A} \int_{z'-(n_A-k') \cdot L'-H'}^{z'-(n_{A-k'}) \cdot L'+H'} e^{-\xi} \cdot K_0'(\xi) \cdot d\xi + \sum\nolimits_{k'=0}^{k'=n_A-1} \frac{q_2 \cdot a}{\pi \cdot \lambda \cdot V_A} \int_{z'-(n_A-k') \cdot L'-H'}^{z'-(n_{A-k'}) \cdot L'+H'} e^{-\xi} \cdot K_0'(\xi) \cdot d\xi + \frac{q_2 \cdot a}{\pi \cdot \lambda \cdot V_A} \int_{-2 \cdot H'}^{0} e^{-\xi} : K_0'(\xi) \cdot d\xi \quad (2)$$

where: $Z = n_A \cdot L'$

$$n_A = \frac{\sqrt{D_{kr} \cdot m + m^2}}{S}$$

Va – the speed of the heat source around the point "A" (on the dividing circle), m/s;

The expression that describes the temperature in the contact zone on the tooth stem (around the point "B") has the form:

$$T* = \sum\nolimits_{k'=0}^{k'=n_A-1} \frac{1.3 \cdot q_1 \cdot a}{\pi \cdot \lambda \cdot V_B} \int_{z'-(n_B-k') \cdot L'-H'}^{z'-(n_{B-k'}) \cdot L'+H'} e^{-\xi} \cdot K_0'(\xi) \cdot d\xi + \sum\nolimits_{k'=0}^{k'=n_B-1} \frac{q_2 \cdot a}{\pi \cdot \lambda \cdot V_B} \int_{z'-(n_B-k') \cdot L'-H'}^{z'-(n_{B-k'}) \cdot L'+H'} e^{-\xi} \cdot K_0'(\xi) \cdot d\xi + \frac{q_2 \cdot a}{\pi \cdot \lambda \cdot V_B} \int_{-2 \cdot H'}^{0} e^{-\xi} : K_0'(\xi) \cdot d\xi \quad (3)$$

V_B – the speed of the heat source on the leg (in the region of point "B"), m/s;

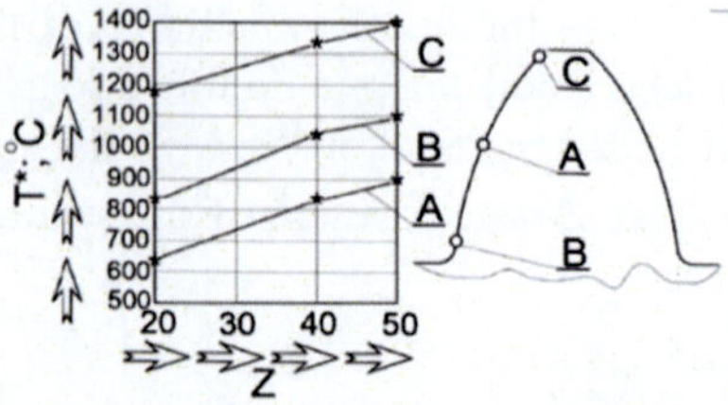

Fig. 2. Calculated temperatures that occur during grinding the gears (m = 8 mm, Z = 20,30,40,50) made of 12x2H4A cemented steel on a MAAG machine with zero settings of circles in the mode: t = 0.02 mm, S = 3 mm/motion; n = 112 motion in the minute.

Figure 2 shows the dependence of the grinding temperature on the number of processed wheel teeth, that is calculated using the formulas (1), (2), (3), respectively, for points "C", "A", "B". From Fig. 2, it can be seen that the lowest temperatures are formed at the midpoint of the tooth profile, where The V_{obk} run-in speed is maximum.

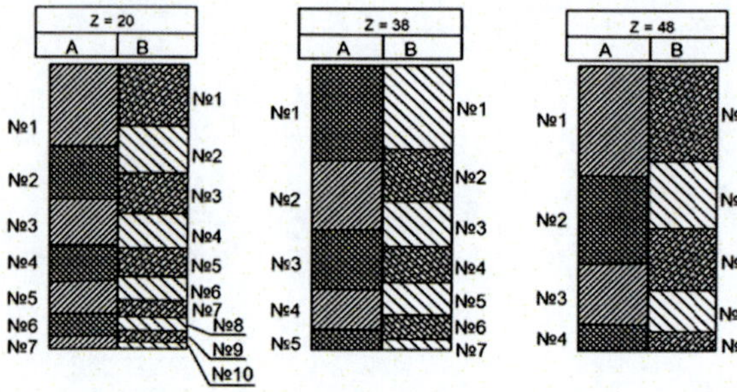

Fig. 3. The distribution of the allowance along with the passes during grinding of the involute tooth profile for wheels with the number of teeth: z = 20 (a), z = 38 (b), z = 48 (c) on the MAAG machine in the middle part (point A) and at the beginning (point B).

Figure 3 shows the distribution of allowances along with the passes during grinding by two-disc circles of involute profiles for gears with different numbers of teeth and the same module m = 2mm according to the zero scheme at the beginning (point B) and in the middle (point A). The division of the allowance along the grinding passes was implemented according to the method based on the formula (4)

$$\chi = \left(\frac{T_{pr}^* \cdot q \cdot H \cdot \left(1.6 \cdot \sqrt{U} - 0.16 + 0.32 \cdot U\right)}{T_{sp}^* \cdot \lambda \cdot U \cdot 2 \cdot T_{kr}^*} \right)^{0.25} \cdot \frac{H}{2 \cdot \left(1.74 \cdot \sqrt{U} - 0.61\right)} \tag{4}$$

$$\frac{T_{pr}^*}{T_{sp}^*} = \frac{1}{0.036 \cdot n + 0.64 \cdot N + 0.74} \tag{5}$$

$$10 \le n \le 20; 0.25 \le N \le 0.50$$

$T^*_{кр}$ – critical temperature, $T^*_{кр}$ = 400 °C. H – contact area width of the grinding wheel and the part, m; $U = \frac{V_{obk} \cdot H}{2 \cdot a}$ – the dimensionless complex of the heat source movement speed; V_{obk} – the run-in speed, m/s; T^*_{pr}, T^*_{sp} – temperatures that are generated by grinding circles with, respectively, intermittent and continuous work surfaces, °C; q – the intensity of the heat flow, W/m^2; a – temperature conductivity of the processed material, J/(m•s• °C).

From Fig. 3, it can be seen that the increase of the teeth number on the processed wheel leads to a cut depth increase assigned to individual grinding passes, and as a result, the number of passes decreases. On the one hand, the number of teeth increases leads to an increase in the temperature of the cutting zone (Fig. 2), which should lead to assigned grinding depths decrease. On the other hand, the number of teeth increase is accompanied by the V_{obk} run-in speed increase (Fig. 1), which leads to a decrease in the penetration depth of the critical temperature into the part, which makes it possible to increase the depth of grinding and, as a result, reduce the number of passes. The calculations have shown that the decrease in the depth of temperature penetration into the processed material is mostly determined by an increase in the Vobk run-in speed than by an increase in the grinding temperature T*.

Besides, in all of these three cases (Fig. 3) the cutting depths assigned in the area of the dividing circle (at the point "A") exceed the cutting depths calculated for a point "B" (near the base of the tooth). It can be explained by the fact that the run-in speed at the point "A" is much higher than at the point "B" and the run-in speed increase slows down the process of heat spreading into the deep layers of the part. The most important criterion for evaluating the performance of an abrasive wheel is the ability to maintain its cutting capacity over time. Grinding wheels on a ceramic bundle lose their cutting ability due to the rapid wear of the abrasive grains tops and "salting" of the working surface. It leads to a temperature increase in the grinding zone and, as a result, to undesirable structural and phase transformations in the surface layer of the processed parts. The cutting capacity of discontinuous circles is estimated by the specific volume of metal removed per unit of time (mm^3\(S* N)) or by the specific volume wear of the circle, determined by the ratio of the volume wear of the circle Qkr (mm^3\min) for a certain period to the volume of the processed material Qm, ground by the circle for the same period:

$$Q = Qkr/Qm \tag{6}$$

The greater the value of this ratio, the better the cutting capacity of the grinding wheel is, due to better self-sharpening of its working surface.

Figure 4(a, b) shows the graphical dependence of specific volumetric wear of intermittent abrasive discs 24A 25P CM2 ПП250x20x76 7 K5 after a 10-min period of grinding the samples of steel 45 (HRC 50–54) on the surface grinding machine model 3Г71M with the cutting depth t = 0.05 mm (a) and t = 0.10 mm (b) cross-feed table S = 1 mm/stroke, the peripheral speed of the circle Vkr = 30 m/s on the speed of

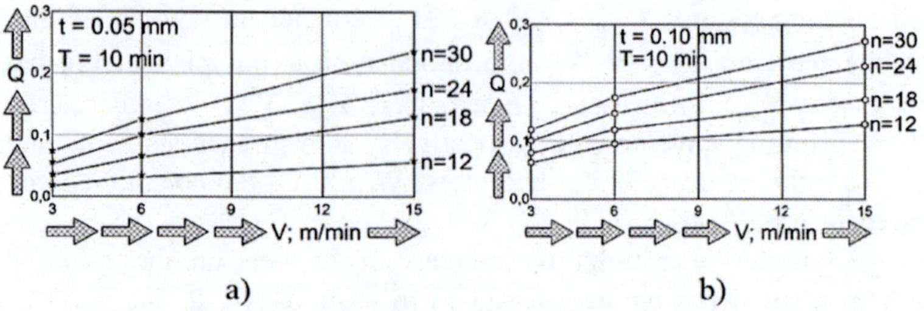

Fig. 4. Dependences of specific volume wear Q of intermittent abrasive wheels with the number of cutting protrusions on working surfaces n = 12; 18;24;30 on the table longitudinal feed V of the flat grinding machine for cutting depths t = 0.5 mm (a) t = 0.10 mm (b).

longitudinal displacement of table V and the number of cutting edges of abrasive tools on the working surfaces n. Working surfaces of all circles have the same ratio of discontinuity N = 0.6. From Fig. 4 (a, b) it can be seen that increasing the speed of longitudinal movement of the flat grinding table **V** and increasing the number of slots on the working surface of the circle **n** lead to an increase in the abrasive tool specific volume wear, i.e., contribute to its self-sharpening and, as a result, increase the cutting capacity of the circle. Figure 5 shows the dependence of the vibrations amplitude that occur in the elastic system of the machine while grinding by intermittent circles on the cutting depth. Calculations of the spindle oscillation amplitude were made using the formula (6).

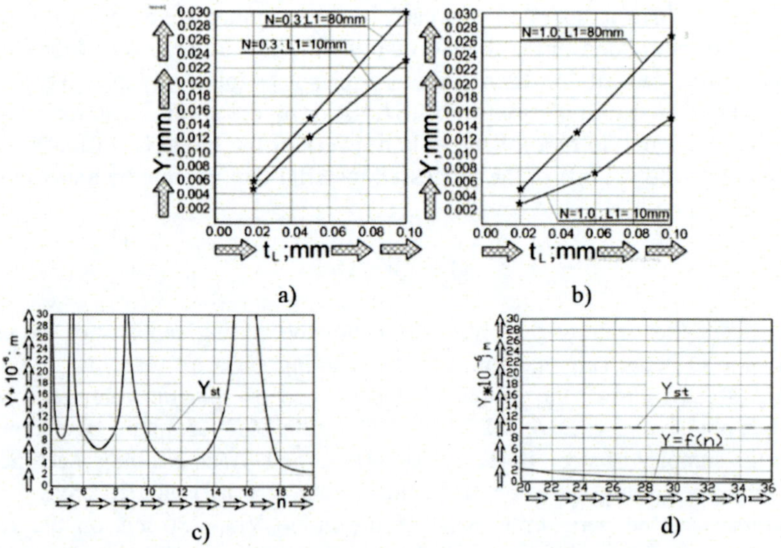

Fig. 5. The effect of the cutting depth t, set on the plane-grinding machine limb, (a, b) and the number of cutting protrusions N on the working surface of the circle (c, d, e) on the amplitude of forced vibrations Y.

$$Y = \frac{Y_{st} \cdot L_1 \cdot k}{V_{kr} \cdot C \cdot 2 \cdot \left|\sin\left(\frac{k \cdot (L_2 + L_1)}{2 \cdot V_{kr}}\right)\right|}, \quad (7)$$

$Y_{st} = \frac{F_{y.o.}}{C}$ – the value of grinding wheel pressing with a continuous working surface under the action of the $F_{y.o.}$, m; C – static rigidity, N/m; k –oscillations frequency 1/s;

V_{kr} – circumferential circle speed, m/s; L_1, L_2 – the length of the depression, and the cutting protrusion of the circle, m.

The size increase of the protrusions L1 and the depressions L2 with the constant discontinuity coefficient N is accompanied by a decrease in the amplitude of the forced oscillations Y (Fig. 5 a, b). The intensity of the amplitude decrease rises with the increase in the discontinuity coefficient and the cutting depth set by the t_l limit of the flat grinding machine. Reducing the size of protrusions and depressions and, as a result, increasing their number on the working surface of the abrasive tool leads to a decrease in the amplitude of forced vibrations (Fig. 5 d, e). It is attributable to the spindle with grinding wheel does not have enough time to do push-ups under the influence of impulse force numerically equal to the product of the force F_{yo} and the time of one cutting ledge ($\tau_1 = L_1/V_{kr}$), and the value equal to the value of extraction that occurs when statistical annex of total cutting forces F_{yo}. At the same time, an increase in the number of cutting protrusions on the circle (i.e., with a decrease in their length L_1), the force impulse decreases, which leads to a reducing of the abrasive tool pressing value from the processed material and, as a result, to an increase in the actual cutting depth t_F. It leads to an increase in the thickness of the slices contributed to individual cutting grains, which favors the self-formation of the cutting microrelief of the working surface and the abrasive tool. Besides, the increase in the number of cutting protrusions on the circle leads to an increase in the frequency of individual vibrations of the elastic system, which facilitates the chip formation process, prevents the processed material from sticking to the cutting grains and thus increases the time between two edits of the grinding wheel. Graphical data (Fig. 5 c, d, e) perfectly fit with the results of research [22, 23], under which, with an increase in the number of cutting protrusions on an intermittent circle, the amplitude of forced oscillations in the resonant mode decreases, and the grinding process proceeds in a stable mode. During the process of grinding by the circles with a small number of cutting protrusions, there is a risk of intense vibrations in the elastic system of the machine. While grinding by the intermittent circles, the cutting process is discrete and, therefore, is accompanied by periodic changes in the machine rigidity and elastic system. It can lead to parametric resonance under certain conditions. The condition of parametric instability of the elastic system of a flat grinding machine can be written as:

$$|L| > (1 + M)/2 \quad (8)$$

$$L = \left[k_1^2 \cdot sink_1 \cdot \tau_1 \cdot sink_2 \cdot \tau_2 - k_2^2 \cdot sink_1\tau_1 \cdot \sin(2 \cdot k_2 \cdot (\tau_1 + \tau_2)) - 2 \cdot k_1 \cdot k_2 \cdot cosk_2 \cdot \tau_1 \cdot cosk_2(\tau_1 + \tau_2)\right] \cdot \frac{1}{F \cdot E} \quad (9)$$

$$M = k_1 \cdot k_2 \cdot \cos(2 \cdot k_2 \cdot (\tau_1 + \tau_2)) \cdot \frac{1}{F^2 \cdot E} \tag{10}$$

$$E = h \cdot (k_2 + h \cdot sin2 \cdot k_2 \cdot \tau_1) \tag{11}$$

$$F = e^{h(\tau_1 + \tau_2)} \tag{12}$$

$$k_1 = \sqrt{\frac{C_o}{m'} + \frac{K_o}{2 \cdot m'} - h^2} \tag{13}$$

$$k_2 = \sqrt{\frac{C_o}{m'} - \frac{K_o}{2 \cdot m'} - h^2} \tag{14}$$

m' – reduced mass, $(N \cdot s^2)/m$; Ko – the hardness of the cutting, kg/m; Co – reduced rigidity of the elastic system of the grinding machine, kg/m; h – parameter that characterizes the degree of decrease in the intensity of oscillations over time;

$$K_o = C_o \cdot \left(\frac{t_L}{t_F} - 1 \right) \tag{15}$$

t_L – cutting depth without the pressing of the grinding wheel from the workpiece, mm; t_F – actual cutting depth, mm; τ_1, τ_2 – a time of contact and interruption duration of contact of the discontinuous circle's working surface respectively with the processed material, s.

$$\tau_1 = \frac{\pi \cdot D_{kr}}{n \cdot (1 + N) \cdot V_{kr}} \tag{16}$$

$$\tau_2 = \frac{\pi \cdot D_{kr}}{n \cdot \left(1 + \frac{1}{N}\right) \cdot V_{kr}} \tag{17}$$

$$N = \frac{L_2}{L_1}$$

L_2, L_1 – the lengths of the working sections of the circle surface and respectively the sections that do not take part in cutting, m; n – the number of sections of the abrasive discontinuous circle working surface that participate in the removal of the processed material; V_{kr} – circumferential rotation speed of the abrasive tool, m/s; D_{kr} – diameter of the grinding wheel, m. Calculations performed using formulas (7)–(16) allowed us to determine the boundaries of the elastic system parametric instability regions of the flat grinding machine in the coordinate system (Ko, N, n). Figure 6 shows the zones of parametric instability calculated for the range of values of the depression length ratio to the length of the ledge $0.30 \leq N \leq 0.60$ and in Fig. 7 – for the interval $1 \leq N \leq 4$.

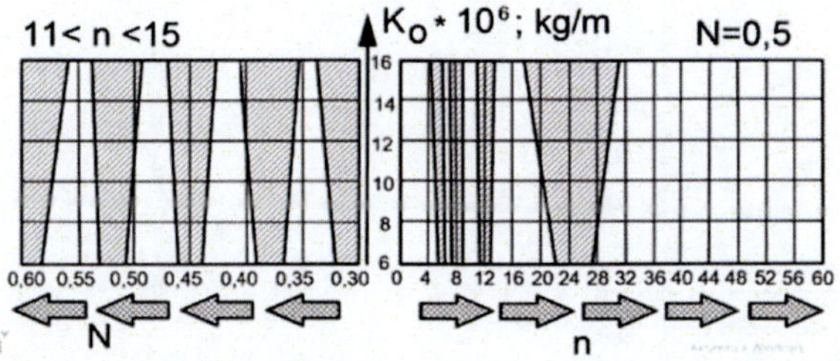

Fig. 6. Parametric instability zones constructed in two-dimensional coordinate systems.

The same pattern is observed for each studied range of variation N: an increase in the cutting rigidity of the K_0 (i.e., a deterioration in the cutting ability of the circles) is accompanied by an expansion of the parametric instability areas of the elastic system of the flat grinding machine.

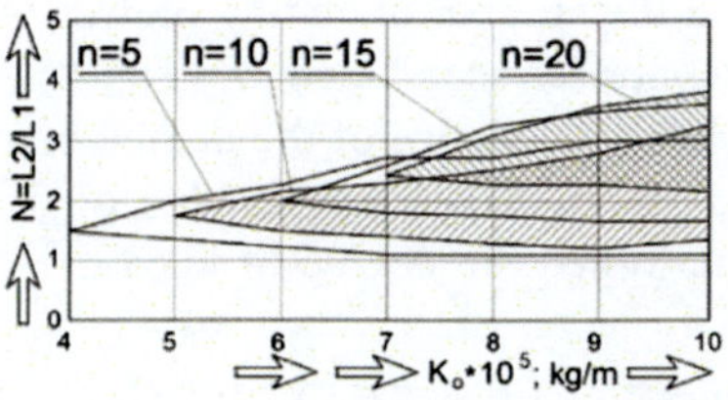

Fig. 7. Areas of parametric instability of the elastic system, for flat grinding by abrasive discontinuous circles, with the number of cutting protrusions on the working surface n = 5,10,15,20.

An increase in the number of cutting protrusions n on the circle leads to a shift in the zones of parametric instability to high values of cutting rigidity K_0 (Fig. 7). It means that while grinding by circles with a small number of slots on the working surfaces, parametric resonance occurs even in cases where the cutting capacity of the circles is high (i.e., at low values of the dynamic rigidity of the K_0). In the process of grinding by circles with a large number of slots, the parametric resonance occurs only when they are heavily salted, i.e., a significant decrease in their cutting capacity. Taking into account, that circles with a large number of cutting protrusions operate in the mode of intensive self-sharpening and, as a result, has good cutting properties, the probability of parametric perturbation of the elastic machine system while grinding these circles is much lower than while grinding circles with a small number of cutting protrusions. It can also be seen in Fig. 6 (B): there is no parametric resonance in the range of cutting protrusions $30 \leq n \leq 60$.

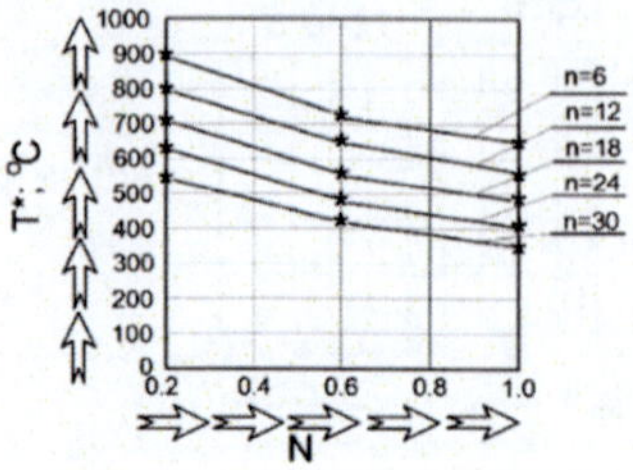

Fig. 8. The effect of the cutting protrusions number and the discontinuity coefficient of the grinding wheel on the grinding temperature.

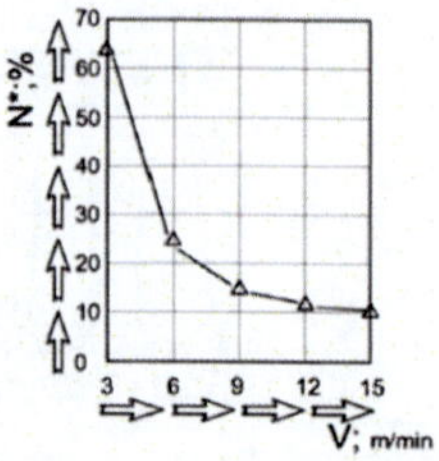

Fig. 9. The effect of the part speed V on the quality of the treated surface.

Figure 8 shows the impact of the design parameters of an intermittent abrasive wheel on the heat stress of the grinding process. The grinding temperature T * decreases with an increase in the number of cutting protrusions n on the circle, and also with an increase in the discontinuity coefficient N (Fig. 8).

In Fig. 9 [22], the graphic dependence of the degree of surface layer release of the processed detail on the longitudinal feed of the table of the flat grinding machine is presented. The formula determined the degree of release:

$$N^* = \left(H_m - H_m^{'}\right) \cdot 100/H_m,$$

where H_m - microhardness of the starting material before grinding (kg/mm^2); $H^{'}_m$ - microhardness of the layer lying at a depth of 20 ÷ 30 μm after grinding (kg/mm^2).

Samples made of 12x2H4A steel were grinded without cooling by a solid circle 24A 25 CM2 7 K5 in the following modes: Vkr = 22 m/s; t = 0.03 mm; V = 3 m/min; 6 m/min; 9 m/min; 12 m/min; 15 m/min. The degree of release of the treated surface decreases with longitudinal feed increasing (Fig. 9).

5 Conclusions

The conditions for the occurrence of the elastic system resonant states of a flat grinding machine are theoretically determined when the cutting process is interrupted periodically due to the absence of an abrasive circle of cutting grains on some parts of the working surface. The conditions of a significant decrease of the forced oscillations intensity and a probability decrease of parametric resonance in the elastic system of a flat grinding machine are revealed. It is shown that it is possible to exclude resonant states by reducing the duration of operation of the cutting protrusion of the discontinuous circle by creating conditions that ensure its high cutting capacity of the abrasive tool during grinding. It is Theoretically justified and experimentally verified that it is possible to reduce the depth of extension of structural and phase transformations in the processed material during intermittent grinding by increasing the speed of heat source movement and reducing the time connection of the abrasive tool and the part and the time between contacts. It is found that the self-sharpening of cutting grains during grinding can be achieved by increasing the number of cutting protrusions on the working surface of an intermittent abrasive wheel. It is established that applying the circles with a large number of slots on the working surface we can significantly reduce the amplitude of forced oscillations, reduce the temperature in the cutting area of the high cutting ability of the abrasive tool and, as a consequence, improve the quality and efficient processing. The purpose of our further research is to identify possibilities of using abrasive wheels with a large number of breaks in their working surfaces for grinding large-sized gears made of hardened and case-hardened steel with a long processing cycle.

References

1. Jackson, M., Davim, P.: Machining with Abrasives. Springer Science, New York (2011)
2. Marinescu, I., Hitcher, M.: Handbook of Machining with Grinding Wheels, 2nd edn. CRC Press, USA (2016)
3. Lebedev, V., Tonkonogyi, V., Yakimov, A., Bovnegra, L., Klymenko, N: Provision of the quality of manufacturing gear wheels in energy engineering. In: Ivanov, V., et al. (eds) Advances in Design, Simulation and Manufacturing. DSMIE-2018. LNME, pp. 89–96. Springer, Cham (2019). /https://doi.org/10.1007/978-3-319-93587-4_10
4. Davim, P. (ed.): Modern Machining Technology. A Practical Guide. Woodhead Publishing, UK (2011)
5. Torrubia, P., Billingham, J., Axinte, D.: Stochastic simplified modelling of abrasive waterjet footprints. Proc. Roy. Soc. **472**, 400–40 (2016)
6. Lebedev, V., Tonkonogyi, V., Chumachenko, T., Klymenko, N., Frolenkova, O.: Experimental and analytical study of CBN grinding of welded martensitic aging steel. In: Ivanov, V., et al. (eds.) Advances in Design, Simulation and Manufacturing II. DSMIE-2019. LNME, pp. 180–187. Springer, Cham (2020). https://doi.org/10.1007/978-3-030-22365-6_18
7. Wang, W., Li, J.: Characteristic quantitative evaluation and stochastic modeling of surface typography for zirconia aluminia abrasive belt. Int. J. Adv. Manuf. Tech. **87**, 111–115 (2016)
8. Lukyanchuk, Y., Denisyuk, V., Mikhalevich, V.: The use of intermittent grinding wheels on operations of centerless grinding of bearing rollers working surfaces. Bull. Khmelnitsky National Univ. **2**(211), 12–16 (2014)
9. Usov, A., Tonkonogyi, V., Dašic, P., Rybak, O.: Modelling of temperature field and stress-strain state of the workpiece with plasma coatings during surface grinding. Machines, Switzerland **7**(1), 20 (2019). https://doi.org/10.3390/machines7010020
10. Tonkonogyi, V., Sidelnykova, T., Dašić, P., Yakimov, A., Bovnegra, L.: Improving the performance properties of abrasive tools at the stage of their operation. In: Karabegovich, I. (eds) New Technologies: Development and Application. NT-2019. LNNS, vol. 67, pp. 136–145. Springer, Cham (2020). https://doi.org/https://doi.org/10.1007/978-3-030-18072-0_15
11. Lischenko, N., Larshin, V.: Determination of temperature while grinding by intermittent and highly porous circles. Sci. Notes Intercollegiate Collect. **40**, 150–158 (2013)

12. Tonkonogyi, V., Yakimov, A., Bovnegra, L., Sidelnykova, T., Dašić, P.: The use of intermittent wheels, impregnated by the contact method to reduce the thermal stress of the grinding process. In: IOP Conference Series: Materials Science and Engineering, vol. 708 (2019). https://doi.org/10.1088/1757-899X/708/1/012034
13. Yakimov, A.V., Yakimov, A.A.: Efficiency evaluation of intermittent grinding. Physical and computer technologies: In: Proceedings of the 19th International Scientific and Technological Conference, pp. 61–66. Kharkiv (2014)
14. Bogutsky, V., Shron, L., Bogutsky, B., Shron, B.: Analysis of structural features of grinding wheels with intermittent working surface. Educational notes of the Crimean engineering and pedagogical University. Tech. Sci. **35**, 60–64 (2012)
15. Tonkonogyi, V., Yakimov, A., Bovnegra, L.: Increase of performance of grinding by plate circles. In: Karabegovich, I. (eds) New Technologies: Development and Application. NT-2018. LNNS, vol. 42, pp. 121–127. Springer, Cham (2019). https://doi.org/10.1007/978-3-319-90893-9_14
16. Bogutsky, V.: Features of using an abrasive tool with an intermittent working surface for sharpening tools from high- speed steels. Rostov Sci. J., 223–231 (2019)
17. Bezpalova, A., Lebedev, V., Tonkonogyi, V., Morozov, Y., Frolenkova, O.: Cutting stone building materials and ceramic tiles with diamond disc. In: Ivanov, V., et al. (eds.) Advances in Design, Simulation and Manufacturing II DSMIE-2019. LNME, pp. 510–521. Springer, Cham (2020). https://doi.org/10.1007/978-3-030-22365-6_51
18. Bogutsky, V., Shron, L.: Ensuring stability of quality parameters while grinding and broaches sharpening. Scientific notes of the Crimean engineering and pedagogical University. Tech. Sci. **49**, 77–82 (2014)
19. Bogutsky, V., Bratan, S.: Analysis of the process of grinding chip grooves of broaches. Progressive technologies in mechanical engineering: Int. Coll. Sci. Papers **3**(49), 15–22 (2014)
20. Khudobin, L., Husanov, A.: Thermophysics of grinding thin-boned and wedge-shaped blanks. Physical and mathematical theory of processing materials technology and engineering. In ten volumes, vol. 2(6). Thermophysics of cutting materials. ONPU, Odessa (2003)
21. Yakimov, A.V., Novikov, F., Lynchevsky, P., Larshin, V., Grischenko, E., Fadeev, A., Novikov, G.: Engineering technology: textbook (for students.of higher.academic institute). ONPU, Odessa (2012)
22. Novikov, F., Polyansky, V.: Modern technologies and technical re-equipment of enterprises (monograph). LIRA, Dnipro (2018)
23. Tonkonogyi, V., Dašić, P., Rybak, O., Lysenko, T.: Application of the modified genetic algorithm for optimization of plasma coatings grinding process. In: Karabegovich, I. (eds) New Technologies: Development and Application. NT-2019. LNNS, vol. 67, 199–211. Springer, Cham (2020). https://doi.org/10.1007/978-3-030-18072-0_23

Influence of Technological Methods of Processing on Wear Resistance of Conjugated Cylindrical Surfaces

Valentyn Zablotskyi, Anatolii Tkachuk(✉), Serhii Moroz, Stanislav Prystupa, and Kostiantyn Svirzhevskyi

Lutsk National Technical University, 75, Lvivska Street, Lutsk 43018, Ukraine
a.tkachuk@lntu.edu.ua

Abstract. Wear of any tribological pair leads to malfunction or failure of the entire mechanism. It is known that the process of wear of friction pairs occurs in three periods: running-in, normal wear, accelerated wear (destruction). Working surfaces of parts that are formed in the manufacturing process receive micro geometric characteristics (roughness, undulation), according to the regulated technological norms. These characteristics describe the properties of the technological surface topography. However, quantitative micro geometric parameters change as they evolve during running-in, and the surface receives a new operational relief that is stable for a long time during normal wear. It is an operational relief of conjugate surfaces that characterizes the qualitative properties of parts for a long period of operation. The running-in process, which is based on complex mechanical, physical, and chemical processes, determines the overall wear resistance of parts. After this period, the physical, mechanical, and geometric characteristics of the surface acquire optimal values corresponding to the operating conditions. Rational performance characteristics during normal wear can self-sustain, and they are continuously reproduced independently in the same values. This state of the surface layer is observed before the beginning of the third stage of the life cycle. Traditionally, to ensure regulated micro geometric characteristics of the working surfaces used grinding, smoothing, and lapping diamond pastes. At the present stage of development of equipment and technology in a row with the mentioned finishing operations, there is an operation of high-speed turning, which allows reaching high purity of surfaces with the lowest cost of the technological process.

Keywords: Analysis · Micro relief · Running-in · Abbott–firestone curve · Dimensionless complex

1 Introduction

For the effective operation of machines and mechanisms, it is necessary that in the process of machining, the surfaces of parts acquire a set of characteristics that occur during the running-in period [1]. Then the friction pair in the manufacturing process will acquire the properties that are inherent in the conjugate surfaces of lapped parts, minimizing the running-in period [2].

V. Tonkonogyi et al. (Eds.): InterPartner 2020, LNME, pp. 477–487, 2021.
https://doi.org/10.1007/978-3-030-68014-5_47

However, the acquisition of optimal surface characteristics and compensation of deviations in the properties of the treated surfaces, during the operation of the part can take place only within certain limits [3]. Thus, during the study of the inner rings of roller bearings found that the roughness decreases particularly intensively in the first 2 h of operation. After 8 h of operation, the surfaces acquire optimal roughness Ra = 0.08–0.04 μm. High projections microrelief rather reduce its height and lower-slower. It is noted that the surfaces with Ra = 0.08–0.04 μm, which had an initial roughness of Ra = 1.25–0.32 μm, during the operation suddenly acquired a "coarse" surface roughness and re-started proportionately [4]. This phenomenon is explained by the following: in the process of burnishing "rough" output surfaces, the optimum value of roughness is reached due to the plastic flow of the metal, particularly due to the formation of a surface film of the metal. Thus, remain unfilled metal sharp corners of hollows of a microrelief of a surface. During operation (normal wear and tear), the metal layer is destroyed, and the surface with the initial roughness is exposed under it. This phenomenon is harmful because it increases the running-in time and scratches the mating surfaces with metal particles. Thus, it can be noted that rough initial surfaces pass the stage of "false burn-in" [1].

Surfaces with initial roughness Ra = 0.32–0.08 μm have a normal break-in period. Here the optimum roughness value Ra = 0.08 is achieved 0.04 μm in about 1.5 h of operation, and a surface with a roughness close to operational has a short running-in period (about 5 min). Normal running-in is characterized by the fact that under the microrelief, which was formed is a dense, well-filled metal base. Changes in microrelief parameters occur mainly because of abrasion and minor plastic deformation of the vertices. Rolling cavities microrelief surface and subsequent failure of the bulk layer does not occur here.

Thus, with the aim of the study of the formation of the technological surface during running, it is proposed to undertake a study on the character of changes of the working surfaces of mating parts to establish as close machining operation to this process.

2 Literature Review

The period of intensive reduction of roughness coincides with the period of intensive increase in the degree of work hardening and, consequently, the value of micro hardness. By the end of the running-in period, its value is stabilized and takes an optimal value.

In [5, 6], it is shown that the process of burnishing the working surfaces of machine parts is accompanied by a change in the initial technological roughness and micro-hardness obtained after machining and the formation of operational values of these characteristics. However, in papers [7, 8], only the change in the height characteristics of the roughness Ra or Rmax is considered. However, the change in the shape of the micro-irregularities determined by the plane and volume characteristics is insufficiently investigated. This issue is especially important in the study of the influence of technological processing methods on the formation in the process of burnishing the operational microrelief [9–11], as well as for ensuring the accuracy and the optimal configuration of technological systems [12, 13]. During the study, the analysis of changes in the main characteristics of the microgeometry and microhardness of the surface layer during running-in.

3 Research Methodology

The process of stable work of parts, which is based on complex mechanical, physical, and chemical processes, determines its overall wear resistance. Before this process, the physical and geometric characteristics of the surface, such as roughness, microhardness, magnitude and sign of residual stresses, structure metal, friction coefficient, and others, acquire optimum values, respectively, of operating conditions and wear. Optimum performance characteristics during normal wear and tear are supported, evolutionary, and they are continuously reproduced in the same values that are manifested by technological heredity [3].

As the object of study was a pair that operates under boundary friction: P_d = 0.8 MPa; v_d = 0.9 m/s; oil – 10W-40. The material of mating parts – the 100Cr6 steel (hardness HRC = 58–60). Wear tests were carried out according to the scheme of friction of the liner on the roller that rotates. The roller was processed in a special cartridge to eliminate the manifestations of technological heredity as much as possible [10, 14]. Two series of samples with different hardness and consisting of five samples were investigated. Samples underwent finishing by grinding (v_{ctrl} – 30 m/s; v_d = 25 m/min, s = 0.03 mm/rev; t = 0.02 mm). The change in the profile of the rollers' microroughness during running-in is shown in Fig. 1.

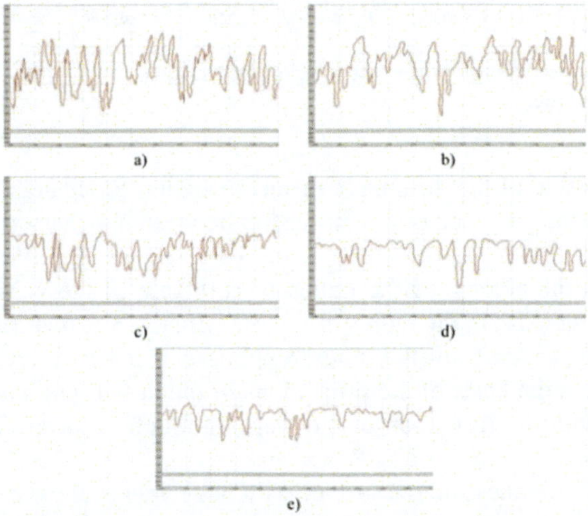

Fig. 1. The changing profile of asperities rollers (steel 100Cr6) during running-in depending on the traversed path friction: a) initial profile; b) S_T = 0.406 km; c) S_T = 0.812 km; d) S_T = 1.62 km; e) S_T = 2.85 km.

The profilograms were shot on the same part of the roller surface in the same direction after passing a different friction path S_T.

For Fig. 2 curves of change of average values of roughness characteristics Ra, R_{max}, b', v, r, β in the process of samples wear (HRC = 58) are shown). Obtained after grinding the characteristics of surface quality (Ra = 0.68 μm, R_{max} = 4.2 μm; b' = 1.55 V; v = 1.9; r = 72 μm; β = 9°30′; $N\mu$ = 7900 MPa) during running we observed a decrease of the height irregularities R_{max}. And the main change of R_{max} occurred in the first hours of work. So, within the first hour (S_T = 0.406 km) output height of the irregularities was reduced by 0.32 $R_{max\ int}$, but the next 10 h of operation (S_T = 4.46 km), it has declined by only 0.42$R_{max\ int}$.

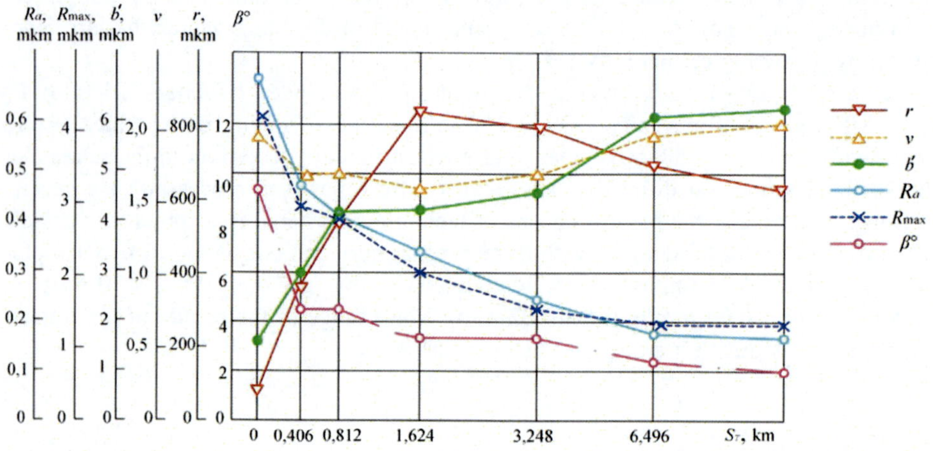

Fig. 2. Characteristics of the roughness for the polished surface in the process of burnishing the roller on the distance traveled.

Similarly, the value of the arithmetic means deviation Ra changes during running-in, which indicates the presence of a relationship between the characteristics of Ra and R_{max}.

Interesting was the change of the radius of curvature of the vertices of the irregularities r. a Sharp increase in the radius to r = 850 μm (S_T = 1.624 km) shows that the removal of the upper part of the microroughness reaches a level at which the inequalities represent the basis of the projections of the initial roughness. In the future, this surface ends with the formation of the working relief. A lower value of r characterizes the last one.

The creation of roughness on the surface with large values of the radius of curvature r can contribute to an accelerated transition to the operational microrelief and reduce the overall value of wear.

The peculiarity of changing the angle of inclination of micro-irregularities β is its rapid stabilization with a total change from 9° to 2°. Thus, during the first two hours of operation (S_T = 0.812 km), there was a 70% change in the angle β.

The most complex changes are the parameters of the curve of the support surface b' and v. The rapid erasure of the individual most prominent irregularities in the first two hours (S_T = 0.812 km) leads to an increase in *b'* more than three times. After the formation of roughness with significant values of the bearing surface area, the value of linear wear, changes in the height of the microroughness, and other characteristics are stabilized, which leads to changes in the following hours of operation of the parameter *b'*. With the further formation of operational roughness (S_T > 2.85 km), this parameter is qualitatively different from the initial roughness, i.e., the value of *b'* increases. This is confirmed by the curves of the support surfaces (Fig. 3).

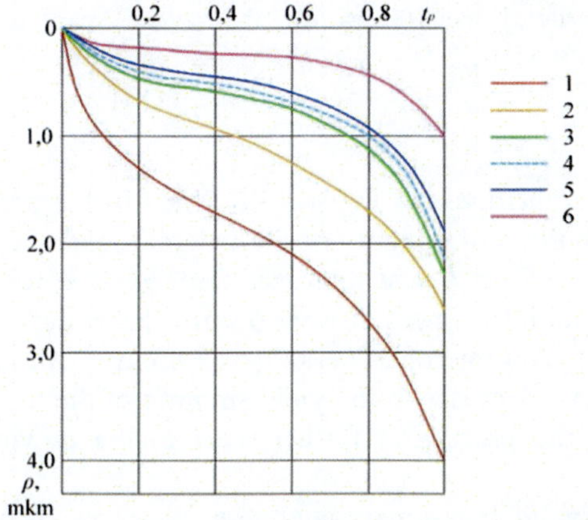

Fig. 3. The dependence of the support surface from the distance *d* after different periods of running rollers made of steel 100Cr6 (HRC = 58–60): 1 – after grinding (S_T = 0 km); 2 – S_T = 0.406 km; 3 – S_T = 0.812 km; 4 – S_T = 1.624 km; 5 – S_T = 2.85 km; 6 – S_T = 4.46 km.

Curves 3–5, corresponding to the friction path 0.812; 1.624; 2.85 km, respectively, are almost identical, changing their shape occurs during the first two hours of operation (S_T = 0.812 km).

The results of a study of changes in the hardening of the surface layer during running, which was carried out on two series of samples of steel 100Cr6 after finishing sanding with the initial microhardness of 7000 MPa (HRC = 51–52) and 7900 MPa (HRC = 58–60), and the corresponding curves are shown in Fig. 4.

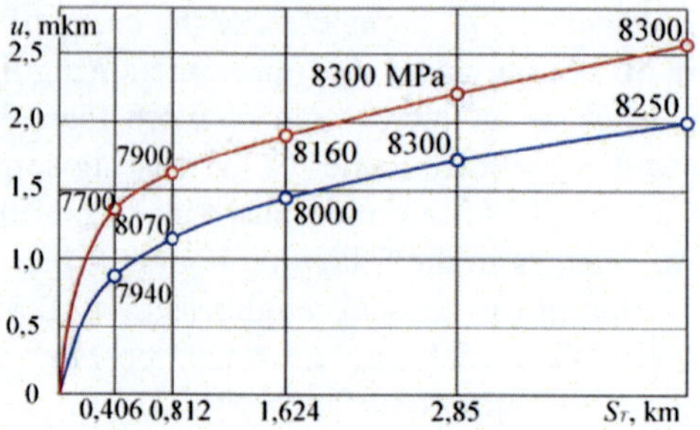

Fig. 4. The dependence of wear and microhardness of the surface of the conjugate surface during running rollers made of steel 100Cr6 from the traversed path of friction.

4 Results

The change in the microhardness of the surface layers in the process of burnishing is determined by the condition of equality of the external load and the yield strength of the metal by the value of the actual reference surface. Due to the small size of the bearing surface at the initial moment of wear occurs plastic deformation of the surface layers, which increases their microhardness ($S_T = 2.85$ km). An increase in the bearing surface further leads to a decrease in the yield strength of the metal, because of which the microhardness in the process of further wear of the previously hardened layer reduces its value.

After the formation of functional roughness, wear occurs without changing its characteristics. This leads to the formation of the optimal microhardness of the surface layer [15–17].

Abbott-Firestone curves (Fig. 3) show that the wear resistance of the samples depends on compliance with all processing conditions. Even minor changes in the technological process lead to an increase in the value of wear by about 25%. This fact indicates the influence of technological heredity on the wear resistance of the studied parts.

From the point of view of technological control of the burnishing process, it is important to know how the processing methods finally form the surface quality, that is, affect the change in the characteristics of the surface layer. With this purpose we studied diamond grinding ($v_{cntr} = 30$ m/s; $v_d = 35$ m/min, $P = 80$ N; $s = 0.15$ mm/rev) diamond lapping pastes, smoothing ($v = 62$ m/min; $s = 0.07$ mm/rev, $P = 1200$ N) and high-speed turning ($v = 200$ m/min; $s = 0.2$ mm/rev; $t = 1.5$ mm).

As in the previous case, a series of five samples were processed and tested for wear by each method. Pre-working the surface of samples of hardened steel 100Cr6 processed round grinding ($v_{cntr} = 30$ m/s; $v_d = 25$ m/min, $S_{pr} = 0.03$ mm/rev; $t = 0.02$ mm).

Change of surface microgeometry characteristics Ra and R_{max}. For samples treated with diamond grinding wheel end, smoothing and high-speed turning (Fig. 5), has a similar character as for round grinding.

According to another algorithm, changes the values of Ra and R_{max} during burnishing for samples treated with lapping diamond pastes. This method obtained roughness with a value of Ra < 0.08 μm, which in its magnitude, is close to the operational roughness. In this regard, it can be argued that the surface will run faster than others, having 2…3 big roughness Ra. It turned out that on the samples ground with diamond pastes, there is an increase in the characteristics Ra and R_{max} to a certain value (S_T = 1.22 km), after which there is a decrease in these characteristics to the final values of the operational roughness.

The phenomenon under consideration, similar to the process of false burnishing, in this case, can be explained by different height microgeometry of the sample and counterbody, which was processed by internal grinding (v_{cntr} = 30 m/s; v_d = 30 m/min, S_{gr} = 0.02 mm/rev; t = 0.01 mm) with a roughness Ra = 0.63–0,32 μm.

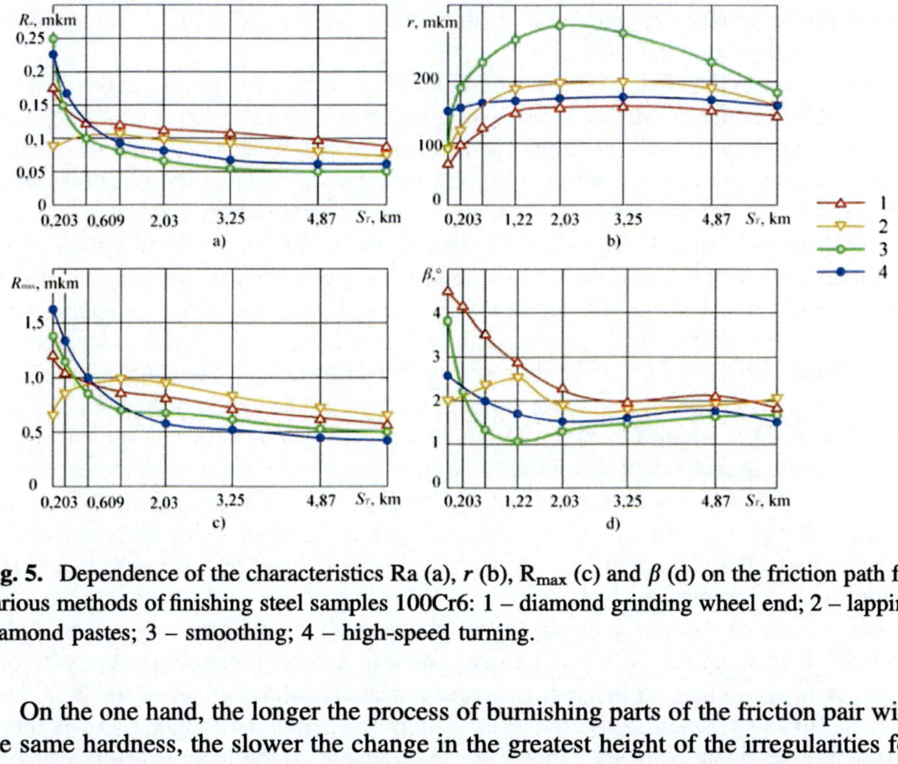

Fig. 5. Dependence of the characteristics Ra (a), r (b), R_{max} (c) and β (d) on the friction path for various methods of finishing steel samples 100Cr6: 1 – diamond grinding wheel end; 2 – lapping diamond pastes; 3 – smoothing; 4 – high-speed turning.

On the one hand, the longer the process of burnishing parts of the friction pair with the same hardness, the slower the change in the greatest height of the irregularities for one of the conjugate parts. At the same time, the lower the height of the sample microroughness, the slower the counterbody roughness changes.

On the other hand, roughness value 0.63 of the hardened counterbody interacting with the surface of the sample leads to an increase in the height of the microroughness. The resulting new roughness of the sample leads to faster wear of the counterbody surface and the pair as a whole.

The desire to reduce the amount of wear in the running-in process by obtaining microgeometry, equal only to the height of the operational roughness, both for the sample and the counterbody (processing was carried out by lapping diamond pastes to Ra = 0.08–0.04 μm) led to intense wear by setting. This example proves that obtaining during the treatment of friction surfaces roughness with a height of R_{max} equal to the operational one cannot be a reliable indicator of the optimality of microgeometry, which ensures minimal wear during running-in.

It is obvious that the term "optimal micro geometry" should be understood as the optimal values of all characteristics of microgeometry or complex expression of surface properties that have the maximum impact on wear resistance.

The change in the radius of curvature of the vertices of micro-irregularities for different processing methods during burnishing is shown in Fig. 5b. Similarly, as for round grinding in general, there is an increase in the index r, and then a decrease to a certain value, which characterizes the operational roughness. However, high-speed turning creates a microgeometry in which there is a slight change in *r*. It can also be assumed that the decrease in microroughness R_{max} (which determines the value of the radii of the protrusions *r*) leads to a smaller change in the radius of curvature of the irregularities during running-in.

The change in the angle of inclination of the side of the micro-irregularities β in the process of burnishing for different processing methods is shown in Fig. 5d. Changes in the microgeometry of samples lapped with diamond pastes leads to wear in the first hours to an increase in the angle β. A significant increase in the radius of curvature of irregularities r in samples treated with smoothing, to a certain decrease in the values β, smaller than established at the end of the running-in period. As for round grinding, the most complex changes in the running-in process are parameters of the support surface curve *b*' and *v*. For example of samples with a minimum value of R_{max} (lapping with diamond pastes), after a slight increase in the value of *b*′, it decreases due to an increase in the height of the micro-roughness due to the influence of the roughness of the counterbody. After burnishing, these surfaces increase in the values of *b*' (the period of formation of the working relief). From this example, it can be seen that the change in the parameters *b*′ and *v* depends not only on their magnitude but also on the change in the process of wear for other characteristics of the surface. Thus, more favorable values for the running-in conditions of the radius of curvature of the vertices r, associated angles β, as well as characteristics Ra and R_{max} lead to smaller and more monotonic changes in the parameters *b*′ and *v*.

An analysis of the data given in Fig. 2–5 allows concluding that regardless of the methods of finishing for all studied characteristics, there is a certain tendency to the formation of working relief in a narrow range, especially for the values of Ra, $R_{max,}$ and angle β. However, when studying the formation of operational roughness, it is more expedient to evaluate complex expressions that include all the main characteristics of the surface microgeometry. As such expression, it is necessary to use the complex dimensionless characteristic of the roughness of a bearing surface $\Delta = R_{max} / r\, b^{I\ 1/\upsilon}$ which is widely applied in settlement dependences for determination of the size of wear [8, 18, 19]. According to the results of the experiment, the values of Δ were calculated. As a result, the dependencies of the change of the dimensionless complex Δ in the process of running-in were built (Fig. 6).

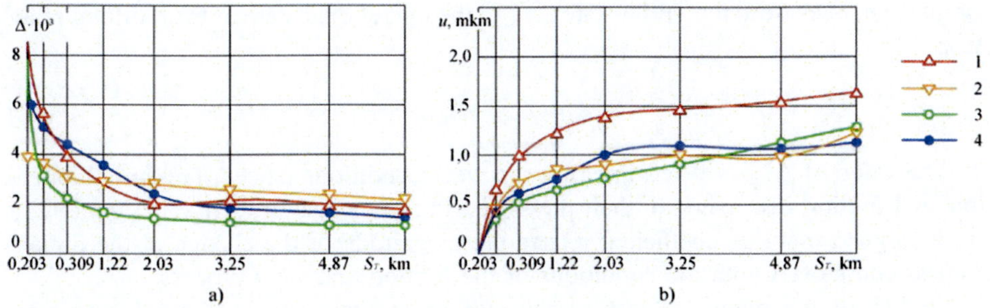

Fig. 6. The dependence of the dimensionless complex Δ (a) and the wear of the samples (b) from the traversed path of friction during running for different methods of finishing processing of samples of steel 100Cr6: 1 – diamond grinding; 2 – lapping with diamond pastes; 3 – smoothing; 4 – high-speed turning.

These dependencies show: despite significant fluctuations in individual characteristics of microgeometry (Fig. 1), there is a certain regularity of monotonic reduction of the Δ complex in the process of burnishing, which is typical for all the studied processing methods.

5 Conclusions

According to the results obtained, it is possible to make an important assumption that the deviation of one of the characteristics from the general pattern should be compensated by a corresponding change in other characteristics included in the dimensionless complex. As a result, the general pattern of change Δ should remain constant for a particular processing method. For example, for samples after lapping diamond smoothing treatment, the increase in the height of the microroughness R_{max} does not lead to an increase in Δ due to compensating changes in the characteristics of r, b' and v.

The obtained data confirm the correctness of the use of dimensionless complex Δ as the main characteristic of surface microgeometry in the study of friction and wear processes. To predict the change in microgeometry during running-in, an empirical dependencies of the change in the dimensionless complex Δ as a function of the number of wear cycles N or the friction path S_T are obtained.

The equation, which describes the studied dependences, was found with three arbitrary constants and had the form:

$$\Delta = \Delta_{out} - k(1 + e^{-C \cdot x}), \tag{1}$$

where x – the number of wear cycles N or the friction path S_T, km; Δ_{out} – the value of the dimensionless complex after finishing; C – the coefficient depending on the quality of the surface layer, which is determined by technological methods of the processing; k – the coefficient showing how much the value of the dimensionless complex changes from the initial state Δ_{out} to the operational state of the working relief Δ_{exp}:

$$k = \Delta_{out} - \Delta_{exp}. \quad (2)$$

The value of Δ_{exp} is determined by the wear conditions (speed, pressure, lubrication, and friction pair material, their physical and mechanical properties). Therefore, it can be argued that this coefficient relates the magnitude of the change of the dimensionless complex Δ with the conditions of the friction and wear process.

Studies of the microhardness of the surface in the process of burnishing after various methods of finishing showed that the pattern of its change has a similar character as in round grinding, with the formation of optimal values of microhardness. Thus, the proposed method of analyzing the wear resistance of the conjugate surfaces of parts to ensure the maximum duration of the period of stable operation by assigning technological modes of machining is confirmed by experimental data. It can be used in the design of technological processes for the manufacture of parts.

References

1. Dykha, A., Sorokatyi, R., Makovkin, O., Babak, O.: Calculation-experimental modeling of wear of cylindrical sliding bearings. East.-Eur. J. Enterp. Technol. **51**(1(89)), 51–59 (2017)
2. Kindrachuk, M., Radionenko, O., Kryzhanovskyi, A., Marchuk, V.: The friction mechanism between surfaces with regular micro grooves under boundary lubrication. Aviation **8**(2), 64–71 (2014)
3. Tkachuk, A., Zablotskyi, V., Kononenko, A., Moroz, S., Prystupa, S.: Directed formation of quality, as a way of improving the durability of conjugated parts of friction pairs. In: Ivanov, V., et al. (eds.) Advances in Design, Simulation and Manufacturing II. DSMIE 2019. Lecture Notes in Mechanical Engineering, pp. 370–377. Springer, Cham (2020)
4. Marchuk, V., Kindrachuk, M., Kryzhanovskyi, A.: System analysis of the properties of discrete and oriented structure surfaces. Aviation **18**(4), 161–165 (2014)
5. Abboud, E., Shi, B., Attia, H., Thomson, V., Mebrahtu, Y.: Finite element-based modeling of machining-induced residual stresses in Ti-6Al-4V under finish turning conditions publication: In: 14th Conference on Modeling of Machining Operations, vol. 8, pp. 63–68. Elsevier, Turin (2013)
6. Jia, X., Zuo, X., Liu, Y., Chen, N., Rong, Y.: High wear resistance of white cast Iron treated by novel process: principle and mechanism. Metall. Mater. Trans. A **46**(12), 5514–5525 (2015)
7. Koval, A.D., Efremenko, V.G., Brykov, M.N., Andrushchenko, M.I., Kulikovskii, R.A., Efremenko, A.V.: Principles of development of grinding media with increased wear resistance. Part 2. Optimization of steel composition to suit conditions of operation of grinding media. J. Frict. Wear **33**(2), 153–159 (2012)
8. Kun, H., Yang, W., Ye, X.: Adjustment of machining-induced residual stress based on parameter inversion. Int. J. Mech. Sci. **135**, 43–52 (2018)
9. Gold, P., Loos, J.: Wear resistance of PVD-coatings in roller bearings. Wear **253**(3–4), 465–472 (2002)

10. Mohammadpour, M., Razfar, M.R., Saffar, R.J.: Numerical investigating the effect of machining parameters on residual stresses in orthogonal cutting. Simul. Model. Pract. Theory **18**(3), 378–389 (2010)
11. Teja, P.S., Kumar, M.D., Krishna, R., Sreenivasan, M.: Simulation and optimization studies on the ring rolling process using steel and aluminum alloys. J. Eng. Sci. **6**(2), E36–E40 (2019). https://doi.org/10.21272/jes.2019.6(2).e6
12. Karpus, V.E., Ivanov, V.A.: Locating accuracy of shafts in V-blocks. Russ. Eng. Res. **32**, 144–150 (2012). https://doi.org/10.3103/S1068798X1202013X
13. Karpus, V.E., Ivanov, V.A.: Choice of the optimal configuration of modular reusable fixtures. Russ. Eng. Res. **32**, 213–219 (2012). https://doi.org/10.3103/S1068798X12030124
14. Redko, R., Zabolotnyi, O., Redko, O., Savchuk, S., Kovalchuk, V.: Improvement of manufacturing technology and recovery of clamping collets for lathe automats. In: Ivanov, V., et al. (eds.) Advances in Design, Simulation and Manufacturing II. DSMIE 2019. Lecture Notes in Mechanical Engineering, pp. 290–301. Springer, Cham (2020)
15. Pawara, S., Jhab, A.K., Mukhopadhyaya, G.: Effect of different carbides on the wear resistance of Fe-based hardfacing alloys. Int. J. Refract. Metal. Hard Mater. **78**, 288–295 (2019)
16. Sanjay, K.T., Gopal Krishnan, R.: Advances in Applied Surface Engineering. Research Publishing Services, Singapore (2011)
17. Yuan, M., Pingfa, F., Zhang, J., Wu, Z., Yu, D.: Prediction of surface residual stress after end milling based on cutting force and temperature. J. Mater. Process. Technol. **235**, 41–48 (2016)
18. Stephens, R., Fatemi, A., Stephens, R., Fuchs, O.H.: Metal Fatigue in Engineering, 2nd edn. Wiley, Hoboken (2001)
19. Zablotskyi, V., Moroz, S., Tkachuk, A., Prystupa, S., Zabolotnyi, O.: Influence of diamond smoothering treatment power parameters on microgeometry of working surfaces of conjugated parts. In: Tonkonogyi, V., et al. (eds.) Advanced Manufacturing Processes. InterPartner 2019. Lecture Notes in Mechanical Engineering, pp. 372–381. Springer, Cham (2020)

Machining Process

Multicriteria Optimization of the Part's Finishing Turning Process Working in the Conditions of Alternating Loadings

Viktor Antonyuk, Kateryna Barandych(✉), and Sergii Vysloukh

National Technical University of Ukraine "Igor Sikorsky Kyiv Polytechnic Institute", 37, Peremohy Avenue, Kiev 03056, Ukraine
victor.antoniuk@gmail.com, {barandichk,vspl}@ukr.net

Abstract. The paper evaluates issues of technological parameters optimization of the machining process of parts operating under alternating loads. It is noted that high cyclic loads on the part during operation often lead to their failure. It is suggested that the finishing of such parts should be carried out by a turning tool, the cutting part of which is made of superhard materials. The main task of this work is formulated, which is the determination of the optimal modes of turning for the part made of the corresponding structural material, which provides the specified quality parameters. To solve this problem, we created a mathematical model of the turning process, which is multicriteria, in which the criteria of optimality selected as maximum values of cyclic durability and productivity of the process. The area of feasible solutions to the optimization problem is provided by the necessary values of the quality parameters of the workpiece and the technical capabilities of the equipment used. Practical testing of the proposed method of optimization of finishing turning with a tool made of cubic boron nitride of 37Cr4(DIN) steel parts, working in difficult operating conditions, showed its great efficiency.

Keywords: Finishing turning · Cyclic durability · Processing performance · Complex objective function · Multicriteria optimization · Cubic boron nitride

1 Introduction

In the manufacture of responsible parts of machines and mechanisms that operate under conditions of high cyclic loading of variables in magnitude and direction, the most common and dangerous cause of failure is fatigue failure, which often leads to serious consequences, as it occurs suddenly [1, 2]. Dynamic load parameters such as speed, cycle frequency, alternation, fracture occur at stresses much smaller than the ultimate strength characteristics because all processes leading to fracture occur in the surface layer [3, 4].

The goal of the work is the development of a problem-solving method for multicriteria optimization of finishing turning process for parts, which are operating under cyclic loading, by creating a comprehensive mathematical model with criteria for maximum process performance and maximum cyclic durability.

V. Tonkonogyi et al. (Eds.): InterPartner 2020, LNME, pp. 491–501, 2021.
https://doi.org/10.1007/978-3-030-68014-5_48

2 Literature Review

Researches [5, 6] have established that for providing the set values of surface layer characteristics, it is necessary to determine the appropriate combination of elements of the cutting mode with the parameters of the quality of the surface layer, in particular surface roughness, depth and degree of slander, residual stresses. In order to achieve this goal, it is necessary to establish dependencies between the parameters of the surface layer and the processing conditions, including the cutting mode [7, 8]. The solution to this problem is facilitated if generalized theoretical dependencies are known, which establishes the relationship between the quality criteria of the surface layer and the parameters of the cutting process [9, 10]. It allows you to assign cutting modes based on the required parameters of roughness, depth, and degree of slander, the level of residual stresses that will provide fatigue resistance [11, 12]. During machining of important parts, there was a tendency to replace grinding by finishing the surfaces with sharpening using a tool whose cutting part is made of superhard materials [13, 14]. Therefore, there is a need to scientifically substantiate the technological support for the required quality parameters of the surface layer of important parts and the reliability of their further operation during such finishing. In this regard, the work is aimed at creating a mathematical model of the process of turning parts at high cyclic loads in order to determine the optimal modes. It will allow for the stages of technological preparation of production to realize the technological provision of the necessary values of the cyclic durability of the parts during their finishing turning [15, 16]. It is advisable to use the maximum cyclic durability and the highest productivity of the machining process to solve the problem of technological assurance of cyclic longevity of components operating under alternating loads. To solve such problems, it is advisable to use multicriteria optimization applied to the processing of the blade tool (turning, boring, etc.), as well as to grinding, which is presented in [17, 18].

3 Research Methodology

The technique of solving the optimization problem involves the use of a mathematical model of the process of finishing a workpiece, the objective function of which will be a complex, consisting of two partial criteria [3, 17].

The first partial criterion is presented in the form of mathematical dependence of cyclic durability on the modes of turning and stress cycle amplitude for parts:

$$\mathrm{N(X)} = f_1(S, V, \sigma) \tag{1}$$

where $\boldsymbol{N}$ is cyclic durability, number of cycles;

$\boldsymbol{X}$ is vector optimizing variables that provide the maximum value of the cyclic durability of parts;
$\boldsymbol{S}$ is tool feed for one spindle rotation, mm/rot;
$\boldsymbol{V}$ is cutting speed, m/min;
$\boldsymbol{\sigma}$ is stress cycle amplitude, GPa.

The second partial criterion of optimality is the maximum productivity of the machining process at finishing [3]. This criterion is a mathematical model that defines the time spent on cutting, which can be represented as follows:

$$P(X) = f_2(\tau_0) = \frac{1}{\tau_0} \tag{2}$$

where τ_0 – is the main time in minutes, which is calculated by the formula:

$$\tau_0 = \frac{Lh}{nSt} = \frac{\pi DLh}{1000VSt} \tag{3}$$

where D is the diameter of the processed surface, mm;

L is the estimated length of machining, that is, the total length of the passage of the tool in the feed direction, mm;
H is the amount of overmeasure, mm;
T is cutting depth, mm.

Considering (2) the dependence (3) of the turning process productivity takes the following form:

$$\mathrm{P(S, V)} = \frac{1000VSt}{\pi DLh} \tag{4}$$

Then, according to the multicriteria optimization technique, a complex objective function that takes into account the specified partial criteria will look like this:

$$C(S, V, \sigma) = \left(\alpha_1 \left(\frac{N(S, V, \sigma) - N_{min}(S, V, \sigma)}{N_{max}(S, V, \sigma) - N_{min}(S, V, \sigma)}\right) + \alpha_2 \left(\frac{P(S, V) - P_{min}(S, V)}{P_{max}(S, V) - P_{min}(S, V)}\right)\right) \tag{5}$$

where α_1 and α_2 are coefficients that determine the importance of each partial criterion when solving an optimization problem, the values of which are expert estimates;

$N(S,V,\sigma)$ is cyclic durability, number of cycles;

N_{min}, N_{max} are minimum and maximum value of cyclic durability, number of cycles;

$P(S, V)$ is a productivity of the finishing process, 1/min;

P_{min}, P_{max} are minimum and maximum performance values, 1/min;

The optimum values of the speed, feed rate, and tension in the part's material, which is resulted by operating conditions, are determined from the area of acceptable solutions.

4 Results

When solving an optimization problem, the range of feasible solutions is formed by many constraints on the turning process. Here is a general view of the mathematical dependencies that form the area of the problem. The used dependencies and corresponding values of coefficients decisions are given in [19, 20].

1. Feed limit:

$$S_{min} \le S \le S_{max} \tag{6}$$

where S_{min} and S_{max} are minimum and maximum allowable feed values determined by the capabilities of the machine tool, mm/rot.

2. Cutting speed limits:

$$V_{min} \le V \le V_{max} \tag{7}$$

$$V_{min} = \frac{\pi D_0 n_{min}}{1000}, V_{max} = \frac{\pi D_0 n_{max}}{1000} \tag{8}$$

where D_0 is the diameter of the workpiece before its processing, mm;

V_{min} and V_{max} are minimum and maximum cutting speed, m/min;
n_{min} and n_{max} are minimum and a maximum speed of the machine spindle, rpm

3. Cutting force limits:

$$P_x = 10C_{P_x} t^{x_{Px}} S^{y_{Px}} V^{n_{Px}} K_{P_x} \le P_{\text{max}a.f.} \tag{9}$$

where P_x is an axial component of cutting force, N;

$P_{max\ a.f.}$ is the maximum axial force of the machine, N;
C_{Px} is constant;
x_{Px}, y_{Px}, n_{Px} are power indicators;
K_{Px} is the correction factor, which is the multiplication of the coefficients K_{mr}, $K_{\varphi Px}$, $K_{\gamma Px}$, $K_{\lambda Px}$, K_{rPx}, with $K_{mr} = \left(\frac{\sigma_v}{750}\right)^n$, where σ_v is the temporary resistance (tensile strength) of the material, MPa; n is a power indicator.

4. Cutting capacity limitations:

$$N_c \le N_{sp} \tag{10}$$

where N_c is effective cutting power, kW; N_{sp} is power developed by the machine on a spindle, kW:

$$N_{sp} = N_{en}\eta \tag{11}$$

where N_{en} is the power of the engine of the main drive of the machine, kW;
η is the efficiency of the main drive of the machine.

The formula calculates the effective power required for turning:

$$N_c = \frac{P_z V}{1020 \cdot 60} \tag{12}$$

where P_z is the tangential component of the force is defined as follows

$$P_z = 10 C_{P_z} t^{x_{P_z}} S^{y_{P_z}} V^{n_{P_z}} K_{P_z} \tag{13}$$

where C_{Pz} is constant;

x_{Pz}, y_{Pz}, n_{Pz} are power indicators;
K_{Pz} is the correction factor, which is the coefficients multiplication K_{mr}, $K_{\varphi Pz}$, $K_{\gamma Pz}$, $K_{\lambda Pz}$, K_{rPz}.

Then, given formulas (11–13), inequality (10) takes the following form:

$$N_{en}\eta \geq \frac{10 C_{P_z} t^{x_{P_z}} S^{y_{P_z}} V^{n_{P_z}} K_{P_z}}{1020 \cdot 60} \tag{14}$$

5. Limitations on machining accuracy:

$$\Delta_\Sigma \leq 0.5 \cdot TD \tag{15}$$

where Δ_Σ is a total error of processing, μm;
TD is the tolerance value corresponding to the qualification of the precision of the maintained size, μm.

The total error includes the following errors:

Δ_y is the error due to oscillation of elastic deformations of the technological system under the action of load instability (cutting forces, inertia forces, etc.) operating in the system of variable rigidity, μm;

$\Delta_{\varepsilon y}$ is the error in mounting the workpiece into the fixture, μm;

Δ_{ad} is the error in adjusting the technological system to the size that is maintained, μm;

Δ_t is the error due to the dimensional wear of the cutting tool, μm. So,

$$\Delta_\Sigma = \Delta_y + \Delta_{\varepsilon y} + \Delta_{ad} + \Delta_t \tag{16}$$

With

$$\Delta_y = y_{max} - y_{min} = W_{max} P_{max} - W_{min} P_{min} \tag{17}$$

where W_{max}, W_{min} are the highest and lowest system flexibility, μm/N;

P_{max}, P_{min} are maximum and minimum value of the component of the cutting force, which coincides with the direction of the size being maintained, N.

That is, the radial component of the cutting force, determined by the formula:

$$P_y = 10C_{P_y} t^{x_{Py}} S^{y_{Py}} V^{n_{Py}} K_{P_y} \tag{18}$$

where C_{Py} is constant;
x_{Py}, y_{Py}, n_{Py} are power indicators;
K_{Py} is the correction factor, which is the product of the coefficients' multiplication K_{mr}, $K_{\varphi Py}$, $K_{\gamma Py}$, $K_{\lambda Py}$, K_{rPy}.

The setup errors of the workpiece in fixture $\Delta_{\varepsilon\ y}$ depending on the type of workpiece and sizes, as well as adjusting the technological system to the size that is maintained Δ_{ad}, that depends on the error of adjustment of the tool's position and the error of measuring the size of workpiece is given in reference literature.

The formula calculates the component of the total error resulting from dimensional wear of the cutting tool:

$$\Delta_t = \frac{L_f}{1000} u_0 \tag{19}$$

where L_f is the full length of cutting path for the batch of parts, mm;
u_0 is the relative dimensional wear of the cutter for these conditions of operation.

The length of the cutting path when turning one workpiece $L_p = \frac{\pi D l_p}{1000S}$,
l_p is the length of the treated part's surface, mm.

The length of the cutting path L_N for a batch of blanks N that are processed in between machine adjustments – $L_N = L_p N$.

Then, in order to take into account the intensive initial wear, the calculated cutting length is increased by $L_{in} = 1000$ mm, so $L_f = L_N + L_{in}$. Therefore

$$\Delta_t = \left(\frac{\pi D l_p N}{1000S} + L_{in}\right) u_0/1000 \tag{20}$$

6. Cutting tool stability limits:

$$T \geq T_r \tag{21}$$

where T_r has required tool stability, min.

The formula can determine the life of the tool, the cutting part of which is made of superhard materials:

$$T = \left(\frac{C_v K_V}{V t^x S^y}\right)^{1/m} \tag{22}$$

where C_v is constant;

x, ***y***, ***m*** are power indicators;
K_v is a correction factor that takes into account the impact of the workpiece material K_{mv}, the state of the surface layer K_{sv} and tool material K_{tv}:

$$K_v = K_{mv}K_{sv}K_{tv} \tag{23}$$

with

$$K_{mv} = K_r\left(\frac{750}{\sigma_v}\right)^{n_v} \tag{24}$$

where K_r is a coefficient;
n_v is a power indicator.

7. Restriction on surface roughness:

$$R_a \leq R_{areq} \tag{25}$$

where $R_{a\ req}$ is the required roughness value;
R_a is estimated roughness after machining the workpiece surface.

The average height of the roughness profile in the general case for all machining methods is determined by equality:

$$R_z = h_1 + h_2 + h_3 + h_4 \tag{26}$$

where h_1 is a component of the roughness profile, caused by geometry and kinematics of movement of the tool's working part;

h_2 is the component of the roughness profile, which depends on the vibrations of the tool relative to the work surface;

h_3 is the component of the roughness profile, caused by plastic deformation in the area of contact between the tool and the workpiece;

h_4 is a component of the roughness profile, which is determined by the roughness of the workpiece of the tool.

So, in the most common cases with $\varphi \geq arcsin\frac{S}{2r}$ and $\varphi_1 < arcsin\frac{S}{2r}$:

$$R_a = 0.2\left(\begin{array}{c}\dfrac{r(1-\cos\varphi_1) + \sin\varphi_1\left[S\cdot\cos\varphi_1 - \sqrt{S\cdot\sin\varphi_1(2\mathrm{r} - \mathrm{S}\cdot\sin\varphi_1)}\right]}{\cos\gamma} + \\ \dfrac{C_Y S^{y_{Py}} V^{z_{Py}}\left[HB^n_{max}t^{x_{Py}} - HB^n_{min}(t - R_{z_{in}})^{x_{Py}}\right]}{HB^n_{av}j_{ts}} + \dfrac{0.5\rho\left(1 - \dfrac{\tau_{slip}}{\sqrt{\tau^2_{slip} + \sigma^2_{ys}}}\right)}{\frac{1}{tg\varphi_1} + \frac{2r}{S}} + R_{Ztc}\end{array}\right) \tag{27}$$

where φ and φ_1 are the main and auxiliary angles of the cutting tool in the plan;

γ is the front angle;
r is the radius at the top of the cutting part of the tool;

C_Y is coefficient;
x_{Py}, y_{Py}, z_{Py}, n are power indicators;
$R_{z\ in}$ is the initial average height of the surface roughness profile, μm;
HB_{max} and HB_{min} are the highest and lowest value of the workpiece hardness;
HB_{av} is the average value of the workpiece hardness;
j_{ts} is the rigidity of the technological system, N/μm;
τ_{slid} is the strength of the treated material to shear, MPa;
$\sigma_{y\ s}$ is the yield strength of the treated material, MPa;
ρ is the radius of rounding of the auxiliary cutting edge, mm;
$R_{z\ tc}$ is the average height of the roughness profile at the top of the cutter.

Considering the dependencies of partial optimization criteria (1) and (3) and the maximum permissible values of cyclic durability and productivity of the machining process, the complex optimality criterion takes the following form:

$$C(S,V,\sigma) = \left(\alpha_1\left(\frac{N(S,V,\sigma) - N_{min}(S,V,\sigma)}{N_{max}(S,V,\sigma) - N_{min}(S,V,\sigma)}\right) + \alpha_2\left(\frac{\frac{1000VSt}{\pi DLh} - P_{min}(S,V)}{P_{max}(S,V) - P_{min}(S,V)}\right)\right) \quad (28)$$

Given the technological constraints presented by dependencies (6)–(27), the mathematical model for solving the task of optimizing the modes of turning the workpiece, which provides the highest cyclic durability of the workpiece during its operation and the maximum productivity of its quality will be as follows:

$$C(S,V,\sigma) \Rightarrow max\,\text{with}\begin{cases} S_{min} \le \mathrm{S} \le S_{max} \\ \frac{\pi D_0 n_{min}}{1000} \le \mathrm{V} \le \frac{\pi D_0 n_{max}}{1000} \\ P_{\mathrm{maxa}.f.} \ge \mathrm{P}_x = 10C_{P_x} t^{x_{P_x}} S^{y_{P_x}} V^{n_{P_x}} K_{P_x} \\ N_{en}\eta \ge \frac{10C_{P_z} t^{x_{P_z}} S^{y_{P_z}} V^{n_{P_z}} K_{P_z}}{1020\cdot 60} \\ 0.5\cdot\mathrm{TD} \ge \left(\begin{array}{c}(W_{max}P_{max} - W_{min}P_{min}) + \Delta_{\varepsilon y} \\ +\Delta_{ad} + \left(\frac{\pi D l_p N}{1000S} + L_{in}\right)u_0/1000\end{array}\right) \\ T \le \left(\frac{C_v K_V}{V t^x S^y}\right)^{1/m} \\ R_a \le R_{areq} \end{cases} \quad (29)$$

As an example of application, the offered technique of the multicriteria optimization of a detail's "drive shaft" working in the conditions of alternating loadings finishing turning is solved. This part is made of 37Cr4(DIN).

Turning was performed on a turning center HAAS ST20 with cutter tool PVVNN 2525 M-16Q without cooling [11].

Tests of steel 37Cr4(DIN) samples for cyclic durability were performed based on $N = 2\cdot 10^7$ cycles at a temperature of 20°C and a speed of 2000 rpm on the test machine MUI-6000.

After conducting experimental studies and processing their results, a mathematical model of the cyclic durability of the material steel 37Cr4(DIN) from the modes of turning and stress cycle amplitude was made:

$$N(S, V, \sigma) = e^k \quad (30)$$

where $k = 14.437 + 0.0048V + 13.006S - 13.19\sigma + 0.002VS - 0.002V\sigma - 5.941S\sigma + 0.0000004V^2 + 2.929S^2 + 3.013\sigma^2$.

The obtained dependence, according to the performed experimental researches, is valid within the following limits of parameters change V = 80–180 m/min; S = 0.08–0.12 mm/rev; σ = 225–670 MPa.

Then, taking into account the dependence (4), a mathematical model of the part's finishing turning process is formed taking into account its cyclic load during operation, which allows determining the turning modes that give the maximum value to the complex objective function:

$$C(S, V, \sigma) = \left(1.15\left(\frac{e^k - N_{min}(S, V, t)}{N_{max}(S, V, t) - N_{min}(S, V, t)}\right) + 0.85\left(\frac{\frac{1000VSt}{\pi DLh} - P_{min}(S, V)}{P_{max}(S, V) - P_{min}(S, V)}\right)\right) \quad (31)$$

in the area of allowable solutions, which is specified by the system of constraints:

$$\begin{cases} 0.004 \leq S \leq 8 \\ \frac{\pi D_0 350}{1000} \leq V \leq \frac{\pi D_0 3500}{1000} \\ 0.2 \geq P_x = 10C_{P_x} t^{x_{P_x}} S^{y_{P_x}} V^{n_{P_x}} K_{P_x} \\ 12.4 \cdot 0.85 \geq \frac{10C_{P_z} t^{x_{P_z}} S^{y_{P_z}} V^{n_{P_z}} K_{P_z}}{1020 \cdot 60} \\ 0.25 \geq \left(\begin{array}{c} (W_{max}P_{max} - W_{min}P_{min}) + \Delta_{\varepsilon y} \\ + \Delta_{ad} + \left(\frac{\pi D l_p N}{1000 S} + L_{in}\right) u_0 / 1000 \end{array}\right) \\ 60 \leq \left(\frac{C_v K_V}{V t^x S^y}\right)^{1/m} \\ 0.8 \leq R_{areq} \end{cases} \quad (32)$$

The optimization problem for the finishing turning process of a part, represented by the given multicriteria model, is a problem of multidimensional nonlinear mathematical programming, which is solved by the method of sliding tolerance.

5 Conclusions

It is defined that the technological support of cyclic durability of details, which are working in the conditions of alternating loads during operation, is possible by the creation of complex mathematical model with two optimization criteria – the maximum values of productivity of detail manufacturing and durability of its operation. The optimal modes of machining of the part are determined from the range of acceptable solutions, which determined by a set of restrictions on the quality parameters of the machined surfaces and the technological system characteristics. A technique based on multicriteria optimization is proposed to solve the problem at the process planning stage.

The given example of the finishing process multicriteria optimization shows that the proposed technique allows determining the optimal conditions of turning, a tool made of cubic boron nitride, parts made of 37Cr4 (DIN) steel, which provides maximum values of cyclic durability and productivity. According to this method, it is possible to predict the number of load cycles during the operation of the part before its destruction, which will let avoid accidents.

As a result of its solution, the optimal modes for turning of the part "Shaft" are defined (S = 0.106 mm/rev; V = 139.6 mm/min; t = 0.1 mm), providing maximum cyclic durability (N = 89602 cycles) when its operation and maximum manufacturing productivity (P = 0.383 1/min).

The method of technological support is universal. It is generally recommended without minor adjustment of the mathematical model for use for any methods of machining parts, which are operating in hard operating conditions, regardless of the processed and tooling materials.

References

1. Christ, H.-J.: Fatigue of Materials at Very High Numbers of Loading Cycles. Springer, Heidelberg (2018)
2. M'Saoubi, J.C., Outeiro, H., Chandrasekaran, O.W., Dillon Jr., Jawahir, I.S.: A review of surface integrity in machining and its impact on functional performance and life of machined products. Int. J. Sustain. Manuf. **1**(1/2), 203–236 (2008)
3. Barandych, K.S., Vysloukh, S.P., Antonyuk, V.S.: Ensuring fatigue life of parts during finish turning with cubic boron nitride tools. J. Superhard Mater. **40**, 206–215 (2018). https://doi.org/10.3103/S1063457618030085
4. Robinson, J.S., Tanner, D.A., Truman, C.E., Wimpory, R.C.: Measurement and prediction of machining induced redistribution of residual stress in the aluminium alloy 7449. Exp. Mech. **51**(6), 981–993 (2011). https://doi.org/10.1007/s11340-010-93
5. Belgasim, O., El-Axir, M.H.: Modeling of residual stresses induced in machining aluminum magnesium alloy (Al-3 Mg). In: Proceedings of the World Congress on Engineering 2010, WCE 2010, London, U.K., pp. 148–156 (2010)
6. Sasahara, H.: The effect on fatigue life of residual stress and surface hardness resulting from different cutting conditions of 0.45% steel. Int. J. Mach. Tools Manuf. **45**(2), 131–136 (2005). https://doi.org/10.1016/j.ijmachtools.2004.08.002
7. Zhou, J., Bushlya, V., Peng, R.L., Chen, Zh., Johansson, S., Stahl, J.E.: Analysis of subsurface microstructure and residual stresses in machined Inconel 718 with PCBN and Al2O3-SiCw tools. In: 2nd CIRP Conference on Surface Integrity (CSI), pp. 150–155 (2014). https://doi.org/10.1016/j.procir.2014.04.026
8. Guo, Y.B., Li, W., Jawahir, I.S.: Surface integrity characterization and prediction in machining of hardened and difficult-to-machine alloys: a state-of-art research and analysis. Mach. Sci. Technol. **13**, 437–470 (2009). https://doi.org/10.1016/j.procir.2013.03.046
9. Zawada-Tomkiewicz, A.: Analysis of surface roughness parameters achieved by hard turning with the use of PCBN tools. Est. J. Eng. **17**(1), 88–99 (2011). https://doi.org/10.3176/eng.2011.1.09
10. Kevin, Y., Evans, C.J., Barashb, M.M.: Experimental investigation on cubic boron nitride turning of hardened AISI 52100 steel. J. Mater. Process. Technol. **134**(1), 1–9 (2003). https://doi.org/10.1016/s0924-0136(02)00070-5

11. Klymenko, S., Kopeikina, M.: Improvement of technologies for edge cutting machining with tools equipped with superhard structured composites. Modern manufacturing process and systems: collective monograph. SaTCIP Publisher Ltd, Moscow (Russia)-Belgrade-Vrnjačka Banja (Serbia) (2020)
12. Athmane Yallese, M., Chaoui, K., Zeghibb, N., Boulanouar, L., Rigal, J.-F.: Hard machining of hardened bearing steel using cubic boron nitride tool. J. Mater. Process. Technol. **209**(2), 1092–1104 (2009). https://doi.org/10.1016/j.jmatprotec.2008.03.014
13. Volkogon, V.M., Antonyuk, V.S.: The effect of grafite-like boron nitride to the formation of residual stresses, strength, and performance of materials based on wurtzitic boron bitride. J. Superhard Mater. **23**(5), 50–53 (2001)
14. Borovskii, G.V., Pini, B.E., Khachikyan, E.A.: High-speed precision machining of hardened steels using small-sized cBN tools. Izv. MGTU "MAMI" **2**(14), 30–38 (2012). (in Russian)
15. Samanta, B.: Surface roughness prediction in machining using soft computing. Int. J. Comput. Integr. Manuf. **22**(3), 257–266 (2009). https://doi.org/10.1080/09511920802287138
16. Kadirgama, K., Noor, M.M., Zuki, N.M., Rahman, M.M., Rejab, M.R.M., Daud, R., Abou-El-Hossein, K.A.: Surface roughness prediction model of 6061-T6 aluminum alloy machining using statistica l method. Eur. J. Sci. Res. **25**(2), 250–256 (2009)
17. Vyslokh, S.P.: Information technology in the tasks of technological preparation of the adjunct of machine-excitation technology: monograph. NTUU "KPI", Kyev (2011). (in Ukrainian)
18. Visloukh, S.P.: Determination of process parameters of new tool. J. Superhard Mater. **23**(5), 65–69 (2001)
19. Dalsky, A.M., Suslov, A.G., Kasilov, A.G., Meshcheryakov, R.K.: Reference technologist-mechanical engineer. V.2. Mechanical Engineering-1, Moscow (2003). (in Russian)
20. Kheifetz, M.L., Vasilyev, A.S., Klimenko, S.A.: Technological control of the heredity of operational quality parameters for machine parts. Adv. Mater. Technol. **2**, 8–18 (2019). https://doi.org/10.17277/amt.2019.02.pp.008-018

Check for updates

Improvement of the Quality for Cutting Tool Monitoring by Optimizing the Features of the State Space

Oleksandr Derevianchenko(✉), Oleksandr Fomin, and Natalia Skrypnyk

Odessa National Polytechnic University,
1, Shevchenko Avenue, Odessa 65044, Ukraine
algdrv@opu.ua

Abstract. The goal of the research is to develop an approach to improve the quality of monitoring cutting tools (for example of mills) based on the feature spaces of their state optimization. The scientific novelty of the work is to create a new approach to improving the quality of monitoring the process of face milling, based on obtaining the optimal sets of indirect features for specified quality parameters – monitoring speed and accuracy of recognition of tool states. The practical value of the developed approach lies in the possibility of using developments in the creation of modern machine tools (flexible manufacturing modules). An approach demonstrated for the option of indirect monitoring of the mills conditions, based on the registration and processing of the active power signals of the asynchronous electric drive of the main motion of the machine. Two parameters used as optimization criteria: training time and recognition quality in monitoring systems. The maximum likelihood method used for constructing the classifier of cutting tools states. An algorithm developed and the corresponding software in the Python programming language created to optimize the feature space with indirect cutting tool control. The high quality of the tools states recognition achieved by using secondary features obtained by wavelet transform of signals. When using 4 secondary features, the classification result of the mill states: monitoring speed is 0.15 s, an error of recognition – 0.1.

Keywords: Indirect control · Cutting tools states · Machining processes · Advanced monitoring systems

1 Introduction

In the conditions of modern industry level "Industry 4.0", there is a continuous increase in the pace of technological processes. Accordingly, in the machining processes by cutting, the modes (primarily, the cutting and feed speeds) were significantly intensified. It leads to an increase in the rate of change of cutting tools (CT) state. In this regard, there is a need to improve the efficiency and quality of tools states monitoring on modern machine tools.

V. Tonkonogyi et al. (Eds.): InterPartner 2020, LNME, pp. 502–512, 2021.
https://doi.org/10.1007/978-3-030-68014-5_49

The main components of the CT monitoring process are:

1. Continuous or periodic CT monitoring or cutting process using sensors of direct, indirect, or combined tools assessment, which ensures the receipt of appropriate signals or patterns (digital images of the cutting part (CP); various signals of the cutting system, machine drives, etc.).
2. Processing of these signals by one or another method and the formation of sets of diagnostic features (DF) that display the current CT state.
3. Training of the monitoring system (building classifiers of CT states) based on the formed DF sets.
4. The classes of CT states recognize in the features space using the constructed classifiers.
5. The CT residual life predicts and determines the optimal moment of its replacement.

The speed of CT monitoring depends on many factors, and first of all – of the DF quantity and quality. The initial space of primary features Y has a dimension n. The CT state in this space, which belongs to one of the classes (Ω_1, Ω_2,… Ω_n), characterized by vector $\boldsymbol{y}$. The result of reducing its dimension, one of the methods of qualitative analysis, is the formation of a vector of secondary features $\tilde{\boldsymbol{y}}$ of m dimension, and the relation: $m \ll n$.

Research Purpose. This research is aimed to develop an approach for improving the quality of CT states monitoring, based on the optimization of feature space of their states. As an optimality criterion, considering the objective function, which depends on the training time of the monitoring system and the quality of tools states recognition. The choice of the optimal method for reducing the dimension of the feature space made using the indirect method of mills monitoring.

2 Literature Review

Improving the quality of cutting processes is one of the main tasks of modern manufacturing, in which more and more intelligent technologies are used. The monitoring of modern machine tools and other technological systems has been the subject of many works, including [1–5]. Mukku V., Lang S. and Reggelin T. [1] note the fact of integration of advanced technologies in an Industry 4.0 Learning Factories. Li B. H., Hou B.C. [2] propose new models, means, and forms of intelligent manufacturing systems architecture and technologies, based on the integration of AI with manufacturing. It improves the quality of the processing. According to P. Oborski's research [3], monitoring is an important part of manufacturing process control. It plays a crucial role in ensuring agility in manufacturing systems: monitoring of manufacturing process, machine tools, cutting tools condition monitoring, processing parameters and efficiency, and other. M. Kumar [4] and W. Mu Lan [5] investigated real-time monitoring systems, a management system for CNC machine tools. Most often, tools used in roughing and impact loads (for example, during milling) fail. Much attention in the literature is devoted to the study of processes and tools for milling. In particular, we note the work [6, 7]. Kang, Y.-G., Wang, Z.-Q. [6] completed researches of peripheral milling [6], Gao, H., Liu, X., Chen, Z. [7] investigated the process of hard milling. The processes of wear and breakage of the cutting elements of mills operating under conditions of periodic shock loads, in the first place, require the creation of continuous monitoring systems. Continuous monitoring of CT provides the ability to change the characteristics of the cutting process directly in the processing. For example, V Tonkonogyi, A. Yakimov [8] proposed to apply solid technological lubricant directly to the working surface of the grinding wheel during the cutting process. It provides a reduction in the thermal stress of the grinding process. The literature analysis indicates the obvious relevance of the development of an approach to improving the quality of cutting tools (and, first of all – mills) monitoring.

3 Research Methodology

For checking the indirect control results' credibility, the system of direct control, which equipped with a special technical vision system, was using. Digital images of wear zones of the mills cutting elements were treating and analyzing with the special program complex [9] using.

In manufacturing, the method of indirect control conditions is widely used. Consider the method of CT states indirect control. Its main positions are set out in [10], and the schemes for forming signals (patterns) of the cutting process (when machining parts with mills) shown in Fig. 1.

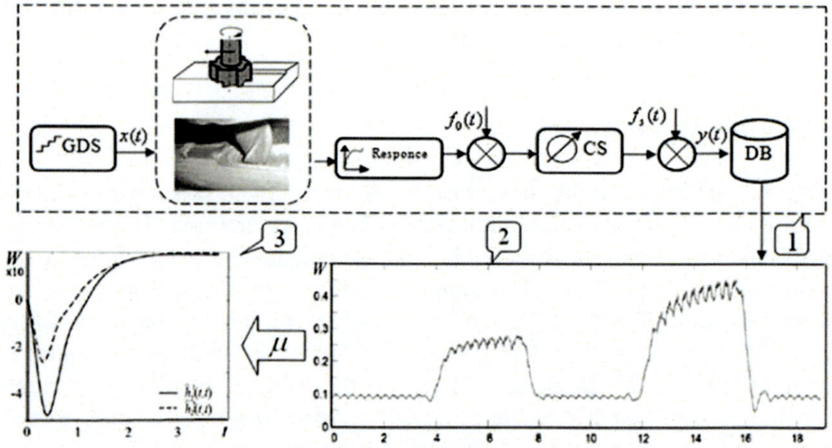

Fig. 1. The general structure of the indirect CT monitoring method: 1 – general organization of the control process [10]; 2 – Filtering the response of the CS: changing the active power of the cutting process *W*, kW in time *t*(*s*); 3 – Estimates for the first-order transition functions $h_2^1(t,t)$ and $h_2^2(t,t)$, for the CT belonging to the classes of states Ω_1 and Ω_2 (on the abscissa axis – processing time *t*(*s*) on the axis of ordinates – active power of the cutting process *W* (kW).

Here, the test action $x(t)$ from the generator of diagnostic signals (GDS) induces a response, which adds the noise $f_o(t)$ and $f_s(t)$ from the diagnostic object and the recording device (CS). The received response $y(t)$ is stored in the database (DB).

The operator μ denotes the process of forming a nonlinear dynamic model of the milling process. As a result of the experiments, a set of signals (patterns of the cutting process with new and worn face milling cutters) was investigated. This set used to form the optimal feature space of CT states. The criteria for optimizing the feature space are:

1. The speed of monitoring (*t*), which directly related to the number of diagnostic features used to recognize the CT state;

2. Monitoring quality (*Q*), which is characterized by the probability of correct recognition (*P*) of CT states, or the probability of recognition error (*e*).

The block diagram of the algorithm for the formation of the optimal CT states features space is shown in Fig. 2. It includes two sequentially implemented modules (blocks) A and C (in the diagram, they indicated by dashed rectangles): the module for the formation of the space of primary features of CT states (A); the module for the formation of the space of secondary features of the CT states (C).

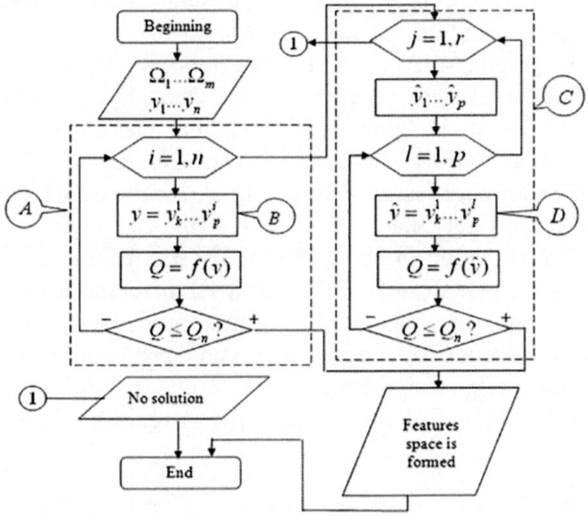

Fig. 2. The algorithm block diagram for the formation of the optimal CT states features space.

Input (source) data in the general case are: many classes of CT states ($\Omega_1 \ldots \Omega_m$) of dimension m to be recognized; sets of primary features ($y_1 \ldots y_n$) of dimension n, formed by processing signals from a sensor of indirect CP control (Fig. 1).

In the analysis cycle of primary features (in module A), their dimension is reduced [11, 12] (by sequentially forming their vectors from the set: $y = y_k^1 \ldots y_p^i$). Each new vector arrives at the input of the CT state classifier, which is constructed using the maximum likelihood method [13].

Next, class recognition is performed with the assessment of quality criteria $Q(t, e)$ and comparing them with predetermined threshold values. When a positive result is achieved ($Q > Q_n$), further analysis is terminated; the optimal feature space is considered formed.

If the result is not achieved (no solution), data is transferred to module C – for the formation of secondary features. Here, the features are the results of its wavelet transform (module D). The corresponding software package implements in the Python programming language.

4 Results

A training sample of the second-order responses of the indirect diagnostic problem [10] used. A fragment of the training sample is presented in Table 1. The research results presented in Fig. 3.

Module A implementation of the process formation of the optimal feature space (using the initial vector **y** in the amount of 10 primary features) allowed obtaining the results in two graphs. It is a graph of the classification error dependence from the number of features ($e = F_1(k)$) and a graph of the corresponding classification time on the number of features ($t = U_1(k)$). The results of their analysis show in Fig. 3. The recognition quality criteria Q were set: $t = 0.12$ s, $e = 0.18$.

For part S of the graph $t = U_1(k)$, the relation holds: $k \leq 3$, i.e., no more than three primary features can be used to recognize the CT states by this criterion (for a time $t < 0.12$ s). For part M of the graph $e = F_1(k)$, the relation holds: $k \geq 6$, i.e., for recognize the CT state according to this criterion (with a recognition error of not more than 0.18), at least six primary features must be used. A primary solution features using was not obtained, because the relation is: $S \cap M = 0$.

The need for the formation of secondary diagnostic features of CT conditions is obvious [10, 14]. As a transformation at the level of compression of information models, families of diagnostic features, based on spectral transformations, frequency characteristics, Karunen-Loev orthogonal expansion coefficients, wavelet transform coefficients, and moments of a signal are using.

Coefficients of signal's wavelet-transform $w_k(t)$ of order k (1) [14, 15]:

$$C(a,b) = \int_0^\infty w_k(t-\tau_1,\ldots,t-\tau_k)a^{-1/2}\psi(\frac{t-b}{a})dt \tag{1}$$

where: $\psi(t)$ – conversion function (mother wavelet); a, b – respectively, the scale and displacement parameters of the wavelet.

To select the wavelet, that provides the best signal recovery, many numerical experiments performed on the conversion and reconstruction of aperiodic and oscillatory signals using *bior*, *coiflet*, *dobechi*, *haar*, *symlet* wavelets. The smallest errors in reconstructing the studied signals are achieved when using the *coiflet* wavelet of order 4, which is accepted as the base when compressing diagnostic models.

Table 1. A fragment of the training sample from the responses of the 2nd order in diagnosing the CT states.

№	Primary features x_i, $i = 1,\ldots,10$										Class Ω_k, $k = 1,2$
	1	2	3	4	5	6	7	8	9	10	
1	0.00	44.59	31.32	16.74	8.21	3.89	1.81	0.84	0.39	0.18	1
2	0.00	44.35	33.89	19.58	10.33	5.25	2.62	1.29	0.64	0.31	1
3	0.00	45.08	32.77	18.08	9.13	4.45	2.13	1.01	0.48	0.22	1
4	0.00	44.88	31.07	16.38	7.93	3.71	1.71	0.78	0.36	0.16	1
5	0.00	44.88	32.77	18.15	9.20	4.50	2.16	1.03	0.49	0.23	1
…	…	…	…	…	…	…	…	…	…	…	…
196	0.00	32.88	19.01	8.50	3.52	1.42	0.57	0.22	0.09	0.03	2
197	0.00	33.62	19.32	8.58	3.54	1.42	0.56	0.22	0.08	0.03	2
198	0.00	33.47	19.31	8.61	3.57	1.44	0.57	0.22	0.09	0.03	2
199	0.00	33.84	19.48	8.67	3.59	1.44	0.57	0.22	0.09	0.03	2
200	0.00	33.72	19.50	8.71	3.61	1.46	0.58	0.23	0.09	0.03	2

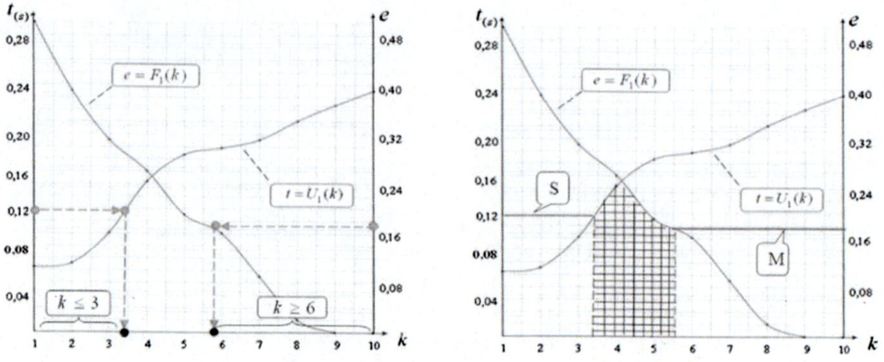

a b

Fig. 3. The results of the analysis of the CT states recognition quality using primary features (a) and module A implementation of the optimal CT states space forming process (b).

The corresponding software package implements in the Python programming language. Here is a listing of one of the modules – module D:

```
import pywt
from pylab import *
from numpy import *
data = numpy.loadtxt("data.txt", delimiter='\t', dtype=
numpy.float)
st='coif4'
(cA, cD) = pywt.dwt(data,st)
```

A training sample of wavelet coefficients of second-order responses of the indirect diagnostic problem was used [10]. A fragment of the training sample presented in Table 2.

The implementation of module C of the process of forming the optimal feature space (when using a vector ў in the amount of 9 secondary features) allowed to obtain the results in the form of two function graphs: e = F2(k) and t = U2(k) (Fig. 4a). The results of their analysis show in Fig. 4b. Obviously, a solution using secondary features, obtains ($S \cap M \neq 0$). The specified quality of the recognition of the CT states is achievable using 2 or 3 features.

Let us consider another variant of the boundary conditions for the features of quality recognition $Q(t, e)$: $t = 0.15$ s, $e = 0.1$. Here, the time for CT state classifying is somewhat longer, but the error is much less.

Table 2. A fragment of a training sample of wavelet coefficients (second-order response) in diagnosing the mill conditions.

№	Secondary features x_i, $i = 1,\dots,9$									Class Ω_k, $k = 1,2$
	1	2	3	4	5	6	7	8	9	
1	21.60	14.75	90.81	129.96	33.00	8.97	2.00	2.39	4.14	1
2	19.66	11.84	91.23	116.32	28.79	3.63	1.38	1.38	1.21	1
3	20.18	12.32	89.56	127.54	34.75	8.10	3.00	1.29	0.05	1
4	16.94	9.79	87.03	121.68	31.15	7.26	2.71	2.14	3.07	1
5	23.66	15.54	95.54	115.72	26.22	6.01	2.86	3.84	5.25	1
…										…
196	22.53	15.85	74.69	69.58	12.06	0.52	1.37	1.98	3.30	2
197	20.51	13.98	70.85	71.03	11.08	1.58	0.07	4.02	5.01	2
198	24.27	17.11	76.11	69.83	14.68	4.45	3.28	0.40	2.65	2
199	19.04	11.39	71.94	71.57	9.59	0.00	0.92	0.89	1.01	2
200	24.46	17.70	76.33	73.83	10.71	0.72	1.68	1.95	1.73	2

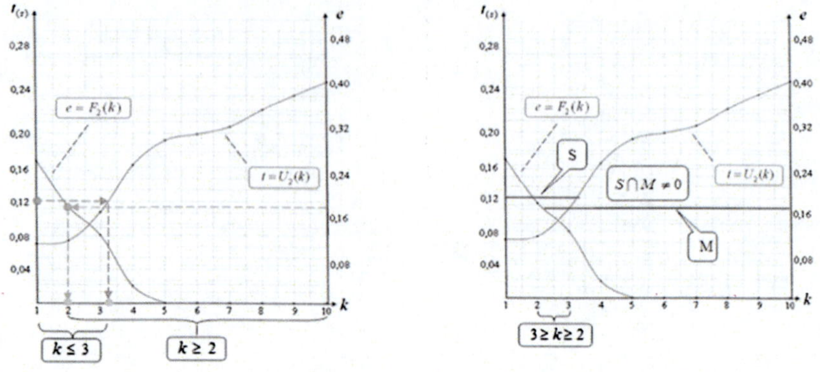

a b

Fig. 4. The analysis results of the quality of recognition of the CT states with the secondary features using.

The beginning graphs are shown in Fig. 5a and Fig. 6a. Results presented in Fig. 5b (primary features – no solution) and Fig. 6b (secondary features – there is a solution, $k = 4$). For part S (Fig. 5b) of the graph $t = U_1(k)$, the relation holds: $k \leq 4$. For part M (Fig. 5b) of the graph $e = F_1(k)$, the relation holds: $k \geq 7$. Obviously, a solution primary features using was not obtained, because the relation is: $S \cap M = 0$.

For part S (Fig. 6b) of the graph $t = U_2(k)$, the relation holds: $k \leq 4$. For part M (Fig. 6b) of the graph $e = F_2(k)$, the relation holds: $k \geq 3$. Obviously, a solution secondary features using was obtained, because the relation is: $S \cap M \neq 0$, $k = 4$.

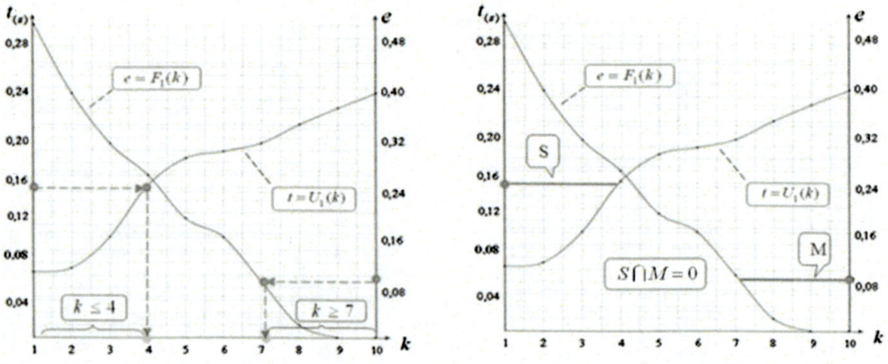

a b

Fig. 5. Analysis results of the CT states recognition quality $Q(t, e)$ when using primary features (specified quality criteria Q: $t = 0.15$ s, $e = 0.1$).

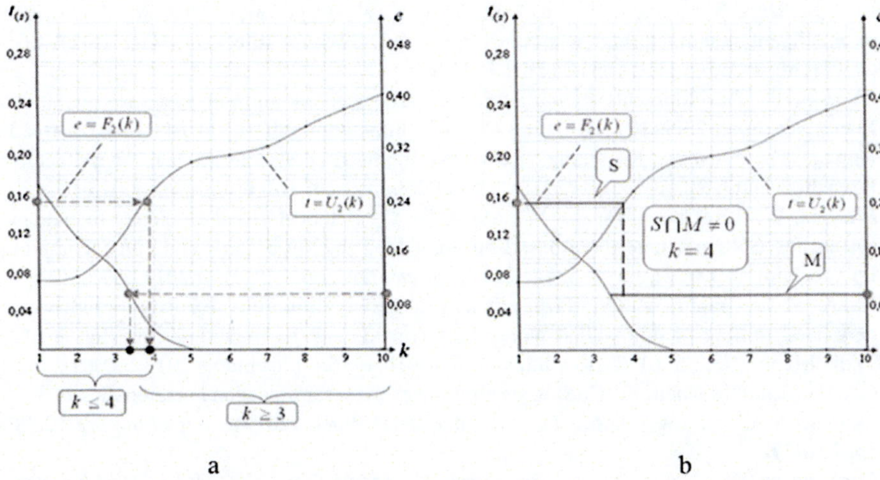

Fig. 6. Analysis results of the CT states recognition quality $Q(t, e)$ when using secondary features (specified quality criteria Q: $t = 0.15$ s, $e = 0.1$).

5 Conclusions

An approach to improve the quality of cutting tools monitoring, based on the optimization of the feature spaces of their states, has been developed. Two parameters used as optimization criteria: time and recognition quality. The approach demonstrated for the option of indirect monitoring of the face mills states, based on the registration and processing of signals of the active power of the asynchronous electric drive of the main motion of the machine.

The method of maximum likelihood used as a method for constructing a classifier of CT states is developed. An algorithm has been developed, and the corresponding software in the Python programming language has been created to optimize the feature space of face mills state with indirect control using. The high quality of the CT states recognition is achieved by using secondary features, obtained as a result of the wavelet transform of signals. Already on preliminary research results, the use of 3 secondary features provides a classification of the CT state within 0.12 s with an error of not more than 0.18. The next result is the classification of the CT state in 0.15 s with a significantly smaller error – 0.1. In future research, it is planned to reduce the classification error of CT states to 0.05 – 0.03.

References

1. Mukku, V.D., Lang, S., Reggelin, T.: Integration of Li Fi technology in an industry 4.0 learning factory. In: 9th Conference on Learning Factories, CLF 2019, Braunschweig; Germany (2019). Procedia Manufacturing, vol. 31, pp. 232–238. https://doi.org/10.1016/j.promfg.2019.03.037

2. Li, B.H., Hou, B.C., Yu, W.T., Lu, X.B., Yang, C.W.: Applications of artificial intelligence in intelligent manufacturing: a review. Front. Inf. Technol. Electron. Eng. **18**(1), 86–96 (2017). https://doi.org/10.1631/FITEE.1601885
3. Oborski, P.: Developments in integration of advanced monitoring systems. Int. J. Adv. Manuf. Technol. **75**(9–12), 1613–1632 (2014). https://doi.org/10.1007/s00170-014-6123-x/
4. Kumar, M., Vaishya, R., Parag: Real-time monitoring system to lean manufacturing. Procedia Manuf. **20**, 135–140 (2018). https://doi.org/10.1016/j.promfg.2018.02.019
5. Mu-Lan, W., Kai-Cheng, F., Bo, W., Meng-Jia, C.: Practical visual efficiency management system for CNC machine tools based on embedded microprocessor and cloud computing. Inf. Technol. J. **12**(22), 6577–6582 (2013). https://doi.org/10.3923/itj.2013.6577.6582
6. Kang, Y.-G., Wang, Z.-Q.: Two efficient iterative algorithms for error prediction in peripheral milling of thin-walled workpieces considering the in-cutting chip. Int. J. Mach. Tools Manuf. **73**, 55–61 (2013). https://doi.org/10.1016/j.ijmachtools.2013.06.001
7. Gao, H., Liu, X., Chen, Z.: Cutting performance and wear/damage characteristics of PCBN tool in hard milling. Appl. Sci. (Switzerland) **9**(4) (2019). https://doi.org/10.3390/app9040772
8. Tonkonogyi, V., Yakimov, A., Bovnegra, L., Sidelnykova, T., Dašić, P.: The use of intermittent wheels, impregnated by the contact method to reduce the thermal stress of the grinding process. IOP Conf. Ser. Mater. Sci. Eng. **708**(1), 012034 (2019). https://doi.org/10.1088/1757-899X/708/1/012034
9. Antoshchuk, S., Derevianchenko, O., Tkachenko, E.: The hierarchical objects analysis on images of cutting tool wear zones. In: International Conference TCSET 2006 Conference Proceedings (2007). https://doi.org/10.1109/TCSET20064404512
10. Fomin, O., Pavlenko, O., Derevyanchenko, O., Ruban, O.: An approach to the construction of a nonlinear dynamic model process cutting for diagnosis condition of tools. Appl. Aspects Inf. Technol. **2**(3), 115–126 (2019). https://doi.org/10.15276/aait.02.2019.3
11. Shahana, A.H., Preeja, V.: Survey on feature subset selection for high dimensional data. In: 2016 International Conference on Circuit, Power and Computing Technologies (ICCPCT), pp. 1–4. IEEE (2016). https://doi.org/10.1109/ICCPCT.2016.7530147
12. Kuhn, M., Johnson, K.: Feature Engineering and Selection: A Practical Approach for Predictive Models. Chapman & Hall/CRC Data Science Series (2019)
13. Fainzilberg, L.S.: Mathematical methods for assessing the utility of diagnostic features. Osvita Ukraine, Kiyv (2010)
14. Medvedew, A., Fomin, O., Pavlenko, V., Speranskyy, V.: Diagnostic features space construction using Volterra kernels wavelet transforms. In: Proceedings of the 2017 IEEE 9th International Conference on Intelligent Data Acquisition and Advanced Computing Systems: Technology and Applications (IDAACS), pp. 1077–1081 (2017)
15. Pavlenko, V.D., Pavlenko, S.V., Speranskyy, V.O.: Identification of systems using Volterra model in time and frequency domain. In: Advanced Data Acquisition and Intelligent Data Processing, chapter 10, pp. 233–270 (2014). https://doi.org/10.5772/58354

Modeling of Tool Surface Dressing with Two-Sided Grinding of the Parts Ends

Vitaliy Kalchenko, Volodymyr Kalchenko, Nataliia Sira(✉), Vladimir Venzhega, and Dmytro Kalchenko

Chernihiv National University of Technology,
95, Shevchenko Street, Chernigiv 14035, Ukraine
nnseraya@gmail.com

Abstract. Spatial modeling is widely used to study parts grinding processes to optimize them. Existing geometric models of the process do not take into account the microgeometry of the abrasive wheel surface after its dressing. This paper presents a general modular three-dimensional model of the rough and finish grinding wheel surfaces dressing with diamond pencil grain with two-sided face grinding. The presented model has been created based on unified modules: instrumental, orientation, and shaped ones. The spatial model of the wheel end face shape takes into account the relative motions of its one and the diamond pencil in the dressing process. Based on the developed model, the researches of the shaping accuracy of the tool face roughing and conical calibration sections after its dressing have been carried out. For the improvement of the shaping accuracy of the parts ends, the straightness ensuring methods of the wheels calibration sections when dressed on CNC and without machines are presented. For the first time, the proposed dressing of abrasive wheels with roughing and gauge sections for two-sided grinding of the crosses ends has been offered. It can be used in the processing of various parts. High precision of details ends forming is achieved by the use of wheels with rough and gauge sections, and quality – by the reduction of process heat stress.

Keywords: Diamond pencil · Cross-ends · Oriented wheel · Calibration sections · Three-dimensional modeling · Wheel dressing

1 Introduction

Mechanical engineering is widely used for parts that have high precision and quality requirements for their end surfaces. It is related to their operating conditions. Such details are, in particular, the crosspieces of the Cardan shafts, the bearing rings, the valves, the pushers of the internal combustion engine, the rods, the end measures, the many-sided non-grinding plates, etc. It is the two-sided face grinding that provides small deviations from the parallelism and flatness of the parts end surfaces with high processing performance.

For increasing the efficiency of machining processes, their optimization is performed using spatial modeling. Existing three-dimensional models of abrasive tools do not take into account their geometric roughness (height of micro-irregularities caused by the

V. Tonkonogyi et al. (Eds.): InterPartner 2020, LNME, pp. 513–523, 2021.
https://doi.org/10.1007/978-3-030-68014-5_50

geometry of the machining process) after dressing. A significant disadvantage of existing models of machined surfaces is that the latter is given by a set of consecutive positions of the workpiece that moves along the guide. Such models do not take into account shaping movements during machining: the movements of the grinding tool and the details.

Development of modular three-dimensional models of the removal allowance and formation process of abrasive tools will allow investigating the further influence of wheels' geometrical roughness on details forming accuracy at two-sided face grinding.

2 Literature Review

Researches of grinding processes were carried out in studies [1–5] using the two-dimensional and three-dimensional geometric models. In the study [1], the influence of the single abrasive grain size on the thermal and mechanical characteristics of the grinding process is considered. In work [2], the influence of grinding wheel wear on the accuracy of the machined surface was determined. Additionally, in the study [3], the influence of the kinematic angles of the cutting tool grain on the surface roughness was investigated. However, the models presented in the works [1–7] do not consider the movements of the forming and the treated surfaces, the processes of removing the allowance, and the influence of the surface tool microgeometry after its dressing.

In the studies [6, 7], models of workpieces machined surfaces taking into account the cutting tool geometry were developed, as well as the influence of the tool surface shape on the thickness of the chip and the contact line length of the workpiece and details. In the study [8], two-dimensional and three-dimensional models of the machined surface are presented on the basis of the model of the mechanism of interaction of the wheel grains and the machined surface in the contact zone. In works [9–12], research results on 3D modeling of the diamond grain are provided. However, the presented papers [6–12] do not contain the models of tool formation and do not take into account the processes of removing the allowance.

In the study [13], modular three-dimensional modeling of the removing allowance processes and forming during grinding without analyzing the geometric roughness of tool surfaces after the dressing is proposed.

There are no standard modular three-dimensional models of tool dressing, the process of removing the seam allowance, and the formation when the two-sided grinding of the parts end surfaces in the well-known papers.

3 Research Methodology

The processing of parts 1 (Fig. 1) on the double-sided grinding machine is carried out by abrasive wheels 2. Improving the accuracy of two-sided end grinding is ensured by the use of grinding wheels, which consist of two sections: 3 – to remove the rough allowance and 4 – the gauge ones [14]. Diamond pencils rule the abrasive tool on the machine where grinding of parts is carried out. Dressing of areas for removal of draft allowance is carried out by the device 5 (Fig. 1) (which is supplied with the machine and is located on the body of the grinding head), and the gauge with a diamond pencil located in the drum of the feed products 6. When dressing the calibration sections, the wheels are oriented in the vertical (angle α) and horizontal (angle β) planes [14].

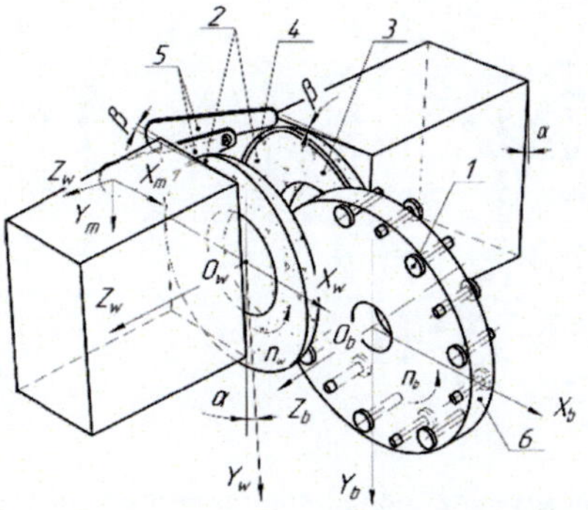

Fig. 1. Scheme of two-sided face grinding with oriented grinding wheels.

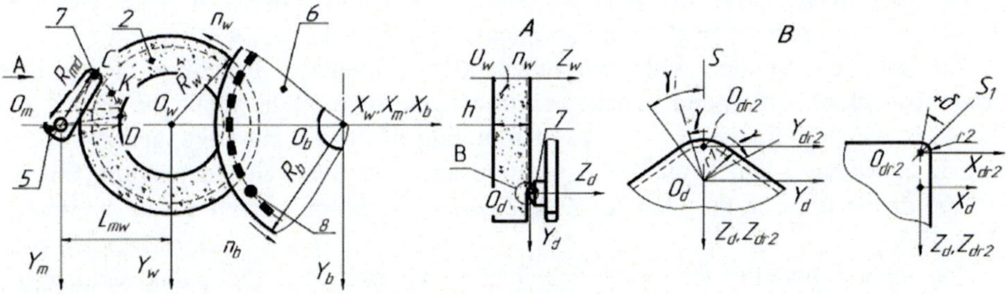

Fig. 2. Calculation scheme for roughing and calibration sections of a grinding wheel with a diamond pencil.

The rough dressing of the end surface 3 of the abrasive wheel 2 (Fig. 2) is carried out with a diamond pencil 7, fixed in the lever 5, with octahedral-shaped cutting grain. The diamond pencil moves along a circular arc *CD*, and the grinding wheel rotates at a frequency of n_w.

Before machining, the grain of diamond pencil 7 must be oriented in such a way that at the point *K* (middle of the arc *CD*) the pencil cutting edge is perpendicular to the wheel machined surface. This will provide a more even distribution of the allowance along the pencil cutting edge and, consequently, less geometric roughness of the abrasive wheel surface.

The clean correction of the calibration conical section 4 of the oriented wheel 2 (Fig. 3) is carried out by a diamond pencil 8, fixed in the product feed drum 6.

The point contact of the diamond pencil 8 (Fig. 3) and the grinding wheel 2 defines the profile of the calibration section in the curve form *II* [14]. For obtaining a straight profile *I*, a diamond tool with a round flat end is used [14, 15]. If the adjustment of the conical calibration section is carried out on a CNC machine tool, its rectilinear profile

I (Fig. 3) is ensured by the movement of the grinding wheel 2 in the axial direction (along the axis O_wZ_w).

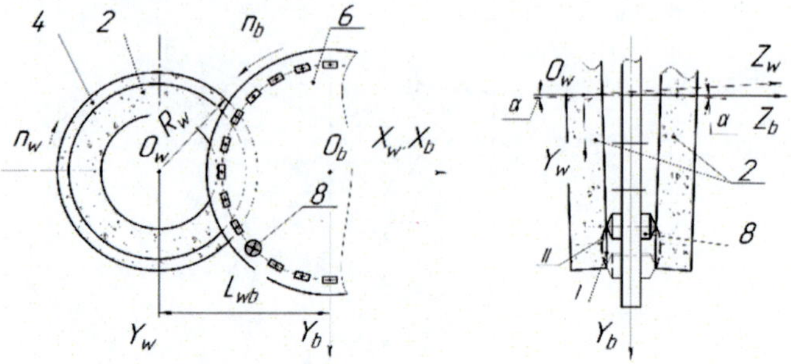

Fig. 3. Scheme of dressing the calibration sections of the grinding wheel.

To develop a common three-dimensional model for removing the allowance and the accuracy of the abrasive wheel end shape when it is dressed, it is necessary to describe the tool surface of the diamond pencil 7 (Fig. 2A).

The spherical module with two independent parameters describes the general surface model of a diamond pencil: δ is the angle of rotation about the axis O_dY_d (Fig. 2B), which defines the radius r2 of rounding of the cutting edge, and γ – is the angular coordinate, which takes the positive and negative value depending on the position of the point in the plane of $O_dY_dZ_d$ diamond pencil relative to the symmetry line O_dS.

The general model of the tool surface can be represented as the radius vector $\overline{r}_d$ of diamond pencil points:

$$\overline{r}_d = S^d_{\gamma \cdot r1 \cdot \delta \cdot r2} \cdot \overline{e}4, \tag{1}$$

$$S^d_{\gamma \cdot r1 \cdot \delta \cdot r2} = M4(\gamma) \cdot M3(r1 - r2) \cdot M5(\delta) \cdot M3(r2), \tag{2}$$

where $S^d_{\gamma \cdot r1 \cdot \delta \cdot r2}$– the spherical module of diamond pencil surface formation represented as a matrix of radius vector transition of the starting point $\overline{e}4$ in the coordinate system of the tool; r1 – the radius that determines the position of the diamond pencil tip (Fig. 1B).

For the construction of modular three-dimensional models one-coordinate matrices *M1*, *M2*, *M3*, *M4*, *M5*, *M6* are used, which describe displacements and rotations about the axes O_dX_d, O_dY_d, O_dZ_d, respectively [9].

In order to study the accuracy of the diamond pencil dressing process, it is necessary to model its surface in the form of two straight sections and a spherical part (Fig. 1B). One-coordinate matrices and the Heaviside function $\overline{r}_d(\gamma, \delta)$ give the universal model of the tool radius vector $\Phi(\gamma)$:

$$\overline{r}_d(\gamma,\delta) = M4(\gamma)\cdot M3(r1-r2)\cdot M5(\delta)\cdot M3(r2)\cdot \overline{e}4\cdot \Phi(|\gamma|) - M4(\gamma)\cdot M3(r1-r2) \\ \times M5(\delta)\cdot M3(r2)\cdot \overline{e}4\cdot \Phi(|\gamma|-\gamma_i) + M4(\gamma_i)\cdot M2(h|\gamma|))\cdot M3(r1-r2)\cdot M5(\delta) \\ \times M3(r2)\cdot \overline{e}4\cdot \Phi(\gamma-\gamma_i) + M4(-\gamma_i)\cdot M2(-h|\gamma|))\cdot M3(r1-r2)\cdot M5(\delta)\cdot M3(r2)\cdot \overline{e}4\cdot \Phi(-\gamma-\gamma_i), \tag{3}$$

where γ_i – an angle that determines the position of the radius edge relative to the symmetry line of the plate O_dS (Fig. 1B); $h(\gamma) = (r1 - r2)\cdot tg(\gamma - \gamma_i)$ – a function that determines the coordinate of a point along the conical section of the cutting edge.

Calculations and construction of three-dimensional models were carried out in a mathematical package MatcCAD. Spatial models of diamond pencil cutting edges with angle at top $\gamma_i = \pi/3$ and rounding radii $r1 = 1$ mm, $r2 = 0{,}05$ mm presents in Fig. 4. In Fig. 4a, a face of the octahedron shaped diamond-shaped grain is shown. Figure 4b shows a cone-shaped grain.

a b

Fig. 4. Model of the diamond pencil forming surface.

The nominal surface of the grinding wheel 2 (Fig. 2) is given by the transition matrix M^t_{dw} from the coordinate tool system to the coordinate system of the workpiece, taking into account the diamond pencil shape:

$$\overline{r}_w = M^t_{dw}\cdot \overline{r}_d = C^{sh}_{a_y\cdot a_x\cdot\theta_{w0}\cdot\theta_w\cdot x_c}\cdot S^{or}_{\alpha_{or}\cdot\gamma_{or}\cdot\varphi_{or}}\cdot \overline{r}_d, \tag{4}$$

$$S^{or}_{\theta_{or}\cdot\gamma_{or}\cdot\varphi_{or}} = M6(\theta_{or})\cdot M5(\delta_{or})\cdot M4(\gamma_{or}), \tag{5}$$

$$C^{sh}_{a_y\cdot a_x\cdot\theta_{w0}\cdot\theta_w\cdot x_c} = M6(\theta_w)\cdot M2(\theta_w\cdot a_y)\cdot M1(\theta_w\cdot a_x)\cdot M6(\theta_{w0})\cdot M1(-z_c), \tag{6}$$

where $S^{or}_{\alpha_{or}\cdot\gamma_{or}\cdot\varphi_{or}}$ – orientation module of the diamond pencil in the circle coordinate system; $C^{sh}_{a_y\cdot a_x\cdot\theta_{w0}\cdot\theta_w\cdot x_c}$ – a cylindrical module that defines the motion of a diamond pencil relative to a grinding wheel; θ_{or}, δ_{or}, γ_{or} – the angles of inclination of the tool surfaces corresponding to the axes O_dZ_d, O_dY_d, O_dX_d respectively; z_c – coordinate that specifies the movement of the diamond pencil coordinate system to the coordinate system of the abrasive wheel (sets the height h of the grinding wheel before machining (Fig. 2A)); θ_{w0} – the angular coordinate of the diamond pencil initial position in the coordinate system of the wheel (Fig. 5), is given by the radius of the R_w tool, the axial distance L_{mw} between the wheel and the lever and the radius R_{md} on which the diamond pencil is located; θ_w – rotation angle of the workpiece coordinate system, models the rotation of the wheel machined surface around its own axis (Fig. 5); a_x, a_y – projections on the corresponding axes of the Archimedes spiral parameter a, which describes the movement of the diamond pencil relative to the wheel.

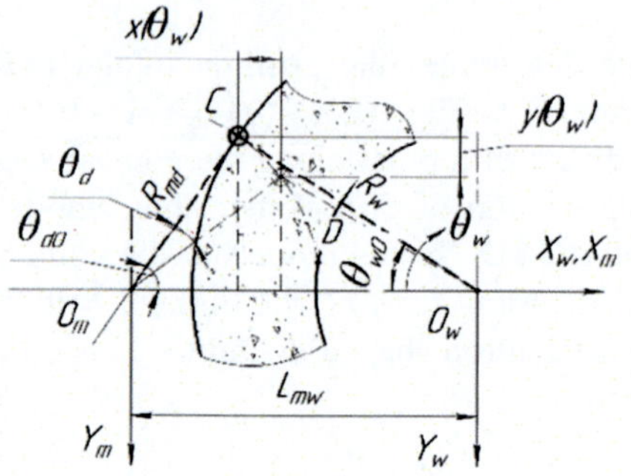

Fig. 5. Calculation scheme for determining the components of the formation module of the end surface of the grinding wheel.

Archimedes spiral parameter projections depend on the projections s_x, s_y of the circular feed of the diamond pencil on the respective axes:

$$a_x = \frac{s_x}{2 \cdot \pi} = \frac{x(\theta_w)}{2 \cdot \pi} = \frac{R_{md} \cdot (\cos(\theta_w \cdot k) - \cos\theta_{d0})}{2 \cdot \pi}, \tag{7}$$

$$a_y = \frac{s_y}{2 \cdot \pi} = \frac{y(\theta_w)}{2 \cdot \pi} = \frac{R_{md} \cdot (\sin\theta_{d0} - \sin(\theta_w \cdot k))}{2 \cdot \pi}, \tag{8}$$

where $x(\theta_w)$, $y(\theta_w)$ – instant coordinates of the diamond pencil along the respective axes in the dressing process (Fig. 5); $\theta_{d0} = \arccos\left(\left(R_{md}^2 + L_{mw}^2 - R_w^2\right)/2R_{md} \cdot L_{mw}\right)$– the angular coordinate of the diamond pencil starting position at point C (Fig. 5); k – coefficient, defined as the ratio of the speed of rotation of the grinding wheel to the speed of circular flow of the diamond pencil.

Given formulas (1), (4) the equation of the machined surface of the grinding wheel has the form:

$$\bar{r}_w(\theta_w, \gamma, \delta) = C^{sh}_{a_y \cdot a_x \cdot \theta_{w0} \cdot \theta_w \cdot x_c} \cdot S^{or}_{\theta_{or} \cdot \gamma_{or} \cdot \varphi_{or}} \cdot S^{d}_{\gamma \cdot r1 \cdot \delta \cdot r2} \cdot \bar{e}4. \tag{9}$$

For the determination of the machining surface of a wheel, it is necessary to write down an equation that determines the contact line of the workpiece and the tool surface:

$$\bar{v} \cdot \bar{n} = \left(\frac{\partial \bar{r}_d}{\partial \gamma} \times \frac{\partial \bar{r}_d}{\partial \delta}\right) \cdot \frac{\partial \bar{r}_d}{\partial \theta_w} = 0, \tag{10}$$

where $\overline{n}$ – a single vector of normal to the diamond pencil surface, is the differential of the radius vector $\overline{r}_d$ of the pencil surface by two independent parameters γ and δ; $\overline{v}$ – velocity vector of the pencil relative motion in the wheel coordinate system (differential of the radius vector $\overline{r}_d$ by parameter θ_w, which models the angle of machined surface rotation per unit time).

For finding the angles γ_{min}, γ_{max}, which determine the initial and final locations of the points on the forming face of a diamond pencil, it is necessary to find the solution of equality (10) with the help of the block (11):

$$angle = \left| \begin{array}{l} \delta \leftarrow 0 \\ for\ k\ \in\ 0..n_i \\ \left| \begin{array}{l} \gamma \leftarrow \gamma_{\min} + \frac{\gamma_{\max}}{10} \cdot k \\ \theta \leftarrow root(n(\gamma,\ \delta,\ 0) \cdot v(\gamma,\ \delta,\ 0), \delta) \\ M^{\langle k+1 \rangle} \leftarrow \begin{pmatrix} \gamma \\ \delta \end{pmatrix} \end{array} \right. \\ M^T \end{array} \right. , \tag{11}$$

where M – a matrix containing the coordinates of the contact line points; n_i – the number of segments that the contact line is conditionally broken.

To facilitate the calculations, we interpolate the discrete dependence into a functional one:

$$angle(\gamma) = \text{interp}\left(cspline\left(angle^{\langle 1 \rangle}, angle^{\langle 2 \rangle}\right), angle^{\langle 1 \rangle}, angle^{\langle 2 \rangle}, \gamma\right). \tag{12}$$

Equation of contact line between the abrasive wheel and pencil:

$$KontLine(\gamma) = \overline{r}_d(\gamma, angle(\gamma),\ 0). \tag{13}$$

The workpiece surface is formed by rotating the contact line around the workpiece axis:

$$PSurf(\gamma, \theta_w) = \overline{r}_d(\gamma, angle(\gamma),\ \theta_w). \tag{14}$$

Equations (9) and (10) describe a common modular three-dimensional model of grinding wheel end surface forming during the dressing.

4 Results

Based on the grinding wheel surface common modular three-dimensional model when straightened with a diamond pencil, the simulated grinding wheel end surface after its dressing (Fig. 6). Grinding wheel radius R_w = 40 mm; height h = 20 mm, angle at top of diamond pencil $\gamma_i = \pi/3$; rotation speed of a grinding wheel of 35 m/s, lever with a diamond pencil – 35 m/min; cutting depth t = 0,1 mm.

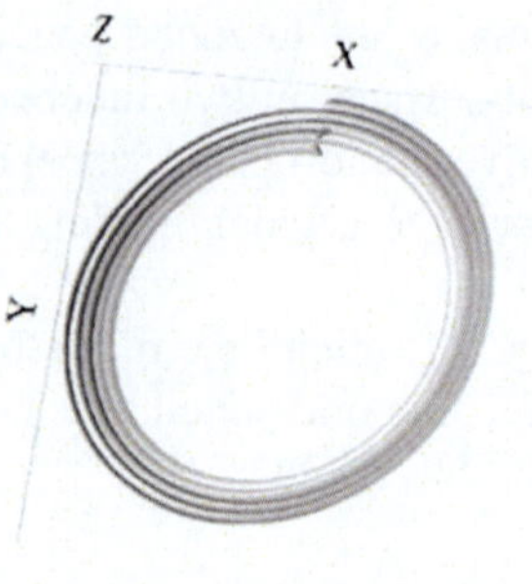

Fig. 6. The profile of the end surface of the grinding wheel after dressing.

As can be seen from Fig. 6, as a result of the dressing on the wheel surface, a geometric roughness *Rz* is formed. Define it as the distance from the nominal workpiece surface to the point *N* (Fig. 7) of the intersection of two consecutive positions 1 and 2. 1 and 2 – the projection of the contact lines on the plane of the wheel.

$$Rz = Poz1(\gamma 1) - z_c, \tag{15}$$

where $Poz1(\gamma 1)$ – the height of the surface profile at the point *N* (intersection of two consecutive cross-sectional positions of the wheel machined surface).

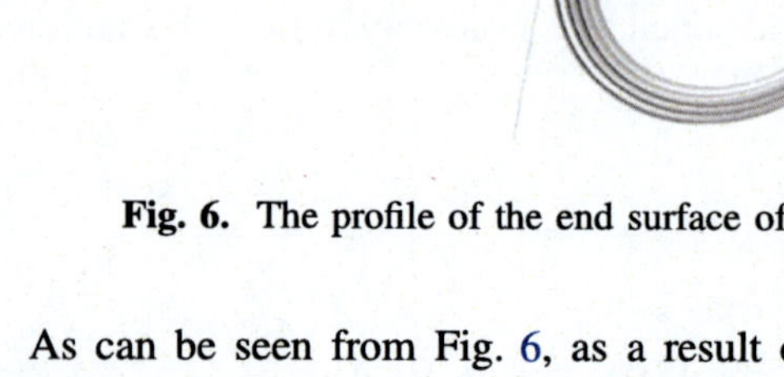

Fig. 7. Geometric roughness of the end surface of the grinding wheel.

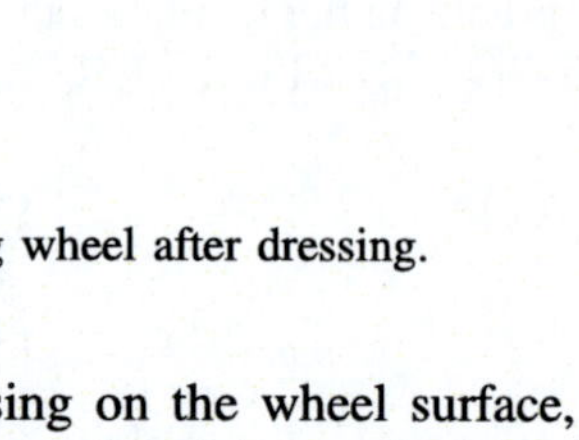

Fig. 8. Determination of the area of the slice layer.

To determine the area of the cut layer, consider the axial section of the abrasive wheel (Fig. 8). Three curves limit the section area: two consecutive positions of the tool surface 1, 2, and line 3 that define the contour of the wheel before dressing. The area of the slice layer is determined from (16):

$$F := \int_{A1}^{A2} Sz(\gamma)\, d\gamma + \int_{A2}^{A3} Poz2(\gamma)\, d\gamma - \int_{A1}^{A3} Poz2(\gamma)\, d\gamma, \tag{16}$$

where $Sz(\gamma)$ – equation of the axial section of the workpiece; A1, A2, A3 – are the points of intersection of the functions limiting the slice area (Fig. 8).

When dressing an abrasive wheel height $h = 20$ mm with a cutting depth $t = 0{,}1$mm, the area of the layer to be cut is 0,35 mm^2.

The analysis of the graph (Fig. 7) shows that for ensuring the high accuracy of the machined parts, the rough section should be corrected with a larger value of the flow value *s*, providing a high cutting capacity of the grinding wheel, and a clean one with a smaller. The smaller feed values on the grinding wheel calibration section provide lower values of the Archimedes spiral parameter *a* (Fig. 6), along which a diamond pencil moves. And smaller values of machining allowance reduce the height of the diamond pen profile involved in the machining process. This adjustment of the grinding wheel finishing section reduces its geometric roughness *Rz* (Fig. 7) and increases the accuracy of the workpiece ends shaping.

In two-sided grinding, the parts are shifted axially by some value Δ. This increases the removable allowance on one end and reduces it on the other. The crosspiece is the representative of the most complicated part in bilateral face grinding. Unlike circular and non-circular parts, which are machined on two-sided grinding machines, the crosspieces 3 (Fig. 9) are fixed axially in the prism of the drum 5 (Fig. 9A). This causes an error of base Δ, which increases the asymmetry of the ends.

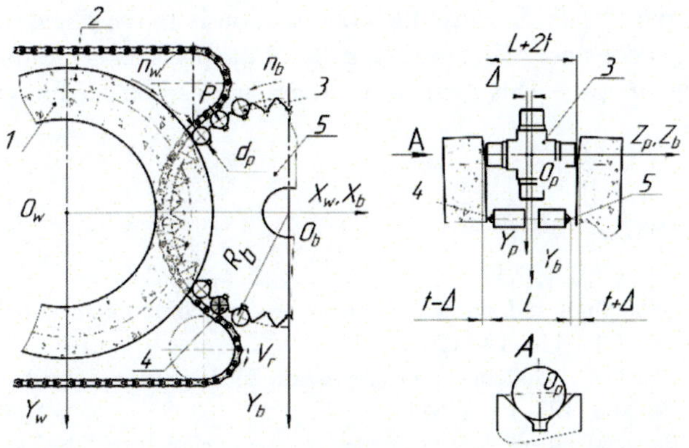

Fig. 9. Cross-sectional machining of grinding wheels with calibration sections.

For the achievement of symmetry and a given roughness (Ra = 0.63 μm) of the ends, their processing is carried out by grinding wheels 1 (Fig. 9) with conical calibration sections 2. In this case, two diamond pencils 4 and 5 are mounted on the product feed drum with the size of the finished part L with tolerances, respectively (t – Δ) and (t + Δ). This will compensate for the asymmetry of the ends by reducing the error of the allowance *t* and the base Δ.

The quality of the machined ends at grinding wheels with rough and finishing sections is improved by reducing the heat stress of the process. When machining a rough section of the wheel, the contact area is equal to the area of the workpiece, and on the calibration section, the treatment occurs with a linear contact on the forming cone. And since the formation of the conical calibration section of the wheel is in the plane of the machined end, reducing its length in the processing process does not affect the accuracy of the workpiece shape and increases the stability of the abrasive tool.

5 Conclusions

A common modular three-dimensional model of grinding wheel dressing was developed based on three unified modules: tooling, orientation, and shaping.

Based on the analysis of the obtained model, it is determined that the high precision of the machined parts is ensured by straightening of the wheel with higher values of feed on its rough section, providing the high cutting ability of the grinding wheel, and on the finishing – with smaller ones. The high quality of the machined ends of the parts is improved by reducing the heat intensity of the process. When machining a rough wheel section, the contact area is equal to the area of the workpiece, and on the calibration section, the treatment occurs with a linear contact on the forming cone.

For the achievement of symmetry and a given roughness (Ra = 0.63 μm) of the crosspieces ends, two diamond pencils are installed on the product feed drum with the size of the workpiece with tolerances for the allowance of error and base. This compensates for the asymmetry of the ends. And using conical gauge wheels ensures high precision of the parts ends. Because the cone section is in the plane of the workpiece, reducing its length does not affect the accuracy of the parts ends molding and increases the stability of the wheels. This method of processing can be used in grinding the ends of various parts.

References

1. Aurich, J.C.: Modelling and simulation of process: machine interaction in grinding. Prod. Eng. Res. Devel. **3**(1), 111–120 (2008)
2. Sharif Ullah, A.M.M.: Modeling and simulation of 3D Surface finish of grinding. Adv. Mater. Res. **126–128**, 128–133 (2011)
3. Zhang, H.: Effects of wheel dressing errors on the accuracy of CNC gear form grinding. Appl. Mech. Mater. **328**, 400–407 (2013)
4. Uhlmanna, E.: Modelling and simulation of grinding processes with mounted points: Part I of II - Grinding tool surface characterization. Procedia CIRP **46**, 599–602 (2016)
5. Choi, J.: Development of the process model for plunge grinding and optimization of grinding process. J. Mech. Eng. Sci. **225**(11), 2628–2637 (2011)
6. Abdalslam, D.: 3D metal removal simulation to determine uncut chip thickness, contact length, and surface finish in grinding. Int. J. Adv. Manuf. Technol. **66**(9–12), 1715–1724 (2012)
7. Changshun, C.: Research about modeling of grinding workpiece surface topography based on real topography of grinding wheel. Int. J. Adv. Manuf. Technol. **93**(5–8), 2411–2421 (2017)
8. Jiang, J.L.: 2D/3D ground surface topography modeling considering dressing and wear effects in grinding process. Int. J. Mach. Tools Manuf **74**, 29–40 (2013)
9. Kundrák, J.: Diamond grinding wheels production study with the use of the finite element method. J. Adv. Res. **7**, 1057–1064 (2016)

10. Markopoulos, A.P.: FEM/AI models for the simulation of precision grinding. Manuf. Technol. **16**(2), 384–390 (2016)
11. Arısoy, Y.M.: Investigations on microstructural changes in machining of Inconel 100 alloy using face turning experiments and 3D finite element simulations. Int. J. Mech. Sci. **107**, 80–92 (2016)
12. Niesłony, P.: Meshing strategies in FEM simulation of the machining process. Arch. Civ. Mech. Eng. **15**(1), 62–70 (2015)
13. Kalchenko, V.: Crossing axes of workpiece and tool at grinding of the circular trough with variable profile. Acta mechanica et automatica **12**(4), 281–285 (2018)
14. Kalchenko, V.I.: Modular 3D modeling of ends bilateral grinding process by wheels with conical calibrating sections. Sci. J. TNTU **4**, 82–92 (2016)
15. Kalchenko, V.: Three-dimensional simulation of machined, tool surfaces and shaping process with two-side grinding of cylindrical parts ends. Adv. Manuf. Process. 118–127 (2020)

Check for updates

Simulation of Metal Transition and Shaping Process by Oriented Turning Tools with Indexable Inserts of a Shaft

Volodymyr Kalchenko, Vitaliy Kalchenko, Olha Kalchenko, Antonina Kolohoida(✉), and Nataliia Sira

Chernihiv National University of Technology, 95, Shevchenko Street, Chernigiv 14035, Ukraine
kolohoida@gmail.com

Abstract. The study of cutting processes is most often carried out using mathematical modeling. A number of mathematical models are proposed that describe the grinding process. For turning, they usually consider a flat cutting scheme. Existing modular spatial models of turning do not provide a description of all known processing schemes. In this research work, the term “space cutting wedge” is proposed. This is the part of the indexable inserts that are involved in the metal transition and shaping process. The mathematical general three-dimensional models of metal transition and shaping process of cylindrical surface and end face by oriented turning tools with indexable inserts is proposed. The three-dimensional model of the wearing spatial cutting wedge for the coordinate of processing is developed. The influence of wear on the accuracy of forming is investigated. The mathematical model takes into account the change in radius of rounding of the variable plaster due to frictional wear. Combined with the use of adaptive control on CNC machines, it improves machining accuracy. On the basis of the offered modular models, the study single-point turning, oblique cutting, and the intermediate position was carried out. It is shown that the geometry roughness of the surface decreases during the treatment of the oriented indexable inserts. Most of the cutting blade of the tool is involved in the work, reducing the local loading along the edge. It is found that machining with an oriented indexable insert reduces the cost of turning operations.

Keywords: Cylindrical surface · End face · Single-point turning · Oblique cutting oriented · Indexable inserts · Three-dimensional modeling

1 Introduction

Turning is one of the most used methods of cutting different metal parts. CNC turning machines can produce considerable nomenclature of surface types. It is possible to provide a curved surface of revolution by using a few turning tools with indexable inserts.

Turning has the highest level of energy efficiency. So it is a fundamental problem of engineering to increased the accuracy of turning. This will give an opportunity of using turning on the finishing operation, for example, instead of grinding.

V. Tonkonogyi et al. (Eds.): InterPartner 2020, LNME, pp. 524–533, 2021.
https://doi.org/10.1007/978-3-030-68014-5_51

The development of general three-dimensional models of metal transition and shaping process by oriented turning tools with indexable inserts and their research will contribute to the creation of new high-precision methods of high quality turning.

2 Literature Review

Complex analysis of metal transition and shaping process for grinding was shown in [1]. Where proposed the three-dimensional schemes of grinding. Simulation of milling processing is proposed [2, 3]. However, existing methodologies for the study of the turning process use a flat forming scheme, with the cutting edge being assumed to be quite sharp. And its wear during processing is not taken into account.

In [3], the authors consider the typical turning of a cylindrical surface. Oblique cutting is described in [4–7]. It mainly uses for finishing operations, due to the formation of the surface with less roughness. Modelling of the turning process is considered [8–11]. Modular modeling of the shaping process for turning is described in [12]. However, the proposed model does not allow to describe the oblique cutting and the method of choosing the orientation of the indexable inserts offered.

Improving processing productivity and reducing its cost is achieved through the use of cutters with indexable inserts. The market offered a wide variety of plates of different sizes and made of different materials. The shape, size, and material of the plate are chosen considering the properties of the workpiece, the available machine equipment, and the requirements for the accuracy.

When exploring the grinding process, the concept of specific productivity is widely used, which describes the amount of abrasive spent on removing a single volume of workpiece material. It is also convenient to use to study the turning process, to determine the effect of fretting wear on the cutting surface on the accuracy of the forming.

3 Research Methodology

In this research work, the mathematical general three-dimensional models of metal transition and shaping process of cylindrical surface and end face by oriented turning tools with indexable inserts is proposed. The angles of orienting turning tools depend on the quality of detail, which is processed.

The general scheme of the turning process is shown in Fig. 1a. In this scheme, pos. 1 is the detail, on which the cylindrical surface and end face between cylindrical surfaces with different diameters are processed; pos. 2 – the turning tool, it is fixed in a special cutter holder, which allows it to be oriented in three planes; pos. 3 – indexable inserts.

For the cylindrical surfaces will be processed, the cutter makes an axial feed *f*, and detail makes the rotational motion with angular velocity ω_p (Fig. 1b, I, II, and III). For the end face will be processed, the cutter is orient thus that one of its work edges coincide end face (Fig. 1b, IV). The rotation of detail provides cutting speed. The feed movement is missing. In this case, a part needs to do not less one full rotation.

The indexable inserts is a plate with defined shape and size, and a certain number of working faces. The working element of it is a vertex with two compatible faces. For convenience, we introduce the notion of a space cutting wedge.

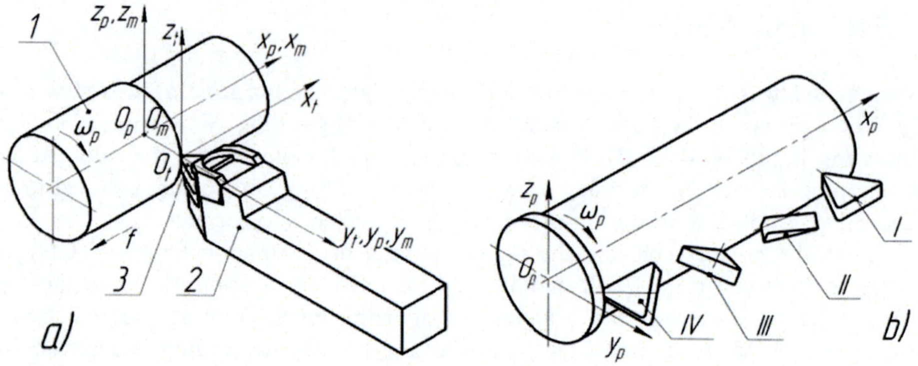

Fig. 1. The scheme of processing of cylindrical surface and end face of the shaft.

The space cutting wedge is part of cutting tools, which take part in the process of metal transition and shaping process. Geometrically represents the result of the motion of an arc, with a radius equal to the radius of curvature of the plate ρ along a curve consisting of two straight sections connected by a radius at the apex of the cutter r (Fig. 2, 3).

For creating a mathematical general three-dimensional model of metal transition and shaping process by oriented turning tools with indexable inserts, we need to write a three-dimensional mathematical model of a cutting surface – the space cutting wedge.

The space cutting wedge of indexable inserts, as a continuous surface of a definite shape, can be written by the equation using the one-coordinate matrices and the Heaviside function.

$$\begin{aligned}\bar{r}_t(\varphi,\theta) = {} & M4(\varphi)\cdot M2(r(i)-\rho)\cdot M6(\theta)\cdot M2(\rho)\cdot \bar{e}4\cdot \Phi(|\varphi|) \\ & - M4(\varphi)\cdot M2(r(i)-\rho)\cdot M6(\theta)\cdot M2(\rho)\cdot \bar{e}4\cdot \Phi(|\varphi|-\varphi_k) \\ & + M4(\varphi_k)\cdot M3(h(\varphi))\cdot M2(r(i)-\rho)\cdot M6(\theta)\cdot M2(\rho)\cdot \bar{e}4\cdot \Phi(\varphi-\varphi_k) \\ & + M4(-\varphi_k)\cdot M3(-h(|\varphi|))\cdot \mathrm{M2(r(i)}-\rho)\cdot \mathrm{M6}(\theta)\cdot \mathrm{M2}(\rho)\cdot \bar{e}4\cdot \Phi(-\varphi-\varphi_k),\end{aligned} \tag{1}$$

where φ_k – an angle that determines the position of the radius edge relative to the symmetry line of the plate (Fig. 2A) (for trihedral indexable inserts $\varphi_k = \pi/3$, tetrahedral – $\varphi_k = \pi/4$, pentahedral – $\varphi_k = \pi/5$, hexahedral – $\varphi_k = \pi/6$), $h(\varphi) = (r(i)-\rho)\cdot tg(\varphi-\varphi_k)$ – a function that determines the coordinate of a point along a straight section of the cutting part; $\Phi(\varphi)$ – Heaviside function, with a positive argument it is one, with a negative – zero; $r(i)$ – the radius at the top of the indexable inserts, it depends on fretting wear; ρ – the radius of the indexable inserts along the cutting edge

(Fig. 2); $\bar{e}_4 = (0,0,0,1)^T$ – radius vector point of the coordinates beginning; one-coordinate matrices M1, M2, M3, M4, M5, M6 [13] – describe the displacements along the X, Y, Z directions and the rotations about the OX, OY, OZ axes, respectively.

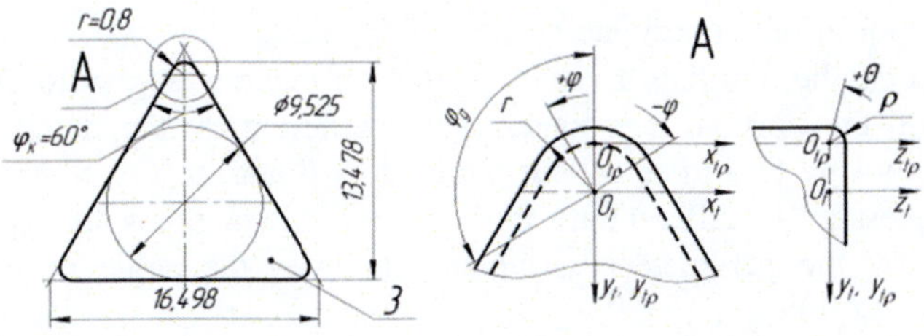

Fig. 2. Overall dimensions of the trihedral indexable inserts.

The mathematical model of the space cutting wedge is created in the special program Mathcad using Eq. (1). The results show in Fig. 3 that there are draws of the mathematical model for trihedral, tetrahedral, and pentahedral indexable inserts.

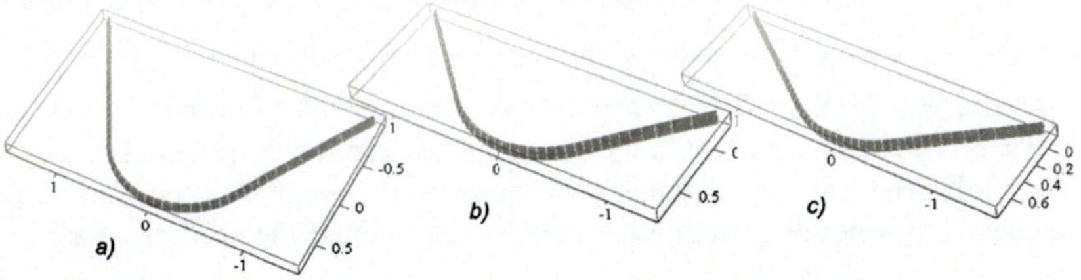

Fig. 3. The space cutting wedge of the indexable inserts a) – trihedral, b) – tetrahedral, c) – pentahedral.

For determining the nominal surface of the workpiece, it is necessary to transition the tool to the coordinate system of the workpiece and to describe the shaping motions (Fig. 1a, b)

$$\begin{aligned}\bar{r}_p(\varphi,\theta,\theta_p,) = M_{tp}\cdot\bar{r}_t = M6(\theta_p)\cdot M3(\theta_p\cdot p_z)\cdot M2(-y_c(i))\\ \times M6(\theta_r)\cdot M4(\varphi_r)\cdot M5(\psi_r)\cdot\bar{r}_t,\end{aligned} \tag{2}$$

where M_{tp} – the transition matrix from the coordinate system of the tool to the coordinate system of the detail; θ_r, φ_r, ψ_r – the angles of orientation tool surface relative to the axes O_tZ_t, O_tX_t, O_tY_t in accordance; $y_c(i)$ – specifies the movement of the tool coordinate system to the workpiece coordinate system, that is, determines the radius of the cylindrical surface; θ_p – rotation angle of the workpiece coordinate system, simulates the rotation of the machined surface around its axis; $p_z = \frac{f_z}{2\cdot\pi}$ – parameter of the screw movement of the space cutting wedge along the workpiece surface; f_z – feed per revolution.

The final shape of the workpiece profile is determined by the shape condition

$$\overline{V} \cdot \overline{n} = 0, \quad (3)$$

where $\overline{n}$ – single normal vector to the tool surface; $\overline{V}$ – velocity vector of the relative motion of the tool in the coordinate system of the part.

The normal can be found as a vector product of vectors tangent to the surface of a tool. So, to find the normal, it is necessary to differentiate the radius vector of the surface of the tool by two independent parameters θ and φ. For finding the velocity vector, it is necessary to differentiate the tools radius vector on the workpiece coordinate system by the parameter θ_p, which simulates the angle of rotation of the workpiece per unit time.

Therefore, in a one-parameter bend, the relation between the parameters θ and φ has the appearance of zero equality of the mixed product of the three vectors of partial derivatives of a vector $\overline{r}_t$

$$\left(\frac{\partial \overline{r}_t}{\partial \theta} \times \frac{\partial \overline{r}_t}{\partial \varphi}\right) \cdot \frac{\partial \overline{r}_t}{\partial \theta_p} = 0, \quad (4)$$

where $\frac{\partial \overline{r}_t}{\partial \theta} \times \frac{\partial \overline{r}_t}{\partial \varphi} = \overline{n}$ – the normal vector to the tool surface at the point with coordinates θ and φ; $\frac{\partial \overline{r}_t}{\partial \theta_p} = \overline{V}$ – velocity vector of the tool relative to detail.

For solving Eq. (4) is used as the logical calculation block recorded in the computer-algebra system Mathcad. As a result, the angles that determine the position of the contact line of the tool surface and the details on the space cutting edge is found. The surface of the workpiece is formed by rotating the contact line around the workpiece axis

$$Pov(\varphi, \theta_p) = \overline{r}_t(\varphi, Ang(\varphi), \theta_p). \quad (5)$$

Equations (2) and (4) determine the general three-dimensional models of metal transition and shaping process of cylindrical surface and end face by oriented turning tools with indexable inserts. For describing the production of specific surfaces, it is necessary to take as constants or to equate to zero some parameters of the model. For example, we will make and analyze the possible partial models from the general one.

As a tool, we take a cutter with triangular indexable inserts. In this case, the tools surface equation will look like (1), provided $\varphi_k = \pi/3$ (Fig. 2, 3).

We will model the process of machining cylindrical surfaces and end faces.

For produce cylindrical surfaces, we will accept: $f = 0.4$ mm/rev. – the workpiece feed per revolution; $t = 0.5$ mm – the depth of cutting; $y_c = 15$ mm – the radius of the detail. The orientation of the tool in the coordinate system of the part: $\theta_r = -\frac{\pi}{36}$, $\varphi_r = \frac{11\pi}{180}$, $\psi_r = \frac{\pi}{36}$ (Fig. 4a).

To produce the end face, we need to change the orientation of the tool. The cutting tool is additionally rotated 90° around the O_tX_t, axis. That is, the angles which determine the orientation of the indexable inserts relative to the parts are equal: $\theta_r = -\frac{\pi}{36}$, $\varphi_r = \frac{101\pi}{180}$, $\psi_r = \frac{\pi}{36}$ (Fig. 4b).

The three-dimensional models of the cylindrical surface and end face are shown in Fig. 4a, b.

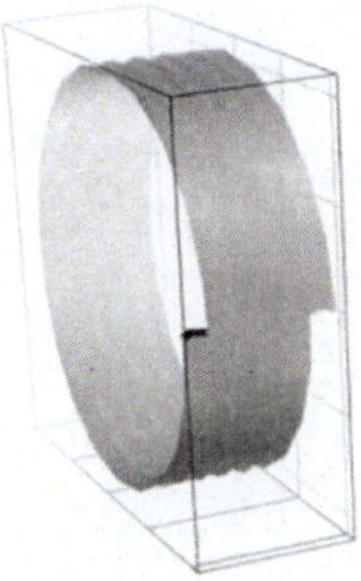
a)

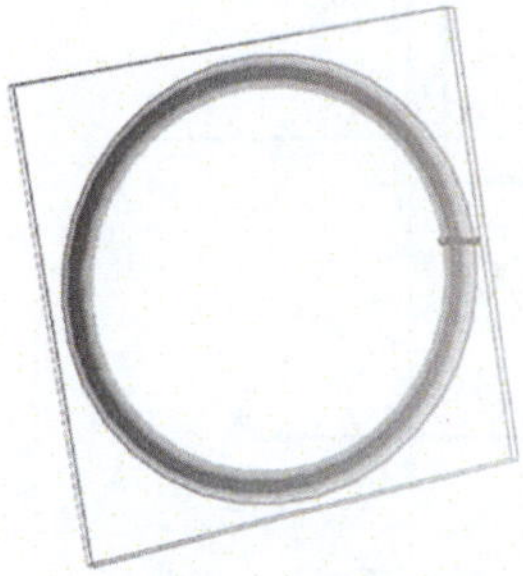
b)

Fig. 4. Shaping of surfaces during turning: a – cylindrical surface; b – end face.

Analyze the main characteristics of the cutting process for the case of turning with different orientations of the indexable inserts relative to the surface of the part were analyzed. For example, choose three angles of placement: the first one it is single-point turning with commonly used values of cutting angles (Fig. 1b, I); the second one – oblique cutting (Fig. 1b, III); and the third provision – the intermediate position (Fig. 1b, III).

The orientation of the indexable inserts in the intermediate position is determined by the condition of the full loading of the cutting blade [14]. The cutter must be rotated so that some point F (Fig. 5), located on the cutting blade, coincides with the point on the surface of the workpiece with radius $R_z = R_d + t$. The chord length H is determined from the equation:

$$H = \sqrt{R_w^2 - R_p^2} = \sqrt{(R_p + t)^2 - R_p^2}, \quad (6)$$

where H – chord length; R_w – workpiece radius; R_p – detail radius; t – cutting depth.

The offset of the extreme working point of the cutting edge x from the plane of symmetry of the indexable inserts determine from the triangle KCP

$$x = (t + t') \cdot tg(0,5 \cdot \varphi_k), \quad (7)$$

where φ_k – the angle at the apex of the indexable inserts of turning tool (Fig. 2).

The value of the edge CP is equal to the sum of t + t', where t' – the shortest distance from the radius of rounding of the indexable inserts to its conditional intersection of the faces (point P) (Fig. 5)

$$t' = \frac{r}{\sin(0,5\varphi_k)} - r, \quad (8)$$

where r – the radius of indexable inserts.

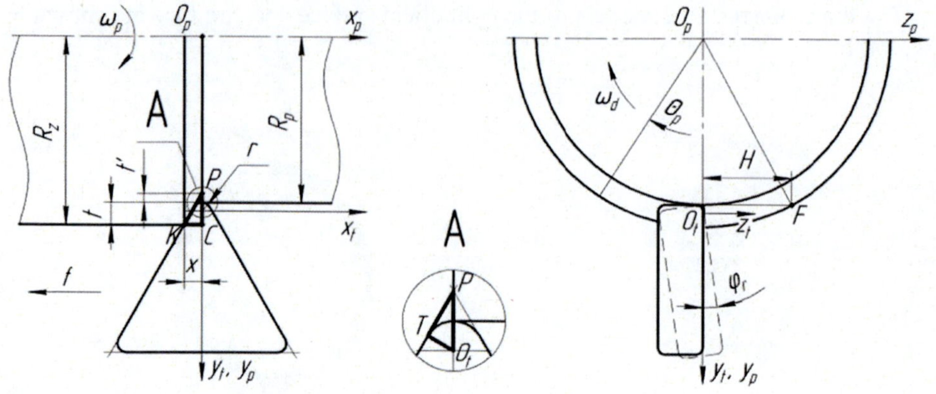

Fig. 5. Determination of the orientation of indexable inserts.

The Eq. (7) can be rewritten with using (8)

$$x = \left(t + r \cdot \left(\frac{1}{\sin(0,5\varphi_k)} - 1\right)\right) \cdot tg(0,5 \cdot \varphi_k), \tag{9}$$

Then the angle of rotation of the cutter relative to the axis Y_u (Fig. 5) is equal to

$$\psi_r = arctg\left(\frac{H}{x}\right) = arctg\left(\frac{\sqrt{(R_p + t)^2 - R_p^2}}{\left[t + r \cdot \left(\frac{1}{\sin(0,5\varphi_k)} - 1\right)\right] \cdot tg(0,5\varphi_k)}\right). \tag{10}$$

4 Results

Figure 6 shows the position of the contact line on the space cutting wedge, where 1 – the space cutting wedge, 2 – the contact line. It can determine with using the general three-dimensional models of metal transition and shaping process, Eqs. (2) and (4).

After turning, the cutter leaves typical lines on the surface of the part, which form the surface roughness. The calculating block can determine the trace equation of the tool profile in the OXY plane

$$Si1 = \left| \begin{array}{l} for\ j \in 0\ldots N_i \\ \left| \begin{array}{l} \varphi \leftarrow \varphi_{\max} - \frac{\varphi_{\max} + |\varphi_{\min}|}{N_i} \cdot j \\ x \leftarrow Pov(\varphi, 0)_{1,1} \\ y \leftarrow Pov(\varphi, 0)_{2,1} \\ M^{\langle j+1 \rangle} \quad \begin{pmatrix} x \\ y \end{pmatrix} \end{array} \right. \\ M^T \end{array} \right. , \tag{11}$$

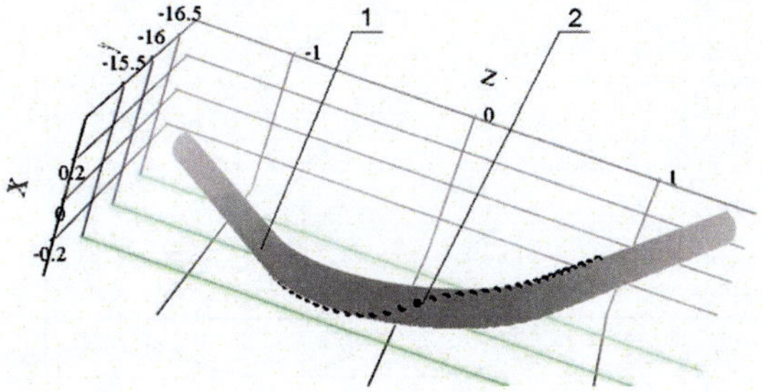

Fig. 6. The position of the contact line on the space cutting wedge.

where φ_{min}, φ_{max} – minimum and maximum angular coordinates of points on the profile of the workpiece; N_i – the number of points on the workpiece which considered to determine the cutting blade trace; M – matrix of coordinates of track points.

The scheme of forming a profile of a part after turning a cylindrical surface by the cutter with oriented indexable inserts is shown in Fig. 7. As can be seen from the figure, the geometric roughness is determined

$$Ra(\varphi 1) = Si1(\varphi 1) - S_p(\varphi 1), \tag{12}$$

where $\varphi 1$ – the coordinate of the point of intersection of two consecutive traces.

Figure 7 shows that as the angle of orientation increases, the surface roughness decreases. In single-point turning, only the radial section of the space cutting wedge is involved in the shaping process. When the cutter is turned at some angle, the cutting process includes a straight section of the cutting blade, which reduces the roughness of the workpiece. The highest value is achieved by oblique cutting by the rectangular part of indexable inserts.

The slice thickness at some point of the space cutting wedge is defined as the shortest distance between two consecutive traces (Fig. 7a). The thickness distribution of the cut layer for different tool orientations is shown in Fig. 8.

The tools orientation provides a more even distribution of the thickness of the cut layer along the cutting blade. Besides, the length of the working section of the indexable inserts is increased. This unloads the working part of the tool and improves machining performance by increasing the velocity. The considered schemes have been analyzed from the point of view of economic efficiency [15], it is found that machining with an oriented indexable insert reduces the cost of turning operation.

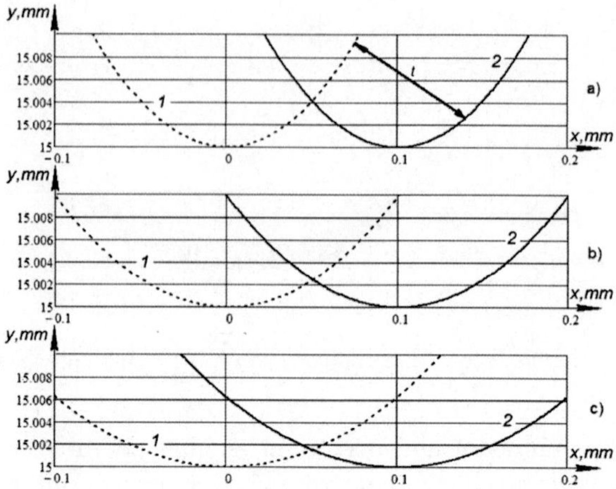

Fig. 7. Formation of profile details: a – single-point turning; b – the intermediate position; c – oblique cutting.

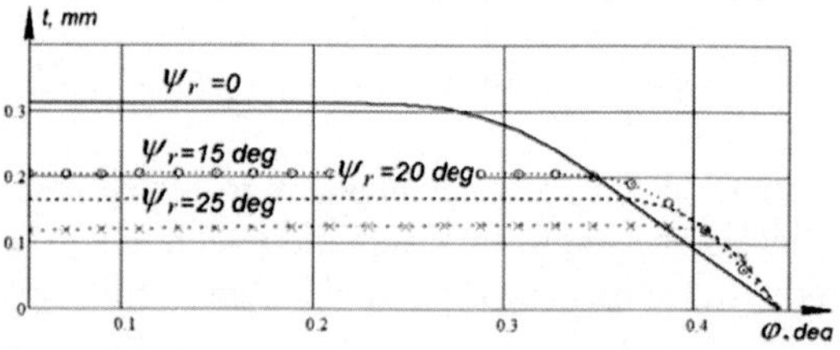

Fig. 8. Slice thickness distribution along the cutting edge for different orientations.

5 Conclusions

The term "space cutting wedge" is proposed. This is the part of the indexable inserts that is involved in the metal transition and shaping process. A three-dimensional model is developed. It describes the dependence of its shape on the cutting allowance and the orientation of the indexable inserts by the processing coordinate.

The mathematical general three-dimensional models of metal transition and shaping process of cylindrical surface and end face by oriented turning tools with indexable inserts is proposed. The proposed general model describes the turning process at different tools orientations. From it can be possible to take a private model for processing cylindrical surface and end face. The analysis of cutting parameters for various turning schemes, such as single-point turning, oblique cutting, and intermediate position, is shown. Research has shown that geometric roughness is improved for turning a cylindrical surface by the cutter with oriented indexable inserts. Besides, the distribution of the thickness of the cut layer becomes more uniform along the cutting edge. It allows increasing the velocity to reach more productivity, without loss of tools life. The three-dimensional model of the wearing spatial cutting wedge for the coordinate of processing is developed. It is based on the removal of a specific allowance. The influence of wear on the accuracy of forming is investigated. Consideration in the general mathematical model of a change in the radius of the indexable inserts as a result of fretting wear allows increasing the accuracy of processing due to adaptive process control using CNC systems.

References

1. Kalchenko, V.: Three-dimensional simulation of machined, tool surfaces and shaping process with two-side grinding of cylindrical parts ends. In: LNME, pp. 118–127. Springer, Cham (2020)
2. Grabchenko, A., Kondusova, E.: 3D modeling of tools, shaping and removal of stock during machining. Kharkiv (2001)
3. Ravska, N., Rodin, P., Nikolenko, T., Melnichuk, P.: The basics of formwork on top with mechanical testing. ZHITI (2000)
4. Venu, V.P.: On a model of oblique cutting. J. Ing. for Ind. **100**, 287–292 (1978)
5. Filippov, A.: Constructing a model of the equivalent wedge oblique cutting edge. Appl. Mech. Mater. **379**, 139–144 (2013)
6. Filippov, A.: Oblique turning skiving cutters. Yurginsky Institute of Technology (branch) of Tomsk Polytechnical University, pp. 236–241 (2014)
7. Grzesik, W.: Investigation of surface textures produced by oblique machining of different workpiece materials. Arch. Mater. Sci. Eng. **52**, 46–53 (2011)
8. Aoki, T., Sencer, B.: Development of a high-performance chip-guiding turning process tool design and chip flow control. Int. J. Adv. Manuf. Technol. **85**, 791–805 (2015)
9. Grzesik, W.: Modelling of heat generation and transfer in metal cutting: a short review. J. Mach. Eng. **20**, 24–33 (2020)
10. Vasilevich, Yu., Dounar, S.: Finite element analysis of centreless-lunette turning of heavy shaft. Sci. Tech. **16**, 196–205 (2017)
11. Tokarev, D., Drozdov, A., Gulyaev, M.: The study of roughness of the PTFE O-rings when turning. Master's J. **1**, 9–13 (2018)
12. Kalchenko, V., Yurchenko, Y., Kalchenko, D.: 3D module modeling tools, shaping and knowing the allowance for turning when machining with non-regrind plastics. Bull. Cherkasy State Technol. Univ. **3**, 8–14 (2012)
13. Reshetov, D., Portman, V.: Precision of Metal Cutting Machines. Machine-Building, Moscow (1986)
14. Kalchenko, V., Kalchenko, V., Kologoida, A., Kalchenko, D.: A way of perfecting with a sacrificial cross with a polyhedral plate. Patent of Ukraine, No. 136841 (2019)
15. Shkarlet, S., Prokopenko, V., Dubyna, M.: Directions of development of the financial services market of Ukraine. Baltic J. Econ. Stud. **4**, 412–420 (2018)

Improvement of a Stochastic Dynamic Model for Grinding of Cylindrical Surfaces with Wear-Resistant Coatings

Maksym Kunitsyn(✉) and Anatoly Usov

Odessa National Polytechnic University,
1, Shevchenko Avenue, Odessa 65044, Ukraine
m.v.kunitsyn@opu.ua

Abstract. Grinding is a finishing method to treat cylindrical surfaces with wear-resistant coatings, which is the micro-cutting of material with many grains that affect each other through the work surface. Since vibrations inevitably occur in the dynamic system when cutting, the accuracy of the shape of the surface obtained by grinding depends on the rigidity of the attachment of the tool and workpiece, as well as on the machining modes. An advanced stochastic model has been developed to study the dynamics of the grinding process. In the model, the machining process is depicted as micro-cutting with abrasive grains that are randomly distributed over the grinding wheel's surface. The geometric parameters of the grains are also random. The simulation results in surface textures after machining, the distribution of cutting forces, and the dynamic deviations of the tool. Their spectral characteristics were constructed, which made it possible to test the influence of processing modes and technological system parameters on vibrations, which leads to a loss of processing quality.

Keywords: Abrasive grains · Frequencies · Dynamic system · Cutting · Vibration

1 Introduction

Grinding is a finishing method to treat cylindrical surfaces with wear-resistant coatings, which is the micro-cutting of material with many grains that affect each other through the work surface. Grinding provides specified levels of surface quality and workpiece accuracy within geometric tolerances. The stochastic geometric characteristics of the grains and their random nature of distribution over the surface layer of the tool create difficulties in analyzing the grinding process [1, 2].

On the surface of the grinding wheel distributed a lot of abrasive grains, which are fastened with a binder [3]. The cutting edge of each abrasive grain of the grinding wheel as a separate element is involved in the grinding process. The complexity of modeling the grinding process lies in the fact that during the interaction of the grinding wheel and the surface of the workpiece, each grain performs microscopic modes, such as cutting, scratching and sliding [4].

V. Tonkonogyi et al. (Eds.): InterPartner 2020, LNME, pp. 534–544, 2021.
https://doi.org/10.1007/978-3-030-68014-5_52

2 Literature Review

The accuracy of the shape of the surface obtained by grinding depends on the rigidity of the attachment of the tool and the workpiece and the machining modes because vibrations inevitably arise in the dynamic system during cutting, which requires additional investigation [1, 2].

There are no full-scale studies on modeling grinding functional-gradient materials, which determine the performance characteristics of the units of mechanisms [3], which include cylinders with wear-resistant coatings. Stochastic geometric characteristics of grains and their random nature of distribution in the surface layer of the tool create difficulties in the grinding's analysis process to ensure the required roughness of the working surface of cylinders with wear-resistant coatings [4].

A review of the literature [5, 6] shows that most of the work is devoted to grinding surfaces from a simple geometric shape based on a model with one freedom. The paper proposes to improve the model of grinding dynamics. Each abrasive grain is considered a distinct cutting edge with random characteristics of shape and location on the grinding wheel's surface. Abrasive grains, crystals of synthetic diamond, are modeled as conical pyramids, randomly embedded in the grinding wheel matrix [7–10].

With small-displacement amplitudes, grinding can be considered as a large-scale milling process with cutting edges [11, 12]. For cases where the amplitude of dimensionless displacements is greater than one, the grinding model should take into account the geometric limitation of the grain height and the contribution of the contact that binds to the workpiece material.

3 Research Methodology

The abrasive grains are evenly distributed along the tracks on the cutting surface of the tool so that the central angle between the axes of the adjacent grains is the same (Fig. 1). This kind of distribution makes it possible to describe in the simulation their motion by a system of differential equations with a constant delay in the angle of rotation $\Delta\varphi$ at time T/N (the time of passing of one grain), while delay will be variable.

All grains have random geometric characteristics (grain height h_0, conical grain angle θ, the angle at the top of the grains α, relative to the radial direction in the plane perpendicular to the track). A radius of 10 μm rounds the tops of the grains.

To simulate the grinding process, taking into account the tool's pliability, we assume that the grinding wheel moves in the plane as a rigid body on elastic supports with a given rigidity and damping, as shown in Fig. 1.

To model the grinding process taking into account the workpiece's flexibility, the grinding process of low-rigidity workpieces is also modeled as a flat system having two orthogonal degrees of freedom, as shown in Fig. 2.

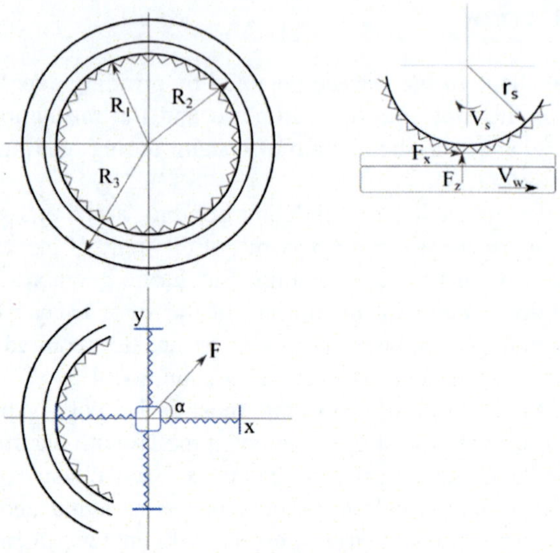

Fig. 1. Schematic of a stochastic grinding model with allowance for tool compliance.

For grinding tools, nominal geometrical parameters of the density and grain size are usually specified, which can be used in the simulation. The distribution of grain volumes is usually [13, 14] random and can be determined by the standard law used in stochastic approaches in cutting force modeling [5].

The distribution function can be written through the Riemann integral as follows [15]:

$$\Phi(h_0) = \frac{1}{\sigma\sqrt{2\pi}} \int_{-\infty}^{\infty} exp\left(-\frac{\left(h_0 - \bar{h}_0\right)^2}{2\sigma^2}\right) dx \tag{1}$$

where h_0 is the mathematical expectation of grain height, $\sigma = \left(h_{0,max} - h_{0,min}\right)/6$ is the standard deviation of the distribution function.

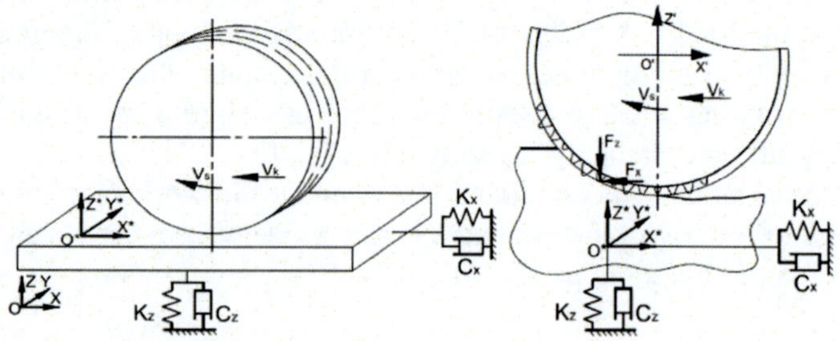

Fig. 2. Dynamic scheme of modeling of processing of a fragment of a cylindrical surface at grinding considering the pliability of a part.

The conical angle of the grain θ is essentially the maximum angle of the cutting wedge γ with the opposite sign. We assume that the anterior angle is a strongly asymmetric distribution with a maximum probability close to the angle –30°, which is well described by the Rayleigh distribution [13]. Some researchers also use Student's [14] or Weibull distribution [16] for these cases. We shall use for the cone angle of the cutting edge of the grains the Rayleigh distribution, for which the distribution function looks like:

$$\Phi(\theta) = 1 - exp\left(-\frac{\theta^2}{2\sigma^2}\right)(\theta \geq 0) \tag{2}$$

where σ is a scale parameter, for $\theta = \sigma$, the probability density is maximal.

Assume that the angle of the vertex of the grain α is distributed by the law, which is described by the function of normal random distribution, and has the form:

$$\Phi(\alpha) = \frac{1}{\sigma\sqrt{2\pi}}\int_{-\infty}^{\infty} exp\left(-\frac{\alpha^2}{2\sigma^2}\right)dx \tag{3}$$

where $\alpha = 0$ is the mathematical expectation of the yaw angle $\sigma = (\alpha_{max} - \alpha_{min})/6$ is the standard deviation of the distribution function.

The distribution function looks like:

$$\Phi(\delta) = \begin{cases} 0, \delta < -\pi/N \\ \frac{\delta + \pi/N}{2\pi/N}, -\pi/N \leq \delta \leq \pi/N \\ 1, \delta > \pi/N \end{cases} \tag{4}$$

where $\delta_{max} = \pi/N$ is the maximum angular offset of the track, $\delta_{min} = -\pi/N$ is the minimum angular offset of the track, $\sigma^2 = (2\pi/N)^2/12$ is the dispersion.

The system of Eqs. (5) describing the formation of new surfaces can be written in the form [13, 17]:

$$D_j(t) = \begin{cases} \left[L_j|_{t-T/N} + V_w \cdot \frac{T}{N} \cdot \sin(\varphi_j(t)) - \Delta x \cdot \sin(\varphi_j(t)) - \right. \\ \left. -\Delta z \cdot \cos(\varphi_j(t)) - (h_{0,j-1} - h_{0,j})\right] \cdot g(\varphi_j(t)) \\ h_{cu,j}(t) = \max(0, D_j(t)) \\ \Pi(t) = \mathrm{Trend}\left[\Pi_j|_{t-T/n} - h_{cu,j}(t)\right] \\ L_j|_0 = 0 \\ g(\varphi_j(t)) = \begin{cases} 1, \text{if } 0 \leq \varphi_j(t) \leq \varphi_{ex} \\ 0, \text{else} \end{cases} \end{cases} \tag{5}$$

where T/N is the period of passage of the grain; $L_j|_{t-T/N}$ is the polar deviation from the workpiece surface at time $t - T/N$ for the *j*-th grain only as a result of rotation; $\Pi(t)$ is the surface coordinates recorded in the information database in Matlab software; Δx, Δz is the dynamic displacement projections on the X, Z-axis; $D_j(t)$ is the distance from the *j*-th cutting edge to the untreated surface, which consists of the static part $L_j|_{t-T/N} + V_w \cdot \frac{T}{N} \cdot \sin(\varphi_j(t))$, the dynamic part of $\Delta x \cdot \sin\varphi_j + \Delta z \cdot \cos\varphi_j$ and the magnitude of the radial height difference of adjacent grains $h_{0,j-1} - h_{0,j}$; $h_{0,j}$ is the height of the *j*-th grain, $h_{0,j-1}$ is the height of the $(j-1)$-th grain; $h_{cu,j}(t)$ is the thickness of the layer cut off by the *j*-th grain at time *t*, $\max(0,\ D_j(t))$ is the function equal to the maximum of 0 and $D_j(t)$, i.e. the cutting thickness cannot be negative; $\Pi_j|_{t-T/N}$ is the position of the point of intersection between the axis of the *j*-th grain at time *t*, and the workpiece surface at time $t - T/N$ in the polar coordinate system; Trend is the function that interpolates values to points specified in an array; $g(\varphi_j(t))$ is the function that determines if the *j*-th grain is in the contact area.

To simulate the process of grinding, taking into account the workability of the workpiece, in order to analyze the geometry of the cut layer, the shape of the treated surface, consider the scheme of embedding adjacent grains into the workpiece material. In Fig. 3, and the current grain positions at a given time *t* and the previous moment $(t - T/N)$ before its displacement is shown due to the feed movement and rotation of one grain.

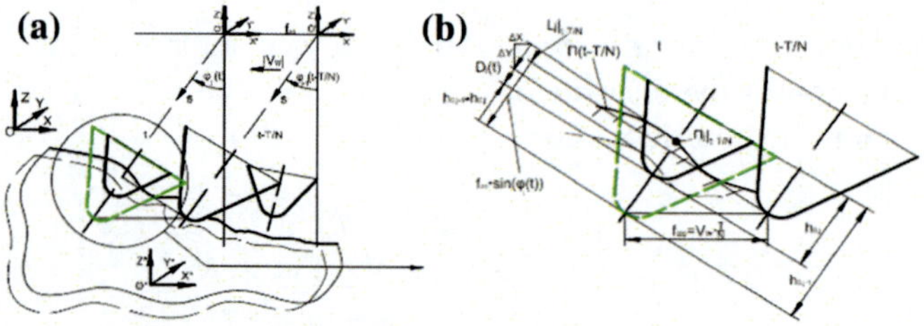

Fig. 3. (a) The coordinate system and the position of the (j − 1)-th and j-th grains during cutting; (b) Geometric analysis of the thickness of the cut layer of the j-th grain.

Suppose that, for $t = 0$, the first grain is in the contact zone, its number is $j = 1$. We will number all the grains moving anticlockwise. That is, at time $t = k \cdot T/N$, the numbering of the grains will correspond to that shown in Fig. 4.

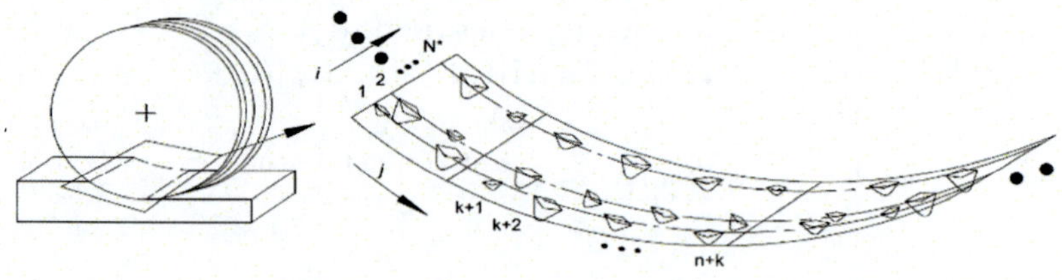

Fig. 4. Scheme of numbering grains per count at the moment of time $t = k \cdot T/N$.

After determining the coordinates of the profile $\Pi_i(t)$ on all tracks from the 1-st to the N″, interpolating the N″ curves with the track deflection in space after this forms the surface of the workpiece (Fig. 5).

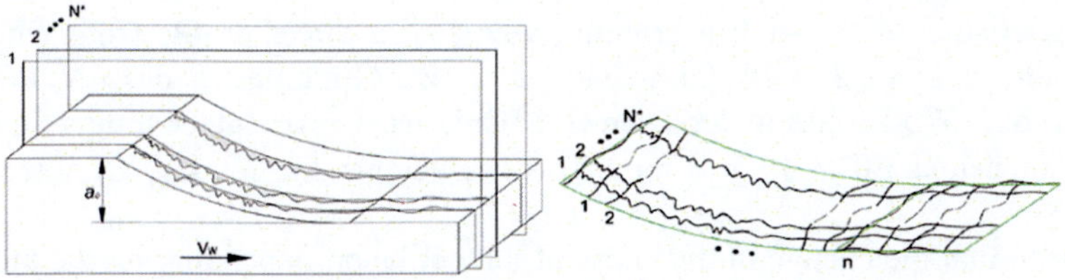

Fig. 5. Scheme of the surface formation when grinding multiple tracks.

We transform the components of the cutting forces for the individual grain $F_{t,ij}$, $F_{n,ij}$ acting on the *i*-th track into the *OXYZ* coordinate system, to the forces $F_{x,ij}$ and $F_{z,ij}$ (projection of cutting forces on the *X*, *Z* axis) using the following relations:

$$\begin{cases} F_{x,ij} = F_{t,ij} \cdot \cos \varphi_{ij} + F_{n,ij} \cdot \sin \varphi_{ij}; \\ F_{z,ij} = -F_{t,ij} \cdot \sin \varphi_{ij} + F_{n,ij} \cdot \cos \varphi_{ij}. \end{cases} \tag{6}$$

Summing up the cutting forces acting on the grain in the area of contact between the circle and the workpiece, we obtain the forces acting on the grinding wheel:

$$\begin{cases} F_x = \sum_{i=1}^{N''} \sum_{j=1}^{n} F_{x,ij}; \\ F_z = \sum_{i=1}^{N''} \sum_{j=1}^{n} F_{z,ij}, \end{cases} \tag{7}$$

where *n* is the number of grains in the contact area on one track ($n = 49$); N″ is the number of tracks ($N'' = 50$).

The differential equations describing the movement of the system during machining, for the case where the system's flexibility is mainly determined by the tool's flexibility and the workpiece is considered rigid, have the form:

$$\begin{cases} m \cdot \ddot{x} + C_x \cdot \dot{x} + K_x \cdot x = F_x \\ m \cdot \ddot{z} + C_z \cdot \dot{z} + K_z \cdot z = F_z \end{cases} \tag{8}$$

and, on the contrary, when the part's elasticity mainly determines the system's durability, and the tool is considered as absolutely rigid, have the form [10]:

$$\begin{cases} m \cdot \ddot{x} + C_x \cdot \dot{x} + K_x \cdot x = -F_x; \\ m \cdot \ddot{z} + C_z \cdot \dot{z} + K_z \cdot z = -F_z, \end{cases} \tag{9}$$

where *m* is the mass of the tool-part system; C_x, C_z – damping coefficients of the tool-part system in the direction of the *X*, *Z* axes. K_x, K_z – given rigidity of the tool-part system in the direction of the *X*, *Z* axes.

4 Results

Figure 6 presents the surface texture of the workpiece in the simulation concerning vibration. The textures of the workpiece's treated surface will be compared with the results without taking into account the vibration.

Characteristics of the tool: a grinding wheel of a direct profile from white electrocorundum 210 × 32 × 76 25A F46 O 7 B. Grinding modes: $N_s = 35\,\mathrm{m/s}$, $a_p = 0.01\,\mathrm{mm}$. Workpiece material steel 37Cr4, wear-resistant coating Ni/Ni-TiO2, sample dimensions 70 × 40 × 4 mm. The coefficients K_x, K_z, C_x, C_z were chosen from [8, 9].

It is seen that the change in thickness of the cut layer, depending on the number of grain in time, also has a wave nature. In the beginning, the tool grains are cut into the workpiece material. With the displacement amplitudes being large, i.e., at these times, the thickness of the cutting layer decreases or increases due to the dynamic displacement of the tool. Further, for the cutting period of all grains that fall into the contact zone, compared to the results obtained without considering the dynamics, the thickness distribution of the cut layer is more sparse.

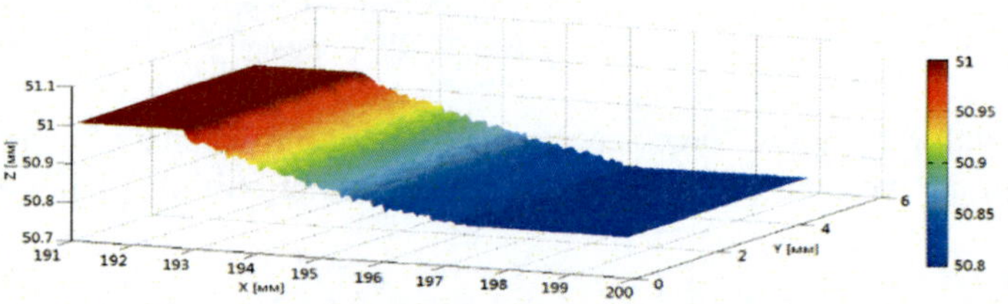

Fig. 6. Three-dimensional textures of the workpiece surface when modeling with vibration at time $t = 100 \cdot T/N$.

Grinding forces are significantly altered due to the influence of dynamics concerning vibration. The change in cutting forces occurs due to the interruption of the cutting of individual grains moving along the uneven surface of the tool and workpiece contact, as well as additional shifts due to the vibration of the tool (Fig. 7).

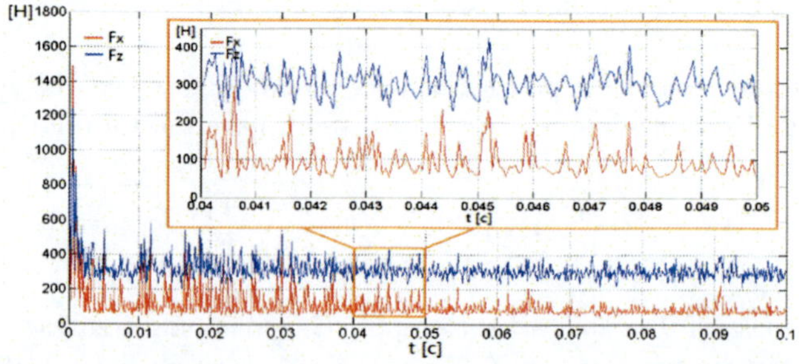

Fig. 7. Dependence of grinding forces F_x, F_z on time t with vibration.

Comparison of the image of the treated area shown in Fig. 8, it is possible to obtain the texture of the machined surface of the workpiece in the simulation without and with taking into account the vibration at time $t = 100 \cdot T/N$ for different variants of randomly selected parameters of the location of the grains makes it possible to establish that the average value of the coordinate *Z* of the machined surface with vibration higher than without. This is explained by the influence of the elastic pressing of the tool with elastic fastening, taking into account vibrations.

The results of modeling a topographic picture of the surface show that there are peaks and troughs of different magnitudes. Based on the surface treatment modeled in this paper, the following parameters are estimated to estimate the surface waviness: Mean is the average coordinate value *Z* of the surface that defines the centerline of the profile; *Max*, *Min* are the maximum and minimum value of the coordinate *Z* of the surface; R_a is the arithmetic mean deviation of the profile, which is defined as the arithmetic mean of the absolute values of the deviations of the profile *z* within the base length *l*. Methods and examples of surface waviness estimation after cutting are presented in [18, 19].

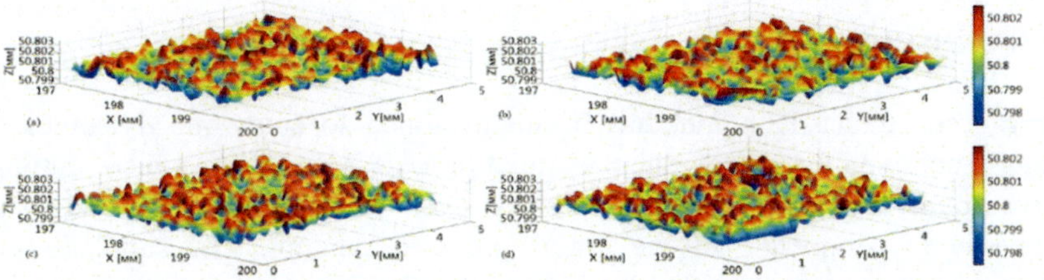

Fig. 8. Textures of the machined surface of the workpiece in the simulation taking into account the vibration at time $t = 100 \cdot T/N$.

The results show that for the case of K_x with large stiffness values ($K_x > 45$ kN/mm), the oscillations of the system damp off, and the behavior is stable. And for cases of K_x with smaller values, oscillations with large amplitudes exceeding the size of the grains are observed, which shows the inability of such modes. Between them, there are modes for which the amplitude of oscillations increases at the initial stage, but the system is stable. After a short transition, it enters a steady-state, and such a regime can be satisfactory.

(a) (b)

Fig. 9. View of the surface after passing the tool obtained using an improved model (a), and the final microrelief of the surface (b) after grinding.

When comparing the natural frequencies with the spectra's peaks, it can be observed that in transient modes, there is a tendency for individual peaks to coincide or multiple natural frequencies. And in unstable modes, there are pronounced peaks in the values of frequencies, multiple natural frequencies of the system. Analyzing the temporal characteristics of the solution using the "chatter" detection method well described in [20, 21], one can capture the appearance of "chatter" in the simulation. This behavior is characteristic of the case for $K_x/K_z = 4$, $K_x = 30$ kN/mm (in this case $f_{nx} = 2 \cdot f_{nz}$), for which the frequency of the highest peak is a multiple of 3 to the natural frequency f_{nx}. Here, the frequency of "chatter" becomes dominant, and the system becomes unstable with the dominant frequency of "chatter".

For comparing the simulation results with the experimental results, the treatment of the flat surface of the cylinder material with a wear-resistant Ni/Ni-TiO_2 coating with a grinding wheel 25A F46 O 7 B was considered. Figure 9 shows the surface's appearance after passing the tool, obtained using an improved model (a), and the final microrelief of the surface (b) after grinding $R_a = 40\ \mu m$.

5 Conclusions

An advanced stochastic model has been developed to study the dynamics of the grinding process. In the model, the machining process is depicted as micro-cutting with abrasive grains that are randomly distributed over the surface of the grinding wheel. The geometric parameters of the grains are also random. The simulation results includes the surface textures after machining, the distribution of cutting forces, and the dynamic deviations of the tool.

Their spectral characteristics were constructed, which made it possible to test the influence of processing modes and technological system parameters on vibrations. It is shown that vibrations are excited in the system at the frequencies of external excitation (the frequency of passage of grains) and the frequencies of the natural vibrations of the flexible, dynamic system characteristic of the regenerative excitation source. This behavior is especially characteristic of the case at $K_x/K_z = 4$, $K_x = 30$ kN/mm, for which the frequency of the highest peak is a multiple of 3 to the natural frequency f_{nx}. It is established that at low rigidity of the technological system, self-oscillations of the "chatter" type with large amplitude are disrupted because of the mechanism of regenerative excitation from grain to grain at frequencies multiple of the natural frequencies of oscillation, which lead to the loss of quality of processing and increase of tool wear.

References

1. Voronov, S.A., Ma, V.: The influence of the geometry of abrasive grain on cutting forces during grinding. Bull. MSTU. N.E. Bauman. Ser. Eng. **5**, 52–63 (2017)
2. Voronov, S.A., Ma, V.: Mathematical modeling of the cylindrical grinding process. J. Mach. Manuf. Reliab. **46**(4), 394–403 (2017)
3. Komanduri, R.: Machining and grinding: a historical review of the classical papers. Appl. Mech. Rev. **46**, 80–132 (1993)
4. Malkin, S., Guo, C.: Grinding Technology: Theory and Applications of Machining with Abrasives. Industrial Press Publisher, New York (2008)
5. Zhen, B.H., Ranga, K.: On the mechanics of the grinding process – Part I. Stochastic nature of the grinding process. Int. J. Mach. Tools Manuf. **43**, 1579–1593 (2003)
6. Maslov, E.N.: Theory of Grinding Materials. Engineering, Moscow (1974). (in Russian)
7. Kashcheev, V.N.: Abrasive Destruction of Solids. Science, Moscow (1970). (in Russian)
8. Grabchenko, A.I., Dobroskok, V.L., Fedorovich, V.A.: 3D modeling of diamond abrasive tools and grinding processes. National Technical University "Kharkiv Polytechnic Institute", Kharkiv (2006)
9. Grabchenko, A.I., Fedorovich, V.A.: 3D diamond abrasive processes. National Technical University "Kharkiv Polytechnic Institute", Kharkiv (2008)
10. Oborskiy, G.A., Daschenko, A.F., Usov, A.V., Dmitrishin, D.V.: System Modeling. Astroprint, Odessa (2013)
11. Altintas, Y.: Manufacturing Automation: Metal Cutting Mechanics, Machine Tool Vibrations, and CNC Design. Cambridge University Press, Cambridge (2000)
12. Brecher, C., Esser, M., Witt, S.: Interaction of manufacturing process and machine tool. CIRP Ann. Manuf. Technol. **58**, 588–607 (2009)
13. Hecker, R., Liang, S.Y.: Predictive modeling of surface roughness in grinding. Int. J. Mach. Tools Manuf. **43**, 755–761 (2003)
14. Stepien, P.: A probabilistic model of the grinding process. Appl. Math. Model. **33**(10), 3863–3884 (2009)
15. Stephen, H.C., William, D.M.: Random Vibration in Mechanical Systems. Academic Press, New York (1963)
16. Holtermann, R., Schumann, S., Menzel, A.: Modelling simulation and experimental investigation of chip formation in internal traverse grinding. Prod. Eng. Res. Dev. **7**, 251–263 (2013)
17. Ioan, D.M., Brian, R.W., Dimitrov, B.: Tribology of Abrasive Machining Processes. William Andrew Inc. (2004)
18. Li, H.N., Yu, T.B., Zhu, L.D.: Analytical modeling of ground surface topography in monocrystalline silicon considering the ductile-regime effect. Arch. Civ. Mech. Eng. **17**(4), 880–893 (2017)
19. Young, P.L., Brackbill, T.P., Kandlikar, S.G.: Estimating roughness parameters resulting from various machining techniques for fluid flow applications. In: Proceedings of the fifth International Conference on Nanochannels, Microchannels and Minichannels, pp. 827–836 (2007)
20. Altintas, Y., Stepan, G., Merdol, D., Dombovari, Z.: Chatter stability of milling in frequency and discrete time domain. CIRP J. Manuf. Sci. Technol. **1**, 35–44 (2008)
21. Riviére, E., Stalon, V., Van den Abeele, O., Filippi, E., Dehombreux, P.: Chatter detection techniques using microphone. In: 7th National Congress on Theoretical and Applied Mechanics (2006)

22. Han, Z., Jin, H., Li, M., Fu, H.: An open modular architecture controller based online chatter suppression system for CNC milling. Math. Probl. Eng. (2015)
23. Janez, G., Andreas, B., Edvard, G., Klocke, F., Igor, G.: Automatic chatter detection in grinding. Int. J. Mach. Tools Manuf. **43**, 1397–1403 (2003)
24. Kovac, P., Gostimirovic, M., Mankova, I., Soos, L., Savkovic, B.: Overview of experimental investigation of cutting process dynamic. J. Prod. Eng. **16**(2), 1–4 (2013)
25. Caixu, Y., Fenghuize, C.C., Haitao, Z., Cui, H.: Research on dynamic characteristics of hard cutting system. Int. J. u- e- Serv. Sci. Technol. **8**(3), 311–320 (2015)
26. Niels, J., Nathan, W., Doppenberg, J.: Robust active chatter control in the high-speed milling process. IEEE Trans. Control Syst. Technol. **20**(4), 901–917 (2012)

Modeling of the Machining Process by PCBN Tool with Self-adaptive Coating

Andrey Manokhin(✉), Sergiy Klymenko, Maryna Kopeikina, Sergiy Klymenko, and Yuriy Melniychuk

Bakul Institute for Superhard Materials of the National Academy of Sciences of Ukraine, 2, Avtozavodskaya Street, Kyiv 04074, Ukraine
the.manokhin@gmail.com

Abstract. The paper discusses the analytical calculation's results of the effect of the self-adaptive coating on such parameters of the contact interaction at the cutting zone as average contact temperature and cutting force during machining of hardened steel with PCBN. The calculation procedure was realized in the form of an iteration algorithm. It was assumed that the formation of thing tribofilms takes place at a specific temperature, and friction factor rapidly drops down in a narrow temperature diapason. The other hypothesis is that coating is not fractured, and friction is taking place on its surface. The results of the calculations show that the effect of the self-adaptive coating implies the growth of the shear angle while cutting temperature and force decline.

Keywords: Self-adaptive coating · PCBN cutting tool · Machining · Average temperature · Wear mechanism

1 Introduction

A promising direction for controlling the performance of cutting tools made of superhard materials is the use of protective coatings obtained by vacuum-plasma sputtering technology. Such coatings obtained by the PVD or CVD method can significantly change the characteristics of the contact interaction in the cutting area and thereby improve the operational properties of the tool. As the peculiarities of the influence of the protective coating on the cutting process have not been sufficiently studied, research in this area seems to be relevant. This problem is especially challenging in the case of the application of PcBN cutting tools with so-called self-adaptive coatings because an impact of such type of films on the cutting forces and cutting temperature has not been researched previously. Taking into account that self-adaptive films are potentially very effective, the quantitative estimation of these parameters during hard machining is very important in order to get a theoretical basement for the development of self-adaptive coatings, optimized for PcBN tools.

V. Tonkonogyi et al. (Eds.): InterPartner 2020, LNME, pp. 545–554, 2021.
https://doi.org/10.1007/978-3-030-68014-5_53

2 Literature Review

In the development of protective coatings, several approaches can be distinguished that include the endowment of coatings with various functional properties. The most common type of coating is a film, the hardness of which should substantially exceed the substrate's hardness. Such coatings, as a rule, consist of carbides, nitrides, and carbonitrides of refractory metals: (TiN, TiCN, TiAlN, AlCrN, TiSiN, etc.) [1–3] and can increase the tool life of a cemented carbide cutting tool by reducing the intensity of its abrasive wear. When applied to PCBN tools, such films are usually ineffective, which requires the development of more complex structures: nanolayer or nanocomposite systems [4, 5].

Another mechanism for increasing tool life is the use of a coating as an intermediate substance, which directly affects the parameters of the cutting process [6]. By suppressing the chemically determined wear mechanism of the cutting tool, such coatings increase the stability of PCNB during high-speed cutting or processing of nickel alloys.

An alternative for these approaches is to optimize the tribological properties of the surface layer of the instrument using the so-called solid lubricants. It is known that MoS2, WS2 [7, 8] coatings, or more complex compositions based on them (MoS_2/Zr, MoS_2/Ti) are used as a solid lubricant; however, they do not have sufficient thermal stability and mechanical properties required for tool coatings. Significantly higher thermal stability is intrinsic to a coating based on amorphous crystalline boron nitride [9]. The low coefficient of friction, in this case, helps to reduce the thermobaric load on the working sections of the tool.

The mechanism for increasing tool life in which the effect of films with a low coefficient of friction is applied is the creation of self-adaptive coatings [10]. The main idea is not to prevent oxidation, but to use this phenomenon to control the parameters of contact interaction during cutting to reduce the friction coefficient and, therefore, the thermobaric load in the contact zone. Such coatings are oxidized during cutting, forming a polyoxide secondary structure (tribofilm), which plays the role of a solid lubricant [11].

Such coatings as TiAlN/VN are known as self-adaptive systems. Moreover, in contrast to films of the previous types, they have a nanostructured state. High mechanical properties characterize nano-layer coatings: hardness, often exceeding 40 GPa [12, 13], and increased strength compared to monoblock systems [14]. It determines the high efficiency of such structures for the cutting tool performance. The uniqueness of the tribological properties of TiAlN/VN coatings is explained by the formation of friction V2O5, which has high lubricating properties [15–17]. At a certain temperature, oxide films are formed during the formation process, the thickness of which is comparable with the parameters of the nanolayer coating structure [18]. A sharp decrease in the intensity of the frictional interaction reduces the load on the tool, and the formation of new films compensates for the periodic destruction of oxide structures due to the layered structure of the coating.

Self-adaptive coatings have a high potential for use, especially when machining hard materials. In this work, model studies of the contact effects that occur during the cutting process with a tool with self-adaptive coatings are performed to quantitatively evaluate the effect of this type of coating on the mechanics of the cutting process.

3 Research Methodology

In order to evaluate the characteristics of the machining process, the one needs to know three groups of parameters, which depend on the process conditions: - the strength of the processed material; - parameter characterizing the deformation of the material being processed (shear angle); and the coefficient of friction between the tool material and the machined material. All these parameters affect the contact temperature and, in turn, depend on it.

When modeling the deformation processes of a material, it is necessary to take into account the influence of conditions in the cutting zone on its mechanical characteristics. Currently, the Johnson-Cook model has become widespread in the modeling of machining processes. The coefficients of the constitutive equation of workpiece material for the hardened ShH15 steel, accepted for the calculations in this study, are: $A = 3538$ MPa; $B = -1965$ MPa, $C = 0.061$, $n = -0.062$, $m = 3.17$. The shear flow stresses in the zones of plastic deformation depend on flow stress $\sigma = \sigma_s / \sqrt{3}$.

The value of the shear angle can be determined using the following iterative algorithm. For the arbitrary initial value of the temperature and the shear angle β, the shear stresses in the shear plane τ_d, and the contact zone on the front surface of the tool q_f is determined as a function of the temperature in expression (1). The values of the strength characteristics change the temperatures in the zones of deformation, which in turn again affects q_f and τ_d. The calculation for this cycle of block A in the scheme (Fig. 1) continues until the difference in temperatures in the zone of primary deformation at the previous and next iterative course will not be less than the accepted error (assume ε = 0.5%).

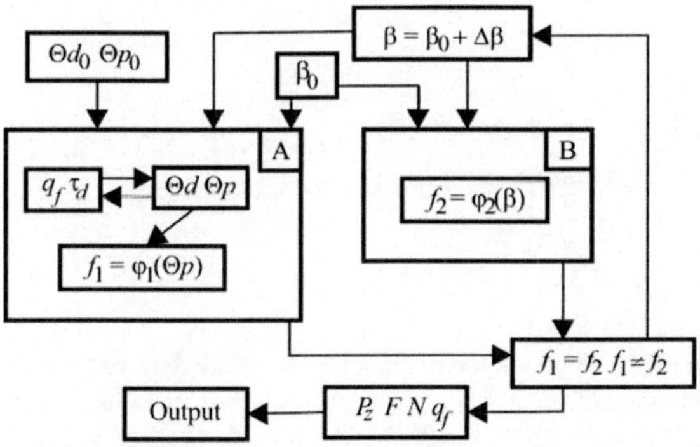

Fig. 1. Calculation scheme for modeling the cutting process.

The coefficient of friction in block A is calculated, depending on the average temperature at the surface of the tool contact with the chips. The value of this temperature is calculated by the method of S. Silin [19]. In block B, the coefficient of friction is calculated as a function of the shear angle and geometric parameters of the cutting tool. This calculation can be carried out following the hypothesis of the minimum main cutting force [19] or under the Oxley's predictive machining theory [20]. After appropriate calculations, the values of the friction coefficient defined in blocks A and B of the calculation scheme are compared. In the following, an increase in the magnitude of the shear angle by $\Delta\beta$ is performed until the condition of the allowable error of calculation is fulfilled: $\frac{f_1 - f_2}{f_1} < \varepsilon = 0,5\%$ after which the external cycle of calculation is terminated. The main characteristics of the cutting process are determined (the length of the contact of the chip with the front surface of the cutter, the physical components of the cutting force on the front surface of the tool), and output the results. The calculations were performed for discrete values of the cutting cross-section thickness from 0 to a_{sr} (mean values of undeformed chip thickness) to account for the increase of this parameter during the initial phase of the machining.

The influence of the self-adaptive coating on the characteristics of contact interaction in the proposed algorithm is taken into account by the parametric relationship of the coefficient of friction on the contact temperature:

$$f(\theta) = \begin{cases} b_{01} + b_{02}e^{-k_1(\theta-\theta_0)}, \theta < \theta_{cr} \\ b_{11} + b_{12}e^{-k_2(\theta-\theta_0)}, \theta \geq \theta_{cr} \end{cases}$$

The general view of the dependence of the friction coefficient on the contact temperature is adopted under the expression given in [21]. A basic view of this relationship is presented in the graph (Fig. 2).

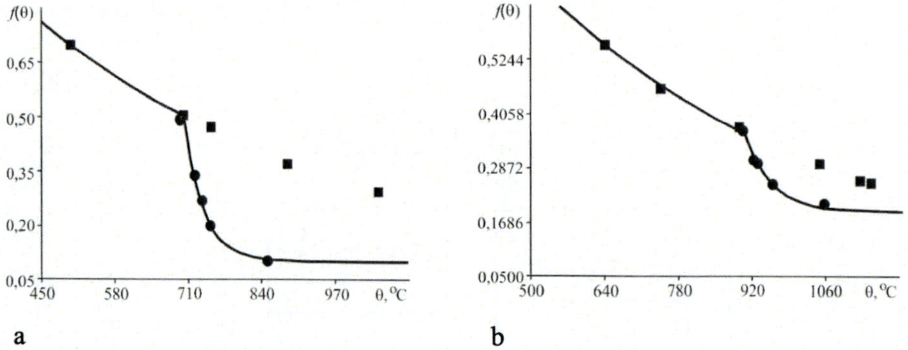

Fig. 2. General view of the dependence of the average coefficient of friction on the temperature, intrinsic for self-adaptive coatings *a* – θ_{cr} = 700 °C, f_{min} = 0,1; *b* – θ_{cr} = 900 °C, f_{min} = 0,2. -■- Experimental data for uncoated PCBN. -●- Model data for PCBN with self-adaptive coating.

The inflection point on the graph corresponds to the beginning of the phase transformation in the upper layers of the coating with the formation of polyoxide secondary structures. This process begins to occur at a critical temperature θ_{cr}. As a hypothesis, we assume that in the temperature range $\Delta\theta$, the ratio of the areas covered by the secondary structures and the total contact area of the chip and the tool surface is equal to 1, and the friction coefficient f is equal to a constant value of $f_{\min}$ due to the tribological properties of the secondary structures. We assume that the coefficient of friction at the temperature range, lower than the critical temperature, corresponds to the tribological characteristics of the cutting process with the application of standard PCBN inserts.

There are no empirical data, which describes the effect of temperature on the friction coefficient of the self-adaptive coating, in which phase transformations from the formation of secondary structures began to occur. So we will vary the coefficients of Eqs. (1) to satisfy the conditions of numerical experiments with different combinations of θ_{cr} and minimal value of friction coefficient $f_{\min}$.

Carrying out such calculations will allow estimating the effects that will arise in extreme cases: the minimum possible under practical conditions of cutting the coefficient of friction and different temperatures of phase transformation in the coating. The temperature range $\Delta\theta$ in which the friction coefficient decreases at the critical point θ_{cr} is assumed to be 200 °C. We determine the coefficients in the system (1) by the least-squares method for all combinations of conditions: θ_{cr} = (600; 700; 800; 900; 1000 °C), $f_{\min}$ = (0.1; 0.2). For the calculations of the contact stresses on the cutting surface front face and the main cutting force according to standard expressions of the mechanics of the cutting process, we assume that the length of contact of the chips with the front surface of the tool is equal to $l_c = a_{sr}(2,05\zeta - 0,55)$ and the specific friction force on the front surface $q_f \frac{4}{3} F/(l_c b)$.

An important role in the proposed model belongs to the issue of determining the coefficient of friction as a function of the shear angle in the shear zone. We compared the results of the calculations of this parameter with the use of the S. Silin and Oxley methods.

S. Silin suggested that the chip formation cutting force has an angle η = 45° with the shear plane, from which it follows from geometric considerations that $\mu = \pi / 4 - \beta + \gamma$. The Oxley chip model assumes that the expression determines the friction angle:

$$\mu = a\tan\left[\frac{1}{2} + \frac{\pi}{4} - \beta + \frac{\cos 2(\beta - \gamma)}{2tg(\beta)} - \frac{\sin 2(\beta - \gamma)}{2}\right]$$

The relationship between the angle of action of the chip formation cutting force, the shear angle and the friction angle is $\eta = \beta + \mu - \gamma$. The numerical solution of the last two equations for the rake angle $\gamma = -10°$ allows us to obtain a simplified dependence: $\mu = -1.35\beta + 47.185$ (Fig. 3). The determination of the friction angle according to the Oxley equation better corresponds to the value of this parameter, determined by experimental our data, so in the subsequent calculations, we will use the expression $\mu = -1,35\beta + 47,185$.

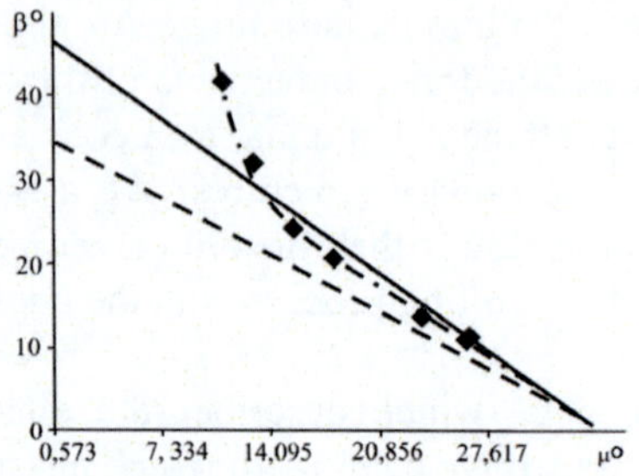

Fig. 3. Shear angle as a function of friction angle defined by Silin (- - -) i Oxley (—), as well as points calculated from the experimental data (-♦-).

4 Results

The results of calculations according to the proposed algorithm are shown in Figs. 4, 5 and 6. These graphs demonstrate differences in shear angle, average contact temperature, and main cutting force for coated and uncoated tools as a function of the average value of undeformed chip thickness a_{sr}. This parameter is growing from 0 to its nominal value to illustrate conditions of the initial stage of the machining process. The nominal value of the average thickness of the undeformed chip when turning with round cutting insert (insert $d = 7$ mm and a feed rate $S = 0.14$ mm/rev) is equal to 27 μm.

The parameter characterizing the degree of deformation of the workpiece at a constant value of the rake angle of the tool is the shear angle, the growth of which indicates a decrease in the stress-strain state of removed material. Accordingly, achieving the maximum value of the shear angle is desirable when optimizing the processing conditions and properties of the tool. As follows from the graphs in Fig. 4, the maximum increase in shear angle, compared to the cutting tool without coating (solid curve), can be achieved if the phase transformation in the surface layers of coatings implicates the formation of secondary structures-solid lubrication (dashed lines), at as low temperature as possible (600–700 °C). A significant increase in the shear angle is observed at $\theta_{cr} = 800$, and 700 °C, a further decrease in θ_{cr} to 600 °C does not significantly affect the increase in the shear angle. High-temperature phases with $\theta_{cr} = 900$–1000 °C will affect the cutting process only at high processing speed, as illustrates the graph calculated for $v = 2.5$ m/s.

Cutting temperature is a characteristic directly related to the magnitude of the shear angle. The reduction of the friction coefficient on the contact surface at the rake face and the heat release in the zone of primary deformation compensates for the increase in temperature associated with the increase in the thickness of the undeformed chip thickness.

As a result, the rate of the temperature growth at the contact areas of the tool with self-adaptive coated is significantly lower in comparison to the uncoated tool (Fig. 5). In the case of machining with cutting speeds of 1.5 and 2.5 m/s, the declination of the temperature in the contact zone will be 11 and 17%, respectively ($\theta_{cr} = 700$ °C). High temperature compositions (such as coatings with $\theta_{cr} \geq 900$ °C) are not effective at $v = 1.5$ m/s and reduce contact temperature θ by 10% at $v = 2.5$ m/s.

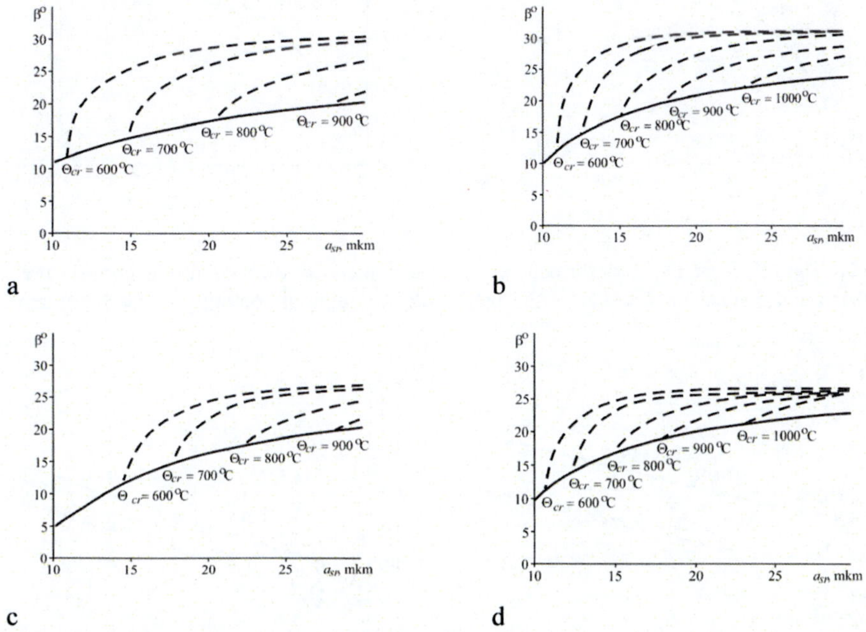

Fig. 4. The effect of the undeformedchip thickness increase on the shear angle: *a*, *b* – *f* = 0,1; *a* – *v* = 1,5 м/с; *b* – *v* = 2,5 м/с; c, *d* – *f* = 0,2; *c* – *v* = 1,5 м/с; *d* – *v* = 2,5 м/с (tool: — without coating; - - - with coating).

The intense decrease of the friction coefficient with the beginning of the formation of secondary structures when the contact temperature reaches θ_{cr} causes a sharp drop in the calculated values of the cutting force. The graphs of these dependencies (Fig. 6) are a mirror image of the graphs of the functions β(asr). A significant reduction in force loads on the tool is observed if $\theta_{cr} \approx 700$ °C at $v = 1.5$ m/s and $\theta_{cr} \approx 800$ °C at $v = 2.5$ m/s. This decrease, according to the calculation results, is 28 and 20%, respectively.

The proposed model is idealized, assuming, in particular, the possibility of maintaining coating over the entire surface of the contact surface of the tool at least during the period of the initial cutting phase. Also, the effect of contact interaction on the clearance face of the tool is not taken into account. However, the analysis allows us to draw some conclusions based on which it is possible to make recommendations on the composition of self-adaptive coatings.

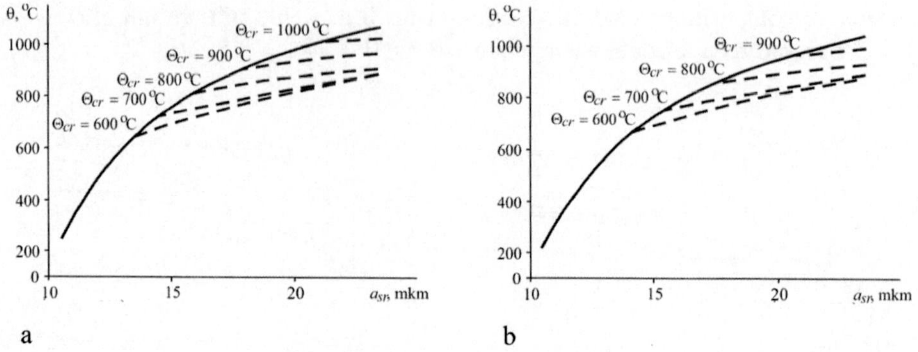

Fig. 5. The effect of the undeformedchip thickness increase on the cutting temperature: *a* – $f = 0{,}1$; $v = 1{,}5$ m/s; *б* – $f = 0{,}1$;$v = 2{,}5$ m/s (tool: —— without coating; - - - with coating).

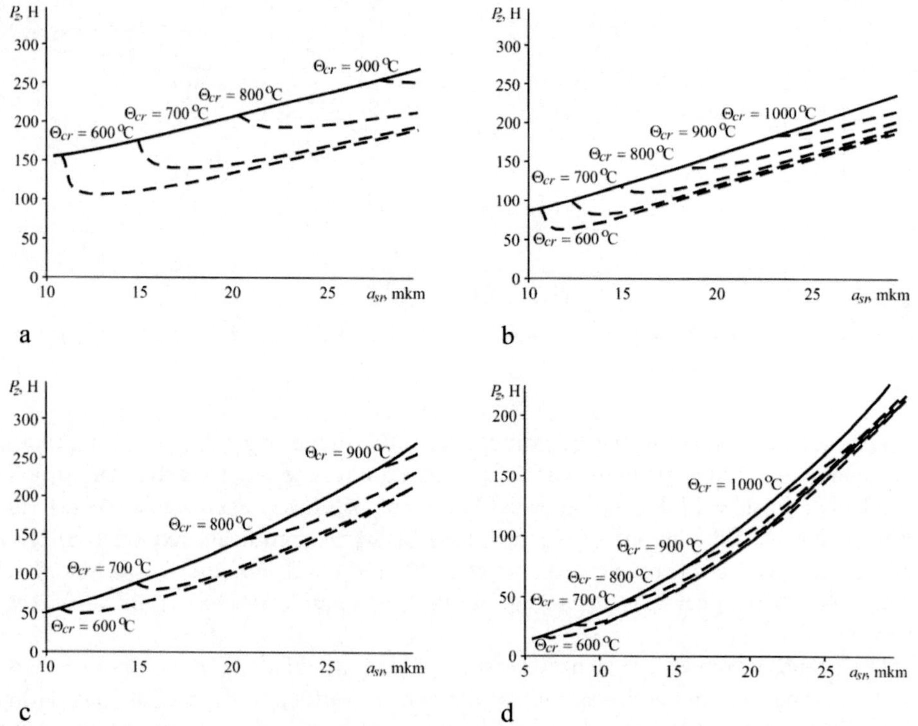

Fig. 6. The effect of the undeformedchip thickness increase on the main cutting force P_z: *a, b* – $f = 0{,}1$; *a* – $v = 1{,}5$ м/c; *b* – $v = 2{,}5$ м/c; *c, d* – $f = 0{,}2$; *c* – $v = 1{,}5$ м/c; *d* – $v = 2{,}5$ м/c (tool: —— without coating; - - - with coating).

5 Conclusions

For reducing the thermobaric load on the tool, the formation of secondary structures on the contact surfaces will be the most effective at temperatures of 600–700 °C. This condition can be fulfilled if the coating will include nitrides of refractory metals, which form the Magnelli phases. In particular, VN forms at a temperature of ~ 500 °C vanadium oxide V_2O_5, which is a solid lubricant up to its melting point of 685 °C.

Higher temperature stability (up to 1300 °C) has solid lubricants formed by amorphous glass films based on solid solutions of oxides like Fe_2O_3-Al_2O_3, Al_2O_3-B_2O_3, Al_2O_3-Ti_2O, Al_2O_3-AlN. However, the formation temperature of these compounds is quite high. In a system containing AlN and TiB_2 the oxidation reactions with formation of titanium diboride at ~ 700 °C are possible, and the formation of aluminum oxynitride in the temperature range 1000–1200 °C is taking place: $Ti(Cr)B_2 + 2{,}5O_2 = (TiCr)O_2 + B_2O_3 \quad \theta_{cr} \sim 700$ °C; $Ti(Cr)B_2 + O_2 \rightarrow (TiCr)B + B_2O_3 \theta_{cr} \sim 700$ °C; $AlN + O_2 \rightarrow Al_xO_yN_z \theta_{cr} \sim 1000$ °C (1200 °C in an equilibrium state).

Considering the results of modeling, the self-adaptive coating of such a system will have limited efficiency due to the high formation temperature of their major component (Al_2O_3, $Al_xO_yN_z$). That is why it is advisable to use such coatings, the phase transformations of which will occur at temperatures of 700–800 °C, and at the same time, they will be characterized by high values of temperature resistance ($\geq$ 1300 °C).

References

1. Selinder, T., Sjöstrand, M., Nordin, M.: Performance of PVD TiN/TaN and TiN/NbN superlattice coated cemented carbide tools in stainless steel machining. Surf. Coat. Technol. **105**(1–2), 51–55 (1998)
2. Caliskan, H., Altas, E., Panjan, P.: Study of nanolayer AlTiN/TiN coating deposition on cemented carbide and its performance as a cutting tool. J. NanoRes. **47**, 1–0 (2017)
3. Vereschaka, A.A., Bublikov, J.I., Sitnikov, N.N.: Influence of nanolayer thickness on the performance properties of multilayer composite nano-structured modified coatings for metal-cutting tools. Int. J. Adv. Manuf. Technol. **95**, 2625–2640 (2018)
4. Uhlmann, E., Riemer, H., Schröter, D.: Investigation of wear resistance of coated PcBN turning tools for hard machining. Int. J. Refract. Met. Hard Mat. **72**, 270–275 (2018)
5. Coelho, R.T., Ng, E.G., Elbestawi, M.A.: Tool wear when turning hardened AISI 4340 with coated PCBN tools using finishing cutting conditions. Int. J. Mach. Tools Manufact. **47**, 263–272 (2007)
6. Kopeikina, M.Y., Klimenko, S.A., Mel'niichuk, Y.A.: Efficiency of cutting tools equipped with cBN-based polycrystalline superhard materials having vacuum-plasma coating. J. Superhard Mat. **30**, 355–362 (2008)
7. Renevier, N.M., Fox, V.C., Teer, D.G., Hampshire, J.: Coating characteristics and tribological properties of sputter-deposited MoS_2/metal composite coatings deposited by closed fifield unbalanced magnetron sputter ion plating. Surf. Coat. Technol. **127**, 24–37 (2000)
8. Jianxin, D., Wenlong, S., Hui, Z., Jinlong, Z.: Performance of PVD MoS_2/Zr-coated carbide in cutting processes. Int. J. Mach. Tools Manufact. **48**(14), 1546–1552 (2008)

9. Klimenko, S.A., Klimenko, S.A., Manokhin, A.S., Beresnev, V.M.: Special features of the applications of cutting tools from polycrystalline cubic boron nitride with protective coatings. J. Superhard Mat. **39**(4), 288–297 (2017)
10. Fox-Rabinovich, G.S., Yamamoto, K., Veldhuis, S.C.: Self-adaptive wear behavior of nano-multilayered TiAlCrN/WN coatings under severe machining conditions. Surf. Coat. Technol. **201**(3–4), 1852–1860 (2006)
11. Podchernyaeva, I.A., Klimenko, S.A., Beresnev, V.M.: Formation of a tribofilm in the surface layer of Al–Ti–Cr–N–B magnetron coating on boron nitride during turning of hardened steel. Powder Metall. Metal Ceram. **54**(¾), 140–150 (2015)
12. Lopez, S., Wong, M., Sproul, W.D.: Thermal behavior of carbon nitride and TiN/NbN superlattice films. J. Vac. Sci. Technol. A Vac. Surf. Films **13**(3), 1644–1648 (1995)
13. Shinn, M., Hultman, L., Barnett, S.A.: Growth, structure, and microhardness of epitaxial TiN/NbN superlattices. J. Mat. Res. **7**(4), 901–911 (1992)
14. Hahn, R., Bartosik, M., Soler, R.: Superlattice effect for enhanced fracture toughness of hard coatings. Scripta Mater. **124**, 67–70 (2016)
15. Zhou, Z., Rainforth, W.M., Lewis, D.B., Crsisy, S., Forsyth, J.J., Clegg, F., Ehiarsarin, A.P., Hoverspain, P.E., Munz, W.D.: Oxidation behavior on nanjscale TiAlN/VN multilayer coating. Surf. Coat. Technol **177–178**, 198–2004 (2004)
16. Zhou, Z., Rainforth, W., Rodenburg, C.: Oxidation behavior and mechanisms of TiAlN/VN coatings. Met. Mat. Trans. **A38**, 2464–2478 (2007)
17. Rainforth, W., Luo, Q., Hovsepian, P.: Wear and friction of TiAlN/VN coatings against Al_2O_3 in air at room and elevated temperatures. Acta Mater. **58**, 2912–2925 (2010)
18. Hovsepian, P.E., Lewis, D.B., Luo, Q.: TiAlN based nanoscale multilayer coatings designed to adapt their tribological properties at elevated temperatures. Thin Solid Films **485**(1–2), 160–168 (2005)
19. Silin, S.S.: Similarity Method when Cutting Materials. Engineering, Moscow (1979). [in Russian]
20. Oxley, P.L.B.: Mechanics of Machining an Analytical Approach to Assessing Machinability. Ellis Horwood, West Sussex (1989)
21. Kragelsky, I.V.: Friction and Wear. Engineering, Moscow (1968). [in Russian]

Kinematics of the Tapered Thread Machining by Lathe: Analytical Study

Iuliia Medvid, Oleh Onysko(✉), Vitalii Panchuk, Lolita Pituley, and Iryna Schuliar

Ivano-Frankivsk National Technical University of Oil and Gas, 15, Karpatska Street, Ivano-Frankivsk 76019, Ukraine
onysko.oleg@nung.edu.ua

Abstract. The accuracy of tapered thread, especially those used in oil and gas production, largely determines the productivity of drilling processes and environmental friendliness of it. This accuracy depends on the accuracy of the movement of the tool cutter regarding the workpiece. The article deals with the study of the vector components of the kinematics of the surface forming of the tapered thread by turning machining. The obtained differential equations describe theoretically-exact necessary motions and really-possible machine-tool motions that provide the formation of tapered thread surfaces by turning machining. These equations and their comparison make it possible to find the radial and tangential deviations of the points of the real-possible surface relatively to the designed surface of the standard tapered thread. Among the parameters of the obtained analytical formulas, there are the parameters of the tapered thread for the drill pipes on the one hand, as well as the geometrical parameters of the tool and the rotate speed of the workpiece on the other one.

Keywords: Threading process by lathe · Circular vector functions · Helical curve on tapered surface · Machine feed · Differential equations of velocity · Radial deviations

1 Introduction

Tapered thread is an important element of oil and gas drill string connections. For improving the durability of the drilling tool, drill pipes of advanced structures are used [1], and drill bits are made using advanced technologies [2]. The accuracy of the tapered joint connection very strong influence on the tightness of the drill string associated with the emission of the drilling fluid into the outside and pollution of the environment as a result [3, 4]. Today these are made using special cutters on high-precision lathes. However, the influence of the kinematics on the accuracy of the surface forming process of the obtained tapered thread has not been investigated. This is explained by the fact that most tool makers produce cutters with zero value of the rake angle at the nose, which greatly simplifies the kinematics of the turning process and does not require its detailed investigation. Considering that in the long run, the need for the use of hard-machined alloy steels for the inclined and horizontal drilling is increasing, the need for using cutters with greater cutting capabilities, i.e., with

V. Tonkonogyi et al. (Eds.): InterPartner 2020, LNME, pp. 555–565, 2021.
https://doi.org/10.1007/978-3-030-68014-5_54

non-zero values of the rake angle increases. [5]. Therefore, the study of the influence of the kinematics of the conical turning process on the accuracy of the obtained thread is a topical problem.

2 Literature Review

Investigation of the accuracy of the tapered thread depending on the geometrical parameters of the cutting edge of the lathe tools is analyzed in the work [5]. The authors investigate the effect of precision and deviation of threading machining on the stress-strain state of the tapered thread connections [6]. Analytical studies of static mechanical tensions in connectors of the oil and gas pipe assortment, depending on the accuracy of the thread profile, are shown in the articles [7–10]. The dynamics of the chip formation process during threading machining by lathe is considered in the articles [11, 12]. The parametric model of the cutting process, whose parameters are the kinematic geometric parameters of the cutter, is proposed in the article [13]. In [14], the research of cylindrical thread joints by finite element method, depending on the accuracy of the thread profile and the clamping strength is presented. Article [15] writes about the dependence of the process of cutting hard-machining steels, depending on the static and kinematic geometric parameters, in particular, the rake angle of the cutter. The article [16] deals with the modelling of the wear resistance of the cutter depending on the change of cutting force in the turning process. The experimental investigations of the cutting process using workpieces from hard-machining high-alloy chromium steels are carefully considered in work [17]. In [18], the results of studies of the vortex turning kinematics of cylindrical thread have been offered. The researches of the accuracy of the cylindrical thread obtained by applying the cutting process are shown in [19]. Of all these reports, the work [20] is the closest to the subject of the article. However, it mainly focuses on the study of the distribution of the values of the kinematic rake angles along the cutting edge of the cutter. It thus investigates the kinematics of the formation of conical helical lines without taking into account the real capabilities of the lathe.

3 Research Methodology

Thus, there is a necessity to analyze the movement of an arbitrary point of the cutting edge of the cutter, which is provided with the kinematic capabilities of the lathe in the process of turning a tapered thread and to identify its difference from the movement of an arbitrary point of a theoretical conical helix. In this case, the placement of cutting edge points should be determined by the parameters of the tapered thread and the geometric parameters of the cutter used for its machining. This will make it possible to identify the functional dependencies of the accuracy of the screw guides of the thread on its parameters and the parameters of the points of the cutting edge of the cutter. The theory of circular vector functions is the most suitable method for the realization of this purpose [21]. According to this theory, the radius vector $\bar{r}$ describes the movement of an arbitrary point M along a helical curve with a constant pitch on the tapered surface using this equation [20] (look at Fig. 1):

$$\vec{r} = a(\vartheta) \cdot \vec{e}(\vartheta) + {}^{\vec{k}P\vartheta}/_{2\pi}, \tag{1}$$

where:

P – thread pitch; υ – value of the rotate angle arbitrary point M relatively thread axis; a(υ) – the value of the tapered radius at arbitrary point M; $\overline{e}$(υ), and (υ) – two unit vectors, which are located relatively axis $0X$ form angles u i u + $90°$.

The circular vector function $\vec{g}$ (υ) can be expressed as follows [20, 21]:

$$\vec{g}(\vartheta) = \vec{e}(\vartheta + {}^{\pi}/_{2})$$

On the other hand, since this vector lies on the tangent line to the circuit in which the vector $\vec{e}$(υ) is placed radially, then [15, 16]:

$$\vec{g}(\vartheta) = \frac{d\vec{e}}{d\vartheta}$$

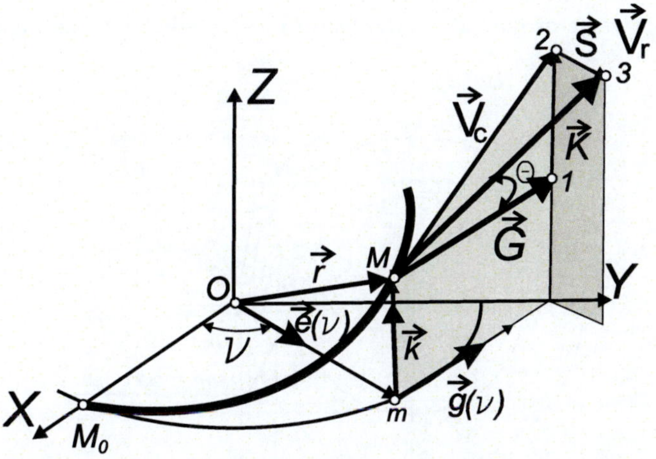

Fig. 1. Scheme of the moving an arbitrary point M along a helical curve.

The differential equation that defines the derivative of the position vector $\vec{r}$ with respect to its rotate angle υ around the axis Ox is offered in the article [15]:

$$\frac{d\vec{r}}{d\vartheta} = \left(\frac{Ptg(\varphi)}{2\pi}\right) \cdot \vec{e}(\vartheta) + \left(P \cdot tg(\varphi) \cdot \frac{\vartheta}{2\pi} + \left(\frac{d_3}{2} - \Delta\right)\right) \cdot \vec{g}(\vartheta) + \vec{k} \cdot {}^{P}/_{2\pi} \tag{2}$$

On the Fig. 1 the vector $\bar{S}$ (between points *2* and *3*) corresponds the first addend of the formula 2, the vector $\bar{G}$ (between points *M* and *1*) corresponds the second addend of the same equation and the vector $\bar{K}$ (between points *1* and *2*) corresponds the third addend of the formula. Thus, formula 2 is presented as follows [20]:

$$\frac{d\vec{r}}{d\vartheta} = \vec{V}_r = \vec{S} + \vec{G} + \vec{K} \tag{3}$$

Vector $\bar{V}_r$ (between points *M* and *3*) is the tangent line to the helical curve at a point *M*, but it doesn't correspond summary speed of its motion $\vec{V}_r$ as the derivative of $\vec{r}$ with respect to angle υ but it isn't the derivative of $\vec{r}$ with respect to time *t*. So:

- equation (2) does not describe the speed of the motion of the point M;
- because the vector $\bar{S}$ in formula 3 lies individually for every arbitrary point on the helical curve, it is impossible to provide it by the machine tool.

Figure 2 shows a scheme of machining of the thread with a taper angle φ and constant pitch P by lathe machine. Kinematical process of the tapered thread turning is realized due to such movements:

- the rotating movement of the workpiece n;
- the constant value of the cutter's longitudinal travel Fa parallel to workpiece centerline providing constant pitch P;
- the constant value of the cutter crossfeed Fi to provide the taper angle φ.

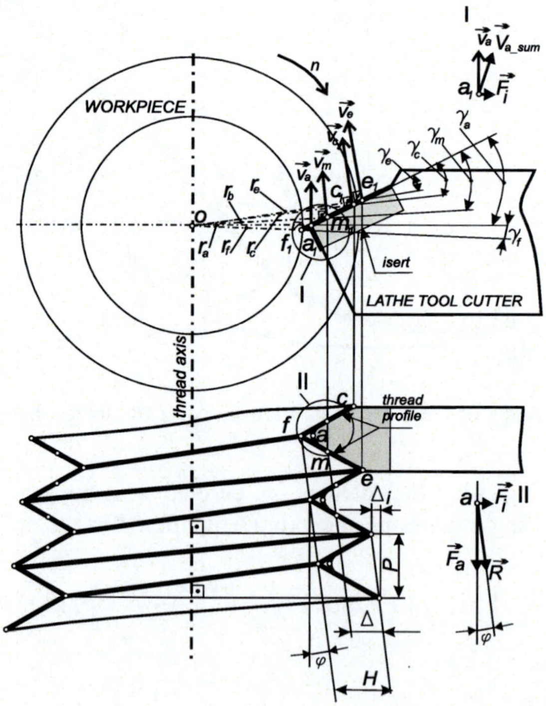

Fig. 2. Scheme of machining of the thread with a taper angle φ and constant pitch P by lathe.

According to the methods adopted in modern orthogonal drawings [22], Fig. 2 shows the horizontal projection of the displacement Δ_i of the cutter due to the crossfeed F_i of machine tool. The real magnitude of the specified displacement coincides with this projection magnitude and is determined using the formula:

$$\Delta_i = Ptg(\varphi) \tag{4}$$

The rotating movement of the workpiece with frequency n provides the main cutting movement at each point of the cutting edge of the insertion of the *CEAM* cutter, the horizontal projection of which coincides with the profile of the thread (Fig. 2). On the vertical projection of the cutting edge represented by the line segment $c_1e_1a_1m_1$ vectors of the velocity of the main cutting movement $\bar{V}_a$, $\bar{V}_m$, $\bar{V}_c$, $\bar{V}_e$ of the corresponding points are set. These vectors lie perpendicular to the radii r_a, r_m, r_c, r_e. The efficiency of the threading process by lathe depends on the magnitude of the static rake angle at the cutter nose – γ_a. At other points, the static rake angles are determined between r_a and the radius at which the corresponding point is placed [7, 20]. The value of the static rake angle at an arbitrary point of the cutting edge of the threading cutter can be determined as follows:

$$\gamma_m = arcsin\left(\gamma_a - \frac{H - \Delta}{r_m cos(\gamma_a)} sin(180^\circ - \gamma_a)\right), \tag{5}$$

where at the example of tool joint tapered thread:

$$r_m = \frac{d_3}{2} + Ptg\varphi \cdot \frac{\vartheta}{2\pi} - \Delta + b,$$

according to standard API7: d_3 – major diameter of the tapered thread at its minor side; P – the pitch of thread; f – taper angle; f – a truncation of thread crest; H – the height of the fundamental triangle (Fig. 2, 3).

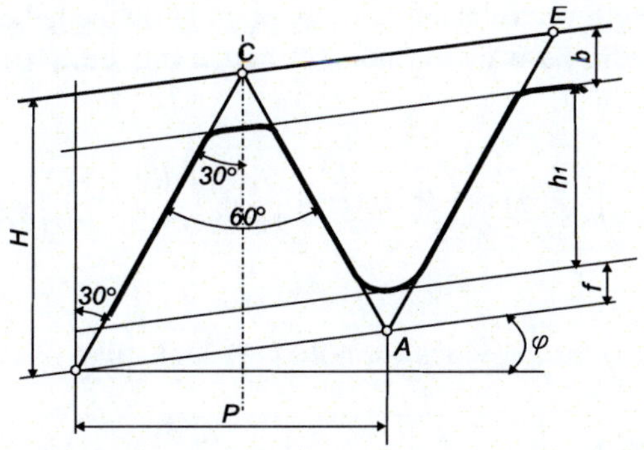

Fig. 3. Scheme of the tool joint tapered thread for drill string.

and also: υ – the value of the angular displacement of the arbitrary point at the cutting edge relative axis of the thread (Fig. 1); *D* – a radial distance of an arbitrary point m of the cutting edge from the crest of thread (Fig. 2).

Using the techniques of differential geometry [21] it is possible to determine the rate of change of the length of the radius vector using its dependence on the angle of its rotating υ (Fig. 1). So, because this one is a function of time:

$$\vartheta = 2\pi nt, \tag{6}$$

the derivative of the angle change with respect to time change has the following expression:

$$\frac{d\vartheta}{dt} = 2\pi n \tag{7}$$

Therefore the differential of the angle function looks like this:

$$d\vartheta = 2\pi ndt \tag{8}$$

So then:

$$\frac{d\vec{r}}{d\vartheta} = \frac{d\vec{r}}{dt} \cdot \frac{1}{2\pi n}, \tag{9}$$

so the rate of position vector can be defined as follows:

$$\frac{d\vec{r}}{dt} = 2\pi n \frac{d\vec{r}}{d\vartheta} \tag{10}$$

4 Results

Using differential geometry techniques [21], it is possible to determine the rate of change of position length vector at an arbitrary point *M* on the helical line with constant pith. In that way on the base of formula (2) and using Eq. (10) the velocity of the *M* point is received:

$$\frac{d\vec{r}}{dt} = 2\pi n \left[\left(\frac{Ptg(\varphi)}{2\pi} \right) \cdot \vec{e}(\vartheta) + \left(P \cdot tg(\varphi) \cdot \frac{\vartheta}{2\pi} + \left(\frac{d_3}{2} - \Delta \right) \right) \cdot \vec{g}(\vartheta) + \vec{k} \cdot {}^{P}\!/_{2\pi} \right] \tag{11}$$

Considering the Eq. (6) the previous formula (11) looks like this:

$$\frac{d\vec{r}}{dt} = (Ptg(\varphi)n) \cdot \vec{e}(\vartheta) + \left(P \cdot tg(\varphi) \cdot n \cdot t \cdot 2\pi n + \left(\frac{d_3}{2} - \Delta \right) 2\pi n \right) \cdot \vec{g}(\vartheta) + \vec{k} \cdot Pn \tag{12}$$

So to determine the velocity of the arbitrary moving point of the tapered helix with a constant pitch on the tapered surface, the equation can be used:

$$\frac{d\vec{r}}{dt} = \left|\overrightarrow{S_m}\right| \cdot \vec{e}(\vartheta) + \left|\overrightarrow{V_m}\right| \cdot \vec{g}(\vartheta) + \left|\overrightarrow{F_a}\right| \cdot \vec{k}, \tag{13}$$

where:

$$\overrightarrow{S_m} = (Ptg(\varphi)n) \cdot \vec{e}(\vartheta) \tag{14}$$

$$\overrightarrow{V_m} = 2\pi n\left(P \cdot tg(\varphi) \cdot n \cdot t + \left(\frac{d_3}{2} - \Delta\right)\right) \cdot \vec{g}(\vartheta) \tag{15}$$

$$\overrightarrow{F_a} = \overrightarrow{k} \cdot Pn \tag{16}$$

Figure 4 presents the 3D scheme that shows the position of the vectors $\bar{V}_r$, $\overrightarrow{F_a}$ and $\bar{S}_m$, moreover the vectors $\bar{V}_r$ and $\bar{S}_m$ for convenience of visualization are placed not at point M as a vector $\overrightarrow{F_a}$, but at its projection on the XOY plane – point m.

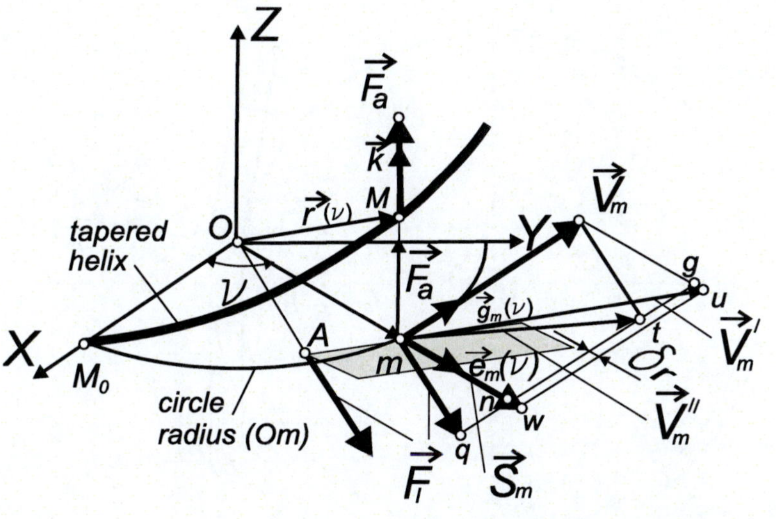

Fig. 4. Scheme of the velocity and feed vectors an arbitrary point M along a helical curve on the tapered surface.

At point A (the projection of the cutting edge nose) the vector of the crossfeed of the machine $\overrightarrow{F_i}$ is placed. Its direction coincides with the direction of the segment line OA, that is, the radius r_a. Because crossfeed is used to provide a taper angle. So, its absolute value is determined according to formula 4:

$$\left|\overrightarrow{F_i}\right| = Ptg(\varphi)n \tag{17}$$

That is the absolute value of the vectors $\overrightarrow{F_i}$ and $\bar{S}_m$ are the same, but their directions aren't. Herewith the vector $\overrightarrow{F_i}$ is the same at any point of the cutting edge, while the vector $\bar{S}_m$ coincides in the direction with the radius Om (r_m) corresponding to the arbitrary point m. The sum of the vectors $\bar{V}_r$ and $\bar{S}_m$, i.e., vector $\overrightarrow{V'_m}$ between the points m and g theoretically provides motion of the arbitrary point on the cutting edge M on a given trajectory with constant pitch Δ_i in radial direction – Archimedean spiral (Fig. 5):

$$\overrightarrow{V'_m} = \overrightarrow{V_m} + \overrightarrow{S_m} \tag{18}$$

Real machine kinematic ensures sum vector $\overrightarrow{V''_m}$ between the points m and t (Fig. 4):

$$\overrightarrow{V''_m} = \overrightarrow{V_m} + \overrightarrow{F_i} \tag{19}$$

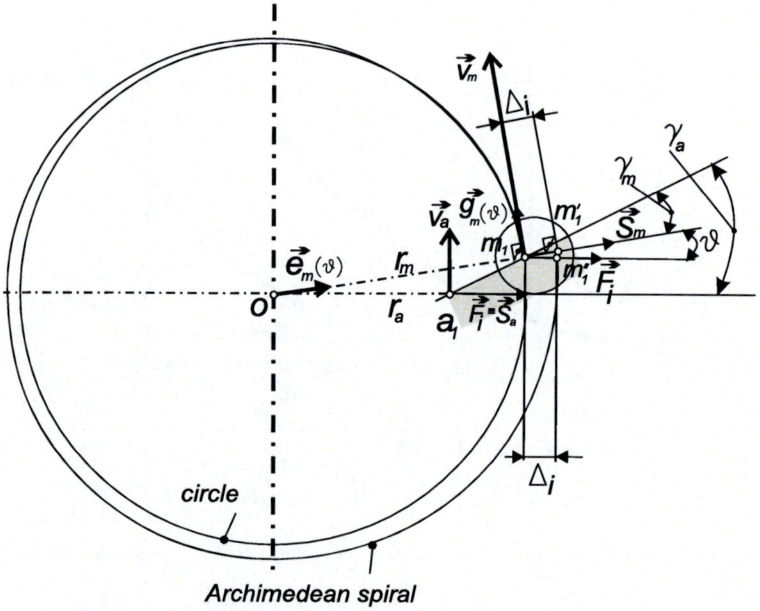

Fig. 5. Scheme to explain the receiving of the Archimedean spiral and deviation from it.

Thus due to the real kinematic capabilities of the machine, there are deviations from the exact theoretical surface of the taper thread in radial δ_r (Fig. 4, 6) and tangential δ_t directions (Fig. 6).

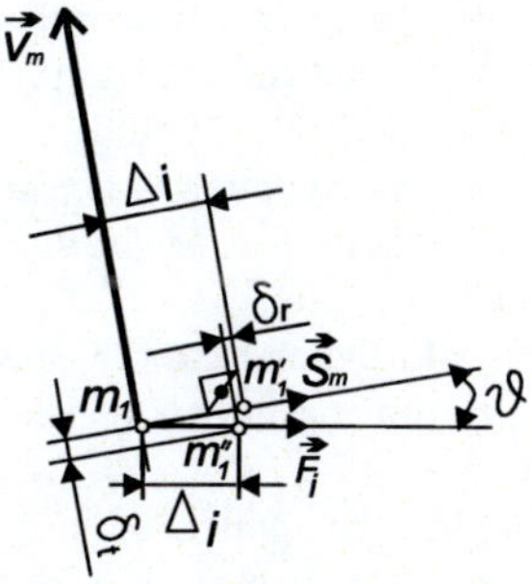

Fig. 6. Scheme for defining the radial and tangential deviation from the Archimedean spiral.

So the deviations from the exact theoretical surface of the taper thread in radial δ_r and tangential δ_t directions are determined as follows:

$$\delta_r = \Delta_i(1 - cos\vartheta) \tag{20}$$

$$\delta_t = \Delta_i sin\vartheta, \tag{21}$$

where:

$$\vartheta = \gamma_a - \gamma_m$$

5 Conclusions

Based on the application of circular vector functions and methods of differential geometry, an analytical description of the theoretically given kinematics of the process of tapered thread and its comparison with the analytical kinematic description of the real tapered threading process by lathe are obtained for the first time.

As a result, the equations that make it possible to calculate the probable tangential and radial deviations of the points on the thread surface with respect to their given theoretical position are obtained for the first time too.

In the nearest studies, based on the analytical expressions obtained here, it is planned to make a study of the accuracy of the tapered thread obtained depending on their parameters and the value of the static rake angle at the nose of the cutter.

References

1. Vlasiy, O., Mazurenko, V., Ropyak, L., Rogal, O.: Improving the aluminum drill pipes stability by optimizing the shape of protector thickening. Eastern-Eur. J. Enterp. Technol. **1** (7–85), 25–31 (2017). https://doi.org/10.15587/1729-4061.2017.65718

2. Ropyak, L., Schuliar, I., Bohachenko, O.: Influence of technological parameters of centrifugal reinforcement upon quality indicators of parts. Eastern-Eur. J. Enterp. Technol. **1** (5), 53–62 (2016). https://doi.org/10.15587/1729-4061.2016.59850
3. Shkitsa, L., Yatsyshyn, T., Lyakh, M., Sydorenko, O.: Means of atmospheric air pollution reduction during drilling wells. IOP Conf. Ser. Mater. Sci. Eng. **144** (2016). https://doi.org/10.1088/1757-899X/144/1/012009
4. Skitsa, L., Yatsyshyn, T., Liakh, M., Sydorenko, O.: Ways to improve safety of pumping-circulatory system of a drilling rig. Min. Miner. Deposits **12**(3), 71–79 (2018). https://doi.org/10.15407/mining12.03.071
5. Kopei, V., Onysko, O., Panchuk, V.: The application of the uncorrected tool with a negative rake angle for tapered thread turning. In: Ivanov, V. (eds.) Advances in Design, Simulation and Manufacturing II. DSMIE 2019. LNME, pp. 149–158. Springer, Cham (2020). https://doi.org/10.1007/978-3-030-22365-6_15
6. Pryhorovska, T., Ropyak, L.: Machining error influence on stress state of conical thread joint details. In: Proceedings of the International Conference on Advanced Optoelectronics and Lasers, CAOL 2019, pp. 493–497 (2019). https://doi.org/10.1109/caol46282.2019.9019544. Article no. 9019544
7. Onysko, O., Kopey, V., Panchuk, V.: Theoretical investigation of the tapered thread joint surface contact pressure in the dependence on the profile and the geometric parameters of the threading turning tool. IOP Conf. Ser. Mater. Sci. Eng. **749**, 012007 (2020). https://doi.org/10.1088/1757-899X/749/1/012007
8. Xu, H., Shi, T., Zhang, Z., Shi, B.: Loading and Contact Stress Analysis on the Thread Teeth in Tubing and Casing Premium Threaded Connection. Mathematical Problems in Engineering. https://www.hindawi.com/journals/mpe/2014/287076/. Accessed 03 Jan 2020
9. Luo, S., Wu, S.: Effect of stress distribution on the tool joint failure of internal and external upset drill pipes. Mater. Des. **52**, 308–314 (2013)
10. Shats'kyi, I., Lyskanych, O., Kornuta, V.: Combined deformation conditions for fatigue damage indicator and well-drilling tool joint. Strength Mater. **48**, 469–472 (2016). https://doi.org/10.1007/s11223-016-9786-8
11. Rezayi, M., Altinta, K.Y.: Generalized modeling of chip geometry and cutting forces in multi-point thread turning. Int. J. Mach. Tools Manuf. **98**, 21–32 (2015). https://doi.org/10.1016/j.ijmachtools.2015.08.005
12. Khoshdareggi, M.R., Altintas, Y.: Generalized modeling of chip geometry and cutting forces in multi-point thread turning. Int. J. Mach. Tools Manuf. **98**, 21–32 (2015)
13. Kopei, V., Onysko, O., Panchuk, V.: Computerized system based on FreeCAD for geometric simulation of the oil and gas equipment thread turning. IOP Conf. Ser. Mater. Sci. Eng. **477**, 012032 (2019). https://doi.org/10.1088/1757-899x/477/1/012032
14. Chen, S., An, Q., Zhang, Y., Gao, L., Li, Q.: Loading analysis on the thread teeth in cylindrical pipe thread connection. J. Pressure Vessel Technol. **3**(132), 0312021–0312028 (2010). https://doi.org/10.1115/1.4000729
15. Baizeau, T., Campocasso, S., Fromentin, G., Rossi, F., Poulachon, G.: Effect of rake angle on strain field during orthogonal cutting of hardened steel with c-BN tools. Procedia CIRP **31**, 166–171 (2015). https://doi.org/10.1016/j.procir.2015.03.089
16. Zhang, G., Guo, C.: Modeling flank wear progression based on cutting force and energy prediction in turning process. Procedia Manuf. **5**, 536–545 (2016). https://doi.org/10.1016/j.promfg.2016.08.044
17. Wang, J., Mathew, P., Li, X., Huang, C., Zhu, H.: Experimental study on cutting characteristics for buttress thread turning of 13% Cr stainless steel. Key Eng. Mater. **443**, 262–267 (2010). https://doi.org/10.4028/www.scientific.net/KEM.443.262

18. Zanger, F., Sellmeier, V., Klose, J., Bartkowiak, M., Schulze, V.: Comparison of modeling methods to determine cutting tool profile for conventional and synchronized whirling. Procedia CIRP **58**, 222–227 (2017). https://doi.org/10.1016/j.procir.2017.03.216
19. Slătineanu, L., Radovanovic, M., Coteaţă, M., Beşliu, I., Dodun, O., Coman, I., Olaru, S.C.: Requirements in designing a device for experimental investigation of threading accuracy. In: 21st Innovative Manufacturing Engineering & Energy International Conference – IManE&E 2017 (2017). https://doi.org/10.1051/matecconf/201711201005.MATEC. Web of Conference vol. 112, 01005
20. Onysko, O., Kopey, V., Panchuk, V., Medvid, I., Lukan, T.: Analytical study of the kinematic rake angles of the cutting edge of the lathe tool for the tapered thread manufacturing. In: Tonkonogyi, V., et al. (eds.) Advanced Manufacturing Processes. InterPartner 2019. LNME, pp. 236–245. Springer, Cham (2020). https://doi.org/10.1007/978-3-030-40724-7_24
21. Norden, A.P.: Short Course of Differential Geometry. State Publishing House of Physical and Mathematical Literature, Moscow (1958)
22. Taras, I.: Features of teaching the graphic discipline courses for different specialty students. In: Cocchiarella, L. (eds.) 18th International Conference on Geometry and Graphics. ICGG 2018. Advances in Intelligent Systems and Computing, vol. 809, pp. 2257–2261. Springer, Cham (2019). https://doi.org/10.1007/978-3-319-95588-9_215

Mathematical Modeling of the Device for Radial Vibroturning

Roman Obertyukh, Andrii Slabkyi(✉), Oleksandr Petrov, and Vitalii Kudrash

Vinnytsia National Technical University, 95, Khmelnytsky Highway, Vinnytsia 21021, Ukraine
slabkiyandrey@gmail.com

Abstract. The structural and design scheme of the hydropulse device for radial vibration cutting with a built-in pulse generator (PPG) is considered. Based on the structural scheme of the device, scientifically substantiated structure of assumptions and representation of a hydraulic link (hereinafter HL) in the form of a Kelvin-Voigt body dynamic models of the hydropulse drive of the device for direct and return moves of the consolidated masses which interact with an HL through transfer numbers are constructed. Four simple dynamic models are presented, based on which D'Alembert principle is based on the mathematical model of the hydro-impulse drive of the device in the form of differential equations of motion of masses, conditions of unambiguity, which cause restrictions on the displacement of this energy carrier. By replacing the variables in the original differential equations of mass with new variables, these differential equations are reduced to the form of classical nonlinear differential equations of the second order, describing the forced oscillations of the masses under the action of variable oscillations of the amplitude of linear deformations of the HL during the operating cycle, and also the natural circular frequencies of the hydropulse drive of the device are established and analyzed.

Keywords: Mathematical model · Dynamic model · Hydropulse device · Ring spring · Frequency · Amplitude

1 Introduction

One of the modern progressive ways of material machining is the process of vibration cutting [4, 5], in particular vibratory turning [1–3, 18]. The most effective devices for low-frequency vibration cutting are devices built based on a hydraulic actuator. In particular, it is a kind of hydropulse, which has several advantages over other known vibration actuators. In order to develop a scientifically sound methodology for design calculation and optimize the design parameters of new devices developed [1], theoretical studies of the dynamic processes that accompany the operation of these devices are needed, for example, by examining their dynamic and mathematical models with subsequent experimental verification of the adequacy of these models.

V. Tonkonogyi et al. (Eds.): InterPartner 2020, LNME, pp. 566–576, 2021.
https://doi.org/10.1007/978-3-030-68014-5_55

2 Literature Review

The methods of a theoretical study of the dynamics of various vibration systems [3, 6–8, 17, 19, 20] by analyzing their mathematical models, in particular hydropulse machines, are usually based on the mathematical model of executive units of machines in the form of differential equations of motion and energy expenditure equations distributed by the PPG during the duty cycle [9]. In order to simplify the mathematical models of the hydropulse drive, sometimes the dynamic processes are neglected, considering the triggering of the PPG instantaneous ("relay"), which, as confirmed by the results of experimental studies, does not allow to describe the dynamics of this drive adequately and causes a significant difference between the results of its theoretical and experimental researches.

In the traditionally constructed mathematical models of hydropulse machines and devices, the differential equations of motion of locking and distribution elements are an algebraic sum of forces of inertia, motion, positional (force of weight, and elastic elements) and friction (dry and viscous). For the uniqueness of the solution of the system of differential equations of movement of the locking and distribution elements, it adds a system of energy consumption equations through the corresponding passage sections of the PPG, which, as a rule, consists of a differential equation of the first order describing the law of pressure changes in the system of PPG, and equations of the expenses balance (OR cost balance) between the cavities of the hydropulse drive in the process of carrying out the duty cycle. The mathematical model also contains uniqueness in the form of certain constraints for the changes in the pressure of the working fluid (energy), the moves and speeds of the locking and distributing elements of the PPG, etc.

Simplified mathematical models describe the cycle of operation of the hydropulse drive as a single-act process, without the allocation of individual phases [10–12], which, in our opinion, does not allow adequately reflect the dynamics of the hydropulse drive and create the correct methods of design calculation of the device or machine.

In hydropulse actuators, in order to obtain high frequencies of pressure pulses passage, they limit the volume of the pressure cavity [1, 13] and use, for the construction of dynamic and mathematical models of such actuators (devices, machines, etc.), an elastically-concentrated model [1] energy carrier, neglecting its mass in pressure cavities. The hydraulic link in the dynamic model of such a hydraulic impulse actuator (hereinafter referred to as the PPG) is represented in the form of a Kelvin-Voigt body [1, 15, 16], composed of parallel inertial spring and dissipative elements, which allows creating mathematical models of hydraulic impulses, devices, and machines based on them, take into account both elastic and viscous properties of the energy carrier. Modeling of the HL in the form of a Kelvin-Voigt body does not contradict the laws of static and dynamic hydraulics; it allows taking into account the properties of the energy carrier (elastic and viscous) during the calculation of the PPG and to drive elastic elements to create the driving forces caused by the energy pressure to the type characteristic of forces.

3 Research Methodology

The schematic diagram of the hydroimpulse device for radial vibration turning shown in Fig. 1. The device consists of single-stage PPG 1 increased throughput and powerhead 2 in the vibration motion of the cutter 3, which performs the processing of vibration cutting part *D*. PPG 1 consists of a housing 1.1, the shut-off element 1.2 which in the form of a conical valve combined with a slotted spring (hereinafter SS1), located in the sleeve 1.3 installed in the housing 1.1, contacts the conical chamfer with the seat 1.4. The saddle 1.4 is located in the same outlet as the sleeve 1.3. In the axial direction of the sleeve 1.3 and the saddle 1.4 are fixed by the cover 1.11 PPG 1. In cover 1.11, a plunger 1.10 is installed, the left spherical (according to the drawing) end of which rests on the support ring SS1, and the right end contact with the screw of the regulator of pre-deformation SS1 (pressure "opening" p_1 PPG).

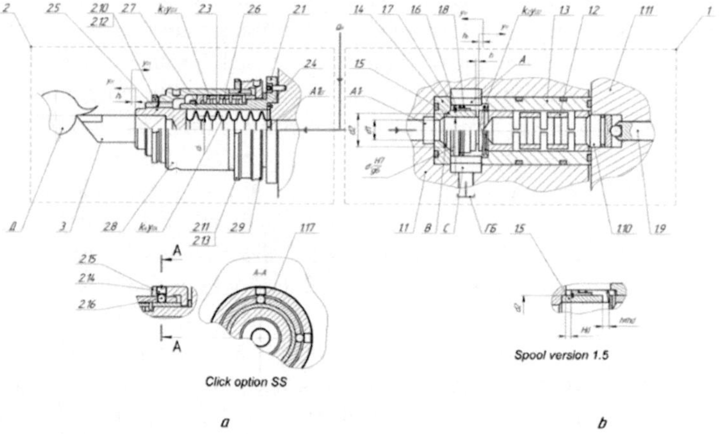

Fig. 1. Schematic diagram of a hydropulse device for radial vibrating turning.

The first stage of sealing PPG1 is carried out on the average diameter d_1 of the contact of the locking element's 1.2 conical part with the saddle 1.4. The second degree of sealing PPG1 on the average diameter d_2 is implemented by the sleeve-valve 1.5, which has an internal hole in the diameter of the cylindrical valve, according to the exact running landing indicated in Fig. 1. Length of conjugation the surfaces of the sleeve-valve 1.5 and the cylindrical part of the conical valve of the shut-off element 1.2 are developed (not less than (0,8… 1,0) d_1'), which ensures high tightness at the moment of opening PPG1.

The initial contact pressure at the interface of the ground grindings of the sleeve-valve 1.5 and the saddle 1.4 is provided by the force of a twisted spring 1.8 acting on the sleeve-valve 1.5 via a stepped sleeve 1.6 located on the outer surface of the sleeve-valve 1.5, and a spring lock ring (washer) 1.7. A gap $h < h_c$ is formed between the flat (right-hand side) face of the valve sleeve 5 and the spool of the SS (here h_c is the stroke of the shut-off element 1.2, which is appropriate to take equal to the stroke of the valve sleeve 1.5).

The working fluid pressure (energy carrier) during the opening of p_1 and closing of p_2 PPG and the initial contact pressure p_k in the chamfer coupling of the sleeve-valve 1.5 with the saddle 1.4 is calculated according to the known dependencies:

$$p_1 \geq 4k_1y_{01}/\left(\pi d_1^2\right) \approx 0,785k_1y_{01}d_1^{-2}; \tag{1}$$

$$p_2 \leq p_1 d_1^2 d_2^{-2} + 0,785k_1h_{в}d_2^{-2}; \tag{2}$$

$$p_{\kappa} = 4k_2y_{02}/\left(\pi d_2^2\right) \approx 0,785k_2y_{02}d_2^{-2}, \tag{3}$$

where k_1, k_2, y_{01}, y_{02} are, respectively, the rigidity and previous deformation of PP1 and the twisted spring 1.8.

The formula can estimate the working force Fp on the cutting edge of the cutter 3 at the time of opening PPG1:

$$F_p = p_1 \cdot \pi d_3^2/4 - k_3 \cdot y_{03} - k_4y_{04} - \overline{F_y}, \tag{4}$$

where d_3 – the average diameter of the cross-section in the connection of the bellows 2.3 with the bush-holder 2.5; k_3, k_4, y_{03}, y_{04} – respectively, the rigidity of SS 2.6, bellows 2.3 and their previous deformations; $\overline{F_p}$ – the average value of the cutting force component.

The pressure in the pressure cavities $A1_g$ and $A1_{cg}$ (in the device is essentially one pressure cavity) $p_r \geq p_1$ (here rr the current pressure in the pressure cavity PPG) acts on the sleeve-valve 1.5, which, moving quickly, passes the distance h, rests on the bush SS, opens the locking element 2 to the value of the negative overlap $h_{в}$, and locks it in this position. The pressure cavities $A1_g$ and $A1_{cg}$ through the intermediate cavity B, groove "a" on the sleeve 1.3 are connected to the drainage cavity C (hydraulic tank (HT). The pressure of the energy carrier in the drive system decreases to level p_2, which causes the return stroke of the cutter 3 under the action of SS 2.6 and bellows 2.3 (since the stiffness $k_3 >> k_4$, the share in the total effort of the return stroke of the cutter from the bellows is not significant). Valve sleeve 1.5 and shut-off element 1.2 to the starting position by the force of SS 1 and the twisted spring 1.8.

Cutter 3 reverses the cutting process, resulting in the shredding of the chip. After locking the locking element 1.2 PPG 1 and the sleeve 1.5 in the initial position (closing PPG 1) the operating cycle of the device is repeated. Pressure pulses are generated in the hydraulic system of its drive – $\Delta p = p_1 - p_2$ and the frequency ν and the vibrational motion of the cutter 3 with the same frequency and amplitude, the level of which is determined by the magnitude of the pressure p1 and the force F_p (see 4).

In the dynamic process, the locking element 1.2 PPG1 performs two types of movement - dynamic and deformation. The valve portion of element 1.2 moves kinematically, and SS1 is mostly involved in kinematic and deformation displacement. The fixed part of SS1 can be considered only its support ring, which rests on the plunger 1.10 (Fig. 1).

An approximate cycle diagram of the duty cycle of a device with corresponding time dependencies can be obtained as described in [1, 13, 14].

Due to the small volume of the intermediate cavity B (see Fig. 1), at the moment of separation of the locking element 1.2 PPG 1, the pressure of the energy carrier in the cavities $A1_g$ (and $A1_{Cg}$) and B is balanced almost instantly, which causes rapid movement of the sleeve-valve 1.5 on the path h under the action of the middle forces (excluding friction forces):

$$\overline{F_{1.5}} = (p_1 - p_K) \cdot A_{1.5}, \quad (5)$$

where $A_{1.5} \approx \pi[d_2^2 - (d_1')^2]/4 \approx 0,785[d_2^2 - (d_1')^2]$ – the annular cross-sectional area of the sleeve-valve 1.5.

After passing the sleeve-valve 1.5 gap h and pushing it into the collar SS1, the locking element 1.2 is moved under the action of energy pressure p_1 on the cross-sectional area $A_2 = \pi d_2^2/4 \approx 0,785 d_2^2$ of the second stage of sealing PPG 1.

During the reverse stroke of shut-off element 1.2 under the action of SS1, it can move with greater acceleration than sleeve-valve 1.5, since in case of proportional deformations the force of the SS1 is much greater than the force of the twisted spring 1.8 ($k_1 \gg k_2$), but $m_2 \ll m_1$ which allows for almost simultaneous movement of the shut-off element 1.2 and the valve sleeve 1.5 on the return path $t_{K3} \approx t_{K3}'$. The phase shift between the movement of the shut-off element 1.2 and the sleeve-valve 1.5 can be more or less accurately determined by mathematical modeling of the dynamic processes occurring in the hydroimpulse drive of the device and the experimental study of its dynamics.

4 Results

Factors describing the real dynamics of the hydraulic drive (HD) device for radial vibration turning with single-stage high bandwidth PPG are dissipative and variable, which also often change randomly. Taking into account, all these factors in the mathematical modeling of the PPG would lead to an overly complex mathematical model of the HD device and would create insurmountable mathematical difficulties in its analysis and study works [1, 13].Given the low mass m_2 compared to the mass m_1 and the short duration of self-movement of the bushing-valve 1.5 on the way h_1 since on the way $h_{g} + h$ the shut-off element 1.2 and the bushing-valve 1.5 (masses m_1 and m_2) move during forward and backward moves (see Fig. 2) as a whole, it is permissible not to consider a separate differential equation of motion, but taking into account the peculiarities of the operation of PPG 1 in the uniqueness of the mathematical model of the device for radial vibration turning. Taking into account the orientation cycle of the device operating cycle (due to the size limit of the article in the paper is not presented), the accepted structure of assumptions and the above observations, the dynamic models of the forward (Fig. 2a) and reverse (Fig. 2b) moves consist of two concentrated masses $m_1 + m_2$ and m_3 and interact with the HL in the form of parallel non-inertial elastic k_{0r} and dissipative C_0 elements through gear relations $U_{01(02)}$ and U_{03}.

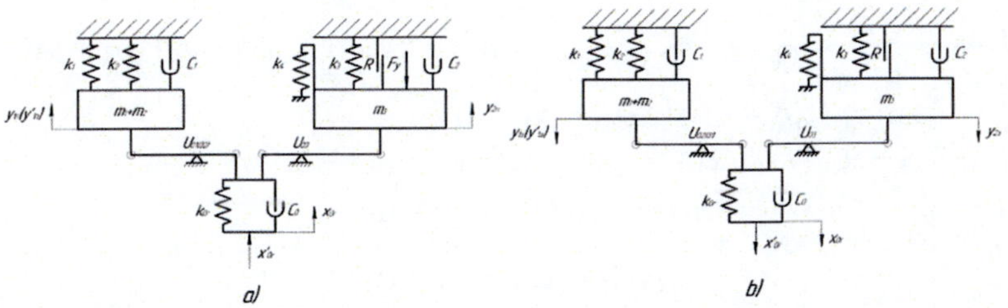

Fig. 2. Dynamic models of forward (a) and reverse (b) moves of masses $m_1 + m_2$.

The HL is deformed at variable speeds $\dot{x}_{0r}$ in directions x_{0r} during the operating cycle of the device. Motor masses $m_1 + m_2$ and m_3 during their forward stroke $(y_{1n}(y'_{1n}), y_{2n})$ and reverse $(y_{13}(y'_{13}), y_{23})$ strokes counteract the positional forces of elastic resistance, which are characterized by the rigidities of SS1 k_1, twisted spring 1.8 k_2, bellows 2.3 k_4, SS 2.6. k_3 (see Fig. 1), the viscous friction, the level of which is determined by the coefficients C_1 and C_2 and velocities $\dot{y}_{1n}(\dot{y}_{1n}), \dot{y}_{2n}, y_{13}(\dot{y}_{13})$, and $\dot{y}_{23}$ the dry friction force R (see assumption structure) and the average cutting force $\overline{F_y}$, which is valid only during direct mass m_3 travel. The coefficient of viscous friction C_1 is summarized for the locking element 1.2 and the bushing-valve 1.5, and the coefficient C_2 takes into account the possible viscous friction in case of lubrication of the guides SS 2.6 and the bush-holder.

The rigidities of SS1 k_1 and SS 2.6 k_3 for the masses m_1 and m_3 assumptions taken during the calculation can be determined by the simplified dependence [1, 9]

$$k_i = \left(1,035Ea_i^4\right) / \left(R_i^3 \cdot n_i\right), \tag{6}$$

where $i = 1-$ order number for SS1; $i = 2$ – serial number of software SS 2.6; $E = 2,15 \cdot 10^5$ MPa modulus of material elasticity SS1 and SS 2.6 (steel 60C2A). The rigidity k_2 of the twisted spring 1.8 (see Fig. 1) can be calculated by the standard [1] method and the rigidity of the bellows 2.3 k_4, which in the powerhead acts as a tension spring, is loaded by tensile force.

It is advisable to simplify the initial dynamic models of the forward and reverse masses $m_1 + m_2$ and m_3 to construct a mathematical model of a hydropulse vibration device by dividing HL [9] to masses $m_1 + m_2$ and m_3. As a result of this reduction, we obtain four simple dynamic models of the straight (Fig. 3a, b) and reverse (Fig. 3c, d) masses $m_1 + m_2$ and m_3.

Using the D'Alembert principle [1, 9], based on simplified dynamic models (Fig. 3), we make the differential equations of motion of the locking element 1.2 PPG 1 and the plug-valve 1.5 (mass $m_1 + m_2$) and the cutter 3 (mass m_3) (Fig. 1) during moves:

$$\text{direct } (x_{01} \geq x_{0r} > x_{02})\_$$

$$\begin{cases} m_{\Sigma 1} \cdot \ddot{y}_{1\Pi} = U_{01(02)} \cdot k_{0r}\left(x_{A_{1(2)}} - y_{1\Pi}\right) - k_1(y_{1\Pi} + y_{01}) - k_2(y_{1\Pi} + y_{02}) \\ -C_1\dot{y}_{1\Pi} - U_{01(02)}^{0,25} \cdot C_0\left(\dot{x}_{A_{1(2)}} - \dot{y}_{1\Pi}\right); \\ m_3\ddot{y}_{2\Pi} = U_{03} \cdot k_{0r}(x_{A_3} - y_{2\Pi}) - k_3(y_{2\Pi} + y_{03}) - k_4(y_{2\Pi} + y_{04}) \\ -U_{03}^{0,25} \cdot C_0(\dot{x}_{A_3} - \dot{y}_{2\Pi}) - R - \overline{F_y} - C_2 \cdot \dot{y}_{2\Pi}; \end{cases} \tag{7}$$

reverse $(\boldsymbol{x}_{02} \geq \boldsymbol{x}_{0r} \geq 0)$_

$$\begin{cases} m_{\Sigma 1} \cdot \ddot{y}_{13} = k_1(y_{01} + h_{\text{B}} - y_{23}) + k_2(y_{02} + h_{\text{B}} + h - y_{23}) \\ -U_{02(01)}k_{0r}\left[x_{A_{2(1)}} - (h_{\text{B}} - y_{13})\right] - U_{01(02)}^{0,25} \cdot C_0\left(\dot{x}_{A_{2(1)}} - \dot{y}_{23}\right) - C_1\dot{y}_{13}; \\ m_2 \cdot \ddot{y}_{23} = k_3(y_{03} + h_p - y_{23}) + k_4(y_{04} + h_p - y_{23}) - U_{03} \cdot k_{0r} \\ \cdot\left[x_{A_3} - (h_p - y_2)\right] - R - C_2 \cdot \dot{y}_{23} - U_{03}^{0,25} \cdot C_0(\dot{x}_{A_3} - \dot{y}_{23}), \end{cases} \tag{8}$$

where

$$x_{01} = p_1 \cdot A_0 \cdot k_{0r}^{-1}; \tag{9}$$

$$x_{02} = p_2 \cdot A_0 k_{0r}^{-1} = x_{01} \cdot U_{21}^{0,5} + k_1 h_{\text{B}} U_{02}^{-0,5} \tag{10}$$

respectively, the boundary deformations of the HL (formulas (9) and (10) are obtained by comparing dependence (1) and (2); $U_{21} = A_1^2 \cdot A_2^{-2}$ – the internal gear ratio in PPG 1 between the first and second sealing stages of the shut-off element 1.2 and the bushing-valve 1.5 (see Fig. 1); $A_1 = \pi d_1^2/4 \approx 0,785 d_1^2$, $A_2 = \pi d_2^2/4 \approx 0,785 d_2^2$ - cross-sectional areas, respectively, of the first and second stages of sealing PPG 1 (the second stage of sealing is fully realized after passing the sleeve-valve 1.5 gap h and pushing it into the collar SS1 and moving these parts as a whole); for: $0 < y_{1\Pi} < h - U_{01(02)} = A_1^2 \cdot A_0^{(-2)}$; $m_{\Sigma 1} = m_1$; $h \leq y_{1n} \leq h_в - U_{01(02)} = A_2^2 \cdot A_0^{-2}$; $m_{\Sigma 1} = m_1 + m_2$; $\dot{x}_{A_{1(2)}} = A_2^2 \cdot A_0^{-2}$; ($x_{A_{1(2)}}$ and $\dot{x}_{A_{1(2)}}$ determined according to the above changes $y_{1\Pi}$ and $U_{01(02)}$); for: $0 \leq y_{13} < h_в - h - U_{02(01)} = A_2^2 \cdot A_0^{-2}$; $m_{\Sigma 1} = m_1 + m_2$; $h_{\text{B}} - h < y_{13} \leq h_{\text{B}} - U_{02(01)} = A_1^2 \cdot A_0^{-2}$; $m_{\Sigma 1} = m_1$; $x_{A_{2(1)}} = x_{0r} \cdot U_{02(01)}$; $\dot{x}_{A_{2(1)}} = \dot{x}_{0r} U_{02(01)}$ ($x_{A_{2(1)}}$ and $\dot{x}_{A_{2(1)}}$ are determined by changes y_{13} and $U_{02(01)}$); $U_{03} = A_{сеф}^2 \cdot A_0^{-2}$; $x_{A_3} = x_{0r} \cdot U_{03}$; $x_{A_3} = \dot{x}_{0r} \cdot U_{03}$; $y_{1n}, y_{13}, 0 \leq y_{23} \leq h_p, 0 \leq y_{23} \leq h_p$, $\dot{y}_{1n}, \dot{y}_{2n}, \dot{y}_{13}, \dot{y}_{23}, \ddot{y}_{1n}, \ddot{y}_{2n}, \ddot{y}_{23}$ - respectively the current coordinates and accelerations of masses $m_1 + m_2$ and m_3; $x_{A_{2(1)}}, x_{A_3}, \dot{x}_{A_{1(2)}}, \dot{x}_{A_{2(1)}}, \dot{x}_{A_3}$ - respectively, the deformations and deformation rates of the HL are reduced to the corresponding cross-sectional areas of the sealing stages of the PPG and the power head 2 (see Fig. 1).

Differential equations of systems (7) and (8), in order to exclude free terms, can lead to the form of equations that describe the forced vibrations of masses $m_1 + m_2$ and m_3 in the form and content under the action of the variable amplitude of the linear

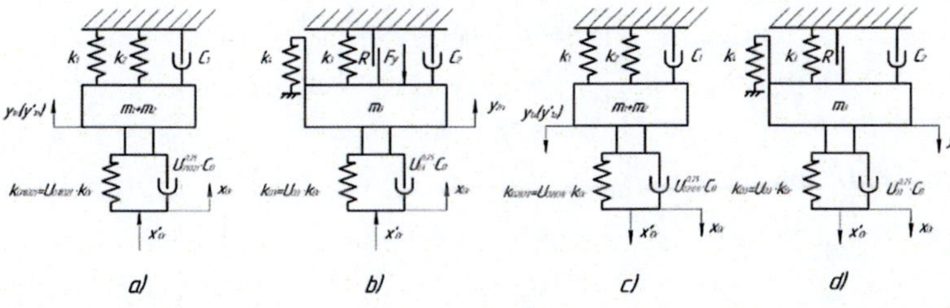

Fig. 3. Simplified dynamic models of forward (a, b) and reverse (c, d) masses

deformation of the HL. This adjustment is accomplished y_{1n}, y_{2n}, y_{13} and y_{23} by replacing other variables z_{1n}, z_2, z_{13} and z_{23} not changing the nature and dynamics of mass movement $m_\Sigma = m_1 + m_2$ and m_3:

$$\begin{cases} z_{1n} = y_{1n} + \omega_{\Sigma 1}^{-2}\left(\omega_{01}^2 \cdot y_{01} + \omega_{02}^2 \cdot y_{02}\right); z_{2n} = y_{2n} + \omega_{\Sigma 2}^{-2}\begin{bmatrix} \omega_{03}^2 \cdot y_{03} + \omega_{04}^2 \cdot y_{04} \\ + (R + \overline{F}_y) m_3^{-1} \end{bmatrix}; \\ z_{13} = y_{13} - \omega_{\Sigma 1}^{-2}\left[\omega_{01}^2 (y_{01} + h_B) + \omega_{02}^2 (y_{02} + h_B + h) + \omega_{01}^2 \cdot U_{01(02)} \cdot h_B\right]; \\ z_{23} = y_{23} - \omega_{\Sigma 2}^{-2}\left[\omega_{03}^2 (y_{03} + h_p) + \omega_{04}^2 (y_{04} + h_p) + \omega_{p_2}^2 \cdot U_{03} \cdot h_p + (R + \overline{F}_y) m_3^{-1}\right], \end{cases} \tag{11}$$

where $\omega_{\Sigma 1} = \sqrt{\omega_{p_1}^2 U_{01(02)} + \omega_{01}^2 + \omega_{02}^2}, \omega_{p_1} = \sqrt{k_{0r} \cdot m_{\Sigma 1}^{-1}}; \omega_{01} = \sqrt{k_1 \cdot m_{\Sigma 1}^{-1}};$ $\omega_{02} = \gamma^{-1}\delta \cdot \omega_{01};$ $\omega_{\Sigma 2} = \alpha \omega_{\Sigma 1};$ $\omega_{p_2} = \gamma^{-1}\omega_{p1};$ $\omega_{03} = \gamma_1^{-1}\delta_1 \omega_{01};$ $\omega_{04} = \gamma_2 \cdot \delta_2 \omega_{01};$ $\alpha = \left\{U_{\Gamma\Pi}\left[1 + \omega_{01}^2 \omega_{\Sigma 1}^{-2}\left(\gamma^{-1}\delta U_{\Gamma\Pi}^{-1} - 1\right)\right]\right\}^{0,5};$ $\gamma = m_{\Sigma 1}/m_3;$ $\delta = k_1/k_3;$ $\delta_1 = k_1/k_2;$ $\delta_2 = k_1/k_4;$ $\gamma_1 = m_{\Sigma 1}/m_3;$ $\gamma_1 = \gamma_2$ - accordingly, the natural frequencies of the PPG device, defined relative to the mass: PPGI1 - HL system; HL reduced to a mass $m_{\Sigma 1}$, locking element 1.2; valve sleeve 1.5 (see Fig. 1), powerhead system 2 - HL; HL reduced to a mass m_3; SS 2.6 relative to mass $m_{\Sigma 1}$; bellows 2.3 relative to the mass $m_{\Sigma 1}$; $\gamma_1, \delta, \delta_1, \delta_2, \gamma_2$ - the ratio between the relevant parameters; $U_{\Gamma\Pi} = U_{03}/U_{01(02)}$ – is the internal gear ratio between the powerhead 2 and the PPG 1.

Substituting in (7) and (8) the variables y_{1n}, y_{2n}, y_{13} and y_{23} the variables z_{1n}, z_2, z_{13} and z_{23}, after algebraic transformations, we obtain for the masses $m_{\Sigma 1}$ and m_3:

$$\text{direct}(x_{02} \geq x_{0r} \geq 0)\_$$

$$\begin{cases} \ddot{z}_{1n} + 2\beta_{1n}\dot{z}_{1n} + \omega_{\Sigma 1}^2 z_{1n} = \omega_{p_1}^2 U_{01(02)}^{0,5} x_{0r}; \\ z_{2n} + 2\beta_{2n}\dot{z}_{2n} + \alpha\omega_{\Sigma 1}^2 z_{2n} = \gamma^{-1}\omega_{p_1}^2 U_{03}^{0,5} x_{0r}, \end{cases} \tag{12}$$

$$\text{reverse}(x_{02} \geq x_{0r} \geq 0)\_$$

$$\begin{cases} \ddot{z}_{13} + 2\beta_{13}\dot{z}_{13} + \omega_{\Sigma 1}^2 z_{13} = -\omega_{p_1}^2 U_{02(01)}^{0,5} x_{0r}; \\ z_{23} + 2\beta_{23}\dot{z}_{23} + \alpha\omega_{\Sigma 1}^2 z_{23} = -\gamma^{-1}\omega_{p_1}^2 U_{03}^{0,5} x_{0r}, \end{cases} \quad (13)$$

where $\beta_{1\Pi} = 0,5m_{\Sigma 1}^{-1}\left[C_1 + U_{01(02)}^{0,25} \cdot C_0\left(\dot{x}_{A_{1(2)}} - \dot{y}_{1\Pi}^{-1} - 1\right)\right]$; $\beta_{2\Pi} = 0,5m_3^{-1}[C + U_{03}^{0,25} \cdot C_0\left(\dot{x}_{A_3} - \dot{y}_{2\Pi}^{-1} - 1\right)]$;

$$\beta_{13} = 0,5m_{\Sigma 1}^{-1}\left[C_1 + U_{02(01)}^{0,25} \cdot C_0\left(\dot{x}_{A_{2(1)}} - \dot{y}_{13}^{-1} - 1\right)\right];$$

$\beta_{23} = 0,5m_3^{-1}\left[C_2 + U_{03}^{0,25} \cdot C_0\left(\dot{x}_{A_3} - \dot{y}_{23}^{-1} - 1\right)\right]$; - variable damping factors during the mass movement $m_{\Sigma 1}$ and m_3. During the operating cycle of the device PPG, the natural circular frequency changes as the gear ratio $\omega_{\Sigma 1}$ changes $U_{01(02)}$ and $U_{02(01)}$ at different time intervals of movement of the links of the PPG 1 and the powerhead 2 (see description of parameters to systems (7) and (8)).

To complete the mathematical model of the PPG device for radial vibration turning to the system of differential equations of motion of the mass $m_{\Sigma 1}$ and m_3 it is necessary to add the equation of connection - the equation of energy flow to move the locking element 1.2 and the sleeve-valve 1.5 PPG 1, cutter 3 (see Fig. 1) and overflow energy in tank *B* through open PPG 1, and conditions of uniqueness that describe the movement of units PPG 1 and cutter 3 at characteristic intervals (see Fig. 2).

Typically, systems of nonlinear equations of type (12) and (13) are solved and investigated by numerical methods using various computer applications, such as MATLAB, where process time is the main explicit argument and is divided into a certain step. In this case, it is advisable to represent the deformation change of the HL in the form of simple dependencies [9] for the masses $m_{\Sigma 1}$ and m_3:

$$\text{direct}: x_{0r} = x_{01} - Q_0^{-1}{}_{\Sigma ln} \quad (14)$$

$$\text{reverse}: x_{0r} = x_{02} \pm Q_{\Sigma l3} \cdot t' \cdot A_0^{-1} \quad (15)$$

where $Q_{\Sigma ln}, Q_{\Sigma l3}$ – respectively, the flow of energy through the hydroline and open slit PPG 1, which corresponds to the first movement of the links of the HD device on the path of forward (l_n) and reverse (l_3) moves of masses $m_{\Sigma 1}$ and m_3; t, t'- the current time of change of a certain stage "l_n" (or "l_3") $Q_{\Sigma ln}$ or $Q_{\Sigma l3}$ (Fig. 2).

5 Conclusions

Based on the design scheme of the device, scientifically sound structure of assumptions and representation of the HL in the form of Kelvin-Voigt body, dynamic models of the PPG device for the forward and backward moves of the masses $m_{\Sigma 1}$ and m_3 interacting with the HL through the transmission numbers $U_{01(02)}$ and U_{03} are constructed.

Using the principle of the dismemberment of the model of forwarding and backward mass moves, four simple dynamic models are presented, based on which

D'Alembert principle is based on the mathematical model of the hydroimpulse drive of the device in the form of differential equations of motion of masses $m_{\Sigma 1}$ and m_3, conditions of unambiguity, which cause restrictions on the displacement of this energy carrier.

The proposed models can be installed and value the relationship between the individual time intervals characterizing pulse pressure, displacement PPG units 1 and of the tool 3 (see. Fig. 1). This makes it possible to adjust the design parameters of the device in accordance with the requirements of the cutting process.

The analysis and research of the proposed models of the hydroimpulse device for radial vibration turning, followed by experimental verification of the degree of adequacy of these models to the real system of the device, will allow to create a scientifically-grounded method for designing similar structures of devices with hydropulse drive.

References

1. Obertyukh, R., Slabkyi, A.: Devices for vibroturning on the basis of a hydropulse drive. VNTU, Vinnitsa (2015)
2. Danilchik, S.: Vibration turning of structural steels. BNTU, Minsk (2018)
3. Overcash, J.: Tunable, ultrasonic, vibration assisted diamond turning: Charlotte (2006)
4. Haidong, Z., Ping, Z., Wenbin, M., Zhongming, Z.: A Study on ultrasonic elliptical vibration cutting of Inconel 718. Shock Vib. **2016**, 3638574 (2016). https://doi.org/10.1155/2016/3638574
5. Tong, F.K., Zhang, Y., Zhang, F.H., Lu, J.Y., Gu, L.Z., Chen, Y.S.: Investigation of the mechanism of vibration cutting. Key Eng. Mater. **375–376**, 42–46 (2008). https://doi.org/10.4028/www.scientific.net/kem.375-376.42
6. Gursky, V., Kuzio, I.: Dynamic analysis of a rod vibro-impact system with intermediate supports. Acta mechanica et automatic **12**(2), 127–134 (2018)
7. Polishchuk, L.: Dynamics of the built in hydraulic drive of conveyors of mobile cars. VNTU, Vinnitsa (2015)
8. Povstyanoy, O.: Theoretical and practical background of computer modeling implementation for solving of problems of powder materials forming. In: The 2018 E-MRS Fall Meeting, Warshaw, Poland (2018)
9. Arkhipchuk, M.: Speed pressure impulson generator for controlling hydropulse drivers of vibration technological machines, Vinnytsia (2006)
10. Iskovich-Lototskiy, R., Ivanchuk, Y., Veselovsky, Y.: Modeling on the process of crushing other woodworking materials under the guidance of a new collection. Sci. Bull. NLTU Ukraine (2018)
11. Iskovich-Lototskiy, R., Ivanchuk, Y., Sevostyanov, I.: Determination of the working parameters of the hydropulse vibration press for flow vibration shock dehydration of moist disperse materials. Ind. Hydraul. Pneumatics **4** (2012)
12. Berezyuk, O.: Mathematical modeling of the hydraulic drive dynamics of the working bodies of the container reversal during the loading of solid household waste in the garbage truck. Bull. Khmelnytsky Natl. Univ. **5** (2013)
13. Obertyukh, R., Slabkyi, A., Marushchak, M., et al.: Dynamic and mathematical models of the hydraulic-pulse device for deformation strengthening of materials. In: Proceedings Volume 10808, Photonics Applications in Astronomy, Communications, Industry, and High-Energy Physics Experiments, Poland (2018). https://doi.org/10.1117/12.2501519

14. Obertyukh, R., Slabkyi, A., Marushchak, M.: etc: Method of design calculation of a hydropulse device for strain hardening of materials. Przeglagd elektrotechniczny **2019**(4), 65–73 (2019). https://doi.org/10.15199/48.2019.04.12
15. Neittaanmaki, P., Repin, S., Yuvinen, T.: Mathematical Modeling and Optimization of Complex Structures. Springer, Switzerland (2016). https://doi.org/10.1007/978-3-319-23564-6
16. Bulíček, M., Málek, J., Rajagopal, K.: On Kelvin-Voigt model and its generalizations. Evol. Equ. Control Theory **1**(1), 17–42 (2012). https://doi.org/10.3934/eect.2012.1.17
17. Gursky, V., Kuzio, I., Lanets, O., et al.: Implementation of dual-frequency resonant vibratory machines with pulsed electromagnetic drive. Przeglagd elektrotechniczny **2019**(4) (2019). https://doi.org/10.15199/48.2019.04.08
18. Xu, Y., Wan, Z., Zou, P., Zhang, Q.: Experimental study on chip shape in ultrasonic vibration–assisted turning of 304 austenitic stainless steel. Adv. Mech. Eng. **11**(8) (2019). https://doi.org/10.1177/1687814019870896
19. Śniegulska-Grądzka, D., Nejman, M., Jemielniak, K.: Cutting force coefficients determination using vibratory cutting. In: 10th CIRP Conference on Intelligent Computation in Manufacturing Engineering, Poland (2017). https://doi.org/10.1016/j.procir.2016.06.091
20. Zhang, C., Liu, W., Wang, S., et al.: Dynamic modeling and trajectory measurement on vibratory finishing. Int. J. Adv. Manuf. Technol. **106**, 253–263 (2020). https://doi.org/10.1007/s00170-019-04644-8

Dynamics of Fine Boring with Multicutting Console Drilling Rods

Gennadiy Oborskyi[1], Alexandr Orgiyan[1], Vladimir Tonkonogyi[1], Anna Balaniuk[1](✉), and Iryna Muraviova[2]

[1] Odessa National Polytechnic University, 1, Shevchenko Avenue, Odessa 65044, Ukraine
annabalanyuk24@gmail.com
[2] Odessa State Academy of Civil Engineering and Architecture, 4, Didrihson Street, Odessa 65029, Ukraine

Abstract. Operations concentration and overlapping is one of the basic reserves for precision enhancement and performance raising in machine-building technology. In case of fine and finishing boring of stepped holes, using multicutting console drilling rods allows raising machining quality, while ensuring holes alignment, processing from one setup, and maintaining axis linearity. In production conditions, two- and three stepped drilling rods are used more often. Alongside with that, regularities of changes in stepped console drilling rods oscillations, exerting a negative influence on longitudinal and transversal sections shapes, are still insufficiently studied. Dynamic features of such machining are determined based on examples of two-stepped holes boring, regularities of forced oscillations amplitudes changes were established. The dynamic system design model was provided that describes oscillations of spindle-drilling rod dual-mass elastic system limited to two cutting processes, and that includes coefficients of impact between cutters, as well as static and dynamic parameters of the elastic system and cutting modes. Design programs were developed, comparative results of calculations and experiments were provided, and further studies lines were discussed.

Keywords: Fine boring · Process dynamics · Multicutting stepped drilling rod · Oscillations · Closed dynamic system · Vibration resistance

1 Introduction

Fine hole boring, external surfaces grinding, groove milling, and butts cutting are performed mostly using high-performance bore finishing machines (HPBFM), ensuring high-precision of shape and location of machined surfaces. The basic portion of these machines park is special machines with setup for a specific part. The most widely spread are horizontal HPBFM with a movable table. Bridges are installed on these machines stands on which spindle heads are fixed. High dynamic quality is achieved based on rigidity studies and oscillations of the machine's bearing system and its spindle assemblies. It is established that spindle assemblies influence machining

V. Tonkonogyi et al. (Eds.): InterPartner 2020, LNME, pp. 577–587, 2021.
https://doi.org/10.1007/978-3-030-68014-5_56

precision with tools (console drilling rods) and tools for setting and workpiece clamping. These two subsystems rigidity and oscillations measuring was carried out on different models machines with improved spindle heads UAR-1 – UAR-4.

2 Literature Review

Console drilling rods are widely used for HPBFM-based drilling, although their application causes cutter bending compliance (process compliance), which leads to a reduction in vibration resistance. Alongside with that, console drilling rods ensure high performance when boring is performed on special machines. These features of console drilling rods led to the necessity of their oscillations study, which manifested itself in a large number of the works published [1–5]. Many works are oriented to vibrations measuring while specifying the quality of surface machined, which allows detecting cutting process problematic parameters [6].

The fine boring process is characterized by many features differing it from other methods of precision holes machining. Fine boring corrects deviation from workpiece hole axis linearity. Processing is performed on high speeds at which there is no build-up on the cutter front surface, which ensures the reduction in surface roughness. Precise holes boring is performed at light feeds ($\sim$up to 0.12 mm/r) and small cutting depths ($\sim$up to 0.8 mm; in this case, cutting section occurs extremely small (0.002–0.04 mm).

The features of the above described fine boring process were studied in experiments performed using single-cutter console drilling rods [7, 8]. Step boring results are virtually not available in reference literature; special drilling rods are to be specially used for step boring. The manufacturing practice testifies that 15% of total HPBFM numbers are set up for workpieces machining with two-stepped holes, whereas for three-stepped holes machining – about 5%. It should be noted that stepped holes geometrical parameters and their design features should manifest themselves in machining process flowsheets.

Fine boring performance can be significantly raised based on operations concentration and overlapping. For the implementation of this reserve, multi-stepped console drilling rods can be introduced into the fine drilling practices. It should be noted that step boring dynamics are insufficiently studied, which leads to insufficient use of stepped drilling rods in metal processing. That is why, to enhance the quality of stepped holes processing, not only static but dynamic methods shall be used in order to raise vibration resistance and fine boring precision, which is a rather urgent problem. In case of a fine step, boring dynamic interactions occur between process system elements that require considering the cutters inter influence, lengths and diameters ratio of console drilling rods steps and considering cutting process parameters that determine machining modes.

Under current conditions subject to ever-growing stiff requirements for bored holes quality (for instance, increase in dimensional accuracy, and for errors of transversal and longitudinal sections shapes and surface layer condition) solution of dynamics problems in machining process manifests itself in the line – process dynamics which is a part of machine-building technology [9].

Thus, when researching processes of multicutting fine boring and during dynamic models formations for calculation methods development, potentially unstable oscillations modes should be studied, their physical nature, as well as time and spatial forms, should be taken into account. Oscillation occurring in a general case, contain a combination of free, forced, parametrical, and self-exciting oscillations (self-oscillations). For different boring conditions, each of these oscillation types can occur dominating. This work investigates oscillations in the elastic-dissipative inertial system (EDIS), within the framework of process dynamics, while taking into account return coupling of cutting, friction processes in electrical motors, etc., which are called working processes (WP) in machines dynamics [9–11].

We should also note that the use of multicutter console drilling rods allows to perform boring at one workpieces setup, which leads to precision enhancement and reduction in the changeover number. The number of publications in technical literature dedicated to fine step boring is relatively small, and in most research works, designs of multicutter boring heads (or special drilling rods) are discussed [7, 12]. In some works, vibration resistance of the boring process is characterized by resultant limit cutting depth in the absence of higher vibrations [13]. Dynamic features of boring two- and three-stepped holes are just mentioned in the majority of works published, or not analyzed at all, oscillations excitation sources are not studied, nor regularities of their amplitudes change, liquids changes, own frequencies, and drilling rods geometry parameters that affect vibration resistance.

In work [14] a study of dynamic effects impact while cutting on hole shape errors under the effect of oscillations during the cutting process was carried out; longitudinal and transversal cross-section errors were formally described; the need for research in the field of multicutter boring and the method for calculating the amplitudes of tool's forced oscillations was emphasized.

The development of a hole boring quality enhancement method in non-rotational parts by a targeted combination of rigidity parameters of the processing system is described in work [15].

In the research study [16] the method for increasing the bending stiffness of the cantilever boring bar during fine boring of long holes with small diameter is proposed and confirmed by experimental studies.

Summarizing results of the published research on step boring, we should note that the absence of necessary scientific substantiation for stepped drilling rods designing hinders their introduction into metal processing practice.

3 Research Methodology

The main aim of the work is the experimental study of forced oscillations amplitudes regularities and their main excitation sources, as well as the development of a closed dynamic model of multicutting fine boring process for the theoretical calculation of forced oscillations amplitudes.

Results of experiments for boring with two-stepped console drilling rods are provided. Fig. 1 demonstrates a draft of a two-stepped drilling rod.

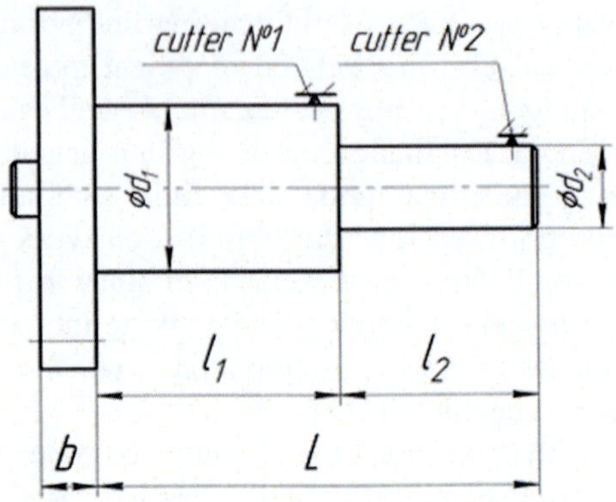

Fig. 1. A draft of two-stepped drilling rod.

Drilling rods of different sizes were used in experimental studies with changes in total length, diameters and lengths ratios; cutting depth and speed were changed as well. Testing bench and instrumentation for oscillations research using non-rotational drilling rods are provided in Fig. 2.

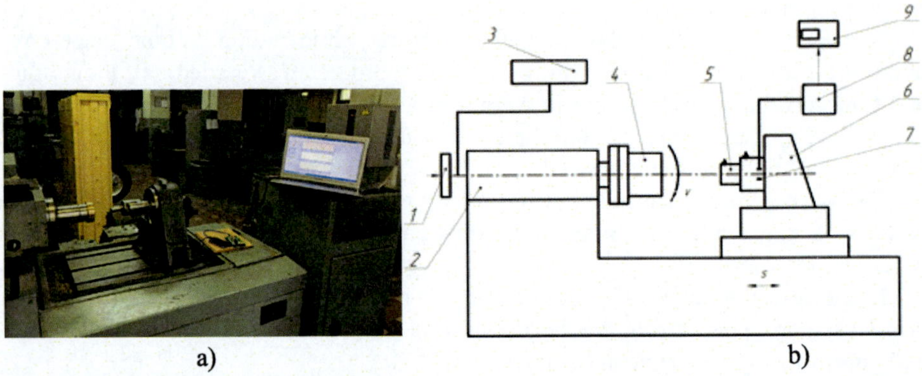

Fig. 2. General arrangement (a) and experimental testing bench diagram (b) for oscillations study using non-rotational drilling rods: 1 – tachometer; 2 – spindle head; 3 – frequency converter; 4 – workpiece; 5 – drilling rod; 6 – device; 7 – tensoresistors; 8 – oscillations recording device (ORD); 9 – personal computer (PC).

The processed workpiece (4) with two stepped holes was installed on the spindle flange coaxially with its axis. Two-stepped console drilling rod (5) for holes drilling in a workpiece, was fixed in the device (6) also coaxially with the spindle axis. Two pairs of tensoresistors (7) were glued on drilling rod (5) next to the attachment. Tensoresistors and ORD were used for measuring drilling rod oscillations in two planes. Two pairs of tensoresistors were turned on in a differential scheme and ensured a reliable process measuring within the band of 0–7 kHz. The above frequency band is quite sufficient, as characteristic frequencies of drilling rods bending oscillations are within 0.4–3.0 kHz band.

ORD (8) consists of 4 tensometric sensors that are connected between each other using bridge connection, Discovery STM32L0538-DISCO microcontroller, tensoamplifier, microCD memory card, and Bluetooth radio module.

The device design distinctive feature is that signal is recorded not by PC, but by Discovery STM32L0538-DISCO microcontroller, which is governed by PC using Bluetooth radio module. ORD calibration was carried out concerning drilling rod static displacement in the section of the cutter, which is most distant from the flange.

In the experiment described below, regularities of changing of forced oscillations amplitudes of drilling rods having equal length L were determined for constant value d_1 and changing d_2 and l_2 (ref. to Fig. 2). The experiment described allows us to determine the effect of d_2 diameter change and increase in l_2 console length effect on regularities of forced oscillations level changes when cutters are operated simultaneously.

Radial stiffness values and own frequencies of tested drilling rods are determined whose values are provided in Fig. 3a.

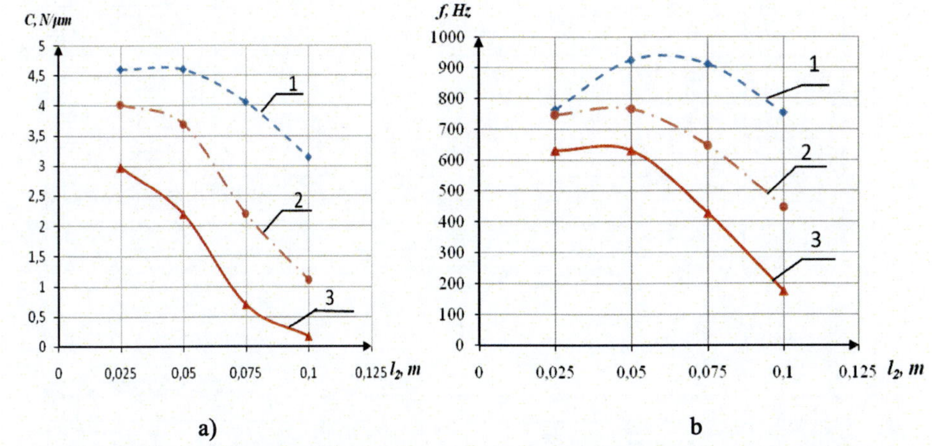

Fig. 3. Experimental values of radial stiffnesses (a) and own frequencies (b) of tested drilling rods when steps lengths are changed; 1 – d_1 = 0,06 m, d_2 = 0,04 m, L = 0,15 m; 2 – d_1 = 0,06 m, d_2 = 0,025 m, L = 0,15 m; 3 – d_1 = 0,06 m, d_2 = 0,015 m, L = 0,15 m.

Experimental results indicate the presence of extremums in the graph of their own frequencies change. Frequencies of drilling rods' own oscillations were determined by damped oscillations oscillograms. Drilling rods free oscillations were excited by an impact in cutter's section.

4 Results

Figure 4 shows experimental results for the described drilling rod in case of cutters No. 1, 2 separate operation, in case of their joint operation.

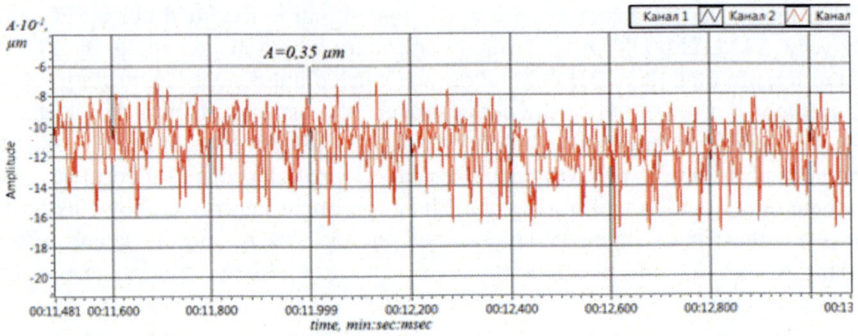

a

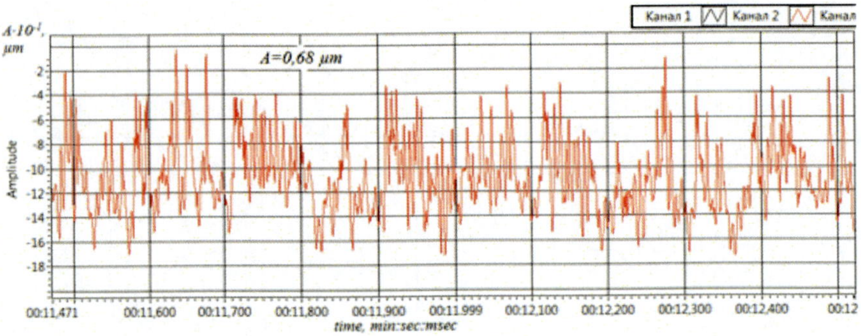

b

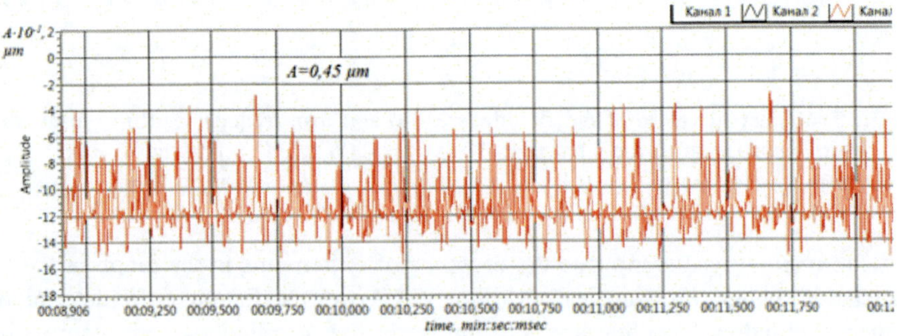

c

Fig. 4. Oscillograms of oscillations amplitudes for cutter No. 1 operation (a), No. 2 – (b) and for cutters No. 1 and No. 2 simultaneous operation (c); L = 0,15 m, l_1 = 0,075 m, l_2 = 0,075 m, d_1 = *0,06* m, d_2 = *0,04* m, processed material is steel 45, cutting modes: n = 1,250 rpm, s = 0,09 mm/r, $t_1 = t_2$ = 0,05 mm.

It should be noted that in this series of experiments, nontrivial results were found which are related to a feature of dynamic interactions when boring is performed with two-stepped drilling rod: at the ratio of a certain length of the steps (in this case equal to 1), the joint operation of two cutters leads to a decrease in the amplitudes of the forced oscillations as compared to the operation of one cutter No. 2. This fact testifies to mutual damping of cutting processes for drilling rod's some parameters and the cutters' joint operation. Alongside that, we detected the cases of increase in oscillation levels in case of cutters No. 1 and No. 2 joint operation compared with the operation of one cutter No. 2 (Table 1). The experiment described confirms the necessity of dynamic calculations when designing two-stepped drilling rods.

Table 1. Experimental values of forced amplitudes oscillations for cutters simultaneous and separate operation.

Drilling rod parameters		$L = 0.15\ m,\ d_1 = 0.06\ m,\ d_2 = 0.04\ m,$		
Steps lengths		Oscillations amplitude		
$l_1,\ m$	$l_2,\ m$	For cutter No. 1 operation $A,\ \mu m$	For cutter No. 2 operation $A,\ \mu m$	For simultaneous operation of cutters No. 1 and No. 2 $A,\ \mu m$
0.1	0.05	0.15	0.35	0.25
0.075	0.075	0.35	0.68	0.45
0.05	0.1	0.15	0.25	0.35

Experimental results testify to non-monotonic changes in forced oscillations amplitudes and on the significant influence of elastic system rigidity on vibration resistance: decrease in d_2 diameter leads to an increase in oscillation levels.

For consummating forced oscillations amplitude calculations for multicutting fine boring, spindle assembly design model was developed, which takes into account spindle compliance contribution to cutters displacement, and spindle inertia characteristics effect on the system 39 s own frequency [17].

Experimental data relating to boring (turning) testify to the fact that damping in the spindle-drilling rod subsystem is structural by nature and proportional to the rate of elastic displacements. Assuming that all oscillations occur at a specific ratio of nodalization diagram different points (that is when the spatial form is preserved) when building a model, the damping element can be associated with arbitrary spindle point. Therefore, the model was chosen in which the damping element is applied to the mass normalization point m_0. Adjusted for dynamic characteristics of cutting processes, forced oscillations in the closed system when turning is performed by two-stepped drilling rod and cutters No. 1 and No. 2 simultaneous operation, are described by Eqs. (1):

$$\alpha_{00} m_0 \ddot{y}_0 + \alpha_{00} b \dot{y}_0 + y_0 + \alpha_{02} m \ddot{y}_2 = \alpha_{00} P_0 \sin \omega t + \alpha_{02} \mu P_{z2} + \alpha_{01} \mu P_{z1},$$

$$\alpha_{22} m \ddot{y}_2 + y_2 + \alpha_{02} m_0 \ddot{y}_0 + \alpha_{02} b \dot{y}_0 = \alpha_{02} P_0 \sin \omega t + \alpha_{12} \mu P_{z1} + \alpha_{22} \mu P_{z2},$$

$$T_p \dot{P}_{z1} + P_{z1} = -K_{p1} y_2 \frac{\alpha_{12}}{\alpha_{22}},$$
$$T_p \dot{P}_{z2} + P_{z2} = -K_{p2} y_2,$$

$$m = m_{red} = 0,243 \cdot \left[m_1 \frac{l_1^3}{(l_1 + l_2)^3} + m_2 \right], \quad (1)$$

where: y_o, y_2 – elastic displacements of masses m_o and m_2; $\ddot{y}_2, \ddot{y}_0, \dot{y}_0$ – second- and first-time derivatives; P_{z1}, P_{z2} – components of cutting force on cutters; T_p – the inertial constant of chip formation; K_{p1}, K_{p2} – cutting coefficients on respective cutters; α_{ik}– movement in *i*-section under the single force applied at κ-section; μ – coefficient of chips friction on the cutter's front edge.

Equations (1) system solution we find in the form of (2):

$$y_0 = a_{11} \cos \omega t + a_{12} \sin \omega t,$$
$$y_2 = a_{21} \cos \omega t + a_{22} \sin \omega t,$$
$$P_{z1} = a_{31} \cos \omega t + a_{32} \sin \omega t,$$

$$P_{z2} = a_{41} \cos \omega t + a_{42} \sin \omega t, \quad (2)$$

where a_{11}, a_{12}, a_{21}, a_{22}, a_{31}, a_{32}, a_{41}, a_{42} - are unknown coefficients calculated using the determinant.

Having substituted (2) into (1), we obtain the algebraic equation system whose solution at a variation from ω 0 to 2,000π allows to build elastic system's frequency-amplitude characteristic:

$$A(\omega) = \sqrt{a_{21}^2 + a_{22}^2}.$$

The given equations system allows not only to calculate oscillations for the cutters No. 1 and No. 2 simultaneous operation but for their individual operation. Motion equations are composed of three-stepped console drilling rods' oscillations calculation.

A program was developed, ensuring calculation of forced oscillations resonance amplitudes. Figure 5 provides some comparative results of the calculations and experiments described above. Calculations are performed based on the following parameters: $T_p = 4,6 \cdot 10^{-4} c$;$в = 13,1 \cdot 10N \cdot c/m$; $K_{p1} = K_{p2} = 2 \cdot 10^{-6}$N/m; $L =$ 0,15 m; $P_0 = 4N, = 2,5N, = 1N$; $\mu = 0,6$, $C_{ang} = 8 \cdot 10^7$N · m.

Theoretical calculation supports experimental results (Fig. 5) and testifies to amplitudes values lessening for the cutters' joint operation due to mutual damping of cutting processes at specific ratios of step lengths. Analysis of oscillation amplitudes confirms the known regularity when an increase in steps rigidity at cutting leads to a decrease in oscillation levels. The experimental results and calculation results difference lies within the 20% range.

Results analysis testifies that amplitudes values decrease proportionally to diameter d_2 increase due to an increase in the step rigidity. Alongside that, in all cases, the regularity of amplitudes change is non-monotonic, maximums, and minimums values are alternating due to complex dynamic interactions in the elastic system.

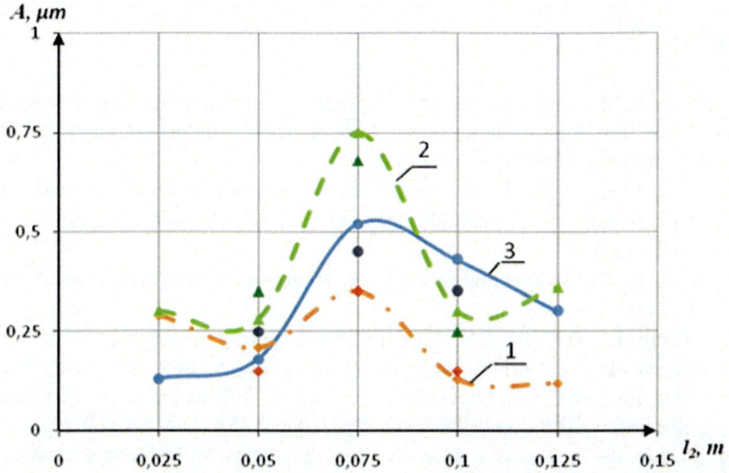

Fig. 5. Comparative results of design and experimental amplitudes of forced oscillations for the cutters joint operation, $L = 0{,}15$ m, $d_1 = 0{,}06$ m, $d_2 = 0{,}04$ m, material processed – steel 45, cutting modes $n = 1{,}250$ rpm, $s = 0{,}09$ mm/r, $t_1 = t_2 = 0{,}05$ mm.; calculation: 1 - oscillation amplitudes for cutter No.1 operation; 2 – for cutter No. 2 operation; 3 – for the cutters No. 1 and No. 2 joint operation; experiment – individual points.

5 Conclusion

Calculations and experimental results reflect complex and dynamic interactions occurring in HPBFM elastic system, limited to two cutting processes, where the interactions are determined by joint changes in influence coefficients, own frequencies, console drilling rods parameters, and by cutting processes characteristics. In all experiments conducted, these interactions lead to a non-monotonic change in oscillations amplitudes and are characterized by the alternation of higher and lower levels, and the presence of extremums. Thus, when designing multicutting console drilling rods and forecasting amplitudes values in specific processing conditions, not only static but dynamic calculations as well, should be performed. The following is supposed to be done in further research:

1. To establish correlation relationships between the level of forced oscillations and fine boring errors (shape precision of longitudinal and transversal section, by surfaces roughness and surface layer characteristics).
2. It is expedient to carry out a number of studies aimed at reduction in bending oscillations of two- and three-stepped console drilling rods using effective vibration dampers that allow reducing not only nonmonotonicity of amplitudes changes but also to determine the achievable processing precision.
3. For the purpose of introduction of special stepped boring drilling rods, it is supposed that vibration resistance criteria should be worked out based on their limit compliance determining.

References

1. Bansal, A., Law, M.: A receptance coupling approach to optimally tune and place absorbers on boring bars for chatter suppression. In: 8th CIRP Conference on High Performance Cutting, pp. 167–170 (2018)
2. Suyama, D.I., Diniz, A.E., Pederiva, R.: The use of carbide and particle-damped bars to increase tool overhang in the internal turning of hardened steel. Int. J. Adv. Manuf. Technol. **86**, 2083–2092 (2016)
3. Bansal, A., Law, M.: A receptance coupling approach to design damped boring bars. In: COPEN (2017)
4. Lijia, L., Xianli, L., Yuanhong, L.: Non-uniform sampling finite-time control for networked control systems via event-driven transmission. Adv. Mech. Eng. **8**(4), 1–0 (2016)
5. Gulyaev, V.I., Lugovoi, P.Z., Glushkova, O.V., et al.: Self-excitation of torsional vibrations of long drillstrings in a viscous fluid. Int. Appl. Mech. **52**, 155–164 (2016)
6. Östling, D., Jensen, T., Tjomsland, M., Standal, O., Mugaas, T.: Cutting process monitoring with an instrumented boring bar measuring cutting force and vibration. In: 8th CIRP Conference on High Performance Cutting, pp. 235–238 (2018)
7. Sørby, K., Østling, D.: Precision turning with instrumented vibration-damped boring bars. In: 8th CIRP Conference on High Performance Cutting, pp. 666–669 (2018)
8. Sørby, K., Sundseth, E.: High-accuracy turning with slender boring bars. Adv. Manuf. **3**(2), 105–110 (2015)
9. Oborsky, G.A., Orgiyan, A.A., Minchev, R.M., Balanyuk, A.V.: Dynamics problems in machine engineering technology. Cutting Tools Technol. Syst. **87**, 3–11 (2017). (in Russian)
10. Zhang, S.J., To, S., Zhang, G.Q.: Diamond tool wear in ultra-precision machining. Int. J. Adv. Manuf. Technol. **88**, 613–641 (2017)
11. Liu, C., Hu, J.: A magnetorheological hydrostatic guideway system for machining vibration control. J. Braz. Soc. Mech. Sci. Eng. **41**, 12 (2019)
12. Xianming, C., Tieliu, W., Mingming, D., Jing, W., Chen, J., Jun, X.: Analysis and prediction on the cutting process of constrained damping boring bars based on PSO-BP neural network model. J. Vibroengineering **19**, 878–893 (2017)
13. Guo, Y., Dong, H., Wang, G., Ke, Y.: Vibration analysis and suppression in robotic boring process. Int. J. Mach. Tools Manuf. **101**, 102–110 (2016)
14. Shen, N., Guo, Z., Li, J., et al.: A practical method of improving hole position accuracy in the robotic drilling process. Int. J. Adv. Manuf. Technol. **96**, 2973–2987 (2018)
15. Kalistru, V., Firsu, A.: Methodology for predicting deviations in the accuracy of the shape of the openings of thin-walled body parts. Actual Probl. Mech. Eng. **3**, 117–122 (2016)

16. Orgiyan, A., Kobelev, V., Ivanov, V., Balaniuk, A., Aymen, A.: Ensuring the bending stiffness of pre-compressed cantilever boring bars during fine boring. In: Tonkonogyi, V., et al. (eds.) Advanced Manufacturing Processes. Inter-Partner 2019. Lecture Notes in Mechanical Engineering, pp. 315–324. Springer, Cham (2020). https://doi.org/10.1007/978-3-030-50794-7_31
17. Balanyuk, A.V.: Oscillations in two-stepped console drilling rods at fine boring. Collection of scientific works (industry machine building, construction), 131–139 (2014). (in Russian)

Development of Calculation of Statistical and Dynamic Errors upon Fine Boring with Console Boring Bars

Alexandr Orgiyan[1(✉)], Gennadiy Oborskyi[1], Anna Balaniuk[1], Vladimir Tonkonogyi[1], and Predrag Dasic[2]

[1] Odessa National Polytechnic University, 1, Shevchenko Avenue, Odessa 65044, Ukraine
annabalanyuk24@gmail.com
[2] SaTCIP Publisher Ltd, Vrnjačka Banja 36210, Serbia

Abstract. Assembly and operational reliability of machines and mechanisms are significantly depending on the accuracy of sizes, forms, and positioning of parts. Systematic or accidental errors from set accuracy indexes define the error of edge metalwork. Related elastic displacements of edge and blank unit plus set movements of flexible system elements lead to change in the set trajectory of edge movement, causing shaping errors. In this paper, there are statistical and dynamic parts of errors resulting from reversible deformations. In this paper, there is provided calculation methodology for errors caused by reversible deformations under the impact of cutting forces. In this paper, there are considered the errors caused by uneven radial flexibility by the turn angle of the system spindle – boring bar caused by preliminary out-of-roundness of the bored hole concerning the spindle axis. There are provided the correlations for calculation of total errors upon boring stepped holes. In this paper, there are calculated dynamical errors based on the determination of amplitude of forced oscillations upon boring smooth and stepped holes. Computer programs were created to calculate separate component parts and also the total reversible error. This paper studies the dependency of influence indexes from geometrical parameters of console boring bars and also comparative results of statistical and dynamic errors.

Keywords: Statistical and dynamic errors · Reversible deformations · Stepped boring · Oscillations amplitude · Frequency · Spindle head

1 Introduction

Finishing boring machines (FBM) are efficiently used for the final processing of highly precise holes. It is noteworthy that the efficiency of the application of separate methods of increasing shaping accuracy depends on specific features of the type of blank unit mechanical processing and machine structure. On these machines, the following details different by shape and deformation properties: bullet casings, piston rods, connectors, pistons, and also hull parts, such as cylinder blocks, crankcases, etc. Horizontal FBMs with transversely and axially movable tables are the most common. On their body,

V. Tonkonogyi et al. (Eds.): InterPartner 2020, LNME, pp. 588–597, 2021.
https://doi.org/10.1007/978-3-030-68014-5_57

bridges are installed, on which spindle heads are mounted. The tasks of increasing dynamic quality and processing accuracy are solved by way of an increase in rigidity and oscillations of machine load-carrying elements and spindle heads. The main impact on processing accuracy is carried out by spindle assemblies with boring console bars fixed on them (spindle – boring bar subsystem) and also clamping devices (part-device subsystem). For providing the high technological reliability of FBM alignment, it is necessary to research the dynamic interaction of these subsystems to define optimal machine parameters.

The tasks of increasing the holes' machining accuracy are of high relevance that are certified with numerous results of published researches [1–3].

The increase in processing accuracy is linked with a decrease in cutting force and, more specifically, with a decrease in its radial content and also with an increase in oscillation amplitudes from the normal axis to the processed surface.

2 Literature Review

Despite numerous researches on the accuracy of mechanical processing using FBM, upon designing boring finishing operations, there is a common approach to the calculation of shaping expected errors even upon processing smooth holes. It is even more hard to estimate the accuracy of multi-edge boring of stepped holes.

Console boring bars, while providing for alignment of the bored hole and high productivity, have increased bending compliance, often leading to loss of vibration resistance [4, 5]. Noteworthy that bending compliance becomes variable upon rotation due to bumps in spindle bearing supports or resulting from the blank unit fixation conditions and its rigidity characteristics. Summarizing the results of the research, we should note that upon accuracy calculations, the following have been taken into account:

- impact of FBM general layout solutions [6];
- impact of edges geometry and material, cutting modes [7, 8];
- quality of spindle assemblies [9];
- FBM vibrational resistance and dynamic quality [10, 11].

In the machine-building technology such characteristic sources of tooling errors are distinguished: elastic deformations of the technological system under the influence of cutting and fastening forces [12]; errors of the blank unit and residual stresses in it [13]; geometrical and kinematic inaccuracies and machine setting errors; dimensional wear and tool manufacturing inaccuracies; temperature deformations of the technological system.

In this paper, the main attention is paid to elastic deformations of the technological system under the influence of cutting forces.

3 Research Methodology

Let us present the methods of calculation of errors and the main results obtained. Static and dynamic components of errors are distinguished when calculating the accuracy of machined holes at fine boring. Static components are characterized by a lower frequency of lower own frequency of the elastic system and a multiple to the rotation speed. Vibrations determine dynamic components at frequencies close to the machine's closed dynamic system frequencies.

These errors lead to errors from the roundness of the cross-sections and conicality in the longitudinal sections of the boring holes. Comparison of the minimum possible natural frequencies of boring bars vibrations with the maximum spindle speeds shows that the smallest number of periods per revolution can reach 5–6. It means that static errors of the cross-section form can have the oval shape (two periods per revolution), triangle (three periods per revolution), or four-angled (four periods per revolution). At the same time, the analysis of round diagrams of boring holes confirms that an oval most often characterizes low-frequency errors of the cross-sectional form. Therefore, static errors leading to hole oval-shaping have been studied.

In our work, static errors are divided into three components: Π_k – diameter error of the processed hole form due to anisotropy of the radial compliance of the system near the edge:

$$\Pi_K = 2(Kyy_{min\,max}) \tag{1}$$

where P_y – radial cutting force,

$K_{\min}, K_{\max}$ – minimal and maximal spindle compliance under the turning angle; Π_0 – diameter error of the processed hole form, caused by oval-shape of the hole in the blank unit:

$$\Pi_o = 2 \cdot K\left[P_y(t) - P_y(t - H_o)\right] \tag{2}$$

where H_o – the largest difference of hole radiuses in the blank unit (oval-shape). Π_e – diameter error of the form of the processed hole, caused by the displacement of the hole axis in the blank unit in relation to the spindle axis:

$$\Pi_e = 2 \cdot K \cdot P_y(t) - K\left[P_y(t + e) + P_y(t - e)\right]. \tag{3}$$

where K – radial compliance of the system close to the edge; e – eccentricity.

While passing from the diameter error measure to the radial, complete static error from roundness should be calculated using the formula:

$$\Delta R_{st} = 0,5 \cdot \sqrt{\Pi_e^2 + \Pi_0^2 + \Pi_k^2} \tag{4}$$

Taking into account that the cutting force upon fine boring can be presented as [8]:

$$P_y = C_{py} \cdot t^{X_{py}} \cdot y^{Y_{py}}, \tag{5}$$

where the attribute C_{py} and the exponents X_{py} and Y_{py} depend on the processed material, and while processing the correlations 1, 2, and 3, we shall represent the calculated error in the short form:

$$\Pi_j = C_{py} \cdot t^{X_{py}} \cdot s^{Y_{py}} \cdot \lambda_j, \tag{6}$$

Where $\lambda_j = K \cdot \varepsilon_j$, where ε_j corresponds to the source of statistical error:

$$\varepsilon_e = 2 - (1 + e/t)^{X_{py}} - (1 - e/t)^{X_{py}},$$
$$\varepsilon_o = 2[1 - (1 - H_o/t)^{X_{py}}],$$
$$\varepsilon_K = 2\Delta K/K. \tag{7}$$

A special program has been developed to calculate static errors.

Determining static errors upon multi-edge fine boring, we note that the total processing error for each edge is the sum of errors – own and additional from all working edges. The additional errors are calculated using influence coefficients [14].

The total error from the roundness of the cross-sectional shape, determined for this pick by the contribution of individual static errors, can be found as:

$$\Delta R_{ci} = 0,5\sqrt{\left[\sum_{i=1}^{n} \Pi_{ei} \cdot \frac{\delta_{ni}}{\delta_{ii}}\right]^2 + \left[\sum_{i=1}^{n} \Pi_{Ki} \cdot \frac{\delta_{ni}}{\delta_{ii}}\right]^2 + \left[\sum_{i=1}^{n} \Pi_{oi} \cdot \frac{\delta_{ni}}{\delta_{ii}}\right]^2} \tag{8}$$

where n – the quantity of edges; i – edge number.

Calculation of dynamic errors.

Out-of-roundness of the cross-section caused by forced oscillations of the edge installed on the boring bars are defined by the oscillations amplitude "A":

$$\Delta R_{\partial} = 2 \cdot A$$

hence upon oscillations, the radius of the incircle is decreased on "A", and the excircle is increased on "A". The sum of static and dynamic components of out-of-roundness errors is equal to:

$$\Delta R = \Delta R_{st} + \Delta R_{\partial}$$

Let us see the calculation of forced oscillations amplitudes on the example of boring two-stepped holes. For that purpose, we should make the calculation model and reconfirmed equations of closed dynamic system movement to calculate the forced oscillations amplitudes. Calculation model is shown in Fig. 1.

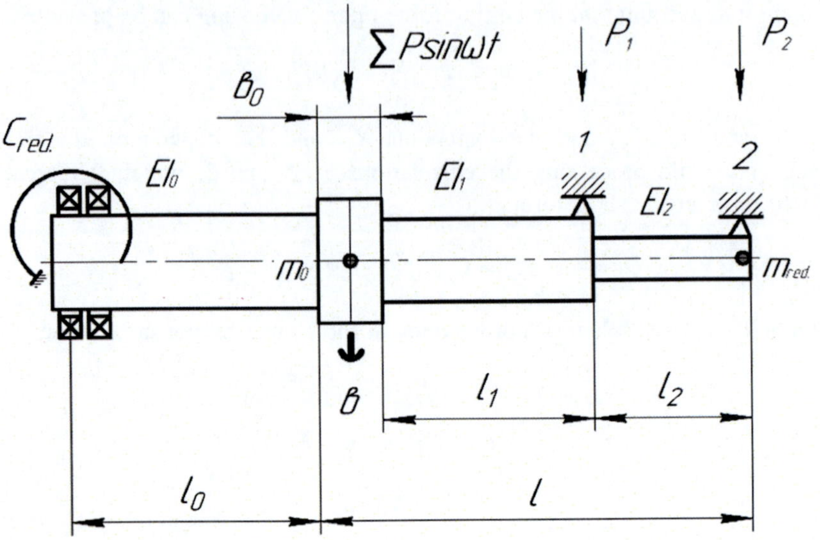

Fig. 1. Spindle – boring bar calculation model.

Where the elastic characteristics of the spindle are shown using the following parameters:

EI_0 (N·m^2) – bending stiffness of spindle section; l_0 (m) – length of the spindle console part, measured from the first support; $C_{red.}$ (N·m/rad) – reduced angulated stiffness of fixation of the frontal support in the middle section, reflecting the supports and spindle beam part bending resistance; m_0 (kg) – spindle inertial property resulting from mass, reduced to flange section.

Boring bar properties:

EI_1 (N·m^2) – cross-section stiffness of the first step; EI_2 (N·m^2) – cross-section stiffness of the second step; b_0, (m) – a sum of thicknesses of spindle flanges and boring bar (for spindle head UAR-2 – b_0= 0.04 m); m_{red} (kg) – inertial property of the boring bar, resulting from mass, reduced to second edge section; b – damping property in a flexible system.

Movement differential equations upon simultaneous work of two edges taking into account force harmonics from bumps in spindle bearing supports and also designing indexes, shall make out [14]:

$$\begin{cases} m_0\ddot{y}_0 + b\dot{y}_0 + \frac{\delta_{22}}{\delta}y_0 - \frac{\delta_{02}}{\delta}y_2 = 4\sin\frac{\omega_c}{9}t + 2\sin\frac{\omega_c}{18}t + 1,5\sin\frac{\omega_c}{24}t + \frac{\delta_{01}\cdot\delta_{22}-\delta_{02}\cdot\delta_{12}}{\delta}K_1P_{z1}; \\ m\ddot{y}_2 + \frac{\delta_{00}}{\delta}y_2 + \frac{\delta_{02}}{\delta}y_0 = K_2P_{z2} + \frac{\delta_{00}\cdot\delta_{12}-\delta_{01}\cdot\delta_{02}}{\delta}K_1P_{z1}; \\ T_p\dot{P}_{z1} + P_{z1} = -K_{p1}y_2\frac{\delta_{12}}{\delta_{22}}; \\ T_p\dot{P}_{z2} + P_{z2} = -K_{p2}y_2, \end{cases} \quad (9)$$

$$\delta = \delta_{00}\cdot\delta_{22} - \delta_{02}^2,$$

$$m = m_{np} = 0,243\cdot\left[m_1\frac{l_1^3}{l^3} + m_2\right].$$

where: y_0 – oscillating mass movement m_0; y_2 - oscillating movement of the edge the furthest to the flanges (edge No. 2); $\ddot{y}_2$ - second derivative by time; $\ddot{y}_0, \dot{y}_0$ – second and first derivative by time; P_{z1}, P_{z2} – component cutting forces on edges; T_p – chip formation permanent inertion; K_{p1}, K_{p2} – curring indexes on respective edges; δ_{iK} – movement in i-section under the single force applied at κ-section; K_1, K_2 – designing indexes; μ – index friction of chip to edge frontal cut.

Impact indexes for all cases of edges work shall be:

$$\delta_{00} = \frac{l_0^3}{3EI_0} + \frac{l_0^2}{c};$$

$$\delta_{01} = \frac{l_0^3}{3EI_0} + \frac{l_0^2}{c} + \left(\frac{l_0}{c} + \frac{l_0^2}{2EI_0}\right) \cdot L_1;$$

$$\delta_{11} = \frac{l_0^3}{3EI_0} + \frac{(l_0 + L_1)^2}{c} + \frac{L_1 \cdot l_0^2}{EI_0} + \frac{L_1^2 \cdot l_0}{EI_0} + \frac{l_1^3}{3EI_1};$$

$$\delta_{02} = \frac{1}{EI_0} \cdot \left[\frac{l_0^2}{2} \cdot \left(l_2 + L_1 + \frac{2}{3} l_0\right)\right] + \frac{l_0(l_0 + L_1 + l_2)}{c};$$

$$\delta_{22} = \frac{1}{EI_0} \cdot \left[(L_1 + l_2) \cdot l_0 \cdot \left(L_1 + l_2 + \frac{l_0}{2}\right) + \frac{l_0^2}{2} \cdot \left(L_1 + l_2 + \frac{2}{3} l_0\right)\right]$$

$$+ \frac{1}{EI_1}\left[l_2 \cdot l_1 \left(l_2 + \frac{l_1}{2}\right) + \frac{l_1^2}{2}\left(l_2 + \frac{2}{3} l_1\right)\right] + \frac{l_2^3}{3EI_2} + \frac{(l_0 + L_1 + l_2)^2}{c};$$

$$\delta_{12} = \frac{1}{EI_0}\left[L_1 \cdot l_0 \left(l_2 + L_1 + \frac{l_0}{2}\right) + \frac{l_0^2}{2}\left(l_2 + L_1 + \frac{2}{3} l_0\right)\right]$$

$$+ \frac{1}{EI_1}\left[\frac{l_1^2}{2}\left(l_2 + \frac{2}{3} l_1\right)\right] + \frac{(l_0 + L_1) \cdot (l_0 + L_1 + l_2)}{c}.$$

Indexes of impact indexes respond to edges numeration.

Calculated the values of two own oscillation frequencies of system $\omega_{1,2}$ and resonance values of forces oscillations amplitudes in a closed dynamic system:

$$A(\omega_{1,2}) = \sqrt{a_{21}^2(\omega) + a_{22}^2(\omega)} \tag{10}$$

4 Results

With the purpose of calculation of the dynamic processing deviations and estimation of the parameters of the stepped console boring bars, the impact of changes in the influence indexes on the force transfer distribution between the edges is researched.

Figure 2 shows the calculated dependencies for all six influence coefficients on the step length l_1 of the processed hole at the specified diameters and the total length of the boring bar.

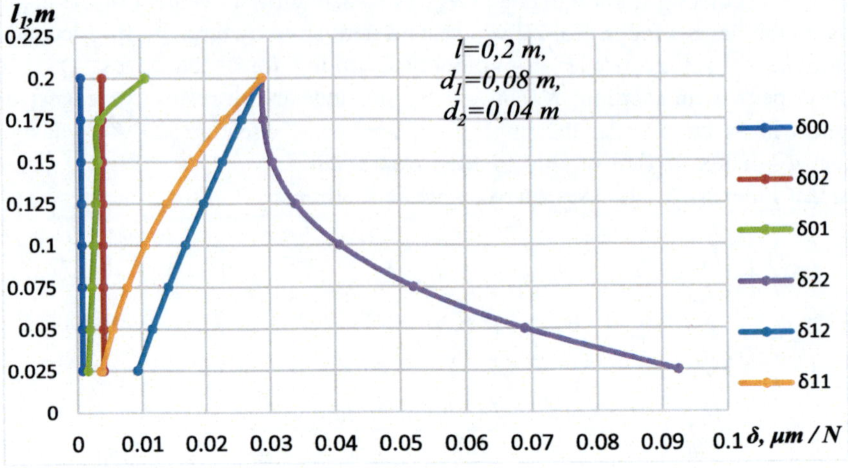

Fig. 2. Estimated impact indexes dependency δ_{ik} on boring bar steps length proportion.

From Eqs. (3), it is clear that the impact index δ_{00} does not depend on the design factors of the boring bar. Changes δ_{01} and δ_{02} are impacted only by the lengths of steps l_1 and l_2, changes δ_{11} and δ_{12} are impacted only by diameter d_1 and steps lengths l_1 and l_2, and impact index δ_{22} is influenced only by step lengths l_1 and l_2 and diameters d_1 and d_2.

The given values of influence coefficients characterize the flexibility of the boring bar. We can see that the impact index δ_{00} does not depend on the design parameters of the boring bar, but takes into account only the flexibility of the spindle assembly, so this coefficient remains constant for the selected spindle head size. The index δ_{02}, which characterizes the transmission of disturbances from the flange to the edge No. 2, depends only on the total length of the boring bar and, therefore, also has a constant value for each individual boring bar. Similarly, the index δ_{01} characterizes the perturbation transfer from the flange of the spindle to edge No. 1 and depends only on the length of the first stage - l_1. Changes in the value of this coefficient occur in a narrow range of values of small pliability and are located almost between the values of the coefficients δ_{00} and δ_{02}. The graphs of the δ_{11}, δ_{12} и δ_{22} indexes changes and should be considered separately. These three graphs come from a single point, as in this case, only one edge, the furthest from the flange, works. As already mentioned, the change of the coefficient values is determined by the value of EI_1 and the lengths of steps l_1 and l_2. Their dependence on the step lengths is monotonic, and the absolute values are close to each other. The coefficient of influence depends on the values of EI_1 and EI_2, as well as the lengths of steps l_1 and l_2. Its dependence on the lengths l_1 and l_2 varies monotonically, and the absolute values increase significantly with the length of the second stage. It should also be noted that the graphs δ_{22}, δ_{11} and δ_{12} are of different nature: based on the common point of the index δ_{22} values with the increasing length of the second step increases, and the values δ_{11} and δ_{22} monotonically decreases. The mutual influence of picks 1 and 2 is characterized by changes in the index value δ_{12}, which is between the coefficient values δ_{11} and δ_{22}. The study of the dependence of the influence indexes δ_{ik} on the geometrical parameters of the boring bar shows that an increase in pliability on the most distant pick leads to a reduction in the mutual influence of the picks.

The complete boring hole error from the roundness considering the static and dynamic components is shown in Fig. 3.

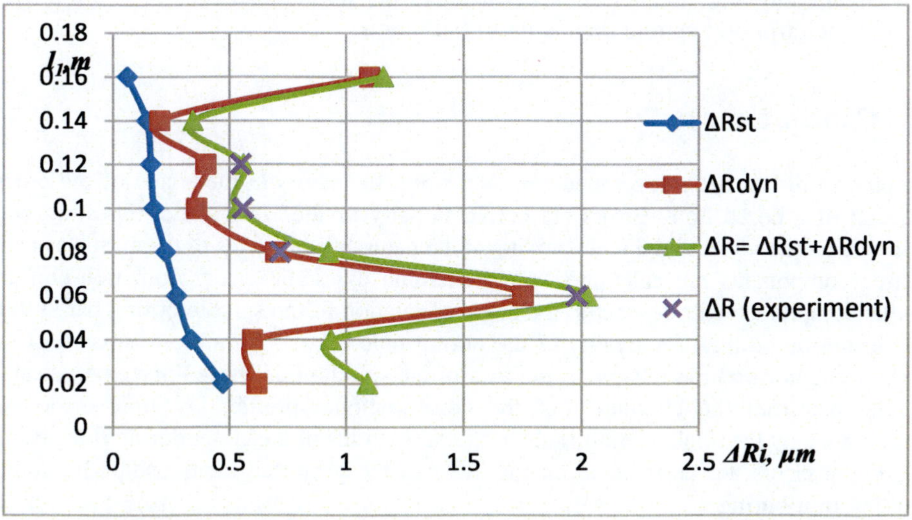

Fig. 3. Changes of error amounts upon multi-edge boring: processed material DIN C45 steel: do_1= 0,085 m, do_2 = 0,045 m, l = 0,16 m, s = 0,06 mm/r, n = 1500 min^{-1}, t_1= t_2= 0,05 mm.

It is notable that upon multi-edge boring, the relations of values of the minimum dynamic errors to values of total static errors are in a range from 5 to 8. It becomes obvious that upon multi-edge fine boring dynamic interactions influence errors in a decisive way. In absolute terms, static errors range from 0.03 to 0.24 μm, and dynamic errors range from 0.4 to 2 μm. A distinctive feature of changes in the total errors in the cross-section roundness upon multi-edge boring is the shifting of their increased and decreased values in accordance with the alternation of the amplitudes of the forced oscillations.

Calculations of total errors at the final reaming show that dynamic errors can be 5–6 times higher than static elastic errors. That is, dynamic interactions designing of console boring bars and for forecasting of vibration levels should be performed not only static but also dynamic calculations.

5 Conclusions

The developed calculation methodology that takes into account dynamic boring interactions clarifies and develops general ideas and methods of determining cross-section accuracy during multi-edge boring: static errors from the inaccuracy of the blank unit installation, the anisotropy of compliance with the rotation angle and errors from the roundness of the blank unit hole, as well as dynamic disturbances. Π_o and Π_e errors are caused by changes in cutting forces as a result of the variable cut-off allowance, while the properties of the spindle assembly determine Π_k error.

A distinctive feature of changes in the total errors in the cross-section roundness upon multi-edge boring is the shifting of their increased and decreased values in accordance with the alternation of the amplitudes of the forced oscillation.

Upon multi-edge fine boring the dynamic errors are dominant and 5-6 times higher than static errors (at cutting depths up to 0.05 mm).

6 Conclusion

A procedure has been developed for calculating the errors in the shape of the cross-section of a bored hole. Errors are conventionally divided into static, repeating with rotation frequency, and dynamic, arising at the natural frequencies of the elastic system. Three components are calculated that determine the static error: from non-uniform radial compliance with respect to the angle of rotation of the spindle-boring bar system Π_k, from the preliminary ovality of the boring hole Π_o and from the mismatch of the axis of the boring hole relative to the axis of the spindle Π_e. These three errors lead to deviations from the circularity of the cross-sectional profile. Dynamic errors are determined by the doubled amplitude of the oscillations of the cutter during fine boring. The total errors are determined on the basis of the proposed relationships for multi-cutting thin boring.

The software has been developed to design multi-cut thin boring operations and increase the efficiency of error prediction.

In the future, the influence of parts fixation efforts on forming errors due to changes in internal stresses during and after processing will be studied.

References

1. Chen, X., Dai, Y., Hu, H., Tie, G., Guan, C.: Influence of installation error on roundness error measurement. In: IOP Conference Series: Mater. Sci. Eng. **612**(3), paper № 032033 (2019)
2. Katsuki, A., Onikura, H., Sajima, T., Murakami, H., Sato, T., Caetano, T., Ali, H., Ohnishi, O.: Study on high-speed on-machine measurement of deep-hole accuracy. Seimitsu Kogaku Kaishi/J. Japan Soc. Precision Eng. **77**(7), 681–687 (2011)
3. Bian, X., Cui, J., Lu, Y., Tan, J. Ultraprecision diameter measurement of small holes with large depth-to-diameter ratios based on spherical scattering electrical-field probing. Appl. Sci. **9**(2), paper № 242 (2019)
4. Grossi, N., Croppi, L., Scippa, A., Campatelli, G.: A dedicated design strategy for active boring bar. Appl. Sci. **9**(17), 3541 (2019)
5. Ren, Y., Zhao, Q., Liu, Y., Ma, J.: Analysis of bending vibration characteristics of rotating composite boring bar. J. Phys. Conf. Series **1** (2019)
6. Linchevskyi, P.A., Dzhuhurian, T.H., Orgiyan, O.A.: Processing parts with processing-boring machinesedited. Technics (2000). (in Russian)

7. Rodyna, A.A., Fydarov, V.Kh, Yermakov, Ye.S.: Research of cutters wearing upon boring. Mach. Build. Relevant Issues **3**, 224–228 (2016). (in Russian)
8. Suslov, A.H., Dalskyi A.M.: Scientific basics of machine building technology. Mach. Build. (2002). (in Russian)
9. Kopieliev, YuF., Orgiyan, A.A., Kobieliev, V.M.: Parametric oscillations of metal cutting machines. Pechatnyi Dim, ONPU (2008). [in Russian]
10. Zaloga, W., Shapoval, Y., Kolesnyk, V. Increasing of efficiency of parts turning al spindle speed from 5,000 to 10.000 rpm by controlling the dynamics of machining. In: Goroshko, A., Royzman, V., Zembytska, M. (eds.) JVE International Ltd., Lithuania, vol. 2, pp. 90–102. Naugarduko, Vilnius (2018)
11. Wang, R., Gao, G., Hao, R.: Chatter depression of boring bod based on friction energy dissipation. J. China Mech. Eng. **26**(6), 2143–2148 (2015)
12. Brust, S., Röttger, A., Kimm, J., Usta, E., Theisen, W.: Manufacturing of hard composite materials on Fe-base with oxide particles. Key Eng. Mater. **742** KEM, 106–112 (2017)
13. Hong, R.J., Yeh, S.S.: Calculation and compensation method for fixture errors in five-axis CNC machine tools. Comput.-Aided Design Appl. **17**(2), 312–324 (2020)
14. Balaniuk, A.: Two-stepped console boring bars oscillations upon fine boring. Academic Papers Collection, pp. 131–139 (2014)

Efficiency and Performance of Milling Using Cutting Tools with Plates of a New Class

Gennadiy Kostyuk[1], Viktor Popov[2], Yurii Shyrokyi[1], and Hanna Yevsieienkova[1(✉)]

[1] National Aerospace University Named By N.Ye. Zhukovsky «KhAI», 17, Chkalov Street, Kharkiv 61070, Ukraine
Yu.shirokiy@gmail.com

[2] Joint Stock Company "FED", 132, Sumska Street, Kharkiv 61023, Ukraine

Abstract. Issues of technological processes' intensification and increase of cutting tools efficiency are the first-priority tasks in engineering. Taking into account the theoretical forecasts, two modifications were created based on the hard alloy BK10: the first with the addition of chromium 0.51% and aluminum 0.76%; the second with chromium 1.21% and aluminum 6.94%. Tested first modification when applied to hard alloy in the first modification in the mode of 200 in (application time was 30 min) and 450 V (application time was 25 min), and with the second modification in the mode to 250, application time was 25 min has tested the efficiency of morphogenesis (removal of the volume of material during the period of resistance of cutting tool) depending on the wear of the cutting tool to the rear surface and from the plates. It is shown that the second version of the modification is more effective when milling for minimal wear and cutting speed when milling without cooling. The volume to be removed at 300 rpm is twice as large as when the cutter is working with plates in the first modification mode. The proposed method for selecting the type of modification and application modes of nanostructures allows us to obtain coating modes and cutting conditions, thereby implementing effective forming, high productivity, and durability of the cutting tool when milling without cooling. The practical value of the research is in the studying of ways of cutting tool efficiency increasing through the possibility of technological process intensification.

Keywords: Nanostructures · Design of chemical composition · Grain volume · Hard alloy

1 Introduction

Currently, the aircraft industry, especially aviation aggregate construction, remains an open question titanium alloy, the use of already become traditional TiN coating does not affect due to the high adhesion to titanium alloy. It is, therefore, necessary to look for a coating having minimum adhesion to the titanium, in addition to ensuring the minimum adhesion should create a low-wear coating, which can be realized by forming nanostructures in the surface coating layer on the cutting tool. All this shows the importance of research on the effectiveness and efficiency of forming a coated cutting

V. Tonkonogyi et al. (Eds.): InterPartner 2020, LNME, pp. 598–608, 2021.
https://doi.org/10.1007/978-3-030-68014-5_58

tool. Based on studies of the grain size and composition of the coating 0.18 HfN + 0.82 ZrN before and after treatment inflicted on the substrate at a voltage of 200 and 250 V, were conducted studies wear, resistance CT with uncoated and coated. Dependences remove material volume during the period of wear resistance of the inserts of the rear surface. It is shown that the efficiency (amount of removable material over a period resistance) is higher for the plates, which was treated at a voltage on the substrate 250 – 2.5 (1.54 times), both at 200 – 1.46 (1.21 times), this value smaller for medium to rough machining (raw plate). It is related to grain size, which in the first case is 93.9 nm, i.e., nanostructures implemented, whereas in the second case above grain size due to a longer coating (30 min instead of 25 min in the first case), which increases with increasing temperature. It is shown that a more effective forming is realized for the second case of wafer BK10 modification (modification of Al2O3 and CrN).

Method of hardening the cutting tool (CT) by applying nano-hardened layer and formation of nanostructures devoted the significant number of studies, the results of which are summarized in works [1, 2], where both theoretical and experimental work, but the machining of hard alloys (e.g., titanium and nickel-chromium) are not considered, especially milling. However, it should be noted that it is milling today that is taken as one of the most promising methods for machining parts in the aviation industry [3–5]. Therefore, carrying out research on the effectiveness of formation in milling and turning these alloys is a relevant and timely task.

2 Literature Review

The results of theoretical studies, which are presented in [6–8] have shown that the nanostructure can be obtained by the action of ions of different varieties, charge, and energy. At the same time, there is the prospect of nanostructures under the action of laser radiation, especially under the action of femtosecond lasers, which provide high performance and low power during the processing of CT [9]. Experimentally it has been shown that one can obtain high microhardness with a small grain size [10] and also improve operability - resistance CT and removes material volume during the period resistance CT [11, 12]. Similar studies have been carried out by the action of the laser that revealed the possibility of increasing efficiency and effectiveness of the CT of high-speed steel and carbide single-carbon [13].

The paper [14] reviews experimental research on nanocomposite protective coatings of various chemical compositions and structures. For adaptive multielement and multilayer systems with specific phase composition, structure, substructure, stress state, and high functional properties, formation conditions are considered. The behavior of such systems under extreme operating conditions and in tribological applications is examined. The structural, phase and chemical composition are discussed as well as the hardness, friction, and wear at elevated temperatures, and the adhesive strength of hierarchical protective coatings are analyzed.

The formation of the (TiZrNbHfTa)N/WN multicomponent coating is considered in the paper [15]. The structural investigations showed the formation of a simple disordered solid solution in (TiZrNbHfTa)N layer, b-W_2N phase in WN layer with the fcc crystal structure and highly disordered bcc (1 1 0) and (2 2 0) -oriented high-entropy alloy phases, regardless of the applied bias potential. It was shown that with increasing of substrate bias from -90 to -280 V, there is a slight decrease of hardness from 34 to 31 GPa and increase of Young's modulus from 325 to 337 GPa, which can be explained by the annihilation of point defects and precipitation of relatively softer metallic phase.

The authors of work [16] believe that one of the most perspective directions of the development of surface engineering concerns hard multicomponent coatings prepared using PVD technologies. The authors present the results of the analysis of the transmission rate of the chemical composition of cathodes composed based on elements with different melting points (Al, Ti, Cr). It was presented the influence of the chemical composition of two and three-component cathodes on the chemical composition of the obtained coating in this work. The study was carried out with the EDS method using a scanning electron microscope with a chemical composition analyzer.

Multicomponent coatings with layers containing different functionality are of interest for a variety of applications, including electronic devices, energy storage, and biomaterials, according to the authors of the paper [17]. A comprehensive geometrical behavior against the multicomponent nano-activating flux, multi-walled carbon nanotubes-titanium oxide (MWCNTs-TiO_2), was configured in the study [18].

The paper [19] presents some of the directions of developing the plasma electrolytic oxidation technique to form the coatings with magnetic, catalytic, biocompatible, or biocidal properties on the valve metals and alloys. It reflects the relationships between the structure, composition, and functional properties of plasma electrolytic oxidation coatings. The data presented suggest that plasma electrolytic oxidation is an effective method of physicochemical synthesis on metals and alloys of the surface layers with different chemical composition and certain characteristics.

3 Research Methodology

Wear of the cutting tool was measured instrumentally on a microscope on which determined the resistance CT at the finishing (flank wear h3 $\leq$ 0,25 mm) semi-finishing (h3 $\leq$ 0,4 mm) and the blister (h3 $\leq$ 0,6 mm) processing. A baseline test for resistance CT was performed using an optical microscope. An example of measurement is shown in Fig. 1. The grain size was determined by scanning electron microscope SEM 10^8, the composition of analyzer DRON-3M, the measurements were carried out on the CT uncoated, coated before treatment, and after treatment – BT-22 alloy. The hardness and microhardness were measured by the device QNESS hardness testing (Vickers EN ISO 6507 Q60M 0.25 g, 62.5 kg. For high accuracy, the automated experiment used a system for measuring cutting forces, which is described in the next section. Similar studies were carried out and classical BK10 and hafnium nitride after upward, and zirconium nitride - reduction.

Given the high adhesion to titanium alloy coatings, most of the calculations were carried out adhesion characteristics of our method of [20], and the coating has a rating of 7 and 9 in the event of contact with the titanium alloy.

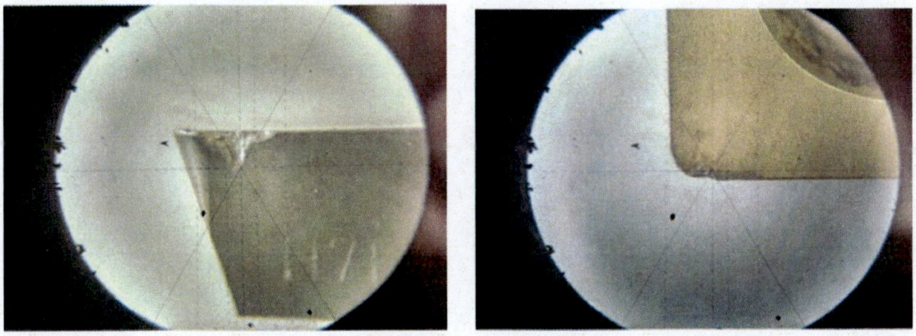

Fig. 1. Photomicrographs destruction (thermoelastic spalling) of the cutting edge of the coated CT.

The coatings were applied to the modernized Bulat-6, allowing to obtain a constant diameter covering at potentials on substrate 200 and 250 and a coating time of 30–25 min, respectively, at nitrogen pressure in the chamber $3 \cdot 10^{-3}$ mmHg (TOR).

4 Results

Pre-checked material composition CT plate treated solid alloy BK10, wherein additionally introduced aluminum oxide and titanium chromium nitride), from panels whose composition, photomicrograph of the cutting part and with a grain size of a = 253 nm are shown in Fig. 2.

A study of the spectral composition of (a) elemental composition, (b) fracture zone, (c) grain size, and (d) a modified hard alloy is presented in Fig. 2. It is seen that solid modified aluminum alloy (aluminum nitride), titanium (titanium nitride), chromium (chromium nitride), i.e., BK10 modified significantly different from the classic BK10, which obviously will affect its efficiency and effectiveness.

Research conducted on the grain size is not operating and operating plate (SEM-106), spectrograms, and composition of the surface layer in the working area and the working plate. Such studies have been conducted for carbide inserts of the modified solid BK10 alloy coated 0.2 HfN + 0.8 ZrN. For example, in Fig. 3 shows a spectrogram (a), the coating (b) is a micrograph of the grain size before treatment (c) and after milling BT22 alloy (2). Analysis of these figures shows that in the process the proportion of zirconium (zirconium nitride) decreases, increasing the proportion of hafnium nitride and hafnium, and the grain size due to thermal effects is increased for all three tested plates with 93.9 to 169 nm, 332 to 373 nm. Thermal effects on the coating material cause grain growth and deterioration of cutting properties of the coating by reducing the microhardness.

Analysis results of the study ended working CT uncoated shown (Fig. 3), which as a result of substantially reduced titanium (TiN) (Fig. 4a) and the alumina (Fig. 3b), which reduces the resistance of the plate.

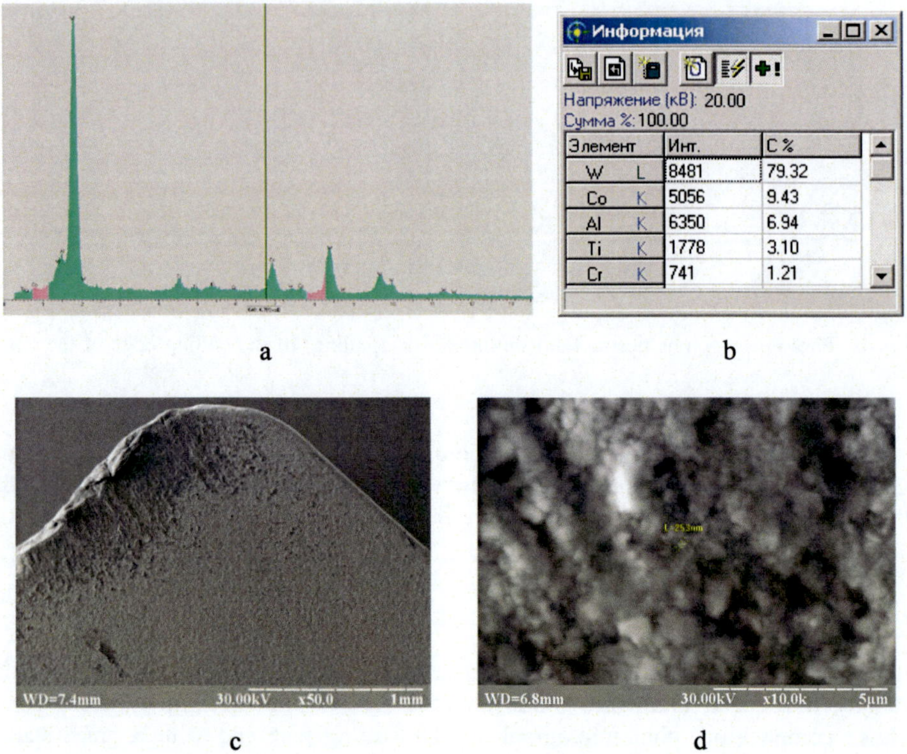

Fig. 2. Spectrogram (a), composition (b) is a photomicrograph of fracture zone (c), a micrograph of the modified solid Al_2O_3 and TiN alloy BK10 uncoated not worked plate (d) - and grain size = 253 nm.

Then we carried out according to the above algorithm for studying the second material; for the correlation result, the research results are shown in Fig. 3. It can be seen that the solid modified aluminum alloy or AlN and Al_2O_3.

The research was conducted on the crystal grain size is not operating and operating plate (SEM-106), the spectrogram, and the composition of the surface layer in the working area and the working plate. Such studies have been conducted for carbide inserts of the modified solid BK10 alloy coated 0.2 HfN + 0.8 ZrN.

Thus in Fig. 3 is spectrogram (a), the coating (b) is a micrograph of the grain size before treatment (c), a micrograph ruined blade (d), and the grain size after milling (e) BT22 alloy (3). Analysis of these figures shows that in the process the proportion of zirconium (zirconium nitride) decreases, increasing the proportion of hafnium nitride and hafnium, and the grain size due to thermal effects is increased for all three tested plates with 93.9 to 169 nm, 332 to 373 nm. Thermal effects on the coating material cause grain growth and deterioration of cutting properties of the coating by reducing the microhardness.

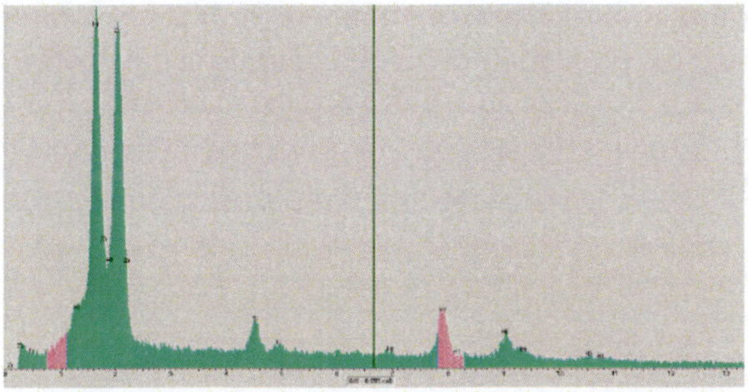

a

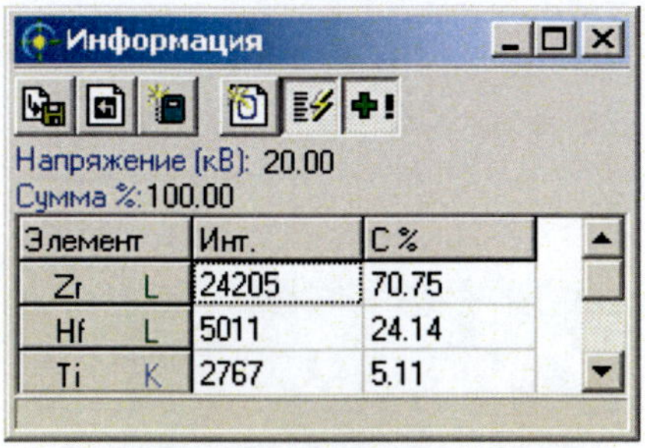

b

c

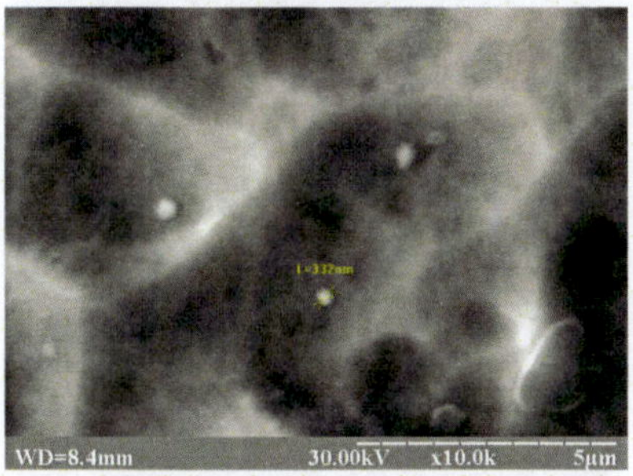

d

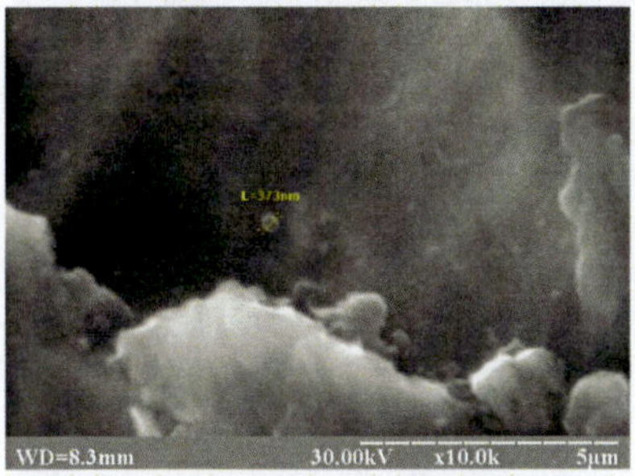

e

Fig. 3. Spectrogram stage (a), composition (b) is a photomicrograph of fracture zone (c) and the modified solid micrograph BK10 alloys coated 0.2 HfN + 0.8 ZrN not worked of the plate (d) – and = 332 nm and a spent (e) – a = 378 nm.

We also performed the study of the spectral composition of (a) elemental composition, (b) fracture zone, (c) grain size, and (d) a modified cemented carbide (Fig. 4). It can be seen that the solid modified aluminum alloy.

The research was conducted on the crystal grain size is not operating and operating plate (SEM-106), the spectrogram, and the composition of the surface layer in the working area and the working plate. Such studies have been conducted for carbide inserts of the modified solid BK10 alloy coated 0.2 HfN + 0.8 ZrN.

It was shown that in the process the proportion of zirconium (zirconium nitride) decreases, increasing the proportion of hafnium nitride and hafnium, and the grain size due to thermal effects is increased for all three tested plates with 93.9 to 169 nm, 332 to 373 Nm. Thermal effects on the coating material cause grain growth and deterioration of cutting properties of the coating by reducing the microhardness.

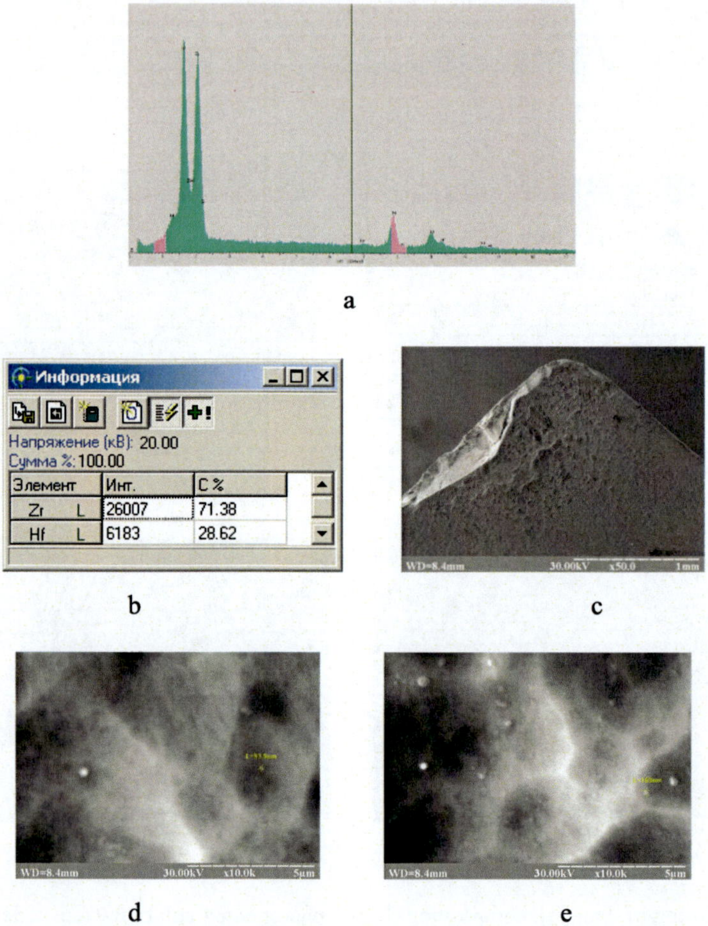

Fig. 4. The spectrogram of the stage (a), composition (b) is a photomicrograph of fracture zone (c) and modified micrograph BK10 carbide coated 0,2HfN + 0.8ZrN not worked Plate (d) – and = 93.9 nm and a spent (e) – and = 169 nm.

At the same time, the formation of nanostructures in the second modified BK10 layer also affects the efficiency of positive plates, and as we have seen in the future, even more significant than the coating.

The effect of the mode of application of the multi-component coating 0.2 HfN + 0.8 ZrN the possibility of its use in finishing (flank wear less than h3 = 0.25 mm) at semi-finishing (h3 $\leq$ 0.4 mm) and a roughing (h3 $\leq$ 0.6 mm). When coating two modes: first - the voltage at the substrate is 200 V, and time τnap = 30 min; second – Un = 250 V, and τnap = 25 min. The dynamics of the fracture BK10 modified carbide coated (Fig. 5), and it is shown that it is realized thermoelastic chipping blades in action maximum stress zone relatively far from the top cutter.

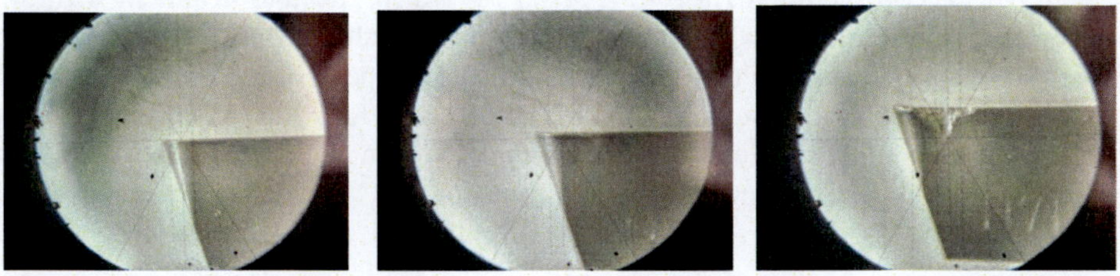

a b c

Fig. 5. Photomicrographs dynamics destruction (thermoelastic spalling) of the cutting edge of the coated CT modified solid alloy BK10.

Ion cleaning was carried out at Un = 1.2 kW and τnap = 5 min. We were obtained depending removes material volume during the period resistance against wear and flank wear on the rear surface of the operating time, it possible to find the volume of removable material during roughing, finishing and at various cutting speeds: 1.07; 1.63 and 2.2 m/s using a coating applied in a mode Un = 250V and τnap = 25 min; second - Un = 200V and τnap = 30 min. Thus, for a surface speed of 1.07 m/s (Fig. 6) can increase the volume of the material removes the period resistance more than an order of magnitude, while the cutting speed of 2.2 m/s it is possible to increase only 2.1 times.

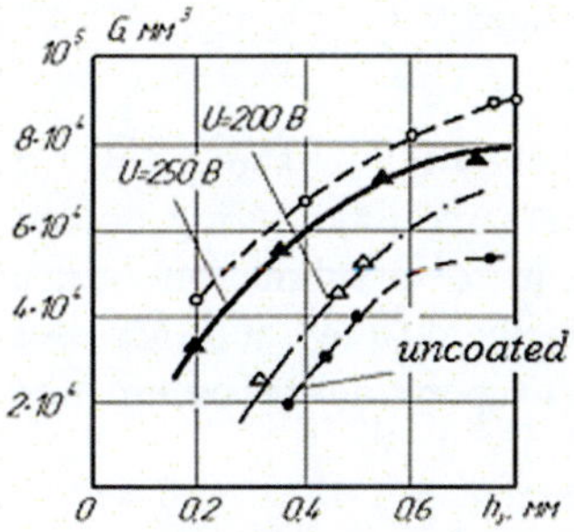

Fig. 6. The dependences remove material volume during the period of the resistance CT BK10 modified coated 0.2 HfN + 0.8 ZrN (Un = 250 - ▲), (Un = 200 -Δ) and uncoated – • at a cutting speed 1.07 m/s from wear and second rear surface modified plate - o.

Minimal volume value removes at 200 can be associated with a longer duration of plasma ion processing plate (30 min instead of 25 min), increasing the average temperature and the grain growth size, and hence to reduce the physical and mechanical characteristics of the coating. Primarily microhardness of the coating is reduced, and hence the abrasion increases, which is particularly important at low speeds.

A study of the influence of the mill revolutions on the stock removal rate during the tool life during semi-finishing milling was carried out (Fig. 7). It is shown that at 205 rpm, the maximum of this value is realized, which for the 1st modified layer of the plate allows using this value for efficient processing of BT-22 titanium alloy. This CT, as well as a second modified BK10-o (Fig. 7) for which effective treatment with the number of revolutions n = 300 rev/min. When the removable volume of material over the period resistance approaching 10^5 mm^3, the cutting speed V = 2,05 m/min.

This CT, as well as a second modified BK10-o(Fig. 7) for which effective treatment with the number of revolutions n = 300 rev/min. When the removable volume of material over the period resistance approaching 10^5 mm^3, the cutting speed V = 2,05 m/min.

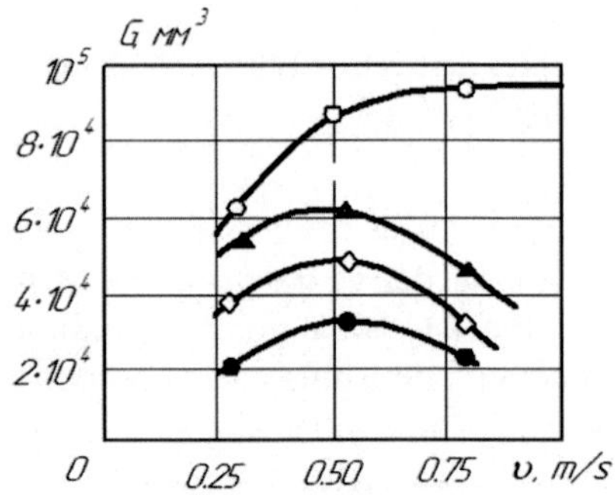

Fig. 7. Dependencies removes material volume during the period when the resistance value of the semi-finishing cutter rpm for BK10 plates (modified) coated 0,2HfN + 0,8ZrN at substrate potential Un = 250 B – ▲ (t = 25 min) at Un = 200 B – ♦ (T = 30 min) and uncoated plates ◊, for a second coating with modified V10 - o.

Milling Research titanium alloy BT-22 showed that operate most efficiently plates with nanostructures, but at the same time during operation of the cutter is observed grain growth of the nanostructured to submicron structure, thus decreasing the proportion of zirconium nitride in the coating, and increases the proportion of nitride of hafnium that due to a large evaporation zirconium nitride compared with hafnium nitride.

5 Conclusions

As a result of studies on the influence of tool wear on the rear surface of a plate of modified carbide VC-10 (modified with nitrides of aluminum and chromium) to remove the volume of material during the period of resistance (G) when milling set:

- with an ion energy of 250 V and an action time of 25 min at the maximum value of removing the volume of material during the period of resistance (G);
- at an energy of 200 V and an exposure time of 30 min for lower values (G) are realized by increasing the temperature during longer processing. It is due to grain growth at higher temperatures

A possible thermoelastic fracture in the zone of maximum temperature stresses was detected for the modified BK10 coating.

The conducted research allows you to choose the most effective processing modes for finishing, semi-finishing and roughing.

It is shown that at a cutting speed of 0.52 m/s, the maximum value is realized, allowing this value to be used for effective processing of the titanium alloy in the case of plates made of the 1st modified BK10 alloy. At the same time, for the second modified BK10 at 300 rpm (0.77 m/s), you can increase the volume of the removed material for the period of resistance to 8.7·104 mm^3.

Acknowledgement. The authors would like to acknowledge financing of National Research Foundation of Ukraine under grant agreement No. 190/02.2020.

References

1. Kostyuk, G., Popov, V., Kostyk, K.: Computer modeling of the obtaining nanostructures process under the action of laser radiation on steel. In: CEUR Workshop Proceedings, vol. 2353, pp. 729–743 (2019)
2. Kostyuk, G., Nechyporuk, M., Kostyk, K.: Determination of technological parameters for obtaining nanostructures under pulse laser radiation on steel of drone engine parts. In: 10th International Conference on Dependable Systems, Services and Technologies (DESSERT), pp. 208–212. IEEE, Leeds, United Kingdom (2019)
3. Dobrotvorskiy, S., Basova, Y., Kononenko, S., Dobrovolska, L., Ivanova, M.: Numerical deflections analysis of variable low stiffness of thin-walled parts during milling. In: Ivanov, V. (ed.) DSMIE 2019, LNME, pp. 43–53. Springer, Cham (2020)
4. Kononenko, S., Dobrotvorskiy, S., Basova, Y., Gasanov, M., Dobrovolska, L.: Deflections and frequency analysis in the milling of thin-walled parts with variable low stiffness. Acta Polytechnica **59**, 283–291 (2019)
5. Dobrotvorskiy, S., Basova, Y., Ivanova, M., Kotliar, A., Dobrovolska, L.: Forecasting of the productivity of parts machining by high-speed milling with the method of half-overlap. Diagnostyka **19**(3), 37–42 (2018)
6. Lu, Y., Yu, L., Lou, X.W.D.: Nanostructured conversion-type anode materials for advanced lithium-ion batteries. Chem **4**(5), 972–996 (2018)
7. Wang, J., Tang, H., Zhang, L., Ren, H., Yu, R., Jin, Q., Liu, P.: Multi-shelled metal oxides prepared via an anion-adsorption mechanism for lithium-ion batteries. Nature Energy **1**(5), 16050 (2016)
8. Wang, J., Tang, H., Wang, H., Yu, R., Wang, D.: Multi-shelled hollow micro-/nanostructures: Promising platforms for lithium-ion batteries. Mater. Chem. Front. **1**(3), 414–430 (2017)

9. Qi, W., Shapter, J.G., Wu, Q., Yin, T., Gao, G., Cui, D.: Nanostructured anode materials for lithium-ion batteries: principle, recent progress and future perspectives. J. Mater. Chem. **A5** (37), 19521–19540 (2017)
10. Bi, R., Ma, Q., Mo, H., Olivo, M., Pu, Y.: Optical-resolution photoacoustic microscopy of brain vascular imaging in small animal tumor model using nanosecond solid-state laser. Neurophotonics Biomed. Spectrosc. **2019**, 159–187 (2019)
11. Pelleg, J.: Grain size effect on mechanical properties. In: Mechanical Properties of Silicon Based Compounds, 1st edn. Springer, Cham (2019)
12. Igouchkine, O., Zhang, Y., Ma, K.L.: Multi-material volume rendering with a physically-based surface reflection model. IEEE Trans. Visual Comput. Graph. **24**(12), 3147–3159 (2017)
13. Davenport, M.S., Parikh, K.R., Mayo-Smith, W.W., Israel, G.M., Brown, R.K., Ellis, J.H.: Effect of fixed-volume and weight-based dosing regimens on the cost and volume of administered iodinated contrast material at abdominal CT. J. Am. Coll. Radiol. **14**(3), 359–370 (2017)
14. Pogrebnjak, A.D., Bagdasaryan, A.A.E., Pshyk, A., Dyadyura, K.: Adaptive multicomponent nanocomposite coatings in surface engineering. Phys. Usp. **60**(6), 586–607 (2017)
15. Bagdasaryan, A.A., Pshyk, A.V., Coy, L.E., Kempiński, M., Pogrebnjak, A.D., Beresnev, V. M., Jurga, S.: Structural and mechanical characterization of (TiZrNbHfTa) N/WN multilayered nitride coatings. Mater. Lett. **229**, 364–367 (2018)
16. Kacprzyńska-Gołacka, J., Słomka, Z., Osuch-Słomka, E., Rydzewski, M., Mazurkiewicz, A., and Smolik, J.: The dependence of the chemical composition of Al-Ti-Cr multi-component coatings on parameters of the arc-evaporation process. Journal of Machine Construction and Maintenance. Problemy Eksploatacji **3**, 53–57 (2017)
17. Liu, X., Liu, W., Carr, A.J., Vazquez, D.S., Nykypanchuk, D., Majewski, P.W., Bhatia, S. R.: Stratification during evaporative assembly of multicomponent nanoparticle films. J. Colloid Interface Sci. **515**, 70–77 (2018)
18. Muzamil, M., Wu, J., Akhtar, M., Patel, V., Majeed, A., Yang, J.: Multicomponent enabled MWCNTs-TiO2 nano-activating flux for controlling the geometrical behavior of modified TIG welding joint process. Diam. Relat. Mater. **97**, 107442 (2019)
19. Rudnev, V.S., Lukiyanchuk, I.V., Vasilyeva, M.S., Medkov, M.A., Adigamova, M.V., Sergienko, V.I.: Aluminum-and titanium-supported plasma electrolytic multicomponent coatings with magnetic, catalytic, biocide or biocompatible properties. Surf. Coat. Technol. **307**, 1219–1235 (2016)
20. Kostyuk, G.: Prediction of the microhardness characteristics, the removable material volume for the durability period, cutting tools durability and processing productivity depending on the grain size of the coating or cutting tool base material. In: Gapiński B., Szostak M., Ivanov V. (eds.) Advances in Manufacturing II. MANUFACTURING 2019. Lecture Notes in Mechanical Engineering, pp. 300–316. Springer, Cham (2019)

Performance and Relative Consumption of Diamond Grains During High-Speed Diamond Sharpening of Superhard Materials

Dmitry Romashov(✉), Vladimir Fedorovich, Vladimir Dobroskok, Ivan Pyzhov, and Yevgeniy Ostroverkh

National Technical University "Kharkiv Polytechnic Institute", 2, Kirpichova Street, Kharkiv 61002, Ukraine
romashov86@ukr.net

Abstract. The most important output indicator of the grinding process is its productivity and relative consumption of the wheel achieved with satisfying the requirements for the quality of processing. This work proposes a methodology for calculating the productivity and relative consumption of the wheel for high-speed diamond grinding of superhard materials (SHM). The actual productivity of the processing process is a function of cutting ability of the wheel - maximum volume of material that the working surface can remove. A fundamental feature of the process of diamond grinding of superhard materials is the practical equality of the hardness of the cutting tool (diamond grains) and the processed material. In this process, the classical concept of the cutting process is not realized, and in the grinding zone, there is a controlled mutual micro destruction of two equally hard materials. The processing of superhard materials is accompanied by low productivity, and high values of relative consumption of diamond wheels, the coefficient of use of diamond grains in these processes does not exceed 5–10%. In this paper, we study the possibility of using high-speed modes to increase productivity processing and the definition of the parameter (speed), the control of which allows ensuring the maximum productivity of the process without compromising the quality of the processed surface. Increasing speed is one of the radical ways to enhance performance during grinding. Industry produces a range of circles designed for these purposes. High-speed grinding required the use of more precise grinding spindles in machines and the requirements for balancing and dressing increased.

Keywords: High-speed diamond grinding · Productivity · Efficiency · Finite element method · Stress-Strain state · Tool sharpening · Surface roughness · Amount of destroyed diamond · Relative consumption

1 Introduction

Grinding performance, surface layer quality, wheel resistance, cutting forces, and temperature in the cutting zone depend on the grain size of the wheel, type of binder, wheel width, concentration, the properties of the processed material, and cutting conditions [1–6].

V. Tonkonogyi et al. (Eds.): InterPartner 2020, LNME, pp. 609–618, 2021.
https://doi.org/10.1007/978-3-030-68014-5_59

When sharpening diamond - the hardest material in nature, it becomes difficult to sharp and fine-tune such tools since the classical requirement of the cutting theory is not fulfilled - exceeding the hardness of the tool material over the workpiece. Discussion on the advantages of grinding or blade processing comes down to the efficiency of abrasive machining since the performance of blade cutting is largely determined by the quality of the tool sharpening. So, for example, with the appropriate quality of sharpening the SHM tool, it can successfully replace natural diamonds.

High-speed grinding increases the productivity of the grinding machine. The reason for this is the possibility to increase the grinding depth and part feed. The use of a high speed grinding wheel improves the quality of the polished surface [6, 7].

Binder of diamond wheels largely determines their operational properties. It not only keeps cutting grains in the working layer of the tool but also affects the processing performance and the quality of polished surfaces, helps to reduce the coefficient of friction with the surface being machined, ensures the tool operates in self-sharpening mode, determines the strength, rigidity and wear resistance of the working wheel, participates in the formation and removal the heat from the treatment zone.

From the previous, we can conclude that the method of sequentially determining the optimal conditions for unchanged other parameters and processing conditions is laborious and requires a large number of preliminary experiments. Grinding is a multi-factor process. The same results in terms of productivity and quality can be obtained with various combinations of grinding modes.

2 Literature Review

The efficiency of the diamond grinding process using wheels on various binders is largely determined by the stability of the parameters of their working surface. It is known [7, 8] that the processing of superhard polycrystalline materials is accompanied by high values of relative consumption of diamond wheels. So the coefficient of use of diamond grains in these processes does not exceed 5-10%. As studies have shown, this indicator decreases with a decrease in the graininess of the wheel. Therefore, at the final stages of processing, the problem of increasing the utilization rate of diamond grains becomes even more relevant [8–10]. In this regard, there is a need for theoretical and experimental research aimed at its effective solution.

Also, the widespread use of SHM blade tools is constrained by the relatively low reliability of its operation. This drawback is due to the imperfection of SHM synthesis (production) processes: the presence of residual stresses, internal defects that appear even after the sharpening of the tool [11].

An increase in cutting speed during grinding is mainly carried out to increase processing productivity. The productivity of a technological operation, including grinding, is determined by the number of workpieces processed per unit of the time on this machine, or by calculation of piece-time for processing.

Improving the productivity sharpening the tool from SHM due to increased transverse and longitudinal feeds is not possible without increasing the cutting speed. Increasing the circumferential speed of the grinding wheel from 45 to 270 (m/s) (6 times), with a longitudinal feed, allows to increase productivity up to 3–4 times.

Grinding performance is determined by the relative removal of the material. Increasing the cutting speed during sharpening in one time allows increasing feeds and thereby to reduce process time. However, in such processes, we should take into account many negative phenomena that accompany the process of high-speed grinding and limiting the possibility of obtaining positive results. Growing centrifugal forces of unbalanced masses lead to an increase in the intensity of vibrations, which also increases the wear of the grinding wheel and worsens the quality of processing [12–16].

That allows us to conclude that there is an undeniable trend towards the need to increase the peripheral speed of the grinding wheel significantly.

3 Research Methodology

Based on the features of superhard materials from synthetic diamond, as well as a large number of factors and their combination, which affect the processing, such methodological approach was substantiated, based on the widespread use of model experiments, which greatly reduces facilitates research, time, cost and allows you to vary a wide range of research parameters.

All elements of the systems under study (diamond grain, metal phase, bond, coating of diamond grains, ground SHM) and processing conditions are considered in the interaction.

Dynamic 3D modeling is widely used to determine performance parameters and consider the processes of destruction of diamond grains and a binder component during processing with high speed grinding wheels.

We will dwell on the algorithmic foundations of dynamic modeling in more detail.

Regardless of the software used, an effective algorithm for creating a model and solving problems using research packages based on finite element method (FEM) contains the following steps (Fig. 1):

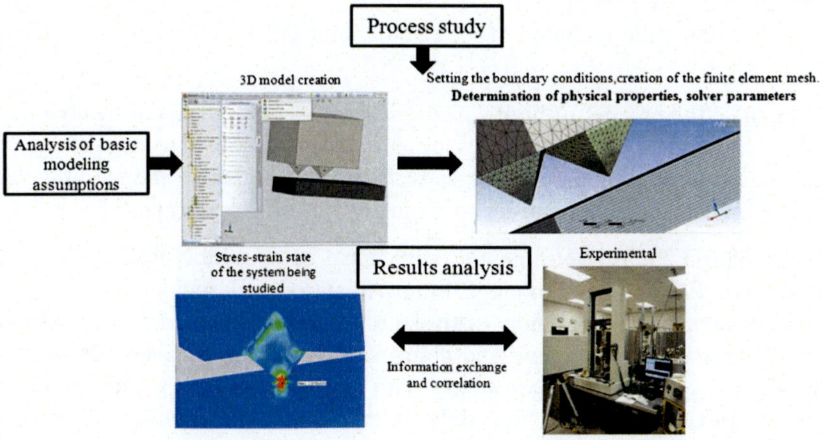

Fig. 1. General concept of research on improving the efficiency of diamond-abrasive tools in the processing of superhard materials.

1. Construction of a geometric model of a structure (or export of this model from a design module). The construction elements used in this case are ranked in increasing order of priority: $pt \rightarrow cr \rightarrow sf \rightarrow vl$ (point → line → surface → volume). In our case, using VL elements, a three-dimensional (3D) model of the "diamond grain - coating - metal phase – binder – processed material" system is created.
2. The choice of the type of finite elements from the package library for each design zone. Since the most common form of diamond and its destruction products is the octahedron [16].
3. Setting material properties for each construction zone. All properties of materials depending on temperature, as well as plastic behavior: elastic modulus, Poisson's ratio, thermo-physical properties, 4. Assignment of geometric characteristics of elements by system zones (depending on the selected element type).
4. Construction of the finite element model (partition of the structure into finite elements). At this stage, based on a geometric model consisting of points, lines, surfaces, and volumes (pt, cr, sf, vl), a finite element model is created consisting of nodes (nd) and elements (el).
5. Verification of the construction of the finite element model and its correction (merging matching nodes, renumbering nodes, and elements) (MERGE, COMPRESS).
6. Fixing the finite element model (setting zero or other given displacements). It is produced in nodes but can be produced using attributes of a geometric model (points, lines, surfaces) if they are associated with the corresponding nodes.
7. Loading of the model (concentrated, distributed loads) over nodes, lines, and surfaces. Specific loading options need to be considered for a specific package version.
8. Export of finite element model data to the calculation module. This stage is performed if the model is built and calculated in different modules.
9. Calculation of the created finite element model in a specific calculation module, depending on the type of task.
10. Conclusion and interpretation of the calculation results (on the screen, in the output file).

To simulate the process of high-speed grinding, we used material behavior models that, with appropriate assumptions, took into account the behavior of super-hard materials based on synthetic diamond, as well as the materials of the binding elements of grinding wheels, coatings and metal inclusions of diamond grains:

1. Elastic, Elastic_Fluid (Isotropic elasticity, Ideal compressible fluid).
2. Orthotropic_Elastic, Anisotropic_Elastic (Anisotropic elasticity).
3. Plastic_Kinematic (Flow theory with isotropic or translational hardening with the effect of high-speed hardening and deformation criterion of destruction).
4. Elastic_Plastic_Thermal (Flow theory with isotropic hardening and temperature)
5. For the behavior of brittle materials of the studied diamond-based system, the Johnson-Holmquist fracture model was first used:

The Johnson-Holmquist material behavior model was first proposed to describe the reaction of brittle materials under large deformations. The second version of this model, known as JH-2 [17], where the strength and damage of the material are considered as functions of representing variables. Moreover, the material deals with the development of damage.

4 Results

With an increase of the cutting speed during sharpening, the cutting processes change significantly: the brittle fracture zone of the material is localized, the friction forces and the conditional stress of cutting are reduced. Chips are formed at smaller thicknesses of the cut, etc. In the process of forming grinding marks, decreases the ratio of the chip thickness to the rounding radius of the top of the diamond grains, which characterizes the moment of the start of chip formation. A decrease in this ratio leads to the fact that the treated surface is the result of cutting, with fewer traces of rude destruction. It is also facilitated by a decrease in the friction coefficient on the rear surface of the diamond grain's cutting part. The transition to a more pure brittle fracture, which is characteristic of high grinding speeds, determines the localization of fracture for the surface layer [18].

Using the LS-Dyna calculation complex in combination with ANSYS allowed us to theoretically calculate the volume of destroyed working diamond grains and the volume of destroyed processed SHM being sharpened (Fig. 2).

A large number of series of model experiments make it possible to establish trends and mathematical relationships in high-speed diamond grinding.

Model experiments were carried out under the theory of experimental design [19, 20]. The values of the factors are encoded by linearly transforming the coordinates of the factor space with the transfer of the origin to the zero point and the choice of scales along the axes in units of the intervals of variation of the factors. The mathematical model of the process description (general view) was used:

$$Y = b_0 + \sum b_i X_i + \sum b_{i,j} X_i X_j + \sum b_{ii} X_i^2 + \ldots \tag{1}$$

A wide range of materials of the system "binder-diamond grains-metal phase-coatings-processed SHM" was used. The main input factors were accepted: the coefficient of thermal expansion (CTE) of the metal phase, the strength properties of the binder (elastic modulus), the properties of the coating of grains (elastic modulus); processing conditions (speed).

As factor levels for the materials of the binder component, metal-phase and coatings were used: alloy steel, nickel alloys, and copper alloys. Such a choice makes it possible to study a wide range of the most suitable materials for these purposes with the help of one 3D model, without spending time on its redefinition.

The obtained mathematical model of the process of high-speed diamond sharpening of an SHM tool has the form:

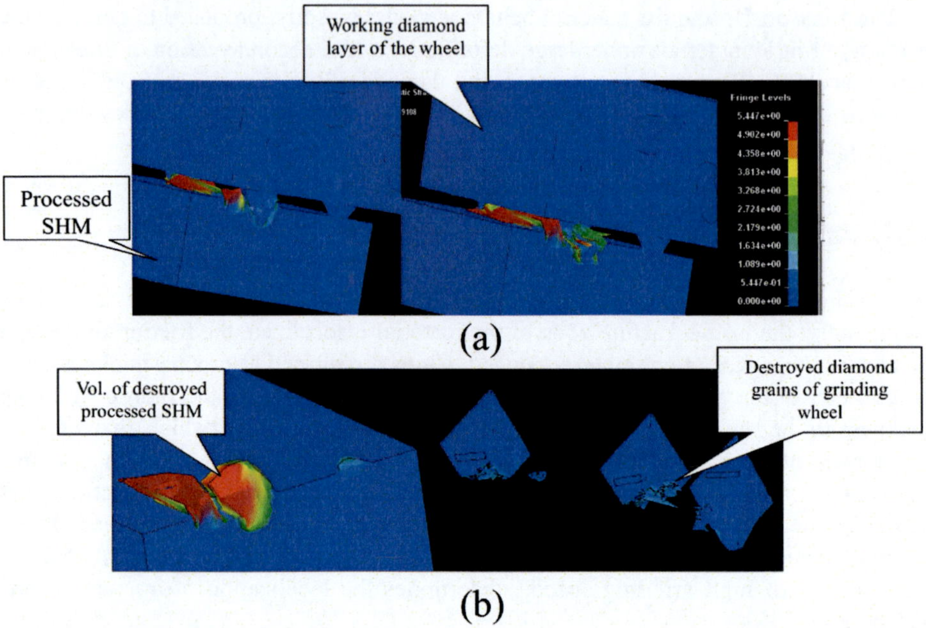

Fig. 2. Visualization of the results of calculating the volume of the destroyed SHM material and diamond grains at high-speed sharpening of a tool from SHM - integration step number 2, time T = 0.0033 s, processing speed V = 150 m/s. a - model "grain-metal phase-binder-sharpened tool"; b - shows the destruction zone, sharp edges of the cutting grains and the volume treated for this time is highlighted

$$\begin{aligned} Y = {} & 23.12 + 0.04X_1 - 0.17X_2 + 3.66X_3 + 0,56X_4 - 0.001X_1X_2 + 0.006X_1X_3 \\ & + 0.023X_1X_4 - 0.03X_2X_3 + 0.014X_2X_4 + 0.031X_3X_4 + 1.28X_1^2 + 1.50X_2^2 + 1.25X_3^2 \\ & - 2.36X_4^2 \end{aligned} \tag{2}$$

where Y – a ratio of the volume of the destroyed working diamond grain and the volume of the destroyed processed SHM (relative wheel consumption); X_1 – coefficient of thermal expansion (CTE) of the metal phase; X_2 - strength properties of the binder (elastic modulus); X_3 – strength properties of the coating of grains (elastic modulus); X_4- processing conditions (speed).

At the same time, we can make conclusions about the relative consumption of the wheel when varying a large number of factors (speed, the material of the binder, grade of grain, feed, etc.).

In the general case, the performance of the diamond abrasive processing is determined by the volume of material being processed per unit of time [21]:

$$Q = Vm/t, \tag{3}$$

where Vm – a volume of material being processed over time t, t - processing time in which the micro-cutting process takes place.

The volume of material being processed during flat grinding is determined by the formula [21]:

$$Vm = abc, \tag{4}$$

where a, b - length and width of the treated surface, mm; c - thickness of being processed layer, mm.

Moreover, the relative consumption of the wheel is the ratio of the volume of the destroyed diamond grains (Vzer) to the volume of the removed SHM:

$$G = V_{zer}/V_{SHM} \tag{5}$$

The increase of the volume of processed material per unit of time with an increase in the processing speed is shown in Fig. 3. Analysis of the calculation results showed that when the processing speed is increased to 130 m/s using a diamond wheel with an iron-based binder, the volume of processed diamond increases by 3 times.

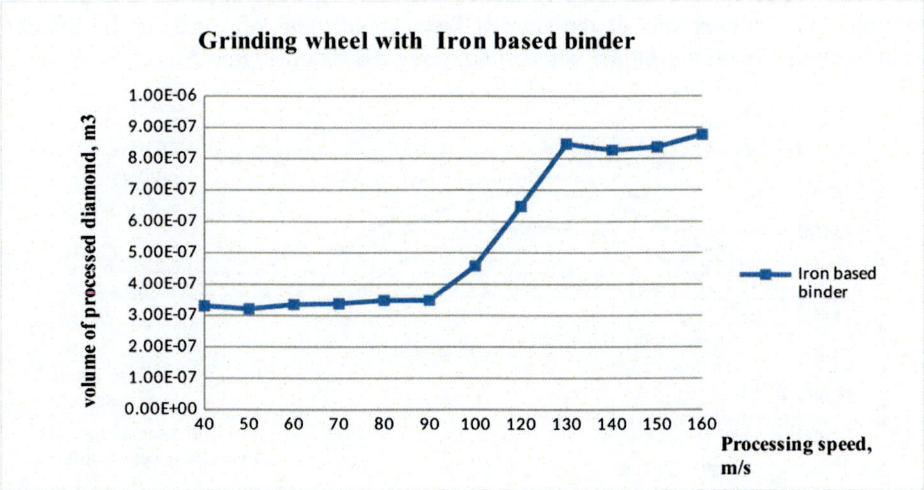

Fig. 3. Dependence of the volume of the processed diamond per unit of time vs. grinding speed (grinding performance).

Accordingly, it seems possible to theoretically calculate the ratio of the volume of spent (worn out) grains to the volume of the processed SHM, i.e., relative consumption (Fig. 4).

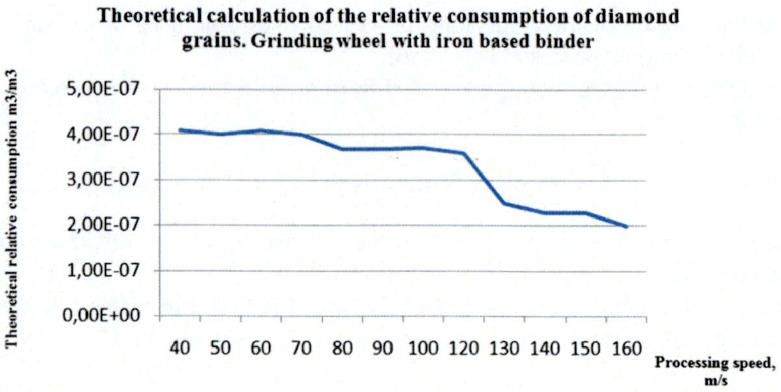

Fig. 4. Theoretical calculation of the relative consumption of diamond grains of a grinding wheel.

The generated database is one of the components of a scientifically based expert system for the selection of optimal processing modes, a combination of grinding wheel parameters when sharpening SHM tools, proposed in the conclusion of the research data. So, using these data and the mathematical model (1), it seems possible to track the most optimal combinations of the studied factors and their influence on the relative consumption of diamond grains when sharpening SHM tools (Fig. 5).

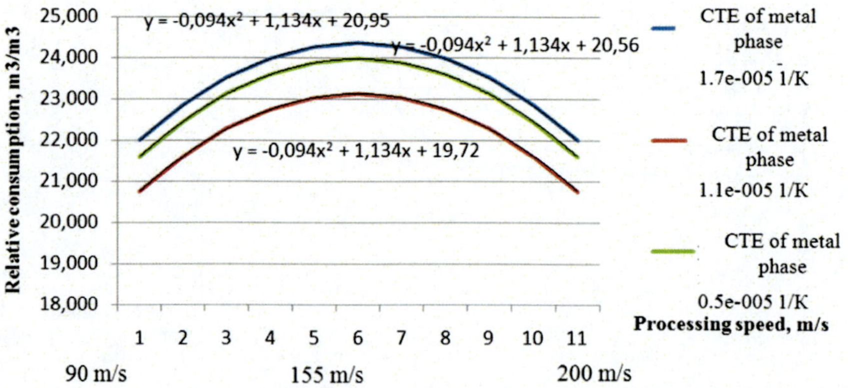

Fig. 5. Dependence of the relative consumption of grains of a wheel during high-speed sharpening SHM tools vs. processing speed when varying the CTE of the metal phase of diamond grain (of processing wheel).

5 Conclusions

A methodology for calculating the processing productivity and relative consumption of a wheel for high-speed diamond grinding of superhard materials was proposed. The accepted methodological approach is based on the widespread use of modeling, taking into account the characteristic features of the studied processes, with the maximum involvement of experimental data. The technical result consists in the fact that with the help of model experiments it was confirmed that in the first stage of sharpening (high-speed mode) with an increase the processing speed to 130 m/s, a wheel with an iron-based binder increases grinding performance by 3 times, and the relative consumption of grains is reduced about 1.5 times. The proposed methodology can become the basis for the development of a scientifically based expert system for choosing the optimal machining modes, combining grinding wheel parameters when sharpening SHM tools within the framework of the 4th scientific and technological revolution (Industry 4.0).

References

1. Fedorovich, V.A.: Methods of grinding with combined control of the cutting relief of the wheels. Bull. KSPU **45**, 26–28 (1999)
2. Li, W., Liu, M., Ren, Y., Chen, Q.: A high-speed precision micro-spindle use for mechanical micro-machining. Int. J. Adv. Manuf. Technol. **102**, 9–12 (2019)
3. Li, W., Ren, Y., Li, C., Li, Z.: Investigation of machining and wear performance of various diamond micro-grinding tools. Int. J. Adv. Manuf. Technol. **106**(3–4), 10–15 (2020)
4. Mao, C., Zhou, F., Hu, Y., Cai, P.: Tribological behavior of cBN-WC-10Co composites for dry reciprocating sliding wear. Ceram. Int. **45**(5), 6447–6458 (2018)
5. Pham, T.-H., Nguyen, D.-T., Banh, T.-L., Van Canh, T.: Experimental study on the chip morphology, tool–chip contact length, workpiece vibration, and surface roughness during high-speed face milling of A6061 aluminum alloy. J. Eng. Manuf. **234**, 610–620 (2019)
6. Li, B., Dai, C., Ding, W., Yang, C., Li, C., Kulik, O., Shumyacher, V.: Prediction on grinding force during grinding powder metallurgy nickel-based superalloy FGH96 with electroplated CBN abrasive wheel. Chin. J. Aeronaut. (2020)
7. Webster, J., Tricard, M.: Innovations in abrasive products for precision grinding. Ann. CIRP **53**(2), 597–617 (2004)
8. Barlow, N., Jackson, M., Hitchiner, M.: Mechanical design of high-speed vitrified cBN grinding wheels. In: Proceedings of IMEC, pp. 568–570. University of Connecticut, USA (1996)
9. Jackson, M.J., Davis, C.J., Hitchiner, M.P., Mills, B.: High-speed grinding with c.B.N. Grinding wheels – applications and future developments. J. Mater. Process. Technol. **110**, 78–88 (2001)
10. Taubert, M., Püschel, A.: High speed grinding passes the test in Germany. Int. Railway J. 31–33 (2009)
11. Kundrak, J., Varga, G., Deszpoth, I., Molnar, V.: Some aspects of the hard machining of bore holes. Appl. Mech. Mater. **309**, 126–132 (2013)
12. Li, W., Wang, Y., Fan, S.H., Xu, J.F.: Wear of diamond grinding wheels and material removal rate of silicon nitrides under different machining conditions. Mater. Lett. **61**(54), 4–8 (2007)

13. Kopač, J., Krajnik, P.: High-performance grinding - a review. J. Mater. Process. Technol. **175**(1–3), 278–284 (2006)
14. Xu, C., Dou, J., Chai, Y., Li, H.: The relationships between cutting parameters, tool wear, cutting force and vibration. Adv. Mech. Eng. **10**(1) (2018)
15. Bakul, V.N.: Optimal grades of diamonds for wheels on an organic binder. Synthetic Diamonds **4**, 4–9 (1970)
16. Konig, W., Klocke, F., Stuff, D.: High speed grinding with cBN wheels – boundary conditions, applications and prospects of a future oriented technology. In: 1st French and German Conference on High Speed Machining, pp. 207–218. Metz (1997)
17. Cronin, D.S., Bui, K., Kaufmann, C., McIntosh, G., Berstad, T.: Implementation and validation of the johnson-holmquist ceramic material model in LS-dyna. In: 4th European LS-DYNA Users Conference, pp. 47–60 (2008)
18. Oliveira, J.F.G., Silva, E.J., Guo, C., Hashimoto, F.: Industrial challenges in grinding. Ann. CIRP **58**, 663–680 (2009)
19. Krasovsky, G.I., Filaretov, G.F.: Experiment Planning. Publishing House of BSU, Minsk (1982)
20. Katsev, P.G.: Statistical research methods of a cutting tool. 2nd edn., rev. Mechanical Engineering, Moscow (1974)
21. Siradze, A.M.: Studies of diamond machinability. GPI them. Lenin, Tbilisi (1975)

Mechanics of Micro-cutting Using FANT

Ihor Shepelenko[1(✉)], Yuri Tsekhanov[2], Michael Storchak[3], Yakiv Nemyrovskyi[1], and Vitalii Cherkun[4]

[1] Central Ukrainian National Technical University, 7, Universytetskyi Ave., Kropyvnytskyi 25006, Ukraine
kntucpfzk@gmail.com

[2] Voronezh State Technical University, 84, 20 let Oktyabrya St., Voronezh 394026, Russia

[3] University of Stuttgart, 7, Keplerstraße, 70174 Stuttgart, Germany

[4] Dmytro Motornyi, Tavria State Agrotechnological University, 18, B. Khmelnytsky Ave., Melitopol 72312, Ukraine

Abstract. The analysis of the antifriction material brass L63 surface layer formation mechanics during finishing antifriction non-abrasive treatment (FANT) at the stage of micro-cutting was performed experimentally by the method of microhardness measurement and computer simulation using the finite element method (FEM). The analysis of the micro-cutting processes simulation results during FANT using the FEM allowed us to identify the main technologically important patterns. It was found that large tensile stresses arise behind the contact surface in the surface layer of the brass tool, which leads to the formation of surface microcracks to facilitate micro-cutting during subsequent passages of the FANT. On the back surface of the contact zone, there are extreme contact pressures for brass, which create the conditions for the formation of an adhesive durable brass antifriction layer. The thickness of the deformation-hardened surface layer is very small and does not significantly affect subsequent technological transitions during coating.

Keywords: Finishing antifriction non-abrasive treatment · Hardness measurement method · Finite element method · Brass · Contact interaction · Back surface

1 Introduction

A characteristic feature of modern engineering is a constant technical update and exceptional mobility to introduce innovative technologies based on the achievements of science and technology. Its level is also determined by the creation of contact operational coatings that increase the service life of products [1]. A significant influence on their formation is exerted by the properties of an intermediate medium through which the interaction of microroughness of contact surfaces occurs. Therefore, an important reserve for improving the performance of parts during their manufacture and repair is the modification of their working surfaces by the creation of antifriction coatings.

The advanced technologies for applying such coatings include the finishing antifriction non-abrasive treatment (FANT), which is realized due to the frictional

V. Tonkonogyi et al. (Eds.): InterPartner 2020, LNME, pp. 619–628, 2021.
https://doi.org/10.1007/978-3-030-68014-5_60

interaction of a copper-containing tool with the surface of the workpiece in the presence of process fluid. It ensures the transfer of tool material and the formation of an antifriction coating with a thickness of up to 5 μm to the surface of the part, as well as hardening of the surface layer of the base material to a depth of 70–80 μm [2].

Existing FANT technologies are characterized by low productivity, uneven coating properties in thickness, high tool loads, and significant heat dissipation. Also, the lack of consensus on the formation mechanism of such a coating shows the need for studies on the interaction of contacting surfaces. Such studies should be carried out using modern methods of engineering analysis.

Currently, the finite element method (FEM) has been widely used to study cutting processes. Its use allows us to minimize physical experiments and identify important patterns of cutting mechanics occurring close to the tip of the cutter in the contact zone, the dimensions of which are comparable to the radius of the tip rounding. The latter is very relevant for the study of the micro-cutting process that occurs during FANT, which largely determines the quality of the antifriction coating.

Another method for studying the mechanics of the formation of a hardened surface layer is experimental - by measuring its microhardness. Both of these methods were used in this work.

2 Literature Review

The FANT method [3–5], is widely used in various branches of manufacturing engineering, including the parts machining (cylinder liners, crankshafts, etc.) that limit the life of internal combustion engines.

The formation of the FANT antifriction coating largely depends on the conditions of the contact interaction of the tool with the surface being treated, and the shape and size of microroughness determine the quality of the obtained coating, its continuity, and adhesive strength [6].

Previously, the authors of [7] simulated the contact interaction of a single microroughness in the form of a cutter from the processed material—SCh20 cast iron and a surface made of antifriction material—L63 brass. This model allowed us to get an idea of the contact interaction of these elements and highlight the main stages of the FANT:

- micro-cutting of the source material with microroughness peaks;
- adhesive sticking and seizure of particles formed as a result of micro-cutting, with the surface onto which the transfer occurs and subsequent micro-smoothing.

It should be noted that a clear time frame cannot demarcate these processes. They exist in close interaction in the unified temporal, geometric, mechanical limits of the process parameters, and therefore the separation is very arbitrary.

It was established [7] that, during micro-cutting, the antifriction material that flows onto the cutting wedge of microroughness is divided into two flows, one of which moves along its front surface, and the second layer with a thickness approximately equal to the cutting wedge rounding radius is deformed by the rear surface of the cutting wedge. It was noted that the cutting blade of a cast-iron micro-cutter wears out intensively in the process of interaction with a brass surface, and this phenomenon occurs already at the very beginning of its operation. Consequently, the micro-cutting process during FANT occurs at small thicknesses of cuts comparable with the cutting edge radius. A distinctive feature of the process is the prevailing influence of the work of friction forces on the rear face [8, 9]. In this case, rounding of the cutting edge is one of the reasons causing the formation of growth on the back surface. This fact must be taken into account in the process of studying contact interaction and constructing an adequate model. The presence of such a model will allow us to investigate the possibility of the formation of areas with adhesive stuck brass, thereby intensifying the FANT coating process, and also to evaluate the stress-deformed state (SDS) and strain hardening of the surface layer of brass L63 from the positions of the process mechanics. In our opinion, this is carried out most objectively with the help of a technique for simulating the micro-cutting process using FEM. It should be mentioned that the software developers who use FEM for modeling develop the quality of their research and expand their capabilities [10–12].

The present work aims to study the features of contact interaction with FANT at the stage of micro-cutting from the standpoint of process mechanics.

3 Research Methodology

An SDS analysis of the surface layer of antifriction material made of brass L63 was carried out by the FEM using the DEFORM-3D software package.

The DEFORM-3D program provides step-by-step restructuring of the entire finite element mesh (FE), taking into account their deformation in the previous step. The values of all parameters calculated in the nodes of the previous step are recalculated to the nodes of the newly rebuilt grid. It must be done so that, due to large deformation gradients, a significant calculated deformation error does not accumulate.

Trial calculations showed that it is advisable at the beginning of the calculations to immediately thicken the FE grid in the area of the cutting edge since very large plastic deformations with significant gradients appear already in the first steps of the calculation.

Next, the pattern of friction was set. When calculating the processes of cutting and deformation, we most often use the Coulomb law of friction [13]:

$$\tau_n = f \cdot \sigma_n, \tag{1}$$

or the Prandtl law of friction:

$$\tau_n = 2\mu \cdot \tau_s, \tag{2}$$

where τ_n – friction stress; σ_n – contact pressure; τ_s – yield strength on the shear of the processed material; f, μ – Coulomb and Prandtl friction coefficient.

As experiments have shown, brass sticking immediately occurs on the back contact surface (Fig. 1).

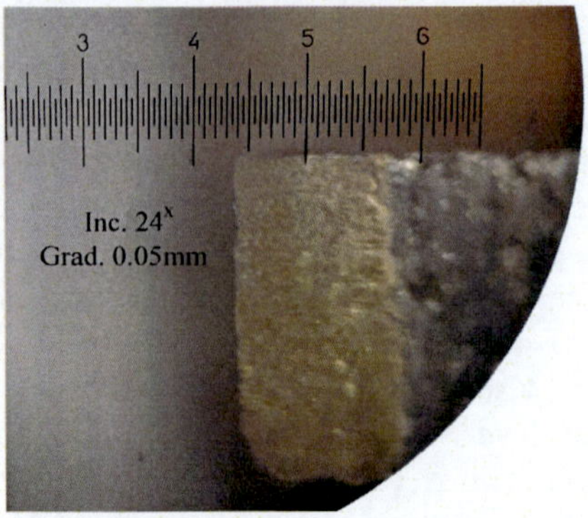

Fig. 1. A back surface region with adhesive stuck brass.

Therefore, in the calculations for this contact zone, the Prandtl law of friction with a value of $\mu = 0.5$ was applied, which corresponds to the conditions of complete sticking. In [13], for deformation processes with high contact pressures, a relationship was established between the coefficients f and μ when the rear angle changes from 0 to 5°:

$$f = \frac{2\mu}{\left[2\alpha + \arcsin 2\mu - \sqrt{1 - 4\mu^2} - \pi - 1\right]}. \tag{3}$$

Substituting the limiting value $\mu = 0.5$ into formula (3), we obtain $f = 0.41$. For the cutting front surface with contact with the chips, where σ_n is much smaller, the value $f = 0.15$ was used, which was determined experimentally by the method described in [7].

The micro-cutting simulation process was stopped when all the SDS parameters in the cutting zone acquired a steady-state value.

To verify the contact interaction simulation data obtained according to the developed model, the calculation data were compared with experimental data obtained from the flow curve and graded graph hardness - stress for brass L63. For this, flow curves were experimentally constructed: stress intensity – microhardness, as well as stress intensity – deformation intensity, which, with large plastic deformations, were approximated by the expressions:

$$\sigma_i = 0,33HV, \tag{4}$$

$$\sigma_i = 820 + 120e_i. \tag{5}$$

The calibration curve hardness – stress (Fig. 2, 3) was constructed according to the methodology [8, 13]. The microhardness was measured on a PMT-3 microhardness meter with a load of 10 g.

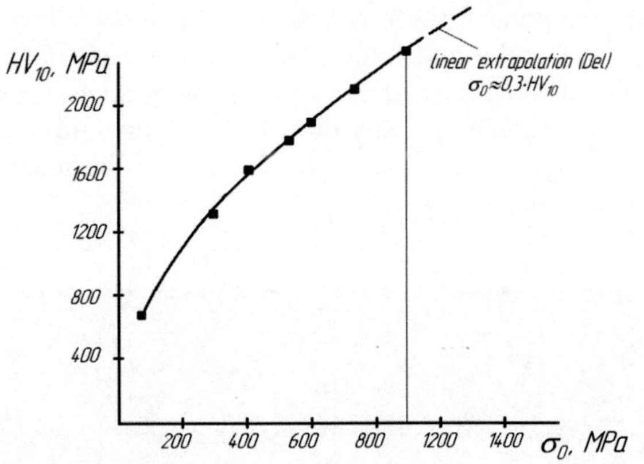

Fig. 2. Dependence between the stress intensity σ_0 and the microhardness HV_{10} during compression of brass L63.

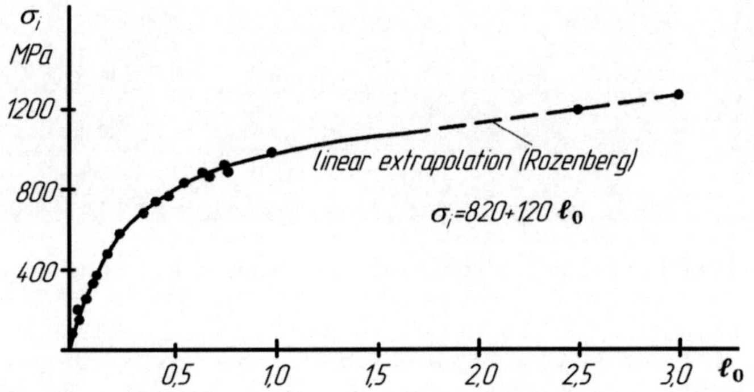

Fig. 3. Dependence between stress intensity σ_0 and deformation intensity ℓ_0 during brass compression L63.

4 Results

For practical use in designing the initial stage of FANT - micro-cutting, the greatest interest is the degree of hardening of the surface layer of the brass sample, which is determined by the accumulated deformation value e_0 (Strain Effective). Moreover, it indicates the degree of hardening of the deformable material. Its distribution in the micro-cutting zone is shown in Fig. 4.

The performed calculation meets the following conditions:

- front angle $\gamma = 0°$ as the most effective for chip formation, and technological cutting capabilities of the cutting microprofile of a cast-iron surface;
- the total depth of micro-cutting t_p is approximately two times greater than the initial rounding radius of the micro-cutter tip.

Moreover, a very important phenomenon in this process is that the micro-cutting process is accompanied at the initial stage by intensive wear of the cutting edge. It occurs until the moment when the entire system "wearing micro-cutter - brass sample" does not come into the established state of system self-organization [14]. In the calculation process, the experimentally determined ratio of the rounding radius to the actual cutting depth was taken: $\frac{r}{t_f} \approx 2$.

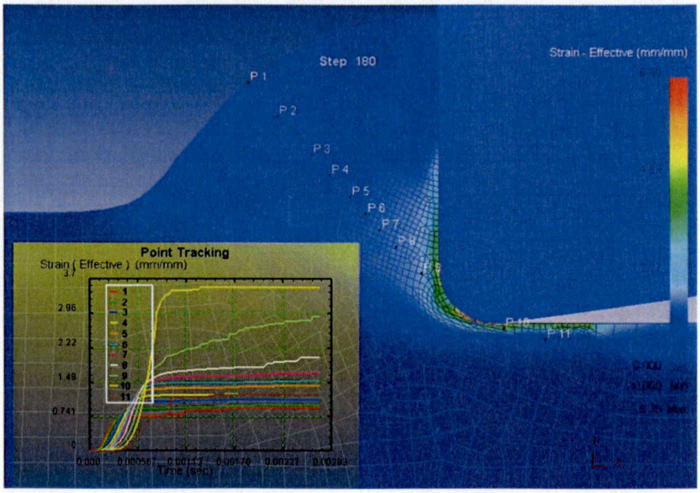

Fig. 4. Field of distribution of accumulated deformation e_0 in the micro-cutting zone.

In Fig. 4, it is seen that the depth of the hardened layer of the brass surface is several times smaller than the blunting radius of the microroughness $\boldsymbol{r} \approx 0.05$ mm. In this case, a very large gradient $\boldsymbol{e_0}$ is observed over the depth of the surface layer. Its maximum value lies at a depth of about 0.01 mm. Therefore, this nature of hardening cannot significantly affect the micro-cutting process during subsequent passes of the first stage of the FANT, and even more so, its power characteristics. It means that during subsequent micro-cutting (subsequent passages of the tool), the influence of deformation hardening on the strength characteristics of the passage is insignificant, which is confirmed by the almost constant $\boldsymbol{P_y}$ force in each of the subsequent cycles. Therefore, this hardening can not be taken into account in technological calculations.

The distribution of accumulated deformation and microhardness over the depth of the hardened surface layer is shown in Fig. 5. They were obtained in two ways: theoretically using the FEM (Fig. 4) and experimentally by measuring the microhardness by constructing the flow curves of brass L63.

Knowing the distribution of microhardness based on the calibration curve, which connects it with the value of the stress intensity σ_i, we obtain the depth distribution e_o using the experimental dependence $\sigma_i = f(e_o)$. As can be seen (Fig. 6), the coincidence of the degree of hardening of the surface layer, determined theoretically and experimentally, is satisfactory.

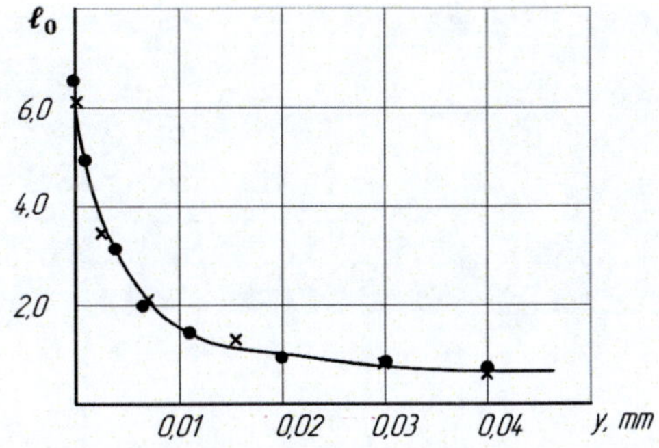

Fig. 5. Distribution of accumulated deformation e_o along y – a depth of the hardened surface layer of a brass sample: ●- according to FEM; ×- according to the distribution of microhardness HV_{10} at a load of 10 g.

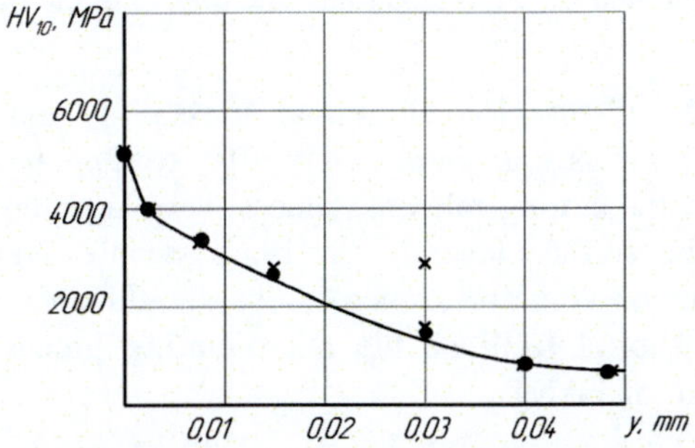

Fig. 6. Distribution of microhardness HV_{10} along y – a depth of the hardened surface layer of a brass sample: ●– according to FEM; ×– according to the distribution of microhardness HV_{10} at a load of 10 g.

Additionally, the stress-deformed state in the micro-cutting zone was also studied by the following parameters: deformation rate intensity (Strainerate), stress intensity (StressEffective), normal stresses σ_x (SX), and σ_y (SY); material flow rate (TotalVelocity).

The value σ_y determines the magnitude of contact stresses on the back surface of the tool – a micro-cutter.

An analysis of the micro-cutting processes simulation results revealed the following patterns.

Figure 7 shows the intensity distribution of deformation rates. A very small region of the most intense plastic deformation is located at the bottom of the rounded surface of the incisor tip. It is here, where the intensive adhesion of brass to cast iron begins, and deformation hardening of the surface of the brass sample occurs.

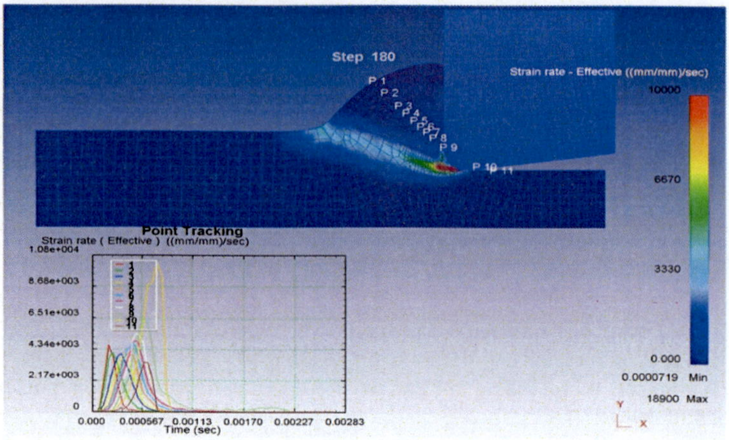

Fig. 7. Distribution field of deformation rate intensity ξ; in the micro-cutting zone.

Figure 8 shows the distribution of normal stresses σ_x and σ_y. A characteristic feature of this distribution is that immediately after coming out of contact with the rounded radius part of the cutting microroughness, very large **tensile** stresses of $\sigma_x \approx$ 2000 MPa arise in the surface layer of the brass sample. Naturally, brass cannot withstand such a stretch. Therefore, microcracks should appear in this ultra-small surface zone, which should facilitate the micro-cutting process during subsequent passes at the first stage of FANT.

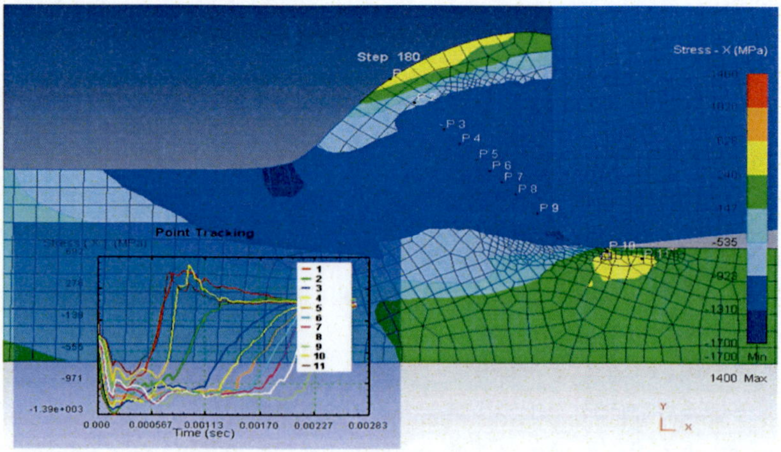

Fig. 8. Distribution of the normal stress σ_x and σ_y in the micro-cutting zone.

As can be seen from Fig. 8 in the contact zone on the back surface, there are very large contact pressures of about 1.4 GPa. Such a high contact pressure ensures the adhesion of brass to the surface of the cast iron part in this area at the first stage of FANT - micro-cutting.

The obtained result is confirmed experimentally by the data shown in Fig. 1, which shows the back surface of a cast-iron sample, where the area of adhesive stuck brass is visible. A similar fact occurred during the deforming broaching of brass pipes with a steel element [15]. On its surface in the zone of maximum contact loads, comparable with the hardness of the material being processed, we observed high strength adhesive sticking with brass.

In subsequent traditionally used rubbing modes, the contact pressure is significantly (by order of magnitude) less than this value. It shows the importance of the brass sticking foci formation, which occurs precisely at the first stage of FANT. Later they become the basis for the formation of the required antifriction brass skin on the cast iron surface.

5 Conclusions

During the analysis of the deformation mechanics of a brass specimen surface layer performed using the FEM, we took into account experimentally established patterns, which allowed us to assert the following:

- strong strain hardening of the brass surface which has a small penetration depth does not affect the micro-cutting process during subsequent passes of the FANT technological operation;
- large tensile stresses arise in the surface layer immediately behind the contact surface. It leads to the formation of surface microcracks, and hence to the facilitation of micro-cutting during subsequent passages of FANT;
- contact loads of about 1.4 GPa (which is extreme for brass) can be observed on the back surface of the contact zone near the lower part of the rounding. It means that the material is in the ultimate state. This state ensures the adhesive sticking of brass on the surface of cast iron and the formation of local areas of adhesively fixed brass on the microrelief of the cast-iron surface. The appearance of these areas with maximum adhesive strength form the basis for the formation of the subsequent antifriction coating.

References

1. Solovykh, E.K.: Trends in the development of surface hardening technologies in mechanical engineering. Kirovograd, 92 (2012). [in Russian]
2. Balabanov, V.I., Bolgov, V.Ju., Ishhenko, S.A.: Friction application of nanoscale antifriction coatings on parts. Nanotechnol. Ecol. Prod. **1**(3), 104–107 (2010). [in Russian]
3. Garkunov, D.N.: Finishing antifriction non-abrasive treatment (FABO) of friction surfaces of parts. RVM (Repair. Restorat. Mod.) **3**, 36–41 (2009). [in Russian]

4. Bugaev, A.M.: FANT as a technological method of increasing the internal combustion engine resource. Int. Res. J. **55**, 36–38 (2017). https://doi.org/10.23670/irj.2017.55.063. [in Russian]
5. Ragutkin, A.V., Sidorov, M.I., Stavrovskij, M.E.: Some aspects of antifriction coatings application efficiency by means of finishing nonabrasive antifriction treatment. J. Min. Inst. **236**, 239–244 (2019). https://doi.org/10.31897/pmi.2019.2.239
6. Pogonyshev, V.A., Panov, M.V.: Theoretical and experimental basis for increasing the wear resistance of machine parts. Mechanics and Phys. Process. Surf. Contact solids Parts Technol. Energy Equip. **4**, 78–84 (2011). [in Russian]
7. Shepelenko, I., Tsekhanov, Y., Nemyrovskyi, Y., Posviatenko, E.: Improving the efficiency of antifriction coatings by means of finishing the antifriction non-abrasive treatment. In: Tonkonogyi, V., et al. (eds.) Advanced Manufacturing Processes. InterPartner 2019. Lecture Notes in Mechanical Engineering, pp. 289–298. Springer, Cham (2020). https://doi.org/10.1007/978-3-030-40724-7_30
8. Rozenberg, Ju.A.: The Mechanics of the Cutting Process, p. 193. Kurgan University Publishing House (2005). [in Russian]
9. Belozerov, V.A., Uteshev, M.Kh., Kaliev, A.N.: Mechanics of deformation and fracture during cutting, p. 128. Tyumen (2012). [in Russian]
10. Stupnytskyy, V., Hrytsay, I.: Simulation study of cutting-induced residual stress. In: Advances in Design, Simulation and Manufacturing. DSMIE-2019. Lecture Notes in Mechanical Engineering, pp. 341–350 (2020)
11. Dobrotvorskiy, S., Balog, M., Basova, Y., Dobrovolska, L., Zinchenko, A.: Concept of the software for materials selection using.NET technologies. In: Tonkonogyi, V., et al. (eds.) Advanced Manufacturing Processes. InterPartner 2019. Lecture Notes in Mechanical Engineering, pp. 32–43. Springer, Cham (2020). https://doi.org/10.1007/978-3-030-40724-7_4
12. Hrytsay, I., Stupnytskyy, V., Topchii, V.: Improved method of gear hobbing computer aided simulation. Arch. Mech. Eng. **66**(4), 475–494 (2019). https://doi.org/10.24425/ame.2019.131358
13. Tsekhanov, Yu.A., Sheikin, S.E.: Mechanics of workpieces forming during deforming broaching, p. 200. Voronezh (2001). [in Russian]
14. Jakubov, F.Ja., Kim, V.A.: Structural and Energetic Aspects of Hardening and Increasing the Durability of the Cutting Tool, p. 300. Crimean Educational and Pedagogic State Publishing House (2005). [in Russian]
15. Nemyrovskyi, Ya.B., Krivosheya, V. V., Sardak, S.E., Shepelenko, I.V., Tsekhanov, Yu.A.: The use of deforming broaching for enhancing the efficiency of cutter chisels. Naukovyi Visnyk Natsionalnoho Hirnychoho Univ. **2**, 61–66 (2020). https://doi.org/10.33271/nvngu/2020-2/061

Finite Element Analysis of Thermal State and Deflected Mode During Titanium Alloys Machining

Vadym Stupnytskyy(✉), Ihor Hrytsay, and She Xianning

Lviv Polytechnic National University, 12, Bandera Street, Lviv, Ukraine
stupn@i.ua

Abstract. Titanium alloys are difficult to machining due to their physical and mechanical properties. An effective method of research the cutting process is to study using simulation models. The article describes the results of rheological simulation of titanium alloys cutting processes using the DEFORM software. The results of the stress-strain and thermodynamic state of the workpiece and tool, cutting force studies in the chip formation zone depend on the machining parameters are given. It has been proven that the process of chip formation in the machining of titanium and nickel alloys takes place under conditions of unstable cutting, which contributes to the generation of residual stresses, which arise synchronously with the increase of radial cutting force. The result of this process is an intense tool wear and dynamic instability in the cutting zone. The machinability of the titanium alloys essentially depends on the cutting parameters. The main causes of low workability are high cutting temperature, saw-tooth type of chip (as a result of an asynchronous change in longitudinal and transverse cutting forces), the adiabatic process of the chip formation, tool vibration and intense tool wear.

Keywords: Functional-oriented technological process · Titanium alloys · Simulation study · Finite element analysis · Cutting parameters

1 Introduction

Titanium alloys are characterized by high specific strength (the ratio of strength to density reaches 30–35 and more), which is almost twice the specific strength of the most common alloyed steels in mechanical engineering [1]. Besides, at elevated temperatures, titanium alloys are superior in strength to high-strength alloys of aluminum and magnesium [2]. Thus, titanium alloys are the main material for products of modern aircraft engineering, rocket construction, and the military-industrial complex. Increased corrosion resistance causes the use of titanium alloys in chemical, power engineering, manufacturing of medical equipment, and other fields. Currently, 41% of titanium is used in the aerospace industry, including 33% in the civilian industry and 8% in the military industry; 47% in the chemical industry and energy; and 12% - in other areas (sports goods – 8%, armor – 2%, etc.). In the United States, 45% of titanium is used in civil aircraft engineering, 15% in military aviation and space engineering, and 40% in other industries, including chemical, oil and gas industries, shipbuilding,

V. Tonkonogyi et al. (Eds.): InterPartner 2020, LNME, pp. 629–639, 2021.
https://doi.org/10.1007/978-3-030-68014-5_61

and medicine. Among the titanium alloys used, Ti-6A1-4V alloy (80–85%) predominates as well as Ti-10V-2Fe-3Al and Ti-6Al-2Sn-4Zr-2Mo-Si (10–15%) alloys [2]. According to forecasts, in the next 3–5 years, despite the crisis in the economies of industrialized countries, titanium consumption in the world economy will increase (due to the increase in the demand for automobile disks and aircraft engines) [1]).

When the structure and parameters of the machining process are planned, it is necessary to take into account the provision of wear resistance, residual stresses, and deformations, fatigue strength, but not only the dimensional accuracy and roughness of the surface layer. These indicators are especially considering conditions of large cyclic force and thermodynamic loads during product operation. Implementation of such technologies should be carried out on the basis of functional-oriented technologies (FOT) [3–5]. The scientific and information base of the FOT is carried out based on the research of the workpiece's stress-strain and thermodynamic state. Given publication addresses such problems.

2 Literature Review

Titanium alloys are hard-to-cut materials due to the high ratio of yield stress to temporal fracture resistance [1, 2]. For example, this ratio for titanium alloys is in the range 0.85–0.95 and for steels it is only 0.65–0.75. At the same time, mechanical characteristics of titanium alloys (relative elongation δ and reduction of the area ψ) compared to heat-resistant steels are significantly less. Low plastic properties of titanium alloys during their machining contribute to micro- and macro-cracks in front of the tool blade [6].

Chips formed during titanium alloys machining have defects of cyclic saw-tooth shape, which divide it into very weakly deformed elements connected by a thin and highly deformed contact layer. The features and reasons for the formation of such chips will be discussed in this article. Thus, in the machining of titanium, large specific loads occur, which predeterminants the presence of high temperatures in the cutting zone due to the low thermal conductivity of the compacted zones in the chip forming region [7, 8]. As a result of strong adhesion and high temperatures, the material to be treated adheres to the cutting tool, and this contributes to a large increase in friction force. The adhesion of titanium onto the contact surfaces of the cutting tool also results in a change in its kinematic and edge-geometric parameters. The shear angle when cutting titanic alloys reaches 38–44. When cutting speed exceeds 60 m/min, it is possible to form chips with a chip thickness ratio of more than 1, that is, the chips have a less length than the cutting path. The temperature in the cutting zone increases with increasing cutting speed, and to a lesser extent, with increasing feed [1].

The machinability of the titanium alloys is 3-4 times less than for carbon steels and 5–7 times less than for aluminum alloys [1, 11]. For example, the ratio of relative workability of titanium alloys Ti-6Al-4V relative to steel AISI 1045 is 0.22–0.26 [2]. It is recommended to use low cutting speeds at small feeds with an intensive supply of lubricating-cooling liquid during machining titanium alloys. Therefore, the cutting tool should be made of more wear-resistant titanium alloy machining tool materials than for carbon steels, giving preference to hard alloys. However, even if these conditions are ensured, the cutting speed should be reduced by 3–4 times as compared to the machining of the steels to provide sufficient tool life, especially for CNC machines [9, 12].

For knowledge-intensive mechanical engineering production, an essential stage of its technological planning is the study of the process of cutting hard-to-cut materials under conditions for which technological systems or their elements are only created. For example, cutting with tools from new tool materials, with new coatings and with new blade geometry, nanoprocessing, and the like are studied [12]. These tasks can be solved by simulating cutting processes. Existing mathematical models of machining processes require data on-chip type, the structure of contact stress distribution, average friction coefficient, etc. Such data can be obtained using complex experimental studies that require significant time and costs. Creating simulation models of cutting processes is a more efficient method of investigation. Therefore, theoretical study of power, stress-strain and thermal forming processes, the use of adequate deformation and fracture criteria to describe the behavior of metals during chip formation taking into account the actual state of the tool contact surfaces, the conditions of various types of chips formation is an important scientific task [1].

The most promising are models of the cutting process, based on numerical methods, which make it possible to solve the problem of deformation and destruction of the shear layer based on fundamental equations of body thermodynamics by discrete smoothing. One such method is the Finite Element Analysis (FEA) [10, 13]. Given the above, by the finite element method, the DEFORM software product implements models of various materials cutting processes (including titanium alloys). Methods have been developed for calculating the shape of chips, stress-strain and thermodynamic state of the workpiece and tool blade, cutting force, etc. DEFORM is a specialized engineering software product developed by the Scientific Forming Technologies Corporation (SFTC), designed for analysis of metal forming processes, heat, and mechanical treatment [13].

Simulation modeling of cutting processes is based on physical laws on destruction, as loss of material resistance to deformation due to disruption of internal bonds, achieved as a result of a critical concentration of microcracks (brittle material) or reaching the energy threshold by a certain grid cell (elastic material) [13]. It is important to note that the ideal conditions of destruction do not exist in nature. Any material can only, with a certain assumption, be attributed to brittle or plastic materials for which there are characteristic processing conditions and a mathematical apparatus that describes or mimics them. It is accepted in engineering practice that the compression strength of a perfectly brittle material is eight times its tensile strength [14]. To date, there is also no single concept for the analytical description of cutting mechanisms, nor is there a universal criterion by which machining processes can be adequately described. At a sufficient level, only selected cases of analytical models of destruction have been theoretically developed and experimentally confirmed, and their generalizations have been carried out [1]. Studies of loading, deformation, and energy criteria of destruction and based on comparison with experimental data and conclusions about their adequacy to real processes were described in the article [13].

3 Research Methodology

The initial data for the cut simulation in the DEFORM are following: 2D or 3D model of machining part (surface); cutting parameters; 2D or a 3D model of the tool, geometry of the cutting edge, material and a covering; durability, mechanical, thermophysical characteristics of the processed material; model of tool's wear; re-mesh criteria (such as error of compliance of modeling on a power vector, vector of speed and admissible geometrical error; strain type (Lagrangian Incremental or Steady-State machining); iteration method (Direct Iteration or Newton-Raphson), type of the deformation and temperature solver (the Sparce or the Skyline methods).

Simulation results – graphs of cutting forces during titanium alloy Ti6Al4V machining (feed rate S = 0.25 mm; cutting depth t = 1 mm; cutting speed V = 120 m/min) are shown in Fig. 1. From the graphs, it can be concluded that the cutting forces in the transverse and longitudinal direction fluctuate in the dissonance at a high frequency (approximately 1.6 kHz).

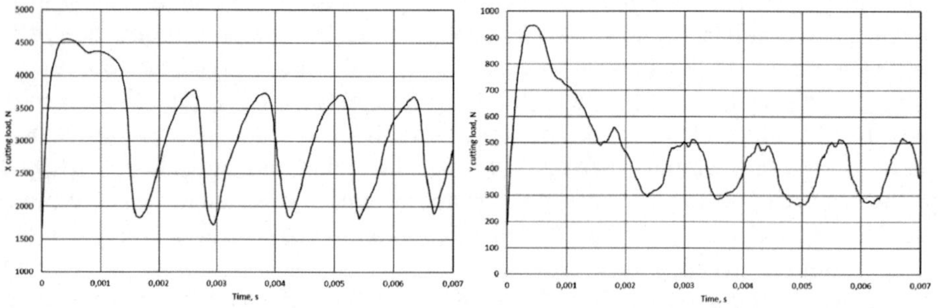

a b

Fig. 1. Components of cutting forces during machining of titanium alloy Ti6Al4V (S = 0.25 mm; t = 1 mm; V = 100 m/min), which are determined based on simulation model analysis in DEFORM 2D: a – in the transverse direction; b – in the longitudinal direction.

The McClintock fracture criterion was used in the simulation rheological model according to the recommendations [13] because the break-up of material periodically on the rake and relief surfaces dominate in the titanium alloy machining.

Under conditions of high-speed deformation, there is a phenomenon of adiabatic shear - formation of local zones of increased plastic strain state, the energy of which turns to thermal, which reduces yield stress. Titanium alloys have extremely low thermal conductivity (significantly lower than heat resistant carbon steels [2]). For example, for Ti6Al4V alloy, the parameter of heat conductivity is 11 W/(m·K) whereas for steel AISI1045 of 45–40 W/(m·K). Thus, the cutting of titanium alloys produces a temperature that is more than two times the temperature level of the steel machining. Heat is concentrated near the strips or sliding planes, which contributes to the formation of so-called adiabatic shear strips when the heat is released at a higher rate than the rate of its withdrawal into the cooling environment due to thermal conductivity.

In shear deformation, the transition from uniform deformation to adiabatic shear is determined by the magnitude and rate of strain. On the example given in Fig. 2, it is possible to define precisely the geometry of excess strain zone (ε = 4.8 mm/mm) (Fig. 2a). The zone of the increased strain rate $\dot{\varepsilon} = (1.04–1.30) \cdot 10^5$ (mm/mm)/s. (Fig. 2b) is even more localized. The temperature in the field of adiabatic shear is 1350...1390 °C and depends, except value and rate of strain, same on thermophysical characteristics of the material (Fig. 2c). The formation of adiabatic shear bands can lead to breakage across these bands and the formation of segment chips. Besides, polymorphic transformations can occur in high-temperature regions. Under metal cutting conditions, a dynamic equilibrium is established between the crystalline and amorphous structural phases of the titanium alloy Ti6Al4V.

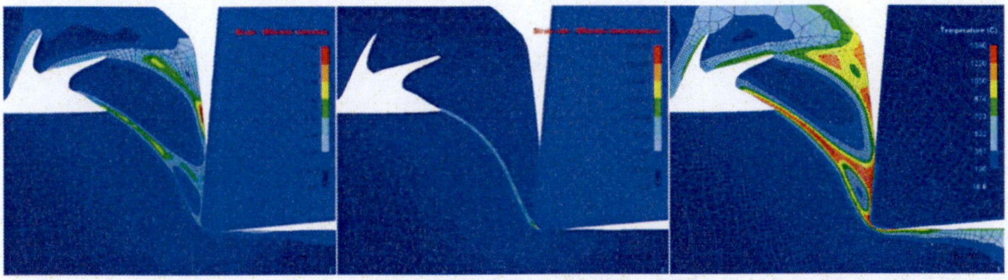
a b c

Fig. 2. Demonstration of strain state (a), strain rate (b), and thermodynamic state (c) in the adiabatic shear zone when machining titanium alloy Ti6Al4V.

Adiabatic shear-type instability is generally thought to develop at the point of stress concentration in a solid body whose stress state is usually uniform. The narrow strip of large plastic strain propagates along the planes of maximum shear stresses until unloading occurs or a crack is formed in the material [5]. Adiabatic shear is one of the most important mechanisms of destruction under dynamic load, in which destruction can occur in a fragile scenario, despite the prevalence of plastic properties of the material being machined. Fig. 3 shows the mechanism of adiabatic shear zone occurrence. This is the case: at a certain point in the simulation study (t = 1.16 ms), the transverse cutting force of the maximum value ($P_X \approx 3.8$ kN) occurs. At the same time, the longitudinal cutting force has a minimum value ($P_Y \approx 300$ N). This chip-forming step occurs with the prevailing process of compressing the chip formation zone with the chip root convexity over the outer surface. At the moment of cutting (t = 1.35 ms) opposite phenomena occur: transverse cutting force takes minimum value ($P_X \approx 1.7$ kN), and longitudinal cutting force - maximum value ($P_Y \approx 520$ N). This step corresponds to the complete prevalence of the shear mechanism with characteristic chip concavity over the outer surface. At the moment of cutting (t = 1.6 ms), the step of dominant compression of the chip formation zone is repeated, etc.

Thus, the following physical cutting process is confirmed. First, the tool cuts into the material of the workpiece and is subjected to plastic deformation of the metal layer, which is accompanied by the absorption of external energy. At the same time, the

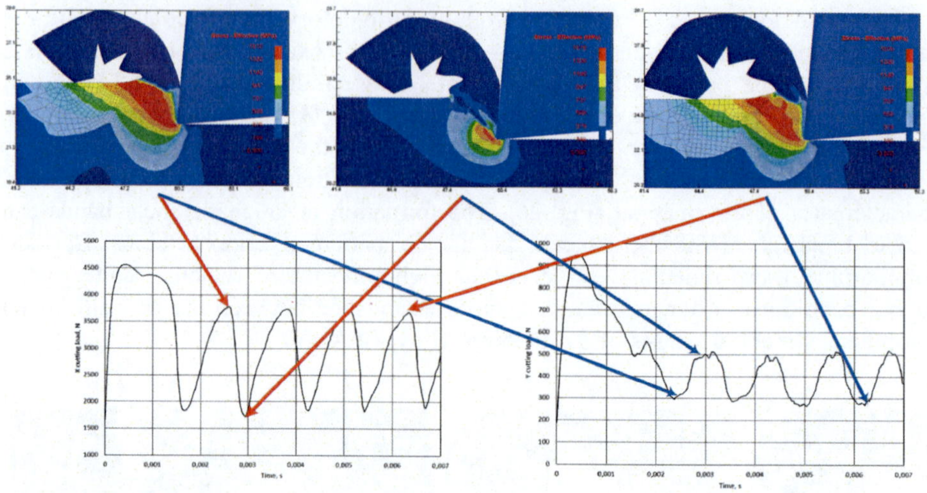

Fig. 3. The relationship between the rheological pattern of chip formation and the loading parameters of cutting in dynamics during titanium alloy machining as a result of DEFORM 2D simulation analysis.

shearing layer of metal is strengthened and becomes brittle. Then the metal is shifted, and the chip element is formed. As a result of the low thermal conductivity of the material being processed, the cutting heat is concentrated in the chip separation zone and contributes to the activation of the adhesion and diffusion processes, causing the cutting edges of the tool to be destroyed. These phenomena, together with the improved abrasive and mechanical properties of titanium-based alloys, cause vibrations and intensify the wear process of the cutting blade of the tool. These features of alloys significantly impair their workability compared to carbon steels and other alloys. Therefore, some intermediate conclusions can be drawn from the machinability of titanium alloys, namely: great viscosity and high-temperature generation can be avoided if titanium alloy is machined at low cutting speed; high-frequency change of the cutting forces in the longitudinal and transverse directions contributes to the vibrations of the tool and workpiece during machining; low cutting speed and high plasticity of the alloy contributes to the formation of a build-up on the rake surface of the tool, and this impairs the quality of the treatment high cutting temperature and intense vibrations lead to rapid wear of the cutting tool; high cutting-induced residual stress appears as a result of the high machining temperature, adiabatic chip making conditions and changeable loading.

4 Results

The logical question arises: what machining parameters need to be used in order to achieve maximum productivity and sufficient quality of the treated layer? Studies carried out in the field of cutting dynamics can be carried out by two methods:

analytical analysis of the cutting pattern based on the study of the destruction process as a partial case of the plastic deformation process; rheological simulation of the stress-strain and thermal state of the workpiece using the finite element analyze methods. Moreover, the simulation type of research is carried out not only for establishing the adequacy of the theoretical foundations of forming the surfaces to be machined but mainly for the effective study of the dynamic stress-strain state of the workpiece in different chip formation zones using the variable simulation data (tool geometry, materials, cutting parameters, etc.) for next applying in optimization models. These structural-parametric models can be used for effective implementation of a functionally-oriented technological process.

Studies of the dependence of the loading, thermodynamic state, and deflected mode during the machining process on the cutting depth (Fig. 4), speed (Fig. 5) and tool's geometrical parameters (Fig. 6) have been simulated in Deform 2D.

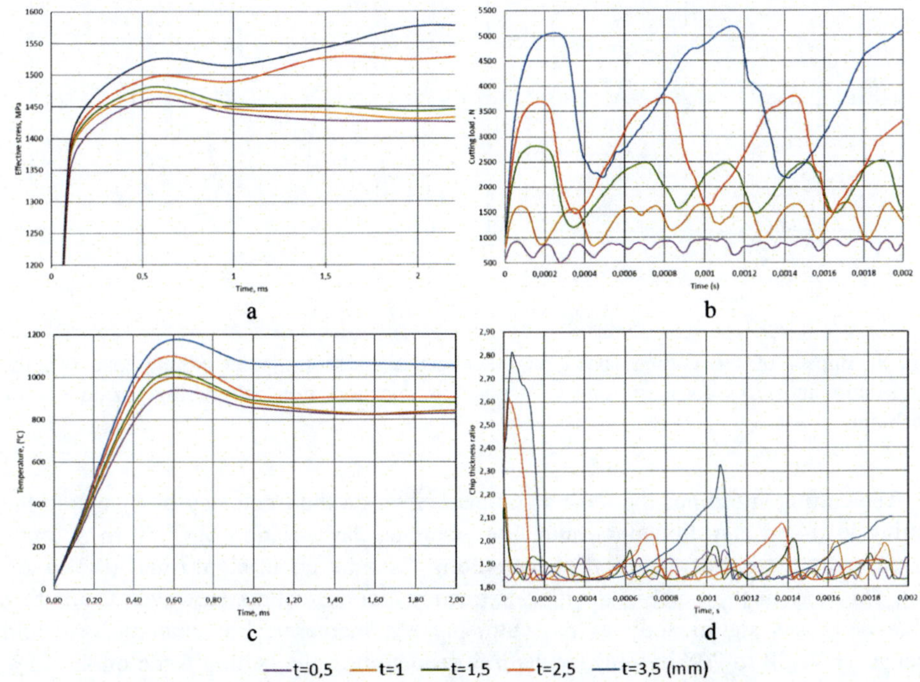

Fig. 4. Impact of the cutting depth on the effective stress (a), total cutting load (b), the temperature in ship's forming zone (c), and chip thickness ratio (d) during titanium alloy machining.

The cutting depth strongly affects the cutting force and chip thickness ratio (Fig. 4). For example, varying the cutting depth from 1.0 to 2.5 mm increases the cutting force by about two times, and increases the thickness ratio by about 1.6 times. A particular change in these parameters occurs if the cutting depth increases beyond 2 mm. However, changing the cutting depth does not significantly affect the effective stress and cutting temperature (increasing t by seven times (from 0.5 to 3.5 mm) results in changing these parameters by 5–7% only).

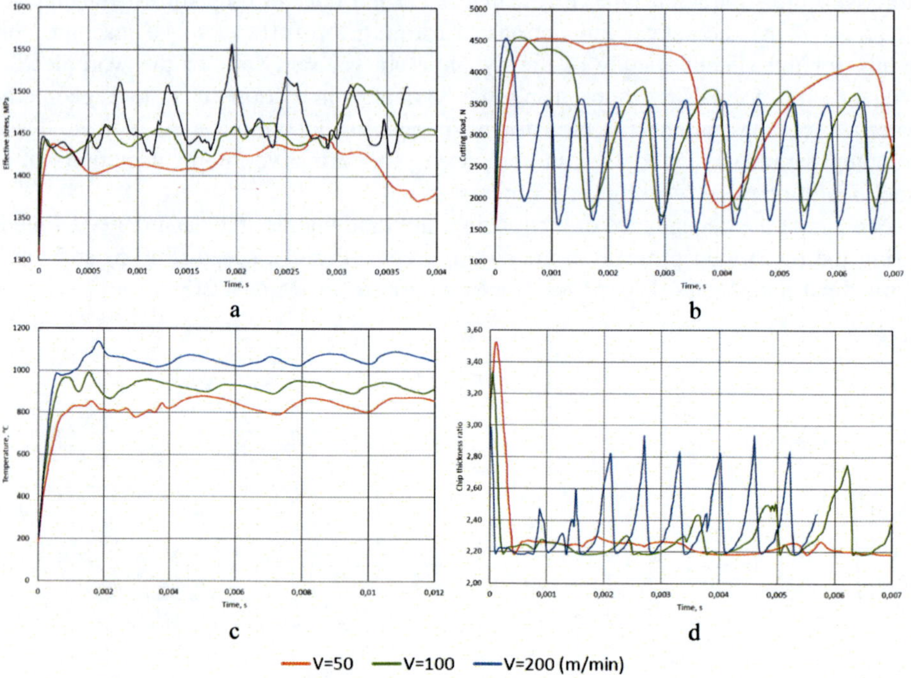

Fig. 5. Impact of the cutting speed on the effective stress (a), total cutting load (b), the temperature in ship's forming zone (c), and chip thickness ratio (d) during titanium alloy machining.

The cutting speed has the most significant effect on the frequency of change in the cutting load and chip thickness ratio. For example, the variation V of 50 to 200 m/s increases the period of variation of the cutting force by about eight times (0.3 to 2.5 ms). This may be the cause of the high-frequency tool oscillation. However, the cutting force decreases significantly as the cutting speed increases. For example, a 4-fold change in V (50 to 200 m/min) results in a reduction in the cutting force on the 25% (from 3.2 to 2.4 kN). Effective stress and cutting temperature do not vary much changing with the cutting speed increasing (7–10%).

The geometry of the tool's cutting edge has a significant effect on the cutting load, chip thickness ratio, and processing temperature. For example, when changing the rake angle from +10° to –5° increases the average cutting force by about 80% (from 2 to 3.6 kN), increases the chip thickness ratio by 27% (from 1.8 to 2.3), and increases the temperature by 32% (from 830 °C to 1100 °C). Besides, an increase in the rake angle results in a large variation in the frequency of the cutting force, especially for a tool with a negative rake angle. However, changing the geometric parameters of the cutting edge has an insignificant effect on the effective stress.

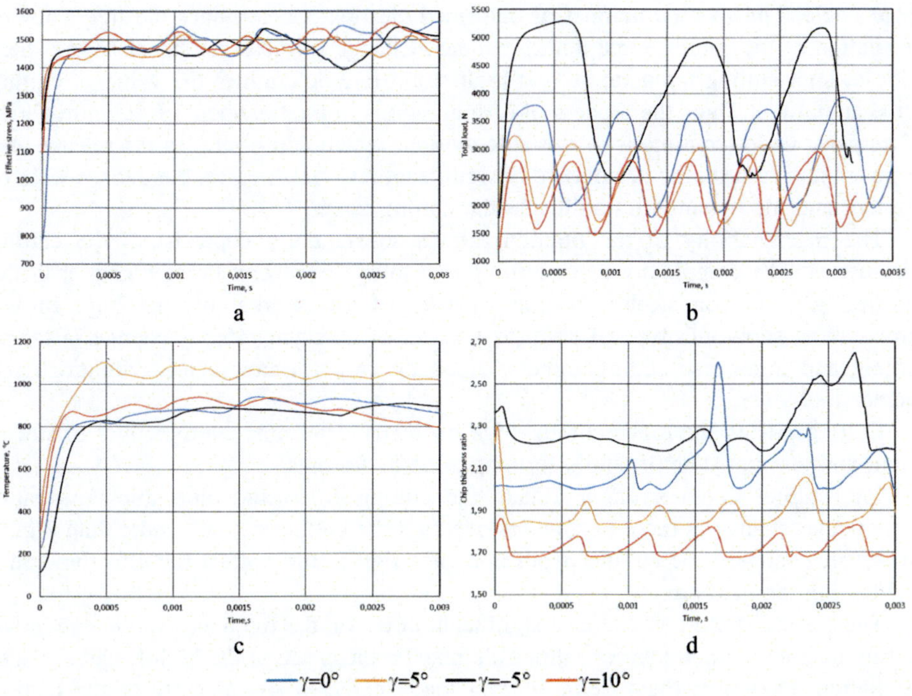

Fig. 6. Impact of the tool geometry (rake angle) on the effective stress (a), total cutting load (b), the temperature in ship's forming zone (c), and chip thickness ratio (d) during titanium alloy machining.

We compared the results of theoretical studies with the results of simulation of the cutting process in the DEFORM 2D, as well as with the results of experimental studies. Statistical analysis of the results showed a satisfactory error of geometric parameters of chips obtained experimentally (up to 10%), and full correspondence of the location of the zone of increased deformation (determined according to [11] in the direction of metal texture) with the results of theoretical and simulation studies.

Considering that the dynamics of the chip thickness ratio and the variation of the load, heat and deflected modes of the titanium alloy cutting can be determined in the simulation rheological model, using the Computer-Aided Forming System described in [3, 4] is the most efficient solution for the process control of the cyclic machining processes.

5 Conclusions

The results of simulation and experimental studies of the chip formation process in the titanium alloys machining show that the geometric shape of the chips, which is formed during the adiabatic shear process, is determined by the conditions of plastic deformation resistance, the cyclicity of the cutting forces and the geometric change of the shear angle. The specific number of deformed chip parts determines the frequency of oscillation of the cutting force during contact. Change of metal resistance in the shear zone at high cutting temperature and high strain rate determines the value of cutting force amplitude. Thus, the process of forming chips in the treatment of titanium alloys takes place under conditions of unstable cutting, which contributes to the release of a large amount of heat, the occurrence of significant residual stresses, the intense wear of the tool and the dynamic instability in the cutting zone.

The machinability of the titanium alloys substantially depends on the cutting parameters. The complexity of titanium-based alloys is several times higher than the machining of carbon steels. The main causes of low workability are high cutting temperature, saw-tooth type of chip (as a result of an asynchronous change in longitudinal and transverse cutting forces), adiabatic chip formation, tool vibration, and intense tool wear.

Cutting modes and tool's cutting-edge geometry determine the dynamics of stress-strain and thermal state of chip's forming zone as follows:

The cutting depth strongly affects the cutting force and chip thickness ratio. A particular change in these parameters occurs if the cutting depth is more than 2 mm. However, changing the cutting depth does not significantly affect the effective stress and cutting temperature.

The cutting speed has the most significant effect on the frequency of change in the cutting load and chip thickness ratio. This may be the cause of the high-frequency tool oscillation. However, the specific cutting load decreases significantly as the cutting speed increases.

The geometry of the tool's cutting edge has a significant effect on the cutting load, chip thickness ratio, and processing temperature. Besides, an increase in the rake angle results in a considerable variation in the frequency of the cutting force, especially for a tool with a negative rake angle. However, changing the geometric parameters of the cutting edge has an insignificant effect on the effective stress.

Subsequent studies will be directed to the analysis of the effect of the obtained results of the stress-strain and thermodynamic state of the titanium workpiece on the formation of parameters that affect the operational characteristics of the product (residual stresses, surface roughness, structural-phase transformations, hardening of the machined surface layer, etc.). However, this research will be a separate investigation that will be a follow-up to the studies described above.

References

1. Davim, J.P.: Machining of Titanium Alloys. Materials Forming, Machining and Tribology. Springer, Heidelberg (2014)
2. Parry, J.: Titanium Alloys: Types, Properties, and Research Insights. Nova Science Publishers, Hauppauge, New York (2017)
3. Stupnytskyy, V.: Features of functionally-oriented engineering technologies in concurrent environment. Int. J. Eng. Res. Technol. (IJERT) **2**(9), 1181–1186 (2013)

4. Stupnytskyy, V., Hrytsay, I.: Simulation study of cutting-induced residual stress. In: Advances in Design, Simulation and Manufacturing. DSMIE-2019. Lecture Notes in Mechanical Engineering, 2019, pp. 341–350 (2020)
5. Davim, J.P.: Surface Integrity in Machining. Springer, London (2010)
6. Ulutan, D., Ozel, T.: Machining induced surface integrity in titanium and nickel alloys: A review. Int. J. Mach. Tools Manuf. **51**, 250–280 (2011)
7. Klocke, F., König, W., Gerschwiler, K. Advanced machining of titanium- and nickel-based alloys. In: Kuljanic, E. (ed.) Advanced Manufacturing Systems and Technology. International Centre for Mechanical Sciences (Courses and Lectures), vol. 372. Springer, Vienna (1996)
8. Nouari, M., Makich, H.: On the physics of machining titanium alloys: interactions between cutting parameters, microstructure and tool wear. Metals **4**, 335–358 (2014)
9. da Silva, R.B., Machado, Á.R., Ezugwu, E.O., Bonney, J., Salesc, W.F.: Tool life and wear mechanisms in high speed machining of Ti–6Al–4V alloy with PCD tools under various coolant pressures. J. Mater. Process. Technol. 213(8), 1459–1464 (2013).
10. Ali, M.H., Ansari, M.N.M., Khidhir, B.A., et al.: Simulation machining of titanium alloy (Ti-6Al-4V) based on the finite element modeling. J. Braz. Soc. Mech. Sci. Eng. **36**, 315–324 (2014)
11. Arrazola, P.-J., Garay, A., Iriarte, L.-M., Armendia, M., Marya, S., Maître, F.L.: Machinability of titanium alloys (Ti6Al4V and Ti555.3). J. Mater. Process. Technol. 209 (5), 2223–2230 (2009).
12. Jawaida, A., Che-Harona, C.H., Abdullah, A.: Tool wear characteristics in turning of titanium alloy Ti-6246. J. Mater. Process. Technol. **92–93**, 329–334 (1999)
13. Stupnytskyy, V.: Analysis and selection of the criterion of local destruction during the simulation cutting process with deform 2D. Bull. "Lviv Polytech." Natl. Univ. "Optim. Prod. Process. Tech. Control Mech. Eng. Instr." **729**, 107–115 (2012).
14. Ginting, A., Nouari, M.: Surface integrity of dry machined titanium alloys. Int. J. Mach. Tools Manuf. **49**(3–4), 325–332 (2009)

Process Engineering

Parameter Identification of the Heat Supply System in a Coach

Serhii Khovanskyi[1], Ivan Pavlenko[1(✉)], Jan Pitel[2], Oleg Bogdaniuk[1], and Vitalii Ivanov[1]

[1] Sumy State University, 2, Rymskogo-Korsakova St., Sumy 40007, Ukraine
i.pavlenko@omdm.sumdu.edu.ua
[2] Technical University of Kosice, 1, Bayerova St., 08001 Presov, Slovak Republic

Abstract. The article presents the issue of providing comfortable conditions for passengers in coaches. At the same time, the main goal is to increase the efficiency of thermal energy use with the combined heating supply system. The research considers the analysis of thermal modes using the developed, comprehensive approach. This approach is based on both the numerical simulation and the regression analysis. The use of CFD analysis allowed analyzing the fields of temperature and air velocity in a coach. The simulation established that the use of the combined heating supply system causes turbulent airflow. A multifactor experiment was conducted to evaluate the impact of the radiator temperature on the internal one at different external temperatures. For the radiator temperature in a range of 20 ÷ 70 °C and the external temperature of –8 ÷ 24 °C, coefficients of the developed regression model were established. The obtained values allow estimating the thermal condition of the coach. Besides, the analytical dependence for the control of the radiator temperature, depending on the external temperature was obtained. The corresponding regulating curves and the transfer function were built. Overall, practical recommendations for compliance with sanitary and hygienic standards and providing comfortable conditions for coaches were formulated.

Keywords: Convective heat transfer · Comfortable condition · CFD modeling · Regression analysis · Regulating curve · Transfer function

1 Introduction

The issue of energy-efficient use of fuel and energy resources has become one of the most important world problems. This issue is particularly acute in Europe due to the scarcity of natural resources, high-energy intensity, and the gradual increase in energy consumption. The main solution to these problems is to carry out energy-saving measures aimed at improving the current situation. An important area of energy conservation is rail transport, which is widespread in Europe. It is also a significant consumer of energy resources.

European railway networks play an important role in cargo and passenger turnover. Soon, given the requirement of the European Union, which is the reduction in

V. Tonkonogyi et al. (Eds.): InterPartner 2020, LNME, pp. 643–653, 2021.
https://doi.org/10.1007/978-3-030-68014-5_62

emissions of harmful substances into the atmosphere by motor transport, the popularity, and importance of rail transport will increase. Improving the energy efficiency of rail transport involves the use of innovative technologies, as well as the introduction of modern energy efficiency standards for the design of new and reconstruction of existing coaches. It is also necessary to take into account the energy consumption of all their technical systems.

The analysis of the condition of life support systems of coaches allowed formulating the purpose of this research, which is to increase the efficiency of thermal energy use of coaches with a combined heating supply system based on the analysis of their heating modes.

To achieve this goal, the research tasks were set and solved. They are as follows: a mathematical and numerical model of the thermal condition of a coach was developed; modeling of aerodynamics and heat and mass transfer processes in a coach for the analysis of its thermal condition; based on the developed model, the assessment of a thermal condition of the coach is carried out; the influence of non-stationary processes in the internal volume of the coach on its general thermal condition is investigated.

2 Literature Review

The temperature mode of carriages is most influenced by external (ambient temperature, solar radiation, precipitation) and internal (different types of heat release) parameters. Calculating the heat balance of stationary conditions in comparison with non-stationary conditions is not a difficult task, because non-stationary processes of thermal conductivity are those characterized by changes in surface temperature not only in space but in time as well. In calculations based on the differential equations of heat and mass transfer, mathematical modeling is often used to analyze the formation of microclimate parameters in a non-stationary process.

The general significance of the problem in terms of carrying out experimental and computational studies of flow modes and thermal patterns in coaches was substantiated in the papers [1, 2]. The improvement of the cooling/heating cogeneration systems using the parametric analysis and experimental research was carried out in the articles [3, 4]. Analytical models of heat transfer from the heating supply system in a premise were proposed in the papers [5, 6] considering the heat transfer through external constructions [7]. Ways for increasing the efficiency of the equipment by the intensification of technological processes are presented in research works [8, 9].

The engineering methodology developed by Bogoslovskyi V., which considers the impact of non-stationary parameters on the thermal balance of the premise, should be highlighted. It allows considering the influence of harmonic and intermittent receipts. Based on this technique, the mathematical model in which features of enclosing designs and change of external factors were considered is constructed. Based on the developed model, the research of the influence of a periodic change of the ambient temperature on the temperature of a zone of service of the premise with various thermal inertness of furnace skin and indoor equipment is carried out. The model provides an understanding of the delay time of the temperature wave from the outside to the middle of the room and the amplitude of fluctuations in internal air temperature.

The approaches and methods mentioned above involve a large number of calculations. The application of this approach to such work involves the availability of expensive specialized equipment for measurements, during which you may encounter some difficulties. Therefore, today numerical computer modeling is widely used, which allows conducting a large number of experiments with different variations of parameters and choosing the most optimal version of the project. Besides, one of the advantages of this method is the ability to model an existing object, which allows assessing the effectiveness of the work and finding ways to modernize.

3 Research Methodology

3.1 CFD Modeling

A three-dimensional model of the coach model 48–060 was created (Fig. 1) in order to conduct a numerical experiment on the thermal condition of the coach according to the method proposed by the authors in [10], using SolidWorks software. The overall dimensions of the coach model (11.6 m × 2.1 m × 2.6m) are approximate to real conditions. When creating a three-dimensional model, radiators with heating pipes behind the protective covers were designed as heating devices. To simplify further calculations, the model was idealized. Namely, it was decided to ignore the influence of some objects of the interior of the coach, and the calculations will be carried out only for half of the coach.

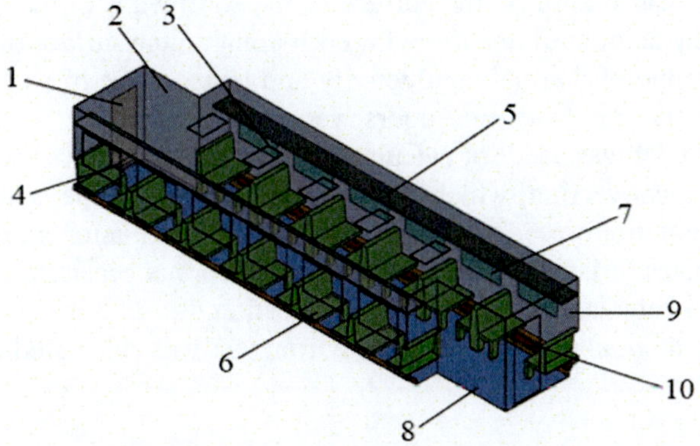

Fig. 1. Three-dimensional model of a coach model 48–060: 1 - doors; 2 - inner wall; 3 - ventilation (inlet); 4 - ventilation (outflow); 5 - luggage racks; 6 - berths; 7 - window; 8 - floor; 9 - outer wall; 10 - heating device.

The calculation area in this task is the volume of the coach, which consists of air, heaters, berths for seating, and luggage racks. When creating the calculation area of the coach, several zones were identified: windows, doors, a ceiling, a floor, exterior, and interior walls. It is because, in the output data, each zone has different parameters.

A calculation grid was constructed (Fig. 2) to perform a numerical study using the ICEM CFD software product [11]. This calculation grid is block-structured, i.e., it consists only of hexagonal (volume) elements – hexahedra. This type of grid was chosen because this geometric model of the calculation area is a set of regular elementary volumes. The main parameters of the air area of this calculation grid are the number of elements $-3.3 \cdot 10^6$; the number of nodes $-3.2 \cdot 10^6$.

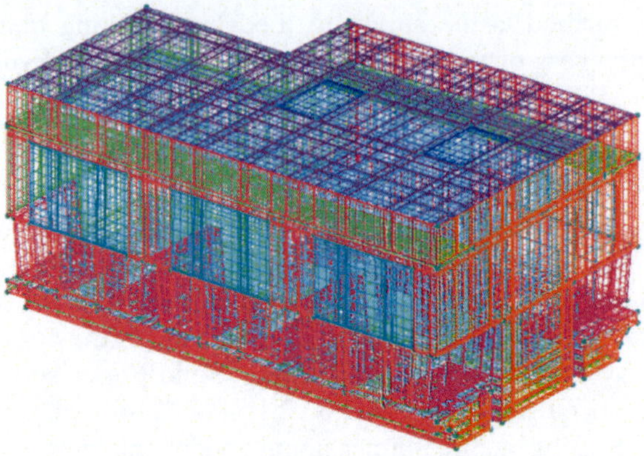

Fig. 2. Calculation grid in the product ICEM CFD.

For further research, the calculation grid was transferred to a special software product called CFX-Pre. The boundary conditions were the boundary conditions of the first (temperature distribution on the surface of the solid walls of the calculation area) and the second (the amount of heat flow for each point on the surface of the body at any time) kind. The value of thermal conductivity and temperature of surfaces of ceilings, floors, walls, heaters, windows, and doors were set.

The gravity model was used for calculations. This model allows taking into account the process of free convection, which occurs in the volume of the coach; this refers to the action of volumetric forces. The construction of the calculation area was carried out in spatial coordinates. It is necessary to set the gravitational acceleration of the vertical coordinate axis to calculate the natural air convection.

The spread of thermal radiation in the environment was calculated using the Monte Carlo model.

3.2 Parameter Identification

In this research, a combination of an artificial neural network and a factorial experiment was used. In particular, for a factorial experiment, a quadratic form was proposed [12] to describe the effect of temperature on the heating radiator T_r and the temperature of the filler structures T_e on the air temperature in the coach T_i:

$$T_i = \{A\}^T\{T\} + [B]^T\{T\}[B], \tag{1}$$

where $\{T\} = \{T_r, T_e\}^T$ is a vector-column of temperatures; $\{A\}$, $[B]$ – vector-column and matrix of estimated parameters, which for a two-factor experiment [13] take the following form:

$$\{A\} = \begin{Bmatrix} a_1 \\ a_2 \end{Bmatrix}; [B] = \begin{bmatrix} b_{11} & \frac{b_{12}}{2} \\ \frac{b_{12}}{2} & b_{22} \end{bmatrix}, \qquad (2)$$

where a_0, a_1, b_{11}, b_{12}, and b_{22} – estimated parameters.
In this case, formula (1) can be rewritten as follows:

$$T_i = a_1 T_r + a_2 T_e + b_{11} T_r^2 + b_{12} T_r T_e + b_{22} T_e^2. \qquad (3)$$

Mathematical model (1) allows using the results of CFD analysis to calculate the numerical values of the estimated coefficients of formula (3). The least-squares method [14] is used as a condition for minimizing the next error functional

$$R(\{X\}) = \sum\nolimits_{k=1}^{N} \left\langle \left(\{A\}^T \{T\}\right)_k + \left([B]^T \{T\}[B]\right)_k - T_i^{<k>} \right\rangle^2 \rightarrow min, \qquad (4)$$

where k is the number of the experimental point *(k* = 1, 2,..., *N); N* is the total number of virtual experiments; $\{X\} = \{a_1, a_2, b_{11}, b_{12}, b_{22}\}^T$ is the combined vector-column of the estimated parameters, which is determined by the linear regression formula [15]:

$$\{X\} = \left([C]^T [C]\right)^{-1} [C]^T \{Y\}, \qquad (5)$$

where $[C]$, $\{Y\}$ is the rigidity matrix of the system and the vector-column of external influence, respectively, the elements of which are determined by the following dependencies:

$$[C]_{m,n} = \sum\nolimits_{k=1}^{N} \left(T_r^k\right)^{\alpha_{m,n}} \left(T_e^k\right)^{\beta_{m,n}}; \{Y\}_m = \sum\nolimits_{k=1}^{N} T_i^k \left(T_r^k\right)^{\gamma_m} \left(T_e^k\right)^{\delta_m}, \qquad (6)$$

which degree is elements of the following symmetric matrices:

$$[\alpha] = \begin{bmatrix} 2 & 1 & 3 & 2 & 1 \\ 1 & 0 & 2 & 1 & 0 \\ 3 & 2 & 4 & 3 & 2 \\ 2 & 1 & 3 & 2 & 1 \\ 1 & 0 & 2 & 1 & 0 \end{bmatrix}; [\beta] = \begin{bmatrix} 0 & 1 & 0 & 1 & 2 \\ 1 & 2 & 1 & 2 & 3 \\ 0 & 1 & 0 & 1 & 2 \\ 1 & 2 & 1 & 2 & 3 \\ 2 & 3 & 2 & 3 & 4 \end{bmatrix}; \gamma = \begin{bmatrix} 1 \\ 0 \\ 2 \\ 1 \\ 0 \end{bmatrix}; \delta = \begin{bmatrix} 0 \\ 1 \\ 0 \\ 1 \\ 2 \end{bmatrix}. \qquad (7)$$

To develop practical recommendations for compliance with sanitary and hygienic standards and provide comfortable conditions [16] for passengers in coaches, it is necessary to establish the dependence of radiator temperature T_r as a controlled value on ambient temperature Te to ensure the constant comfortable temperature in car T_i. In particular, we can obtain the following analytical dependence from Eq. (3):

$$T_r = \frac{1}{2b_{11}}\left[\sqrt{(a_1 + b_{12}T_e)^2 + 4b_{11}\left(T_i - a_2T_e - b_{22}T_e^2\right)} - a_1 - b_{12}T_e\right]. \quad (8)$$

4 Results

4.1 Numerical Simulation

In the process of the calculation, the main thermophysical, hydromechanical, thermodynamic, and optical parameters of the calculation areas were obtained when reaching the stationary calculation mode [17] at the coolant temperature in heating devices 70 °C and filler structures –24 °C. These parameters were averaged over the volume of the calculation area and taking into account the steady-state process of complex heat transfer [18, 19].

Temperature distribution by volume with the installed system of combined heating of the closed coach with various set boundary conditions, namely with the stable temperature of heating devices with a range of variation from 70 °C and to 20 °C and ambient temperature from −24 °C to 8 °C indicated in Fig. 3.

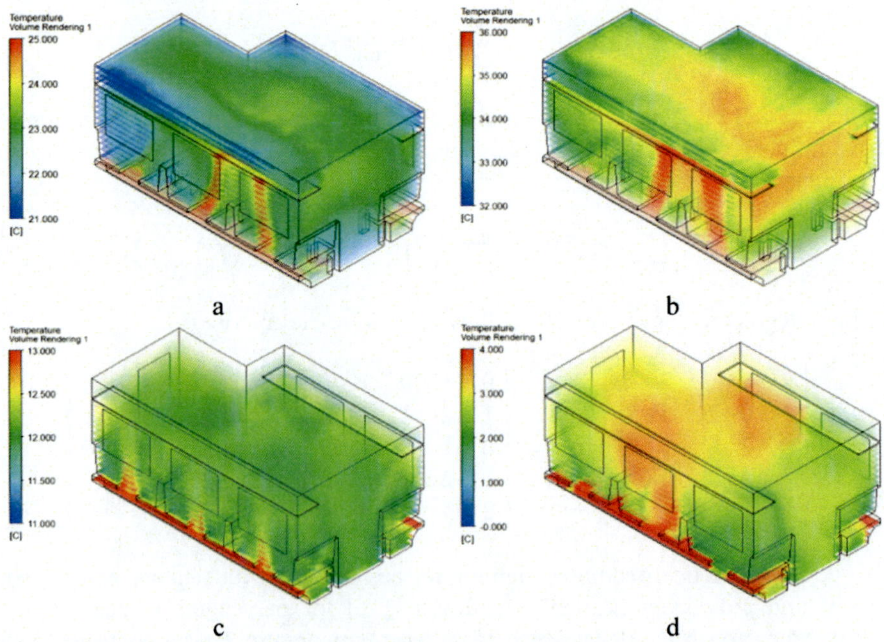

Fig. 3. Temperature distribution on the volume of the coach: a – 70 °C on the radiator and –24 °C of filler structures; b – 70 °C on the radiator and 8 °C of filler structures c – 20 °C on the radiator and 8 °C of filler structures; d – 20 °C on the radiator and –24 °C of filler structures.

Visualization of the air velocity in the volume and plane in the middle of the coach in the stationary mode is shown in Fig. 4.

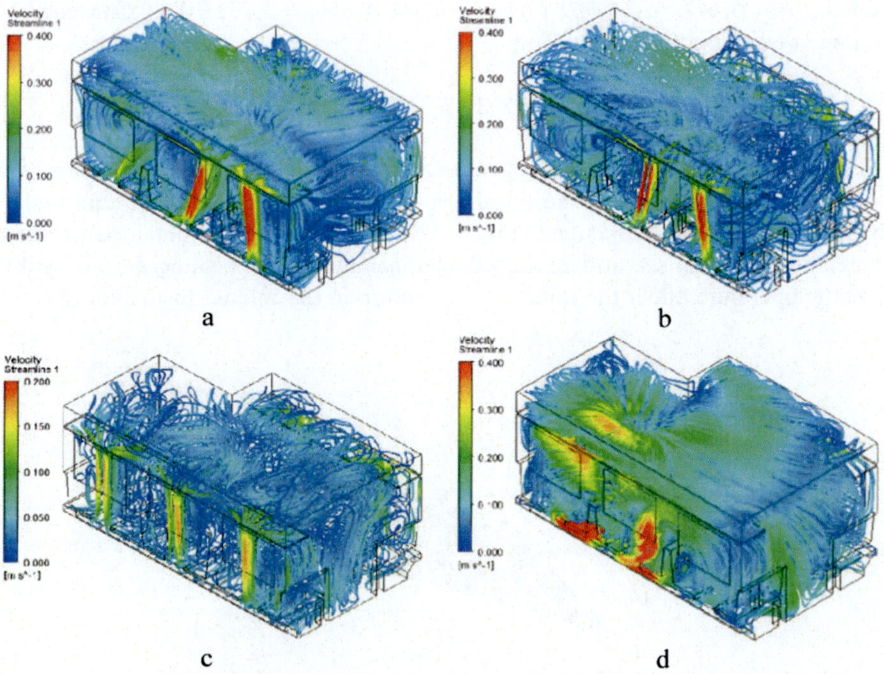

Fig. 4. The air movement velocity in the volume of the coach at a – 70 °C on the radiator and –24 °C of filler structures; b – 70 °C on the radiator and 8 °C of filler structures; c – 20 °C on the radiator and 8 °C of filler structures; d – 20 °C on the radiator and –24 °C of filler structures.

Figure 3, 4 show that as a result of the use of the combined heating type, where the radiators are heating pipes covered with a protective casing, there is a turbulent movement of air in the coach [20] with the formation of several vortices. Taking into account the set different temperatures on the radiator and filler structures, the formed vortices in each case behave differently.

Having analyzed the results of the calculations visually, we can say that the air heated by the radiator moves up the walls, envelops the berths and luggage racks uncounted on the way, then rises to the ceiling and falls to the bottom and this process is repeated.

In the premise, there is a circular movement of air, which causes vortices [21]. In most cases, the air movement velocity is higher near the outer walls of the coach due to the radiators located there.

4.2 Regression Analysis

As a result of the regression analysis [22] of using the proposed mathematical model describing formulas (1)–(7), the following values of estimated parameters have been obtained: $a_1 = 0.442$, $a_2 = 0.251$, $b_{11} = b_{22} = 0$, $b_{12} = 1.731 \cdot 10^{-3}$. Consequently, formula (3) takes the following form:

$$T_i = 0.442 \cdot T_r + 0.251 \cdot T_e + 1.731 \cdot 10^{-3} \cdot T_r \cdot T_e. \tag{9}$$

A graphical representation of the geometric interpretation of the dependence of the temperature in the middle of the coach on the temperature of the heaters and the ambient temperature was obtained (Fig. 5). Having analyzed the obtained geometric interpretation, one can see how exactly the temperature of the heating devices and the ambient temperature affect the internal temperature in the middle of the coach.

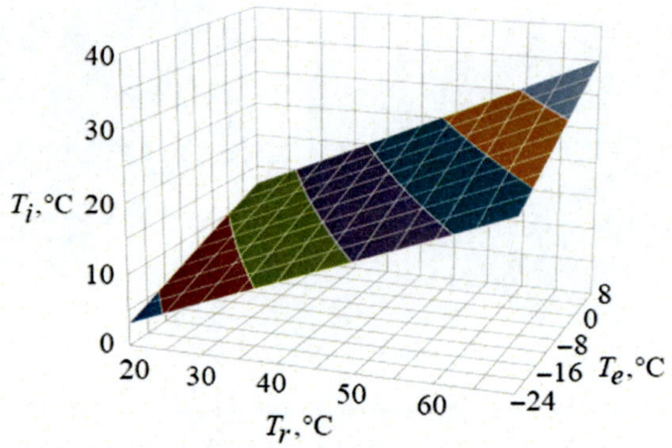

Fig. 5. A regression surface.

For the obtained numerical values of estimated parameters of the model of ensuring the thermal condition of the coach, comfortable conditions for passengers in the coaches will be provided with proper regulation of the radiator temperature (8), which can be determined by Eq. (9):

$$T_r = \frac{T_i - a_2 T_e}{a_1 + b_{12} T_e}. \tag{10}$$

In this case, the effect of changes in the ambient temperature on the change of the regulated temperature of the radiator is determined by the following transfer function:

$$\varphi(T_e) = \frac{\Delta_r / T_r}{\Delta_e / T_e} = \frac{\partial T_r}{\partial T_e} \frac{T_e}{T_r} = \frac{T_e}{a_1 + b_{12} T_e} \frac{a_1 a_2 + b_{12} T_i}{a_2 T_e - T_i}, \tag{11}$$

where Δ_r is the variation of the radiator temperature as a result of the change of the ambient temperature Δ_e.

The corresponding radiator temperature control curves are shown in Fig. 6, taking into account the values mentioned above of the estimated coefficients. Particularly, for the case study, when the ambient temperature varies in a range of –10 ÷ 25 °C, the radiator temperature is regulated in the range of 25 ÷ 55 °C.

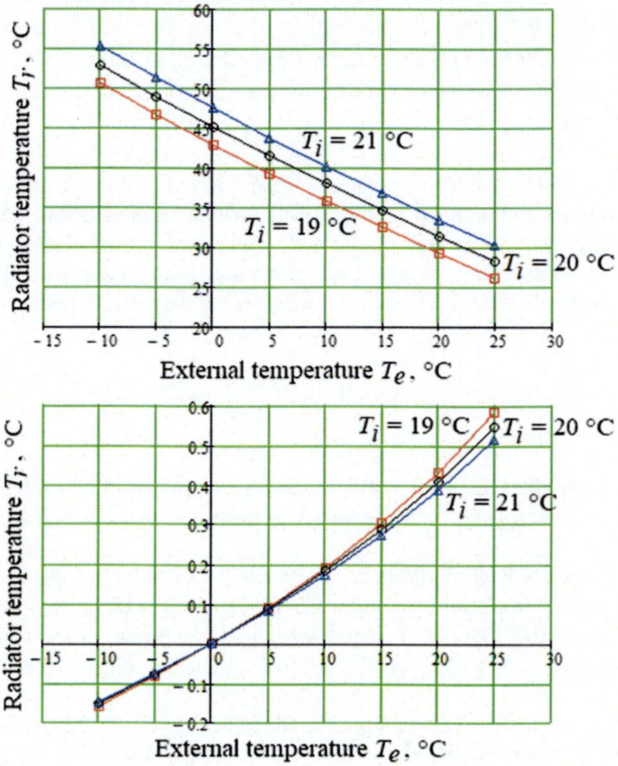

Fig. 6. Regulation curves (a) and the transfer function (b).

5 Conclusions

In this paper, a three-dimensional model of a coach model 48-060 with a combined heating system has been created. A calculated CFD model has been developed that allows analyzing the temperature distribution and air velocity in a coach. The numerical simulation established that the use of the combined heating type in the coach causes a turbulent movement of air with the formation of two vortices. It has also been determined that the air heated by the radiator moves up along the walls, enveloping the berths and luggage racks. At the same time, the air movement velocity is higher near the outer walls of the car due to the location of the radiators.

A two-factor experiment has been performed to evaluate the effect of temperature on the radiator on the internal temperature of the coach at different ambient temperatures. Thus, for the radiator temperature in the range of 20 ÷ 70 °C, as well as the ambient temperature of –10 ÷ 25 °C, the coefficients of the regression model have been found to be a_1 = 0.442, a_2 = 0.251, and b_{12} = $1.731 \cdot 10^{-3}$. Due to the proposed regression dependences, the coefficients of the corresponding mathematical model for estimating the thermal condition of the coach have been obtained.

The proposed approach, based on the integrated application of numerical modeling and regression analysis, is useful for developing practical recommendations for compliance with sanitary and hygienic standards and providing comfortable conditions for passengers in the coaches. In particular, the dependence of the radiator temperature on the ambient temperature to ensure the required comfortable temperature in the coach has been established.

Acknowledgments. The numerical simulations were realized at the Faculty of Manufacturing Technologies with a seat in Presov of Technical University of Kosice (Presov, Slovak Republic) within the research projects "Research of Unsteady Temperature Condition of Premises Heated by Low-grade Renewable Energy Sources" and "Identification of Parameters for Technological Equipment using Artificial Neural Networks" funded by the National Scholarship Programme of the Slovak Republic.

References

1. El-Bialy, E.M., Khalil, E.E.: Flow regimes and thermal patterns in a subway station. In: 9th Annual International Energy Conversion Engineering Conference, IECEC 2011, 98128 (2011)
2. Khalil, E.E., El-Bialy, E.M.: Experimental and computational investigation of flow regimes and thermal patterns in a subway station. ASHRAE Trans. **118**(1), 199–206 (2012)
3. Yue, C., Han, D., Pu, W., He, W.: Parametric analysis of a vehicle power and cooling/heating cogeneration system. Energy **115**, 800–810 (2016). https://doi.org/10.1016/j.energy.2016.09.072
4. Jianbo, L., Shiming, X., Xiangqiang, K., Kai, L., Fulin, C.: Experimental study on absorption/compression hybrid refrigeration cycle. Energy **168**, 1237–1245 (2019). https://doi.org/10.1016/j.energy.2018.11.093
5. Mizakova, J., Pitel, J.: An analytical dynamic model of heat transfer from the heating body to the heated room. In: MATEC Web of Conferences, vol. 125, p. 02047 (2017). https://doi.org/10.1051/matecconf/201712502047
6. Mizakova, J., Pitel, J., Hrehova, S.: Some simulation results of heat transfer through the wall model. Int. J. Math. Models Methods Appl. Sci. **8**(1), 1–8 (2014)
7. Pitel, J., Khovanskyi, S., Pavlenko, I., Mizakova, J.: Dynamic simulation of heat transfer through external building constructions. J. Eng. Sci. **6**(1), E33–E38 (2019). https://doi.org/10.21272/jes.2019.6(1).e6
8. Fesenko, A., Basova, Y., Ivanov, V., Ivanova, M., Yevsiukova, F., Gasanov, M.: Increasing of equipment efficiency by intensification of technological processes. Periodica Polytechnica Mech. Eng. **63**(1), 67–73 (2019). https://doi.org/10.3311/PPme.13198
9. Liaposhchenko, O., Pavlenko, I., Monkova, K., Demianenko, M., Starynskyi, O.: Numerical simulation of aeroelastic interaction between gas-liquid flow and deformable elements in modular separation devices. In: Ivanov, V., et al. (eds.) Advances in Design, Simulation and Manufacturing II. DSMIE-2019. Lecture Notes in Mechanical Engineering. Springer, Cham, pp. 765–774 (2020). https://doi.org/10.1007/978-3-030-22365-6_76

10. Rimar, M., Kulikov, A., Fedak, M., Khovanskyi, S., Pavlenko, I.: Application of the CFD software for modeling thermal comfort in sport hall. MM Sci. J. **2020**(March), 3723–3727 (2020). https://doi.org/10.17973/MMSJ.2020_03_2019023
11. Wei, X., Duan, B., Zhang, X., Zhao, Y., Yu, M., Zheng, Y.: Numerical simulation of heat and mass transfer in air-water direct contact using computational fluid dynamics. Procedia Eng. **205**, 2537–2544 (2017). https://doi.org/10.1016/j.proeng.2017.10.218
12. Al-khlyleh, M., Hajja, M.: Copositive and positive quadratic forms on matrices. RM **74**(4), 155 (2019). https://doi.org/10.1007/s00025-019-1079-7
13. Ozawa, K., Mejza, I., Mejza, S., Kuriki, S.: Two-factor experiments with split units constructed by cyclic designs and square lattice designs. Revstat Stat. J. **16**(3), 279–294 (2018)
14. Zhang, Z., Yuan, S., Xu, Z., Fang, L.: A prediction method about central heating parameters based on method of least square. In: 2nd International Conference on Measurement, Information and Control, ICMIC 2013, vol. 2, pp. 1163–1166 (2013). https://doi.org/10.1109/MIC.2013.6758165
15. Sieres, J., Campo, A.: Uncertainty analysis for the experimental estimation of heat transfer correlations combining the Wilson plot method and the Monte Carlo technique. Int. J. Therm. Sci. **129**, 309–319 (2018). https://doi.org/10.1016/j.ijthermalsci.2018.03.019
16. Ampofo, F., Maidment, G., Missenden, J.: Underground railway environment in the UK. Part 1: review of thermal comfort. Appl. Thermal Eng. **24**(5–6), 611–631 (2004). https://doi.org/10.1016/j.applthermaleng.2003.10.017
17. Khovanskyi, S., Pavlenko, I., Pitel, J., Mizakova, J., Ochowiak, M., Grechka, I.: Solving the coupled aerodynamic and thermal problem for modeling the air distribution devices with perforated plates. Energies **12**(18), 3488 (2019). https://doi.org/10.3390/en12183488
18. Petinrin, M.O., Dare, A.A.: Numerical investigation of the concave-cut baffles effect in shell-and-tube heat exchanger. J. Eng. Sci. **6**(1), E1–E9 (2019). https://doi.org/10.21272/jes.2019.6(1).e1
19. Petinrin, M.O., Towoju, O.A., Ajiboye, S.A., Zebulun, O.E.: Numerical study of the effect of changing tube pitches on heat and flow characteristics from tube bundles in cross flow. J. Eng. Sci. **6**(2), E1–E10 (2019). https://doi.org/10.21272/jes.2019.6(2).e1
20. Lin, F.K.T., Hwang, G.J., Wong, S.-C., Soong, C.Y.: Numerical computation of turbulent flow and heat transfer in a radially rotating channel with wall conduction. Int. J. Rotating Mach. **7**(3), 209–222 (2001). https://doi.org/10.1155/S1023621X01000197
21. Beghein, C., Jiang, Y., Chen, Q.Y.: Using large eddy simulation to study particle motions in a room. Indoor Air **15**(4), 281–290 (2005). https://doi.org/10.1111/j.1600-0668.2005.00373.x
22. Pavlenko, I., Trojanowska, J., Ivanov, V., Liaposhchenko, O.: Parameter identification of hydro-mechanical processes using artificial intelligence systems. Int. J. Mech. Appl. Mech. **2019**(5), 19–26 (2019)

Optimal Sizing of the Evaporation Chamber in the Low-Flow Aerothermopressor for a Combustion Engine

Dmytro Konovalov[1], Halina Kobalava[1(✉)], Mykola Radchenko[1], Vyacheslav Sviridov[1], and Ionut Cristian Scurtu[2]

[1] Admiral Makarov National University of Shipbuilding, 9, Heroes of Ukraine Avenue, Mykolayiv 54025, Ukraine
g.lavamay@gmail.com
[2] Naval Academy "Mircea cel Batran", Constanta, Romania

Abstract. The efficiency of gas turbine plants will be improved by cooling the cyclic air with an aerothermopressor. Constructive and technological factors affecting the aerothermopressor work were considered in this paper. CFD simulation and calculation of the water evaporation process in the evaporation chamber (working chamber) of the aerothermopressor was carried out. The Eulerian–Lagrangian approach was used to simulate the interaction of water droplets injected and airflow. The developed software was used in order to determine the characteristics of the aerothermopressor workflow. It is based on the methods of calculating thermogasdynamic compression and pressure losses due to the aerodynamic resistance of the two-phase flow in the aerothermopressor flow part. An empirical equation has been determined for calculating the optimal relative length of the aerothermopressor evaporation chamber depending on the initial droplet diameter at the inlet and the mass water concentration in the airflow. Analytical determination of the optimal relative length of the aerothermopressor evaporation chamber allows determining the length section of the evaporation chamber in order to obtain the maximum value of pressure increase, as a result of thermogasdynamic compression.

Keywords: CFD simulation · Pressure increase · Two-phase flow

1 Introduction

The energy-saving technologies are used to improve the auxiliary systems work of gas turbine plants (GTP) [1–6]. They ensure the low-grade heat utilization of secondary energy resources and contribute to the improvement of the GTP fuel and energy efficiency. Such technologies include contact cooling of the gas turbine plant cyclic air, which is based on the process of thermogasdynamic compression. A feature of this process is an increase in pressure as a result of the instantaneous evaporation of injected liquid into the airflow accelerated to the speed of sound. The process of contact cooling cyclic air GTP is taken place in a two-phase jet device.

V. Tonkonogyi et al. (Eds.): InterPartner 2020, LNME, pp. 654–663, 2021.
https://doi.org/10.1007/978-3-030-68014-5_63

Many design and technological factors are affecting the aerothermopressor efficiency, namely, the flow part design and the method of injected water in the apparatus, the optimal relative length of the evaporation chamber, the aerodynamic resistance influence of the injection system, etc. The corresponding development of the flow part design and the workflow organization is relevant in the development of aerothermopressor technologies. At the same time, it is necessary to be able to determine the relative length of the aerothermopressor evaporation chamber analytically, at the output of which the maximum pressure increase will be obtained as a result of thermogasdynamic compression.

2 Literature Review

Contact air cooling by instantaneous evaporation of water droplets is widely used in the cooling systems of working fluids of power plants [1–4]. One of the effective methods of contact air cooling is to use a two-phase jet device. The process of thermogasdynamic compression at a flow speed closed to sound is taken place in this apparatus. And besides deep air cooling, there is also an effectively dispersed spraying of liquid in the airflow and the pressure of this flow is increased. This device was named the aerothermopressor [7].

Achieving more efficient spraying of a liquid in the aerothermopressor, and accordingly, a more productive process of water droplets evaporation is one of the main tasks in designing a jet apparatus. The water droplet diameter at the inlet rather highly affects the length of the evaporation section and the velocity mode: with a decrease in the initial droplet size, the length of the evaporation section and friction losses is decreased significantly. Water injection increases the friction coefficient by 10–20%; therefore, in order to avoid a further increase in losses, it is necessary to reduce the aerodynamic resistance of the system structural elements. It is desirable to place the injection system devices in the flow with a low gas velocity (in front of the nozzle) and to perform them of a more streamlined shape [7]. Another negative factor affected the work of the aerothermopressor is the surface friction effect, which occurs with a gradual decrease in the evaporation rate due to a decrease in the droplet's surface area.

The chamber length should ensure full evaporation of the droplet at maximum pressure increase. At the aerothermopressor designing, it is necessary to choose a length of the working chamber that would ensure the full evaporation of liquid droplets. It should be noted that there are no methods for analytically determining the length of the evaporation section and, accordingly, the working chamber. The length of the working chamber is chosen experimentally or under the practice of operating jet devices of different types (ejectors, injectors) [7, 8]. Moreover, to ensure full evaporation, the length is chosen with a reserve. However, this design approach in practice increases the length of the working chamber section, which corresponds to the dry operating mode of the aerothermopressor. Hence, surface friction becomes the predominant factor, and the pressure loss increases significantly.

3 Research Methodology

Simulation and calculation of the process of liquid evaporation in the aerothermopressor evaporation chamber were carried out using computer CFD modeling techniques. The CFD-modeling will expand the range of changes of input parameters and boundary conditions for obtaining the results of a gas-dynamic analysis of two-phase flow.

Many models and analytical solutions are describing the processes of droplet evaporation in the airflow (taking into account the transfer of mass, energy, and momentum between the droplets and air) [9–12]. The airflow was modeled as a compressible, turbulent, and continuous flow. The airflow model is based on a system of differential equations of mass conservation, momentum conservation, and energy conservation [11]. Among existing turbulence models Reynolds-Averaged Navier–Stokes (RANS), the k-ε Realizable turbulence model is preferred. The k-ε Realizable model satisfies the mathematical constraint on the Reynolds stresses and makes it possible to predict the behavior of the two-phase flow propagation velocity [13].

The calculation of the two-phase flow is carried out in the following sequence: first of all, the airflow is calculated without taking into account the influence of the injected water droplets. After that, the discrete phase (trajectories, size, velocity, position, and temperature changes of the droplets) is calculated in the Lagrangian approach based on the calculated airflow field. At the stage of coupling calculation, the mass, energy, and momentum transfer exchanges of the two phases are calculated [11, 14–16].

Due to the interaction between the continuous phase (airflow) and the discrete phase (droplets), changes of thermodynamic characteristics and motion of injected water droplets are done. The following equation determines the trajectory of the droplet:

$$\frac{dw_{\mathrm{d}}}{dt} = F_{\mathrm{D}}(w - w_{\mathrm{d}}) + \frac{g(\rho_{\mathrm{d}} - \rho)}{\rho_{\mathrm{d}}}, \tag{1}$$

where F_{D} – aerodynamic drag force is determined by:

$$F_{\mathrm{D}} = \frac{18\mu}{\rho_{\mathrm{p}}\delta_{\mathrm{d}}^{2}}\frac{C_{\mathrm{D}}Re}{24}, \tag{2}$$

where w, w_{d} – air and water droplet velocity; δ_{p} – droplet diameter; C_{D} – aerodynamic drag coefficient; μ – dynamic viscosity.

It is accepted that the injection is carried out in the confuser of the aerothermopressor (the droplet diameter is up to 20 μm). In the confuser, the droplets are additionally crushed, and the flow is aligned. Therefore, it is assumed that at the inlet to the evaporation chamber, droplets are getting the same diameter with the same initial parameters for each size of the evaporation chamber.

The processes of transfer and evaporation of water droplets in the evaporation chamber for air cooling are considered. The aerothermopressor model and the evaporation chamber are shown in Fig. 1. The main standard size of the chamber with a diameter of 50 mm was adopted, at which the air mass flow is 1.0–1.5 kg/s. The chamber length is optimal for determining the maxima of the pressure increase in the process of thermogasdynamic compression. The site for water injection is located at the inlet to the chamber and occupies the entire cross-section for uniform droplets distribution.

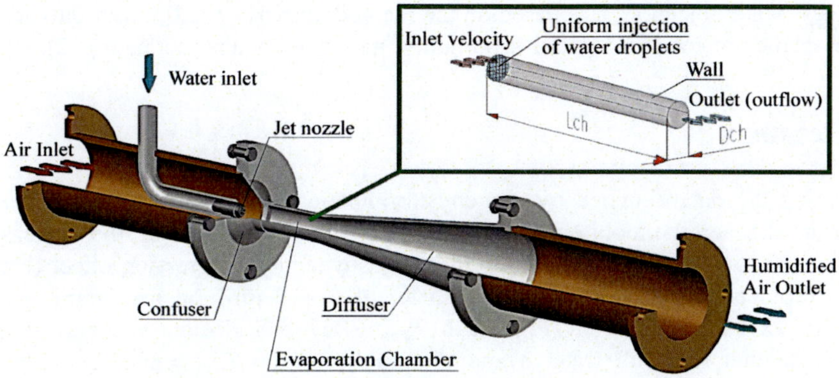

Fig. 1 Aerothermopressor model and boundary conditions for the evaporation chamber

The computational mesh is a hybrid mesh. It was constructed using the Automatic Method. The main elements are tetrahedrons. The boundary wall layers were constructed using Inflation; the mesh elements of wall layers are prisms.

To carry out a numerical simulation of the transfer and evaporation of water droplets in the airflow in the aerothermopressor evaporation chamber, the calculation method was determined in the ANSYS Fluent software package. The calculation was carried out, taking into account the convergence of the results, and the output of streamlines for workflow parameters was processed and visualized in the CFD-Post.

The parameters of the airflow at the inlet (velocity, pressure, temperature) correspond to the parameters of the cyclic air between the compressors of the gas turbine plant. The boundary conditions at the outlet (Fig. 1) correspond to Outflow (outflow to the atmosphere, without indicating the pressure at the outlet).

Water droplets (discrete phase) are taken the same spherical shape. The collision and merging of droplets were not taken into account in the simulation due to the moderate concentration of water in the airflow (up to 10%) and a rapid evaporation process (relative velocity of injected water $w_w/w_{air1} = 0.3$). Water droplets were injected in the same direction as flow.

The study was conducted for a number of values of the water droplet diameter ($\delta_p = 3, 4, 5, 6, 7, 10, 15, 20$ μm) injected into the airflow at the evaporation chamber inlet. The main inlet parameters of the airflow and water injection: evaporation chamber diameter $D_{ch} = 50$ mm; evaporation chamber length $L_{ch} = 2500$ mm; inlet air pressure $P_1 = 3{,}01 \cdot 10^5$ Pa; inlet air temperature $T_{atp1} = 500; 550$ K, inlet air velocity $w_{air1} = 331.7; 350.0$ m/s; air mass flow $G_{air} = 1.352; 1.298$ kg/s; Mach number at the inlet evaporation chamber $M = 0.50; 0.60; 0.74$; water temperature $T_w = 300$ K; water velocity $w_w = 100$ m/s; water mass flow $G_w = 0.09$ kg/s.

To determine the characteristics of the aerothermopressor workflow, the developed software was used. It is based on the methods of calculating thermogasdynamic compression, as well as pressure losses due to the aerodynamic resistance of the two-phase flow in the aerothermopressor flow part [9, 17]. The software takes into account the influence of local resistances in the flow part, the droplet drag in the flow, the effect of excess water injection. It is based on the methodology [2, 7, 18], taken into account the properties of moist air (relative humidity, moisture content, enthalpy) [2].

4 Results

Analyzing the data obtained by numerical simulation of the evaporation process of water in the aerothermopressor evaporation chamber, maxima of the increase in pressure was found as a result of the thermogasdynamic compression effect (Fig. 2). When droplets are injected, with a diameter of $\delta_p = 3$ μm, the maximum pressure increase value shifts to the left (Fig. 2 a, b) $P_{abs} = 337$–338 kPa, and the relative length of the chamber with full water evaporation is $l_{ev} = 8$ (at $T_{air1} = 500$ K), $l_{ev} = 4$ (at $T_{air1} = 550$ K). With a droplet diameter increase of $\delta_p = 7$ μm, the relative length of the section with full water evaporation increases more than 5 times and is $l_{ev} = 50$ (at $T_{air1} = 500$ K), $l_{ev} = 20$ (at $T_{air1} = 550$ K). For low-flow aerothermopressors, with air mass flow to $G_{air} = 1.5$ kg/s, with an initial droplet diameter $\delta_p = 10$–20 μm, there is no full evaporation of water droplets. Consequently, the evaporation rate depends on the diameter of the initial droplet of the injected liquid, and the better the spray (the diameter of the droplets does not exceed 7 μm), the more efficient the evaporation process. It is primarily due to the increase in the total surface area of water evaporated droplets.

Depending on the inlet air temperature T_{air1}, Mach number at the inlet to the evaporation chamber $M = 0.74$ and the initial droplet diameter δ_p, the pressure increase at $T_{air1} = 500$ K is $\Delta P_{abs} = 0.5$–10.5% (2–31 kPa), at $T_{air1} = 550$ K is $\Delta P_{abs} = 1.0$–12.5% (4–37 kPa) (Fig. 2a, b). The maximum pressure increase occurs with the injection of the most finely sprayed water, with an initial droplet diameter $\delta_p = 3$ μm, as a result of thermogasdynamic compression, the total pressure increases by 31–37 kPa.

The pattern of absolute pressure distribution along the evaporation chamber length for different initial droplet diameters has the same tendency. Then the droplet's diameter is larger; the effect of thermogasdynamic compression is smaller. It is associated with an increase in the droplet's aerodynamic resistance and a decrease in the evaporation rate.

Depending on the air temperature at the evaporation chamber inlet T_{air1} and the initial droplet diameter δ_p, the depth of cooling of the humidified air leaving the evaporation chamber is: at $T_{air1} = 500$ K – $\Delta T_{air} = 116$–153 K; at $T_{air1} = 550$ K – $\Delta T_{air} = 140$–163 K (Fig. 2 c, d) and the amount of injected water in the airflow is $\mu_{H2O} = 5.8$–6.1%.

The main characteristics of the water evaporation process in the low-flow aerothermopressor evaporation chamber (G_{air} = 1.0–1.5 kg/s) obtained by CFD simulation are shown in the following chart (Fig. 3).

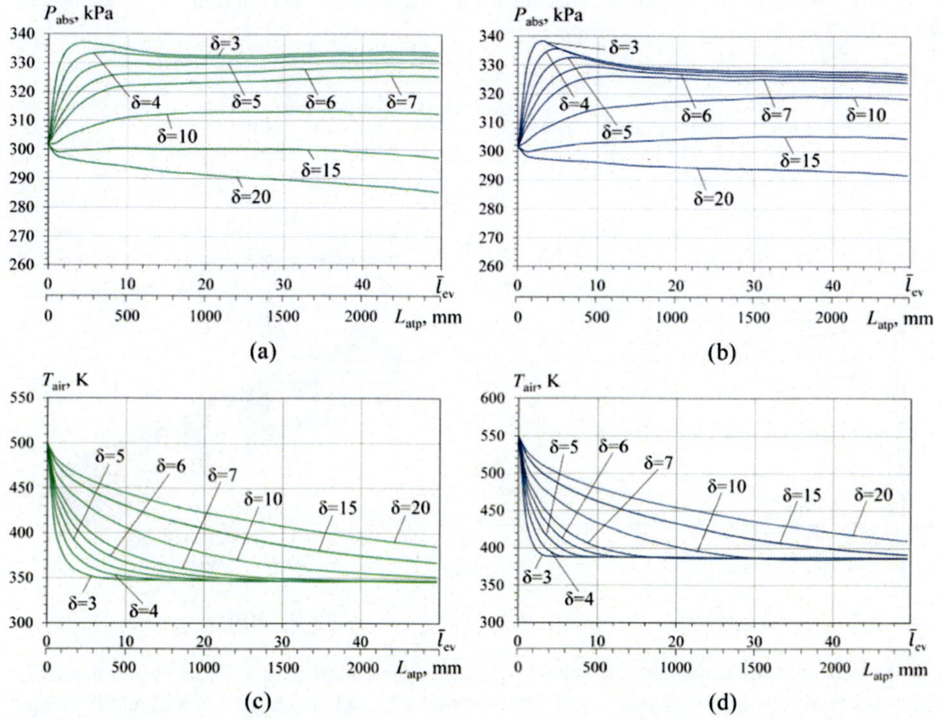

Fig. 2 Dependences of absolute pressure P_{abs} at T_{air1} = 500 K (a), at T_{air1} = 550 K (b), two-phase flow temperature T_{air} at T_{air1} = 500 K (c), at T_{air1} = 550 K (d) on the length L_{atp} and relative length l_{ev} of the evaporation chamber.

To compare the calculation results, the minimum and maximum initial droplet diameters δ_p = 3 μm and δ_p = 15 μm were taken, at which the positive effect of thermogasdynamic compression was obtained. Thus, at δ_p = 3 μm, the maximum absolute pressure is P_{abs} = 338 kPa, the maximum pressure increase is ΔP_{abs} = 12% (37 kPa). As a result of contact cooling, the air temperature decreases by ΔT_{air} = 153–163 K, and the amount of injected water in the airflow is μ_{H2O} = 6.1%. With an increase in the initial droplet diameter to δ_p = 15 μm, the maximum absolute pressure value significantly decreases P_{abs} = 305 kPa. Consequently, the maximum pressure increase also decreases ΔP_{abs} = 1.2% (4 kPa), as a result of contact cooling the decrease in air temperature is also low ΔT_{air} = 133–159 K, the amount of injected water in the airflow is μ_{H2O} = 5.0–5.9%.

As a result of computer simulations, the values of the main characteristics of the evaporation process were obtained (Fig. 2, 3). A three-dimensional surface was constructed that takes into account the effect of initial droplet diameter at the inlet δ_p and water mass concentration mc_{H2O} on the length of the evaporation chamber at maximum pressure increase. Based on this, the analytical equation was determined for calculating the optimal relative length of the aerothermopressor evaporation chamber L/D (l_{ev}). The determination of this equation was carried out by the method of approximation. In this case, the following equation was chosen (regression coefficient – R = 0.9944; $R^2 = 0.9889$):

$$\frac{L}{D} = 5.57 \cdot 10^4 \cdot e^{-\frac{\left[\left(\frac{\delta - 17.7}{5.96}\right)^2 + \left(\frac{mc_{H2O} + 0.523}{0.145}\right)^2\right]}{2}}. \quad (3)$$

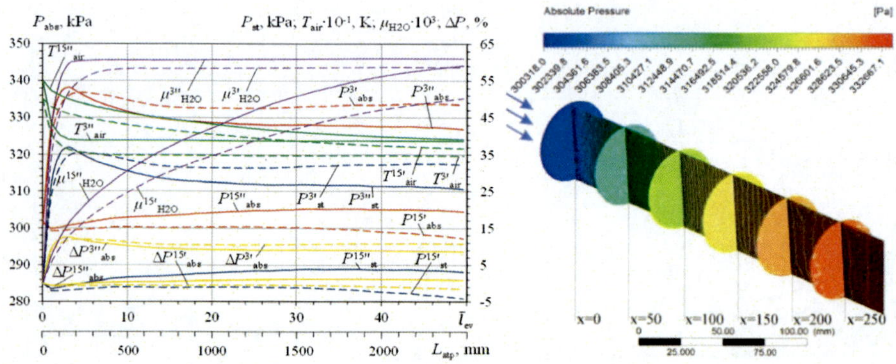

(a) (b)

Fig. 3 Dependences of absolute pressure P_{abs}, static pressure P_{st}, two-phase flow temperature T_{air}, amount of injected water μ_{H2O}, pressure increase ΔP_{abs} on the length L_{atp} and relative length l_{ev} of the evaporation chamber (a): ' – at T_{air1} = 500 K; "– at T_{air1} = 550 K; the contour of the two-phase flow absolute pressure P_{abs} (b) at T_{air1} = 550 K and δ_p = 3 μm

The obtained Eq. (3) was used in the software package to calculate the optimal length of the evaporation chamber. This equation was obtained for the following characteristics of the water evaporation process in the evaporation chamber: δ_p = 3–18 μm; M = 0.74; T_{air1} = 500–550 K; mc_{H2O} = 0–0.25 kg/m^3. The aerothermopressor operation as part of the Rolls Royce WR-21 gas turbine was simulated. The aerothermopressor was used for contact cooling of cyclic air between low and high-pressure compressor stages. The increase in the total pressure in the "ideal" aerothermopressor at the inlet air temperature T_{atp1} = 500 K (friction losses are not taken into account) at M = 0.50 is $\Delta P_{atp.id}$ = 6.5% ($19\cdot 10^3$ Pa) (Fig. 4, a–b). At the inlet temperature increase (T_{atp1} = 550 K), the total pressure increases too $\Delta P_{atp.id}$ = 7.5% ($23\cdot 10^3$ Pa) (Fig. 4, a–b). It is explained by an increase in the relative temperature from (T_{atp1}/T_{atp2}) = 1.466–1.470 (at T_{atp1} = 500 K) to (T_{atp1}/T_{atp2}) = 1.594–1.596 (at T_{atp1} = 550 K). The same behavior is observed with increasing Mach number M. At M = 0.74: T_{atp1} = 500 K – $\Delta P_{atp.id}$ = 14.0% ($42\cdot 10^3$ Pa); T_{atp1} = 550 K – $\Delta P_{atp.id}$ = 17% ($52\cdot 10^3$ Pa) (Fig. 4, c–d). It can be seen than Mach number M and the inlet temperature T_{atp1} are greater. The total pressure increase $P_{atp.id}$ is higher.

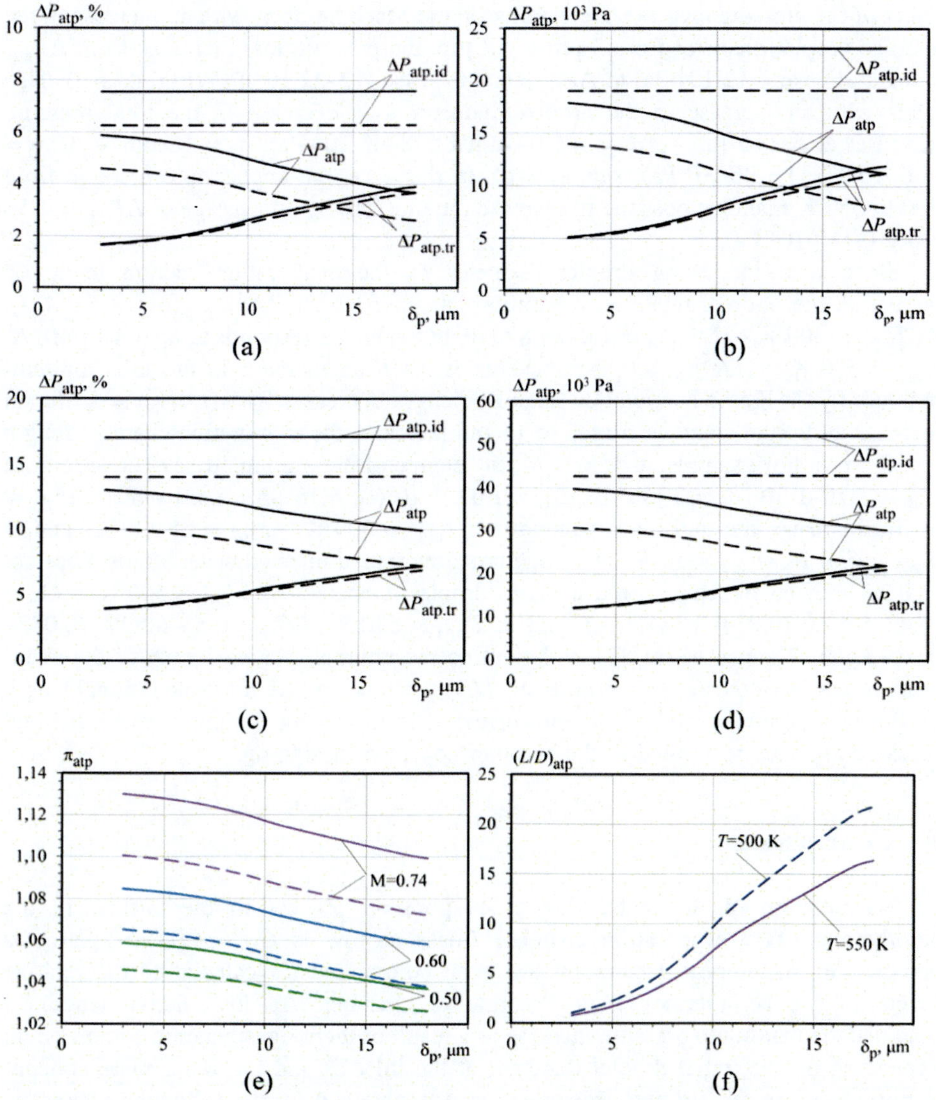

Fig. 4 Dependences of the total pressure increase in the aerothermopressor ΔP_{atp}, in the "ideal" aerothermopressor $\Delta P_{atp.id}$, pressure losses $\Delta P_{atp.tr}$ (a, b, c, d) and the degree of pressure increase in the aerothermopressor π_{atp} (e) and optimal relative evaporation chamber length $(L/D)_{atp}$ (f) of the average droplet δ_p: (a), (b) – $M = 0.50$; (c), (d) – $M = 0.74$; ________ – $T_{atp1} = 550$ K (227 °C); _ _ _ _ _ – $T_{atp1} = 500$ K (277 °C).

There are losses of hydraulic pressure in the aerothermopressor flow part in real conditions. These are local pressure losses in the confuser and diffuser, friction pressure losses in the evaporation chamber, as well as losses due to the droplet drag in the airflow. Thus, the total pressure loss in the aerothermopressor can determine the final effect of the pressure increase. Depending on the Mach number, with an increase in the average droplet diameter from 3 μm to 18 μm, the pressure loss increases from $\Delta P_{atp.tr}$ = 1.8–4.0% (5–12·10^3 Pa) to $\Delta P_{atp.tr}$ = 4.0–7.0% (12–22·10^3 Pa) (Fig. 4, a–d). This leads with an increase in the droplet diameter to a decrease in the total pressure increase: at δ_p = 3 μm – ΔP_{atp} = 6.0–13.0% (18–40·10^3 Pa); δ_p = 18 μm – ΔP_{atp} = 4.0–10.0% (12–30·10^3 Pa) (Fig. 4, a–d). In this case, the temperature increase from 500 to 550 K makes it possible to increase the pressure by an average of ΔP_{atp} = 1.5–3.0% (5–10·10^3 Pa).

Since with increasing droplet diameter δ_p the evaporation section length is increased, the optimal relative evaporation chamber length $(L/D)_{atp}$ grows (Fig. 4, f). At T_{atp1} = 500 K – $(L/D)_{atp}$ = 1.0–21.8 (a higher value corresponds to δ_p = 18 μm). At T_{atp1} = 550 K – $(L/D)_{atp}$ = 0.8–16.4. That is, with an increase in the inlet air temperature by 50 K, the evaporation chamber length decreases by $\Delta(L/D)_{atp}$ = 25%. An increase in droplet diameter from 3 to 18 μm increases the evaporation chamber length by 20 times. For example, at M = 0.74, the degree of pressure increase decreases from π_{atp} = 1.10–1.30 (δ_p = 3 μm) to π_{atp} = 1.07–1.10 (δ_p = 18 μm), i.e. by 30% (Fig. 4, e). In this case, the outlet air temperature: T_{atp1} = 500 K – t_{atp2} = 65–68 °C, and at T_{atp1} = 550 K – t_{atp2} = 70–72 °C. The mass flow rate of injected water almost does not change with an increase in the average droplet diameter. At T_{atp1} = 500 K – G_w = 5.40–5.43% (0.073–0.074 kg/s), and at T_{atp1} = 550 K – G_w = 6.55–6.59% (0.088–0.089 kg/s). The use of Eq. (3) to determine the optimal relative length of the evaporation section of the aerothermopressor $(L/D)_{atp}$ allows to determine the rational length of the evaporation chamber, which corresponds to the maximum increase in total pressure and can be applied in the aerothermopressor designing.

5 Conclusions

CFD simulation of the water droplet evaporation process in the airflow in the aerothermopressor evaporation chamber (mass airflow is G_{air} = 1.0–1.5 kg/s) was carried out. An empirical equation has been determined for calculating the relative length of the aerothermopressor evaporation chamber. At this relative length, it obtained the maximum pressure increase as a result of thermogasdynamic compression, depending on the initial droplet diameter at the inlet δ_p and the water mass concentration mc_{H2O} in the airflow. This equation was obtained for the following characteristics: δ_p = 3–18 μm; M = 0.74; T_{air1} = 500–550 K; mc_{H2O} = 0–0.25 kg/m^3. The aerothermopressor design parameters were obtained for the Rolls Royce WR-21 gas turbine cooling system. In this case, the maximum pressure increase ΔP_{abs} = 13% (40 kPa) corresponds to the lowest initial droplet diameter δ_p = 3 μm with a relative length of the evaporation chamber $(L/D)_{atp}$ = 0.8–1.0 at the inlet air temperature T_{atp1} = 500–550 K.

The equation for determining the optimal relative length of the evaporation section can be recommended for use in the design methodology for low-flow aerothermopressors.

References

1. Chaker, M.: Key Parameters for the performance of impaction-pin nozzles used in inlet fogging of gas turbine engines. In: Proceedings of GT2005 ASME Turbo Expo: Power for Land, Sea and Air, pp. 1–7. Reno-Tahoe, Nevada, USA (2005)
2. Radchenko, A., Bohdal, L., Zongming, Y., Portnoi, B., Tkachenko, V.: Rational designing of gas turbine inlet air cooling system. In: Tonkonogyi, V., et al. (eds.) Advanced Manufacturing Processes. InterPartner-2019. Lecture Notes in Mechanical Engineering, pp. 591–599. Springer, Cham (2020)
3. Lin, A., et al.: Evaluation of mass injection cooling on flow and heat transfer characteristics for high-temperature inlet air in a MIPCC engine. Int. J. Heat Mass Transf. **135**, 620–630 (2019)
4. Konovalov, D., Trushliakov, E., Radchenko, M., Kobalava, H., Maksymov, V.: Research of the aerothermopressor cooling system of charge air of a marine internal combustion engine under variable climatic conditions of operation. In: Tonkonogyi, V. et al. (eds.) Advanced Manufacturing Processes. InterPartner 2019. Lecture Notes in Mechanical Engineering, pp. 520–529. Springer, Cham (2020)
5. Bohdal, T., Sikora, M., Widomska, K., Radchenko, A.M.: Investigation of flow structures during HFE-7100 refrigerant condensation. Arch. Thermodyn. **36**(4), 25–34 (2015)
6. Jonsson, M., Yan, J.: Humidified gas turbines – a review of proposed and implemented cycles. Energy **30**, 1013–1078 (2005)
7. Stepanov, I.R., Chudinov, V.I.: Some Problems of Gas and Liquid Motion in the Channels and Pipelines of Power Plants. The Science Publ., Leningrad (1977)
8. Fowle, A.: An experimental investigation of an aerothermopressor having a gas flow capacity of 25 pounds per second. Massachusetts Institute of Technology, USA (1972)
9. Bergman, T.L., et al.: Fundamentals of Heat and Mass Transfer, 7th edn. Wiley, New Jersey (2011)
10. Sirignano, W.A.: Fluid Dynamics and Transport of Droplets and Sprays, 2nd edn. Cambridge University Press, New York (2010)
11. ANSYS Fluent Tutorial Theory Guide Release 17.0.: ANSYS. Inc. Canonsburg (2016)
12. Chen, Z., Xie, Q., Chen, G., Yu, Y., Zhao, Z.: Numerical simulation of single-nozzle large scale spray cooling on drum wall. Thermal Sci. **22**(1A), 359–370 (2018)
13. Korkodinov, Y.A.: An overview of the k-ε family of models for modeling turbulence. Mech. Eng. Mater. Sci. **5**(2), 5–16 (2013)
14. Nijdam, J.J., et al.: Lagrangian and Eulerian models for simulating turbulent dispersion and coalescence of droplets within a spray. Appl. Math. Modeling **30**, 1196–1211 (2006)
15. Jafarnadar, S., Jahangiramini, A.: Numerical simulation of flash boiling effect in a 3-dimensional chamber using computational fluid dynamic techniques. Int. J. Eng. **29**(5), 87–95 (2016)
16. Montazeri, H., Blocken, B., Hensen, J.: Evaporative cooling by water spray systems: CFD simulation, experimental validation and sensitivity analysis. Build. Environ. **83**, 129–141 (2015)
17. Konovalov, D., Kobalava, H., Radchenko, M., Scurtu, I.C., Radchenko, R.: Determination ofhydraulic resistance of the aerothermopressor for gas turbine cyclic air cooling. In: 9th InternationalConference on Thermal Equipments, Renewable Energy and Rural Development (TE-RE-RD 2020),Constanta, Romania. E3S Web Conference, vol. 180, article no. 01012 (2020)
18. Vulis, L.A.: Gas Flow Thermodynamics. Gosenergoizdat, Moscow (1950). [in Russian]

Improvement of Characteristics of Water-Fuel Rotary Cup Atomizer in a Boiler

Victoria Kornienko[1(✉)], Roman Radchenko[2], Dariusz Mikielewicz[3], Maxim Pyrysunko[1], and Andrii Andreev[1]

[1] Kherson Branch of Admiral Makarov National University of Shipbuilding, 44, Ushakova Avenue, Kherson 73022, Ukraine
kornienkovika1987@gmail.com
[2] Admiral Makarov National University of Shipbuilding, 9, Heroes of Ukraine Avenue, Mykolayiv 54025, Ukraine
[3] Gdansk University of Technology, Gdańsk, Poland

Abstract. Two types of its atomization can be distinguished when water-fuel emulsion (WFE) is burned in boiler furnaces: decomposition of the fuel stream in atomizer and secondary self-atomization during a micro explosion of WFE droplets. It allows reducing the size of fuel fragments, increases the area of contact with the oxidizing agent, and intensifies the burning process. During WFE combustion, the deposits on heating surfaces are reduced or even fully absent due to a decrease of soot and coke generation. The size of fuel fragments formed after secondary flow crushing can be determined only on base experimental data. The research aims to investigate rotary cup atomizer characteristics. Experimental researches of rotary cup atomizer characteristics were carried out on the experimental setup with atomized liquid of fuel oil, water, and WFE. The dependence of diameter droplets of atomized liquid from revolution number of atomizer cup, the relative distribution of atomized liquid over the radius of a torch with different atomizer diameter and the specific flow fields of atomized liquid at various distances from the cutoff of atomizer cup, atomizer diameter, and fluid flow have been investigated developed based on the experimental data. A criterial dependence is obtained for determining the average diameter of droplets when WFE are sprayed. Based on the experimental and theoretical data, a nozzle with atomizer diameter 25 mm was selected, which satisfactorily atomizes the fuel at a flow rate of 1…3 kg/h and provides the required diameter of emulsion droplets.

Keywords: Water-fuel emulsion · Exhaust gases · Rotary cup atomizer

1 Introduction

Recently, fuel components in the cost of production began to increase sharply. As a result, cheaper high-viscosity heavy sulfur fuels, which require modernization of fuel preparation systems and furnace devices, began to be burned in boiler plants: instead of centrifugal, it is necessary to use rotary cup atomizer.

V. Tonkonogyi et al. (Eds.): InterPartner 2020, LNME, pp. 664–674, 2021.
https://doi.org/10.1007/978-3-030-68014-5_64

The use of alternative fuels and water-fuel emulsions (WTE) is quite promising. When WFE is burned in boiler furnaces, two types of its atomization can be distinguished: decomposition of the fuel stream in an atomizer and secondary self-atomization during microexplosion of WFE droplets. It assists in reducing the size of fuel fragments, increase the area of contact with the oxidizing agent, and intensify the burning process. During WFE combustion, the deposits on heating surfaces are reduced or even fully absent due to a decrease of soot and coke generation. To generalize the experimental data on the dispersed characteristics of boiler atomizers, the relationship between power-law complexes characterizing the average diameter of dry fuel droplets was used. To calculate the burning of WFE, in addition to the initial diameter of the droplets, the size of the fuel fragments formed after the secondary in-line crushing is necessary. The essence of the WFE droplet crushing phenomenon is the boiling of the dispersed water phase during the heating of emulsion and the following break of the fuel shell into fragments under the influence of the generated vapor pressure. In this case, an explosive effect appears, provided by a significant difference in the boiling points of dispersed water and fuel.

The hydrodynamics of the fluid flow determines the formation of the film on the surface of rotary cup. Such a flow is characterized by the uncertainty of the boundary conditions on the free surface of the film, which does not allow obtaining the equation of fluid motion in explicit form. Therefore, complex analysis is needed for obtaining the necessary data for designing the rotary cup atomizer.

2 Literature Review

Atomizer characteristics of nozzle influence considerably on the efficiency of any thermal power plant from cogenerative internal combustion engines [1] and gas turbines [2] to trigeneration plants [3] generating heat, electricity power [4] and cooling (refrigeration [5, 6] and air conditioning [7]), i.e., practically in all waste heat recovery technics [8, 9]. The use of alternative fuels is quite a promising trend in stationary [10] as well as in marine energetics for diesel engines [11] and gas turbine power plants [12].

The paper [13] presents the results of experimental studies on the atomization of the emulsions flowing through twin-fluid atomizers obtained by applying the digital microphotography method. The analysis of photos shows that the droplets being formed during the liquid atomization have very different sizes. The effect of the shape of the orifice and the injection pressure on the droplet size distributions and mean droplet diameters were analyzed [14]. The results of experimental studies of the spraying of Chinese fuel oil with water additives by a Saacke burner with a rotary cup atomizer are presented in [15]. Water does not worsen, but improves combustion processes due to additional crushing of fuel droplets due to microexplosions, an increase in the combustion surface of particles [16, 17], and good mixing of fuel with air. With an increase in the water content in WFE up to 25%, a decrease in the concentration of solid residues of fuel combustion, carbon monoxide, and hydrocarbons is observed, which indicates the efficiency of WFE combustion [18, 19]. However, in [15], data on the operation of rotary cup atomizer are not presented. According to the literature, the droplet diameter obtained by the atomization of fuel affects NO_x emissions and particulate matter. In [20], Marshall's experimental data on droplet diameter at high flow rates of a liquid sprayed by a rotating disk are published. In [20], a criterial dependence is given for determining the relative mass-average diameter of droplets when spraying of liquid with rotary cup atomizer with a flow rate of 400… 1100 kg/h. The rotary cup atomizer is a special atomizer used for improved automotive painting [21]. A model to predict the droplet size was developed by combining two already-reported models [22, 23]. Studies [24, 25] present study investigates liquid flow patterns on the inner surface of a rotary cup atomizer and the effect of these flow patterns on the breakup patterns at the edge of the rotary cup. It also estimates the atomization characteristics based on the diameter of liquid ligaments issuing from the cup edge. In [26, 27] behaviors of liquid films scattering from a disk-type or cup-type rotary atomizer are studied using computations based on the three-dimensional Smoothed Particle Hydrodynamics method. Authors' [28, 29] experimentally investigated the effects of operational conditions and liquid properties on the formation and breakup of ligaments from a high-speed rotary cup atomizer.

Experimental and literature data on the rotary cup atomizer are small, which does not allow obtaining criteria equations and to determine the average droplet size. The research aims to investigate rotary cup atomizer characteristics. The research tasks:

- determine the effect of revolution number of rotary cup atomizer on average diameter droplets of atomized fuel;
- determine the relative distribution of atomized liquid over the radius of the torch;
- determine the specific flow fields of atomized liquid at various distances from the cutoff of atomizer cup, atomizer diameter, and fluid flow.

3 Research Methodology

Rotary cup atomizer is insensitive to fuel oil viscosity since the fuel is exposed to a high temperature before atomization, washing the inner surface of the atomized cone, as a result of which the viscosity of the fuel oil is quickly brought to 2° or less. For ensuring the required dispersion during the atomization of fuel and WFE, sustainable combustion of fuel at a flow rate of about 1…3 kg/h was studied the operation of rotary cup atomizer on experimental installation, taking into account the above provisions.

The layout of the experimental setup for studying the operation of the rotary cup atomizer is shown in Fig. 1.

The design of setup provides for mounting the rotary cup atomizer *1*, drive motor *2*, installation of equipment and accessories necessary for examining operation atomizer - a circle collector *7* and sector collector *8* with measuring beakers *9*, amplifier 5. The test liquid is supplied by gravity: water is supplied from supply tank *3*, and fuel or WFE is supplied from tank *4*.

In [15] the average diameter of droplets d during of rotary cup atomizer operation depends on such values:

$$d_{\kappa}/\delta = f(We, La, M, N) \quad (1)$$

$We = \upsilon_2^2 \rho_2 \delta/\sigma$ – Weber criterion, which represents the ratio of pressure head to surface tension pressure;

$La = \rho_1 v_1^2/\sigma\delta$ – Laplace criterion, which determines the ratio of viscosity forces and liquid surface tension;

$M = \rho_2/\rho_1$ – a criterion, which characterizes the inertial properties of a gaseous and a liquid medium;

$N = \mu_2/\mu_1$ – a criterion, which represents the ratio of the viscosity forces of a gaseous medium and a sprayed liquid.

In the above formulas, it is accepted: d – droplet diameter; δ – fluid film thickness; σ – surface tension coefficient; μ_1, μ_2 – dynamic viscosity coefficient of the liquid and gaseous medium; ρ_1, ρ_2 – fluid and gas density.

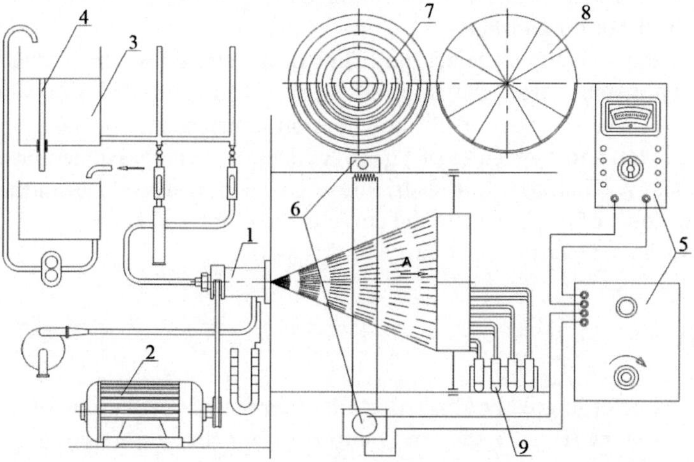

Fig. 1. Scheme of an experimental setup for investigating afterburning installation characteristics: 1 – rotary cup atomizer; 2 – drive motor; 3 – tank of water; 4 – tube; 5 – amplifier; 6 – collector; 7 – circle collector; 8 – sector collector; 9 – measuring beakers.

The distribution of the atomized liquid in the torch created by the atomizer is an essential characteristic that determines the perfection of the process (technological or the process of burning liquid fuel in the combustion chamber of engine or boiler).

The field of specific fluid most fully characterizes the distribution of atomized liquid in the torch flows at various points of the torch. The specific fluid flow is equal to the ratio of its second flow rate through the area, perpendicular to the direction of droplets flight, to the value of this area:

$$q_n = \Delta G/f_n \quad (2)$$

With this definition, the measurement of a specific flow is not difficult. It is necessary only to position the collection cells of the collection in a plane perpendicular to the axis of an atomizer.

When studying the distribution of atomized liquid along the radius of the torch, a circle collector is also used. Summing up the volumes of liquid that fell into the collector cells sequentially located from the center to the periphery, and relating the obtained values of volumes to the total volume of liquid that fell into the collector, the total distribution curve similar to the total distribution curve of droplets by diameter is obtained. The equation can approximate this curve:

$$y = 1 - e^{-ax\chi} \tag{3}$$

$$y = \sum\nolimits_{i=N}^{i} q_i / \sum\nolimits_{i=1}^{N} q_i;\ x = r/\bar{r}; \tag{4}$$

Where N – the number of circle cells in collector; $\bar{r}$ – size constant; r – radial coordinate; χ – distribution constant.

As the constant r, you can select the radius of the circle for which 50% of the fluid flow at a given distance from the atomizer passes inside and 50% - outside it. In this case, $a = 0.693$. The meaning of the distribution constant can be easily found if we differentiate the left and right sides of Eq. (4) concerning x and assume that $x = 1$. Then $dy/dx_i = 0{,}3465\ \chi$. Therefore, the distribution constant is directly proportional to the tangent of the angle of inclination to the curve at the point $x = 1$, $y = 0.5$. The larger the value of χ is, the steeper the total curve at this point.

4 Results

The droplet diameter is greatly affected by the number of revolutions of the atomizer. As experiments show (Fig. 2), the rotary cup atomizer stably provides a good spray in the range of 20…35 μm starting from n = 5000 rpm.

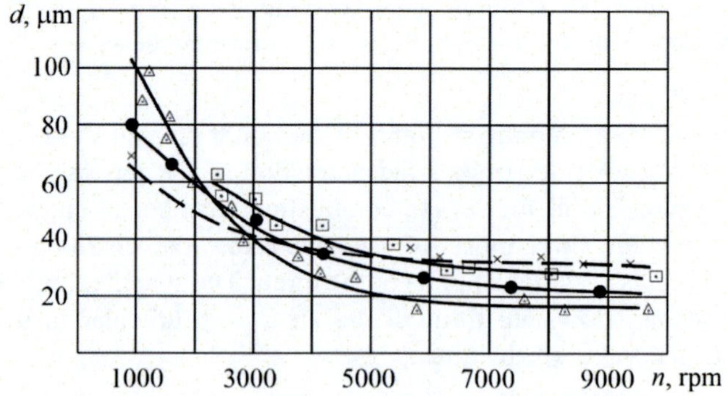

Fig. 2. Dependences of diameter droplets of atomized liquid from revolution number of atomizer cup (air pressure h = 250 mm w. col., atomizer diameter d_a = 25 mm): ⊡ - fuel oil consumption 1 kg/h; × - WFE consumption 1 kg/h; ● - water consumption 4 kg/h; △ - water consumption 1,2 kg/h.

With decreasing *n*, the size of the fraction of the atomized liquid sharply increases and is 30…100 μm. Their greater viscosity explains the large size of the drops of emulsion and fuel oil compared with the size of the droplets of water.

Figure 3 shows the total curves of the flow distribution of atomized liquid over the radius of the torch with different atomizer diameters.

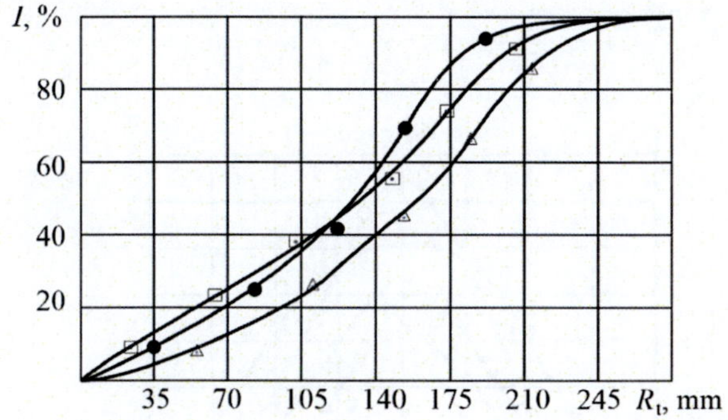

Fig. 3. Dependences of relative distribution of atomized liquid over the radius of the torch with different atomizer diameter d_a (air pressure h = 250 mm w. col., revolution number of atomizer cup n = 6200 rpm): ● – d_a = 25 mm; ⊡ – d_a = 26 mm; ⚠ – d_a = 28 mm

These curves are obtained based on data using a ring collector, which was installed perpendicular to the axis of the torch. These curves allow determining the radius R_{99}, including 99% of the atomized liquid. For atomizer with a large spray radius, there is a correspondingly larger R_{99} (for atomizer d_a = 25 mm - R_{99} = 180 mm, for d_a = 26 mm - R_{99} = 200 mm, for d_a = 28 mm - R_{99} = 268 mm.

In Fig. 4–5 the specific flow fields of atomized liquid are shown. These fields are measured using a selective tube. The maximum specific flow is located on the axis of the torch, and as the distance from the axis increases, the values of specific flows decrease. As the distance from the atomizer opening increases, the specific flux fields are aligned in the cross-section of the torch.

As the gas back pressure increases, the specific flow fields also become more uniform. If the rate of outflow (pressure drop) from the atomizer increases, then this also leads to an equalization of the specific flow fields. An increase in the atomizer diameter (with other constant conditions of outflow) causes an increase in the values of specific flows and the expansion of the boundaries of the torch. Moreover, the atomizer with a small diameter (d_a = 25 mm) at a flow rate of 1.2 kg/h has a more uniform field of specific flows of atomized liquid.

Based on the experiments and conclusions made, a nozzle with an atomizer diameter d_a = 25 mm was selected with an air pressure h = 250 mm water column and an atomizer speed 5000 rpm, which satisfactorily atomizes the fuel at a flow rate of 1… 3 kg/h and provides the required diameter emulsion droplets for targeted research.

Processing the obtained experimental data made it possible to obtain criteria equation for determining the average diameter of WFE droplets in the form of:

$$d = 5.8We^{-0.5}La^{0.2}\delta \quad (5)$$

Experimental verification of the obtained criterion dependence showed that the proposed dependence is recommended for *We* and *La* values within: $0.125 < We < 1.2$; $107 < La < 920$.

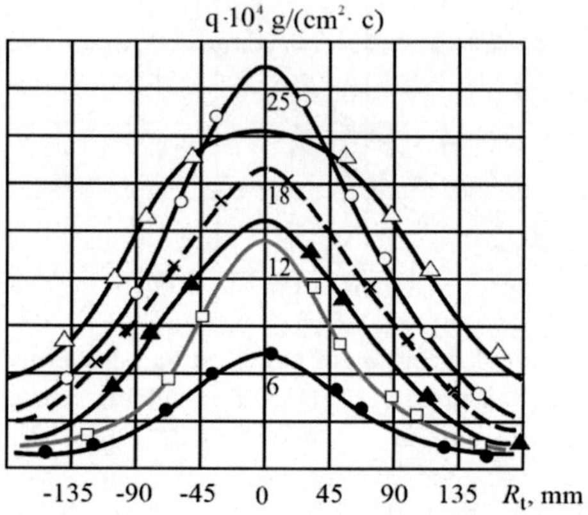

Fig. 4. Specific flow fields of atomized liquid with different atomizer diameter d_a and fluid flow: ● – d_a = 25 mm; ▲ – d_a = 26 mm; ⊙ – d_a = 28 mm; d_a = 26 MM: ⊡ – water consumption 1,0 kg/h; × – WFE consumption 1,0 kg/h; ◬ – fuel oil consumption 1,0 kg/h.

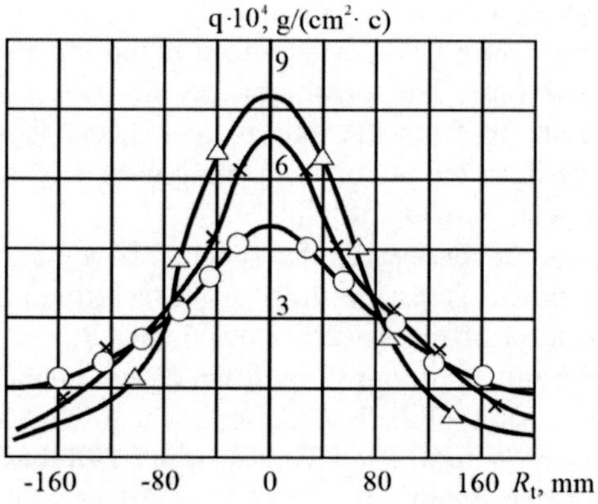

Fig. 5. Specific flow fields of atomized liquid at various distances *l* from the cutoff of atomizer cup (water consumption 1.2 kg/h, atomizer diameter d_a = 25 mm, air pressure *h* = 250 mm w. col.): ⊙ – *l* = 150 mm; × – *l* = 250 mm; ◬ – *l* = 350 mm.

A comparison of the mean droplet diameter calculated from the proposed dependence with the experimental data is shown in Fig. 6.

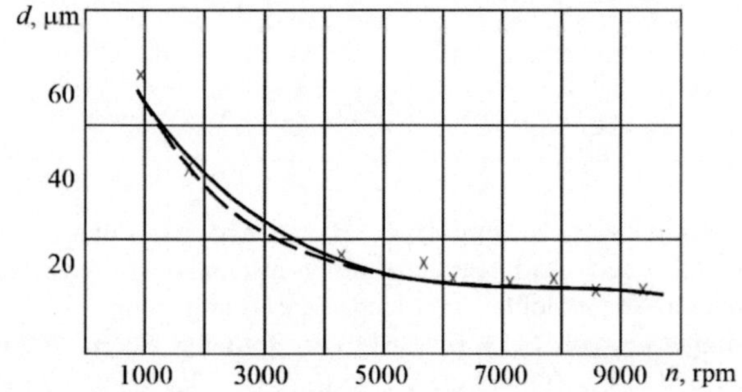

Fig. 6. Dependences of diameter droplets of atomized liquid from revolution number of atomizer cup (atomizer diameter d_a = 25 mm, WFE consumption 1,0 kg/h): --- - experimental data; — - calculation data.

The presented dependencies show a complete coincidence of the calculated data with the experimental ones at rotation frequencies above 5000 rpm and a slight discrepancy in the range from 2000 to 5000 rpm.

5 Conclusions

The requirements concerning the development of high-performance equipment with low environmental impact and high flexibility have increased lately. Therefore, complex analysis is needed for obtaining the necessary data for designing the rotary cup atomizer. There are few experimental and literature data on a rotary cup atomizer, that does not allow to obtain criteria equations and especially to determine the average droplet size.

Experimental researches rotary cup atomizer performance was carried out on the experimental setup with atomized liquid of fuel oil, water, and WFE.

The dependence of diameter droplets of atomized liquid on revolution number of atomizer cup, the relative distribution of atomized liquid over the radius of a torch with different atomizer diameter and the specific flow fields of atomized liquid at various distances from the cutoff of atomizer cup, atomizer diameter and fluid flow have been investigated based on the experimental data.

The obtained criteria dependence for determining the average diameter of droplets when WFE sprayed are quite adequate to experimental data with an accuracy of 5%.

Issuing from the experimental and calculation data, a nozzle with atomizer diameter d_a = 25 mm was selected, which satisfactorily atomizes the fuel at a flow rate of 1… 3 kg/h and provides the required diameter of emulsion droplets for targeted research.

The mathematical model for determining the diameters of droplets of WFE atomized by a rotary cup atomizer is presented. The results of the research in this direction proved the prospects for further theoretical studies of ranges of particular flows of atomized liquid for different atomizer diameters, fluid flow rates, and different distances from an atomizer.

References

1. Radchenko, A., Mikielewicz, D., Forduy, S., Radchenko, M., Zubarev, A.: Monitoring the fuel efficiency of gas engine in integrated energy system. In: Nechyporuk, M. et al. (eds.) Integrated Computer Technologies in Mechanical Engineering. Advances in Intelligent Systems and Computing, vol. 1113, pp. 361–370. Springer, Cham (2020)
2. Radchenko, A., Bohdal, L., Zongming, Y., Portnoi, B., Tkachenko, V.: Rational designing of gas turbine inlet air cooling system. In: Tonkonogyi, V. et al. (eds.) Grabchenko's International Conference on Advanced Manufacturing Processes. InterPartner-2019. Lecture Notes in Mechanical Engineering, pp. 591–599. Springer, Cham (2020)
3. Radchenko, A., Radchenko, M., Konovalov, A., Zubarev, A.: Increasing electrical power output and fuel efficiency of gas engines in integrated energy system by absorption chiller scavenge air cooling on the base of monitoring data treatment. In: HTRSE-2018, 6 p. E3S Web of Conferences, vol. 70, p. 03011 (2018)
4. Forduy, S., Radchenko, A., Kuczynski, W., Zubarev, A., Konovalov, D.: Enhancing the fuel efficiency of gas engines in integrated energy system by chilling cyclic air. In: Tonkonogyi, V., et al. (eds.) Grabchenko's International Conference on Advanced Manufacturing Processes. InterPartner-2019. Lecture Notes in Mechanical Engineering, pp. 500–509. Springer, Cham (2020)
5. Konovalov, D., Trushliakov, E., Radchenko, M., Kobalava, G., Maksymov, V.: Research of the aerothermopresor cooling system of charge air of a marine internal combustion engine under variable climatic conditions of operation. In: Tonkonogyi, V., et al. (eds.) Grabchenko's International Conference on Advanced Manufacturing Processes. InterPartner-2019. Lecture Notes in Mechanical Engineering, pp. 520–529. Springer, Cham (2020)
6. Bohdal, T., Sikora, M., Widomska, K., Radchenko, A.M.: Investigation of flow structures during HFE-7100 refrigerant condensation. Arch. Thermodynamics: Polish Acad. Sci. **36**(4), 25–34 (2015)
7. Trushliakov, E., Radchenko, A., Forduy, S., Zubarev, A., Hrych, A.: Increasing the operation efficiency of air conditioning system for integrated power plant on the base of its monitoring. In: Nechyporuk, M., et al. (eds.) Integrated Computer Technologies in Mechanical Engineering. Advances in Intelligent Systems and Computing, vol. 1113, pp. 351–360. Springer, Cham (2020)
8. Radchenko, R., Kornienko, V., Pyrysunko, M., Bogdanov, M., Andreev, A.: Enhancing the efficiency of marine diesel engine by deep waste heat recovery on the base of its simulation along the route line. In: Nechyporuk, M., et al. (eds.) Integrated Computer Technologies in Mechanical Engineering. Advances in Intelligent Systems and Computing, vol. 1113, pp. 337–350. Springer, Cham (2020)
9. Kornienko, V., Radchenko, M., Radchenko, R., Konovalov, D., Andreev, A., Pyrysunko, M.: Improving the efficiency of heat recovery circuits of cogeneration plants with combustion of water-fuel emulsions. Thermal Sci. **00**, 154 (2020)

10. Radchenko, N.: A concept of the design and operation of heat exchangers with change of phase. Arch. Thermodynamics: Polish Acad. Sci. **25**(4), 3–19 (2004)
11. Pham, V.V.: Advanced technology solutions for treatment and control noxious emission of large marine diesel engines: A brief review. J. Mech. Eng. Res. Dev. **42**(5), 21–27 (2019)
12. Cherednichenko, O., Serbin, S., Dzida, M.: Investigation of the combustion processes in the gas turbine module of an FPSO operating on associated gas conversion products. Polish Maritime Res. **26**(4), 149–156 (2020)
13. Broniarz-Press, L., Ochowiak, M., Rozanski, J., Woziwodzki, S.: The atomization of water–oil emulsions. Experimental Thermal Fluid Sci. **33**(6), 955–962 (2009)
14. Broniarz-Press, L., Włodarczak, S., Matuszak, M., Ochowiak, M., Idziak, R., Sobiech, Ł, Szulc, T., Skrzypczak, G.: The effect of orifice shape and the injection pressure on enhancement of the atomization process for pressure-swirl atomizers. Crop Protection **82**, 65–74 (2016)
15. Lozitskii, N.G., Kotler, V.R.: Use of burners with a rotational nozzle for burning fuel oil with water additive. Ind. Energy **3**, 37–40 (2002). [in Russian]
16. Baskar, P.K., Senthil, A.: Experimental investigation on performance characteristics of a diesel engine using diesel-water emulsion with oxygen enriched air. Alexandria Eng. J. **56** (1), 37–146 (2017)
17. Patel, K.R., Dhiman, V.: Research study of water-diesel emulsion as alternative fuel in diesel engine – an overview. Int. J. Latest Eng. Res. Appl. **2**(9), 37–41 (2017)
18. Kornienko, V., Radchenko, R., Stachel, A., Andreev, A., Pyrysunko, M.: Correlations for pollution on condensing surfaces of exhaust gas boilers with water-fuel emulsion combustion. In: Tonkonogyi, V., et al. (eds.) Grabchenko's International Conference on Advanced Manufacturing Processes. InterPartner-2019. Lecture Notes in Mechanical Engineering, pp. 530–539. Springer, Cham (2020)
19. Kornienko, V., Radchenko, R., Konovalov, D., Andreev, A., Pyrysunko, M.: Characteristics of the rotary cup atomizer used as afterburning installation in exhaust gas boiler flue. In: Ivanov, V., et al. (eds.) Advances in Design, Simulation and Manufacturing III (DSMIE 2020). Lecture Notes in Mechanical Engineering, pp. 302–311. Springer, Cham (2020)
20. Sumenkov, V.M., Sen', L.I., Blinnikov, O.V.: Liquid atomization model by rotary nozzles. Shipbuilding **6**, 53–55 (2008). [in Russian]
21. Soma, T., Katayama, T., Tanimoto, J., Saito, Y., Matsushita, Y., et al.: Liquid film flow on a high speed rotary bell-cup atomizer. Int. J. Multiphase Flow **70**, 96–103 (2015)
22. Ray, R., Henshaw, P., Biswas, N.: Characteristics of spray atomization for liquid droplets formed using a rotary bell atomizer. J. Fluids Eng. **141**(8), 081303 (2019)
23. Ray, R.: Evaporation of spray from a rotary bell atomizer. Electronic Theses and Dissertations, 5704 (2015)
24. Guettler, N., Paustian, S., Ye, Q., Tiedje, O.: Numerical and experimental investigations on rotary bell atomizers with predominant air flow rates. In: 28th European Conference on Liquid Atomization and Spray Systems, ILASS 2017. Valencia, Spain (2017)
25. Ogasawara, S., Daikoku, M., Shirota, M., Inamura, T., Saito, Y., Yasumura, K., Shol, M., Aoki, H., Miura, T.: Liquid atomization using a rotary bell cup atomizer. J. Fluid Sci. Technol. **5**(3), 464–474 (2010)
26. Petrone, G., Cammarata, G., Caggia, S., Anastasi, M.: Reacting flows in industrial duct-burners of a heat recovery steam generator. In: Excerpt from the Proceedings of the COMSOL Conference. Hannover (2008)
27. Izawa, S., Iso, T., Nishio, Y., Fukunishi, Y.: Ligament formation and droplet breakup on disk-type and cup-type rotary atomizers. Trans. JSME **84**(862) (2018)

28. Stevenin, C., Béreaux, Y., Charmeau, J.-Y., Balcaen, J.: Shaping air flow characteristics of a high-speed rotary-bell sprayer for automotive painting processes. J. Fluids Eng. **137**(11), 111304 (2015)
29. Hatayama, Y., Haneda, T., Shirota, M., Inamura, T., Daikoku, M., Soma, T., Saito, Y., Aoki, H.: Formation and breakup of ligaments from a high speed rotary bell cup atomizer (Part 1: Observation and quantitative evaluation of formation and breakup of ligaments). Trans. Japan Soc. Mech. Eng. Ser. B **79**(802), 1081–1094 (2013)

Substantiation of Pressure Compensator Construction for Nuclear Power Plants in Emergency Situations

Igor Kozlov[1(✉)], Vladimir Skalozubov[1], Vladislav Spinov[1], Dmitriy Spinov[1], and Predrag Dasic[2]

[1] Odessa National Polytechnic University, 1, Shevchenko Av., Odessa 65044, Ukraine
kozlov_i.l_@ukr.net
[2] SaTCIP Publisher Ltd., 36210 Vrnjačka Banja, Serbia

Abstract. This paper presents the results of the process to qualify the pressure compensator system design as a result of the reactor coolant flow thermohydrodynamic instability for the management of accidents with a complete continued loss of power supply. Exposed are the features of transonic flow regimes at two-phase flows in the pipeline valves' flow part. Using computational modeling, the water hammer occurrence conditions have been determined for managing accidents with a continued blackout during the emergency opening of the pressure compensator safety valves due to its volume overflow. It has been found that an effective way to eliminate the water hammer emergence conditions is to increase the pressure compensator by upper part hydrodynamic resistance installing the grating structures. Based on computational modeling, functional dependencies are determined, and a method for qualifying the pressure compensator system design under accident with a complete continued loss of power supply is proposed. The conditions for ensuring the safety functions of the pressure compensator system for managing accidents with continued blacking-out shall be satisfied by maintaining the required level of reactor feed water, preventing the hydrodynamic shocks emergence, and determining the pressure compensator effective action delay of the to manage accidents.

Keywords: Qualification · Hydrodynamic shock · Continued loss of power supply accident · Pipeline fittings · Safety valve · Pulse safety device

1 Introduction

The main cause of the severe accident with damage to nuclear fuel and destructive steam-gas explosions at the Fukushima-Daiichi Nuclear Power Plant (NPP) in 2011 refers to the complete loss of long-term power supply (CLPS) [1]. CLPS led to the failure of active safety systems with electric pumps, at the same time that passive safety systems also failed to prevent severe accidents and explosions [2].

One of the relevant lessons drawn from the major accident at the Fukushima-Daiichi NPP reveals the need to review safety at beyond design basis accidents with

V. Tonkonogyi et al. (Eds.): InterPartner 2020, LNME, pp. 675–684, 2021.
https://doi.org/10.1007/978-3-030-68014-5_65

multiple equipment failures (including the complete loss of continued power supply) [3].

Studying those lessons of the Fukushima accident also identified the necessity to modernize the requirements and develop effective strategies for managing accidents and "mitigating" the consequences of severe accidents in extreme conditions [4, 5].

The creation of promising passive safety systems for nuclear power plants is associated with the need to qualify new/prospective and existing passive safety systems managing the accident that involves complete continued blackout (CCB). Under the IAEA Glossary, the equipment qualification definition refers to the collection and justification of operability guarantees for structures, systems, and equipment components under loads arising when normal operation conditions, as well as during transition and emergency modes [6, 7].

Thus, the issues of pressure compensator (PC) system compliance with the requirements imposed by the CLPS emergency at a nuclear power plant determine the relevance of the presented study.

This study's purpose is to elaborate a calculation method for PC attestation when a CLPS occurs; another goal of such research is to determine the PC effective response in order to manage CLPS emergencies.

2 Literature Review

The issues of computational modeling of accidents with nuclear power plants (NPP) complete continued blackout (CCB) and the analysis of emergency events with multiple failures of systems important to safety have been considered in works [8, 9].

These studies' results determine the justification of time delay chosen as required to restore power supply and/or to actuate additional tools of accident management (mobile diesel generators, mobile fire extinguishing systems, etc.). The Fukushima accident experience shows the insufficient capacities and effectiveness of such measures to ensure the performance of safety functions during the accident.

This research authors' recent works [10, 11] considered the water hammers (WH) area boundaries at pressure compensator VS-99 (producer Sempell) pulse protection device (SV PSD) safety valves' opening through the section as represented in the format of determining the criteria:

$$\mathrm{K}_5 = \frac{\rho_0 v_0^2}{P_{\max}};\quad \mathrm{K}_6 = \frac{L_{ik}}{\Pi_{iK}} \cdot \mathrm{grad}_z(\Pi_{iK})$$

where L_{iK} – safety valves confuser length; $\mathrm{grad}_z(\Pi_{iK})$ – the average gradient of the change in the area of the bore along the longitudinal coordinate z in the confuser part of the safety valves.

However, the WH regions boundaries inside the pressure compensator system given in [10, 11], never take into account the dynamic behavior of the coolant in the hydraulic system immediately in the reactor when CLSP accidents [12], that circumstance determines this work relevance.

3 Research Methodology

3.1 Basic Principles and Assumptions of the Pressure Compensator System Qualification Method for CLSP Accident Management

Main conservative assumptions and basic principles of this method are represented in a chronologically structured sequence of CLSP accident [10, 11]:

1. Disabling of the functions for removing the reactor-released residual heat and the steam generator feedwater required level maintained in the event of failure of the active safety systems (ASS) electric pumps.
2. A decrease in the coolant flow rate through the reactor in the event of ASS failure leads to the onset of steamization in the core and an increase in the reactor's internal steam pressure.
3. The function of residual heat-safe removal and maintaining the required level of steam generator feedwater at the expense of the stopped main circulation pump "run-out" and natural circulation in the 1st circuit conservatively is not taken into account.
4. An increase in steam volume pressure at the pressure compensator above the permissible values (P_{max}) causes the automatic opening of the pulse-safety device safety valves SV PSD, thus a pressure drop below P_{max}.
5. When opening/closing the pressure compensator SV PSD, three WH types are possible to occur, critical for the CLSP accident control reliability [10–12]:

- WH to the pressure compensator body due to its full volume overflow with the coolant (WH type WH1);
- WH at the PC SV PSD closure, initiated by pressure pulses when condensation under two-phase flows transonic flow regimes conditions in the PC through-flow path (WH type WH2);
- WH when safety valves closure caused by their unacceptable closing speed (WH type WH3).

The design scheme of the pressure compensator system qualification for CLSP accident conditions [10] is shown in Fig. 1. The structural and technical data of pressure compensator SVPSD are given in [13].

Mass balance equations for steam and coolant volumes in pressure compensator [14]:

$$\frac{d(\rho_v V_{vK})}{dt} = \rho_v \cdot \frac{dV_{vK}}{dt} + V_{vK} \cdot \frac{d\rho_v}{dP_{vK}} \cdot \frac{dP_{vK}}{dt} = -\rho_l \Pi_K \frac{dH}{dt} - G_{iK} \quad (1)$$

$$\rho_l \Pi_K \frac{dH}{dt} = -G_{iK} \quad (2)$$

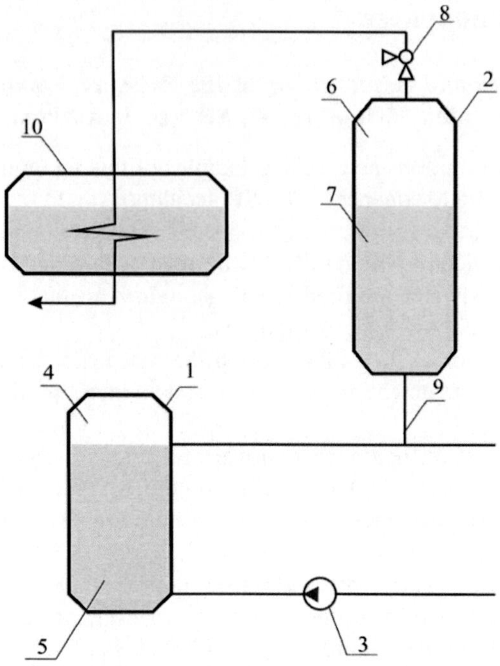

Fig. 1. Calculation scheme of qualification of the pressure compensator system: 1 – reactor; 2 – pressure compensator; 3 – reactor main coolant pump; 4 – vapor volume in the reactor; 5 – the volume of coolant in the reactor; 6 – steam volume in pressurizer; 7 – the amount of coolant in the pressurizer; 8 – safety valves of the pulse-safety device of pressure compensator (SV PSD of pressurizer); 9 – a connecting line of the pressure compensator with the 1st circuit; 10 – tank bubbler.

Initial conditions:

$$V_{vK}(t=0) = V_{vK_0}; H(t=0) = H_0(V_{vK_0}) \tag{3}$$

$$P_{vK}(t=0) = P_{vR_0} - \rho_l g(H_0 + L) \tag{4}$$

Mass flow from pressure compensator to the 1st circuit:

$$G_{iK} = \mu_K \Pi_T \sqrt{2\rho_l[P_{vK} - P_{vR} + g\rho_l(H+L)]} \tag{5}$$

where ρ_v, ρ_l – density of steam and coolant, respectively; V_{vK} – pressure compensator steam volume; t – time; P_{vK}, P_{vR} – pressure in the steam volume of the pressure compensator and the reactor, respectively; H, L – the height of the coolant level in the pressure compensator and the connecting pipe (Fig. 1), respectively; Π_K, Π_T– respectively, the pressure compensator and the connecting pipeline flow area; g – gravity acceleration; μ_K– total minimum coefficient of flow from the pressure compensator to the 1st circuit (see Table 2 [10]); G_{iK} – flow through the pressure compensator SV PSD system:

$$G_K = \begin{cases} \mu_{iK}\Pi_{iK}\sqrt{2\rho(P_{vK} - P_0}\text{, for opened SV PSD of pressurizer}(P_{vK} \geq P_{\max}) \\ 0,\ \text{for closed SV PSD of pressurizer} \end{cases} \tag{6}$$

where μ_{iK} – flow rate through SV PSD of pressure compensator ($P_{vK} < P_{\max}$); Π_{iK} – minimum flow area of pressure compensator SV PSD; ρ - density of the medium; P_0 – containment pressure; $P_{\max}$ – maximum pressure in the steam volume of pressure compensator.

Necessary conditions for water hammer occurrence in the pressure compensator body and pressure compensator SV PSD [11, 12]:

$$\text{WH1} - H = H_K \tag{7}$$

$$\text{for WH2} - \text{Mach criterion} \cdots = \frac{v(\Pi_{iK_0})}{a_{TF}(\Pi_{iK_0})} \geq 1 \tag{8}$$

for WH3 – closing speed of pressure compensator SV PSD

$$\frac{d\Pi_{iK}}{dt} \geq \frac{2 \cdot \Pi_{iK_0}}{t_0}\left[\frac{2(P_{vK} - P_0)\rho\Pi^2_{iK_0}}{G^2_{iK}} - \xi_0\right] \tag{9}$$

where H_K – high of pressure compensator; $v(\Pi_{iko})$ – two-phase flow rate when pressure compensator SV PSD fully open with a minimum flow area Π_{iko}; a_{TF} – speed of sound in a two-phase flow; ξ_0 – coefficient of hydraulic resistance when pressure compensator SV PSD fully open; t_0 – pressure compensator design opening/closing time.

The condition for the effective influence of the pressure compensator on ensuring the residual heat-safe removal function:

$$dH/dt \leq 0\ \ \Pi\text{pN}\ \ t \leq t_K \tag{10}$$

Mass balance and thermal energy equations for steam and coolant volumes in the reactor:

$$\frac{d(\rho_v V_{vR})}{dt} = \rho_v \cdot \frac{dV_{vR}}{dt} + V_{vR} \cdot \frac{d\rho_v}{dP_{vR}} \cdot \frac{dP_{vR}}{dt} = G_{iv} \tag{11}$$

$$\rho_l \Pi_R \frac{dh}{dt} = G_K - G_{iv} + G_{gP}(t) \tag{12}$$

$$G_{iv} \cdot i_v(P_{vR}) + \rho_l \cdot \Pi_R \cdot (i_v - i_l)\frac{dh}{dt} = N(t) \tag{13}$$

Under the initial conditions:

$$V_{vR}(t=0) = V_{vR_0}; P_{vR}(t=0) P_{vR_0}; h(t=0) = h_0; i_v(t=0) = i_{v0}; i_l(t=0) = il_0 \tag{14}$$

$$G_{gP} = G_0(1 - t/t_B) \tag{15}$$

where V_{vR}, P_{vR} – steam volume and pressure in the reactor; G_{iv} – vaporization flow rate in the reactor core; Π_R, h – flow area and coolant level in the reactor core, respectively; i_v, i_l – specific (per unit mass) enthalpy of steam and coolant, respectively; $N(t)$ - residual heat power; $G_{gP}(t)$ – a run-out flow of the stopped main coolant pump; t_B – a full run-down time of the main recirculation pump.

The maximum water hammer amplitude (ΔP_{gm}) on the pressure compensator body at $H = H_K$ can be determined from the energy conservation equation when the kinetic energy of braking of the coolant level is converted into the energy of the pressure water hammer pulse in the isometric approximation:

$$\frac{d}{dt}\left[\frac{\rho_l}{2}\cdot\left(\frac{d}{dt}\right)^2 + i_l\right] = 0 \tag{16}$$

After the transformations, it follows from (16):

$$\Delta P_{gm} = \int_0^{t_g} \frac{dP}{d\tau} d\tau = -\frac{\rho_l}{di_l dP} \int_0^{t_g} \frac{dH}{d\tau} \cdot \frac{d^2H}{d\tau^2} d\tau \tag{17}$$

Where $t_g = H_K/a_l$; a_l– the speed of sound in the coolant.

In the criteria (dimensionless) format of the equation of mass balance and thermal energy:

$$\rho_v \cdot \frac{dV_{vK}}{dt} + V_{vR} \cdot \frac{d\rho_v}{dP_{vR}} \cdot \frac{dP_{vR}}{dt} = \sqrt{K_7 \cdot (P_{vK} - P_{vR} + K_8 \cdot H + K_9)} - G_{iK} \tag{18}$$

$$\rho_v \cdot \frac{dV_{vR}}{dt} + V_{vR} \cdot \frac{d\rho_v}{dP_{vR}} \cdot \frac{dP_{vR}}{dt} = K_{10} \cdot G_{lv} \tag{19}$$

$$K_{11} \cdot \frac{dh}{dt} = G_K - G_{lv} + G_{gP} \tag{20}$$

$$K_{12} \cdot G_{lv} \cdot i_v + \frac{dh}{dt} = N(t) \tag{21}$$

Initial conditions:

$$\mathrm{V}_{vK_0} = 1; \mathrm{H}_0 = 1; P_{vK_0} = \mathrm{K}_{13}; V_{vR_0} = \mathrm{K}_{10}^{-1}; \mathrm{h}_0 = 1; \mathrm{P}_{vR_0} = P_{vR_0}/P_{vK_0}; \mathrm{i}_{v0} = i_{l0}/i_{v0} \tag{22}$$

where are the similarity criteria:

$$\left.\begin{array}{l} \mathrm{K}_7 = \frac{2\rho_{l_0}^3 P_{vK_0}\mu_S^2\Pi_{gP}^2\Pi_R^2 h_0^2 i_{v_0}^2 (1-i_{l_0}/i_{v_0})^2}{\rho_{v_0}^2 V_{vK_0}^2 N_0^2} \\ \mathrm{K}_8 = \frac{\rho_l g H_0}{P_{vK_0}}; \mathrm{K}_9 = \frac{\rho_l g L}{P_{vK_0}}; \mathrm{K}_{10} = \frac{V_{vK_0}}{V_{vR_0}}; \\ \mathrm{K}_{11} = \frac{\rho_{l_0}\Pi_R h_0}{\rho_{v_0}\cdot V_{vK_0}}; \mathrm{K}_{12} = \frac{\rho_{v_0}\cdot V_{vK_0}}{\rho_{l_0}\cdot \Pi_R \cdot h_0(1-i_{l_0}/i_{v_0})}; \\ \mathrm{K}_{13} = \left[P_{vR_0} - \rho_{l_0} g(H_0 + L)\right]/P_{vK_0} \end{array}\right\} \tag{23}$$

μ_S – total flow rate at the inlet to the reactor; Π_{Gp} – flow area of the coolant of the main circulation pipe.

The system of Eqs. (18)–(23) is nonlinear and in the general case can be solved by the numerical Runge-Kutta method [15].

4 Results

4.1 Analysis of the Results of Computational Modeling

Following (7), (8), (9) and the developed method, the criteria and qualification conditions of pressurizer system for water hammers in the process of an accident with LLPS:

$$\text{for WH1} - \mathrm{K}_{K1} = \frac{H}{H_K}(\mathrm{K}_7, \ldots, \mathrm{K}_{13}) < 1 \tag{24}$$

$$\text{for WH2 at } P_{vK} \geq P_{\max} - \mathrm{K}_{K2} = \mathrm{M}(\Pi_{iK_0}, \mathrm{K}_5, \ldots, \mathrm{K}_{13}) < 11 \tag{25}$$

$$\text{for WH3 at } P_{vK} \geq P_{\max} - \mathrm{K}_{K3} = \frac{1}{2}\left[\frac{2(P_{\max} - P_0)\rho\Pi_{iK_0}^2}{G_{iK}^2} - \xi_0\right]^{-1} < 1 \tag{26}$$

The feasibility of qualification conditions (24), (25) is determined by the results of integrating the system of nonlinear equations by the Runge-Kutta method.

To verify the proposed method for determining the conditions and parameters of water hammer in a pressure compensator, the well-known experimental data of A.V. Korolev obtained on the model of a pressure compensator VVER-440 were used [16].

Figure 2 shows the experimental data [16] on the relative maximum amplitude of WH 1 $\Delta P_{gm} = \Delta P_{gm}/P_0$ for various shutter diameters of the pressurizer VVER-440 model (SV PSD of pressurizer simulator). From the presented results it follows that the calculations according to the well-known formula N.E. Zhukovsky [17] have

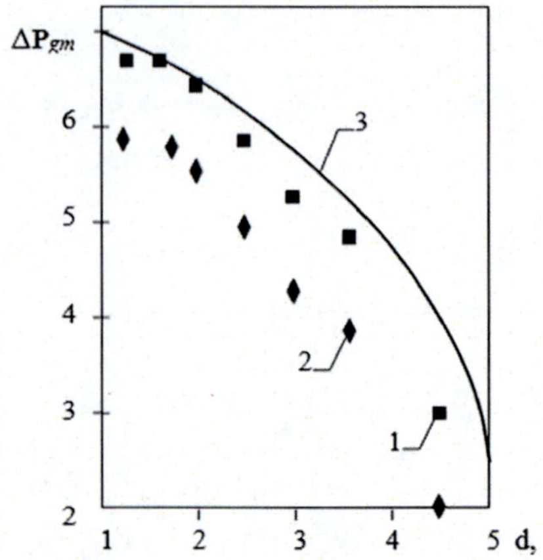

Fig. 2. The maximum amplitudes of the pressure of hydroblow when filling out the experimental model of the VVER-440 pressure compensator depending on the shutter diameter d (SV PSD of pressurizer simulator): 1 - experiment [16]; 2 - calculation by the formula N.E. Zhukovsky; 3 - calculation by formulas (17), (18)… (23).

underestimated ΔP_{gm} values with respect to experimental data, and solutions of Eqs. (17), (18)… (23) have quite satisfactory conservative estimates.

At the time of the opening of the SV PSD of pressurizer, the qualification condition (24) for the absence of a water hammer due to the overflow of the full volume of the pressure compensator (WH type WH1) with the coolant is not provided.

The qualification results for the conditions of the WH type WH2 are shown in Fig. 3 in the format of criteria K_5 and K_6. From the obtained results of computational modeling, it follows that the qualification condition for WH2 (25) is also not provided.

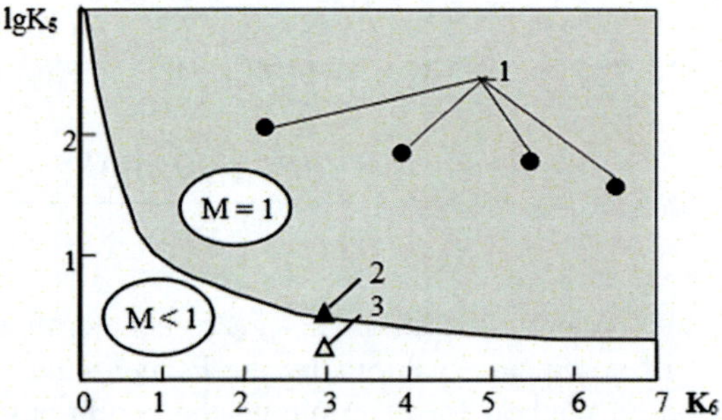

Fig. 3. The range of conditions for the occurrence of water hammer as a consequence of aperiodic instability in transonic flows of two-phase vapor-liquid flows: 1 - experiment 2; 2 - VS-99 at the rated power of the reactor; 3 - VS-99 during tests at the "hot" shutdown of the reactor.

The feasibility of qualification conditions (26) for the absence of a water hammer due to the accelerated closure of the SV PSD of pressurizer (WH type WH3) is ensured.

Computational models of emergency codes (e.g. 9–15) shall account for the supposed passive security system operation.

5 Conclusions

The conservative estimate of the pressure compensator system effective action time for managing accidents with a long-term total loss of power supply makes about 900 s from the moment of the emergency process start.

In the developed method framework the criteria, conditions and issuing consequences of water hammers occurrence due to the pressure compensator coolant overflow, two-phase flow transonic flow regimes in the pressure compensator safety valves' flow-through part and unacceptably accelerated closing of the safety valves when the pressure compensator steam volume pressure is less than the maximum allowable values. The obtained criteria, conditions, and consequences of the pressure compensator system water hammer do completely agree with already known experimental data.

As a result of carried out calculation analysis, it was found that during an accident with a complete continued loss of power supply accident, possible is the occurrence of water hammers due to the pressure compensator overflow when the safety valves are opened under transonic modes of the two-phase flow in the open safety valves' flow part.

An effective measure to prevent water hammer in the pressure compensator system shall be to increase the pressure compensator upper part hydrodynamic resistance by installing distance gratings.

References

1. International Fact Finding Expert Mission of the Fukushima-Daiichi NPP Accident Following The Great East Japan Earthquake and Tsunami. IAEA Mission Report (2011)
2. Gauntt, R., Kalinich, D., Cardoni, J., et al.: Fukushima Daiichi Accident Study Report. Sandia National Laboratories (2012)
3. Nielsen, N.: EU nuclear reactors fall short on safety. EUObserver (2012). https://euobserver.com/justice/117755.
4. Naffea, H., Gerliga, V., Shevelev, D., Balashevsky, A.: Assessing the effectiveness of passive heat removal from the containment of a VVER RP reactor under prolonged blackout conditions. Nucl. Radiat. Saf. **2**(58), 27–31 (2013)
5. Gromov, G., Dybach, A., Zelenyi, O., et al.: Results from Review of Stress Tests for Operating NPPs of Ukraine in the Light of the Fukushima-1AccidentinJapan. Nucl. Radiat. Saf. **1**(53), 3–9 (2012)
6. IAEA Safety Glossary: 2018 Edition, IAEA, Vienna (2019)
7. Safety Standards for protecting people and the environment General Safety No. GSR Part 4 (Rev. 1) Safety Assessment for Facilities and Activities, IAEA, Vienna (2016)
8. Skalozubov, V., Klyuchnikov, A., Komarov, Y., Shavlakov, A.: Scientific and technical foundations of measures to improve the safety of nuclear power plants with VVER. Institute of NPP Safety Problems, Chernobyl, Ukraine (2010). [In Russian]

9. INSAG Series No. 12, Basic Safety Principles for Nuclear Power Plants 75-INSAG-3, Rev.1. Report by the International Nuclear Safety Group. IAEA, Vienna (2015)
10. Skalozubov, V., Spinov, V., Pirkovskiy, D., Gablaya, T., Rafalskyi, R.: Qualification of the pressure compensator system for the management of accidents with complete loss of long power supply from VVER power plant. Odes'kyi Politechnichnyi Universytet Pratsi **2**(58), 60–68 (2019). https://dspace.opu.ua/jspui/handle/123456789/10518.
11. Skalozubov, V., Bilous, N., Pirkovsky, D., Kozlov, I., Komarov, Y., Chulkin, O.: Water hammers in transonic modes of steam-liquid flows in NPP equipment. Nucl. Radiat. Saf. **2** (82), 46–49 (2019)
12. Skalozubov, V., Chulkin, O., Pirkovsky, D., Kozlov, I., Komarov, Y.: Method for determination of water hammer conditions & consequences in VVER pressurizer. Turk. J. Phys. (2019)
13. IPU KD || ChEM. Specifications Pulse safety device of the pressure compensator. https://chzem.nt-rt.ru/images/manuals/IPUKD.pdf
14. Kljuchnikov, A., Sharaevskij, I., Fialko, N., Zimin, L., Sharaevskaja, N.: Thermophysics of nuclear reactor accidents. ISP NPP NASU, Chornobyl (2012)
15. Galkin, A., Dyatchina, D.: Numerical decision of mathematical models of objects given by component systems of the differential equations. Mod. Prob. Sci. Educ. **6** (2011). https://science-education.ru/ru/article/view?id=5196, last accessed 2020/05/30.
16. Korolev, A., Ischenko, A., Ishchenko, O.: The study of water hammers when filling the pressure compensation system in water-water power reactors, News of higher educational institutions and energy associations of the CIS. Energy **5**, 459–469 (2017). [in Russian]
17. Korolev, A.: Analysis and modeling of heat and power equipment, working with the two-phase flow. Astroprint, Odessa (2010). [in Russian]

High-Octane Fuel Compositions Based on Petroleum and Biocomponents

Nina Merezhko[1], Valentyna Tkachuk[2(✉)], Viktoria Romanchuk[3], Oksana Rechun[2], and Victor Zagoruiko[2]

[1] Kyiv National University of Trade and Economics, 19, Kioto St., Kyiv 02156, Ukraine
[2] Lutsk National Technical University, 75, Lvivska St., Lutsk 43018, Ukraine
v.tkachuk@lntu.edu.ua
[3] Lviv Polytechnic National University, 12, Bandera St., Lviv 79013, Ukraine

Abstract. The problem of the environmental safety of road transport has become part of the security of Ukraine. The annual increase in emissions of auto-transport into the atmosphere requires stricter environmental requirements for commercial vehicles and exhaust gases of internal combustion engines. World experience shows that the use of 10–15% bioadditives in a gasoline mixture does not harm the technical and operational performance of the internal combustion engine. Besides, there are more advantages in using this type of fuel, since it is an environmentally friendly type of fuel, and it generates fewer emissions during combustion. Biofuels can be adapted to existing engine designs that will be well used in all conditions. At the same time, this fuel is better for engines; it reduces the overall cost of controlling engine pollution and, consequently, its use requires fewer maintenance costs. At the same time, modern requirements for gasoline consist in the fact that they must ensure the creation of a homogeneous fuel-air mixture of the necessary composition under any temperature conditions. This article investigates the production of high-octane gasoline A-92 and A-95 by using high-octane applications. When conducting the research used the following BodogAri: methyl tert-butyl ether (MTBE), ethanol, and isobutanol. As a basis for gasoline, catalytic cracking gasoline, catalytic reforming reformate, benzene production raffinate, straight-run gasoline, and petroleum solvent were used.

Keywords: Gasoline · Operational properties · Quality · Ecological purity

1 Introduction

Gasoline fractions obtained at refineries using atmospheric distillation units have low octane numbers, in the range of 65–75, according to the research method [1]. At low octane values, the use of such fuel is fraught with negative consequences for the engine due to fuel detonation. Among the most common negative phenomena: the front wear of valves and seats, as well as the residue of carbon on the walls and floors. That is, the

V. Tonkonogyi et al. (Eds.): InterPartner 2020, LNME, pp. 685–694, 2021.
https://doi.org/10.1007/978-3-030-68014-5_66

octane number must be suitable for a particular engine, so it is relevant to use various methods to increase it for gasoline.

In the traditional scheme of refineries, a catalytic reforming process is used to increase the octane number of straight-run gasoline, and its final product-reformate is added to gasoline as the main source of aromatic hydrocarbons and benzene (the aromatics content in reformat is 60–70%, including benzene 2–7% wt.) [2]. The reforming process itself is quite energy-intensive using expensive platinum catalysts, and the output of the target products is 80–84% by weight. At the same time, the standard content of aromatic hydrocarbons in commercial gasoline is not more than 35%, and benzene is not more than 1%, according to ISO 6246:2017 for Euro-5 gasoline. For reducing the share of aromatic hydrocarbons and benzene, various options for processing reformation products are proposed, but they lead to significant additional costs [3].

A drastic reduction of aromatic hydrocarbons, including benzene, can be achieved by reducing the proportion of reformate and using compounds that increase the octane number and do not contain aromatic carbohydrates to maintain the octane number of gasoline at the required level [2].

In this regard, there is a growing interest in using of bioadditives, which would cover the environmental and operational properties of the fuel and make it economically feasible when added to gasoline with a low octane number [4].

World experience shows that the use of 10–15% bioadditives in a gasoline mixture does not have a negative impact on the technical and operational performance of the internal combustion engine. In addition, there are more advantages in using this type of fuel, since it is an environmentally friendly type of fuel, and it generates fewer emissions during combustion.

Biofuels can be adapted to existing engine designs that will be well used in all conditions. At the same time, this fuel is better for engines. It reduces the overall cost of controlling engine pollution and, consequently, its use requires fewer maintenance costs. At the same time, modern requirements for gasoline consist in the fact that they must ensure the creation of a homogeneous fuel-air mixture of the necessary composition under any temperature conditions. At the same time, the fuel should evaporate well, have a small surface tension, good starting properties, and provide a quick warm-up of the cold engine and its high pick-up rate. For this purpose, the fuel must have a high heat of combustion, burn completely, without detonation in all modes of engine operation.

2 Literature Review

The works of such scientists as P. Topilnytskyi, B. Buhai, S. Zubenko, O. Hrynyshyn, V. Brostov, S. Boichenko, and others are devoted to certain aspects of the functioning of the petrochemical industry in Ukraine [5–7]. P. Topilnytskyi's works are devoted to the development and implementation of demulsifiers for oil dehydration, neutralizers, and corrosion inhibitors of oil-water equipment, the study of the action of depressor additives in diesel fuels and liquefied gases. The works of S. Boichenko on the problems of effective and rational use of fuels and lubricants and technical liquids. The problems of using alternative fuels, in particular, the effect of additives based on nanocarbon clusters on the performance properties of ethanol-gasoline are investigated in works of O. Haidai, Ya. Bereznytskyi, N. Himach, V. Pishevskyi, et al. Issues of long-term storage of gasoline are covered in the works of B. Kochirko, I. Budzynska, and N. Kharchenko [6]. In our previous publications, the quality of light oil products, problems, and prospects of the market for these products are studied [8–10].

Many bioadditives and fuel compositions based on them are patented and used in the world [2]. The experience of using high-octane compounds of various classes has shown that the most promising among them are oxygen-containing applications, or oxygenates [4, 11, 12]. The use of bioalcohols as a fuel or additive to fuel has long been known since in 1914 it was proved that the transition from gasoline to alcohol is possible, and in 1934 in Europe already produced more than $2.65 \cdot 10^6$ m^3 of alcohol-zine mixtures [13].

The advantage of alcohol-containing fuels is to reduce the amount of carbon monoxide, nitrogen oxides, and soot in engine exhaust gases. Also, alcohols have high anti-knock properties and are the most effective anti-knock additive. Alcohol-gasoline mixtures are not inferior to traditional petroleum fuels by motor properties. Despite the lower heat of combustion, such compositions provide work on impoverished mixtures, so the increase in fuel consumption, in this case, is insignificant: the average consumption of ethanol-containing fuel is 5%. Because of the use of alcohols, not only carbon monoxide and hydrocarbon emissions are reduced but also the emission of nitrogen oxides from car exhaust gases. Bioethers are a promising addition to motor fuel. Their advantages in comparison with alcohols are that the oxygen content in the molecules is two times lower. The lower heat of combustion is much higher than in alcohols. Esters are corrosion-active or low active, practically insoluble in water, environmentally safer, and the detonation resistance is not inferior to alcohols.

The problems of efficient and economical use of energy resources and expansion in the structure of the total energy consumption of renewable energy sources have been widely covered in scientific works of foreign scientists: Ajanovic A. Bentivoglio D., Rasetti M., Popp J., Lakner Z., Harangi-Rákos M., Fári M., Tyner W. and others [14–17].

Additionally, the influence of fuel properties to performance of the compression ignition for engines, as well as quantification of execution and emission efficiency are presented in the research works [18, 19].

The purpose of the work is to obtain high-octane gasoline A-92 and A-95 by using high-octane applications. The object of the research is as follows: the following bioadditives were used in the studies: methyl tert-butyl ether (MTBE), ethanol, and isobutanol.

As a basis for gasoline, catalytic cracking gasoline, catalytic reforming reformate, benzene production raffinate, straight-run gasoline, and petroleum solvent were used. These are products that are manufactured at a modern oil refinery and are available on the market for oil products.

Catalytic cracking gasoline (CC gasoline) is produced at the appropriate installation, which is designed to increase the depth of oil refining and obtain high-quality gasoline. The industrial process is carried out at a temperature of 450–510 °C, a pressure of 0.14–0.18 MPa, using zeolite catalysts. The catalytic reforming reformat (reformat KR) is produced at the catalytic reforming station, which is designed to produce a high-octane component of gasoline from low-octane gasoline by flavoring them. Industrial processes of catalytic reforming are carried out on platinum catalysts in a water-based gas environment at a temperature of 470–520 °C, a pressure of 1.5–4.0 MPa, and a volume feed rate of 1–2 h^{-1}. To reduce the benzene content of reformate two ways are possible – selection of raw faction reforming resolutory or separation from the reformed benzene fraction. Benzene production refining (BP refining) is a by-product of the extraction of aromatic hydrocarbons from the catalytic reforming reformate. Straight-run gasoline (SG) is obtained by atmospheric distillation of oil. Solvent oil is a mixture of xylenes, which is obtained in the process of extraction and catalytic reforming. The characteristics of the oil products that were used as the basis for the production of a-92 and a-95 gasolines are shown in Table 1.

Table 1. Characteristics of oil products that were used as the basis for the production of A-92 and A-95 gasoline.

Name	CC gasoline hydrotreated	Reformat KR without the benzene fraction	BP refining	Gasoline straight-run	Oil solvent
Sulfur content, ppm	10	0,1	0,1	0,1	0,2
The content of aromatic hydrocarbons,% vol.	19,7	64,2	0,4	3,5	99,9
Benzene content,% vol.	0,6	0,7	0,1	0,3	–
Octane number by experimental method	89,3	97,7	69,5	72	105
Motor octane number	81,5	87,5	61,2	66	100
Density at 20 ˚C, kg/m^3	715	803	674	685	862
Paraffin hydrocarbon content,%	57,0	33,7	92,8	62,6	–
The content of naphthenic hydrocarbons,%	11,3	1,5	6,4	33,9	0,1
Olefin hydrocarbons content,%	12,0	0,6	0,4	–	–
Fractional composition:					
beginning of boiling	30	41,5	44,5	31	137
end of boiling	210	189,5	109,8	85	142

3 Researches Methodology

At the first stage of research, the basic bases of gasoline A-92 and a-95 were obtained by mixing oil components: catalytic cracking gasoline, catalytic reforming reformate, refined gasoline production and oil solvent in such a ratio as to provide an octane number in the mixture of about 90 for gasoline A-92 and 93 for gasoline A-95. As a result, the basic bases of gasoline A-92 and gasoline A-95 were obtained. In them, octane numbers were determined using the motor method (ON by MM) and the research method (ON by RM) according to the methodology described in the ISO 5164:2005 SSTU [4].

At the second stage, the influence of bioapplications on the change in the octane number of the base gasoline base was determined. To do this, bioavailables were added to the basic bases in the amount of 5, 7, and 10% by volume each. In each mixture, the octane number was determined by the motor and test method, according to SSTU ISO 5164: 2005 [4].

Based on the data obtained, we recommend the formulation for obtaining A-92 and A-95 gasoline with bioadditives. In these gasolines, the octane number was determined by the experimental method, density, sulfur content, aromatic hydrocarbons content, fractional composition, benzene and oxygen particles, the concentration of actual resins, and the data were tested for compliance with the requirements of DSTU 7687: 2015. All these indicators were determined by standard methods [2, 20]. The processing of experimental results was carried out on the software [21]. Physico-chemical parameters of bioadditives to gasoline are shown in Table 2.

Table 2. Physico-chemical parameters of bioadditives to gasoline.

Indicators	Bioadditives		
	Bioethanol	Bioisobutanol	MTBE
Density at 20 C, kg/m^3	790	802	742
Heat, kJ / kg	26945	35690	33200
Evaporation combustion	839	562	–
Flash point, (in a closed crucible), °C	12	27	−28
Octane number			
- by experimental method	121	108	115
- by the motor method	97	91	97
Saturated vapor pressure, kPa	17	9,7	53,0
Oxygen content,% vol	34,78	21,6	16,18

4 Results

The composition of the basic basis for obtaining gasoline A-92 and A-95, which are used to study the impact of bioadditives, are presented in Table 3.

Table 3. The composition of the base for obtaining gasoline A-92 and A-95.

The oil component	Composition of the base for gasoline production, %	
	A-92	A-95
CC пasoline hydrotreated	53	50
KR reformate without benzene fraction	35	44
BP refining	12	–
Oil solvent	–	6

In the course of the research, the influence of applications on the octane characteristics of the base of gasoline A-92 and gasoline A-95 was established (Table 4). The octane number was determined using the motor method (ON per MM) and the experimental method (ON per EM).

Table 4. The influence of additives on the octane characteristics of the base for A-92 and A-95 gasoline.

Basic framework + additive	A-92		A-95	
	ON by MM	ON by EM	ON by MM	ON by EM
100% vol. Basic basis	82,2	90	84,4	94,3
95% base + 5% vol. MTBE	83,4	91,3	85,0	95,3
93% base + 7% vol. MTBE	83,8	92,9	85,3	95,8
90% base + 10% vol. MTBE	84,4	92,7	85,5	96,4
95% base + 5% vol. Bioethanol	83,5	92,1	85,0	96,0
93% base + 7% vol. Bioethanol	83,8	92,6	85,4	96,4
90% base + 10% vol. Bioethanol	84,3	93,8	85,8	97,1
95% base + 5% vol. Bio IBS	82,8	90,5	84,5	94,8
93% base + 7% vol. IBS	83,3	91,4	84,8	95,1
90% base + 10% vol. IBS	83,9	91,8	85,2	95,7

From the above data, it can be seen that the largest increase in the octane number is given by adding ethanol in an amount of 10% for gasoline A-92 and a-95, to a lesser extent MTBE, and even less than Isobutanol. However, considering the cost of ethanol, which is currently 35,000 UAH/t and the cost of bioizobutanol-26,000 UAH/t [22], to reduce the cost of gasoline, we recommend Isobutanol as a high-octane addition to gasoline. Besides, to reduce the amount of aromatic hydrocarbons in gasoline, it is recommended to reduce the content of catalytic cracking gasoline in gasoline A-92 from 53 to 30%, leave the content of catalytic reforming to reformat at the level of 35%, and reduce the refined gasoline production from 12 to 10%.

For A-95 gasoline, the content of catalytic reforming reformat should be reduced from 44 to 37%, and catalytic cracking gasoline from 50 to 28%. Since each refinery has a fraction of straight-run gasoline of - 85°C with a low octane number, it is advisable to add it to gasoline. In our case, it is added in an amount of 10% to the a-92 gasoline and 9% to the A-95. It should be noted that methyl tert-butyl ether belongs to the hazard class 4 [23] and is used by us within 5–9%, and is therefore safe for environmental impact. Scientific studies [24] have shown that MTBE has a low soil accumulation capacity and is found in the soil only at high concentrations and partly migrates into the air due to its high volatility. The amount of Isobutanol is added in an amount of 10%. To increase the octane number of gasoline A-92, we recommend adding MTBE in an amount of 5% as the component with the highest octane number for gasoline A-95-in an amount of 9%. Besides, for high-octane gasoline A-95 is given a solvent as a high-octane component in an amount of 5%. The recommended formulation and characteristics of the obtained A-92 and A-95 gasolines are presented in Tables 5 and 6.

Table 5. The recommended formula for gasoline A-92.

Additives	Characteristic of the additive		Characteristics of the additive in the mixture		
	ON per EM	Aromatic hydrocarbons content %	Additive content,%	ON per EM *	Aromatic hydrocarbons content,%
CC gasoline	89,3	19,7	30	26,8	5,91
Reformat KR	97,7	64,2	35	34,2	22,47
BP refining	69,2	0,4	10	6,9	0,04
Benzene production	72,0	3,5	10	7,2	0,35
MTBE	115,0	0	5	5,8	0
Bioisobutanol	108,0	0	10	10,8	0

The indicators of the obtained gasolines presented in these tables are obtained by the calculation method. The composition of gasoline is selected in such a way as to obtain gasoline with an octane number of 92, while the content of aromatic hydrocarbons in the resulting gasoline should not exceed 35%. As in the A-92 version of gasoline, the gasoline composition was selected so that its octane number was close to 95, and the content of aromatic hydrocarbons did not exceed 35%. The resulting mixtures were analyzed in the laboratory for compliance of all quality indicators with the requirements of State standard of technical conditions 7687: 2015. Comparative data are presented in Table 7.

Analysis of the data from Table 7 allows establishing that gasoline with an oil base and with the addition of bioadditives meet the requirements of State standard of technical conditions according to the State standard 7687: 2015.

Table 6. The recommended formula for gasoline A-95.

Additives	Characteristic of the additive		Characteristics of the additive in the mixture		
	ON per EM	Aromatic hydro-carbons content, %	Additive content,%	ON per EM *	Aromatic hydrocarbons content,%
Gasoline catalytic cracking	89,3	19,7	28	25,00	5,52
Reformed catalytic reforming	97,7	64,2	37	36,15	23,75
Refined benzene production	69,2	0,4	7	4,84	0,03
Oil solvent	105	100	5	5,25	5,00
Straight-run gasoline	72	3,5	9	6,48	0,32
Methyl tert-butyl ether	115	0	9	10,35	0,00
Bioisobutanol	108	0	5	5,40	0,00

Table 7. Compliance of gasoline performance indicators with State standard of technical conditions requirements 7687:2015.

Name of indicator	The norm value for A-92/A-95	Defined indicators for A-92	Defined indicators for A-95
Octane number by the experimental method	92/95	91,7	93,8
The saturated vapor pressure in summer, kPa	45-80	69	70
Density at temperatures of 15 °C, kg/m^3, within	720–775	721	729
Fractional composition:			
- is the volumetric fraction of evaporation at 70 ° C,%, within	20,0–50,0	34,0	33,0
- the volumetric fraction of evaporation at 100 ° C,%, within	46,0–71,0	66,0	67,2
- volumetric volume of evaporation at 150 °C,%, not less	75,0	78,0	81,0
- boiling point final, C, not higher	210	198	201
- the volume fraction of the balance after the outburst, % (not more than)	2	1,5	1,5
Sulfur content, mg/kg	10	4	4
Volume fraction of aromatic hydrocarbons, %	35	29,5	33,6
Benzene volume fraction, %	1	0,6	0,7
Oxygen mass fraction, %	2,7	2,3	2,7
The concentration of the actual resin, mg/(100 cm^3)	5	4	4
Corrosion on copper plate (3 h at 50 °C) class	1	1	1

5 Conclusions

The influence of bioadditives – ethanol, bioisobutanol, MTBE on the increase of octane number of gasoline prepared based on gasoline-catalytic cracking, reform of catalytic reforming, benzene production refinement, solvent, and straight-run gasoline is investigated. An optimal formulation of gasoline has been developed with the use of all refinery components and high-octane bioadditives.

The largest increase in octane is found to be the addition of 10% ethanol for gasoline A-92 and A-95, to a lesser extent MTBE, in less isobutanol. However, given the cost of ethanol, to reduce the cost of gasoline, we recommend adding isobutanol as a high-co-octane additive to gasoline. For increasing the octane number of gasoline A-92, we recommend the addition of MTBE in the amount of 5% as the component with the highest octane, for gasoline A-95 – in the amount of 9%. Besides, solvent for high-octane A-95 gasoline is added as a high-octane component, in the amount of 5%.

The result of these studies is to establish the optimal ratio of components in fuels, which are different from the existing counterparts of the new recipe, obtained from oil refining water to obtain a composition that meets the Euro-5 gasoline standard in terms of octane number, aromatic hydrocarbon content, benzene content, sulfur content, fractional composition, concentration of actual resins, and would not be corrosive. The indicators of refined gasoline meet all the regulatory requirements for high-octane gasoline A-95 Euro-5 according to the State standard of technical conditions requirements 7687: 2015.

References

1. Topilnytskyi, P., Hrynyshyn, O., Machynskyi, O.: Primary oil and gas processing technology: a textbook. Lviv Polytechnic Publishing House, Lviv, Ukraine (2014)
2. Boichenko, S., Pushak, A., Topilnytskyi, P., Leida, K: Motor fuels: properties and quality. Center of educational literature, Kyiv, Ukraine (2017)
3. Kapustin, V.M., Rudyn, M.H.: Chemistry and oil refining technology: a textbook. Russia, Moscow (2013)
4. Kozak, F.V., Melnyk, V.M., Prunko, I.B., Voitsechivska, Y.I.: Cost-effective use of bioethanol on internal combustion engines. Road Transp. **42**, 22–28 (2018)
5. Topilnytskyi, P.I., Holych, Y., Boichenko, S.V., Romanchuk, V.V.: Dependence of oil dehydration on their physical and chemical characteristics. Oil Gas Ind. Ukraine **1**, 25–30 (2015)
6. Haidai, O.O., Zubenko, S.O., Polynkin, Ye.V., Pyliavskyi, V.S.: Environmental and operational characteristics of fuel biological motor E-85. In: Proceedings of the scientific articles of the third All-Ukrainian Congress of Ecologists, pp. 308–310. Ukraine (2011)
7. Boichenko, S.V., Ivanov, S.V., Burlaka, V.H.: Motor fuels and oils for modern technology: monograph, Ukraine (2005)
8. Tkachuk, V.V.: Research of modern problems of production of alternative fuels for gasoline engines in Ukraine. Merchand. Bull. Collection Sci. Works **12**, 249–256 (2019)

9. Tkachuk, V. Rechun, O., Merezhko, N., Bozhydarnik, T., Karavaiev, T.: Assessment of the quality of alternative fuels for gasoline engines. In: Advances in Design, Simulation and Manufacturing II Proceedings of the 2nd International Conference on Design, Simulation, Manufacturing: The Innovation Exchange, DSMIE-2019, p. 871–881. Springer, Cham (2020). https://doi.org/10.1007/978-3-030-22365-6_46
10. Merezhko, N., Tkachuk, V., Zinchenko, O.: Performance of gasoline with multifunctional additives. Goods Mark. **4**, 50–61 (2019)
11. Nguen, Van Ty.: Coisonerization of light fractions of straight-run gasoline and reformate. Chem. Technol. Fuels Oils **4**, 5–9 (2008)
12. Markov, V.A., Vallejo Maldnado, P.R., Biriukov, V.V.: Alcohol fuels for diesel engines. News of higher educational institutions. Engineering **11**(668), 39–51 (2015)
13. Markov, V.V., Petrykin, A.A., Emelianov, V.E., Shamonina, A.V., Varannik, V.P., Onoichenko, S.N.: Alcohols as additives to gasoline. Autom. Ind. **8**, 24 (2005)
14. Veloso, A., José, M., Machado, C., Márcia, Y., Pasa, D.: The effect of temperature on the properties of the NBR/PVC blend exposed to ethanolfuel and different gasolines. Fuel **113**, 679–689 (2013)
15. Ajanovic, A.: Biofuels versus food production: Does biofuels production increase food prices? Energy **36**, 2070–2076 (2011)
16. Bentivoglio, D., Rasetti, M.: Biofuel sustainability: review of implications for land use and food price. Rivista di Economia Agraria, Anno LXX, 7–31 (2015)
17. Popp, J., Lakner, Z., Harangi-Rákos, M., Fári, M.: The effect of bioenergy expansion: Food, energy, and environment. Renew. Sustain. Energy Rev. **32**, 559–578 (2014)
18. Towoju, O.A., Jekayinfa, S.O.: Compression ignition engine performance as a function of the fuel properties. J. Eng. Sci. **6**(1), G1–G5 (2019). https://doi.org/10.21272/jes.2019.6(1).g1
19. Nagendra, S., Kumar, B.S., Kiran, A.V., Ramanjaneyalu, C., Nagamani, K.: Quantification of execution and emission efficiency of a fueled diesel engine. J. Eng. Sci. **7**(1), G15–G20 (2020). https://doi.org/10.21272/jes.2020.7(1).g3
20. Zinchenko, O., Topilnytskyy, P., Romanchuk, V.: Equipmentof PJSC "Ukrtatnafta" laboratories in the aspect of the quality of petroleum products, Poland (2018)
21. Kuts, Y., Povstyanoy, O.: Computer-informative software for research of the new materials of constructional application. Funct. Mater. **1**, 175–178 (2017)
22. Wholesale prices for alcohols. https://ua.all.biz/uk/spyrt-izobutylovyj-izobutanol-bgg1084909. Accessed 29 Nov 2019
23. Yavorsky, O.P., Paustovsky, Y., Anisimova, I.G., Zaprivoda, L.P.: Experimental study of metyl tertiary-butyl ether effect on the embryogenesis in white rats. Environ. Health **2**, 9–13 (2015)
24. Garkavyi, S.S., Omelchuk, S.T., Brozgina, T.S.: Experimental study of metyl tertiary-butyl ether effect in the system "soil-groundwater" as as a stage of establishing its hygiene standard in the soil. Prev. Med. **2**, 105–109 (2011)

Method for Measuring the Temperature in the Elements of a Wind Turbine Multiplier

Boris Morgun[1], Raul Turmanidze[2], Julya Morgun[1], Pavlo Shvahirev[1], and Oleksandr Levynskyi[1(✉)]

[1] Odessa National Polytechnic University, 1, Shevchenko Ave., Odessa 65044, Ukraine
Rikokilller.rc@gmal.com
[2] Georgian Technical University (GTU), 77, Kostava Street, 0175 Tbilisi, Georgia

Abstract. The possibilities of using a new type of device to evaluate the temperature parameters of the design of the planetary-friction multiplier developed by the authors for a wind power installation are considered. Unlike gear-based animators, planetary-friction gears have compactness, manufacturability, and overload protection, which is especially important for a wind power installation, but they are at risk of high temperatures due to frictional torque transmission. The work presents the design of the multiplier in which, in contrast to the known designs, the contacts between the balls and cages are removed from the bearing interference system, which reduces the risk of heating. The idea of introducing lubricant into the multiplier, which reduces friction in the contacts between the balls and separators, where the specific pressure is low, is squeezed out between the balls and bearing rings, and also using the spring, creates an estimated high pressure, which allows semi-dry friction. For estimating the temperature of multiplier construction elements with small dimensions, the authors developed a contact artificial thermocouple method that allows using its two thermoelectrodes closely located and fixed in the probe to determine its temperature quickly and accurately with contact with the conductive surface of the element. The article provides an analysis of ensuring high measurement accuracy and the result of evaluating the temperature parameters of the multiplier elements during its long-term operation, both with and without lubricant. It is established that the proposed design of the multiplier with lubricant provides heating not higher than the established standards.

Keywords: Thermocouple · Multiplier · Temperature

1 Introduction

Known disadvantages of gearboxes and gear-based multipliers are difficulty in manufacturing, high cost, large dimensions, increased levels of vibration, and noise. Therefore, original developments and scientific studies of the designs of ball planetary-friction mechanisms (BPFM) in wind turbines, which act as planetary mechanisms and are distinguished by high manufacturability, compact structures reduced vibration activity, and drive protection from overloads, are relevant [1–4].

V. Tonkonogyi et al. (Eds.): InterPartner 2020, LNME, pp. 695–703, 2021.
https://doi.org/10.1007/978-3-030-68014-5_67

All the basic requirements for the quality indicators of structures are reduced to ensuring the normal thermal regime of their operation [5–8]. For planetary-friction mechanisms, where the transmission of torques is carried out by friction, the temperature of structural elements is a fundamental requirement. It should be carefully evaluated in the study of experimental samples.

The purpose of the article is the development of a method for measuring the temperature of conductive surfaces with high accuracy and speed and allowing to measure the temperature of small parts located in hard to reach places; measuring the temperature of the elements proposed by the authors of the planetary-friction multiplier [9, 10] at nominal operating conditions to assess its performance and durability and confirm its practical expediency; confirmation of the hypothesis about the possible use of lubricant in the friction gears of the multiplier to reduce friction between the balls and the separators.

2 Literature Review

In [11], the construction of a planetary-friction mechanism was proposed, which makes it possible to use it as a multiplier for wind power plants. However, the interference mechanisms of the bearings of the proposed design do not provide adjustment of contact forces in the bearings, and spurious dry friction between the balls of bearings and cages is not eliminated, which significantly increases the temperature of the planetary-friction mechanism. Temperature measurement by known methods [12] does not allow quickly and with high accuracy to determine the surface temperature, especially in hard-to-reach places of the structure. Measuring the temperature of the elements of the planetary-friction multiplier is a difficult task since its parts are in close contact with each other and access to them is difficult. The metrological support of the tests largely depends on the accuracy and speed of the used measuring equipment.

At present, thermocouples, pyrometers, and thermal imagers are widely used to measure the temperature of body surfaces [13]. However, pyrometers and thermal imagers based on the analysis of radiation emanating from the object under study [12] cannot be used to solve the problem. Firstly, the area of the investigated object must be significant. Secondly, at the same body temperature, different amounts of heat are emitted from different metals, that is, there is a different emissivity, which excludes the possibility of accurate temperature measurement.

Known thermocouples having a junction are usually attached to the surface to be measured, which increases the heat sink with the small dimensions of the elements under study and leads to temperature estimation errors.

A probe for measuring temperature is also known, a junction of thermoelectrodes can be quickly contacted with a given surface. But the thermocouple junction in contact with the surface is pressed by a leaf spring, which causes heat removal, leading to a loss in the accuracy of irradiation [14].

3 Research Methodology

The authors developed the "Probe for measuring the surface temperature of the body" [15], which has high measurement accuracy, speed, and the ability to measure temperature in hard-to-reach places. In the probe, there are two thermoelectrodes made in the form of compensation spirals creating a thermocouple and fixed in an insulated core. The working end of the thermocouple has no junction, and the thermoelectrodes freely contact the surface of the body, the temperature of which is measured.

The technical effect achieved by applying the proposed method is that the probe design eliminates the high inertia of heating the thermocouple junction, which reduces the measurement error. Fig. 1 shows a probe that has thermoelectrodes 1 and 2, freely located at the outlet and protruding beyond the plane of the probe end face. The leads of the thermoelectrodes are fixed in an insulated core 3 built into the metal cup 4. A holder 6 with a cable 7 is laid through a ferrule 5 made of plastic, the wires of which are connected to the leads of the thermoelectrodes 1 and 2. Each thermoelectrode has a length compensator made in the form of a spiral 8.

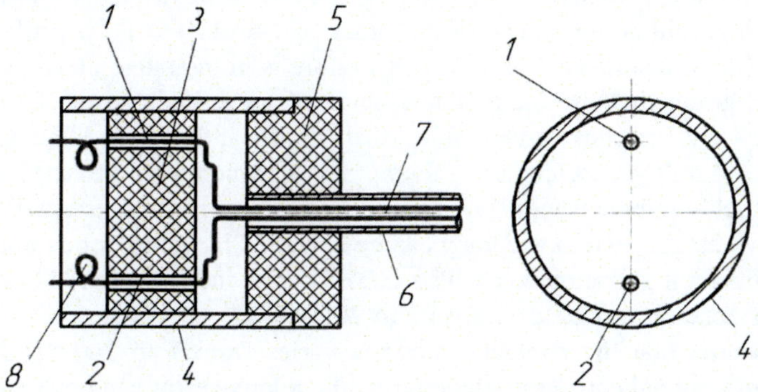

Fig. 1. Temperature probe for the proposed measurement method surface temperatures.

For measuring temperature, the probe is pressed against the test surface. In this case, the thermoelectrodes come into contact with the metal surface and create a thermocouple in which there is a third intermediate electrode, which does not affect the thermo-emf, since the temperature of the contacts is [12, 13].

For achieving high accuracy in determining the temperature using this method, it is obvious that it seeks to bring the thermocouple conductors closer together, as well as to their minimum diameter. In the above studies, chromel-drop thermocouple conductors with a diameter of 0.2 mm were used. Thermo-EMF was recorded by a multi-limit light galvanometer type 167311 from GOERZ.

The error of the contact artificial thermocouple method can be estimated by considering the well-known formula [15].

$$I = \frac{E(t, t_0)}{R_M + R_n + R_m}, \tag{1}$$

where R_M – is the resistance of the millivoltmeter; R_n – the resistance of connecting wires; R_m - thermocouple resistance.

In this case, it depends on several factors, namely: on the transition resistance between the contacting surfaces; the resistance of the contact body, and the resistance of the films that may be on the contact surface. As practice has shown, contact resistance can be expressed by the following formula [16]:

$$R_m = \frac{d}{P_k^b}, \tag{2}$$

where P_K –the contact pressure; d – constant, which is depended on the contact material and temperature; b – constant, depending on the form of contact.

From formula 2 it follows that the contact temperature is affected by the contact temperature and its shape, as well as the magnitude of the applied force. At high temperatures, oxide films with a thickness of 10–20 A always appear on the surface of metals with high ohmic resistance of the order of 140–170 Ω [17], while the contact resistance of these materials at normal temperature is in the range of 1–5 Ω. It follows from the foregoing that the internal resistance of an artificial contact thermocouple varies widely and reaches large values that significantly exceed the resistance of commonly used millivoltmeters and galvanometers, which is in the range of 20–60 Ω. The use of such equipment with low internal resistance, in this case, will not provide the necessary accuracy. High accuracy can be achieved by using, for example, a highly sensitive multi-limit galvanometer with a GOERZ type indicator 167311, which has an internal resistance on a scale of 100 mV to 200,000 Ω.

If we assume that the contact resistance varies from 1 to 200 Ω, then for two contacts of an artificial contact thermocouple, the relative error can be determined from the ratio

$$\begin{gathered}\Delta = 2\frac{R^{max} - R^{min}}{R^{min}}100\% = 2\frac{\left(\mathrm{R}_M + \mathrm{R}_K^{max}\right) - \left(\mathrm{R}_M + \mathrm{R}_K^{min}\right)}{\mathrm{R}_M + \mathrm{R}_K^{min}}100\% == \\ 2\frac{(200000 + 400) - (200000 + 2)}{200000 + 2}100\% = 0,004 \cdot 100\% == 0,4\%\end{gathered} \tag{3}$$

The test developed by the method for assessing the temperature regime was subjected to the design of a two-stage SHFM for wind power plants [9, 10], which consists of an output shaft 1, housing 2, in which bearings 4 and 6 are installed, leads 3 and 5, which engage with bearing cages 3 and 5, a cover 7, which compresses the outer ring of the bearing 6, the output shaft 8, on which the inner ring of the bearing 6 sits rigidly, and a thrust bearing 9, which abuts the spring 10 (Fig. 2, 3). The outer rings of bearings 4 and 6 are secured against rotation by screws. When turning the cover 7, the outer ring of the bearing 6 moves along the axis of the multiplier and compresses the spring 10 through the balls and the inner ring together with the output shaft 8.

Since the outer ring of the bearing is fixed, the spring 10 unclenches the inner rings of the bearings 4 and 6 with a certain force, creating the necessary adjustable tension in them, which makes it possible to transmit torque due to the friction of the balls on the ring.

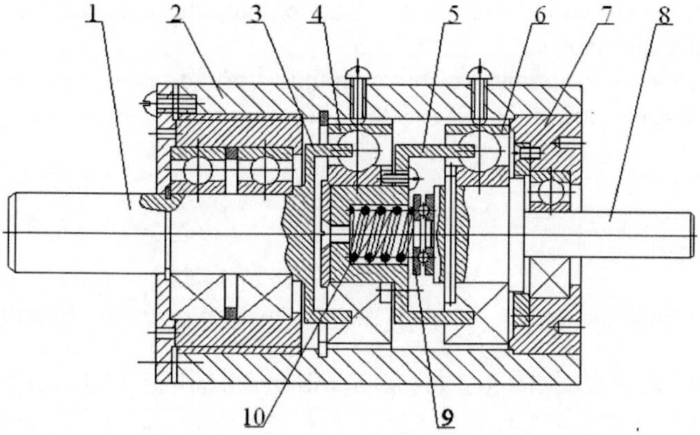

Fig. 2. Scheme of the experimental design of BPFM.

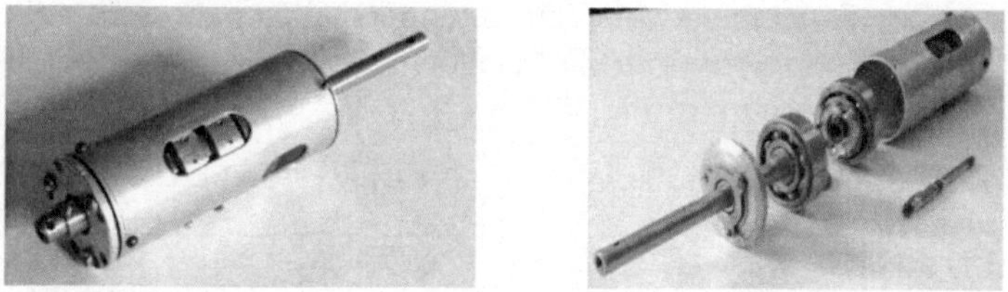

a b

Fig. 3. An experimental sample of BPFM in the assembled state (a) and disassembled (b).

When the input shaft 1 rotates, the leash 3 rotates the separator with the balls of the bearing 4. The rotation is transmitted to the inner ring of the bearing 4 with the leash 5, from it to the bearing cage 6 and its inner ring, mounted on the output shaft. Bearings 4 and 6 in this design act as planetary mechanisms, with a gear ratio $i_M = i_1 i_2$, where i_1, i_2 are determined by the formula:

$$i_{KC_n} = \left(\frac{D_{\mathrm{B}} + D_{\mathrm{H}}}{D_{\mathrm{B}}}\right)^{n-1} \tag{4}$$

where D_{B} and D_H are the diameters of the raceways, respectively, of the inner and outer rings of the bearing.

Next, the rotation is transmitted to the bearing cage 6 and transmitted through its inner ring to the output shaft. Since the inner rings of the bearings 4 and 6 rotate at different speeds, a thrust bearing 9 is installed between the output shaft and the spring 10, which creates the tension.

The developed BPFM design allows to exclude the transmission of sliding force through the separators, reduces the corresponding friction between the separators and balls in the axial direction, and, as a result, provides a significant increase in the efficiency of the mechanism.

For a preliminary assessment of the operability of the original proposed design of BPFM, a prototype based on bearings 306 with a gear ratio $i_M = 6{,}8$ was manufactured and tested.

4 Results

Research is carried out twice: without the use of grease in the bearings and using grease.

The first study was carried out using grease-free bearings in the BPFM. The input shaft BPFM was fixed in the cartridge of the machine 1K62. The machine spindle revolutions were: $n_1 = 160\ \text{min}^{-1}$. The output shaft of the BPFM was fixed using the rotating cone located in the tailstock and received a value of revolutions $n_2 = 1088\ \text{min}^{-1}$.

The study lasted 80 min with fixing temperature indicators every 10 min. According to the results, a graph of the temperature fluctuations of bearings without lubrication is obtained, shown in Fig. 4.

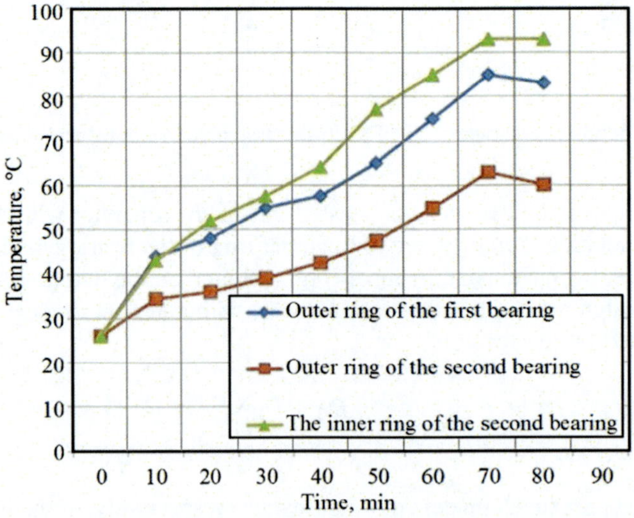

Fig. 4. Graph of fluctuations in temperature of bearings without lubrication.

According to the data obtained, a significant increase in the temperature of the BPFM bearing rings without lubrication becomes apparent even at relatively low rotational speeds.

The second study was carried out using grease-lubricated bearings in BPFM. The input shaft BPFM was fixed in the cartridge of the machine 1K62. The machine spindle received revolutions $n_1 = 160\ \text{min}^{-1}$. The BPFM output shaft was fixed in a 550 W generator used as a loading device and received revolutions $n_2 = 4284\ \text{min}^{-1}$.

The temperature of the multiplier elements was determined using an artificial contact thermocouple and was also recorded by a highly sensitive multi-target galvanometer with a light pointer type 167311 from GOERZ.

The study lasted 35 min with fixing temperature indicators every 10 min. Based on the results, we obtained a graph of the temperature fluctuations of bearings with lubrication, shown in Fig. 5.

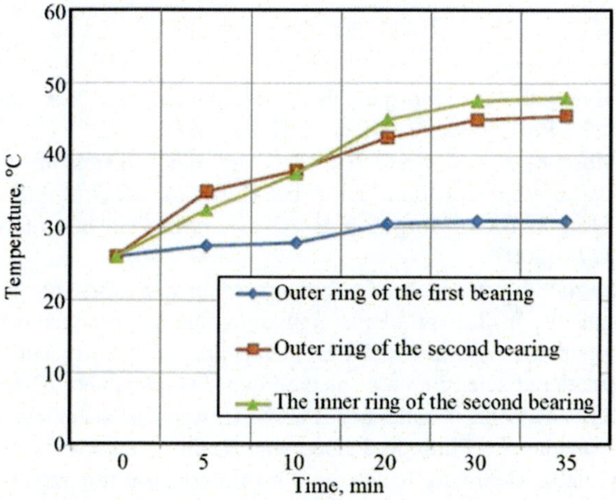

Fig. 5. Graph of fluctuations in temperature of bearings with lubricant.

According to the obtained graph, the temperature of the lubricated bearing rings does not increase by more than 20 °C even at input shaft $n_1 = 630\ \text{min}^{-1}$ speeds and loads for 35 min.

Thus, this experiment confirms the hypothesis put forward on the possibility of using lubricant in planetary friction mechanisms to significantly reduce friction losses and reduce the heating of the structure. Fig. 5 confirms the presence of slight heating of structural elements even with prolonged use.

5 Conclusions

The method developed by the authors for measuring the surface temperature of conductive bodies by two contact thermoelectrodes has a high measurement accuracy, eliminates the inertia of heating the thermocouple junction, and ensures speed.

The proposed method for measuring the surface temperature of conductive materials can be used in systems with such features as inaccessibility, movement of system elements, the small surface area for measurements, and the inability to use standard temperature sensors.

The use of lubricant in the planetary-friction multiplier with a simultaneous increase in the interference of the bearings causes crushing of the oil film in the spots of contact of the balls with the rings and leads to the appearance of semi-dry friction, which provides torque transmission. At the same time, the friction between the balls and the separator in contact on a large surface is carried out under lubrication conditions and does not cause significant friction losses.

Tests made it possible to establish the operability of the proposed design of the planetary-friction multiplier and its operation at permissible temperatures in the case of bearings with grease.

References

1. Oborsky, G., Bundyuk, A., Morgun, B., Prokopovich, I.: New and Non-traditional Technologies in the Energy Sector. Phoenix, Odessa (2016)
2. Oborsky, G., Morgun, B., Morgun, Y.: Planetary mechanisms of wind power plants. Information technology in education. Sci. Prod. Coll. Sci. BC **3**(8), 102–109 (2015)
3. Gutyrya, S.S., Morgun B.O., Morgun Y.B.: Planetary-friction box of tweezers. Patent of Ukraine, No. 87778 (2009)
4. Gutyrya, S., Morgun, B., Morgun, Y.: Increase of power efficiency of the ball friction-planetary mechanisms. In: Les problems contemporains du technosphere et de formations des cardes d'ingenieurs. CONFERENCE 2010, pp. 65–74. Recueil d'ovrages du IV Conference international scientifique et methodique a Hammamet (2010)
5. DSTU ISO 81400 - 4. Wind installations. Part 4. Design and specifications of the gearbox
6. Turmanidse, R., Dadone, L., Philippe, J., Demaret, B.: Investigation, Development and Tests Results of the Variable Geometry Rotor. In: 33rd European Rotorcraft Forum, Kazan, p. 11 (2007)
7. Vanyeyev, S.M., Miroshnichenko, D.V., Rodymchenko, T.S., Protsenko, M., Smolenko, D. V.: Data measuring system for torque measurement on running shafts based on a non-contact torsional dynamometer. J. Eng. Sci. **6**(2), E15–E23 (2019). https://doi.org/10.21272/jes.2019.6(2).e3
8. Kanwal, T., Altaf, S., Javed, M.K.: Environmental monitoring smart system with self-sustaining wireless sensor network using data validation algorithms. J. Eng. Sci. **7**(1), E10–E19 (2020). https://doi.org/10.21272/jes.2020.7(1).e3
9. Oborsky, G.O., Morgun, B.O., Guturya, S.S., Morgun, Y.B.: Planetary-frictional multiplier. Patent of Ukraine, No. 112370 (2016)
10. Gutyrya, S., Morgun, Y., Lyashevsky, A., Belozerov A.: Ball planetary-friction animator for wind turbines. In: Material International Scientific-Practical Conference New and Non-traditional Technologies in Resource and Energy Conservation. CONFERENCE 2014, ATM Ukraine, pp. 44–47 (2014)
11. Guturya, S.S, Morgun, B.O., Morgun Y.B.: Planetary-friction gearbox. Patent of Ukraine, No. 87778 (2009)
12. Oborsky, G., Slobodianyk, P., Kostenko, V., Antoshuk, S.: Measurement of Physical Quantities. Astroprint, Odessa (2012)

13. Lutsyk, Y., Guk, O., Lakh, O., Stadnyk, B.: Temperature Measurement: Theory and Practice. Beskid Bit Publishing House, Lviv (2006)
14. Serheev, S.S.: Probe for measuring the surface temperature of bodies. Patent RU, No. 2393442 (2010)
15. Oborsky, G.O., Morgun, B.O., Morgun, Y.B., Prokopovich, I.V.: Probe for measuring the surface temperature of bodies. Patent of Ukraine, No. 104319 (2016)
16. Sotskov, B.: Reference on Elements of Automation and Telemechanics. Gosenergoizdat, Moskov (1967). [in Russian]
17. Slutskaya, V.: Thin Films using Microwave Technology. Tekhnika, Kyiv (1967)

Check for updates

The Study of Dynamic Processes of Mechatronic Systems with Planetary Hydraulic Motors

Anatolii Panchenko[1(✉)], Angela Voloshina[1], Olena Titova[1], Igor Panchenko[1], and Andrii Zasiadko[2]

[1] Tavria State Agrotechnological University, 18, B. Khmelnytskyi Ave., Melitopol 72310, Ukraine
tia_tgatu@ukr.net

[2] Berdyansk College of the Tavria State Agrotechnological University, 23, Eastern Ave., Berdyansk 71108, Ukraine

Abstract. The study of the dynamic processes of mechatronic systems with planetary hydraulic motors is a solution to one of the urgent problems associated with improving the quality of functioning of self-propelled vehicles. A structural-functional diagram of a dynamic model of a mechatronic system with a planetary hydraulic motor is developed. The initial data and initial conditions for the simulation of transients occurring in the mechatronic system with a planetary hydraulic motor, as well as the design parameters of the rotor system of the planetary hydraulic motor, which affect the change in its output characteristics, are substantiated. The dynamics of changes in the output characteristics of the mechatronic system with serial and modernized hydraulic motors is studied, taking into account the design features of the rotor system. It has been established that fluctuations in the diametrical clearance in the rotor system of a serial hydraulic motor cause significant pulsations of pressure and torque. At the same time, stabilization of the diametrical clearance in the rotor system of the modernized hydraulic motor allows eliminating pressure and torque pulsations throughout the study of the acceleration process. It was established that the magnitude of the diametrical clearance in the rotor system does not affect the nature of the change in the flow rate of the working fluid and the shaft speed of the serial and upgraded hydraulic motors.

Keywords: Planetary hydraulic motor · Dynamic model · Rotor system · Diametrical clearance · Output Characteristics

1 Introduction

In hydraulic drives of mechatronic systems of self-propelled machinery, gerotor [1, 2], orbital [3–5] and relatively new planetary [6, 7] hydraulic motors are widely used. Planetary hydraulic motors are similar to orbital [8, 9], and gerotor [10] motors, the principle of a planetary gearbox is laid in them. The rotor system is one of the main components of a planetary hydraulic motor that determines its performance. The design of the rotor system is based on the principle of operation of a gear pair (movable and

V. Tonkonogyi et al. (Eds.): InterPartner 2020, LNME, pp. 704–713, 2021.
https://doi.org/10.1007/978-3-030-68014-5_68

fixed rotor) with internal hypocycloidal gearing [11]. Under the influence of the hydraulic field [6, 7], which acts as a crank of a planetary gearbox, the movable rotor rolls along with stationary one.

Planetary hydraulic motors ensure the operation of the mechatronic system of the self-propelled machine at low (50…75 min^{-1}) speeds of rotation of the working body with high (4000…4500 N·m) torque, at a working pressure of 20…25 MPa [12]. Their disadvantage is the uneven output characteristics, especially during transients, due to the error in the shape of the elements of the rotor system [8, 9]. The study of dynamic processes occurring in mechatronic systems with planetary hydraulic motors, in order to predict changes in their output characteristics, is the solution to one of the urgent scientific problems to improve the quality of self-propelled vehicles.

2 Literature Review

A justified assessment of the efficiency of hydrodynamic machines by dissipative power is presented [13]. A multi-criteria choice of the optimal configuration of the device [14] is proposed. A mathematical model of rotor vibrations on non-linear bearings for multistage centrifugal machines [15] is considered, the dynamics of a hydrostatic unit with a low rotation speed is studied generated by the epitrochoid of a rotary piston machine [16]. A parametric optimization of parts machining is considered in [17] Issues related to the methodology for calculating, designing, and operating gerotor, orbital and planetary hydraulic machines have not been investigated.

The forces and moments acting on the rotor of an orbital hydraulic motor are considered [5, 18], their effect on a gear pair is studied [19]. Physical [20], mathematical [21, 22], dynamic [12, 23], and structural-functional models [8, 24] are proposed that describe the relationship between the design features of the rotor system of the orbital hydraulic motor. A methodology for designing and manufacturing rotors has been developed [25]. A program was developed [9], which allows simulating the coupling of a system of rotors (external and internal). The proposed program allows you to get a three-dimensional image of the zones with valid interfaces that provide efficient and reliable operation of orbital engines. The pressure distribution in the special chambers of the hydraulic gerotor engine was determined [10]. The solution for the orbital hydraulic engine was examined from tribology and the environment [26]. A numerical model has been developed that can quantitatively measure the wear of an orbital engine under various operating conditions [27]. An algorithm for generating a gerotor gear is presented, which evaluates key performance targets that need to be minimized or maximized, and then an optimization algorithm is applied to determine the best option [28]. Methods for increasing the load capacity of cycloidal gears are considered [29]. Based on the geometry and working mechanism of the orbital gerotor, the algorithm of a moving deforming grid was introduced and implemented in the CFD software package [30]. Complex studies of a hydrostatic installation with a low rotational speed of the orbital rotor in a hydrostatic transmission system were carried out [31]. Transient processes occurring in orbital hydraulic machines are not considered considering the design features of their rotors.

To date, planetary hydraulic machines deserve much attention. To obtain the planetary motion of the rotor system, a distribution system is needed that creates a rotating hydraulic field [6, 7, 24]. The hydraulic field rotates parallel to the surface of the stationary rotor with rotation. The movable rotor rolls in motionless at the same speed as the hydraulic field, while rotating in the opposite direction. The hydraulic field, in this hydraulic motor, acts as a crank of a planetary gearbox. Design schemes [24], a mathematical apparatus [7], and a calculation algorithm [6] are developed, which make it possible to justify the angular arrangement of the working and unloading windows of the movable distributor. The influence of changes in the arrangement of windows on the output characteristics of the planetary hydraulic motor was determined [6]. The influence of the design features of the rotors on the output characteristics of the planetary hydraulic motor has not been studied.

An analysis of the studies performed showed that the dynamic processes of mechatronic systems with planetary hydraulic motors had not been studied in order to predict changes in their output characteristics. Hydraulic elements and working fluid are not considered as a whole when studying mechatronic systems with a hydraulic drive. The mutual influence of all hydraulic elements and the working fluid on the change in the dynamic characteristics of the mechatronic system of self-propelled equipment with rotary hydraulic machines has not been studied. Thus, the study of the dynamic processes of mechatronic systems with planetary hydraulic motors, in order to predict changes in their output characteristics, is a solution to one of the urgent problems associated with improving the quality of functioning of self-propelled vehicles.

3 Research Methodology

To solve the problem of studying the dynamic processes of mechatronic systems with planetary hydraulic motors, in order to predict changes in their output characteristics, it is necessary:

- to develop a structural-functional diagram of a dynamic model of a mechatronic system with a planetary hydraulic motor;
- substantiate the initial data and initial conditions for the simulation of transients occurring in a mechatronic system with a planetary hydraulic motor, as well as the design parameters of the rotor system of the planetary hydraulic motor, affecting the change in its output characteristics;
- to study the dynamics of changes in the output characteristics of the mechatronic system with serial and upgraded hydraulic motors, taking into account the design features of the rotor system.

Theoretical and parametric studies performed earlier [6–9, 12, 24] show that the developed mathematical models allow studies of the dynamics of changes in the output characteristics of the mechatronic system with a planetary hydraulic motor in order to predict changes in their output characteristics. The studies were carried out based on the developed universal model of the mechatronic system [12], taking into account the design features of the rotors of the serial and modernized hydraulic motors, as well as the features of their movement [8].

The developed universal model adequately describes the processes occurring in mechatronic systems with a hydraulic actuator, with a probability of more than 95% [8] and consists of a pump, a hydraulic motor, and a safety valve. It is supposed to use the parameters of a serial and upgraded planetary type hydraulic motors as modeling parameters of a hydraulic motor.

The dynamic processes of the mechatronic system with a planetary hydraulic motor were studied using the Vissim simulation package.

When studying the mechatronic system, the following initial data and initial conditions were adopted [8, 12]:

- *pump*: the pump flow is constant and equal to $Q_{p.g}(t)$ = 1770 cm^3/s; the angular velocity of the pump shaft is equal to ω_p = 125 s^{-1}; for an unregulated pump, the control parameter is equal to $e = 1$; pressure in the drain line is $p_s = 0$;
- *hydraulic motor*: the working volume of the hydraulic motor is V_{0hm} = 160 cm^3; the moment of resistance is constant and equal to M_r = 365 N·m; the moment of inertia of the rotating mass is J = 3.6 N·m; volumetric efficiency of the hydraulic motor – $\eta_{vol} = 0.95$; changes in the diametrical clearance are G = 0.055…0.21 mm for a serial one and G = 0.05…0.06 mm for a modernized hydraulic motor; the total error in the manufacturing form of the rotors of a serial hydraulic motor is E = 0.15 mm, for a modernized one – E = 0.01 mm. Hydromechanical efficiency of the hydraulic motor – $\eta_{hm} = 0.9$;
- *valve*: safety, spring stiffness equal to C = 200 kg/cm; spring pre-compression value x_0 = 0.125 cm; positive gap overlap is x_z = 0.53 cm;
- *working fluid*: the polytropic index is $K = 1.2$; the parameters of the working fluid, depending on the type of oil and the working temperature of the hydraulic system are $A = 12.62$ and $B = 1740$; the content of undissolved air in the working fluid in relative units is $m_0 = 0.925$.

When modeling, the initial data are set by block 1 of the structural-functional diagram of the dynamic model of the mechatronic system with a planetary hydraulic motor (Fig. 1). Changing the mechanical efficiency, taking into account the design features of the rotor system (diametrical clearance G and the manufacturing form error of the rotors E) of the serial and upgraded hydraulic motors, is described by block 2 and is determined from the expression [8]:

$$\eta_{hm} = \frac{(1-S)\cdot 2\cdot\left(e\pm\dfrac{G\pm E}{2}\right)\cdot b\cdot(z_1+1)\cdot h_i}{V_{0hm}\cdot\mu\cdot b\cdot S'\sqrt{\dfrac{2\Delta p}{\rho}}}$$

The change in torque depending on the change in mechanical efficiency is described by block 3.

The change in the flow rate of the working fluid supplied to the hydraulic motor and the geometric flow rate, considering the design features of the rotor system, is presented in block 4. The change in the shaft speed of the serial and upgraded hydraulic motors is described in block 5. Block 6 allows you to display graphical dependencies of the pressure in the mechatronic system, torque, shaft speeds of serial and modernized hydraulic motors, as well as costs through the safety valve and hydraulic motor at the current time.

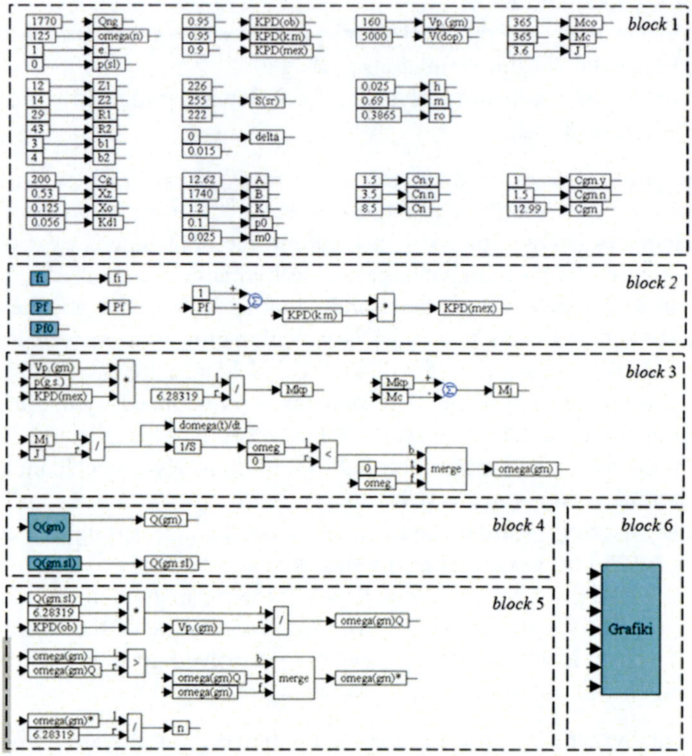

Fig. 1. Structural and functional diagram of a dynamic model of a mechatronic system with a planetary hydraulic motor.

4 Results

The corresponding dependencies represent the simulation results:

- the dynamics of pressure changes in the discharge line of the mechatronic system and the flow rate of the working fluid through the hydraulic motor and safety valve (Fig. 2);
- torque and frequency of rotation of the motor shaft (Fig. 3).

An analysis of the dependence of the pressure change in the mechatronic system shows that during acceleration at time t = 0…0.02 s there is a rather large pressure surge up to 88 MPa, 5.5 times higher than the nominal value, both for serial and modernized (Fig. 2, curves 1) of hydraulic motors. Further, over a period of time

0.04…0.8 s, the pressure stabilizes, and its value gradually decreases to 28.5… 27.5 MPa, exceeding the nominal value by 1.8 times (Fig. 2, curves 1). With further acceleration, during a period of 0.8…0.9 s, the pressure decreases quite sharply, reaching its nominal value. In a period of time t > 0.9 s, there is a steady motion of the motor shaft.

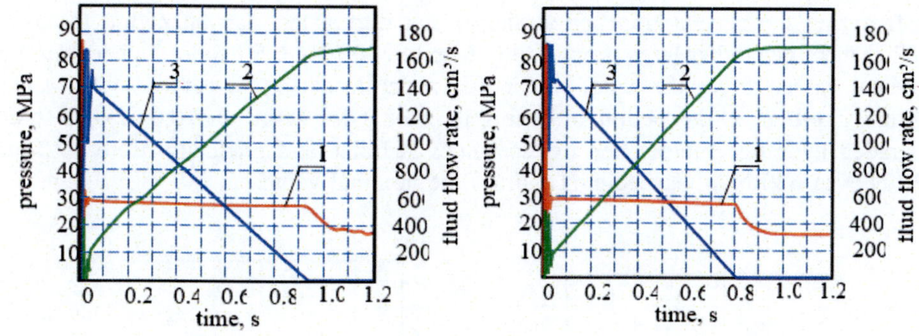

a b

Fig. 2. Dependences of pressure and flow rate of the fluid in the acceleration mode of the mechatronic system: a – serial hydraulic motor; b – modernized hydraulic motor; 1 – a curve of the pressure; 2 – a curve of the flow of the fluid through the hydraulic motor; 3 – a curve of the flow of fluid through the safety valve

The pressure pulsation (Fig. 2, a – curve 1), caused by the design features of the rotor system of a serial hydraulic motor (change in the diametrical clearance $G = 0.055\ldots0.21$ mm), although implicitly expressed, reaches 9…10% of the nominal value (vibration amplitude pressure is 1.5 MPa). There are no pressure pulsations in the mechatronic system with a modernized hydraulic motor at the change in the diametrical clearance $G = 0.05\ldots0.06$ mm (Fig. 2b, curve 1).

An analysis of the dependence of the change in the flow rate of the working fluid passing through the hydraulic motors shows (Fig. 2, curves 2) that at the time t = 0… 0.04 s the flow rate value of the flowing fluid has a rather significant pulsation up to 430 cm^3/s, caused by the opening of the safety valve. Over a period of time, 0.04… 0.9 s for serial and 0.04…0.82 s for upgraded hydraulic motors (Fig. 2, curves 2), the flow rate of the working fluid increases uniformly, and there are no ripples. In a period of time t > 0.9 s, steady motion of the hydraulic motor shaft is observed at a nominal flow rate of 1660 cm^3/s.

It should be noted that a change of the diametrical gap in the rotors system does not affect the nature of the change in the flow rate of the working fluid, both in serial and in modernized hydraulic motors.

Analysis of the dependence characterizing the change in the flow rate of the working fluid through the safety valve during acceleration (t = 0…0.04 s) shows that the flow rate of the working fluid through the safety valve reaches a maximum value of 1660 cm^3/s and has a rather significant pulsation up to 640 cm^3/s – for a serial hydraulic motor and up to 500 cm^3/s – for a modernized one (Fig. 2, curves 3). Over a period of time, 0.04…0.84 s for serial and 0.04…0.82 s for modernized hydraulic motors, the flow rate of the working fluid through the valve is uniformly reduced, and there are practically no ripples (Fig. 2, curves 3). Then the valve is completely closed, characterizing the steady motion of the shaft of the serial and modernized hydraulic motors.

An analysis of the dependences of the change in the torque of the hydraulic motor shaft during the acceleration period shows that during the start-up period (t = 0… 0.02 s) there is a rather large peak of torque up to 2000 N·m, 5.7 times higher than the nominal value, as for the serial so for the modernized (Fig. 3, curves 1) hydraulic motors. Further, in the period of time 0.04…0.8 s for serial and 0.04…0.72 s for upgraded hydraulic motors, the torque values stabilize and gradually decrease from 630 N·m to 620 N·m, exceeding 1.7 times the nominal value.

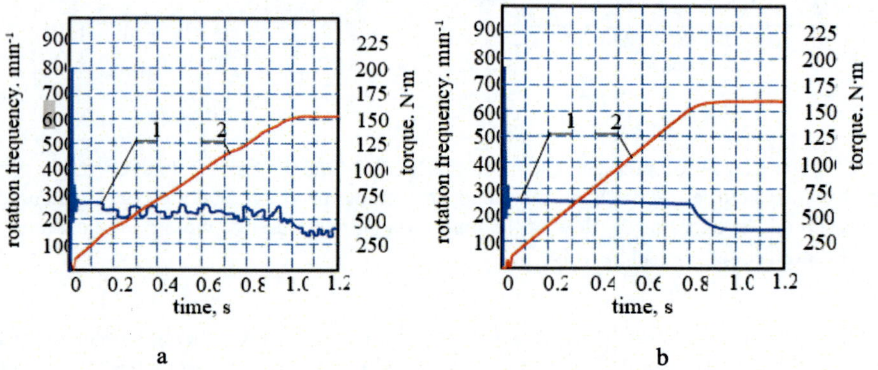

Fig. 3. Dependences of torque and angular velocity in the acceleration mode of the mechatronic system: a – serial hydraulic motor; b – modernized hydraulic motor; 1 – a curve of the torque; 2 – a curve of the frequency of rotation of the shaft of the hydraulic motor

The torque values on the shaft of a serial hydraulic motor have significant pulsations of up to 34%, the amplitude of which is up to 120…130 N·m (Fig. 3a, curve 1), caused by a change in the diametrical clearance in the rotors system. With further acceleration, in the period of time 0.8…0.9 s for serial and 0.72…0.8 s for modernized hydraulic motors, the torque value decreases quite sharply, reaching its nominal value (Fig. 3, curves 1). In a period of time t > 0.9 s, steady motion of the shaft of serial and modernized hydraulic motors is observed. The pulsations caused by the rotor system of a serial hydraulic motor are reduced to 18% with an oscillation amplitude of up to 63 N·m (Fig. 3, a – curve 1). It should be noted that throughout the studies of the acceleration process, there are no torque pulsations on the shaft of the modernized hydraulic motor (Fig. 3b, curve 1).

An analysis of the dependence of the change in the shaft speed of a serial and modernized hydraulic motor during acceleration shows (Fig. 3, curves 2) that the shaft speed values reach their nominal value of 600 min^{-1} in a period of time t = 0.9 s for serial and t = 0.8 s - for the upgraded hydraulic motors, this is due to the operation of the safety valve (Fig. 2, curves 3).

It should be noted that the design features of the rotor system (the value of the diametrical clearance) do not affect the change in the rotational speed of the shaft of the serial and modernized hydraulic motors.

The studies of dynamic processes occurring in mechatronic systems with planetary hydraulic motors make it possible to predict a change in the output characteristics, due to their design features, at the design and modernization stages.

5 Conclusions

As a result of the studies, a structural-functional diagram of a dynamic model of a mechatronic system with a planetary hydraulic motor was developed. The initial data and initial conditions for the simulation of transients occurring in the mechatronic system with a planetary hydraulic motor are justified, as well as the design parameters of the rotor system of the modernized planetary hydraulic motor, which affect the change in its output characteristics. The dynamics of changes in the output characteristics of the mechatronic system with serial and modernized hydraulic motors is studied, taking into account the design features of the rotor system.

It has been established that fluctuations in the diametrical clearance (0.055… 0.21 mm) in the rotor system of a serial hydraulic motor cause significant pulsations of pressure (9…10%) and torque (18…34%). At the same time, the stabilization of the diametrical gap (0.05…0.06 mm) in the rotor system of the modernized hydraulic motor eliminates pressure and torque pulsations throughout the acceleration research. It was established that the magnitude of the diametrical clearance in the rotor system does not affect the nature of the change in the flow rate of the working fluid and the shaft speed of the serial and upgraded hydraulic motors.

The conducted studies allow us to predict a change in the output characteristics of mechatronic systems with planetary hydraulic motors at the design and modernization stages.

References

1. Chang, Y., Kim, J., Jeon, C., Chul, K., Jung, S.: Development of an integrated system for the automated design of a gerotor oil pump. J. Mech. Des. **129**(10), 1099–1105 (2006). https://doi.org/10.1115/1.2757629
2. Gamez-Montero, P.J., Garcia-Vilchez, M., Raush, G., Freire, J., Codina, E.: Teeth clearance and relief grooves effects in a trochoidal-gear pump using new modules of GeroLAB. J. Mech. Des. **134**(5), 054502–054507 (2012). https://doi.org/10.1115/1.4006440
3. Yang, D., Yan, J., Tong, S.: Flowrate formulation of deviation function based gerotor pumps. J. Mech. Des. **132**(6), 064503–064505 (2010). https://doi.org/10.1115/1.4001595
4. Choi, T., Kim, M., Lee, G., et al.: Design of rotor for internal gear pump using cycloid and circular-arc curves. J. Mech. Des. **134**(1), 011005–011012 (2012). https://doi.org/10.1115/1.4004423

5. Xing, K., Zhang, Y., Jin, X.: Performance analysis and experimental research of the orbital motor. In: Proceedings of the Seventh International Conference on Fluid Power Transmission and Control, pp. 131–135 (2009)
6. Voloshina, A., Panchenko, A., Panchenko, I., Titova, O., Zasiadko, A.: Improving the output characteristics of planetary hydraulic machines. In: IOP Conference Series: Materials Science and Engineering, vol. 708, no. 1, p. 012038 (2019). https://doi.org/10.1088/1757-899X/708/1/012038
7. Voloshina, A., Panchenko, A., Boltyansky, O., Panchenko, I., Titova, O.: Justification of the kinematic diagrams for the distribution system of a planetary hydraulic motor. Int. J. Eng. Technol. **7**(4.3), 6–11 (2018). https://doi.org/10.14419/ijet.v7i4.3.19544
8. Panchenko, A., Voloshina, A., Milaeva, I., Panchenko, I., Titova, O.: The Influence of the form error after rotor manufacturing on the output characteristics of an orbital hydraulic motor. Int. J. Eng. Technol. **7**(4.3), 1–5 (2018). https://doi.org/10.14419/ijet.v7i4.3.19542
9. Panchenko, A., Voloshina, A., Panchenko, I., Titova, O., Pastushenko, A.: Reliability design of rotors for orbital hydraulic motors. In: IOP Conference Series: Materials Science and Engineering, vol. 708, no. 1, p. 012017 (2019). https://doi.org/10.1088/1757-899X/708/1/012017
10. Strmcnik, E., Majdic, F.: The pressure and efficiency characteristic of hydraulic gerotor motor with the floating outer ring. Tehnicki vjesnik-Tech. Gaz. **25**(2), 609–615 (2018)
11. Stryczek, J.: Characteristic parameters of hydraulic displacement machines with cycloidal gearing. Mech. Mach. Theory **1**(28), 97–112 (1993). https://doi.org/10.1016/0094-114x(93)90050-6
12. Panchenko, A., Voloshina, A., Kiurchev, S., et al.: Development of the universal model of mechatronic system with a hydraulic drive. East.-Eur. J. Enterp. Technol. **4**(7(94)), 51–60 (2018). https://doi.org/10.15587/1729-4061.2018.139577
13. Fesenko, A., Basova, Y., Ivanov, V., Ivanova, M., Yevsiukova, F., Gasanov, M.: Increasing of equipment efficiency by intensification of technological processes. Periodica Polytech. Mech. Eng. **63**(1), 67–73 (2019). https://doi.org/10.3311/PPme.13198
14. Karpus, V.E., Ivanov, V.A.: Choice of the optimal configuration of modular reusable fixtures. Russ. Eng. Res. **32**(3), 213–219 (2012). https://doi.org/10.3103/S1068798X12030124
15. Verbovyi, A., Neamtu, C., Sieryk, M., Vashyst, B., Pavlenko, V., Simonovskiy, V., Pavlenko, I.: Ensuring the vibration reliability of rotors connected by spline joints. J. Eng. Sci. **6**(2), D14–D19 (2019). https://doi.org/10.21272/jes.2019.6(2).d3
16. Dasgupta, K., Mukherjee, A., Maiti, R.: Modeling and dynamics of epitrochoid generated orbital rotary piston LSHT hydraulic motor. J. Manuf. Sci. Eng. **118**(3), 415–421 (1996)
17. Ivanov, V., Dehtiarov, I., Pavlenko, I., Liaposhchenko, O., Zaloga, V.: Parametric optimization of fixtures for multiaxis machining of parts. In: Hamrol, A., Kujawińska, A., Barraza, M. (eds.) Advances in Manufacturing II, MANUFACTURING 2019, Lecture Notes in Mechanical Engineering, pp. 335–347 (2019). https://doi.org/10.1007/978-3-030-18789-7_28
18. Ivanovic, L., Miric, N., Devedzic, G., Ćuković, S.: Analysis of forces and moments in gerotor pumps. J. Mech. Eng. Sci. **10**(224), 2257–2269 (2010). https://doi.org/10.1243/09544062JMES2041
19. Ivanović, L., Blagojević, M., Devedžić, G., Assoul, Y.: Analytical and numerical analysis of load gerotor pumps. Sci. Tech. Rev. **60**(1), 30–38 (2010)
20. Altare, G., Rundo, M.: Computational fluid dynamics analysis of gerotor lubricating pumps at high-speed: geometric features influencing the filling capability. J. Fluids Eng. **38**(11), FE-15-1757 (2016). https://doi.org/10.1115/1.4033675

21. Van de Ven, J.D.: On fluid compressibility in switch-mode hydraulic circuits – part I: modeling and analysis. J. Dyn. Syst. Meas. Control **135**(2), 021013–021013–13 (2012). https://doi.org/10.1115/1.4023062
22. Van de Ven, J.D.: On fluid compressibility in switch-mode hydraulic circuits – part II: modeling and analysis. J. Dyn. Syst. Meas. Control **135**(2), 021014-021014-7 (2012). https://doi.org/10.1115/1.4023063
23. Chiu-Fan, H.: Flow characteristics of gerotor pumps with novel variable clearance designs. J. Fluids Eng. **137**(4), FE-14-1137 (2015). https://doi.org/10.1115/1.4029274
24. Voloshina, A., Panchenko, A., Boltyansky, O., Titova, O.: Improvement of manufacture workability for distribution systems of planetary hydraulic machines. In: Ivanov, V., et al. (eds.) Advances in Design, Simulation and Manufacturing II. DSMIE 2019. Lecture Notes in Mechanical Engineering, pp. 732–741. Springer, Cham (2020). https://doi.org/10.1007/978-3-030-22365-6_73
25. Furustig, J., Almqvist, A., Bates, C.A., Ennemark, P., Larsson, R.: A two scale mixed lubrication wearing-in model, applied to hydraulic motors. Tribol. Int. **90**, 248–256 (2015). https://doi.org/10.1016/j.triboint.2015.04.033
26. Strmcnik, E., Majdic, F., Kalin, M.: Water-lubricated behaviour of AISI 440C stainless steel and a DLC coating for an orbital hydraulic motor application. Tribol. Int. **131**, 128–136 (2019)
27. Furustig, J., Almqvist, A., Pelcastre, L.: A strategy for wear analysis using numerical and experimental tools, applied to orbital type hydraulic motors. Proc. Inst. Mech. Eng. Part C-J. Mech. Eng. Sci. **230**(12), 2086–2097 (2016)
28. Robison, A., Vacca, A.: Kinematic multi-objective optimization of circular-toothed gerotor pumps by genetic algorithm. In: Proceedings of the ASME/Bath Symposium on Fluid Power and Motion Control, UNSP V001701A016 (2017)
29. Biernacki, K.: Methods of increasing loadability for the plastic cycloidal gears. Fundam. Mach. Des. **490**, 156–165 (2012)
30. Ding, H., Lu, X.J., Jiang, B.: A CFD model for orbital gerotor motor. In: IOP Conference Series: Earth and Environmental Science, vol. 15, no. 6, p. 062006 (2012)
31. Dasgupta, K., Mukherjee, A., Maiti, R.: Estimation of critical system parameters that affect orbit motor performance-combining simulation and experiments. J. Manuf. Sci. Eng.-Trans. ASME **121**(2), 300–306 (1999)

Improvement of the Refrigeration Capacity Utilizing for the Ambient Air Conditioning System

Andrii Radchenko(✉), Eugeniy Trushliakov, Veniamin Tkachenko, Bohdan Portnoi, and Alexandr Prjadko

Admiral Makarov National University of Shipbuilding, 9, Heroes of Ukraine Ave., Mykolayiv 54025, Ukraine
nirad50@gmail.com

Abstract. One of the most reasonable reserves to improve the efficiency of ambient air conditioning systems is to enable the operation of refrigeration compressors in close to nominal modes by selecting a rational design refrigeration capacity and its distribution in response to the current thermal load according to the actual variable climatic conditions to provide closed to maximum annual cooling production and to match current conditioning duties at the same time. The approach to improve the efficiency of utilizing the refrigeration capacity of the air conditioning system is based on shearing a rational design refrigeration capacity in two ranges concerning current cooling consumption. The first booster range of ambient air precooling to a certain intermediate threshold temperature is characterized by considerable fluctuations in cooling load, whereas the second range of subsequent subcooling air to a target leaving temperature is characterized by the comparatively stable cooling load. The first booster range of cooling load requires regulation of the cooling capacity, for instance, by the application of a variable speed compressor, whereas the second range of subsequent subcooling air can be covered by the operation of a conventional compressor in a mode closed to a nominal value.

Keywords: Annual refrigeration energy production · Design refrigeration capacity · Current cooling load

1 Introduction

Energy consumption for treating the ambient air in air conditioning systems (ACS) depends on the ambient air temperature t_a and relative humidity φ_a, which changes considerably during the day. It is obvious that the production of refrigeration by refrigeration machine (RM) according to ambient air conditioning duties over a certain period of time, for instance a year (annual refrigeration) $\sum(Q_0 \cdot \tau)$, depends on the current thermal loading Q_0, caused by ambient air processing, as well as on the hour duration τ of the ACS operation. Issuing from this the efficiency of using the refrigeration capacity for ambient air processing in ACS can be estimated by annual refrigeration $\sum(Q_0 \cdot \tau)$ as a primary criterion in further complex thermal and economical optimization of the whole ACS, including indoor air processing about actual cooling

V. Tonkonogyi et al. (Eds.): InterPartner 2020, LNME, pp. 714–723, 2021.
https://doi.org/10.1007/978-3-030-68014-5_69

loads of building and indoor thermal-humidity environments in targeted climatic conditions.

The existing approaches to improve the efficiency of ambient air processing in ACS are aimed at determining a design refrigeration capacity to cover the maximum yearly cooling load, which leads to oversizing the RM and other equipment and enlarged their cost as a consequence. The application of various controlling systems and variable speed compressors provides lowering energy consumption, but the problem of determining a rational design refrigeration capacity, providing closed to maximum annual effect without overestimating the installed refrigeration capacity, needs further solving.

2 Literature Review

Many publications are devoted to improving air processing in ACS by the intensification of heat transfer processes in air coolers [1, 2], evaporators [3, 4], alternate safe refrigerants [5, 6], application of various refrigerant circulation contours [7, 8], waste heat recovery technics, including combined cooling, heating and power generation [9, 10], modeling [11, 12], analysis [13, 14], optimization [15, 16], experimental and monitoring [17, 18] methods to match current cooling demands. Some of the principal technical innovations and methodological approaches in waste heat recovery refrigeration might be successfully applied for air conditioning, in particular, two-stage air cooling [19], evaporative cooling [20].

Numerous researchers have studied the energy efficiency of the Variable Refrigerant Flow (VRF) systems [21, 22] and proposed some practical recommendations. Mainly the studies have been conducted on solutions of efficient operation of the VRF system in buildings and control strategies of the systems [23]. A control algorithm of the supply air temperature as a set temperature in the outdoor air processing (OAP) unit to run the VRF–OAP system more efficiently for buildings was developed in [24]. The control algorithm was conducted with adjusting the refrigerant flow supply to the OAP and the indoor unit appropriately through supporting the supply air temperature according to the outdoor ambient and indoor temperature and humidity conditions. Results [25] show that ACS have a great potential for energy saving, and the adjustability of the VRF system is better than of centralized ACS.

The VRF system with heat recuperative ventilation [26] and outdoor ACS was introduced [27]. The evaluation of indoor thermal-humidity environments and energy consumption of the VRF system with a heat pump desiccant was conducted [28].

The author [29] proposes the method of calculating the thermal load of the building. The VRF systems operate with high part-load efficiency, that results into high daily and seasonal energy efficiency, so as ACS typically spend most of their operating hours within the range of 40% to 80% of maximum cooling capacity [30].

Despite the existence of a lot of control algorithms and devices to minimize the power consumption and provide indoor thermal-humidity environments in all the central ACS, the problem of choosing a rational design cooling capacity to provide annually efficient operation at changeable current loadings and to exclude oversizing RM and cooling equipment and its cost as a consequence needs a solution.

The study aims to develop an approach to improve the efficiency of utilizing the refrigeration capacity of ACS and method to determine a rational design cooling capacity that provides closed to maximum annual effect without oversizing and its distribution concerning current climatic conditions.

3 Research Methodology

The efficiency of ACS and their RM performance depends on their current loading Q_0 and an hour duration τ of yearly operation. The efficiency of using the refrigeration capacity for covering the ambient air processing needs is estimated by annual refrigeration energy production $\sum(Q_0{\cdot}\tau)$ as a primary criterion to determine a rational design refrigeration capacity.

The proposed method to determine the rational design refrigeration capacity is based on the yearly loading characteristic cumulative curve of annual refrigeration production dependence on the design refrigeration capacity of the RM. The current refrigeration, generated by RM at any time period for ambient air cooling down to the target temperature t_{a2} = 10 °C, has been summarized over the year. The rational design refrigeration capacity is selected according to the yearly loading characteristic cumulative curve to provide closed to maximum annual refrigeration production.

In order to generalize the results and simplify calculations for any total refrigeration capacities Q_0, it is convenient to present the refrigeration capacity of the RM ACS, not in absolute Q_0. However, in relative (specific) values per unit airflow rate (G_a = 1 kg/s) – in the form of specific refrigeration capacity, $q_0 = Q_0/G_a$, kW/(kg/s), or kJ/kg, where Q_0 is the total refrigeration capacity when cooling the air with the flow rate G_a: $Q_0 = (c_a\,\xi{\cdot}\Delta t_a)G_a$, where $\Delta t_a = t_a - t_{a2}$ – decrease in air temperature.

The specific annual production of refrigeration:

$$\sum(q_0\cdot\tau)=\sum(\xi c_{ma}(t_a-t_{a2})\tau) \tag{1}$$

where: q_0 – specific refrigeration capacity [kJ/kg] or [kW/(kg/s)]; $\sum(q_0{\cdot}\tau)$ – specific annual production of refrigeration [kJ/(kg/h)] or [kW•h/(kg/s)]; ξ – specific heat ratio of total heat, including sensible and latent heat, to sensible heat of humid air; t_a – ambient air temperature [°C]; t_{a2} – air temperature at the air cooler outlet [°C]; c_{ma} – specific heat of moist air [kJ/(kg·K)]; τ – time interval [h].

The specific refrigeration capacity is calculated as:

$$q_0=\xi c_{ma}(t_a-t_{a2}) \tag{2}$$

A rational specific refrigeration capacity $q_{0.rat}$ is determined to exclude unproductive expenses of refrigeration capacity q_0 caused by oversizing RM without obtaining a noticeable effect in increasing the annual refrigeration production $\sum(q_0{\cdot}\tau)$.

The further improved approach to enhance the efficiency of utilizing the ACS refrigeration capacity is based on shearing the overall refrigeration capacity, spent for ambient air processing, in two ranges with regard to current cooling loads. The first, booster, range of ambient air precooling to a certain intermediate threshold temperature, is characterized by considerable fluctuations in cooling load, whereas the second range of subsequent subcooling air to a target leaving temperature is characterized by the comparatively stable cooling load. The first booster range of cooling load requires regulation of the cooling capacity by the application of a variable speed compressor, whereas the second range of subsequent subcooling air can be covered by the operation of a conventional compressor in a mode closed to a nominal value.

Nomenclature		Δt_a	Ambient air temperature reduction
$\sum(q_0 \cdot \tau)$	Specific annual production of refrigeration energy, $\sum(q_0 \cdot \tau) = \sum(Q_0 \cdot \tau)/G_a$	φ_a	Ambient air relative humidity
AC	Air cooler	c_{ma}	Specific heat of moist air
ACS	Air conditioning system	*Subscripts*	
c	Specific heat of the air	10, 15	Parameters of air-cooled to temperatures 10 °C and 15 °C
G_a	Air mass flow rate	10–15	The difference in the parameters of air-cooled from 10 °C to 15 °C
OAP	Outdoor air processing	a	Ambient air at the inlet of cooler
q_0	Specific refrigeration capacity, $q_0 = Q_0/G_a$	$a2$	Cooled air at the outlet of the cooler
Q_0	Refrigeration capacity	b	Booster stage of air cooler
$q_{0.10}$, $q_{0.15}$	Specific refrigeration capacity for cooling ambient air to 10 °C and 15 °C	def	Deficit
$q_{0.10\text{–}15}$	Specific refrigeration capacity for cooling air from 10 °C to 15 °C	ma	Moist air
$q_{0.b}$	Booster stage specific refrigeration capacity	rat	Rational
$q_{0.bdef}$	Booster stage specific refrigeration capacity deficit	*Greek letters*	
$q_{0.rat}$	Rational specific refrigeration capacity	Δ	Difference
RM	Refrigeration machine	ξ	Specific heat ratio of total heat (sensible and latent) to sensible heat of moist air
t	Temperature	Σ	Sum
t_a	Ambient air temperature	τ	Time interval
t_{a2}	Air temperature at the air cooler outlet	φ	Relative humidity
VRF	Variable refrigerant flow		

4 Results

Specific annual refrigeration energy production $\sum(q_0 \cdot \tau)$ required for cooling the ambient air to t_{a2} = 10 and 15 °C against a design specific refrigeration capacity q_0, calculated for 2017 and 2019 in Mykolaiv region, south of Ukraine, is presented in Fig. 1.

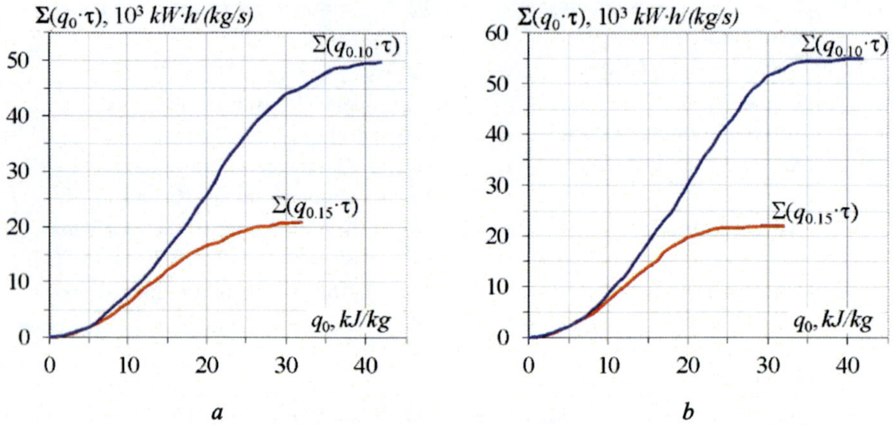

Fig. 1. Specific annual refrigeration energy production $\sum(q_0 \cdot \tau)$ required for cooling the ambient air to t_{a2} = 10 and 15 °C against a design specific refrigeration capacity q_0 for 2017 (*a*) and 2019 (*b*).

For the considered climatic conditions when the ambient air is cooled to the temperature t_{a2} = 10 °C a design specific (at G_a = 1 kg/s) refrigeration capacity of RM $q_{0.10rat}$ = 35 kJ/kg provides close to the maximum annual refrigeration production $\sum(q_0 \cdot \tau) \approx 48 \cdot 10^3$ kW·h/(kg/s) while maintaining its increment with a noticeable high rate. A design specific refrigeration capacity $q_{0.10rat}$ = 35 kJ/kg for cooling the ambient air to t_{a2} = 10 °C is assumed as rational one to provide closed to a maximum annual refrigeration production $\sum(q_0 \cdot \tau)$. However, with decreased specific refrigeration capacity $q_{0.10rat}$ = 35 kJ/kg versus its maximum value $q_{0.10}$ = 42 kJ/kg, id est. by more than 15%. Similarly, for cooling ambient air to t_{a2} = 15 °C the rational value of specific refrigeration capacity is $q_{0.15rat}$ = 25 kJ/kg is lower than its maximum value $q_{0.15}$ = 32 kJ/kg by about 20% with the corresponding reduction in sizes of RM and other equipment.

In order to prove the approach to the analysis of the efficiency of using design refrigeration capacities of ACS chillers, taking into account the change in cooling loads by actual climatic conditions, the current values of specific refrigeration capacity q_0 of RM ACS when cooling the ambient air from the current temperature t_a to t_{a2} = 10 and 15 °C, respectively $q_{0.10}$ and $q_{0.15}$ for July of 2017 and 2019 year, Mykolaiv region, south of Ukraine, have been considered (Fig. 2).

As calculation results in Fig. 1 shows, when the ambient air is cooled from its current temperatures t_a to t_{a2}, the cooling load fluctuations of $q_{0.10}$ and $q_{0.15}$ are very significant. The almost equidistant trend lines of the specific cooling load $q_{0.10}$ and $q_{0.15}$ indicate that these fluctuations are caused primarily by changes in the specific cooling load $q_{0.15}$ for precooling the ambient air to the temperature t_{a2} = 15 °C, within which there is practically damping of the fluctuations of the current cooling load. The intermediate temperature t_{a2} = 15 °C is assumed as threshold temperature for the rational distribution of design overall cooling capacity of ACS between two ranges with different characters of cooling load behavior.

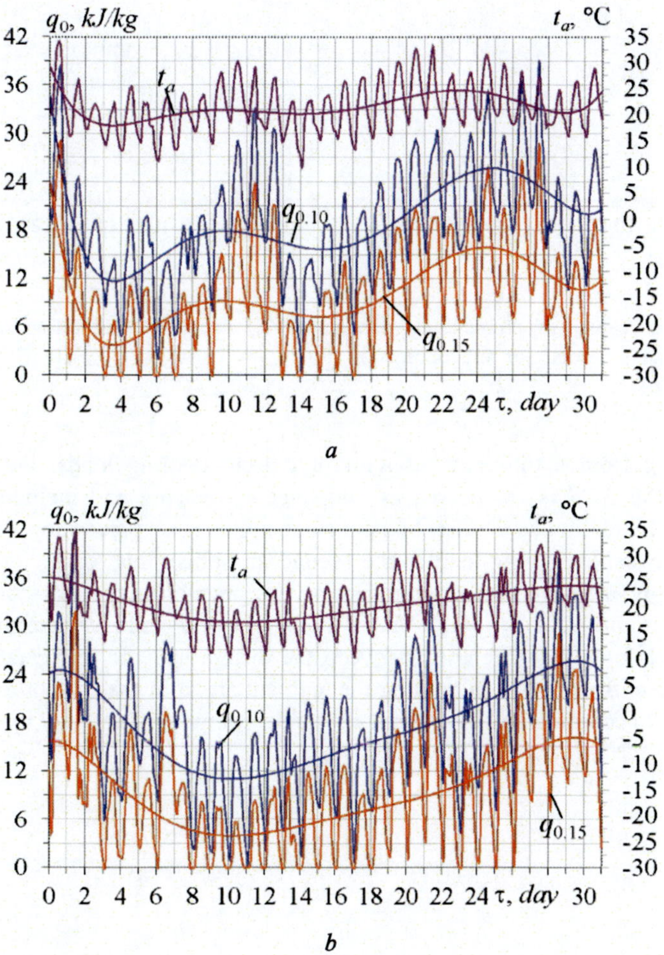

Fig. 2. Current values of ambient air temperature t_a, specific refrigeration capacity $q_{0.10}$, needed for cooling ambient air from t_a to t_{a2} = 10 °C, specific refrigeration capacity $q_{0.15}$, needed for cooling ambient air from t_a to intermediate temperature t_{a2} = 15 °C for July 2017 (*a*) and 2019 (*b*).

At the same time, within further cooling of the air from the intermediate temperature t_{a2} = 15 °C to t_{a2} = 10 °C the fluctuations of the specific cooling loads on the ACS $q_{0.10–15} = q_{0.10} - q_{0.15}$ are comparatively small (within a range of 6–10 kJ/kg in Fig. 3a versus 24–2 kJ/kg for $q_{0.15}$ in Fig. 2a and within a range of 4–10 kJ/kg in Fig. 3b versus 24–0 kJ/kg for $q_{0.15}$ in Fig. 2b for a bit lower monthly ambient air temperature) without taking into account 3–5 short-term bursts-drops, caused by a decrease in the current values of the ambient air temperature below 15 °C (Fig. 3).

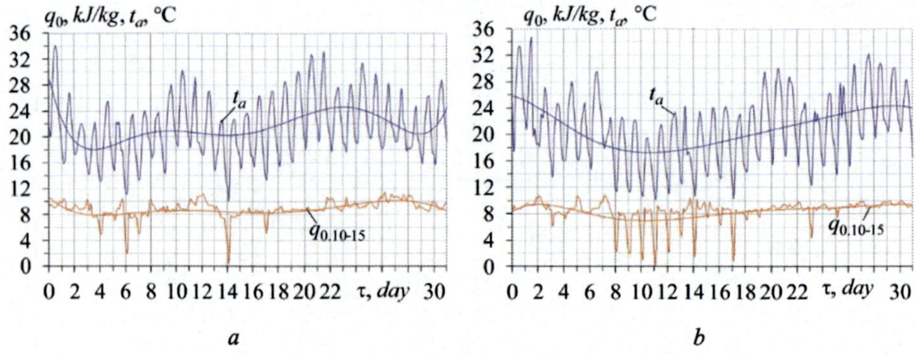

Fig. 3. Current values of ambient air temperature t_a, specific refrigeration capacity $q_{0.10–15} = q_{0.10} - q_{0.15}$, needed for subcooling air from t_{a2} = 15 °C to t_{a2} = 10 °C for July 2017 (*a*) and 2019 (*b*).

Proceeding from a different behavior of current cooling loads, the ambient air treatment in the ACS is considered as two-stage processing and includes a range of cooling load fluctuation as the first booster stage ambient air precooling and a range of comparatively stable cooling load as the second air subcooling stage.

Accordingly, a design refrigeration capacity for precooling the ambient air from the current temperature t_a to t_{a2} = 15 °C, as booster component, is determined by the residual principle as the difference between the design specific refrigeration capacity $q_{0.10rat}$ for the entire process of cooling the ambient air from the current temperature t_a to t_{a2} = 10 °C according to Fig. 1, and its stable component $q_{0.10–15}$ for further subcooling air: $q_{0b} = q_{0.10rat} - q_{0.10–15}$ (Fig. 4).

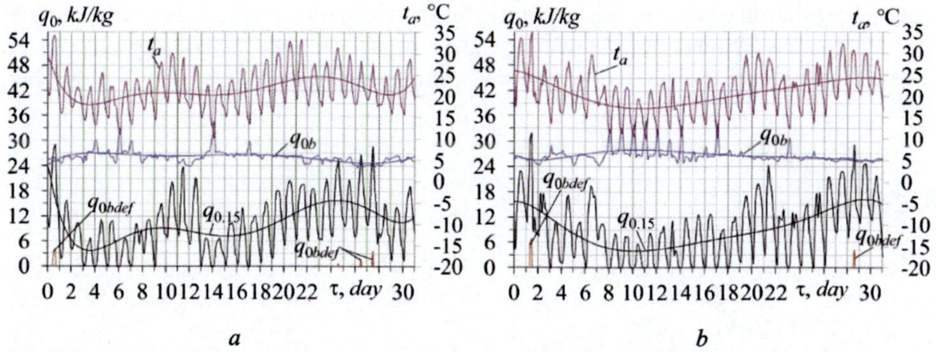

Fig. 4. Current values of ambient air temperature t_a, specific refrigeration capacity $q_{0.15}$, needed for cooling ambient air from t_a to intermediate temperature t_{a2} = 15 °C, booster specific refrigeration capacity $q_{0b} = q_{0.10rat} - q_{0.10\text{–}15}$ and booster specific refrigeration capacity deficit $q_{0bdef} = q_{0.15} - q_{0b}$ within subcooling air from t_{a2} = 15 °C to t_{a2} = 10 °C for July 2017 (*a*) and 2019 (*b*).

Since the fluctuations of the current refrigeration capacity, spent for cooling the ambient air from t_a to t_{a2} = 10 °C, are caused mainly by its booster part $q_{0.15}$, which corresponds to precooling the ambient air from t_a to t_{a2} = 15 °C, at elevated current cooling loads $q_{0.15}$ there is some deficit of the booster component q_{0bdef} of refrigeration capacity calculated by the residual principle, $q_{0bdef} = q_{0b} - q_{0.15}$, compared to the current specific thermal loads $q_{0.15}$. However, the booster refrigeration capacity deficit q_{0bdef} occurs very seldom 1, 27, 28 July of 2017 (Fig. 4a) and 2 and 29 July of 2019 (Fig. 4b). It proves the appropriate results of the application of the proposed method for analyzing the efficiency and rational distribution of a design refrigeration capacity used for ambient air processing in ACS according to current cooling loading.

5 Conclusions

The approach to analyzing the efficiency of utilizing a refrigeration capacity used for ambient air processing in ACS and method of determining its rational values and their distribution during the operation in actual changeable current climatic conditions is presented. The annual refrigeration production is assumed as a primary criterion for estimation of utilizing a refrigeration capacity, and its dependence on a design refrigeration capacity is used as a basic yearly loading characteristic curve for determining a rational design refrigeration capacity.

The method to determine a rational design cooling capacity to provide a closed to maximum annual refrigeration production and, at the same time, to avoid oversizing a refrigeration machine and its enlarged cost has been developed.

The ambient air treatment in ACS is proposed to consider as two-stage processing, which includes a range of cooling load fluctuation as the first booster stage of air cooler for ambient air precooling and a range of comparatively stable cooling load as the second stage for further air deep cooling. The first booster range of cooling load requires regulation of the cooling capacity, for instance, by the application of a variable speed compressor, whereas the second range of subsequent subcooling air can be covered by the operation of a conventional compressor in a mode closed to a nominal value.

References

1. Li, H., Liu, E., Zhou, G., Yang, F., Su, Z., Xu, Y., Jian, C., Shi, S.: Influence of fin thickness on heat transfer and flow performance of a parallel flow evaporator. Therm. Sci. **23**(4), 2413–2419 (2019)
2. Saab, R., Ali, M.I.H.: Variable-refrigerant-flow cooling-systems performance at different operation-pressures and types-of-refrigerants. Energy Procedia **119**, 426–432 (2017)
3. Dąbrowski, P., Klugmann, M., Mikielewicz, D.: Channel blockage and flow maldistribution during unsteady flow in a model microchannel plate heat exchanger. J. Appl. Fluid Mech. **12**, 1023–1035 (2019)
4. Dąbrowski, P., Klugmann, M., Mikielewicz, D.: Selected studies of flow maldistribution in a minichannel plate heat exchanger. Arch. Thermodyn. **38**, 135–148 (2017)
5. Saeed, M.U., Qureshi, S.R., Hashmi, K.J., Khan, M.A., Danish, S.N.: Performance assessment of alternate refrigerants for retrofitting R22 based air conditioning system. Therm. Sci. **22**(2), 931–941 (2018)
6. Konovalov, D., Trushliakov, E., Radchenko, M., Kobalava, H., Maksymov, V.: Research of the aerothermopressor cooling system of charge air of a marine internal combustion engine under variable climatic conditions of operation. In: Tonkonogyi V., et al. (eds.) Advanced Manufacturing Processes. InterPartner-2019, LNME, pp. 520–529. Springer, Cham (2020)
7. Trushliakov, E., Radchenko, M., Bohdal, T., Radchenko, R., Kantor, S.: An innovative air conditioning system for changeable heat loads. In: Tonkonogyi, V., et al. (eds.) Advanced Manufacturing Processes. InterPartner-2019, LNME, pp. 616–625. Springer, Cham (2020)
8. Radchenko, M., Radchenko, R., Tkachenko, V., Kantor, S., Smolyanoy, E.: Increasing the operation efficiency of railway air conditioning system on the base of its simulation along the route line. In: Nechyporuk, M., Pavlikov, V., Kritskiy, D. (eds.) Integrated Computer Technologies in Mechanical Engineering. Advances in Intelligent Systems and Computing, vol. 1113, pp. 461–467. Springer, Cham (2020)
9. Forduy, S., Radchenko, A., Kuczynski, W., Zubarev, A., Konovalov, D.: Enhancing the gas engines fuel efficiency in integrated energy system by chilling cyclic air. In: Tonkonogyi V., et al. (eds.) Advanced Manufacturing Processes. InterPartner-2019, LNME, pp. 500–509. Springer, Cham (2020)
10. Radchenko, A., Mikielewicz, D., Forduy, S., Radchenko, M., Zubarev, A.: Monitoring the fuel efficiency of gas engine in integrated energy system. In: Nechyporuk, M., Pavlikov, V., Kritskiy, D. (eds.) Integrated Computer Technologies in Mechanical Engineering. Advances in Intelligent Systems and Computing, vol. 1113, pp. 361–370. Springer, Cham (2020)
11. Volintiru, O.N., Scurtu, I.C., Stefănescu, T.M.: Modeling and optimization of HVAC system for special ships. J. Phys: Conf. Ser. **1122**(1), 012004 (2018)
12. Mikielewicz, D., Mikielewicz, J.: A thermodynamic criterion for selection of working fluid for subcritical and supercritical domestic micro CHP. Appl. Therm. Eng. **30**, 2357–2362 (2010)
13. Bohdal, L., Kukielka, L., Świłło, S., Radchenko, A.M., Kułakowska, A.: Modelling and experimental analysis of shear-slitting process of light metal alloys using FEM, SPH and vi-sion-based methods. In: AIP Conference Proceedings, vol. 2078, p. 020060 (2019)

14. Bohdal, L., Kukielka, L., Radchenko, A.M., Patyk, R., Kułakowski, M., Chodór, J.: Modelling of guillotining process of grain oriented silicon steel using FEM. In: AIP Conference Proceedings, vol. 2078, p. 020080 (2019)
15. Kornienko, V., Radchenko, M., Radchenko, R., Konovalov, D., Andreev, A., Pyrysunko, M.: Improving the efficiency of heat recovery circuits of cogeneration plants with combustion of water-fuel emulsions. Therm. Sci. **00**, 154 (2020)
16. Ortiga, J., Bruno, J.C., Coronas, A.: Operational optimisation of a complex trigeneration system connected to a district heating and cooling network. Appl. Therm. Eng. **50**, 1536–1542 (2013)
17. Preda, A., Scurtu, I.C.: Thermal image building inspection for heat loss diagnosis. In: 5th International Scientific Conference SEA-CONF 2019, Journal of Physics: Conference Series, vol. 1297, no. 012004, 6 p. (2019)
18. Balasubramanian, A., Jayaraman, V., Sivan, S., Vairavan, M.: Experimental analysis of triple fluid vapour absorption refrigeration system driven by electrical energy and engine waste heat. Therm. Sci. **23**(5B), 2995–3001 (2019)
19. Radchenko, A., Bohdal, L., Zongming, Y., Portnoi, B., Tkachenko, V.: Rational designing of gas turbine inlet air cooling system. In: Tonkonogyi V., et al. (eds.) Advanced Manufacturing Processes. InterPartner-2019, LNME, pp. 591–599. Springer, Cham (2020)
20. Eidan, A.A., Alwan, K.J.: Enhancement of the performance characteristics for air conditioning system by using direct evaporative cooling in hot climates. Energy Procedia **142**, 3998–4003 (2017)
21. Im, P., Malhotra, M., Munk, J.D., Lee, J.: Cooling season full and part load performance evaluation of variable refrigerant flow (VRF) system using an occupancy simulated research building. In: 16th International Refrigeration and Air Conditioning Conference, Purdue, West Lafayette, USA (2016)
22. Khatri, R., Joshi, A.: Energy performance comparison of inverter based variable refrigerant flow unitary AC with constant volume unitary AC. Energy Procedia **109**, 18–26 (2017)
23. Zhang, L., Wang, Y., Meng, X.: Qualitative analysis of the cooling load in the typical room under continuous and intermittent runnings of air-conditioning. Procedia Eng. **205**, 405–409 (2017)
24. Lee, J.H., Yoon, H.J., Im, P., Song, Y.-H.: Verification of Energy Reduction Effect Through Control Optimization of Supply Air Temperature in VRF-OAP System. Energies **11**(1), 15 (2018)
25. Liu, C., Zhao, T., Zhang, J.: Operational electricity consumption analyze of VRF air conditioning system and centralized air conditioning system based on building energy monitoring and management system. Procedia Eng. **121**, 1856–1863 (2015)
26. Fengxia, H., Zhongbin, Z., Hu, H., Zemin, C.: Experimental study on the all-fresh-air handling unit with exhaust air energy recovery. Energy Procedia **152**, 431–437 (2018)
27. Park, D.Y., Yun, G., Kim, K.S.: Experimental evaluation and simulation of a variable refrigerant-flow (VRF) air-conditioning system with outdoor air processing unit. Energy Build. **146**, 122–140 (2017)
28. Aynur, T.N., Hwang, Y., Radermacher, R.: Integration of variable refrigerant flow and heat pump desiccant systems for the cooling season. Appl. Therm. Eng. **30**, 917–927 (2010)
29. Sait, H.H.: Estimated thermal load and selecting of suitable air-conditioning systems for a three story educational building. Procedia Comput. Sci. **19**, 636–645 (2013)
30. Enteria, N., Yamaguchi, H., Miyata, M., Sawachi, T., Kuwasawa, Y.: Performance evaluation of the variable refrigerant flow (VRF) air-conditioning system subjected to partial and unbalanced thermal loadings. J. Therm. Sci. Technol. **11**(1), 1–11 (2016)

Rational Thermal Loading the Engine Inlet Air Chilling Complex with Cooling Towers

Mykola Radchenko[1(✉)], Bohdan Portnoi[1], Serhiy Kantor[1], Serhiy Forduy[2], and Dmytro Konovalov[3]

[1] Admiral Makarov National University of Shipbuilding, 9, Heroes of Ukraine Avenue, Mykolayiv, Ukraine
nirad50@gmail.com
[2] PepsiCo, Inc., Kyiv, Ukraine
[3] Admiral Makarov National University of Shipbuilding (Kherson Branch), Kherson, Ukraine

Abstract. The processes of cooling air at the inlet of energy installations by exhaust heat conversion chillers with heat removal from them by cooling towers of the circulating cooling system are studied for the gas turbine. Air cooling is conducted by using combined type exhaust heat conversion chillers, which utilize the exhaust gas heat of gas turbine and which include absorption lithium-bromide and refrigerant ejector chillers as stages to convert waste heat into cold. The data on current thermal loads on air coolers and cooling towers under actual climatic conditions of their operation with a different distribution of thermal loads between the cooling towers were obtained on the base of the results of modeling the operation of the gas turbine cooling complex during the year. It was shown the possibility to increase the fuel-saving due to turbine inlet air cooling. A novel approach to analyzing the efficiency of using the installed thermal capacity of the cooling towers has been proposed which can be applied both to choose the rational distribution of thermal capacity depending on the change in actual thermal load and to determine the ranges of thermal loads for effective application (of various methods of controlling the operation) of cooling towers.

Keywords: Turbine · Inlet air cooler · Combined chiller · Thermal load distribution · Climatic condition

1 Introduction

The efficiency of combustion engines decreases because of the rising temperature of the ambient air at the inlet [1, 2]. There are innovative contact air cooling technologies proposed [3, 4]. Energy-saving technologies are developed to increase the fuel efficiency of engines by utilization of exhaust heat [5, 6]. The use of exhaust gases heat for GTU inlet air cooling provides an increase in GTU efficiency at increased ambient air temperatures t_{amb} [7, 8] as well as in gas engines [9]. The application of exhaust heat conversion chillers (EHCCh), using the GTU exhaust gas heat for inlet air cooling is the most widespread method to increase turbine efficiency [10, 11].

V. Tonkonogyi et al. (Eds.): InterPartner 2020, LNME, pp. 724–733, 2021.
https://doi.org/10.1007/978-3-030-68014-5_70

The effect of turbine inlet air cooling depends on its depth, which in turn depends on the type of chiller. So, in absorption lithium-bromide chiller (ACh) it is possible to cool air to temperatures 15–20 °C [12, 13], with a coefficient of performance (COP) is ζ_A = 0.7…0.8 and in refrigerant ejector chiller (ECh) – to temperatures about 10 °C [14, 15], but its COP is much less: ζ_E = 0.2… 0.25. Therefore, a combined absorption-ejector chiller (AECh) complex with ACh for cooling ambient air to a temperature t_{a2} ≈ 15 °C in a high-temperature stage AC_{HT} and ECh for deeper air cooling to t_{a2} 10 °C in a low-temperature stage AC_{LT} of a two-stage air cooler at the inlet of GTU has been proposed [16, 17].

Since heat is removed from the AECh by cooling towers (CT) of the reverse cooling system, and their thermal load, in turn, depends on the AECh COP, then we have to solve the problem of determining the specific rational thermal loads $q_{0.AC}$ on AC depending on the target air cooling temperature, respectively thermal loads for chillers, which removed in CT, which poses problems of determining and distribution of design specific load q_{CT} on CT depending on the target air cooling temperature.

The study aims to develop an approach to analyzing the efficiency of using the installed capacity of cooling towers taking into account changes in thermal loads following the current climatic conditions. The following tasks to be solved: determining of the rational design thermal load on two-stage air coolers (AC) at the inlet of GTU (rational design cooling capacity of the AECh) to provide close to maximum annual fuel saving; determining of the rational design thermal load on cooling towers; determining of actual current thermal loads on cooling towers and their distribution depending on the target air cooling temperature.

2 Literature Review

In combustion engines, a cyclic air is an ambient air; therefore, their performance is strongly affected by climatic conditions [18, 19]. With an increase in temperature at the GTU inlet, ambient temperature, and relative humidity, the fuel consumption increases, the GTU efficiency and effective power drops [20, 21]. The simplest and cheapest method of cooling air is conducted by spraying water directly into the airstream at the GTU inlet [22], but its efficiency is limited by wet bulb temperature [23]. Also, saturated evaporative cooling (inlet fogging) relies heavily on the climatic data and the design capacity of inlet air cooling systems. Generally, GTU inlet air cooling is preferable in the zones with high ambient air temperature and low humidity [24, 25]. Furthermore, it is more appropriate for units with lower design economic efficiency [11]. An enhancement of GTU fuel efficiency at high temperatures of ambient air is possible by its cooling in chillers using the exhaust gas heat [26, 27]. Integrating cooling and heating subsystems into a conventional plant could increase its efficiency to 80% [28, 29]. The available waste heat potential of engines used by ACh can be enhanced through deep utilization [31].

The ACh provides cooling GTU inlet air to the temperature of about 15 °C with the use of exhaust gas heat. In a high efficient ACh, producing water with temperatures of around 7 °C [32], the cooling of air is possible to the temperature 15 °C [31], whereas in ECh – down to 10 °C, but the efficiency of heat conversion in ECh is low compared to ACh: the ECh coefficient of performance is ζ_E = 0.2–0.25 versus ζ_A = 0.7–0.8 for ACh [29, 30]. So it is reasonable to apply ACh for ambient air cooling to 15 °C in a AC_{HT} at the GTU inlet and further air subcooling in an AC_{LT} to 10 °C by ECh [31].

The CT is the most effective for heat extraction in the circulating water supply systems and got a widespread use in air conditioning and combustion engines [33].

3 Research Methodology

For a generalization of received results over a wide range of GTU a design cooling load, i.e., the cooling capacity of a chiller, required for cooling the ambient air to the target temperatures 10 °C and 15 °C in different climatic conditions (ambient air temperature t_{amb}, relative humidity φ_{amb}) varying during a year, is evaluated as specific cooling capacity q_0, related to the unit of air mass flow G_a = 1 kg/s in the AC.

The specific cooling capacity:

$$q_{0.AC} = \xi \cdot c_{ma} \cdot (t_{amb} - t_{a2}),\ kJ/kg, \tag{1}$$

where: $q_{0.AC}$ – specific cooling capacity [kJ/kg]; ξ – specific heat ratio of total heat, including sensible and latent heat, to sensible heat of moist air; t_{amb} – ambient air temperature [°C]; t_{a2} – air temperature at the AC outlet [°C]; c_{ma} – moist air specific heat [kJ/(kg·K)].

The GTU with two-stage cooling inlet air to 15 °C in the AC_{HT} by ACh and further subcooling air down to 10 °C in the AC_{LT} by ECh are considered for evaluation of the fuel-saving due to inlet air cooling and rational design cooling capacities q_0 of chillers in different climatic conditions. The rational values of specific thermal load on the AC that is the AECh cooling capacity, required for cooling the ambient air to the temperatures 10 and 15 °C to ensure the maximum specific annular fuel saving $\Sigma\Delta b_e$ taking into account on-site climatic conditions are determined [34].

The specific annular fuel saving $\Sigma\Delta b_e$:

$$\Sigma\Delta b_e = \sum(\Delta t_a \cdot \tau) \cdot b_{et}, \tag{2}$$

where: $\Sigma\Delta b_e$ – specific annual fuel consumption economy [kg/kW]; Δt_a – current air temperature drops [°C]; τ – time interval [h]; b_{et} – specific fuel consumption economy for 1 °C air temperature drop [kg/(kW·h)].

The thermal load on the CT was calculated as:

$$q_{CT} = (q_{0.HT}/\zeta_A + q_{0.HT}) + (q_{0.LT}/\zeta_E + q_{0.HT}) \tag{3}$$

where: $q_{0.HT}$ – specific thermal load on AC_{HT} [kJ/kg]; $q_{0.LT}$ – specific thermal load on AC_{LT} [kJ/kg]; ζ_A – the ACh thermal coefficient; ζ_E – the ECh thermal coefficient.

4 Results

The ambient air parameters change considerably during the GTU operation. Changes in ambient air temperature t_{amb}, relative humidity φ_{amb} [35, 36], and absolute humidity d_{amb} during 2018 and July month for Pivdennobuz'ka compressor station, Mykolaiv region, Ukraine are presented in Fig. 1.

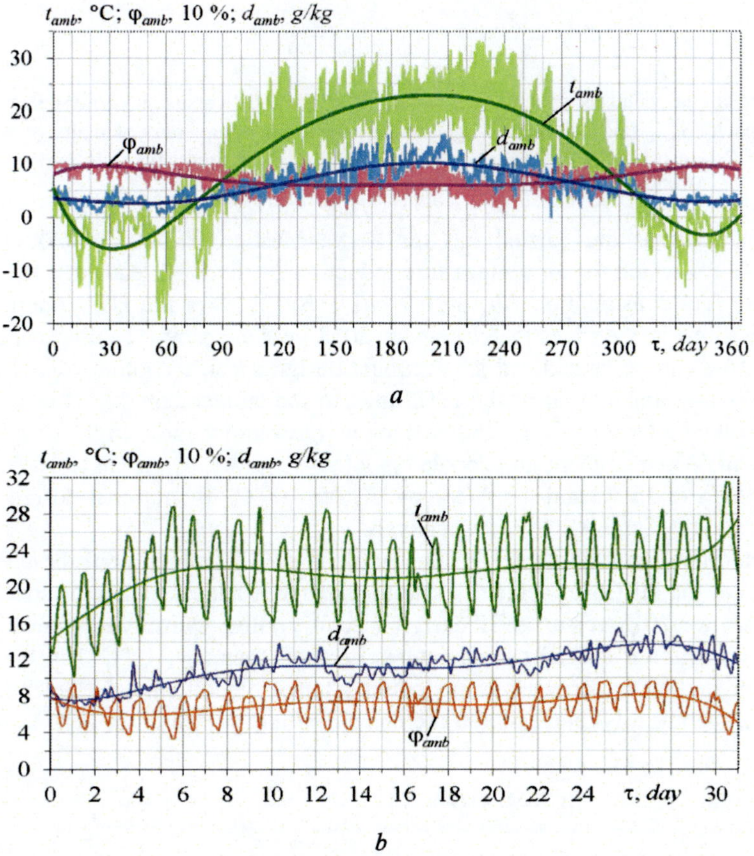

Fig. 1. Current values of ambient air temperature t_{amb}, absolute humidity d_{amb}, and relative humidity φ_{amb} for a year (*a*) and July month (*b*) 2018 for Pivdennobuz'ka compressor station, Mykolaiv region.

To determine the rational design thermal load on two-stage AC at the inlet of GTU (the AECh rational design cooling capacity) that provides closed to maximum annual values of specific fuel saving $\Sigma\Delta b_e$ for 2018 versus an air cooler design specific cooling capacity $q_{0.AC}$ (airflow G_a = 1 kg/s) at different temperatures of cooled air (10 °C – in the AECh; 15 °C – in the ACh) for various operation climatic conditions of Pivdennobuz'ka compressor station, Mykolaiv region are calculated and presented in Fig. 2.

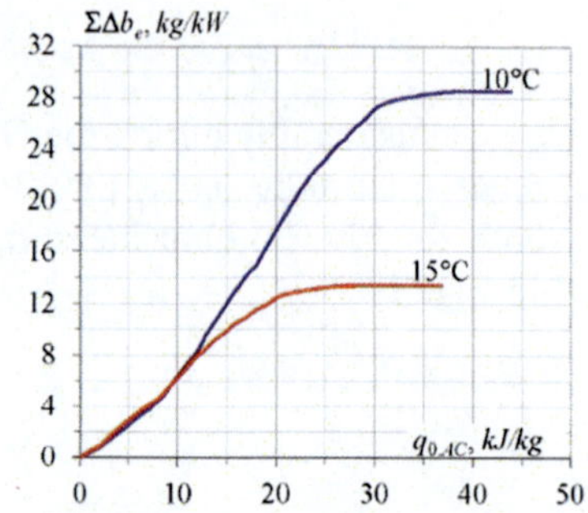

Fig. 2. The annual values of specific fuel saving $\Sigma\Delta b_e$ for 2018 versus air cooler design specific cooling capacity $q_{0.AC}$ (airflow G_a = 1 kg/s) at different temperatures of cooled air t_{a2}: 10 °C – in AECh; 15 °C – in the ACh for Pivdennobuz'ka compressor station, Mykolaiv region.

The calculations are carried out for Zorya–Mashproekt UGT 10000 (N_{eISO} = 10,5 MW), for which the air temperature reduction Δt_a by 1 °C leads to a decrease in the specific fuel consumption Δb_e by 0,7 g/(kW·h) [37]. For the GTU operation climatic conditions in the investigated region, 2018 year, a specific cooling capacity $q_{0.AC}$ = 24 kJ/kg can be considered as a rational design specific cooling capacity of the ACh for cooling ambient air at the GTU inlet to the temperature 15 °C at which the annual specific fuel saving $\Sigma\Delta b_e$ is closed to the maximum value at high enough rate of its increment. When ambient air is cooled in a two-stage AC by AECh to 10 °C (in the AC_{HT} by ACh to 15 °C and the AC_{LT} by ECh to 10 °C), the rational design specific cooling capacity of the AECh is respectively $q_{0.AC}$ = 34 kJ/kg.

To determine the rational design thermal load on CT, which cool the circulating water to cool the AECh to achieve close to maximum annual values of specific fuel saving $\Sigma\Delta b_e$ for 2018 versus a CT design specific cooling capacity q_{CT} (airflow from GTU G_a = 1 kg/s) at different temperatures of cooled air t_{a2} (10 °C – in the AECh; 15 °C – in the ACh) for various operation climatic conditions of Pivdennobuz'ka compressor station, Mykolaiv region are calculated and presented in Fig. 3.

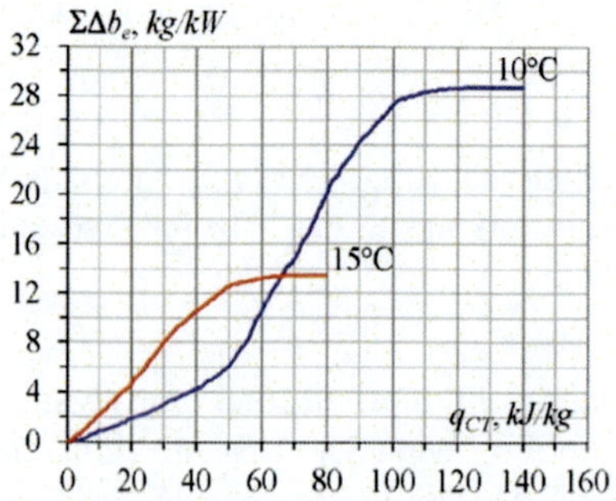

Fig. 3. The annual values of specific fuel saving $\Sigma\Delta b_e$ for 2018 versus cooling tower design specific thermal load q_{CT} (airflow G_a = 1 kg/s) at different temperatures of cooled air t_{a2}: 10 °C – in AECh; 15 °C – in the ACh for Pivdennobuz'ka compressor station, Mykolaiv region.

The thermal load on the CT was calculated q_{CT} in dependence on thermal load $q_{0.HT}$ on the ACh and $q_{0.LT}$ on the ECh and on their thermal coefficients $\zeta_A = 0{,}7$ and $\zeta_E = 0{,}25$.

For the GTU operation climatic conditions in the investigated region, 2018 year, a specific cooling capacity $q_{CT} = 55$ kJ/kg can be considered as a rational design specific cooling capacity of the CT for cooling circulating water for cooling the ACh when ambient air cooling at the GTU inlet to the temperature 15 ºC at which the annual specific fuel saving $\Sigma\Delta b_e$ is closed to the maximum value at high enough rate of its increment. When ambient air is cooled in a two-stage by the AECh to 10 °C, the rational design specific cooling capacity of the CT is respectively $q_{CT} = 105$ kJ/kg.

The current values of specific thermal loads q_{CT} and its rational value $q_{CT.rat}$ of cooling towers when ambient air cooling from the current temperature t_{amb} to $t_{a2} = 10$ and 15, respectively $q_{CT.10}$ and $q_{CT.15}$ for the 2018 year Pivdennobuz'ka compressor station, Mykolaiv region, Ukraine (Fig. 4) are presented.

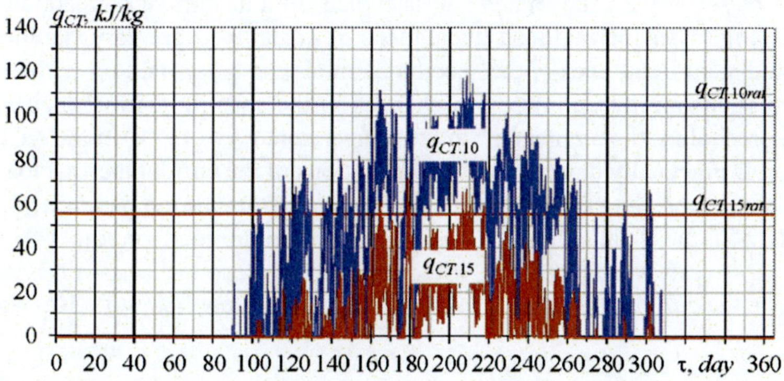

Fig. 4. Current values of specific thermal loads $q_{CT.15}$ on cooling towers and rational value of specific thermal load $q_{CT.15rat}$, when ambient air cooling from t_a to $t_{a2} = 15$ °C in ACh, specific thermal loads $q_{CT.10}$ on cooling towers and rational value of specific thermal load $q_{CT.10rat}$, when ambient air cooling from t_a to $t_{a2} = 10$ °C in the AECh in 2018 for Pivdennobuz'ka compressor station, Mykolaiv region.

The proposed method to determine a CT rational design thermal load and its rational distribution according to the behavior of the actual thermal loads following current climatic conditions is quite useful to determine the effective application of energy-saving methods of maintenance of capacity, in particular, through the accumulation of excessive capacity at lowered current thermal loads to cover increased loads.

To justify the approach to the analysis of the efficiency of using the rational design cooling capacity of CT taking into account the change in thermal loads following the current climatic conditions, the current values of specific thermal loads q_{CT} of CT when ambient air cooling from the current temperature t_{amb} to $t_{a2} = 10$ and 15, respectively $q_{CT.10}$ and $q_{CT.15}$ 2018 July for Pivdennobuz'ka compressor station, Mykolaiv region, Ukraine (Fig. 5) are presented.

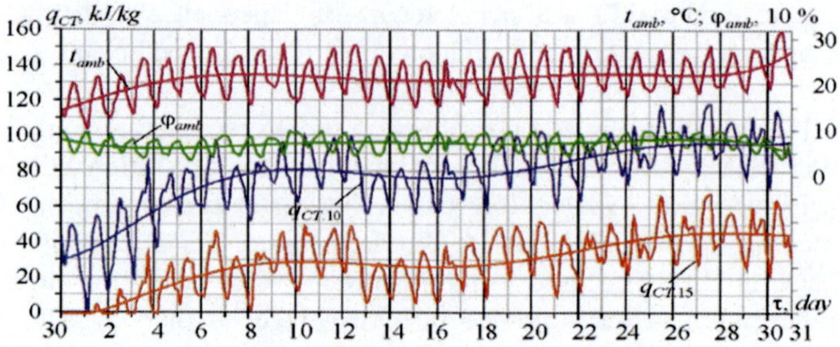

Fig. 5. Current values of ambient air temperature t_{amb} and relative humidity φ_{amb}, specific thermal loads $q_{CT.15}$ on cooling towers, when ambient air cooling from t_{amb} to t_{a2} = 15 °C in ACh, specific thermal loads $q_{CT.10}$ on cooling towers, when ambient air cooling from t_a to t_{a2} = 10 °C in AECh in July 2018 for Pivdennobuz'ka compressor station, Mykolaiv region.

As can be seen, when the ambient air is cooled from its current temperatures t_{amb} to t_{a2}, the cooling tower thermal load fluctuations $q_{0.10}$ and $q_{0.15}$ are very significant. The almost equidistant trend lines of the specific thermal load $q_{CT.10}$ and $q_{CT.15}$ indicate that these fluctuations are due primarily to changes in the specific thermal load $q_{CT.15}$ for providing precooling the ambient air to the temperature t_{a2} = 15 °C in the ACh, within which there is practically damping of the fluctuations of the current thermal load on CT.

At the same time, with the further cooling of the air from the temperature t_{a2} = 15 °C to t_{a2} = 10 °C, the fluctuations of the specific thermal load on the CT $q_{CT.10\text{–}15} = q_{CT.10} - q_{CT.15}$ are relatively small, without taking into account bursts-drops, caused by a decrease in the current values of the ambient air temperature below 15 °C (Fig. 6).

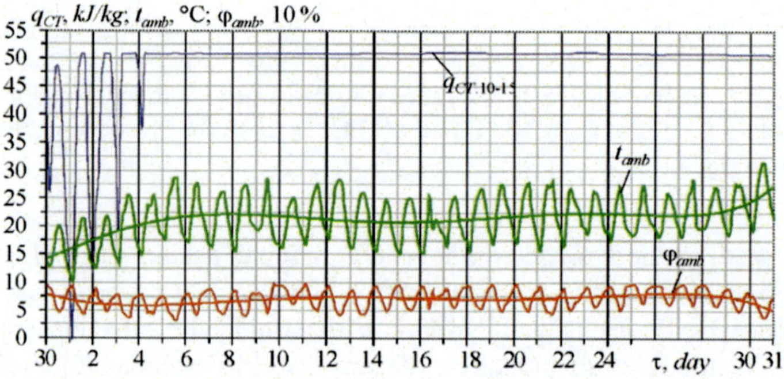

Fig. 6. Current values of ambient air temperature t_{amb} and relative humidity φ_{amb}, specific thermal loads $q_{CT.10\text{–}15} = q_{CT.10} - q_{CT.15}$ on cooling towers, needed for cooling air from t_{a2} = 15 °C to t_{a2} = 10 °C in July 2018 for Pivdennobuz'ka compressor station, Mykolaiv region.

Taking into account a relatively stable behavior of the specific thermal load on the cooling tower within the range $q_{CT.10-15} = q_{CT.10} - q_{CT.15}$ when the air is subcooled air cooler from t_{a2} = 15 °C to t_{a2} = 10 °C compared to large fluctuations of the CT capacity $q_{CT.15}$ within the range of ambient air pre-cooling in the air cooler from t_a to t_{a2} = 15 °C; the thermal load $q_{CT.10-15}$ is taken as a stable design component of thermal load on cooling towers when the air is deeply cooled from t_{a2} = 15 °C to t_{a2} = 10 °C in the air cooler. Accordingly, the CT design thermal load $q_{CT.10rat} - q_{CT.10-15}$ that necessary to ensure the operation of the ACh when precooling the ambient air from the current temperatures t_{amb} to t_{a2} = 15 °C is determined by a residual principle as the difference between a design specific thermal load $q_{CT.10rat}$ for providing the entire process of cooling the ambient air from the current temperatures t_{amb} to t_{a2} = 10 °C and its stable component q_{010-15}.

5 Conclusions

The data on actual current thermal loads while two-stage cooling ambient air at the inlet of GTU to the temperature of about 15 °C in a high-temperature stage AC_{HT} fed by chilled water from ACh and further subcooling to 10 °C and lower in a low-temperature stage AC_{LT} fed by refrigerant from ECh to determine the heat removed from ACh and ECh through CT was obtained by modeling the operation of the GTU cooling complex for climatic conditions of a compressor station in Mykolaiv region, South Ukraine.

The rational design thermal loads on two-stage air coolers (AC) at the inlet of GTU (rational design cooling capacity of AECh) to provide closed to maximum annual fuel-saving, taking into account the current variable thermal loads according to actual site climatic conditions, were calculated with using the method developed by authors.

This method was modified to determine the rational load on the whole GTU inlet air cooling complex, including CT removing heat from AECh. For this aim, the approach has been proposed to divide the overall design thermal load on the CT into two parts: a variable part of the load providing the GTU inlet ambient air cooling to 15 °C and a relatively stable part providing the further air cooling to 10 °C and lower.

The method is reasonable for designing ambient air cooling complexes to provide their minimum sizes due to the rational distribution of thermal loads regarding the changes in actual thermal loads. It might be easily implemented into more complex thermodynamical and economical optimization as the primary stage.

References

1. Kumar, T.A., Muzaffarul, H.M., Islam, M.: Effect of ambient temperature on the performance of a combined cycle power plant. Trans. Can. Soc. Mech. Eng. **37**(4), 1177–1188 (2013)
2. Günnur, ŞG., et al.: The effect of ambient temperature on electric power generation in natural gas combined cycle power plant-a case study. Energy Rep. **4**, 682–690 (2018)

3. Sohani, A., Farasati, Y., Sayyaadi, H.: A systematic approach to find the best road map for enhancement of a power plant with dew point inlet air pre-cooling of the air compressor. Energy Convers. Manag. **150**, 463–484 (2017)
4. Saghafifar, M., Gadalla, M.: Analysis of Maisotsenko open gas turbine power cycle with a detailed air saturator model. Appl. Energy **149**, 338–353 (2015)
5. Radchenko, M., Radchenko, R., Kornienko, V., Pyrysunko, M.: Semi-empirical correlations of pollution processes on the condensation surfaces of exhaust gas boilers with water-fuel emulsion combustion. In: Ivanov, V., et al. (eds.) Advances in Design, Simulation and Manufacturing II. DSMIE 2019, Lecture Notes in Mechanical Engineering, pp. 853–862. Springer, Cham (2020)
6. Kornienko, V., Radchenko, R., Stachel, A., Pyrysunko, M.: Correlations for pollution on condensing surfaces of exhaust gas boilers with water-fuel emulsion combustion. In: Tonkonogyi, V., et al. (eds.) Advanced Manufacturing Processes. InterPartner-2019, Lecture Notes in Mechanical Engineering, pp. 530–539. Springer, Cham (2020)
7. Sadighi Dizaji, H., Hu, E.J., Chen, L., Pourhedayat, S.: Using novel integrated Maisotsenko cooler and absorption chiller for cooling of gas turbine inlet air. Energy Convers. Manag. **195**, 1067–1078 (2019)
8. Shukla, A.K., Singh, O.: Thermodynamic investigation of parameters affecting the execution of steam injected cooled gas turbine based combined cycle power plant with vapor absorption inlet air cooling. Appl. Therm. Eng. **122**, 380–388 (2017)
9. Radchenko, A., Mikielewicz, D., Forduy, S., Radchenko, M., Zubarev, A.: Monitoring the fuel efficiency of gas engine in integrated energy system. In: Nechyporuk, M. (ed.) Integrated Computer Technologies in Mechanical Engineering. Advances in Intelligent Systems and Computing, vol. 1113, pp. 361–370. Springer, Cham (2020)
10. Zhang, T., Liu, Z., Hao, H., Chang, L.J.: Application research of intake-air cooling technologies in gas-steam combined cycle power plants in China. J. Power Energy Eng. **2**, 304–311 (2014)
11. Yang, C., Yang, Z., Cai, R.: Analytical method for evaluation of gas turbine inlet air cooling in combined cycle power plant. Appl. Energy **86**, 848–856 (2009)
12. Elberry, M.F., et al.: Performance improvement of power plants using absorption cooling system. Alexandria Eng. J. **57**, 2679–2686 (2018)
13. Mahmoudi, S.M., Zare, V., Ranjbar, F., Farshi, L.: Energy and exergy analysis of simple and regenerative gas turbines inlet air cooling using absorption refrigeration. J. Appl. Sci. **9**(13), 2399–2407 (2009)
14. Butrymowicz, D., et al.: Investigations of prototype ejection refrigeration system driven by low grade heat. In: HTRSE-2018, E3S Web of Conferences, vol. 70, no. 03002, 7 p. (2018)
15. Elbel, S., Lawrence, N.: Review of recent developments in advanced ejector technology. Int. J. Refrig. **62**, 1–8 (2016)
16. Konovalov, D., Trushliakov, E., Radchenko, M., Kobalava, H., Maksymov, V.: Research of the aerothermopressor cooling system of charge air of a marine internal combustion engine under variable climatic conditions of operation. In: Tonkonogyi, V., et al. (eds.) Advanced Manufacturing Processes. InterPartner-2019. Lecture Notes in Mechanical Engineering, pp. 520–529. Springer, Cham (2020)
17. Forduy, S., Radchenko, A., Kuczynski, W., Zubarev, A., Konovalov, D.: Enhancing the gas engines fuel efficiency in integrated energy system by chilling cyclic air. In: Tonkonogyi, V., et al. (eds.) Advanced Manufacturing Processes. InterPartner-2019. Lecture Notes in Mechanical Engineering, pp. 500–509. Springer, Cham (2020)
18. Popli, S., Rodgers, P., Eveloy, V.: Trigeneration scheme for energy efficiency enhancement in a natural gas processing plant through turbine exhaust gas waste heat utilization. Appl. Energy **93**, 623–636 (2012)

19. Popli, S., Rodgers, P., Eveloy, V.: Gas turbine efficiency enhancement using waste heat powered absorption chillers in the oil and gas industry. Appl. Therm. Eng. **50**, 918–931 (2013)
20. De Sa, A., Al Zubaidy, S.: Gas turbine performance at varying ambient temperature. Appl. Therm. Eng. **31**(14–15), 2735–2739 (2011)
21. Baakeem, S.S., Orfi, J., AlAnsary, H.: Performance of a typical simple gas turbine unit under Saudi weather conditions. Int. J. Fluid Mech. Therm. Sci. **1**, 59–71 (2015)
22. Lin, A.Q., Sun, Y.G., Zhang, H., Lin, X., Yang, L., Zheng, Q.: Fluctuating characteristics of air-mist mixture flow with conjugate wall-film motion in a compressor of gas turbine. Appl. Therm. Eng. **142**, 779–792 (2018)
23. Nixdorf, M., Prelipceanu, A., Hein, D.: Thermo-economic analysis of inlet air conditioning methods of a cogeneration gas turbine plant. In: ASME TURBO EXPO, 10 p. (2002)
24. Ameri, M., Hejazi, S.H.: The study of capacity enhancement of the Chabahar gas turbine installation using an absorption chiller. Appl. Therm. Eng. **24**, 59–68 (2004)
25. Baakeem, S.S., Orfi, J., Al-Ansary, H.: Performance improvement of gas turbine power plants by utilizing turbine inlet air-cooling (TIAC) technologies in Riyadh Saudi Arabia. Appl. Therm. Eng. **138**, 417–432 (2018)
26. Al-Ibrahim, A.M., Varnhamm, A.: A review of inlet air-cooling technologies for enhancing the performance of combustion turbines in Saudi Arabia. Appl. Therm. Eng. **30**, 1879–1888 (2010)
27. Ehyaei, M.A., Hakimzadeh, S., Enadi, N., Ahmadi, P.: Exergy, economic and environment (3E) analysis of absorption chiller inlet air cooler used in gas turbine power plants. Int. J. Energy Res. **36**(4), 486–498 (2012)
28. Tom, K.: Combined heating and power and emissions trading: options for policy makers. Int. Energy Agency, 15 p. (2008)
29. International Energy Agency: Combined heat and power: evaluating the benefits of greater global investment (2008)
30. Radchenko, R., Kornienko, V., Pyrysunko, M., Bogdanov, M., Andreev, A.: Enhancing the efficiency of marine diesel engine by deep waste heat recovery on the base of its simulation along the route line. In: Nechyporuk, M., et al. (eds.) Integrated Computer Technologies in Mechanical Engineering. Advances in Intelligent Systems and Computing, vol. 1113, pp. 337–350. Springer, Cham (2020)
31. Kornienko, V., Radchenko, M., Radchenko, R., Konovalov, D., Andreev, A., Pyrysunko, M.: Improving the efficiency of heat recovery circuits of cogeneration plants with combustion of water-fuel emulsions. Therm. Sci. 154 (2020)
32. Trushliakov, E., Radchenko, M., Bohdal, T., Radchenko, R., Kantor, S.: An innovative air conditioning system for changeable heat loads. In: Tonkonogyi, V., et al. (eds.) Advanced Manufacturing Processes. InterPartner-2019. Lecture Notes in Mechanical Engineering, pp. 616–625. Springer, Cham (2020)
33. Mulyandasari, V.: Cooling tower selection and sizing (engineering design guideline). KLM Techn. Group. Practical Engineering Guidelines for Processing Plant Solutions (2011)
34. Radchenko, A., Bohdal, L., Zongming, Y., Portnoi, B., Tkachenko, V.: Rational designing of gas turbine inlet air cooling system. In: Tonkonogyi, V., et al. (eds.) Advanced Manufacturing Processes. InterPartner-2019, Lecture Notes in Mechanical Engineering, pp. 591–599. Springer, Cham (2020)
35. Meteomanz Homepage. https://www.meteomanz.com/. Accessed 15 Jan 2020
36. Reference and Information Portal “Weather and Climate”. https://www.pogodaiklimat.ru/. Accessed 12 Jan 2020
37. SE SPCGTC “Zorya”–“Mashproekt”: Nikolaev gas turbines of industrial application, Publ. 20 p. (2004)

Ship Engine Intake Air Cooling by Ejector Chiller Using Recirculation Gas Heat

Roman Radchenko[1], Maxim Pyrysunko[2], Andrii Radchenko[1(✉)], Andrii Andreev[2], and Victoria Kornienko[2]

[1] Admiral Makarov National University of Shipbuilding, 9, Heroes of Ukraine Avenue, Mykolayiv 54025, Ukraine
nirad50@gmail.com

[2] Kherson Branch of Admiral Makarov National University of Shipbuilding, 44, Ushakova Street, Kherson 73000, Ukraine

Abstract. One of the promising ways in environmental protection of the marine internal combustion engines (ICE) is the neutralization of harmful substances in exhaust gas through engine gas recirculation (EGR) technology. However, the use of such techniques conflicts with the engine's energy efficiency and leads to increased fuel consumption. It is promising to use technologies that would increase fuel and energy efficiency of ICE with EGR systems and combine high environmental efficiency with engine fuel efficiency. The technology of precooling intake air at the suction of turbocharger by waste heat using chiller (WHUCh) was developed for ICE with the EGR system. The advantage of this solution is the possibility of using the waste heat of recirculation gases to reduce the heat load on the scrubber recycling system and fuel consumption. The scheme-design solution of the EGR system using the heat of recirculation gas by an ejector chiller (ECh) for cooling the air at the intake of the main ship engine is proposed. The effect of using the heat of recirculation gas for cooling engine intake air is analyzed, taking into account the changing climatic conditions on a vessel's route line. It is shown that using the heat of recirculation gas for cooling engine intake air by ejector refrigeration machine reduces the air temperature at the entrance of the main engine by 5–15 °C, which decreases the specific fuel consumption by 0.5–1.4 g/(kW•h). It reduces emissions of harmful substances (NOx by 26–38%; SOx by 9–14%) when the engine is running with recirculation of gas.

Keywords: Harmful emissions · Exhaust gas recirculation · Ejector chiller · Specific fuel consumption · Ecology

1 Introduction

The most sensitive environmental impact comes from ship power plants, in which the main source of energy (thermal, mechanical, electrical) is internal combustion engines (ICE) [1]. During the operation of internal combustion engines, the greatest harm is caused by toxic substances contained in the exhaust gas [2]. The formation of harmful gases, such as carbon dioxide CO_2, nitrogen oxides NOx, carbon monoxide CO, sulfur

V. Tonkonogyi et al. (Eds.): InterPartner 2020, LNME, pp. 734–743, 2021.
https://doi.org/10.1007/978-3-030-68014-5_71

oxides SOx, etc., depends on the organization of work processes in the internal combustion engine. A very effective way of greening marine ICE is the artificial neutralization of harmful substances in exhaust gases, for example, by exhaust gas recirculation (EGR technology). However, the use of such technology conflicts with the energy efficiency of the ICE and causes increased fuel consumption compared with conventional engine operation without exhaust gas recirculation.

It is reasonable to use the technologies that would ensure an increase in fuel efficiency of ICE with EGR systems, id est. Would combine high environmental efficiency with fuel efficiency. These technologies would provide fuel saving due to precooling air at the suction of the turbocharger by waste heat using chillers (WHUCh).

2 Literature Review

Exhaust gas recirculation is a method that significantly reduces the formation of NOx in marine diesel engines. Using this method fully meet Tier III requirements for NOx [3, 4]. In the EGR system, after the cooling and cleaning process, a part of the exhaust gas is recirculated to the scavenge air receiver. Thus, part of the oxygen in the air that is used in the combustion process is replaced by CO_2 oxide [5, 6]. The oxygen content of O_2 and the burning rate is reduced, thereby reducing the maximum burning temperature, and thus reducing the intensity of NOx formation [7]. NOx emission reduction is almost linear to exhaust gas recirculation [8, 9].

The increased amount of recirculation gases causes under-burning of the fuel and an increase in emissions of CO, CO_2, and soot [10]. Therefore, the quantity (part in the total volume of exhaust gases) of the recirculation gases must be limited. That is, a compromise must be found between the effect of NO recirculation and combustion efficiency (in the form of low-grade hydrocarbons). It imposes restrictions on the upper limit of the recycling rate [11, 12].

When using EGR, there is an increased fuel consumption caused by the slowed rate of heat release [13, 14]. In the upper engine load range, fuel consumption gradually increases with the recycling rate. At recirculation ratio, Kr = 20%, the fuel consumption increases by 10–20 g/(kWh), in the average engine load range, fuel consumption increases linearly by the Kr, but in the lower load range, it slightly affects the fuel consumption. Using recirculation with a Kr = 10% can reduce NOx by about 30% without a significant increase in fuel consumption, although the exhaust gases smoking increases slightly. When Kr = 20%, the reduction in emissions of nitrogen oxides can reach 60%. However, already at Kr > 10–15%, the fuel economy deteriorates by 4–7%. One advantage of the EGR is the low cost compared with other methods of reducing NOx. In the case of recycling, there is no need to use complex and expensive devices, the manufacture of which for large-sized engines causes great technical difficulties [15, 16].

Since the CO_2 and water molecules have higher heat capacity, the combustion temperature somewhat decreases [17]. According to the data of [7, 12], an increased mass flow rate gives about 93% of the effect of reducing the flue gas temperature [18, 19], while an increase in the specific heat gives about 7% [20, 21].

The cooling of recirculation gases leads to a decrease in NOx and particulate emissions with comparable recirculation ratio Kr. This effect is more significant, with a large recirculation ratio [22].

The work [9] shows the results of using exhaust gas recirculation for the engine MAN 12K90ME9, which is applied as the main engine on the container vessel. The recycling rate at nominal and partial engine loads was Kr = 30–40%, while NOx emissions decreased to 80%, SOx emissions decreased by 25%, particulate matter (PM) by 50%, but this led to an increase in specific fuel oil consumption by 5–6 g/(kWh). The increase in specific fuel oil consumption of low-speed engines up to 5 g/(kWh) is also confirmed in [23]. It is also indicated that it is a technology for EGR at a concentration of fuel sulfur level S < 3.5%.

A typical EGR system for a ship's main low-speed engine includes a scrubber, a cooler, a water separator, a fan, and a support system for NaOH solution with a pump and tank [24]. It should be noted that the components of the system are quite dimensional, the pumps of the gas cleaning system, cooling system and fan (or electric compressor) require a small amount of electrical energy [25, 26].

Today, it is promising to use technologies that would ensure an increase in fuel and energy efficiency of combustion engines with EGR systems, i.e., would combine high environmental efficiency with engine fuel efficiency due to deep waste heat utilization [27, 28], using alternative fuels [29, 30] and water-fuel emulsion [31], engine cyclic air cooling by applying refrigeration [32] and jet technologies [33], modern modeling methods [34, 35] and data statistics for the influence of actual climatic conditions on performance characteristics of engines [36, 37]. Such technologies provide engine cyclic air cooling by waste heat using chillers (WHUCh): ejector chillers (ECh) as the most simple in design [38] and absorption chillers (ACh) as high efficient heat transformer with a coefficient of performance $\zeta = 0.6-0.7$[39, 40].

The study aims at the assessment of the environmental and energy efficiency of marine diesel engine due to cooling the cyclic air of a ship's main engine by an ejector chiller using the waste heat of recirculation gases, (the heat of hot water, obtained from the heat of the exhaust gases of the EGR system,) leading to reduce the heat load on the scrubber recycling system, concerning variable climatic conditions during the vessel route.

3 Research Methodology

The effectiveness of the application of the proposed technical solution was analyzed based on the EGR system typical for MAN low-speed two-stroke diesel engines following the Tier III environmental conditions. Recirculation is provided by bypassing part of the exhaust gases purified from harmful gases in the scrubber after cooling in the heat exchanger-gas cooler. The EGR system comprises a scrubber, chilling unit, condensate trap, the fan, and the system is cleaned from the solution of NaOH.

Circuit solution with the use of the heat-using circuit of the ECh is considered for the ship's low-speed two-stroke diesel engine MAN B&W 6G50ME-C9.6. The CEAS software package [11] was used to analyze the parameters of the recirculation system, as well as the characteristics of the main engine. The calculation was made for the following initial data: the performance characteristics of the main engine (under ISO conditions) – engine load – NMCR = 90%; power – N_e = 9288 kW; speed – n_e = 96.5 rpm; specific fuel oil consumption (SFOC) – g_e = 166.0 g/(kWh); exhaust gas recirculation system (EGR) – as bypass with scrubber and gas cooler, responsible for Tier III environmental conditions.

The calculation of the characteristics of the engine was carried out on the operating mode during the voyage of the dry-cargo ship from Odessa to Yokohama. The variation of climatic conditions (ambient air temperature t_a, absolute humidity d_a, and relative humidity φ_a, the temperature of seawater t_w) during the vessel's voyage is presented in Fig. 1.

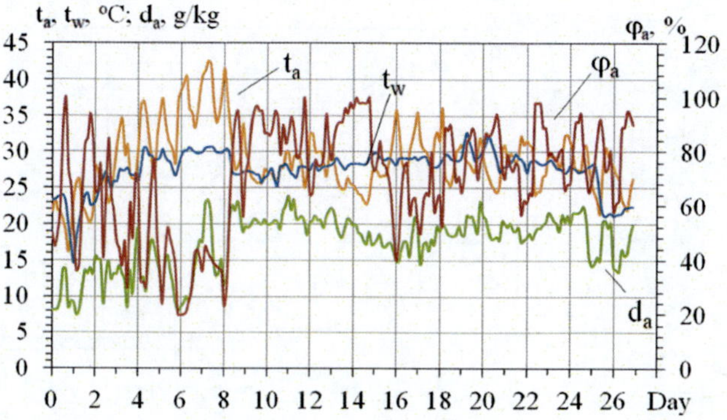

Fig. 1. The variation of temperature t_a, absolute humidity d_a, and relative humidity φ_a of ambient air and temperature of seawater t_w during the vessel trade route Odessa-Yokohama.

The operating parameters of the heat-recovery contour based on the ECh were calculated using the well-known equations, applied in the software complex developed at the Department of Conditioning and Refrigeration.

The following characteristics of the ECh for cooling the ship engine intake air were chosen: refrigerant – R142b; refrigerant evaporation temperature in the evaporator-air cooler t_0 = 5 °C and in the generator t_g = 80–120 °C, refrigerant condensing temperature in the condenser t_c = 25–45 °C.

The values of coefficient of performance for WHUCh: ζ = 0.30; 0.35.

4 Results

The solution to using the ECh was developed and analyzed (Fig. 2). The bypassing recirculation system runs as follows: exhaust gases from 10 to 40% in quantity are fed to the scrubber, where they are partially cooled and cleaned by spraying water with special nozzles. Then the exhaust gases are cooled in the heat exchanger - gas cooler (heater of water for a refrigerant generator of ECh), condensed vapor from exhaust gases is drained through condensate trap and cooled gases are fed by the fan to the scavenge air receiver, where gases are mixed with the scavenge air coming from the turbocharger.

It is proposed to use the heat of the recirculating gases for high-pressure liquid refrigerant evaporation in the generator of ECh with the generation of high-pressure refrigerant vapor as motive fluid for the ejector to suck a low-pressure refrigerant vapor from refrigerant evaporator - air cooler (AC-RE) at the intake of the turbocharger. Thus, the cooling capacity of ECh is used for cooling air at the intake of the engine turbocharger.

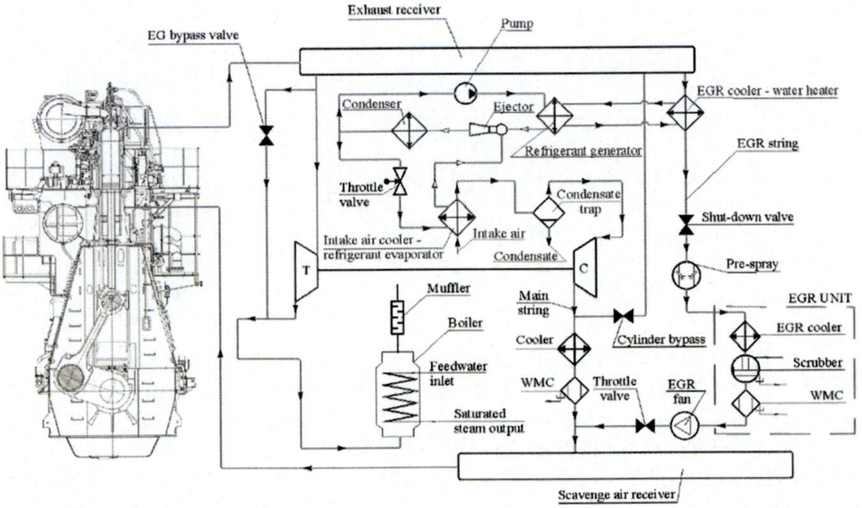

Fig. 2. Scheme of EGR-technology with bypass for the MAN marine diesel engine with ejector chiller: WMC – water mist catcher.

For MAN B & W 6G50ME-C9.6 engine considered, the specific fuel consumption decreases by about 0.5–1.6 g/(kWh) due to engine inlet air cooling during a voyage (Fig. 3). The number of recirculation during a voyage is K_p = 14–20%, the flow of recirculating flue gases is $G_{g.r}$ = 3–4 kg/s and a total exhaust gas flow G_g = 14–16 kg/s. The flow of "fresh" air to the engine turbocharger is $G_{a.egr}$ = 13–16 kg/s with exhaust gas recirculation and G_a = 17–19 kg/s - without recirculation of exhaust gases.

For the 6G50ME-C9.6 engine, according to the data of the MAN company (according to the calculations using the CEAS software package), when cooling intake air for every 10 °C a reduction in specific fuel consumption is 1.09 g/(kWh) or 0.109 g/(kWh·K) for every 1 °C air temperature drop.

The results of analyzing the operation efficiency of recirculation gas heat-recovery chiller with different coefficients of performance ζ = 0.30; 0.35 show the following cooling capacities received in chiller (Fig. 4): $Q_{0(0.3)}$ = 160–280 kW (ζ = 0.30) and $Q_{0(0.35)}$ = 180–245 kW (ζ = 0.35). The heat load on the ECh generator (heat consumption of chiller) is Q_g = 520–820 kW (Fig. 4) with appropriate cooling of the recirculation gas in the gas cooler (before the scrubber) from the temperature t_{g2} = 370–440 °C to the temperature t_{g1} = 180 °C (limited taken into account the danger of corrosion) provides decrease in air temperature at the inlet of the engine turbocharger, respectively: $\Delta t_{a(0.3)}$ = 4.1–10.8 °C (ζ = 0.30); $\Delta t_{a(0.35)}$ = 4.8–12.6 °C (ζ = 0.35).

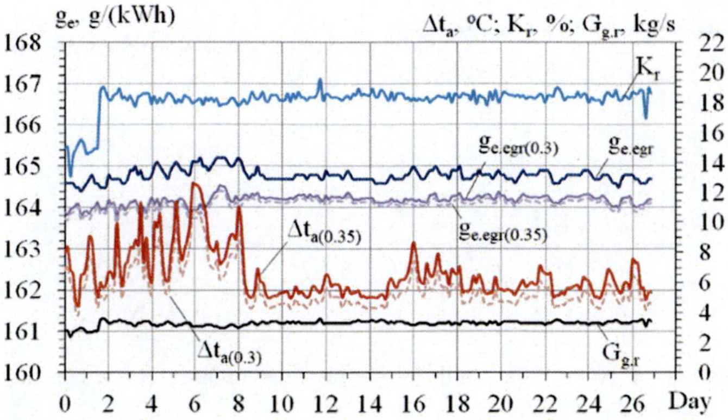

Fig. 3. Variation of specific fuel consumption g_e, number of recirculation K_r, decrease in air temperature at the engine intake Δt_a and recirculation gas flow $G_{g.r}$ during the vessel route Odessa-Yokohama at different coefficients of performance of ECh ζ = 0.30; 0.35.

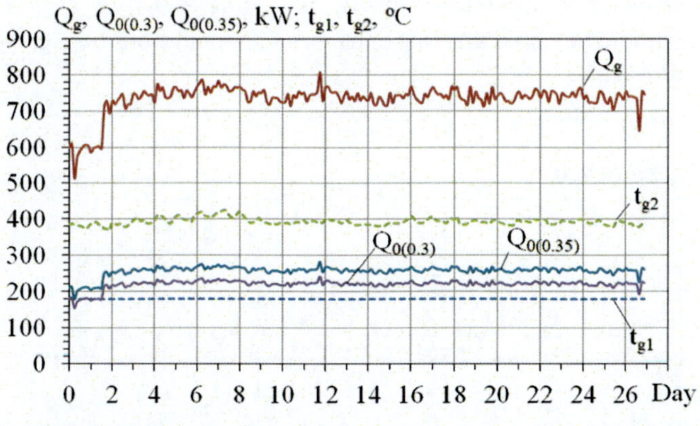

Fig. 4. The temperature of recirculation gases cooled by removing the heat for ECh generator (at the exit from heat exchanger- gas cooler) t_{g1}, the temperature of the gases at the inlet of heat exchanger-gas cooler t_{g2}, thermal load on the generator of the chiller Q_g, the cooling capacity of the chiller at different coefficients of performance $Q_{0(0.3)}$ and $Q_{0(0.35)}$ during the route of the vessel.

A decrease in the engine intake air temperature ensures a reduction in the specific fuel consumption (Fig. 3) in accordance with: $\Delta g_{e(0.3)}$ = 0.4–1.2 g/(kWh) (ζ = 0.30); $\Delta g_{e(0.35)}$ = 0.5–1.4 g/(kWh) (ζ = 0.35). The maximum efficiency of engine intake air cooling through recirculation exhaust gas heat-recovery corresponds to the coefficient of performance of the chiller $\zeta = 0.35$ and is $\Delta g_e = 0.8–2.2\%$, while the total specific fuel consumption will be $g_{e(0.35)} = 164–165$ g/(kWh), and without engine intake air cooling - $g_e = 163.7–164.4$ g/(kWh).

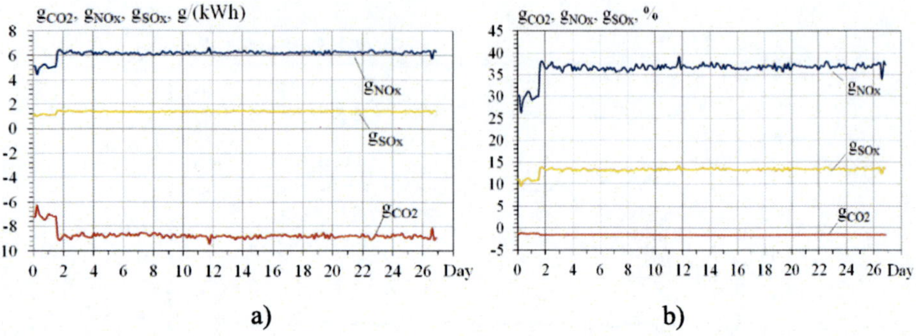

Fig. 5. The values of NOx, SOx, and CO_2 emissions in absolute values (a) and relative (b) during the route of the vessel.

Reduction of emissions due to lowering the engine intake air temperature when using the heat of recirculation gases is insignificant and amounts to no more than 0.2–0.3% for NOx and SOx, but for the system with gas recirculation and $\zeta = 0.35$ (Fig. 5) is: $\Delta g_{NOx(0.35)} = 26–38\%$ (4.6–6.8 g/(kWh)); $\Delta g_{SOx(0.35)} = 9–14\%$ (1.1–1.5 g/(kWh)).

However, it should be noted that this enhances CO_2 emissions by $\Delta g_{CO2(0.35)} = 1.3–1.7\%$ (6.6–9.1 g/(kWh)).

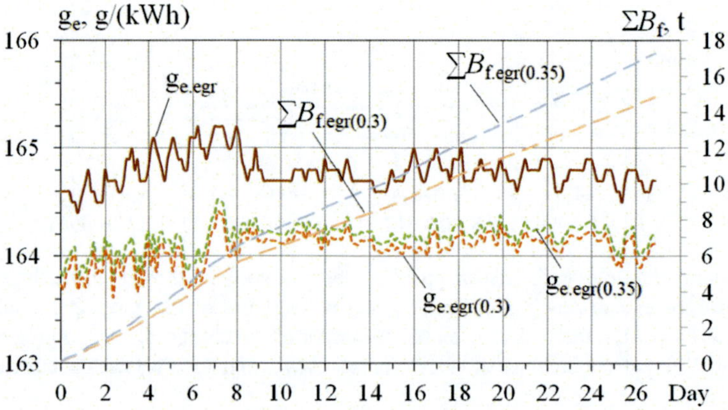

Fig. 6. Variation of specific fuel consumption ge.egr, a total of fuel economy ∑Bf.egr during the vessel route Odessa-Yokohama at different coefficients of performance of chiller $\zeta = 0.30$; 0.35.

The total fuel economy $\sum B_f$ due to a decrease in recirculation gas temperature during the vessel route Odessa-Yokohama at different coefficients of performance of chiller is (Fig. 6): $\Sigma B_f(0.3) = 14.8$ t ($\zeta = 0.3$); $\Sigma B_f(0.35) = 17.3$ t ($\zeta = 0.35$).

5 Conclusions

The technology of precooling intake air at the suction of turbocharger by waste heat using ECh is implemented in ICE with the EGR system. The scheme solution of the ecological exhaust gas recirculation system with the use of its heat by ECh for cooling the air at the intake of the ship's main engine is developed.

The effect of using the heat of recirculating gases for cooling air at the turbocharger intake was analyzed for the engine MAN 6G50ME-C9.6 with the account of changing climatic conditions on the vessel route line Odesa-Yokohama.

It is shown that the use of ecological recirculation gas heat in ECh allows reducing the air temperature at the intake of the ship's main engine by 5–15 °C, which provides a reduction of the specific fuel consumption by 0.5–1.4 g/(kW•h). The emissions of harmful substances are reduced: of NOx by 26–38%; SOx by 9–14%.

In the future, it is planned to analyze the environmental and energy efficiency of the cooling of the cyclic air of a ship's main engine by an absorption chiller using the heat of hot water obtained from the heat of the exhaust gases of the EGR system.

References

1. Kumar, J., Thirumala, S., et al.: Effect of reformed EGR on the performance and emissions of a diesel engine: a numerical study. Alexandria Eng. J. **57**, 517–525 (2018)
2. Ghosh, S., Dutta, D.: The effects of EGR on the performance and exhaust emissions of a diesel engine operated on diesel oil and pongamia pinata Methyl ester (PPME). Int. J. Eng. Inventions **12**(1), 39–44 (2012)
3. MAN Diesel & Turbo. https://marine.man-es.com/two-stroke/project-guides. Accessed 12 Feb 2020
4. Nielsen, K.V.: Exhaust recirculation control for reduction of NOx from large two-stroke diesel engines. Technical University of Denmark, Department of Electrical Engineering (2016)
5. Hussain, J., Palaniradja, K., Alagumurthi, N., Manimaran, R.: Effect of exhaust gas recirculation (EGR) on performance and emission characteristics of a three cylinder direct injection compression ignition engine. Alexandria Eng. J. **51**(4), 241–247 (2012)
6. Pandhare, A.P., Zende, K.C., Joglekar, A.S., Bhave, S.C., Padalkar, A.S.: Effect of EGR on the exhaust gas temperature and exhaust opacity in compression ignition engines using Jatropha oil as fuel. Applied mechanics and materials. In: 2nd International Conference on Mechanical and Aerospace Engineering, vol. 110–116, pp. 431–436 (2012)
7. Agarwal, D., Singh, S.K., Agarwal, A.K.: Effect of exhaust gas recirculation (EGR) on performance, emissions, deposits and durability of a constant speed compression ignition engine. Appl. Energy **88**(8), 2900–2907 (2011)
8. Diesel engine control systems. 1st edn. KGI Behind the wheel, Moscow (2004). [in Russian]
9. Wei, H., Zhu, T., Shu, G., Tan, L., Wang, Y.: Gasoline engine exhaust gas recirculation – a review. Appl. Energy **99**, 534–544 (2012)

10. Kumar, J.T.S., Sharma, T.K., Murthy, K.M., Rao, G.A.P.: Effect of reformed EGR on the performance and emissions of a diesel engine: A numerical study. Alexandria Eng. J. **57**, 517–525 (2018)
11. Wartsila Homepage. https://wartsila.com. Accessed 12 Feb 2020
12. MAN Diesel Turbo: CEAS Engine Calculations. https://marine.man-es.com/two-stroke/ceas. Accessed 22 Jun 2019
13. Hadjab, R.: Computational fluid dynamics simulation of heat transfer performance of exhaust gas recirculation coolers for heavy-duty diesel engines. Thermal Sci. **22**(6B), 2733–2745 (2018)
14. Grünenwald, B., Knödler, W., Saumweber, C.: A methodology for evaluation of charge air coolers for low pressure EGR systems with respect to corrosion. Vehicle Thermal Manage. Syst. Conf. Exhib. (VTMS) **10**, 57–67 (2011)
15. Ramakrishna Reddy, E., Keerthana, B.V.S., Dhana Raju, V., Sai Mahith, A., Dhanush, B.: Effect of EGR on diverse characteristics of diesel engine operated with corn seed biodiesel blend. Int. J. Ambient Energy **41**, 1–8 (2020)
16. De Serio, D., De Oliveira, A., Sodré, J.R.: Application of an EGR system in a direct injection diesel engine to reduce NOx emissions. J. Phys.: Conf. Series **74**(3), 032001 (2016).
17. Larsenb, U., Pierobona, L., Baldib, F., Haglinda, F., Ivarssona, A.: Development of a model for the prediction of the fuel consumption and nitrogen oxides emission trade-o for large ships. Energy **80**, 545–555 (2015)
18. Llamas, X.: Modeling and Control of EGR on Marine Two-Stroke Diesel Engines. Linköping Studies in Science and Technology. Dissertations, vol. 1904 (2018)
19. Peilong, X., Hongzong, S., Yiqian, W., Pin, W.: Effects of exhaust gas recirculation on combustion and emission of a marine diesel engine. Adv. Mater. Res. **926–930**, 905–908 (2014)
20. Asad, U., Ming, Z.: Exhaust gas recirculation for advanced diesel combustion cycles. Appl. Energy **123**, 242–252 (2014)
21. Bermúdez, V., Serrano, J.R., Piqueras, P., Gómez, J., Bender, S.: Analysis of the role of altitude on diesel engine performance and emissions using an atmosphere simulator. Int. J. Engine Res. **18**(1–2), 105–117 (2017)
22. Guan, B., Zan, R., Lin, H., Huang, Z.: Review of state of the art technologies catalytic reduction of NOx from diesel engine exhaust. Appl. Therm. Eng. **66**, 395–414 (2014)
23. Pham, V.V.: Advanced technology solutions for treatment and control noxious emission of large marine diesel engines: a brief review. J. Mech. Eng. Res. Dev. **42**(5), 21–27 (2019)
24. Thangaraja, J., Kannan, C.: Effect of exhaust gas recirculation on advanced diesel combustion and alternate fuels-a review. Appl. Energy **180**, 169–184 (2016)
25. Radchenko, N.: A concept of the design and operation of heat exchangers with change of phase. Arch. Thermodyn.: Polish Acad. Sci. **4**(25), 3–19 (2004)
26. Radchenko, N.: On reducing the size of liquid separators for injector circulation plate freezers. Int. J. Refrig. **5**(8), 267–269 (1985)
27. Radchenko, M., Radchenko, R., Kornienko, V., Pyrysunko, M.: Semi-empirical correlations of pollution processes on the condensation surfaces of exhaust gas boilers with water-fuel emulsion combustion. In: Ivanov, V., et al. (eds.) Advances in Design, Simulation and Manufacturing II. DSMIE 2019. Lecture Notes in Mechanical Engineering, pp. 853–862. Springer, Cham (2020)
28. Radchenko, M., Radchenko, R., Ostapenko, O., Zubarev, A., Hrych, A.: Enhancing the utilization of gas engine module exhaust heat by two-stage chillers for combined electricity, heat and refrigeration. In: 5th International Conference on Systems and Informatics, ICSAI 2018, Jiangsu, Nanjing, China, pp. 240–244 (2019)

29. Pourkhesalian, A.M., Shamekhi, A.H., Salimi, F.: Alternative fuel and gasoline in an SI engine: a comparative study of performance and emissions characteristics. Fuel **5**(89), 1056–1063 (2010)
30. Bae, C., Kim, J.: Alternative fuels for internal combustion engines. Proc. Combust. Inst. **3** (36), 3389–3413 (2017)
31. Kornienko, V., Radchenko, R., Stachel, A., Andreev, A., Pyrysunko, M.: Correlations for pollution on condensing surfaces of exhaust gas boilers with water-fuel emulsion combustion. In: Tonkonogyi, V., et al. (eds.) Grabchenko's International Conference on Advanced Manufacturing Processes. InterPartner-2019. Lecture Notes in Mechanical Engineering, pp. 530–539. Springer, Cham (2020)
32. Chen, X., Omer, S., Worall, M., Riffat, S.: Recent developments in ejector refrigeration technologies. Renew. Sustain. Energy Rev. **19**, 629–651 (2013)
33. Konovalov, D., Kobalava, H.: Efficiency analysis of gas turbine plant cycles with water injection by the aerothermopressor. In: Ivanov, V., et al. (eds.) Advances in Design, Simulation and Manufacturing II. DSMIE 2019. Lecture Notes in Mechanical Engineering, pp. 581–591. Springer, Cham (2020)
34. Konovalov, D., Trushliakov, E., Radchenko, M., Kobalava, G., Maksymov, V.: Research of the aerothermopresor cooling system of charge air of a marine internal combustion engine under variable climatic conditions of operation. In: Tonkonogyi, V., et al. (eds.) Grabchenko's International Conference on Advanced Manufacturing Processes. InterPartner-2019. Lecture Notes in Mechanical Engineering, pp. 520–529. Springer, Cham (2020)
35. Radchenko, R., Kornienko, V., Pyrysunko, M., Bogdanov, M., Andreev, A.: Enhancing the efficiency of marine diesel engine by deep waste heat recovery on the base of its simulation along the route line. In: Nechyporuk, M. (ed.) ICTME. Advances in Intelligent Systems and Computing, vol. 1113, pp. 337–350. Springer, Cham (2020)
36. Trushliakov, E., Radchenko, M., Radchenko, A., Kantor, S., Zongming, Y.: Statistical approach to improve the efficiency of air conditioning system performance in changeable climatic conditions. In: 5th International Conference on Systems and Informatics: ICSAI 2018, Jiangsu, Nanjing, China, pp. 1303–1307 (2018)
37. Radchenko, A., Radchenko, M., Trushliakov, E., Kantor, S., Tkachenko, V.: Statistical method to define rational heat loads on railway air conditioning system for changeable climatic conditions. In: 5th International Conference on Systems and Informatics: ICSAI 2018, Jiangsu, Nanjing, China, pp. 1308–1312 (2018).
38. Kornienko, V., Radchenko, R., Konovalov, D., Andreev, A., Pyrysunko, M.: Characteristics of the rotary cup atomizer used as afterburning installation in exhaust gas boiler flue. In: Ivanov, V., et al. (eds.) Advances in Design, Simulation and Manufacturing III. DSMIE 2020. Lecture Notes in Mechanical Engineering, vol. 2, pp. 302–311. Springer, Cham (2020)
39. Radchenko, R., Radchenko, A., Serbin,S., Kantor, S., Portnoi, B.: Gas turbine unite inlet air cooling by using an excessive refrigeration capacity of absorption-ejector chiller in booster air cooler. In: HTRSE-2018, E3S Web of Conferences, vol. 70, p. 03012, 6 p. (2018)
40. Radchenko, A., Radchenko, M., Konovalov, A., Zubarev, A.: Increasing electrical power output and fuel efficiency of gas engines in integrated energy system by absorption chiller scavenge air cooling on the base of monitoring data treatment. In: HTRSE-2018, E3S Web of Conferences, vol. 70, p. 03011, 6 p. (2018)

Prediction of Changes in the Output Characteristics of the Planetary Hydraulic Motor

Angela Voloshina[1](✉), Anatolii Panchenko[1], Olena Titova[1], Irina Milaeva[1], and Andrey Pastushenko[2]

[1] Tavria State Agrotechnological University, 18, B. Khmelnytsky Ave., Melitopol 72310, Ukraine
voloshinaa2012@gmail.com

[2] Mykolayiv National Agrarian University, 9, G. Gongadze Street, Mykolaiv 54020, Ukraine

Abstract. In recent years, particular attention is paid to the development of low-speed hydraulic motors, the working elements of which should have a low speed of movement and develop high torques. Planetary hydraulic machines meet these requirements. The disadvantages of planetary hydraulic machines can be attributed to the unevenness of the output characteristics due to the error in the shape of the elements of its rotor system and distribution system, which causes a pulsation of the flow of the working fluid. Thus, conducting research aimed at predicting a change in the output characteristics of a planetary hydraulic motor during operation, for a given range of changes in its operating parameters, is an urgent task. The various parameters that determine the difference in the output characteristics of the planetary hydraulic motor depending on the design features of its rotor system and distribution system are substantiated. The regression equations are obtained that describe the change in the output characteristics of the planetary hydraulic motor during operation, for a given range of changes in its operating parameters. The obtained regression equations adequately describe the difference in the functional parameters of the planetary hydraulic motor during its operation. The change in the efficiency of the planetary hydraulic motor in a wide range of operating parameters changes is substantiated. The performed studies make it possible to predict changes in the output parameters of planetary hydraulic motors in operating conditions.

Keywords: Rotor system · Distribution system · Full-factor experiment · Regression equations · Adequacy

1 Introduction

In recent years, there has been an intense search and improvement of the hydraulic machines used to create hydraulic machines with the most uncomplicated design, manufacturing, and repair technology at the lowest cost. Particular attention is paid to the development of low-speed hydraulic motors, the working elements of which must

V. Tonkonogyi et al. (Eds.): InterPartner 2020, LNME, pp. 744–754, 2021.
https://doi.org/10.1007/978-3-030-68014-5_72

have a low speed of movement and have a high load capacity (i.e., develop high torques). The developed hydraulic machines must be reliable in operation, have high efficiency, relatively small specific gravity, and dimensions.

At present, gerotor [1, 2], orbital [3, 4] and planetary [5, 6] hydraulic machines are widely used in hydraulic drives of self-propelled machinery. For self-propelled equipment, planetary hydraulic machines deserve the greatest attention. The main components of planetary hydraulic machines are the rotor system and distribution system. The design of the rotor system is based on the principle of operation of a gear pair (movable and fixed rotor) with internal hypocycloidal gearing [7, 8]. In these rotor systems, the separation of the discharge zone from the discharge zone is ensured simultaneously with the break-in condition. The distribution system of planetary hydraulic machines is formed by movable and fixed distributors and creates a rotating hydraulic field necessary to obtain the planetary motion of the rotor system [5, 6]. The hydraulic field rotates parallel to the surface of the stationary rotor with rotation. The movable rotor revolves around a still motionless at the same speed as the hydraulic field, while rotating in the opposite direction. The hydraulic field in this hydraulic motor acts as a crank of a planetary gearbox.

The main disadvantages of planetary hydraulic machines are the unevenness of the output characteristics due to the error in the shape of the elements of the rotor system [7], as well as the pulsation of the working fluid pressure in their distribution system [9].

Thus, today the issue of predicting changes in the output characteristics of a planetary hydraulic motor in a given range of changes in the geometric and operating parameters of its rotor system and distribution system is significant for increasing the efficiency of using planetary hydraulic motors.

2 Literature Review

The implementation of a mathematical model of rotor vibrations [10], the study of the rotor dynamics of multistage centrifugal machines [11], the basic principles of energy exposure when using a rotor with radial channels [12] are proposed. Multicriteria selection [13] and optimizing hydrodynamic profiling [14] were made. An analysis of these studies showed that today there is very little literature on the design methodology of gerotor, orbital, and planetary hydraulic machines [15].

The features of epicycloidal and hypocycloidal engagement are considered [16]. The interaction of the rotors of orbital hydraulic machines is considered as a gapless (theoretical) compound [4]. The influence of forces and moments in the gearing of a rotary trochoidal pump is investigated [17]. The work does not take into account the absence of a rigid kinematic connection between the rotors of the orbital hydraulic motor, which allows the internal rotor to move within the diametrical gap (self-install) arbitrarily. The forces acting in the gearing of the gerotor pump are considered, taking into account the compression state of the working fluid. Recommendations have been developed for the design of gerotor pumps [18]. Much attention is paid to the design of hypocycloidal surfaces of the rotors of gerotor hydraulic machines. A program has been developed for designing hypocycloidal surfaces, considering the equations of hydrodynamics of fluid motion in working chambers [19]. The technological process of manufacturing rotors is proposed [20]. Experimental studies of prototypes of gerotor pumps were carried out [21]. The marginal deviations of the error in the shape of the gear surfaces of the rotors are justified [7]. In the paper [8], the proposed method for the determination of hydraulic motor reliability by simulating changes in the technical state of the rotors. Issues related to the design of elements of the rotor system of planetary hydraulic machines were not considered.

A design scheme has been developed, and a mathematical apparatus has been proposed [5], which allows one to study the influence of the design parameters of a distribution system on the output characteristics of a planetary hydraulic motor. The influence of the shape of the distribution windows [6] on the output characteristics of the planetary hydraulic motor is investigated. The kinematic schemes of distribution systems are justified [9]. Issues related to the design of elements of the distribution systems of planetary hydraulic machines were not considered.

A universal model of a mechatronic system with a hydraulic drive has been developed [22]. Geometric [23], mathematical [24, 25], and hydrodynamic [26] models, cavitation processes [27] are investigated [28], are considered, theoretical studies of the influence of the geometric parameters of the flowing parts of the gerotor pump on its output characteristics are performed. The full three-dimensional transition model CFD for the gerotor engine is presented [29]. One of the key modeling technologies for such a machine is mesh processing for a dynamically changing fluid volume. No experimental studies were confirming the adequacy of the developed models.

Thus, conducting studies to predict a change in the output characteristics of a planetary hydraulic motor in a given range of changes in its geometric and operating parameters is an urgent task.

3 Research Methodology

To solve the problem of predicting changes in the output characteristics of the planetary hydraulic motor during operation, it is necessary:

- substantiate the various parameters that determine the difference in the output characteristics of the planetary hydraulic motor, depending on the design features of its rotor system and distribution system;
- conduct a full-factor experiment to obtain regression equations describing the change in the output characteristics of the planetary hydraulic motor in a given range of changes in the design parameters of the rotor system;
- conduct a full-factor experiment to obtain the regression equations describing the change in the output characteristics of the planetary hydraulic motor in a given range of changes in the design parameters of the distribution system;
- conduct comparative tests of serial and upgraded hydraulic motors.

Theoretical studies performed earlier [5, 6, 8, 9] show that the developed mathematical models, parametric studies, etc. allow predicting a change in the output characteristics of planetary hydraulic machines at the design stage. For developing a mathematical model to predict the change in output characteristics, as well as to determine the effectiveness of the developed planetary hydraulic machines in operating conditions, it is necessary to conduct experimental studies of serial and modernized hydraulic motors.

At this stage of the research, it is planned to conduct a full-factor experiment to obtain a mathematical model in the form of regression equations that describe the change in the output characteristics of a planetary hydraulic motor under operating conditions.

The tests were carried out with two planetary hydraulic motors of the PRG-22 series, with a power of 22 kW and a working volume of 630 cm^3 (with a serial and modernized rotor system and distribution system).

The nominal (current) value of the flow rate was established by changing the flow of the working fluid when the hydraulic motor was idling (without load), respectively, equal to 50, 70, 90, and 110 l/min.

The required load was set using the brake device of the test bench in the range of 0.1–1.5 kN·m in increments of 100 N·m.

To study the process of changing the output characteristics of a planetary hydraulic motor depending on the design features of the rotor system and the distribution system, the mathematical planning of the factor experiment was used.

The main input factors x_i, determining changes in the output characteristics of the planetary hydraulic motor, in the study of the rotor system, are the follows: the differential pressure of the working fluid $\Delta p - x_1$; working fluid flow $Q - x_2$; rotor shape error $E - x_3$.

The following parameters y_i were chosen as the response functions: torque on the motor shaft: $M_{tor} - y_1$; frequency of rotation of the hydraulic motor shaft $n - y_2$; hydromechanical efficiency $\eta_{mech} - y_3$; volumetric efficiency $\eta_{vol} - y_4$; total efficiency of the hydraulic motor $\eta - y_5$.

The constant factors z_i are as follows: temperature of the working fluid $t - z_1$; kinematic viscosity of the working fluid $\nu - z_2$; fineness of filtration of the working fluid $F - z_3$.

The main input factors x_i, determining changes in the output characteristics of the planetary hydraulic motor when studying the distribution system, are the follows: the differential pressure of the working fluid $\Delta p - x_1$; rotation frequency of the hydraulic motor shaft $n - x_2$; the gap between the distribution windows of the movable and fixed distributors $\delta - x_3$.

The following parameters were chosen as the response functions y_i: working fluid flow $Q - y_1$; torque on the shaft of the hydraulic motor $M_{tor} - y_2$; hydromechanical efficiency $\eta_{mech} - y_3$; volumetric efficiency $\eta_{vol} - y_4$; total efficiency of the hydraulic motor $\eta - y_5$.

The constant factors z_i are the follows: temperature of the working fluid $t - z_1$; kinematic viscosity of the working fluid $\nu - z_2$; fineness of filtration of the working fluid $F - z_3$.

4 Results

For determining the relationship between the geometric and functional parameters of a planetary hydraulic motor, it is necessary to study the process of changing its output characteristics depending on the design features of the rotors (shape error production of serial and modernized rotors).

As a result of the studies, the regression Eqs. (1)–(5) were obtained taking into account significant coefficients that adequately describe the investigated process of changing the output characteristics of the planetary hydraulic motor depending on the pressure drop, flow rate and the shape error of the displacers:

$$y_1(M_{\mathrm{tor}}) = 1170,5 + 495 \cdot \mathrm{x}_1 - 60,25 \cdot \mathrm{x}_2 + 69,5 \cdot \mathrm{x}_3 - 28,25 \cdot \mathrm{x}_1 \cdot \mathrm{x}_2 + 23 \cdot \mathrm{x}_1 \cdot \mathrm{x}_3 + 5,75 \cdot \mathrm{x}_2 \cdot \mathrm{x}_3; \quad (1)$$

$$y_2(n) = 111 + 75 \cdot x_2; \quad (2)$$

$$y_3(\eta_{mech}) = 0,843 - 0,037 \cdot x_2 - 0,056 \cdot x_3 + 0,0051 \cdot x_1 \cdot x_2 + 0,012 \cdot x_2 \cdot x_3; \quad (3)$$

$$y_4(\eta_{vol}) = 0,942 + 0,015 \cdot x_1 + 0,0345 \cdot x_2 - 0,01 \cdot x_1 \cdot x_2; \quad (4)$$

$$y_5(\eta) = 0,792 + 0,013 \cdot x_1 - 0,0054 \cdot x_2 - 0,053 \cdot x_3 + 0,013 \cdot x_2 \cdot x_3. \quad (5)$$

An analysis of Eqs. (1)–(5) confirms that a change in the shape error of the rotors affects the change (increase) in the torque M_{tor} on the motor shaft and does not affect the difference in the rotational speed n.

Having decoded Eqs. (1)–(5), we determine the change in the output characteristics of the planetary hydraulic motor, taking into account the design features of its rotor system under operating conditions:

- change in torque $M_{tor} = f\,(\Delta p,\, Q,\, E)$:

$$M_{tor} = 83,84 \cdot \Delta p - Q \cdot (0,094 \cdot \Delta p + 0,04) + E \cdot (1,21 \cdot \Delta p + 0,036 \cdot Q + 2,46) - 2,91; \quad (6)$$

- change in speed $n = f\,(Q)$:

$$n = 6 + 1,5 \cdot Q; \quad (7)$$

- change in hydromechanical efficiency $\eta_{mech} = f\,(\Delta p,\, Q,\, E)$:

$$\eta_{mech} = 0,858 + 0,013 \cdot \mathrm{E} - 0,0012 \cdot \Delta \mathrm{p} - Q \cdot (0,0013 - 0,000017 \cdot \Delta \mathrm{p} - 0,000075 \cdot \mathrm{E}); \quad (8)$$

– change in volumetric efficiency $\eta_{vol} = f\,(\Delta p,\, Q)$:

$$\eta_{vol} = 0,831 + 0,0048 \cdot \Delta \mathrm{p} + Q \cdot (0,00115 - 01000033 \cdot \Delta \mathrm{p}); \quad (9)$$

– change in overall efficiency $\eta = f\,(\Delta p,\, Q,\, E)$:

$$\eta = 0,723 + 0,0022 \cdot \Delta p - 0,00047 \cdot Q + E \cdot (0,000082 \cdot Q + 0,011). \quad (10)$$

The obtained dependences (6)–(10) describe the change in the functional parameters of the planetary hydraulic motor under operating conditions when the pressure drop Δp changes in the range from 8 MPa to 20 MPa; the flow rate of the working fluid Q is from 20 l/min to 120 l/min, and the shape error of the rotors is from 0.02 mm to 0.21 mm [7].

For determining the relationship between the geometric and functional parameters of the planetary hydraulic motor, it is necessary to study the process of changing its output characteristics depending on the design features of the distribution system (with a gap between the distribution windows and without a gap).

As a result of the studies, the regression Eqs. (11)–(15) were obtained taking into account significant coefficients that adequately describe the investigated process of changing the output characteristics of the planetary hydraulic motor under operating conditions depending on the pressure drop, flow rate of the working fluid and the gap between the distribution windows:

$$y_1(Q) = 54,7 + 44,6 \cdot x_2 + 0,1475 \cdot x_2 \cdot x_3; \quad (11)$$

$$y_2(M_{tor}) = 251 + 81 \cdot x_1 - 4 \cdot x_2 - 20,75 \cdot x_3 + 3 \cdot x_1 \cdot x_2 - 6,25 \cdot x_1 \cdot x_3; \quad (12)$$

$$y_3(\eta_{mech}) = 0,851 - 0,022 \cdot x_2 - 0,061 \cdot x_3 + 0,013 \cdot x_1 \cdot x_2 - 0,005 \cdot x_1 \cdot x_3 - 0,00325 \cdot x_2 \cdot x_3; \quad (13)$$

$$y_4(\eta_{vol}) = 0,945 - 0,00325 \cdot x_1 + 0,00625 \cdot x_2 - 0,00275 \cdot x_3; \quad (14)$$

$$y_5(\eta) = 0,804 - 0,00363 \cdot x_1 - 0,0161 \cdot x_2 - 0,0609 \cdot x_3 + 0,0129 \cdot x_1 \cdot x_2 - 0,00387 \cdot x_1 \cdot x_3 - 0,00387 \cdot x_2 \cdot x_3. \quad (15)$$

The adequacy of the obtained regression Eqs. (11)–(15) was determined by the standard method using the Fisher criterion.

An analysis of Eqs. (11)–(15) confirms that a change in the gap δ between the distribution windows of the movable and fixed distributors affects the difference in the output characteristics of the planetary hydraulic motor under operating conditions. Changing the pressure drop, Δp affects the change in the torque M_{tor} and the efficiency η of t hydraulic motor. A change in the rotational speed n affects the difference in the flow rate of the working fluid Q, torque M_{tor}, and efficiency η of the hydraulic motor.

The change in the torque M_{tor}, the rotation speed n and the efficiency η of the hydraulic motor was determined when the flow rate of the working fluid Q was varied in the range from 20 l/min 120 l/min, the pressure drop Δp was from 8 MPa to 20 MPa and the gap δ between the distribution windows was from 0 up to 0.015 rad.

To determine the dependence $M_{tor} = f(\Delta p, Q, \delta)$ from Eq. (11) we determine the change in the rotational speed $n = f(Q)$, having previously decoded it

$$n = \frac{270 \cdot Q - 14769}{44,45 + 19,4 \cdot \delta} + 330. \tag{16}$$

Having decoded Eq. (12) and substituting into it (16), we determine the change in torque on the motor shaft:

$$\begin{aligned} \mathrm{M}_{tor} &= 8,99 - 263,15 \cdot \delta + (21,82 - 205,59 \cdot \delta) \\ &\cdot \Delta\mathrm{p} + \frac{(0,75 \cdot \Delta\mathrm{p} - 11,99) \cdot \mathrm{Q} - 41,06 \cdot \Delta\mathrm{p} + 655,74}{44,45 + 19,4 \cdot \delta} \end{aligned} \tag{17}$$

Having decoded Eqs. (13)–(15), we determine the change in the output characteristics of the planetary hydraulic motor, taking into account the design features of its distribution system:

– change in the hydromechanical efficiency of the hydraulic motor $\eta_{mech} = f(n, \Delta p, \delta)$:

$$\begin{aligned} \eta_{mech} &= 0,967 - n \cdot (0,0002 - 0,000012 \cdot \Delta p + 0,0016 \cdot \delta) - \delta \\ &\cdot (5,53 + 0,16 \cdot \Delta p) - 0,0003 \cdot \Delta p. \end{aligned} \tag{18}$$

– change in volumetric efficiency of the hydraulic motor $\eta_{vol} = f(n, \Delta p, \delta)$:

$$\eta_{vol} = 0,95 - 0,0008 \cdot \Delta p + 0,00002 \cdot n - 0,36 \cdot \delta. \tag{19}$$

– change in the overall efficiency of the hydraulic motor $\eta = f(n, \Delta p, \delta)$:

$$\begin{aligned} \eta = 0,926 - n \cdot (0,0002 - 0,000012 \cdot \Delta p + 0,0019 \cdot \delta) - \delta \cdot (5,81 + 0,13 \cdot \Delta p) \\ - 0,004 \cdot \Delta p. \end{aligned} \tag{20}$$

The obtained dependences (16)–(20) describe the operation of a planetary hydraulic motor under operating conditions from an upgraded distribution system [5, 9] with a 7/6 kinematic scheme using two additional discharge windows of a movable distributor.

The implementation of the obtained regression equations to determine the overall efficiency of the upgraded hydraulic motor under operating conditions is shown in Fig. 1.

It should be noted that under operating conditions, planetary hydraulic motors of the PRG series have sufficiently high and stable efficiency in a wide range of output characteristics. At the same time, the overall efficiency of the upgraded hydraulic motor (Fig. 1a) is 11% higher than the serial one (Fig. 1b).

It should be noted that the overall efficiency of the upgraded hydraulic motor (Fig. 1a) varies in the range of 0.80–0.88 depending on the change in the output parameters. Moreover, the best value of efficiency ($\eta = 0.88$) remains stable in the range of changes in torque from 0.3 to 1.3 kN·m and rotation speed from 20 to 110 rpm. The efficiency of a serial hydraulic motor (Fig. 1b) varies in the range 0.70–0.78 and is maximum ($\eta = 0.78$) with a change in torque from 0.5 to 1.2 kN·m and rotation speed from 30 to 90 rpm.

As a result of the experimental studies, mathematical models are obtained in the form of regression equations that adequately describe the change in the functional parameters of the planetary hydraulic motor in the entire range of changes in the parameters of its operation. The performed studies make it possible to predict changes in the output characteristics of planetary hydraulic motors in operating conditions.

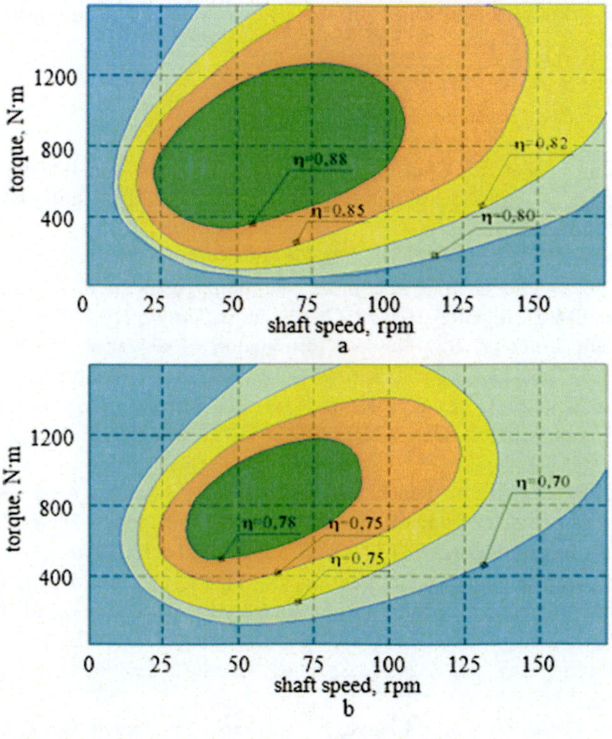

Fig. 1. External characteristics of planetary hydraulic motors of the PRG-22 series under operating conditions: a – modernized and b – serial.

5 Conclusions

As a result of the studies, mathematical models in the form of regression equations are obtained that describe the change in the functional parameters of the planetary hydraulic motor of the PRG-22 series over the entire range of operating parameters.

The pressure drop varied from 8 MPa to 20 MPa, the flow rate of the working fluid was from 20 to 120 l/min, the shape error of the rotor system was from 0.02 to 0.21 mm, and the gap between the distribution windows was from 0 rad up to 0.015 rad. The obtained mathematical models adequately describe the change in the functional parameters of the planetary hydraulic motor in the entire range of operating parameters.

The implementation of the obtained models to determine the overall efficiency of the modernized hydraulic motor shows that the total efficiency in operating conditions varies in the range of 0.80–0.88 depending on the change in the output parameters. The best value of the efficiency (η = 0.88) remains stable in the range of changes in torque from 0.3 to 1.3 kN·m and rotation speed – from 20 to 110 rpm. At the same time, the efficiency of a serial hydraulic motor varies in the range 0.70–0.78 and has a maximum value (η = 0.78) when the torque changes from 0.5 to 1.2 kN·m and the rotation speed is from 30 to 90 rpm.

Thus, the studies performed make it possible to predict changes in the output characteristics of planetary hydraulic motors in a wide range of operational parameters.

References

1. Chang, Y., Kim, J., Jeon, C., Chul, K., Jung, S.: Development of an integrated system for the automated design of a gerotor oil pump. J. Mech. Des. **129**(10), 1099–1105 (2006). https://doi.org/10.1115/1.2757629
2. Gamez-Montero, P.J., Garcia-Vilchez, M., Raush, G., Freire, J., Codina, E.: Teeth clearance and relief grooves effects in a trochoidal-gear pump using new modules of GeroLAB. J. Mech. Des. **134**(5), 054502–105457 (2012). https://doi.org/10.1115/1.4006440
3. Yang, D., Yan, J., Tong, S.: Flowrate formulation of deviation function based gerotor pumps. J. Mech. Des. **132**(6), 064503–106455 (2010). https://doi.org/10.1115/1.4001595
4. Choi, T., Kim, M., Lee, G., et al.: Design of rotor for internal gear pump using cycloid and circular-arc curves. J. Mech. Des. **134**(1), 011005–0011012 (2012). https://doi.org/10.1115/1.4004423
5. Voloshina, A., Panchenko, A., Panchenko, I., Titova, O., Zasiadko, A.: Improving the output characteristics of planetary hydraulic machines. IOP Conf. Series: Mater. Sci. Eng. **708**(1), 012038 (2019). https://doi.org/10.1088/1757-899X/708/1/012038
6. Voloshina, A., Panchenko, A., Boltyansky, O., Titova, O.: Improvement of manufacture workability for distribution systems of planetary hydraulic machines. In: Ivanov, V., et al. (ed.) Advances in Design, Simulation and Manufacturing II. DSMIE-2019. Lecture Notes in Mechanical Engineering, pp. 732–741.Springer, Cham (2019). https://doi.org/10.1007/978-3-030-22365-6_73
7. Panchenko, A., Voloshina, A., Milaeva, I., Panchenko, I., Titova, O.: The Influence of the form error after rotor manufacturing on the output characteristics of an orbital hydraulic motor. Int. J. Eng. Technol. **7**(4.3), 1–5 (2018). https://doi.org/10.14419/ijet.v7i4.3.19542.
8. Panchenko, A., Voloshina, A., Panchenko, I,, Titova, O., Pastushenko, A.: Reliability design of rotors for orbital hydraulic motors. IOP Conf. Ser.: Mater. Sci. Eng. **708**(1), 012017 (2019). https://doi.org/10.1088/1757-899X/708/1/012017
9. Voloshina, A., Panchenko, A., Boltyansky, O., Panchenko, I., Titova, O.: Justification of the kinematic diagrams for the distribution system of a planetary hydraulic motor. Int. J. Eng. Technol. **7**(4.3), 6–11 (2018). https://doi.org/10.14419/ijet.v7i4.3.19544.

10. Pavlenko, I.: Static and dynamic analysis of the closing rotor balancing device of the multistage centrifugal pump. Appl. Mech. Mater. **630**, 248–254 (2014). https://doi.org/10.4028/www.scientific.net/amm.630.248
11. Pavlenko, I., Trojanowska, J., Gusak, O., Ivanov, V., Pitel, J., Pavlenko, V.: Estimation of the reliability of automatic axial-balancing devices for multistage centrifugal pumps. Periodica Polytechnica Mech. Eng. **63**(1), 52–56 (2019). https://doi.org/10.3311/PPme.12801
12. Fesenko, A., Basova, Y., Ivanov, V., Ivanova, M., Yevsiukova, F., Gasanov, M.: Increasing of equipment efficiency by intensification of technological processes. Periodica Polytechnica Mech. Eng. **63**(1), 67–73 (2019). https://doi.org/10.3311/PPme.13198
13. Karpus, V.E., Ivanov, V.A.: Choice of the optimal configuration of modular reusable fixtures. Russian Eng. Res. **32**(3), 213–219 (2012). https://doi.org/10.3103/S1068798X12030124
14. Pavlenko, I., Liaposhchenko, A., Ochowiak, M., Demyanenko, M.: Solving the stationary hydroaeroelasticity problem for dynamic deflection elements of separation devices. Vibr. Phys. Syst. **2018026**, 1–7 (2018)
15. Gamez-Montero, P., Codina, E., Castilla, R.: A review of gerotor technology in hydraulic machines. Energies **12**, 2423 (2019). https://doi.org/10.3390/en12122423
16. Stryczek, J.: Characteristic parameters of hydraulic displacement machines with cycloidal gearing. Mech. Mach. Theory **1**(28), 97–112 (1993). https://doi.org/10.1016/0094-114x(93)90050-6
17. Ivanović, L., Blagojević, M., Devedžić, G., Assoul, Y.: Analitycal and numerical analysis of load gerotor pumps. Sci. Tech. Rev. **60**(1), 30–38 (2010)
18. Ivanovic, L., Miric, N., Devedzic, G., Ćuković, S.: Analysis of forces and moments in gerotor pumps. J. Mech. Eng. Sci. **224**(10), 2257–2269 (2010). https://doi.org/10.1243/09544062JMES2041
19. Stryczek, J., Bednarczyk, S., Biernacki, K.: Gerotor pump with POM gears: Design, production technology, research. Arch. Civil Mech. Eng. **14**(3), 391–397 (2014). https://doi.org/10.1016/j.acme.2013.12.008
20. Furustig, J., Almqvist, A., Bates, C.A., Ennemark, P., Larsson, R.: A two scale mixed lubrication wearing-in model, applied to hydraulic motors. Tribol. Int. **90**, 248–256 (2015). https://doi.org/10.1016/j.triboint.2015.04.033
21. Stryczek, J., Bednarczyk, S., Biernacki, K.: Strength analysis of the polyoxymethylenecycloidal gears of the gerotor pump. Arch. Civil Mech. Eng. **14**(4), 647–660 (2014). https://doi.org/10.1016/j.acme.2013.12.005
22. Panchenko, A., Voloshina, A., Kiurchev, S. et al.: Development of the universal model of mechatronic system with a hydraulic drive. Eastern-Eur. J. Enterprise Technol. **4**(7(94)), 51–60 (2018). https://doi.org/10.15587/1729-4061.2018.139577
23. Altare, G., Rundo, M.: Computational fluid dynamics analysis of gerotor lubricating pumps at high-speed: geometric features influencing the filling capability. J. Fluids Eng. **38**(11), FE15–1757 (2016). https://doi.org/10.1115/1.4033675
24. Van de Ven, J.D.: On fluid compressibility in switch-mode hydraulic circuits. – Part I: modeling and analysis. J. Dyn. Syst. Meas. Control **135**(2), 021013–021013–13 (2012). https://doi.org/10.1115/1.4023062
25. Van de Ven, J.D.: On Fluid Compressibility in switch-mode hydraulic circuits. – Part II: modeling and analysis. J. Dyn. Syst. Meas. Control **135**(2), 021014–021014–7 (2012). https://doi.org/10.1115/1.4023063
26. Chiu-Fan, H.: Flow characteristics of gerotor pumps with novel variable clearance designs. J. Fluids Eng. **137**(4), FE-14–1137 (2015). https://doi.org/10.1115/1.4029274

27. Moloshnyi, O.M., Szulc, P., Sotnyk, M.I.: Influence of an inlet rotating axial device on the cavitation processes in a low specific speed centrifugal pump. J. Eng. Sci. **6**(1), E25–E32 (2019). https://doi.org/10.21272/jes.2019.6(1).e5
28. Velev, E.: Study cavitation gerotor motors, using computer simulation. In: XV International Scientific Conference: Renewable Energies and Innovative Technologies, pp. 64–66 (2016)
29. Ding, H., Lu, J.X., Jiang, B.A.: CFD model for orbital gerotor motor. IOP Conf. Ser.: Earth Environ. Sci. **15**(6), 062006 (2012). https://doi.org/10.1088/1755-1315/15/6/062006

Cavitational Impact on Electrical Conductivity in the Beet Processing Industry

Marija Zheplinska[1], Mikhailo Mushtruk[1](✉), and Oksana Salavor[2]

[1] National University of Life and Environmental Sciences of Ukraine, 15, Heroes of Defense Street, Kyiv 03041, Ukraine
mixej.1984@ukr.net

[2] National University of Food Technologies, 68, Volodymyrska Street, Kyiv 016041, Ukraine

Abstract. The paper presents the results of studies the influence cavitation effects arising from the juices steam treatment on cellular and diffusion juices in sugar production. In conducting parallel studies, the treatment of the juice by steam and their usual heating and cooling structural transformations in the constituent parts of juices have been established, which leads to an increase in electrical conductivity. Thus ion carriers are releasing, which will allow increasing the juice purity and the sugar-sand production amount. When the temperature rises above 65 °C, electrical conductivity increases dramatically due to an increase in the mobility of ions during the increase in temperature, as well as a decrease in the solution viscosity. During the cooling of the juice, the electrical conductivity decreases smoothly, and the characteristic schedule curvature is not present. Thus, it is established that due to the steam-condensation cavitation effects, the structural transformations of the diffusion juice macromolecular compounds and colloidal dispersion substances are occurring. This effect leads to an increase in the electrical conductivity of juice due to the colloidal dispersion substances disaggregation and the ion carriers components releasing.

Keywords: Cellular juice · Sugar production · Electrical conductivity · Macromolecular compounds · Colloidal dispersion substances

1 Introduction

Diffusion juice is known to be a complex sucrose solution with a variety of organic and non-organic compounds. According to one of the hypotheses, most of them are contained in the associated and complex compounds form varying stability degrees, and the processes occurring during the diffusion juice purification by lime-carbon dioxide, are mainly the complexes transformations processes. The sucrose also takes part in this and eventually becomes a free state [1].

Classical methods of intensifying the processes of calc-carbon dioxide purification of diffusion juice of sugar beet production by changing the temperature, alkalinity, or duration of individual stages do not provide an increase in the purification efficiency

V. Tonkonogyi et al. (Eds.): InterPartner 2020, LNME, pp. 755–762, 2021.
https://doi.org/10.1007/978-3-030-68014-5_73

[2]. In particular, analyzing the conditions of the processes at the previous defecation, it is found that the limiting factor at this stage is the rate of delivery of reagents to the reaction zone [3]. In this regard, the authors propose a method of intensifying preliminary bowel movements using hydrodynamic (HD) cavitation, which significantly accelerates the mixing of reagents due to shock-wave action during the collapse of cavitation bubbles [4]. But under these conditions, along with positive signs, there is also a slight deterioration in the sedimentation-filtration properties of juice of I saturation, which can most likely be explained by cavitation destruction (grinding) of sediment particles formed during the previous defecation [5].

There is also information about the application of the method of blowing steam into the juice stream to intensify the purification and sedimentation processes, which helps to reduce the content of calcium salts and the color of the juice of II saturation [6, 7].

To destroy and remove part of non-sugars from solution by coagulation, deposition, or adsorption, it is necessary to use heat or chemical reaction energy, which is carried out according to the typical technological scheme of juice purification in the conditions of the previous and main defecation [8].

The heat energy is brought to the juice at all stages of purification, mainly through recuperative heat exchange equipment [9]. However, as practice suggests, the heating of juice through the heat exchange surface is not able to induce radical changes in the hydrated substances of the most common associate in the juice system, or in real complex compounds formed with the metals and other, mostly organic compounds participation which are always present in juice [10]. This led to the fact that the radical destruction of natural associative in known purification methods is realized through the chemical reaction energy use, which is from reagent juice processing [11].

But this method is quite long and requires much chemical reagent consumption. At the same time, as the experiment proves, the destruction of associated structures is possible by another method, namely by the water steam introduction to the juice [12]. This is especially noticeable in the method proposed to clean the raw cane juice "mesclado" that is in the production where the consumption for cleaning are 0.03% to the mass of raw materials (while in sugar production they are 3.0%) [13]. During the method application, an increase in the purity of juice in more than two units, and the sedimentation rate of the coagulated particles colloidal dispersion substances were noted [14].

But in the mentioned works (especially on the injection of water vapor), the physical essence of the effects and the mechanisms of their action on the physico-chemical transformations of non-curved diffusion juice is not considered, but only the positive consequences taking place, in this case, are given [15]. Besides, the literature does not contain data comparing the effects of hydrodynamic (HD) or vapor condensation (VC) cavitation processing of juices in terms of improving the purification efficiency and research to establish the optimal place in the technological scheme of purification and has become the subject of our study [16].

During condensation of high-potential vapor (160–180 °C), energy is released, which causes the dehydration of colloidal dispersion substances. This is especially true for protein and pectin substances. This provides them with complete coagulation, which is why the physic-chemical properties of beet juice are improved. Sugar beet juices are considered as a difficult mixture of solutes. Some of the sugar sugars may be associated because they are in the form of not very stable complexes. Our scientific hypothesis is that under the influence of a strongly dispersed pair, the schedule of some associates will pass. They will have a simpler structure and will become anionic and can be precipitated with calcium ion in the form of lime milk. This should change the electrical conductivity of the juice. Therefore, the purpose of our studies was to determine the electrical conductivity in cell and diffusion juices using the conductivity method and to establish different values of conductivity in the conventional heating of diffusion juice and diffusion juice treated with steam.

2 Literature Review

Positive results were obtained in case of the diffusion juice treatment with water steam and simultaneous lime milk introduction. Using this method allowed not only to increase the purity of the juice and to reduce its color but also to reduce the lime milk consumption for cleaning [17].

The works mentioned above cover mainly the final technological result. Still, the physical nature of the effects arising during hydrodynamic or steam-condensation cavitation treatment and the mechanisms of their manifestation for the diffusion juice compounds physic-chemical transformations, which are necessarily taking place, are not considered. This became the subject of our further research.

Physical phenomena that cause intense destructive or intensifying actions, despite the manifestation mechanism and forms diversity, unite the general pattern: they arise in liquid environments during a sudden change in external pressure and are accompanied by intense grows or splashes the formed bubbles if they are contained in a liquid [18]. A distinctive feature of these phenomena is the space-time localization of energy, which makes it possible to generate directed pulses of high power at a relatively low energy level [19].

The hydrodynamic cavitation effect intensity depends on the energy released by the collapse of the cavitation bubbles of appropriate sizes and concentrations. As a result, the reaction of the treated medium and other physical and chemical properties is changing [20].

The information about the steam-condensation cavitation effects influences on the treated medium is not found in literary sources. It is possible that by analogy with hydrodynamic cavitation, the energy released by the steam bubbles destruction is sufficient to destroy part of the complex and associated compounds. In this case, the components released under these conditions will be ion-carriers and will be able to participate in reactions with the calcium ion. Indirect evidence of such destruction can be the change of the electrical conductivity of juice, which was determined in the laboratory of the Processes and Equipment for Processing of Agricultural Production Department the National University of Life and Environmental Sciences of Ukraine.

3 Research Methodology

The objects of research are diffuse juice, juice of preliminary defecation, juice of the first and second saturation. Diffuse juice is processed in a hydrodynamic cavitation installation under optimal conditions. Samples of diffusion juice according to standard methods in laboratory conditions are processed before the juice of preliminary defecation and the first and second saturation.

For preliminary defecation juice and juice of the first saturation, the sedimentation rate and juice sedimentation volume after settling are determined after 20 min; for the second saturation juice, the content of calcium salts in percent by weight of the juice and its purity is determined.

It is well-known that the shock-wave effect of a hydrodynamic cavitation field on the medium is processed, determined by the stage of cavitation. In this case, the effective regime of cavitation mixing and dispersion corresponds to the maximum of the shock-wave action of cavitation field bubbles on the medium, is being processed. Therefore, the cavitation stage λ is used by us as a determining parameter for characterizing the operation mode of a hydrodynamic cavitation device and ranged from 0.62 to 4.0. This choice is due to the fact that at $\lambda = 1.0$ a flow regime arises, which is characterized by a bubble form of cavitation, which can be considered transitional from turbulent to cavitation. At $\lambda = 4.0$, which is characterized by a mixed form of cavitation, a super-cavitation regime begins to form.

Processing of the medium in the working area of the hydrodynamic cavitation installation occurs under the influence of cavitation bubbles, splashing. Therefore, the number and size of the formed cavitation bubbles are the determining factors of the technological efficiency of the cavitation treatment. In turn, the structure of the field of cavitation bubbles depends on the hydrodynamic parameters of the process, the main of which are the flow rate in the gap between the cavitator and the wall of the working section of the cavitation device and the stage of cavitation λ [20]. Therefore, the establishment of the hydrodynamic operating conditions of the installation was first carried out on barometric water with a temperature of 50 °C, which is similar for diffusion juice (diffusion column). The compression ratio of the flow in the working area of the laboratory setup is changed by establishing cavitators of different diameters with a Reynolds number of (23.6–25.4) 10^4. The cavitation stage was also determined on the water by visual measurement of the cavitation length in its characteristic radius.

The electrical conductivity of juices was measured using an AC bridge P5021 with zero indicators and signal generator GZ-33. For the control of the values, parallel correctness measurements were performed using the KHL-1M conductivity meter. The difference between a bridge and a conductivity meter lies in the fact that while determining the electrical conductivity of the AC bridge, both the active and reactive resistance components are taken into account, and the voltage is applied to the electrodes at the frequency of 1 kHz. This minimizes the electrodes polarization result effect and hence destruction or formation of new substances in solution. Measurements were made in a particular electrochemical cell in which the temperature of the juice was maintained constant using the thermostat.

4 Results

Cellular and diffusion beet juices were used for the electrical conductivity study. Samples were taken in a 40 ml volume of each juice, and electrical conductivity was measured. In Fig. 1, it is evident that the electrical conductivity of diffusion juice is higher than the electrical conductivity of cellular juice. This can be explained by the fact that a part of the cell juice associate collapses during extraction in a diffusion apparatus, which causes the release certain of its constituent elements, which are electric charge carriers. The same dependence was obtained for juice treated with different amounts of water steam (Fig. 2).

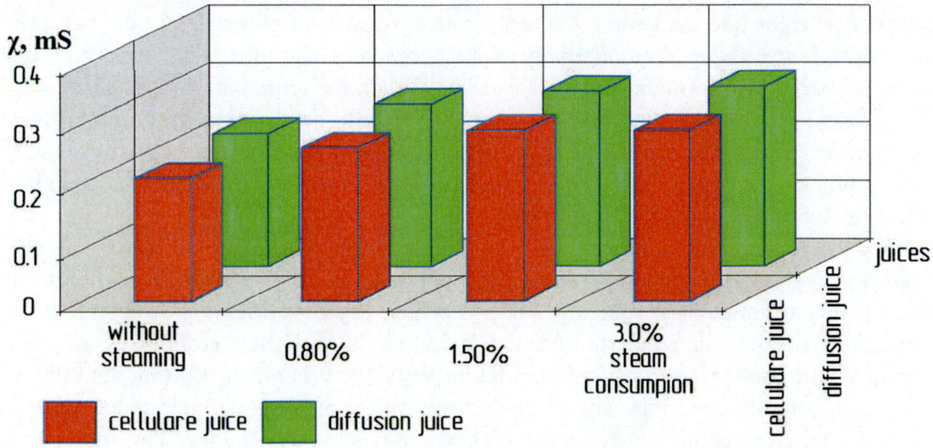

Fig. 1. The dependence of juices electrical conductivity on steaming them.

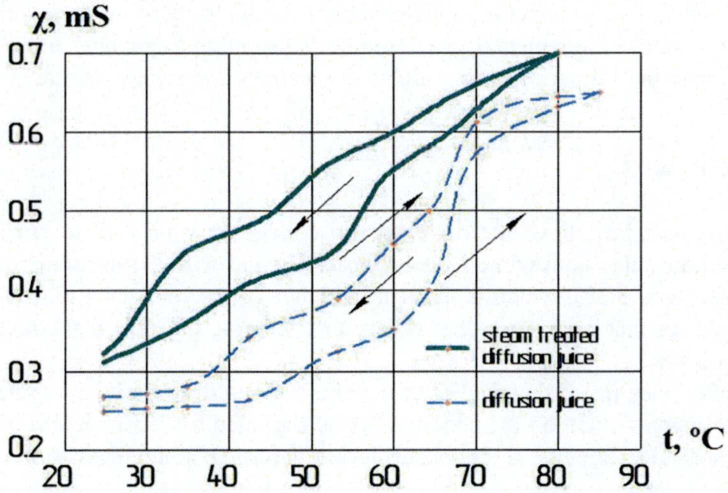

Fig. 2. The dependence of diffusion juice electrical conductivity on temperature during its heating and cooling.

Despite the dilution caused by the steam condensation, the electrical conductivity of the processed diffusion juice is always higher than the electrical conductivity of the raw juice.

To ensure that the steam-condensation cavitation effects lead to the above-described phenomena rather than heating, we conducted studies that determined the impact of juice heating through the heat exchange surface on the electrical conductivity of the treated and untreated diffusion juice.

The diffusion juice obtained in industrial conditions was heated to 85 °C in a water bath and then cooled to an initial temperature, and electrical conductivity was measured for each 5 °C. The data presented in Fig. 2 shows that during the heating of juice, the electrical conductivity dependence on heating at different temperatures has a different nature. Thus, when the juice is heated to 45 °C, this dependence is characterized by almost a straight line, and when the temperature rises from 45 to 65 °C, the nature of this dependence changes qualitatively, which indicates complex transformations that occur primarily with protein-pectin substances. When the temperature rises above 65 ° C, electrical conductivity increases dramatically due to an increase in the mobility of ions during temperature rising, as well as a decrease in the solution viscosity. During the cooling of the juice, the electrical conductivity decreases smoothly, and the characteristic schedule curvature is not present.

For juice with steam processing, the schedule characterizing the electrical conductivity dependence on temperature does not have clearly expressed bends during heating in the temperature range 45–65 °C which again verifies the structural transformations of the juice macromolecular compounds and their complexes structure during steam-condensation cavitation juice processing. When heated more than 65 °C, the dependence is quantitatively similar to the diffusion juice without processing. In general, electrical conductivity in the second case is always higher. The more rapid growth of electrical conductivity in the first stage can be explained by the high-molecular compounds associative collapse in diffusion juice under the steam-condensation cavitation effects influence with the release of ion carriers. This is evidenced by the less rapid increase in the electrical conductivity of the same juice at high temperatures as the fundamental processes of physical and chemical transformations occurred under the influence of the effects of steam-condensation cavitation.

5 Conclusions

The results of studies show that the electrical conductivity for diffuse juice is higher than for cellular juice both without steam treatment and with the introduction of water vapor from 0.8 to 3.3%, which is indirect evidence of the release of ok-rem elements from cells juices that have an electric charge and increase the electrical conductivity of the diffusion juice.

It has also been found that studies of the effect of heating the diffusion juice through a heat exchanger surface on the conductivity of the juice treated with and treated with water vapor lead to the effects that occur during vapor-cavitation cavitation rather than heating.

It is established that the cavitation treatment of diffusion juice before liming causes a significant acceleration of the course of the main reactions to the previous and main bowel movements and an increase in the overall purification effect. But the mechanism of the influence of cavitation phenomena on the components of diffusion juice remains unknown and will be the subject of our further research since without this, it is impossible to improve or optimize the processes of sugar beet production.

Thus it is established that due to the steam-condensation cavitation effects, the structural transformations of the diffusion juice macromolecular compounds and colloidal dispersion substances are occurring, which leads to an increase in the electrical conductivity of juice due to the colloidal dispersion substances disaggregation and the ion carriers components releasing.

References

1. Matyyashchuk, A., Khomichak, L., Nemirovich, P., et al.: Theoretical substantiation of steam injection for the diffusion juice purification. Express-news: Sci. Technol. Prod. 21–22, 9–10 (1997)
2. Dhar, B., Elbeshbishy, E., Hafez, H., Lee, H.: Hydrogen production from sugar beet juice using an integrated bio hydrogen process of dark fermentation and microbial electrolysis cell. Bioresour. Technol. **198**, 223–230 (2015)
3. Palamarchuk, I., Mushtruk, M., Vasyliv, V., Zheplinska, M.: Substantiation of regime parameters of vibrating conveyor infrared dryers. Potravinarstvo Slovak J. Food Sci. **13**(1), 751–758 (2019)
4. Khomichak, L.: Improvement of the technique and the device for the surface properties of saturation sediment determination. Sci. Works USUFT **4**(2), 79–81 (1998)
5. Sukhenko, Yu., Mushtruk, M., Vasyliv, V., et al.: Production of pumpkin pectin paste. In: Ivanov, V., et al. (ed.) Advances in Design, Simulation and Manufacturing II. DSMIE-2019. Lecture Notes in Mechanical Engineering, pp. 805–812. Springer, Cham (2020)
6. Lee, K., Morad, N., Teng, T., Poh, B.: Development, characterization and the application of hybrid materials in coagulation/flocculation of waste water: a review. Chem. Eng. J. **203**, 370–386 (2012)
7. Zheplinska, M., Mushtruk, M., Vasyliv, V., Deviatko, O.: Investigation of the process of production of crafted beer with spicy and aromatic raw materials. Potravinarstvo Slovak J. Food Sci. **13**(1), 806–814 (2019)
8. Kozelová, D., Mura, L., Matejková, E., et al.: Organic products, consumer behavior on market and European organic product market situation. Potravinarstvo Slovak J. Food Sci. **5** (3), 20–26 (2011)
9. Kim, J., Ghafoor, K., Ahn, J., et al.: Kinetic modeling and characterization of a diffusion-based time-temperature indicator (TTI) for monitoring microbial quality of non-pasteurized angelica juice. LWT-Food Sci. Technol. **67**, 143–150 (2016)
10. Lebovka, N., Shynkaryk, M., El-Belghiti, K., et al.: Plasmolysis of sugar beet: pulsed electric fields and thermal treatment. J. Food Eng. **80**(2), 639–644 (2007)
11. Matyyashchuk, A., Khomichak, L., Nemirovich, P., et al.: Hydrodynamic cavitation as one of the methods for intensification previous defecation. Sci. Works USUFT **4**(2), 83–85 (1998)
12. Ozerov, D., Sapronov, A.: Coagulation and aggregation of colloidal dispersion substances in pre-defecation. Sugar Ind. **8**, 24–27 (1985)

13. Sheiko, T., Tkachenko, S., Mushtruk, M., et al.: The Studying the processing of food dye from beet juice. Potravinarstvo Slovak J. Food Sci. **13**(1), 688–694 (2019)
14. Jiang, J.: The role of coagulation in water treatment. Curr. Opinion Chem. Eng. **8**, 36–44 (2018)
15. Guo, S., Luo, J., Yang, Q., et al.: Decoloration of sugarcane molasses by tight ultra-alteration: filtration behavior and fouling control. Sep. Purif. Technol. **204**, 66–74 (2018)
16. Johnson, B., Zhou, X., Wangersky, P.: Surface coagulation in sea water. Neth. J. Sea Res. **20** (2, 3), 201–210 (1986)
17. Loginova, K., Loginov, M., Vorobiev, E., Lebovka, N.: Quality and filtration characteristics of sugar beet juice obtained by "cold" extraction assisted by pulsed electric field. J. Food Eng. **106**(2), 144–151 (2011)
18. Almohammed, F., Mhemdi, H., Grimi, N., Vorobiev, E.: Alkaline pressing of electro orated sugar beet tissue: process behavior and qualitative characteristics of raw juice. Food Bioprocess Technol. **8**(9), 1947–1957 (2015)
19. Dalfré Filho, J., Assis, M., Genovez, A.: Bacterial inactivation in artificially and naturally contaminated water using a capitating jet apparatus. J. Hydro-Environ. Res. **9**(2), 259–267 (2015)
20. Luo, J., Guo, S., Qiang, X., et al.: Sustainable utilization of cane molasses by an integrated separation process: interplay between adsorption and nano filtration. Separ. Purif. Technol. **219**, 16–24 (2019)

Quality Assurance

Ensuring the Quality of Training Engineers in a Virtual Environment

Olena Bilous, Tetiana Hovorun, Kristina Berladir(✉), and Marina Dunaeva

Sumy State University, 2, Rymskogo-Korsakova Street, Sumy 40007, Ukraine
kr.berladir@pmtkm.sumdu.edu.ua

Abstract. The introduction of information technology in the educational process provides many opportunities for the development and implementation of modern methods of training future engineering professionals with the use of electronic educational content. Paper considers the peculiarities of the introduction of blended and distance learning technologies in Sumy State University into the educational process system. The stages of formation and application of virtual laboratory work in the educational process are considered, taking into account the formation of the professional competences of the engineer, the shortcomings and advantages of their use in training are revealed. It should be noted that the paper examines the introduction of a virtual simulator in the learning process to improve the efficiency of the learning process, improve the quality of acquisition of skills and knowledge, development of professional engineering competencies. The analysis of the research results suggests that the virtual objects of distance learning are somewhat imperfect in the question of forming the communicative competence of the future engineer. However, at the same time, they allow to expand the student's ability to prepare and analyze complex engineering problems, to evaluate and select the necessary information, apply the required theoretical and practical methods for the analysis of complex engineering developments. This study provides recommendations for improving the effectiveness of learning through the introduction of virtual objects in the study of engineering disciplines.

Keywords: Selective training content · Virtual laboratory work · Electronic simulator · Virtual training facility · Engineer training

1 Introduction

The transition to a new level of the education system becomes relevant due to the introduction of modern methods of informatization of the learning process. They provide an opportunity to move from traditional paper media, textbooks, books to electronic learning complexes, virtual simulators, multi-level tests, electronic laboratory work, from the visual audience to the multimedia.

Dissemination of information technology in the educational process creates conditions for the development of creative skills, forms the ability to analyze and forecast processes in the study of professionally oriented disciplines. Distance learning

V. Tonkonogyi et al. (Eds.): InterPartner 2020, LNME, pp. 765–774, 2021.
https://doi.org/10.1007/978-3-030-68014-5_74

technologies allow us to implement the tasks of quality training of future engineers effectively.

Blended and distance learning is relevant today in the training of engineers. Information training technologies solve several problems related to the provision of laboratories with modern technical equipment, the presentation of efficient production technologies, the providing of relevant information, and the ability to study the latest developments and research. These aspects underlie the formation of professional competencies in engineering students. It should be noted that one of the areas of modernization of the educational process is the transition to competence training of future professionals.

Techniques and methods of organizing the cognitive activity of students form the effectiveness of the educational process, which can be assessed by examining the relative change in learning outcomes over time [1].

Of course, when organizing training with the use of information educational technologies, special attention should be paid to electronic learning objects, such as, for example, interactive tests, simulators, virtual laboratory works, etc. [2, 3]. A well-designed distance course should provide conditions for the organization of the learning process, the effectiveness of which is at least comparable to the efficacy of traditional education.

2 Literature Review

The issue of the introduction of electronic learning objects in the educational process is continuously in the field of view of scientists-teachers [4–6]. The possibility of intensification and increase the efficiency of the educational process through the introduction of distance technologies in Ukraine are studied by O. Bondarenko, V. Zabolotny, O. Pinchuk, etc.

Authors [7] presented a study of the effectiveness of the introduction of virtual learning objects (VLO) in the study of differential calculus. An analysis of learning outcomes showed the improvement of theoretical knowledge acquired in lessons with the use of such technologies. Paper [8] is devoted to the didactic principles for the construction of laboratory workshops on physical disciplines. Article [9] aims at the modernization of online education as one of the most perspective ways for the development of modern educational systems.

Features of developing the virtual practical and interactive tools are considered in the works by P. Mazur, S. Petrovsky, M. Yanovsky, A. Kalashnikov, L. Polischuk, D. Rodkin, V. Evstifeev, and O. Chorny investigate the problems of creating and using the universal virtual complexes for carrying out laboratory classes according to the cycles of technical disciplines.

In the system of the university, it is necessary to solve the tasks of vocational education by the competence approach, which emphasizes the formation of future specialist readiness for the practical application of knowledge, skills, and abilities in the context of solving real production problems.

It was investigated that interactive educational materials attract special attention and interest to both students and teachers. Such objects include virtual laboratory work

(VLW), interactive simulators, and test tasks. Also, such objects are the main components in the organization of blended and distance learning [10, 11]. It should be noted that modern students have information technology since school, which significantly facilitates the introduction of new elements of learning in the educational process. Students are prepared to work in a virtual laboratory, or to perform tests, simulators.

The analysis of existing pedagogical studies indicates that the question of forming the professional qualities of a specialist in the engineering profile using information technology during the study of distance learning has not been sufficiently studied. The problem of research of features of the formation of professional competencies at students of engineering specialties at the organization of training on the application of remote technologies remains actual. Also relevant is the question of researching the process of forming exactly those qualities of the future engineer, which can be formed only when applied in the training of virtual electronic laboratory work, simulators, etc.

The purpose of the paper is to analyze the peculiarities of the introduction of distance learning information technologies and to investigate the effectiveness of the formation of engineering competencies in future professionals.

3 Research Methodology

There are e-Learning platforms to support distance learning. They are developed with the support of various database management systems and application servers [12, 13]. To work in distance courses and to communicate with students, it is possible to use a variety of electronic learning tools and online resources, both for individual and joint learning.

E-Learning platform at Sumy State University includes modern remote technologies, for example, designer of educational materials Lecture.ED, which allows you to create interactive web pages, tests, forums and tasks for collaboration; the MiX learning platform, which is an online environment for all types of communication between students and teachers, allows you to monitor the progress of each student and the discipline as a whole; open resource of educational materials "OCW SSU".

The methodical model of online courses stimulates a high level of interactivity of educational content on a wide application of virtual simulators based on Java, JS, Flash, Unity3D, including the use of technologies of virtual and augmented reality [14–19].

4 Results

The teaching staff of Sumy State University is actively developing and studying the peculiarities of the organization of distance learning and mixed technologies. Thus, several electronic primary objects have been introduced into the educational process, including virtual laboratory works (Fig. 1).

Laboratory work plays a significant role in training technical specialists. The work in the laboratory promotes the formation and development of the relevant competencies in the engineer. In the laboratory, students get the experience in studying the properties

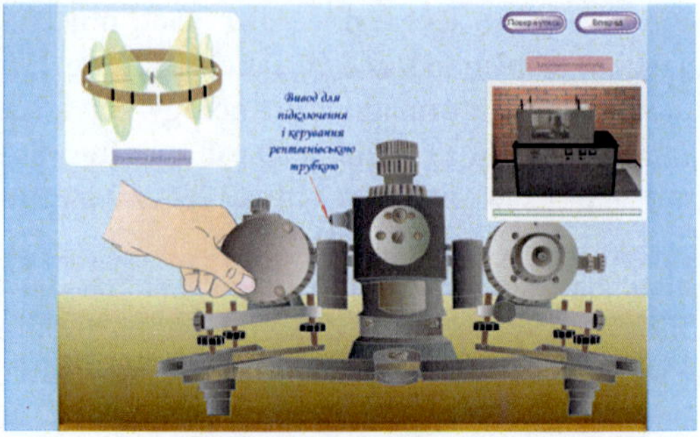

Fig. 1. Virtual laboratory work of the course "Methods of Structural Analysis of Materials".

of the materials, studying the operation of devices, learn to process and interpret the results. Carrying out the work requires a student's ability to focus attention, use the instruments in the research, the ability to find the necessary information. Carrying out the laboratory work allows developing such competencies in future specialists: research, search, information-organizing, appraisal-analytical. When performing a laboratory practice, a worldview and communicative competence develop. Besides, the projective, constructive, research, search, heuristic, information-organizing, assessment-analytical, and other competencies necessary for the successful work of a graduate at an institution of higher education, regardless of which professional direction he chooses for himself in the field of modern engineering, will become in demand.

In determining the competence of a university graduate as a future specialist, it is necessary to consider his personal qualities that manifest themselves in the approach to performing professional duties. Personal qualities of the graduate are needed in the modern market, and this is not only a conscientious execution of the regulated duties of a specialist but also innovative thinking, creative inclinations, a manifestation of communicative, organizational, projective, prognostic and other professional and personal competences.

The creation and maintenance of modern working laboratory equipment is a time-consuming and costly process that many educational institutions cannot afford. Some of them are forced to refuse such work on real physical stands, replacing them with virtual laboratory workshops. However, they can exist in addition to the development of real workshops, but not completely replace them. Complete replacement of the physical laboratory to the virtual one will not ensure the acquiring skills and abilities of the future engineer. Combining virtual work with "physical" one, the advantages of student contact with real laboratory equipment and capabilities of using computer tools are united (multimedia technologies, Internet, etc.).

On the other hand, if the virtual one at the universities replaces all real laboratory work, it will lead to separating students from real situations and equipment. When performing, the VLW does not develop practical skills in measuring physical quantities with the use of devices and equipment, the skills of experimenting, the assembly of electrical circuits, etc. Unacceptable is the training of a specialist who cannot work with real objects. Often, being able to work well with the VLW, a modern student practically does not have practical skills to work with real models, and we consider this a significant problem in the education of a modern student.

Computer laboratory installations in virtual laboratories, as a rule, represent a computer model of a real experimental installation. Since virtual laboratory work (Fig. 2) in the distance course is only a model, it does not always reflect the specific properties of the phenomenon or object under study. It can be considered as some limitation of presenting the real environment through electronic capabilities. Also, the lack of virtual work, to some extent, can be considered that it is an individual performance of laboratory work by a student at home or in a laboratory Specialist who graduated from an institution of higher education must live and work in a society. It is necessary to teach students to work in a group, a team, to be able to design and set experimental tasks and translate them into reality. Furthermore, this is possible only when performing real work. It is impossible, however, not to mention several advantages of the VLW against traditional methods of laboratory work. Laboratory equipment in educational laboratories is not updated as often as we would like. Some works cannot be carried out in training laboratories; for example, many works on quantum, atomic and nuclear physics. Computer simulation allows you to do without expensive equipment and enable you to recreate almost any real physical model.

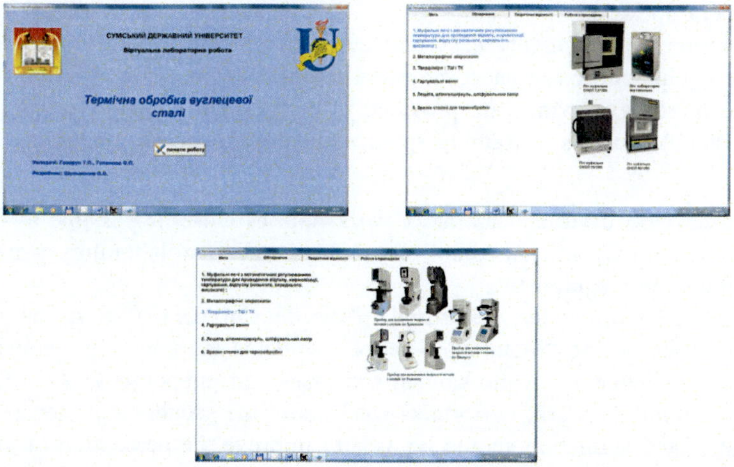

Fig. 2. Virtual laboratory work "Thermal Treatment of Carbon Steel" of the course "Thermal Treatment".

Realization of laboratory work, real or virtual, consists of several stages. When performing both real and virtual laboratory work, virtually all the stages mentioned above coincide.

However, the main difference is that the VLW is performed individually, and real laboratory work is performed in groups of 2–3 people. Work in the group forms the ability for students to work together, a sense of collectivism, and responsibility, which is a necessary factor in their further professional activity. We also note the distinction when performing the experimental part of the work.

For example, the current laboratory practice on the discipline "Materials Science" is a traditional approach to laboratory work. Nevertheless, the state of the equipment used and its quantity leads to the difficulties or impossibility of a personal experiment in the laboratory on the specific features of the equipment operation (increased danger of thermal installations, the toxicity of cooling media, considerable time of conducting thermal treatments, high probability of failure of equipment elements, which is exploited); significant energy costs, lack of a sufficient number of options for conducting experiments; the complexity of carrying out phased control over the execution of work.

The advantages of using simulators (even compared to working on real laboratory stands) are:

- intensification of learning without losing the quality of mastering the material;
- the possibility of conducting laboratory work by the frontal method (all students simultaneously perform one task), which significantly increases the efficiency of this type of training;
- the possibility of a wide variation of the experimental conditions;
- the possibility of modeling and safe research of extreme and emergency modes of equipment operation;
- the possibility of relatively easy and quick modification of the elements of the studied equipment to the latest industrial designs;
- significant energy savings are provided in comparison with the use of real laboratory stands, savings in training space, reduction in capital, operating, and other costs.

A well-developed virtual laboratory workshop, as close as possible to the conditions of a real experiment, can completely replace traditional laboratory work with the use of real complex equipment.

The following principles were used in the development of a virtual laboratory workshop on the course "Materials Science": students must have previously formed theoretical knowledge about the studied processes and phenomena; according to the purpose of laboratory work, performers can choose the sequence of steps; video and photo images and animation should be used to improve the perception of processes.

For example, consider the process of virtual laboratory work on heat treatment of steels, which includes the following steps:

Step 1 – the student determines the practical purpose of heat treatment of steels, its types, and basic characteristics, gets acquainted with the necessary equipment for heat treatment, can conduct a demonstration process of one of the options for the heat treatment of steels;

Step 2 – the student randomly chooses one of the options of the material and the proposed heat treatment, determines the heating temperature of the steel for the recommended type of heat treatment and the kind of medium for cooling;

Step 3 – the student virtually on the simulator conducts the heat treatment process, builds graphs for heat treatment and performs identification of the structural-phase state of the material after heat treatment, at each stage in parallel determining the hardness;

Step 4 – the student analyzes and explains the results, compiles an electronic report, which indicates the purpose of the work, the graphs of heat treatment, provides images of structures after heat treatment.

The logic of presenting material in a virtual laboratory work differs from real work by a more detailed description of the research process, an abundance of tips and links, and the presence of animation.

Virtual work requires greater clarity in describing the sequence of actions; therefore, it is methodologically justified to present this kind of work in the form of a certain number of sections – tabs, each of which has its semantic load:

1. Theoretical material – for the successful completion of any laboratory work, the student must carefully work out the theoretical material on the research topic; therefore, in virtual laboratory work, a section with a similar title should be presented in more detail than in a classical workshop.
2. Description of work – the purpose of the laboratory work is formulated, a diagram of the installation, calculation formulas are provided, and work with graphs is described.
3. The order of work – students receive step-by-step instructions for performing laboratory work.
4. Laboratory equipment – the student independently goes through all the stages of heat treatment.
5. Report – the student completes the required items and draws graphs.

Virtual thermal devices identical to the real ones are given for the maximum approximation of virtual work performed on real equipment (namely, models and photos of real heating furnaces and measuring equipment (Fig. 2)) and equipment control algorithm, temperature control, and measurement, the transmission of the color change of the heated sample, equipment for loading into furnaces, hardness testers, etc.).

Yes, the simulator allows you to create the properties of the investigated processes, objects, and images that do not exist, perform virtual actions (Fig. 3).

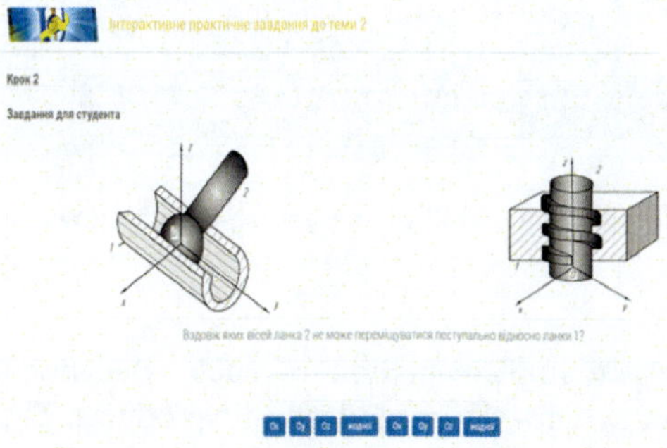

Fig. 3. Interactive practical task "Kinematic Pairs and Kinematic Chains" of the course "Theory of Mechanisms and Machines".

Simulators give the chance to use various forms for presentation of training material, control of the performance of operations, and estimation, depending on the results of work.

With the help of a virtual simulator, it is possible to:

- set the algorithm for actions and operations;
- to implement an individual approach in the education of each student;
- rationally present the theoretical and practical material of the discipline;
- motivate students to study;
- to expand the possibilities of visual presentation of educational content with the help of video blocks.

The virtual simulator consists of several blocks. Each of them has a task. The transition to the next unit of the simulator is possible only if the previous task is performed correctly. While working with the simulator, the student has the opportunity to ask questions to the teacher, view the theoretical material of the discipline. As a rule, typical tasks in the simulator are enough for the organization of individual work of students of the group. It is especially effective to use simulators to perform tasks consisting of several stages or steps of work. The introduction of simulators in the educational process forms (Fig. 2–4) the skills and abilities of theoretical and practical direction. It contributes to the development of the relevant competencies of future engineers.

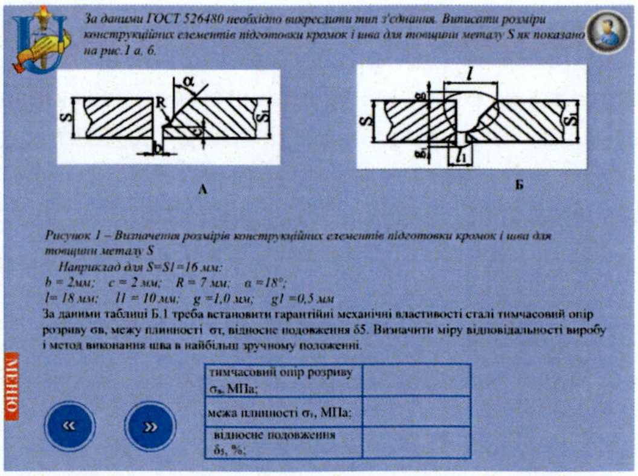

Fig. 4. Simulator "Welding Production" of the course "Technology of Construction Materials and Materials Science".

The development and implementation of such educational content make high demands on the teacher as a developer of electronic products. The teacher must have an idea of the capabilities of the simulators, form the stages of development of events in the performance of the tasks of the simulator, present the criteria for evaluating the results of the student's work.

5 Conclusions

Distance learning technologies are becoming an integral part of the modern training of engineers. The analysis of the research results suggests that the virtual objects of distance learning are somewhat imperfect in the question of forming the communicative competence of the future engineer. However, at the same time, they allow to expand the student's ability to prepare and analyze complex engineering problems, to evaluate and select the necessary information, apply the necessary theoretical and practical methods for the analysis of complex engineering developments.

Analysis of the study of the problem allows us to draw the following conclusions:

- introduction of virtual laboratory work promotes the development of exploratory, philosophical, constructive and projective competence;
- interactive simulator increases the effectiveness of training, develops professional-oriented skills and abilities;
- electronic content allows you to implement an individual approach to the organization of education of each student.

At the same time, some of the shortcomings of applying virtual laboratory work are highlighted, namely:

- virtual laboratory work somewhat limits getting the skills and competencies of the research, creative plan because the work is carried out within the programmed algorithm of actions;
- the implementation of only virtual laboratory workshops within any engineering discipline complicates the stage of adaptation of the graduate engineer in real production and communication in the absence of skills in working with real engineering equipment, limit the formation of functional competence;
- student's training in the virtual space does not allow the development of communicative skills, lack of experience in the team.

In further research, it will be useful to expand the list of e-learning objects of distance and blended learning technologies and study their influence on the formation of relevant competencies of future engineers.

References

1. Kuzmina, N.: The effectiveness of the process of teaching and learning. Eastern Eur. Sci. J. **5**, 121–126 (2014)
2. Burke, R.D., Jonge, N.D., Avola, C., Forte, B.: A virtual engine laboratory for teaching powertrain engineering. Comput. Appl. Eng. Educ. **25**(6), 948–960 (2017). https://doi.org/10.1002/cae.21847
3. Grodotzki, J., Ortelt, R.T., Tekkaya A.E.: Remote and virtual labs for engineering education 4.0: achievements of the ELLI project at the TU Dortmund university. Proc. Manuf. **26**, 1349–1360 (2018). https://doi.org/10.1016/j.promfg.2018.07.126.
4. Fiialka, S., Onkovych, H., Baliun, O.: The use of modern teaching methods in editor education in Ukraine. Adv. Educ. **7**, 57–63 (2017). https://doi.org/10.20535/2410-8286.88019.

5. Zhai, G., Wang, Y., Liu, L.: Design of electrical online laboratory and e-learning. In: International Conference on Future Computer Supported Education, vol. 2, pp. 325–330 (2012)
6. Hongjun, Z., Zhi, Y., Youming, X., Yongyou, W., Lu, K.: Virtual emulation laboratories for teaching offshore oil and gas engineering. Comput. Appl. Eng. Educ. **26**(5), 1603–1613 (2018). https://doi.org/10.1002/cae.21977
7. Arango, J., Gaviria, D., Valencia, A.: Differential calculus teaching through virtual learning objects in the field of management sciences. Proc. – Soc. Behav. Sci. **176**, 412–418 (2015). https://doi.org/10.1016/j.sbspro.2015.01.490.
8. Kurilovas, E., Kubilinskiene, S., Dagiene, V.: Web 3.0 – based personalisation of learning objects in virtual learning environments. Comput. Hum. Behav. **30**, 654–662 (2014). https://doi.org/10.1016/j.chb.2013.07.039.
9. Salaimeh, S.A., Hjouj, A.A.: Visual object-oriented technology and case-tools of developing the Internet/Intranet-oriented training courses. J. Eng. Sci. **4**(2), H9–H11 (2017). https://doi.org/10.21272/jes.2017.4(2).h9.
10. Onime, C., Uhomoibhi, J., Zennaro, M.: Demonstration of a low cost implementation of an existing hands-on laboratory experiment in electronic engineering. In: 11th International Conference on Remote Engineering and Virtual Instrumentation (REV), pp. 195–197. Inst Super Engn Porto, Porto, Portugal (2014)
11. Prasetya, D., Wibawa, A., Hirashima, T.: An interactive digital book for engineering education students. World Trans. Eng. Technol. Educ. **16**(1), 54–59 (2018)
12. Švač, V., Cagáňová, D.: Managerial skills for innovation support. Mobile Netw. Appl. (2020). https://doi.org/10.1007/s11036-020-01517-3
13. Cagáňová, D., Čambál, M., Weidlichová Luptáková, S.: Intercultural management - trend of contemporary globalized world. Electron. Electr. Eng. **6**(102), 51–54 (2010)
14. Ivanov, V., Pavlenko, I., Trojanowska, J., Zuban, Y., Samokhvalov, D., Bun, P.: Using the augmented reality for training engineering students. In: Bruzzone A.G., et al. (eds.) Proceedings of the 4th International Conference of the Virtual and Augmented Reality in Education, VARE 2018, pp. 57–64 (2018)
15. Ivanov, V., Pavlenko, I., Liaposhchenko, O., Gusak, O., Pavlenko, V.: Determination of contact points between workpiece and fixture elements as a tool for augmented reality in fixture design. Wirel. Netw. (2019). https://doi.org/10.1007/s11276-019-02026-2
16. Bilous, O., Govorun, T., Berladir, Ch., Dynaeva, M.: Peculiarities of implementation in educational process of electronic interactive materials for students of engineering specialties. Eng. Educ. Technol. **2**(22), 40–49 (2018). https://doi.org/10.30929/2307-9770-2018-22-40-49
17. Bun, P., Trojanowska, J., Ivanov, V., Pavlenko, I.: The use of virtual reality training application to increase the effectiveness of workshops in the field of lean manufacturing. In: Bruzzone A.G. et al. (eds.) Proceedings of the 4th International Conference of the Virtual and Augmented Reality in Education, VARE 2018, pp. 65–71 (2018)
18. Denysenko, Yu., Ivanov, V., Ivchenko, O.: Quality assessment of teaching the disciplines in the e-learning environment of sumy state university. In: CEUR Workshop Proceedings. 13th International Conference on ICT in Education, Research and Industrial Applications. Integration, Harmonization and Knowledge Transfer, ICTERI 2017, vol. 1844, pp. 166–175 (2017).
19. Korotun, M., Denysenko, Yu., Malovana, N., Dutchenko, O.: Improvement of the effectiveness of general engineering courses using trainers. In: Ivanov, V., et al. (ed.) Advances in Design, Simulation and Manufacturing. DSMIE-2020. Lecture Notes in Mechanical Engineering (2020). https://doi.org/10.1007/978-3-030-50794-7_3.

Ring Laser for Angle Measurement Devices

Irina Cherepanska[1(✉)], Olena Bezvesilna[2], Artem Sazonov[2], Petro Melnychuk[3], and Valerii Kyrylovych[3]

[1] Zhytomyr National Agroecological University, 7, Staryi Blvd, Zhytomyr 10008, Ukraine
cherepanskairina@gmail.com

[2] National Technical University of Ukraine "Igor Sikorsky Kyiv Polytechnic Institute", 37, Prosp. Peremohy, Kyiv 03056, Ukraine

[3] Zhytomyr Polytechnic State University, 103, Chudnivska St., Zhytomyr 10005, Ukraine

Abstract. Angle measurement means are one of the advanced directions for the application of gas ring lasers. The requirements of ring lasers used in angle measurement devices are different, in many respects, from those used in navigation. A simple ring laser developed for implementation in high-precision angle measurement instruments is described. The main specifications are presented. Theoretical and experimental studies and computer simulation of ring laser parameters are presented. The studies were carried out, taking into account changes in the Earth's rotation speed around the axis and without taking into account changes in the Earth's rotation speed. The obtained results indicate that the change in the speed of rotation of the Earth affects the accuracy of a ring laser. In order to reduce this error, it is necessary to fulfill the requirements of the rotational axis of the ring laser relative to the rotational axis of the Earth. Methods of the accelerated tests are imperfect for determining a term of storage. Therefore, tests during a real storage term remain the most reliable. The operation of the ring lasers in angle measuring instruments for more than 20 years has demonstrated their high stability. Using the ring laser, the angle measurement means of accuracy that exceeds the accuracy of the existing National Standards for a plane angle can be developed.

Keywords: Error · Earth rotation velocity · Measurement accuracy · Design of measuring system

1 Introduction

In high precision angle measurement means, photoelectric angle converters, inductive or capacity converters, as well as limbs, are used as angle sensors. Manufacturing of the sensitive elements (scales) for such sensors is performed on special equipment such as dividing machines. In this case, an error of the sensors includes an error of manufacturing equipment.

Laser Goniometric Systems provide the high accuracy of the angles contactless measurements. They use the fundamental features of the Ring Semiconductor Lasers (RL). Many semiconductor lasers have been developed for the practical purpose within

V. Tonkonogyi et al. (Eds.): InterPartner 2020, LNME, pp. 775–784, 2021.
https://doi.org/10.1007/978-3-030-68014-5_75

last few decades such as Fabry–Pérot interferometer, lasers with distributed feedback, also with distributed Bragg mirrors as well as Fiber Bragg grating as a part of fiber optic gyro [1], and external hybrid mirrors, also vertical-cavity surface-emitting lasers (VCSEL). Thus, laser characteristics are explored by scientists [2, 3].

A ring laser comprises an angular scale set by the wavelength of laser radiation. This angular scale is a qualitatively different one in which the errors of dividing machines are absent. The use of such a scale improves the parameters of angle measurement, essentially means such as accuracy, operating speed, and measurement authenticity.

Ring lasers can be used in angle measurement devices for different purposes. For example, it is used as a basic component of an intelligent precise goniometric system [4]. The last one allows measuring with accuracy about 0.1″ in single measurements and 0.08″ in multiple ones.

However, during angles measuring by the transducers, which are based on the RL, some errors are occurred in instability of RL rotation velocity and caused by the influence of external magnetic fields and primarily the Earth's magnetic field.

The research is aimed to review main construction principles of the RL 3.970.029 which is developed by ARSENAL SDP SE (Kyiv, Ukraine), to show error analysis of the RL as well as results of the experimental examination and computer simulation of its parameters (drift of scale factor and zero of output characteristic) which are affected by Earth rotation velocity instability, to define the ability of the RL 3.970.029 use as a precise angles transmitter in the systems of their contactless measuring with high accuracy, to develop the recommendations for errors decreasing and accuracy of angles measuring improving by use of that ring laser.

2 Literature Review

Significant attention is paid to the development of goniometers. For example, in [5–13], the design of goniometers, as well as test results, are described. In [5], the concept of constructing precision laser goniometric systems based on the combination of a ring laser and an optical angle sensor with the holographic principle of recording the angular scale is considered. In [6], this paper presents the results of experimental and modeling metrological studies on a digital goniometer, measuring angular orientations of an image of an optical mark using a digital camera array are presented in [6] of this paper. Fully digital technology is also described for measuring angles using virtual angular-scale carriers (digital images). By using a mathematical model for the goniometer, it is shown that the error in angle measurements can be reduced to the level of thousandths of an Arcsec in principle. In [7] presents the main characteristics of a new kind of laser goniometer being developed as the result of a collaboration between INRIM and INFN. The target is an absolute accuracy of 10 nrad, being the present accuracy of the most precise angular encoders at the level of some 100 nrad. Our key idea is to setup a square laser cavity of 0.5 m in the side, making use of the last generation dielectric super-mirrors applied in the larger gyroscopes for seismology and geophysics. It [8] became necessary for proposing a remote non–contact method to measure angular positions and movement of objects using Laser Dynamic Goniometer (LDG) as compared with the usual Photo-electrical autocollimators with a narrow range of about 1° in order to propose a remote non-contact method. This article presents the analysis, the errors as well as the experimental results of using Laser Dynamic Goniometer to measure wide range with an accuracy of approximately 0.1 arcs and a possibility of measuring constant angles with an accuracy of 0.005 ...0.1 arcs in the range of possible angles of 15...30°. We consider the fundamental physical principles of Ring Gas Lasers (RGL) in the mode of laser gyroscope in [9]. The impact of nonreciprocal effects on the parameters of RGL generation and the methods of RGL parameter control in the mode of laser gyroscope is discussed. It is performed by applying the Zeeman non-reciprocal magneto-optic effect. The first commercial automatic angle measurement system (goniometer-spectrometer) GS1L is described in [10]. The system is designed for the measurement of plane angles, and pyramidally of the prism faces, as well as the refractive index of optical media. This system is produced in lots and widely used at many plants and metrological centers.

The ring lasers designed for navigation were used in the first angle measurement devices. Ring laser gyroscopes are one of the most accurate sensors for angles measuring. However, due to the instability of their parameters (drift of scale factor and zero of output characteristic), they are still under research. In [11] is shown that the problem of instability of ring laser parameters is minor with proper analysis of data. Those results allow not only improving their functional parameters but also pave the way for the development of small-sized In [12] described the experimental and theoretical results of the development of solid ring laser gyroscope (RLG), which are already obtained as well as perspectives of their application in future. In [13] shown the method of ring laser error calibration in frequency shift mode.

3 Research Methodology

3.1 Features of Ring Laser Applications in Angle Measurement Means

It is known that the majority of gas ring lasers are used in military and civil navigation systems. The requirements of these ring lasers are high. For example, they must have a minimum lock-in zone, a special frequency separation unit for low angular rate measurements, they also must provide the operation in a wide temperature range, in conditions of shocks, vibrations, radiation exposure, etc. The most important characteristic of the ring laser is scale factor stability during a long period of time. There are rigid weight and dimension restrictions for such devices. Therefore, the ring lasers used in navigation systems are expensive.

The requirements of parameters of ring lasers used in industrial angle measurement means are different in many aspects. Generally, such devices operate in a narrow temperature range. They are not exposed to vibrations and shocks during the measurement process. Generally, they do not require a frequency separation unit. The implementation of the self-calibration method [7] in angle measurement devices such as goniometers allows decreasing the requirements for the long-term stability of the ring laser scale factor. For most applications, the requirements of minimum dimensions and weight are not the basic ones. Therefore, the design of such lasers is simpler and cheaper. Furthermore, their cost falls due to a decrease in tests number.

At the same time, additional requirements are imposed on such lasers. For example, in many cases, such as expensive, high-precision angle measurement devices operate for several decades. Therefore, the ring lasers implemented must have a service life of the order of 20–30 years. An operating life must be from several thousand to tens of thousands of hours. To obtain high precision, the ring laser must have a high angular resolution.

The precision of modern RL is quite high, but more significant errors characterize it. The precision of angle measurement by measuring means based on RL is determined by many factors conditioned with the parameters of subsystems of such a device. For example, in the case of a goniometer, it is non-stability of rotation velocity of a rotating device, inaccuracy of registration the beginning and the end of counting the controlled angle, the discretion of information signal quantization, etc. While measuring the general error of angles, RL introduces its components conditioned by nonlinearity and output characteristic zero shift, non-stability of scale coefficient, ratio signal/noise of information signal, etc. The structural scheme of fluctuation processes and errors of RL is presented in Fig. 1.

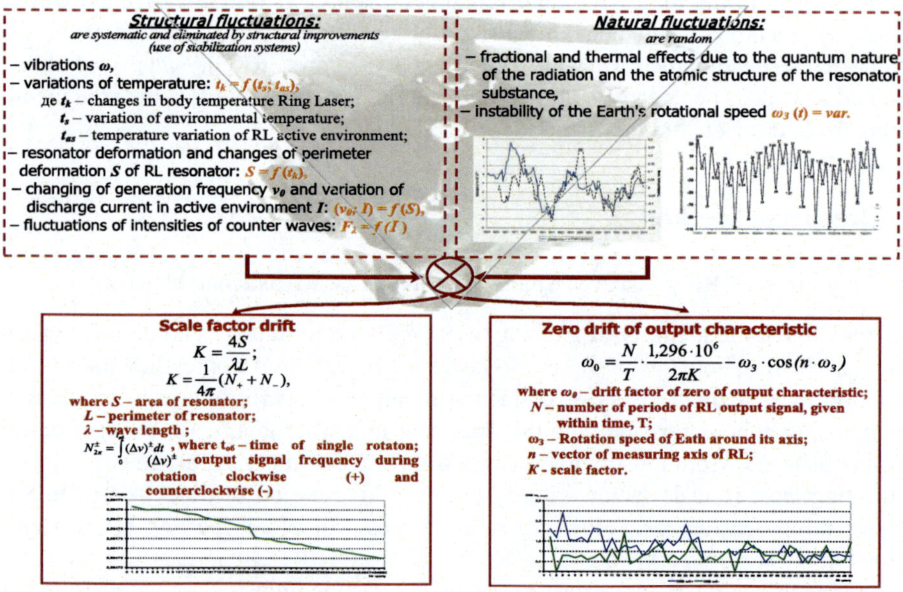

Fig. 1. The structure scheme of fluctuation processes and errors of RL.

3.2 Ring Laser Design

The design of the RL 3.970.029 is given in Fig. 2. It was developed in the ARSENAL SDP SE (Kyiv, Ukraine) and made following a classical scheme.

RL resonator consists of three mirrors and is performed in the form of an equilateral triangle. One mirror is placed on the piezo converter. It is used to adjust the size of the perimeter P of the resonator. The mixing optics with four-section photo receiver of information circuit is on the second mirror. The prism with a photoreceiver of radiation power setting is placed on the third mirror.

The batteries are arranged in such a way that they provide compensation of Langmuir drift (according to the scheme: two anodes, one cathode). The diaphragm is performed in the passive channel.

Vacuum-tight connectors achieve pressurization of the working volume. Mirrors are mounted on the optical contact, and input connections of anode and cathode are performed using metal soldering.

The monoblock is glued to the metal plate with its landing area to the metal plate of material with a low coefficient of linear expansion. There are three holes in the plate designed for fastening in angle measuring devices.

In order to provide a larger volume of working gas mixture (He-Ne), sinuses are made in the monoblock.

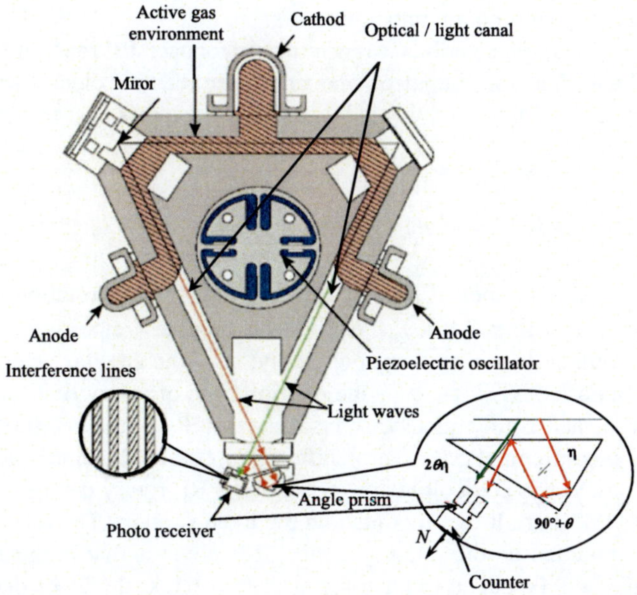

Fig. 2. RL3.970.029 is developed by enterprise “Arsenal” in collaboration with the Department of Instrument Engineering of the National Technical University of Ukraine “Kyiv Polytechnic Institute named after Igor Sikorsky”.

The zero drift of the original characteristic and the drift of the scale factor are the RL parameters that determine its precision characteristics and have to be distinguished.

4 Results

The estimations of the scale factor and zero shift of the output characteristic have been carried out in the following way.

The estimation of scale factor was conducted with counting and without taking into account the changes in rotation velocity of the Earth around its axis. Change in rotation velocity of the Earth influences the work of the RL and leads to significant values of errors while measuring angles. Therefore, RL was installed in such a way to set its axis in a parallel position to the Earth's rotation axis. It was done in order to achieve a partial reduction of the error, which arises to the impact of changes in the velocity of Earth rotation at the measurement. It means that the deviation angle of the RL rotation axis and the axis of rotation of the Earth was heading to zero. In this case, the linear drift of the RL scale factor is about 0.01361 per/s and causes an error in the angle of measurement not more than 0.1″. The scale factor was measured in cycles. Each cycle made 5 turns per angle 2π either clockwise or counterclockwise. The total number of cycles is 50. The angular rotation velocity of the rotary device with RL is ω = 90°/c.

Therefore, the measurements were performed on the geographic latitude of the measured location Ψ = 50°27′ of northern latitude (Kyiv), the rotation time of one rotation of clockwise and counterclockwise 4 s:

The number of RL signal periods was determined for each RL rotation. The number of periods was obtained at the output of one ring laser rotation clockwise and counterclockwise at angle 2π. Moreover, this is by the known expressions: $t^{+} = t^{-} = t = 4$ s.

$$N_{2\pi}^{+} = K\omega_{2\pi}^{+} + K\omega_3 \cos\Psi_{2\pi} + v_0 t^{+};$$
$$N_{2\pi}^{-} = K\omega_{2\pi}^{-} - K\omega_3 \cos\Psi_{2\pi} - v_0 t^{-}, \text{ c} \tag{1}$$

where $N_{2\pi}^{+}$, $N_{2\pi}^{-}$ – the number of periods of RL signal for one rotation clockwise and counterclockwise at angle 2π; $\omega_{2\pi}^{+}$, $\omega_{2\pi}^{-}$ – the angular velocity of rotary device clockwise and counterclockwise correspondingly; ω_3 – the angular velocity of rotation of the Earth around its axis i; t+, t− – the rotation time of rotary device at angle 2π at rotation clockwise and counterclockwise respectively; $\Psi_{2\pi}$ – the geographic latitude of the location of measurement; v0 – zero shift of RL output characteristic.

The slit photoelectric autocollimator recorded the RL rotary device rotations with a focal length of 1000 mm. It was mounted on the fixed support. The reflecting elements (mirrors) were mounted on the rotating device. It is obvious that in this case, the error of autocollimator will be added to the error of drift of RL scale factor determination as well as non-stability of rotary device rotation due to friction in supports, non-stable functioning of electric drive, etc. Therefore, it is recommended to calculate the average value of periods of signal $\overline{N_{2\pi}^{+}}$ and $\overline{N_{2\pi}^{-}}$ from RL for 5 rotations of the rotary device clockwise and counterclockwise, respectively. The total average value of the periods of the signal is:

$$N_{\Sigma} = \overline{N_{2\pi}^{+}} + \overline{N_{2\pi}^{-}}. \tag{2}$$

The obtained values NΣ of signal periods for every among 50 cycles at the rotary device with RL rotation ω = 90°/s taking into account and ignoring the velocity of the Earth's rotation around its axis, are presented in Fig. 3.

The expression estimates a scale factor of RL:

$$K = \frac{\overline{N}_{+} + \overline{N}_{-}}{4\pi}. \tag{3}$$

The estimation of scale factor was conducted, taking into account and ignoring the velocity of the Earth's rotation around its axis. The obtained results are presented in Fig. 4, Fig. 5.

The constant component of drift ω0 is ω0 ≈ 0.3389795 GHz. The scale factor has suddenly changed to the relative value of 3 • 10–6 between measurement cycles 24 and 25. Changes of the ring laser scale factor before and after the jump are represented as a linear drift, combined with the random fluctuations. Its linear drift was about 0.002 rad−1/s before and after the scale factor jump.

To improve an angle measurement accuracy, the angular rate of the rotary table can be increased; cumulative processing of several measurement cycles allows the estimation of the scale factor drift and using corresponding calculations for decreasing a component of the error caused by this drift.

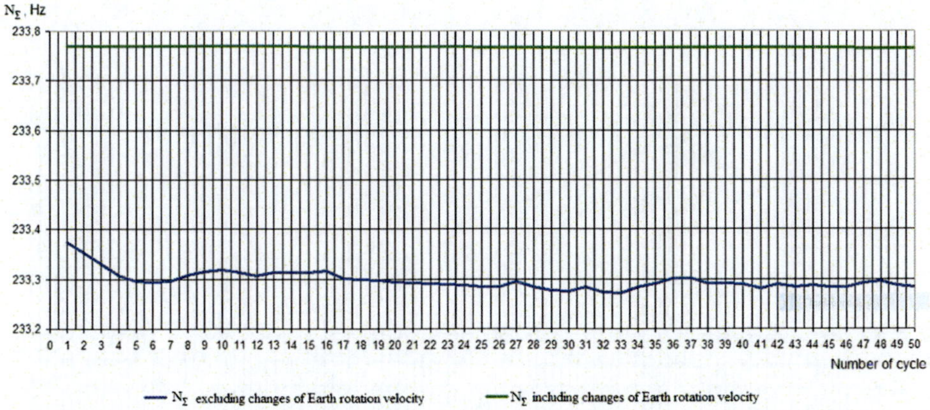

Fig. 3. The summarizing chart of values of the total average value of signal periods per 50 cycles at the velocity of the rotary device rotation with RL ω = 90°/s taking into account and ignoring the velocity of the Earth's rotation around its axis.

Fig. 4. The drift of scale factor per 50 cycles at the velocity of the rotary device rotation with RL ω = 90°/s taking into account and ignoring the velocity of the Earth's rotation around its axis.

Except for a transient process at the beginning of measurements (cycles 1.2) and one cycle before and one cycle after the jump, the average value ΔN of cycles from 3 to 22 is 187.51, and that of cycles from 26 to 50 is 185.74.

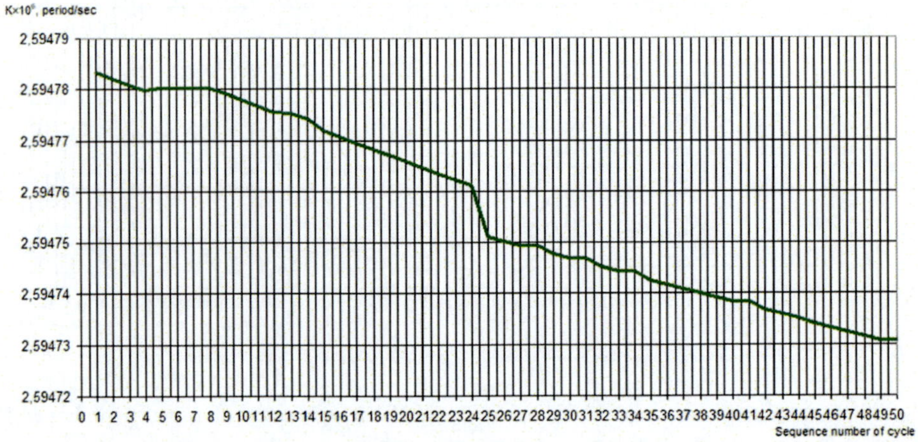

Fig. 5. The drift of scale factor per 50 cycles at the velocity of the rotary device rotation of RL ω = 90 rot/s taking into account the changes in the velocity of the Earth's rotation around its axis.

Thus, the changes of ring laser parameters between measurements cycle 24 and 25 results in a change of both a scale factor and a zero shift of the output characteristic.

Thus, the obtained results confirm the possibility of practical use of the RL 3.970.029 as the component of high-precision angle measurement systems as well as a precision angle sensor. Therefore, RL that is used as a component of the proposed PHSAM as a precision angle converter has the perimeter stabilization system as well as the system of radiation power stability. It allows supporting RL precision parameters within reasonable limits and providing high precision of measurements up to 0.1″.

5 Conclusions

The research and simulation results of the RL parameters confirmed the possibility of its use as a precise angle transmitter in the contactless measuring of angles with high accuracy. Received results show the influence of Earth rotational velocity on the RL parameters. It is recommended to follow the requirements of RL axes installation regarding rotation axes of the Earth. Further improvements of the angles measuring by use of RL can be achieved first of all due to improving parameters of the angles measuring subsystems as well as the use of new methods of data processing, particularly neural networks [14] and algorithmic correction of the measuring results [15].

References

1. Mou, J., Pang, B., Huang, T., Ying, G., Shu, X.: A new method to eliminate the misalignment angle in dynamic goniometer based on fiber optic gyro. Optik **193** (2019). https://doi.org/10.1016/j.ijleo.2019.162998. Accessed 21 May 2020
2. Sasan, M., Ghanbarisabagh, M., Golmohammadi, S.: Quantum effect on characteristics of semiconductor ring laser. Opt. Laser Technol. **108**, 136–141 (2018)
3. Yang, R., Sun, H., Hu, P., Yang, H., Fu, H., Fan, Z., Tan, J.: Experimental exploration of mode-locking evolution mechanism in dual-ring fiber laser. Optik **208** (2020). https://doi.org/10.1016/j.ijleo.2019.163899. Accessed 21 May 2020
4. Cherepanska, I., Bezvesilna, O., Koval, Y., Sazonov, A.: Intelligent precise goniometric system of analysis of spectral distribution intensities for definition of chemical composition of metal-containing substances. Metallofiz. Noveishie Tekhnol. **41**(2), 263–278 (2019)
5. Burnashev, M.N., Pavlov, P.A., Filatov, Y.V.: Development of precision laser goniometer systems. Quantum Electron. **43**(2), 130–138 (2013)
6. Bokhman, E.D., Venediktov, V.Yu., Korolev, A.N., Lukin, A.Ya.: Digital goniometer with a two-dimensional scale. J. Opt. Technol. **85**(5), 269–274 (2018)
7. Astrua, M., Belfi, J., Beverini, N., Virgilio, A., Carelli, G., Maccioni, E., Ortolan, A., Pisani, M., Santiano, M.: The INRIM - INFN ring laser gyroscope for planar angle metrology application (2015). https://doi.org/10.1049/cp.2015.0159
8. Eno, N.A., Pavlov, P.A., Filatov, Y.: Laser goniometer used for remote measurement of angular position and movement for metrology. LASE (2018). https://doi.org/10.1117/12.2287491
9. Azarova, V.V., et al.: Ring gas lasers with magneto-optical control for laser gyroscopy. Quantum Electron. **30**(2), 96–104 (2000)
10. Angle Measurement System GS1L. Technical Specification and Operating Instructions. Ukraine, Kyiv, Central Design Bureau “Arsenal”. https://old.arsenal.co.ua/index.phtml?id=_3_4#3. Accessed 21 Nov 2019
11. Di Virgilio, A.D.V., Beverini, N., Carelli, G., et al.: Analysis of ring laser gyroscopes including laser dynamics. Eur. Phys. J. C **79**, 573 (2019)

12. Badaoui, N., Morbieu, B., Martin, P., Rouchon, P., et al.: Towards a solid-state ring laser gyroscopeVers un gyrolaser à état solide. Comptes Rendus Physique **15**(10), 841–850 (2014)
13. Liu, Z., Wang, L., Li, K., Ban, J., Wang, M.: A calibration method for the errors of ring laser gyro in rate-biased mode. Sensors (Basel) **19**(21), 47–54 (2019)
14. Cherepanska, I., Bezvesilna, E., Sazonov, A., Nechai, S., Pidtychenko, O.: Development of artificial neural network for determining the components of errors when measuring angles using a goniometric software–hardware complex. East.-Eur. J. Enterp. Technol. **95**(9), 43–51 (2018)
15. Cherepanska, I., Bezvesilna, O., Sazonov, A.: Algorithmic correction of measuring results of the precise instrument goniometric system. Sci. Notes Taurida Natl. V.I. Vernadsky Univ. Ser.: Tech. Sci. **30**(69), 6–11 (2019). (in Ukrainia)

Check for updates

Optical Inspection Software for a Selected Product on the Smart Factory Production Line

Magdalena Diering(✉) and Jan Kacprzak

Poznan University of Technology, Piotrowo 3, 60-965 Poznan, Poland
Magdalena.Diering@put.poznan.pl

Abstract. The article describes research aimed at the determination of parameters of an optical inspection station on the Smart Factory automatic production line. This work proposes a fully automated approach for vision-based quality control. The starting point for achieving the set objective was to perform a concise analysis of literature on quality control and to become familiar with the functioning of software provided by the hardware producer. The optical inspection software was developed for a product consisting of blocks. The developed software consists of two modules used for two cameras. As part of the design work, an optical inspection algorithm for a specific finished product was developed, parameters were determined, and the place for the introduction of an optical assessment station in the existing production line was indicated. The article describes an algorithm that enables the introduction of an optical inspection station in the production line. The developed algorithm was evaluated under real (in a laboratory) inspection conditions.

Keywords: AOI · Quality control · OMRON · Inspection program · Algorithm

1 Introduction

The control itself does not create added value because it does not affect (change) the assessed product. However, it is one of the auxiliary processes that, thanks to collecting information about the finished product, allows gaining knowledge which elements can be improved and corrected [1, 2]. In this way, it indirectly allows receiving the added value that is significant both from a process and customer perspective. From quality control in the production process, the key element is the quality of workmanship, which determines the compliance of the finished product with its original design.

It is currently believed that quality is noticed by the customer only when it "does not appear" in the product. Therefore, nowadays, it is required that the methods and techniques of quality control and also appraisers competencies [3] are as accurate as possible, and at the same time, do not extend the production time of a single end product. This requirement may be fulfilled by Automatic Optical Inspection (AOI). It involves replacing the employee's subjective visual assessment with control using advanced vision systems and computer software. Depending on the standard used in the inspection, AOI based on image analysis and AOI using algorithm analysis can be distinguished.

V. Tonkonogyi et al. (Eds.): InterPartner 2020, LNME, pp. 785–796, 2021.
https://doi.org/10.1007/978-3-030-68014-5_76

The paper aims to develop and describe the parameters of an optical inspection station on the Smart Factory automatic production line. The optical inspection software was developed for a product consisting of blocks. The developed software consists of two modules used for two cameras. The starting point for achieving the set objective was to perform a concise analysis of literature on quality control and to become familiar with the functioning of software provided by the hardware producer.

2 Literature Review

Today, machine vision and optical inspection applications are frequently used in many different areas such as automotive, food, and electronics. They are more and more popular as they allow contactless control. In contactless measurement systems, products are usually controlled by one or more specialized devices in the form of vision systems (cameras) [4]. Automated Optical Inspection is an advanced form of visual inspection, i.e., it consists of replacing the employee's subjective visual assessment with inspection using cameras and computer equipment with dedicated software [5, 6].

The use of automatic optical inspection is part of the automation of the production process [7–9]. The vision system automatically carries out activities performed traditionally by an employee. The human task is to supervise the operation of devices as well as to create and implement appropriate control programs using modules (subprograms) supplied by the software producer [5, 10]. This approach is consistent with modern I4.0 industry assumptions [11–13].

Vision systems allow performing [14, 15]: object geometry checks; object dimensions and structure control; object color assessment; text control; code reading; control of the presence and position of object elements.

3 Research Methodology

The research object was an OMRON automatic optical inspection station. The research aimed to determine the parameters of the station using the finished product (Fig. 1), which consisted of blocks of different shapes and colors. The diversity of colors and shapes of individual elements allowed for the assessment of the correctness of the optical inspection algorithm presented in the next part of the work.

The OMRON automatic optical inspection station is located in the Smart Factory laboratory housed in the building of the Faculty of Mechanical Engineering at Poznan University of Technology. The laboratory belongs to the Institute of Materials Technology. The elements of the optical inspection station are (Fig. 2):

- Computer (FH-1050-10 box type controller; Intel® Celeron® dual-core processor allowing for up to four cameras to be connected);
- Two cameras 3Z4S-LEVS-0814H1 (each with 8 mm lens and aperture from F/1.4 to F/16);

- Two LED light sources (light intensity can be adjusted in the range from 1 to 400 units for both sources simultaneously or by selecting the appropriate channel for each one individually);
- Monitor;
- Omron S8VK-C24024 power supply;
- Leads.

Fig. 1. Research object.

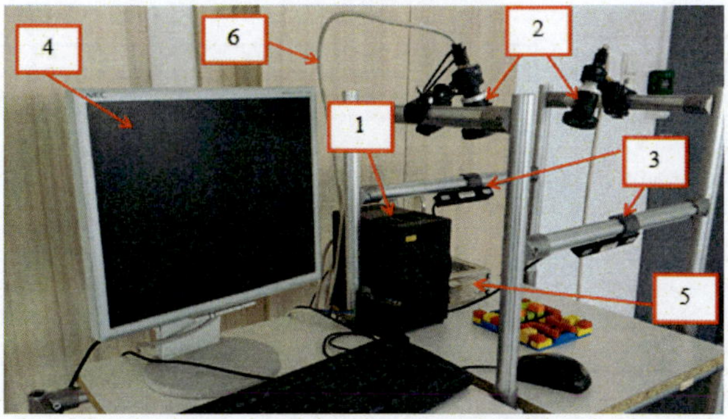

Fig. 2. Optical inspection station.

The second aim of the research was to indicate the place of implementation of the optical inspection station in the production line in the Smart Factory laboratory (built for scientific and didactic purposes). The line, built by FlexLink, allows manufacturing finished products consisting of blocks in diverse variants. The production line consists of four loops enabling the product to be transported to subsequent stations arranged along with it. Each of the loops has crossovers enabling the product to be redirected to any transport loop. The product moves along the line on a pallet marked with an RFID tag, which allows it to be identified. There are RFID reading heads in front of each site and workstation that allows changing the direction of the route. A control panel controls the entirety. Assembly stations are independent organizational units responsible for the production of components of finished products. Each station is equipped with flow racks that allow storing containers with parts and components for assembly. Managing the production system is carried out using the 4Factory IT system, and communication between system elements is carried out via the Internet of Things (IoT).

4 Results

Vision system software includes numerous control methods, the so-called subprograms, and enables the creation and saving of own control programs [5].

The first step to properly create an optical inspection algorithm using software is to place the tested object within the range of cameras. Then, the next step is to configure the image captured by the vision system at the site. The range of the basic configuration is shown in the scheme (Fig. 3).

The optical inspection software was developed for a product consisting of blocks. The assembly of this product is the basis for work in the laboratory – this type of product is made on the production line in the Smart Factory lab. The proposed software uses modules and subprograms of the OMRON hardware and software manufacturer. The main purpose of the inspection software is to check several critical elements of the product. Yellow (dashed line) indicates critical elements tested for the correct color of the blocks, while white for elements tested for the correct shape (Fig. 4).

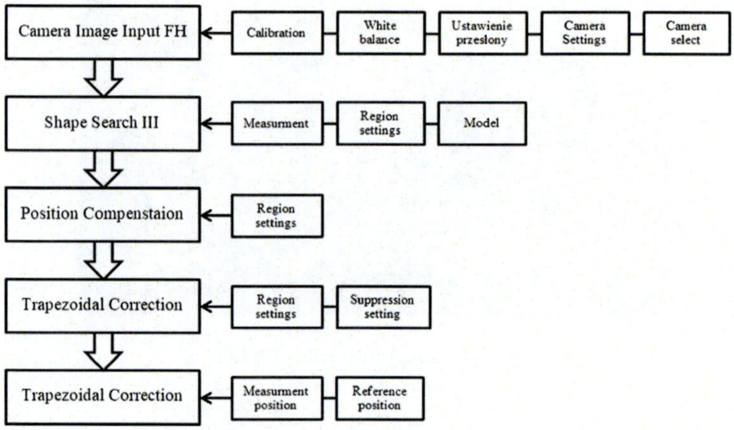

Fig. 3. Basic configuration scheme.

The developed software consists of two modules used for two cameras with the *Camera Switching* option (Fig. 5). The first module is the basic module (Fig. 5, 0–4) responsible for proper image capturing and subsequent processing, i.e., it is responsible for performing basic software preparatory functions.

Fig. 4. Tested object.

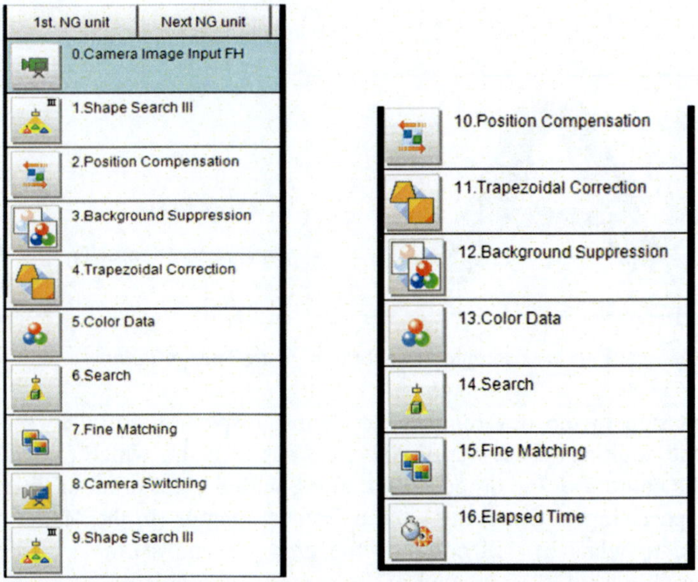

Fig. 5. Inspection software implemented in the OMRON environment.

The second module (Fig. 5, 5–7) aims to carry out a proper optical inspection of the object. It compares and assesses the color of specific blocks using *Colour Data*, checks the shape of the marked critical elements of the product, and compares the captured image with the reference model pixel by pixel using *Fine Matching*.

Camera Image Input FH forms the basis of the entire software. After proper configuration, it captures the image of an object, ensuring adequate focus and exposure. It also allows narrowing the captured image to a minimum, which speeds up further work of the entire software. The incorrect setting of capture parameters can lead to distorting the entire test results.

Configuring both cameras is the basis for creating each inspection program using software provided by the producer. Each camera requires a separate parameter setting, including:

- Manual setting of the aperture and image focus using the moving parts of the camera;
- *Camera settings-Camera shutter* – camera shutter speed and *Gain* – granularity;
- Number of lines to be read – an option that allows to reduce the captured image.
- *White Balance*;
- Calibration – allows the measurement data to be converted into real dimensions, e.g., using the three-point method (using the *Specify Points* option – one should mark three points selected by the user on the screen (Fig. 6), and then provide their actual position; the software will calculate the calibration parameters).

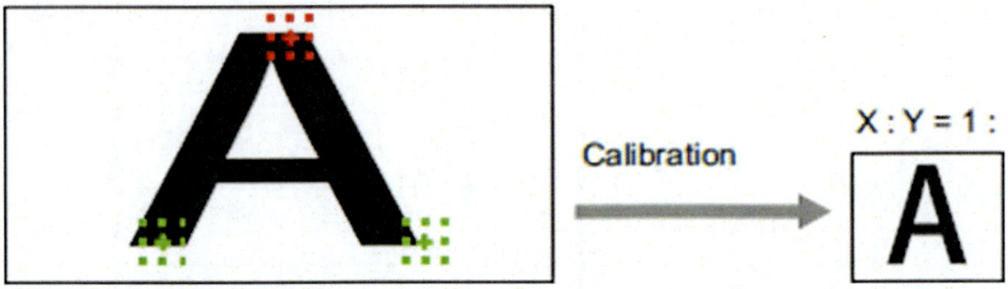

Fig. 6. Example of calibration using *Specify Points.*

The next step in the development of optical inspection software for a selected product is the addition of the *Shape Search 3* subprogram, which allows to quickly search for the contour of the tested object, along with the detection of interferences. It consists in specifying in the settings the reference contour (in the *Model* and *Region Settings* tabs) to which it will compare the products and setting a detection point, which, if possible, should be located in the center of the object.

Shape Search III is a function that allows registering the object based on its contour, taking into account interferences. It allows specifying the degree of acceptable correlation. It enables fast and accurate detection of the tested object.

Position Compensation allows for correcting the image of an object when it is turned to the reference image. In more precise terms, it compares the measured coordinates of the object and compares it with the proper coordinates, and in the case of displacement, it corrects the position of the object by the difference observed. It should be emphasized that if the object exceeds the frames set in the *Camera Image Input FH*, the inspection result may be negative.

Background Suppression allows additional narrowing of the tested area by "muting" the part whose brightness is lower or higher than assumed when configuring this option. Similarly to narrowing the tested area using *Camera Image Input FH*, this option also reduces the work time of the entire program.

A *Trapezoidal Correction* is an option that allows the correction of shape errors of the examined object. These errors occur when the lens captures the image. They mainly result from the fact that the cameras are not placed directly above the product in a perpendicular way to the product base. They are located on a metal structure, thanks to which they capture the image at an angle, making the object illusively resemble a trapezoid, not a rectangle or square. It is an important option for examining the correctness of the object's shape critical features.

As it was mentioned, the second proper part of the created program is the object inspection module (Fig. 5, 5–7). The task of this module is to perform the optical inspection of a given object. It uses subprograms that allow controlling the colors and shape of individual blocks and the entire product.

The *Colour Data* function allows assessing the correctness of the color of the blocks. The assessment is carried out by calculating the average color of a given area (block) and comparing it to the color of the reference element. In this case, lighting and a variety of marked blocks have a significant impact on the measurement and assessment of color. If two blocks of different colors are marked as one area, then the permissible color can be averaged in such a way that the measurement result and assessment will be false.

The *Search* function allows searching for the desired shapes/components of an object in the previously marked area of interest. It allows determining if an element of a certain shape is in the right place. The inspection result depends on the degree of correlation specified by the user.

Fine matching is a subprogram that compares the reference image with the image captured by the camera pixel by pixel. The assessment of the object's correctness is carried out by comparing half of the surface of the object's elements that differ from the reference pattern to the allowable size of the different fields. It is worth noting that the subprogram is very sensitive to shifts in the position of the examined object.

Table 1 presents the parameters and their values that were adopted (based on own empirical research in the Smart Factory laboratory) and set during the configuration of subprograms, i.e., functions of the developed inspection software for the OMRON inspection station.

In order to estimate the time necessary to be reserved on the production line for quality control, many measurements of the duration of a single inspection were made. Forty measurements of the duration of optical inspection were made depending on different light intensity (Fig. 7), and basic statistics were calculated for the duration of inspection (average value, median, standard deviation, and the maximum and minimum values were determined).

The measurement results, presented in Fig. 7, show that the duration of inspection is not affected by the change in light intensity. The longest inspection time was 518 ms, the shortest 460. It is worth noticing that the times are significantly exceeding the average value, i.e., 476.14 ms were obtained after prolonged program inactivity. The standard deviation is 6.37 ms, and compared to the average value is very small. The most common measurement time is 473 ms, and the median value is 475 ms.

Table 1. Parameters of subprograms of the developed inspection software.

Subprogram	Tab	Option	Parameters
Camera input	Camera setting	Shutter speed	2000
		Gain	56
		Number of line to be read	132-1677
	White balance	R	1.067
		G	0.890
		B	1.423
	Calibration	A	0.999284
		B	0.046880
		C	-84,062632
		D	0.269777
		E	0.988920
		F	-70,121197
		Field of view	2591.084714
Shape Search III	Detection point	Detection coordinate	1267.5000 911.000
		x	1267.5
		y	911.0
	Measurement	Angle	0
		Angle range	-180 - 180
	Judgment	Count	0 -32
Position compensation	Scroll method	Default	
		Last image	
	Reference position	Upper left	493.266
		Lower left	501.1556
		Lower right	2082.1544
		Upper right	2034.258
	Method:	Figure	
	Measurement position	Upper left	545.237
		Lower left	484.1527
		Lower right	2051.1552
		Upper right	1985.280
Background suppression	RGB common		15 - 202
Colour data	Measurement	Static mash	
		Correction condition	Normalization
		R	161
		G	75
		B	42
	Judgment	RGB judgment condition	
		Colour difference	0 - 5

(Continued)

Table 1. *(Continued)*

Subprogram	Tab	Option	Parameters
		Colour deviation	0 - 85
	HSV judgment condition	H	16 - 68
		S	47 - 255
		V	139 - 255
	Difference image display	Compensation processing	Normalization
		Difference parameter	30
Fine matching	Measurement	Binary	
	Judgment condition	Quantity	0 - 9999
		Area	0 - 1000.000
		Defect pos x:	-9999 - 9999
		Defect pos y:	- 9999 - 9999
Background suppresion 2	RGB		14 - 196
		Static mash	
			Normalization
	Measurement	R	172
		G	94
Colour data 2		B	56
	Judgment	Colour difference	0 - 5
		Colour deviation	0 - 85
		H	16 - 68
	HSV	S	47 - 255
		V	139 - 255
Search	Measurement	Default	
	Judgment	Default	

By implementing the OMRON station in the Smart Factory production line, it can be assumed that approx. 476 ms should be spent on performing an inspection in the program. It should be kept in mind that the measurements presented in the tables do not include the time during which the production line was stopped.

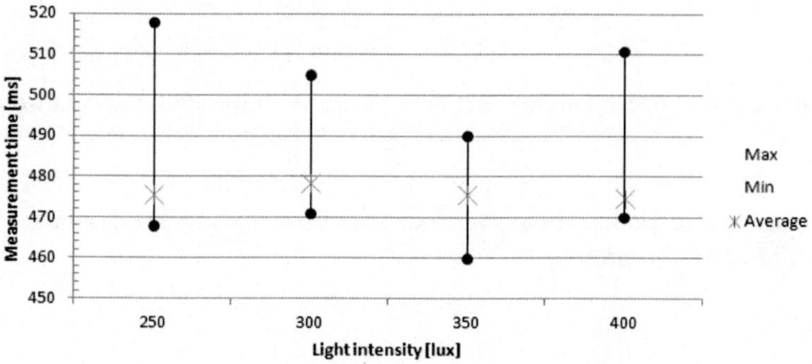

Fig. 7. Measurement results – duration of optical inspection depending on light intensity.

The line for scientific and didactic purposes in the Smart Factory laboratory was built by FlexLink. The line allows the manufacturing of finished products consisting of blocks in diverse variants. The production line (Fig. 8) consists of four loops enabling the product to be transported to subsequent stations arranged along with it.

Each of the loops in Fig. 8 has crossovers enabling the product to be redirected to any transport loop. The product moves along the line on a pallet marked with an RFID tag, which allows it to be identified. There are RFID reading heads in front of each site and workstations that allow changing the direction of the route. A control panel controls the entirety. Assembly stations are independent organizational units responsible for the production of components of finished products. Each station is equipped with flow racks which allow to store containers with parts and components for assembly. Managing the production system is carried out using the 4Factory IT system, and communication between system elements is carried out via the Internet of Things (IoT).

This place was chosen due to the fact that if the station is moved to one of the workstations, there is a significant risk of creating an unwanted buffer.

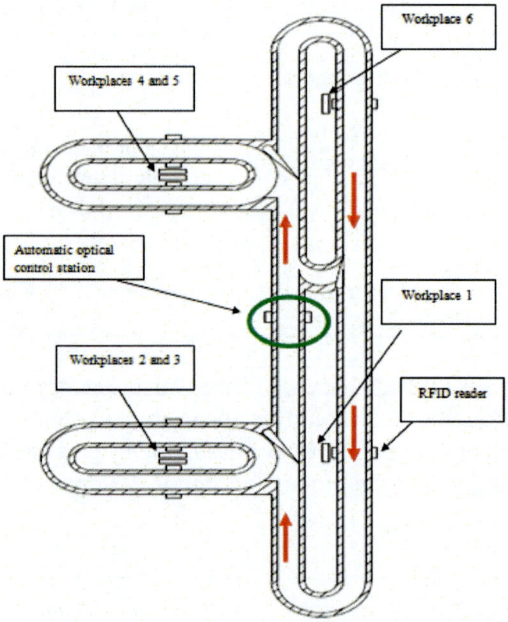

Fig. 8. Smart Factory production line with the place for an inspection station (area outlined in green).

Moreover, the indicated area contains an RFID reading head which allows to control and identify the product.

5 Conclusions

The aim of the work was, above all, to determine the parameters of an optical inspection station on the automatic production line in the Smart Factory laboratory. The starting point for achieving the set objective was to perform a concise analysis of literature on quality control and to become familiar with the functioning of software provided by the hardware producer. Eventually, an inspection program was created and described in the article.

The most important conclusions, useful for physically introducing an optical inspection station in the production line:

- Inspection time for a single product is approximately 480 ms and should be extended by the time needed to stop the product. However, this should not be a problem because the production line software makes it possible to manually set the rhythm of individual operations and the speed of product movement.
- The Fine Matching option used in the program, which analyzes the image pixel by pixel, is very susceptible to changes in the position of the examined product. This problem should be kept in mind, especially due to the malfunctioning of transfer pallets, which often rotate when moving on the production line. It is a mechanical problem that needs a solution to ensure the proper functioning of the entire line because robots are unable to assemble the finished product properly.

It is worth emphasizing that after introducing the station in the production line, the parameters adopted during the software configuration need to be checked and may need to be corrected due to the change in the environment (measurement conditions).

Acknowledgments. The paper is prepared and financed by scientific statutory research conducted by the Division of Production Engineering, Faculty of Mechanical Engineering, Poznan University of Technology, Poznan, Poland, supported by the Polish Ministry of Science and Higher Education from the financial means in 2019–2020 (0613/SBAD/8727).

References

1. Kujawińska, A., Vogt, K.: Analysis of the impact of selected work factors on the efficiency of visual inspection. Mach. Eng. **18**(1), 40–51 (2013). [in Polish]
2. Hamrol, A., Kujawińska, A., Bożek, M.: Quality inspection planning within a multistage manufacturing process based on the added value criterion. Int. J. Adv. Manuf. Technol. **108**, 1–14 (2020)
3. Goliński, M., Spychała, M., Szafranski, M., Graczyk-Kucharska, M.: Competency management as the direction of the development of enterprises-based on research. In: 3rd International Conference on Social Science, Shanghai, China, pp. 391–399. Published by DEStech Publications, Inc. (2016)
4. Baygini, M., Aygin, M.: Deep learning based approaches for machine vision inspection applications. In: International Conference on Advanced Technologies, ICAT 2018, p. 63 (2018)
5. Materials from the company OMRON

6. Garbacz, P., Giesko, T.: Multi-camera vision system for the inspection of metal shafts. In: Szewczyk, R., Zieliński, C., Kaliczyńska, M. (eds.) Challenges in Automation, Robotics and Measurement Techniques, ICA 2016. Advances in Intelligent Systems and Computing, vol. 440. Springer, Cham (2016)
7. Wojciechowski, J., Suszynski, M.: Optical scanner assisted robotic assembly. Assem. Autom. **37**, 4 (2017)
8. Klos, S., Patalas-Maliszewska, J.: Using a simulation method for intelligent maintenance management. In: Burduk, A., Mazurkiewicz, D. (eds.) Advances in Intelligent Systems and Computing, vol. 637, pp. 85–95. Springer International Publishing (2018)
9. Vieira, G.G., Varela, M.L.R., Putnik, G.D., Machado, J.M., Trojanowska, J.: Integrated platform for real-time control and production and productivity monitoring and analysis. Rom. Rev. Precis. Mech. Opt. Mechatron. **50**, 119–127 (2016)
10. Hawary, A.F., Hoe, Y.H., Bakar, E.A., Othman, W.A.F.W.: A study of gauge repeatability and reproducibility of the back-end semiconductor lead inspection system. Robotika **1**(2), 1–6 (2019)
11. Varela, M.L.R., Putnik, G.D., Manupati, V.K., Rajyalakshmi, G., Trojanowska, J., Machado: Collaborative manufacturing based on cloud, and on other I4.0 oriented principles and technologies: a systematic literature review and reflections. Manag. Prod. Eng. Rev. **9**(3), 90–99 (2018)
12. Krenczyk, D., Skolud, B., Olender, M.: Semi-automatic simulation model generation of virtual dynamic networks for production flow planning. IOP Conf. Ser.: Mater. Sci. Eng. **145**, 042021 (2016)
13. Pavlenko, I., Trojanowska, J., Gusak, O., Ivanov, V., Pitel, J., Pavlenko, V.: Estimation of the Reliability of Automatic Axial-balancing Devices for Multistage Centrifugal Pumps. Periodica Polytechnica Mechanical Engineering **63**(1), 52–56 (2019)
14. Batchelor, B.G.: Machine Vision for Industrial Applications. Springer (2012)
15. Malamas, E.N., et al.: A survey on industrial vision systems, applications and tools. Image Vis. Comput. **21**(2), 171–188 (2003)

Standardization Issues of Test Methods for Engineering Nanomaterials

Kostiantyn Dyadyura(✉), Tatyana Ivakhniuk, Liudmyla Hrebenyk, Uriy Ivakhniuk, and Leonid Sukhodub

Sumy State University, 2, Rymskogo-Korsakova Street, Sumy 40007, Ukraine
dyadyura@pmtkm.sumdu.edu.ua

Abstract. The paper relates to the field of standardization of methods for testing by ecotoxicity of engineering nanomaterials using microorganisms. The validity of the methodological approach is discussed in the context of the widespread use of nanomaterials and the high uncertainty of measurement of characteristics to assess their safe application in industry and everyday life. Defining the environmental toxicity of examined samples of ZnS-Alg containing various concentrations of nanoparticles was done in the raw of experiments in vitro according to the level of suppression of growth of test-cultures of Aspergillus niger (1119). Studying the toxic effects of nanoparticles indicated that the medium with ZnS-Alg at concentrations 0.15 μg/ml and 0.10 μg/ml had the same average III-d class of toxicity; 0.05 μg/ml ZnS-Alg had a low toxicity. The results of the study have shown that in the standardization of new engineering nanomaterials, the nanosafety issue can be addressed by using Aspergillus niger as a test-culture. The obtained data confirm the possibility of using unicellular organisms in the creation of universal express test systems for assessing the safety of nanomaterials. The methodology tested in the work has good prospects and can be included in the protocol in the development of standards, taking into account the ethical aspects of the introduction of modern nanomaterials.

Keywords: Standardization · Nanotechnology · Engineered nanomaterials · Nanosafety · Nanotoxicology · Ecotoxicity

1 Introduction

The development of nanotechnologies and nanomaterials is a priority direction for technological innovation, contributing to sustainable and comprehensive economic growth. With the expansion of the production and use of engineering nanomaterials (ENMs) in biomedical, textile, aerospace, cosmetics, petroleum, agricultural and electronic industries, the public's interest is increasingly emphasized in providing risk assessments of the environmental and technological impact of nanotechnological products [1]. It primarily concerns the effects of nano-sized objects, their agglomerates, and aggregates (NOAA) [2–4]. The development of standards and infrastructure of quality is crucial to maintain the highest quality products and services based on nanotechnology. A responsible approach to the development and commercialization of

V. Tonkonogyi et al. (Eds.): InterPartner 2020, LNME, pp. 797–805, 2021.
https://doi.org/10.1007/978-3-030-68014-5_77

nanotechnologies requires the standardization of methods that allow assessing the impact of nanomaterials on the environment [5]. Despite a large number of NOAA research findings and the development of new approaches to the complex analysis of the physicochemical parameters of nanoparticles, current toxicological and ecotoxicological methods do not allow to evaluate all risks fully and to determine the potentially dangerous effects of nanomaterials [1, 4, 6]. The lack of a common methodology makes it difficult to systematize the results of different laboratories on the safety of new, diverse nanomaterials. In this regard, there is a need to develop effective methods for testing nanomaterials to identify potential negative impacts on living organisms and ecosystems [2, 5]. The development of international standards for the correct study of the toxicity and ecotoxicity of NOAA aims to reduce the potential risks of adverse effects, which can significantly simplify the introduction of new technologies and stimulate the development of scientific and technological progress [7, 8]. Therefore, standardized procedures of ENMs toxicity and ecotoxicity studies need to be further improved and implemented.

Scientific and methodological providing a risk assessment of the impact of innovative nanotechnological products on health and the environment, formation, and progress of technological and regulatory framework is necessary for the development of international standards and recommendations on occupational safety, production, use, and disposal of nanomaterials.

2 Literature Review

Standardization in the field of nanotechnology provides the following [9–14]:

- detailed study and control of the substance and processes at the nanoscale level (below 100 nm in one or more dimensions) for further possible use;
- creation of new, improved materials and devices, taking into account the properties of nano-sized objects, which allow optimizing their practical use.

Existing specific challenges in the development of standards relate to terminology and nomenclature, instrumentation and metrology, testing methods, modeling, and nanoethics standards, including nanosafety [8]. For example, ASTM standards contain terminology in this area, provide recommendations for the practical use of nanotechnology and nanomaterials in various industries, as well as algorithms for properties testing and testing procedures, taking into account health and safety issues.

Furthermore, ISO/IES and CEN standards address nanosafety and nanotoxicology issues. They ensure an assessment of the impact of nanomaterials on living objects during the release of nanoparticles during the life cycle of the device or process. Environmental, health, and safety (EHS) standards require measurement protocols, standard tests, and techniques that can achieve clear comparability of nanomaterial research. In the ISO/TC229 Technical Committee, the Working Group WG 3 (Health, Safety and Environment Scope) is responsible for the development of science-based standards in the field of health, safety and environmental aspects of nanotechnology, the main task of which is to develop standardized methods for:

- the monitoring of professional exposure on nanomaterials;
- the identification of relative toxicity and hazard potential of nanomaterials;
- toxicological testing of nanomaterials;
- environmentally sound nanomaterials application;
- making product safety of nanomaterials products.

Nowadays, the regulations to govern the use and implementation of nanomaterials and nanotechnologies have been published. The following standards are used to prepare samples in various appropriate medium for toxicological studies: ISO 10993–18 [9], ISO 14971 [10], ISO/TR 13014 [11], ISO/TR 16197 [12]. ISO/TR 12885:2018 [13]. Information on recommendations for assessment and decision-making for risk management, in the context of incomplete and/or uncertain information on the use of nanomaterials, is contained in the following normative instruments [13, 14]. For better harmonization in the field of regulation and control of nanotechnologies, new standardization projects are planned in various industries at the national, regional, and international levels. The actual revision of the ISO / TC 229 documentation allows us to confidently assert the need for a more intensive discussion (Fig. 1) in this direction with more effective use of the results of scientific research of nanoscale objects.

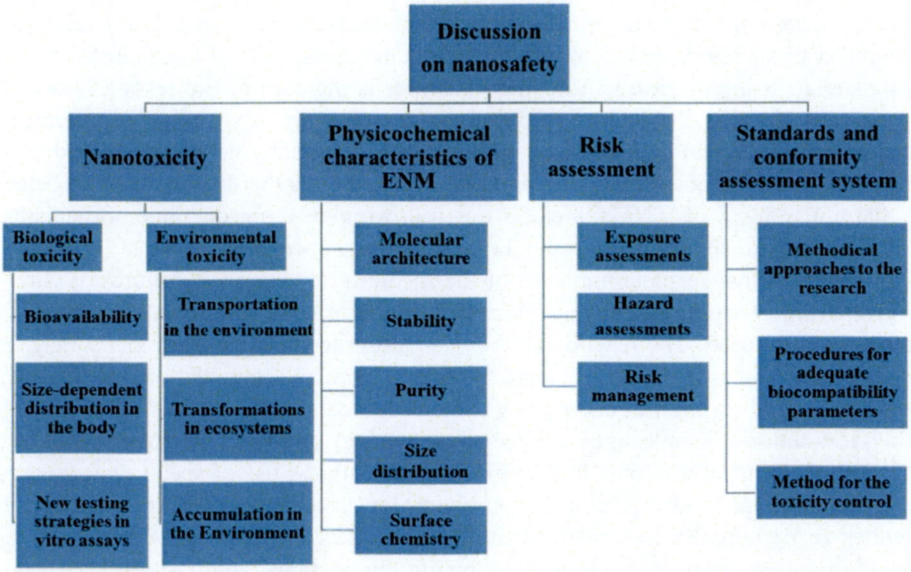

Fig. 1. The discussion scheme of science-based EHS standards for nanotechnologies.

When studying nanomaterials, methodological approaches that take into account the behavior of nanoparticles in the process of their interaction with biological objects deserve special attention. Among them are the following four groups:

- epidemiology/occupational medicine;
- in vivo methods with animals;

- in vitro and in silico methods;
- methods for determining physicochemical properties.

Unfortunately, to date, there is virtually no experimentally proven evidence of the effects of nanoobjects on populations of living organisms as part of an ecosystem. Currently, test-cultures of biological objects are finding to assess the medium toxicity. Strains of bacteria and unicellular fungi, cell cultures of plants and animals actively react to the presence of toxicants in the environment in which they are located.

An analysis of the methods for determining the biosafety of nanomaterials has shown that toxicological screening tests are useful for sound decision-making at the initial stages of research and when developing products with NOAA [2, 5]. Also, nanoparticle assessment should include toxicokinetic studies that can provide information on potential "target organs" for nanoparticles, their transport, and resistance in tissues [7, 8]. Also, there is the offer to evaluate the dangerous effects of nanomaterials through the study of the possibility of the formation of reactive oxygen species (ROS) in cells, the excess of which provokes the development of oxidative stress with harmful consequences for the whole body. Quantitative estimation of ROS is possible using the electron-spin resonance method [15]. Following current ethical and scientific norms for experimental research, there is a global tendency to replace traditional in vivo studies (using laboratory animals) with improved in vitro model systems (using bacteria, tissue culture, tissue sections) and computer modeling - in silico. Each year, the number of standardized techniques that allow the assessment of the nanotoxicity of nanomaterials using in vitro and in silico methods is increasing. Bio testing is one of the research methods for determination of the degree of the toxic effect of chemical, physical and biological factors of the environment, potentially dangerous for humans and components of ecosystems. The biological test systems that may be used to detect the harmful effects of ENMs include the following test objects: microorganisms, hydrations, organisms of higher animals, cellular and subcellular elements [16].

It is known that for determination of the medium toxicity such cultures as *Alternaria alternata, Aspergillus niger, Fusarium moniliforme, Penicillium chrysogenum* are successfully used. The reaction of the test-culture to the medium toxicity may be expressed in changes in the quantity and the rate of biomass accumulation, the intensity of excretion or absorption of energy (e.g., luminescence for luminescent bacteria), etc. Using test-cultures increasingly displaces a chemical analysis, which involves the detection of chemical toxicants in the environment with the use of specific reagents to the detection of particular substances. Applying bio testing has another advantage. It provides an opportunity to determine the integral effect on living organisms of substances of the environment without determining the chemical composition. The analysis of the literature suggests that the use of unicellular fungi that are common in the environment may be the first step in determining the toxicity of nanoparticles for an ecosystem for further certification of nanoproducts. The aim of this work was studying the possibility of using the test-culture of fungi of the genus *Aspergillus*, which are involved in the processes of biodegradation and destruction in the environment, for bio testing of synthesized hybrid nanoparticles of zinc sulfide and alginate (ZnS-Alg) with proven antibacterial properties [17].

3 Research Methodology

Zinc sulfide nanoparticles with alginate (ZnS-Alg) were synthesized by aqueous chemical solution deposition [17]. As a precursor, a mixture of zinc nitrate and thiourea was used (a quantitative ratio was 1: 1). pH level was raised to 8 by adding an ammonia solution. The optimum temperature of the reactor for chemical synthesis was 90 °C. For making a composite of zinc sulfide with alginate (ZnS-Alg) 3% sodium alginate solution was supplemented to the mixture. The mineral composition of the synthesized nanomaterial was investigated by X-ray fluorescence analysis (XRF) with using XRF-spectrometer «ElvaX Light SDD» (Ukrainė). The morphology of the ZnS-Alg nanoparticles was studied by transmission electron microscopy (TEM). TEM analyses were performed by using "PEM-125k" (Ukraine).

The definition of environmental toxicity, containing various concentrations of examined samples, was done in the raw of experiments in vitro according to the level of suppression of growth of test-cultures of *Aspergillus niger* (1119) according to patent [12]. Before the inoculating, the number of spores in one drop of the prepared suspension was defined according to the help of the Horiayev camera. For inoculation, the obtained standard suspension of the *Aspergillus niger* test culture (1119) was used, which is an amount of 100 mkl was inoculated on the Saburo agar medium (LLC "Pharmaktiv", Ukraine) containing various concentrations of the tested substances (Table 1). To control the cultivated of *Aspergillus niger* (1119) test-culture in an isotonic solution, NaCl inoculated in sterile Saburo agar without testing-samples was used.

Table 1. The distribution of test cultures Aspergillus niger (1119) in the experimental series.

Experience number	In vitro test design: *A. niger* (1119) + Saburo agar with	The number of repeat experiment (n)	Experience number	In vitro test design: *A. niger* (1119) + Saburo agar with	The number of repeat experiment (n)
1	0.15 μg/ml ZnS	5	6	0.05 μg/ml Alg	5
2	0.10 μg/ml ZnS	5	7	0.15 μg/ml ZnS-Alg	5
3	0.05 μg/ml ZnS	5	8	0.10 μg/ml ZnS-Alg	5
4	0.15 μg/ml Alg	5	9	0.05 μg/ml ZnS-Alg	5
5	0.10 μg/ml Alg	5	Control	*A. niger* (1119) + Saburo agar without testing-samples	5

In the period of incubation on the solid growing medium, the diameter of the control and experimentally grown colonies of test-cultures was measured with the help of ruler in tow perpendicular directions and on five surfaces (every 24 h) and was fixed on the 3, 5, 7, 10 and 14 day. The number of measurements depended on the speed of growth. The changes in the cultural features of the colonies (nature of edge, change of color, and other) were taken into account, the start of spore-formation was marked, etc.

Grounding of the received data, the level of suppression of test-culture growth affected by the toxicant in percentage rate to the control group (ΔR, %) was defined [17]. Based on the size of index ΔR the integral estimation of growth for each test-culture compared to the control-culture with the further raging of growth range percentage was done: 0% – 0 points – no growth; 1 – 29% – 1 point – highly suppressed growth; 30–60% – 2 points – moderately suppressed growth; 61–90% – 3 points – weakly suppressed growth; 91– $\geq$ 100% – 4 points – normal growth/stimulation of growth [17]. Depending on the received results of the integral estimation of toxicity (in points) of the examined medium the distribution of the testing samples into the classes of danger was done: I – extremely high toxicity – 0 – 5 points; II – high toxicity – 6 – 29 points; III – moderate toxicity – 30 – 64 points; IV – weak toxicity – 65 – 95 points; V – no toxicity – 96 – 100 points [17].

4 Results

An analysis of the results obtained for synthesized ZnS-Alg nanoparticles by using transmission electron microscopy showed that the optimal conditions for chemical synthesis made it possible to obtain particles of the proper spherical shape with the same diameter of spheres. The images obtained at various magnifications made it possible to evaluate the size and structure of nanoparticles (Fig. 2). The average particle size is in the range of from 50 to 80 nm. Each particle has a core formed by ZnS, and a loose shell of the polymer - sodium alginate (Alg). XRF spectrograms of the ZnS-Alg nanocomposite give the opportunity to estimate the chemical composition of the nanoparticles and to prove the presence of ZnS in the test sample. The presence of sodium alginate confirms the presence of a peak in the fluorescence of Na ions in the "light" spectrum.

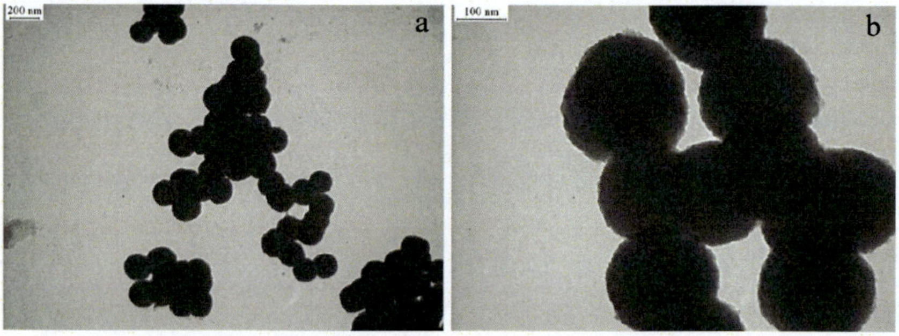

Fig. 2. A TEM images of ZnS-Alg nanoparticles.

Previous studies of the biological properties of synthesized hybrid ZnS-Alg nanoparticles have demonstrated their ability to inhibit the growth of conditionally pathogenic and pathogenic bacteria [16]. One of the most pressing problems of today is the development of antibiotic resistance in bacterial cultures. That is why the presence of antibacterial properties in the new nanomaterial gives optimistic predictions for its further use and dictates the need for studies of the ecotoxicity of these nanoparticles with using test-cultures involved in biodegradation and biodestruction in the environment. Studying the toxicity of ZnS-Alg nanoparticles with using test-cultures of *Aspergillus niger* (1119) allows to evaluate the degree of exposure to organisms and the ecofunctional characteristics of individual populations of living organisms, their communities, and entire ecosystems, and to predict the effects of their use in order to protect environmental effects of anthropogenic pollution and biodegradation. Based on the data obtained to the determination of the medium toxicity with different concentrations of the tested compounds (ZnS, Alg, ZnS-Alg), we found that the medium containing ZnS in concentrations of 0.15 μg / ml; 0.10 μg / ml; 0.05 μg / ml (experiment 1–3) had a weak toxic effect to the test-culture of *Aspergillus niger* (1119) with an exposure of 14 days. So, the total indicator ΔR of the medium with a content of 0.15 μg / ml ZnS was 92.4% (total growth indicator (TGI - 85 points); for 0.10 μg / ml ZnS - 93.8% (TGI - 95 points), for 0.05 μg / ml ZnS - 93.6% (TGI - 90 points), which is typical for class IV of environmental toxicity. Moreover, when evaluating the growth of the test culture colonies on a medium with 0.15 μg / ml ZnS, a decrease in the toxicity of the medium was revealed during the entire experiment: the culture growth criteria compared to the control on the third - the seventh day of the experiment corresponded to 3 points (slightly inhibited growth) and on the 10th and 14th day - 4 points (normal growth). In determining the toxic effect of alginate (experiment 3–6) against *Aspergillus niger* (1119) at a 14-day exposure, it was found that media with low alginate concentrations of 0.10 μg / ml and 0.05 μg / ml (experiment 4 - 6) do not have toxicity to the test culture. TGI of the test culture was 100 points (V class of toxicity); medium with a content of 0.15 μg/ml Alg was classified as slightly toxic (ΔR - 87.6%; TGI - 80 points), and visually the test-culture had slightly inhibited growth (TGI on 7–10 days was 3 points). Based on comparative data obtained in the study of the toxic effects of ZnS-Alg nanoparticles, it can be concluded that the medium with ZnS-Alg at concentrations 0.15 μg/ml and 0.10 μg/ml (experiment 7–8) had the same average III-d toxicity class (TGI was 50 and 60 points respectively); 0.05 μg/ml ZnS-Alg (experiment 9) had low toxicity (TGI for a 14-day exposure was 75 points).

5 Conclusions

The obtained results suggest that the synthesized hybrid ZnS-Alg nanoparticles with proven antibacterial properties do not have a toxic effect on *Aspergillus niger* micromycetes, which are involved in biodegradation and biodestruction processes in the environment. Although the biological and environmental impact of the new ZnS-Alg nanoparticles requires further study, the results of the study have shown that in the standardization of new engineering nanomaterials, the nanosafety issue can be addressed by using *Aspergillus niger* as a test-culture. The research results are promising in the development of technical protocols for studying the environmental toxicity of new nanomaterials but require comparative procedures with already implemented testing methods. Verification of nanomaterials using unicellular organisms is express for decision-making on continuing the development of products containing nanomaterial; the expediency of costs for performance of the following stages in need of multilevel testing; availability of appropriate control for the continuation of laboratory studies of nanomaterials. The results of the research can be used to develop international norms on the testing, use, and disposal of nanomaterials for the health and safety of the public, consumers, workers, and the environment using reasonable assumptions and appropriate risk management practices.

References

1. Gubala, V., Johnston, L.J., Krug, H.F., Moore, C.J., Ober, C.K., Schwenk, M., Vert, M.: Engineered nanomaterials and human health: part 2. Applications and nanotoxicology (IUPAC Technical Report). Pure Appl. Chem. **90**(8), 1325–1356 (2018)
2. Jeevanandam, J., Barhoum, A., Chan, Y.S., Dufresne, A., Danquah, M.K.: Review on nanoparticles and nanostructured materials: history, sources, toxicity and regulations. Beilstein J. Nanotechnol. **9**, 1050–1074 (2018)
3. Jahan, S., Yusoff, I.B., Alias, Y.B., Bakar, A.F.B.A.: Reviews of the toxicity behavior of five potential engineered nanomaterials (ENMs) into the aquatic ecosystem. Toxicol Rep. **4**, 211–220 (2017)
4. Park, H., Yeo, M.: Nanomaterial regulatory policy for human health and environment. Mol. Cell. Toxicol. **12**, 223–236 (2016)
5. Hodoroaba, V., Ghanem, A.: Nanoscale advances a technique-driven materials categorisation scheme to support regulatory identification. Nanoscale Adv. **1**, 781–791 (2019)
6. Wu, T., Tang, M.: Review of the effects of manufactured nanoparticles on mammalian target organs. J. Appl. Toxicol. **38**, 25–40 (2018)
7. Kaur, J., Khatri, M., Puri, S.: Toxicological evaluation of metal oxide nanoparticles and mixed exposures at low doses using zebra fish and THP1 cell line. Environ. Toxicol. **34**(4), 375–387 (2019)
8. Rauscher, H., Rasmussen, K., Sokull-Klüttgen, B.: Regulatory aspects of nanomaterials in the EU. Chem. Ing. Tech. **89**, 224–231 (2017)
9. ISO 10993–18:2020 Biological evaluation of medical devices – Part 18: Chemical characterization of materials
10. ISO 14971:2019 Medical devices – Application of risk management to medical devices
11. ISO/TR 13014:2012 Nanotechnologies – Guidance on physico-chemical characterization of engineered nanoscale materials for toxicologic assessment
12. ISO/TR 16197:2014 Nanotechnologies – Compilation and description of toxicological screening methods for manufactured nanomaterials
13. ISO/TR 12885:2018 Nanotechnologies – Health and safety practices in occupational settings
14. ISO/TS 12901–2:2014 Nanotechnologies – Occupational risk management applied to engineered nanomaterials – Part 2: Use of the control banding approach
15. Fadeel, B., Fornara, A., Toprak, M.S., Bhattacharya, K.: Keeping it real: the importance of material characterization in nanotoxicology. Biochem. Biophys. Res. Commun. **468**, 498–503 (2015)

16. Hrebenyk, L.I., Ivakhniuk, T.V., Sukhodub, L.F.: ZnS quantum dots encapculated with alginate: Synthesis and antibacterial properties. In: Pogrebnjak, A. (eds.) International Conference on Nanomaterials: Applications & Properties 2017, NAP-2017, pp. 04NB07-1–04NB07-7. IEEE, Piscataway (2017)
17. Smirnov, V.F.: Method for determining the toxicity of the environment by the degree of inhibition of growth of test cultures of microorganisms. Patent of RU, No. 2570637 (2015)

The Role of Information Quality in Energy Management Systems

Łukasz Grudzień(✉) and Filip Osiński

Poznan University of Technology, 3, Piotrowo Street, 61-138 Poznan, Poland
filip.osinski@put.poznan.pl

Abstract. The paper presents research results and conclusions arisen during the implementation of Energy Management Systems based on ISO 50001 standard in production and service enterprises. EnMS should be primarily data-driven and fact-based; however, the availability of data on energy consumption and its quality in many cases is a significant problem. One of the basic elements of such a system is Energy Performance Indicator (EPI), which by definition, should be based on data that an organization can monitor, measure, and analyze. The quality of data and information about EPI depend the correctness and credibility of whole EnMS. The paper presents the most common problems associated with basic information attributes. Also, article shows areas, which should be identified, in which data, static factors and relevant variables, can be found that may affect final energy performance indicators. Ensuring the accuracy, correctness, and currency of data will affect, firstly, the ability to manage individual areas of the system and, secondly, the effectiveness of decisions taken.

Keywords: ISO 50001 · Data quality · Energy performance

1 Introduction

Energy efficiency is described in the literature, mostly with the production process [1–6]. Any organization that wants to remain competitive must pay attention to the costs of energy used, not only during its production processes but in its entire scope. It is related to both Energy efficiency and indirectly with the number of greenhouse gases issued by the organization, which is increasingly the area of interest of its clients and principals [7]. It is the most important in the case of production enterprises, where energy costs can absorb 15–30% of all production costs [8]. The generation of energy is a source of most pollutions in the environment. Due to the increasing awareness of clients, entrepreneurs try to manage the energy efficiency of their companies to improve the company's image. Energy efficiency is not only part of the general trend of taking care of the environment, especially in the EU, but it is also an element of the wider issue of CSR (Corporate Social Responsibility) [9]. Achieving a high level of energy efficiency is not only a manifestation of concern for reducing the internal costs of the company but also shows that the interest of society and welfare of the environment is understood by the company. The reduction in energy use results in a reduction of raw materials consumption, as well as the reduction of greenhouse gas emissions [10]. This problem was recognized by the EU commission, which is implementing legal solutions to

V. Tonkonogyi et al. (Eds.): InterPartner 2020, LNME, pp. 806–812, 2021.
https://doi.org/10.1007/978-3-030-68014-5_78

accelerate the process of improving energy efficiency. The 2020 climate & energy package can be an example of a legal requirement [11]. It assumes a reduction of emissions of greenhouse gases by 20% with a simultaneous increase in energy efficiency by 20% and use of renewable energy sources (RES) in the total energy generated at the level of 20%. These are the EU's main objectives for achieving sustainable growth by 2020. The problem of energy efficiency was also noticed by the International Organization for Standardization, which issued in 2011 ISO 50001 - a standard, which contains requirements for planning, implementation, and maintenance of the Energy Management System (EnMS) in organizations [12].

The first edition of the international standard ISO 50001 in 2011 was in line with the standard picture of requirements for management systems, describing elements such as energy policy, operational planning, and control, internal audits and management review. In 2018, an updated edition of the standard inscribed in the High Level Structure system was published. The main difference was the application of a standardized 10 point layout, allowing for easy integration of individual systems. The purpose of the standard is to improve company energy performance. ISO 50001 provides a range of requirements that support organizations in:

- development of energy policy for more efficient use of energy,
- setting goals, targets, and objectives to meet the policy,
- use of data to improve the decision-making process,
- measuring of results,
- review the effectiveness of the energy policy,
- continuous improvement of energy management and energy performance.

The next part of the paper presents conclusions related to the use of data in Energy Management Systems, drawn based on practical experience of the authors gathered during their work with organizations implementing these systems. The purpose of the research is to determine a universal method of data collection that can be used in Energy Management Systems.

2 Literature Review

The ISO 9000 standard uses term information to refer to data that are significant to the management system [13]. Information is one of the most valuable assets of modern enterprises. Information significantly affects the efficiency and effectiveness of an organization. The information is contained in practically every activity carried out, starting from organizations know-how, patents, technology, through information about clients and competitors, ending with the instructions issued by superiors to employees in order to perform operational activities. Therefore, that information is also a basic resource management systems. Decisions are made with the use of information, without which it would not be possible for the company to function efficiently. It reflects one of the basic functions of quality management that talks about making decisions based on facts [14]. Thus, the information concerning business enterprises and their business processes plays a key role. In this approach, the basic functions of information are: notification (informative), decision-making, and controlling. Without information, it is not possible to make decision-making processes. It determines, as the choice of the decision-maker [15]. In the decision-making process, the task of information is to provide an appropriate description of a given situation so that the decision is made in the state of least uncertainty. The accuracy of decisions made depends on the quality of information of decision-maker.

3 Research Methodology

Due to the specificity of information (immateriality), its quality is often assessed using the so-called information attributes. The most frequently mentioned may include comprehensiveness, accuracy, clarity, and currency [16, 17]. Table 1 shows their brief characteristics.

Table 1. Information attributes selected for evaluating the quality of information [18].

Information criteria	Description
Comprehensiveness	Is the scope of information adequate? (not too much nor too little)
Accuracy	Is the information precise enough and close enough to reality?
Clarity	Is the information understandable or comprehensible to the target group?
Applicability	Can the information be directly applied? Is it useful?
Conciseness	Is the information to the point, void of unnecessary elements?
Consistency	Is the information free of contradictions or convention breaks?
Correctness	Is the information free of distortion, bias, or error?
Currency	Is the information up-to-date and not obsolete?

The situation in energy management systems (EnMS) is similar to that described above. Under the requirements of ISO 50001: 2018, EnMS should be primarily data-driven and fact-based [12]. One of the basic elements of such a system is Energy Performance Indicator (EPI), which by definition, should be based on data that an organization can monitor, measure, and analyze [19]. Examples of data that can be collected in organizations, along with their description and relevant variables are given in Table 2.

Table 2. Examples of energy data in individual organization processes.

Energy use	Data collected	Static factors	Relevant variable
Maintenance of infrastructure	Building Heating [kWh], Lighting of buildings and infrastructure [kWh], Air conditioning [kWh]	Building volume, Lighting area	Outdoor temperature, Duration of sunshine, Energy performance of buildings

(continued)

Table 2. (*continued*)

Energy use	Data collected	Static factors	Relevant variable
Internal transport	Quantity of fuels consumed - oil, gas, gasoline [Mg, m^3]	Euro emissions standard, Technical condition of vehicles,	Distance traveled by vehicles
External transport	Quantity of fuels consumed - oil, gas, gasoline [Mg, m^3]	Euro emissions standard Technical condition of vehicles	Distance traveled by vehicles
Production	The consumption of: Production gas (including welding gases) [m^3] Coal [Mg] Heating oil [Mg] Electric current [kWh] Steam Compressed air	Number of work shifts, Energy consumption of machines and equipment	Production volume, Amount of waste, Number of manufacturing defects
Provision of services	Electricity consumption [kWh]	Work time	Order amount
Office work and social facilities	Electricity consumption [kWh]	Number of employees, Number of working shifts	Work time

Relevant variables are understood as characteristic variables that may affect the amount of energy used in specific areas of energy use. These variables, such as the outdoor temperature, often do not depend directly on the organization, so their correct determination and monitoring may be the key to developing authoritative EPIs. The issue of static factors, which should also be monitored, looks similar, but the situation is so much easier because they do not have so much volatility as significant variables.

In order to correctly collect the data necessary to determine and monitor the current energy consumption of an organization, each company should carry out an energy review on its own. During this review, the commitment organization is required not only to identify current types of energy and its use but also to evaluate its consumption in the past and forecast future changes.

Despite the apparent good availability of data on energy use in individual areas of the company, a large part of these data is subject to considerable uncertainty, and their utility in the management of energy is limited.

4 Results

As stated above, the base values for EnMS, such as areas of significant energy use, energy performance indicators, etc., are identified and calculated based on data collected from processes implemented in the scope of the system. The quality of data and information will depend on the correctness and credibility of EnMS. Concerning the qualitative features of information, it can be stated that data used in EnMS must have, above all, an appropriate level of: accuracy, correctness, and currency. The following observations related to the listed attributes of information result from the practical experience of the authors gathered during the implementation of EnMS in organizations.

Data accuracy is related to the ability to measure data for specific areas of energy use. For example, the company has 3 buildings of different sizes, in which various processes are carried out. These are manufacturing processes, storage processes, and auxiliary processes (maintenance workshop). In each of these buildings, electricity is used for the processes. The company has an electricity meter installed that allows reading only data on the total amount of electricity consumed. In this case, the accuracy of data is insufficient, as it makes it impossible to read how much electricity was consumed by the production machines, how much to illuminate the production or storage area and how much was used for the work of power tools used in the workshop. Data accuracy is also related to parametrized static factors. In both cases, the obvious priority comes from ensuring adequate measurement accuracy. Due to the lack of adequate data accuracy, it is not possible to properly manage energy use areas in the context of energy efficiency.

The second attribute of information - correctness, is mainly related to the reliability of the data. Reading data from a non-calibrated device may be affected by an error, which will affect the lack of comparability and reproducibility of data obtained. The reliability of information is also affected by the use of analytical methods. An example of the lack of information correctness can be the calculation of EnMS performance indicators. For the calculation of EPI of individual buildings, related to the consumption of heat energy produced from gas, the ratio of the amount of energy consumed (GJ) to the area (m2) of heated rooms was assumed. In the company, there are, among others, 2 buildings of similar area and in which similar processes are carried out. EPIs differ significantly for both objects. It is due to incorrect assumptions for calculating the indicator. The height of the rooms is not taken into account, which is important because not a certain area is heated, but the volume. The adoption of the volume unit (m^3) in the denominator caused the EPI indicators in the monitored rooms became more similar.

The third attribute, the currency of data, is of particular importance in the case of highly dynamic data. These data are most often associated with areas for which relevant variables have been identified. The need to monitor and provide current data applies to both measured values and relevant variables. In particular, the identification and consideration of the variables is a task that is difficult for organizations and therefore often overlooked. For example, the consumption of electricity for the operation of air-conditioning devices will be different in different seasons of the year for a moderate climate, where is an issue with large differences in external temperatures in winter and summer. Additionally, it can be compared differently for periods with different average temperatures like cold and hot summer (e.g. average temperature in Wroclaw, Poland in July 2011 to 18.0 degrees C, and in July 2014 - 22.0 degrees C [20]). The absence of these variables or estimated approach to data may lead to erroneous conclusions drawn as to the unjustified increase in electricity consumption.

Table 3 presents which areas or elements of EnMS are associated with the individual attributes of information quality.

Table 3. Relationship of information attributes with areas EnMS.

Information criteria	Relation with EnMS area
Accuracy	The ability to read data from all areas of energy use, Parameterization and measurement of static factors and relevant variables Adopted accuracy of measurement
Correctness	Reliability of measuring devices Unit comparability Correct use of analytical methods
Currency	Frequency of data collection, Dynamics of relevant variables

5 Conclusions

Due to the increasing problems of modern civilization, often related to dwindling energy resources, energy management systems are now becoming one of the most popular types of management systems. One of the basic elements of the system, which allows its proper management is data from processes, among other relevant variables, energy use, or EPI. The quality of these data and information is crucial for the credibility and effectiveness of EnMS. Ensuring the accuracy, correctness, and currency of data will affect, firstly, the ability to manage individual areas of the system and, secondly, the effectiveness of decisions taken. Failure to ensure an adequate level of the three attributes of information can lead to situations where areas of significant energy use will be identified incorrectly. It is, therefore, important to provide the right tools for data collection, use of appropriate analytical methods, and analysis tools. It will allow making decisions that will be adequate to the current state of energy efficiency. It will significantly reduce the uncertainty of decisions taken and thus enable improvement of energy efficiency, as well as improvement of the whole EnMS. The assumptions described in this paper may serve as guidelines for building such a mechanism for collecting data of appropriate quality.

References

1. Antosz, K., Chandima, R.M.: Machinery classification and prioritization: empirical models and AHP based approach for effective preventive maintenance. In: IEEE International Conference on Industrial Engineering and Engineering Management, pp. 1380–1386 (2016)
2. Antosz, K., Stadnicka, D.: Lean philosophy implementation in SMEs - study results. In: 7th International Conference on Engineering, Project and Production Management, Procedia Engineering, vol. 182, pp. 25–32 (2017)
3. Kujawińska, A., Diering, M., Rogalewicz, M., Żywicki, K., Hetman, Ł.: Soft modelling-based methodology of raw material waste estimation. In: Burduk, A., Mazurkiewicz, D. (eds.) Intelligent Systems in Production Engineering and Maintenance – ISPEM 2017, Advances in Intelligent Systems and Computing, vol. 637, pp. 407–417. Springer (2018)

4. Rewers, P., Trojanowska, J., Diakun, J., Rocha, A., Reis, L.P.: A study of priority rules for a levelled production plan. In: Hamrol, A., Ciszak, O., Legutko, S., Jurczyk, M. (eds.) Advances in Manufacturing. Lecture Notes in Mechanical Engineering, pp.111–120. Springer (2018)
5. Trojanowska, J., Kolinski, A., Varela, M.L.R., Machado, J.: The use of theory of constraints to improve production efficiency–industrial practice and research results. DEStech Transactions on Engineering and Technology Research (ICPR 2017), pp. 537–542 (2017)
6. Trojanowska, J., Kolinski, A., Galusik, D., Varela, M.L.R., Machado, J.: A methodology of improvement of manufacturing productivity through increasing operational efficiency of the production process. In: Hamrol, A., Ciszak, O., Legutko, S., Jurczyk, M. (eds.) Advances in Manufacturing. Lecture Notes in Mechanical Engineering, pp. 23–32. Springer (2018)
7. McKanne, A.: Thinking Globally: How ISO 50001 – Energy Management can make Industrial Energy Efficiency Standard Practice. Ernest Orlando, Lawrence (2009)
8. Giacone, E., Mancò, S.: Energy efficiency measurement in industrial processes. Energy **38** (1), 331–345 (2012)
9. Lindgreen, A., Swaen, V.: Corporate social responsibility. Int. J. Manag. Rev. **12**, 1–7 (2010)
10. Finnerty, N., Sterling, R., Coakley, D., Contreras, S., Coffey, R., Keane, M.M.: Development of a global energy management system for non-energy intensive multi-site industrial organisations: a methodology. Energy **136**, 16–31 (2017)
11. Limiting Global Climate Change to 2 degrees Celsius – The way ahead for 2020 and beyond, Communication by the Commission to the European Council. https://eur-lex.europa.eu/LexUriServ/LexUriServ.do?uri=COM:2007:0002:FIN:EN:PDF. Accessed 07 Feb 2019
12. EN ISO 50001: Energy management systems – Requirements with guidance for use. Geneva (2018)
13. ISO 9000: Quality Management Systems. Requirements. Geneva (2018)
14. Materska, K.: Information in Organisations of Knowledge Society. SBP, Warsaw (2007).[in Polish]
15. Hamrol, A.: Quality Management with Examples. PWN, Warsaw (2015)
16. Eppler, M.J.: Managing Information Quality. Springer, Heidelberg (2006)
17. Floridi, L.: Information ethics, its nature and scope. Comput. Soc. **34**(5), 3 (2005)
18. Grudzień, Ł., Hamrol, A.: Information quality in design process documentation of quality management systems. Int. J. Inf. Manag. **36**(4), 599–606 (2016)
19. Osiński, F., Grudzień, Ł.: Polish SME energy efficiency in the years 2014–2016. In: Machado, J., Soares, F., Veiga, G. (eds.) Innovation, Engineering and Entrepreneurship. HELIX 2018. Lecture Notes in Electrical Engineering, vol. 505. Springer, Cham (2019)
20. Climate-data. https://pl.climate-data.org/europa/polska/lower-silesian-voivodeship/wroc%C5%82aw-4531/. Accessed 27 Jan 2020

Automated Attestation of Metrics for Industrial Robots' Manipulation Systems

Valerii Kyrylovych(✉), Anton Kravchuk, Petro Melnychuk, and Liudmyla Mohelnytska

Zhytomyr Polytechnic State University, 103, Chudnivska Street, Zhytomyr 10005, Ukraine
kiril_va@yahoo.com

Abstract. A new approach to determine the technological capabilities of industrial robots (IR) using the developed methodology on automated attestation of IR manipulation systems metrics is described. Briefly, the essence of the process of IR metric attestation is a determination of the geometric characteristics of the IR working area, in which the user-defined orientation of the analyzed attestation parameter H is ensured. H is determined by the construction and kinematic parameters of the MS links, the clamping device (CD), and the object of manipulation (OM), if present in the CD. Procedurally, the IR metric attestation involves multiple solutions of the kinematics inverse problem on many IR MS links, while maintaining the user-defined orientation of the CD with/without OM in it in the IR coordinate system using the SolidWorks and RoboDK. The IR metric evaluation outcome is the geometric characteristics of the IR work areas, in which the attestation parameter H retains the predetermined user orientation. Visually, the attestation results are presented as a part of the analyzed IR work area with geometrically defined characteristics of the parameter H. The advantages of the proposed approach are the reduction of financial, time and intellectual resources for carrying out the attestation due to the absence of the need for real IR. The efficiency of the developed methodology is illustrated by the attestation of the MS metric of the Tinker Kit Braccio robot. The results obtained were visualized, commented on, and they confirmed the efficiency of the proposed approach and the methodology that implements it.

Keywords: Working area · Clamping device · Geometric characteristic · Parameter · Visualization

1 Introduction

The widespread use of industrial robots (IR) in modern mechanical assembly manufactures (RMAM) is a characteristic feature of these industries' automation. The annual production and implementation of IR increase on average by 15% [1]. The design and operation of robotic technological structures, such as flexible manufacturing cells (FMC), requires the availability of technical information that facilitates and/or provides for the solution of certain tasks in the technological preparation (TP) of RMAM.

V. Tonkonogyi et al. (Eds.): InterPartner 2020, LNME, pp. 813–822, 2021.
https://doi.org/10.1007/978-3-030-68014-5_79

It refers, for example, to the need to have information about maintaining a certain orientation in the coordinate system (CS) of the IR clamping devices (CD), regardless of their design. It refers primarily to the determination of the geometric characteristics of the IR work areas, in which the CD maintains its certain orientation, for example, vertical, horizontal, and other IR CD intermediate orientations. Herewith, the "one-handed" and "one-gripper" robots are analyzed.

These tasks can be solved by specific tools and approaches that are designed to improve the efficiency of solving TP RMAM tasks. One such approach, based on the so-called MS metric attestation of the analyzed IR, is presented in this paper.

In this case, the attestation of the IR MS metric refers to the definition of certain dependencies and regularities of the IR investigated parameter values, which is due to the construction and kinematic characteristics of the MS links of the investigated IR. In this case, the parameter under study is the orientation of the CD with/without the object of manipulation (OM) in it in the coordinate system (CS) of the IR.

The concept of construction and kinematic characteristics should be understood as the size of the links, their shape, type, and magnitudes of their mutual movements, which for the compactness and brevity of this term is called the IR MS metric [2]. For the attestation of the IR MS metric, it is necessary to have data about the IR and the construction and geometric characteristics of the end-effector type, first of all, a CD or a tool, taking into account the overall dimensions of the OM if present in the CD.

Advanced manufacturers of such IR models as Mitsubishi [3], ABB [4], Fanuc [5], Kuka [6], do not specify the above parameters in the technical documentation. The information available in the technical documentation of the above mentioned IRs is not sufficient to solve many tasks of TP RMAM.

Thus, on the one hand, the enterprises producing IR do not provide the information needed to improve the solution of the TP RMAM tasks; no existing regulatory document either in Ukraine or in other countries provides for its mandatory availability. It considerably complicates the solution of such practical problems of TP RMAM, which is usually quite laborious, such as placement and planning of technological equipment in the IR work areas, determination of coordinates of trajectories anchor points, synthesis, and planning of trajectories themselves, etc.

2 Literature Review

The following analysis of information sources indicates that, in general, the subject of attestation is not new to IR. At the same time, in one way or another, similar parameters, considered in this paper, or similar procedures of attestations for similar parameters of IR are highlighted. A considerable amount of research is focused on determining, in one way or another, such parameters of IR as accuracy and geometric characteristics of IR working areas [7–11].

Accuracy is the parameter of attestation in the paper [7]. Its use is possible in TP RMAM, taking into account different values of the precision parameter in different parts of the IR working areas. It can be used for attestation regarding the parameter of the vector of approach to the analyzed point, in determining the accuracy of the positioning the CD in it, but not in any way for the attestation of the IR MS metric.

The result of the study [8] is the determined dependence of the IR CD positioning accuracy and the geometry of the analyzed IR working area, where the specified accuracy is provided. Fragmentation and lack of attestation parameter determine the impossibility of using this work for attestation of the IR MS metric.

Metric determination elements are used indirectly in research [9]. They are determined by a limited movement of IR links, which implicitly correlates with accuracy. However, this is not indicated, and in fact, the attestation of the IR working areas is not performed either by the IR MS metric or by the parameter of accuracy. Therefore, the value of this work for disclosing the purpose of this paper is doubtful.

The result of [10] is its approach to the method of estimating the presence of inverse kinematics, based on the Denavit-Hartenberg (DH) model. It cannot be used to solve the mentioned problems because the description of the IR kinematics by DH matrices involves the use of only thread models of MS links and angles between them using directional cosines. In order to conduct the metric attestation, it is necessary to have 3D models of IR MS links.

The process of IR MS metric attestation in the above interpretation was first considered in [11]. Nevertheless, the content of these works is rather declarative and does not give a systematic and detailed account of the methodological sequence of the automated IR MS metric attestation, which is the content of the work proposed here.

Available object-oriented software products, namely RoboMaster [12], ROS [13], RobotAnalyzer [14], do not directly solve similar problems in this formulation, but according to the detailed analysis of their capabilities, they can be used as a tool for such research. For example, RoboDK software, as one of the most versatile simulators for IR and their off-line programming, allows you to integrate a 3D model of a robot and to simulate the processes of moving MS links in the 3D space.

So, the analysis of information sources indicates that the results obtained in them are fragmentarily useful and not focused on implementation at TP of RMAM.

Thus, the purpose of this work is to develop a new approach to investigate the technological capabilities of IR by conducting the automated IR MS metric attestation as an integral part of TP RMAM, which increases the efficiency and reduces the complexity of its (TP) execution when designing new and/or modernizing the existing machinery and instrument making FMC.

3 Research Methodology

The content of the proposed methodology for IR metrics attestation is a methodological basis of the proposed approach to the study of IR technological capabilities. The process is performed by the following chain of actions: structuring, parameterization, implementation, research, and visualization. It is a sequence of actions, that is the basis of the developed step (St)-by-step methodology for IR MS metric attestation. It is implemented in a sequence of stages SI, SII, SIII (Fig. 1).

Considering the content and essence of the IR MS metric attestation process proposed here, solving the tasks of automated IR MS metric attestation is reduced to the implementation of a set of computational procedures $\left(\varphi_k | k = \overline{1,\, n_\varphi}\right)$ with a total

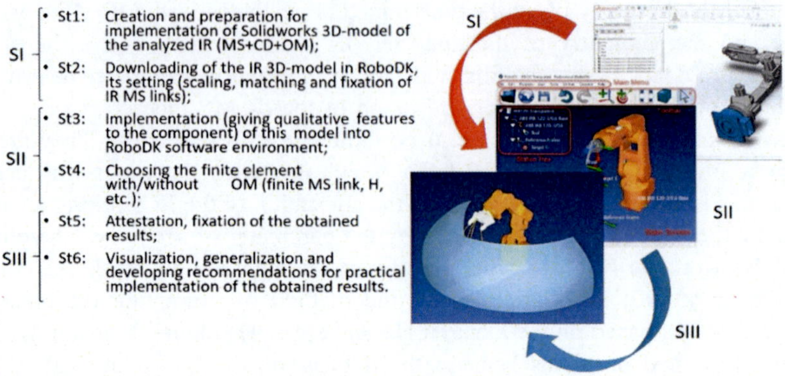

Fig. 1. General scheme of the developed methodology for automated IR MS metrics attestation.

number n_φ, that converts IR MS metric parameters of many links $\left(l_i | i = \overline{1,\, n_l}\right)$ with the total number n_l, sets of generalized coordinates $\left(q_j | j = \overline{1,\, n_q}\right)$ with total number n_q, where $n_l > n_q$, into the attestation parameter P. In this case, P is a vector V_p of parameter P, which is analyzed for its position in the IR work area (WA), i.e., in the IR CS. V_p can be directed and therefore investigated at any angle to the vertical V and/or the horizontal H, i.e. $\widehat{VH}$.

The mathematical model of the IR MS metric attestation process (expression (1)) in terms of set theory [15] is a surjective mapping (symbol $\rightarrow$) of the Cartesian product (symbol x) of the above IR MS metric parameters (l_i) and (q_j) onto the studied attestation parameter P:

$$\left(\varphi_k | k = \overline{1,\, n_\varphi}\right) : \left(\left(l_i | i = \overline{1,\, n_l}\right) \times \left(q_j | j = \overline{1,\, n_l}\right) | n_l \geq n_q\right) \rightarrow \\ \rightarrow \left(P = (V_P \in WA); V_P = \widehat{VH}\right). \quad (1)$$

The first stage SI (St1, St2) presupposes the creation of the IR 3D model in SolidWorks using the stl format (St1). St2 is downloading, one by one, 3D-components of IR parts models into RoboDK and setting them, namely, scaling, matching, locking, etc.

The second stage SII, implemented by St3 and St4, involves the final tuning of the IR as a single mechanical system of downloaded parts. At this stage, the IR CS and the number of degrees of freedom are chosen, as well as qualitative and quantitative parameters of the MS links mobility ratio are provided, and avoidance of various collisions in the RoboDK (St3) is ensured. Step St4 is the choice of an MS finite element, that is, the CD or tool with/without OM in the CD with their respective geometric dimensions.

At the final stage SIII, the process of the analyzed IR MS metric attestation is performed, that is, the stepwise determination of the extreme boundary points in the IR working area, in which the specified orientation of the attestation parameter is preserved, as components of the IR finite element. This operation has a certain discreteness, which is determined by the step of moving between the *i*-th and *j*-th points, which in turn are determined by the accepted steps of linear and angular sampling (St5). The theoretical and methodological basis of the St5 implementation is the solution of the inverse kinematics problem [2], taking into account the geometric parameters of the 3D model of the IR metric components. These components determine the attestation parameter H (Fig. 2, Fig. 3, Fig. 5).

The final step St6 is to visualize the information received, summarize it, and develop recommendations for practical use. SolidWorks provides visualization.

4 Results

The above method of IR MS metric attestation was tested using automated attestation of the Tinker Kit Braccio model [16]. The attestation was carried out in the absence of OM in the CD with the value of the attestation parameter H = 65 mm (Fig. 2, 3 and 4) and for the case of OM fixed in the CD at H = 77 mm (Fig. 5, 6 and 7). The obtained data of the performed MS metric attestation for the analyzed robot processed and visualized.

In Fig. 2, the obtained results are presented in the form of a part's cross-section of the analyzed robot workspace, where the orientation of the finite element H without OM in the CD is kept in the vertically oriented position of the parameter H with the angle ($\widehat{VH} = 90°$ by expression (1)), i.e., CD pointing down. Green indicates the area in which the specified position of the attestation parameter H is guaranteed.

Here (Fig. 2), the markings of the parameter H at its position in the green zone are indicated, namely with the characteristic designated here (and hereinafter with the

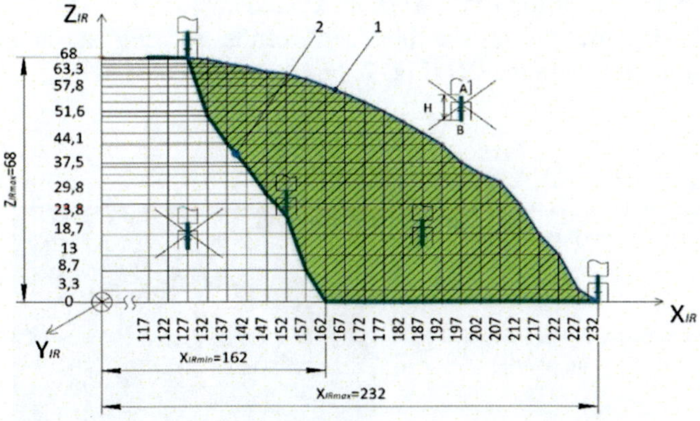

Fig. 2. Graphical visualization of the coordinates Z_{IR} and X_{IR} dependences with the vertical orientation of the parameter H = 65 mm without OM in the CD directed downward relative to the basis of IR – H-Axis Vertical Down (WAVD): 1 – the curve of the attestation zone outer contour when changing the coordinates of the parameter H in the interval [X_{IRmin}...X_{IRmax}] = [162...232] mm; 2 – the curve of the attestation zone inner contour when changing the coordinates of the parameter H in the interval [Z_{IRmin}...Z_{IRmax}] = [0...68] mm.

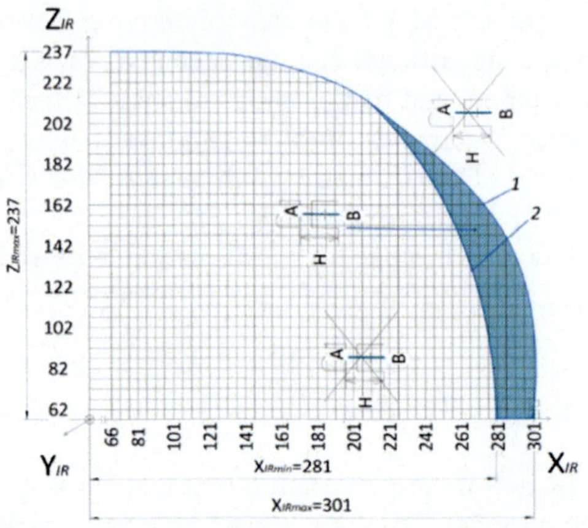

Fig. 3. Graphical visualization of the coordinates Z_{IR} and X_{IR} dependences with the horizontal orientation of the parameter H (without OM in the CD) is directed forward (to the right) from the robot stand (Z_{IR} axis) in the direction of increasing the radius of the IR working area - H-Axis Horizontal Forward (WAHF): 1 – the curve of the attestation zone outer contour when changing the coordinates in the interval $[X_{IRmin}\ldots X_{IRmax}]$ = [281…301] mm; 2 – the curve of the attestation zone inner contour when changing coordinates in the interval $[Z_{IRmin}\ldots Z_{IRmax}]$ = [62…242] mm.

corresponding adjustments) as (H-Axis Vertical Down) (WAVD). Such designations are further considered as characteristics of the parameter H. The fact of the non-fixed investigated orientation of the parameter H in the non-darkened zone is indicated by the correspondingly crossed images of the CD outside the green zone.

Similar attestation studies were performed for H = 65 mm without OM with the direction of the WAHF characteristic (Fig. 3), that is, the vector H is directed towards the increase (Forward) of the robot work area radius.

The final visualization of the obtained results for the attestation parameter H = 65 mm without OM in the CD (Fig. 2, Fig. 3) is presented in Fig. 4. 3D work area

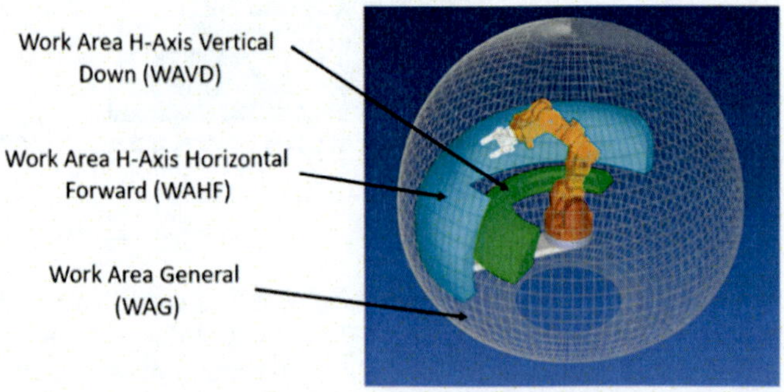

Fig. 4. WAVD and WAHF as the parts of WAG (Working Area General) with H = 65 mm without OM.

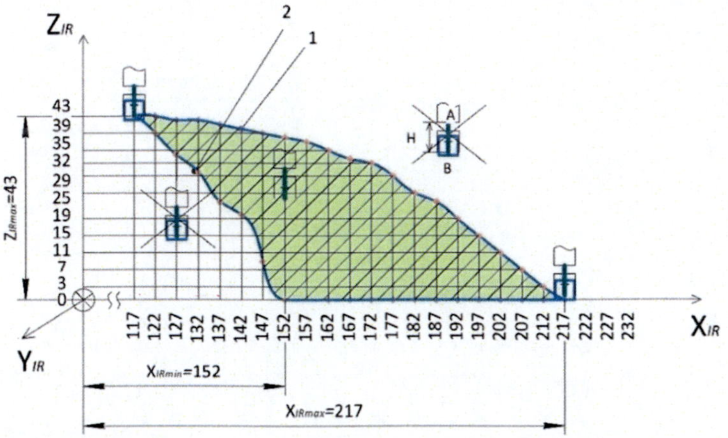

Fig. 5. Graphical visualization of the coordinates Z_{IR} and X_{IR} dependences with the vertical orientation of the parameter H = 77 mm (with OM in the CD), directed downward relative to the IR basis – H-Axis Vertical Down + OM, i.e. (WAVD+OM): 1 – the curve of the attestation zone outer contour when changing the parameter H coordinates in the interval $[X_{IRmin}...X_{IRmax}]$ = [152...217] mm; 2 – the curve of the attestation zone inner contour when changing the parameter H coordinates in the interval $[Z_{IRmin}...Z_{IRmax}]$ = [0...43] mm.

models of the Tinker Kit Braccio robot are created in SolidWorks specialized software environment.

Relevant parts of the IR work area with a guaranteed parameter H position of the final element (CD) without OM are depicted. The blue zone reflects the part of the robot work area, where the orientation of the finite element of the CD without OM in it, i.e., with WAHF characteristic. The green robot work area stores the orientation of the attestation parameter H as a part of the finite element of the CD without the OM in CD in a robot with WAVD characteristic.

Similar studies were performed with the fixed OM in the CD of this robot, with the parameter H = 77 mm (Fig. 5, 6 and 7).

The results of the studies showed that practically determined working area, in this case, has shifted to robot CS and decreased in the area for both positions of the clamping device with H = 77 mm with fixed OM in the CD. Graphical visualization of the results is presented in Fig. 5. Comparison of the areas of coordinate sections of orientation H working areas in Fig. 4 and Fig. 5 (CD without OM) is approximately 40%.

Similar results are shown by the analysis of the study of the attestation parameter H coordinates and orientation to the plane on which the basis of the robot is fixed (Fig. 6). There is a characteristic difference in the size of the intersection between the parameter H features (WAVD) and (WAVD+OM). In this case, this difference is 75%. That is, the area of intersection of the robot working space for H = 77 mm with the characteristic (WAVD+OM) decreased nearly 4 times relative to the area of the zone without OM in the CD with the characteristic (WAVD) (Fig. 2).

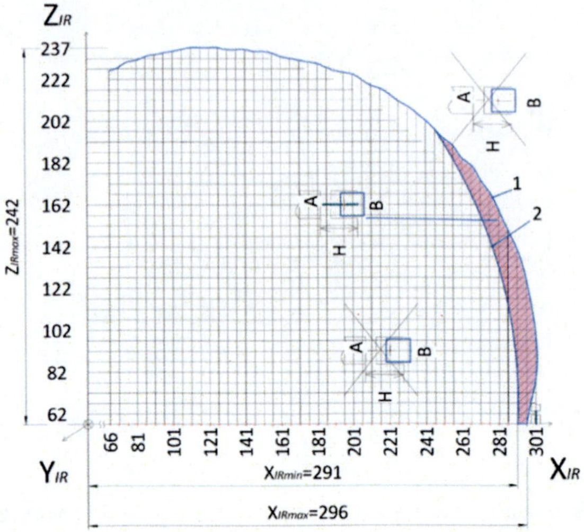

Fig. 6. Graphical visualization of the coordinates Z_{IR} and X_{IR} dependences with the horizontal orientation of the parameter H (with OM), directed forward (to the right) from the robot stand (Z_{IR} axis) in the direction of increasing the radius of the IR working area - H-Axis Horizontal Forward+OM, i.e. with the OM fixed in CD (WAHF+OM): 1 – the curve of the attestation zone outer contour when changing the parameter H coordinates in the interval $[X_{IRmin}...X_{IRmax}]$ = [291…296] mm; 2 – the curve of the attestation zone inner contour when changing attestation parameter H coordinates in the interval $[Z_{IRmin}...Z_{IRmax}]$ = [62…242] mm.

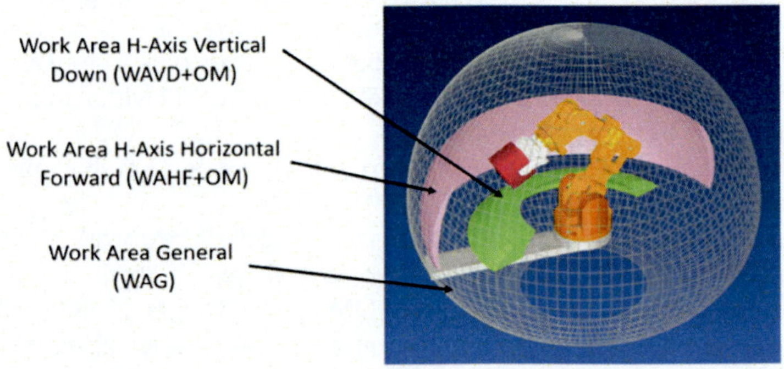

Fig. 7. WAVD and WAHF as the parts of WAG (Working Area General) with H = 77 mm and with OM.

Figure 7 presents the visualization of the obtained data for the attestation parameter H = 65 mm with OM in the CD (see Fig. 5, Fig. 6), namely the work areas in which the position of the parameter H of the finite part (CD) with the OM is guaranteed. The pink zone represents the robot work area, where the orientation of the finite element with OM in the position with (WAHF+OM) characteristics is stored, and the green robot work area stores the orientation of the attestation parameter H as a part of the CD finite element with the OM in CD in a robot with (WAVD+OM) characteristics.

Figure 8 presents a generalized visualization of the results obtained. This figure characterizes the difference in the size of robot work areas with OM and without OM in the CD, which is determined by the values of the parameter H.

Fig. 8. WAVD and WAHF as the parts of WAG with OM (H = 77 mm) and without OM (H = 65 mm).

5 Conclusions

A new approach to identify the technological capabilities of IR is proposed. It is the automated attestation of the IR MS metric. Its main component is the multiple solutions of the inverse kinematics problem to provide the necessary orientation of the attestation parameter. It is a geometrical and construction part of the CD with/without OM in it. Process automation is provided through the use of Solid Work and RoboDK software. The developed methodology does not require the presence of a real IR.

The results of the proposed approach and IR MS metric attestation allow solving certain technological problems of industrial robotics, both in the analysis of the possibility to use existing mechanical assembly FMCs and a priori they are the basis for solving the tasks of planning, placement of technological equipment in the IR work areas and other tasks of TP RMAM when designing new FMC of machine building and tool industry.

References

1. International Federation of Robotics. https://ifr.org/. Accessed 2 Dec 2020
2. Korendyasev, A.I., Salamander, L.I.: Theoretical foundations of robotics. Science, Moscow (2006). (In Russian)
3. Mitsubishi Electric Homepage. https://www.mitsubishielectric.com/. Accessed 10 Feb 2020
4. ABB Homepage. https://new.abb.com/. Accessed 10 Feb 2020
5. Fanuc Homepage. https://www.fanuc.com/. Accessed 10 Feb 2020

6. Kuka Homepage. https://www.kuka.com/. Accessed 10 Feb 2020
7. Kyrylovych., V., Sazonov, A.: Impositions of the brush loam of the industrial robot. Patent of Ukraine, No. 58988 (2011)
8. Tian, W., Mei, D., Li, P., Zeng, Y., Hong, P., Zhou, W.: Determination of optimal samples for robot calibration based on error similarity. Chin. J. Aeronaut. **28**, 946–953 (2015)
9. Bader, A.M., Maciejewski, A.A.: Maximizing the failure-tolerant workspace area for planar redundant robots. Mechanism and Machine Theory 143, (2020). https://www.sciencedirect.com/science/article/pii/S0094114X19310055. Accessed 10 Feb 2020
10. Wang, S., Luo, X., Luo, Q., Han, B.: Existence Conditions and General Solutions of Closed-form Inverse Kinematics for Revolute Serial Robots. Applied Sciences 9 (2019). https://www.mdpi.com/2076-3417/9/20. Accessed 10 Feb 2020
11. Kyrylovych., V., Kravchuk, A.: The automated certification of working areas of industrial robots' geometrical characteristics: necessity and essence. In: 4th International Proceedings on Theoretical and Applied Aspects in Radio Engineering, Instrument Making and Computer Technologies, pp. 190–191. Ternopil (2019). (In Ukrainian)
12. RobotMaster Homepage. https://www.robotmaster.com/en/. Accessed 11 Feb 2020
13. ROS Homepage. https://www.ros.org/. Accessed 12 Feb 2020
14. RobotAnalyzer Homepage. http://www.roboanalyzer.com/. Accessed 12 Feb 2020
15. Sigorskiy, V.P.: Math for Engineer. Technika, Kyiv (1975). (In Russian)
16. Tinker Kit Braccio Homepage. https://store.arduino.cc/tinkerkit-braccio-robot. Accessed 10 Feb 2020

Using the Specific Molarity Indicator of the Chemical Parameters of Mineral Waters in Assessing Their Biological Effects

Alona Kysylevska[1(✉)], Konstantin Babov[2], Sergey Gushcha[2], Igor Prokopovich[1], and Boris Nasibullin[2]

[1] Odessa National Polytechnic University, 1, Shevchenko Avenue, Odessa 65044, Ukraine
kisilevskaya07@gmail.com

[2] State Institute «Ukrainian Scientific-Research Institute of Medical Rehabilitation and Balneology of the Ministry of Health of Ukraine», 6, Lermontovsky Lane, Odessa 65014, Ukraine

Abstract. The article presents materials on the scientific justification for the use of the specific molarity index of mineral water components in assessing their biological effects. The authors studied 18 types of mineral waters of Ukraine with a salinity of 0.14 g/l to 6.86 g/l and the content of metasilicic acid from 1.0 mg/l to 226.0 mg/l. The effectiveness of their biological action was studied by the example of an excretory function in animals (daily diuresis and glomerular filtration rate). The experiments proved that the features of the biological action of mineral waters (urination) depend on the content and ratio of all components. Microcomponents, in particular, organic substances, most affect the diuretic effect of mineral waters. An inverse relationship was established between the processes of urination and the specific molarity of chlorides and metasilicic acid. The possible biological activity of mineral waters is mostly reflected in the particular molarity of the components. Mineralization does not have a significant effect on urination; therefore, it cannot be a criterion for the differentiation of mineral waters by their biological outcome. The results make it possible to carry out the next stage of work to create methodological foundations for the biomedical classification of mineral waters, depending on the effectiveness of their biological effects.

Keywords: Natural mineral waters · Biological effect · Specific molarity · Daily diuresis · Glomerular filtration rate

1 Introduction

Water is a necessary factor in human existence. About 70% of a person's body weight consists of water [1–3]. Water is involved in the transport of nutrients, the elimination of metabolites and toxins, in maintaining water-electrolyte metabolism, which plays a vital role in maintaining homeostasis of the body. The body needs a constant supply of water [3]. A unique role in maintaining human life is played by mineral waters (MW), the consumed volume of which is increasing every year. MW is used for drinking,

V. Tonkonogyi et al. (Eds.): InterPartner 2020, LNME, pp. 823–832, 2021.
https://doi.org/10.1007/978-3-030-68014-5_80

treatment, and prevention of diseases, because they have an effect on homeostatic processes, contribute to the correction of micronutrient deficiencies, remove harmful substances from the body. The digestibility of minerals and biologically active substances from MW is better than from food [3]. Given Ukraine's European integration course, there is a need for new approaches to the development and implementation of regulatory requirements for MW to harmonize them with European requirements, which should be carried out, taking into account the broader spectrum of Ukrainian MW and their specific action [4].

2 Literature Review

Today, in Ukraine, the main criterion for assessing the presence of the biological effect of MW is mineralization (M), and for weakly mineralized MW– the presence of biologically active components in concentrations exceeding the balneological norm [5]. The biomedical assessment of MW in Ukraine provides for preclinical studies to determine their safety and quality. Experimental studies in animals (white rats) include physiological, immunological, biochemical, and morphological methods [5]. The state of the blood system, the structural and functional characteristics of the stomach, liver, pancreas, kidneys, central nervous, cardiovascular, and respiratory systems for the use of MW in healthy animals, and then in animals with a model of pathological conditions, are evaluated. In the presence of biological action after clinical trials, MW for packing acquire the status of natural medicinal-canteen, and for use in resorts–medicinal. The remaining MW are related to mineral table water [5].

Such a classification of MW in Ukraine is different from the European. According to the Directive [6], MW are food products. There is no limitation of mineralization. Any designation of MW properties associated with the prevention and treatment of human diseases is prohibited. It is allowed to put on the marking designations related to the content of components and effects of influence on the body according to the results of clinical and pharmacological analyzes of MW [6, 7].

Most of the studies on the biological action of MW are devoted to the role of their macro-components. For example, bicarbonate MW have a positive effect on the digestive tract [1, 8] and lower cholesterol [1]. Sulphate MW improve the motility and functioning of the intestine and the hepatobiliary system [1, 8]. Chloride MW stimulate motility of the gastrointestinal tract, excretion of electrolytes, and have a choleretic effect [1, 8]. Calcium MW increase the mineral density of bone tissue [1, 8, 9]. Magnesium MW improve motility and function of the intestines and biliary tract [1, 8]. The role of other trace elements in the biological activity of MW, except for fluorine and iron [1], is little considered. However, it is noted that the diuretic effect of MW with TDS < 50 mg/l also depends on their components, acting as catalysts for the enzymatic reaction of biochemical processes [8].

Trace elements in the chemical composition of MW can be considered from two sides– as safety indicators and as biologically active components, in the presence of which in significant concentrations, MW can exert a biological effect (therapeutic). Directive [10] establishes requirements for the safety criteria of MW. For each indicator, except boron, standards are set. The boron content in MW is normalized by regulatory documents of Ukraine [5] and Poland [11]– 5 mg/l. The same criterion is established as a balneological norm for natural healing MW. They significantly stimulate metabolic processes, activate the excretion of nitrogen metabolism products (creatinine and urea), and the function of the organs of the gastrointestinal tract [12].

Analyzing the types of packaged MW in Europe [13, 14], we noted that among them, there are practically no MW with organic substances (naphthenes, humins, bitumen, etc.) (> 5.0 mg/l [5]). In Ukraine, most of these MW are used at health resorts as therapeutic ones. Nevertheless, many MW are packaged in bottles (medical and canteen MW) (TDS < 1 g/l). The biological effect of mentioned waters on the organism of animals and humans is proved, in particular, stimulation of the functional state of the kidneys. The authors showed the healing effect of these MW when used in laboratory animals with experimental pathologies of carbohydrate metabolism and kidneys [12, 15].

In Europe, silicon in MW is also not standardized. In Ukraine, a balneological norm for the content of metasilicic acid for natural healing waters (50.0 mg/l [5]) has been established, as well as in many European countries. Silicon-containing MWs activate the processes of urination, excretion of nitrogenous substances [11, 16], positively affect the course of gastritis [17].

The beneficial effect of MW on the human body is a consequence of the synergistic effect of the components contained in the MW [18]. Trace elements, in this case, have a decisive, sometimes leading role in the implementation of biological actions. The existing classification of MW in Ukraine and Europe does not fully reflect the direction of their natural activity. It takes into account only the types of MW, using only dominant macrocomponents, and the presence of some biologically active components. There are differences in the approach to the differentiation of MW in Ukraine and Europe, both in terms of the degree of biological action and the content of components.

The effectiveness of the biological action of MW, which contain the same biologically active components, depends not only on the level of mineralization of MW but also on the concentration and ratio of ingredients [15]. According to [16], MW of one mineralization level, which differ only in the content of metasilicic acid, have different biological effects. MW with a lower concentration of metasilicic acid caused a more pronounced effect on the function of urination, and MW with a higher concentration affected a carbohydrate metabolism [16].

It was established in [17] that the use of MW with different content of metasilicic acid has a unidirectional, positive effect on the course of gastritis. MW with greater salinity, molarity, and a higher content of metasilicic acid has a more significant corrective impact on the development of gastritis. It was established in [19] that MW with higher molarity and content of metasilicic acid has a more pronounced restoring effect on the urinary function of the kidneys and the excretion of chloride ions.

The body's response to the intake of a complex of macro- and microelements is due to the need to maintain a constant water-electrolyte balance since even slight changes in the osmotic characteristics of extracellular water can disrupt the course of homeostatically critical metabolic processes [20, 21]. The body regulates the maintenance of water balance not by the amount of water entering the body, but by the osmolarity of extracellular fluid. Therefore, in assessing the effect of water on the body, several studies use indicators of osmolarity and molarity.

The authors of [22] proposed not using absolute indicators of mineralization and osmolarity, but relative ones, that is, attached to 1 g of salts, when assessing the biological effect. However, the authors of [22] believed that the sum of moles of dissolved substances in MW is osmolarity, although osmolarity can only be determined using an osmometer. Presumably, the calculated total molarities of all salts of MW were given in the works. Nevertheless, at the same time, the molarity of the micro-components should also be taken into account.

Based on the analysis of literature data and previous studies, it can be concluded that the macro and micro component composition, as well as the molarity of the MW and their components, play a significant role in the effectiveness of the biological action of MW.

Therefore, the goal of our work was to determine the relationship between the characteristics of the ratio of the concentration of macrocomponents, biologically active components (for instance, metasilicic acid and organic carbon), and the biological effect of MW (for example, excretory function).

3 Research Methodology

We evaluated 32 MW of Ukraine of different chemical composition. The concentrations of calcium, magnesium, hydro carbonates and chlorides were determined by the titrimetric method, metasilicic acid, organic carbon by the photometric method, sulfates by the gravimetric method, sodium and potassium by the calculation method. Relative values of the content of MW components were also expressed as specific molarity (y_i) according to formula (1):

$$y_i = \frac{M_i}{M} \cdot 100, \% \tag{1}$$

where *Mi* is the molar mass of the component MW, *M* is the total molar mass of MW.

To assess the biological effect of MW on the functional state of the urinary system–daily diuresis (DD) and glomerular filtration rate (GFR), we experimented on 40 white Wistar female rats weighing 180–200 g according to the Directive [23]. MW were administered to rats in the esophagus by a probe, once a day, at a dose of 1% of the animal's body weight for seven days. Rats were divided into 2 equal groups: the first group (20 pcs.) – intact animals (control group), the second group (20 pcs.) – healthy animals that received a course of introduction of MW. We used techniques [24]. GFR was determined by the clearance of endogenous creatinine:

$$GFR = \frac{V \cdot U_{cr} \cdot}{S \cdot P_{cr}}, \tag{2}$$

where *GFR* is the glomerular filtration rate, ml/(dm^2•min);
V is urine output in 1 min;
U_{cr} is urinary creatinine concentration;
S is the surface of the body;
P_{cr} is the plasma creatinine concentration [24].

Since the weight of the rats is not the same, GFR was assigned to the unit surface of the body (*S*) (3):

$$S = 9\sqrt[3]{w^2}, \tag{3}$$

where *w* is the body weight, g [24].

Statistical processing of the obtained data was carried out by the method of indirect differences and direct regression dependence of the change in the indicator on the duration of the course of water loads [25]. Significant changes were considered those in which the student coefficient was < 0.05. We got 18 MW with significant shifts in both DD and GFR. To calculate the ratios of the components of MW and the effectiveness of their biological action, the change in the values of DD and GFR was calculated to assess the biological activity of MW in percent relative to the control group.

4 Results

In 18 of 32 MW, both DD and GFR were significantly changed. Their mineralization (M) was from 0.14 g/l to 6.86 g/l with the content of metasilicic acid from 1.0 mg/l to 226.0 mg/l. DD compared with the control group, ranged from 23.9% to 128.4%. GFR ranged from 25% to 70%. To identify patterns of distribution of MW components depending on their chemical composition and biological action, we performed a correlation analysis (Table 1, 2).

Table 1. The results of the correlation analysis of the chemical composition and functional state of the urinary system of the MW (Spearmen correlation coefficients, r) (absolute values, mg).

Variable	M	H_2SiO_3	C_{org}	(Na + K)	Ca	Mg	HCO_3	Cl	SO_4	DD	GFR
M	1.0000	−0.0155	0.2839	**0.9334**	**0.7381**	**0.6035**	**0.7539**	**0.7667**	0.2446	0.2447	−0.0062
H_2SiO_3	−0.0155	1.0000	**−0,5996**	−0.1298	0.0052	−0.1493	−0.0922	−0.1487	**−0.4744**	0.0522	−0.0942
C_{org}	0.2839	−0.1298	1,0000	0.3394	0.2928	0.1880	0.4311	0.2440	0.4130	0.1955	**0.5711**
(Na + K)	**0.9334**	−0.1298	0,3394	1.0000	**0.6111**	**0.4837**	**0.7305**	**0.8147**	0.2870	0.2056	0.0702
Ca	**0.7381**	0.0052	0,2928	**0.6111**	1.000	**0.6163**	**0.6390**	**0.5049**	0.3199	**0.5070**	0.1652
Mg	**0.6035**	−0.1493	0,1880	**0.4837**	**0.6163**	1.0000	**0.6234**	0.2595	**0.5021**	0.3055	0.0281
HCO_3	**0.7539**	−0.0922	0,4311	**0.7305**	**0.6390**	**0.6234**	1.0000	0.3608	0.0424	0.4097	0.3103
Cl	**0.7667**	−0.1487	0,2440	**0.8147**	**0.5049**	0.2595	0.3608	1.0000	0.2845	0.0352	0.0740
SO_4	0.2446	**−0.4744**	0,4130	0.2870	0.3199	**0.5021**	0.0424	0.2845	1.0000	0.0547	0.0717
DD *	0.2447	0.0522	0,1955	0.2056	**0.5070**	0.3055	0.4097	0.0352	0.0547	1.0000	0.2152
GFR *	−0.0062	−0.0942	**0,5711**	0.0702	0.1652	0.0281	0.3103	0.0740	0.0717	0.2152	1.0000

* – % change. ** – Bold font shows statistically significant correlation coefficients. *** M – mineralization. **** DD – daily diuresis. ***** GFR – glomerular filtration rate.

From Table 1 it can be seen that positively significant r bind mineralization with sodium and potassium ($r = 0.9334$), chlorides ($r = 0.7667$) with hydrogen carbonates ($r = 0.7539$), and calcium ($r = 0.7381$). Significant positively significant sodium r with chlorides and bicarbonates are due to the good solubility of $NaHCO_3$ and NaCl salts. The negatively significant relationship between metasilicic acid and sulfates ($r = -0.744$) is explained by the high values of r sulfates with magnesium (r = 0.5021) and calcium ($r = 0.3199$) and the presence of a genetic bond between these ions. The greatest r connection between the change in DD is with calcium ($r = 0.5070$), less, but significant, with bicarbonates ($r = 0.4097$), even less with magnesium (r = 0.3055) and mineralization ($r = 0.2447$). A significant r is the relationship between the changes in GFR and C_{org} ($r = 0.5711$), less is with bicarbonates ($r = 0.3103$).

Table 2. The results of the correlation analysis of the chemical composition and functional state of the urinary system of the MW (relative values, millimoles).

Variable	H_2SiO_3	C_{org}	(Na+K)	Ca	Mg	HCO_3	Cl	SO_4	DD	GFR
H_2SiO_3	1,0000	−0,3950	**−0,6074**	**0,5909**	0,2230	−0,0134	0,0610	−0,1349	−0,2448	−0,0109
C_{org}	−0.3950	1,0000	−0.0032	−0,1197	−0,0235	0,2732	−0,1530	−0,1537	0,2969	**0,6330**
(Na+K)	**−0,6074**	−0,0032	1,0000	**−0,8404**	**−0,5803**	−0,3738	0,1560	0,4491	−0,0579	−0,2640
Ca	**0,5909**	−0,1197	**−0,8404**	1,0000	0,4130	0,0981	0,0145	−0,2362	0,0579	0,0400
Mg	0,2230	−0,0235	**−0,5803**	0,4130	1,0000	0,2136	−0,2075	0,1395	0,0671	0,0145
HCO_3	−0,0134	0,2732	−0,3738	0,0981	0,2136	1,0000	**−0,8745**	−0,2293	0,3490	0,3574
Cl	0,0610	−0,1530	0,1560	0,0145	−0,2075	**−0,8745**	1,0000	0,0036	−0,4236	−0,1850
SO_4	−0,1349	−0,1537	0,4491	−0,2362	0,1395	−0,2293	0,0036	1,0000	−0,2388	−0,1763
DD*	−0,2448	0,2969	−0,0579	0,0579	0,0671	0,3490	−0,4236	−0,2388	1,0000	0,2152
GFR*	−0,0109	**0,6330**	−0,2640	0,0400	0,0146	0,3574	−0,1850	−0,1763	0,2152	1,0000

* – % change. ** – Bold font shows statistically significant correlation coefficients. *** M – mineralization. **** DD – daily diuresis. ***** GFR – glomerular filtration rate.

From Table 2, it can be seen that r, calculated for specific values of the molarity of the components of MW, differ from their absolute values. So, the r of the relation between the specific molarity of Ca and the change in DD is much lower than the absolute value ($r = 0.0579$) (Fig. 1, 2). If there was practically no relationship between the absolute values of Cl and a change in DD, then there is a negatively significant relationship between the specific molarity of Cl and the change in DD ($r = -0.4236$). Positive correlations of the absolute values of sulfates and other macrocomponents of MW, and alterations in DD change to negative for specific molarities.

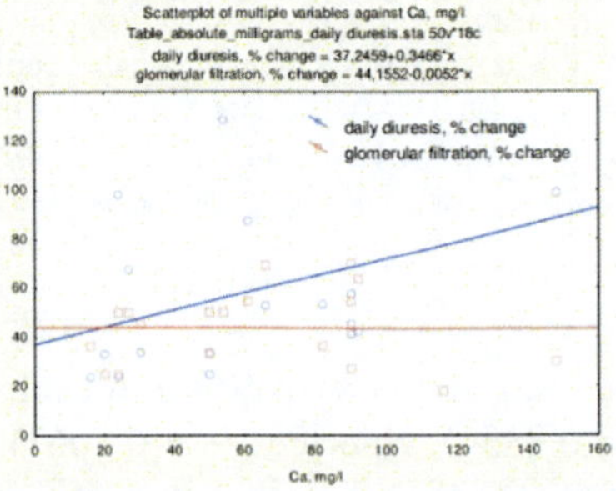

Fig. 1. The scattering diagram of the relationship of Ca with DD and GFR (absolute values).

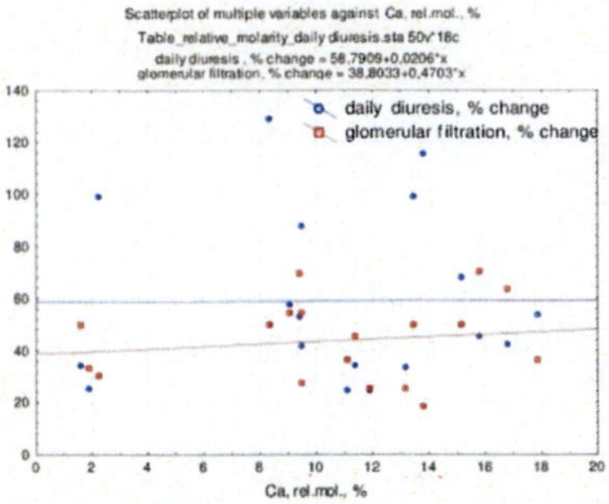

Fig. 2. The scattering diagram of the relationship of Ca with DD and GFR (specific molarity).

The difference between the relationship between the changes in DD and hydrocarbonates in the absolute and relative values is insignificant ($r = 0.4097$ and $r = 0.3490$). However, the relationship between the changes in DD and chlorides in absolute and positively insignificant $r = 0.0352$ significantly increases in negatively significant ($r = -0.4236$) for specific molarity. To a lesser extent, this happens with sulfates (changes from r = 0.0547 to r = –0.2388) and metasilicic acid (changes from $r = 0.0522$ by $r = -0.2448$) (Fig. 3, 4). Concerning bonds, changes in GFR, and specific molarity of all components, bond strengthening is observed. The r bond of specific C_{org} molarity increases for both DD and GFR changes. For GFR, these relationships are very significant ($r = 0.6330$).

To study the fundamental processes of urination, we examined the quantitative ratio of GFR to DD, the data of the obtained regression equations, and significant r. It was determined that glomerular filtration plays a considerable part in the diuretic effect of MW if the dominant role of organic substances and bicarbonates in the composition of MW is. DD most increases with the action of MW with the presence of organic substances. The data obtained confirmed the results obtained previously [23].

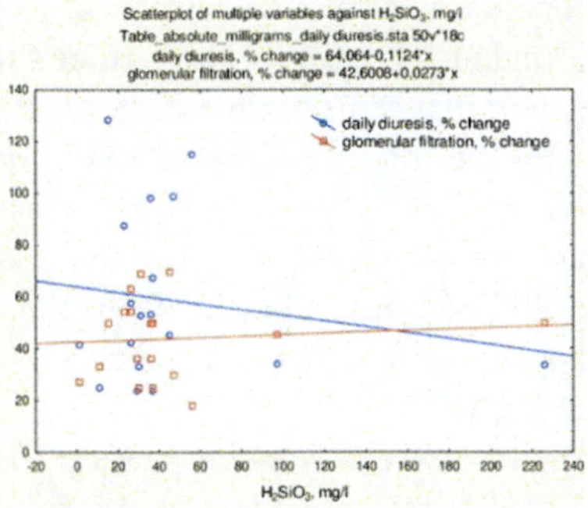

Fig. 3. The scattering diagram of the connection H_2SiO_3 with DD and GFR (absolute values).

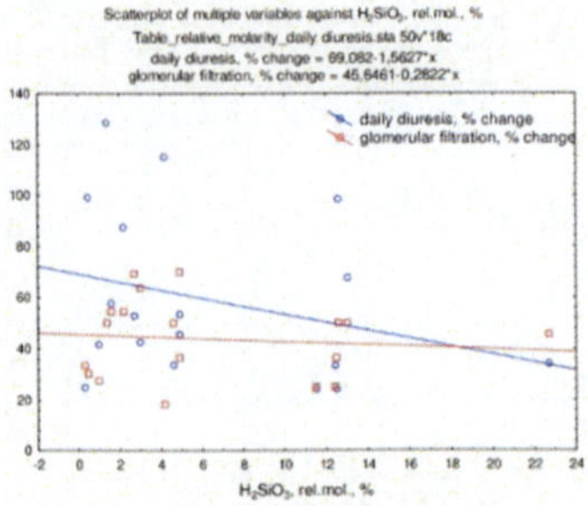

Fig. 4. The scattering diagram of the connection H_2SiO_3 with DD and GFR (specific molarity).

An inverse relationship in MW was established between the processes of urination and the specific molarity of chlorides and H_2SiO_3. With a lower content of H_2SiO_3, an increase in DD occurs due to the stimulation of glomerular filtration. A more significant amount of H_2SiO_3 in the MW, liquid reabsorption in the tubules is also activated, as was shown earlier [23], which leads to a smaller effect of DD growth. Our studies have proved the previously revealed correcting effect of artificial model solutions with different concentrations of H_2SiO_3 on the course of the pathological state of gastritis [20] and the hypothesis of an increase in its effectiveness on the presence of macro-components of MW. Namely, the efficacy of the biological effect depends on the ratio of macrocomponents, in particular, their specific molarity.

5 Conclusions

Features of the biological effect of MW, in particular, the process of urination, depending on the content of both macro- and microcomponents, as well as on their ratio, were investigated. The systemic biological activity of MW is caused not only by the content of a specific agent (trace element, organic substance) but also by the presence of a macro component component.

Among the other components of MW, the formation of urine is most affected by organic matter due to a significant increase in glomerular filtration rate.

The results obtained suggest that the possible biological activity of MW waters is reflected to a greater extent by the total molarity of MW and the specific molarity of their components.

Having data on the content of components, using the criterion of specific molarity of the components, and the obtained dependencies from the regression equations, it is possible to calculate the approximate range of efficiency and the direction of the biological action of a particular MW.

For a full study of the role of MV components in their biological action, additional studies of the effect of MV of different chemical composition on all indicators of the state of the functional systems of the body are necessary.

Mineralization does not have a significant effect on urination; therefore, it cannot be a criterion for the differentiation of MW by the effectiveness of their activities without the availability of experimental studies.

To harmonize the regulatory requirements for MW in Ukraine and Europe to differentiate the biological effect of MW (impact on the body or proven therapeutic effect), additional studies are needed both on healthy animals and with simulated pathological conditions with subsequent clinical studies in humans. At the next stage of our research, it is planned to study the relationship between the peculiarities of the ratio of the concentration of the components of MW and the strength of their biological effect on the example of other body functions, in particular, the functional state of the liver and pancreas. It will make it possible to develop "markers" of the biological effect of MW (the most informative indicators of the state of the functional systems of the body) that change under the influence of MW of different chemical composition. It will provide an opportunity to create methodological foundations of the biomedical classification of MW, depending on the effectiveness of their biological action.

Acknowledgment. The research study was partially funded by the Ministry of Health of Ukraine for state budget funds.

References

1. Quattrini, S., Pampaloni, B., Luisa, M.: Brandi Natural mineral waters: chemical characteristics and health effects. Clin. Cases Min. Bone Metab. **13**(3), 173–180 (2016)
2. Petraccia, L., et al.: Water, mineral waters and health. Clin. Nutr. **25**(3), 377–85 (2006)
3. Casado, Á., et al.: Types and characteristics of drinking water for hydration in the elderly. Crit. Rev. Food Sci. Nutr. **55**(12), 633–641 (2015)
4. Kysylevska, A., et al.: Harmonization of the EU and ukrainian normative documentation: case study on determination of barium content in mineral waters to develop quality and safety criteria. In: Tonkonogyi, V. et al. (eds) Advanced Manufacturing Processes. InterPartner-2019. Lecture Notes in Mechanical Engineering. Springer, Cham (2020). https://doi.org/10.1007/978-3-030-40724-7_16
5. Order of Ministry of Public Health of the Ukraine from 02.06.2003, № 243. https://zakon.rada.gov.ua/laws/show/z0752-03. Accessed 12 Feb 2020
6. Directive 2009/54/EC of the European Parliament and of the Council of 18 June 2009. Off J Eur Union L 164/45-58, 26.06.2009 (2009). https://eur-lex.europa.eu/legal-content/EN/TXT/?uri=uriserv:OJ.L_.2009.164.01.0042.01.ENG&toc=OJ:L:2009:164:TOC. Accessed 12 Feb 2020
7. Unione Geotermica Italiana. Geotermia, notizario dell' Unione Geotermica Italiana. VIII. Aprile. 26 (2010). https://www.unionegeotermica.it/notiziari/UgiNotiziario26.pdf. Accessed 12 Feb 2020
8. Albertini, M., Dacha, M., Teodori, L., Conti, M.: Drinking mineral waters: biochemical effects and health implications – the state-of-the-art. Int. J. Environ. Health **1**(1), 153–169 (2007)
9. Wynn, E., Krieg, M.A., Aeschlimann, J.M., Burckhardt, P.: Alkaline mineral water lowers bone resorption even in calcium sufficiency: alkaline mineral water and bone metabolism. Bone **44**(1), 120–124 (2009). https://doi.org/10.1016/j.bone.2008.09.007

10. Commission Directive 2003/40/EC of 16 May 2003. Off J Eur Union L 126/34-39, 22.05.2003. https://eur-lex.europa.eu/legal-content/EN/TXT/?qid=1590141046254&uri = CELEX:32003L0040, https://eur-lex.europa.eu/legal-content/EN/TXT/PDF/?uri=CELEX:32003L0040&from=en. Accessed 12 Feb 2020
11. Regulation of the Minister of Health from 31st March (2011). Accessed 12 Feb 2020
12. Babov, K.D., et al.: Features of the Biological Effect of Mineral Waters of Different Mineralization: Monograph. KIM, Kiev (2009). [in Russian]
13. Diduch, M., Polkowska, Z., Namiesnik, J.: Chemical quality of bottled waters: a review. J. Food Sci. **76**(9), 178–196 (2011). https://doi.org/10.1111/j.1750-3841.2011.02386.x
14. Carstea, E.M., et al.: Quality assessment of Romanian bottled mineral water and tap water. Environ. Monit. Assess. **188**, 521 (2016). https://doi.org/10.1007/s10661-016-5531-9
15. Gushcha, S.G.: To the mechanisms of correcting influence of mineral waters of different osmolyarity and microelement composition on the structural-functional state of kidneys of rats with experimental nephritis. Bull. Prob. Biol. Med. **2**(144), 301–306 (2018). https://doi.org/10.29254/2077-4214-2018-2-144-301-306
16. Jurkić, L.M., Cepanec, S.K., Pavelić, K.: Biological and therapeutic effects of ortho-silicic acid and some ortho-silicic acid-releasing compounds: new perspectives for therapy. Nutriton Metab. **10**(2), 1–12 (2013)
17. Nasibullin, B.A., et al.: Application of silicon mineral waters of Ukraine and their artificial analogues in the correction of experimental gastritis. Water Hyg. Ecol. **1–4**(16), 40–45 (2018). [in Russian]
18. Salomon, A., Regulska-Ilow, B.: Polish bottled mineral and healing water - characterization and application. Bromatologia I Chem. Toksykologiczna **46**(1), 53–65 (2013)
19. Gushcha, S.G., Nasibullin, B.A., et al.: Pathogenetic and sanogenetic mechanisms of the influence of mineral waters (siliconed and with increased organic substances) of different osmularity on the exposure of toxic nephritis. J. Biotechnol. Bioeng. **2**(2), 7–12 (2018)
20. Gozhenko, A.I., Shumilova, P.A., Dolomatov, S.I.: The effect of osmotic loads on the functional state of the kidneys of healthy people. Nefrologiya **8**(2), 44–48 (2004). [in Russian]
21. Shvets, V.I.: Interconnection of regulation of water-salt metabolism of homeostasis. Med. Hydrol. Rehabil. **4**(1), 66–69 (2006). [in Ukrainian]
22. Gozhenko, A.I.: Physiological bases of the optimal water consumption. Actual Probl. Transp. Med. **4**(14), 14–21 (2008). [in Russian]
23. Council Directive 2010/63/EU of the European Parliament and of the council of 22 september 2010. Official Journal of the European Communities L 276, 33–79 (2010). https://eur-lex.europa.eu/legal-content/EN/TXT/?qid=1590141272990&uri=CELEX:32010L0063. Accessed 02 Dec 2020
24. Command of Ministry of Public Health of the Ukraine № 692 dated 28.09.2009 (2009). http://search.ligazakon.ua/l_doc2.nsf/link1/MOZ10099.html. Accessed 02 Dec 2020
25. McDonald, J.H.: Handbook of Biological Statistics, 3rd edn. Sparky House Publishing, Baltimore (2014)

Compilation of the Best Practices for Auditing the Sustainable Development of Organizations

Beata Starzyńska[1(✉)] and Mariusz Bryke[2]

[1] Poznan University of Technology,
5, Maria Skłodowska-Curie Square, 60-965 Poznan, Poland
beata.starzynska@put.poznan.pl
[2] Kaizen Institute Poland, 13, Koreańska St., 52-121 Wrocław, Poland

Abstract. Contemporary organizations tend to be moving away from the traditional model of management and head towards the idea of sustainable development. The article puts forward the idea of sustainable development concerning a company's operations. The compiled catalog of best practices provides an operationalized concept of sustainable development and assumes the claim to achieve all well-balanced corporate economic, environmental, and social goals simultaneously. The catalog includes collections of best practices in three areas – Human, Lean and Green. According to the HLG concept, an enterprise is supposed to enhance its performance in the three areas: production (services) in terms of waste elimination in processes (Lean), the impact on the natural environment (Green), and workplace quality (Human). The catalog was developed to design and implement an audit tool aimed to measure sustainable development in manufacturing and service companies. Assessment of corporate sustainability is carried out by measuring indices obtained from the analysis of answers given to the questions regarding the application of good practices in the audited organization.

Keywords: Sustainability · Production · Manufacturing · Services · Good practices · Audit questions

1 Introduction

Contemporary organizations tend to be moving away from the traditional model of management and head towards the idea of sustainable development. The integration of economic, environmental, and social issues has become an overriding objective for modern organizations that strive to gain competitive advantage and fulfill the expectations of their clients, both existing and prospective ones.

– Sustainable development was defined as "the growth that ensures that humanity meets the needs of the present generations without compromising the ability of future ones to meet their own needs" [1, 2].
– To run responsible business in compliance with the idea of sustainable development means to address economic, social, and ecological issues in each life phase of the offered products (provided services).

V. Tonkonogyi et al. (Eds.): InterPartner 2020, LNME, pp. 833–842, 2021.
https://doi.org/10.1007/978-3-030-68014-5_81

- So-called sustainable production involves economically viable manufacturing goods, yet the systems used in the production do not lead to the degradation of the ecosystem. At the same time, implementation processes ensure a friendly and safe work environment. As a result, the products are manufactured to meet the client's expectations.
- Each phase – from design to the end of the product life cycle should be focused on reducing the adverse impact on the environment, not only within the manufacturing phase but also when it is delivered and used [3]. Designing phase seems to be gaining importance – it this stage the model and its technical specification are made, and environment-friendly materials (i.e., recycled or degradable ones) may be introduced [4, 5].

Sustainable development involves the implementation of state-of-the-art technologies that facilitate the effective exploitation of natural resources and energy and also reduce the emission of greenhouse gases to the atmosphere.

Sustainable enterprises voluntarily take measures to support their employees and local communities as well. Internally, social responsibility is based on respecting labor market rights, increasing the sense of stability, and creating safe, ergonomic workplaces. Externally, it should be perceived as respect for the rules of fair competition, basing on reliable information and commitment to projects dedicated to the development of a company's public environment. Corporate Social Responsibility (CSR) is also all about compliance with the environment-related regulations and reasonable exploitation of natural resources [6, 7].

The key management issue is to measure and monitor the execution of business obligations for operational sustainability. In order to review a business' sustainable development, one usually needs a dedicated set of indicators and special measurement tools. It seems quite obvious that a proper form of the audit would certainly work for the goal [8, 9].

The title catalog of best practices was developed to design and implement an audit tool aimed to measure sustainable development in manufacturing and service companies. The catalog includes collections of best practices in three areas – human, lean and green.

2 Literature Review

The implementation of the idea of sustainable development involves activities embracing each of the three following aspects – economic, social, and ecological. Being this the reason, methods, and tools to measure and assess business activities in the areas [10–12] have been gaining popularity over the past few years.

The methods and tools facilitate the process of establishing measurable goals, help to assess the progress, and monitor the changes across the organization as well as to evaluate the impact of decisions made and measures taken [11–13]. Contrary to traditional economic measures, sustainable development indicators show the economy, society, and environmental issues concerning the relevant business factors.

So-called indicators of sustainable production may measure an organization's progress on the way towards more sustainable manufacturing practices.

Why good practices?

Each management approach and related strategies have to be "armed" with appropriate operational tools [14]. Mere declaration of Lean Manufacturing or Lean Green implementation has little consequence, while effective practices have the power to translate strategies into results. Such practices may be defined as regular activities that bring about expected improvements.

The idea of best practices was recognized when firsts steps were taken to introduce scientific management, i.e., F.W. Taylor used the idea of the one best way to perform certain activities.

Subject literature provides numerous definitions and classifications of "good practices" [15–17]. Most of them perceive good practices as the best, verified in practice way of performing "any activity", the way that guarantees that the goals are achieved. At the same time, such a universal activity may be applied by any entity in comparable conditions.

Over time some verified practices get the status of approved standards, such as good manufacturing practice (GMP) or good hygiene practice (GHP). Several management approaches, also in the area of quality, environment or health and safety, assume specific rules, methods, and tools that are considered to be best practices. One of the iconic solutions of the Lean Manufacturing concept takes its name after the name of one of the practices (5S Practices). Environmental management concepts postulate, among others, the practice of applying closed circuits to reduce water consumption, using packaging made of recycled materials or increased commitment in pro-ecological campaigns.

Practices of different nature (organizational, technical, informational) share a particular characteristic – they are applied to ensure, maintain, and improve processes (relevant to get a finished product). Effectiveness of dedicated practices in selected areas of manufacturing enterprises (and also service providing businesses) has been the subject of extensive research [15, 18, 19]. The results indicate clearly that the regular application of such practices is a prerequisite for long-lasting effects.

Well described and documented, they may provide a model for the implementation and development of corresponding solutions in other organizations. On a small scale, the activities are aimed to increase compctitivc advantage and improve the financial performance of an enterprise; further, well-balanced growth of the whole organization should be recognized as well [20].

It should be noticed, though, that the literature presents some opinions deflating the legitimacy of good practice catalogs. Admittedly, some practices are not universal enough to face the challenges of each environment. Addressees, their needs, and potential (e.g., microenterprises, SMEs, large organizations), specifics (e.g., manufacturing enterprise or service provider), conditions for the application of selected practices (e.g., obligatory or discretionary) must be taken into consideration when the catalog is being built.

Best practices in the catalog prepared to facilitate auditing for sustainable development in organizations are dedicated to manufacturing and service processes in the organizations.

3 Research Methodology

In order to manufacture a product compliant with the customers' demand with short lead time, low cost, and reduced resource consumption, contemporary enterprises tend to take advantage of the concept of Lean Manufacturing. Lean management (applicable also for service providers) assumes the idea of continuous improvement and elimination of all the activities that fail to provide added value for the clients. The most frequently observed forms of waste include excessive motion and transport, inventory, overproduction, defects, over-processing, and waiting. Lean tools and methodology are considered effective in recognizing waste and eliminating unnecessary costs, so organizations are willing to incorporate the practices to their standard effective management strategies, e.g., in the area of environment protection. As a result of the dynamic development of a systematic approach, new hybrid solutions, such as Human Sigma came into being [21, 22]. The integrated approach referred to in the literature as Lean-Green has won broad practical applications [23–27].

The Integrated Lean-Green approach was extended [3, 28] to incorporate the idea of corporate business responsibility focused around a human being. Its name, Human Lean Green (HLG), reflects the combination of three concepts – supporting the interests of employees (Human), waste elimination (Lean), and environmental protection (Green). The new approach involves three equally important areas [29]:

– Human – concern about employees and friendly work environment;
– Lean – improving processes (through waste elimination);
– Green – protecting the natural environment.

The proposed idea assumes first and foremost that each model organization is supposed to pursue the state of organizational balance [20] in the areas mentioned above [3].

However, the question is how to assess strategy preparation and progress made in an enterprise which declares the implementation of HLG strategy and compliance with HLG criteria.

Naturally, an organization should consider performing an audit – one of the most popular solutions used to assess management systems. It gained importance and recognition upon the implementation of standardized systems for quality management in the 80s of the previous century. In the broad sense, an audit is perceived as systematic and independent assessment of an institution, system, process, project, product, etc. An audited entity will be assessed in terms of its compliance with particular criteria, e.g., provisions of legal regulations, norms, standards, procedures, or instructions.

Audit assumptions are compatible with the assumptions used for indicators of sustainable production:

– the development of the sustainable production indicators is a continuous process that requires systematic goals and relevant measures,
– each company, depending on the industry, starts the process at a different stage,
– developing sustainable production systems involves the cooperation of enterprises, communities, and the government at all possible levels – local, regional, national, and international [11].

The catalog of good practices was prepared for the project and implementation of an audit tool designed to measure sustainable development in production enterprises/service providers.

According to the HLG concept, an enterprise is supposed to enhance its performance in the three areas: production (services) in terms of waste elimination in processes (Lean), the impact on the natural environment (Green), workplace quality (Human). The HLG concept is, therefore, compatible with the idea of sustainable development discussed in chapter 1, with its three major pillars. Figure 1 presents the model of organizational effectiveness as postulated by the authors of the concept [3, 28, 29]. The Human Lean Green model of the organizational effectiveness includes five components necessary for the successful management in the production enterprise (to be defined separately for service providers) – management, materials, machines, employees, methods. They are presented at the bottom of the model (Fig. 1).

Fig. 1. Components defined for the human lean green method.

The assessment criteria for the corporate effectiveness in the Lean area were selected in compliance with the classification of waste in the production and service processes as postulated by [30, 31]. The selection of Green criteria was based on the review of popular standards related to environmental reporting. The review of Human criteria focused on the conditions reported for the work environment, including ergonomics as well as safety and hygiene conditions.

The next step was to define the assessment criteria for each category (Fig. 1). The following were recognized as waste:

- Human – no attention for health, poor ergonomics, poor health, and safety conditions, lost human potential, declining biodiversity;
- Lean – excessive motion and transport, inventory, overproduction, defects, overprocessing and waiting;
- Green – scrap, excessive consumption of energy, water, raw materials, and other materials, air pollution.

4 Results

Audit questions were formulated based on the selected criteria for assessing a company's performance towards sustainable development. They were grouped about components identified above and presented in the form of matrices marked:

- horizontally by constant components of corporate assessment, regardless of the area to be assessed. The components are subject to the designed audit: *management, materials, machines, employees, and methods;*
- vertically by waste categories identified in one of the three selected areas: Human, Lean, and Green (Fig. 2).

PILLARS	CORE COMPONENTS OF THE MODEL - HUMAN				
	MANAGEMENT	MATERIALS	MACHINES	EMPLOYEES	METHODS
NO ATTENTION FOR HEALTH	Does the organisation take measures to increase efficiency and thus eliminate overtime work?	Are workers that come into contact with hazardous materials and substances properly protected?	Are the machines and other equipment inspected, maintained, repaired and modernised according to a schedule?	Is staff absenteesm caused by diseases of non-occupational origin only?	Does the organisation take measures to prevent accidents?
POOR ERGONOMICS	Does the corporate policy address workplace ergonomics?	Can workers access all support equipment when manipulating materials? If so, do they use it?	Does the work/ operations at workplace cause excessive physical/ mental burden?	Are staff members willing to put forward initiatives to improve and reorganise their workplaces?	Are employees effectively educated how to adjust their workplaces to make them comfortable and friendly?
POOR HEALTH AND SAFETY CONDITIONS	Are visits at Gemba an opportunity to show deep concern about the safety at workplaces?	Do workers use personal protective equipment when performing their duties?	Do all the machines and equipment work in compliance with health and safety regulations?	Do employees follow health and safety regulations and fire protection requirements when performing their routine duties?	Are employees urgently required to know health and safety regulations and fire protection system?
LOST HUMAN POTENTIAL	Does the management bring extra value by encouraging staff creativity?	Does the company management organise professional trainings for staff and provides materials to encourage their carrier development?	Are all the ideas and initiatives postulated by the staff recorded in the IT system and discussed?	Are staff members willing to participate in initiatives and events proposed by their superiors?	Can employees address some of their problems together with their team mates?
DECLINING BIODIVERSITY	Is the company involved in any initiatives aimed at restoring/ creating "green" areas (biologically active) within the company/ in its immediate surrounding?	Are the resources in place to introduce initiatives aimed at restoring/ creating "green" areas within the company/ in its immediate surrounding?	Are employees aware of the environmental impact of the company`s operations/ processes/ applied technologies?	Do employees engage in proecological initiatives?	Are employees encouraged (motivated) to engage in some proecological initiatives? Are they effective?

Fig. 2. Audit questions regarding good practices in the human area.

The complete catalog of good practices includes three dedicated matrices with successive audit questions regarding Green and Lean good practices. Questions have been formulated in line with the best practices applied in modern organizations [15–18, 25, 26]. The method provides for two sets of audit questions (separately for a production enterprise and for a service provider).

The unit called "Good Practices" provides a set of questions focused on the company's activities dedicated to waste elimination, process optimization, increasing effectiveness, cost reduction, improving work conditions, and natural environmental protection. A general assessment of corporate sustainability is carried out by measuring indices obtained from the analysis of answers given to the questions about good practices.

Apart from the good practice catalog, the audit tool also includes the unit "Figures" with numerical parameters related to the activities in terms of its effectiveness in the three areas highlighted above.

The audit follow-up report presents the current picture of the company and provides recommendations for the application/further implementation of good practices in compliance with the idea of sustainable development. The recommendations are based on globally approved models and focus on waste elimination in processes, increasing process effectiveness, natural environment protection, improving workplace ergonomics as well as health and safety regulations.

As part of the research on the developed method, several HLG audits were carried out in various enterprises. In order to check the functionality of an audit tool for validation, companies representing both the manufacturing and service sectors were selected.

The final report discusses key activities in terms of the organization's sustainable development. The first part of the report includes basic information about the organization, and the audit generated automatically based on the data put in the organization's survey. The second part of the report includes a universal introduction defining a proposed tool and the report application in the three sustainable development perspectives to improve the organization's effectiveness. The third part describes the employed approach in each of the three perspectives, the auditor's comments, and the audit follow-up recommendations.

The company's current condition is reflected by equally calibrated quantitative metrics (for each area) calculated based on the replies to the auditor's questions. They directly indicate the level of a company's sustainability in each area in question. Additional (supporting) metrics are possible as well. They are calculated according to quantitative data acquired from the organization.

The final element of the report presents the so-called aggregated indicator in each of the three perspectives of the Human Lean Green approach (Fig. 3).

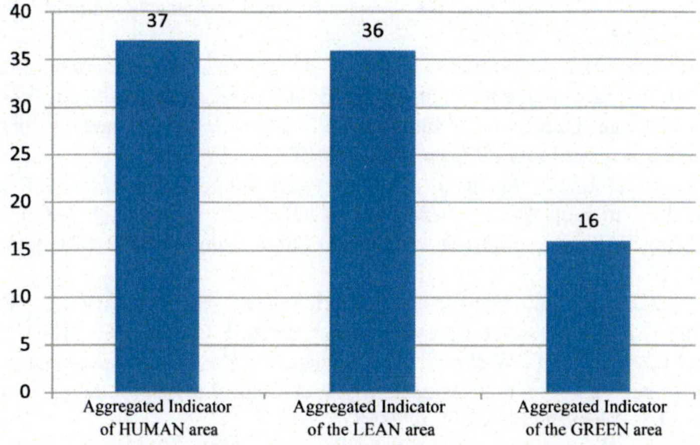

Fig. 3. Part of the HLG report.

The aggregated HLG indicator presents a graphic and quantitative picture of organizational sustainability in terms of the three examined perspectives. It also works as a starting point to introduce improvement activities. Taking the common elements of the organization's surroundings as a denominator for the proposed model (Fig. 1), one can compare employed practices in the area of processes improvement, natural environment protection and working environment optimization.

5 Conclusions

The article discussed the idea of compiling the best practices catalog for auditing the sustainable development in production enterprises and service companies.

The complete catalog of good practices includes three dedicated matrices with successive audit questions regarding Human, Lean, and Green areas. A general assessment of corporate sustainability is carried out by measuring indices obtained from the analysis of answers given to the questions regarding the application of good practices in the audited organization.

Audits carried out so far were well evaluated by the surveyed enterprises. The following features of the tool were emphasized – ease of use, its universal character suitable for various industries, a complete report, user-friendly interface (dialog boxes, drop-down lists), potential for development, reusability, the ability to study selected areas within HLG approach. A suggested audit may be widely used as a measuring tool to examine organizational maturity in terms of sustainable development.

References

1. World Commission on Environment and Development: Our Common Future. Oxford University Press, New York (1987)
2. Verrier, B., Rose, B., Caillaud, E., Remita, H.: Combining organizational performance with sustainable development issues: the lean and green project benchmarking repository. J. Clean. Prod. **85**, 83–93 (2014)
3. Bryke, M., Starzyńska, B.: Human Lean Green conception as the instrument of sustainability of organizational development oriented towards the increase of its effectiveness. Research Papers of Wroclaw University of Economics, Sustainable development of organization – environmental responsibility 377, pp. 119–136 (2015)
4. Rojek, I., Dostatni, E.: Artificial neural network-supported selection of materials in ecodesign. In: Trojanowska, J., Ciszak, O., Machado, J., Pavlenko, I. (eds.) Advances in Manufacturing II. Lecture Notes in Mechanical Engineering, pp. 422–431. Springer, Cham (2019)
5. Dostatni, E., Diakun, J., Grajewski, D., Wichniarek, R., Karwasz, A.: Functionality assessment of ecodesign support system. Manag. Prod. Eng. Rev. **6**(1), 10–15 (2015)
6. Asharavi, M., Adams, M., Walker, T.R., Magnan, G.: How corporate social responsibility can be integrated into corporate sustainability: a theoretical review of their relationships. Int. J. Sustain. Dev. World Ecol. **25**(8), 672–682 (2018)
7. Nielsen, A.E., Andersen, S.E.: Corporate social responsibility. Int. Encycl. Strate. Commun. (2018). https://doi.org/10.1002/9781119010722.iesc0051

8. Alic, M., Rusjan, B.: Contribution of the ISO 9001 internal audit to business performance. Int. J. Qual. Reliab. Manag. **27**(8), 916–937 (2010)
9. Lenning, J., Gremyr, I.: Turning internal audits into business audits that drive business relevant improvements. In: Proceedings of 9th QMOD-ICQSS International Conference on Quality and Service Sciences, pp. 712–725 (2016)
10. Finkbeiner, M., Schau, E.M., Lehmann, A., Traverso, M.: Towards life cycle sustainability assessment. Sustainability **2**, 3309–3322 (2010)
11. Veleva, V., Ellenbecker, M.: Indicators of sustainable production: framework and methodology. J. Clean. Prod. **9**, 519–549 (2001)
12. Harik, R., El Hachem, W., Medini, K., Bernard, A.: Towards a holistic sustainability index for measuring sustainability of manufacturing companies. Int. J. Prod. Res. **53**(13), 4117–4139 (2015)
13. Wilson, M.C., Wu, J.: The problems of weak sustainability and associated indicators. Int. J. Sustain. Dev. World Ecol. **24**(1), 44–51 (2017)
14. Hamrol, A.: Quality Management with Examples. PWN, Warszawa (2008). (in Polish)
15. Eswaramoorthi, M., Kathiresam, G.R., Prasad, P.S.S., Mohanram, P.V.: A survey on lean practices in Indian machine tool industries. Int. J. Adv. Manuf. Technol. **52**, 1091–1101 (2011)
16. Shah, R., Ward, P.T.: Lean manufacturing: context, practice bundles, and performance. J. Oper. Manag. **21**, 129–149 (2003)
17. Salleh, N.A.M., Kasolang, S., Jaffar, A.: Green lean TQM human resource management practices in malaysian automotive companies. Int. Sch. Sci. Res. Innov. **6**(10), 2065–2069 (2012)
18. Tortorella, G.L., Fogliatto, F.S.: Method for assessing human resources management practices and organisational learning factors in a company under lean manufacturing implementation. Int. J. Prod. Res. **52**(15), 4623–4645 (2014)
19. Garetti, M., Taisch, M.: Sustainable manufacturing: trends and research challenges. Prod. Plan. Control **23**(2–3), 83–104 (2012). https://doi.org/10.1080/09537287.2011.591619
20. Koźmiński, A.K., Obłój, K.: Outline of Organizational Balance Theory. PWE, Warszawa (1989). (in Polish)
21. Caiado, R., Nascimento, D., Quelhas, O., Tortorella, G., Rangel, L.: Toward sustainability through green, lean and six sigma integration at service industry: review and framework. Technol. Econ. Dev. **24**, 1659–1678 (2018)
22. Shah, J., Deshpande, V.: Lean six sigma: an integrative approach of lean and six sigma methodology. Int. J. Curr. Eng. and Technol. **5**, 3528–3534 (2015)
23. Galeazzo, A., Furlan, A., Vinelli, A.: Lean and green in action: interdependencies and performance of pollution prevention project. J. Clean. Prod. **85**, 191–200 (2014)
24. Garza-Reyes, J.A.: Lean and green – a systematic review of the state of the art literature. J. Clean. Prod. **102**, 18–29 (2015)
25. Bryke, M.: Lean Green. Will kaizen tools leading to lean help achieve the organization's green status? KAIZEN Magazine, Medialog, Poznań **3**(4) (2012). (in Polish)
26. Bryke, M.: Lean Green. Environmental activities improve organization and allow you to control the costs of operation. KAIZEN Magazine, Medialog, Poznań **4**(5) (2012). (in Polish)
27. Bryke, M.: Lean Green. Why is it worth being green? Green value stream mapping. KAIZEN Magazine, Medialog, Poznań **2**(7) (2013). (in Polish)
28. Bryke, M., Kanus, M.: Human Lean Green. We are building an organizational culture ensuring sustainable development and effectiveness of the organization. KAIZEN Magazine, Medialog, Poznań **3**(12) (2014). (in Polish)

29. Bryke, M., Hamrol, A., Starzyńska, B.: Developing human lean green tool as an instrument of measuring sustainable organization development. In: 19th QMOD – ICQSS International Conference on Quality and Science Services, pp. 1035–1046 (2016)
30. Imai, M.: Kaizen: The Key To Japan's Competitive Success. McGraw-Hill Education, New York (1986)
31. Imai, M.: Gemba Kaizen: A Commonsense Approach to a Continuous Improvement Strategy. McGraw-Hill Education, New York (2012)

The Selection of Lithological Layers According to Measurements of Drilled Wells

Alexandr Shpinkovski(✉) and Maria Shpinkovska

Odessa National Polytechnic University,
1, Shevchenko Avenue, Odesa 65044, Ukraine
csonpu@ukr.net

Abstract. The paper states that according to the results of measurements when drilling wells, for the industrial development of wells, it is important to interpret the lithological layers reliably. To solve this problem, the manual lithology assessment is often used. At the same time, concomitant errors are not excluded, which in the future can lead to economic losses. Therefore, it becomes relevant to automate the process of assessment and accurate determination of lithological layers under measurements. The paper used data from twelve wells. In them, manually identified five main lithological layers. It is proposed to carry out a classification (allocation of layers) in order to check the quality of lithology assessment and to be able to predict the type of layers for new data sets. An analysis of the measurement data was carried out and their preliminary processing. Based on the results of descriptive statistics, outliers in the measurement data were eliminated. After that, the measurement data is divided into training and test samples. Several methods classify the lithological layers. Quality criteria showed acceptable results in the four main categories of lithology and incomplete category information with the least number of measurements. Ways of further improvement of the methodology for determining lithological strata of wells are proposed.

Keywords: Measurement data · Well log · Quality metrics · Correlation matrix

1 Introduction

Understanding the location and composition of rocks and fluids in the bowels is the main task of finding oil, gas, and fossil fuels. The resulting measurement information is critical to improving the efficiency of mineral exploration. Well logs with measurement data provide direct information about what is in the bowels.

Such economic prospects as the cost of drilling wells and the work of returning mineral deposits directly depend on the data of well logging measurements. It also affects the decision to develop new wells in the future [1].

In the process of drilling wells, different logs are formed for each characteristic. Using a special cable, high-precision measuring instruments capture the necessary data. After raising the tools from the depth, the measurement data is read and recorded.

V. Tonkonogyi et al. (Eds.): InterPartner 2020, LNME, pp. 843–852, 2021.
https://doi.org/10.1007/978-3-030-68014-5_82

Well drilling data can also be obtained with the support of devices at the bottom of the wells. Logging data include those that describe the near-wellbore fluid together with the collected material.

Many well logs contain the results of measuring the properties of the rock and fluids depending on the types of minerals, the number and types of fluids, porosity, and rock properties. Among others, they include such measurement results as gamma radiation intensity, electrical resistance, porosity and density, sound velocity [1].

Well logging data may have gaps or missing values. They occur for several reasons, such as temporary inoperability or breakdown of tools, inaccuracies or errors of operators, inaccurate determination of the measurement interval. Gaps in the logging data lead to a decrease in useful information and affect the accuracy of modeling for subsequent measurements, studies, and works [2].

Given the above factors, when interpreting measurement data from well logs, there is a risk of losing some of the measurement information. Firstly, due to technological and technical imperfection of the measurement procedure or miscalculations of the operators, and secondly, due to inconsistencies or inaccuracies in the expert interpretation of logging data.

From a technical point of view, modern technologies are impossible without machine learning - research in the field of computer science, artificial intelligence, and statistics [3]. The focus of machine learning is the development of algorithms for studying patterns and data prediction [4]. Its special value lies in the fact that it allows you to use computers to automate and increase the efficiency of decision-making procedures, including in the field of interpretation of large data from production measurements [3, 5].

Therefore, it is proposed to use machine learning methods with a teacher to automate the interpretation of measurement data. Classification methods allow you to train the algorithm based on the known correspondence of the measurement data to the rock type. A trained model can predict the type of rock or make up for missing values from other new values for well logging.

The purpose of this paper is to use machine learning to explore the possibility of using geophysical measurements to highlight the types (classes) of mineral layers. It will allow in the future to automate the decision-making process on mining and replace manual, labor-intensive expert assessment.

2 Literature Review

The paper uses a data set to evaluate the effectiveness of lithological identification using the cross-hole method. There are 12 wells with the same logs and 5 lithologies: mudstones, siltstones, muddy siltstones, silt mudstones, and oil shales [6].

Each data well is represented by the following sets of log curves:

_CAL (Caliper Log). The support log is a well logging tool that provides continuous measurement of the size and shape of a well along with its depth and is commonly used in hydrocarbon exploration for well drilling. Recorded measurements can be an important indicator of hollows or swelling of oil shale in a well, which can affect the results of other well logs. The caliper log provides a view of the diameter of the well along with its depth. Well trunks tend to have the wrong diameter, so it is important to have a tool like a log that measures the diameter in several places. Caliper logs can also provide some insight into the types of formation and whether the rocks are hard and compact or loose gravel [7].

_GR (Gamma Ray log). Gamma logging is a method of measuring natural gamma radiation to characterize a rock or sediment in a well or borehole. It is a logging method used in mining, mineral exploration, water well drilling, for evaluating formations when drilling oil and gas wells and for other related purposes. Different types of rocks emit different amounts and different spectra of natural gamma radiation. The presence of altered zones, core rocks and shale layers can also be detected [7].

_SP (Spontaneous Potential). SP, also called Self-Potential, is a method of passive electrical resistance. For SP logging, the electrode descends the wellbore, and the other electrode is grounded at the surface. The voltage difference between the two electrodes is measured. Natural electric currents in the bowels create voltages. SP logging requires very simple tools, and the SP electrode is usually combined with other well logging tools [8].

_LLD, _LLS. Resistivity logging involves the measurement of the electrical properties of the rock. Changes in the rock's electrical properties can be due to the clay mineral content, water content and porosity, temperature, and water conductivity. The Dual Laterolog tool (DLL) used in deep thermal water exploration and allows the estimation of fracture parameters in hard rocks quantitatively as well. By using the Dual Laterolog, based on resistivity anomalies and separation between shallow (LLS) and deep laterolog (LLD) and mud conductivity, fracture zones can be detected; in addition, the fracture porosity and the fracture aperture of horizontal and vertical fractures can be estimated [9].

_AC (Acoustic Logs). Displays the travel time of acoustic waves as a function of depth in the well. The term is commonly used as a synonym for a sound journal. Some acoustic logs display speed. An acoustic logging technique includes those methods that use a transducer to transmit an acoustic wave through a fluid in a well and surrounding elastic material. Acoustic logs are used to determine the lithology and porosity of the rocks surrounding the well. This information may be useful for determining future well locations and potential areas for stimulation of the wellbore. The journal, in combination with other journals, provides the basis for a detailed analysis of lithologies, changes, stratigraphy, etc.

_DEN (Density Log). Density logging is a well logging tool that can provide continuous recording of the bulk density of a formation along the length of a well. In geology, bulk density is a function of the density of the minerals that make up the rock and the fluid contained in the pore space. A modern density measurement tool emits gamma rays from a source at the bottom of the instrument. The gamma rays emitted by the source penetrate the rocks when the instrument passes by, where some are absorbed. This information can be used to calculate parameters such as rock porosity and rock type. It is calculated by measuring the difference in fluxes between the source and the detector, which are at a given distance.

_PEF (Photoelectric Factor). The Photoelectric Factor was introduced as an additional measure to the bulk density measurement and registered the absorption of low-energy gamma rays when they form barns per electron in units. The recorded value is a direct function of the total atomic number of elements in the formation, and therefore is a sensitive indicator of mineralogy. This relationship and the fact that the value slightly depends on the size of the pores or the content of the liquid or gas means that the index is an excellent indicator of mineralogy [10].

In addition to measuring the physical values of the log curves, the set of source data contains:

Top_Depth. Logging depth of measured logging values (upper bound);

Bot_Depth. Logging depth of measured logging values (lower bound);

Lith_Section. Interpretation of the lithological layer according to the results of preliminary expert assessment.

It indicates one of the five classes (mudstones, siltstones, muddy siltstones, silt mudstones, and oil shales).

Paper with material on the use of intelligent processing of measurement data of well logging parameters is considered. The paper shows that any classification method uses a set of features to characterize each object, where these features should be relevant to the task at hand. In supervised classification, there are two phases to constructing a classifier: the training phase and the testing phase. In the training phase, the training set is used to decide how the features ought to be weighted and combined in order to separate the various classes of objects. In the testing phase, the weights determined in the training set are applied to a set to calculate the overall classification error of the solution. This error is used to adjust some parameters [11].

This article compares the results of supervised learning algorithms, non-supervised learning algorithms, and a neural network machine learning algorithm. An integrated approach to data set processing and function selection is also proposed. The well log data used in this article is for wells in the Anadarko Basin, Kansas. The data set is divided into training, testing, and evaluating wells used to test the model. The goal is to evaluate the algorithms and constraints of each algorithm. It is assumed that a simple controlled learning algorithm can give a higher estimate than the neural network algorithm, depending on the selected model parameter. An analysis of the selection of parameters was made for all models. The optimal parameter was used for the corresponding classifier [12]. In this paper, incomplete information on the accuracy of classifiers, except for the neural network. There is also no information about the logging data, on which the marginal accuracy of the calculations depends.

The paper notes that currently, shale gas is one of the focuses of the unconventional reservoir. Well logs play an important role in shale gas production, and it is the bridge connecting geology, geophysics and petroleum engineering. In the exploration stage, well logs are used to identify lithology, evaluate the parameters of mineral types and compositions, total organic carbon, porosity, permeability, gas content, and the quantity of the potential resources. An analysis is made of the response characteristics of the logging and calculation of parameters for a shale gas well using regression models [3, 13].

3 Research Methodology

To create lithology layer classification models, we use Google Colaboratory online environment as a Jupyter notebook shell [14].

It is used Python programming libraries:

- Python Standard Library: built-in Python module
- Numpy: Python scientific computing library
- Matplotlib: graphics library
- Seaborn: a data visualization module
- Pandas: data analysis toolkit
- Scikit-learn: machine learning tools.

The procedure for classifying logging measurement data is carried out in several stages [15].

The data set consists of ten elements (eight logging measurements, and two variable depths), lithology labels with depth intervals, a column with borehole names (Table 1). A set of measurements at each depth interval contains a vector of features, each of which is associated with a class of facies (a type of lithology). We will use the pandas library to load data into the data frame, which provides a convenient data structure for working with log data.

The measurement data consists of 7469 rows, which contain the results of geophysical measurements of 12 wells. Object vectors consist of the variables listed in Sect. 2.

Table 1. Logging data from twelve wells.

	TopDepth	_CAL	_GR	…	Lith_Section	Well Name
0	999.4	8.4541	90.9256	…	siltstone	G39
1	1000.6	8.8013	108.8170	…	mudstone	G39
2	1002.0	8.5569	62.4191	…	siltstone	G39
3	1010.2	9.7062	114.2539	…	mudstone	G39
4	1017.6	9.2788	81.7255	…	siltstone	G39
…	…	…	…	…	…	…
7465	2052.33	8.8402	95.6815	…	mudstone	X78
7466	2052.89	8.8407	95.6815	…	mudstone	X78
7467	2053.33	8.8389	95.6815	…	muddy siltstone	X78

A brief overview of the statistical distribution (descriptive statistics) of training data is given in Fig. 1. I most columns, the standard deviation values are large. Also, the minimum and maximum values are very noticeably different from the 25th and 75th percentiles. It indicates the presence of large outliers in the data.

	TopDepth	_CAL	_GR	_SP	_LLD	_LLS	_AC	_DEN	_PEF
count	7470.000000	7470.000000	7470.000000	7470.000000	7470.000000	7470.000000	7470.000000	7470.000000	7470.000000
mean	1803.702829	9.316698	108.038393	-9.207900	12.720103	12.768499	81.348994	2.424630	0.387226
std	287.782026	0.737949	16.775717	8.011636	9.399133	9.350825	13.483099	0.123808	0.304100
min	999.400000	-0.558118	-86.809700	-57.263100	-243.385000	-3.535470	42.333400	1.433920	-1.644140
25%	1637.700000	8.827845	98.725500	-14.602025	8.108377	7.771933	71.242200	2.359450	0.241269
50%	1829.260000	9.080895	109.278500	-10.450600	11.132300	10.619900	77.633650	2.448410	0.399115
75%	2013.632500	9.740285	117.812000	-3.745073	14.226575	14.865875	90.544325	2.518820	0.593645
max	2384.850000	13.165500	281.861000	16.304700	154.435000	166.230000	134.654000	2.738280	1.432340

Fig. 1. A descriptive statistical overview of the measured data (Jupyter Notebook cell result).

For further analysis and work with numerical data, textual (categorical) values of lithological layers in the "Well Name" column are transformed into numerical ones. The result is shown in Table 2.

For further analysis, it is preferable to remove the uninformative parameters "Formation" (duplicates the name of the well) and "BotDepth" (strongly correlates with "TopDepth" and is one step behind it).

Table 2. Change the value of the Well Name column.

Number	Categorical text value	Numerical equivalent
1.	mudstone	1.0
2.	siltstone	2.0
3.	muddy siltstone	3.0
4.	silty mudstone	4.0
5.	oil shale	5.0

Also, we estimate the number of outliers that go beyond two standard deviations for each of the curves.

A correlation matrix is created, and a heat map (Fig. 2) is built for more assess the relationships between the curves (columns).

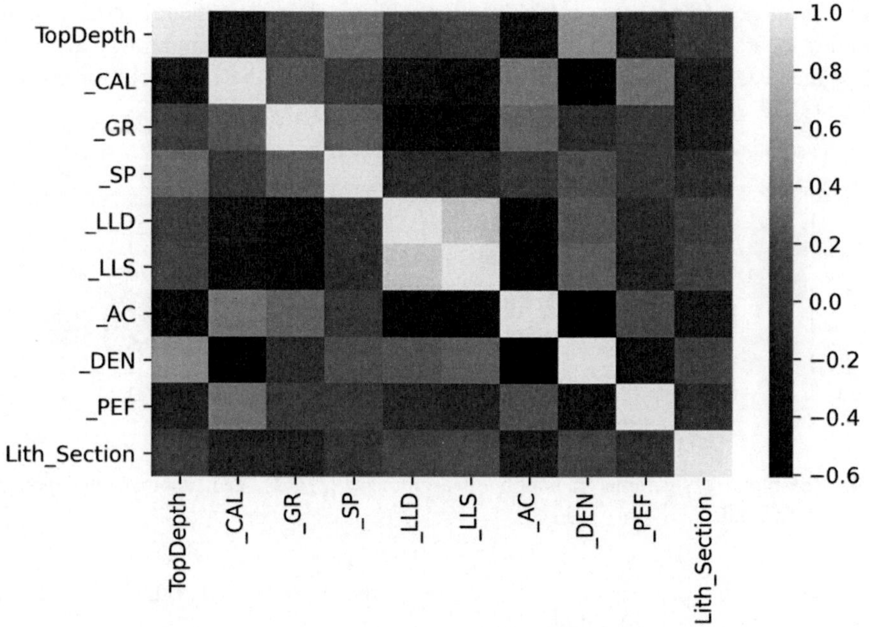

Fig. 2. Correlation heat map between measurement data.

From the heat map, you can see the positive tendency of density (_DEN) to depth (TopDepth). Depth is weakly or negatively correlated with other instrument data.

That is, it is not yet advisable to remove depth from the characteristics. The highest correlation occurs between later logs _LLD and _LLS. That is, it makes sense to remove one of them to eliminate possible overfitting.

We construct box-plots using geophysical curves to estimate the distribution of values within classes. Figure 4 box-plots of the two characteristics are presented (_GR, _SP). Outliers are visible in the figure.

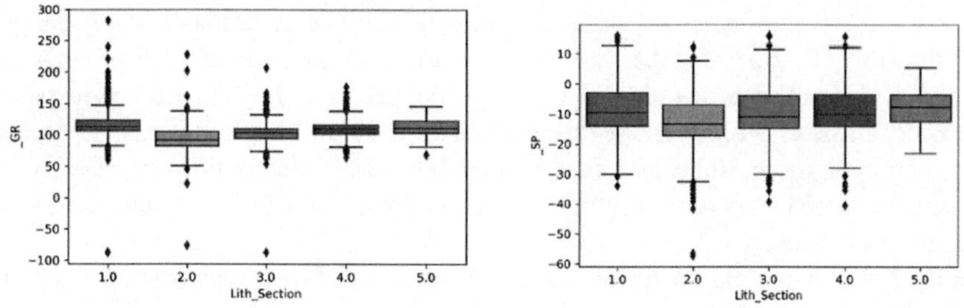

Fig. 3. Box-plot of gamma radiation measurements (_GR) and spontaneous potential (_SP).

It is proposed to process outliers of all geophysical measurement curves, equating all values located beyond the 5th or 95th percentile to the corresponding values.

For the data specified in the previous section, we get the box-plots for the measurement data _GR and _SP (Fig. 4).

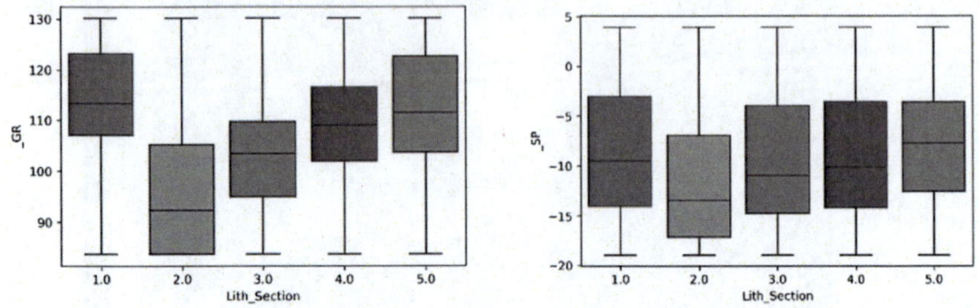

Fig. 4. Box-plots of gamma radiation measurements (_GR) and spontaneous potential (_SP) without outliers.

Even after the processing of the outliers, the data is still difficult to separate according to the classes of lithologies.

It is proposed to divide the data into parts of training and testing, which will be used to train the model and configure parameters. It would be correct to break the entire data set into separate wells, and not just to divide all the corresponding values of the set randomly. Different ratios of data sets for training and validation were tested. The selected best values for separation are 0.5 for the training part and 0.5 for the test part.

4 Results

Since we are dealing with classification methods, the choice of a metric for assessing the quality of separation of lithological layers can be made from the following list of the most common metrics for a similar problem: accuracy, precision, recall, F1, ROC-AUC [14, 15].

The accuracy metric is not suitable, because there is an imbalance of classes (a small selection of values for the "oil shale" layer), and the forecast will be erroneous. Precision and recall metrics should not be used because they do not correspond to unbalanced classes. Moreover, in this case, it is difficult to set the task of choosing which metric is more important in this problem. The best option in case of class imbalance is the F1 and AUC metrics because they are indifferent to the number and presentation of classes.

Table 3 shows the classification results in the form of quality metrics F1 and AUC for the six applied methods.

The best quality indicators for determining the types of lithological layers of a well of six methods were shown by the method of Logistic Regression. Nearby are the value of Ridge Classifier and other methods and algorithms. The low value of the quality metric F1 is compensated by more acceptable values of the AUC metric.

Table 3. Values of the metric of quality classification.

Number	Classification method	Metric F1	Metric ROC-AUC
1	RidgeClassifier	0,6217	0,7606
2	LogisticRegression	0,6402	0,7763
3	DecisionTreeClassifier	0,5609	0,7116
4	KNeighborsClassifier	0,5826	0,7345
5	GaussianNB	0,6173	0,7520
6	SVM classifier	0,5667	0,7149

The best option in case of class imbalance is indicators F1 and AUC because they are indifferent to the number and presentation of classes. The table shows the average values. Class imbalances cause low rates.

5 Conclusions

The paper examined ways to improve the classification of mineral layers based on well logging measurements that provide information on the characteristics of the measured data. The methodology of the application of machine learning and data mining for the interpretation and classification of logging data is described. A comparison of several classification methods is proposed, and under the F1 and ROC-AUC metrics, the best result in logistic regression.

The goal of the work is generally achieved. However, overall, the results can be improved. Other classification methods may be considered. It can lead to higher quality indicators, but not more than a few percents.

In the future, you can try using more complex models to test their quality. One of the areas of work may be the construction of additional log data based on existing ones since additional functions can improve quality. It is also advisable to build a machine learning model based on a neural network, which can significantly improve the result, although it will require additional configuration and computer resources.

References

1. Lopes, R., Jorge, A.: Mind the Gap: A Well Log Data Analysis (2017)
2. Lopes, R., Jorge, A.: Assessment of predictive learning methods for the completion of gaps in well log data. J. Pet. Sci. Eng. **162**, 873–886 (2018). https://doi.org/10.1016/j.petrol.2017.11.019
3. Prokopovych, I., Shpinkovski, O.: The use of intelligent technologies in diagnosing the disease. In: 1st International Scientific and Practical Conference: Information Systems and Technologies in Medicine, pp. 127–129 (2018). [in Ukrainian]

4. Denysenko, Y., Kysylevska, A., Panchenko, O., Zaloga, V., Dynnyk, O.: Decision-making based on prediction of oil quality indicators in the enterprise's information system. In: Tonkonogyi, V. et al. (eds) Advanced Manufacturing Processes. Inter Partner 2019. LNME. Springer, Cham (2020). https://doi.org/10.1007/978-3-030-40724-7_3
5. Kuric, I., Kandera, M., Klarák, J., Ivanov, V., Więcek, D.: Visual product inspection based on deep learning methods. In: Tonkonogyi, V. et al. (eds) Advanced Manufacturing Processes. Inter Partner 2019. LNME. Springer, Cham (2020). https://doi.org/10.1007/978-3-030-40724-7_15
6. Zhimin, C.: Cross-well Lithology Identification (2018). https://doi.org/10.6084/m9.figshare.6667646.v1
7. Energy Information and Data| OpenEI.org (2020). https://openei.org/wiki/Main_Page
8. Corwin, R., Hoover, D.: The self-potential method in geothermal exploration. Geophysics **44** (2), 226–245 (1979). https://doi.org/10.1190/1.1440964
9. Vasvári, V.: On the applicability of dual Laterolog for the determination of fracture parameters in hard rock aquifers. Austrian J. Earth. Sci. **104**(2), 80–89 (2011)
10. Kansas Geological Survey - Geological Log Analysis - Nuclear Porosity Logs. http://www.kgs.ku.edu/Publications/Bulletins/LA/04_nuclear.html. Accessed 03 May 2020
11. Nisbet, R., Miner, G., Yale, K.: Handbook of Statistical Analysis and Data Mining Applications, 2nd edn. Academic Press, London (2018)
12. Mohamed, I., Mohamed, S., Mazher, I., Chester, P.: Formation lithology classification: insights into machine learning methods. In: SPE Annual Technical Conference and Exhibition. Society of Petroleum Engineers. Calgary, Canada (2019). https://doi.org/10.2118/196096-MS
13. Zhang, Y., Jin, S., Jiang, H., Wang, Y., Jia, P.: Review of well logs and petrophysical approaches for shale gas in sichuan basin. Chin. TOPEJ **8**, 316–324 (2015). https://doi.org/10.2174/1874834101508010316
14. Welcome to Colaboratory. A free Jupyter notebook environment. https://colab.research.google.com/notebooks/welcome.ipynb. Accessed 03 May 2020
15. Hall, B.: Facies classification using machine learning. Lead. Edge **35**, 906–909 (2016). https://doi.org/10.1190/tle35100906.1

Author Index

V. Tonkonogyi et al. (Eds.): InterPartner 2020, LNME, pp. 853–856, 2021.
https://doi.org/10.1007/978-3-030-68014-5